T0305980

ORDINARY DIFFERENTIAL EQUATIONS

An Introduction to the Fundamentals

Second Edition

Textbooks in Mathematics

Series editors:

Al Boggess and Ken Rosen

CRYPTOGRAPHY: THEORY AND PRACTICE, FOURTH EDITION
Douglas R. Stinson and Maura B. Paterson

GRAPH THEORY AND ITS APPLICATIONS, THIRD EDITION
Jonathan L. Gross, Jay Yellen and Mark Anderson

COMPLEX VARIABLES: A PHYSICAL APPROACH WITH APPLICATIONS, SECOND EDITION
Steven G. Krantz

GAME THEORY: A MODELING APPROACH
Richard Alan Gillman and David Housman

FORMAL METHODS IN COMPUTER SCIENCE
Jiacun Wang and William Tepfenhart

SPHERICAL GEOMETRY AND ITS APPLICATIONS
Marshall A. Whittlesey

AN INTRODUCTION TO MATHEMATICAL PROOFS
Nicholas A. Loehr

COMPUTATIONAL PARTIAL DIFFERENTIAL EQUATIONS USING MATLAB®
Jichun Li and Yi-Tung Chen

AN ELEMENTARY TRANSITION TO ABSTRACT MATHEMATICS
Gove Effinger and Gary L. Mullen

MATHEMATICAL MODELING WITH EXCEL, SECOND EDITION
Brian Albright and William P. Fox

PRINCIPLES OF FOURIER ANALYSIS, SECOND EDITION
Kenneth B. Howell

https://www.crcpress.com/Textbooks-in-Mathematics/book-series/CANDHTEXBOOMTH

TEXTBOOKS IN MATHEMATICS

ORDINARY DIFFERENTIAL EQUATIONS

An Introduction to the Fundamentals

Second Edition

Kenneth B. Howell

University of Alabama in Huntsville, USA

CRC Press is an imprint of the
Taylor & Francis Group, an **informa** business
A CHAPMAN & HALL BOOK

Second edition published 2020
by CRC Press
6000 Broken Sound Parkway NW, Suite 300, Boca Raton, FL 33487-2742

and by CRC Press
2 Park Square, Milton Park, Abingdon, Oxon, OX14 4RN

[First edition published by CRC Press 2015]

CRC Press is an imprint of Taylor & Francis Group, LLC

ISBN: 978-1-138-60583-1 (hbk)
ISBN: 978-0-429-34742-9 (ebk)

Typeset in NimbusRomNo9L
by Author

Contents

Preface
(With Important Information for the Reader)

This textbook reflects my personal views on how an "ideal" introductory course in ordinary differential equations should be taught, tempered by such practical constraints as the time available, and the level and interests of the students. It also reflects my beliefs that a good text should both be a useful resource beyond the course, and back up its claims with solid and clear proofs (even if some of those proofs are ignored by most readers). Moreover, I hope, it reflects the fact that such a book should be written to engage those normally taking such a course; namely, students who are reasonably acquainted with the differentiation and integration of functions of one variable, but who might not yet be experts and may, on occasion, need to return to their elementary calculus texts for review. Most of these students are not mathematicians and probably have no desire to become professional mathematicians. Still, most are interested in fields of study in which a fundamental understanding of the mathematics and applications of differential equations is valuable.

Do note that this text only assumes the student reader is acquainted with single-variable calculus. It is not assumed that the reader has had courses in multi-variable calculus or linear algebra. If you, the reader, have had (or are taking) these and other more advanced courses, then great. You can delve into a few more topics. In particular, if you've had a course in linear algebra or real analysis, then you can (and should) be on the lookout for points where the theory from those more advanced courses can be applied to simplify some of the discussion.

Of course, while I wrote this text for the students, the needs of the instructors were kept firmly in mind. After all, this is the text my colleagues and I have been using for the last several years.

Whether you are a student, instructor or just a casual reader, there are a number of things you should be aware of before starting the first chapter:

1. *Extra material*: There is more material in this text than can be reasonably covered in a "standard" one-semester introductory course. In part, this is to provide the material for a variety of "standard" courses which may or may not cover such topics as Laplace transforms, series solutions, systems and numerical methods. Beyond that, though, there are expanded discussions of topics normally covered, as well as topics rarely covered, but which are still elementary enough and potentially useful enough to merit discussion. There are also proofs that are not simple and illuminating enough to be included in the basic exposition, but should still be there to keep the author honest and to serve as a reference for others. Because of this extra material, there is an appendix, *Author's Guide to Using This Text*, with advice on which sections must be covered, which are optional, and which are best avoided by the first-time reader. It also contains a few opinionated comments.

2. *Computer math packages*: At several points in the text, the use of a "computer math package" is advised or, in exercises, required. By a "computer math package", I mean one of those powerful software packages such as *Maple* or *Mathematica* that can do symbolic calculations, graphing and so forth. Unfortunately, software changes over time, new products emerge, and companies providing this software can be bought and sold. In addition, you may be able to find other computational resources on the Internet (but be aware that websites can be much more fickle and untrustworthy than major software providers). For these reasons, details on using such software are not included in this text. You will have to figure that out yourself (it's not that hard). I will tell you this: Extensive use of *Maple* was made in preparing this text. In fact, most of the graphs were generated in *Maple* and then cleaned up using commercial graphics software.

   On the subject of computer math packages: Please become reasonably proficient in at least one. If you are reading this, you are probably working in or will be working in a field in which this sort of knowledge is invaluable. But don't think this software can replace a basic understanding of the mathematics you are using. Even a simple calculator is useless to someone who doesn't understand just what $+$ and $\times$ mean. Mindlessly using this software can lead to serious and costly mistakes (as discussed in Section 10.3).

3. *Additional chapters*: By the way, I do not consider this text complete. Additional chapters on systems of differential equations and boundary-value problems are being written for a possible follow-up text. As these chapters become written (and rewritten), they will become available at the website for this text (see below).

4. *Text website*: While this edition remains in publication, I intend to maintain a website for this text containing at least the following:

   - A lengthy solution manual
   - The aforementioned chapters extending the material in this text
   - A list of known errata discovered since the book's publication

   At the time I was writing this, the text's website was at http://howellkb.uah.edu/DEtext/. With luck, that will still be the website's location when you need it. Unfortunately, I cannot guarantee that my university will not change its website policies and conventions, forcing you to search for the current location of the text's website. If you must search for this site, I would suggest starting with the website of the Department of Mathematical Sciences of the University of Alabama in Huntsville.

Those acquainted with the first edition may wonder how this, the second edition, differs from the first. The answer: Not much — known typos have been corrected, some of the discussion has been cleaned up, and many more exercises have been added. In fact, two new "chapters" consist of nothing more than review exercises for Parts II and III. Aside from that, two chapters on numerical methods (extending the original discussion on the Euler method) have been added.

Finally, I must thank the many students and fellow faculty who have used earlier versions of this text and have provided the feedback that I have found invaluable in preparing this edition. Those comments are very much appreciated. And, if you, the reader, should find any errors or would like to make any suggestions or comments regarding this text, please let me know. That, too, would be very much appreciated.

Dr. Kenneth B. Howell
(howellkb@uah.edu)

Part I

The Basics

1

The Starting Point: Basic Concepts and Terminology

Let us begin our study of "differential equations" with a few basic questions — questions that any beginner should ask:

What are "differential equations"?

What can we do with them? Solve them? If so, what do we solve for? And how?

and, of course,

What good are they, anyway?

In this chapter, we will try to answer these questions (along with a few you would not yet think to ask), at least well enough to begin our studies. With luck we will even raise a few questions that cannot be answered now, but which will justify continuing our study. In the process, we will also introduce and examine some of the basic concepts, terminology and notation that will be used throughout this book.

1.1 Differential Equations: Basic Definitions and Classifications

A *differential equation* is an equation involving some function of interest along with a few of its derivatives. Typically, the function is unknown, and the challenge is to determine what that function could possibly be.

Differential equations can be classified either as "ordinary" or as "partial". An *ordinary differential equation* is a differential equation in which the function in question is a function of only one variable. Hence, its derivatives are the "ordinary" derivatives encountered early in calculus. For the most part, these will be the sort of equations we'll be examining in this text. For example,

$$\frac{dy}{dx} = 4x^3$$

$$\frac{dy}{dx} + \frac{4}{x}y = x^2$$

$$\frac{d^2y}{dx^2} - 2\frac{dy}{dx} - 3y = 65\cos(2x)$$

$$4x^2 \frac{d^2y}{dx^2} + 4x \frac{dy}{dx} + [4x^2 - 1]y = 0$$

and

$$\frac{d^4y}{dx^4} = 81y$$

are some differential equations that we will later deal with. In each, y denotes a function that is given by some, yet unknown, formula of x. Of course, there is nothing sacred about our choice of symbols. We will use whatever symbols are convenient for the variables and functions, especially if the problem comes from an application and the symbols help remind us of what they denote (such as when we use t for a measurement of time).[1]

A *partial differential equation* is a differential equation in which the function of interest depends on two or more variables. Consequently, the derivatives of this function are the partial derivatives developed in the later part of most calculus courses.[2] Because the methods for studying partial differential equations often involve solving ordinary differential equations, it is wise to first become reasonably adept at dealing with ordinary differential equations before tackling partial differential equations.

As already noted, this text is mainly concerned with ordinary differential equations. So let us agree that, unless otherwise indicated, the phrase "differential equation" in this text means "ordinary differential equation". If you wish to further simplify the phrasing to "DE" or even to something like "Diffy-Q", go ahead. This author, however, will not be so informal.

Differential equations are also classified by their "order". The *order* of a differential equation is simply the order of the highest order derivative explicitly appearing in the equation. The equations

$$\frac{dy}{dx} = 4x^3 \quad , \quad \frac{dy}{dx} + \frac{4}{x}y = x^2 \quad \text{and} \quad y\frac{dy}{dx} = -9.8x$$

are all first-order equations. So is

$$\frac{dy}{dx} + 3y^2 = y\left(\frac{dy}{dx}\right)^4 \quad ,$$

despite the appearance of the higher powers — dy/dx is still the highest *order* derivative in this equation, even if it is multiplied by itself a few times.

The equations

$$\frac{d^2y}{dx^2} - 2\frac{dy}{dx} - 3y = 65\cos(2x) \quad \text{and} \quad 4x^2\frac{d^2y}{dx^2} + 4x\frac{dy}{dx} + [4x^2 - 1]y = 0$$

are second-order equations, while

$$\frac{d^3y}{dx^3} = e^{4x} \quad \text{and} \quad \frac{d^3y}{dx^3} - \frac{d^2y}{dx^2} + \frac{dy}{dx} - y = x^2$$

are third-order equations.

?▶Exercise 1.1: *What is the order of each of the following equations?*

$$\frac{d^2y}{dx^2} + 3\frac{dy}{dx} - 7y = \sin(x)$$

---

[1] On occasion, we may write "$y = y(x)$" to explicitly indicate that, in some expression, y denotes a function given by some formula of x with $y(x)$ denoting that "formula of x". More often, it will simply be understood that y is a function given by some formula of whatever variable appears in our expressions.

[2] A brief introduction to partial derivatives is given in Section 3.7 for those who are interested and haven't yet seen partial derivatives.

$$\frac{d^5y}{dx^5} - \cos(x)\frac{d^3y}{dx^3} = y^2$$

$$\frac{d^5y}{dx^5} - \cos(x)\frac{d^3y}{dx^3} = y^6$$

$$\frac{d^{42}y}{dx^{42}} = \left(\frac{d^3y}{dx^3}\right)^2 \quad .$$

In practice, higher-order differential equations are usually more difficult to solve than lower-order equations. This, of course, is not an absolute rule. There are some very difficult first-order equations, as well as some very easily solved twenty-seventh-order equations.

Solutions: The Basic Notions*

Any function that satisfies a given differential equation is called a *solution* to that differential equation. "Satisfies the equation", means that, if you plug the function into the differential equation and compute the derivatives, then the result is an equation that is true no matter what real value we replace the variable with. And if that resulting equation is not true for some real values of the variable, then that function is not a solution to that differential equation.

!►Example 1.1: *Consider the differential equation*

$$\frac{dy}{dx} - 3y = 0 \quad .$$

If, in this differential equation, we let $y(x) = e^{3x}$ (i.e., if we replace y with e^{3x}), we get

$$\frac{d}{dx}\left[e^{3x}\right] - 3e^{3x} = 0$$

$$\hookrightarrow \qquad 3e^{3x} - 3e^{3x} = 0$$

$$\hookrightarrow \qquad 0 = 0 \quad ,$$

which certainly is true for every real value of x. So $y(x) = e^{3x}$ is a solution to our differential equation.

On the other hand, if we let $y(x) = x^3$ in this differential equation, we get

$$\frac{d}{dx}\left[x^3\right] - 3x^3 = 0$$

$$\hookrightarrow \qquad 3x^2 - 3x^3 = 0$$

$$\hookrightarrow \qquad 3x^2(1-x) = 0 \quad ,$$

which is true only if $x = 0$ or $x = 1$. But our interest is not in finding values of x that make the equation true; our interest is in finding functions *of x (i.e., $y(x)$) that make the equation true for all values of x. So $y(x) = x^3$ is not a solution to our differential equation. (And it makes no sense, whatsoever, to refer to either $x = 0$ or $x = 1$ as solutions, here.)*

---

* Warning: The discussion of "solutions" here is rather incomplete so that we can get to the basic, intuitive concepts quickly. We will refine our notion of "solutions" in Section 1.3 starting on page 14.

Typically, a differential equation will have many different solutions. Any formula (or set of formulas) that describes all possible solutions is called a *general solution* to the equation.

!▶Example 1.2: *Consider the differential equation*

$$\frac{dy}{dx} = 6x \quad .$$

All possible solutions can be obtained by just taking the indefinite integral of both sides,

$$\int \frac{dy}{dx}\,dx = \int 6x\,dx$$

$$\hookrightarrow \qquad y(x) + c_1 = 3x^2 + c_2$$

$$\hookrightarrow \qquad y(x) = 3x^2 + c_2 - c_1$$

where c_1 and c_2 are arbitrary constants. Since the difference of two arbitrary constants is just another arbitrary constant, we can replace the above $c_2 - c_1$ with a single arbitrary constant c and rewrite our last equation more succinctly as

$$y(x) = 3x^2 + c \quad .$$

This formula for y describes all possible solutions to our original differential equation — it is a general solution *to the differential equation in this example. To obtain an individual solution to our differential equation, just replace c with any particular number. For example, respectively letting $c = 1$, $c = -3$, and $c = 827$ yield the following three solutions to our differential equation:*

$$3x^2 + 1 \quad , \quad 3x^2 - 3 \quad \text{and} \quad 3x^2 + 827 \quad .$$

As just illustrated, general solutions typically involve arbitrary constants. In many applications, we will find that the values of these constants are not truly arbitrary but are fixed by additional conditions imposed on the possible solutions (so, in these applications at least, it would be more accurate to refer to the "arbitrary" constants in the general solutions of the differential equations as "yet undetermined" constants).

Normally, when given a differential equation and no additional conditions, we will want to determine all possible solutions to the given differential equation. Hence, "solving a differential equation" often means "finding a general solution" to that differential equation. That will be the default meaning of the phrase "solving a differential equation" in this text. Notice, however, that the resulting "solution" is not a single function that satisfies the differential equation (which is what we originally defined "a solution" to be), but is a formula describing all possible functions satisfying the differential equation (i.e., a "general solution"). Such ambiguity often arises in everyday language, and we'll just have to live with it. Simply remember that, in practice, the phrase "a solution to a differential equation" can refer either to

any single function that satisfies the differential equation,

or

any formula describing all the possible solutions (more correctly called a general solution).

In practice, it is usually clear from the context just which meaning of the word "solution" is being used. On occasions where it might not be clear, or when we wish to be very precise, it is standard to call any single function satisfying the given differential equation a *particular solution*. So, in the last example, the formulas

$$3x^2 + 1 \quad , \quad 3x^2 - 3 \quad \text{and} \quad 3x^2 + 827$$

describe particular solutions to

$$\frac{dy}{dx} = 6x \quad .$$

Initial-Value Problems

One set of auxiliary conditions that often arises in applications is a set of "initial values" for the desired solution. This is a specification of the values of the desired solution and some of its derivatives at a single point. To be precise, an *N^{th}-order set of initial values* for a solution y consists of an assignment of values to

$$y(x_0) \quad , \quad y'(x_0) \quad , \quad y''(x_0) \quad , \quad y'''(x_0) \quad , \quad \ldots \quad \text{and} \quad y^{(N-1)}(x_0)$$

where x_0 is some fixed number (in practice, x_0 is often 0) and N is some nonnegative integer.[3] Note that there are exactly N values being assigned and that the highest derivative in this set is of order $N - 1$.

We will find that N^{th}-order sets of initial values are especially appropriate for N^{th}-order differential equations. Accordingly, the term *N^{th}-order initial-value problem* will always mean a problem consisting of

1. an Nth-order differential equation, and
2. an Nth-order set of initial values.

For example,

$$\frac{dy}{dx} - 3y = 0 \qquad \text{with} \quad y(0) = 4$$

is a first-order initial-value problem. "$dy/dx - 3y = 0$" is the first-order differential equation, and "$y(0) = 4$" is the first-order set of initial values. On the other hand, the third-order differential equation

$$\frac{d^3y}{dx^3} + \frac{dy}{dx} = 0$$

along with the third-order set of initial conditions

$$y(1) = 3 \quad , \quad y'(1) = -4 \quad \text{and} \quad y''(1) = 10$$

makes up a third-order initial-value problem.

A *solution* to an initial-value problem is a solution to the differential equation that also satisfies the given initial values. The usual approach to solving such a problem is to first find the general solution to the differential equation (via any of the methods we'll develop later), and then determine the values of the 'arbitrary' constants in the general solution so that the resulting function also satisfies each of the given initial values.

---

[3] Remember, if $y = y(x)$, then

$$y' = \frac{dy}{dx} \quad , \quad y'' = \frac{d^2y}{dx^2} \quad , \quad y''' = \frac{d^3y}{dx^3} \quad , \quad \ldots \quad \text{and} \quad y^{(k)} = \frac{d^ky}{dx^k} \quad .$$

We will use the 'prime' notation for derivatives when the d/dx notation becomes cumbersome.

!▶Example 1.3: *Consider the initial-value problem*

$$\frac{dy}{dx} = 6x \qquad \text{with} \quad y(1) = 8 \quad .$$

From Example 1.2, we know that the general solution to the above differential equation is

$$y(x) = 3x^2 + c$$

where c is an arbitrary constant. Combining this formula for y with the requirement that $y(1) = 8$, we have

$$8 = y(1) = 3 \cdot 1^2 + c = 3 + c \quad ,$$

which, in turn, requires that

$$c = 8 - 3 = 5 \quad .$$

So the solution to the initial-value problem is given by

$$y(x) = 3x^2 + c \qquad \text{with} \quad c = 5 \quad ;$$

that is,

$$y(x) = 3x^2 + 5 \quad .$$

By the way, the terms "initial values", "initial conditions", and "initial data" are essentially synonymous and, in practice, are used interchangeably.

1.2 Why Care About Differential Equations? Some Illustrative Examples

Perhaps the main reason to study differential equations is that they naturally arise when we attempt to mathematically describe "real-world" processes that vary with, say, time or position. Let us look at one well-known process: the falling of some object towards the earth. To illustrate some of the issues involved, we'll develop two different sets of mathematical descriptions for this process.

By the way, any collection of equations and formulas describing some process is called a *(mathematical) model* of the process, and the process of developing a mathematical model is called, unsurprisingly, *modeling*.

The Situation to Be Modeled:

Let us concern ourselves with the vertical position and motion of an object dropped from a plane at a height of 1,000 meters. Since it's just being dropped, we may assume its initial downward velocity is 0 meters per second. The precise nature of the object — whether it's a falling marble, a frozen duck (live, unfrozen ducks don't usually fall) or some other familiar falling object — is not important at this time. Visualize it as you will.

The first two things one should do when developing a model is to sketch the process (if possible) and to assign symbols to quantities that may be relevant. A crude sketch of the process is in Figure 1.1 (I've sketched the object as a ball since a ball is easy to sketch). Following ancient traditions, let's make the following symbolic assignments:

m = the mass (in grams) of the object

t = time (in seconds) since the object was dropped

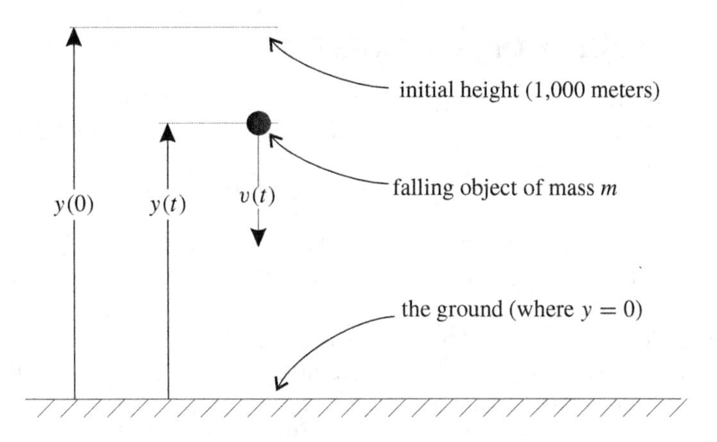

Figure 1.1: Rough sketch of a falling object of mass m.

$$\begin{aligned} y(t) &= \text{vertical distance (in meters) between the object and the ground at time } t \\ v(t) &= \text{vertical velocity (in meters/second) of the object at time } t \\ a(t) &= \text{vertical acceleration (in meters/second}^2\text{) of the object at time } t \end{aligned}$$

Where convenient, we will use y, v and a as shorthand for $y(t)$, $v(t)$ and $a(t)$. Remember that, by the definition of velocity and acceleration,

$$v = \frac{dy}{dt} \qquad \text{and} \qquad a = \frac{dv}{dt} = \frac{d^2y}{dt^2} \quad .$$

From our assumptions regarding the object's position and velocity at the instant it was dropped, we have that

$$y(0) = 1{,}000 \qquad \text{and} \qquad \left.\frac{dy}{dt}\right|_{t=0} = v(0) = 0 \quad . \tag{1.1}$$

These will be our initial values. (Notice how appropriate it is to call these the "initial values" — $y(0)$ and $v(0)$ are, indeed, the initial position and velocity of the object.)

As time goes on, we expect the object to be falling faster and faster downwards, so we expect both the position and velocity to vary with time. Precisely how these quantities vary with time might be something we don't yet know. However, from Newton's laws, we do know

$$F = ma$$

where F is the sum of the (vertically acting) forces on the object. Replacing a with either the corresponding derivative of velocity or position, this equation becomes

$$F = m\frac{dv}{dt} \tag{1.2}$$

or, equivalently,

$$F = m\frac{d^2y}{dt^2} \quad . \tag{1.2$'$}$$

If we can adequately describe the forces acting on the falling object (i.e., the F), then the velocity, $v(t)$, and vertical position, $y(t)$, can be found by solving the above differential equations, subject to the initial conditions in line (1.1).

The Simplest Falling Object Model

The Earth's gravity is the most obvious force acting on our falling object. Checking a convenient physics text, we find that the force of the Earth's gravity acting on an object of mass m is given by

$$F_{\text{grav}} = -gm \qquad \text{where} \qquad g = 9.8 \quad \left(\text{meters/second}^2\right) \quad .$$

Of course, the value for g is an approximation and assumes that the object is not too far above the Earth's surface. It also assumes that we've chosen "up" to be the positive direction (hence the negative sign).

For this model, let us suppose the Earth's gravity, F_{grav}, is the only significant force involved. Assuming this (and keeping in mind that we are measuring distance in meters and time in seconds), we have

$$F = F_{\text{grav}} = -9.8m$$

in the "$F = ma$" equation. In particular, equation (1.2$'$) becomes

$$-9.8m = m\frac{d^2y}{dt^2} \quad .$$

The mass conveniently divides out, leaving us with

$$\frac{d^2y}{dt^2} = -9.8 \quad .$$

Taking the indefinite integral with respect to t of both sides of this equation yields

$$\int \frac{d^2y}{dt^2}\,dt = \int -9.8\,dt$$

$$\hookrightarrow \qquad \int \frac{d}{dt}\left(\frac{dy}{dt}\right) dt = \int -9.8\,dt$$

$$\hookrightarrow \qquad \frac{dy}{dt} + c_1 = -9.8t + c_2$$

$$\hookrightarrow \qquad \frac{dy}{dt} = -9.8t + c$$

where c_1 and c_2 are the arbitrary constants of integration and $c = c_2 - c_1$. This gives us our formula for ${}^{dy}/_{dt}$ up to an unknown constant c. But recall that the initial velocity is zero,

$$\left.\frac{dy}{dt}\right|_{t=0} = v(0) = 0 \quad .$$

On the other hand, setting t equal to zero in the formula just derived for ${}^{dy}/_{dt}$ yields

$$\left.\frac{dy}{dt}\right|_{t=0} = -9.8 \cdot 0 + c \quad .$$

Combining these two expressions for $y'(0)$ yields

$$0 = \left.\frac{dy}{dt}\right|_{t=0} = -9.8 \cdot 0 + c \quad .$$

Thus, $c = 0$ and our formula for ${}^{dy}/_{dt}$ reduces to

$$\frac{dy}{dt} = -9.8t \quad .$$

Again, we have a differential equation that is easily solved by simple integration,

$$\int \frac{dy}{dt}\,dt = \int -9.8t\,dt$$

$$\hookrightarrow \qquad y(t) + C_1 = -9.8\left[\frac{1}{2}t^2\right] + C_2$$

$$\hookrightarrow \qquad y(t) = -4.9t^2 + C$$

where, again, C_1 and C_2 are the arbitrary constants of integration and $C = C_2 - C_1$.[4] Combining this last equation with the initial condition for $y(t)$ (from line (1.1)), we get

$$1{,}000 = y(0) = -4.9 \cdot 0^2 + C \quad .$$

Thus, $C = 1{,}000$ and the vertical position (in meters) at time t is given by

$$y(t) = -4.9t^2 + 1{,}000 \quad .$$

A Better Falling Object Model

The above model does not take into account the resistance of the air to the falling object — a very important force if the object is relatively light or has a parachute. Let us add this force to our model. That is, for our " $F = ma$ " equation, we'll use

$$F = F_{\text{grav}} + F_{\text{air}}$$

where F_{grav} is the force of gravity discussed above, and F_{air} is the force due to the air resistance acting on this particular falling body.

Part of our problem now is to determine a good way of describing F_{air} in terms relevant to our problem. To do that, let us list a few basic properties of air resistance that should be obvious to anyone who has stuck their hand out of a car window:

1. The force of air resistance does not depend on the position of the object, only on the relative velocity between it and the surrounding air. So, for us, F_{air} will just be a function of v, $F_{\text{air}} = F_{\text{air}}(v)$. (This assumes, of course, that the air is still — no up- or downdrafts — and that the density of the air remains fairly constant throughout the distance this object falls.)

2. This force is zero when the object is not moving, and its magnitude increases as the speed increases (remember, speed is the magnitude of the velocity). Hence, $F_{\text{air}}(v) = 0$ when $v = 0$, and $|F_{\text{air}}(v)|$ gets bigger as $|v|$ gets bigger.

3. Air resistance acts *against* the direction of motion. This means that the direction of the force of air resistance is opposite to the direction of motion. Thus, the sign of $F_{\text{air}}(v)$ will be opposite that of v.

While there are many formulas for $F_{\text{air}}(v)$ that would satisfy the above conditions, common sense suggests that we first use the simplest. That would be

$$F_{\text{air}}(v) = -\gamma v$$

[4] Note that slightly different symbols are being used to denote the different constants. This is highly recommended to prevent confusion when (and if) we ever review our computations.

where γ is some positive value. The actual value of γ will depend on such parameters as the object's size, shape, and orientation, as well as the density of the air through which the object is moving. For any given object, this value could be determined by experiment (with the aid of the equations we will soon derive).

?►Exercise 1.2: *Convince yourself that*

a: *this formula for $F_{\text{air}}(v)$ does satisfy the above three conditions, and*

b: *no simpler formula would work.*

We are now ready to derive the appropriate differential equations for our improved model of a falling object. The total force is given by

$$F = F_{\text{grav}} + F_{\text{air}} = -9.8m - \gamma v \quad .$$

Since this formula explicitly involves v instead of dy/dt, let us use the equation (1.2) version of "$F = ma$" from page 9,

$$F = m\frac{dv}{dt} \quad .$$

Combining the last two equations,

$$m\frac{dv}{dt} = F = -9.8m - \gamma v \quad .$$

Cutting out the middle and dividing through by the mass gives the slightly simpler equation

$$\frac{dv}{dt} = -9.8 - \kappa v \qquad \text{where} \quad \kappa = \frac{\gamma}{m} \quad . \tag{1.3}$$

Remember that γ, m and, hence, κ are positive constants, while $v = v(t)$ is a yet unknown function that satisfies the initial condition $v(0) = 0$. After solving this initial-value problem for $v(t)$, we could then find the corresponding formula for height at time t, $y(t)$, by solving the simple initial-value problem

$$\frac{dy}{dt} = v(t) \qquad \text{with} \quad y(0) = 1{,}000 \quad .$$

Unfortunately, we cannot solve equation (1.3) by simply integrating both sides with respect to t,

$$\int \frac{dv}{dt}\,dt = \int [-9.8 - \kappa v]\,dt \quad .$$

The first integral is not a problem. By the relation between derivatives and integrals, we still have

$$\int \frac{dv}{dt}\,dt = v(t) + c_1$$

where c_1 is an arbitrary constant. It's the other side that is the problem. Since κ is a constant, but $v = v(t)$ is an unknown function of t, the best we can do with the right-hand side is

$$\int [-9.8 - \kappa v]\,dt = -\int 9.8\,dt - \kappa \int v(t)\,dt = -9.8t + c_2 - \kappa \int v(t)\,dt \quad .$$

Again, c_2 is an arbitrary constant. However, since $v(t)$ is an unknown function, its integral is simply another unknown function of t. Thus, letting $c = c_2 - c_1$ and "integrating the equation" simply gives us the rather unhelpful formula

$$v(t) = -9.8t + c - (\kappa \cdot \textit{some unknown function of } t) \quad .$$

Fortunately, this is a text on differential equations, and methods for solving equations such as equation (1.3) will be discussed in Chapters 4 and 5. But there's no need to rush things. The main goal here is just to see how differential equations arise in applications. Of course, now that we have equation (1.3), we also have a good reason to continue on and learn how to solve it.

By the way, if we replace v in equation (1.3) with ${}^{dy}/_{dt}$, we get the second-order differential equation

$$\frac{d^2y}{dt^2} = -9.8 - \kappa\frac{dy}{dt} \quad .$$

This can be integrated, yielding

$$\frac{dy}{dt} = -9.8t - \kappa y + c$$

where c is an arbitrary constant. Again, this is a first-order differential equation that we cannot solve until we delve more deeply into the various methods for solving these equations. And if, in this last equation, we again use the fact that $v = {}^{dy}/_{dt}$, all we get is

$$v = -9.8t - \kappa y + c \tag{1.4}$$

which is another not-very-helpful equation relating the unknown functions $v(t)$ and $y(t)$.[5]

Summary of Our Models and the Related Initial Value Problems

For the first model of a falling body, we had the second-order differential equation

$$\frac{d^2y}{dt^2} = -9.8 \quad .$$

along with the initial conditions

$$y(0) = 1{,}000 \qquad \text{and} \qquad y'(0) = 0 \quad .$$

In other words, we had a second-order initial-value problem. This problem, as we saw, was rather easy to solve.

For the second model, we still had the initial conditions

$$y(0) = 1{,}000 \qquad \text{and} \qquad y'(0) = 0 \quad ,$$

but we found it a little more convenient to write the differential equation as

$$\frac{dv}{dt} = -9.8 - \kappa v \qquad \text{where} \qquad \frac{dy}{dt} = v$$

and κ was some positive constant. There are a couple of ways we can view this collection of equations. First of all, we could simply replace the v with ${}^{dy}/_{dt}$ and say we have the second-order initial problem

$$\frac{d^2y}{dt^2} = -9.8 - \kappa\frac{dy}{dt}$$

with

$$y(0) = 1,000 \qquad \text{and} \qquad y'(0) = 0 \quad .$$

[5] Well, not completely useless — see Exercise 1.10 b on page 20.

Alternatively, we could (as was actually suggested) view the problem as two successive first-order problems:

$$\frac{dv}{dt} = -9.8 - \kappa v \qquad \text{with} \quad v(0) = 0 \quad ,$$

followed by

$$\frac{dy}{dt} = v(t) \qquad \text{with} \quad y(0) = 1{,}000 \quad .$$

The first of these two problems can be solved using methods we'll develop later. And once we have the solution, $v(t)$, to that, the second can easily be solved by integration.

Though, ultimately, the two ways of viewing our second model are equivalent, there are advantages to the second. It is conceptually simple, and it makes it a little easier to use solution methods that will be developed relatively early in this text. It also leads us to finding $v(t)$ before even considering $y(t)$. Moreover, it is probably the velocity of landing, not the height of landing, that most concerns a falling person with (or without) a parachute. Indeed, if we are lucky, the solution to the first, $v(t)$, may tell us everything we are interested in, and we won't have to deal with the initial-value problem for y at all.

Finally, it should be mentioned that, together, the two equations

$$\frac{dv}{dt} = -9.8 - \kappa v \qquad \text{and} \qquad \frac{dy}{dt} = v$$

form a "system of differential equations". That is, they comprise a set of differential equations involving unknown functions that are related to each other. This is an especially simple system since it can be solved by successively solving the individual equations in the system. Much more complicated systems can arise that are not so easily solved, especially when modeling physical systems consisting of many components, each of which can be modeled by a differential equation involving several different functions (as in, say, a complex electronic circuit). Dealing with these sorts of systems will have to wait until we've become reasonably adept at dealing with individual differential equations.

1.3 More on Solutions

Intervals of Interest

When discussing a differential equation and its solutions, we should include a specification of an interval (of nonzero length) over which the solution(s) is (are) to be valid. The choice of this interval, which we may call the *interval of solution*, the *interval of the solution's validity*, or, simply, the *interval of interest*, may be based on the problem leading to the differential equation, on mathematical considerations, or, to a certain extent, on the whim of the person presenting the differential equation. One thing we will insist on, in this text at least, is that **solutions must be continuous over this interval**.

!► *Example 1.4:* *Consider the equation*

$$\frac{dy}{dx} = \frac{1}{(x-1)^2} \quad .$$

Integrating this gives

$$y(x) = c - \frac{1}{x-1}$$

where c is an arbitrary constant. No matter what value c is, however, this function cannot be continuous over any interval containing $x = 1$ because $(x-1)^{-1}$ "blows up" at $x = 1$. So we will only claim that our solutions are valid over intervals that do not include $x = 1$. In particular, we have valid (continuous) solutions to this differential equation over the intervals $[0, 1)$, $(-\infty, 1)$, $(1, \infty)$, and $(2, 5)$; but not over $(0, 2)$ or $(0, 1]$ or $(-\infty, \infty)$.

Why should we make such an issue of continuity? Well consider, if a function is not continuous at a point, then its derivatives do not exist at that point — and without the derivatives existing, how can we claim that the function satisfies a particular differential equation?

Another reason for requiring continuity is that the differential equations most people are interested in are models for "real-world" phenomena, and real-world phenomena are normally continuous processes while they occur — the temperature of an object does not instantaneously jump by fifty degrees nor does the position of an object instantaneously change by three kilometers. If the solutions do "blow up" at some point, then

1. some of the assumptions made in developing the model are probably not valid, or
2. a catastrophic event is occurring in our process at that point, or
3. both.

Whatever is the case, it would be foolish to use the solution derived to predict what is happening beyond the point where "things blow up". That should certainly be considered a point where the validity of the solution ends.

Sometimes, it's not the mathematics that restricts the interval of interest, but the problem leading to the differential equation. Consider the simplest falling object model discussed earlier. There we had an object start falling from an airplane at $t = 0$ from a height of 1,000 meters. Solving the corresponding initial-value problem, we obtained

$$y(t) = -4.9t^2 + 1{,}000$$

as the formula for the height above the earth at time t. Admittedly, this satisfies the differential equation for all t, but, in fact, it only gives the height of the object from the time it starts falling, $t = 0$, to the time it hits the ground, T_{hit}.[6] So the above formula for $y(t)$ is a valid description of the position of the object only for $0 \leq t \leq T_{\text{hit}}$; that is, $[0, T_{\text{hit}}]$ is the interval of interest for this problem. Any use of this formula to predict the position of the object at a time outside the interval $[0, T_{\text{hit}}]$ is just plain foolish.

In practice, the interval of interest is often not explicitly given. This may be because the interval is implicitly described in the problem, or because determining this interval is part of the problem (e.g., determining where the solutions must "blow up"). It may also be because the person giving the differential equation is lazy or doesn't care what interval is used because the issue at hand is to find formulas that hold independently of the interval of interest.

In this text, if no interval of interest is given or hinted at, assume it to be any interval that makes sense. Often, this will be the entire real line, $(-\infty, \infty)$.

Solutions Over Intervals

In introducing the concept of the "interval of interest", we have implicitly refined our notion of "a (particular) solution to a differential equation". Let us make that refinement explicit: A *solution* to a differential equation over an interval of interest is a function that is both continuous and satisfies the differential equation over the given interval.

[6] T_{hit} is computed in Exercise 1.9 on page 19.

Recall that the *domain* of a function is the set of all numbers that can be plugged into the function. Naturally, if a function is a solution to a differential equation over some interval, then that function's domain must include that interval.[7]

Since we've refined our definition of particular solutions, we should make the corresponding refinement to our definition of a general solution. A *general solution* to a differential equation over an interval of interest is a formula or set of formulas describing all possible particular solutions over that interval.

Describing Particular Solutions

Let us get somewhat technical for a moment. Suppose we have a solution y to some differential equation over some interval of interest. Remember, we've defined y to be a "function". If you look up the basic definition of "function" in your calculus text, you'll find that, strictly speaking, y is a mapping of one set of numbers (the domain of y) onto another set of numbers (the range of y). This means that, for each value x in the function's domain, y assigns a corresponding number which we usually denote $y(x)$ and call "the value of y at x". If we are lucky, the function y is described by some formula, say, x^2. That is, the value of $y(x)$ can be determined for each x in the domain by the equation

$$y(x) = x^2 \quad .$$

Strictly speaking, the function y, its value at x (i.e., $y(x)$), and any formula describing how to compute $y(x)$ are different things. In everyday usage, however, the fine distinctions between these concepts are often ignored, and we say things like

consider the function x^2 or *consider* $y = x^2$

instead of the more correct statement

consider the function y *where* $y(x) = x^2$ *for each* x *in the domain of* y.

For our purposes, "everyday usage" will usually suffice, and we won't worry that much about the differences between y, $y(x)$, and a formula describing y. This will save ink and paper, simplify the English, and, frankly, make it easier to follow many of our computations.

In particular, when we seek a particular solution to a differential equation, we will usually be quite happy to find a convenient formula describing the solution. We will then probably mildly abuse terminology by referring to that formula as "the solution". Please keep in mind that, in fact, any such formula is just one *description* of the solution — a very useful description since it tells you how to compute $y(x)$ for every x in the interval of interest. But other formulas can also describe the same function. For example, you can easily verify that

$$x^2 \quad , \quad (x+3)(x-3)+9 \quad \text{and} \quad \int_{t=0}^{x} 2t\,dt$$

are all formulas describing the same function on the real line.

There will also be differential equations for which we simply cannot find a convenient formula describing the desired solution (or solutions). On those occasions we will have to find some alternative way to describe our solutions. Some of these will involve using the differential equations to sketch approximations to the graphs of their solutions. Other alternative descriptions will involve formulas that approximate the solutions and allow us to generate lists of values approximating a solution at different points. These alternative descriptions may not be as convenient or as accurate as explicit formulas for the solutions, but they will provide usable information about the solutions.

---

[7] In theory, it makes sense to restrict the domain of a solution to the interval of interest so that irrelevant questions regarding the behavior of the function off the interval have no chance of arising. At this point of our studies, let us just be sure that a function serving as a solution makes sense at least over whatever interval we have interest in.

Additional Exercises

1.3. *For each differential equation given below, three choices for a possible solution $y = y(x)$ are given. Determine whether each choice is or is not a solution to the given differential equation. (In each case, assume the interval of interest is the entire real line $(-\infty, \infty)$.)*

a. $\dfrac{dy}{dx} = 3y$

i. $y(x) = e^{3x}$ **ii.** $y(x) = x^3$ **iii.** $y(x) = \sin(3x)$

b. $x\dfrac{dy}{dx} = 3y$

i. $y(x) = e^{3x}$ **ii.** $y(x) = x^3$ **iii.** $y(x) = \sin(3x)$

c. $\dfrac{d^2y}{dx^2} = 9y$

i. $y(x) = e^{3x}$ **ii.** $y(x) = x^3$ **iii.** $y(x) = \sin(3x)$

d. $\dfrac{d^2y}{dx^2} = -9y$

i. $y(x) = e^{3x}$ **ii.** $y(x) = x^3$ **iii.** $y(x) = \sin(3x)$

e. $x\dfrac{dy}{dx} - 2y = 6x^4$

i. $y(x) = x^4$ **ii.** $y(x) = 3x^4$ **iii.** $y(x) = 3x^4 + 5x^2$

f. $\dfrac{d^2y}{dx^2} - 2x\dfrac{dy}{dx} - 2y = 0$

i. $y(x) = \sin(x)$ **ii.** $y(x) = x^3$ **iii.** $y(x) = e^{x^2}$

g. $\dfrac{d^2y}{dx^2} + 4y = 12x$

i. $y(x) = \sin(2x)$ **ii.** $y(x) = 3x$ **iii.** $y(x) = \sin(2x) + 3x$

h. $\dfrac{d^2y}{dx^2} - 6\dfrac{dy}{dx} + 9y = 0$

i. $y(x) = e^{3x}$ **ii.** $y(x) = xe^{3x}$ **iii.** $y(x) = 7e^{3x} - 4xe^{3x}$

1.4. For each initial-value problem given below, three choices for a possible solution $y = y(x)$ are given. Determine whether each choice is or is not a solution to the given initial-value problem.

a. $\dfrac{dy}{dx} = 4y$ *with* $y(0) = 5$

i. $y(x) = e^{4x}$ **ii.** $y(x) = 5e^{4x}$ **iii.** $y(x) = e^{4x} + 1$

b. $x\dfrac{dy}{dx} = 2y$ *with* $y(2) = 20$

i. $y(x) = x^2$ **ii.** $y(x) = 10x$ **iii.** $y(x) = 5x^2$

c. $\dfrac{d^2y}{dx^2} - 9y = 0$ *with* $y(0) = 1$ *and* $y'(0) = 9$

i. $y(x) = 2e^{3x} - e^{-3x}$ **ii.** $y(x) = e^{3x}$ **iii.** $y(x) = e^{3x} + 1$

d. $x^2\dfrac{d^2y}{dx^2} - 4x\dfrac{dy}{dx} + 6y = 36x^6$ *with* $y(1) = 1$ *and* $y'(1) = 12$

i. $y(x) = 10x^3 - 9x^2$ **ii.** $y(x) = 3x^6 - 2x^2$ **iii.** $y(x) = 3x^6 - 2x^3$

1.5. *For the following, let*

$$y(x) = \sqrt{x^2 + c}$$

where c *is an arbitrary constant.*

a. *Verify that this* y *is a solution to*

$$\frac{dy}{dx} = \frac{x}{y}$$

no matter what value c *is.*

b. *What value should* c *be so that the above* y *satisfies the initial condition*

i. $y(0) = 3$? **ii.** $y(2) = 3$?

c. *Using your results for the above, give a solution to each of the following initial-value problems:*

i. $\dfrac{dy}{dx} = \dfrac{x}{y}$ *with* $y(0) = 3$

ii. $\dfrac{dy}{dx} = \dfrac{x}{y}$ *with* $y(2) = 3$

1.6. *For the following, let*

$$y(x) = Ae^{x^2} - 3$$

where A *is an arbitrary constant.*

a. *Verify that this* y *is a solution to*

$$\frac{dy}{dx} - 2xy = 6x$$

no matter what value A *is.*

b. *In fact, it can be verified (using methods that will be developed later) that the above formula for* y *is a general solution to the above differential equation. Using this fact, finish solving each of the following initial-value problems:*

i. $\dfrac{dy}{dx} - 2xy = 6x$ *with* $y(0) = 1$

ii. $\dfrac{dy}{dx} - 2xy = 6x$ *with* $y(1) = 0$

1.7. *For the following, let*

$$y(x) = A\cos(2x) + B\sin(2x)$$

where A *and* B *are arbitrary constants.*

a. *Verify that this* y *is a solution to*

$$\frac{d^2y}{dx^2} + 4y = 0$$

no matter what values A *and* B *are.*

b. *Again, it can be verified that the above formula for* y *is a general solution to the above differential equation. Using this fact, finish solving each of the following initial-value problems:*

i. $\dfrac{d^2y}{dx^2} + 4y = 0$ *with* $y(0) = 3$ *and* $y'(0) = 8$

ii. $\dfrac{d^2y}{dx^2} + 4y = 0$ *with* $y(0) = 0$ *and* $y'(0) = 1$

1.8. *It was stated (on page 7) that "N^{th}-order sets of initial values are especially appropriate for N^{th}-order differential equations." The following problems illustrate one reason this is true. In particular, they demonstrate that, if* y *satisfies some N^{th}-order initial-value problem, then it automatically satisfies particular higher-order sets of initial values. Because of this, specifying the initial values for* $y^{(m)}$ *with* $m \geq N$ *is unnecessary and may even lead to problems with no solutions.*

a. *Assume* y *satisfies the first-order initial-value problem*

$$\frac{dy}{dx} = 3xy + x^2 \qquad \text{with} \qquad y(1) = 2 \quad .$$

i. *Using the differential equation along with the given value for* $y(1)$, *determine what value* $y'(1)$ *must be.*

ii. *Is it possible to have a solution to*

$$\frac{dy}{dx} = 3xy + x^2$$

that also satisfies both $y(1) = 2$ *and* $y'(1) = 4$ *? (Give a reason.)*

iii. *Differentiate the given differential equation to obtain a second-order differential equation. Using the equation obtained along with the now known values for* $y(1)$ *and* $y'(1)$, *find the value of* $y''(1)$.

iv. *Can we continue and find* $y'''(1)$, $y^{(4)}(1)$, $\ldots$?

b. *Assume* y *satisfies the second-order initial-value problem*

$$\frac{d^2y}{dx^2} + 4\frac{dy}{dx} - 8y = 0 \qquad \text{with} \qquad y(0) = 3 \quad \text{and} \quad y'(0) = 5 \quad .$$

i. *Find the value of* $y''(0)$ *and of* $y'''(0)$

ii. *Is it possible to have a solution to*

$$\frac{d^2y}{dx^2} + 4\frac{dy}{dx} - 8y = 0$$

that also satisfies all of the following:

$$y(0) = 3 \quad , \quad y'(0) = 5 \quad \text{and} \quad y'''(0) = 0 \quad ?$$

1.9. *Consider the simplest model we developed for a falling object (see page 10). In that, we derived*

$$y(t) = -4.9t^2 + 1{,}000$$

as the formula for the height y *above ground of some falling object at time* t.

a. *Find* T_{hit}, *the time the object hits the ground.*

b. *What is the velocity of the object when it hits the ground?*

c. *Suppose that, instead of being dropped at $t = 0$, the object is tossed up with an initial velocity of 2 meters per second. If this is the only change to our problem, then:*

i. *How does the corresponding initial-value problem change?*

ii. *What is the solution $y(t)$ to this initial value problem?*

iii. *What is the velocity of the object when it hits the ground?*

1.10. *Consider the "better" falling object model (see page 11), in which we derived the differential equation*

$$\frac{dv}{dt} = -9.8 - \kappa v \tag{1.5}$$

for the velocity. In this, κ is some positive constant used to describe the air resistance felt by the falling object.

a. *This differential equation was derived assuming the air was still. What differential equation would we have derived if, instead, we had assumed there was a steady updraft of 2 meters per second?*

b. *Recall that, from equation (1.5) we derived the equation*

$$v = -9.8t - \kappa y + c$$

relating the velocity v to the distance above ground y and the time t (see page 13). In the following, you will show that it, along with experimental data, can be used to determine the value of κ.

i. *Determine the value of the constant of integration, c, in the above equation using the given initial values (i.e., $y(0) = 1{,}000$ and $v(0) = 0$).*

ii. *Suppose that, in an experiment, the object was found to hit the ground at $t = T_{hit}$ with a speed of $v = v_{hit}$. Use this, along with the above equation, to find κ in terms of T_{hit} and v_{hit}.*

1.11. *For the following, let*

$$y(x) = Ax + Bx \ln|x|$$

where A and B are arbitrary constants.

a. *Verify that this y is a solution to*

$$x^2 \frac{d^2y}{dx^2} - x\frac{dy}{dx} + y = 0 \qquad \text{on the intervals } (0,\infty) \text{ and } (-\infty,0) \quad ,$$

no matter what values A and B are.

b. *Again, we will later be able to show that the above formula for y is a general solution for the above differential equation. Given this, find the solution to the above differential equation satisfying $y(1) = 3$ and $y'(1) = 8$.*

c. *Why should your answer to* **1.11 b** *not be considered a valid solution to*

$$x^2 \frac{d^2y}{dx^2} - x\frac{dy}{dx} + y = 0$$

over the entire real line, $(-\infty,\infty)$?

2

Integration and Differential Equations

Often, when attempting to solve a differential equation, we are naturally led to computing one or more integrals — after all, integration is the inverse of differentiation. Indeed, we have already solved one simple second-order differential equation by repeated integration (the one arising in the simplest falling object model, starting on page 10). Let us now briefly consider the general case where integration is immediately applicable, and also consider some practical aspects of using both the indefinite integral and the definite integral.

2.1 Directly-Integrable Equations

We will say that a given first-order differential equation is *directly integrable* if (and only if) it can be (re)written as

$$\frac{dy}{dx} = f(x) \tag{2.1}$$

where $f(x)$ is some known function of just x (no y's). More generally, any N^{th}-order differential equation will be said to be *directly integrable* if and only if it can be (re)written as

$$\frac{d^N y}{dx^N} = f(x) \tag{2.1$'$}$$

where, again, $f(x)$ is some known function of just x (no y's or derivatives of y).

!►Example 2.1: *Consider the equation*

$$x^2 \frac{dy}{dx} - 4x = 6 \quad . \tag{2.2}$$

Solving this equation for the derivative:

$$x^2 \frac{dy}{dx} = 4x + 6$$

$$\hookrightarrow \qquad \frac{dy}{dx} = \frac{4x + 6}{x^2} \quad .$$

Since the right-hand side of the last equation depends only on x, we do have

$$\frac{dy}{dx} = f(x) \qquad \left(\text{with } f(x) = \frac{4x+6}{x^2}\right) \quad .$$

So equation (2.2) is directly integrable.

!▶ Example 2.2: *Consider the equation*

$$x^2 \frac{dy}{dx} - 4xy = 6 \quad . \tag{2.3}$$

Solving this equation for the derivative:

$$x^2 \frac{dy}{dx} = 4xy + 6$$

$$\hookrightarrow \qquad \frac{dy}{dx} = \frac{4xy+6}{x^2} \quad .$$

Here, the right-hand side of the last equation depends on both x *and* y *, not just* x *. So equation (2.3) is not directly integrable.*

Solving a directly-integrable equation is easy. First solve for the derivative to get the equation into form (2.1) or (2.1 ′), then integrate both sides as many times as needed to eliminate the derivatives, and, finally, do whatever simplification seems appropriate.

!▶ Example 2.3: *Again, consider*

$$x^2 \frac{dy}{dx} - 4x = 6 \quad . \tag{2.4}$$

In Example 2.1, we saw that it is directly integrable and can be rewritten as

$$\frac{dy}{dx} = \frac{4x+6}{x^2} \quad .$$

Integrating both sides of this equation with respect to x *(and doing a little algebra):*

$$\int \frac{dy}{dx}\,dx = \int \frac{4x+6}{x^2}\,dx \tag{2.5}$$

$$\hookrightarrow \qquad y(x) + c_1 = \int \left[\frac{4}{x} + \frac{6}{x^2}\right] dx$$

$$= 4\int x^{-1}\,dx + 6\int x^{-2}\,dx$$

$$= 4\ln|x| + c_2 - 6x^{-1} + c_3$$

where c_1 *,* c_2 *and* c_3 *are arbitrary constants. Rearranging things slightly and letting* $c = c_2 + c_3 - c_1$ *, this last equation simplifies to*

$$y(x) = 4\ln|x| - 6x^{-1} + c \quad . \tag{2.6}$$

This is our general solution to differential equation (2.4). Since both $\ln|x|$ *and* x^{-1} *are discontinuous just at* $x = 0$ *, the solution can be valid over any interval not containing* $x = 0$ *.*

?▶ Exercise 2.1: *Consider the differential equation in Example 2.2 and explain why the* y *, which is an unknown function of* x *, makes it impossible to completely integrate both sides of*

$$\frac{dy}{dx} = \frac{4xy+6}{x^2}$$

with respect to x *.*

2.2 On Using Indefinite Integrals

This is a good point to observe that, whenever we take the indefinite integrals of both sides of an equation, we obtain a bunch of arbitrary constants — c_1, c_2, ... (one constant for each integral) — that can be combined into a single arbitrary constant c. In the future, rather than note all the arbitrary constants that arise and how they combine into a single arbitrary constant c that is added to the right-hand side in the end, let us agree to simply add that c at the end. Let's not explicitly note all the intermediate arbitrary constants. If, for example, we had agreed to this before doing the last example, then we could have replaced all that material from equation (2.5) to equation (2.6) with

$$\int \frac{dy}{dx}\,dx = \int \frac{4x+6}{x^2}\,dx$$

$$\hookrightarrow \qquad y(x) = \int \left[\frac{4}{x} + \frac{6}{x^2}\right] dx$$

$$= 4\int x^{-1}\,dx + 6\int x^{-2}\,dx$$

$$= 4\ln|x| - 6x^{-1} + c \quad .$$

This should simplify our computations a little.

This convention of "implicitly combining all the arbitrary constants" also allows us to write

$$y(x) = \int \frac{dy}{dx}\,dx \tag{2.7}$$

instead of

$$y(x) + \text{some arbitrary constant} = \int \frac{dy}{dx}\,dx \quad .$$

By our new convention, that "some arbitrary constant" is still in equation (2.7) — it's just been moved to the right-hand side of the equation and combined with the constants arising from the integral there.

Finally, like you, this author will get tired of repeatedly saying "where c is an arbitrary constant" when it is obvious that the c (or the c_1 or the A or ...) that just appeared in the previous line is, indeed, some arbitrary constant. So let us not feel compelled to constantly repeat the obvious, and agree that, when a new symbol suddenly appears in the computation of an indefinite integral, then, yes, that is an arbitrary constant. Remember, though, to use different symbols for the different constants that arise when integrating a function already involving an arbitrary constant.

!►Example 2.4: *Consider solving*

$$\frac{d^2y}{dx^2} = 18x^2 \quad . \tag{2.8}$$

Clearly, this is directly integrable and will require two integrations. The first integration yields

$$\frac{dy}{dx} = \int \frac{d^2y}{dx^2}\,dx = \int 18x^2\,dx = \frac{18}{3}x^3 + c_1 \quad .$$

Cutting out the middle leaves

$$\frac{dy}{dx} = 6x^3 + c_1 \quad .$$

Integrating this, we have

$$y(x) = \int \frac{dy}{dx}\,dx = \int \left[6x^3 + c_1\right] dx = \frac{6}{4}x^4 + c_1x + c_2 \quad .$$

So the general solution to equation (2.8) is

$$y(x) = \frac{3}{2}x^4 + c_1 x + c_2 \quad .$$

In practice, rather than use the same letter with different subscripts for different arbitrary constants (as we did in the above example), you might just want to use different letters, say, writing

$$y(x) = \frac{3}{2}x^4 + ax + b$$

instead of

$$y(x) = \frac{3}{2}x^4 + c_1 x + c_2 \quad .$$

This sometimes prevents dumb mistakes due to bad handwriting.

2.3 On Using Definite Integrals

Basic Ideas

We have been using the *indefinite* integral to recover $y(x)$ from dy/dx via the relation

$$\int \frac{dy}{dx}\, dx = y(x) + c \quad .$$

Here, c is some constant (which we've agreed to automatically combine with other constants from other integrals).

We could just about as easily have used the corresponding *definite* integral relation

$$\int_a^x \frac{dy}{ds}\, ds = y(x) - y(a) \tag{2.9}$$

to recover $y(x)$ from its derivative. Note that, here, we've used s instead of x to denote the variable of integration. This prevents the confusion that can arise when using the same symbol for both the variable of integration *and* the upper limit in the integral. The lower limit, a, can be chosen to be any convenient value. In particular, if we are also dealing with initial values, then it makes sense to set a equal to the point at which the initial values are given. That way (as we will soon see) we will obtain a general solution in which the undetermined constant is simply the initial value.

Aside from getting it into the form

$$\frac{dy}{dx} = f(x) \quad ,$$

there are two simple steps that should be taken before using the definite integral to solve a first-order, directly-integrable differential equation:

1. Pick a convenient value for the lower limit of integration a. In particular, if the value of $y(x_0)$ is given for some point x_0, set $a = x_0$.

2. Rewrite the differential equation with s denoting the variable instead of x (i.e., replace x with s),

$$\frac{dy}{ds} = f(s) \quad . \tag{2.10}$$

After that, simply integrate both sides of equation (2.10) with respect to s from a to x:

$$\int_a^x \frac{dy}{ds}\,ds = \int_a^x f(s)\,ds$$

$$\hookrightarrow \qquad y(x) - y(a) = \int_a^x f(s)\,ds \quad .$$

Then solve for $y(x)$ by adding $y(a)$ to both sides,

$$y(x) = \int_a^x f(s)\,ds + y(a) \quad . \tag{2.11}$$

This is a general solution to the given differential equation. It should be noted that the integral here is a definite integral. Its evaluation does not lead to any arbitrary constants. However, the value of $y(a)$, until specified, can be anything; so $y(a)$ is the "arbitrary constant" in this general solution.

!►Example 2.5: *Consider solving the initial-value problem*

$$\frac{dy}{dx} = 3x^2 \qquad \text{with} \qquad y(2) = 12 \quad .$$

Since we know the value of $y(2)$, we will use 2 as the lower limit for our integrals. Rewriting the differential equation with s replacing x gives

$$\frac{dy}{ds} = 3s^2 \quad .$$

Integrating this with respect to s from 2 to x:

$$\int_2^x \frac{dy}{ds}\,ds = \int_2^x 3s^2\,ds$$

$$\hookrightarrow \qquad y(x) - y(2) = s^3\Big|_2^x = x^3 - 2^3 \quad .$$

Solving for $y(x)$ (and computing 2^3) then gives us

$$y(x) = x^3 - 8 + y(2) \quad .$$

This is a general solution to our differential equation. To find the particular solution that also satisfies $y(2) = 12$, as desired, we simply replace the $y(2)$ in the general solution with its given value,

$$\begin{aligned} y(x) &= x^3 - 8 + y(2) \\ &= x^3 - 8 + 12 = x^3 + 4 \quad . \end{aligned}$$

Of course, rather than go through the procedure just outlined to solve

$$\frac{dy}{dx} = f(x) \quad ,$$

we could, after determining a and $f(s)$, just plug these into equation (2.11),

$$y(x) = \int_a^x f(s)\,ds + y(a) \quad ,$$

and compute the integral. That is, after all, what we derived for any choice of f.

Advantages of Using Definite Integrals

By using definite integrals instead of indefinite integrals, we avoid dealing with arbitrary constants and end up with expressions explicitly involving initial values. This is sometimes convenient.

A much more important advantage of using definite integrals is that they result in concrete, computable formulas even when the corresponding indefinite integrals can*not* be evaluated. Let us look at a classic example.

!▸Example 2.6: *Consider solving the initial-value problem*

$$\frac{dy}{dx} = e^{-x^2} \qquad \text{with} \qquad y(0) = 0 \quad .$$

In particular, determine the value of $y(x)$ when $x = 10$.

Using indefinite integrals yields

$$y(x) = \int \frac{dy}{dx}\,dx = \int e^{-x^2}\,dx \quad .$$

Unfortunately, this integral was not one you learned to evaluate in calculus.[1] And if you check the tables, you will discover that no one else has discovered a usable formula for this integral. Consequently, the above formula for $y(x)$ is not very usable. Heck, we can't even isolate an arbitrary constant or see how the solution depends on the initial value.

On the other hand, using definite integrals, we get

$$\int_0^x \frac{dy}{ds}\,ds = \int_0^x e^{-s^2}\,ds$$

$$\hookrightarrow \qquad y(x) - y(0) = \int_0^x e^{-s^2}\,ds$$

$$\hookrightarrow \qquad y(x) = \int_0^x e^{-s^2}\,ds + y(0) \quad .$$

This last formula explicitly describes how $y(x)$ depends on the initial value $y(0)$. Since we are assuming $y(0) = 0$, this reduces to

$$y(x) = \int_0^x e^{-s^2}\,ds \quad .$$

We still cannot find a computable formula for this integral, but, if we choose a specific value for x, say, $x = 10$, this expression becomes

$$y(10) = \int_0^{10} e^{-s^2}\,ds \quad .$$

The value of this integral can be very accurately approximated using any of a number of numerical integration methods such as the trapezoidal rule or Simpson's rule. In practice, of course, we'll just use the numerical integration command in our favorite computer math package (Maple, Mathematica, etc.). Using any such package, you will find that

$$y(10) = \int_0^{10} e^{-s^2}\,ds \approx 0.886 \quad .$$

[1] Well, you could expand e^{-x^2} in a Taylor series and integrate the series.

In one sense,

$$y(x) = \int f(x)\,dx \tag{2.12}$$

and

$$y(x) = \int_a^x f(s)\,ds + y(a) \tag{2.13}$$

are completely equivalent mathematical expressions. In practice, either can be used just about as easily, *provided* a reasonable formula for the indefinite integral in (2.12) can be found. If no such formula can be found, however, then expression (2.13) is much more useful because it can still be used, along with a numerical integration routine, to evaluate $y(x)$ for specific values of x. Indeed, one can compute $y(x)$ for a large number of values of x, plot each of these values of $y(x)$ against x, and thereby construct a very accurate approximation of the graph of y.

There are other ways to approximate solutions to differential equations, and we will discuss some of them. However, if you can express your solution in terms of definite integrals — even if the integral must be computed approximately — then it is usually best to do so. The other approximation methods for differential equations are typically more difficult to implement, and more likely to result in poor approximations.

Important "Named" Definite Integrals with Variable Limits

You should be familiar with a number of "named" functions (such as the natural logarithm and the arctangent) that can be given by definite integrals. For the two examples just cited,

$$\ln(x) = \int_1^x \frac{1}{s}\,ds \qquad \text{for} \quad x > 0$$

and

$$\arctan(x) = \int_0^x \frac{1}{1+s^2}\,ds \quad .$$

While $\ln(x)$ and $\arctan(x)$ can be defined independently of these integrals, their alternative definitions do not provide us with particularly useful ways to compute these functions by hand (unless x is something special, such as 1). Indeed, if you need the value of $\ln(x)$ or $\arctan(x)$ for, say, $x = 18$, then you are most likely to "compute" these values by having your calculator or computer or published tables[2] tell you the (approximate) value of $\ln(18)$ or $\arctan(18)$. Thus, for computational purposes, we might as well just view $\ln(x)$ and $\arctan(x)$ as names for the above integrals, and be glad that their values can easily be looked up electronically or in published tables.

It turns out that other integrals arise often enough in applications that workers dealing with these applications have decided to "name" these integrals, and to have their values tabulated. Two noteworthy "named integrals" are:

- The *error function*, denoted by erf and given by

$$\operatorname{erf}(x) = \int_0^x \frac{2}{\sqrt{\pi}} e^{-s^2}\,ds \quad .$$

- The *sine-integral function*, denoted by Si and given by[3]

$$\operatorname{Si}(x) = \int_0^x \frac{\sin(s)}{s}\,ds \quad .$$

[2] if you are an old-timer

[3] This integral is clearly mis-named since it is not the integral of the sine. In fact, the function being integrated, $\sin(x)/x$, is often called the "sinc" function (pronounced "sink"), so Si should really be called the "sinc-integral function". But nobody does.

Both of these are considered to be well-known functions, at least among certain groups of mathematicians, scientists and engineers. They (the functions, not the people) can be found in published tables and standard mathematical software (such as Maple or Mathematica) alongside such better-known functions as the natural logarithm and the trigonometric functions. Moreover, using tables or software, the value of $\operatorname{erf}(x)$ and $\operatorname{Si}(x)$ for any real value of x can be accurately computed just as easily as can the value of $\arctan(x)$. For these reasons, and because "$\operatorname{erf}(x)$" and "$\operatorname{Si}(x)$" take up less space than the integrals they represent, we will often follow the lead of others and use these function names instead of writing out the integrals.

!▶Example 2.7: *In Example 2.6, above, we saw that the solution to*

$$\frac{dy}{dx} = e^{-x^2} \qquad \text{with} \qquad y(0) = 0$$

is

$$y(x) = \int_0^x e^{-s^2}\,ds \quad .$$

Since this integral is the same as the integral for the error function with $2/\sqrt{\pi}$ *divided out, we can also express our answer as*

$$y(x) = \frac{\sqrt{\pi}}{2}\operatorname{erf}(x) \quad .$$

2.4 Integrals of Piecewise-Defined Functions
Computing the Integrals

Be aware that the functions appearing in differential equations can be piecewise defined, as in

$$\frac{dy}{dx} = f(x) \qquad \text{where} \quad f(x) = \begin{cases} x^2 & \text{if } x < 2 \\ 1 & \text{if } 2 \le x \end{cases} \quad .$$

Indeed, two such functions occur often enough that they have their own names: the *step function*, given by

$$\operatorname{step}(x) = \begin{cases} 0 & \text{if } x < 0 \\ 1 & \text{if } 0 \le x \end{cases} \quad ,$$

and the *ramp function*, given by

$$\operatorname{ramp}(x) = \begin{cases} 0 & \text{if } x < 0 \\ x & \text{if } 0 \le x \end{cases} \quad .$$

The reasons for these names should be obvious from their graphs (see Figure 2.1)

Such functions regularly arise when we attempt to model things reacting to discontinuous influences. For example, if $y(t)$ is the amount of energy produced up to time t by some light-sensitive device, and the rate at which this energy is produced depends proportionally on the intensity of the light received by the device, then

$$\frac{dy}{dt} = \operatorname{step}(t)$$

models the energy production of this device when it's kept in the dark until a light bulb (of unit intensity) is suddenly switched on at $t = 0$.

Computing the integrals of such functions is simply a matter of computing the integrals of the various "pieces", and then putting the integrated pieces together appropriately. While this can be done using indefinite integrals, each such integral introduces a new constant of integration which must then be related to the others to ensure that the final result describes a continuous function with a minimum number of arbitrary constants. On the other hand, the intelligent use of a definite integral eliminates the extra bookkeeping arising from an excessive number of "arbitrary" constant, and also ensures that the result is a continuous function (as required for solutions). Let's illustrate this by solving the differential equation given at the start of this section.

!►Example 2.8 (using definite integrals): *We seek a general solution to*

$$\frac{dy}{dx} = f(x) \quad \text{where} \quad f(x) = \begin{cases} x^2 & \text{if} \quad x < 2 \\ 1 & \text{if} \quad 2 \leq x \end{cases} .$$

Taking the definite integral (starting, for no good reason, at 0), we have

$$y(x) = \int_0^x f(s)\,ds + y(0) \qquad \text{where} \quad f(s) = \begin{cases} s^2 & \text{if} \quad s < 2 \\ 1 & \text{if} \quad 2 \leq s \end{cases} .$$

Now, if $x \leq 2$, then $f(s) = s^2$ for every value of s in the interval $(0, x)$. So, when $x \leq 2$,

$$\int_0^x f(s)\,ds = \int_0^x s^2\,ds = \left.\frac{1}{3}s^3\right|_{s=0}^{x} = \frac{1}{3}x^3 .$$

(Notice that this integral is valid for $x = 2$ even though the formula used for $f(s)$, s^2, was only valid for $s < 2$.)

On the other hand, if $2 < x$, we must break the integral into two pieces, the one over $(0, 2)$ and the one over $(2, x)$:

$$\begin{aligned} \int_0^x f(s)\,ds &= \int_0^2 f(s)\,ds + \int_2^x f(s)\,ds \\ &= \int_0^2 s^2\,ds + \int_2^x 1\,ds \\ &= \left.\frac{1}{3}s^3\right|_{s=0}^{2} + \left. s \right|_{s=2}^{x} \\ &= \left[\frac{1}{3}\cdot 2^3 - 0\right] + [x - 2] = x + \frac{2}{3} . \end{aligned}$$

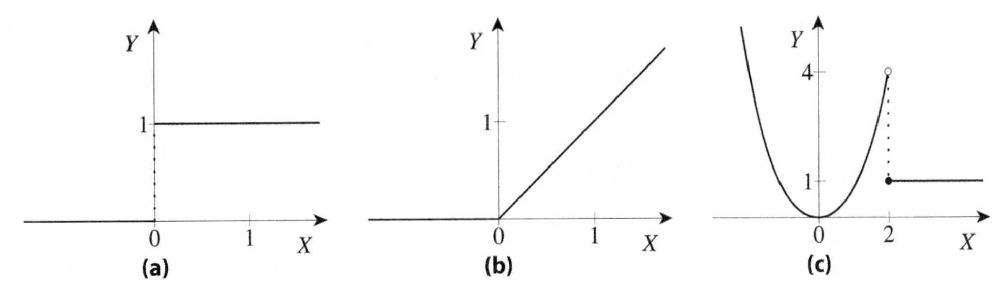

Figure 2.1: Three piecewise defined functions: **(a)** the step function, **(b)** the ramp function, **(c)** $f(x)$ from Example 2.8.

Thus, our general solution is

$$y(x) = \int_0^x f(s)\,ds + y(0) = \begin{cases} \frac{1}{3}x^3 + y(0) & \text{if } x \leq 2 \\ x + \frac{2}{3} + y(0) & \text{if } 2 < x \end{cases} .$$

Keep in mind that solutions to differential equations are required to be continuous. After checking the above formulas, it should be obvious that the $y(x)$ obtained in the last example is continuous everywhere except, possibly, at $x = 2$. With a little work we could also verify that, in fact, we also have continuity at $x = 2$. But we won't bother because, in the next subsection, it will be seen that solutions so obtained via definite integration are guaranteed to be continuous, provided the discontinuities in the function being integrated are not too bad.

In practice, a given piecewise defined function may have more than two "pieces", and the differential equation may have order higher than one. For example, you may be called upon to solve

$$\frac{d^2y}{dx^2} = f(x) \qquad \text{where} \quad f(x) = \begin{cases} 0 & \text{if } x < 1 \\ 1 & \text{if } 1 \leq x < 2 \\ 0 & \text{if } 2 \leq x \end{cases}$$

or even something involving infinitely many pieces, such as

$$\frac{d^4y}{dx^4} = \text{stair}(x) \qquad \text{where} \quad \text{stair}(x) = \begin{cases} 0 & \text{if } x < 0 \\ 1 & \text{if } 0 \leq x < 1 \\ 2 & \text{if } 2 \leq x < 3 \\ 3 & \text{if } 3 \leq x < 4 \\ \vdots \end{cases} . \tag{2.14}$$

The method illustrated in the last example can still be applied; you just have more integrals to keep track of.

Continuity of the Integrals

Theorem 2.1

Let f be a function on an interval (α, β) and let a be a point in that interval. Suppose, further, that f is continuous at all but, at most, a finite number of points in (α, β), and that, at each such point x_0 of discontinuity, the left- and right-hand limits

$$\lim_{x \to x_0^-} f(x) \qquad \text{and} \qquad \lim_{x \to x_0^+} f(x)$$

exist (and are finite).[4] *Then the function given by*

$$g(x) = \int_a^x f(s)\,ds$$

is continuous on (α, β).

---

[4] Such discontinuities are said to be *finite-jump* discontinuities.

PROOF: First of all, note that the two requirements placed on f ensure

$$g(x) = \int_a^x f(s)\,ds$$

is well defined for any x in (α, β) using any of the definitions for the integral found in most calculus texts (check this out yourself, using the definition in your calculus text). They also prevent $f(x)$ from "blowing up" on any closed subinterval $[\alpha', \beta']$ of (α, β). Thus, for each such closed subinterval $[\alpha', \beta']$, there is a corresponding finite constant M such that[5]

$$|f(s)| \leq M \qquad \text{whenever} \quad \alpha' \leq s \leq \beta' \quad .$$

Now, to verify the claimed continuity of g, we must show that

$$\lim_{x \to x_0} g(x) = g(x_0) \tag{2.15}$$

for any x_0 in (α, β). But by the definition of g and well-known properties of integration,

$$\begin{aligned}
\lim_{x \to x_0} g(x) &= \lim_{x \to x_0} \int_a^x f(s)\,ds \\
&= \lim_{x \to x_0} \left[\int_a^{x_0} f(s)\,ds + \int_{x_0}^x f(s)\,ds \right] \\
&= \lim_{x \to x_0} \left[g(x_0) + \int_{x_0}^x f(s)\,ds \right] = g(x_0) + \lim_{x \to x_0} \int_{x_0}^x f(s)\,ds \quad .
\end{aligned}$$

So, to show equation (2.15) holds, it suffices to confirm that

$$\lim_{x \to x_0} \int_{x_0}^x f(s)\,ds = 0 \quad ,$$

which, in turn, is equivalent to confirming that

$$\lim_{x \to x_0^+} \left| \int_{x_0}^x f(s)\,ds \right| = 0 \qquad \text{and} \qquad \lim_{x \to x_0^-} \left| \int_{x_0}^x f(s)\,ds \right| = 0 \quad . \tag{2.16}$$

To do this, pick any two finite values α' and β' satisfying $\alpha < \alpha' < x_0 < \beta' < \beta$. As noted, there is some finite constant $M \geq |f(s)|$ on $[\alpha', \beta']$. So, if $x_0 \leq x \leq \beta'$,

$$0 \leq \left| \int_{x_0}^x f(s)\,ds \right| \leq \int_{x_0}^x |f(s)|\,ds \leq \int_{x_0}^x M\,ds = M[x - x_0] \quad .$$

Similarly, if $\alpha' < x < x_0$, then

$$\begin{aligned}
0 \leq \left| \int_{x_0}^x f(s)\,ds \right| = \left| - \int_x^{x_0} f(s)\,ds \right| &= \left| \int_x^{x_0} f(s)\,ds \right| \\
&\leq \int_x^{x_0} |f(s)|\,ds \leq \int_x^{x_0} M\,ds = M[x_0 - x] \quad .
\end{aligned}$$

Hence,

$$0 \leq \lim_{x \to x_0^+} \left| \int_{x_0}^x f(s)\,ds \right| \leq \lim_{x \to x_0^+} M[x - x_0] = M[x_0 - x_0] = 0$$

and

$$0 \leq \lim_{x \to x_0^-} \left| \int_{x_0}^x f(s)\,ds \right| \leq \lim_{x \to x_0^-} M[x_0 - x] = M[x_0 - x_0] = 0 \quad ,$$

which, of course, means that equation set (2.16) holds.

[5] The constant M can be the maximum value of $|f(s)|$ on $[\alpha', \beta']$, provided that maximum exists. It may change if either endpoint α' or β' is changed.

Additional Exercises

2.2. *Determine whether each of the following differential equations is or is not directly integrable:*

a. $\frac{dy}{dx} = 3 - \sin(x)$

b. $\frac{dy}{dx} = 3 - \sin(y)$

c. $\frac{dy}{dx} + 4y = e^{2x}$

d. $x\frac{dy}{dx} = \arcsin(x^2)$

e. $y\frac{dy}{dx} = 2x$

f. $\frac{d^2y}{dx^2} = \frac{x+1}{x-1}$

g. $x^2\frac{d^2y}{dx^2} = 1$

h. $y^2\frac{d^2y}{dx^2} = 8x^2$

i. $\frac{d^2y}{dx^2} + 3\frac{dy}{dx} + 8y = e^{-x^2}$

j. $x^2\frac{d^2y}{dx^2} + 3x\frac{dy}{dx} = 0$

2.3. *Find a general solution for each of the following directly integrable equations. (Use indefinite integrals on these.)*

a. $\frac{dy}{dx} = 4x^3$

b. $\frac{dy}{dx} = 20e^{-4x}$

c. $x\frac{dy}{dx} + \sqrt{x} = 2$

d. $\sqrt{x+4}\,\frac{dy}{dx} = 1$

e. $\frac{dy}{dx} = x\cos\left(x^2\right)$

f. $\frac{dy}{dx} = x\cos(x)$

g. $x = \left(x^2 - 9\right)\frac{dy}{dx}$

h. $1 = \left(x^2 - 9\right)\frac{dy}{dx}$

i. $1 = x^2 - 9\frac{dy}{dx}$

j. $\frac{d^2y}{dx^2} = \sin(2x)$

k. $\frac{d^2y}{dx^2} - 3 = x$

l. $\frac{d^4y}{dx^4} = 1$

2.4. *Solve each of the following initial-value problems (using the indefinite integral). Also, state the largest interval over which the solution is valid (i.e., the maximal possible interval of interest).*

a. $\frac{dy}{dx} = 4x + 10e^{2x}$ *with* $y(0) = 4$

b. $\sqrt[3]{x+6}\,\frac{dy}{dx} = 1$ *with* $y(2) = 10$

c. $\frac{dy}{dx} = \frac{x-1}{x+1}$ *with* $y(0) = 8$

d. $x\frac{dy}{dx} + 2 = \sqrt{x}$ *with* $y(1) = 6$

e. $\cos(x)\frac{dy}{dx} - \sin(x) = 0$ *with* $y(0) = 3$

f. $\left(x^2 + 1\right)\frac{dy}{dx} = 1$ *with* $y(0) = 3$

g. $x\dfrac{d^2y}{dx^2} + 2 = \sqrt{x}$ with $y(1) = 8$ and $y'(1) = 6$

2.5 a. *Using definite integrals (as in Example 2.5 on page 25), find the general solution to*

$$\frac{dy}{dx} = \sin\left(\frac{x}{2}\right)$$

with $y(0)$ *acting as the arbitrary constant.*

b. *Using the formula just found for* $y(x)$ *:*

i. *Find* $y(\pi)$ *when* $y(0) = 0$.

ii. *Find* $y(\pi)$ *when* $y(0) = 3$.

iii. *Find* $y(2\pi)$ *when* $y(0) = 3$.

2.6 a. *Using definite integrals (as in Example 2.5 on page 25), find the general solution to*

$$\frac{dy}{dx} = 3\sqrt{x+3}$$

with $y(1)$ *acting as the arbitrary constant.*

b. *Using the formula just found for* $y(x)$ *:*

i. *Find* $y(6)$ *when* $y(1) = 16$.

ii. *Find* $y(6)$ *when* $y(1) = 20$.

iii. *Find* $y(-2)$ *when* $y(1) = 0$.

2.7. *Using definite integrals (as in Example 2.5 on page 25), find the solution to each of the following initial-value problems. (In some cases, you may want to use the error function or the sine-integral function.)*

a. $\dfrac{dy}{dx} = x\,e^{-x^2}$ with $y(0) = 3$

b. $\dfrac{dy}{dx} = \dfrac{x}{\sqrt{x^2+5}}$ with $y(2) = 7$

c. $\dfrac{dy}{dx} = \dfrac{1}{x^2+1}$ with $y(1) = 0$

d. $\dfrac{dy}{dx} = e^{-9x^2}$ with $y(0) = 1$

e. $x\dfrac{dy}{dx} = \sin(x)$ with $y(0) = 4$

f. $x\dfrac{dy}{dx} = \sin\left(x^2\right)$ with $y(0) = 0$

2.8. *Using an appropriate computer math package (such as Maple or Mathematica), graph each of the following over the interval* $0 \le x \le 10$ *:*

a. *the error function,* $\operatorname{erf}(x)$.

b. *the sine integral function,* $\operatorname{Si}(x)$.

c. *the solution to*

$$\frac{dy}{dx} = \ln\left|2 + x^2\sin(x)\right| \quad \text{with} \quad y(0) = 0 \quad .$$

2.9. *Each of the following differential equations involves a function that is (or can be) piecewise defined. Sketch the graph of each of these piecewise defined functions, and find the general solution of each differential equation. If an initial value is also given, then also solve the given initial-value problem:*

a. $\dfrac{dy}{dx} = \operatorname{step}(x)$ with $y(0) = 0$ and $\operatorname{step}(x)$ as defined on page 28

b. $\dfrac{dy}{dx} = f(x)$ with $y(0) = 2$ and $f(x) = \begin{cases} 0 & \text{if } x < 1 \\ 1 & \text{if } 1 \le x \end{cases}$

c. $\frac{dy}{dx} = f(x)$ *with* $y(0) = 0$ *and* $f(x) = \begin{cases} 0 & \textit{if} \quad x < 1 \\ 1 & \textit{if} \quad 1 \le x < 2 \\ 0 & \textit{if} \quad 2 \le x \end{cases}$

d. $\frac{dy}{dx} = |x - 2|$

e. $\frac{dy}{dx} = \text{stair}(x)$ *for* $x < 4$ *with* $y(0) = 0$ *and* $\text{stair}(x)$ *as defined on page 30*

Part II

First-Order Equations

3

Some Basics about First-Order Equations

For the next few chapters, our attention will be focused on first-order differential equations. We will discover that these equations can often be solved using methods developed directly from the tools of elementary calculus. And even when these equations cannot be explicitly solved, we will still be able to use fundamental concepts from elementary calculus to obtain good approximations to the desired solutions.

But first, let us discuss a few basic ideas that will be relevant throughout our discussion of first-order differential equations.

3.1 Algebraically Solving for the Derivative

Here are some of the first-order differential equations that we have seen or will see in the next few chapters:

$$x^2\frac{dy}{dx} - 4x = 6 \quad ,$$

$$\frac{dy}{dx} - x^2y^2 = x^2 \quad ,$$

$$\frac{dy}{dx} + 4xy = 2xy^2$$

and

$$x\frac{dy}{dx} + 4y = x^3 \quad .$$

One thing we can do with each of these equations is to algebraically solve for the derivative. Doing this with the first equation:

$$x^2\frac{dy}{dx} - 4x = 6$$

$$\hookrightarrow \qquad x^2\frac{dy}{dx} = 6 + 4x$$

$$\hookrightarrow \qquad \frac{dy}{dx} = \frac{4x+6}{x^2} \quad .$$

For the second equation:

$$\frac{dy}{dx} - x^2y^2 = x^2$$

$$\hookrightarrow \qquad \frac{dy}{dx} = x^2 + x^2y^2 \quad .$$

Solving for the derivative is often a good first step towards solving a first-order differential equation. For example, the first equation above is directly integrable — solving for the derivative yielded

$$\frac{dy}{dx} = \frac{4x+6}{x^2} \quad ,$$

and $y(x)$ can now be found by simply integrating both sides with respect to x.

Even when the equation is not directly integrable and we get

$$\frac{dy}{dx} = \text{“a formula of both } x \text{ and } y\text{”} \quad ,$$

— as in our second equation above,

$$\frac{dy}{dx} = x^2 + x^2y^2$$

— that formula on the right can still give us useful information about the possible solutions and can help us determine which method is appropriate for obtaining the general solution. Observe, for example, that the right-hand side of the last equation can be factored into a formula of x and a formula of y,

$$\frac{dy}{dx} = x^2\left(1+y^2\right) \quad .$$

In the next chapter, we will find that this means the equation is “separable” and can be solved by a procedure developed for just such equations.

For convenience, let us say that a first-order differential equation is in *derivative formula form* if it is written as

$$\frac{dy}{dx} = F(x,y) \tag{3.1}$$

where $F(x,y)$ is some (known) formula of x and/or y. Remember, to convert a given first-order differential equation to derivative form, simply use a little algebra to solve the differential equation for the derivative.

?►Exercise 3.1: *Verify that the derivative formula forms of*

$$\frac{dy}{dx} + 4y = 3y^3 \quad \text{and} \quad x\frac{dy}{dx} + 4xy = 2y^2$$

are

$$\frac{dy}{dx} = 3y^3 - 4y \quad \text{and} \quad \frac{dy}{dx} = \frac{2y^2 - 4xy}{x} \quad ,$$

respectively.

Keep in mind that the right side of equation (3.1), $F(x,y)$, need not always be a formula of *both* x and y. As we saw in an example above, the equation might be directly integrable. In this case, the right side of the above derivative formula form reduces to some $f(x)$, a formula involving only x,

$$\frac{dy}{dx} = f(x) \quad .$$

Alternatively, the right side may end up being a formula involving only y, $F(x,y) = g(y)$. We have a word for such differential equations; that word is “autonomous”. That is, an *autonomous* first-order differential equation is a differential equation that can be written as

$$\frac{dy}{dx} = g(y)$$

where $g(y)$ is some formula involving y but not x. The first equation in the last exercise is an example of an autonomous differential equation. Autonomous equations arise fairly often in applications, and the fact that dy/dx is given by a formula of just y will make an autonomous equation easier to graphically analyze in Chapter 9. But, as we'll see in the next chapter, they are just special cases of "separable" equations, and can be solved using the methods that will be developed there.

You should also be aware that the derivative formula form is not the only way we will attempt to rewrite our first-order differential equations. Frankly, much of the theory for first-order differential equations involves determining how a given differential equation can be rewritten so that we can cleverly apply tricks from calculus to further reduce the equation to something that can be easily integrated. We've already seen this with directly-integrable differential equations (for which the "derivative formula" form is ideal). In the next few chapters, we will see this with other equations for which other forms are useful.

By the way, there are first-order differential equations that cannot be put in derivative formula form. Consider

$$\frac{dy}{dx} + \sin\left(\frac{dy}{dx}\right) = x \quad .$$

It can be safely said that solving this equation for dy/dx is beyond the algebraic skills of most mortals. Fortunately, first-order differential equations that cannot be rewritten in the derivative formula form rarely arise in real-world applications.

3.2 Constant (or Equilibrium) Solutions

There is one type of particular solution that is easily determined for many first-order differential equations using elementary algebra: the "constant" solution.

A *constant solution* to a given differential equation is simply a constant function that satisfies that differential equation. Remember, y is a *constant function* if its value, $y(x)$, is some fixed constant for all x; that is, for some single number y_0,

$$y(x) = y_0 \qquad \text{for all} \quad x \quad .$$

Such solutions are also sometimes called *equilibrium solutions*. In an application involving some process that can vary with x, these solutions describe situations in which the process does not vary with x. This often means that all the factors influencing the process are "balancing out", leaving the process in a "state of equilibrium". As we will later see, this sometimes means that these solutions — whether called constant or equilibrium — are the most important solutions to a given differential equation.[1]

!►Example 3.1: *Consider the differential equation*

$$\frac{dy}{dx} = 2xy^2 - 4xy$$

and the constant function

$$y(x) = 2 \qquad \textit{for all} \quad x \quad .$$

[1] According to mathematical tradition, one only refers to a constant solution as an "equilibrium solution" if the differential equation is autonomous.

Since the derivative of a constant function is zero, plugging in this function, $y = 2$ *into*

$$\frac{dy}{dx} = 2xy^2 - 4xy$$

gives

$$0 = 2x \cdot 2^2 - 4x \cdot 2 \quad ,$$

which, after a little arithmetic and algebra, reduces further to

$$0 = 0 \quad .$$

Hence, our constant function satisfies our differential equation, and, so, is a constant solution to that differential equation.

On the other hand, plugging the constant function

$$y(x) = 3 \qquad \text{for all} \quad x$$

into

$$\frac{dy}{dx} = 2xy^2 - 4xy$$

gives

$$0 = 2x \cdot 3^2 - 4x \cdot 3 \quad .$$

This only reduces to

$$0 = 6x \quad ,$$

which is not valid for all values of x *on any nontrivial interval. Thus,* $y = 3$ *is not a constant solution to our differential equation.*

Admittedly, constant functions are not usually considered particularly exciting. The graph of a constant function,

$$y(x) = y_0 \qquad \text{for all} \quad x$$

is just a horizontal line (at $y = y_0$), and its derivative (as noted in the above example) is zero. But the fact that its derivative is zero is what simplifies the task of finding all possible constant solutions to a given differential equation, especially if the equation is in derivative formula form. After all, if we plug a constant function

$$y(x) = y_0 \qquad \text{for all} \quad x$$

into an equation of the form

$$\frac{dy}{dx} = F(x, y) \quad ,$$

then, since the derivative of a constant is zero, this equation reduces to

$$0 = F(x, y_0) \quad .$$

We can then determine all values y_0 that make $y = y_0$ a constant solution for our differential equation by simply determining every constant y_0 that satisfies

$$F(x, y_0) = 0 \qquad \text{for all} \quad x \quad .$$

!►Example 3.2: *Suppose we have a differential equation that, after a bit of algebra, can be written as*

$$\frac{dy}{dx} = (y - 2x)\left(y^2 - 9\right) \quad .$$

If it has a constant solution,

$$y(x) = y_0 \qquad \text{for all} \quad x \quad ,$$

then, after plugging this simple formula for y into the differential equation (and remembering that the derivative of a constant is zero), we get

$$0 = (y_0 - 2x)(y_0{}^2 - 9) \quad , \tag{3.2}$$

which is possible if and only if either

$$y_0 - 2x = 0 \qquad \text{or} \qquad y_0{}^2 - 9 = 0 \quad .$$

Now,

$$y_0 - 2x = 0 \iff y_0 = 2x \quad .$$

This gives us a value for y_0 that varies with x, contradicting the original assumption that y_0 was a constant. So this does not lead to any constant solutions (or any other solutions, either!). If there is such a solution, $y = y_0$, it must satisfy the other equation,

$$y_0{}^2 - 9 = 0 \quad .$$

But

$$y_0{}^2 - 9 = 0 \iff y_0{}^2 = 9 \iff y_0 = \pm\sqrt{9} = \pm 3 \quad .$$

So there are exactly two constant values for y_0, 3 and -3, that satisfy equation (3.2). And thus, our differential equation has exactly two constant (or equilibrium) solutions,

$$y(x) = 3 \qquad \text{for all} \quad x$$

and

$$y(x) = -3 \qquad \text{for all} \quad x \quad .$$

Keep in mind that, while the constant solutions to a given differential equation may be important, they rarely are the only solutions. And in practice, the solution to a given initial-value problem will typically *not* be one of the constant solutions. However, as we will see later, one of the constant solutions may tell us something about the long-term behavior of the solution to that particular initial-value problem. That is one of the reasons constant solutions are so important.

You should also realize that many differential equations have no constant solutions. Consider, for example, the directly-integrable differential equation

$$\frac{dy}{dx} = 2x \quad .$$

Integrating this, we get the general solution

$$y(x) = x^2 + c \quad .$$

No matter what value we pick for c, this function varies as x varies. It cannot be a constant.

In fact, it is not hard to see that no directly-integrable differential equation,

$$\frac{dy}{dx} = f(x) \quad ,$$

can have a constant solution (unless $f \equiv 0$). Just consider what you get when you integrate the $f(x)$. (That's why we did not mention such solutions when we discussed directly-integrable equations.)

3.3 On the Existence and Uniqueness of Solutions

Unfortunately, not all problems are solvable, and those that are solvable sometimes have several solutions. This is true in mathematics just as it is true in real life.

Before attempting to solve a problem involving some given differential equation and auxiliary condition (such as an initial value), it would certainly be nice to know that the given differential equation actually has a solution satisfying the given auxiliary condition. This would be especially true if the given differential equation looks difficult and we expect that considerable effort will be required in solving it (effort which would be wasted if that solution did not exist). And even if we can find a solution, we normally would like some assurance that it is the only solution.

The following theorem is the standard theorem quoted in most elementary differential equation texts addressing these issues for fairly general first-order initial-value problems.

Theorem 3.1 (on existence and uniqueness)
Consider a first-order initial-value problem

$$\frac{dy}{dx} = F(x, y) \qquad \text{with} \quad y(x_0) = y_0$$

in which both F and $\partial F/\partial y$ are continuous functions on some open region of the XY–plane containing the point (x_0, y_0).[2] *The initial-value problem then has exactly one solution over some open interval (α, β) containing x_0. Moreover, this solution and its derivative are continuous over that interval.*

This theorem assures us that, if we can write a first-order differential equation in the derivative formula form,

$$\frac{dy}{dx} = F(x, y) \quad ,$$

and that $F(x, y)$ is a 'reasonably well-behaved' formula on some region of interest, then our differential equation has solutions — with luck and skill, we will be able to find them. Moreover, if we can find a solution to this equation that also satisfies some initial value $y(x_0) = y_0$ corresponding to a point at which F is 'reasonably well-behaved', then that solution is unique (i.e., it is the only solution) — there is no need to worry about alternative solutions — at least over some interval (α, β). Just what that interval (α, β) is, however, is not explicitly described in this theorem. It turns out to depend in subtle ways on just how well behaved $F(x, y)$ is. More will be said about this in a few paragraphs.

!►*Example 3.3:* *Consider the initial-value problem*

$$\frac{dy}{dx} - x^2 y^2 = x^2 \qquad \text{with} \quad y(0) = 3 \quad .$$

As derived earlier, the derivative formula form for this equation is

$$\frac{dy}{dx} = x^2 + x^2 y^2 \quad .$$

So

$$F(x, y) = x^2 + x^2 y^2$$

[2] The $\partial F/\partial y$ is a "partial derivative". If you are not acquainted with partial derivatives see the appendix on page 59.

and

$$\frac{\partial F}{\partial y} = \frac{\partial}{\partial y}\left[x^2 + x^2y^2\right] = 0 + x^2 2y = 2x^2 y \quad .$$

It should be clear that these two functions are continuous everywhere on the XY*–plane. Hence, we can take the entire plane to be that "open region" in the above theorem, which then assures us that the above initial-value problem has one (and only one) solution valid over some interval* (a, b) *with* $a < 0 < b$ *. Unfortunately, the theorem doesn't tell us what that solution is nor what that interval* (a, b) *might be. We will have to wait until we develop a method for solving this differential equation.*

The proof of the above theorem is nontrivial and can be safely skipped by most beginning readers. In fact, despite the importance of the above theorem, we will rarely explicitly refer to it in the chapters that follow. The main explicit references will be a "graphical" discussion of the theorem in Chapter 9 using methods developed there,[3] and to note that analogous theorems can be proven for higher-order differential equations. Nonetheless, it is an important theorem whose proof should be included in this text if only to assure you that the author is not making it up. Besides, the basic core of the proof is fairly accessible to most readers and contains some clever and interesting ideas. We will go over that basic core in the next section (Section 3.4), leaving the more challenging details for the section after that (Section 3.5).

Part of the proof will be to identify the interval (α, β) mentioned in the above theorem. In fact, the interval (α, β) can be easily determined if F and ${}^{\partial F}/_{\partial y}$ are sufficiently well behaved. That is what the next theorem gives us. Its proof requires just a few modifications of the proof of the above, and will be briefly discussed after that proof.

Theorem 3.2
Consider a first-order initial-value problem

$$\frac{dy}{dx} = F(x, y) \qquad \text{with} \quad y(x_0) = y_0$$

over an interval (α, β) *containing* x_0 *, and with* $F = F(x, y)$ *being a continuous function on the infinite strip*

$$\mathcal{R} = \{\, (x, y) : \alpha < x < \beta \text{ and } -\infty < y < \infty \,\} \quad .$$

Further suppose that, on $\mathcal{R}$ *, the partial derivative* ${}^{\partial F}/_{\partial y}$ *is continuous and is a function of* x *only.*[4] *Then the initial-value problem has exactly one solution over* (α, β) *. Moreover, this solution and its derivative are continuous on that interval.*

In practice, many of our first-order differential equations will not satisfy the conditions described in the last theorem. So this theorem is of relatively limited value for now. However, it leads to higher-order analogs that will be used in developing the theory needed for important classes of higher-order differential equations. That is why Theorem 3.2 is mentioned here.

[3] which you may find more illuminating than the proof given here

[4] More generally, the theorem remains true if we replace the phrase "a function of x only" with "a bounded function on $\mathcal{R}$". Our future interest, however, will be with the theorem as stated.

3.4 Confirming the Existence of Solutions (Core Ideas)

So let us consider the first-order initial-value problem

$$\frac{dy}{dx} = F(x, y) \qquad \text{with} \quad y(x_0) = y_0 \quad ,$$

assuming that both F and ${}^{\partial F}/_{\partial y}$ are continuous on some open region in the XY–plane containing the point (x_0, y_0). Our goal is to verify that a solution y exists over some interval. (This is the existence claim of Theorem 3.1. The uniqueness claim of that theorem will be left as an exercise using material developed in the next section — see Exercise 3.2 on page 56.)

The gist of our proof consists of three steps:

1. Observe that the initial-value problem is equivalent to a corresponding integral equation.
2. Derive a sequence of functions — ψ_0, ψ_1, ψ_2, ψ_3, ... — using a formula inspired by that integral equation.
3. Show that this sequence of functions converges on some interval to a solution y of the original initial-value problem.

The "hard" part of the proof is in the details of the last step. We can skip over these details initially, returning to them in the next section.

Two comments should be made here:

1. The ψ_k's end up being approximations to the solution y, and, in theory at least, the method we are about to describe can be used to find approximate solutions to an initial-value problem. Other methods, however, are often more practical.
2. This method was developed by the French mathematician Emile Picard and is often referred to as the *(Picard's) method of successive approximations* or as *Picard's iterative method* (because of the way the ψ_k's are generated).

To simplify discussion let us assume $x_0 = 0$, so that our initial-value problem is

$$\frac{dy}{dx} = F(x, y) \qquad \text{with} \quad y(0) = y_0 \quad . \tag{3.3}$$

There is no loss of generality here. After all, if $x_0 \neq 0$, we can apply the change of variable $s = x - x_0$ and convert our original problem into problem (3.3) (with x replaced by s).

Converting to an Integral Equation

Suppose $y = y(x)$ is a solution to initial-value problem (3.3) on some interval (α, β) with $\alpha < 0 < \beta$. Renaming x as s, our differential equation becomes

$$\frac{dy}{ds} = F(s, y(s)) \qquad \text{for each } s \text{ in } (\alpha, \beta) \quad .$$

Integrating this from 0 to any x in (α, β) and remembering that $y(0) = y_0$, we get

$$\int_0^x \frac{dy}{ds}\, ds = \int_0^x F(s, y(s))\, ds$$

$$\hookrightarrow \qquad y(x) - y(0) = \int_0^x F(s, y(s))\, ds$$

$$\hookrightarrow \qquad y(x) - y_0 = \int_0^x F(s, y(s))\, ds \quad .$$

That is, y satisfies the integral equation

$$y(x) = y_0 + \int_0^x F(s, y(s))\,ds \qquad \text{whenever} \quad \alpha < x < \beta \quad .$$

On the other hand, if y is any continuous function on (α, β) satisfying this integral equation, then basic calculus tells us that, on this interval, y is differentiable with

$$\frac{dy}{dx} = \frac{d}{dx}\left[y_0 + \int_0^x F(s, y(s))\,ds\right] = 0 + \frac{d}{dx}\int_0^x F(s, y(s))\,ds = F(x, y(x)) \quad .$$

and

$$y(0) = y_0 + \underbrace{\int_0^0 F(s, y(s))\,ds}_{0} = y_0 \quad .$$

Thus, y also satisfies our original initial-value problem.

We should note that, in the above, we implicitly assumed $F(x, y)$ was a reasonably behaved function at each point (x, y) where $\alpha < x < \beta$ and $y = y(x)$. In particular, if F is continuous at each of these points, then this continuity, the continuity of y, and the fact that $y' = F(x, y)$ ensures that y is not only differentiable on (α, β) but that y' is continuous on (α, β).

In summary, we have the following theorem:

Theorem 3.3

Let y be any continuous function on some interval (α, β) containing 0, and assume F is a function of two variables continuous at every (x, y) with $\alpha < x < \beta$ and $y = y(x)$. Then y has a continuous derivative on (α, β) and satisfies the initial-value problem

$$\frac{dy}{dx} = F(x, y) \qquad \text{with} \quad y(0) = y_0 \quad \text{on} \quad (\alpha, \beta)$$

if and only if y satisfies the integral equation

$$y(x) = y_0 + \int_0^x F(s, y(s))\,ds \qquad \text{whenever} \quad \alpha < x < \beta \quad .$$

Generating a Sequence of "Approximate Solutions"

Begin with any continuous function ψ_0. For example, we could simply choose ψ_0 to be the constant function

$$\psi_0(x) = y_0 \qquad \text{for all} \quad x \quad .$$

(Later, we will place some additional restrictions on ψ_0, but the above constant function will still be a valid choice for ψ_0.)

Next, let ψ_1 be the function constructed from ψ_0 by

$$\psi_1(x) = y_0 + \int_0^x F(s, \psi_0(s))\,ds \quad .$$

Then construct ψ_2 from ψ_1 via

$$\psi_2(x) = y_0 + \int_0^x F(s, \psi_1(s))\,ds \quad .$$

Continue the process, defining ψ_3, ψ_4, ψ_5, ... by

$$\psi_3(x) = y_0 + \int_0^x F(s, \psi_2(s))\,ds \quad ,$$

$$\psi_4(x) = y_0 + \int_0^x F(s, \psi_3(s))\,ds \quad ,$$

$$\vdots$$

In general, once ψ_k is defined, we define ψ_{k+1} by

$$\psi_{k+1}(x) = y_0 + \int_0^x F(s, \psi_k(s))\,ds \quad . \tag{3.4}$$

Since we apparently can continue this iterative process forever, we have an infinite sequence of functions

$$\psi_0 \quad , \quad \psi_1 \quad , \quad \psi_2 \quad , \quad \psi_3 \quad , \quad \psi_4 \quad , \quad \ldots \quad .$$

In the future, we may refer to this sequence as the *Picard sequence* (based on ψ_0 and F). Note that, for $k = 1, 2, 3, \ldots$,

$$\psi_k(0) = y_0 + \underbrace{\int_0^0 F(s, \psi_{k-1}(s))\,ds}_{0} = y_0 \quad .$$

So each of these ψ_k's satisfies the initial condition in our initial-value problem. Moreover, since each of these ψ_k's is a constant added to an integral from 0 to x, each of these ψ_k's should be continuous at least over the interval of x's on which the integral is finite.

(Naively) Taking the Limit

Now suppose there is an interval (α, β) containing 0 on which this sequence of ψ_k's converges to some continuous function. Let y denote this function,

$$y(x) = \lim_{k\to\infty} \psi_k(x) \qquad \text{for} \quad \alpha < x < \beta \quad .$$

Now let x be any point in (a, b). Blithely (and naively) taking the limit of both sides of equation (3.4), we get

$$\begin{aligned}
y(x) = \lim_{k\to\infty} \psi_k(x) &= \lim_{k\to\infty} \psi_{k+1}(x) \\
&= \lim_{k\to\infty} \left[y_0 + \int_0^x F(s, \psi_k(s))\,ds \right] \\
&= y_0 + \lim_{k\to\infty} \int_0^x F(s, \psi_k(s))\,ds \\
&= y_0 + \int_0^x \lim_{k\to\infty} F(s, \psi_k(s))\,ds \\
&= y_0 + \int_0^x F(s, y(s))\,ds \quad .
\end{aligned}$$

Thus (assuming the above limits are valid) we see that y satisfies the integral equation

$$y(x) = y_0 + \int_0^x F(s, y(s))\,ds \qquad \text{for} \quad a < x < b \quad .$$

As noted in Theorem 3.3, this means the function y is a solution to our original initial-value problem, thus verifying the claimed existence of such a solution.

That was the essence of Picard's method of successive approximations.

3.5 Details in the Proof of Theorem 3.1

Confirming the Existence of Solutions

What Are the Remaining Details?

Before proclaiming that we have rigorously verified the existence of a solution to our initial-value problem via the Picard method, we need to rigorously verify the assumptions made in the last section. If you check carefully, you will see that we still need to rigorously confirm the following three statements concerning the functions ψ_1, ψ_2, ... generated by the Picard iteration method:

1. There is an interval (α, β) containing 0 such that

$$\lim_{k\to\infty} \psi_k(x)$$

exists for each x in (α, β).

2. The function given by

$$y(x) = \lim_{k\to\infty} \psi_k(x)$$

is continuous on the interval (α, β).

3. The above defined function y satisfies

$$y(x) = y_0 + \int_0^x F(s, y(s))\, ds \qquad \text{whenever} \quad \alpha < x < \beta \quad .$$

Confirming these claims under the assumptions in Theorem 3.1 on page 42 will be the main goal of this section.[5]

Some Preliminary Bounds

In carrying out our analysis, we will make use of a number of facts normally discussed in standard introductory calculus courses. For example, we will use without comment that fact that, for any summation,

$$\left|\sum_k c_k\right| \leq \sum_k |c_k| \quad .$$

This is the triangle inequality. Recall, also, that "the absolute value of an integral is less than or equal to the integral of the absolute value". We will need to be a little careful about this because the lower limits on our integrals will not always be less than our upper limits. If $\sigma < \tau$, then we do have

$$\left|\int_\sigma^\tau g(s)\, ds\right| \leq \int_\sigma^\tau |g(s)|\, ds \quad .$$

[5] Some of the analysis in this section can be shortened considerably using tools from advanced real analysis. Since the typical reader is not expected to have yet had a such a course, we will not use those tools. However, if you have had such a course and are acquainted with such terms as "uniform convergence" and "Cauchy sequences", then you should look to see how your more advanced mathematics can shorten the analysis given here.

On the other hand, if $\tau < \sigma$, then

$$\left| \int_{\sigma}^{\tau} g(s)\, ds \right| = \left| - \int_{\tau}^{\sigma} g(s)\, ds \right| = \left| \int_{\tau}^{\sigma} g(s)\, ds \right| \leq \int_{\tau}^{\sigma} |g(s)|\, ds \quad .$$

Suppose that, in addition, $|g(s)| \leq M$ for all s in some interval containing σ and τ . Then, if $\sigma < \tau$,

$$\left| \int_{\sigma}^{\tau} g(s)\, ds \right| \leq \int_{\sigma}^{\tau} |g(s)|\, ds \leq \int_{\sigma}^{\tau} M\, ds = M[\tau - \sigma] = M\,|\tau - \sigma| \quad ,$$

while, if $\tau < \sigma$,

$$\left| \int_{\sigma}^{\tau} g(s)\, ds \right| \leq \int_{\tau}^{\sigma} |g(s)|\, ds \leq \int_{\tau}^{\sigma} M\, ds = M[-(\tau - \sigma)] = M\,|\tau - \sigma| \quad .$$

So, in general, we have the following little lemma:

Lemma 3.4
If $|g(s)| \leq M$ for all s in some interval containing σ and τ , then

$$\left| \int_{\sigma}^{\tau} g(s)\, ds \right| = M\,|\tau - \sigma| \quad .$$

Other facts from calculus will be used and slightly expanded as needed. These facts will include material on the absolute convergence of summations and the Taylor series for the exponentials.

The next lemma establishes the interval (α, β) mentioned in the existence theorem (Theorem 3.1) along with some function bounds that will be useful in our analysis.

Lemma 3.5
Assume both $F(x, y)$ and ${}^{\partial F}/_{\partial y}$ are continuous on some open region $\mathcal{R}$ in the XY–plane containing the point $(0, y_0)$. Then there are positive constants M and B , a closed interval $[\alpha, \beta]$ and a finite distance ΔY such that all the following hold:

1. $\alpha < 0 < \beta$.

2. *The open region $\mathcal{R}$ contains the closed rectangular region*

$$\mathcal{R}_1 - \{ (x, y) : \alpha \leq x \leq \beta \text{ and } |y - y_0| \leq \Delta Y \} \quad .$$

3. *For each (x, y) in $\mathcal{R}_1$,*

$$|F(x, y)| \leq M \qquad \text{and} \qquad \left| \left. \frac{\partial F}{\partial y} \right|_{(x,y)} \right| \leq B \quad .$$

4. $0 < -\alpha M \leq \Delta Y$ *and* $0 < \beta M \leq \Delta Y$.

5. *If ϕ is a continuous function on (α, β) satisfying*

$$|\phi(x) - y_0| \leq \Delta Y \qquad \text{for} \quad \alpha \leq x \leq \beta \quad ,$$

then

$$\psi(x) = y_0 + \int_0^x F(s, \phi(s))\, ds$$

defines the function ψ on the interval $[\alpha, \beta]$. Moreover, ψ is continuous on $[\alpha, \beta]$ and satisfies

$$|\psi(x) - y_0| \leq \Delta Y \qquad \text{for} \quad \alpha \leq x \leq \beta \quad .$$

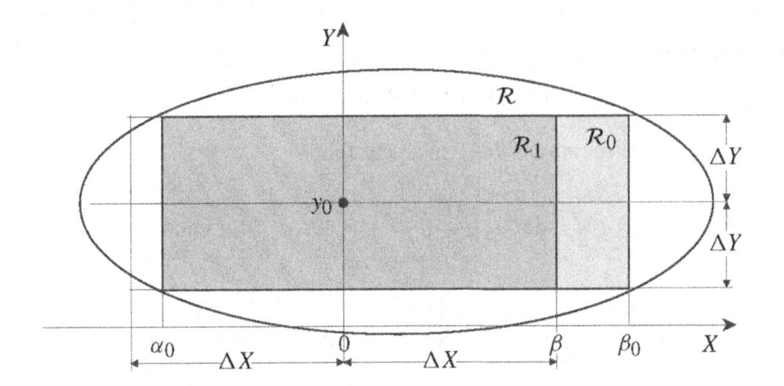

Figure 3.1: Rectangles contained in region $\mathcal{R}$ for the proof of Lemma 3.5 (with $|\alpha_0| < \Delta X$ and $\Delta X < \beta_0$).

PROOF: The goal is to find a rectangle $\mathcal{R}_1$ on which the above holds. We start by noting that, because $\mathcal{R}$ is an open region containing the point $(0, y_0)$, that point is not on the boundary of $\mathcal{R}$, and we can pick a negative value α_0 and two positive values β_0 and ΔY so that the closed rectangular region

$$\mathcal{R}_0 = \{ (x, y) : \alpha_0 \le x \le \beta_0 \text{ and } |y - y_0| \le \Delta Y \}$$

is contained in $\mathcal{R}$, as in Figure 3.1.

Since F and ${}^{\partial F}/_{\partial y}$ are continuous on $\mathcal{R}$, they (and their absolute values) must be continuous on that portion of $\mathcal{R}$ which is $\mathcal{R}_0$. But recall that a continuous function of one variable on a closed finite interval will always have a maximum value on that interval. Likewise, a continuous function of two variables will always have a maximum value over a closed finite rectangle. Let M and B be, respectively, the maximum values of $|F|$ and $\left|{}^{\partial F}/_{\partial y}\right|$ on $\mathcal{R}_0$. Then, of course,

$$|F(x, y)| \le M \quad \text{and} \quad \left| \left.\frac{\partial F}{\partial y}\right|_{(x,y)} \right| \le B \qquad \text{for each } (x, y) \text{ in } \mathcal{R}_0 \quad .$$

Now let us further restrict the possible values of x by first setting

$$\Delta X = \frac{\Delta Y}{M} \qquad \left(\text{so } M = \frac{\Delta Y}{\Delta X}\right) \quad ,$$

and then defining the endpoints of the interval (α, β) by

$$\alpha = \begin{cases} \alpha_0 & \text{if } -\Delta X < \alpha_0 \\ -\Delta X & \text{if } \alpha_0 \le -\Delta X \end{cases} \quad \text{and} \quad \beta = \begin{cases} \Delta X & \text{if } \Delta X < \beta_0 \\ \beta_0 & \text{if } \beta_0 \le \Delta X \end{cases}$$

(again, see Figure 3.1).

By these choices,

$$\alpha_0 \le \alpha < 0 < \beta \le \beta_0 \quad ,$$

$$|x| \le \Delta X \qquad \text{whenever} \quad \alpha \le x \le \beta \quad ,$$

$$0 < -\alpha M \le \Delta X M = \Delta Y \quad ,$$

$$0 < \beta M \le \Delta X M = \Delta Y \quad ,$$

and the closed rectangle

$$\mathcal{R}_1 = \{ (x, y) : \alpha \leq x \leq \beta \text{ and } |y - y_0| \leq \Delta Y \}$$

is contained in the closed rectangle $\mathcal{R}_0$, ensuring that

$$|F(x, y)| \leq M \quad \text{and} \quad \left| \left. \frac{\partial F}{\partial y} \right|_{(x,y)} \right| \leq B \qquad \text{for each } (x, y) \text{ in } \mathcal{R}_1 \quad .$$

This takes care of the first four claims of the lemma.

To confirm the lemma's final claim, let ϕ be a continuous function on (α, β) satisfying

$$|\phi(x) - y_0| \leq \Delta Y \qquad \text{for} \quad \alpha \leq x \leq \beta \quad .$$

Then, $(s, \phi(s))$ is a point in $\mathcal{R}_1$ for each s in the interval $[\alpha, \beta]$. This, in turn, means that $F(s, \phi(s))$ exists and is bounded by M over the interval $[\alpha, \beta]$. Moreover, it is easily verified that the continuity of both F over $\mathcal{R}$ and ϕ over (α, β) ensures that $F(s, \phi(s))$ is a bounded continuous function of s over $[\alpha, \beta]$. Consequently, the integral in

$$\psi(x) = y_0 + \int_0^x F(s, \phi(s))\, ds$$

exists (and is finite) for each x in $[\alpha, \beta]$.

To help confirm the claimed continuity of ψ, take any two points x and x_1 in (α, β). Using Lemma 3.4 and the fact that F is bounded by M on $\mathcal{R}_1$, we have that

$$\begin{aligned} |\psi(x_1) - \psi(x)| &= \left| y_0 + \int_0^{x_1} F(s, \phi(s))\, ds - y_0 - \int_0^x F(s, \phi(s))\, ds \right| \\ &= \left| \int_x^{x_1} F(s, \phi(s))\, ds \right| \\ &\leq M\, |x_1 - x| \quad . \end{aligned}$$

Hence,

$$\lim_{x \to x_1} |\psi(x_1) - \psi(x)| \leq \lim_{x \to x_1} M\, |x_1 - x| = M \cdot 0 = 0 \quad ,$$

which, in turn, means that

$$\lim_{x \to x_1} \psi(x) = \psi(x_1) \quad ,$$

confirming that ψ is continuous at each x_1 in (α, β). By almost identical arguments, we also have

$$\lim_{x \to \alpha^+} \psi(x) = \psi(\alpha) \qquad \text{and} \qquad \lim_{x \to \beta^-} \psi(x) = \psi(\beta) \quad .$$

Altogether, these limits tell us that ψ is continuous on the closed interval $[\alpha, \beta]$.

Finally, let $\alpha \leq x \leq \beta$. Again using Lemma 3.4 and the boundedness of F, along with the definition of ΔX, we see that

$$|\psi(x) - y_0| = \left| \int_0^x F(s, \phi(s))\, ds \right| \leq M\, |x| \leq M \Delta X = M \cdot \frac{\Delta Y}{M} = \Delta Y \quad . \qquad \blacksquare$$

Convergence of the Picard Sequence

Let us now look more closely at the Picard sequence of functions,

$$\psi_0 \ , \ \psi_1 \ , \ \psi_2 \ , \ \psi_3 \ , \ \ldots$$

with ψ_0 being "some continuous function" and

$$\psi_{k+1}(x) = y_0 + \int_0^x F(s, \psi_k(s))\, ds \qquad \text{for} \quad k = 0, 1, 2, 3, \ldots \quad .$$

Remember, F and $\partial F/\partial y$ are continuous on some open region containing the point $(0, y_0)$. This means Lemma 3.5 applies. Let $[\alpha, \beta]$, M, B and ΔY be the interval and constants from that lemma. Let us also now impose an additional restriction on the choice for ψ_0: Let us insist that ψ_0 be any continuous function on $[\alpha, \beta]$ such that

$$|\psi_0(x) - y_0| \leq \Delta Y \qquad \text{for} \quad \alpha < x < \beta \quad .$$

In particular, we could let ψ_0 be the constant function $\psi_0(x) = y_0$ for all x.

We now want to show that the sequence of ψ_k's converges to a function y on $[\alpha, \beta]$. Our first step in this direction is to observe that, thanks to the additional requirement on ψ_0, Lemma 3.5 can be applied repeatedly to show that ψ_1, ψ_2, ψ_3, ... are all well-defined, continuous functions on the interval $[\alpha, \beta]$ with each satisfying

$$|\psi_k(x) - y_0| \leq \Delta Y \qquad \text{for} \quad \alpha \leq x \leq \beta \quad .$$

Next, we need to establish useful bounds on the sequence

$$|\psi_1(x) - \psi_0(x)| \quad , \quad |\psi_2(x) - \psi_1(x)| \quad , \quad |\psi_3(x) - \psi_2(x)| \quad , \quad \ldots$$

when $\alpha \leq x \leq \beta$. The first is easy:

$$\begin{aligned} |\psi_1(x) - \psi_0(x)| &= |\psi_1(x) - y_0 - \psi_0(x) + y_0| \\ &= |[\psi_1(x) - y_0] + (-[\psi_0(x) - y_0])| \\ &\leq |\psi_1(x) - y_0| + |\psi_0(x) - y_0| \leq 2\Delta Y \quad . \end{aligned}$$

To simplify the derivation of useful bounds on the others, let us observe that, if $k \geq 1$,

$$\begin{aligned} |\psi_{k+1}(x) - \psi_k(x)| &= \left|\left[y_0 + \int_0^x F(s, \psi_k(s))\, ds\right] - \left[y_0 + \int_0^x F(s, \psi_{k-1}(s))\, ds\right]\right| \\ &= \left|\int_0^x [F(s, \psi_k(s)) - F(s, \psi_{k-1}(s))]\, ds\right| \\ &\leq \int_0^x |F(s, \psi_k(s)) - F(s, \psi_{k-1}(s))|\, ds \quad . \end{aligned}$$

Now recall that, if f is any continuous and differentiable function on an interval $\mathcal{I}$, and t_1 and t_2 are two points in $\mathcal{I}$, then there is a point τ between t_1 and t_2 such that

$$f(t_2) - f(t_1) = f'(\tau)\,[t_2 - t_1] \quad .$$

This was the mean value theorem for derivatives. Consequently, if

$$\left|f'(t)\right| \leq B \qquad \text{for each } t \text{ in } \mathcal{I} \quad ,$$

then

$$|f(t_2) - f(t_1)| = \left|f'(\tau)\,[t_2 - t_1]\right| = \left|f'(\tau)\right| |t_2 - t_1| \le B\,|t_2 - t_1| \quad .$$

The same holds for partial derivatives. In particular, for each pair of points (x, y_1) and (x, y_2) in the closed rectangle

$$\mathcal{R}_1 = \{\,(x, y) : \alpha \le x \le \beta \text{ and } |y - y_0| \le \Delta Y\,\} \quad ,$$

we have a γ between y_1 and y_2 such that

$$|F(x, y_2) - F(x, y_1)| = \left|\left.\frac{\partial F}{\partial y}\right|_{(x,\gamma)} \cdot [y_2 - y_1]\right| \le B\,|y_2 - y_1| \quad .$$

Thus, for $0 \le x \le \beta$ and $k = 1, 2, 3, \ldots$,

$$\begin{aligned} |\psi_{k+1}(x) - \psi_k(x)| &\le \int_0^x |F(s, \psi_k(s)) - F(s, \psi_{k-1}(s))|\, ds \\ &\le \int_0^x B\,|\psi_k(s) - \psi_{k-1}(s)|\, ds \quad . \end{aligned}$$

Repeatedly using this (with $0 \le x \le \beta$), we get

$$\begin{aligned} |\psi_2(x) - \psi_1(x)| &\le \int_0^x B\,|\psi_1(s) - \psi_0(s)|\, ds \\ &\le \int_0^x B \cdot 2\Delta Y\, ds = 2\Delta Y B x \quad , \end{aligned}$$

$$\begin{aligned} |\psi_3(x) - \psi_2(x)| &\le \int_0^x B\,|\psi_2(s) - \psi_1(s)|\, ds \\ &\le \int_0^x B \cdot 2\Delta Y B\, s\, ds = 2\Delta Y \frac{(Bx)^2}{2} \quad , \end{aligned}$$

$$\begin{aligned} |\psi_4(x) - \psi_3(x)| &\le \int_0^x |\psi_3(s) - \psi_2(s)|\, ds \\ &\le \int_0^x B \cdot 2\Delta Y B^2 \frac{s^2}{2}\, ds \le 2\Delta Y \frac{(Bx)^3}{3 \cdot 2} \quad , \end{aligned}$$

$$\begin{aligned} |\psi_5(x) - \psi_4(x)| &\le \int_0^x |\psi_4(s) - \psi_3(s)|\, ds \\ &\le \int_0^x B \cdot 2\Delta Y B^3 \frac{s^3}{3 \cdot 2}\, ds \le 2\Delta Y \frac{(Bx)^4}{4!} \quad , \end{aligned}$$

$$\vdots$$

Continuing, we get

$$|\psi_{k+1}(x) - \psi_k(x)| \le 2\Delta Y \frac{(Bx)^k}{k!} \qquad \text{for} \quad 0 \le x \le \beta \quad \text{and} \quad k = 1, 2, 3, \ldots \quad .$$

Virtually the same arguments give us

$$|\psi_{k+1}(x) - \psi_k(x)| \le 2\Delta Y \frac{(-Bx)^k}{k!} \qquad \text{for} \quad \alpha \le x \le 0 \quad \text{and} \quad k = 1, 2, 3, \ldots \quad .$$

More concisely, for $\alpha \leq x \leq \beta$ and $k = 1, 2, 3, \ldots$,

$$|\psi_{k+1}(x) - \psi_k(x)| \leq 2\Delta Y \frac{(B|x|)^k}{k!} \quad . \tag{3.5}$$

At this point it is worth recalling that the Taylor series for e^X is

$$\sum_{k=0}^{\infty} \frac{X^k}{k!}$$

and that this series converges for each real value X. In particular, for any x,

$$2\Delta Y e^{B|x|} = \sum_{k=0}^{\infty} 2\Delta Y \frac{(B|x|)^k}{k!} \quad .$$

Now consider the infinite series

$$S(x) = \sum_{k=0}^{\infty} \left[\psi_{k+1}(x) - \psi_k(x)\right] \quad .$$

According to inequality (3.5), the absolute value of each term in this series is bounded by the corresponding term in the Taylor series for $2\Delta Y e^{B|x|}$. The comparison test then tells us that $S(x)$ converges absolutely for each x in $[\alpha, \beta]$. And this means that the limit

$$S(x) = \lim_{N\to\infty} \sum_{k=0}^{N} \left[\psi_{k+1}(x) - \psi_k(x)\right]$$

exists for each x in the interval $[\alpha, \beta]$. But

$$\begin{aligned}
\sum_{k=0}^{N} \left[\psi_{k+1}(x) - \psi_k(x)\right] &= [\psi_1(x) - \psi_0(x)] + [\psi_2(x) - \psi_1(x)] + [\psi_3(x) - \psi_2(x)] \\
&\quad + \cdots + [\psi_N(x) - \psi_{N-1}(x)] + [\psi_{N+1}(x) - \psi_N(x)] \\
&= -\psi_0(x) + \psi_1(x) - \psi_1(x) + \psi_2(x) - \psi_2(x) + \psi_3(x) \\
&\quad + \cdots - \psi_{N-1}(x) + \psi_N(x) - \psi_N(x) + \psi_{N+1}(x) \quad .
\end{aligned}$$

Most of the terms cancel out, leaving us with

$$\sum_{k=0}^{N} \left[\psi_{k+1}(x) - \psi_k(x)\right] = \psi_{N+1}(x) - \psi_0(x) \quad . \tag{3.6}$$

So

$$\lim_{k\to\infty} \psi_k(x) = \lim_{k\to\infty} \left[\psi_0(x) + \sum_{k=0}^{k-1} \left[\psi_{k+1}(x) - \psi_k(x)\right]\right] = \psi_0(x) + S(x) \quad .$$

This shows that the limit

$$y(x) = \lim_{N\to\infty} \psi_N(x)$$

exists for each x in $[\alpha, \beta]$, confirming the first statement we wished to confirm at the beginning of this section (see page 47).

At this point, let us observe that, for $\alpha \leq x \leq \beta$, we have the formulas

$$\psi_N(x) = \psi_0(x) + S(x) = \psi_0(x) + \sum_{k=0}^{N-1} \left[\psi_{k+1}(x) - \psi_k(x)\right] \tag{3.7a}$$

and

$$y(x) = \psi_0(x) + S(x) = \psi_0(x) + \sum_{k=0}^{\infty} \left[\psi_{k+1}(x) - \psi_k(x)\right] \quad . \tag{3.7b}$$

Let us also observe what we get when we combine the above formula for ψ_N with inequality (3.5) and the observations regarding the Taylor series of the exponential:

$$\begin{aligned} |\psi_N(x)| &\leq |\psi_0(x)| + \sum_{k=0}^{N-1} |\psi_{k+1}(x) - \psi_k(x)| \\ &\leq |\psi_0(x)| + \sum_{k=0}^{N} 2\Delta Y \frac{(B\,|x|)^k}{k!} = |\psi_0(x)| + \Delta Y e^{B|x|} \quad . \end{aligned} \tag{3.8a}$$

Likewise

$$|y(x)| \leq |\psi_0(x)| + \Delta Y e^{B|x|} \quad . \tag{3.8b}$$

These observations may later prove useful.

Continuity of the Limit

Now to confirm the continuity of y claimed by the second statement from the beginning of this section. We start by picking any two points x_1 and x in $[\alpha, \beta]$, and any positive integer N, and then observe that, because F is bounded by M,

$$\begin{aligned} |\psi_N(x_1) - \psi_N(x)| &= \left|\left[y_0 + \int_0^{x_1} F(s, \psi_{N-1}(s))\,ds\right] - \left[y_0 + \int_0^{x} F(s, \psi_{N-1}(s)\,ds\right]\right| \\ &= \left|\int_x^{x_1} F(s, \psi_{N-1}(s))\,ds\right| \\ &\leq M\,|x_1 - x| \quad . \end{aligned}$$

Combined with the definition of y and some basic facts about limits, this gives us

$$|y(x_1) - y(x)| = \lim_{N\to\infty} |\psi_N(x_1) - \psi_N(x)| \leq M\,|x_1 - x| \quad .$$

As demonstrated at the end of the proof of Lemma 3.5, this immediately tells us that y is continuous on $[\alpha, \beta]$.

The Limit as a Solution

Finally, let us verify the third statement made at the beginning of this section, namely that the above defined y satisfies

$$y(x) = y_0 + \int_0^x F(s, y(s))\,ds \qquad \text{whenever} \quad \alpha < x < \beta \quad .$$

This, according to Theorem 3.3 on page 45, is equivalent to showing that y satisfies the differential equation in our initial-value problem over the interval (α, β).[6]

[6] Yes, we've already shown that y is defined and continuous on $[\alpha, \beta]$, not just (α, β). However, the derivative of a function is ill-defined at the endpoints of the interval over which it is defined, and that is why we are now limiting x to being in (α, β).

We start by assuming $\alpha \leq x \leq \beta$. Using equation set (3.7) and inequality (3.5), we see that

$$\begin{aligned}
|y(x) - \psi_N(x)| &= |[y(x) - \psi_0(x)] - [\psi_N(x) - \psi_0(x)]| \\
&= \left|\sum_{k=0}^{\infty}[\psi_{k+1}(x) - \psi_k(x)] - \sum_{k=0}^{N-1}[\psi_{k+1}(x) - \psi_k(x)]\right| \\
&= \left|\sum_{k=N}^{\infty}[\psi_{k+1}(x) - \psi_k(x)]\right| \\
&\leq \sum_{k=N}^{\infty}|\psi_{k+1}(x) - \psi_k(x)| \\
&\leq \sum_{k=N}^{\infty} 2\Delta Y \frac{(B\,|x|)^k}{k!} \quad .
\end{aligned}$$

Under the change of index $k = N + n$, this becomes

$$|y(x) - \psi_N(x)| \leq 2\Delta Y \sum_{n=0}^{\infty} \frac{(B\,|x|)^{N+n}}{(N+n)!} \quad . \tag{3.9}$$

But

$$\begin{aligned}
(N+n)! &= \underbrace{(N+n)}_{\geq N}\underbrace{(N+n-1)}_{\geq N-1}\underbrace{(N+n-2)}_{\geq N-2}\cdots\underbrace{(N+n-[N-1])}_{\geq 1}\underbrace{n(n-1)\cdots 2\cdot 1}_{=n!} \\
&\geq N!\,n! \quad .
\end{aligned}$$

Thus,

$$\frac{1}{(N+n)!} \leq \frac{1}{N!\,n!}$$

and

$$\sum_{n=0}^{\infty}\frac{(B\,|x|)^{N+n}}{(N+n)!} \leq \sum_{n=0}^{\infty}\frac{(B\,|x|)^{N+n}}{N!\,n!} \leq \frac{(B\,|x|)^N}{N!}\sum_{n=0}^{\infty}\frac{(B\,|x|)^n}{n!} = \frac{(B\,|x|)^N}{N!}e^{B|x|} \quad .$$

Combining this with inequality (3.9) yields

$$|y(x) - \psi_N(x)| \leq 2\Delta Y \frac{(B\,|x|)^N}{N!}e^{B|x|} \quad .$$

Consequently,

$$\begin{aligned}
&\left|\psi_{N+1}(x) - y_0 - \int_0^x F(s, y(s))\,ds\right| \\
&\qquad = \left|\left[y_0 + \int_0^x F(s, \psi_N(s))\,ds\right] - y_0 - \int_0^x F(s, y(s))\,ds\right| \\
&\qquad \leq \int_0^x |F(s, \psi_N(s)) - F(s, y(s))|\,ds \\
&\qquad \leq \int_0^x B\,|\psi_{N-1}(s) - y(s)|\,ds
\end{aligned}$$

$$\leq \int_0^x B \cdot 2\Delta Y \frac{(B\,|s|)^{N-1}}{(N-1)!} e^{B|s|}\, ds$$

$$= 2\Delta Y \frac{B^N e^{B|x|}}{(N-1)!} \int_0^x |s|^{N-1}\, ds \quad .$$

Computing the last integral leaves us with

$$\left| \psi_{N+1}(x) - y_0 - \int_0^x F(s, y(s))\, ds \right| \leq 2\Delta Y \frac{(B\,|x|)^N}{N!} e^{B|x|} \quad .$$

But, as is well known,

$$\frac{(B\,|x|)^N}{N!} \to 0 \quad \text{as} \quad N \to \infty$$

for any finite value $B\,|x|$. Hence

$$\left| \psi_N(x) - y_0 - \int_0^x F(s, y(s))\, ds \right| \to 0 \quad \text{as} \quad N \to \infty \quad .$$

That is

$$0 = \lim_{N\to\infty} \left[\psi_N(x) - y_0 - \int_0^x F(s, y(s))\, ds \right]$$

$$= \lim_{N\to\infty} \psi_N(x) - y_0 - \int_0^x F(s, y(s))\, ds$$

$$= y(x) - y_0 - \int_0^x F(s, y(s))\, ds \quad ,$$

verifying that

$$y(x) = y_0 + \int_0^x F(s, y(s))\, ds \qquad \text{whenever} \quad \alpha < x < \beta \quad ,$$

as desired.

Where Are We?

Let's stop for a moment and review what we have done. We have just spent several pages rigorously verifying the three statements made at the beginning of this section under the assumptions made in Theorem 3.1 on page 42. By verifying these statements, we've rigorously justified the computations made in the previous section showing that the limit of a Picard sequence is a solution to the initial-value problem in Theorem 3.1. Consequently, we have now rigorously verified the claim in Theorem 3.1 that a solution to the given initial-value problem exists on at least some interval (α, β) .

We now need to show that this y is the only solution on that interval.

The Uniqueness Claim in Theorem 3.1

If you've made it through this section up to this point, then you should have little difficulty in finishing the proof of Theorem 3.1 by doing the following exercises. Do make use of the work we've done in the previous several pages.

?►Exercise 3.2: *Consider a first-order initial-value problem*

$$\frac{dy}{dx} = F(x, y) \qquad \text{with} \quad y(0) = y_0 \quad ,$$

and with both F and ${}^{\partial F}/_{\partial y}$ being continuous functions on some open region containing the point $(0, y_0)$. Since Lemma 3.5 applies, we can let $[\alpha, \beta]$ be the interval, and M, B and ΔY the positive constants from that lemma. Using this interval and these constants:

a i: Verify that

$$0 \leq M|x| \leq \Delta Y \qquad \text{for} \quad \alpha \leq x \leq \beta \quad .$$

ii: Also verify that any solution y to the above initial-value problem satisfies

$$|y(x) - y_0| \leq M|x| \qquad \text{for} \quad a < x < b \quad .$$

Now observe that the last two inequalities yield

$$|y(x) - y_0| \leq M|x| \leq \Delta Y \qquad \text{for} \quad \alpha \leq x \leq \beta$$

whenever y is a solution to the above initial-value problem.

b: For the following, let y_1 and y_2 be any two solutions to the above initial-value problem on (α, β), and let

$$\psi_0 \ , \ \psi_1 \ , \ \psi_2 \ , \ \psi_3 \ , \ \ldots \qquad \text{and} \qquad \phi_0 \ , \ \phi_1 \ , \ \phi_2 \ , \ \phi_3 \ , \ \ldots$$

be the two Picard sequences of functions on (α, β) generated by setting

$$\psi_{k+1}(x) = y_0 + \int_0^x F(s, \psi_k(s))\, ds$$

and

$$\phi_{k+1}(x) = y_0 + \int_0^x F(s, \phi_k(s))\, ds$$

with

$$\psi_0(x) = y_1(x) \qquad \text{and} \qquad \phi_0(x) = y_2(x) \quad .$$

i: Using ideas similar to those used above to prove the convergence of the Picard sequence, show that, for each x in (α, β) and each positive integer k,

$$|\psi_{k+1}(x) - \phi_{k+1}(x)| \leq \int_0^x B\,|\psi_k(s) - \phi_k(s)|\, ds \quad .$$

ii: Then verify that, for each x in (α, β),

$$|\psi_0(x) - \phi_0(x)| \leq 2\Delta Y \quad ,$$

and

$$\lim_{k\to\infty} |\psi_{k+1}(x) - \phi_{k+1}(x)| = 0 \quad .$$

(Hint: This is very similar to our showing that $|\psi_{k+1}(x) - \psi_{k+1}(x)| \to 0$ as $k \to \infty$.)

iii: Verify that, for each x in (α, β) and positive integer k,

$$\psi_k(x) = y_1(x) \qquad \text{and} \qquad \phi_k(x) = y_2(x) \quad .$$

iv: Combine the results of the last two parts to show that

$$y_1(x) = y_2(x) \qquad \text{for} \quad \alpha < x < \beta \quad .$$

The end result of the above set of exercises is that there cannot be two *different* solutions on the interval (α, β) to the initial-value problem. That was the uniqueness claim of Theorem 3.1.

3.6 On Proving Theorem 3.2

We could spend several more enjoyable pages redoing the work in the previous section, but under the assumptions made in Theorem 3.2 on page 43 instead of those in Theorem 3.1. To avoid that, let us briefly discuss how you can modify that work, and, thereby, prove Theorem 3.2.

First of all, recall that much of the initial effort in proving the convergence of the Picard sequence,

$$\psi_0 \ , \ \psi_1 \ , \ \psi_2 \ , \ \psi_3 \ , \ \ldots$$

with

$$\psi_{k+1}(x) = y_0 + \int_0^x F(s, \psi_k(s))\,ds \qquad \text{for} \quad k = 0, 1, 2, 3, \ldots \quad ,$$

was in showing that there is an interval (α, β) such that, as long as $\alpha \leq s \leq \beta$, then $\psi_k(s)$ is never so large or so small that $(s, \psi_k(s))$ is outside a rectangular region on which F is "well-behaved" (this was the main result of Lemma 3.5 on page 48). However, if (as in Theorem 3.2) $F = F(x, y)$ is a continuous function on the infinite strip

$$\mathcal{R} = \{ (x, y) : \alpha < x < \beta \text{ and } -\infty < y < \infty \} \quad ,$$

then, for any continuous function ϕ on (α, β), $F(s, \phi(s))$ is a well-defined, continuous function of s over (α, β), and the integral in

$$\psi(x) = y_0 + \int_0^x F(s, \phi(s))\,ds$$

exists (and is finite) whenever $\alpha < x < \beta$. Verifying that ψ is continuous requires a little more thought than was needed in the proof of Lemma 3.5, but is still pretty easy — simply appeal to the continuity of $F(s, \phi(s))$ as a function of s along with the fact that

$$\psi(x_1) - \psi(x) = \int_x^{x_1} F(s, \phi(s))\,ds$$

to show that

$$\lim_{x \to x_1} \psi(x) = \psi(x_1) \qquad \text{for each } x_1 \text{ in } (\alpha, \beta) \quad .$$

Consequently, all the functions in the Picard sequence ψ_0, ψ_1, ψ_2, ... are continuous on (α, β) (provided, of course, that we started with ψ_0 being continuous).

Now choose finite values α_1 and β_1 so that $\alpha < \alpha_1 < 0 < \beta_1 < \beta$; let ΔY be the maximum value of

$$\frac{1}{2} |\psi_1(x) - \psi_0(x)| \qquad \text{for} \quad \alpha_1 \leq x \leq \beta_1 \quad ,$$

and let $\mathcal{R}_0$ be the infinite strip

$$\mathcal{R}_0 = \{ (x, y) : \alpha_1 < x < \beta_1 \text{ and } -\infty < y < \infty \} \quad .$$

By the assumptions in the theorem, we know that, on $\mathcal{R}$, the continuous function $\partial F/\partial y$ depends only on x. So we can treat it as a continuous function on the closed interval $[\alpha_1, \beta_1]$. But such functions are bounded. Thus, for some positive constant B and every point in $\mathcal{R}_0$,

$$\left| \frac{\partial F}{\partial y} \right| \leq B \quad .$$

Using this, the bounds on

$$|\psi_{k+1}(x) - \psi_k(x)| \qquad \text{for} \quad \alpha_1 \leq x \leq \beta_1 \quad \text{and} \quad k = 1, 2, 3, \ldots$$

can now be rederived exactly as in the previous section (leading to inequality (3.5) on page 53 and inequality set (3.8) on page 54), and we can then use arguments almost identical to those used in the previous section to show that the Picard sequence converges on (α_1, β_1) to a solution y of the given initial-value problem. The only notable modification is that the bound M used to show the continuity of y must be rederived. For this proof, let M be the maximum value of $F(x, y)$ on the closed rectangle

$$\{(x, y) : \alpha_1 \leq x \leq \beta_1 \text{ and } |y| \leq H\}$$

where H is the maximum value of

$$|\psi_0(x)| + \Delta Y e^{B|x|} \qquad \text{for} \quad \alpha_1 \leq x \leq \beta_1 \quad .$$

Inequality set (3.8) then tells us that

$$|\psi_k(s)| \leq H \qquad \text{for} \quad \alpha_1 \leq s \leq \beta_1 \quad \text{and} \quad k = 0, 1, 2, 3, \ldots \quad .$$

This, in turn, assures us that

$$|F(s, \psi_k(s))| \leq M \qquad \text{for} \quad \alpha_1 \leq s \leq \beta_1 \quad \text{and} \quad k = 0, 1, 2, 3, \ldots \quad ,$$

which is what we used in the previous section to prove the continuity of y.

Finally, since every point x in the interval (α, β) is also in some such subinterval (α_1, β_1), we must have that the Picard sequence converges at every point x in (α, β), and what it converges to, $y(x)$, is a solution to the given initial-value problem. Straightforward modifications to the arguments outlined in Exercise 3.2 then show that this solution is the only solution.

3.7 Appendix: A Little Multivariable Calculus

There are a few places in our discussions where some knowledge of the calculus of functions of two or more variables (i.e., "multivariable" calculus) is needed. These include the commentary about existence and uniqueness in this chapter (Theorems 3.1 and 3.2), and the use of the multivariable version of the chain rule in Chapter 7. This appendix is a brief introduction to those elements of multivariable calculus that are needed for these discussions. It is for those who have not yet been formally introduced to calculus of several variables, and contains just barely enough to get by.

Functions of Two Variables

At least while we are only concerned with first-order differential equations, the only multivariable calculus we will need involves functions of just two variables, such as

$$f(x, y) = x^2 + x^2 y^2 \quad , \qquad g(x, y) = \frac{x^3 + 4y}{x} \qquad \text{and} \qquad h(x, y) = \sqrt{x^3 + y^2} \quad .$$

These functions will be defined on "regions" of the XY–plane.

Open and Closed Regions

Functions of one variable are typically defined on intervals of the X–axis. For functions of two variables, we must replace the concept of an interval with that of a "region". For our purposes, a *region* (in the XY–plane) refers to the collection of all points enclosed by some curve or set of curves on the plane (with the understanding that this curve or set of curves actually does enclose some collection of points in the plane). If we include the curves with the enclosed points, then we say the region is *closed*; if the curves are all excluded, then we refer to the region as *open*. This corresponds to the distinction between a closed interval $[a, b]$ (which does contain the endpoints), and an open interval (a, b) (which does not contain the endpoints).

!►Example 3.4: *Consider the rectangular region $\mathcal{R}$ whose sides form the rectangle generated from the vertical lines $x = 1$ and $x = 4$ along with the horizontal lines $y = 2$ and $y = 6$. If $\mathcal{R}$ is to be a closed region, then it must include this rectangle; that is,*

$$\mathcal{R} = \{(x, y) : 1 \leq x \leq 4 \text{ and } 2 \leq y \leq 6\} \quad .$$

If $\mathcal{R}$ is to be an open region, then it must exclude this rectangle; that is,

$$\mathcal{R} = \{(x, y) : 1 < x < 4 \text{ and } 2 < y < 6\} \quad .$$

On the other hand, if $\mathcal{R}$ just includes one of its sides, say, its right side,

$$\mathcal{R} = \{(x, y) : 1 < x \leq 4 \text{ and } 2 < y < 6\} \quad ,$$

then it is considered to be neither open or closed.

Limits

The concept of limits for functions of two variables is a natural extension of the concept of limits for functions of one variable.

Given a function $f(x, y)$ of two variables, a point (x_0, y_0) in the plane, and a finite value A, we say that

A is the limit of $f(x, y)$ as (x, y) approaches (x_0, y_0),

equivalently,

$$\lim_{(x,y)\to(x_0,y_0)} f(x, y) = A \qquad \text{or} \qquad f(x, y) \to A \quad \text{as} \quad (x, y) \to (x_0, y_0) \quad ,$$

if and only if we can make the value of $f(x, y)$ as close (but not necessarily equal) to A as we desire by requiring (x, y) be sufficiently close (but not necessarily equal) to (x_0, y_0). More formally,

$$\lim_{(x,y)\to(x_0,y_0)} f(x, y) = A$$

if and only if, for every positive value ϵ there is a corresponding positive distance δ_ϵ such that $f(x, y)$ is within ϵ of A whenever (x, y) is within δ_ϵ of (x_0, y_0). That is, (in mathematical shorthand), for each $\epsilon > 0$ there is a $\delta_\epsilon > 0$ such that

$$\text{distance from } (x, y) \text{ to } (x_0, y_0) < \delta_\epsilon \quad \Longrightarrow \quad |f(x, y) - A| < \epsilon \quad .$$

The rules for the existence and computation of these limits are straightforward extensions of those for functions of one variable, and need not be discussed in detail here.

!►Example 3.5: *"Obviously", if*

$$f(x, y) = x^2 + x^2y^2 \quad ,$$

then

$$\lim_{(x,y)\to(2,3)} f(x,y) = \lim_{(x,y)\to(2,3)} \left[x^2 + x^2y^2\right] = 2^2 + (2^2)(3^2) = 40 \quad .$$

On the other hand

$$\lim_{(x,y)\to(0,3)} g(x,y)$$

does not exist if

$$g(x,y) = \frac{x^3 + 4y}{x}$$

because $(x,y) \to (0,3)$ *leads to* ${}^{12}/_{0}$.

Continuity

The only difference between "continuity for a function of one variable" and "continuity for a function of two variables" is the number of variables involved.

Basically, a function $f(x,y)$ is continuous at a point (x_0, y_0) if and only if we can legitimately write

$$\lim_{(x,y)\to(x_0,y_0)} f(x,y) = f(x_0, y_0) \quad .$$

That function is then continuous on a region $\mathcal{R}$ if and only if it is continuous at every point in $\mathcal{R}$. Note that this does require $f(x,y)$ to be defined at every point in the region.

Partial Derivatives

Recall that the derivative of a function of one variable $f = f(t)$ is given by the limit formula

$$\frac{df}{dt} = \lim_{\Delta t\to 0} \frac{f(t+\Delta t) - f(t)}{\Delta t}$$

provided the limit exists. The simplest extension of this for a function of two variables $f = f(x,y)$ is the "partial" derivatives with respect to each variable:

1. The *(first) partial derivative with respect to* x is denoted and defined by

$$\frac{\partial f}{\partial x} = \lim_{\Delta x\to 0} \frac{f(x+\Delta x, y) - f(x,y)}{\Delta x}$$

provided the limit exists.

2. The *(first) partial derivative with respect to* y is denoted and defined by

$$\frac{\partial f}{\partial y} = \lim_{\Delta y\to 0} \frac{f(x, y+\Delta y) - f(x,y)}{\Delta y}$$

provided the limit exists.

Note the notation, ${}^{\partial f}/_{\partial x}$ and ${}^{\partial f}/_{\partial y}$, in which we use ∂ instead of d.[7]

An important thing to observe about the limit formula for ${}^{\partial f}/_{\partial x}$ is that, in essence, x replaces the variable t in the previous formula for ${}^{df}/_{dt}$ while y does not vary. Consequently, to compute ${}^{\partial f}/_{\partial x}$, simply take the derivative of $f(x,y)$ using x as the variable while pretending y is a constant. Likewise, to compute ${}^{\partial f}/_{\partial y}$ simply take the derivative of $f(x,y)$ using y as the variable while pretending x is a constant. As a result, everything already learned about computing ordinary derivatives applies to computing partial derivatives, provided we keep straight which variable is being treated (temporarily) as a constant.

[7] Some authors prefer using such notation as $D_x f$ and f_x instead of ${}^{\partial f}/_{\partial x}$, and $D_y f$ and f_y instead of ${}^{\partial f}/_{\partial y}$.

!▶Example 3.6: *Let*

$$f(x,y) = x^2 + x^2y^2 \quad .$$

Then

$$\frac{\partial f}{\partial x} = \frac{\partial}{\partial x}\left[x^2 + x^2y^2\right] = \frac{\partial}{\partial x}\left[x^2\right] + \frac{\partial}{\partial x}\left[x^2y^2\right] = 2x + 2xy^2 \quad ,$$

while

$$\frac{\partial f}{\partial y} = \frac{\partial}{\partial y}\left[x^2 + x^2y^2\right] = \frac{\partial}{\partial y}\left[x^2\right] + \frac{\partial}{\partial y}\left[x^2y^2\right] = 0 + x^2 2y \quad .$$

?▶Exercise 3.3: *Let*

$$g(x,y) = x^2y^3 \quad \text{and} \quad h(x,y) = \sin\left(x^2+y^2\right) \quad .$$

Verify that

$$\frac{\partial g}{\partial x} = 2xy^3 \quad \text{and} \quad \frac{\partial g}{\partial y} = 3x^2y^2 \quad ,$$

while

$$\frac{\partial h}{\partial x} = 2x\cos\left(x^2+y^2\right) \quad \text{and} \quad \frac{\partial h}{\partial y} = 2y\cos\left(x^2+y^2\right) \quad .$$

Functions of More than Two Variables

The notation can become a bit more cumbersome, and the pictures even harder to draw, but everything discussed above for functions of two variables naturally extends to functions of three or more variables. For example, we may have a function of three variables $f = f(x,y,z)$ defined on, say, an open box-like region

$$\mathcal{R} = \{\, (x,y,z) : x_{\min} < x < x_{\max} \ , \ y_{\min} < y < y_{\max} \text{ and } z_{\min} < z < z_{\max} \,\}$$

where $x_{\min}$, $x_{\max}$, $y_{\min}$, $y_{\max}$, $z_{\min}$ and $z_{\max}$ are finite numbers. We will then say that, for any given point (x_0, y_0, z_0) and value A,

$$\lim_{(x,y,)\rightarrow(x_0,y_0,z_0)} f(x,y,z) = A$$

if and only if there is a corresponding positive distance δ_ϵ for every positive value ϵ such that $f(x,y,z)$ is within ϵ of A whenever (x,y,z) is within δ_ϵ of (x_0,y_0,z_0). We will also say that this function is continuous on $\mathcal{R}$ if and only if we can legitimately write

$$\lim_{(x,y,)\rightarrow(x_0,y_0,z_0)} f(x,y,z) = f(x_0,y_0,z_0)$$

for every point (x_0,y_0,z_0) in $\mathcal{R}$. Finally, the three (first) partial derivatives of this function are given by

$$\frac{\partial f}{\partial x} = \lim_{\Delta x\rightarrow 0} \frac{f(x+\Delta x,y,z) - f(x,y,z)}{\Delta x} \quad ,$$

$$\frac{\partial f}{\partial y} = \lim_{\Delta y\rightarrow 0} \frac{f(x,y+\Delta y,z) - f(x,y,z)}{\Delta y}$$

and

$$\frac{\partial f}{\partial z} = \lim_{\Delta y\rightarrow 0} \frac{f(x,y,z+\Delta z) - f(x,y,z)}{\Delta z} \quad ,$$

provided the limits exist. Again, in practice, the partial derivative with respect to any one of the three variables is the derivative obtained by pretending the other variables are constants.

Additional Exercises

3.4. Rewrite each of the following in derivative formula form, and then find all constant solutions. (In some cases, you may have to use the quadratic formula to find any constant solutions.)

a. $\frac{dy}{dx} + 3xy = 6x$

b. $\sin(x+y) - y\frac{dy}{dx} = 0$

c. $\frac{dy}{dx} - y^3 = 8$

d. $x^2\frac{dy}{dx} + xy^2 = x$

e. $\frac{dy}{dx} - y^2 = x$

f. $y^3 - 25y + \frac{dy}{dx} = 0$

g. $(x-2)\frac{dy}{dx} = y + 3$

h. $(y-2)\frac{dy}{dx} = x - 3$

i. $\frac{dy}{dx} + 2y - y^2 = -2$

j. $\frac{dy}{dx} + (8-x)y - y^2 = -8x$

3.5. Which of the equations in the above exercise set are autonomous?

3.6. Consider the first-order initial-value problem

$$\frac{dy}{dx} = 2\sqrt{y} \qquad \text{with} \quad y(1) = 0 \quad .$$

a. Verify that each of the following is a solution on the interval $(-\infty, \infty)$, and graph that solution:

i. $y(x) = 0 \quad \text{for} \quad -\infty < x < \infty$.

ii. $y(x) = \begin{cases} 0 & \text{if } x < 1 \\ (x-1)^2 & \text{if } 1 \le x \end{cases}$.

iii. $y(x) = \begin{cases} 0 & \text{if } x < 3 \\ (x-3)^2 & \text{if } 3 \le x \end{cases}$.

b. You've just verified three different functions as being solutions to the above initial-value problem. Why does this not violate Theorem 3.1?

3.7. Let $\psi_0, \psi_1, \psi_2, \psi_3, \ldots$ be the sequence of functions generated by the Picard iterative method (as described in Section 3.4) using the initial-value problem

$$\frac{dy}{dx} = xy \qquad \text{with} \quad y(0) = 2$$

along with

$$\psi_0(x) = 2 \qquad \text{for all} \quad x \quad .$$

Using the formula for Picard's method (formula (3.4) on page 46), compute the following:

a. $\psi_1(x)$

b. $\psi_2(x)$

c. $\psi_3(x)$

3.8. *Let* $\psi_0\,,\ \psi_1\,,\ \psi_2\,,\ \psi_3\,,\ \ldots$ *be the sequence of functions generated by the Picard iterative method (as described in Section 3.4) using the initial-value problem*

$$\frac{dy}{dx} = 2x + y^2 \qquad \text{with} \quad y(0) = 3$$

along with

$$\psi_0(x) = 3 \qquad \text{for all} \quad x \quad .$$

Compute the following:

a. $\psi_1(x)$ **b.** $\psi_2(x)$

4

Separable First-Order Equations

As we will see below, the notion of a differential equation being "separable" is a natural generalization of the notion of a first-order differential equation being directly integrable. What's more, a fairly natural modification of the method for solving directly integrable first-order equations gives us the basic approach to solving "separable" differential equations. However, it cannot be said that the theory of separable equations is just a trivial extension of the theory of directly integrable equations. Certain issues can arise that do not arise in solving directly integrable equations. Some of these issues are pertinent to even more general classes of first-order differential equations than those that are just separable, and may play a role later on in this text.

In this chapter we will, of course, learn how to identify and solve separable first-order differential equations. We will also see what sort of issues can arise, examine those issues, and discuss some ways to deal with them. Since many of these issues involve graphing, we will also draw a bunch of pictures.

4.1 Basic Notions

Separability

A first-order differential equation is said to be *separable* if, after solving it for the derivative,

$$\frac{dy}{dx} = F(x, y) \quad ,$$

the right-hand side can then be factored as "a formula of just x" times "a formula of just y";

$$F(x, y) = f(x)g(y) \quad .$$

If this factoring is not possible, the equation is not separable.

More concisely, a first-order differential equation is *separable* if and only if it can be written as

$$\frac{dy}{dx} = f(x)g(y) \tag{4.1}$$

where f and g are known functions.

!►Example 4.1: *Consider the differential equation*

$$\frac{dy}{dx} - x^2y^2 = x^2 \quad . \tag{4.2}$$

Solving for the derivative (by adding x^2y^2 *to both sides),*

$$\frac{dy}{dx} = x^2 + x^2y^2 \quad ,$$

and then factoring out the x^2 *on the right-hand side gives*

$$\frac{dy}{dx} = x^2\left(1 + y^2\right) \quad ,$$

which is in form

$$\frac{dy}{dx} = f(x)g(y)$$

with

$$f(x) = \underbrace{x^2}_{\text{no } y\text{'s}} \qquad \text{and} \qquad g(y) = \underbrace{\left(1 + y^2\right)}_{\text{no } x\text{'s}} \quad .$$

So equation (4.2) is a separable differential equation.

!►Example 4.2: *On the other hand, consider*

$$\frac{dy}{dx} - x^2y^2 = 4 \quad . \tag{4.3}$$

Solving for the derivative here yields

$$\frac{dy}{dx} = x^2y^2 + 4 \quad .$$

The right-hand side of this clearly cannot be factored into a function of just x *times a function of just* y *. Thus, equation (4.3) is not separable.*

We should (briefly) note that any directly integrable first-order differential equation

$$\frac{dy}{dx} = f(x)$$

can be viewed as also being the separable equation

$$\frac{dy}{dx} = f(x)g(y)$$

with $g(y)$ being the constant 1 . Likewise, a first-order autonomous differential equation

$$\frac{dy}{dx} = g(y)$$

can also be viewed as being separable, this time with $f(x)$ being 1 . Thus, both directly integrable and autonomous differential equations are all special cases of separable differential equations.

Integrating Separable Equations

As just noted, a directly-integrable equation

$$\frac{dy}{dx} = f(x)$$

can be viewed as the separable equation

$$\frac{dy}{dx} = f(x)g(y) \qquad \text{with} \quad g(y) = 1 \quad .$$

We point this out again because the method used to solve directly-integrable equations (integrating both sides with respect to x) is rather easily adapted to solving separable equations. Let us try to figure out this adaptation using the differential equation from the first example. Then, if we are successful, we can discuss its use more generally.

!▶Example 4.3: *Consider the differential equation*

$$\frac{dy}{dx} - x^2y^2 = x^2 \quad .$$

In Example 4.1, we saw that this is a separable equation, and can be written as

$$\frac{dy}{dx} = x^2\left(1+y^2\right) \quad .$$

If we simply try to integrate both sides with respect to x , the right-hand side would become

$$\int x^2\left(1+y^2\right)\,dx \quad .$$

Unfortunately, the y here is really $y(x)$, some unknown formula of x ; so the above is just the integral of some unknown function of x — something we cannot effectively evaluate. To eliminate the y's on the right-hand side, we could, before attempting the integration, divide through by $1+y^2$, obtaining

$$\frac{1}{1+y^2}\frac{dy}{dx} = x^2 \quad . \tag{4.4}$$

The right-hand side can now be integrated with respect to x . What about the left-hand side? The integral of that with respect to x is

$$\int \frac{1}{1+y^2}\frac{dy}{dx}\,dx \quad .$$

Tempting as it is to simply "cancel out the dx's", let's not (at least, not yet). After all, ${}^{dy}/_{dx}$ is not a fraction; it denotes the derivative $y'(x)$ where $y(x)$ is some unknown formula of x . But y is also shorthand for that same unknown formula $y(x)$. So this integral is more precisely written as

$$\int \frac{1}{1+[y(x)]^2}\,y'(x)\,dx \quad .$$

Fortunately, this is just the right form for applying the generic substitution $y = y(x)$ to convert the integral with respect to x to an integral with respect to y . No matter what $y(x)$ might be (so long as it is differentiable), we know

$$\int \underbrace{\frac{1}{1+[y(x)]^2}}_{\frac{1}{1+y^2}}\,\underbrace{y'(x)\,dx}_{dy} = \int \frac{1}{1+y^2}\,dy \quad .$$

Combining all this, we get

$$\int \frac{1}{1+y^2}\frac{dy}{dx}\,dx = \int \frac{1}{1+[y(x)]^2}\,y'(x)\,dx = \int \frac{1}{1+y^2}\,dy \quad ,$$

which, after cutting out the middle, reduces to

$$\int \frac{1}{1+y^2}\frac{dy}{dx}\,dx = \int \frac{1}{1+y^2}\,dy \quad ,$$

the very equation we would have obtained if we had yielded to temptation and naively "cancelled out the dx*'s".*

Consequently, the equation obtained by integrating both sides of equation (4.4) with respect to x *,*

$$\int \frac{1}{1+y^2}\frac{dy}{dx}\,dx = \int x^2\,dx \quad ,$$

is the same as

$$\int \frac{1}{1+y^2}\,dy = \int x^2\,dx \quad .$$

Doing the indicated integration on both sides then yields

$$\arctan(y) = \frac{1}{3}x^3 + c \quad ,$$

which, in turn, tells us that

$$y = \tan\left(\frac{1}{3}x^3 + c\right) \quad .$$

This is the general solution to our differential equation.

Two generally useful ideas were illustrated in the last example. One is that, whenever we have an integral of the form

$$\int H(y)\frac{dy}{dx}\,dx$$

where y denotes some (differentiable) function of x , then this integral is more properly written as

$$\int H(y(x))\,y'(x)\,dx \quad ,$$

which reduces to

$$\int H(y)\,dy$$

via the substitution $y = y(x)$ (even though we don't yet know what $y(x)$ is). Thus, in general,

$$\int H(y)\frac{dy}{dx}\,dx = \int H(y)\,dy \quad . \tag{4.5}$$

This equation is true whether you derive it rigorously, as we have, or obtain it naively by mechanically canceling out the dx's.[1]

The other idea seen in the example was that, if we divide an equation of the form

$$\frac{dy}{dx} = f(x)g(y)$$

by $g(y)$, then (with the help of equation (4.5)) we can compute the integral with respect to x of each side of the resulting equation,

$$\frac{1}{g(y)}\frac{dy}{dx} = f(x) \quad .$$

This leads us to a *basic procedure for solving separable first-order differential equations*:

[1] One of the reasons our notation is so useful is that naive manipulations of the differentials often do lead to valid equations. Just don't be too naive and cancel out the d's in $^{dy}/_{dx}$.

1. Get the differential equation into the form
$$\frac{dy}{dx} = f(x)g(y) \quad .$$

2. Divide through by $g(y)$ to get
$$\frac{1}{g(y)}\frac{dy}{dx} = f(x) \quad .$$
(Note: At this point we've "separated the variables", getting all the y's and derivatives of y on one side, and all the x's on the other.)

3. Integrate both sides with respect to x, making use of the fact that
$$\int \frac{1}{g(y)}\frac{dy}{dx}\,dx = \int \frac{1}{g(y)}\,dy \quad .$$

4. Solve the resulting equation for y.

There are a few issues that can arise in some of these steps, and we will have to slightly refine this procedure to address those issues. Before doing that, though, let us practice with another differential equation for which the above approach can be applied without any difficulty.

!▶Example 4.4: *Consider solving the initial-value problem*
$$\frac{dy}{dx} = -\frac{x}{y-3} \qquad \text{with} \quad y(0) = 1 \quad .$$

Here,
$$\frac{dy}{dx} = f(x)g(y) \qquad \text{with} \quad f(x) = -x \quad \text{and} \quad g(y) = \frac{1}{y-3} \quad ,$$
and "dividing through by $g(y)$" is the same as multiplying through by $y-3$. Doing so, and then integrating both sides with respect to x, we get the following:
$$[y-3]\frac{dy}{dx} = -x$$
$$\hookrightarrow \qquad \int [y-3]\frac{dy}{dx}\,dx = -\int x\,dx$$
$$\hookrightarrow \qquad \int [y-3]\,dy = -\int x\,dx$$
$$\hookrightarrow \qquad \frac{1}{2}y^2 - 3y = -\frac{1}{2}x^2 + c \quad .$$
Though hardly necessary, we can multiply through by 2, obtaining the slightly simpler expression
$$y^2 - 6y = -x^2 + 2c \quad .$$

We are now faced with the less-than-trivial task of solving the last equation for y in terms of x. Since the left-hand side looks something like a quadratic for y, let us rewrite this equation as
$$y^2 - 6y + \left[x^2 - 2c\right] = 0$$
so that we can apply the quadratic formula to solve for y. Applying that venerable formula, we get
$$y = \frac{-(-6) \pm \sqrt{(-6)^2 - 4\left[x^2 - 2c\right]}}{2} = 3 \pm \sqrt{9 - x^2 + 2c} \quad ,$$

which, since $9 + 2c$ is just another unknown constant, can be written a little more simply as

$$y = 3 \pm \sqrt{a - x^2} \quad . \tag{4.6}$$

This is the general solution to our differential equation.

Now for the initial-value problem. Combining the general solution just derived with the given initial value at $x = 0$ yields

$$1 = y(0) = 3 \pm \sqrt{a - 0^2} = 3 \pm \sqrt{a} \quad .$$

So

$$\pm\sqrt{a} = -2 \quad .$$

This means that $a = 4$, and that we must use the negative root in formula (4.6) for y. Thus, the solution to our initial-value problem is

$$y = 3 - \sqrt{4 - x^2} \quad .$$

4.2 Constant Solutions
Avoiding Division by Zero

In the above procedure for solving

$$\frac{dy}{dx} = f(x)g(y) \quad ,$$

we divided both sides by $g(y)$. This requires, of course, that $g(y)$ not be zero — which is often *not* the case.

!►Example 4.5: *Consider solving*

$$\frac{dy}{dx} = 2x(y - 5) \quad .$$

As long as $y \neq 5$, we can divide through by $y - 5$ and follow our basic procedure:

$$\frac{1}{y-5}\frac{dy}{dx} = 2x$$

$$\hookrightarrow \quad \int \frac{1}{y-5}\frac{dy}{dx}\,dx = \int 2x\,dx$$

$$\hookrightarrow \quad \int \frac{1}{y-5}\,dy = \int 2x\,dx$$

$$\hookrightarrow \quad \ln|y-5| = x^2 + c$$

$$\hookrightarrow \quad |y-5| = e^{x^2+c} = e^{x^2}e^c$$

$$\hookrightarrow \quad y - 5 = \pm e^{x^2}e^c \quad .$$

So, assuming $y \neq 5$, we get

$$y = 5 \pm e^c e^{x^2} \quad .$$

Notice that, because $e^c \neq 0$ for every real value c, this formula for y never gives us $y = 5$ for any real choice of c and x.

But what about the case where $y = 5$?

Well, suppose $y = 5$. To be more specific, let y be the constant function

$$y(x) = 5 \qquad \text{for every} \quad x \quad ,$$

and plug this constant function into our differential equation

$$\frac{dy}{dx} = 2x(y - 5) \quad .$$

Recalling (again) that derivatives of constants are zero, we get

$$0 = 2x(5 - 5) \quad ,$$

which is certainly a true equation. So $y = 5$ is a solution. In fact, it is one of those "constant" solutions we discussed in the previous chapter.

Combining all the above, we see that the "general solution" to the given differential equation is actually the set consisting of the solutions

$$y(x) = 5 \qquad \text{and} \qquad y(x) = 5 \pm e^c e^{x^2} \quad .$$

Now consider the general case, where we seek all possible solutions to

$$\frac{dy}{dx} = f(x)g(y) \quad .$$

If y_0 is any single value for which

$$g(y_0) = 0 \quad ,$$

then plugging the corresponding constant function

$$y(x) = y_0 \qquad \text{for all} \quad x$$

into the differential equation gives, after a trivial bit of computation,

$$0 = 0 \quad ,$$

showing that

$$y(x) = y_0 \quad \text{is a constant solution to} \quad \frac{dy}{dx} = f(x)g(y) \quad ,$$

just as we saw (in the above example) that

$$y(x) = 5 \quad \text{is a constant solution to} \quad \frac{dy}{dx} = 2x(y - 5) \quad .$$

Conversely, suppose $y = y_0$ is a constant solution to

$$\frac{dy}{dx} = f(x)g(y)$$

(and f is not the zero function). Then the equation is valid with y replaced by the constant y_0, giving us

$$0 = f(x)g(y_0) \quad ,$$

which, in turn, means that y_0 must be a constant such that

$$g(y_0) = 0 \quad .$$

What all this shows is that our basic method for solving separable equations may miss the constant solutions because those solutions correspond to a division by zero in our basic method.[2]

Because constant solutions are often important in understanding the physical process the differential equation might be modeling, let us be careful to find them. Accordingly, we will insert the following step into our procedure on page 68 for solving separable equations:

- Identify all constant solutions by finding all values y_0 , y_1 , y_2 , ... such that

  $$g(y_k) = 0 \quad ,$$

  and then write down

  $$y(x) = y_0 \quad , \quad y(x) = y_1 \quad , \quad y(x) = y_2 \quad , \quad \ldots \quad .$$

  (These are the constant solutions.)

(And we will renumber the other steps as appropriate.)

Sometimes, the formula obtained by our basic procedure for solving can be 'tweaked' to also account for the constant solutions. A standard 'tweak' can be seen by reconsidering the general solution obtained in our last example.

!▶ **Example 4.6:** *The general solution obtained in the previous example was the set containing*

$$y(x) = 5 \qquad \text{and} \qquad y(x) = 5 \pm e^c e^{x^2} \quad .$$

If we let $A = \pm e^c$, the second equation reduces to

$$y(x) = 5 + Ae^{x^2} \quad .$$

Remember, though, $A = \pm e^c$ can be any positive or negative number, but cannot be zero (because of the nature of the exponential function). So, by our definition of A , our general solution is

$$y(x) = 5 \tag{4.7a}$$

and

$$y(x) = 5 + Ae^{x^2} \quad \text{where } A \text{ can be any nonzero real number} \quad . \tag{4.7b}$$

However, if we allow A to be zero, then equation (4.7b) reduces to equation (4.7a),

$$y(x) = 5 + 0 \cdot e^{x^2} = 5 \quad ,$$

which means the entire set of possible solutions can be expressed more simply as

$$y(x) = 5 + Ae^{x^2}$$

where A is an arbitrary constant with no restrictions on its possible values.

---

[2] Because $g(y_0) = 0$ is a 'singular' value for division, many authors refer to constant solutions of separable equations as *singular* solutions.

In the future, we will usually express our general solutions as simply as practical, with the trick of letting

$$A = \pm e^c \text{ or } 0$$

often being used without comment. Keep in mind, though, that the sort of tweaking just described is not always possible.

?►Exercise 4.1: *Verify that the general solution to*

$$\frac{dy}{dx} = -y^2$$

is given by the set consisting of

$$y(x) = 0 \qquad \text{and} \qquad y(x) = \frac{1}{x+c} \quad .$$

Is there any way to rewrite these two formulas for $y(x)$ as a single formula using just one arbitrary constant?

The Importance of Constant Solutions

Even if we can use the same general formula to describe all the solutions (constant and otherwise), it is often worthwhile to explicitly identify any constant solutions. To see this, let us now solve the differential equation from Chapter 1 describing a falling object when we take into account air resistance.

!►Example 4.7: *Let $v = v(t)$ be the velocity (in meters per second) at time t of some object of mass m plummeting towards the ground. In Chapter 1, we decided that F_{air}, the force of air resistance acting on the falling body, could be described by*

$$F_{\text{air}} = -\gamma v$$

where γ was some positive constant dependent on the size and shape of the object (and probably determined by experiment). Using this, we obtained the differential equation

$$\frac{dv}{dt} = -9.8 - \kappa v \qquad \text{where} \quad \kappa = \frac{\gamma}{m} \quad .$$

This is a relatively simple separable equation. Assuming v equals a constant v_0 yields

$$0 = -9.8 - \kappa v_0 \quad \Longrightarrow \quad v_0 = -\frac{9.8}{\kappa} = -\frac{9.8m}{\gamma} \quad .$$

So, we have one constant solution,

$$v(t) = v_0 \qquad \text{for all } t$$

where

$$v_0 = -\frac{9.8}{\kappa} = -\frac{9.8m}{\gamma} \quad .$$

For reasons that will soon become clear, v_0 is called the terminal velocity *of the object that is falling.*

To find the other possible solutions, we assume $v \neq v_0$ and proceed:

$$\frac{dv}{dt} = -9.8 - \kappa v$$

$$\hookrightarrow \quad \frac{1}{9.8 + \kappa v}\frac{dv}{dt} = -1$$

$$\hookrightarrow \quad \int \frac{1}{9.8 + \kappa v}\frac{dv}{dt}\,dt = -\int 1\,dt$$

$$\hookrightarrow \quad \int \frac{1}{9.8 + \kappa v}\,dv = -\int dt$$

$$\hookrightarrow \quad \frac{1}{\kappa}\ln|9.8 + \kappa v| = -t + c$$

$$\hookrightarrow \quad \ln|9.8 + \kappa v| = -\kappa t + \kappa c$$

$$\hookrightarrow \quad 9.8 + \kappa v = \pm e^{-\kappa t + \kappa c}$$

$$\hookrightarrow \quad v(t) = \frac{1}{\kappa}\left[-9.8 \pm e^{\kappa c}e^{-\kappa t}\right] \quad .$$

Since $v_0 = -9.8\kappa^{-1}$, the last equation reduces to

$$v(t) = v_0 + Ae^{-\kappa t} \qquad \text{where} \quad A = \pm\frac{1}{\kappa}e^{\kappa c} \quad .$$

This formula for $v(t)$ yields the constant solution, $v = v_0$, if we allow $A = 0$. Thus, letting A be a completely arbitrary constant, we have that

$$v(t) = v_0 + Ae^{-\kappa t} \tag{4.8a}$$

where

$$v_0 = -\frac{9.8m}{\gamma} \qquad \text{and} \qquad \kappa = \frac{\gamma}{m} \tag{4.8b}$$

describes all possible solutions to the differential equation of interest here. The graphs of some possible solutions (assuming a terminal velocity of -10 meters/second) are sketched in Figure 4.1.

Notice how the constant in the constant solution, v_0, appears in the general solution (equation (4.8a)). More importantly, notice that the exponential term in this solution rapidly goes to zero as t increases, so

$$v(t) = v_0 + Ae^{-\kappa t} \quad \rightarrow \quad v(t) = v_0 \qquad \text{as} \quad t \rightarrow \infty \quad .$$

This is graphically obvious in Figure 4.1. Consequently, no matter what the initial velocity and initial height were, eventually the velocity of this falling object will be very close to v_0 (provided it doesn't hit the ground first). That is why v_0 is called the terminal velocity. That is also why that constant solution is so important here (and is appropriately also called the equilibrium solution). It accurately predicts the final velocity of any object falling from a sufficiently high height. And if you are that falling object, then that velocity[3] is probably a major concern.

[3] between 120 and 150 miles per hour for a typical human body

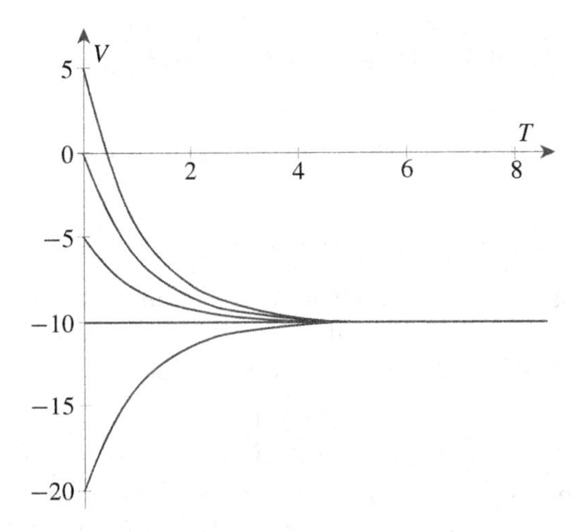

Figure 4.1: Graphs of the velocity of a falling object during the first 8 seconds of its fall assuming a terminal velocity of -10 meters per second. Each graph corresponds to a different initial velocity.

4.3 Explicit Versus Implicit Solutions

Thus far, we have been able to find explicit formulas for all of our solutions; that is, we have been able to carry out the last step in our basic procedure — that of solving the resulting (integrated) equation for y in terms of x — obtaining

$$y = y(x) \qquad \text{where } y(x) \text{ is some formula of } x \text{ (with no } y\text{'s)}.$$

For example, as the general solution to

$$\frac{dy}{dx} - x^2y^2 = x^2 \quad ,$$

we obtained (in Example 4.3)

$$y = \underbrace{\tan\left(\frac{1}{3}x^3 + c\right)}_{y(x)} \quad .$$

Unfortunately, this is not always possible.

!▶Example 4.8: *Consider*

$$\frac{dy}{dx} = \frac{x+1}{8 + 2\pi \sin(\pi y)} \quad .$$

In this case,

$$g(y) = \frac{1}{8 + 2\pi \sin(\pi y)} \quad ,$$

which can never be zero. So there are no constant solutions, and we can blithely proceed with our procedure. Doing so:

$$\frac{dy}{dx} = \frac{x+1}{8 + 2\pi \sin(\pi y)}$$

$$\hookrightarrow \qquad [8 + 2\pi \sin(\pi y)]\frac{dy}{dx} = x + 1$$

$$\hookrightarrow \qquad \int [8 + 2\pi \sin(\pi y)] \frac{dy}{dx}\, dx = \int x + 1\, dx$$

$$\hookrightarrow \qquad \int [8 + 2\pi \sin(\pi y)]\, dy = \int x + 1\, dx$$

$$\hookrightarrow \qquad 8y - 2\cos(\pi y) = \frac{1}{2}x^2 + x + c \quad .$$

The next step would be to solve the last equation for y in terms of x. But look at that last equation. Can you solve it for y as a formula of x? Neither can anyone else. So we are not able to obtain an explicit formula for y. At best, we can say that $y = y(x)$ satisfies the equation

$$8y - 2\cos(\pi y) = \frac{1}{2}x^2 + x + c \quad .$$

Still, this equation is not without value. It does implicitly describe the possible relations between x and y. In particular, the graphs of this equation can be sketched for different values of c (we'll do this later on in this chapter). These graphs, in turn, give you the graphs you would obtain for $y(x)$ if you could actually find the formula for $y(x)$.

In practice, we must deal with both "explicit" and "implicit" solutions to differential equations. When we have an explicit formula for the solution in terms of the variable, that is, we have something of the form

$$y = y(x) \qquad \text{where } y(x) \text{ is some formula of } x \text{ (with no } y\text{'s)} \quad , \tag{4.9}$$

then we say that we have an *explicit solution* to our differential equation. Technically, it is that "formula of x" in equation (4.9) which is the *explicit solution*. In practice, though, it is common to refer to the entire equation as "an explicit solution". For example, we found that the solution to

$$\frac{dy}{dx} - x^2y^2 = x^2$$

is explicitly given by

$$y = \tan\left(\frac{1}{3}x^3 + c\right) \quad .$$

Strictly speaking, the explicit solution here is the formula

$$\tan\left(\frac{1}{3}x^3 + c\right) \quad .$$

That, of course, is what is really meant when someone answers the question

$$\textit{What is the explicit solution to} \qquad \frac{dy}{dx} - x^2y^2 = x^2 \qquad ?$$

with the equation

$$y = \tan\left(\frac{1}{3}x^3 + c\right) \quad .$$

If, on the other hand, we have an equation (other than something like (4.9)) involving the solution and the variable, then that *equation* is called an *implicit solution*. In trying to solve the differential equation in Example 4.8,

$$\frac{dy}{dx} = \frac{x + 1}{8 + 2\pi \sin(\pi y)} \quad ,$$

we derived the equation

$$8y - 2\cos(\pi y) = \frac{1}{2}x^2 + x + c \quad .$$

This equation is an implicit solution for the given differential equation.[4]

Differential equations — be they separable or not — can have both implicit and explicit solutions. Indeed, implicit solutions often arise in the process of deriving an explicit solution. For example, in solving

$$\frac{dy}{dx} - x^2y^2 = x^2 \quad ,$$

we first obtained

$$\arctan(y) = \frac{1}{3}x^3 + c \quad .$$

This is an implicit solution. Fortunately, it could be easily solved for y, giving us the explicit solution

$$y = \tan\left(\frac{1}{3}x^3 + c\right) \quad .$$

As a general rule, explicit solutions are preferred over implicit solutions. Explicit solutions usually give more information about the solutions, and are easier to use than implicit solutions (even when you have sophisticated computer math packages). So, whenever you solve a differential equation,

FIND AN EXPLICIT SOLUTION IF AT ALL PRACTICAL.

Do not be surprised, however, if you encounter a differential equation for which an explicit solution is not obtainable. This is not a disaster; it just means a little more work may be needed to extract useful information about the possible solutions.

4.4 Full Procedure for Solving Separable Equations

In light of the possibility of singular solutions and the possibility of not finding explicit solutions, we should refine our procedure for solving a separable differential equation to:

1. Get the differential equation into the form

$$\frac{dy}{dx} = f(x)g(y) \quad . \tag{4.10}$$

2. Identify all constant solutions by finding all values y_0, y_1, y_2, ... such that

$$g(y_k) = 0 \quad ,$$

and then write down

$$y(x) = y_0 \quad , \quad y(x) = y_1 \quad , \quad y(x) = y_2 \quad , \quad \ldots \quad .$$

(These are the constant solutions.)

3a. Divide equation (4.10) through by $g(y)$ to get

$$\frac{1}{g(y)}\frac{dy}{dx} = f(x)$$

(assuming y is not one of the constant solutions just found).

[4] The fact that an explicit solution is a formula while an implicit solution is an equation may be a little confusing at first. If it helps, think of the phrase "implicit solution" as being shorthand for "an equation implicitly defining the solution $y = y(x)$".

b. Integrate both sides of the equation just obtained with respect to x.

c. Solve the resulting equation for y, if practical (thus obtaining an explicit solution). If not practical, use that resulting equation as an implicit solution, possibly rearranged or simplified if appropriate.

4. If constant solutions were found, see if the formulas obtained for the other solutions can be tweaked to also describe the constant solutions. In any case, be sure to write out all solution(s) obtained.

The above yields the general solution. If initial values are also given, then use those initial conditions with the general solution just obtained to derive the particular solutions satisfying the given initial-value problems.

4.5 Existence, Uniqueness, and False Solutions

On the Existence and Uniqueness of Solutions

Let's consider a generic initial-value problem involving a separable differential equation,

$$\frac{dy}{dx} = f(x)g(y) \qquad \text{with} \quad y(x_0) = y_0 \quad .$$

Letting $F(x,y) = f(x)g(y)$ this is

$$\frac{dy}{dx} = F(x,y) \qquad \text{with} \quad y(x_0) = y_0 \quad ,$$

which was the initial-value problem considered in Theorem 3.1 on page 42. That theorem assures us that there is exactly one solution to our initial-value problem on some interval (a,b) containing x_0 provided

$$F(x,y) = f(x)g(y)$$

and

$$\frac{\partial F}{\partial y} = \frac{\partial}{\partial y}[f(x)g(y)] = f(x)g'(y)$$

are continuous in some open region containing the point (x_0, y_0). This means our initial-value problem will have exactly one solution on some interval (a,b) provided $f(x)$ is continuous on some open interval containing x_0, and both $g(y)$ and $g'(y)$ are continuous on some open interval containing y_0. In practice, this is typically what we have.

Typically, also, one rarely worries about the existence and uniqueness of the solution to an initial-value problem with a separable differential equation, at least not when one can carry out the integration and algebra required by our procedure. After all, doesn't our refined procedure for solving separable differential equations always lead us to "the solution"? Well, here are two reasons to have at least a little concern about existence and uniqueness:

1. After the integration in step 3, the resulting equation may involve a nontrivial formula of y. After applying the initial condition and solving for y, it is possible to end up with more than one formula for $y(x)$. But as long as f, g and g' are sufficiently continuous, the above tells us that there is only one solution. Thus, only one of these formulas for $y(x)$ can be correct. The others are "false solutions" that should be identified and eliminated. (An example is given in the next subsection.)

2. Suppose $g(y_0) = 0$. Our refined procedure tells us that the constant function $y = y_0$, which certainly satisfies the initial condition, is also a solution to the differential equation. So $y = y_0$ is immediately seen to be a solution to our initial-value problem. Do we then need to go through the rest of our procedure to see if any other solutions to the differential equation satisfy $y(x_0) = y_0$? The answer is *No, not if f is continuous on an open interval containing x_0, and both g and g' are continuous on an open interval containing y_0.* If that continuity holds, then the above analysis assures us that there is only one solution. Thus, if we find *a* solution, we have found *the* solution.

It is possible to have an initial-value problem

$$\frac{dy}{dx} = f(x)g(y) \qquad \text{with} \quad y(x_0) = y_0 \quad ,$$

in which the f or g or g' is not suitably continuous. The problem in Exercise 3.6 on page 63,

$$\frac{dy}{dx} = 2\sqrt{y} \qquad \text{with} \quad y(1) = 0 \quad ,$$

is one such problem. Here,

$$y_0 = 0 \quad , \quad f(x) = 1 \quad , \quad g(y) = 2\sqrt{y} \quad \text{and} \quad g'(y) = \frac{1}{\sqrt{y}} \quad .$$

Clearly, g and, especially, g' are not continuous in any open interval containing $y_0 = 0$. So the above results on existence and uniqueness cannot be assumed. Indeed, in this case there is not just the one constant solution $y = 0$, but, as shown in that exercise, there are many different solutions, including

$$y(x) = \begin{cases} 0 & \text{if} \quad x < 1 \\ (x-1)^2 & \text{if} \quad 1 \le x \end{cases} \qquad \text{and} \qquad y(x) = \begin{cases} 0 & \text{if} \quad x < 3 \\ (x-3)^2 & \text{if} \quad 3 \le x \end{cases} \quad .$$

A Caution on False Solutions

It is always a good idea to verify that any 'solution' obtained in solving a differential equation really is a solution. This is even more true when solving separable differential equations. Not only does the extra algebra involved naturally increase the likelihood of human error, this algebra can, as noted above, lead to 'false solutions' — formulas that are obtained as solutions but do not actually satisfy the original problem.

!► Example 4.9: *Consider the initial-value problem*

$$\frac{dy}{dx} = 2\sqrt{y} \qquad \textit{with} \quad y(0) = 4 \quad .$$

The differential equation does have one constant solution, $y = 0$, but since that doesn't satisfy the initial condition, it hardly seems relevant. To find the other solutions, let's divide the differential equation by $\sqrt{y}$ and proceed with the basic procedure:

$$\frac{1}{\sqrt{y}}\frac{dy}{dx} = 2$$

$$\hookrightarrow \qquad \int \frac{1}{\sqrt{y}}\frac{dy}{dx}\,dx = \int 2\,dx$$

$$\hookrightarrow \qquad \int y^{-1/2}\,dy = \int 2\,dx$$

$$\hookrightarrow \qquad 2y^{1/2} = 2x + c \quad .$$

Dividing by 2 *and squaring (and letting* $a = c/2$*), we get*

$$y = (x+a)^2 \quad . \tag{4.11}$$

Plugging this into the initial condition, we obtain

$$4 = y(0) = (0+a)^2 = a^2 \quad ,$$

which means that

$$a = \pm 2 \quad .$$

Hence, we have two formulas for the solution to our initial-value problem,

$$y_+(x) = (x+2)^2 \qquad \text{and} \qquad y_-(x) = (x-2)^2 \quad .$$

Both satisfy the initial condition. Do both satisfy the differential equation

$$\frac{dy}{dx} = 2\sqrt{y} \quad ?$$

Well, plugging

$$y = y_\pm(x) = (x \pm 2)^2$$

into the differential equation yields

$$\frac{d}{dx}(x \pm 2)^2 = 2\sqrt{(x \pm 2)^2}$$

$$\hookrightarrow \qquad 2(x \pm 2) = 2\sqrt{(x \pm 2)^2} \quad .$$

So, for $y = y_\pm(x)$ *to be solutions to our differential equation, we must have*

$$x \pm 2 = \sqrt{(x \pm 2)^2} \tag{4.12}$$

for all values of x *'of interest'. In particular, this equation must be valid at the initial point* $x = 0$.

So, consider what happens to equation (4.12) at the initial point $x = 0$. *With* $y = y_+(x)$ *and* $x = 0$ *equation (4.12) becomes*

$$0 + 2 = \sqrt{(0+2)^2} = \sqrt{4} \quad ,$$

which, of course, simplifies to the perfectly acceptable equation

$$2 = 2 \quad .$$

But with $y = y_-(x)$ *and* $x = 0$ *we get*

$$0 - 2 = \sqrt{(0-2)^2} = \sqrt{4} = 2 \quad ,$$

which, of course, simplifies to

$$-2 = 2 \quad ,$$

which is not acceptable. So we cannot accept $y = y_-(x)$ as a solution to our initial-value problem. It was a false solution.

While we are at it, let's look a little more closely at equation (4.12) with $y = y_+(x)$,

$$x + 2 = \sqrt{(x+2)^2} \quad .$$

Remember, if A is any real number, then

$$\sqrt{A^2} = |A| \quad .$$

So equation (4.12) with $y = y_+$ can be written as

$$x + 2 = |x + 2| \quad ,$$

which is true if and only if $x + 2 \geq 0$ (i.e., $x \geq -2$). This means that our solution, $y = y_+(x)$, is not valid for all values of x, only for those greater than or equal to -2. Thus, the actual solution that we have is

$$y = y_+(x) = (x+2)^2 \quad \text{for} \quad -2 \leq x \quad .$$

There was a lot of analysis done in the last example after obtaining the apparent solutions

$$y = (x \pm 2)^2 \quad .$$

Don't be alarmed. In most of the problems you will be given, verifying that your formula is a solution should be fairly easy. Still, take the moral of this example to heart: It is a good idea to verify that any formulas derived as solutions truly are solutions.

By the way, in a later chapter we will develop some graphical techniques that would have simplified our work in the above example.

4.6 On the Nature of Solutions to Differential Equations

When we solve a first-order directly integrable differential equation,

$$\frac{dy}{dx} = f(x) \quad ,$$

we get something of the form

$$y = F(x) + c$$

where F is any antiderivative of f and c is an arbitrary constant. Computationally, all we have to do is find a single antiderivative F for f and then add an arbitrary constant. Thus, also, the graph of any possible solution is nothing more than the graph of $F(x)$ shifted vertically by the value of c (up if $c > 0$, down if $c < 0$). What's more, the interval for x over which

$$y = F(x) + c$$

is a valid solution depends only on the one function F. If $F(x)$ is continuous for all x in an interval (a, b), then (a, b) is a valid interval for our solution. This interval does not depend on the choice for c.

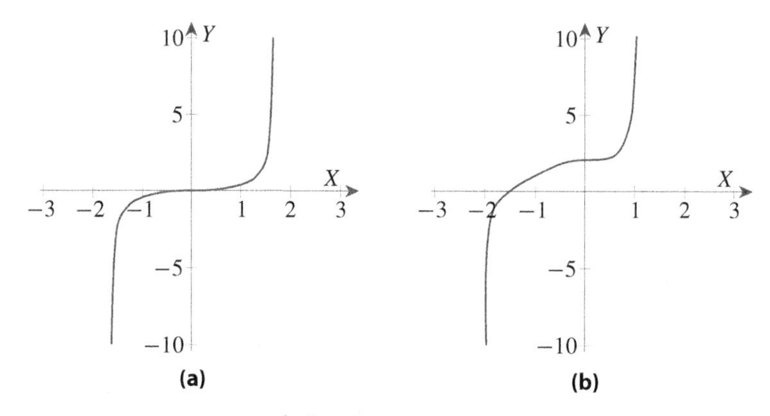

Figure 4.2: The graph of $y = \tan\left(\frac{1}{3}x^3 + c\right)$ **(a)** when $y(0) = 0$ and **(b)** when $y(0) = 2$.

The situation can be much more complicated if our differential equation is not directly integrable. First of all, finding an explicit solution can be impossible. And consider those explicit general solutions we have found,

$$y = \tan\left(\frac{1}{3}x^3 + c\right) \qquad \text{(from Example 4.3 on page 67)}$$

and

$$y = 3 \pm \sqrt{a - x^2} \qquad \text{(from Example 4.4 on page 69)} \quad .$$

In both of these, the arbitrary constants are not simply "added on" to some formula of x. Instead, each solution formula combines the variable, x, with the arbitrary constant, c or a, in a very nontrivial manner. There are two immediate consequences of this:

1. The graphs of the solutions are no longer simply vertically shifted copies of some single function.

2. The possible intervals over which any solution is valid may depend on the arbitrary constant. And since the value of that constant can be determined by the initial condition, the interval of validity for our solutions may depend on the initial condition.

Both of these consequences are illustrated in Figure 4.2, in which the graphs of two solutions to the differential equation in Example 4.3 have been sketched corresponding to two different initial values (namely, $y(0) = 0$ and $y(0) = 2$). In these figures you can see how changing the initial condition from $y(0) = 0$ to $y(0) = 2$ changes the interval over which the solution exists. Even more apparent is that the graph corresponding to $y(0) = 2$ is not merely a 'shift' of the graph corresponding to $y(0) = 0$; there is also a small but clear distortion in the shape of the graph.

The possible dependence of a solution's interval of validity is even better illustrated by the solutions obtained in Example 4.4. There, the differential equation was

$$\frac{dy}{dx} = -\frac{x}{y - 3}$$

and the general solution was found to be

$$y = 3 \pm \sqrt{a - x^2} \quad .$$

The arbitrary constant here, a, occurs in the square root. For this square root to be real, we must have

$$a - x^2 \geq 0 \quad .$$

That is,

$$-\sqrt{a} \leq x \leq \sqrt{a}$$

is the maximal interval over which

$$y = 3 + \sqrt{a - x^2} \qquad \text{and} \qquad y = 3 - \sqrt{a - x^2}$$

are valid solutions.

To properly indicate this dependence of the solution's possible domain on the arbitrary constant or the initial value, we should state the maximal interval of validity along with any formula or equation describing our solution(s). For Example 4.4, that would mean writing the general solution as

$$y = 3 \pm \sqrt{a - x^2} \qquad \text{for all} \qquad -\sqrt{a} \leq x \leq \sqrt{a} \quad .$$

When this is particularly convenient or noteworthy, we will attempt to remember to do so. Even when we don't, keep in mind that there may be limits as to the possible values of x , and that these limits may depend on the values assumed by the arbitrary constants.

By the way, notice also that the above a cannot be negative (otherwise, $\sqrt{a}$ will not be a real number). This points out that, in general, the 'arbitrary' constants appearing in general solutions are not always completely arbitrary.

4.7 Using and Graphing Implicit Solutions

Outside of courses specifically geared towards learning about differential equations, the main reason to solve an initial-value problem such as

$$\frac{dy}{dx} = \frac{x+1}{8 + 2\pi \sin(y\pi)} \qquad \text{with} \quad y(0) = 2$$

is so that we can predict what values $y(x)$ will assume when x has values other than 0 . In practice, of course, $y(x)$ will represent something of interest (position, velocity, promises made, number of ducks, etc.) that varies with whatever x represents (time, position, money invested, food available, etc.). When the solution y is given explicitly by some formula $y(x)$, then those values are relatively easily obtained by just computing that formula for different values of x , and a picture of how $y(x)$ varies with x is easily obtained by graphing $y = y(x)$. If, instead, the solution is given implicitly by some equation, then the possible values of $y(x)$ for different x's , along with any graph of $y(x)$, must be extracted from that equation. It may be necessary to use advanced numerical methods to extract the desired information, but that should not be a significant problem — these methods are probably already incorporated into your favorite computer math package.

!►Example 4.10: *Let's consider the initial-value problem*

$$\frac{dy}{dx} = \frac{x+1}{8 + 2\pi \sin(y\pi)} \qquad \textit{with} \quad y(0) = 2 \quad .$$

In Example 4.8, we saw that the general solution to the differential equation is given implicitly by

$$8y - 2\cos(y\pi) = \frac{1}{2}x^2 + x + c \quad . \tag{4.13}$$

The initial condition $y(0) = 2$ *tells us that* $y = 2$ *when* $x = 0$ *. With this assumed, our implicit solution reduces to*

$$8 \cdot 2 - 2\cos(2\pi) = \frac{1}{2}\left[0^2\right] + 0 + c \quad .$$

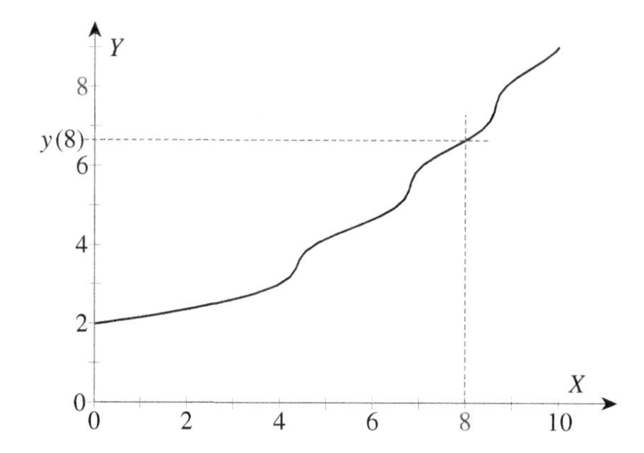

Figure 4.3: Graph of the implicit solution to the initial-value problem of Example 4.10.

So

$$c = 8 \cdot 2 - 2\cos(2\pi) - \frac{1}{2}\left[0^2\right] - 0 = 16 - 2 = 14 \quad .$$

Plugging this back into equation (4.13) gives

$$8y - 2\cos(y\pi) = \frac{1}{2}x^2 + x + 14 \tag{4.14}$$

as an implicit solution for our initial-value problem.

Replacing c with 14 does not make it any easier for us to convert this equation relating y and x into a formula $y(x)$ for y. Still, $y = y(x)$ must satisfy equation (4.14), and the graph of that equation can be generated by invoking the appropriate command(s) in a suitable computer math package. That is how the graph in Figure 4.3 was created. From this graph, we see that the value of $y(8)$ is between 6 and 7. For a more precise determination of $y(8)$, set $x = 8$ in equation (4.14). This gives us

$$8y - 2\cos(y\pi) = \frac{1}{2}8^2 + 8 + 14 \quad ,$$

which, after a little arithmetic, reduces to

$$8y - 2\cos(y\pi) = 54 \quad .$$

Now apply some numerical method (such as the Newton-Raphson method for finding roots[5]) to find, approximately, the corresponding value of y. Again, we need not do the tedious computations ourselves; we can go to our favorite computer math package, look up the appropriate commands, and let it compute that value for y. Doing so, we find that $y(8) \approx 6.642079594$.

Any curve that is at least part of the graph of an implicit solution for a differential equation is called an *integral curve* for the differential equation. Remember, this is the graph of an *equation*. If a *function* $y(x)$ is a solution to that differential equation, then $y = y(x)$ must also satisfy any equation serving as an implicit solution, and, thus, the graph of that $y(x)$ (which we will call a *solution curve*) must be at least a portion of one of the integral curves for that differential equation. Sometimes an integral curve will be a solution curve. That is "clearly" the case in Figure 4.3, because that curve is "clearly" the graph of a function (more on that later).

---

[5] It should be in your calculus text.

Sometimes though, there are two (or more) different functions y_1 and y_2 such that both $y = y_1(x)$ and $y = y_2(x)$ satisfy the same equation for the same values of x. If that equation is an implicit solution to some differential equation, then its graph (the integral curve) will contain the graphs of both $y = y_1(x)$ and $y = y_2(x)$. In such a case, the integral curve is not a solution curve but contains two or more solution curves.

To illuminate these comments, let us look at the solution curves and integral curves for one equation we've already solved. At the same time, we will discover that, at least occasionally, the use of implicit solutions can simplify our work, even when explicit solutions are available.

!►Example 4.11: *Consider graphing all the solutions to*

$$\frac{dy}{dx} = -\frac{x}{y-3}$$

and the particular solution satisfying

$$y(0) = 1 \quad .$$

In Example 4.4 (starting on page 69), we "separated and integrated" this differential equation to get the implicit solution

$$y^2 - 6y = -x^2 + c \quad . \tag{4.15}$$

We were then able to solve this equation for y in terms of x by using the quadratic formula.

This time, rather than attempting to solve for y, let's simply move the x^2 to the left, obtaining

$$x^2 + y^2 - 6y = 2c \quad .$$

This looks suspiciously like an equation for a circle. Writing 6 as $2 \cdot 3$ and adding 3^2 to both sides (to complete the square in the y terms) make it look even more so:

$$x^2 + \underbrace{y^2 - 2 \cdot 3y + 3^2}_{(y-3)^2} = 2c + 3^2$$

$$\hookrightarrow \qquad (x-0)^2 + (y-3)^2 = 2c + 9 \quad .$$

Since the left side is the sum of squares, it cannot be negative; hence, neither can the right side. So we can let $R = \sqrt{2c + 9}$ and write our equation as

$$(x-0)^2 + (y-3)^2 = R^2 \quad . \tag{4.16}$$

You should recognize this implicit solution for our differential equation as also being the equation for a circle of radius R centered at $(0,3)$. One such circle (with $R = 2$) is sketched in Figure 4.4a. These circles are integral curves for our differential equation. In this case, we can find the solution curves by solving our last equation for the explicit solutions

$$y = 3 \pm \sqrt{R^2 - x^2} \quad .$$

The solution curves, then, are the graphs of $y = y_-(x)$ and $y = y_+(x)$ where

$$y_+(x) = 3 + \sqrt{R^2 - x^2} \qquad \text{and} \qquad y_-(x) = 3 - \sqrt{R^2 - x^2} \quad .$$

Since we must have $R^2 - x^2 \geq 0$ for the square roots, each of these functions can only be defined for

$$-R \leq x \leq R \quad .$$

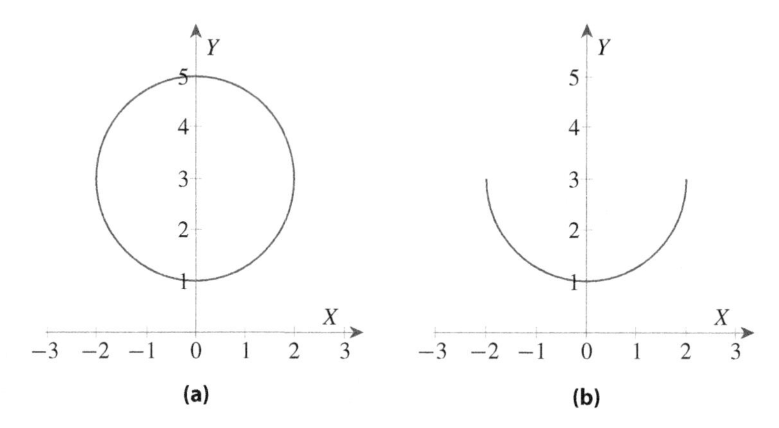

Figure 4.4: **(a)** The integral curve and **(b)** the solution curve for the differential equation in Example 4.11 containing the point $(x, y) = (0, 1)$.

Observe that the graphs of these functions are not the entire circles of the integral curves, but are semicircles, with the graph of

$$y = 3 + \sqrt{R^2 - x^2} \qquad \text{with} \qquad -R \leq x \leq R$$

being the upper half of the circle of radius R about $(0, 3)$, and the graph of

$$y = 3 - \sqrt{R^2 - x^2} \qquad \text{with} \qquad -R \leq x \leq R$$

being the lower half of that same circle.

If we further require that $y(0) = 1$, then implicit solution (4.16) becomes

$$(0 - 0)^2 + (1 - 3)^2 = R^2 \quad .$$

So $R = 2$, and $y = y(x)$ must satisfy

$$(x - 0)^2 + (y - 3)^2 = 2^2 \quad . \tag{4.17}$$

Solving this for y in terms of x, we get the two functions

$$y_+(x) = 3 + \sqrt{2^2 - x^2} \qquad \text{with} \qquad -2 \leq x \leq 2$$

and

$$y_-(x) = 3 - \sqrt{2^2 - x^2} \qquad \text{with} \qquad -2 \leq x \leq 2 \quad .$$

The graph of equation (4.17) (an integral curve) is a circle of radius 2 about $(0, 3)$ (see Figure 4.4a). It contains the point $(x, y) = (0, 1)$ corresponding to the initial value $y(0) = 1$. To be specific, this point, $(0, 1)$, is on the lower half of that circle (the solution curve for $y_-(x)$) and not on the upper half (the solution curve for $y_+(x)$). Thus, the (explicit) solution to our initial-value problem is

$$y = y_-(x) = 3 - \sqrt{2^2 - x^2} \qquad \text{with} \qquad -2 \leq x \leq 2 \quad .$$

This is the solution curve sketched in Figure 4.4b.

Let us now consider things more generally, and assume that we have any first-order differential equation that can be put into derivative formula form. Since what follows does not require "separability", let us simply assume we've managed to get the differential equation into the form

$$\frac{dy}{dx} = F(x, y)$$

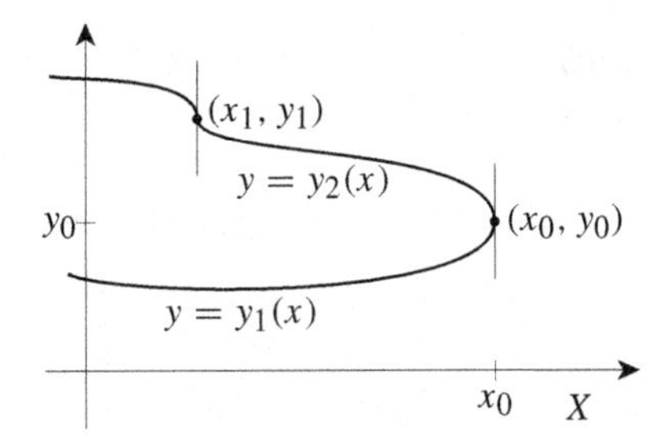

Figure 4.5: An integral curve containing two solution curves, with the portion above y_0 being the graph of $y_2(x)$ and the portion below y_0 being the graph of $y_1(x)$.

where $F(x, y)$ is some formula of x and y. This equation might be a separable differential equation such as

$$\frac{dy}{dx} = -\frac{x}{y-3} \quad ,$$

or it might be one like

$$\frac{dy}{dx} = x^2y^2 + 4 \quad ,$$

which is not separable. Suppose further that, either using methods developed in this chapter or methods that will be developed later, we have found an integral curve for this differential equation.

If no two distinct points on this integral curve have the same x-coordinate, then this curve is the graph of a function $y = y(x)$ that satisfies the differential equation (whether or not we can solve for the formula $y(x)$). So the entire integral curve is a solution curve.

On the other hand, if there are two points on this curve with the same x-coordinate, then the curve has to 'bend back' on itself at some point (x_0, y_0). At this point, the curve changes from being the graph of one solution $y = y_1(x)$ to being the graph of another solution $y = y_2(x)$. Also, at this point, the tangent line to the integral curve must be vertical (i.e., have "infinite slope"), provided that tangent line exists (see Figure 4.5). But the slope of the tangent line to the graph of a differential equation's solution at any point (x, y) is simply the derivative $^{dy}/_{dx}$ of the solution at that point, and that value can be computed directly from the differential equation

$$\frac{dy}{dx} = F(x, y) \quad .$$

Thus, (x_0, y_0), a point at which the integral curve 'bends back on itself', must be a point at which $F(x, y)$ becomes infinite (or, otherwise fails to exist).

Mind you, we cannot say that a curve 'bends back on itself' at a point just because the derivative becomes infinite there. Many functions have isolated points at which their derivative becomes infinite or otherwise fails to exist. Just look at point (x_1, y_1) in Figure 4.5. Or consider

$$y(x) = 3x^{1/3} \quad .$$

This is a well-defined function on the entire real line whose derivative, $y'(x) = x^{-2/3}$, blows up at $x = 0$. So all we can say is that, if the curve does 'bend back on itself' then it can only do so at points where its derivative either becomes infinite or otherwise fails to exist.

Here is a little theorem summarizing some what we have just discussed.

Theorem 4.1

Let $\mathcal{C}$ be a curve contained in the graph of an implicit solution for some first-order differential equation

$$\frac{dy}{dx} = F(x, y) \quad .$$

If $F(x, y)$ is a finite number for each point (x, y) in $\mathcal{C}$, then $\mathcal{C}$ is the graph of a function satisfying the given differential equation (i.e., $\mathcal{C}$ is a solution curve).

?►Exercise 4.2: *Explain why the integral curve graphed in Figure 4.3 is "clearly" a solution curve.*

4.8 On Using Definite Integrals with Separable Equations

Just as with any directly integrable differential equation, a separable differential equation

$$\frac{dy}{dx} = f(x)g(y) \quad ,$$

once separated to the form

$$\frac{1}{g(y)}\frac{dy}{dx} = f(x) \quad ,$$

can be integrated using definite integrals instead of the indefinite integrals we've been using. The basic ideas are pretty much the same as for directly integrable differential equations:

1. Pick a convenient value for the lower limit of integration, a. In particular, if the value of $y(x_0)$ is given for some point x_0, set $a = x_0$.
2. Rewrite the differential equation with s denoting the variable instead of x. This means that we rewrite our separable equation as

$$\frac{dy}{ds} = f(s)g(y) \quad ,$$

which 'separates' to

$$\frac{1}{g(y)}\frac{dy}{ds} = f(s) \quad .$$

3. Then integrate each side with respect to s from $s = a$ to $s = x$.

The integral on the left-hand side will be of the form

$$\int_{s=a}^{x} \frac{1}{g(y)}\frac{dy}{ds}\,ds \quad .$$

Keep in mind that, here, y is some unknown function of s, and that the limits in the integral are limits on s. Using the substitution $y = y(s)$, we see that

$$\int_{s=a}^{x} \frac{1}{g(y)}\frac{dy}{ds}\,ds = \int_{y=y(a)}^{y(x)} \frac{1}{g(y)}\,dy \quad .$$

Do not forget to convert the limits to being the corresponding limits on y, instead of s.

Once the integration is done, we attempt to solve the resulting equation for $y(x)$ just as before.

!►Example 4.12: *Let us solve*

$$\frac{dy}{dx} = \frac{1}{2y}e^{-x^2} \qquad \text{with} \quad y(0) = 3$$

using definite integrals. Proceeding as described above:

$$\frac{dy}{dx} = \frac{1}{2y}e^{-x^2}$$

$$\hookrightarrow \qquad 2y\frac{dy}{dx} = e^{-x^2}$$

$$\hookrightarrow \qquad 2y\frac{dy}{ds} = e^{-s^2}$$

$$\hookrightarrow \qquad \int_{s=0}^{x} 2y\frac{dy}{ds}\,ds = \int_{s=0}^{x} e^{-s^2}\,ds \quad .$$

Since $y(0) = 3$, *we can rewrite the last equation as*

$$\int_{y=3}^{y(x)} 2y\,dy = \int_{s=0}^{x} e^{-s^2}\,ds \quad .$$

The integral on the left is easily evaluated; the one on the right is not. Doing the easy integration and solving for y, *we get*

$$y^2\Big|_{y=3}^{y(x)} = \int_{s=0}^{x} e^{-s^2}\,ds$$

$$\hookrightarrow \qquad [y(x)]^2 - 3^2 = \int_{s=0}^{x} e^{-s^2}\,ds$$

$$\hookrightarrow \qquad [y(x)]^2 = 9 + \int_{s=0}^{x} e^{-s^2}\,ds \quad .$$

So

$$y(x) = \pm\left[9 + \int_{s=0}^{x} e^{-s^2}\,ds\right]^{1/2} \quad .$$

Plugging in the initial value again,

$$3 = y(0) = \pm\left[9 + \int_{s=0}^{0} e^{-s^2}\,ds\right]^{1/2} = \pm[9 + 0]^{1/2} \quad ,$$

we see that the $\pm$ *really should be* $+$, *not* $-$. *Thus, the solution to our initial-value problem is*

$$y = \left[9 + \int_{s=0}^{x} e^{-s^2}\,ds\right]^{1/2} \quad .$$

Going back to the section on "named integrals" in Chapter 2 (see page 27), we see that we can also express this as

$$y = \left[9 + \frac{\sqrt{\pi}}{2}\operatorname{erf}(x)\right]^{1/2} \quad .$$

The advantages of using definite integrals in solving a separable differential equation

$$\frac{dy}{dx} = f(x)g(y)$$

are the same as in solving a directly integrable differential equation:

1. The solution directly involves the initial value instead of a constant to be determined from the initial value, and

2. even if a 'nice' formula for
$$\int_a^x f(s)\,ds$$
cannot be found, the value of this integral can be closely approximated for specific values of x using standard methods (which are already in many computer math packages). Using these values for this integral, it is then often possible to find the corresponding values for $y(x)$ for specific values of x.

Unfortunately, we still have a serious problem if we cannot find a usable formula for
$$\int_{y(a)}^{y(x)} \frac{1}{g(y)}\,dy$$
since the numerical methods for computing this integral require knowing the value of $y(x)$ for the desired choice of x, and that $y(x)$ is exactly what we do *not* know.

Additional Exercises

4.3. *Determine whether each of the following differential equations is or is not separable, and, if it is separable, rewrite the equation in the form*
$$\frac{dy}{dx} = f(x)g(y) \quad .$$

a. $\frac{dy}{dx} = 3y^2 - y^2\sin(x)$

b. $\frac{dy}{dx} = 3x - y\sin(x)$

c. $x\frac{dy}{dx} = (x-y)^2$

d. $\frac{dy}{dx} = \sqrt{1+x^2}$

e. $\frac{dy}{dx} + 4y = 8$

f. $\frac{dy}{dx} + xy = 4x$

g. $\frac{dy}{dx} + 4y = x^2$

h. $\frac{dy}{dx} = xy - 3x - 2y + 6$

i. $\frac{dy}{dx} = \sin(x+y)$

j. $y\frac{dy}{dx} = e^{x-3y^2}$

4.4. *Using the basic procedure, find the general solution to each of the following separable equations:*

a. $\frac{dy}{dx} = \frac{x}{y}$

b. $\frac{dy}{dx} = y^2 + 9$

c. $xy\frac{dy}{dx} = y^2 + 9$

d. $\frac{dy}{dx} = \frac{y^2+1}{x^2+1}$

e. $\cos(y)\frac{dy}{dx} = \sin(x)$

f. $\frac{dy}{dx} = e^{2x-3y}$

4.5. *Using the basic procedure, find the solution to each of the following initial-value problems:*

a. $\dfrac{dy}{dx} = \dfrac{x}{y}$ *with* $y(1) = 3$

b. $\dfrac{dy}{dx} = 2x - 1 + 2xy - y$ *with* $y(0) = 2$

c. $y\dfrac{dy}{dx} = xy^2 + x$ *with* $y(0) = -2$

d. $y\dfrac{dy}{dx} = 3\sqrt{xy^2 + 9x}$ *with* $y(1) = 4$

4.6. *Find all the constant solutions — and only the constant solutions — to each of the following. If no constant solution exists, say so.*

a. $\dfrac{dy}{dx} = xy - 4x$

b. $\dfrac{dy}{dx} - 4y = 2$

c. $y\dfrac{dy}{dx} = xy^2 - 9x$

d. $\dfrac{dy}{dx} = \sin(y)$

e. $\dfrac{dy}{dx} = e^{x+y^2}$

f. $\dfrac{dy}{dx} = 200y - 2y^2$

4.7. *Find the general solution for each of the following. Where possible, write your answer as an explicit solution.*

a. $\dfrac{dy}{dx} = xy - 4x$

b. $\dfrac{dy}{dx} = xy - 3x - 2y + 6$

c. $\dfrac{dy}{dx} = 3y^2 - y^2\sin(x)$

d. $\dfrac{dy}{dx} = \tan(y)$

e. $\dfrac{dy}{dx} = \dfrac{y}{x}$

f. $\dfrac{dy}{dx} = \dfrac{6x^2 + 4}{3y^2 - 4y}$

g. $(x^2 + 1)\dfrac{dy}{dx} = y^2 + 1$

h. $(y^2 - 1)\dfrac{dy}{dx} = 4xy^2$

i. $\dfrac{dy}{dx} = e^{-y}$

j. $\dfrac{dy}{dx} = e^{-y} + 1$

k. $\dfrac{dy}{dx} = 3xy^3$

l. $\dfrac{dy}{dx} = \dfrac{2 + \sqrt{x}}{2 + \sqrt{y}}$

m. $\dfrac{dy}{dx} - 3x^2y^2 = -3x^2$

n. $\dfrac{dy}{dx} - 3x^2y^2 = 3x^2$

o. $\dfrac{dy}{dx} = 200y - 2y^2$

4.8. *Solve each of the following initial-value problems. If possible, express each solution as an explicit solution.*

a. $\dfrac{dy}{dx} - 2y = -10$ *with* $y(0) = 8$

b. $y\dfrac{dy}{dx} = \sin(x)$ *with* $y(0) = -4$

c. $\dfrac{dy}{dx} = 2x - 1 + 2xy - y$ *with* $y(0) = -1$

d. $x\frac{dy}{dx} = y^2 - y$ with $y(2) = 1$

e. $x\frac{dy}{dx} = y^2 - y$ with $y(1) = 2$

f. $\frac{dy}{dx} = \frac{y^2 - 1}{xy}$ with $y(1) = -2$

g. $(y^2 - 1)\frac{dy}{dx} = 4xy$ with $y(0) = 1$

4.9. In Chapter 11, when studying population growth, we will obtain the "logistic equation"

$$\frac{dy}{dx} = \beta y - \gamma y^2$$

with β and γ being positive constants.

a. What are the constant solutions to this equation?

b. Find the general solution to this equation.

4.10. For each of the following initial-value problems, find the largest interval over which the solution is valid. (Note: You've already solved these initial-value problems in exercise set 4.8 or at least found the general solution to the differential equation in 4.7.)

a. $\frac{dy}{dx} - 2y = -10$ with $y(0) = 8$

b. $x\frac{dy}{dx} = y^2 - y$ with $y(1) = 0$

c. $x\frac{dy}{dx} = y^2 - y$ with $y(1) = 2$

d. $\frac{dy}{dx} = e^{-y}$ with $y(0) = 1$

e. $\frac{dy}{dx} = 3xy^3$ with $y(0) = \frac{1}{2}$

5

Linear First-Order Equations

"Linear" first-order differential equations make up another important class of differential equations that commonly arise in applications and are relatively easy to solve (in theory). As with the notion of 'separability', the idea of 'linearity' for first-order equations can be viewed as a simple generalization of the notion of direct integrability, and a relatively straightforward (though, perhaps, not so intuitively obvious) method will allow us to put any first-order linear equation into a form that can be relatively easily integrated. We will derive this method in a short while (after, of course, describing just what it means for a first-order equation to be "linear").

By the way, the criteria given here for a differential equation being linear will be extended later to higher-order differential equations, and a rather extensive theory will be developed to handle linear differential equations of any order. That theory is not needed here; in fact, it would be of very limited value. And, to be honest, the basic techniques we'll develop in this chapter are only of limited use when it comes to solving higher-order linear equations. However, these basic techniques involve an "integrating factor", which is something we'll be able to generalize a little bit later (in Chapter 7) to help solve much more general first-order differential equations.

5.1 Basic Notions

Definitions

A first-order differential equation is said to be *linear* if and only if it can be written as

$$\frac{dy}{dx} = f(x) - p(x)y \tag{5.1}$$

or, equivalently, as

$$\frac{dy}{dx} + p(x)y = f(x) \tag{5.2}$$

where $p(x)$ and $f(x)$ are known functions of x only.

Equation (5.2) is normally considered to be the *standard form* for first-order linear equations. Note that the only appearance of y in a linear equation (other than in the derivative) is in a term where y alone is multiplied by some formula of x. If there are any other functions of y appearing in the equation after you've isolated the derivative, then the equation is not linear.

!►Example 5.1: *Consider the differential equation*

$$x\frac{dy}{dx} + 4y - x^3 = 0 \quad .$$

Solving for the derivative, we get

$$\frac{dy}{dx} = \frac{x^3 - 4y}{x} = x^2 - \frac{4}{x}y \quad ,$$

which is

$$\frac{dy}{dx} = f(x) - p(x)y$$

with

$$p(x) = \frac{4}{x} \quad \text{and} \quad f(x) = x^2 \quad .$$

So this first-order differential equation is linear. Adding $4/x \cdot y$ *to both sides, we then get the equation in standard form,*

$$\frac{dy}{dx} + \frac{4}{x}y = x^2 \quad .$$

On the other hand

$$\frac{dy}{dx} + \frac{4}{x}y^2 = x^2$$

is not linear because of the y^2.

In testing whether a given first-order differential equation is linear, it does not matter whether you attempt to rewrite the equation as

$$\frac{dy}{dx} = f(x) - p(x)y$$

or as

$$\frac{dy}{dx} + p(x)y = f(x) \quad .$$

If you can put it into either form, the equation is linear. You may prefer the first, simply because it is a natural form to look for after solving the equation for the derivative. However, because the second form (the standard form) is more suited for the methods normally used for solving these equations, more experienced workers typically prefer that form.

!►Example 5.2: *Consider the equation*

$$x^2\frac{dy}{dx} + x^3[y - \sin(x))] = 0 \quad .$$

Dividing through by x^2 *and doing a little multiplication and addition convert the equation to*

$$\frac{dy}{dx} + xy = x\sin(x) \quad ,$$

which is the standard form for a linear equation. So this differential equation is linear.

It is possible for a linear equation

$$\frac{dy}{dx} + p(x)y = f(x)$$

to also be a type of equation we've already studied. For example, if $p(x) = 0$ then the equation is

$$\frac{dy}{dx} = f(x) \quad ,$$

which is directly integrable. If, instead, $f(x) = 0$, the equation can be rewritten as

$$\frac{dy}{dx} = -p(x)y \quad ,$$

showing that it is separable. In addition, you can easily verify that a linear equation is separable if $f(x)$ is any constant multiple of $p(x)$.

If a linear equation is also directly integrable or separable, then it can be solved using methods already discussed. Otherwise, a small trick turns out to be very useful.

Deriving the Trick for Solving

Suppose we want to solve some first-order linear equation

$$\frac{dy}{dx} + py = f \tag{5.3}$$

(for brevity, $p = p(x)$ and $f = f(x)$). To avoid triviality, let's assume $p(x)$ is not always 0. Whether $f(x)$ vanishes or not will not be relevant.

The small trick to solving equation (5.3) comes from the product rule for derivatives: If μ and y are two functions of x, then

$$\frac{d}{dx}[\mu y] = \frac{d\mu}{dx}y + \mu\frac{dy}{dx} \quad .$$

Rearranging the terms on the right side, we get

$$\frac{d}{dx}[\mu y] = \mu\frac{dy}{dx} + \frac{d\mu}{dx}y \quad ,$$

and the right side of this equation looks a little like the left side of equation (5.3). To get a better match, let's multiply equation (5.3) by μ,

$$\mu\frac{dy}{dx} + \mu py = \mu f \quad .$$

With luck, the left side of this equation will match the right side of the last equation for the product rule, and we will have

$$\begin{aligned} \frac{d}{dx}[\mu y] &= \mu\frac{dy}{dx} + \frac{d\mu}{dx}y \\ &= \mu\frac{dy}{dx} + \mu py = \mu f \quad . \end{aligned} \tag{5.4}$$

This, of course, requires that

$$\frac{d\mu}{dx} = \mu p \quad .$$

Assuming this requirement is met, the equations in (5.4) hold. Cutting out the middle of that (and recalling that f and μ are functions of x only), we see that the differential equation reduces to

$$\frac{d}{dx}[\mu y] = \mu(x)f(x) \quad . \tag{5.5}$$

The advantage of having our differential equation in this form is that we can actually integrate both sides with respect to x, with the left side being especially easy since it is just a derivative with respect to x.

The function μ is called an *integrating factor* for the differential equation. As noted in the derivation, it must satisfy

$$\frac{d\mu}{dx} = \mu p \quad . \tag{5.6}$$

This is a simple separable differential equation for μ (remember, $p = p(x)$ is a known function). Any nonzero solution to this can be used as an integrating factor (the zero solution, $\mu = 0$, would simplify matters too much!). Applying the approach we learned for separable differential equations, we divide through by μ, integrate, and solve the resulting equation for μ:

$$\int \frac{1}{\mu}\frac{d\mu}{dx}\,dx = \int p(x)\,dx$$

$$\hookrightarrow \qquad \ln|\mu| = \int p(x)\,dx$$

$$\hookrightarrow \qquad \mu = \pm e^{\int p(x)\,dx}$$

Since we only need one function $\mu(x)$ satisfying requirement (5.6), we can drop both the "$\pm$" and any arbitrary constant arising from the integration of $p(x)$. This leaves us with a relatively simple formula for our integrating factor, namely,

$$\mu(x) = e^{\int p(x)\,dx} \tag{5.7}$$

where it is understood that we can let the constant of integration be zero.

5.2 Solving First-Order Linear Equations

As we just derived, the real 'trick' to solving a first-order linear equation is to reduce it to an easily integrated form via the use of an integrating factor. Here is a procedure for actually carrying out the necessary steps. To illustrate these steps, we will immediately use them to find the general solution to the equation from Example 5.1,

$$x\frac{dy}{dx} + 4y = x^3 \quad .$$

The Procedure:

1. Get the equation into the standard form for first-order linear differential equations,

$$\frac{dy}{dx} + p(x)y = f(x) \quad .$$

For our example, we just divide through by x, obtaining

$$\frac{dy}{dx} + \frac{4}{x}y = x^2 \quad .$$

As noted in Example 5.1, this is the desired form with

$$p(x) = \frac{4}{x} \qquad \text{and} \qquad f(x) = x^2 \quad .$$

2. Compute an integrating factor

$$\mu(x) = e^{\int p(x)\,dx} \quad .$$

Remember, since we only need one integrating factor, we can let the constant of integration be zero here.

For our example,

$$\mu(x) = e^{\int p(x)\,dx} = e^{\int 4/x\,dx} = e^{4\ln|x|} \quad .$$

Applying some basic identities for the natural logarithm, we can rewrite this last expression in a much more convenient form:

$$\mu(x) = e^{4\ln|x|} = e^{\ln\left|x^4\right|} = \left|x^4\right| = x^4 \quad .$$

3a. Multiply the differential equation (in standard form) by the integrating factor,

$$\mu\left[\frac{dy}{dx} + p(x)y = f(x)\right]$$

$$\hookrightarrow \qquad \mu\frac{dy}{dx} + \mu py = \mu f \quad ,$$

b. and observe that, via the product rule and choice of μ, the left side can be written as the derivative of the product of μ and y,

$$\underbrace{\mu\frac{dy}{dx} + \mu py}_{\frac{d}{dx}[\mu y]} = \mu f \quad ,$$

c. and then rewrite the differential equation as

$$\frac{d}{dx}[\mu y] = \mu f \quad ,$$

For our example, $\mu = x^4$. *Multiplying our equation by this and proceeding through the three substeps above yields*

$$x^4\left[\frac{dy}{dx} + \frac{4}{x}y = x^2\right]$$

$$\hookrightarrow \qquad \underbrace{x^4\frac{dy}{dx} + 4x^3y}_{\frac{d}{dx}[x^4y]} = x^6$$

$$\hookrightarrow \qquad \frac{d}{dx}[x^4y] = x^6 \quad .$$

4. Integrate with respect to x both sides of the last equation obtained,

$$\int \frac{d}{dx}[\mu y]\,dx = \int \mu(x)f(x)\,dx$$

$$\hookrightarrow \qquad \mu y = \int \mu(x)f(x)\,dx \quad .$$

Don't forget the arbitrary constant here!

Integrating the last equation in our example,

$$\int \frac{d}{dx}[x^4y]\,dx = \int x^6\,dx$$

$$\hookrightarrow \qquad x^4y = \frac{1}{7}x^7 + c \quad .$$

5. Finally, solve for y by dividing through by μ.

   For our example,

$$y = x^{-4}\left[\frac{1}{7}x^7 + c\right] = \frac{1}{7}x^3 + cx^{-4} \quad .$$

Later, we will use the above procedure to derive an explicit formula for computing y from $p(x)$ and $f(x)$. Unfortunately, it is not a particularly simple formula, and those who attempt to memorize it typically make more mistakes than those who simply remember the above procedure.

!►Example 5.3: *Consider*

$$e^x \frac{dy}{dx} = 20 + 3e^x y \quad , \quad y(0) = 7 \quad .$$

Subtracting $3e^x y$ from both sides and then multiplying through by e^{-x} puts this linear differential equation into the desired form,

$$\frac{dy}{dx} - 3y = 20e^{-x} \quad .$$

So $p(x) = -3$, and our integrating factor is

$$\mu = \mu(x) = e^{\int -3\,dx} = e^{-3x} \quad .$$

Multiplying the differential equation by μ and following the rest of the steps in our procedure gives us the following:

$$e^{-3x}\left[\frac{dy}{dx} - 3y = 20e^{-x}\right]$$

$$\hookrightarrow \quad \underbrace{e^{-3x}\frac{dy}{dx} - 3e^{-3x}y}_{\frac{d}{dx}[e^{-3x}y]} = 20e^{-4x}$$

$$\hookrightarrow \quad \frac{d}{dx}[e^{-3x}y] = 20e^{-4x}$$

$$\hookrightarrow \quad \int \frac{d}{dx}[e^{-3x}y]\,dx = \int 20e^{-4x}\,dx$$

$$\hookrightarrow \quad e^{-3x}y = -5e^{-4x} + c$$

$$\hookrightarrow \quad y = e^{3x}\left[-5e^{-4x} + c\right] \quad .$$

So the general solution to our differential equation is

$$y(x) = -5e^{-x} + ce^{3x} \quad .$$

Using this formula for $y(x)$ with the initial condition gives us

$$7 = y(0) = -5e^{-0} + ce^{3\cdot 0} = -5 + c \quad .$$

Thus,

$$c = 7 + 5 = 12 \quad ,$$

and the solution to the given initial-value problem is

$$y(x) = -5e^{-x} + 12e^{3x} \quad .$$

Let us briefly get back to our requirement for $\mu = \mu(x)$ being an integrating factor for

$$\frac{dy}{dx} + py = f \quad .$$

That requirement was equation (5.6),

$$\frac{d\mu}{dx} = \mu p \quad .$$

Now, in computing this μ, you will often get something like

$$\mu(x) = |\mu_0(x)|$$

where $\mu_0(x)$ is a relatively simple continuous function (e.g., $\mu(x) = |\sin(x)|$). Consequently, on any interval over which the graph of $\mu_0(x)$ never crosses the X–axis,

$$\mu_0(x) = \mu(x) \qquad \text{or} \qquad \mu_0(x) = -\mu(x) \quad .$$

Either way,

$$\frac{d\mu_0}{dx} = \frac{d[\pm\mu]}{dx} = \pm\frac{d\mu}{dx} = \pm\mu p = \mu_0 p \quad .$$

So μ_0 also satisfies the requirement for being an integrating factor for the given differential equation. This means that, if in computing μ you do get something like

$$\mu(x) = |\mu_0(x)|$$

where $\mu_0(x)$ is a relatively simple function, then you can ignore the absolute value brackets and just use μ_0 for your integrating factor.

!►Example 5.4: *Consider solving the linear differential equation*

$$\frac{dy}{dx} + \cot(x)y = x\csc(x) \quad .$$

This equation is already in the desired form. In a case like this, it is often a good idea to see what the equation looks like in terms of sines and cosines,

$$\frac{dy}{dx} + \left[\frac{\cos(x)}{\sin(x)}\right]y = \frac{x}{\sin(x)} \quad .$$

To find $\mu = e^{\int p\,dx}$, first observe that, ignoring the constant of integration,

$$\int p(x)\,dx = \int \frac{\cos(x)}{\sin(x)}\,dx = \int \frac{\frac{d}{dx}\sin(x)}{\sin(x)}\,dx = \ln|\sin(x)| \quad .$$

So

$$\mu = \mu(x) = e^{\int p(x)\,dx} = e^{\ln|\sin(x)|} = |\sin(x)| \quad .$$

As discussed above, we can just drop the $|\cdot|$ and use $\sin(x)$ for the integrating factor. Doing so, and stepping through the rest of our procedure, we have

$$\sin(x)\left[\frac{dy}{dx} + \frac{\cos(x)}{\sin(x)}y = \frac{x}{\sin(x)}\right]$$

$$\hookrightarrow \qquad \underbrace{\sin(x)\frac{dy}{dx} + \cos(x)y}_{\frac{d}{dx}[\sin(x)y]} = x$$

$$\hookrightarrow \qquad \int \frac{d}{dx}[\sin(x)y]\,dx = \int x\,dx$$

$$\hookrightarrow \qquad \sin(x)y = \frac{1}{2}x^2 + c_1$$

$$\hookrightarrow \qquad y = \frac{x^2 + c}{2\sin(x)} \quad .$$

5.3 On Using Definite Integrals with Linear Equations

Integration arises twice in our method for solving

$$\frac{dy}{dx} + p(x)y = f(x) \quad .$$

It first arises when we integrate p to get the integrating factor,

$$\mu(x) = e^{\int p(x)\,dx} \quad .$$

It is needed again when we then integrate both sides of the corresponding equation

$$\frac{d}{dx}[\mu y] = \mu f \quad .$$

At either point, of course, we could use definite integrals instead of indefinite integrals.

Let's first look at what happens when we integrate both sides of the last equation using definite integrals. Remember, everything is a function of x, so this equation can be written a bit more explicitly as

$$\frac{d}{dx}[\mu(x)y(x)] = \mu(x)f(x) \quad .$$

As before, to avoid having x represent two different entities, we replace the x's with another variable, say, s, and rewrite our current differential equation as

$$\frac{d}{ds}[\mu(s)y(s)] = \mu(s)f(s) \quad .$$

Then we pick a convenient lower limit a for our integration and integrate each side of the above with respect to s from $s = a$ to $s = x$,

$$\int_a^x \frac{d}{ds}[\mu(s)y(s)]\,ds = \int_a^x \mu(s)f(s)\,ds \quad . \tag{5.8}$$

But

$$\int_a^x \frac{d}{ds}[\mu(s)y(s)]\,ds = \mu(s)y(s)\Big|_a^x = \mu(x)y(x) - \mu(a)y(a) \quad .$$

So equation (5.8) reduces to

$$\mu(x)y(x) - \mu(a)y(a) = \int_a^x \mu(s)f(s)\,ds \quad ,$$

Solving this for $y(x)$ yields

$$y(x) = \frac{1}{\mu(x)}\left[\mu(a)y(a) + \int_a^x \mu(s)f(s)\,ds\right] \quad . \tag{5.9}$$

This is not a simple enough formula to be worth memorizing (especially since you still have to remember what μ is). Nonetheless, it is a formula worth knowing about for at least two good reasons:

1. This formula can automatically take into account an initial value $y(x_0) = y_0$. All we have to do is to choose the lower limit a to be x_0. Then formula (5.9) tells us that the solution to

$$\frac{dy}{dx} + py = f \qquad \text{with} \quad y(x_0) = y_0$$

is

$$y = \frac{1}{\mu(x)}\left[\mu(x_0)y_0 + \int_{x_0}^{x} \mu(s)f(s)\,ds\right] \quad . \tag{5.10}$$

2. Even if we cannot determine a relatively nice formula for the integral of μf (for a given choice of μ and f), the value of the integral in formula (5.9) can, in practice, still be accurately computed for desired values of x using numerical integration routines found in standard computer math packages. Indeed, using any of these packages and formula (5.9), you could probably program a computer to accurately compute $y(x)$ for a number of values of x and use these values to produce a very accurate graph of y.

!►Example 5.5: *Consider solving*

$$\frac{dy}{dx} - 2xy = 4 \qquad \textit{with} \quad y(0) = 3 \quad .$$

The differential equation is clearly linear and in the desired form for the first step of our procedure. Computing the integrating factor, we find that, here,

$$\mu = e^{\int p(x)\,dx} = e^{\int [-2x]\,dx} = e^{-x^2+c} \quad .$$

Choosing, as we may, c to be zero, we then get

$$\mu(x) = e^{-x^2} \quad .$$

Plugging this into formula (5.9) (and choosing $a = 0$ since we have $y(0) = 3$ as the initial condition) yields

$$\begin{aligned} y(x) &= \frac{1}{\mu(x)}\left[\mu(0)y(0) + \int_0^x \mu(s)f(s)\,ds\right] \\ &= \frac{1}{e^{-x^2}}\left[e^{-0^2}\cdot 3 + \int_0^x e^{-s^2}\,4\,ds\right] \\ &= e^{x^2}\left[3 + 4\int_0^x e^{-s^2}\,ds\right] \quad . \end{aligned}$$

This is the solution to our initial-value problem. The integral,

$$\int_0^x e^{-s^2}\,ds \quad ,$$

was left unevaluated because no one has yet found a "nice" formula for this integral. At best, we can 'hide' this integral by using the error function (see page 27), rewriting our formula for y as

$$y(x) = e^{x^2}\left[3 + 2\sqrt{\pi}\,\mathrm{erf}(x)\right] \quad .$$

Still, to find the value of, say, $y(4)$, we would have to either numerically approximate the integral in

$$y(4) = e^{4^2}\left[3 + 4\int_0^4 e^{-s^2}\,ds\right]$$

or look up the value of the error function in

$$y(4) = e^{4^2}\left[3 + 2\sqrt{\pi}\,\mathrm{erf}(4)\right] \quad .$$

Either way, a decent computer math package could be helpful.

As already noted, we could also use a definite integral in determining the integrating factor. This means μ would be given by

$$\mu(x) = \exp\left(\int_a^x p(s)\,ds\right)$$

where a was any appropriate lower limit. Naturally, if we had an initial condition $y(x_0) = y_0$, it would make sense to let $a = x_0$. This would slightly simplify formula (5.10) to

$$y = \frac{1}{\mu(x)}\left[y_0 + \int_{x_0}^x \mu(s)f(s)\,ds\right] \tag{5.11}$$

since

$$\mu(x_0) = \exp\left(\int_{x_0}^{x_0} p(s)\,ds\right) = e^0 = 1 \quad .$$

In practice, there is little to be gained in using a definite integral in the computation of μ *unless* there is not a reasonable formula for the integral of p. Then we are pretty well forced into using a definite integral to compute $\mu(x)$ and to computing this integral numerically for each value of x of interest. That, in turn, would pretty well force us to compute $y(x)$ for each x of interest by using numerical computation of formula (5.10).

5.4 Integrability, Existence and Uniqueness

If you check, you will see that our derivation of the definite integral formula

$$y(x) = \frac{1}{\mu(x)}\left[y_0 + \int_{x_0}^x \mu(s)f(s)\,ds\right] \qquad \text{with} \quad \mu(x) = \exp\left(\int_{x_0}^x p(s)\,ds\right)$$

as a solution to the initial-value problem

$$\frac{dy}{dx} + p(x)y = f(x) \qquad \text{with} \quad y(x_0) = y_0$$

merely required that y be any solution to this problem, and that p and f be 'sufficiently integrable' for the existence of the integrals involving them. That is, every solution to this initial-value problem is given by this *one* formula. Conversely, as long as p and f are 'sufficiently integrable', you can use elementary calculus to differentiate the above definite integral formula and verify that the y defined by this formula is a solution to the above initial-value problem (see problem 5.5). Thus, the above definite integral formula gives us *the one and only* solution to the above initial-value problem, provided p and f are 'sufficiently integrable'.

Just what is 'sufficiently integrable'? Basically, we want the integrals

$$\int_{x_0}^{x} p(s)\,ds \qquad \text{and} \qquad \int_{x_0}^{x} \mu(s) f(s)\,ds$$

to be well-defined, continuous functions of x in whatever interval of interest (α, β) we have. (Note that this ensures

$$\mu(x) = \exp\left(\int_{x_0}^{x} p(s)\,ds\right)$$

is never zero in this interval.) Certainly, p and f will be 'sufficiently integrable' if they are continuous on (α, β). But continuity is not necessary; p and f can have a few discontinuities provided these discontinuities are not too bad. In particular, we can allow the same piecewise-defined functions considered back in Section 2.4. That (along with Theorem 2.1 on page 30) gives us the following existence and uniqueness theorem for initial-value problems involving first-order linear differential equations.

Theorem 5.1 (existence and uniqueness)
Let p and f be functions that are continuous except for, at most, a finite number of finite-jump discontinuities in an interval (α, β). Also let x_0 and y_0 be any two numbers with $\alpha < x_0 < \beta$. Then the initial-value problem

$$\frac{dy}{dx} + p(x)y = f(x) \qquad \text{with} \quad y(x_0) = y_0$$

has exactly one solution over the interval (α, β), and that solution is given by

$$y(x) = \frac{1}{\mu(x)}\left[y_0 + \int_{x_0}^{x} \mu(s) f(s)\,ds\right] \qquad \text{with} \quad \mu(x) = \exp\left(\int_{x_0}^{x} p(s)\,ds\right) \quad .$$

Additional Exercises

5.1. *Determine whether each of the following differential equations is or is not linear, and, if it is linear, rewrite the equation in standard form,*

$$\frac{dy}{dx} + p(x)y = f(x) \quad .$$

a. $x^2\frac{dy}{dx} + 3x^2 y = \sin(x)$

b. $y^2\frac{dy}{dx} + 3x^2 y = \sin(x)$

c. $\frac{dy}{dx} - xy^2 = \sqrt{x}$

d. $\frac{dy}{dx} = 1 + (xy + 3y)^2$

e. $\frac{dy}{dx} = 1 + xy + 3y$

f. $\frac{dy}{dx} = 4y + 8$

g. $\frac{dy}{dx} - e^{2x} = 0$

h. $\frac{dy}{dx} = \sin(x)\,y$

i. $\frac{dy}{dx} + 4y = y^3$

j. $x\frac{dy}{dx} + \cos\left(x^2\right) = 827y$

5.2. *Using the methods developed in this chapter, find the general solution to each of the following first-order linear differential equations:*

a. $\dfrac{dy}{dx} + 2y = 6$

b. $\dfrac{dy}{dx} + 2y = 20e^{3x}$

c. $\dfrac{dy}{dx} = 4y + 16x$

d. $\dfrac{dy}{dx} - 2xy = x$

e. $x\dfrac{dy}{dx} + 3y - 10x^2 = 0$

f. $x^2\dfrac{dy}{dx} + 2xy = \sin(x)$

g. $x\dfrac{dy}{dx} = \sqrt{x} + 3y$

h. $\cos(x)\dfrac{dy}{dx} + \sin(x)\,y = \cos^2(x)$

i. $x\dfrac{dy}{dx} + (5x+2)y = \dfrac{20}{x}$

j. $2\sqrt{x}\dfrac{dy}{dx} + y = 2xe^{-\sqrt{x}}$

5.3. *Find the solution to each of the following initial-value problems using the methods developed in this chapter:*

a. $\dfrac{dy}{dx} - 3y = 6$ *with* $y(0) = 5$

b. $\dfrac{dy}{dx} - 3y = 6$ *with* $y(0) = -2$

c. $\dfrac{dy}{dx} + 5y = e^{-3x}$ *with* $y(0) = 0$

d. $x\dfrac{dy}{dx} + 3y = 20x^2$ *with* $y(1) = 10$

e. $x\dfrac{dy}{dx} = y + x^2\cos(x)$ *with* $y\left(\dfrac{\pi}{2}\right) = 0$

f. $(1+x^2)\dfrac{dy}{dx} = x\left[3 + 3x^2 - y\right]$ *with* $y(2) = 8$

5.4. *Express the answer to each of the following initial-value problems in terms of definite integrals:*

a. $\dfrac{dy}{dx} + 6xy = \sin(x)$ *with* $y(0) = 4$

b. $x^2\dfrac{dy}{dx} + xy = \sqrt{x}\sin(x)$ *with* $y(2) = 5$

c. $x\dfrac{dy}{dx} - y = x^2e^{-x^2}$ *with* $y(3) = 8$

5.5. *Let (α, β) be an interval, and let x_0 and y_0 be any two numbers with $\alpha < x_0 < \beta$. Assume p and f are functions continuous at all but, at most, a finite number of points in (α, β), and that each of these discontinuities is a finite-jump discontinuity. Define $\mu(x)$ and $y(x)$ by*

$$\mu(x) = \exp\left(\int_{x_0}^{x} p(s)\,ds\right)$$

and

$$y(x) = \frac{1}{\mu(x)}\left[y_0 + \int_{x_0}^{x} \mu(s)f(s)\,ds\right] .$$

Compute the first derivatives of μ and y, and then verify that y satisfies the initial condition $y(x_0) = y_0$ as well as the differential equation

$$\frac{dy}{dx} + p(x)y = f(x) \qquad \textit{for} \quad \alpha < x < \beta .$$

6

Simplifying Through Substitution

In previous chapters, we saw how certain types of first-order differential equations (directly integrable, separable, and linear equations) can be identified and put into forms that can be integrated with relative ease. In this chapter, we will see that, sometimes, we can start with a differential equation that is not one of these desirable types and construct a corresponding separable or linear equation whose solution can then be used to construct the solution to the original differential equation.

6.1 Basic Notions

There are many first-order differential equations, such as

$$\frac{dy}{dx} = (x+y)^2 \quad ,$$

that are neither linear nor separable, and which do not yield up their solutions by direct application of the methods developed thus far. One way of attempting to deal with such equations is to replace y with a cleverly chosen formula of x and "u" where u denotes another unknown function of x. This results in a new differential equation with u being the function of interest. If the substitution truly is clever, then this new differential equation will be separable or linear (or, maybe, even directly integrable), and can be be solved for u in terms of x using methods discussed in previous chapters. Then the function of real interest, y, can be determined from the original 'clever' formula relating u, y and x.

Here are the basic steps to this approach, described in a little more detail and illustrated by being used to solve the above differential equation:

1. Identify what is hoped will be a good formula of x and u for y,

$$y = F(x,u) \quad .$$

This 'good formula' is our substitution for y. Here, u represents another unknown function of x (so "$u = u(x)$"), and the above equation tells us how the two unknown functions y and u are related. (Identifying that 'good formula' is the tricky part. We'll discuss that further in a little bit.)

Let's try a substitution that reduces the right side of our differential equation,

$$\frac{dy}{dx} = (x+y)^2 \quad ,$$

to u^2. This means setting $u = x + y$. Solving this for y gives our substitution,

$$y = u - x \quad .$$

2. Replace *every* occurrence of y in the given differential equation with that formula of x and u, *including the y in the derivative.* Keep in mind that u is a function of x, so the ${}^{dy}/_{dx}$ will become a formula of x, u, and ${}^{du}/_{dx}$ (it may be wise to first compute ${}^{dy}/_{dx}$ separately).

   Since we are using $y = u - x$ (equivalently, $u = x + y$), we have

   $$(x + y)^2 = u^2 \quad ,$$

   and

   $$\frac{dy}{dx} = \frac{d}{dx}[u - x] = \frac{du}{dx} - \frac{dx}{dx} = \frac{du}{dx} - 1 \quad .$$

   So, under the substitution $y = u - x$,

   $$\frac{dy}{dx} = (x + y)^2$$

   becomes

   $$\frac{du}{dx} - 1 = u^2 \quad .$$

3. Solve the resulting differential equation for u (don't forget the constant solutions!). If possible, get an explicit solution for u in terms of x. (This assumes, of course, that the differential equation for u is one we can solve. If it isn't, then our substitution wasn't that clever, and we may have to try something else.)

   Adding 1 to both sides of the differential equation just derived for u yields

   $$\frac{du}{dx} = u^2 + 1 \quad ,$$

   which we recognize as being a relatively easily solved separable equation with no constant solutions. Dividing through by $u^2 + 1$ and integrating,

   $$\frac{1}{u^2 + 1}\frac{du}{dx} = 1$$

   $$\hookrightarrow \qquad \int \frac{1}{u^2 + 1}\frac{du}{dx}\,dx = \int 1\,dx$$

   $$\hookrightarrow \qquad \arctan(u) = x + c$$

   $$\hookrightarrow \qquad u = \tan(x + c) \quad .$$

4. If you get an explicit solution $u = u(x)$, then just plug that formula $u(x)$ into the original substitution to get the explicit solution to the original equation,

   $$y(x) = F(x, u(x)) \quad .$$

   If, instead, you only get an implicit solution for u, then go back to the original substitution, $y = F(x, u)$, solve that to get a formula for u in terms of x and y (unless you already have this formula for u), and substitute that formula for u into the solution obtained to convert it to the corresponding implicit solution for y.

Our original substitution was $y = u - x$ *. Combining this with the formula for* u *just obtained, we get*

$$y = u - x = \tan(x + c) - x$$

as a general solution to our original differential equation,

$$\frac{dy}{dx} = (x + y)^2 \quad .$$

The key to this approach is, of course, in identifying a substitution, $y = F(x, u)$, that converts the original differential equation for y to a differential equation for u that can be solved with reasonable ease. Unfortunately, there is no single method for identifying such a substitution. At best, we can look at certain equations and make good guesses at substitutions that are likely to work. We will next look at three cases where good guesses can be made. In these cases the suggested substitutions are guaranteed to lead to either separable or linear differential equations. As you may suspect, though, they are not guaranteed to lead to simple separable or linear differential equations.

6.2 Linear Substitutions

If the given differential equation can be rewritten so that the derivative equals some formula of $Ax + By + C$,

$$\frac{dy}{dx} = f(Ax + By + C) \quad ,$$

where A, B and C are known constants, then a good substitution comes from setting

$$u = Ax + By + C \quad ,$$

and then solving for y. For convenience, we'll call this a *linear* substitution.[1]

We've already seen one case where a linear substitution works — in the example above illustrating the general substitution method. Here is another example, one in which we end up with an implicit solution.

!►Example 6.1: *To solve*

$$\frac{dy}{dx} = \frac{1}{2x - 4y + 7} \quad ,$$

we use the substitution based on setting

$$u = 2x - 4y + 7 \quad .$$

Solving this for y *and then differentiating yields*

$$y = \frac{1}{4}[2x - u + 7] = \frac{x}{2} - \frac{u}{4} + \frac{7}{4}$$

and

$$\frac{dy}{dx} = \frac{d}{dx}\left[\frac{x}{2} - \frac{u}{4} - \frac{7}{4}\right] = \frac{1}{2} - \frac{1}{4}\frac{du}{dx} \quad .$$

[1] because $Ax + By + C = 0$ is the equation for a straight line

So, the substitution based on $u = 2x - 4y + 7$ *converts*

$$\frac{dy}{dx} = \frac{1}{2x - 4y + 7}$$

to

$$\frac{1}{2} - \frac{1}{4}\frac{du}{dx} = \frac{1}{u} \quad .$$

This differential equation for u *looks manageable, especially since it contains no* x*'s. Solving for the derivative in this equation, we get*

$$\frac{du}{dx} = -4\left[\frac{1}{u} - \frac{1}{2}\right] = -4\left[\frac{2}{2u} - \frac{u}{2u}\right] = -4\left[\frac{2-u}{2u}\right] \quad ,$$

which simplifies to

$$\frac{du}{dx} = 2\left[\frac{u-2}{u}\right] \quad . \tag{6.1}$$

Again, this is a separable equation. This time, though, the differential equation has a constant solution,

$$u = 2 \quad . \tag{6.2}$$

To find the other solutions to our differential equation for u*, we multiply both sides of equation (6.1) by* u *and divide through by* $u - 2$*, obtaining*

$$\frac{u}{u-2}\frac{du}{dx} = 2 \quad .$$

After noticing that

$$\frac{u}{u-2} = \frac{u-2+2}{u-2} = \frac{u-2}{u-2} + \frac{2}{u-2} = 1 + \frac{2}{u-2} \quad ,$$

we can integrate both sides of our last differential equation for u*,*

$$\int \frac{u}{u-2}\frac{du}{dx}\,dx = \int 2\,dx$$

$$\hookrightarrow \qquad \int \left[1 + \frac{2}{u-2}\right] du = 2x + c$$

$$\hookrightarrow \qquad u + 2\ln|u-2| = 2x + c \quad . \tag{6.3}$$

Sadly, the last equation is not one we can solve to obtain an explicit formula for u *in terms of* x*. So we are stuck with using it as an implicit solution of our differential equation for* u*.*

Together, formula (6.2) and equation (6.3) give us all the solutions to the differential equation for u*. To obtain all the solutions to our original differential equation for* y*, we must recall the original (equivalent) relations between* u *and* y*,*

$$u = 2x - 4y + 7 \qquad \text{and} \qquad y = \frac{x}{2} - \frac{u}{4} + \frac{7}{4} \quad .$$

The latter with the constant solution $u = 2$ *(formula (6.2)) yields*

$$y = \frac{x}{2} - \frac{2}{4} + \frac{7}{4} = \frac{x}{2} + \frac{5}{4} \quad .$$

On the other hand, it is easier to combine the first relation between u *and* y *with the implicit solution for* u *in equation (6.3),*

$$u = 2x - 4y + 7 \qquad \text{with} \qquad u + 2\ln|u-2| = 2x + c \quad ,$$

obtaining

$$[2x - 4y + 7] + 2\ln|[2x - 4y + 7] - 2| = 2x + c \quad .$$

After a little algebra, this simplifies to

$$\ln|2x - 4y + 5| = 4y + C \quad .$$

which does not include the "constant u *" solution above. So, for* $y = y(x)$ *to be a solution to our original differential equation, it must either be given by*

$$y = \frac{x}{2} + \frac{5}{4}$$

or satisfy

$$\ln|2x - 4y + 5| = 4y + C \quad .$$

Let us see what happens whenever we have a differential equation of the form

$$\frac{dy}{dx} = f(Ax + By + C)$$

(where A, B and C are known constants), and we attempt the substitution based on setting

$$u = Ax + By + C \quad .$$

Solving for y and then differentiating yields

$$y = \frac{1}{B}[u - Ax - C] \qquad \text{and} \qquad \frac{dy}{dx} = \frac{1}{B}\left[\frac{du}{dx} - A\right] \quad .$$

Under these substitutions,

$$\frac{dy}{dx} = f(Ax + By + C)$$

becomes

$$\frac{1}{B}\left[\frac{du}{dx} - A\right] = f(u) \quad .$$

After a little algebra, this can be rewritten as

$$\frac{du}{dx} = A + Bf(u) \quad ,$$

which is clearly a separable equation. Thus, we will always get a separable differential equation for u. Moreover, the ease with which this differential equation can be solved clearly depends only on the ease with which we can evaluate

$$\int \frac{1}{A + Bf(u)}\,du \quad .$$

6.3 Homogeneous Equations

We now consider first-order differential equations in which the derivative can be viewed as a formula of the ratio y/x. In other words, we are now interested in any differential equation that can be rewritten as

$$\frac{dy}{dx} = f\left(\frac{y}{x}\right) \tag{6.4}$$

where f is some function of a single variable. Such equations are sometimes said to be *homogeneous*.[2] Unsurprisingly, the substitution based on setting

$$u = \frac{y}{x} \qquad \text{(i.e., } y = xu \text{)}$$

is often useful in solving these equations. We will, in fact, discover that this substitution will always transform an equation of the form (6.4) into a separable differential equation.

!►Example 6.2: *Consider the differential equation*

$$xy^2\frac{dy}{dx} = x^3 + y^3 \quad .$$

Dividing through by xy^2 and doing a little factoring yields

$$\frac{dy}{dx} = \frac{x^3 + y^3}{xy^2} = \frac{x^3\left[1 + \dfrac{y^3}{x^3}\right]}{x^3\left[\dfrac{y^2}{x^2}\right]} \quad ,$$

which simplifies to

$$\frac{dy}{dx} = \frac{1 + \left[\dfrac{y}{x}\right]^3}{\left[\dfrac{y}{x}\right]^2} \quad . \tag{6.5}$$

That is,

$$\frac{dy}{dx} = f\left(\frac{y}{x}\right) \qquad \text{with} \quad f(\text{whatever}) = \frac{1 + \text{whatever}^3}{\text{whatever}^2} \quad .$$

So we should try letting

$$u = \frac{y}{x}$$

or, equivalently,

$$y = xu \quad .$$

On the right side of equation (6.5), replacing y with xu is just the same as replacing each y/x with u. Either way, the right side becomes

$$\frac{1 + u^3}{u^2} \quad .$$

On the left side of equation (6.5), the substitution $y = xu$ is in the derivative. Keeping in mind that u is also a function of x, we have

$$\frac{dy}{dx} = \frac{d}{dx}[xu] = \frac{dx}{dx}u + x\frac{du}{dx} = u + x\frac{du}{dx} \quad .$$

[2] Warning: Later we will refer to a completely different type of differential equation as being "homogeneous".

So,

$$\frac{dy}{dx} = \frac{1 + \left[\frac{y}{x}\right]^3}{\left[\frac{y}{x}\right]^2} \quad \overset{y=xu}{\Longrightarrow} \quad u + x\frac{du}{dx} = \frac{1+u^3}{u^2} \quad .$$

Solving the last equation for du/dx *and doing a little algebra, we see that*

$$\frac{du}{dx} = \frac{1}{x}\left[\frac{1+u^3}{u^2} - u\right] = \frac{1}{x}\left[\frac{1+u^3}{u^2} - \frac{u^3}{u^2}\right] = \frac{1}{x}\left[\frac{1+u^3-u^3}{u^2}\right] = \frac{1}{xu^2} \quad .$$

How nice! Our differential equation for u *is the very simple separable equation*

$$\frac{du}{dx} = \frac{1}{xu^2} \quad .$$

Multiplying through by u^2 *, integrating, and doing a little more algebra:*

$$\int u^2 \frac{du}{dx}\, dx = \int \frac{1}{x}\, dx$$

$$\hookrightarrow \qquad \frac{1}{3}u^3 = \ln|x| + c$$

$$\hookrightarrow \qquad u^3 = 3\ln|x| + 3c$$

$$\hookrightarrow \qquad u = \sqrt[3]{3\ln|x| + 3c} \quad .$$

Combining this with our substitution $y = xu$ *then gives*

$$y = xu = x\left[\sqrt[3]{3\ln|x| + 3c}\right] = x\sqrt[3]{3\ln|x| + C}$$

as the general solution to our original differential equation.

In practice, it may not be immediately obvious if a given first-order differential equation can be written in form (6.4), but it is usually fairly easy to find out. First, algebraically solve the differential equation for the derivative to get

$$\frac{dy}{dx} = \text{“some formula of } x \text{ and } y\text{”} \quad .$$

With a little luck, you'll be able to do a little algebra (as we did in the above example) to see if that "formula of x and y" can be written as just a formula of y/x, $f(y/x)$.

If it's still not clear, then just go ahead and try the substitution $y = xu$ in that "formula of x and y". If all the x's cancel out and you are left with a formula of u, then that formula, $f(u)$, is the right side of (6.4) (remember, $u = y/x$). So the differential equation can be written in the desired form. Moreover, half the work in plugging the substitution into the differential equation is now done.

On the other hand, if the x's do not cancel out when you substitute xu for y, then the differential equation cannot be written in form (6.4), and there is only a small chance that this substitution will yield an 'easily solved' differential equation for u.

!►Example 6.3: *Again, consider the differential equation*

$$xy^2\frac{dy}{dx} = x^3 + y^3 \quad ,$$

which we had already studied in the previous example. Solving for the derivative again yields

$$\frac{dy}{dx} = \frac{x^3 + y^3}{xy^2} \quad .$$

Instead of factoring out x^3 from the numerator and denominator of the right side, let's go ahead and try the substitution $y = xu$ and see if the x's cancel out:

$$\frac{x^3 + y^3}{xy^2} = \frac{x^3 + [xu]^3}{x[xu]^2} = \frac{x^3 + x^3u^3}{x^3u^2} = \frac{x^3\left(1 + u^3\right)}{x^3u^2} \quad .$$

The x's clearly do cancel out, leaving us with

$$\frac{1 + u^3}{u^2} \quad .$$

Thus, (as we already knew), our differential equation can be put into form (6.4). What's more, getting our differential equation into that form and using $y = xu$ will lead to

$$\frac{1 + u^3}{u^2}$$

for the right side, just as we saw in the previous example.

When employing the substitution $y = xu$ to solve

$$\frac{dy}{dx} = f\left(\frac{y}{x}\right) \quad ,$$

do not forget to treat u as a function of x ! Thus, when we differentiate y, we have

$$\frac{dy}{dx} = \frac{d}{dx}[xu] = \frac{dx}{dx}u + x\frac{du}{dx} = u + x\frac{du}{dx} \quad .$$

This is not a formula worth memorizing — you shouldn't even bother memorizing $y = xu$ — it should be quite enough to remember that $u = u(x)$ with $u = {}^{y}/_{x}$.

However, it is worth noting that, if we plug these substitutions into

$$\frac{dy}{dx} = f\left(\frac{y}{x}\right) \quad ,$$

we always get

$$u + x\frac{du}{dx} = f(u) \quad ,$$

which is the same as

$$\frac{du}{dx} = \frac{f(u) - u}{x} \quad .$$

This confirms that we will always get a separable equation, just as with linear substitutions. This time, the ease with which the differential equation for u can be solved depends on the ease with which we can evaluate

$$\int \frac{1}{f(u) - u}\,du \quad .$$

6.4 Bernoulli Equations

A *Bernoulli equation* is a first-order differential equation that can be written in the form

$$\frac{dy}{dx} + p(x)y = f(x)y^n \tag{6.6}$$

where $p(x)$ and $f(x)$ are known functions of x only, and n is some real number. This looks much like the standard form for linear equations. Indeed, a Bernoulli equation is linear if $n = 0$ or $n = 1$ (and is also separable if $n = 1$). Consequently, our main interest is in solving such an equation when n is neither 0 nor 1.

The above equation can be solved using a substitution, though a good choice for that substitution might not be immediately obvious. You might suspect that setting $u = y^n$ would help, but it doesn't — unless, that is, it leads you to try a substitution based on

$$u = y^r$$

where r is some value yet to be determined. If you solve this for y in terms of u and plug the resulting formula for y into the Bernoulli equation, you will then discover, after a bit of calculus and algebra, that you have a linear differential equation for u if and only if $r = 1 - n$ (see problem 6.8). So the substitution that does work is the one based on setting

$$u = y^{1-n} \quad .$$

In the future, you can either remember this, re-derive it as needed, or know where to look it up.

You should also observe that, if $n > 0$, then the constant function

$$y(x) = 0 \qquad \text{for all} \quad x$$

is a solution to equation (6.6). This particular solution is often overlooked when using the substitution $u = y^{1-n}$ for a reason noted in the next example.

!▶Example 6.4: *Consider the differential equation*

$$\frac{dy}{dx} + 6y = 30e^{3x}y^{2/3} \quad .$$

This is in form (6.6), with $n = 2/3$. Right off, let's note that this Bernoulli equation has a constant solution $y = 0$.

Setting

$$u = y^{1-n} = y^{1-2/3} = y^{1/3} \quad ,$$

we see that the substitution

$$y = u^3$$

is called for. Plugging this into our original differential equation, we get

$$\frac{dy}{dx} + 6y = 30e^{3x}y^{2/3}$$

$$\hookrightarrow \quad \frac{d}{dx}\left[u^3\right] + 6\left[u^3\right] = 30e^{3x}\left[u^3\right]^{2/3}$$

$$\hookrightarrow \quad 3u^2\frac{du}{dx} + 6u^3 = 30e^{3x}u^2 \quad .$$

Dividing this last equation through by $3u^2$ gives

$$\frac{du}{dx} + 2u = 10e^{3x} \quad .$$

(This division assumes $u \neq 0$, corresponding to an assumption that $y \neq 0$. That is why the $y = 0$ solution is often overlooked.)

The last equation is a relatively simple linear equation with integrating factor

$$\mu = e^{\int 2\,dx} = e^{2x} \quad .$$

Continuing as usual with such equations,

$$e^{2x}\left[\frac{du}{dx} + 2u = 10e^{3x}\right]$$

$$\hookrightarrow \qquad e^{2x}\frac{du}{dx} + 2e^{2x}u = 10e^{5x}$$

$$\hookrightarrow \qquad \frac{d}{dx}[e^{2x}u] = 10e^{5x} \quad .$$

Integrating both sides with respect to x then yields

$$e^{2x}u = \int 10e^{5x}\,dx = 2e^{5x} + c \quad ,$$

which tells us that

$$u = e^{-2x}\left[2e^{5x} + c\right] = 2e^{3x} + ce^{-2x} \quad .$$

Finally, after recalling the substitution that led to the differential equation for u (and the fact that $y = 0$ is a solution, we obtain our general solution to the given Bernoulli equation,

$$y(x) = u^3 = \left[2e^{3x} + ce^{-2x}\right]^3 \quad \text{and} \quad y(x) = 0 \quad .$$

Additional Exercises

6.1. *Use linear substitutions (as described in Section 6.2) to find a general solution to each of the following:*

a. $\dfrac{dy}{dx} = \dfrac{1}{(3x+3y+2)^2}$ **b.** $\dfrac{dy}{dx} = \dfrac{(3x-2y)^2+1}{3x-2y} + \dfrac{3}{2}$

c. $\cos(4y - 8x + 3)\dfrac{dy}{dx} = 2 + 2\cos(4y - 8x + 3)$

6.2. *Using a linear substitution, solve the initial-value problem*

$$\frac{dy}{dx} = 1 + (y-x)^2 \quad \text{with} \quad y(0) = \frac{1}{4} \quad .$$

6.3. *Use substitutions appropriate to homogeneous first-order differential equations (as described in Section 6.3) to find a general solution to each of the following:*

a. $x^2\frac{dy}{dx} - xy = y^2$

b. $\frac{dy}{dx} = \frac{y}{x} + \frac{x}{y}$

c. $\cos\left(\frac{y}{x}\right)\left[\frac{dy}{dx} - \frac{y}{x}\right] = 1 + \sin\left(\frac{y}{x}\right)$

6.4. *Again, use a substitution appropriate to homogeneous first-order differential equations, this time to solve the initial-value problem*

$$\frac{dy}{dx} = \frac{x - y}{x + y} \quad \text{with} \quad y(0) = 3 \quad .$$

6.5. *Use substitutions appropriate to Bernoulli equations (as described in Section 6.4) to find a general solution to each of the following:*

a. $\frac{dy}{dx} + 3y = 3y^3 \quad .$

b. $\frac{dy}{dx} - \frac{3}{x}y = \left(\frac{y}{x}\right)^2$

c. $\frac{dy}{dx} + 3\cot(x)y = 6\cos(x)y^{2/3}$

6.6. *Use a substitution appropriate to a Bernoulli equation to solve the initial-value problem*

$$\frac{dy}{dx} - \frac{1}{x}y = \frac{1}{y} \quad \text{with} \quad y(1) = 3 \quad .$$

6.7. *For each of the following, determine a substitution that simplifies the given differential equation, and, using that substitution, find a general solution. (Warning: The substitutions for some of the later equations will not be substitutions already discussed.)*

a. $\frac{dy}{dx} = \frac{y}{x} + \left(\frac{x}{y}\right)^2$

b. $3\frac{dy}{dx} = -2 + \sqrt{2x + 3y + 4}$

c. $\frac{dy}{dx} + \frac{2}{x}y = 4\sqrt{y}$

d. $\frac{dy}{dx} = 4 + \frac{1}{\sin(4x - y)}$

e. $(y - x)\frac{dy}{dx} = 1$

f. $(x + y)\frac{dy}{dx} = y$

g. $\left(2xy + 2x^2\right)\frac{dy}{dx} = x^2 + 2xy + 2y^2$

h. $\frac{dy}{dx} + \frac{1}{x}y = x^2y^3$

i. $\frac{dy}{dx} = 2\sqrt{2x + y - 3} - 2$

j. $\frac{dy}{dx} = 2\sqrt{2x + y - 3}$

k. $x\frac{dy}{dx} - y = \sqrt{xy + x^2}$

l. $\frac{dy}{dx} + 3y = 28e^{2x}y^{-3}$

m. $\frac{dy}{dx} = (x - y + 3)^2$

n. $\frac{dy}{dx} + 2x = 2\sqrt{y + x^2}$

o. $\cos(y)\frac{dy}{dx} = e^{-x} - \sin(y)$

p. $\frac{dy}{dx} = x\left[1 + 2\frac{y}{x^2} + \frac{y^2}{x^4}\right]$

6.8. Consider a generic Bernoulli equation

$$\frac{dy}{dx} + p(x)y = f(x)y^n$$

where $p(x)$ and $f(x)$ are known functions of x and n is any real number other than 0 or 1. Use the substitution $u = y^r$ (equivalently, $y = u^{1/r}$) and derive that the above Bernoulli equation for y reduces to a linear equation for u if and only if $r = 1 - n$. In the process, also derive the resulting linear equation for u.

7

The Exact Form and General Integrating Factors

In the previous chapters, we've seen how separable and linear differential equations can be solved using methods for converting them to forms that can be easily integrated. In this chapter, we will develop a more general approach to converting a differential equation to a form (the "exact form") that can be integrated through a relatively straightforward procedure. We will see just what it means for a differential equation to be in exact form and how to solve differential equations in this form. Because it is not always obvious when a given equation is in exact form, a practical "test for exactness" will also be developed. Finally, we will generalize the notion of integrating factors to help us find exact forms for a variety of differential equations.

The theory and methods we will develop here are more general than those developed earlier for separable and linear equations. In fact, the procedures developed here can be used to solve any separable or linear differential equation (though you'll probably prefer using the methods developed earlier). More importantly, the methods developed in this chapter can, in theory at least, be used to solve a great number of other first-order differential equations. As we will see though, practical issues will reduce the applicability of these methods to a somewhat smaller (but still significant) number of differential equations.

By the way, the theory, the computational procedures, and even the notation that we will develop for equations in exact form are all very similar to that often developed in the later part of many calculus courses for two-dimensional conservative vector fields. If you've seen that theory, look for the parallels between it and what follows.

7.1 The Chain Rule

The exact form for a differential equation comes from one of the chain rules for differentiating a composite function of two variables. Because of this, it may be wise to briefly review these differentiation rules.

First, suppose ϕ is a differentiable function of a single variable y (so $\phi = \phi(y)$), and that y, itself, is a differentiable function of another variable t (so $y = y(t)$). Then the composite function $\phi(y(t))$ is a differentiable function of t whose derivative is given by the (elementary) chain rule

$$\frac{d}{dt}[\phi(y(t))] = \phi'(y(t))\,y'(t) \quad .$$

A less precise (but more suggestive) description of this chain rule is

$$\frac{d}{dt}[\phi(y(t))] = \frac{d\phi}{dy}\frac{dy}{dt} \quad .$$

!▶Example 7.1: *Let*

$$y(t) = t^2 \quad \text{and} \quad \phi(y) = \sin(y) \quad .$$

Then

$$\phi(y(t)) = \sin\left(t^2\right) \quad ,$$

and

$$\frac{d}{dt}\sin\left(t^2\right) = \frac{d}{dt}[\phi(y(t))] = \frac{d\phi}{dy}\frac{dy}{dt}$$

$$= \frac{d}{dy}[\sin(y)] \cdot \frac{d}{dt}\left[t^2\right] = \cos(y) \cdot 2t = \cos\left(t^2\right) 2t \quad .$$

(In practice, of course, you probably do not explicitly write out all the steps listed above.)

Now suppose ϕ is a differentiable function of two variables x and y (so $\phi = \phi(x, y)$), while both x and y are differentiable functions of a single variable t (so $x = x(t)$ and $y = y(t)$). Then the composite function $\phi(x(t), y(t))$ is a differentiable function of t, and its derivative can be computed using a chain rule typically encountered later in the study of calculus, namely,

$$\frac{d}{dt}[\phi(x(t), y(t))] = \frac{\partial\phi}{\partial x}\frac{dx}{dt} + \frac{\partial\phi}{\partial y}\frac{dy}{dt} \quad . \tag{7.1}$$

In practice, it is usually easier to compute this derivative by simply replacing the x and y in the formula for $\phi(x, y)$ with the corresponding formulas $x(t)$ and $y(t)$, and then computing that formula of t to compute the above derivative. Still, this chain rule (and other chain rules involving functions of several variables) can be quite useful in more advanced applications. Our particular interest is in the corresponding chain rule for computing

$$\frac{d}{dx}[\phi(x, y(x))] \quad ,$$

which we can obtain from equation (7.1) by simply letting $x = t$. Then

$$\frac{dx}{dt} = 1 \quad , \quad y = y(t) = y(x) \quad , \quad \frac{dy}{dt} = \frac{dy}{dx} \quad ,$$

and equation (7.1) reduces to

$$\frac{d}{dx}[\phi(x, y(x))] = \frac{\partial\phi}{\partial x} + \frac{\partial\phi}{\partial y}\frac{dy}{dx} \quad . \tag{7.2}$$

For brevity, we will henceforth refer to this formula as *chain rule (7.2)* (not very original, but better than constantly repeating "the chain rule described in equation (7.2)").

Don't forget the difference between $^{d}/_{dx}$ and $^{\partial}/_{\partial x}$. If $\phi = \phi(x, y)$, then

$$\frac{d\phi}{dx} = \text{the derivative of } \phi(x, y) \text{ assuming } x \text{ is the variable and } y \text{ is a function of } x\,.$$

while

$$\frac{\partial\phi}{\partial x} = \text{the derivative of } \phi(x, y) \text{ assuming } x \text{ is the variable and } y \text{ is a constant.}$$

!▶Example 7.2: *Assume y is some function of x (i.e., $y = y(x)$) and*

$$\phi(x, y) = y^2 + x^2 y \quad .$$

Then

$$\frac{\partial \phi}{\partial x} = \frac{\partial}{\partial x}\left[y^2 + x^2 y\right] = 2xy \quad , \quad \frac{\partial \phi}{\partial y} = \frac{\partial}{\partial y}\left[y^2 + x^2 y\right] = 2y + x^2 \quad ,$$

and, by chain rule (7.2),

$$\frac{d}{dx}\left[\phi(x, y(x))\right] = \frac{\partial \phi}{\partial x} + \frac{\partial \phi}{\partial y}\frac{dy}{dx} = 2xy + \left[2y + x^2\right]\frac{dy}{dx} \quad .$$

If, for example, $y = \sin(x)$ *, then the above becomes*

$$\frac{d}{dx}\left[\phi(x, y(x))\right] = 2x\sin(x) + \left[2\sin(x) + x^2\right]\cos(x) \quad .$$

On the other hand, if $y = y(x)$ *is some unknown function, then, after replacing* ϕ *with its formula, we simply have*

$$\frac{d}{dx}\left[y^2 + x^2 y\right] = 2xy + \left[2y + x^2\right]\frac{dy}{dx} \quad . \tag{7.3}$$

In our use of chain rule (7.2), y will be an unknown function of x , and the right side of equation (7.2) will correspond to one side of whatever differential equation is being considered.

7.2 The Exact Form, Defined

Let $\mathcal{R}$ be some region in the XY–plane. We will say that a first-order differential equation is in *exact form* (on $\mathcal{R}$) if and only if both of the following hold:

1. the differential equation is written in the form

$$M(x, y) + N(x, y)\frac{dy}{dx} = 0 \tag{7.4a}$$

   where $M(x, y)$ and $N(x, y)$ are known functions of x and y ,

and

2. there is a differentiable function $\phi = \phi(x, y)$ on $\mathcal{R}$ such that

$$\frac{\partial \phi}{\partial x} = M(x, y) \quad \text{and} \quad \frac{\partial \phi}{\partial y} = N(x, y) \tag{7.4b}$$

   at every point in $\mathcal{R}$.

We will refer to the above $\phi(x, y)$ as a *potential function* for the differential equation.[1]

In practice, the region $\mathcal{R}$ is often either the entire XY–plane or a significant portion of it. There are a few technical issues regarding this region, but we can deal with these issues later. In the meantime, little harm will be done by not explicitly stating the region. Just keep in mind that if we say a certain equation is in exact form, then it is in exact form on some region $\mathcal{R}$, and that the graph of any solution $y = y(x)$ derived will be restricted to being a curve in that region.

[1] Referring to ϕ as a "potential function" comes from the theory of conservative vector fields. In fact, it is not common terminology in other differential equation texts. Most other texts just refer to this function as "ϕ".

!► Example 7.3: *Consider the differential equation*

$$2xy + \left[2y + x^2\right]\frac{dy}{dx} = 0 \quad . \tag{7.5}$$

This equation is in the form

$$M(x,y) + N(x,y)\frac{dy}{dx} = 0$$

with

$$M(x,y) = 2xy \qquad \text{and} \qquad N(x,y) = 2y + x^2 \quad .$$

Moreover, if we glance back at Example 7.2, we immediately see that, letting

$$\phi(x,y) = y^2 + x^2 y \quad ,$$

we have

$$\frac{\partial\phi}{\partial x} = \frac{\partial}{\partial x}\left[y^2 + x^2 y\right] = 2xy = M(x,y)$$

and

$$\frac{\partial\phi}{\partial y} = \frac{\partial}{\partial y}\left[y^2 + x^2 y\right] = 2y + x^2 = N(x,y)$$

everywhere in the XY*–plane. So equation (7.5) is in exact form (and we can take* $\mathcal{R}$ *to be the entire* XY*–plane).*

In the next several sections, we will discuss how to determine when a differential equation is in exact form, how to convert one not in exact form into exact form, how to find a potential function, and how to solve the differential equation using a potential function. However, we will develop the material backwards, starting with solving a differential equation given a known potential function, and working our way towards dealing with equations not in exact form. This ends up being the natural (and least confusing) way to develop the material.

But first, a few general comments regarding exact forms and potential functions:

1. Just being written as

$$M(x,y) + N(x,y)\frac{dy}{dx} = 0$$

does not guarantee that a differential equation is in exact form; there still might not be a $\phi(x,y)$ with

$$\frac{\partial\phi}{\partial x} = M(x,y) \qquad \text{and} \qquad \frac{\partial\phi}{\partial y} = N(x,y) \quad .$$

For example, we will discover that there is no $\phi(x,y)$ satisfying

$$\frac{\partial\phi}{\partial x} = 3y + 3y^3 \qquad \text{and} \qquad \frac{\partial\phi}{\partial y} = xy^2 - x \quad .$$

So

$$3y + 3y^3 + \left[xy^2 - x\right]\frac{dy}{dx} = 0$$

is not in exact form.

2. A single differential equation will have several potential functions. In particular, adding any constant to a potential function yields another potential function. After all, if

$$\frac{\partial\phi_0}{\partial x} = M(x,y) \qquad \text{and} \qquad \frac{\partial\phi_0}{\partial y} = N(x,y)$$

and

$$\phi_1(x, y) = \phi_0(x, y) + c$$

for some constant c, then, since any derivative of any constant is zero,

$$\frac{\partial \phi_0}{\partial x} = \frac{\partial \phi_1}{\partial x} = M(x, y) \quad \text{and} \quad \frac{\partial \phi_0}{\partial y} = \frac{\partial \phi_1}{\partial y} = N(x, y) \quad .$$

Moreover, as you will verify in Exercises 7.2 and 7.3 (starting on page 141), any differential equation that can be written in one exact form will actually have several exact forms, all corresponding to different (but related) potential functions.

Because of the uniqueness results concerning solutions to first-order differential equations, it does not matter which of these potential functions is used to solve the equation. Thus, in the following, we will only worry about finding *a* potential function for any given differential equation.

3. Many authors refer to an equation in exact form as an *exact equation*. We are avoiding this terminology because, unlike linear and separable differential equations, the "exactness" of an equation can be destroyed by legitimate rearrangements of that equation. For example,

$$\frac{dy}{dx} + x^2 y = e^{2x}$$

is a linear differential equation whether it is written as above or written as

$$\frac{dy}{dx} = e^{2x} - x^2 y \quad .$$

Consider, on the other hand,

$$2xy + \left[2y + x^2\right]\frac{dy}{dx} = 0 \quad .$$

As we saw in Example 7.3, this differential equation is in exact form (i.e., is an "exact equation"). However, if we rewrite it as

$$\frac{dy}{dx} = -\frac{2xy}{2y + x^2} \quad ,$$

we have an equation that is not in exact form (i.e., is not an "exact equation").

7.3 Solving Equations in Exact Form Using a Known Potential Function

Observe that, if $\phi = \phi(x, y)$ is any sufficiently differentiable function of x and y, and

$$\frac{\partial \phi}{\partial x} = M(x, y) \quad \text{and} \quad \frac{\partial \phi}{\partial y} = N(x, y) \quad ,$$

then the differential equation

$$M(x, y) + N(x, y)\frac{dy}{dx} = 0$$

can be rewritten as

$$\frac{\partial \phi}{\partial x} + \frac{\partial \phi}{\partial y}\frac{dy}{dx} = 0 \quad .$$

Chain rule (7.2) then tells us that the left side of this equation is just the derivative of $\phi(x, y(x))$. So this last equation can be written more concisely as

$$\frac{d}{dx}[\phi(x, y)] = 0$$

(with $y = y(x)$). Not only is this concise, but it is easily integrated:

$$\int \frac{d}{dx}[\phi(x, y)]\,dx = \int 0\,dx$$

$$\hookrightarrow \qquad \phi(x, y) = c \quad .$$

Think about this last equation for a moment. It describes the relation between x and $y = y(x)$ assuming y satisfies the differential equation

$$M(x, y) + N(x, y)\frac{dy}{dx} = 0 \quad .$$

In other words, the equation $\phi(x, y) = c$ is an implicit solution to the differential equation for which it is a potential function.

All this is important enough to be restated as a theorem.

Theorem 7.1 (importance of a potential function)
Let $\phi(x, y)$ be a potential function for a given first-order differential equation. Then that differential equation can be written as

$$\frac{\partial \phi}{\partial x} + \frac{\partial \phi}{\partial y}\frac{dy}{dx} = 0 \quad .$$

This, in turn, can be rewritten as

$$\frac{d}{dx}[\phi(x, y)] = 0 \qquad \text{with} \quad y = y(x) \quad ,$$

which can be integrated,

$$\int \frac{d}{dx}[\phi(x, y)]\,dx = \int 0\,dx \quad ,$$

to obtain the implicit solution

$$\phi(x, y) = c$$

where c is an arbitrary constant.

The above theorem contains all the steps for finding an implicit solution to a differential equation that can be put in exact form, provided you have a corresponding potential function. Of course, you have probably already noticed that this theorem can be shortened to the following:

Corollary 7.2
Let $\phi(x, y)$ be a potential function for a given first-order differential equation. Then

$$\phi(x, y) = c$$

is an implicit solution to that differential equation.

In solving at least the first few differential equations in the exercises at the end of this chapter, you should use all the steps described in Theorem 7.1 simply to reinforce your understanding of why $\phi(x, y) = c$. Then feel free to cut out the intermediate steps (i.e., use the corollary). And, of course, don't forget to see if the implicit solution can be solved for y in terms of x, yielding an explicit solution.

!►Example 7.4: *Consider the differential equation*

$$2xy + \left[2y + x^2\right]\frac{dy}{dx} = 0 \quad .$$

From Example 7.3, we know this is in exact form and has corresponding potential function

$$\phi(x, y) = y^2 + x^2 y \quad .$$

Since

$$\frac{\partial}{\partial x}\left[y^2 + x^2 y\right] = 2xy \qquad \text{and} \qquad \frac{\partial}{\partial y}\left[y^2 + x^2 y\right] = 2y + x^2 \quad ,$$

our differential equation can be rewritten as

$$\frac{\partial}{\partial x}\left[y^2 + x^2 y\right] + \left(\frac{\partial}{\partial y}\left[y^2 + x^2 y\right]\right)\frac{dy}{dx} = 0 \quad .$$

By chain rule (7.2), this reduces to

$$\frac{d}{dx}\left[y^2 + x^2 y\right] = 0 \qquad \text{with} \quad y = y(x) \quad .$$

Integrating this,

$$\int \frac{d}{dx}\left[y^2 + x^2 y\right] dx = \int 0 \, dx \quad ,$$

yields the implicit solution

$$y^2 + x^2 y = c \quad .$$

Rewriting this as

$$y^2 + x^2 y - c = 0$$

and then solving for y *via the quadratic formula provides the explicit solution*

$$y = \frac{-x^2 \pm \sqrt{x^4 + 4c}}{2} \quad .$$

Finding the Potential Function

Let us now consider a more difficult problem: finding a potential function, $\phi(x, y)$, for a given differential equation that has been written in the form

$$M(x, y) + N(x, y)\frac{dy}{dx} = 0 \quad .$$

This means finding a function $\phi(x, y)$ satisfying both

$$\frac{\partial \phi}{\partial x} = M(x, y) \qquad \text{and} \qquad \frac{\partial \phi}{\partial y} = N(x, y) \quad .$$

A relatively straightforward procedure for finding this ϕ will be outlined in a moment. But first, let us make a few observations regarding this pair of partial differential equations:

1. We are not assuming the given differential equation is in exact form, so there might not be a solution to the above pair of partial differential equations. Our method will find ϕ if it exists (i.e., if the equation is in exact form) and will lead to an obviously impossible equation otherwise.

2. Because the above pair of equations are partial differential equations, we must treat x and y as independent variables. Do *not* view y as a function of x in solving for ϕ.

3. If you are acquainted with methods for "recovering the potential for a conservative vector field", then you will recognize the following as one of those methods.

Now, here is the procedure:

Basic Procedure for Finding a Potential Function

Assume we have a differential equation in the form

$$M(x, y) + N(x, y)\frac{dy}{dx} = 0 \quad .$$

For purposes of illustration, we will use the differential equation

$$2xy + 2 + \left[x^2 + 4\right]\frac{dy}{dx} = 0 \quad .$$

To find a potential function $\phi(x, y)$ for this differential equation (if it exists), do the following:

1. Identify the formulas for $M(x, y)$ and $N(x, y)$, and, using these formulas, write out the pair of partial differential equations to be solved,

$$\frac{\partial \phi}{\partial x} = M(x, y) \qquad \text{and} \qquad \frac{\partial \phi}{\partial y} = N(x, y) \quad .$$

(In doing so, we are naively assuming the given differential equation is in exact form.)

For our example,

$$\underbrace{2xy + 2}_{M(x,y)} + [\underbrace{x^2 + 4}_{N(x,y)}]\frac{dy}{dx} = 0 \quad .$$

So the pair of partial differential equations to be solved is

$$\frac{\partial \phi}{\partial x} = 2xy + 2 \qquad \textit{and} \qquad \frac{\partial \phi}{\partial y} = x^2 + 4 \quad .$$

2. Integrate the first equation in the pair with respect to x, treating y as a constant.

$$\int \frac{\partial \phi}{\partial x}\, dx = \int M(x, y)\, dx \quad .$$

In computing the integral of M with respect to x, you get a "constant of integration". Keep in mind that this "constant" is based on y being a constant, and may thus depend on the value of y. Consequently, this "constant of integration" is actually some yet unknown function of y — call it $h(y)$.

For our example,

$$\int \frac{\partial \phi}{\partial x}\, dx = \int [2xy + 2]\, dx$$

$$\hookrightarrow \qquad \phi(x, y) = x^2 y + 2x + h(y) \quad .$$

(Observe that this step yields a formula for $\phi(x, y)$ involving one yet unknown function of y only. We now just have to determine $h(y)$ to know the formula for $\phi(x, y)$.)

3. Replace ϕ in the other partial differential equation,

$$\frac{\partial \phi}{\partial y} = N(x, y) \quad ,$$

with the formula just derived for $\phi(x, y)$, and compute the partial derivative. Keep in mind that, because $h(y)$ is a function of y only,

$$\frac{\partial}{\partial y}[h(y)] = h'(y) \quad .$$

Then algebraically solve the resulting equation for $h'(y)$.

For our example:

$$\frac{\partial \phi}{\partial y} = N(x, y)$$

$$\hookrightarrow \qquad \frac{\partial}{\partial y}\left[x^2 y + 2x + h(y)\right] = x^2 + 4$$

$$\hookrightarrow \qquad x^2 + h'(y) = x^2 + 4$$

$$\hookrightarrow \qquad h'(y) = 2 \quad .$$

4. Look at the formula just obtained for $h'(y)$. Because $h'(y)$ is a function of y only, its formula must involve only the variable y; no x may appear.

If the x's do not cancel out, then we have an impossible equation. This means the naive assumption made in the first step was wrong; the given differential equation was not in exact form, and there is no $\phi(x, y)$ satisfying the desired pair of partial differential equations. In this case, *Stop! Go no further in this procedure!*

On the other hand, if the previous step yields

$$h'(y) = \text{a formula of } y \text{ only (no } x\text{'s)} \quad ,$$

then integrate both sides of this equation with respect to y to obtain $h(y)$. (Because $h(y)$ does not depend on x, the constant of integration here will truly be a constant, not a function of the other variable.)

For our example, the last step yielded

$$h'(y) = 2 \quad .$$

The right side does not contain x, so we can continue and integrate to obtain

$$h(y) = \int h(y)\,dy = \int 2\,dy = 2y + c_1 \quad .$$

(We will later see that the constant c_1 is not that important.)

5. Combine the formula just obtained for h with the formula obtained for ϕ in step 2.

In our example, combining the results from step 2 and the last step above yields

$$\begin{aligned} \phi(x, y) &= x^2 y + 2x + h(y) \\ &= x^2 y + 2x + 2y + c_1 \end{aligned}$$

where c_1 is an arbitrary constant.

(Because of the arbitrary constant from the integration of $h'(y)$, the formula obtained actually describes all possible ϕ's satisfying the desired pair of partial differential equations. If you look at our discussion above, it should be clear that this formula will always be of the form

$$\phi(x, y) = \phi_0(x, y) + c_1$$

where $\phi_0(x, y)$ is a particular formula and c_1 is an arbitrary constant. But, as noted earlier, we only need to find one potential function $\phi(x, y)$. So you can set c_1 equal to your favorite constant, 0, or keep it arbitrary and see what happens.)

If the above procedure does yield a formula $\phi(x, y)$, then it immediately follows that the given equation is in exact form and $\phi(x, y)$ is a potential function for the given differential equation. But don't forget that the goal is usually to solve the given differential equation,

$$M(x, y) + N(x, y)\frac{dy}{dx} = 0 \quad .$$

The function $\phi = \phi(x, y)$ is *not* that solution. It is a function such that the differential equation can be rewritten as

$$\frac{d}{dx}[\phi(x, y)] = 0 \quad .$$

Integrating this yields the implicit solution

$$\phi(x, y) = c$$

from which, if the implicit solution is not too complicated, we can obtain an explicit solution $y = y(x)$.

!▶Example 7.5: *Consider solving*

$$2xy + 2 + \left[x^2 + 4\right]\frac{dy}{dx} = 0 \quad .$$

As just illustrated, this equation is in exact form, and has a potential function

$$\phi(x, y) = x^2y + 2x + 2y + c_1$$

where c_1 can be any constant. So the differential equation can be rewritten as

$$\frac{d}{dx}\left[x^2y + 2x + 2y + c_1\right] = 0 \quad ,$$

which integrates to

$$x^2y + 2x + 2y + c_1 = c_2 \quad .$$

This is an implicit solution for the differential equation. Solving this for y is easy, and (after letting $c = c_2 - c_1$) gives us the explicit solution

$$y = \frac{c - 2x}{x^2 + 2} \quad .$$

Note that, in the last example, the constants arising from the integration of $h'(y)$ and $^{d\phi}/_{dx}$ were combined at the end. It is easy to see that this will always be possible.

Other Ways to Find a Potential Function

There are other ways to find a function $\phi(x, y)$ satisfying both

$$\frac{\partial \phi}{\partial x} = M(x, y) \qquad \text{and} \qquad \frac{\partial \phi}{\partial y} = N(x, y) \quad .$$

Two, in particular, are worth mentioning:

1. The first is the obvious modification of the one already given in which the roles of ${}^{\partial\phi}/_{\partial x} = M$ and ${}^{\partial\phi}/_{\partial y} = N$ are interchanged. That is, instead of

   integrating ${}^{\partial\phi}/_{\partial x} = M$ with respect to x, and then plugging the result into ${}^{\partial\phi}/_{\partial y} = N$,

   you

   integrate ${}^{\partial\phi}/_{\partial y} = N$ with respect to y, and then plug the result into ${}^{\partial\phi}/_{\partial x} = M$.

   The integration of ${}^{\partial\phi}/_{\partial y} = N$ will yield some formula involving x, y and an unknown function of x, $g(x)$. Plugging this formula into ${}^{\partial\phi}/_{\partial x} = M$ should yield a ordinary differential equation for $g(x)$ which does not contain y. If the y's do not cancel out, the desired ϕ does not exist. Otherwise, $g(x)$ can be obtained by integration and then combined with the formula just obtained for $\phi(x, y)$.

   This is usually the preferred method when $\int N(x, y)\, dy$ is much easier to compute than $\int M(x, y)\, dx$. Indeed, it usually is a good idea to scan these two integrals and, if one looks much easier to compute, compute that one, and plug the result into the partial differential equation corresponding to the other integral. Don't forget to check to see if the equation resulting from that just involves the appropriate variable.

2. The other method is one often attempted by beginners who do not understand why it should not be used: First, independently integrate both ${}^{\partial\phi}/_{\partial x} = M$ and ${}^{\partial\phi}/_{\partial y} = N$. Then stare at the two different formulas obtained for $\phi(x, y)$ (each involving a different unknown function) and try to guess what single formula for $\phi(x, y)$ (without unknown functions) matches the results from the two integrations.

   Fight any temptation to take this approach. Yes, with a little luck and skill, you can get ϕ this way. But it is usually more work, it doesn't easily warn you when $\phi(x, y)$ does not exist, and is more likely to result in errors. Why use a method involving two integrations and two unknown functions when you can use a method involving just one integration and one unknown function with a straightforward way to determine that one function?

7.4 Testing for Exactness — Part I

The procedure just discussed for finding a potential function ϕ for

$$M(x, y) + N(x, y)\frac{dy}{dx} = 0 \tag{7.6}$$

does not tell us whether such a ϕ even exists until step 4, after a possibly tricky integration. Fortunately, there is a simple test that that can often tell us when seeking that ϕ would be futile. This test is based on the fact that, for any sufficiently differentiable $\phi(x, y)$,

$$\frac{\partial^2 \phi}{\partial y\, \partial x} = \frac{\partial^2 \phi}{\partial x\, \partial y} \quad .$$

Now let $\mathcal{R}$ be any region in the XY–plane on which ϕ is sufficiently differentiable and satisfies

$$\frac{\partial \phi}{\partial x} = M(x, y) \quad \text{and} \quad \frac{\partial \phi}{\partial y} = N(x, y) \quad .$$

Then, at every point in $\mathcal{R}$,

$$\frac{\partial M}{\partial y} = \frac{\partial}{\partial y}\left[\frac{\partial \phi}{\partial x}\right] = \frac{\partial^2 \phi}{\partial y \, \partial x} = \frac{\partial^2 \phi}{\partial x \, \partial y} = \frac{\partial}{\partial x}\left[\frac{\partial \phi}{\partial y}\right] = \frac{\partial N}{\partial x} \quad .$$

So, for equation (7.6) to be in exact form over $\mathcal{R}$, we must have

$$\frac{\partial M}{\partial y} = \frac{\partial N}{\partial x} \tag{7.7}$$

at every point in $\mathcal{R}$. If this equation does not hold, differential equation (7.6) is not in exact form over that region — no corresponding potential function ϕ exists.

What we have not shown is that equality (7.7) necessarily implies that differential equation (7.6) is in exact form. In fact, equality (7.7) does imply that, given any point in $\mathcal{R}$, the equation is in exact form over *some* subregion of $\mathcal{R}$ containing that point. Unfortunately, showing that and describing those regions takes more development than is appropriate here. For now, let us just say that, in practice, the equality (7.7) implies that differential equation (7.6) is "probably" in exact form over the given region, and it is worthwhile to seek a corresponding potential function ϕ via the method outlined earlier.

Let us summarize what has just been derived, accepting the term "suitably differentiable" as simply meaning that the necessary partial derivatives can be computed:

Theorem 7.3 (test for probable exactness)
Let $M(x, y)$ and $N(x, y)$ be two suitably differentiable functions of two variables over a region $\mathcal{R}$ in the XY–plane, and consider the differential equation

$$M(x, y) + N(x, y)\frac{dy}{dx} = 0 \quad .$$

1. *If*
$$\frac{\partial M}{\partial y} \neq \frac{\partial N}{\partial x} \quad \text{on} \quad \mathcal{R} \quad ,$$
then the above differential equation is not in exact form on $\mathcal{R}$.

2. *If*
$$\frac{\partial M}{\partial y} = \frac{\partial N}{\partial x} \quad \text{on} \quad \mathcal{R} \quad ,$$
then the above differential equation might be in exact form on $\mathcal{R}$. It is worth seeking a corresponding potential function.

!►Example 7.6: *Consider the differential equation*

$$3y + 3y^3 + \left[xy^2 - x\right]\frac{dy}{dx} = 0 \quad .$$

Here

$$\frac{\partial M}{\partial y} = \frac{\partial}{\partial y}\left[3y + 3y^3\right] = 3 + 9y^2$$

and

$$\frac{\partial N}{\partial x} = \frac{\partial}{\partial x}\left[xy^2 - x\right] = y^2 - 1 \quad .$$

So

$$\frac{\partial M}{\partial y} \neq \frac{\partial N}{\partial x} \quad ,$$

telling us that the given differential equation is not in exact form over any region.

!►Example 7.7: *Consider the differential equation*

$$-\frac{y}{x^2 + y^2} + \frac{x}{x^2 + y^2}\frac{dy}{dx} = 0 \quad .$$

Here

$$M(x, y) = -\frac{y}{x^2 + y^2} \qquad \text{and} \qquad N(x, y) = \frac{x}{x^2 + y^2}$$

are well defined and differentiable at every point on the XY–plane except $(x, y) = (0, 0)$. So let's take $\mathcal{R}$ to be the entire XY–plane with the origin removed. Computing the partial derivatives, we get

$$\frac{\partial M}{\partial y} = \frac{\partial}{\partial y}\left[-\frac{y}{x^2 + y^2}\right] = -\frac{1(x^2 + y^2) - y(2y)}{(x^2 + y^2)^2} = \frac{y^2 - x^2}{(x^2 + y^2)^2}$$

and

$$\frac{\partial N}{\partial x} = \frac{\partial}{\partial x}\left[\frac{x}{x^2 + y^2}\right] = \frac{1(x^2 + y^2) - x(2x)}{(x^2 + y^2)^2} = \frac{y^2 - x^2}{(x^2 + y^2)^2} \quad .$$

So these two partial derivatives are equal throughout $\mathcal{R}$, and our test for probable exactness tells us that the given differential equation might be in exact form on $\mathcal{R}$ — it is worthwhile to try to find a corresponding potential function.

For many, the test described above in Theorem 7.3 will suffice. Those who wish a more complete test should jump to Section 7.7 starting on page 137 (where we will also finish solving the differential equation in Example 7.7).

7.5 "Exact Equations": A Summary

To review:

If you suspect that a given differential equation,

$$M(x, y) + N(x, y)\frac{dy}{dx} = 0 \quad ,$$

is in exact form, then you can quickly check for at least probable exactness by computing $\partial M/\partial y$ and $\partial N/\partial x$ and seeing if

$$\frac{\partial M}{\partial y} = \frac{\partial N}{\partial x} \quad .$$

If the two partial derivatives are equal, then follow the procedure for finding a potential function $\phi(x, y)$ outlined on pages 124 to 126.

If that procedure is successful and yields a $\phi(x, y)$, then finish solving the given differential equation using the fact that the differential equation can be rewritten as

$$\frac{d}{dx}[\phi(x, y)] = 0 \quad ,$$

the integration of which yields the implicit solution

$$\phi(x, y) = c \quad .$$

If this equation can be solved for y in terms of x, do so.

If the given differential equation is not in exact form, then there is a possibility that it can be put into an exact form using appropriate "integrating factors". We will discuss these next.

By the way, don't forget that these equations may be solvable by other means. For example, the equation used to illustrate the procedure for finding ϕ was a linear differential equation, and could have been solved a bit more quickly using the methods from Chapter 5.

7.6 Converting Equations to Exact Form

Basic Notions

Obviously, the first step to converting a given first-order differential equation to exact form is to get it into the form

$$M(x, y) + N(x, y)\frac{dy}{dx} = 0 \quad .$$

Then apply the test for (probable) exactness. With luck, the test result will be positive. More likely, it will not.

To see how we might further convert our equation to exact form, it may help to recall why we want the exact form. It is so that the left side of the differential equation can be identified as an ordinary derivative of some formula of x and $y(x)$,

$$\frac{d}{dx}\phi(x, y(x)) \quad .$$

We had a similar situation with linear equations. Given a linear equation

$$\frac{dy}{dx} + p(x)y = f(x) \quad ,$$

we found that, *after multiplying it by an integrating factor* μ to get

$$\mu\frac{dy}{dx} + \mu py = \mu f \quad ,$$

we could identify the equation's left side as a complete derivative of a formula of x and $y(x)$, namely,

$$\frac{d}{dx}[\mu(x)y(x)] \quad .$$

The same idea can be applied to convert an equation not in exact form to one that is in exact form.

!►Example 7.8: *Consider the differential equation*

$$3y + 3y^3 + \left[xy^2 - x\right]\frac{dy}{dx} = 0 \quad .$$

In Example 7.6, we saw that this equation is not in exact form. But look what happens after we multiply through by $\mu = x^2y^{-2}$ *,*

$$x^2y^{-2}\left(3y + 3y^3 + \left[xy^2 - x\right]\frac{dy}{dx} = 0\right) \quad .$$

We get

$$\underbrace{3x^2y^{-1} + 3x^2y}_{M_{new}(x,y)} + [\underbrace{x^3 - x^3y^{-2}}_{N_{new}(x,y)}]\frac{dy}{dx} = 0$$

with

$$\frac{\partial M_{new}}{\partial y} = \frac{\partial}{\partial y}\left[3x^2y^{-1} + 3x^2y\right] = -3x^2y^{-2} + 3x^2 = 3x^2 - 3x^2y^{-2}$$

and

$$\frac{\partial N_{new}}{\partial x} = \frac{\partial}{\partial x}\left[x^3 - x^3y^{-2}\right] = 3x^2 - 3x^2y^{-2} \quad .$$

So

$$\frac{\partial M_{new}}{\partial y} = \frac{\partial N_{new}}{\partial x} \quad ,$$

telling us that the equation is now (probably) in exact form (over any region where y *never equals* 0 *).*

We will refer to any nonzero function $\mu = \mu(x, y)$ as an *integrating factor* for a first-order differential equation

$$M + N\frac{dy}{dx} = 0$$

if and only if multiplying that equation through by μ ,

$$\mu M + \mu N\frac{dy}{dx} = 0 \quad ,$$

yields a differential equation in exact form.[2] This integrating factor may be a function of x or of y or of both x and y . Notice that, because

$$\underbrace{\mu M}_{\text{"new" } M} + \underbrace{\mu N}_{\text{"new" } N}\frac{dy}{dx} = 0$$

is exact, our test for exactness tells us that

$$\frac{\partial}{\partial y}[\mu M] = \frac{\partial}{\partial x}[\mu N] \quad .$$

[2] You can show that the integrating factors found for linear equations are just special cases of the integrating factors considered here. If there is any danger of confusion, we'll refer to the integrating factors now being discussed as the "more general" integrating factors.

Finding Integrating Factors
General Approach

Finding an integrating factor for an equation

$$M(x, y) + N(x, y)\frac{dy}{dx} = 0$$

starts with the requirement just derived,

$$\frac{\partial}{\partial y}[\mu M] = \frac{\partial}{\partial x}[\mu N] \quad . \tag{7.8}$$

Remember, M and N will be known formulas of x and y. So equation (7.8) is a differential equation in which the unknown function is our integrating factor, μ. Unfortunately, it is a rather nontrivial partial differential equation (unlike the partial differential equations in Section 7.2), and a complete discussion of how to solve it for μ is beyond the scope of any introductory differential equations course. Fortunately, there are some common cases where this partial differential equation reduces to an equation we can handle. The cases we will consider are where there is an integrating factor μ that is either a function of x only, or a function of y only, or a 'simple' formula of x and y. In all cases, the approach is basically the same:

1. Choose the case you think is appropriate, and make the corresponding assumptions on μ.
2. Expand equation (7.8) by computing out the derivatives as far as possible, taking into account the assumptions made on μ.
3. See if the resulting equation can be solved for a μ satisfying the assumptions made. If so, do so. If not, start over using different assumptions on μ (unless you've run out of reasonable options).

We'll illustrate the above approach in a moment. Before that, however, a few more comments should be made:

1. Only one integrating factor is needed. So go ahead and assign convenient values to the arbitrary constants that arise in solving for μ (just as in finding integrating factors for linear equations).
2. Once you have found an integrating factor μ, remember why you wanted it. Use it to rewrite your differential equation in exact form, and then solve the differential equation as described earlier in this chapter.
3. There are tests to determine if there are integrating factors that are functions of just x or just y. Moreover, there are formulas for μ that can be used if any of the tests are satisfied (see Exercise 7.6 on page 142). DO NOT WASTE YOUR TIME TRYING TO USE THESE FORMULAS! They are hard to memorize correctly and are worth learning only if you expect to compute many, many integrating factors over a relatively short time frame. Chances are, you won't have that need, just a need to understand the basic concepts and to compute the occasional integrating factor.

Now, let's look at the common cases:

First Case: μ Being a Function of x Only

If you think the integrating factor μ could be a function of x only, then assume it, and see what happens when you compute out equation (7.8). Under this assumption,

$$\mu = \mu(x) \quad , \quad \frac{\partial \mu}{\partial x} = \frac{d\mu}{dx} \quad \text{and} \quad \frac{\partial \mu}{\partial y} = 0 \quad .$$

Consequently, equation (7.8) should immediately reduce to an ordinary differential equation for μ. Moreover, since μ is supposed to be just a function of x here, this differential equation for μ should not contain any y's. Thus, if the y's do not cancel out, the assumption that μ could be just a function of x is wrong — go to the next case. But if the y's do cancel out, then solve the differential equation just obtained for a $\mu = \mu(x)$.

!▶Example 7.9: *Consider the differential equation*

$$1 + y^3 + xy^2 \frac{dy}{dx} = 0 \quad .$$

Here

$$\frac{\partial M}{\partial y} = \frac{\partial}{\partial y}\left[1 + y^3\right] = 3y^2 \qquad \text{and} \qquad \frac{\partial N}{\partial x} = \frac{\partial}{\partial x}\left[xy^2\right] = y^2 \quad .$$

So $^{\partial M}/_{\partial y} \neq {}^{\partial N}/_{\partial x}$. *The equation is not exact, but, with luck, we can find a function* μ *so that*

$$\mu\left[1 + y^3\right] + \mu\left[xy^2\right]\frac{dy}{dx} = 0$$

is exact. This integrating factor μ *must satisfy*

$$\frac{\partial}{\partial y}\left(\mu\left[1 + y^3\right]\right) = \frac{\partial}{\partial x}\left(\mu\left[xy^2\right]\right) \quad .$$

Let's see if μ *could be a function of just* x. *Assume* $\mu = \mu(x)$. *Then*

$$\frac{\partial \mu}{\partial x} = \frac{d\mu}{dx} \qquad \text{and} \qquad \frac{\partial \mu}{\partial y} = 0 \quad .$$

Using this, we have

$$\frac{\partial}{\partial y}\left(\mu\left[1 + y^3\right]\right) = \frac{\partial}{\partial x}\left(\mu\left[xy^2\right]\right)$$

$$\hookrightarrow \quad \frac{\partial \mu}{\partial y}\left[1 + y^3\right] + \mu\frac{\partial}{\partial y}\left[1 + y^3\right] = \frac{\partial \mu}{\partial x}\left[xy^2\right] + \mu\frac{\partial}{\partial x}\left[xy^2\right]$$

$$\hookrightarrow \quad 0 \cdot \left[1 + y^3\right] + \mu\left[0 + 3y^2\right] = \frac{d\mu}{dx}\left[xy^2\right] + \mu\left[y^2\right]$$

$$\hookrightarrow \quad 3y^2\mu = xy^2\frac{d\mu}{dx} + y^2\mu \quad .$$

The y*'s do cancel out, leaving us with*

$$3\mu = x\frac{d\mu}{dx} + \mu \quad ,$$

which simplifies to

$$\frac{d\mu}{dx} = \frac{2\mu}{x} \quad .$$

So there is an integrating factor μ *that is a function of* x *alone. Moreover, to find such an integrating factor, we need only find one (nonzero) solution to the above simple, separable ordinary differential equation. Proceeding to do so, we get*

$$\int \frac{1}{\mu}\frac{d\mu}{dx}\,dx = \int \frac{2}{x}\,dx$$

$$\hookrightarrow \quad \ln|\mu| = 2\ln|x| + c = \ln|x|^2 + c$$

$$\hookrightarrow \quad \mu = \pm e^{\ln x^2 + c} = Ax^2 \quad .$$

Since only one nonzero integrating factor is needed, we can take $A = 1$, giving us

$$\mu(x) = x^2$$

as our integrating factor.

Unfortunately, there is always the possibility that the y's will not cancel out.

!►Example 7.10: *It is easily verified that*

$$6xy + 5[x^2 + y]\frac{dy}{dx} = 0$$

is not in exact form. To be a corresponding integrating factor, μ must satisfy

$$\frac{\partial}{\partial y}(\mu[6xy]) = \frac{\partial}{\partial x}\left(\mu 5[x^2 + y]\right) .$$

Assume μ is a function of x only. Then

$$\frac{\partial \mu}{\partial x} = \frac{d\mu}{dx} \quad \text{and} \quad \frac{\partial \mu}{\partial y} = 0 ,$$

and, thus,

$$\frac{\partial}{\partial y}(\mu[6xy]) = \frac{\partial}{\partial x}\left(\mu 5[x^2 + y]\right)$$

$$\hookrightarrow \quad \frac{\partial \mu}{\partial y}[6xy] + \mu\frac{\partial}{\partial y}[6xy] = \frac{\partial \mu}{\partial x}5[x^2 + y] + \mu\frac{\partial}{\partial x}\left(5[x^2 + y]\right)$$

$$\hookrightarrow \quad 0 \cdot 6xy + \mu[6x] = \frac{d\mu}{dx}5[x^2 + 5] + \mu[10x]$$

$$\hookrightarrow \quad \frac{d\mu}{dx} = \frac{10x\mu - 6x\mu}{4x^2 + 5y} = \frac{4x\mu}{4x^2 + 5y} .$$

Here, the y's do not cancel out, as they should if our assumption that μ depended only on x were true. Hence, that assumption was wrong. This equation does not have an integrating factor that is a function of x only. We will have to try something else.

Second Case: μ Being a Function of y Only

This is just like the first case, but with the roles of x and y switched. If you think the integrating factor μ could be a function of y only, then assume it, and see what happens when you compute out equation (7.8). Under this assumption, $\mu = \mu(y)$,

$$\frac{\partial \mu}{\partial x} = 0 \quad \text{and} \quad \frac{\partial \mu}{\partial y} = \frac{d\mu}{dy} .$$

Again, equation (7.8) should immediately reduce to an ordinary differential equation for μ, only this time the differential equation for μ should not contain any x's. If the x's do not cancel out, our assumption that μ could be just a function of y is wrong and we can go no further with this hope. But if the x's do cancel out, then solving the differential equation just obtained for a $\mu = \mu(y)$ yields the desired integrating factor.

!►Example 7.11: *As just seen in Example 7.10,*

$$6xy + 5[x^2 + y]\frac{dy}{dx} = 0$$

does not have an integrating factor depending only on x. So instead, let's try to find one that depends on just y. Assuming this,

$$\mu = \mu(y) \quad , \quad \frac{\partial\mu}{\partial x} = 0 \quad \text{and} \quad \frac{\partial\mu}{\partial y} = \frac{d\mu}{dy} \quad .$$

Combining this with equation (7.8):

$$\frac{\partial}{\partial y}(\mu[6xy]) = \frac{\partial}{\partial x}\left(\mu\left[5x^2 + 5y\right]\right)$$

$$\hookrightarrow \quad \frac{\partial\mu}{\partial y}[6xy] + \mu\frac{\partial}{\partial y}[6xy] = \frac{\partial\mu}{\partial x}\left[5x^2 + 5y\right] + \mu\frac{\partial}{\partial x}\left[5x^2 + 5y\right]$$

$$\hookrightarrow \quad \frac{d\mu}{dy}[6xy] + \mu[6x] = 0\cdot\left[5x^2 + 5y\right] + \mu[10x]$$

$$\hookrightarrow \quad \frac{d\mu}{dx} = \frac{10x\mu - 6x\mu}{6xy} = \frac{2\mu}{3y} \quad .$$

The x's cancel out, as hoped, and we have a simple differential equation for $\mu = \mu(y)$. Solving it:

$$\int \frac{1}{\mu}\frac{d\mu}{dy}\,dy = \int \frac{2}{3y}\,dy$$

$$\hookrightarrow \quad \ln|\mu| = \frac{2}{3}\ln|y| + c = \ln|y|^{2/3} + c$$

$$\hookrightarrow \quad \mu = Ay^{2/3} \quad .$$

Taking $A = 1$ gives the integrating factor

$$\mu(y) = y^{2/3} \quad .$$

Third Case: μ Being a 'Simple' Function of Both Variables

Of course, it is quite possible that no function of just x or just y will be an integrating factor for a given equation

$$M(x, y) + N(x, y)\frac{dy}{dx} = 0 \quad .$$

In this case, the best we can usually hope for is that there is a relatively simple function of x and y that will work as an integrating factor. This means making a 'good guess' at $\mu(x, y)$ and verifying that it satisfies equation (7.8). One 'guess' sometimes worth trying is

$$\mu(x, y) = x^\alpha y^\beta$$

where the exponents, α and β, are constants to be determined. To determine the values of the constants so that the guess works (if such constants exist), just plug this formula for μ into equation (7.8) and see if it reduces to an equation that can be solved for α and β. If so, find those values and use $\mu(x, y) = x^\alpha y^\beta$ (with the values just found for α and β) as your integrating factor. If not, keep searching (or consider dealing with the differential equation using the graphical and numerical methods we'll discuss in Chapters 9, 10 and 12).

!► Example 7.12: *Consider the differential equation from Example 7.6,*

$$3y + 3y^3 + \left[xy^2 - x\right]\frac{dy}{dx} = 0 \quad .$$

Plugging

$$\mu(x, y) = x^\alpha y^\beta$$

into equation (7.8) for our differential equation yields:

$$\frac{\partial}{\partial y}(\mu M) = \frac{\partial}{\partial x}(\mu N)$$

$$\hookrightarrow \quad \frac{\partial}{\partial y}\left(x^\alpha y^\beta\left[3y + 3y^3\right]\right) = \frac{\partial}{\partial x}\left(x^\alpha y^\beta\left[xy^2 - x\right]\right)$$

$$\hookrightarrow \quad \frac{\partial}{\partial y}\left(3x^\alpha y^{\beta+1} + 3x^\alpha y^{\beta+3}\right) = \frac{\partial}{\partial x}\left(x^{\alpha+1} y^{\beta+2} - x^{\alpha+1} y^\beta\right)$$

$$\hookrightarrow \quad 3(\beta + 1)x^\alpha y^\beta + 3(\beta + 3)x^\alpha y^{\beta+2} = (\alpha + 1)x^\alpha y^{\beta+2} - (\alpha + 1)x^\alpha y^\beta \quad .$$

Combining like terms then gives

$$[3\beta + \alpha + 4]x^\alpha y^\beta + [3\beta - \alpha + 8]x^\alpha y^{\beta+2} = 0 \quad ,$$

which, in turn, holds if and only if

$$3\beta + \alpha + 4 = 0 \qquad \text{and} \qquad 3\beta - \alpha + 8 = 0 \quad .$$

This last pair of equations constitute a simple system of linear equations,

$$3\beta + \alpha + 4 = 0$$
$$3\beta - \alpha + 8 = 0$$

which can be easily solved by any of a number of ways, yielding

$$\alpha = 2 \qquad \text{and} \qquad \beta = -2 \quad .$$

Thus, the differential equation we started with,

$$3y + 3y^3 + \left[xy^2 - x\right]\frac{dy}{dx} = 0 \quad ,$$

does have an integrating factor of the form $\mu(x, y) = x^\alpha y^\beta$ *, and it is*

$$\mu(x, y) = x^2 y^{-2}$$

(just as was used in Example 7.8 on page 131).

7.7 Testing for Exactness — Part II
Simple Connectivity and the Complete Test for Exactness

A more complete test for exactness than given in Theorem 7.3 can be described if we are more careful about describing our situation. So suppose we have an equation

$$M(x,y) + N(x,y)\frac{dy}{dx} = 0 \quad ,$$

and that, on some open region $\mathcal{R}$ of the XY–plane, all of the following hold:

1. The functions $M(x,y)$ and $N(x,y)$, along with the derivatives $^{\partial M}/_{\partial y}$ and $^{\partial N}/_{\partial x}$, are continuous everywhere in $\mathcal{R}$.
2. At each point (x,y) in $\mathcal{R}$,
$$\frac{\partial M}{\partial y} = \frac{\partial N}{\partial x} \quad .$$

That region $\mathcal{R}$ is said to be *simply connected* if each and every simple closed curve (i.e., loop) in $\mathcal{R}$ encloses only points in $\mathcal{R}$. If any simple closed curve in $\mathcal{R}$ encloses any point not in $\mathcal{R}$, then we will say that $\mathcal{R}$ is not simply connected. If you think about it, you will realize that saying a region is simply connected is just a precise way of saying that the region has no "holes". And if you think a little more about the situation, you will realize that, if our open region $\mathcal{R}$ has "holes" (i.e., is not simply connected), then it is probably because these are points where $M(x,y)$ or $N(x,y)$ or their partial derivatives fail to exist.

!►Example 7.13: *Again, consider the differential equation*

$$-\frac{y}{x^2+y^2} + \frac{x}{x^2+y^2}\frac{dy}{dx} = 0 \quad .$$

As noted in Example 7.7,

$$M(x,y) = -\frac{y}{x^2+y^2} \qquad \text{and} \qquad N(x,y) = \frac{x}{x^2+y^2}$$

are well defined and differentiable and satisfy

$$\frac{\partial M}{\partial y} = \frac{\partial N}{\partial x}$$

everywhere in the region $\mathcal{R}$ consisting of the XY–plane with the origin removed. Removing this point (the origin) creates a "hole" in $\mathcal{R}$. This point is also a point not in $\mathcal{R}$ but which is enclosed by any loop in $\mathcal{R}$ around the origin.

Now we can state the full test for exactness. (Its proof will be briefly discussed at the end of this section.)

Theorem 7.4 (complete test for exactness)
Let $\mathcal{R}$ be a simply-connected open region in the XY–plane, and let $M(x,y)$ and $N(x,y)$ be two continuous functions on $\mathcal{R}$ whose partial derivatives are also continuous on $\mathcal{R}$. Then

$$M(x,y) + N(x,y)\frac{dy}{dx} = 0$$

is in exact form on $\mathcal{R}$ if and only if

$$\frac{\partial M}{\partial y} = \frac{\partial N}{\partial x}$$

at every point in $\mathcal{R}$.

This theorem assures us that, if our region $\mathcal{R}$ is simply connected, then we can (in theory at least) use the procedure outlined on pages 124 to 126 to find the corresponding potential function $\phi(x, y)$ on $\mathcal{R}$, and from that, derive an implicit solution $\phi(x, y) = c$ to our differential equation.

Theorem 7.4 does not definitely say the differential equation is not in exact form if $\mathcal{R}$ is not simply connected. Whether the equation is or is not in exact form over all of $\mathcal{R}$ is still uncertain. What is certain, however, is the following immediate consequence of Theorem 7.4.

Corollary 7.5

Assume $M(x, y)$ and $N(x, y)$ are two continuous functions on some open region $\mathcal{R}$ of the XY–plane. Assume further that, on $\mathcal{R}$, the partial derivatives of M and N are continuous and satisfy

$$\frac{\partial M}{\partial y} = \frac{\partial N}{\partial x} \quad .$$

Then

$$M(x, y) + N(x, y)\frac{dy}{dx} = 0$$

is in exact form on each simply-connected open subregion of $\mathcal{R}$.

Thus, even if our original region is not simply connected, we can at least pick any open, simply connected subregion $\mathcal{R}_1$, and (in theory at least) use the procedure outlined in Section 7.3 to find a corresponding potential function $\phi_1(x, y)$ on $\mathcal{R}_1$, and from that, derive the implicit solution $\phi_1(x, y) = c$ to our differential equation, valid on subregion $\mathcal{R}_1$.

But then, why might there not be a potential function $\phi(x, y)$ valid on the entire region $\mathcal{R}$? Let's go back to an example started earlier to see.

!► Example 7.14: *Let us continue our consideration of the differential equation*

$$-\frac{y}{x^2 + y^2} + \frac{x}{x^2 + y^2}\frac{dy}{dx} = 0$$

from Example 7.7. As just noted in the previous example, the region $\mathcal{R}$ consisting of all the XY–plane except for the origin $(0, 0)$ is not simply connected. But we can partition it into the left and right half-planes

$$\mathcal{R}_+ = \{(x, y) : x > 0\} \quad \text{and} \quad \mathcal{R}_- = \{(x, y) : x < 0\} \quad ,$$

which are simply connected. Theorem 7.4 assures us that our differential equation is in exact form over each of these half-planes, and, indeed, you can easily show that all the corresponding potential functions on these regions for our differential equation are given by

$$\phi_+(x, y) = \text{Arctan}\left(\frac{y}{x}\right) + c_+ \quad \text{on} \quad \mathcal{R}_+$$

and

$$\phi_-(x, y) = \text{Arctan}\left(\frac{y}{x}\right) + c_- \quad \text{on} \quad \mathcal{R}_-$$

where c_+ and c_- are arbitrary constants.

But could there be a potential function ϕ on all of $\mathcal{R}$ corresponding to our differential equation? If so, then ϕ would also be a potential function over the left and right half-planes $\mathcal{R}_+$ and $\mathcal{R}_-$, and, as just noted, there would be constants c_+ and c_- such that

$$\phi(x, y) = \operatorname{Arctan}\left(\frac{y}{x}\right) + c_+ \qquad \text{for} \quad x > 0$$

and

$$\phi(x, y) = \operatorname{Arctan}\left(\frac{y}{x}\right) + c_- \qquad \text{for} \quad x < 0 \quad .$$

Since ϕ must be continuous everywhere except at the origin, it must, in particular, be continuous at any point on the positive Y–axis, $(0, y)$ with $y > 0$. So, letting $x \to 0$ from the positive side, we have

$$\phi(0, y) = \lim_{x \to 0^+} \phi(x, y) = \lim_{x \to 0^+} \operatorname{Arctan}\left(\frac{y}{x}\right) + c_+ \quad .$$

Using the substitution $t = {}^y/_x$ and recalling the limiting values of the Arctangent function, we can rewrite the above as

$$\phi(0, y) = \lim_{t \to +\infty} \operatorname{Arctan}(t) + c_+ = \frac{\pi}{2} + c_+ \quad .$$

Likewise, letting $x \to 0$ from the negative side, we have

$$\begin{aligned} \phi(0, y) &= \lim_{x \to 0^-} \phi(x, y) \\ &= \lim_{x \to 0^-} \operatorname{Arctan}\left(\frac{y}{x}\right) + c_- \\ &= \lim_{t \to -\infty} \operatorname{Arctan}(t) + c_- = -\frac{\pi}{2} + c_- \quad . \end{aligned}$$

Together, the above tells us that

$$-\frac{\pi}{2} + c_- = \phi(0, 1) = \frac{\pi}{2} + c_+ \quad ,$$

which, of course, means that $c_- = \pi + c_+$. Because of the arbitrariness of the constants added to potential functions, we may, for simplicity, let $c_+ = 0$. Then

$$\phi(x, y) = \begin{cases} \operatorname{Arctan}\left(\frac{y}{x}\right) & \text{if} \quad x > 0 \\ \operatorname{Arctan}\left(\frac{y}{x}\right) + \pi & \text{if} \quad x < 0 \\ \frac{\pi}{2} & \text{if} \quad x = 0 \text{ and } y > 0 \end{cases} \quad .$$

But look at what must now happen at a point on the negative Y–axis, say, at $(0, -1)$.

$$\lim_{x \to 0^+} \phi(x, -1) = \lim_{x \to 0^+} \operatorname{Arctan}\left(\frac{-1}{x}\right) = \lim_{t \to -\infty} \operatorname{Arctan}(t) = -\frac{\pi}{2}$$

and

$$\lim_{x \to 0^-} \phi(x, -1) = \lim_{x \to 0^-} \operatorname{Arctan}\left(\frac{-1}{x}\right) + \pi = \lim_{t \to +\infty} \operatorname{Arctan}(t) + \pi = \frac{3\pi}{2} \quad .$$

So

$$\lim_{x \to 0^+} \phi(x, -1) \neq \lim_{x \to 0^-} \phi(x, -1) \quad .$$

Thus, there are points in $\mathcal{R}$ at which $\phi(x, y)$ is not continuous, contrary to the fact that a potential function on $\mathcal{R}$ must be continuous everywhere on $\mathcal{R}$. And thus, the answer to our question "Could there be a potential function ϕ on all of $\mathcal{R}$ corresponding to our differential equation?" is "No".

The above example illustrates that, while we can partition a non-simply connected region into simply-connected subregions and then find all possible potential functions for our differential equation over each subregion, it may still be impossible to "paste together" these regions and potential functions to obtain a potential function that is well defined across all the boundaries between the partitions.

Is this truly a problem for us, whose main interest is in solving the differential equation? Not really. We can still solve the given differential equation. All we need to do is to choose our simply-connected partitions reasonably intelligently.

!► Example 7.15: *Let's solve the initial-value problem*

$$-\frac{y}{x^2+y^2} + \frac{x}{x^2+y^2}\frac{dy}{dx} = 0 \qquad \textit{with} \quad y(1) = 3$$

using the potential functions from the last example.

Since $(x, y) = (1, 3)$ *is in the right half plane, it makes sense to use* ϕ_+ *from the last example. Letting* $c_+ = 0$, *this potential function is*

$$\phi_+(x, y) = \operatorname{Arctan}\left(\frac{y}{x}\right) \qquad \textit{for} \quad x > 0 \quad .$$

So our differential equation has an implicit solution

$$\operatorname{Arctan}\left(\frac{y}{x}\right) = C_+ \qquad \textit{for} \quad x > 0 \quad .$$

Taking the tangent of both sides, letting $A = \tan(C_+)$, *and then solving for* y *yields the general solution*

$$y = Ax \qquad \textit{for} \quad x > 0 \quad .$$

Combined with the initial condition, this is

$$3 = y(1) = A \cdot 1 \quad .$$

So $A = 3$ *and the solution to our initial-value problem is*

$$y = 3x \qquad \textit{for} \quad x > 0 \quad .$$

(We will leave the issue of whether we truly need to restrict x *to being positive as an exercise for the interested.)*

Proving Theorem 7.4

This is one theorem we will not attempt to prove. A good proof would require a review of "integrals over curves in the plane" and "Green's theorem", both of which are subjects you may recall seeing near the end of your calculus course. Moreover, if you replace the equation

$$M(x, y) + N(x, y)\frac{dy}{dx} = 0$$

with

$$\mathbf{F}(x, y) = M(x.y)\mathbf{i} + N(x, y)\mathbf{j}$$

and replace the phrase "in exact form" with "a conservative vector field", then Theorem 7.4 becomes the theorem describing when a two-dimensional vector field is conservative. The proofs of these two theorems are virtually the same, and since the theorem about conservative vector fields is in any good calculus text, this author will save space and writing by directing the interested student to reviewing the relevant chapters in his/her old calculus book.

Additional Exercises

7.1. *For each choice of $\phi(x, y)$ given below, find a differential equation for which the given ϕ is a potential function, and then solve the differential equation using the given potential function.*

a. $\phi(x, y) = 3xy$

b. $\phi(x, y) = y^2 - 2x^3 y$

c. $\phi(x, y) = x^2 y - xy^3$

d. $\phi(x, y) = x \operatorname{Arctan}(y)$

7.2. *The following concern the differential equation*

$$\frac{dy}{dx} = \frac{1}{y} - \frac{y}{2x} \quad . \tag{7.9}$$

a. *Verify that the above differential equation can be rewritten as*

$$\left[y^2 - 2x\right] + 2xy\frac{dy}{dx} = 0 \quad ,$$

and then verify that this is an exact form for equation (7.9) by showing that

$$\phi(x, y) = xy^2 - x^2$$

is a corresponding potential function.

b. *Solve equation (7.9) using the above potential function.*

c. *Note that we can also rewrite equation (7.9) as*

$$e^{xy^2-x^2}\left[y^2 - 2x\right] + e^{xy^2-x^2}2xy\frac{dy}{dx} = 0 \quad .$$

Show that this is also an exact form by showing that

$$\psi(x, y) = e^{xy^2-x^2}$$

is a corresponding potential function.

7.3. *Assume $\phi(x, y)$ is a potential function corresponding to*

$$M(x, y) + N(x, y)\frac{dy}{dx} = 0 \quad .$$

Show that

$$\psi_1(x, y) = e^{\phi(x,y)} \qquad \text{and} \qquad \psi_2(x, y) = \sin(\phi(x, y))$$

are also potential functions for this differential equation, though corresponding to different exact forms.

7.4. *Each of the following differential equations is in exact form. Find a corresponding potential function for each, and then find a general solution to the differential equation using that potential function (even if it can be solved by simpler means).*

a. $2xy + y^2 + \left[2xy + x^2\right]\frac{dy}{dx} = 0$

b. $2xy^3 + 4x^3 + 3x^2y^2\frac{dy}{dx} = 0$

c. $2 - 2x + 3y^2 \frac{dy}{dx} = 0$

d. $1 + 3x^2y^2 + \left[2x^3y + 6y\right]\frac{dy}{dx} = 0$

e. $4x^3y + \left[x^4 - y^4\right]\frac{dy}{dx} = 0$

f. $1 + \ln|xy| + \frac{x}{y}\frac{dy}{dx} = 0$

g. $1 + e^y + xe^y \frac{dy}{dx} = 0$

h. $e^y + \left[xe^y + 1\right]\frac{dy}{dx} = 0$

7.5. *For each of the following differential equations,*

i. *verify that the equation is not in exact form,*

ii. *find an integrating factor, and*

iii. *solve the given differential equation (using the integrating factor just found).*

a. $1 + y^4 + xy^3 \frac{dy}{dx} = 0$

b. $y + \left[y^4 - 3x\right]\frac{dy}{dx} = 0$

c. $2x^{-1}y + \left[4x^2y - 3\right]\frac{dy}{dx} = 0$

d. $1 + [1 - x\tan(y)]\frac{dy}{dx} = 0$

e. $3y + 3y^2 + \left[2x + 4xy\right]\frac{dy}{dx} = 0$

f. $2x(y+1) - \frac{dy}{dx} = 0$

g. $2y^3 + \left[4x^3y^3 - 3xy^2\right]\frac{dy}{dx} = 0$

h. $4xy + \left[3x^2 + 5y\right]\frac{dy}{dx} = 0 \quad \text{for } y > 0$

i. $6 + 12x^2y^2 + \left[7x^3y + \frac{x}{y}\right]\frac{dy}{dx} = 0$

7.6. *The following problems concern the differential equation*

$$M(x, y) + N(x, y)\frac{dy}{dx} = 0 \quad . \tag{7.10}$$

Assume M and N are continuous and have continuous partial derivatives over the entire XY–plane, and let P and Q be the functions given by

$$P = \frac{\dfrac{\partial M}{\partial y} - \dfrac{\partial N}{\partial x}}{N} \quad \text{and} \quad Q = \frac{\dfrac{\partial N}{\partial x} - \dfrac{\partial M}{\partial y}}{M} \quad .$$

a. *Show that, if P is a function of x only (so all the y's cancel out), then*

$$\mu(x) = e^{\int P(x)\,dx}$$

is an integrating factor for equation (7.10).

b. *Show that, if Q is a function of y only (so all the x's cancel out), then*

$$\mu(y) = e^{\int Q(x)\,dx}$$

is an integrating factor for equation (7.10).

8

Review Exercises for Part of Part II

In the last several chapters, we've identified various 'types' (directly integrable, separable, etc.) of first-order differential equations and developed ways to solve them. What follows is a list of first-order differential equations that can be solved by the methods we've developed. For each, identify the equation's type (or types) and then find its general solution. Where possible, express each answer as an explicit solution.

1. $x\dfrac{dy}{dx} = 2y - 6x^3$

2. $x\dfrac{dy}{dx} = 2y^2 - 6y$

3. $4y^2 - x^2y^2 + \dfrac{dy}{dx} = 0$

4. $\dfrac{dy}{dx} = \sqrt{x+y}$

5. $x^2\dfrac{dy}{dx} - \sqrt{x} = 3$

6. $xy\dfrac{dy}{dx} - y^2 = \sqrt{x^4 + x^2y^2}$

7. $\dfrac{dy}{dx} = y^2 - 2xy + x^2$

8. $4xy - 6 + x^2\dfrac{dy}{dx} = 0$

9. $xy^2 - 6 + x^2y\dfrac{dy}{dx} = 0$

10. $x^3 + y^3 - xy^2\dfrac{dy}{dx} = 0$

11. $3y - x^3 + x\dfrac{dy}{dx} = 0$

12. $1 + 2xy^2 + \left[2x^2y + 2y\right]\dfrac{dy}{dx} = 0$

13. $3xy^3 - y + x\dfrac{dy}{dx} = 0$

14. $2 + 2x^2 - 2xy + \left[x^2 + 1\right]\dfrac{dy}{dx} = 0$

15. $\left(y^2 - 4\right)\dfrac{dy}{dx} = y$

16. $\left(x^2 - 4\right)\dfrac{dy}{dx} = x$

17. $\dfrac{dy}{dx} = \dfrac{1}{xy - 3x}$

18. $\dfrac{dy}{dx} = \dfrac{3}{1+x}y - y^2$

19. $\sin(y) + (x + y)\cos(y)\dfrac{dy}{dx} = 0$

20. $\sin(y) + (1 + x)\cos(y)\dfrac{dy}{dx} = 0$

21. $\sin(x) + 2\cos(x)\dfrac{dy}{dx} = 0$

22. $xy\dfrac{dy}{dx} = 2\left(x^2 + y^2\right)$

23. $\dfrac{dy}{dx} = \dfrac{x + 2y}{x + 2y + 3}$

24. $\dfrac{dy}{dx} = \dfrac{x + 2y}{2x - y}$

25. $\dfrac{dy}{dx} = \dfrac{y}{x} + \tan\left(\dfrac{y}{x}\right)$

26. $\dfrac{dy}{dx} = xy^2 + 3y^2 + x + 3$

27. $1 - (x + 2y)\dfrac{dy}{dx} = 0$

28. $\ln|y| + \left[\dfrac{x}{y} + 3\right]\dfrac{dy}{dx} = 0$

29. $y^2 + 1 - \dfrac{dy}{dx} = 0$

30. $\dfrac{dy}{dx} - 3y = 12e^{2x}$

31. $xy\dfrac{dy}{dx} = x^2 + xy + y^2$

32. $(x+2)\dfrac{dy}{dx} - x^3 = 0$

33. $xy^3\dfrac{dy}{dx} = y^4 - x^2$

34. $\dfrac{dy}{dx} = 4y - 16e^{4x}y^{-2}$

35. $(2y - 6x) + (x+1)\dfrac{dy}{dx} = 0$

36. $xy^2 + \left(x^2y + 10y^4\right)\dfrac{dy}{dx} = 0$

37. $y\dfrac{dy}{dx} - xy^2 = 6xe^{4x^2}$

38. $(y - x + 3)^2\left(\dfrac{dy}{dx} - 1\right) = 1$

39. $x + ye^{xy} + xe^{xy}\dfrac{dy}{dx} = 0$

40. $y^2 - y^2\cos(x) + \dfrac{dy}{dx} = 0$

41. $\dfrac{dy}{dx} + 2y = \sin(x)$

42. $\dfrac{dy}{dx} + 2x = \sin(x)$

43. $\dfrac{dy}{dx} = y^3 - y^3\cos(x)$

44. $y^2e^{xy^2} - 2x + 2xye^{xy^2}\dfrac{dy}{dx} = 0$

45. $\dfrac{dy}{dx} = e^{-3x} - 3y$

46. $\dfrac{dy}{dx} = \tan(6x + 3y + 1) - 2$

47. $\dfrac{dy}{dx} = e^{4x+3y}$

48. $\dfrac{dy}{dx} = x\left[6y + e^{x^2}\right]$

49. $x(1 - 2y) + \left(y - x^2\right)\dfrac{dy}{dx} = 0$

50. $x^2\dfrac{dy}{dx} + 3xy = 6e^{-x^2}$

9

Slope Fields: Graphing Solutions Without the Solutions

Up to now, our efforts have been directed mainly towards finding formulas or equations describing solutions to given differential equations. Then, sometimes, we sketched the graphs of these solutions using those formulas or equations. In this chapter, we will do something quite different. Instead of solving the differential equations, we will use the differential equations, directly, to sketch the graphs of their solutions. No other formulas or equations describing the solutions will be needed.

The graphic techniques and underlying ideas that will be developed here are, naturally, especially useful when dealing with differential equations that cannot be readily solved using the methods already discussed. But these methods can be valuable even when we can solve a given differential equation because they yield "pictures" describing the general behavior of the possible solutions. Sometimes, these pictures can be even more enlightening than formulas for the solutions.

9.1 Motivation and Basic Concepts

Suppose we have a first-order differential equation that, for motivational purposes, "just cannot be solved" using the methods already discussed. For illustrative purposes, let's pretend

$$16\frac{dy}{dx} + xy^2 = 9x$$

is that differential equation. (True, this is really a simple separable differential equation. But it is also a good differential equation for illustrating the ideas being developed.)

For our purposes, we need to algebraically solve the differential equation to get it into the derivative formula form, $y' = F(x, y)$. Doing so with the above differential equation, we get

$$\frac{dy}{dx} = \frac{x}{16}\left(9 - y^2\right) \quad . \tag{9.1}$$

Remember, there are infinitely many particular solutions (with different particular solutions typically corresponding to different values for the general solution's 'arbitrary' constant). Let's now pick some point in the plane, say, $(x, y) = (1, 2)$, let $y = y(x)$ be the particular solution to the differential equation whose graph passes through that point, and consider sketching a short line tangent to this graph at this point. From elementary calculus, we know the slope of this tangent line is given by the derivative of $y = y(x)$ at that point. And fortunately, equation (9.1) gives us a formula for computing this very derivative without the bother of actually solving for $y(x)$! So, for the graph of this particular $y(x)$,

$$\begin{aligned}
\text{Slope of the tangent line at } (1,2) &= \frac{dy}{dx} \quad \text{at} \quad (x,y) = (1,2) \\
&= \frac{x}{16}\left(9 - y^2\right) \quad \text{at} \quad (x,y) = (1,2) \\
&= \frac{1}{16}\left(9 - 2^2\right) \\
&= \frac{5}{16} \quad .
\end{aligned}$$

Thus, if we draw a short line with slope $^5/_{16}$ through the point $(1, 2)$, that line will be tangent at that point to the graph of a solution to our differential equation.

So what? Well, consider further: At each point (x, y) in the plane, we can draw a short line whose slope is given by the right side of equation (9.1). For convenience, let's call each of these short lines the *slope line* for the differential equation at the given point. Now consider any curve drawn so that, at each point (x, y) on the curve, the slope line there is tangent to the curve. If this curve is the graph of some function $y = y(x)$, then, at each point (x, y),

$$\frac{dy}{dx} = \text{slope of the slope line at } (x,y) \quad .$$

But we constructed the slope lines so that

$$\text{slope of the slope line at } (x,y) = \text{right side of equation (9.1)} = \frac{x}{16}\left(9 - y^2\right) \quad .$$

So the curve drawn is the graph of a function $y(x)$ satisfying

$$\frac{dy}{dx} = \frac{x}{16}\left(9 - y^2\right) \quad .$$

That is, the curve drawn is the graph of a solution to our differential equation, and we've managed to draw this curve without actually solving the differential equation.

In practice, of course, we cannot draw the slope line at *every* point in the plane. But we can construct the slope lines at the points of any finite grid of points, and then sketch curves that "parallel" these slope lines — that is, sketch curves so that, at each point on each curve, the slope of the tangent line is closely approximated by the slopes of the nearby slope lines. Each of these curves would then approximate the graph of a solution to the differential equation. These curves may not be perfect, but, if we are careful, they should be close to the actual graphs, and, consequently, will give us a good picture of what the solutions to our differential equation look like. Moreover, we can construct these graphs without actually solving the differential equation.

By the way, the phrase "graph of a solution to the differential equation" is a bit long to constantly repeat. For brevity, we will misuse terminology slightly and call these graphs *solution curves* (for the given differential equation).[1]

[1] If you recall the discussion on graphing implicit solutions (Section 4.7 starting on page 83), you may realize that, strictly speaking, the curves being sketched are "integral curves" containing "solution curves". However, we will initially make an assumption that makes the distinction between integral and solution curves irrelevant.

9.2 The Basic Procedure

What follows is a procedure for systematically constructing approximate graphs of solutions to a first-order differential equation using the ideas just developed. We assume that we have a first-order differential equation, possibly with some initial condition $y(x_0) = y_0$, and that we wish to sketch some of the solution curves for this differential equation in some "region of interest" in the XY–plane. To avoid a few complicating issues (which will be dealt with later), an additional requirement will be imposed (in the first step) on the sort of differential equations being considered. Later, we'll discuss what can be done when this requirement is not met. These steps will be illustrated using the initial-value problem

$$16\frac{dy}{dx} + xy^2 = 9x \qquad \text{with} \quad y(0) = 1 \quad . \tag{9.2}$$

The Procedure:

1. Algebraically solve the differential equation for the derivative to get it into the form

$$\frac{dy}{dx} = F(x, y)$$

where $F(x, y)$ is some formula involving x and/or y.

For now, let us limit our discussion to differential equations for which $F(x, y)$ is well defined and continuous throughout the region of interest. In particular, then, we are requiring $F(x, y)$ to be a finite number for each (x, y) in the region we are trying to graph solutions.[2] What may happen when this requirement is not satisfied will be discussed later (in Section 9.4).

Solving equation (9.2) for the derivative, we get

$$\frac{dy}{dx} = \frac{x}{16}\left(9 - y^2\right) \quad .$$

So here,

$$F(x, y) = \frac{x}{16}\left(9 - y^2\right) \quad .$$

There is certainly no problem with computing this for any pair of values x *and* y*; so our differential equation meets the requirement that "*$F(x, y)$ *be well defined" in whatever region we end up using.*

2. Pick a grid of points

$$\begin{array}{l}(x_1, y_1)\,,\ (x_2, y_1)\,,\ (x_3, y_1)\,,\ \ldots,\ (x_J, y_1)\,,\\ (x_1, y_2)\,,\ (x_2, y_2)\,,\ (x_3, y_2)\,,\ \ldots,\ (x_J, y_2)\,,\\ \vdots\\ (x_1, y_K)\,,\ (x_2, y_K)\,,\ (x_3, y_K)\,,\ \ldots,\ (x_J, y_K)\,.\end{array}$$

on which to plot the slope lines. Just which points are chosen is largely a matter of judgment. If the problem involves an initial condition $y(x_0) = y_0$, then the corresponding point, (x_0, y_0), should be one point in the grid. In addition, the grid should:

i. 'cover' the region over which you plan to graph the solutions, and

[2] Since it prevents points with "infinite slope", this requirement ensures that the curves will be "solution curves" in the strict sense of the phrase.

ii. have enough points so that the slope lines at those points will give a good idea of the curves to be drawn.

(More points can always be added later.)

In our example, we have the initial condition $y(0) = 1$*, so we want our grid to contain the point* $(0, 1)$*. For our grid let us pick the set of all points in the region* $0 \leq x \leq 4$ *and* $0 \leq y \leq 4$ *with integral coordinates:*

$$\begin{array}{l} (0,4)\,,\ (1,4)\,,\ (2,4)\,,\ (3,4)\,,\ (4,4)\,, \\ (0,3)\,,\ (1,3)\,,\ (2,3)\,,\ (3,3)\,,\ (4,3)\,, \\ (0,2)\,,\ (1,2)\,,\ (2,2)\,,\ (3,2)\,,\ (4,2)\,, \\ (0,1)\,,\ (1,1)\,,\ (2,1)\,,\ (3,1)\,,\ (4,1)\,, \\ (0,0)\,,\ (1,0)\,,\ (2,0)\,,\ (3,0)\,,\ (4,0)\,. \end{array}$$

Note that this does contain the point $(0, 1)$*, as desired.*

3. For each grid point (x_j, y_k):

(a) Compute $F(x_j, y_k)$, the right side of the differential equation from step 1.

(b) Using the value $F(x_j, y_k)$ just computed, carefully draw a short line at (x_j, y_k) with slope $F(x_j, y_k)$. (As already stated, this short line is called the *slope line* for the differential equation at (x_j, y_k). Keep in mind that the slope line at each point is tangent to the solution curve passing through this point.)

More Terminology: The collection of all the slope lines at all points on the grid is called a *slope field* for the differential equation.[3]

Glancing back at our example from step 1, we see that

$$F(x, y) = \frac{x}{16}\left(9 - y^2\right) \quad .$$

Systematically computing this at each grid point (and noting that these values give us the slopes of the slope lines at these points):

$$\begin{aligned} \textit{slope of slope line at } (0,0) &= F(0,0) = \frac{0}{16}\left(9 - 0^2\right) = 0 \\ \textit{slope of slope line at } (1,0) &= F(1,0) = \frac{1}{16}\left(9 - 0^2\right) = \frac{9}{16} \\ \textit{slope of slope line at } (2,0) &= F(2,0) = \frac{2}{16}\left(9 - 0^2\right) = \frac{9}{8} \\ &\vdots \\ \textit{slope of slope line at } (1,2) &= F(1,2) = \frac{1}{16}\left(9 - 2^2\right) = \frac{5}{16} \\ &\vdots \end{aligned}$$

The results of all these slope computations are contained in the table in Figure 9.1a. Sketching the corresponding slope line at each grid point then gives us the slope field sketched in Figure 9.1b.

[3] Some texts also refer to a slope field as a "direction field". We will use that term for something else in Chapter 39.

	slope values at (x, y)				
$y = 4$	0	$-7/16$	$-7/8$	$-21/16$	$-7/4$
$y = 3$	0	0	0	0	0
$y = 2$	0	$5/16$	$5/8$	$15/16$	$5/4$
$y = 1$	0	$1/2$	1	$3/2$	2
$y = 0$	0	$9/16$	$9/8$	$27/16$	$9/4$
$x =$	0	1	2	3	4

(a)

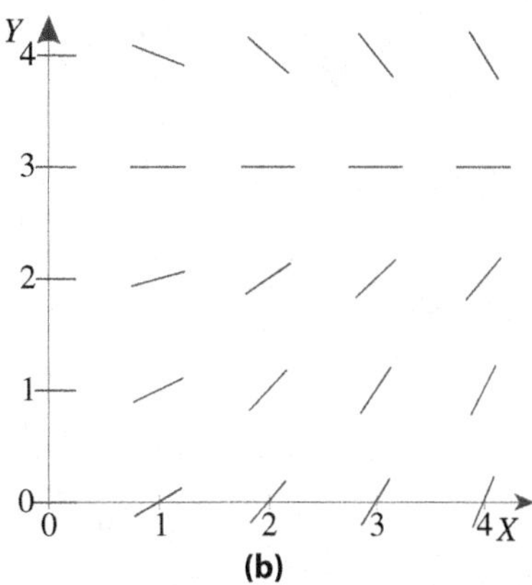

(b)

Figure 9.1: **(a)** The table of slopes of the slope lines at the grid points and **(b)** the corresponding slope lines (and resulting slope field) for $y'(x) = \frac{1}{16}x\left(9 - y^2\right)$.

4. Using the slope field just constructed, sketch curves that "parallel" the slope field. To be precise:
   (a) Pick a convenient grid point as a starting point. Then, as well as can be done freehanded, sketch a curve through that point which "parallels" the slope field. This curve must go through the starting point and must be tangent to the slope line there. Beyond that, however, there is no reason to expect this curve to go through any other grid point — simply draw this curve so that, at each point on the curve, the curve's tangent there closely matches the nearby slope lines. In other words, do *not* attempt to "connect the dots"! Instead, "go with the flow" indicated by the slope field.

   (b) If desired, repeat the previous step and sketch another curve using a different starting point. Continue sketching curves with different starting points until you get as many curves as seem appropriate.

   If done carefully, the curves sketched will be reasonably good approximations of solution curves for the differential equation. If your original problem involves an initial value, be sure that one of your starting points corresponds to that initial value. The resulting curve will be (approximately) the graph of the solution to that initial-value problem.

   Figure 9.2 shows the slope field just constructed, along with four curves sketched according to the instructions just given. The starting points for these curves were chosen to be $(0, 0)$, $(0, 1)$, $(0, 3)$, *and* $(0, 4)$. *Each of these curves approximates the graph of one solution to our differential equation,*

$$\frac{dy}{dx} = \frac{x}{16}\left(9 - y^2\right) \quad ,$$

   with the one passing through $(0, 1)$ *being (approximately) the graph of the solution to the initial-value problem*

$$\frac{dy}{dx} = \frac{x}{16}\left(9 - y^2\right) \qquad \text{with} \quad y(0) = 1 \quad .$$

5. At this point (or possibly some point in the previous step) decide whether there are enough slope lines to accurately draw the desired curves. If not, add more points to the grid, repeat step 3 with the new grid points, and redraw the curves with the improved slope field (but, first, see some of the notes below on this procedure).

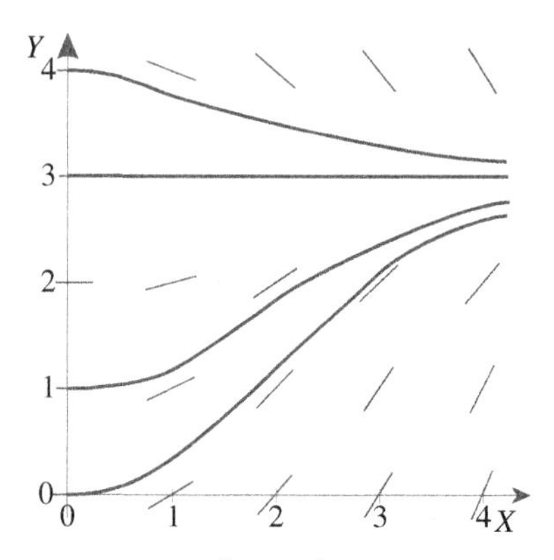

Figure 9.2: A slope field for $y'(x) = \frac{1}{16}x\left(9 - y^2\right)$, along with four curves sketched "parallel" to the field.

It must be admitted that the graphs obtained in Figure 9.2 are somewhat crude and limited. Clearly, we should have used a much bigger grid to cover more area, and should have used more grid points per unit area so we can get better detail (especially in the region where $y \approx 3$). So let us try (sometime in the near future) something like, say, a 19×16 grid covering the region where $0 \leq x \leq 6$ and $0 \leq y \leq 5$. This will give us a field of 504 slope lines instead of the measly 25 used in Figure 9.2.

Though the above procedure took several pages to describe and illustrate, it is really quite simple, and, eventually, yields a picture that gives a fairly good idea of how the solutions of interest behave. Just how good a picture depends on the slope field generated and how carefully the curves are chosen and drawn. Here are a few observations that may help in generating this picture.

1. As indicated in the example, generating a good slope field can require a great deal of tedious computations and careful drawing — if done by hand. But why do it by hand? This is just the sort of tedious, mind-numbing work computers do so well. Program a computer to generate the slope field. Better yet, check your favorite computer math package. There is a good chance that it will already have commands to generate these fields. Use those commands (or find a math package that has those commands). That is how the direction field in Figure 9.3 was generated.

2. As long as $F(x, y)$ is well defined at every point in the region being considered, solution curves cannot cross each other at nonzero angles. This is because any solution curve through any point (x, y) must be tangent to the one and only slope line there, whether or not that slope line is drawn in. Thus, at worst, two solution curves can become tangent to each other at a point. Even this, the merging of two or more solution curves with a common tangent, is not something you should often expect.

3. Just which curves you choose to sketch depend on your goal. If your goal is to graph the solution to an initial-value problem, then it may suffice to just draw that one curve passing through the point corresponding to the initial condition. That curve approximates the graph of the desired solution $y(x)$ and, from it, you can find the approximate value of $y(x)$ for other values of x.

   On the other hand, by drawing a collection of well-chosen curves following the slope field, you can get a fair idea of how all the solutions of interest generally behave and how they depend on the initial condition. Choosing those curves is a matter of judgment, but do

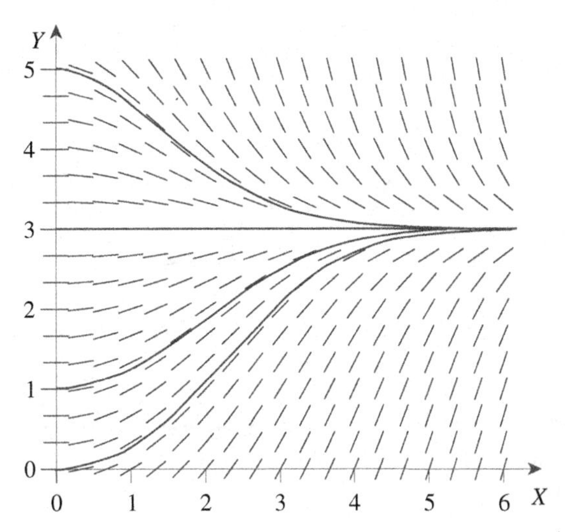

Figure 9.3: A better slope field (based on a 19×16 grid) for $y'(x) = \frac{1}{16}x\left(9 - y^2\right)$, along with four curves sketched "parallel" to the field.

try to identify any curves that are horizontal lines. These are the graphs of constant solutions and are likely to be particularly relevant. In fact, it's often worthwhile to identify and sketch the graphs of all constant solutions in the region, even if they do not pass through any of your grid points.

Consider finding the values of $y(4)$ and $y(6)$ when $y(x)$ is the solution to the initial-value problem

$$\frac{dy}{dx} = \frac{x}{16}\left(9 - y^2\right) \qquad \text{with} \quad y(0) = 1 \quad .$$

Since this differential equation was the one used to generate the slope fields in Figures 9.2 and 9.3, we can use the curve drawn through $(0, 1)$ in either of these figures as an approximate graph for $y(x)$. On this curve in the better slope field of Figure 9.3, we see that $y \approx 2.6$ when $x = 4$, and that $y \approx 3$ when $x = 6$. Thus, according to our sketch, if $y(x)$ satisfies the above initial-value problem, then

$$y(4) \approx 2.6 \qquad \text{and} \qquad y(6) \approx 3 \quad .$$

More generally, after looking at Figure 9.3, it should be apparent that any curve in the sketched region that "parallels" the slope field will approach $y = 3$ when x becomes large. This strongly suggests that, if $y(x)$ is any solution to our differential equation with $0 \leq y(0) \leq 5$, then

$$\lim_{x \to \infty} y(x) = 3 \quad .$$

Do observe that $y = 3$ is a constant solution to our differential equation.

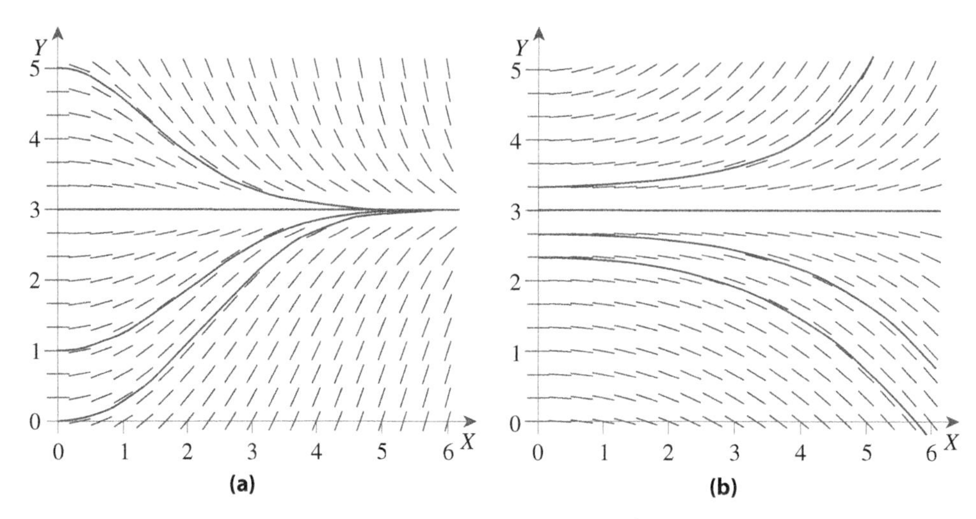

Figure 9.4: **(a)** A slope field and some solutions for $y'(x) = \frac{1}{4}x(3-y)$, and **(b)** a slope field and some solutions for $y'(x) = \frac{1}{3}(y-3)^{1/3}$.

9.3 Observing Long-Term Behavior in Slope Fields

Basic Notions

A slope field of a differential equation gives a picture of the general behavior of the possible solutions to that differential equation, at least in the region covered by that slope field. In many cases, this picture may even give you a good idea of the "long-term" behavior of the solutions.

!►*Example 9.1:* *Consider the differential equation*

$$\frac{dy}{dx} = \frac{x}{4}(3-y) \quad .$$

A slope field (and some solution curves) for this equation is sketched in Figure 9.4a. Now let $y = y(x)$ be any solution to this differential equation, and look at the slope field. It clearly suggests that

$$y(x) \rightarrow 3 \quad \text{as} \quad x \rightarrow \infty \quad .$$

On the other hand, the slope field sketched in Figure 9.4b for

$$\frac{dy}{dx} = \frac{1}{3}(y-3)^{1/3}$$

suggests a rather different sort of behavior for this equation's solutions as x gets large. Here, it looks as if almost no solutions approach any constant value as $x \rightarrow \infty$. Instead, we appear to have

$$\lim_{x\to\infty} y(x) = +\infty \qquad \text{if} \quad y(0) > 3$$

and

$$\lim_{x\to\infty} y(x) = -\infty \qquad \text{if} \quad y(0) < 3 \quad .$$

Of course, one should be cautious about using a slope field to predict the value of $y(x)$ when x is outside the range of x-values used in the slope field. In general, a slope field for a given differential equation sketched on one region of the XY–plane can be quite different from a slope field for that differential equation over a different region. So it is important to be sure that the general pattern of slope lines on which you are basing your prediction does not significantly change as you consider points outside the region of your slope field.

!►Example 9.2: *If you look at the differential equation for the slope field in Figure 9.4a,*

$$\frac{dy}{dx} = \frac{x}{4}(3 - y) \quad ,$$

you can see that the magnitude of the right side

$$\left|\frac{x}{4}(3 - y)\right|$$

becomes larger as either $|x|$ *or* $|y - 3|$ *becomes larger, but the sign of the right side remains*

$$\text{negative if } x > 0 \text{ and } y > 3 \qquad \text{and} \qquad \text{positive if } x > 0 \text{ and } y < 3 \quad .$$

Thus, the slope lines may become steeper as we increase x *or as we go up or down with* y, *but they continue to "direct" all sketched solution curves towards the line* $y = 3$ *as* $x \to \infty$.

There is one class of differential equations whose slope fields are especially suited for making long-term predictions: the *autonomous* first-order differential equations. Remember, such a differential equation can be written as

$$\frac{dy}{dx} = g(y)$$

where $g(y)$ is a known formula of y only. The fact that the right side of this equation does not depend on x means that the vertical column of slope lines at any one value of x is identically repeated at every other value of x. So if the slope field tells you that the solution curve through, say, the point $(x, y) = (1, 4)$ has slope $1/2$, then you are certain that the solution curve through any point (x, y) with $y = 4$ also has slope $1/2$. Moreover, if there is a horizontal slope line at a point (x_0, y_0), then there will be a horizontal slope line wherever $y = y_0$; that is, $y = y_0$ will be a constant solution to the differential equation.

!►Example 9.3: *The differential equation for the slope field sketched in Figure 9.4b,*

$$\frac{dy}{dx} = \frac{1}{3}(y - 3)^{1/3} \quad ,$$

is autonomous since this formula for the derivative does not explicitly involve x. *So the pattern of slope lines in any vertical column in the given slope field will be repeated identically in every vertical column in any slope field covering a larger region (provided we use the same* y*-values). Moreover, from the right side of the equation, we can see that the slopes of the slope lines*

1. *remain positive and steadily increase as* y *increases above* $y = 3$,

and

2. *remain negative and steadily decrease as* y *decreases below* $y = 3$.

Consequently, no matter how large a region we choose for the slope field, we will see that

1. *the slope lines at points above* $y = 3$ *will be "directing" the solution curves more and more steeply upwards as* y *increases*

and

2. *the slope lines at points below $y = 3$ will be "directing" the solution curves more and more steeply downwards as y decreases.*

Thus, we can safely say that, if $y = y(x)$ is any solution to this differential equation, then

$$\lim_{x\to\infty} y(x) = \begin{cases} +\infty & \text{if } y(0) > 3 \\ -\infty & \text{if } y(0) < 3 \end{cases} .$$

Constant Solutions and Stability

The "long-term behavior" of a constant solution

$$y(x) = y_0 \qquad \text{for all} \quad x$$

is quite straightforward: the value of $y(x)$ remains y_0 as $x \to \infty$. What is more varied, and often quite important, is the long-term behavior of the other solutions that are initially "close" to this constant solution. The slope fields in Figures 9.4a and 9.4b clearly illustrate how different this behavior may be.

In Figure 9.4a, the graph of every solution $y = y(x)$ with $y(0) \approx 3$ remains close to the horizontal line $y = 3$ as x increases. Thus, if you know $y(x)$ satisfies the given differential equation, but only know that $y(0) \approx 3$, then it is still safe to expect that $y(x) \approx 3$ for all $x > 0$. In fact, it appears that $y(x) \to 3$ as $x \to \infty$.

In Figure 9.4b, by contrast, the graph of every nonconstant solution $y = y(x)$ with $y(0) \approx 3$ diverges from the horizontal line $y = 3$ as x increases. Thus, if $y(x)$ is a solution to the differential equation for this slope field, but you only know that $y(0) \approx 3$, then you have very little idea what $y(x)$ is for large values of x. This could be a significant concern in real-world applications where, often, initial values are only known approximately.

This leads to the notion of the "stability" of a given constant solution for a first-order differential equation. This concerns the tendency of solutions having initial values close to that of that constant solution to continue having values close to that constant as the variable increases. Whether or not the initially nearby solutions remain nearby determines whether a constant solution is classified as being "stable", "asymptotically stable" or "unstable". Basically, we will say that a constant solution $y = y_0$ to some given first-order differential equation is:

- *stable* if (and only if) every other solution $y = y(x)$ having an initial value $y(0)$ "sufficiently close" to y_0 remains reasonably close to y_0 as x increases.[4]

- *asymptotically stable* if (and only if) it is stable and, in fact, any other solution $y = y(x)$ satisfies

$$\lim_{x\to\infty} y(x) = y_0$$

  whenever that solution's initial value $y(0)$ is "sufficiently close" to y_0.[5] (Typically, this means the horizontal line $y = y_0$ is the horizontal asymptote for these solutions — that's where the term "asymptotically stable" comes from.)

- *unstable* whenever it is not a stable constant solution.

[4] To be more precise: $y = y_0$ is a *stable* constant solution if and only if, for every $\epsilon > 0$, there is a corresponding $\delta > 0$ such that, whenever $y = y(x)$ is a solution to the differential equation satisfying $|y(0) - y_0| < \delta$, then $|y(x) - y_0| < \epsilon$ for all $x > 0$.

[5] More precisely: $y = y_0$ is an *asymptotically stable* constant solution if and only if there is a corresponding $\delta > 0$ such that, whenever $y = y(x)$ is a solution to the differential equation satisfying $|y(0) - y_0| < \delta$, then $\lim_{x\to\infty} y(x) = y_0$.

Of course, the above definitions assume the differential equation is "reasonably well-defined in a region about the constant solution $y = y_0$".[6]

Often, the stability or instability of a constant solution is readily apparent from a given slope field, with rigorous confirmation easily done by fairly simple analysis. Asymptotically stable constant solutions are also often easily identified in slope fields, though rigorously verifying asymptotic stability may require a bit more analysis.

!►Example 9.4: *Recall that the slope field in Figure 9.4a is for*

$$\frac{dy}{dx} = \frac{x}{4}(3 - y) \quad .$$

From our discussions in Examples 9.1 and 9.2, we already know $y = 3$ is a constant solution to this differential equation, and that, if $y = y(x)$ is any other solution satisfying $y(0) \approx 3$, then $y(x) \approx 3$ for all $x > 0$. In fact, because the slope lines are all angled towards $y = 3$ as x increases, it should be clear that, for every $x > 0$, $y(x)$ will be closer to 3 than is $y(0)$. So $y = 3$ is a stable constant solution to the above differential equation.

Is $y = 3$ an asymptotically stable solution? That is, do we have

$$\lim_{x\to\infty} y(x) = 3$$

whenever $y = y(x)$ is a solution with $y(0)$ is sufficiently close to 3 ? The slope field certainly suggests so. Fortunately, this differential equation is a fairly simple separable equation which you can easily solve to get

$$y(x) = 3 + Ae^{-x^2/2}$$

as a general solution. Taking the limit, we see that

$$\lim_{x\to\infty} y(x) = \lim_{x\to\infty} 3 + Ae^{-x^2/2} = 3 + 0 \quad ,$$

no matter what $y(0)$ is. So, yes, $y = 3$ is not just a stable constant solution to the above differential equation; it is an asymptotically stable constant solution.

!►Example 9.5: *Now, again consider the slope field in Figure 9.4b, which is for*

$$\frac{dy}{dx} = \frac{1}{3}(y - 3)^{1/3} \quad .$$

Again, we know $y = 3$ is a constant solution for this differential equation. However, from our discussion in Example 9.3, we also know that, if $y = y(x)$ is any other solution, then

$$\lim_{x\to\infty} y(x) = \pm\infty \quad .$$

Clearly, then, even if $y(0)$ is very close (but not equal) to 3, $y(x)$ will not remain close to 3 as x increases. Thus, $y = 3$ is an unstable constant solution to this differential equation.

In the two examples given so far, all the solutions starting near a stable constant solution converged to that solution, while all nonconstant solutions starting near an unstable solution diverged to $\pm\infty$ as $x \to \infty$. The next two examples show that somewhat different behavior can occur.

---

[6] More precisely: The differential equation can be written as $y' = F(x, y)$ where F is continuous at every (x, y) with $x \geq 0$ and $|y - y_0| < \delta$ for some $\delta > 0$.

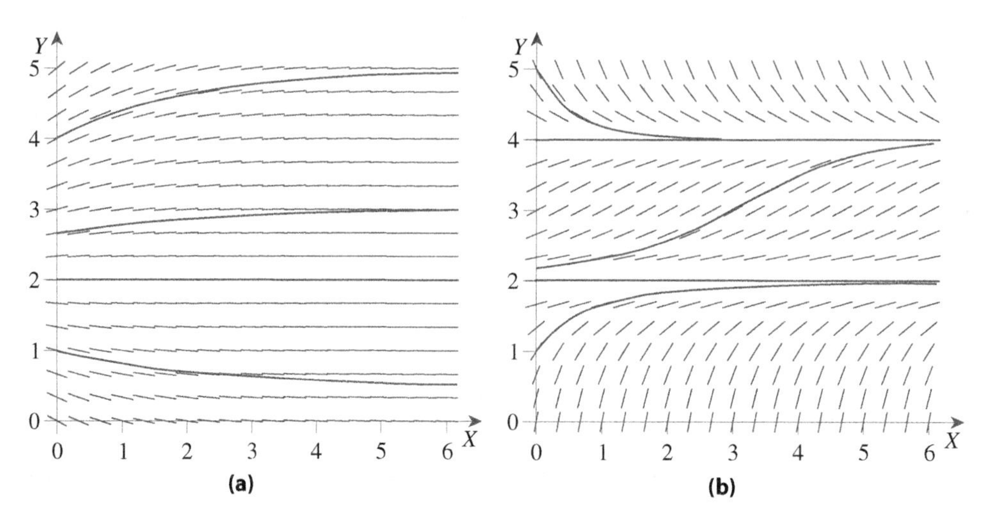

Figure 9.5: Slope fields **(a)** for Example 9.6, and **(b)** for Example 9.7.

!►Example 9.6: *The slope field and solution curves sketched in Figure 9.5a are for*

$$\frac{dy}{dx} = \frac{y-2}{6e^{x/2}-2} \quad .$$

Here, $y = 2$ is the only constant solution. Following the slope lines in this figure, it appears that, although the graph of each nonconstant solution $y = y(x)$ starts at $x = 0$ by moving away from $y = 2$ as x increases, this graph quickly levels out so that $y(x)$ approaches some constant as $x \to \infty$. This behavior can be confirmed by solving the differential equation. With a little work, you can solve this differential equation and show that, if y is any solution to this differential equation, then

$$y(x) - 2 = \left[3 - e^{-x/2}\right][y(0) - 2] \quad .$$

You can also easily verify that

$$\left|3 - e^{-x/2}\right| < 3 \qquad \text{for} \quad x > 0 \quad .$$

So,

$$|y(x) - 2| = \left|3 - e^{-x/2}\right| |y(0) - 2| < 3\,|y(0) - 2| \quad .$$

In other words, the distance between $y(x)$ and $y = 2$ when $x > 0$ is never more than three times the distance between $y(x)$ and $y = 2$ when $x = 0$. So, if we wish $y(x)$ to stay within a certain distance of $y = 2$ for all positive values of x, we merely need to be sure that $y(0)$ is no more than a third of that distance from 2.

This confirms that $y = 2$ is a stable constant solution. However, it is not asymptotically stable because

$$\lim_{x\to\infty} y(x) = 2 + 3[y(0) - 2] \neq 2 \qquad \text{whenever} \quad y(0) \neq 2 \quad .$$

?►Exercise 9.1: *Let $y(x)$ be a solution to the differential equation discussed in the last example. Using the solution formula given above:*

a: *Show that*

$$|y(x) - 2| < 1 \qquad \text{for all} \quad x > 0$$

whenever

$$|y(0) - 2| < \frac{1}{3} \quad .$$

b: *How close should* $y(0)$ *be to* 2 *so that*

$$|y(x) - 2| < \frac{1}{2} \qquad \textit{for all} \quad x > 0 \quad ?$$

In the next example, there are two constant solutions, and the analysis is done without the benefit of having a general solution to the given differential equation.

!►Example 9.7: *The slope field and solution curves sketched in Figure 9.5b are for*

$$\frac{dy}{dx} = \frac{1}{2}(4 - y)(y - 2)^{4/3} \quad .$$

Technically, this separable equation can be solved for an implicit solution by the methods discussed for separable equations, but the resulting equation is too complicated to be of much value here. Fortunately, from a quick examination of the right side of this differential equation, we can see that

1. *There are two constant solutions,* $y = 2$ *and* $y = 4$.
2. *The differential equation is autonomous. So the pattern of slope lines seen in Figure 9.5b continues unchanged throughout the entire horizontal strip with* $0 \leq y \leq 5$.

Following the slope lines in Figure 9.5b, it seems clear that $y = 4$ *is a stable constant solution. In fact, it appears that*

$$\lim_{x \to \infty} y(x) = 4$$

whenever y *is a solution satisfying*

$$2 < y(0) < 5 \quad .$$

This strongly suggests that $y = 4$ *is an asymptotically stable constant solution.*

On the other hand, if

$$\lim_{x \to \infty} y(x) = 4 \qquad \textit{whenever} \quad 2 < y(0) < 5 \quad ,$$

then the constant solution $y = 2$ *cannot be stable. True, it appears that*

$$\lim_{x \to \infty} y(x) = 2 \qquad \textit{whenever} \quad 0 < y(0) \leq 2 \quad ,$$

but, if $y(0)$ *is just a tiny bit larger than* 2, *then* $y(x)$ *does not stay close to* 2 *as* x *increases — it gets close to* 4. *So we must consider this constant solution as being unstable. (We will later see that this type of instability can cause serious problems when attempting to numerically solve a differential equation.)*

In the last example, we did not do the analysis to rigorously verify that $y = 4$ is an asymptotically stable constant solution, and that $y = 2$ is an unstable constant solution. Still, you are probably pretty confident that more rigorous analysis will confirm this. If so, good — you are correct. We'll verify this in Section 9.5 using the more rigorous tests developed there.

Finally, a few comments that should be made regarding, not stability, but our discussion of "stability":

1. Strictly speaking, we've been discussing the stability of solutions to initial-value problems when the initial value of $y(x)$ is given at $x = 0$. To convert our discussion to a discussion of the stability of solutions to initial-value problems with the initial value of $y(x)$ given at some other point $x = x_0$, just repeat the above with $x = 0$ replaced by $x = x_0$. There will be no surprises.

2. Traditionally, discussions of "stability" only involve autonomous differential equations. We did not do so here because there seemed little reason to do so (provided we are careful about taking into account how the differential equation depends on x). Admittedly, limiting discussion to autonomous equations would have simplified things since the slope fields of autonomous differential equations do not depend on x. In addition, constant solutions to autonomous equations are traditionally called equilibrium solutions, and, to this author at least, "stable and unstable equilibriums" sounds more interesting than "stable and unstable constant solutions". Still, that did not justify limiting our discussion to just autonomous equations.

9.4 Problem Points in Slope Fields, and Issues of Existence and Uniqueness

In sketching and using a slope field for

$$\frac{dy}{dx} = F(x, y)$$

we have, up to this point, assumed $F(x, y)$ is well defined and continuous throughout the region of interest. This will not always be the case. So let us look at what can happen when F is not so well behaved at certain points. This, by the way, will naturally lead to a brief continuation of our discussion of "existence" and "uniqueness" that we began in the later part of Chapter 3.

Infinite Slopes

Often, a given $F(x, y)$ becomes infinite at certain points in the XY–plane. This, in turn, means that the corresponding slope lines have "infinite slope", that is, they are vertical. One practical problem is that the software you are using to create your slope fields might object to 'division by zero' and not be able to deal with these points. On a more fundamental level, these infinite slopes may be warning you that something very significant is occurring in the solutions whose graphs include or are near these points.

In particular, these vertical slope lines may be telling you that solutions are, themselves, becoming infinite for finite values of x.

!►***Example 9.8:*** *A slope field for*

$$\frac{dy}{dx} = \frac{1}{3-x}$$

is sketched in Figure 9.6a. Since

$$\lim_{x\to 3} \frac{1}{3-x} = \pm\infty \quad ,$$

there are vertical slope lines at every point (x, y) with $x = 3$. This, along with the pattern of the other nearby slope lines, suggests that the solutions to this differential equation are "blowing

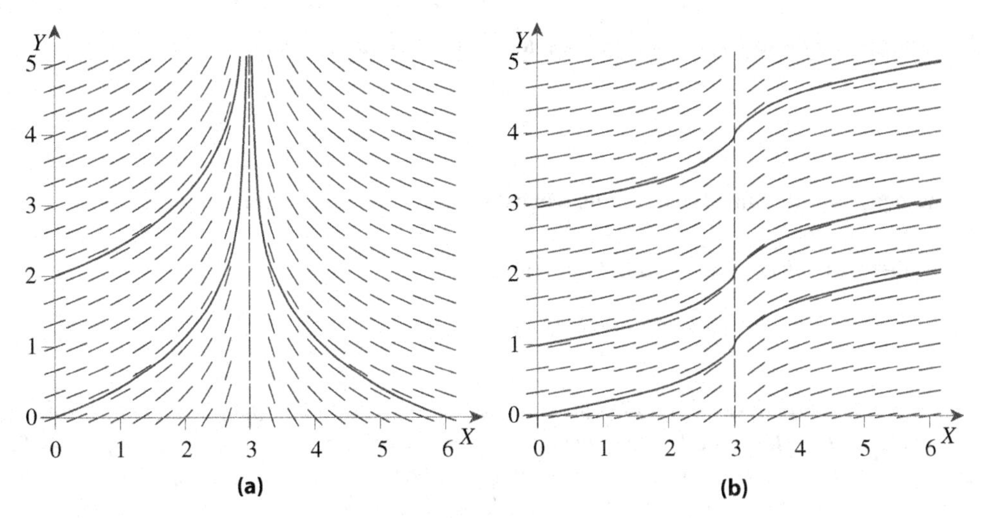

Figure 9.6: Slope fields **(a)** for $y'(x) = (3 - x)^{-1}$ from Example 9.8, and **(b)** for $y'(x) = \frac{1}{3}(x-3)^{-2/3}$ from Example 9.8.

up" as x approaches 3. Fortunately, this differential equation is easily solved — just integrating it yields

$$y = c - \ln|3 - x| \quad ,$$

which does, indeed, "blow up" at $x = 3$ for any choice of c.

Consequently, the vertical slope lines in Figure 9.6a form a vertical asymptote for the graphs of the solutions to the given differential equation. This further means that no solution to the differential equation passes through a point (x, y) with $x = 3$. In particular, if you are asked to solve the initial-value problem

$$\frac{dy}{dx} = \frac{1}{3 - x} \qquad \text{with} \quad y(3) = 2 \quad ,$$

you have every right to respond: "Nonsense, there is no solution to this initial-value problem."

On the other hand, the vertical slope lines might not be harbingers of particularly bad behavior in our solutions. Instead, the solutions may be fairly ordinary functions whose graphs just happen to have vertical tangent lines at a few points.

!▶Example 9.9: *In Figure 9.6b, we have a slope field for*

$$\frac{dy}{dx} = \frac{1}{3(x-3)^{2/3}} \quad .$$

Again, "division by zero" when $x = 3$ gives us vertical slope lines at every (x, y) with $x = 3$. This time, however, integrating the differential equation yields

$$y = (x - 3)^{1/3} + c \quad .$$

For each value of c, this is a continuous function on the entire real line (including at $x = 3$) which just happens to have a vertical tangent when $x = 3$.

In particular, as you can easily verify,

$$y = (x - 3)^{1/3} + 2$$

is the one and only solution on $(-\infty, \infty)$ *to the initial-value problem*

$$\frac{dy}{dx} = \frac{1}{3(x-3)^{2/3}} \qquad \text{with} \quad y(3) = 2 \quad .$$

Another possibility involving infinite slopes is illustrated in the next example.

!►Example 9.10: *The slope field in Figure 9.7a is for*

$$\frac{dy}{dx} = \frac{x-2}{2-y} \quad .$$

This time, the vertical slope lines occur wherever $y = 2$ *(excluding the point* $(2, 2)$*, which we will discuss later). It should be clear that these slope lines do not correspond to asymptotes of the graphs of solutions that "blow up", nor does it appear possible for a curve going from left to right to pass through these points and still parallel the slope lines. Instead, if we carefully sketch the curve that "follows the slope field" through, say, the point* $(x, y) = (0, 2)$*, then we end up with the circle sketched in the figure (which also has a vertical tangent at* $(x, y) = (4, 2)$ *). But such a circle cannot be the graph of a function* $y = y(x)$ *since it corresponds to two different values for* $y(x)$ *for each* x *in the interval* $(0, 4)$*.*

Fortunately, again, our differential equation is a simple separable equation. Solving it (as you can easily do), yields

$$y = 2 \pm \sqrt{A - (x-2)^2} \quad .$$

In particular, if we further require that $y(0) = 2$*, then we obtain exactly two solutions,*

$$y = 2 + \sqrt{4 - (x-2)^2} \qquad \text{and} \qquad y = 2 - \sqrt{4 - (x-2)^2} \quad ,$$

with each defined and continuous on the closed interval $[0, 4]$*. The first satisfies the differential equation on the interval* $(0, 4)$*, and its graph is the upper half of the sketched circle. The second also satisfies the differential equation on the interval* $(0, 4)$*, but its graph is the lower half of the sketched circle.*

Undefined and Indeterminant Slopes

Let's now look at two examples involving points at which slope lines simply cannot be drawn because $F(x, y)$ is neither finite nor infinite at those points.

!►Example 9.11: *Again, consider the slope field in Figure 9.7a for*

$$\frac{dy}{dx} = \frac{x-2}{2-y} \quad .$$

If $(x, y) = (2, 2)$*, this becomes the indeterminant expression*

$$\frac{dy}{dx} = \frac{0}{0} \quad .$$

Moreover, the slopes of the slope lines at points near $(x, y) = (2, 2)$ *range from* 0 *to* $\pm\infty$*. In fact, the point* $(x, y) = (2, 2)$ *appears to be the center of the circles made up of the graphs of the solutions to this differential equation— a fact that can be confirmed using the solution formulas*

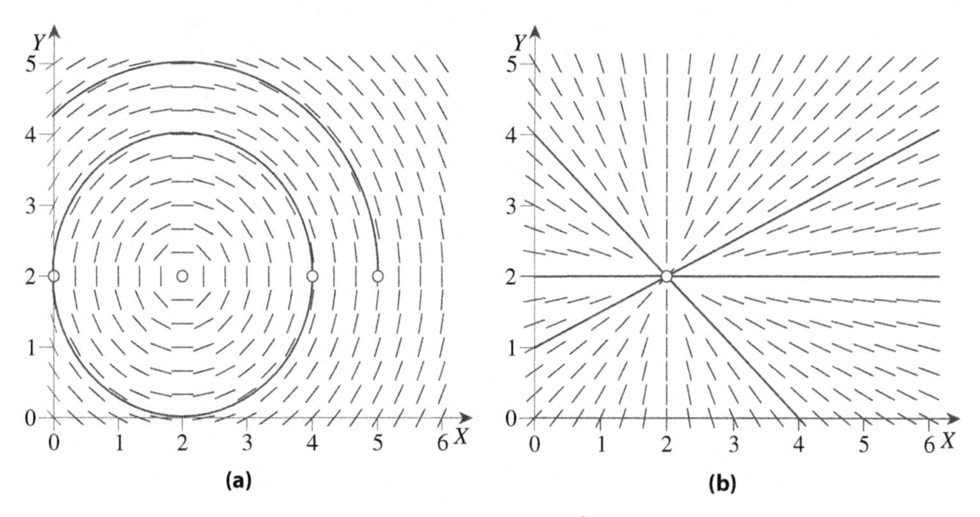

Figure 9.7: Slope fields **(a)** for $y'(x) = (x - 2)(2 - y)^{-1}$ from Examples 9.10 and 9.11, and **(b)** for $y'(x) = (y - 2)(x - 2)^{-1}$ from Example 9.12.

from Example 9.10. Clearly, no real curve can pass through the point $(x, y) = (2, 2)$ and remain parallel to the slope lines near this point. So if we really wanted a solution to

$$\frac{dy}{dx} = \frac{x - 2}{2 - y} \quad \text{with} \quad y(2) = 2 \quad ,$$

which is valid on some interval (α, β), then we would be disappointed. There is no such solution.

!►*Example 9.12:* *We also get*

$$\frac{dy}{dx} = \frac{0}{0}$$

when we let $(x, y) = (2, 2)$ in

$$\frac{dy}{dx} = \frac{y - 2}{x - 2} \quad .$$

This time, however, the slope field (sketched in Figure 9.7b) suggests that every solution curve passes through this point. And, indeed, solving this simple separable equation yields the formula

$$y = 2 + A(x - 2)$$

where A is an arbitrary constant. This formula gives $y = 2$ when $x = 2$ no matter what A is. Consequently, the initial-value problem

$$\frac{dy}{dx} = \frac{y - 2}{x - 2} \quad \text{with} \quad y(2) = 2$$

has infinitely many solutions.

In both of the above examples, the slope lines were all well defined (possibly with infinite slope) at all but one point in the XY–plane. They are fairly representative examples of what can happen when $F(x, y)$ is undefined at isolated points. Of course, we can easily give examples in which $F(x, y)$ is undefined on vast regions of the XY–plane. There isn't much to be said about these cases, but we'll provide one example for the sake of completeness.

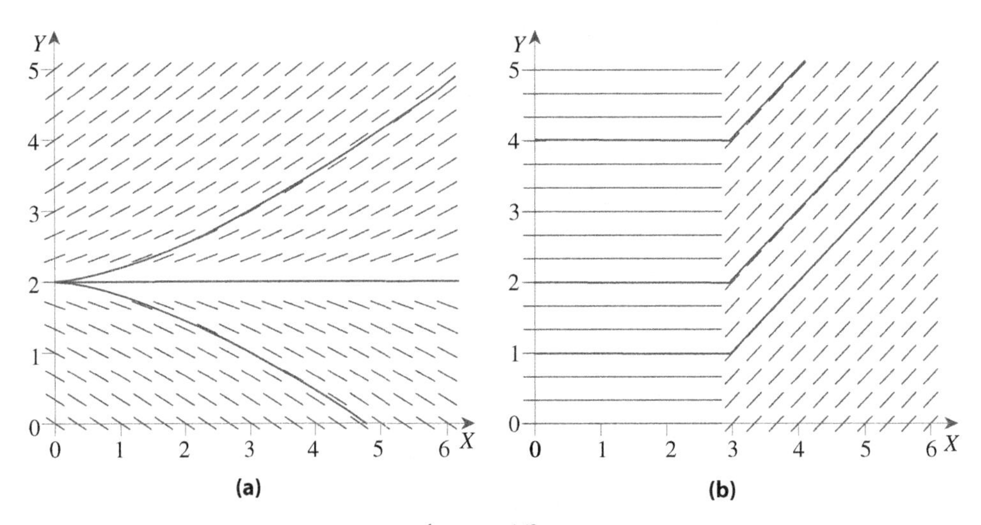

Figure 9.8: Slope fields **(a)** for $y'(x) = \frac{1}{2}(y-2)^{1/3}$ from Example 9.14, and **(b)** for the differential equation in Example 9.15.

!►Example 9.13: *Consider the differential equation*

$$\frac{dy}{dx} = \sqrt{1 - (x^2 + y^2)} \quad .$$

The right side only makes sense if $x^2 + y^2 \leq 1$. Obviously, there can be no "slope field" in any region outside the circle $x^2 + y^2 = 1$ (that's why we didn't attempt to sketch it), and it is just plain silly to ask for a solution to this differential equation satisfying, say, $y(x_0) = y_0$ whenever (x_0, y_0) is a point outside the circle $x^2 + y^2 = 1$.

Curves Diverging From or Converging To a Point

In Example 9.12 (Figure 9.7b), we have solution curves converging to and diverging from the point $(2, 2)$. In that case, $F(x, y)$ was indeterminant at that point. As the next example illustrates, we can have solution curves converging to and diverging from a point even though $F(x, y)$ is a nice well-defined, finite number at that point. Fortunately, for reasons to be explained, this is not very common.

!►Example 9.14: *Consider*

$$\frac{dy}{dx} = \frac{1}{2}(y-2)^{1/3} \quad .$$

A slope field and some solutions for this differential equation are sketched in Figure 9.8a. Note that we've sketched three curves diverging from the point $(0, 2)$. These curves are the graphs of

$$y = 2 \quad , \quad y = 2 + \left(\frac{x}{3}\right)^{3/2} \quad \text{and} \quad y = 2 - \left(\frac{x}{3}\right)^{3/2} \quad ,$$

all of which are solutions on $[0, \infty)$ to the initial-value problem

$$\frac{dy}{dx} = \frac{1}{2}(y-2)^{1/3} \quad \text{with} \quad y(0) = 2 \quad .$$

What distinguishes this from Example 9.12 (Figure 9.7b) is that the right side of the above differential equation is not indeterminant at the point $(0,2)$. Instead, at $(x,y) = (0,2)$ we have

$$\frac{dy}{dx} = \frac{1}{2}(2-2)^{1/3} = 0 \quad ,$$

which is a perfectly reasonable finite value.

On Existence and Uniqueness

Let us return to the issues of the "existence" and "uniqueness" of the solutions to a generic initial-value problem

$$\frac{dy}{dx} = F(x,y) \qquad \text{with} \quad y(x_0) = y_0 \quad . \tag{9.3}$$

We first discussed these issues in the later part of Chapter 3. In particular, you may recall Theorem 3.1 on page 42. That theorem assures us that:

> *If both $F(x,y)$ and $\partial F/\partial y$ are continuous functions on some open region of the XY–plane containing the point (x_0, y_0), then:*
>
> 1. *(existence) The above initial-value problem has at least one solution $y = y(x)$.*
>
> 2. *(uniqueness) There is an open interval (a,b) containing x_0 on which this $y = y(x)$ is the only solution to this initial-value problem.*

Now consider every slope field for this differential equation in some region around (x_0, y_0) on which F is continuous. The continuity of F ensures that the slope lines will be well defined with finite slope at every point, and that these slopes will vary continuously as you move throughout the region. Clearly, there is a curve through the point (x_0, y_0) that is "parallel" to every possible slope field, and this curve will have to be the graph of a function satisfying

$$\frac{dy}{dx} = F(x,y) \qquad \text{with} \quad y(x_0) = y_0 \quad .$$

This graphically verifies the "existence" part of Theorem 3.1. In fact, a good mathematician can take the above argument and construct a rigorous proof that

> *If F is a continuous function on some open region of the XY–plane containing the point (x_0, y_0), then*
>
> $$\frac{dy}{dx} = F(x,y) \qquad \textit{with} \quad y(x_0) = y_0$$
>
> *has at least one solution on some interval (a,b) with $a < x_0 < b$.*

So we can use our slope fields to visually convince ourselves that initial-value problem (9.3) has a solution whenever F is reasonably well behaved. But what about uniqueness? Will the curve drawn be the only possible curve matching the slope fields? Well, in Example 9.14 (Figure 9.8a) we had three different curves passing through the point $(0,2)$, all of which matched the slope field. Thus, we have (at least) three different solutions to the initial-value problem given in that example. And this occurred even though the $F(x,y)$ is a continuous function on all of the XY–plane.

This is where the second part of Theorem 3.1 can help us in using slope fields. It assures us that there is only one solution (over some interval containing x_0) to

$$\frac{dy}{dx} = F(x,y) \qquad \text{with} \quad y(x_0) = y_0$$

provided both $F(x, y)$ and $^{\partial F}/_{\partial y}$ are continuous in a region around (x_0, y_0). In Example 9.14, we had

$$F(x, y) = \frac{1}{2}(y - 2)^{1/3} \quad .$$

While this F is continuous throughout the XY–plane, the corresponding $^{\partial F}/_{\partial y}$,

$$\frac{\partial F}{\partial y} = \frac{1}{2 \cdot 3}(y - 2)^{-2/3} = \frac{1}{6(y - 2)^{2/3}} \quad ,$$

is not continuous at any (x, y) with $y = 2$. Consequently, Theorem 3.1 does not assure us that the initial-value problem given in Example 9.14 has only one solution. And, indeed, we discovered three solutions.

So, what can we say about using slope fields to sketch solutions to

$$\frac{dy}{dx} = F(x, y) \qquad \text{with} \quad y(x_0) = y_0 \quad ?$$

Based on the example we've seen and the discussion above, we can safely make the following three statements:

1. If $F(x, y)$ is reasonably well behaved in some region around the point (x_0, y_0) (i.e., $F(x, y)$ well defined, finite and continuous at each point (x, y) in this region), then we can use slope fields to sketch a curve that will be a reasonable approximation to a solution to the initial-value problem over some interval.

2. If $F(x, y)$ is not reasonably well behaved in some region around the point (x_0, y_0), in particular, if $F(x_0, y_0)$ is not a well-defined finite value, then we may or may not have a solution to the given initial-value problem. The slope field will probably give us an idea of the nature of solution curves passing through points near (x_0, y_0), but more analysis may be needed to determine if the given initial-value problem has a solution, and, if it exists, the nature of that solution.

3. Even if $F(x, y)$ is reasonably well behaved in some region around the point (x_0, y_0), it is worthwhile to see if $^{\partial F}/_{\partial y}$ is also well defined everywhere in that region. If so, then the curve drawn using a decent slope field will be a reasonably good approximation of the graph to the only solution to the initial-value problem. Otherwise, there is a possibility of multiple solutions.

Finally, let us observe that we can have unique, reasonably well-behaved solutions even though both F and $^{\partial F}/_{\partial y}$ have discontinuities. This was evident in Example 9.9 on page 159 (Figure 9.6b), and is evident in the following example.

!► Example 9.15: *The right side of*

$$\frac{dy}{dx} = \begin{cases} 0 & \text{if} \quad x < 3 \\ 1 & \text{if} \quad 3 \leq x \end{cases}$$

is discontinuous at every point (x, y) with $x = 3$. This differential equation yields the simple, yet striking, slope field in Figure 9.8b. And from this slope field, it should be clear that there is exactly one solution to this differential equation satisfying, say, $y(3) = 2$. That is one of the curves sketched, and (as you can verify) that curve is the graph of

$$y(x) = \begin{cases} 2 & \text{if} \quad x < 3 \\ x - 1 & \text{if} \quad 3 \leq x \end{cases} \quad .$$

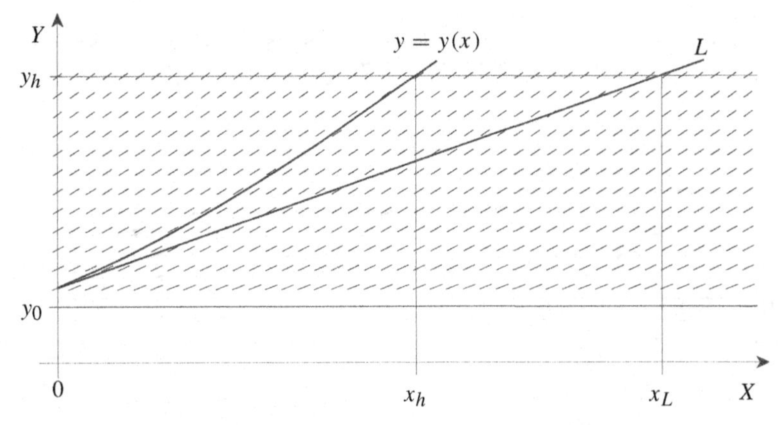

Figure 9.9: A slope field on the strip $y_0 \leq y \leq y_h$ for $y' = g(y)$ when $g(y_0) = 0$ and g is an increasing function on $[y_0, y_h]$.

9.5 Tests for Stability

In Section 9.3, we discussed the stability of constant solutions, using slope fields to visually distinguish between constant solutions that were stable, asymptotically stable or unstable. That was good for developing a basic understanding of stability, but, as we saw in the examples, it is not always possible to determine the stability of a given constant solution from just a slope field. So let us take a closer look at the geometry of the solution curves to a first-order differential equation

$$\frac{dy}{dx} = F(x, y)$$

which start out near the graph of a constant solution $y = y_0$, and see if we can derive some relatively simple "computational" tests for verifying the stability or instability suggested by such slope fields as in Figures 9.9 and 9.10.

Throughout this section, we'll assume we have three finite numbers y_0, y_l and y_h with

$$y_l < y_0 < y_h \quad .$$

The constant solution to our differential equation will be $y = y_0$, and the strips of interest will be those strips bounded by the horizontal lines

$$y = y_0 \quad , \quad y = y_l \quad \text{and} \quad y = y_h \quad .$$

We will also assume $F(x, y)$ is at least a continuous function of both x and y on these strips. This ensures that we need not worry about any truly "bad" problem points in the strips and can safely assume that no solution curve "ends" at a point in one of our strips.

Autonomous Equations

Since the analysis is much easier with autonomous equations, we will start with those. Accordingly, we assume $y = y_0$ is a constant solution to a differential equation of the form

$$\frac{dy}{dx} = g(y)$$

where g is a continuous function on the closed interval $[y_l, y_h]$.

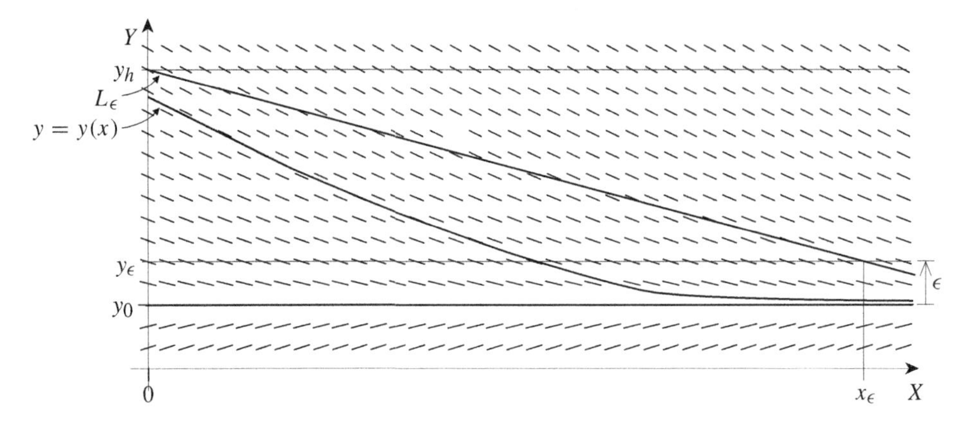

Figure 9.10: Slope field for $y' = g(y)$ when $g(y_0) = 0$ and g is a decreasing function on $[y_l, y_h]$ with $y_l < y_0 < y_h$.

The Single Crossing Point Lemma

We start by observing that no solution curve can cross a horizontal line $y = y_c$ more than once if $g(y_c)$ is a finite, nonzero value. In particular, suppose $g(y_c) > 0$ (as we have for $y_c = y_h$ in Figure 9.9), and suppose $y = y(x)$ is a solution to our autonomous differential equation whose graph crosses the horizontal line $y = y_c$ at the point $(x, y) = (x_c, y_c)$. At this point, the slope of the solution curve is positive, telling us that the solution curve goes from below to above this horizontal line as x goes from the left to the right of x_c. And since $g(y) > 0$ at every point on the horizontal line $y = y_c$, there is no point where the solution curve can come back below this horizontal line as x increases.

Likewise, if $g(y_c) < 0$ (see Figure 9.10), then each solution curve crossing $y = y_c$ goes from above to below $y = y_c$ and can never "come back up" to cross $y = y_c$ a second time.

We'll use this observation several times in what follows, so let us dignify it as a lemma:

Lemma 9.1

Let $y = y(x)$ be a solution to

$$\frac{dy}{dx} = g(y)$$

on some interval $(0, x_{max})$ whose graph crosses a horizontal line $y = y_c$ when $x = x_c$. Suppose, further, that $g(y_c)$ is a finite, nonzero value. Then,

$$g(y_c) > 0 \quad \Longrightarrow \quad y(x) > y_c \quad \text{whenever} \quad x_c < x < x_{max} \quad ,$$

while

$$g(y_c) < 0 \quad \Longrightarrow \quad y(x) < y_c \quad \text{whenever} \quad x_c < x < x_{max} \quad .$$

Instability

Consider the case illustrated in Figure 9.9. Here, $y = y_0$ is a constant solution to

$$\frac{dy}{dx} = g(y) \quad ,$$

and the slope of the slope line at (x, y) (i.e., the value of $g(y)$) increases as y increases from $y = y_0$ to $y = y_h$. So if

$$y_0 < y_1 < y_2 < y_h \quad ,$$

then

$$0 = g(y_0) < g(y_1) < g(y_2) < g(y_h) \quad . \tag{9.4}$$

Now take any solution $y = y(x)$ to

$$\frac{dy}{dx} = g(y) \qquad \text{with} \quad y_0 < y(0) < y_h \quad ,$$

and let L be the straight line tangent to the graph of this solution at the point where $x = 0$ (see Figure 9.9). From inequality set (9.4) (and Figure 9.9), we see that:

1. The slope of tangent line L is positive. Hence, L crosses the horizontal line $y = y_h$ at some point (x_L, y_h) with $0 < x_L < \infty$.

2. At each point in the strip, the slope of the tangent to the graph of $y = y(x)$ is at least as large as the slope of L. So, as x increases, the graph of $y = y(x)$ goes upward faster than L. Consequently, this solution curve crosses the horizontal line $y = y_h$ at a point (x_h, y_h) with $0 < x_h < x_L$.

From this and Lemma 9.1, it follows that, if x is a point in the domain of our solution $y = y(x)$, then

$$y(x) \geq y_h \qquad \text{whenever} \quad x > x_h \quad .$$

That is,

$$y(x) - y_0 > y_h - y_0$$

whenever x is a point in the domain of $y = y(x)$ with $x_h < x$.

This tells us that, no matter how close we pick $y(0)$ to y_0 (at least with $y(0) > y_0$), the graph of our solution will, as x increases, diverge to a distance of at least $y_h - y_0$ from y_0. This means we can*not* choose a distance ϵ with

$$\epsilon < y_h - y_0 \quad ,$$

and find a solution $y = y(x)$ to

$$\frac{dy}{dx} = g(y) \qquad \text{with} \quad y(0) > y_0$$

that remains within ϵ of y_0 for all values of x. In other words, $y = y_0$ is not a stable constant solution.

This, along with analogous arguments when $g(y)$ is an increasing function on $[y_l, y_0]$, gives us:

Theorem 9.2

Let $y = y_0$ be a constant solution to

$$\frac{dy}{dx} = g(y) \quad ,$$

where g is a continuous function on some interval $[y_l, y_h]$ with $y_l < y < y_h$. Then $y = y_0$ is an unstable constant solution if either of the following holds:

1. *$g(y)$ is an increasing function on $[y_l, y_0]$ for some $y_l < y_0$.*

2. *$g(y)$ is an increasing function on $[y_0, y_h]$ for some $y_0 < y_h$.*

Stability

Now consider the case illustrated in Figure 9.10. Here, $y = y_0$ is a constant solution to

$$\frac{dy}{dx} = g(y)$$

when $g(y)$ (the slope of the slope line at point (x, y)) is a decreasing function on an interval $[y_l, y_h]$. So, if

$$y_l < y_{-2} < y_{-1} < y_0 < y_1 < y_2 < y_h \quad ,$$

then

$$g(y_l) > g(y_{-2}) > g(y_{-1}) > 0 > g(y_1) > g(y_2) > g(y_h) \quad .$$

Thus, the slope lines just below the horizontal line $y = y_0$ have positive slope, those just above $y = y_0$ have negative slope, and the slopes become steeper as the distance from the horizontal line $y = y_0$ increases.

The fact that $y = y_0$ is a stable constant solution should be obvious from the figure. After all, the slope lines are all angled toward $y = y_0$ as x increases, "directing" the solutions curves toward $y = y_0$ as x increases.

Figure 9.10 also suggests that, if $y = y(x)$ is any solution to

$$\frac{dy}{dx} = g(y) \qquad \text{with} \quad y_l < y(0) < y_h \quad ,$$

then

$$\lim_{x\to\infty} y(x) = y_0 \quad ,$$

suggesting that $y = y_0$ is also asymptotically stable. To rigorously confirm this, it is convenient to separately consider the three cases

$$y(0) = y_0 \quad , \quad y_0 < y(0) < y_h \quad \text{and} \quad y_l < y(0) < y_0 \quad .$$

The first case is easily taken care of. If $y(0) = y_0$, then our solution $y = y(x)$ must be the constant solution $y = y_0$ (the already noted stability of this constant solution prevents any other possible solutions). Hence,

$$\lim_{x\to\infty} y(x) = \lim_{x\to\infty} y_0 = y_0 \quad .$$

Next, assume $y = y(x)$ is a solution to

$$\frac{dy}{dx} = g(y) \qquad \text{with} \quad y_0 < y(0) < y_h \quad .$$

To show

$$\lim_{x\to\infty} y(x) = y_0 \quad ,$$

it helps to remember that the above limit is equivalent to saying that we can make $y(x)$ as close to y_0 as desired (say, within some small, positive distance ϵ) by simply picking x large enough.

So let ϵ be any small, positive value, and let us show that there is a corresponding "large enough value" x_ϵ so that $y(x)$ is within a distance ϵ of y_0 whenever x is bigger than x_ϵ. And since we are only concerned with ϵ being "small", let's go ahead and assume

$$\epsilon < y_h - y_0 \quad .$$

Now, for notational convenience, let $y_\epsilon = y_0 + \epsilon$, and let L_ϵ be the straight line through the point $(x, y) = (0, y_h)$ with the same slope as the slope lines along the horizontal line $y = y_\epsilon$ (see Figure 9.10). Because $y_h > y_\epsilon > y_0$, the slope lines along the line $y = y_\epsilon$ have negative slope.

Hence, so does L_ϵ. Consequently, the line L_ϵ goes downward from point $(0, y_h)$, intersecting the horizontal line $y = y_\epsilon$ at some point to the right of the Y–axis. Let x_ϵ be the X–coordinate of that point.

Next, consider the graph of our solution $y = y(x)$ when $0 \leq x \leq x_\epsilon$. Observe that:

1. This part of this solution curve starts at the point $(0, y(0))$, which is between the lines L_ϵ and $y = y_0$.

2. The slope at each point of this solution curve above $y = y_\epsilon$ is less than the slope of the line L_ϵ. Hence, this part of the solution curve must go downward faster than L_ϵ as x increases.

3. If $y(x) < y_\epsilon$ for some value of x, then $y(x) < y_\epsilon$ for all larger values of x. (This is from Lemma 9.1.)

4. The graph of $y = y(x)$ cannot go below the horizontal line $y = y_0$ because the slope lines at points just below $y = y_0$ all have positive slope.

These observations tell us that, at least when $0 \leq x \leq x_\epsilon$, our solution curve must remain between the lines L_ϵ and $y = y_0$. In particular, since L_ϵ crosses the horizontal line $y = y_\epsilon$ at $x = x_\epsilon$, we must have

$$y_0 \leq y(x_\epsilon) \leq y_\epsilon = y_0 + \epsilon \quad .$$

From this along with Lemma 9.1, it follows that

$$y_0 \leq y(x) \leq y_0 + \epsilon \qquad \text{for all} \quad x > x_\epsilon \quad .$$

Equivalently,

$$0 \leq y(x) - y_0 \leq \epsilon \qquad \text{for all} \quad x > x_\epsilon \quad ,$$

which tells us that $y(x)$ is within ϵ of y_0 whenever $x > x_\epsilon$. Hence, we can make $y(x)$ as close as desired to y_0 by choosing x large enough. That is,

$$\lim_{x \to \infty} y(x) = y_0 \quad .$$

That leaves the verification of

$$\lim_{x \to \infty} y(x) = y_0$$

when $y = y(x)$ satisfies

$$\frac{dy}{dx} = g(y) \qquad \text{with} \quad y_l < y(0) < y_0 \quad .$$

This will be left to the interested reader (just use straightforward modifications of the arguments in the last few paragraphs — start by vertically flipping Figure 9.10).

To summarize our results:

Theorem 9.3
Let $y = y_0$ be a constant solution to an autonomous differential equation

$$\frac{dy}{dx} = g(y) \quad .$$

This constant solution is both stable and asymptotically stable if there is an interval $[y_l, y_h]$, with $y_l < y_0 < y_h$, on which $g(y)$ is a decreasing continuous function.

Differential Tests for Stability

Recall from elementary calculus that you can determine whether a function g is an increasing or decreasing function by just checking to see if its derivative is positive or negative. To be precise,

$$g'(y) > 0 \quad \text{for} \quad a \leq y \leq b \quad \Longrightarrow \quad g \text{ is an increasing function on } [a, b]$$

and

$$g'(y) < 0 \quad \text{for} \quad a \leq y \leq b \quad \Longrightarrow \quad g \text{ is a decreasing function on } [a, b] \quad .$$

Consequently, we can replace the lines in Theorems 9.3 and 9.2 about g being increasing or decreasing with corresponding conditions on g', obtaining the following:

Theorem 9.4
Let $y = y_0$ be a constant solution to an autonomous differential equation

$$\frac{dy}{dx} = g(y)$$

in which g is a differentiable function on some interval $[y_l, y_h]$ with $y_l < y_0 < y_h$. Then $y = y_0$ is both a stable and asymptotically stable constant solution if

$$g'(y) < 0 \qquad \text{for} \quad y_l \leq y \leq y_h \quad .$$

Theorem 9.5
Let $y = y_0$ be a constant solution to an autonomous differential equation

$$\frac{dy}{dx} = g(y)$$

in which g is a differentiable function on some interval $[y_l, y_h]$ with $y_l < y_0 < y_h$. Then $y = y_0$ is an unstable constant solution if either

$$g'(y) > 0 \qquad \text{for} \quad y_l < y < y_0$$

or

$$g'(y) > 0 \qquad \text{for} \quad y_0 < y < y_h \quad .$$

But now recall that, if a function is sufficiently continuous and is positive (or negative) at some point, then that function remains positive (or negative) over some interval surrounding that point. With this we can reduce the above theorems to the following single theorem:

Theorem 9.6
Let $y = y_0$ be a constant solution to an autonomous differential equation

$$\frac{dy}{dx} = g(y)$$

in which g is differentiable and g' is continuous on some interval $[y_l, y_h]$ with $y_l < y_0 < y_h$. Then:

1. *$y = y_0$ is a stable and asymptotically stable constant solution if $g'(y_0) < 0$.*
2. *$y = y_0$ is an unstable constant solution if $g'(y_0) > 0$.*

!►Example 9.16: *Let us again consider the autonomous differential equation considered earlier in Example 9.7 on page 157,*

$$\frac{dy}{dx} = \frac{1}{2}(4-y)(y-2)^{4/3} \quad ,$$

and whose slope field was sketched in Figure 9.5b on page 156.

Because the right side,

$$g(y) = \frac{1}{2}(4-y)(y-2)^{4/3}$$

is zero when y is either 2 or 4, this differential equation has constant solutions

$$y = 2 \qquad \text{and} \qquad y = 4 \quad .$$

So as to apply any of the above theorems, we compute $g'(y)$:

$$g'(y) = \frac{d}{dy}\left[\frac{1}{2}(4-y)(y-2)^{4/3}\right] = -\frac{1}{2}(y-2)^{4/3} + \frac{2}{3}(4-y)(y-2)^{1/3} \quad .$$

After a bit of algebra, this simplifies to

$$g'(y) = \frac{7}{6}\left(\frac{22}{7} - y\right)(y-2)^{1/3} \quad .$$

Plugging in $y = 4$, we get

$$g'(4) = \frac{7}{6}\left(\frac{22}{7} - 4\right)(4-2)^{1/3} = \frac{7}{6}\left(\frac{22}{7} - \frac{28}{7}\right)\sqrt[3]{2} = -\sqrt[3]{2} < 0 \quad .$$

Theorem 9.6 then tells us that the constant solution $y = 4$ is stable and asymptotically stable, just as we suspected from looking at the slope field in Figure 9.5b.

Unfortunately, we cannot apply Theorem 9.6 to determine the stability of the other constant solution, $y = 2$, since

$$g'(2) = \frac{7}{6}\left(\frac{22}{7} - 2\right)(2-2)^{1/3} = 0 \quad .$$

Instead, we must look a little more closely at the formula for $g'(y)$, and observe that, if

$$2 < y < \frac{22}{7} \quad ,$$

then

$$g'(y) = \frac{7}{6}\underbrace{\left(\frac{22}{7} - y\right)}_{>0}\underbrace{(y-2)^{1/3}}_{>0} > 0 \quad .$$

The test given in Theorem 9.5 (with $[y_0, y_h] = [2, {}^{22}\!/_7]$) applies and assures us that $y = 2$ is an unstable constant solution, just as we suspected from looking at Figure 9.5b.

Nonautonomous Equations

Look again at Figure 9.10, but now imagine that the slope lines are also becoming steeper as x increases. With that picture in your mind, you will realize that the arguments leading to stability Theorems 9.3 and 9.4 remain valid if these slope lines also grow steeper as x increases. In particular, we have the following analog of Theorem 9.4:

Theorem 9.7
Let $y = y_0$ be a constant solution to a differential equation

$$\frac{dy}{dx} = F(x, y)$$

in which $F(x, y)$ is differentiable with respect to both x and y at every point in some strip

$$\{(x, y) : 0 \leq x \text{ and } y_l \leq y \leq y_h\}$$

with $y_l < y_0 < y_h$. Further suppose that, at each point in this strip above the line $y = y_0$,

$$\frac{\partial F}{\partial y} < 0 \quad \text{and} \quad \frac{\partial F}{\partial x} \leq 0 \quad ,$$

and that, at each point in this strip below the line $y = y_0$,

$$\frac{\partial F}{\partial y} < 0 \quad \text{and} \quad \frac{\partial F}{\partial x} \geq 0 \quad .$$

Then $y = y_0$ is both a stable and asymptotically stable constant solution.

We'll leave it to the interested reader to come up with corresponding analogs of the other theorems on stability and instability.

Additional Exercises

9.2. *For each of the following, construct the slope field for the given differential equation on the indicated 2×2 or 3×3 grid of listed points:*

a. $\frac{dy}{dx} = \frac{1}{2}\left(x^2 + y^2\right)$ at $(x, y) = (0, 0), (1, 0), (0, 1)$ and $(1, 1)$

b. $2\frac{dy}{dx} = x^2 - y^2$ at $(x, y) = (0, 0), (1, 0), (0, 1)$ and $(1, 1)$

c. $\frac{dy}{dx} = \frac{y}{x}$ at $(x, y) = (1, 1), \left(3/2, 1\right), \left(1, 3/2\right)$ and $\left(3/2, 3/2\right)$

d. $(2x + 1)\frac{dy}{dx} = x^2 - 2y^2$ at $(x, y) = (0, 1), (1, 1), (0, 2)$ and $(1, 2)$

e. $2\frac{dy}{dx} = (x - y)^2$
at $(x, y) = (0, 0), (0, 1), (0, 2), (1, 0), (0, 1), (1, 2), (2, 0), (2, 1)$ and $(2, 2)$

f. $\frac{dy}{dx} = (1 - y)^3$
at $(x, y) = (0, 0), (0, 1), (0, 2), (1, 0), (0, 1), (1, 2), (2, 0), (2, 1)$ and $(2, 2)$

Several slope fields for unspecified first-order differential equations are given below. For sketching purposes, you may want to use an enlarged photocopy of each given slope field.

9.3. *On the right is a slope field for some first-order differential equation.*

a *Letting $y = y(x)$ be the solution to this differential equation that satisfies $y(0) = 3$:*

i. *Sketch the graph of this solution.*

ii. *Using your sketch, find (approximately) the value of $y(8)$.*

b. *Sketch the graphs of two other solutions to this unspecified differential equation.*

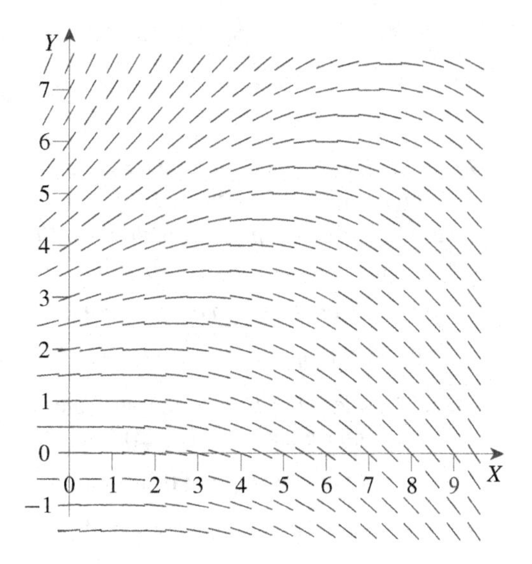

9.4. *On the right is a slope field for some first-order differential equation.*

a. *Sketch the graphs of the solutions to this differential equation that satisfy*

i. $y(0) = 2$

ii. $y(0) = 4$

iii. $y(0) = 4.5$

b. *What, approximately, is $y(4)$ if y is the solution to this unspecified differential equation satisfying*

i. $y(0) = 2$?

ii. $y(0) = 4$?

iii. $y(0) = 4.5$?

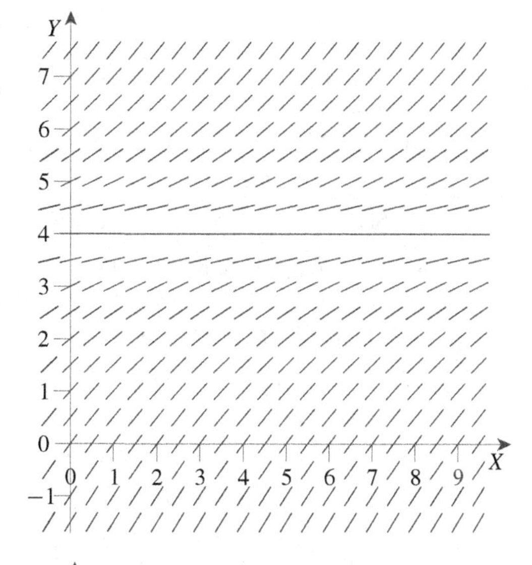

9.5. *On the right is a slope field for some first-order differential equation.*

a. *Let $y(x)$ be the solution to the differential equation with $y(0) = 5$.*

i. *Sketch the graph of this solution.*

ii. *What (approximately) is the maximum value of $y(x)$ on the interval $(0, 9)$, and where does it occur?*

iii. *What (approximately) is $y(8)$?*

b. *Now let $y(x)$ be the solution to the differential equation with $y(0) = 0$.*

i. *Sketch the graph of this solution.*

ii. *What (approximately) is $y(8)$?*

9.6. *On the right is a slope field for some first-order differential equation.*

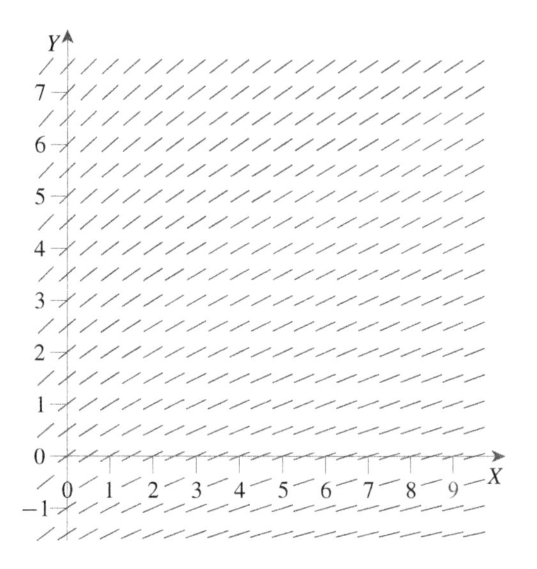

a. *Let* $y(x)$ *be the solution to the differential equation with* $y(0) = 2$.

i. *Sketch the graph of this solution.*

ii. *What (approximately) is* $y(3)$ *?*

b. *Now let* $y(x)$ *be the solution to the differential equation with* $y(3) = 1$.

i. *Sketch the graph of this solution.*

ii. *What (approximately) is* $y(0)$ *?*

9.7. *On the right is a slope field for some first-order differential equation.*

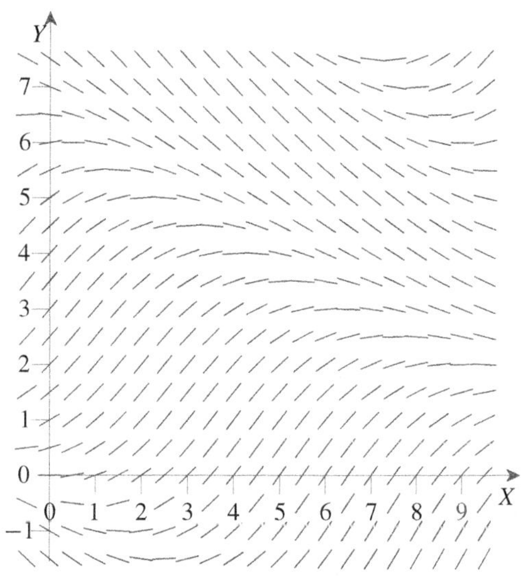

a. *Let* $y(x)$ *be the solution to the differential equation with* $y(0) = 4$.

i. *Sketch the graph of this solution.*

ii. *What (approximately) is the maximum value of* $y(x)$ *on the interval* $(0, 9)$*, and where does it occur?*

b. *Now let* $y(x)$ *be the solution to the differential equation with* $y(2) = 0$.

i. *Sketch the graph of this solution.*

ii. *What (approximately) is the maximum value of* $y(x)$ *on the interval* $(0, 9)$*, and where does it occur?*

9.8. *Look up the commands for generating slope fields for first-order differential equations in your favorite computer math package (they may be the same commands for generating "direction fields"). Then:*

i. *Use the computer math package to sketch the indicated slope field for each differential equation given below,*

ii. *and use the resulting slope field to sketch (by hand) some of the solution curves for the given differential equation.*

a. $\frac{dy}{dx} = \sin(x + y)$ *using a* 25×25 *grid on the region* $-2 \leq x \leq 10$ *and* $-2 \leq y \leq 10$

b. $10\frac{dy}{dx} = y(5 - y)$ *using a* 25×25 *grid on the region* $-2 \leq x \leq 10$ *and* $-2 \leq y \leq 10$

c. $10\dfrac{dy}{dx} = y(y-5)$ *using a* 25×25 *grid on the region* $-2 \leq x \leq 10$ *and* $-2 \leq y \leq 10$

d. $2\dfrac{dy}{dx} = y(y-2)^2$ *using a* 25×17 *grid on the region* $-2 \leq x \leq 10$ *and* $-1 \leq y \leq 3$

e. $3\dfrac{dy}{dx} = (4-y)(y-1)^{4/3}$ *using a* 19×16 *grid on the region* $0 \leq x \leq 6$ *and* $0 \leq y \leq 5$

f. $3\dfrac{dy}{dx} = \sqrt[3]{x-y}$ *using a* 25×21 *grid on the region* $-2 \leq x \leq 10$ *and* $-2 \leq y \leq 8$

9.9. *Slope fields for several (unspecified) first-order differential equations have be sketched below. Assume that each horizontal line is the graph of a constant solution to the corresponding differential equation. Identify each of these constant solutions, and, for each constant solution, decide whether the slope field is indicating that it is a stable, asymptotically stable, or unstable constant solution.*

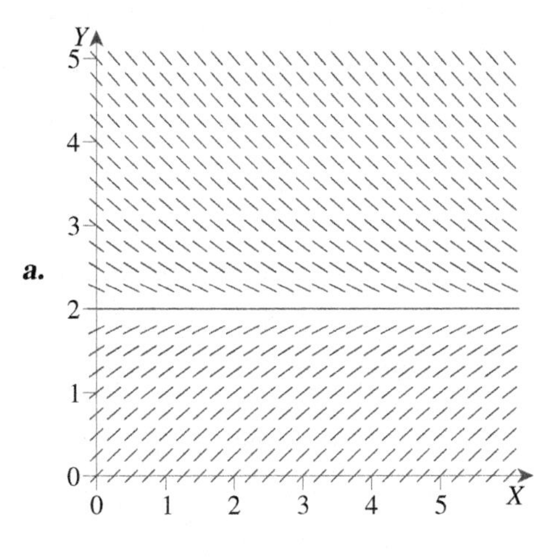

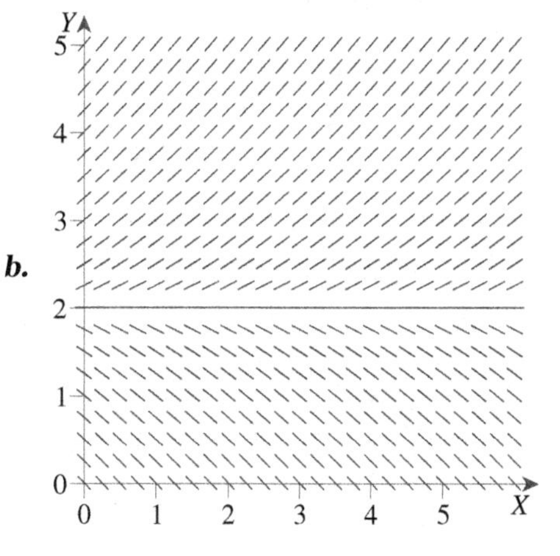

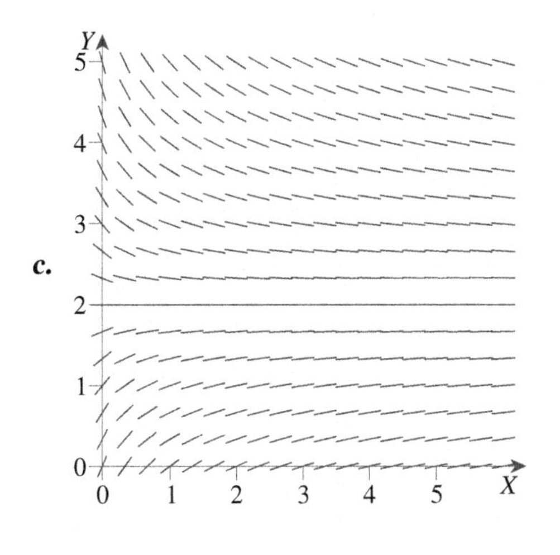

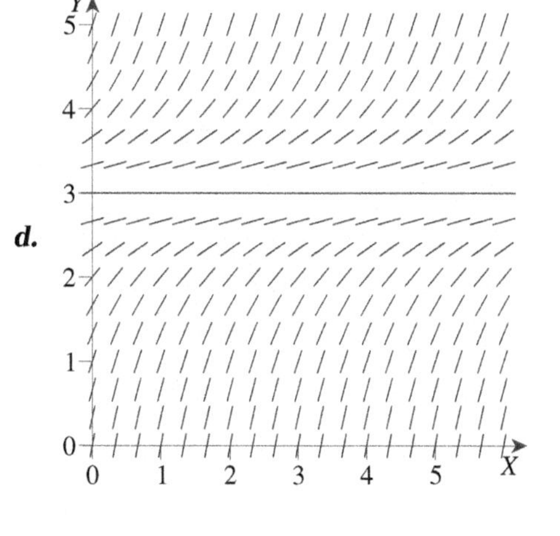

e.

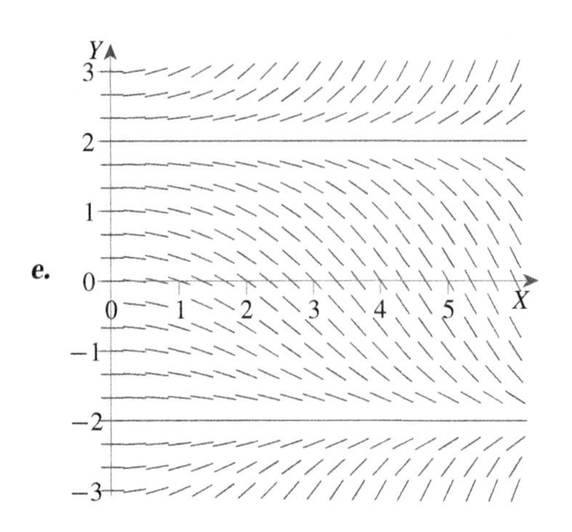

f.

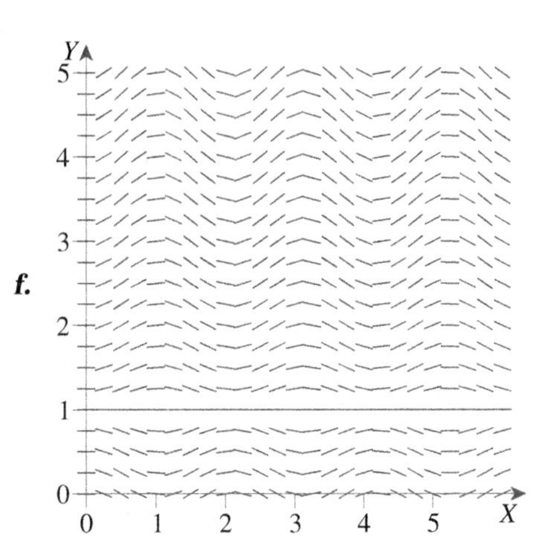

10

Numerical Methods I: The Euler Method

In the last chapter, we saw that a computer can easily generate a slope field for a given first-order differential equation. Using that slope field we can then sketch a fair approximation to the graph of the solution y to a given initial-value problem, and, from that graph, find an approximation to $y(x)$ for any desired x in the region of the sketched slope field. The obvious question now arises: Why not let the computer do all the work and just tell us the approximate value of $y(x)$ for the desired x ?

Well, why not?

In this chapter, we will develop, use and analyze one method for generating a "numerical solution" to a first-order initial-value problem. This type of "solution" is not a formula or equation for the actual solution $y(x)$, but two lists of numbers,

$$\{x_0, x_1, x_2, x_3, \ldots, x_N\} \quad \text{and} \quad \{y_0, y_1, y_2, y_3, \ldots, y_N\}$$

with each y_k approximating the value of $y(x_k)$. Obviously, a nice formula or equation for $y(x)$ would usually be preferred over a list of approximate values, but, when obtaining that nice formula or equation is not practical, a numerical solution is better than nothing.

The method we will study in this chapter is "the Euler method". It is but one of many methods for generating numerical solutions to differential equations. We choose it as the first numerical method to study because it is relatively simple, and, using it, you will be able to see many of the advantages and the disadvantages of numerical solutions. Besides, most of the other methods that might be discussed are refinements of Euler's method, so we might as well learn this method first.

10.1 Deriving the Steps of the Method

Euler's method is based on approximating the graph of a solution $y(x)$ with a sequence of tangent line approximations computed sequentially, in "steps". So, our first task is to derive a useful formula for the tangent line approximation in each step.

The Basic Step Approximation

Let $y = y(x)$ be a solution to some first-order differential equation

$$\frac{dy}{dx} = f(x, y) \quad ,$$

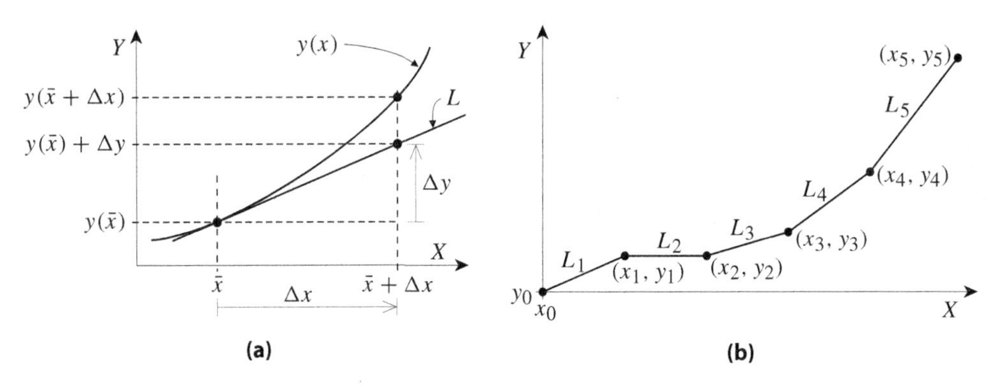

Figure 10.1: **(a)** A single tangent line approximation for the Euler method, and **(b)** the approximation of the solution curve generated by five steps of Euler's method.

where f is a "reasonably smooth" (at least continuous) function of x and y (the precise meaning of "reasonably smooth" will be discussed later). As illustrated in Figure 10.1a, if $\bar{x}$ is a point in the interval of interest, then $(\bar{x}, y(\bar{x}))$ is a point on the graph of $y = y(x)$, and the nearby points on this graph can be approximated by corresponding points on the straight line tangent to the graph at point $(\bar{x}, y(\bar{x}))$ (line L in Figure 10.1a). As with the slope lines in the last chapter, the differential equation can give us the slope of this line:

$$\begin{aligned} \text{the slope of the approximating line} &= \frac{dy}{dx} \quad \text{at} \quad (\bar{x}, y(\bar{x})) \\ &= f(\bar{x}, y(\bar{x})) \quad . \end{aligned}$$

Now let Δx be any positive distance in the X direction. Using our tangent line approximation (again, see Figure 10.1a), we have that

$$y(\bar{x} + \Delta x) \approx y(\bar{x}) + \Delta y$$

where

$$\frac{\Delta y}{\Delta x} = \text{slope of the approximating line} = f(\bar{x}, y(\bar{x})) \quad .$$

So,

$$\Delta y = \Delta x \cdot f(\bar{x}, y(\bar{x}))$$

and

$$y(\bar{x} + \Delta x) \approx y(\bar{x}) + \Delta x \cdot f(\bar{x}, y(\bar{x})) \quad . \tag{10.1}$$

Approximation (10.1) is the fundamental approximation underlying each basic step of Euler's method. However, in what follows, the value of $y(\bar{x})$ will usually only be known by some approximation $\bar{y}$. With this approximation (and the continuity of f), we have

$$y(\bar{x}) + \Delta x \cdot f(\bar{x}, y(\bar{x})) \approx \bar{y} + \Delta x \cdot f(\bar{x}, \bar{y}) \quad ,$$

which, combined with approximation (10.1), yields the actual approximation used in the Euler method,

$$y(\bar{x} + \Delta x) \approx \bar{y} + \Delta x \cdot f(\bar{x}, \bar{y}) \quad . \tag{10.2}$$

The distance Δx in the above approximations is called the *step size*. We will see that choosing a good value for the step size is important.

Generating the Numerical Solution

Now suppose we have a first-order initial-value problem

$$\frac{dy}{dx} = f(x, y) \qquad \text{with} \quad y_0 = y(x_0) \quad ,$$

with $y = y(x)$ as its (possibly unknown) solution. A *numerical solution* for this initial-value problem consists of a pair of sequences

$$\{x_0, x_1, x_2, x_3, \ldots, x_N\} \qquad \text{and} \qquad \{y_0, y_1, y_2, y_3, \ldots, y_N\}$$

where x_0 and y_0 are given by the initial data, and with the other x_k's and y_k's computed so that $y_k \approx y(x_k)$ for each k. It should be emphasized that a numerical solution is not the true solution to the initial-value problem. It's a set of approximations to the true solution on a particular set of points. Clearly, if we can actually find a usable formula for the solution $y = y(x)$, then there is no need to seek a numerical solution. Thus, normally, a numerical solution should be sought only if other methods to solve the problem cannot be applied.

In the Euler method, the two sequences are generated by first choosing a step size Δx, and then successively computing the x_k's and y_k's using

$$x_{k+1} = x_k + \Delta x \tag{10.3a}$$

and

$$y_{k+1} = y_k + \Delta x \cdot f(x_k, y_k) \tag{10.3b}$$

for $k = 0, 1, 2, \ldots, N-1$.

Note the similarity between formula (10.3b) and approximation (10.2). Because of this, it is easily verified (assuming $f(x, y)$ is continuous) that the above generates x_k's and y_k's satisfying $y_k \approx y(x_k)$:

$$\begin{aligned} y_1 = y_{0+1} &= y_0 + \Delta x \cdot f(x_0, y_0) \\ &= y(x_0) + \Delta x \cdot f(x_0, y(x_0)) \approx y(x_0 + \Delta x) = y(x_1) \quad , \end{aligned}$$

$$\begin{aligned} y_2 = y_{1+1} &= y_1 + \Delta x \cdot f(x_1, y_1) \\ &\approx y(x_1) + \Delta x \cdot f(x_1, y(x_1)) \approx y(x_1 + \Delta x) = y(x_2) \quad , \end{aligned}$$

$$\begin{aligned} y_3 = y_{2+1} &= y_2 + \Delta x \cdot f(x_2, y_2) \\ &\approx y(x_2) + \Delta x \cdot f(x_2, y(x_2)) \approx y(x_2 + \Delta x) = y(x_3) \quad , \end{aligned}$$

$$\begin{aligned} y_4 = y_{3+1} &= y_3 + \Delta x \cdot f(x_3, y_3) \\ &\approx y(x_3) + \Delta x \cdot f(x_3, y(x_3)) \approx y(x_3 + \Delta x) = y(x_4) \end{aligned}$$

... *and so on* ...

10.2 Computing via the Euler Method (Illustrated)

Let's now use the Euler method to compute a numerical solution to some first-order differential equation with initial data $y(x_0) = y_0$, say,

$$5\frac{dy}{dx} - y^2 = -x^2 \quad \text{with} \quad y(0) = 1 \quad . \tag{10.4}$$

(As it turns out, this differential equation is not easily solved by any of the methods already discussed. So if we want to find the value of, say, $y(3)$, then a numerical method may be our only choice.)

To use Euler's method, we follow the steps given below. These steps are grouped into two parts: the main part in which the values of the x_k's and y_k's are iteratively computed, and the preliminary part in which the constants and formulas for those iterative computations are determined.

The Steps in the Euler Method
Preliminary Steps

1. Rewrite the initial-value problem with the differential equation in derivative formula form,

$$\frac{dy}{dx} = f(x, y) \quad \text{with} \quad y(x_0) = y_0 \quad ,$$

and, from this, determine the formula for $f(x, y)$ and the values of x_0 and y_0.

For our example, solving for the derivative formula form yields

$$\frac{dy}{dx} = \frac{1}{5}\left[y^2 - x^2\right] \quad \textit{with} \quad y(0) = 1 \quad .$$

So,

$$f(x, y) = \frac{1}{5}\left[y^2 - x^2\right] \quad , \quad x_0 = 0 \quad \textit{and} \quad y_0 = 1 \quad .$$

2. Decide on a distance Δx for the step size, a positive integer N for the maximum number of steps, and a maximum value desired for x, x_{max}. Choose these quantities so that

$$x_{max} = x_N = x_0 + N\Delta x \quad .$$

For the sake of this illustration, let us pick

$$\Delta x = \frac{1}{2} \quad \textit{and} \quad N = 6 \quad .$$

Then

$$x_{max} = x_0 + N\Delta x = 0 + 6 \cdot \frac{1}{2} = 3 \quad .$$

Iterative Steps

In these steps, we iteratively compute x_{k+1} and y_{k+1} from x_k and y_k using formula set (10.3),

$$x_{k+1} = x_k + \Delta x \quad \text{and} \quad y_{k+1} = y_k + \Delta x \cdot f(x_k, y_k) \quad ,$$

using the information from the two preliminary steps.

1. Compute x_1 and y_1 using equation set (10.3) with $k = 0$. (Be sure to save the resulting values of x_1 and y_1!)

For our example, using equation set (10.3) with $k = 0$ gives us

$$x_1 = x_{0+1} = x_0 + \Delta x = 0 + \tfrac{1}{2} = \tfrac{1}{2} \quad ,$$

and

$$\begin{aligned} y_1 = y_{0+1} &= y_0 + \Delta x \cdot f(x_0, y_0) \\ &= 1 + \tfrac{1}{2} \cdot f(0, 1) \\ &= 1 + \tfrac{1}{2} \cdot \tfrac{1}{5}\left[1^2 - 0^2\right] = \tfrac{11}{10} = 1.1 \quad . \end{aligned}$$

2. Compute x_2 and y_2 using equation set (10.3) with $k = 1$ and the values of x_1 and y_1 from the previous step. (Be sure to save the resulting values of x_2 and y_2 !)

   For our example, using equation set (10.3) with $k = 1$ gives us

$$x_2 = x_{1+1} = x_1 + \Delta x = \tfrac{1}{2} + \tfrac{1}{2} = 1 \quad ,$$

and

$$\begin{aligned} y_2 = y_{1+1} &= y_1 + \Delta x \cdot f(x_1, y_1) \\ &= \tfrac{11}{10} + \tfrac{1}{2} \cdot f\left(\tfrac{1}{2}, \tfrac{11}{10}\right) \\ &= \tfrac{11}{10} + \tfrac{1}{2} \cdot \tfrac{1}{5}\left[\left(\tfrac{11}{10}\right)^2 - \left(\tfrac{1}{2}\right)^2\right] = \tfrac{299}{250} = 1.196 \quad . \end{aligned}$$

3. Compute x_3 and y_3 using equation set (10.3) with $k = 2$ and the values of x_2 and y_2 from the previous step. (Be sure to save the resulting values of x_3 and y_3 !)

   For our example, equation set (10.3) with $k = 2$ and the above values for x_2 and y_2 yields

$$x_3 = x_{2+1} = x_2 + \Delta x = 1 + \tfrac{1}{2} = \tfrac{3}{2} \quad ,$$

$$\begin{aligned} y_3 = y_{2+1} &= y_2 + \Delta x \cdot f(x_2, y_2) \\ &= \tfrac{299}{250} + \tfrac{1}{2} \cdot f\left(1, \tfrac{299}{250}\right) \end{aligned}$$

and

$$= \tfrac{299}{250} + \tfrac{1}{2} \cdot \tfrac{1}{5}\left[\left(\tfrac{299}{250}\right)^2 - 1^2\right] = \tfrac{774{,}401}{625{,}000} = 1.239041\ldots \quad .$$

4, 5, ... In each subsequent step, increase k by 1, and compute x_{k+1} and y_{k+1} using equation set (10.3) with the values of x_k and y_k from the previous step. Continue until x_N and y_N are computed. (Be sure to save the resulting values of each x_{k+1} and y_{k+1} !)

For our example (omitting many computational details):

Step 4. With $k = 3$,

$$x_4 = x_{3+1} = x_3 + \Delta x = \tfrac{3}{2} + \tfrac{1}{2} = 2$$

and

$$\begin{aligned} y_4 = y_{3+1} &= y_3 + \Delta x \cdot f(x_3, y_3) \\ &= \tfrac{774{,}401}{625{,}000} + \tfrac{1}{2} \cdot f\left(\tfrac{3}{2}, \tfrac{774{,}401}{625{,}000}\right) \\ &= \tfrac{774{,}401}{625{,}000} + \tfrac{1}{2} \cdot \tfrac{1}{5}\left[\left(\tfrac{774{,}401}{625{,}000}\right)^2 - \left(\tfrac{3}{2}\right)^2\right] = \cdots = 1.167564\ldots \quad . \end{aligned}$$

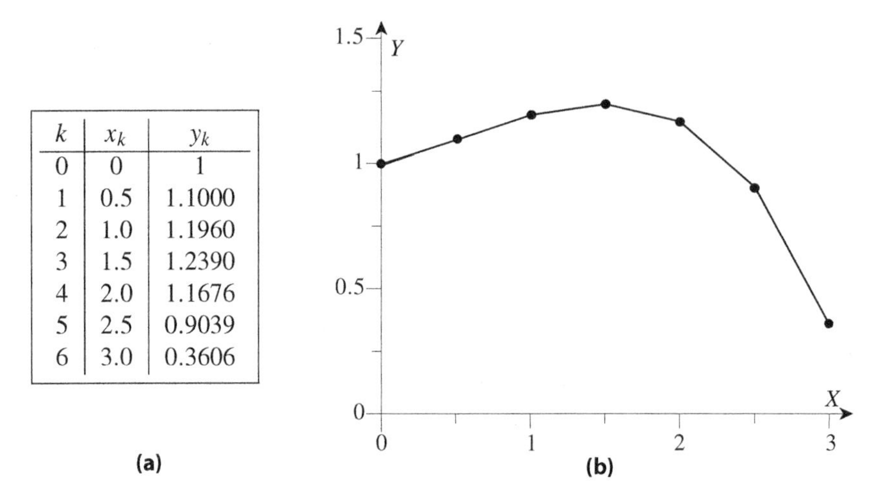

k	x_k	y_k
0	0	1
1	0.5	1.1000
2	1.0	1.1960
3	1.5	1.2390
4	2.0	1.1676
5	2.5	0.9039
6	3.0	0.3606

Figure 10.2: Results of Euler's method to solve $5y' - y^2 = -x^2$ with $y(0) = 1$ using $\Delta x = 1/2$ and $N = 6$: **(a)** The numerical solution in which $y_k \approx y(x_k)$ (for $k \geq 3$, the values of y_k are to the nearest 0.0001). **(b)** The graph of the corresponding piecewise linear approximate to $y = y(x)$.

Step 5. With $k = 4$,

$$x_5 = x_{4+1} = x_4 + \Delta x = 2 + \frac{1}{2} = \frac{5}{2} \quad ,$$

and

$$y_5 = y_{4+1} = y_4 + \Delta x \cdot f(x_4, y_4) = \cdots = 0.903884\ldots \quad .$$

Step 6. With $k = 5$,

$$x_6 = x_{5+1} = x_5 + \Delta x = \frac{5}{2} + \frac{1}{2} = 6 \quad ,$$

and

$$y_6 = y_{5+1} = y_5 + \Delta x \cdot f(x_5, y_5) = \cdots = .360585\ldots \quad .$$

Since we had earlier chosen N*, the maximum number of steps, to be* 6*, we can stop computing.*

The results of these computations are summarized in the table in Figure 10.2a.

On Doing the Computations

The first few times you use Euler's method, attempt to do all the computations by hand. If the numbers become too awkward to handle, use a simple calculator and decimal approximations. This will help you understand and appreciate the method. It will also help you appreciate the tremendous value of programming a computer to do the calculations in the second part of the method. That, of course, is how one should really carry out the computations in the second part of Euler's method.

In fact, the Euler method may already be one of the standard procedures in your favorite computer math package. Still, writing your own version is enlightening and is highly recommended for the good of your soul.

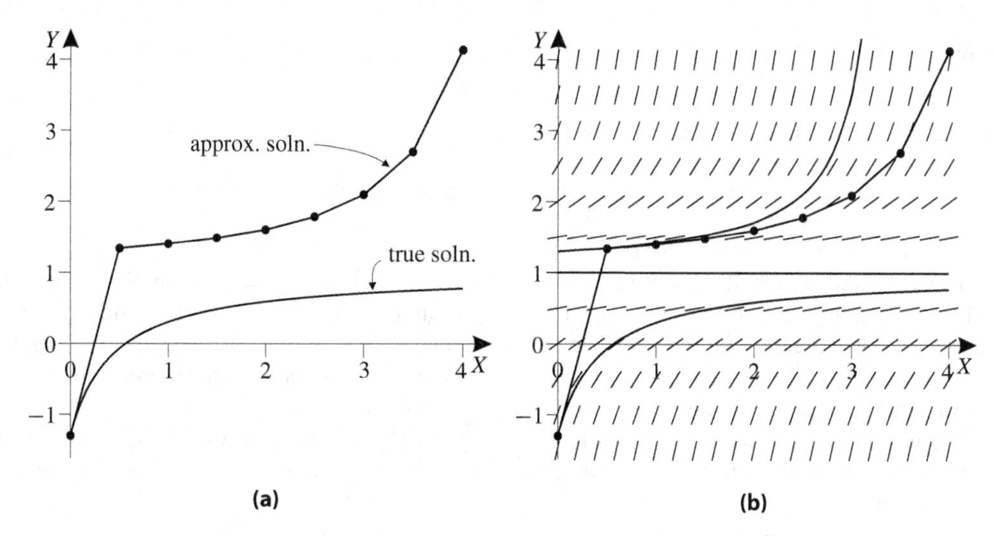

Figure 10.3: Catastrophic failure of Euler's method in solving $y' = (y-1)^2$ with $y(0) = -1.3$: **(a)** Graphs of the true solution and the approximate solution. **(b)** The same graphs with a slope field for the differential equation.

10.3 Using the Results of the Method

Tables and Graphs

What you do with the results of your computations depends on why you are doing these computations. If N is not too large, it is usually a good idea to write the obtained values of

$$\{x_0, x_1, x_2, x_3, \ldots, x_N\} \qquad \text{and} \qquad \{y_0, y_1, y_2, y_3, \ldots, y_N\}$$

in a table for convenient reference (with a note that $y_k \approx y(x_k)$ for each k) as done in Figure 10.2a for our example. And, whatever the size of N, it is always enlightening to graph — as done in Figure 10.2b for our example — the corresponding piecewise linear approximation to the graph of $y = y(x)$ by drawing straight lines between each (x_k, y_k) and (x_{k+1}, y_{k+1}).

What Can Go Wrong

Do not forget that Euler's method does not yield exact answers. Instead, it yields values

$$\{x_0, x_1, x_2, x_3, \ldots, x_N\} \qquad \text{and} \qquad \{y_0, y_1, y_2, y_3, \ldots, y_N\}$$

with

$$y_k \approx y(x_k) \qquad \text{for} \quad k > 0 \quad .$$

What's more, each y_{k+1} is based on the approximation

$$y(x_k + \Delta x) \approx y(x_k) + \Delta x \cdot f(x_k, y(x_k))$$

with $y(x_k)$ being replaced with approximation y_k when $k > 0$. So we are computing approximations based on previous approximations.

Because of this, the accuracy of the approximation $y_k \approx y(x_k)$, especially for larger values of k, is a serious issue. Consider the work done in the previous section: just how well can we trust the approximation

$$y(3) \approx 0.3606$$

obtained for the solution to initial-value problem (10.4)? In fact, it can be shown that

$$y(3) = -.23699 \quad \text{to the nearest } 0.00001 \quad .$$

So our approximation was not very good!

To get an idea of how the errors can build up, look back at Figure 10.1a on page 177. You can see that, if the graphs of the true solutions to the differential equation are generally concave up (as in the figure), then the tangent line approximations used in Euler's method lie below the true graphs and yield underestimates for the approximations. Over several steps, these underestimates can build up so that the y_k's are significantly below the actual values of the $y(x_k)$'s.

Likewise, if the graphs of the true solutions are generally concave down, then the tangent line approximations used in Euler's method lie above the true graphs and yield overestimates for the approximations.

Also keep in mind that most of the tangent line approximations used in Euler's method are not based on lines tangent to the true solution but on lines tangent to solution curves passing through the (x_k, y_k)'s. This can lead to the "catastrophic failure" illustrated in Figure 10.3a. In this figure, the true solution to

$$\frac{dy}{dx} = (y-1)^2 \qquad \text{with} \quad y(0) = -\frac{13}{10} \quad ,$$

is graphed along with the graph of the approximate solution generated from Euler's method with $\Delta x = 1/2$. Exactly why the graphs appear so different becomes apparent when we superimpose the slope field in Figure 10.3b. The differential equation has an unstable equilibrium solution $y = 1$. If $y(0) < 1$, as in the above initial-value problem, then the true solution $y(x)$ should converge to 1 as $x \to \infty$. Here, however, one step of Euler's method overestimated the value of y_1 enough that (x_1, y_1) ended up above equilibrium and in the region where the solutions diverge away from the equilibrium. The tangent lines to these solutions led to higher and higher values for the subsequently computed y_k's. Thus, instead of correctly telling us that

$$\lim_{x \to \infty} y(x) = 1 \quad ,$$

this application of Euler's method suggests that

$$\lim_{x \to \infty} y(x) = \infty \quad .$$

A few other situations where blindly applying Euler's method can lead to misleading results are illustrated in the exercises (see Exercises 10.6, 10.7, 10.8 and 10.9). And these sorts of problems are not unique to Euler's method. Similar problems can occur with all numerical methods for solving differential equations. Because of this, it is highly recommended that Euler's method (or any other numerical method) be used only as a last resort. Try the methods developed in the previous chapters first. Use a numerical method only if the other methods fail to yield usable formulas or equations.

Unfortunately, the world is filled with first-order differential equations for which numerical methods are the only practical choices. So be sure to at least skim the next section on improving the method. Also, if you must use the Euler method (or any other numerical method), be sure to do a reality check. Graph the corresponding approximation on top of the slope field for the differential equation, and ask yourself if the approximations are reasonable. In particular, watch out that your numerical solution does not jump over an unstable equilibrium solution.

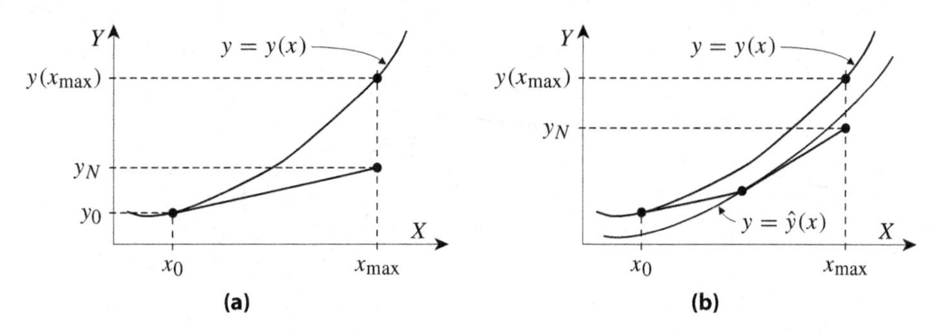

Figure 10.4: Two approximations y_N of $y(x_{max})$ where y is the solution to $y' = f(x, y)$ with $y(x_0) = y_0$: **(a)** Using Euler's method with Δx equaling the distance from x_0 to x_{max} . **(b)** Using Euler's method with Δx equaling half the distance from x_0 to x_{max} (Note: $\hat{y}$ is the solution to $y' = f(x, y)$ with $y(x_1) = y_1$.)

10.4 Reducing the Error

Smaller Step Sizes

Suppose we are applying the Euler method to some initial-value problem on some interval $[x_0, x_{max}]$. The one parameter we can adjust is the step size, Δx (or, equivalently, the number of steps, N , in going from x_0 to x_{max}). By shrinking Δx (increasing N), at least two good things are typically accomplished:

1. The error in the underlying approximation

$$y(x_k + \Delta x) \approx y(x_k) + \Delta x \cdot f(x_k, y(x_k))$$

is reduced.

2. The slope in the piecewise straight line approximation is recomputed at more points, which means that this approximation can better match the bends in the slope field for the differential equation.

Both of these are illustrated in Figure 10.4.

Accordingly, we should expect that shrinking the step size in Euler's method will yield numerical solutions that more accurately approximate the true solution. We can experimentally test this expectation by going back to our initial-value problem

$$5\frac{dy}{dx} - y^2 = -x^2 \qquad \text{with} \quad y(0) = 1 \quad ,$$

computing (as you'll be doing for Exercise 10.5) the numerical solutions arising from Euler's method using, say,

$$\Delta x = 1 \quad , \quad \Delta x = \frac{1}{2} \quad , \quad \Delta x = \frac{1}{4} \quad \text{and} \quad \Delta x = \frac{1}{8} \quad ,$$

and then graphing the corresponding piecewise linear approximations over the interval $[0, 3]$ along with the graph of the true solution. Do this, and you will get the graphs in Figure 10.5.[1] As expected, the graphs of the approximate solutions steadily approach the graph of the true solution as Δx gets smaller. It's even worth observing that the distance between the true value for $y(3)$ and the approximated value appears to be cut roughly in half each time Δx is cut in half.

[1] The graph of the "true solution" in Figure 10.5 is actually the graph of a very accurate approximation. The difference between this graph and the graph of the true solution is less than the thickness of the curve used to sketch it.

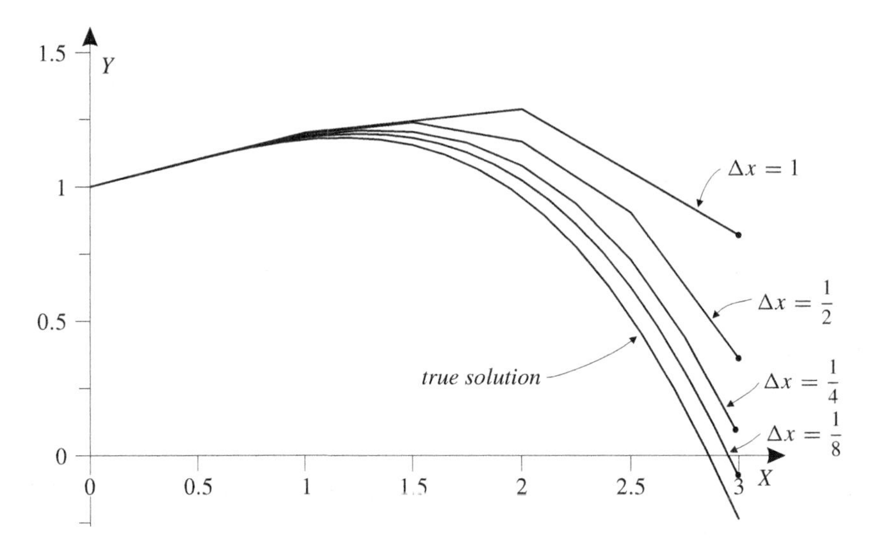

Figure 10.5: Graphs of the different piecewise linear approximations of the solution to $5y' - y^2 = -x^2$ with $y(0) = 1$ obtained using Euler's method with different values for the step size Δx. Also graphed is the true solution.

In fact, our expectations can be rigorously confirmed. In the next section, we will analyze the error in using Euler's method to approximate the solution y to a first-order initial-value problem

$$\frac{dy}{dx} = f(x, y) \qquad \text{with} \quad y(x_0) = y_0 \tag{10.5}$$

on an interval $[x_0, x_0 + L]$. We will see that, assuming f is a "reasonably smooth and bounded" function of x and y, there is a constant M such that

$$|y(x_k) - y_k| < M \cdot \Delta x \tag{10.6}$$

whenever $k = 0, 1, \ldots, N$ and

$$\{x_0, x_1, x_2, x_3, \ldots, x_N\} \qquad \text{and} \qquad \{y_0, y_1, y_2, y_3, \ldots, y_N\}$$

is a numerical solution to initial-value problem 10.5 on $[x_0, x_0 + L]$ generated by the Euler method.

Inequality (10.6) is an *error bound.* It describes the worst theoretical error in using y_N for $y(x_{max})$. In practice, the error may be much less than suggested by this bound, but it cannot be any worse (unless there are other sources of error). Since this bound shrinks to zero as Δx shrinks to zero, we are assured that the approximations to $y(x_{max})$ obtained by Euler's method will converge to the correct value of $y(x_{max})$ if we repeatedly use the method with step sizes shrinking to zero. In fact, if we know the value of M and wish to keep the error below some small positive value, we can use error bound (10.6) to pick a step size, Δx, that will ensure the error is below that desired value. Unfortunately,

1. M can be fairly large.
2. In practice (as we will see), M can be difficult to determine.
3. Error bound (10.6) does not take into account the round-off errors that normally arise in computations.

Let's briefly consider the problem of round-off errors. Inequality (10.6) is only the error bound arising from the theoretically best implementation of Euler's method. In a sense, it is an "ideal error bound" because it is based on all the computations being done with infinite precision. This is rarely practical, even when using a computer math package that can do infinite precision arithmetic — the expressions for the numbers rapidly become too complicated to be usable, even by the computer math packages themselves. In practice, the numbers must be converted to approximations with finite precision, say, decimal approximations accurate to the nearest 0.0001 as done in the table on page 181.

Don't forget that the computations in each step involve numbers from previous steps, and these computations are affected by the round-off errors from those previous steps. So the ultimate error due to round-off will increase as the number of steps increases. With modern computers, the round-off error resulting from each computation is usually very small. Consequently, as long as the number of steps N remains relatively small, the total error due to round-off will usually be insignificant compared to the basic error in Euler's method. But if we attempt to reduce the error in Euler's method by making the step size very, very small, then we must take many, many more steps to go from x_0 to the desired x_{max}. It is quite possible to reach a point where the accumulated round-off error will negate the theoretic improvement in accuracy of the Euler method described by inequality (10.6).

Better Methods

Be aware that Euler's method is a relatively primitive method for numerically solving first-order initial-value problems. Refinements on the method can yield schemes in which the approximations to $y(x_{max})$ converge to the true value much faster as the step size decreases. For example, instead of using the tangent line approximation in each step,

$$y_{k+1} = y_k + \Delta x \cdot f(x_k, y_k) \quad ,$$

we might employ a "tangent parabola" approximation that better accounts for the bend in the graphs. (However, writing a program to determine this "tangent parabola" can be tricky.)

In other approaches, the $f(x_k, y_k)$ in the above equation is replaced with cleverly chosen weighted averages of values of $f(x, y)$ computed at cleverly chosen points near (x_k, y_k). The idea is that this yields a straight a line approximation with the slope adjusted to reduce the over- or undershooting noted a page or two ago. At least two of the more commonly used methods, the "improved Euler method" and the "classic Runge-Kutta method", take this approach. We will examine these methods in Chapter 40.

10.5 Error Analysis for the Euler Method*

The Problem and Assumptions

Throughout this section, we will be concerned with the accuracy of numerical solutions to some first-order initial-value problem

$$\frac{dy}{dx} = f(x, y) \qquad \text{with} \quad y(x_0) = y_0 \quad . \tag{10.7}$$

The precise results will be given in Theorem 10.1, below. For this theorem, L is some finite length, and we will assume there is a corresponding rectangle in the XY–plane

$$\mathcal{R} = \{(x, y) : x_0 \leq x \leq x_0 + L \quad \text{and} \quad y_{min} < y < y_{max}\}$$

* A somewhat advanced discussion for the interested reader.

such that all of the following hold:

1. f and its first partial derivatives are continuous, bounded functions on $\mathcal{R}$. This "boundedness" means there are finite constants A, B and C such that, at each point in $\mathcal{R}$,

$$|f| \leq A \quad , \quad \left|\frac{\partial f}{\partial x}\right| \leq B \quad \text{and} \quad \left|\frac{\partial f}{\partial y}\right| \leq C \quad . \tag{10.8}$$

2. There is a unique solution, $y = y(x)$, to the given initial-value problem valid over the interval $[x_0, x_0 + L]$. (We'll refer to $y = y(x)$ as the "true solution" in what follows.)

3. The rectangle $\mathcal{R}$ contains the graph over the interval $[x_0, x_0 + L]$ of the true solution.

4. If $x_0 \leq x_k \leq x_0 + L$ and (x_k, y_k) is any point generated by any application of Euler's method to solve our initial-value problem, then (x_k, y_k) is in $\mathcal{R}$.

The rectangle $\mathcal{R}$ may be the entire vertical strip

$$\{(x, y) : x_0 \leq x \leq x_0 + L \quad \text{and} \quad -\infty < y < \infty\}$$

if f and its partial derivatives are bounded on this strip. If f and its partial derivatives are not bounded on this strip, then finding the appropriate upper and lower limits for this rectangle is one of the challenges in using the theorem.

Theorem 10.1 (Error bound for the Euler method)
Let f, x_0, y_0, L and $\mathcal{R}$ be as above, and let $y = y(x)$ be the true solution to initial-value problem (10.7). Then there is a finite constant M such that

$$|y(x_k) - y_k| < M \cdot \Delta x \tag{10.9}$$

whenever $k = 0, 1, \ldots, N$ and

$$\{x_0, x_1, x_2, x_3, \ldots, x_N\} \quad \text{and} \quad \{y_0, y_1, y_2, y_3, \ldots, y_N\}$$

is a numerical solution to initial-value problem (10.7) obtained using the Euler method with step spacing Δx and total number of steps N satisfying

$$0 < \Delta x \cdot N \leq L \quad . \tag{10.10}$$

This theorem is only concerned with the error inherent in Euler's method. Inequality (10.9) does not take into account errors arising from rounding off numbers during computation. For a good discussion of round-off errors in computations, the interested reader should consult a good text on numerical analysis.

To prove this theorem, we will derive a constant M that makes inequality (10.9) true. (The impatient can look ahead to equation (10.17) on page 193.) Accordingly, for the rest of this section, $y = y(x)$ will denote the true solution to our initial-value problem, and

$$\{x_0, x_1, x_2, x_3, \ldots, x_N\} \quad \text{and} \quad \{y_0, y_1, y_2, y_3, \ldots, y_N\}$$

will be an arbitrary numerical solution to initial-value problem (10.7) obtained from Euler's method with step spacing Δx and total number of steps N satisfying inequality (10.10).

Also, to simplify discussion, let us agree that, in all the following, k always denotes an arbitrary nonnegative integer less than N.

Preliminary Bounds

Our derivation of a value for M will be based on several basic inequalities and facts from calculus. These include the inequalities

$$|A + B| \leq |A| + |B| \qquad \text{and} \qquad \left| \int_a^b \psi(s)\,ds \right| \leq \int_a^b |\psi(s)|\,ds$$

when $a < b$. Of course, if $|\psi(s)| \leq K$ for some constant K, then, whether or not $a < b$,

$$\int_a^b |\psi(s)|\,ds \leq K\,|b - a| \quad .$$

Also remember that, if $\phi = \phi(x)$ is continuous and differentiable, then

$$\phi(a) - \phi(b) = \int_a^b \frac{d\phi}{ds}\,ds \quad .$$

Combining the above, we get:

Corollary 10.2
Assume ϕ is a continuous differentiable function on some interval. Assume further that $\phi' \leq K$ on this interval for some constant K. Then, for any two points a and b in this interval,

$$|\phi(a) - \phi(b)| \leq K\,|b - a| \quad .$$

We will use this corollary twice.

First, we apply it with $\phi(x) = f(x, y(x))$. Recall that, by the chain rule in Chapter 7,

$$\frac{d}{dx} f(x, y(x)) = \frac{\partial f}{\partial x} + \frac{\partial f}{\partial y}\frac{dy}{dx} \quad ,$$

which we can rewrite as

$$\frac{d}{dx} f(x, y(x)) = \frac{\partial f}{\partial x} + \frac{\partial f}{\partial y} f(x, y)$$

whenever $y = y(x)$ is a solution to $y' = f(x, y)$. Applying bounds (10.8), we then get

$$\left|\frac{df}{dx}\right| \leq \left|\frac{\partial f}{\partial x}\right| + \left|\frac{\partial f}{\partial y}\right| |f(x, y)| \leq B + CA \qquad \text{at every point in } \mathcal{R} \quad .$$

The above corollary (with $\phi(x) = f(x, y(x))$ and $K = B + CA$) then tells us that

$$|f(a, y(a)) - f(b, y(b))| \leq (B + CA)(b - a) \tag{10.11}$$

whenever $x_0 \leq a \leq b \leq x_0 + L$.

The second application of the above corollary is with $\phi(y) = f(x_k, y)$. Here, y is the variable, x remains constant, and $\phi' = {}^{\partial f}/_{\partial y}$. Along with the fact that $\left|{}^{\partial f}/_{\partial y}\right| < C$ on rectangle $\mathcal{R}$, this corollary immediately gives us

$$|f(x_k, b) - f(x_k, a)| \leq C\,|b - a| \tag{10.12}$$

whenever a and b are any two points in the interval $[y_{min}, y_{max}]$.

Maximum Error in the Underlying Approximation

Now consider the error in the underlying approximation

$$y(x_k + \Delta x) \approx y(x_k) + \Delta x \cdot f(x_k, y(x_k)) \quad .$$

Let ϵ_{k+1} be the difference between $y(x_k + \Delta x)$ and the above approximation,

$$\epsilon_{k+1} = y(x_k + \Delta x) - [y(x_k) - \Delta x \cdot f(x_k, y(x_k))] \quad .$$

Note that this can be rewritten both as

$$y(x_{k+1}) = y(x_k) + \Delta x \cdot f(x_k, y(x_k)) + \epsilon_{k+1} \tag{10.13}$$

and as

$$\epsilon_{k+1} = [y(x_k + \Delta x) - y(x_k)] - f(x_k, y(x_k)) \cdot \Delta x \quad .$$

From basic calculus, we know that

$$f(x_k, y(x_k)) \cdot \Delta x = f(x_k, y(x_k)) \int_{x_k}^{x_k+\Delta x} dx = \int_{x_k}^{x_k+\Delta x} f(x_k, y(x_k))\, dx \quad .$$

We also know $y = y(x)$ satisfies $y' = f(x, y)$. Hence,

$$y(x_k + \Delta x) - y(x_k) = \int_{x_k}^{x_k+\Delta x} \frac{dy}{dx}\, dx = \int_{x_k}^{x_k+\Delta x} f(x, y(x))\, dx \quad .$$

Taking the absolute value of ϵ_{k+1} and applying the last three observations yields

$$\begin{aligned}
|\epsilon_{k+1}| &= |[y(x_k + \Delta x) - y(x_k)] - f(x_k, y(x_k)) \cdot \Delta x| \\
&= \left| \int_{x_k}^{x_k+\Delta x} f(x, y(x))\, dx - \int_{x_k}^{x_k+\Delta x} f(x_k, y(x_k))\, dx \right| \\
&= \left| \int_{x_k}^{x_k+\Delta x} f(x, y(x)) - f(x_k, y(x_k))\, dx \right| \\
&\leq \int_{x_k}^{x_k+\Delta x} |f(x, y(x)) - f(x_k, y(x_k))|\, dx \quad .
\end{aligned}$$

Remarkably, we've already found an upper bound for the integrand in the last line (inequality (10.11), with $a = x$ and $b = x_k$). Replacing this integrand with this upper bound, and then doing a little elementary integration, yields

$$|\epsilon_{k+1}| \leq \int_{x_k}^{x_k+\Delta x} (B + CA)(x - x_k)\, dx = \frac{1}{2}(B + CA)(\Delta x)^2 \quad .$$

This last inequality combined with equation (10.13) means that we can rewrite the underlying approximation more precisely as

$$y(x_{k+1}) = y(x_k) + \Delta x \cdot f(x_k, y(x_k)) + \epsilon_{k+1} \tag{10.14a}$$

where

$$|\epsilon_{k+1}| \leq \frac{1}{2}(B + CA)(\Delta x)^2 \quad . \tag{10.14b}$$

Ideal Maximum Error in Euler's Method

Now let E_k be the difference between $y(x_k)$ and y_k,

$$E_k = y(x_k) - y_k \quad .$$

Because $y_0 = y(x_0)$,

$$E_0 = y(x_0) - y_0 = 0 \quad .$$

More generally, using formula (10.14a) for $y(x_k + \Delta x)$ and the formula for y_{k+1} from Euler's method, we have

$$\begin{aligned} E_{k+1} &= y(x_{k+1}) - y_{k+1} \\ &= y(x_k + \Delta x) - y_{k+1} \\ &= \left[y(x_k) + \Delta x \cdot f(x_k, y(x_k)) + \epsilon_{k+1}\right] - [y_k + \Delta x \cdot f(x_k, y_k)] \quad . \end{aligned}$$

Cleverly rearranging the last line and taking the absolute value leads to

$$\begin{aligned} |E_{k+1}| &= |\epsilon_{k+1} + [y(x_k) - y_k] + \Delta x \cdot [f(x_k, y(x_k)) - f(x_k, y_k)]| \\ &= |\epsilon_{k+1} + E_k + \Delta x \cdot [f(x_k, y(x_k)) - f(x_k, y_k)]| \\ &\le |\epsilon_{k+1}| + |E_k| + |\Delta x \cdot [f(x_k, y(x_k)) - f(x_k, y_k)]| \quad . \end{aligned}$$

Fortunately, from inequality (10.14b), we know

$$|\epsilon_{k+1}| \le \frac{1}{2}(B + CA)(\Delta x)^2 \quad ,$$

and from inequality (10.12) and the definition of E_k, we know

$$|f(x_k, y(x_k)) - f(x_k, y_k)| \le C\,|y(x_k) - y_k| = C\,|E_k| \quad .$$

Combining the last three inequalities, we get

$$\begin{aligned} |E_{k+1}| &\le |\epsilon_{k+1}| + |E_k| + |\Delta x \cdot [f(x_k, y(x_k)) - f(x_k, y_k)]| \\ &\le \frac{1}{2}(B + CA)(\Delta x)^2 + |E_k| + \Delta x \cdot C\,|E_k| \\ &\le \frac{1}{2}(B + CA)(\Delta x)^2 + (1 + \Delta x \cdot C)\,|E_k| \quad . \end{aligned}$$

This is starting to look ugly. So let

$$\alpha = \frac{1}{2}(B + CA) \qquad \text{and} \qquad \beta = 1 + \Delta x \cdot C \quad ,$$

just so that the above inequality can be written more simply as

$$|E_{k+1}| \le \alpha(\Delta x)^2 + \beta\,|E_k| \quad .$$

Remember, $E_0 = 0$. Repeatedly applying the last inequality, we then obtain the following:

$$|E_1| = |E_{0+1}| = \alpha(\Delta x)^2 + \beta\,|E_0| = \alpha(\Delta x)^2 \quad ,$$

$$\begin{aligned} |E_2| = |E_{1+1}| &\le \alpha(\Delta x)^2 + \beta\,|E_1| \\ &\le \alpha(\Delta x)^2 + \beta\alpha(\Delta x)^2 \le (1 + \beta)\,\alpha(\Delta x)^2 \quad , \end{aligned}$$

$$\begin{aligned}|E_3| = |E_{2+1}| &\leq \alpha(\Delta x)^2 + \beta\,|E_2| \\ &\leq \alpha(\Delta x)^2 + \beta\,(1+\beta)\,\alpha(\Delta x)^2 \\ &\leq \alpha(\Delta x)^2 + \left(\beta+\beta^2\right)\alpha(\Delta x)^2 \leq \left(1+\beta+\beta^2\right)\alpha(\Delta x)^2 \quad , \\ &\vdots\end{aligned}$$

Continuing, we see that

$$|E_k| \leq S_k\,\alpha(\Delta x)^2 \qquad \text{where} \quad S_k = 1+\beta+\beta^2+\cdots+\beta^{k-1} \quad . \tag{10.15}$$

You may recognize S_k as a partial sum for a geometric series. Whether you do or not, we have

$$\begin{aligned}(\beta-1)S_k &= \beta S_k - S_k \\ &= \beta\left[1+\beta+\beta^2+\cdots+\beta^{k-1}\right] - \left[1+\beta+\beta^2+\cdots+\beta^{k-1}\right] \\ &= \left[\beta+\beta^2+\cdots+\beta^k\right] - \left[1+\beta+\beta^2+\cdots+\beta^{k-1}\right] = \beta^k - 1 \quad .\end{aligned}$$

Dividing through by β and recalling what α and β represent then give us

$$\begin{aligned}S_k\alpha &= \frac{\beta^k - 1}{\beta - 1}\alpha \\ &= \frac{(1+\Delta x\cdot C)^k - 1}{1+\Delta x\cdot C - 1}\cdot\frac{B+CA}{2} = \frac{\left[(1+\Delta x\cdot C)^k - 1\right](B+CA)}{\Delta x\cdot 2C} \quad .\end{aligned}$$

So inequality (10.15) can be rewritten as

$$|E_k| \leq \frac{(1+\Delta x\cdot C)^k - 1}{\Delta x\cdot C}\alpha(\Delta x)^2 \quad .$$

Dividing out one Δx leaves us with

$$|E_k| \leq M_{k,\Delta x}\cdot\Delta x \qquad \text{where} \quad M_{k,\Delta x} = \frac{[(1+\Delta x\cdot C)^k - 1](B+CA)}{2C} \quad . \tag{10.16}$$

The claim of Theorem 10.1 is almost proven with inequality (10.16). All we need to do now is to find a single constant M such that $M_{N,\Delta x} \leq M$ for all possible choices of M and Δx. To this end, recall the Taylor series for the exponential,

$$e^X = \sum_{n=0}^{\infty}\frac{1}{n!}X^n = 1 + X + \frac{1}{2}X^2 + \frac{1}{6}X^3 + \cdots \quad .$$

If $X > 0$ then

$$1 + X < 1 + X + \frac{1}{2}X^2 + \frac{1}{6}X^3 + \cdots = e^X \quad .$$

Cutting out the middle and letting $X = \Delta x\cdot C$, this becomes

$$1 + \Delta x\cdot C < e^{\Delta x\cdot C} \quad .$$

From this and the fact that $N \geq k$, we have

$$(1+\Delta x\cdot C)^k < \left[e^{\Delta x\cdot C}\right]^k = e^{k\Delta x\cdot C} \leq e^{N\Delta x\cdot C} \leq e^{LC}$$

where L is that constant with $N\Delta x \leq L$. So

$$M_{k,\Delta x} = \frac{\left[(1+\Delta x \cdot C)^N - 1\right](B+CA)}{2C} < M$$

where

$$M = \frac{(e^{LC} - 1)(B+CA)}{2C} \quad . \tag{10.17}$$

And this (finally) completes our proof of Theorem 10.1 on page 188.

Additional Exercises

10.1. *Several initial-value problems are given below, along with values for two of the three parameters in Euler's method: step size Δx, number of steps N and maximum variable of interest x_{max}. For each, find the corresponding numerical solution using the Euler method with the indicated parameter values. Do these problems without a calculator or computer.*

a. $\dfrac{dy}{dx} = \dfrac{y}{x}$ with $y(1) = -1$; $\Delta x = \dfrac{1}{3}$ and $N = 3$

b. $\dfrac{dy}{dx} = -8xy$ with $y(0) = 10$; $x_{max} = 1$ and $N = 4$

c. $2x + 5\dfrac{dy}{dx} = y$ with $y(0) = 2$; $x_{max} = 2$ and $\Delta x = \dfrac{1}{2}$

d. $\dfrac{dy}{dx} + \dfrac{y}{x} = 4$ with $y(1) = 8$; $\Delta x = \dfrac{1}{2}$ and $N = 6$

10.2. *Again, several initial-value problems are given below, along with values for two of the three parameters in the Euler method: step size Δx, number of steps N and maximum variable of interest x_{max}. For each, find the corresponding numerical solution using Euler's method with the indicated parameter values. Do these problems with a (nonprogrammable) calculator.*

a. $\dfrac{dy}{dx} = \sqrt{2x+y}$ with $y(0) = 0$; $\Delta x = \dfrac{1}{2}$ and $N = 6$

b. $(1+y)\dfrac{dy}{dx} = x$ with $y(0) = 1$; $N = 6$ and $x_{max} = 2$

c. $\dfrac{dy}{dx} = y^x$ with $y(1) = 2$; $\Delta x = 0.1$ and $x_{max} = 1.5$

d. $\dfrac{dy}{dx} = \cos(y)$ with $y(0) = 0$; $\Delta x = \dfrac{1}{5}$ and $N = 5$

10.3 a. *Using your favorite computer language or computer math package, write a program or worksheet for finding the numerical solution to an arbitrary first-order initial-value problem using the Euler method. Make it easy to change the differential equation and the computational parameters (step size, number of steps, etc.).*[2,3]

[2] If your computer math package uses infinite precision or symbolic arithmetic, you may have to include commands to ensure your results are given as decimal approximations.

[3] It may be easier to compute all the x_k's first, and then compute the y_k's.

b. *Test your program/worksheet by using it to re-compute the numerical solutions for the problems in Exercise 10.2, above.*

10.4. *Using your program/worksheet from Exercise 10.3 a with each of the following step sizes, find an approximation for $y(5)$ where $y = y(x)$ is the solution to*

$$\frac{dy}{dx} = \sqrt[3]{x^2 + y^2 + 1} \quad \text{with} \quad y(0) = 0 \quad .$$

a. $\Delta x = 1$ **b.** $\Delta x = 0.1$ **c.** $\Delta x = 0.01$ **d.** $\Delta x = 0.001$

10.5. *Let y be the (true) solution to the initial-value problem considered in Section 10.2,*

$$5\frac{dy}{dx} - y^2 = -x^2 \quad \text{with} \quad y(0) = 1 \quad .$$

For each step size Δx given below, use your program/worksheet from Exercise 10.3 a to find an approximation to $y(3)$. Also, for each, find the magnitude of the error (to the nearest 0.0001) in using the approximation for $y(3)$, assuming the correct value of $y(3)$ is -0.23699.

a. $\Delta x = 1$ **b.** $\Delta x = \frac{1}{2}$ **c.** $\Delta x = \frac{1}{4}$ **d.** $\Delta x = \frac{1}{8}$

e. $\Delta x = 0.01$ **f.** $\Delta x = 0.001$ **g.** $\Delta x = 0.0001$

10.6. *Consider the initial-value problem*

$$\frac{dy}{dx} = (y - 1)^2 \quad \text{with} \quad y(0) = -\frac{13}{10} \quad .$$

This is the problem discussed in Section 10.3 in the illustration of a "catastrophic failure" of Euler's method.

a. *Find the exact solution to this initial-value problem using methods developed in earlier chapters. What, in particular, is the exact value of $y(4)$?*

b. *Using your program/worksheet from Exercise 10.3 a, find the numerical solution to the above initial-value problem with $x_{max} = 4$ and step size $\Delta x = 1/2$. (Also, confirm that this numerical solution has been properly plotted in Figure 10.3 on page 182.)*

c. *Find the approximation to $y(4)$ generated by Euler's method with each of the following step sizes (use your answer to the previous part or your program/worksheet from Exercise 10.3 a). Also, compute the magnitude of the error in using this approximation for the exact value found in the first part of this exercise.*

i. $\Delta x = 1$ **ii.** $\Delta x = \frac{1}{2}$ **iii.** $\Delta x = \frac{1}{4}$ **iv.** $\Delta x = \frac{1}{10}$

d. *Observe (from your answer to Exercise 10.6 c i) that the approximate value for $y(4)$ obtained using $\Delta x = 1$ is a very poor approximation to the true value of $y(4)$. Why did the Euler method fail so badly in this one case? (Hint: Consider the slope field.)*

10.7. *Consider the following initial-value problem*

$$\frac{dy}{dx} = -4y \quad \text{with} \quad y(0) = 3 \quad .$$

The following will illustrate the importance of choosing appropriate step sizes.

a. *Find the numerical solution using Euler's method with* $\Delta x = 1/2$ *and* N *being any large integer (this will be more easily done by hand than using a calculator!). Then do the following:*

i. *There will be a pattern to the* y_k*'s. What is that pattern? What happens as* $k \to \infty$ *?*

ii. *Plot the piecewise straight approximation corresponding to your numerical solution along with a slope field for the above differential equation. Using these plots, decide whether your numerical solution accurately describes the true solution, especially as* x *gets large.*

iii. *Solve the above initial-value problem exactly using methods developed in earlier chapters. What happens to* $y(x)$ *as* $x \to \infty$ *? Compare this behavior to that of your numerical solution. In particular, what is the approximate error in using* y_k *for* $y(x_k)$ *when* x_k *is large?*

b. *Now find the numerical solution to the above initial-value problem using Euler's method with* $\Delta x = 1/10$ *and* N *being any large integer (do this by hand, looking for patterns in the* y_k*'s)). Then do the following:*

i. *Find a relatively simple formula describing the pattern in the* y_k*'s.*

ii. *Plot the piecewise straight approximation corresponding to this numerical solution along with a slope field for the above differential equation. Does this numerical solution appear to be significantly better (more accurate) than the one found in part 10.7 a?*

10.8. *In this problem we'll see one danger of blindly applying a numerical method to solve an initial-value problem. The initial-value problem is*

$$\frac{dy}{dx} = \frac{3}{7 - 3x} \quad \text{with} \quad y(0) = 0 \quad .$$

a. *Find the numerical solution to this using Euler's method with step size* $\Delta x = 1/2$ *and* $x_{max} = 5$. *(Use your program/worksheet from Exercise 10.3 a).*

b. *Sketch the piecewise straight approximation corresponding to the numerical solution just found.*

c. *Sketch the slope field for this differential equation, and find the exact solution to the above initial-value problem by simple integration.*

d. *What happens in the true solution as* $x \to 7/3$ *?*

e. *What can be said about the approximations to* $y(x_k)$ *obtained in the first part when* $x_k > 7/3$ *?*

10.9. *What goes wrong with attempting to find a numerical solution to*

$$(2y - 1)^2 \frac{dy}{dx} = 1 \quad \text{with} \quad y(0) = 0$$

using Euler's method with step size $\Delta x = 1/2$ *?*

11

The Art and Science of Modeling with First-Order Equations

For some, "modeling" is the building of small plastic replicas of famous ships; for others, "modeling" means standing in front of cameras wearing silly clothing; for us, "modeling" is the process of developing sets of equations and formulas describing some process of interest. This process may be the falling of a frozen duck, the changes in a population over time, the consumption of fuel by a car traveling various distances, the accumulation of wealth by one individual or company, the cooling of a cup of coffee, the electronic transmission of sound and images from a television station to a home television, or any of a huge number of other processes affecting us. A major goal of modeling, of course, is to predict "how things will turn out" at some point of interest, be it a point of time in the future or a position along the road. Along with this, naturally, is often a desire to use the model to determine changes we can make to the process to force things to turn out as we desire.

Of course, some things are more easily modeled mathematically than others. For example, it will certainly be easier to mathematically describe the number of rabbits in a field than to mathematically describe the various emotions of these rabbits. Part of the art of modeling is the determination of which quantities the model will deal with (e.g., "number of rabbits" instead of "emotional states").

Another part of modeling is the balancing between developing as complete a model as possible by taking into account all possible influences on the process as opposed to developing a simple and easy to use model by the use of simplifying assumptions and simple approximations. Attempting to accurately describe all possible influences usually leads to such a complicated set of equations and formulas that the model (i.e., the set of equations and formulas we've developed) is unusable. A model that is too simple, on the other hand, may lead to wildly inaccurate predictions, and, thus, would also not be a useful model.

Here, we will examine various aspects of modeling using first-order differential equations. This will be done mainly by looking at a few illustrative examples, though, in a few sections, we will also discuss how to go about developing and using models with first-order differential equations more generally.

11.1 Preliminaries

Suppose we have a situation in which some measurable quantity of interest (e.g.: velocity of a falling duck, number of rabbits in a field, gallons of fuel in a vehicle, amount of money in a bank, temperature of a cup of coffee) varies as some basic parameter (such as time or position) changes. For convenience, let's assume the parameter is time and denote that parameter by t, as is traditional.

Recall that, if

$$Q(t) = \text{the } \textit{amount} \text{ of that measurable quantity at time } t \quad ,$$

then

$$\frac{dQ}{dt} = \text{the } \textit{rate} \text{ at which } Q \text{ varies as } t \text{ varies} \quad .$$

If we can identify what controls this rate, and can come up with a formula $F(t, Q)$ describing how this rate depends on t and the Q, then

$$\frac{dQ}{dt} = F(t, Q)$$

gives us a first-order differential equation for Q which, with a little luck, can be solved to obtain a general formula for Q in terms of t. At the very least, we will be able to construct this equation's slope field and sketch graphs of $Q(t)$.

Our development of the "improved falling object model" in Section 1.2 is a good example of this sort of modeling. Go back to page 11 and take a quick look at it. There, the 'measurable quantity' is the velocity v (in meters/second); the rate at which it varies with time, dv/dt, is the acceleration, and we were able to determine a formula F for this acceleration by determining and adding together the accelerations due to gravity and air resistance,

$$\begin{aligned} F(t, v) &= \text{total acceleration} \\ &= \text{acceleration due to gravity} + \text{acceleration due to air resistance} \\ &= (-9.8) + (-\kappa v) \end{aligned}$$

where κ is some positive constant that would have to be determined by experiment. This gave us the first-order differential equation

$$\frac{dv}{dt} = F(t, v) = -9.8 - \kappa v \quad ,$$

which we were later able to solve and analyze.

In what follows, we will develop models for several other situations. We will also, in Section 11.5, give further advice on developing your own models with first-order differential equations. Be sure to observe how we develop these models and to read the notes in Section 11.5. You will be developing more models in the exercises and, maybe later, in real life.

11.2 A Rabbit Ranch

The Situation to Be Modeled

Pretend we've been given a breeding pair of rabbits along with acres and acres of prime rabbit range with no predators. Let us further assume this rabbit range is fenced in so that no rabbits can escape or come in, and so that no predators can come in. We release the rabbits, planning to return in a few years (say, five) to harvest rabbits for the Easter trade.

An obvious question is *How many rabbits will we have on our rabbit ranch in five years?*

Setting Up the Model

Experienced modelers typically begin by drawing a simple, enlightening picture of the process (if appropriate) and identifying the relevant variables (as we did for the "falling object model" — see page 8). Since the author could not think of a particularly appropriate and enlightening picture, we will skip the picture and go straight to identifying the obvious variables of interest. They are 'time' and 'the number of rabbits,' which we will naturally denote using the symbols t and R, respectively. The time t can be measured in seconds, days, months, years, centuries, etc. We will use months, with $t = 0$ being the time the rabbits were released. So,

$$t = \text{number of months since the rabbits were released}$$

and

$$R = R(t) = \text{number of rabbits at time } t \quad .$$

Since we started with a pair of rabbits, the initial condition is

$$R(0) = 2 \quad . \tag{11.1}$$

Now, since t is being measured in months,

$$\begin{aligned} \frac{dR}{dt} &= \text{rate } R \text{ varies as } t \text{ varies} \\ &= \text{change in the number of rabbits per month} \quad . \end{aligned}$$

Because the fence prevents rabbits escaping or coming in, the change in the number of rabbits is due entirely to the number of births and deaths in our rabbit population. Thus,

$$\begin{aligned} \frac{dR}{dt} &= \text{change in the number of rabbits per month} \\ &= \text{number of births per month} - \text{number of deaths per month} \quad . \end{aligned} \tag{11.2}$$

Now we need to model the "number of births per month" and the "number of deaths per month." Starting with the birth process, and assuming that half the population are females, we note that

$$\begin{aligned} &\text{number of births per month} \\ &\qquad = \text{number of births per female rabbit per month} \\ &\qquad\qquad \times \text{number of female rabbits that month} \\ &\qquad = \text{number of births per female rabbit per month} \times \frac{1}{2}R \quad . \end{aligned}$$

(We are also assuming that all the females are capable of having babies, no matter their age. Well, these are rabbits; they marry young.)

It seems reasonable to assume the average number of births per female rabbit per month is a constant. For future convenience, let

$$\beta = \frac{1}{2} \times \text{number of births per female rabbit per month} \quad .$$

This is the "monthly birth rate per rabbit" and allows us to write

$$\text{number of births per month} = \beta R \quad . \tag{11.3}$$

Checking a reliable reference on rabbits (any decent encyclopedia will do), it can be found that, on the average, each female rabbit has 6 litters per year with 5 bouncy baby bunnies in each litter. Hence, since there are 12 months in a year,

$$\begin{aligned} \beta &= \frac{1}{2} \times \text{number of births per female rabbit per month} \\ &= \frac{1}{2} \times \frac{1}{12} \times \text{number of births per female rabbit per year} \\ &= \frac{1}{2} \times \frac{1}{12} \times 6 \times 5 \quad . \end{aligned}$$

That is,

$$\beta = \frac{5}{4} \quad . \tag{11.4}$$

What about the death rate? Since there are no predators and plenty of food, it seems reasonable to assume old age is the main cause of death. Again checking a reliable reference on rabbits, it can be found that the average life span for a rabbit is 10 years. Clearly, then, the number of deaths per month will be negligible compared to the number of births. So we will assume

$$\text{number of deaths per month} = 0 \quad . \tag{11.5}$$

Combining equations (11.2), (11.3) and (11.5), we obtain

$$\begin{aligned} \frac{dR}{dt} &= \text{number of births per month} - \text{number of deaths per month} \\ &= \beta R - 0 \quad . \end{aligned}$$

That is,

$$\frac{dR}{dt} = \beta R \tag{11.6}$$

where β is the average monthly birth rate per rabbit.

Of course, equation (11.6) does not just apply to the situation being considered here. The same equation would have been obtained for the changing population of any creature having zero death rate and a constant birth rate β per unit time per creature. But the problem at hand involves rabbits, and for rabbits, we derived $\beta = 5/4$. This, the above differential equation, and the fact that we started with two rabbits means that $R(t)$ must satisfy

$$\frac{dR}{dt} = \frac{5}{4}R \qquad \text{with} \quad R(0) = 2 \quad .$$

This is our "model".

Using Our Model

Our differential equation is

$$\frac{dR}{dt} = \beta R \qquad \text{with} \quad \beta = \frac{5}{4} \quad .$$

This is a simple separable and linear differential equation. You can easily show that its general solution is

$$R(t) = Ae^{\beta t} \quad .$$

Applying the initial condition,

$$2 = Ae^{\beta \cdot 0} = A \quad .$$

So the number of rabbits after t months is given by

$$R(t) = 2e^{\beta t} \quad \text{with} \quad \beta = \frac{5}{4} \quad . \tag{11.7}$$

Five years is 60 months. Using a calculator, we find that the number of rabbits after 5 years is

$$R(60) = 2e^{\frac{5}{4}\cdot 60} = 2e^{75} \approx 7.47 \times 10^{32} \quad .$$

That is a lot of rabbits. At about 3 kilograms each, the mass of all the rabbits on the ranch will then be approximately

$$2.2 \times 10^{33} \text{ kilograms} \quad .$$

By comparison:

$$\text{the mass of the Earth} \approx 6 \times 10^{24} \text{ kilograms}$$

and

$$\text{the mass of the Sun} \approx 2 \times 10^{30} \text{ kilograms} \quad .$$

So our model predicts that, in five years, the total mass of our rabbits will be over a thousand times that of our nearest star.

This does not seem like a very realistic prediction. Later in this chapter, we will derive a more complete (but less simple) model.

But first, let us briefly discuss a few other modeling situations involving differential equations similar to the one derived here (equation (11.6)).

11.3 Exponential Growth and Decay

Whenever the rate of change of some quantity $Q(t)$ is directly proportional to that quantity, we automatically have

$$\frac{dQ}{dt} = \beta Q$$

with β being the constant of proportionality. Since this simple relationship is inherent in many processes of interest, it, along with with its general solution

$$Q(t) = Ae^{\beta t} \quad ,$$

arise in a large number of important applications, most of which do not involve rabbits. If $\beta > 0$, then $Q(t)$ increases rapidly as t increases, and we typically say we have *exponential growth.* If $\beta < 0$, then $Q(t)$ shrinks to 0 fairly rapidly as t increases, and we typically say we have *exponential decay*. And if $\beta = 0$, then $Q(t)$ remains constant — we just have equilibrium solutions.

Simple Population Models

Suppose we are interested in how the population of some set of creatures or plants varies with time. These may be rabbits on a ranch (as in our previous example) or yeast fermenting a vat of grape juice or the people in some city or the algae growing in a pond. They may even be the people in some country that are infected with and are helping spread some contagious disease. Whatever the individuals of this population happen to be, we can let $P(t)$ denote the total number of these individuals at time t, and ask how $P(t)$ varies with time (as we did in our "rabbit ranch" example). If we further assume (as we did in our previous "rabbit ranch" example) that

1. the change in $P(t)$ over a unit of time depends only on the number of "births" and "deaths" in the population;[1]

2. the "average birth rate per individual per unit of time" β_0 is constant,

and

3. the "average death rate per individual per unit of time" δ_0 is constant (i.e., a constant fraction δ_0 of the population dies off during each unit of time);

then

$$\begin{aligned}\frac{dP}{dt} &= \text{change in the number of individuals per unit time}\\ &= \text{number of births per unit time} \;-\; \text{number of deaths per unit time}\\ &= \beta_0 P(t) \;-\; \delta_0 P(t) \quad .\end{aligned}$$

Letting β be the "net birth rate per individual per unit time",

$$\beta = \beta_0 - \delta_0 \quad ,$$

the above equation reduces to

$$\frac{dP}{dt} = \beta P(t) \quad , \tag{11.8}$$

the solution of which, as we already know, is

$$P(t) = P_0 e^{\beta t} = P_0 e^{(\beta_0 - \delta_0)t} \qquad \text{where} \quad P_0 = P(0) \quad .$$

If $\beta_0 > \delta_0$, then the model predicts that the population will grow exponentially. If $\beta_0 < \delta_0$, then the model predicts that the population will decline exponentially. And if $\beta_0 = \delta_0$, then the model predicts that the population remains static.

This is a simple model whose accuracy depends on the validity of the three basic assumptions made above. In many cases, these assumptions are often reasonably acceptable during the early stages of the process, and, initially, we do see exponential growth of populations, say, of yeast added to grape juice or of a new species of plants or animals introduced to a region where it can thrive. As illustrated in our "rabbit ranch" example, however, this is too simple a model to describe the long-term growth of most biological populations.

Natural Radioactive Decay

The effect of radioactive decay on the amount of some radioactive isotope can be described by a model completely analogous to the general population model just discussed. Assume we start with some amount (say, a kilogram) of some radioactive isotope of interest (say, uranium–235). During any given length of time, there is a certain probability that any given atom of that material will spontaneously decay into a smaller atom along with associated radiation and other atomic and subatomic particles. Thus, the amount we have of that particular radioactive isotope will decrease as more and more of the atoms decay (provided there is not some other material that decays into the isotope of interest.)

Let's assume we have some radioactive isotope of interest, and that there is no other radioactive material decaying into that isotope. For convenience, let $A(t)$ denote the amount of that radioactive material at time t, and let δ be the fraction of the material that decays per unit time. In essence, the

[1] Precisely what "birth" or "death" means may depend on the creatures/plants in the population. For a microbe, "birth" may be when a parent cell divides into two copies of itself. If the population is the set of people infected with a particular disease, then "birth" occurs when a person contracts the disease.

decay of an atom is the death of that atom, and this δ is essentially the same as the δ_0 in the above population growth discussion. Virtually the same analysis done to obtain equation (11.8) (but using P instead of A, and noting that $\beta_0 = 0$ since no new atoms of the isotope are being "born") then yields

$$\frac{dA}{dt} = -\delta A(t) \quad .$$

Solving this differential equation then gives us

$$A(t) = A_0 e^{-\delta t} \qquad \text{with} \quad A_0 = A(0) \quad . \tag{11.9}$$

Because radioactive decay is a probabilistic event, and because there are typically huge numbers of atoms in any sample of radioactive material, the laws of probability and statistics ensure that this is usually a very accurate model over long periods of time (unlike the case with biological populations).

The positive constant δ, called the *decay rate*, is different for each different isotope. It is large if the isotope is very unstable and a large fraction of the atoms decay in a given time period, and it is small if the isotope is fairly stable and only a small fraction of the atoms decay in the same time period. In practice, the decay rate δ is usually described indirectly through the *half-life* $\tau_{1/2}$ of the isotope, which is the time it takes for half of the original amount to decay. Using the above formula for $A(t)$, you can easily verify that $\tau_{1/2}$ and δ are related by the equation

$$\delta \times \tau_{1/2} = \ln 2 \quad . \tag{11.10}$$

?►Exercise 11.1: *Derive equation (11.10). Use formula (11.9) for $A(t)$ and the fact that, by the definition of $\tau_{1/2}$,*

$$A(\tau_{1/2}) = \frac{1}{2}A(0) \quad .$$

!►Example 11.1: *Cobalt-60 is a radioactive isotope of cobalt with a half-life of approximately 5.27 years.*[2] *Using equation (11.10), we find that its (approximate) decay constant is given by*

$$\delta = \frac{\ln 2}{\tau_{1/2}} = \frac{\ln 2}{5.27 \text{ (years)}} \approx 0.1315 \quad \text{(per year)} \quad .$$

Combining this with formula (11.9) gives us

$$A(t) \approx A_0 e^{-0.1315t} \qquad \text{with} \quad A_0 = A(0)$$

as the formula for the amount of undecayed cobalt remaining after t years.

Suppose we initially have 10 grams of cobalt-60. At the end of one year, those 10 grams would have decayed to approximately

$$(10 \text{ gm.}) \times e^{-0.1315 \times 1} \approx 8.77 \text{ grams of cobalt-60} \quad .$$

At the end of two years, those 10 grams would have decayed to approximately

$$(10 \text{ gm.}) \times e^{-0.1315 \times 2} \approx 7.69 \text{ grams of cobalt-60} \quad .$$

And at the end of ten years, those 10 grams would have decayed to approximately

$$(10 \text{ gm.}) \times e^{-0.1315 \times 10} \approx 2.68 \text{ grams of cobalt-60} \quad .$$

[2] Cobalt-60 has numerous medical applications, as well as having the potential as an ingredient in some particularly nasty nuclear weapons. It is produced by exposing cobalt-59 to "slow" neutrons, and decays to a stable nickel isotope after giving off one electron and two gamma rays.

11.4 The Rabbit Ranch, Again

Back to wrangling rabbits.

The Situation (and Problem)

Recall that we imagined ourselves having a fenced-in ranch enclosing many acres of prime rabbit range. We start with a breeding pair of rabbits, and plan to return in five years. The question is *How many rabbits will we have then?*

In Section 11.2, we attempted to answer this question using a fairly simple model we had just developed. However, the predicted number of rabbits after five years (which had a corresponding mass a thousand times greater than that of the Sun) was clearly absurd. That model did not account for the problems arising when a population of rabbits grows too large. Let us now see if we can derive a more realistic model.

A Better Model

Again, we let

$$R(t) = \text{number of rabbits after } t \text{ months}$$

with $R(0) = 2$. We still have

$$\begin{aligned} \frac{dR}{dt} &= \text{change in the number of rabbits per month} \\ &= \text{number of births per month} - \text{number of deaths per month} \quad . \end{aligned} \tag{11.11}$$

However, the assumptions that

$$\text{number of deaths per month} = 0 \quad ,$$

and

$$\text{number of births per month} = \beta R$$

where

$$\beta = \text{monthly birth rate per rabbit} = \frac{5}{4}$$

are too simplistic. As the population increases, the amount of range land (and, hence, food) per rabbit decreases. Eventually, the population may become too large for the available fields to support all the rabbits. Some will starve to death, and those female rabbits that survive will be malnourished and will give birth to fewer bunnies. In addition, overcrowding is conducive to the spread of diseases which, in a population already weakened by hunger, can be devastating.

Clearly, we at least need to correct our formula for the number of deaths per month, because, once overcrowding begins, we can expect a certain fraction of the population to die each month. Letting δ denote that fraction,

$$\begin{aligned} \text{number of deaths per month} &= \text{fraction of the population that dies each month} \\ &\qquad \times \text{number of rabbits} \\ &= \delta R \quad . \end{aligned}$$

Keep in mind that this fraction δ, which we can call the monthly death rate per rabbit, will not be constant. It will depend on just how overcrowded the rabbits are. In other words, δ will vary with

R, and, thus, should be treated as a function of R, $\delta = \delta(R)$. Just how δ varies with R is yet unknown, but it should be clear that

> if R is small, then overcrowding is not a problem and $\delta(R)$ should be close to zero,

and

> as R increases, then overcrowding increases and more rabbits start dying. So, $\delta(R)$ should increase as R increases.

The simplest function of R for δ satisfying the two above conditions is

$$\delta = \delta(R) = \gamma_D R$$

where γ_D is some positive constant. This gives us

$$\text{number of deaths per month} = \delta R = \left[\gamma_D R\right] R = \gamma_D R^2 \quad .$$

What sort of "correction" should we now consider for

$$\text{number of births per month} = \beta R \quad ?$$

Well, as with the monthly death rate δ, above, we should expect the monthly birth rate per rabbit, β, to be a function of the number of rabbits, $\beta = \beta(R)$. Moreover:

> If R is small, then overcrowding is not a problem and $\beta(R)$ should be close to its ideal value $\beta_0 = 5/4$,

and

> as R increases, then more rabbits become malnourished, and females have fewer babies each month. So, $\beta(R)$ should decrease from its ideal value as R increases.

A simple formula describing this is obtained by subtracting from the ideal birth rate a simple correction term proportional to the number of rabbits,

$$\beta = \beta(R) = \beta_0 - \gamma_B R$$

where $\beta_0 = 5/4$ is the ideal monthly birth rate per rabbit and γ_B is some positive constant.[3] This then gives us

$$\text{number of births per month} = \beta R = \left[\beta_0 - \gamma_B R\right] R = \beta_0 R - \gamma_B R^2 \quad .$$

As with our simpler model, the one we are developing can be applied to populations of other organisms by using the appropriate value for the ideal birth rate per organism, β_0. For rabbits, we have $\beta_0 = 5/4$.

Combining the above formulas for the monthly number of births and deaths with the generic differential equation (11.11) yields

$$\begin{aligned}
\frac{dR}{dt} &= \text{number of births per month} - \text{number of deaths per month} \\
&= \beta R - \delta R \\
&= \beta_0 R - \gamma_B R^2 - \gamma_D R^2 \quad ,
\end{aligned}$$

which, letting $\gamma = \gamma_B + \gamma_D$, simplifies to

$$\frac{dR}{dt} = \beta_0 R - \gamma R^2 \tag{11.12}$$

where $\beta_0 = 5/4$ is the ideal monthly birth rate per rabbit and γ is some positive constant. Presumably, γ could be determined by observation (if this new model does accurately describe the situation).

[3] Yes, the birth rate becomes negative if R becomes large enough, and negative birth rates are not realistic. But we are still trying for as simple a model as feasible — with luck, R will not get large enough that the birth rate becomes negative.

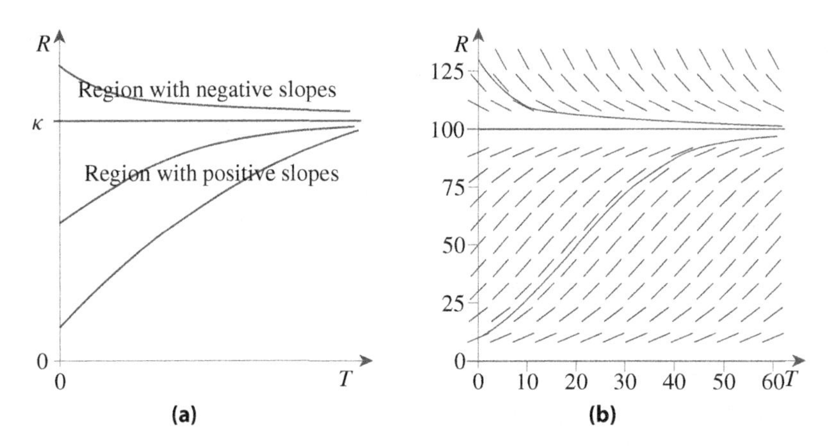

Figure 11.1: Slope Fields for Logistic Equation (11.13): **(a)** A minimal field for a generic logistic equation and **(b)** a slope field for the logistic equation with $\kappa = 100$ and $\beta_0 = {}^{1}/_{10}$. The graphs of a few solutions have been roughly sketched on each.

Using the Better Model

Equation (11.12) is called the *logistic equation* and was an important development in the study of population dynamics. It is a relatively simple separable equation that can be solved without too much difficulty. But let's not, at least, not yet. Instead, let us first rewrite this equation by factoring out γR,

$$\frac{dR}{dt} = \gamma R \left(\frac{\beta_0}{\gamma} - R \right) \quad .$$

From this it is obvious that our differential equation has two constant solutions,

$$R = 0 \qquad \text{and} \qquad R = \frac{\beta_0}{\gamma} \quad .$$

The first tells us that, if we start with no rabbits, then we get no rabbits in the future (no surprise there). The second is more interesting. For convenience, let $\kappa = {}^{\beta_0}/_{\gamma}$. Our differential equation can then be written as

$$\frac{dR}{dt} = \gamma R (\kappa - R) \qquad \text{with} \quad \gamma = \frac{\beta_0}{\kappa} \quad , \tag{11.13}$$

and the two constant solutions are

$$R = 0 \qquad \text{and} \qquad R = \kappa \quad .$$

(We probably should note that κ, as a ratio of positive constants, is a positive constant.)

While we are at it, let's further observe that, if $0 < R < \kappa$, then

$$\frac{dR}{dt} = \underbrace{\gamma R}_{>0} \underbrace{(\kappa - R)}_{>0} > 0 \quad .$$

In other words, if $0 < R < \kappa$, then the population is increasing.

On the other hand, if $\kappa < R$, then

$$\frac{dR}{dt} = \underbrace{\gamma R}_{>0} \underbrace{(\kappa - R)}_{<0} < 0 \quad .$$

That is, the population will be decreasing if $\kappa < R$.

We can graphically represent these observations using the crude slope field sketched in Figure 11.1a. This figure suggests that, over time, the number of rabbits will stabilize around κ. If there are initially fewer than κ rabbits (but at least some), then the rabbit population will increase towards a total of κ rabbits. If there are initially more than κ rabbits, then the population will decrease towards a total of κ rabbits. This prediction is reflected in the more carefully constructed slope field in Figure 11.1b. Because κ is the maximum number of rabbits that can exist in the long term given the resources available, κ is often called the *carrying capacity* of the 'system' consisting of the rabbits and their environment. (Of course, if the carrying capacity is too low, say, $\kappa \approx 0.5$ then, realistically, all the rabbits will die.)

Finding the precise formula for $R(t)$ will be left to you (Exercise 11.7). What you will show is that, in terms of the carrying capacity κ, ideal birth rate β_0, and initial population $R_0 = R(0)$,

$$R(t) = \frac{\kappa R_0}{R_0 + (\kappa - R_0)e^{-\beta_0 t}} \quad . \tag{11.14}$$

This formula reflects the fact that there are three basic parameters in our model: the ideal monthly birth rate β_0, the initial number of rabbits $R(0)$, and the carrying capacity of the system κ. The first two we know or can figure out from basic biology. The last, κ, will have to be determined "from experiment". For example, we might return a year after releasing the original pair (at $t = 12$), count the number of rabbits on the ranch, $R(12)$, and then use this value along with the known values for R_0 and β_0 in formula (11.14) to create an equation for κ. Solving that equation will then give us κ. This, of course, assumes that the model is fairly accurate — an assumption that would require further experiment to verify or disprove. But, at least the model's prediction regarding the population growth seems a good deal more reasonable than that made by the simpler model in Section 11.2.

11.5 Notes on the Art and Science of Modeling

Our current interest is in modeling situations in which the rate at which some quantity varies is fairly well understood. In these sorts of problems, it is often "relatively easy" to develop a first-order differential equation to serve as the basis for a mathematical model for the situation. We've seen several examples already and will see more in the next few sections. But now, let us pause to discuss some of the steps and issues in developing and using such models.

First Steps in the Modeling Process

Naturally, one of your very first steps in modeling something should be to learn whatever you believe is needed for developing the model. Then identify and label the significant basic variables and decide on the units associated with these variables. In our rabbit ranch problems, those variables were R and t (with associated units *rabbits* and *months*, respectively); in the following, we'll use Q for the generic quantity of interest and assume it varies over time t.

Next, write out everything you know using these variables. This includes any initial values you may have for any of the variables. In our rabbit ranch problem, we did not know much at first, only the initial value for R, $R(0) = 2$. If you can draw an illuminating picture representing the situation, do so and label it for easy reference.

Then turn your attention to deriving a differential equation that accurately models the *rate* at which Q varies with t, ${}^{dQ}/_{dt}$. Do *not* attempt to directly derive a formula for $Q(t)$, at least, not with the sort of problems being considered here. We are now dealing with problems for which it is much easier to first find a formula $F(t, Q)$ for ${}^{dQ}/_{dt}$, and then find $Q(t)$ by solving the resulting differential equation.

Developing the Differential Equation for the Model

Coming up with a usable differential equation

$$\frac{dQ}{dt} = F(t, Q)$$

that accurately models how Q's rate of change depends on t and Q is the most important, and, for many, the hardest part of the of the modeling process. After all, as you now know, anyone can solve a first-order differential equation (or, at least, construct a slope field for one). Coming up with the right differential equation can be much more challenging.

Here are a few things you can do to make it less challenging:

Identify and Describe the Processes Driving the Model

Keep in mind that $^{dQ}/_{dt}$ is the *rate* at which $Q(t)$ changes as t changes. This rate depends on the processes driving the situation, not (usually) on the particular value of Q at some particular time. In particular, the actual value of $Q(0)$ is (usually) irrelevant in setting up the differential equation.

Once you've determined your variables and drawn your illuminating pictures, write out

$$\frac{dQ}{dt} = \text{the change in } Q \text{ per unit time}$$

and then identify the different processes that cause that Q to change. In our examples with rabbits, these processes were "birth" and "death", and we initially observed that

$$\begin{aligned}\frac{dR}{dt} &= \text{change in the number of rabbits per month}\\ &= \text{number of births per month} - \text{number of deaths per month} \quad .\end{aligned}$$

Then we worked out how many births and deaths should be expected each month given that we had R rabbits that month.

In general, you want to identify the different processes that cause $Q(t)$ to increase (e.g., births and immigration into the region) and to decrease (e.g., deaths and emigration out of the region). Each of these processes typically corresponds to a different term in $F(t, Q)$ (remember to *add* those terms corresponding to processes that increase Q, and *subtract* those terms corresponding to processes that decrease Q). For example, if $Q(t)$ is the number of people in a certain region at time t, we may have

$$\frac{dQ}{dt} = F(t, Q)$$

where

$$F(t, Q) = F_{\text{birth}} + F_{\text{immig}} - F_{\text{death}} - F_{\text{emig}}$$

with

$$\begin{aligned}F_{\text{birth}} &= \text{number of births per unit time} \quad ,\\ F_{\text{immig}} &= \text{number of number of people immigrating into the region per unit time} \quad ,\\ F_{\text{death}} &= \text{number of deaths per unit time}\end{aligned}$$

and

$$F_{\text{emig}} = \text{number of number of people emigrating out of the region per unit time} \quad .$$

Once you've identified the different terms making up $F(t, Q)$ (e.g., the above F_{birth}, F_{immig}, etc.), take each term, consider the process it is supposed to describe, and try to come up with a reasonable formula describing the change in Q during a unit time interval due to that process alone.

Often, that formula will involve just Q, itself. For example, in our first "rabbit ranch" example (with $R = Q$),

$$\begin{aligned} F_{\text{birth}} &= \text{number of births per month} \\ &= \text{number of births per female rabbit per month} \\ &\quad \times \text{number of female rabbits that month} \\ &= \cdots \\ &= \beta R \qquad \text{with} \quad \beta = \frac{5}{4} \quad . \end{aligned}$$

Often, you will make 'simplifications' and 'assumptions' to keep the model from becoming too complicated. In the above formula for F_{birth}, for example, we did not attempt to account for seasonal variations in birth rate, and we assumed that half the rabbit population were breeding females. We also assumed a constant monthly birth rate and death rate per rabbit, no matter how many rabbits we had.

Balance the Units

As already noted, one of the first steps in modeling a situation is to decide on the main variables and to choose the units for measuring these variables. The subsequent computations and derivations are all in terms of these units, and we can often avoid embarrassing mistakes by just keeping track of our units and being sure that their use is consistent. In particular, the units implicit in any equation must remain balanced; that is, each term in any equation must have the same associated units as every other term in that equation.

For example, the basic quantities in our rabbit ranch models are R and t. Even though we treated these as numbers, we knew that

$$R = \text{number of } \textit{rabbits} \qquad \text{and} \qquad t = \text{time in } \textit{months} \quad .$$

So the units associated with R and t are, respectively, *rabbits* and *months*. Consequently, the units associated with

$$\frac{dR}{dt} = \lim_{\Delta t \to 0} \frac{R(t + \Delta t) - R(t)}{\Delta t} = \lim_{\Delta t \to 0} \frac{\text{change in the number of } \textit{rabbits}}{\text{change in time as measured in } \textit{months}}$$

are $^{rabbits}/_{month}$ (i.e., rabbits per month), and every term in any formula for $^{dR}/_{dt}$ must also have $^{\text{rabbits}}/_{\text{month}}$ as the associated units. If someone suggested a term that was not "rabbits per month", then that term would be wrong and should be immediately rejected. Thus, for example,

$$\frac{dR}{dt} = Rt$$

is clearly wrong because the right side is describing *rabbits* × *months*, not $^{rabbits}/_{month}$.

The constants used in our derivations also have associated units. The monthly birth rate per rabbit,

$$\begin{aligned} \beta &= \text{number of rabbits born per month per rabbit on the ranch} \\ &= \left(\frac{\text{number of } \textit{rabbits} \text{ born}}{\textit{month}} \right) \text{ per } \textit{rabbit} \text{ on the ranch} \\ &= \frac{\text{number of } \textit{rabbits} \text{ born}}{\textit{month} \times \textit{rabbit}} \quad , \end{aligned}$$

has associated units $^{1}/_{\text{month}}$ (since the units of *rabbits* cancel out), and if we had wanted to be a bit more explicit, we would have written equation (11.4) on page 200 as

$$\beta = \frac{5}{4} \left(\frac{1}{\text{month}} \right)$$

instead of just

$$\beta = \frac{5}{4} \quad .$$

Often, you will not see the units being explicitly noted throughout the development and use of a model. There are several possible reasons for this:

1. If the formulas and equations are correctly developed, then the units naturally remain balanced. The modeler knows this and trusts their skill in modeling.

2. The writer assumes the readers can keep track of the units themselves.

3. The writer is lazy or needs to save space.

There is much to be said in favor of explicitly giving the units associated with every element of every formula and equation. It helps prevent stupid mistakes and may help clarify the meaning of some of the formulas for the reader (and for the model builder). We will do this, somewhat, in the next major example. Beginning modelers are strongly encouraged to keep track of the units in every step as they develop their own equations. At the very least, stop every so often and check that the units in the equations balance. If not, you did something wrong — go back, find your error, and correct it.

Oh yes, one more thing about "units": Be sure that anyone who is going to read your work or use any model you've developed knows the units you are using.[4]

Testing and Using the Model

Once you've developed a differential equation modeling the way a quantity of interest $Q(t)$ varies, you will normally want to solve that differential equation or otherwise analyze it to see what it says about $Q(t)$ for various choices of t. If the situation being modeled is fairly simple and straightforward, and your modeling skills are adequate, then your model can probably be trusted to give fairly accurate predictions.

In practice, it is usually wise to check and see if predictions based on this model are reasonable before announcing your new model to the world. After all, it is quite possible that some of your 'simplifications' and 'assumptions' overly simplified your model and caused important issues to be ignored. That certainly happened with our first rabbit ranch model in which assuming constant birth and death rates resulted in a model predicting far more rabbits in five years than possible.

If your predictions are not reasonable, go back, revisit your derivations, and see where a more careful modeling of the individual processes leads. If necessary, learn more about the processes themselves.[5] This should lead to a refined model for $Q(t)$ that, in turn, leads to more reasonable projections as to the behavior of $Q(t)$. The differential equation will probably be more complicated, but that is the price you pay for a better, more accurate model.

Of course, you should not automatically assume that 'apparently reasonable' predictions are accurate. If possible, compare results predicted by the model to real-world data. You may need to do this anyway to determine the values of some of the constants in your model. Hopefully, the results predicted and the real-world data will agree well enough that you can feel confident that your model is sufficiently accurate for the desired applications. If not, refine your model further.

By the way, in using your model, keep in mind the simplifications and assumptions made in deriving it so that you have some idea as to the limitations of this model.

[4] In 1999, the Mars Climate Orbiter crashed into Mars instead of orbiting the planet because the Orbiter's software gave instructions in terms of the imperial system (which measures force in pounds) while the hardware assumed the metric system (which measures force in newtons – with 1 pound $\approx$ 4.45 newtons). This failure to communicate the units being used caused an embarrassing end to a space project costing more than $300 million.

[5] This author once read a paper describing a 'new' model for "laser interaction with a solid material". Using this model, you could then show that any solid can be chilled to absolute zero by suitably heating it with a laser — a rather dubious result. That paper's author should have better tested his model and learned more about thermodynamics.

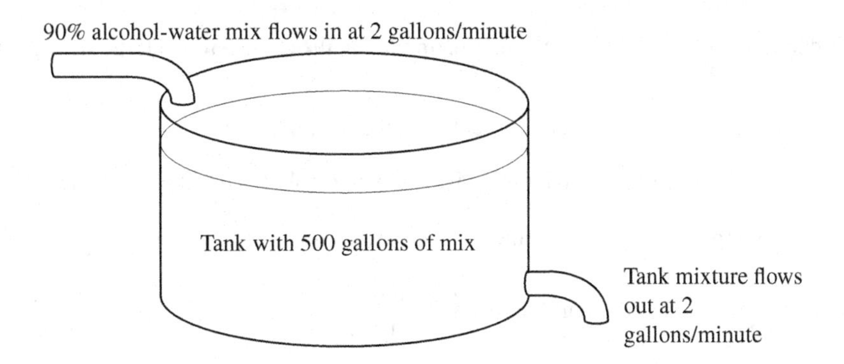

Figure 11.2: Figure illustrating a simple mixing problem.

11.6 Mixing Problems

In a "mixing problem", some substance is continually being added to some container in which the substance is mixed with some other material, and with the resulting mixture being constantly drained off at some rate. This container may be a large tank, a lake or the system of veins and arteries in a body; and the substance being added may be some chemical, pollutant or medicine being added to the liquid in the tank, the water in the lake or the blood in the body. These problems are favorites of authors of differential equation texts because they can be modeled fairly easily using the basic observation that (usually)

$$\begin{aligned} &\text{the rate the amount of substance in the container changes} \\ &\qquad = \text{the rate the substance is added} - \text{the rate the substance is drained off} \quad . \end{aligned}$$

We will do one simple mixing problem, and then briefly mention some possible variations.

A Simple Mixing Problem

The Situation to Be Modeled

We start out with a large tank containing 500 gallons of pure water. Each minute thereafter, two gallons of an alcohol-water mix are added, and two gallons of the mixture in the tank are drained. The alcohol-water mix being added is 90 percent alcohol. Throughout this entire process, we assume the mixture in the tank is thoroughly and uniformly mixed. The problem is to develop a formula describing the amount of alcohol in the tank at any given time. In particular, let's determine if and when the mixture in the tank is 50 percent alcohol.

Setting Up the Model

In this case, a simple, illustrative picture for the process is easily drawn. Just see Figure 11.2. We will let

$$t = \text{number of minutes since we started adding the alcohol-water mix}$$

and

$$y = y(t) = \text{gallons of pure alcohol in the tank at time } t \quad .$$

Since we started with a tank containing pure water (no alcohol), the initial condition is

$$y(0) = 0 \quad .$$

Our derivation of the differential equation modeling the change in y starts with the observation that

$$\begin{aligned}\frac{dy}{dt} &= \text{change in the amount of alcohol in the tank per minute}\\ &= \text{rate alcohol is added to the tank} - \text{rate alcohol is drained from the tank} \quad .\end{aligned}$$

Since we are adding 2 gallons per minute of a 90 percent alcohol mix,

$$\begin{aligned}\text{rate alcohol is added to the tank} &= 2\left(\frac{\text{gallons of input mix}}{\text{minute}}\right) \times \frac{90}{100}\left(\frac{\text{gallons of alcohol}}{\text{gallons of input mix}}\right)\\ &= \frac{9}{5}\left(\frac{\text{gallons of alcohol}}{\text{minute}}\right) \quad .\end{aligned}$$

In determining how much is being drained away, we must determine the concentration of alcohol in the tank's mixture at any given time, which is simply the total amount of alcohol in the tank at that time (i.e., $y(t)$ gallons) divided by the total amount of the mixture in the tank (which, because we drain off as much as we add, remains constant at 500 gallons). So,

$$\begin{aligned}\text{rate alcohol is drained from the tank} &= 2\left(\frac{\text{gallons of tank mix}}{\text{minute}}\right)\\ &\quad \times \text{amount of alcohol per gallon of tank mix}\\ &= 2\left(\frac{\text{gallons of tank mix}}{\text{minute}}\right) \times \frac{y(t)\,(\text{gallons of alcohol})}{500\,(\text{gallons of tank mix})}\\ &= \frac{y(t)}{250}\left(\frac{\text{gallons of alcohol}}{\text{minute}}\right) \quad .\end{aligned}$$

Combining the above gives us

$$\begin{aligned}\frac{dy}{dt} &= \text{rate alcohol is added to the tank} - \text{rate alcohol is drained from the tank}\\ &= \frac{9}{5} - \frac{y(t)}{250} \quad \left(\frac{\text{gallons of alcohol}}{\text{minute}}\right) \quad .\end{aligned}$$

Thus, the initial-value problem that $y = y(t)$ must satisfy is

$$\frac{dy}{dt} = \frac{9}{5} - \frac{y}{250} \quad \text{with} \quad y(0) = 0 \quad . \tag{11.15}$$

Using the Model

Factoring out $1/250$ on the right side of our differential equation yields

$$\frac{dy}{dt} = \frac{1}{250}(450 - y) \quad .$$

From this we see that

$$y = 450$$

is the only constant solution. Moreover,

$$\frac{dy}{dt} = \frac{1}{250}(450 - y) > 0 \qquad \text{if} \quad y < 450$$

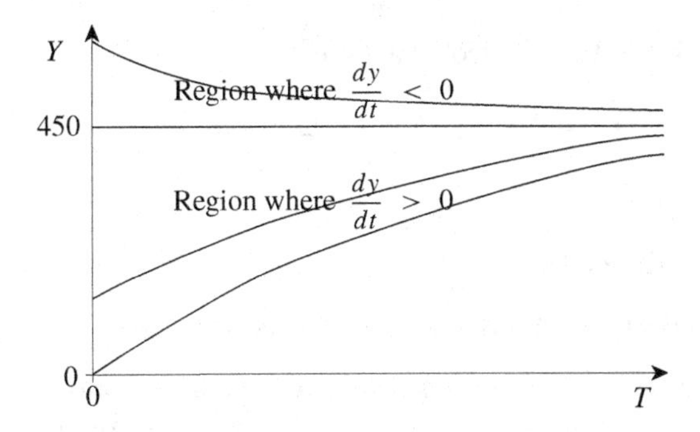

Figure 11.3: Crude graphs of solutions to the simple mixing problem from Figure 11.2 based on the sign of $^{dy}/_{dt}$.

and

$$\frac{dy}{dt} = \frac{1}{250}(450 - y) < 0 \quad \text{if} \quad y > 450 \quad .$$

So, we should expect the graphs of the possible solutions to this differential equation to be something like the curves in Figure 11.3. In other words, no matter what the initial condition is, we should expect $y(t)$ to approach 450 as $t \to \infty$.

Fortunately, the differential equation at hand is fairly simple. It (the differential equation in initial-value problem (11.15)) is both separable and linear, and, using either the method we developed for separable equations or the method we developed for linear equations, you can easily show that

$$y(t) = 450 - Ae^{-t/250}$$

is the general solution. Note that, as $t \to \infty$,

$$y(t) = 450 - Ae^{-t/250} \to 450 - A \cdot 0 = 450 \quad ,$$

just as Figure 11.3 suggests. Consequently, no matter how much alcohol is originally in the tank, eventually there will be nearly 450 gallons of alcohol in the tank. Since the tank holds 500 gallons of mix, the concentration of the alcohol in the mix will eventually be nearly $^{450}/_{500} = {}^{9}/_{10}$ (i.e., 90 percent of the liquid in the tank will be alcohol).

For our particular problem, $y(0) = 0$. So,

$$0 = y(0) = 450 - Ae^{-0/250} = 450 - A \quad .$$

Hence, $A = 450$ and

$$y(t) = 450 - 450e^{-t/250} \quad .$$

Finally, recall that we wanted to know when the mixture in the tank is 50 percent alcohol. This will be the time when half the liquid in the tank (i.e., 250 gallons) is alcohol. Letting τ denote this time, we must have

$$250 = y(\tau) = 450 - 450e^{-\tau/250}$$

$$\hookrightarrow \quad 450e^{-\tau/250} = 450 - 250$$

$$\hookrightarrow \quad e^{-\tau/250} = \frac{200}{450} = \frac{4}{9}$$

$$\hookrightarrow \quad -\frac{\tau}{250} = \ln\left(\frac{4}{9}\right) = -\ln\left(\frac{9}{4}\right) \quad .$$

So the mixture in the tank will be half alcohol at time

$$\tau = 250 \ln\left(\frac{9}{4}\right) \approx 202.7 \text{ (minutes)} \quad .$$

Other Mixing Problems

All sorts of variations of the problem just discussed can be visualized:

1. Instead of adding an alcohol-water mix, we may be adding a mixture of so many ounces of some chemical (such as sugar or salt) dissolved in the water (or other solvent).
2. The flow rate into the tank may be different from the drainage flow rate. In this case, the volume of the mixture in the tank will be changing, and that will affect how the concentration in the tank is computed.
3. We may have the problem considered in our simple mixing problem, but with some of the drained flow being diverted to a machine that magically converts a certain fraction of the alcohol to water, and the flow from that machine being dumped back into the tank. (Think of that machine as the tank's 'liver'.)
4. Instead of adding an alcohol-water mix, we may be adding a certain quantity of some microorganism (yeast, E.coli bacteria, etc.) in a nutrient solution. Then we would have to consider a mixture/population dynamics model to also account for the growth of the microorganism in the tank, as well as the in-flow and drainage.
5. And so on ... .

11.7 Simple Thermodynamics

Bring a hot cup of coffee into a cool room, and, in time, the coffee cools down to room temperature. Put a similar hot cup of coffee into a refrigerator, and you will discover that the coffee cools down faster. Let's try to describe this cooling process a little more precisely.

To be a little more general, let us simply assume we have some object (such as a hot cup of coffee or a cold glass of water) that we place in a room in which the air is at temperature T_{room}. To keep matters simple, assume T_{room} remains constant. Let $T = T(t)$ be the temperature at time t of the object we placed in the room. As time t goes on, we expect T to approach T_{room}. Now consider

$$\frac{dT}{dt} = \text{rate at which } T \text{ approaches } T_{\text{room}} \text{ as time } t \text{ increases} \quad .$$

It should seem reasonable that this rate at any instant of time t depends just on the difference between the temperature of the object and the temperature of the room, $T - T_{\text{room}}$; that is

$$\frac{dT}{dt} = F(T - T_{\text{room}}) \quad . \tag{11.16}$$

for some function F. Moreover, we can make the following observations:

1. If $T - T_{\text{room}} = 0$, then the object is the same temperature as the room. In this case, we do not expect the object's temperature to change. Hence, we should have ${}^{dT}/_{dt} = 0$ when $T = T_{\text{room}}$.

2. If $T - T_{\text{room}}$ is a large positive value, then the object is much warmer than the room. We then expect the object to be rapidly cooling; that is, T should be a rapidly decreasing function of t. Hence $^{dT}/_{dt}$ should be large and negative.

3. If $T - T_{\text{room}}$ is a large negative value, then the object is much cooler than the room. We then expect the object to be rapidly warming; that is, T should be a rapidly increasing function of t. Hence $^{dT}/_{dt}$ should be large and positive.

In terms of the function F on the right side of equation (11.16), these three observations mean

$$T - T_{\text{room}} = 0 \quad \Longrightarrow \quad F(T - T_{\text{room}}) = 0 \quad ,$$

$$T - T_{\text{room}} \text{ is a large positive value} \quad \Longrightarrow \quad F(T - T_{\text{room}}) \text{ is a large negative value}$$

and

$$T - T_{\text{room}} \text{ is a large negative value} \quad \Longrightarrow \quad F(T - T_{\text{room}}) \text{ is a large positive value} \quad .$$

The simplest choice of F satisfying these three conditions is

$$F(T - T_{\text{room}}) = -\kappa(T - T_{\text{room}})$$

where κ is some positive constant. Plugging this into equation (11.16) yields

$$\frac{dT}{dt} = -\kappa(T - T_{\text{room}}) \quad . \tag{11.17}$$

This equation is often known as *Newton's law of heating and cooling*. The positive constant κ describes how easily heat flows between the object and the air, and must be determined by experiment.

Equation (11.17) states that the change in the temperature of the object is proportional to the difference in the temperatures of the object and the room. It's not exactly the same as equation (11.8) on page 202 (unless $T_{\text{room}} = 0$), but it is quite similar in spirit. We'll leave its solution and further discussion as exercises for the reader.

Additional Exercises

11.2. *Do the following using formula (11.7) on page 201 from the simple model for the rabbit population on our rabbit ranch:*

a. *Find the approximate number of rabbits on the ranch after one year.*

b. *How long does it take for the number of rabbits to increase*

i. *from* 2 *to* 4 ? **ii.** *from* 4 *to* 8 ? **iii.** *from* 8 *to* 16 ?

c. *How long does it take for the number of rabbits to increase*

i. *from* 2 *to* 20 ? **ii.** *from* 5 *to* 50 ? **iii.** *from* 10 *to* 100 ?

d. *Approximately how long does it take for the mass of the rabbits on the ranch to equal the mass of the Earth?*

11.3. (Epidemiology) *Imagine the following situation:*

> *A stranger infected with a particularly contagious strain of the sniffles enters a city. Let $I(t)$ be the number of people in the city infected with the sniffles t days after the stranger entered the city. Assume that only the stranger has the sniffles on day 0, and that the number of people with the sniffles increases exponentially thereafter (as derived in the simple population growth model in Section 11.3). Assume further that 50 people have the sniffles on the tenth day after the stranger entered the city.*

Let $I(t)$ be the number of people in the city with sniffles on day t.

a. *What is the formula for $I(t)$?*

b. *How many people have the sniffles on day 20 ?*

c. *Approximately how long until 250,000 people in the city have the sniffles?*

11.4. *Assume that $A(t) = A_0 e^{-\delta t}$ is the amount of some radioactive substance at time t having a half-life $\tau_{1/2}$.*

a. *Verify that, for each value of t (not just $t = 0$),*

$$A(t + \tau_{1/2}) = \frac{1}{2} A(t) \quad .$$

b. *Verify that the formula $A(t) = A_0 e^{-\delta t}$ can be rewritten as*

$$A(t) = A_0 \left(\frac{1}{2}\right)^{t/\tau_{1/2}} \quad .$$

11.5. *Cesium-137 is a radioactive isotope of cesium with a half-life of about 30 years.*

a. *Find the corresponding decay constant δ for cesium-137.*

b. *Suppose we have a bottle (which we never open) containing 20 grams of cesium-137. Approximately how many grams of cesium-137 will still be in the bottle*

i. *after 10 years?* **ii.** *after 25 years?* **iii.** *after 100 years?*

11.6. (Carbon-14 dating) *A little background:*

> *Most of the carbon in living tissue comes, directly or indirectly, from the carbon dioxide in the air. A tiny fraction (about one part per trillion) of this carbon is the radioactive isotope carbon-14 (which has a half-life of approximately 5,730 years). The rest of the carbon is not radioactive. As a result, about one trillionth of the carbon in the tissues of a living plant or animal is that radioactive form of carbon. This ratio of carbon-14 to nonradioactive carbon in the air and living tissue has remained fairly constant[6] because the rate at which carbon-14 is created (through an interaction of cosmic radiation with the nitrogen in the upper atmosphere) matches the rate at which it decays.*
>
> *At death, however, the plant or animal stops absorbing carbon, and the tiny amount of carbon-14 in its tissues begins to decrease due to radioactive decay. By measuring the current ratio of carbon-14 to the nonradioactive carbon in a tissue sample (say, a piece of old bone or wood), and then comparing this ratio to the*

[6] but not perfectly constant — see a good article on carbon-14 dating.

ratio in comparable living tissue, a good estimate of the fraction of the carbon-14 that has decayed can be made. Using that and our model for radioactive decay, the age of the bone or wood can then be approximated.

Using the above information:

a. *Find the (approximate) decay constant* δ *for carbon-14.*

b. *Suppose a piece of wood came from a tree that died* t *years ago. Approximately what percentage of the carbon-14 that was in the piece of wood when the tree died still remains undecayed if*

i. $t = 10$ *years?* **ii.** $t = 100$ *years?* **iii.** $t = 1{,}000$ *years?*

iv. $t = 5{,}000$ *years?* **v.** $t = 10{,}000$ *years?* **vi.** $t = 50{,}000$ *years?*

c. *Suppose a skeleton of a person found in an ancient grave contains* 30 *percent of the carbon-14 normally found in (equally sized) skeletons of living people. Approximately how long ago did this person die?*

d. *The wood in the ornate funeral mask of the ancient fictional ruler Rootietootiekoomin is found to contain* 60 *percent of the carbon-14 originally in the wood. Approximately how long ago did Rootietootiekoomin die?*

e. *Let* A *be the amount of carbon-14 measured in a tissue sample (e.g., an old bone or piece of wood), and let* A_0 *be the amount of carbon-14 in the tissue when the plant or creature died. Derive a formula for the approximate length of time since that plant's or creature's demise in terms of the ratio* A/A_0 *.*

11.7. *Consider the "better model" for the rabbit population in Section 11.4.*

a. *Solve the logistic equation derived there (equation (11.13) on page 206), and verify that the solution can be written as given in formula (11.14) on page 207.*

b. *Assume the same values for the initial number of rabbits and ideal birth rate as assumed in Section 11.4,*

$$R(0) = 2 \qquad \text{and} \qquad \beta_0 = \frac{5}{4} \quad .$$

Also assume that our rabbit ranch has a carrying capacity κ *of* 10,000,000 *rabbits (it's a big ranch). How many rabbits (approximately) does our "better model" predict will be on our ranch*

i. *at the end of the first* 6 *months?*

ii. *at the end of the first year? (Compare this to the number predicted by the simple model in Exercise 11.2 a, and to the carrying capacity.)*

iii. *at the end of the second year? (Compare this to the carrying capacity.)*

c. *Solve formula (11.14) on page 207 for the carrying capacity* κ *in terms of* R_0 *,* $R(t)$ *,* β *and* t *.*

d. *Using the formula for the carrying capacity just derived (and assuming the ideal birth rate* $\beta_0 = 5/4$ *, as before), determine the approximate carrying capacity of a rabbit ranch under each of the following conditions:*

i. *You have* 1,000 *rabbits* 6 *months after starting with a single breeding pair.*

ii. *You have* 2,000 *rabbits* 6 *months after starting with a single breeding pair.*

11.8. *Suppose we have a rabbit ranch and have begun harvesting rabbits. Let*

$$R(t) = \text{number of rabbits on the ranch } t \text{ months after beginning harvesting}$$

and assume the following:

1. *The monthly birth rate per rabbit,* β *, is* $5/4$ *(as we derived).*

2. *We have no problems with overpopulation (i.e., for all practical purposes, we can assume the natural death rate is* 0 *).*

3. *Each month we harvest* 500 *rabbits. (Assume this is done "over the month", so the rabbits are still reproducing as we are harvesting.)*

a. *Derive the differential equation for* $R(t)$ *based on the above assumptions.*

b. *Find any equilibrium solutions to your differential equation (this may surprise you), and analyze how the rabbit population varies over time based on how many we had when we first began harvesting. (Feel free to use a crude slope field as done in Section 11.4.)*

c. *Solve the differential equation. Get your final answer in terms of* t *and* $R_0 = R(0)$ *.*

11.9. *Repeat the previous problem, only, instead of harvesting* 500 *rabbits a month, assume we harvest* 50 *percent of the rabbits on the ranch each month.*

11.10. *Again, assume we have a rabbit ranch, and let*

$$R(t) = \text{number of rabbits on the ranch after } t \text{ months.}$$

Taking into account the problems that arise when the population is too large, we obtained the differential equation

$$\frac{dR}{dt} = \beta R - \gamma R^2$$

where β *is the monthly birth rate per rabbit (which we figured was* $5/4$ *) and* γ *was some positive constant that would have to be determined later.*

This differential equation was obtained assuming we were not harvesting rabbits. Assume, instead, that we are harvesting h *rabbits each month. How do we change the above differential equation to reflect this if*

a. *we harvest a constant number* h_0 *of rabbits each month?*

b. *we harvest one fourth of all the rabbits on the ranch each month?*

11.11. *Consider the following situation:*

> *Mullock the Barbarian begins a campaign of self-enrichment with a horde of 200 vicious warriors. Each week he loses 5 percent of his horde to the unavoidable accidents that occur while sacking and pillaging. Fortunately (for Mullock), the horde's lifestyle of wanton violence and mindless destruction attracts 50 new warriors to the horde each week.*

Let $y(t)$ *be the number of warriors in Mullock's horde* t *weeks after starting the campaign.*

a. *Derive the differential equation describing how* $y(t)$ *changes each week. Is there also an initial value given?*

b. *To what size does the horde eventually grow? (Don't solve the initial-value problem to answer this question. Instead, use equilibrium solutions and graphical methods.)*

c. *Now solve the initial-value problem from the first part.*

d. *How long does it take Mullock's horde to reach* 90 *percent of its final size?*

11.12. (Mixing) *Consider the following mixing problem:*

> *We have a large tank initially containing 1,000 gallons of pure water. We begin adding an alcohol-water mix at a rate of 3 gallons per minute. This alcohol-water mix being added is 75 percent alcohol. At the same time, the mixture in the tank is drained at a rate of 3 gallons per minute. Throughout this entire process, the mixture in the tank is thoroughly and uniformly mixed.*

Let $y(t)$ *be the number of gallons of pure alcohol in the tank* t *minutes after we started adding the alcohol-water mix.*

a. *Find the differential equation for* $y(t)$.

b. *Sketch a crude slope field for the differential equation just obtained, and find any equilibrium solutions.*

c. *Using the differential equation just obtained, find the formula for* $y(t)$.

d. *Approximately how many gallons of alcohol are in the tank at*

i. $t = 10$? **ii.** $t = 60$? **iii.** $t = 1000$?

e. *When will the mixture in the tank be half alcohol?*

11.13. *Redo Exercise 11.12, but assume the tank initially contains* 900 *gallons of pure water and* 100 *gallons of alcohol.*

11.14. *Consider the following mixing problem:*

> *We have a tank initially containing 200 gallons of pure water, and start adding saltwater (containing 3 ounces of salt per gallon of water) at the rate of* $\frac{1}{2}$ *gallon per minute. At the same time, the resulting mixture in the tank is drained at the rate of* $\frac{1}{2}$ *gallon per minute. As usual, the mixture in the tank is thoroughly and uniformly mixed at all times.*

Let $y(t)$ *be the number of ounces of salt in the tank at* t *minutes after we started adding the saltwater.*

a. *Find the differential equation for* $y(t)$.

b. *Sketch a crude slope field for the differential equation just obtained, and find any equilibrium solutions.*

c. *Using the differential equation just obtained along with any given initial values, find the formula for* $y(t)$.

d. *Approximately how many ounces of salt are in the tank at*

i. $t = 10$? **ii.** $t = 60$? **iii.** $t = 100$?

e. *What does the concentration of the salt in the tank approach as* $t \to \infty$?

f. *When will the concentration of the salt in the tank be* 2 *ounces of salt per gallon of water?*

11.15. *Redo Exercise 11.14, but assume that a device has been attached to the tank that, each minute, filters out half the salt in a single gallon from the mixture in the tank.*

11.16. *Consider the following variation of the mixing problem in Exercise 11.12:*

> *We have a large tank initially containing 500 gallons of pure water, and start adding saltwater (containing 2 ounces of salt per gallon of water) at the rate of 2 gallons per minute. At the same time, the resulting mixture in the tank is drained at the rate of 3 gallons per minute. As usual, assume the mixture in the tank is thoroughly and uniformly mixed at all times.*

Note that the tank is being drained faster that it is being filled.

Let $y(t)$ be the number of ounces of salt in the tank at t minutes after we started adding the saltwater.

a. *What is the formula for the volume of the liquid in the tank t minutes after we started adding the saltwater?*

b. *Find the differential equation for $y(t)$. (Keep in mind that the concentration of salt in the outflow at time t will depend on both the amount of salt and the volume of the liquid in the tank at that time.)*

c. *Using the differential equation just obtained along with any given initial values, find the formula for $y(t)$.*

d. *How many ounces of salt are in the tank at*

i. $t = 10$? **ii.** $t = 60$? **iii.** $t = 100$?

e i. *When will there be exactly 1 gallon of saltwater in the tank?*

ii. *How much salt will be in that gallon of saltwater?*

11.17. (Heating/cooling) *Consider the following situation:*

> *At 2 o'clock in the afternoon, the butler reported discovering the dead body of his master, Lord Hakky d'Sack, in the Lord's personal wine cellar. The Lord had apparently been bludgeoned to death with a bottle of* Rip'le 04. *At 4 o'clock, the forensics expert arrived and measured the temperature of the body. It was 90 degrees at that time. One hour later, the body had cooled down to 80 degrees. It was also noted that the wine cellar was maintained at a constant temperature of 50 degrees.*

Should the butler be arrested for murder? (Base your answer on the time of death as determined from the above information, Newton's law of heating and cooling and the fact that a reasonably healthy person's body temperature is about 98.2 degrees.)

12

Numerical Methods II: Beyond the Euler Method

The Euler method developed in Chapter 10 is a good starting point for those learning about numerical methods for differential equations, but is a poor ending point for those computing numerical solutions for initial-value problems arising in actual applications. In this chapter, we will remedy that situation by expanding our repertoire of numerical methods. Our main goals will be to develop the "improved Euler method" and the "classic Runge-Kutta method", both of which typically deliver much more accurate approximations, and both of which are used much more often in real-world applications than the Euler method. It should be noted that this improved accuracy comes with a cost of requiring a greater number of computations per step. This cost, however, is relatively modest, especially when the computations are carried out by a reasonably well-programmed computer.

Before starting, let's recall just what we are talking about. A numerical solution for a first-order initial-value problem

$$\frac{dy}{dx} = f(x, y) \qquad \text{with} \quad y(x_0) = y_0$$

consists of a pair of sequences

$$\{ x_0 , x_1 , x_2 , x_3 , \ldots , x_N \} \qquad \text{and} \qquad \{ y_0 , y_1 , y_2 , y_3 , \ldots , y_N \}$$

where x_0 and y_0 are given by the initial data, and with the other x_k's and y_k's computed so that $y_k \approx y(x_k)$ for each k, with $y = y(x)$ being the (possibly unknown) solution to the initial-value problem. In all of the discussion that follows, Δx is some positive distance (the *step size*), and the x's are related by $x_{k+1} = x_k + \Delta x$.[1]

12.1 Forward and Backward Euler Methods

Before looking at improving the Euler method for numerically solving

$$\frac{dy}{dx} = f(x, y) \qquad \text{with} \quad y(x_0) = y_0 \quad , \tag{12.1}$$

we should note that there are actually two "basic" Euler methods for numerically solving this initial-value problem. The one developed in Chapter 10 started with the approximation

$$y'(x) \approx \frac{y(x + \Delta x) - y(x)}{\Delta x} \quad ,$$

[1] Though we will not discuss them, there are methods in which the step size is allowed to vary between the x_k's.

which, combined with the differential equation, yielded the approximation

$$y(x + \Delta x) \approx y(x) + \Delta x \cdot f(x, y(x))$$

which, in turn, led to the iterative formula

$$y_{k+1} = y_k + \Delta x \cdot f(x_k, y_k)$$

for computing y_{k+1} from y_k. This is the standard Euler method and the one usually meant when the term "Euler method" is used. However, on rare occasions it is also called the *forward Euler method* to distinguish from the *backward Euler method.*

The backward Euler method starts with the approximation

$$y'(x) \approx \frac{y(x) - y(x - \Delta x)}{\Delta x} \quad .$$

Combined with the differential equation, this becomes

$$f(x, y(x)) \approx \frac{y(x) - y(x - \Delta x)}{\Delta x} \quad .$$

Solving for $y(x - \Delta x)$ then yields

$$y(x - \Delta x) \approx y(x) - \Delta x \cdot f(x, y(x)) \quad ,$$

which, in turn, leads to the equivalent iterative formulas

$$y_{k-1} = y_k - \Delta x \cdot f(x_k, y_k) \tag{12.2a}$$

and

$$y_k = y_{k-1} + \Delta x \cdot f(x_k, y_k) \quad . \tag{12.2b}$$

There are at least two ways of using these equations:

1. Equation (12.2a) can be treated as an explicit formula for computing y_{k-1} from y_k. This would be useful if, instead of having an initial-value problem, we have a *final-value problem* in which we know what the final value of y at some point in the future should be, and we wish to determine the initial value that results in that desired final value.
2. Either equation (12.2a) or (12.2b) can be treated as an equation implicitly defining y_k. "Simply" plug in the values for y_{k-1}, Δx and x_k, and then solve the equation for y_k. This, obviously, requires more work on the part of the user, and the practicality of the method depends on ease at which this equation can then be solved for y_k.

Both the forward and the backward Euler methods are based on simple tangent line approximations of the graph of $y = y(x)$ passing through (x, y). Because of this, the error in using the backward Euler method is basically the same as for the forward Euler method. That is:

> *If $y = y(x)$ is the true solution to either initial-value problem (12.1) on an interval $(x_0, x_0 + L)$ (or is the solution to the analogous final-value problem on the same interval), and if f is "reasonably smooth", then there is a constant M_E such that*
>
> $$|y(x_k) - y_k| < M_E \cdot \Delta x$$
>
> *whenever $k = 0, 1, \ldots, N$ and*
>
> $$\{ x_0 , x_1 , x_2 , x_3 , \ldots , x_N \} \qquad \text{and} \qquad \{ y_0 , y_1 , y_2 , y_3 , \ldots , y_N \}$$
>
> *is a numerical solution to initial-value problem (12.1) (or the analogous final-value problem) obtained from the forward (or backward) Euler method with step size Δx and total number of steps N satisfying*
>
> $$0 < \Delta x \cdot N \leq L \quad .$$

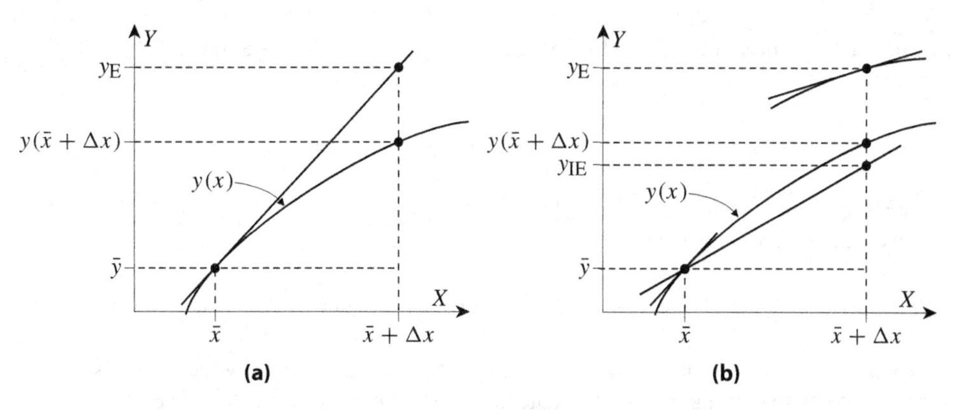

Figure 12.1: Two approximations, y_E and y_{IE}, of $y(\bar{x} + \Delta x)$ where y is the solution to $y' = f(x, y)$ passing through $(\bar{x}, \bar{y})$: In **(a)** (basic Euler approximation) y_E is on the line through $(\bar{x}, \bar{y})$ with slope $f(\bar{x}, \bar{y})$ (i.e., the tangent line). In **(b)** (improved Euler approximation) y_{IE} is on the line through $(\bar{x}, \bar{y})$ whose slope is the average of $f(\bar{x}, \bar{y})$ and $y' = f(\bar{x} + \Delta x, y_E)$.

Since our main interest in this chapter is in methods having error bounds that shrink to zero as $\Delta x \to 0$ faster than the error bound for the Euler method, let us not spend any more time on the backward Euler method, and proceed to an actual improvement on that method. Be aware that, even if we do not mention it later, these methods will have analogous "backward" variations.

Let us also agree that, in future sections, "the Euler method" refers to the forward Euler method.

12.2 The Improved Euler Method

Derivation

The Euler method developed in Chapter 10 is based on the approximation

$$y(\bar{x} + \Delta x) \approx y_E \tag{12.3a}$$

where

$$y_E = \bar{y} + \Delta x \cdot f(\bar{x}, \bar{y}) \tag{12.3b}$$

and $y = y(x)$ is the solution to

$$\frac{dy}{dx} = f(x, y) \quad \text{with} \quad y(\bar{x}) = \bar{y} \quad . \tag{12.3c}$$

This approximation, y_E, came from the simple tangent line approximation illustrated in Figure 12.1a. As can be seen from this figure, this tangent line is likely to overshoot or undershoot the graph of y at $\bar{x} + \Delta x$ by a small but significant amount, meaning that y_E is likely to be a good, but not outstanding, approximation for $y(\bar{x} + \Delta x)$.

Looking at Figure 12.1a, it should be clear that a better approximation probably would be given by a straight line through $(\bar{x}, \bar{y})$ but whose slope S is some "average" of the slopes of tangent lines to $y = y(x)$ over the interval $[\bar{x}, \bar{x} + \Delta x]$, say, the average of the slopes found at the two end points

$$S = \text{Average Slope} = \frac{1}{2}\left[\text{slope at } (\bar{x}, \bar{y}) + \text{slope at } (\bar{x} + \Delta x, y(\bar{x} + \Delta x))\right] \quad .$$

Keeping in mind that the slope is given by the derivative of y and that y is a solution to $y' = f(x, y)$, we see that, in fact,

$$S = \frac{1}{2}\left[y'(\bar{x}) + y'(\bar{x}+\Delta x)\right] = \frac{1}{2}\left[f(\bar{x},\bar{y}) + f(\bar{x}+\Delta x, y(\bar{x}+\Delta x))\right] \quad .$$

Using the straight line through $(\bar{x}, \bar{y})$ with this slope gives us the fundamental approximation for the improved Euler method,

$$y(\bar{x}+\Delta x) \approx \bar{y} + \Delta x \cdot S = \bar{y} + \Delta x \cdot \frac{1}{2}\left[f(\bar{x},\bar{y}) + f(\bar{x}+\Delta x, y(\bar{x}+\Delta x))\right] \quad . \tag{12.4}$$

Unfortunately, the right side of this approximation involves $y(\bar{x}+\Delta x)$, which is not known (otherwise, why would be trying to approximate it?). The best we have at this point is the Euler approximation

$$y(\bar{x}+\Delta x) \approx y_{\text{E}}$$

from equation set (12.3). But then, assuming f is a reasonably smooth function of x and y, we should also have

$$f(\bar{x}+\Delta x, y(\bar{x}+\Delta x)) \approx f(\bar{x}+\Delta x, y_{\text{E}}) \quad .$$

Combining this with approximation (12.4) then yields our improved Euler approximation for solutions to $y' = f(x, y)$:

$$y(\bar{x}+\Delta x) \approx y_{\text{IE}} \tag{12.5a}$$

where

$$y_{\text{IE}} = \bar{y} + \Delta x \cdot \frac{1}{2}\left[f(\bar{x},\bar{y}) + f(\bar{x}+\Delta x, y_{\text{E}})\right] \tag{12.5b}$$

with

$$y_{\text{E}} = \bar{y} + \Delta x \cdot f(\bar{x},\bar{y}) \quad . \tag{12.5c}$$

This approximation is illustrated in Figure 12.1b, which certainly suggests that, typically, y_{IE} is a significantly better approximation to $y(\bar{x}+\Delta x)$ than the first approximation, y_{E}.

The improved Euler method is based on the above approximation for computing $y(x+\Delta x)$. It generates a numerical solution to the initial-value problem

$$\frac{dy}{dx} = f(x, y) \qquad \text{with} \quad y(x_0) = y_0$$

consisting of a pair of sequences

$$\{x_0, x_1, x_2, x_3, \ldots, x_N\} \qquad \text{and} \qquad \{y_0, y_1, y_2, y_3, \ldots, y_N\}$$

with

$$y_0 = y(x_0)$$

and the subsequent x_k's and y_k's computed successively by

$$x_{k+1} = x_k + \Delta x \tag{12.6a}$$

and

$$y_{k+1} = y_k + \Delta x \cdot \frac{1}{2}\left[f(x_k, y_k) + f(x_{k+1}, y_{\text{E},k})\right] \tag{12.6b}$$

where

$$y_{\text{E},k} = y_k + \Delta x \cdot f(x_k, y_k) \tag{12.6c}$$

and with the step size, Δx, being some prechosen distance.

Observe that equations (12.6b) and (12.6c) are equations (12.5b) and (12.5c) with minor changes in notation (and using the fact that $x_k + \Delta x = x_{k+1}$ in equation (12.6b)). Also note that, from approximation (12.5a), it easily follows that, for each k,

$$y_k \approx y(x_k) \quad .$$

Just how good an approximation this is will be discussed later.

Note also that each $y_{E,k}$ is only used to compute y_{k+1}. In any implementation of this method, we can replace these $y_{E,k}$'s with a single symbol, say, p, recomputing it with each new value of the index k.

Implementing the Improved Euler Method (Illustrated)

Let us outline the steps in using the improved Euler Method to numerically solve some first-order differential equation with initial data $y(x_0) = y_0$, say,

$$5\frac{dy}{dx} - y^2 = -x^2 \qquad \text{with} \quad y(0) = 1 \quad . \tag{12.7}$$

(This is the same problem used in Chapter 10 to illustrate the basic Euler method.)

As with the basic Euler method in Chapter 10, the implementation consists of two preliminary steps in which the constants and formulas to be used are determined, followed by the iterative steps in which the values of the x_k's and y_k's are computed.

Preliminary Steps

1. Rewrite the initial-value problem with the differential equation in derivative formula form,

$$\frac{dy}{dx} = f(x, y) \qquad \text{with} \quad y(x_0) = y_0 \quad ,$$

and, from this, determine the formula for $f(x, y)$ and the values of x_0 and y_0.

For our example, solving for the derivative formula form yields

$$\frac{dy}{dx} = \frac{1}{5}\left[y^2 - x^2\right] \qquad \textit{with} \quad y(0) = 1 \quad .$$

So,

$$f(x, y) = \frac{1}{5}\left[y^2 - x^2\right] \quad , \qquad x_0 = 0 \qquad \textit{and} \qquad y_0 = 1 \quad .$$

2. Decide on a distance Δx for the step size, a positive integer N for the maximum number of steps, and a maximum value desired for x, x_{max}. Choose these quantities so that

$$x_{max} = x_N = x_0 + N\Delta x \quad .$$

As with our illustration for the basic Euler method, let us pick

$$\Delta x = \frac{1}{2} \qquad \textit{and} \qquad N = 6 \quad .$$

Then

$$x_{max} = x_0 + N\Delta x = 0 + 6 \cdot \frac{1}{2} = 3 \quad .$$

Iterative Steps

1, 2, ... For each integer $k \geq 0$, successively compute x_{k+1}, p and y_{k+1} using the values of x_k and y_k found in a previous step and the following version of equation set (12.6):

$$x_{k+1} = x_k + \Delta x \quad ,$$

$$p = y_k + \Delta x \cdot f(x_k, y_k)$$

and

$$y_{k+1} = y_k + \Delta x \cdot \frac{1}{2}\left[f(x_k, y_k) + f(x_{k+1}, p)\right]$$

where $f(x, y)$, Δx, x_0 and y_0 are as determined in the preliminary steps.

Continue repeating these computations until x_N and y_N are computed. (Be sure to save the x_k and y_k values computed!)

For our example (omitting many computational details):

Step 1. With $k = 0$*:*

$$x_1 = x_{0+1} = x_0 + \Delta x = 0 + \frac{1}{2} = \frac{1}{2} \quad ,$$

$$\begin{aligned} p &= y_0 + \Delta x \cdot f(x_0, y_0) \\ &= 1 + \frac{1}{2} \cdot f(0, 1) \\ &= 1 + \frac{1}{2} \cdot \frac{1}{5}\left[1^2 - 0^2\right] = \frac{11}{10} \end{aligned}$$

and

$$\begin{aligned} y_1 = y_{0+1} &= y_0 + \Delta x \cdot \frac{1}{2}[f(x_0, y_0) + f(x_1, p)] \\ &= 1 + \frac{1}{2} \cdot \frac{1}{2}\left[f(0, 1) + f\left(\frac{1}{2}, \frac{11}{10}\right)\right] \\ &= 1 + \frac{1}{2} \cdot \frac{1}{2} \cdot \frac{1}{5}\left[1^2 - 0^2 + \left(\frac{11}{10}\right)^2 - \left(\frac{1}{2}\right)^2\right] \\ &= \frac{1{,}098}{1{,}000} = 1.098 \quad . \end{aligned}$$

Step 2. With $k = 1$*:*

$$x_2 = x_{1+1} = x_1 + \Delta x = \frac{1}{2} + \frac{1}{2} = 1 \quad ,$$

$$\begin{aligned} p &= y_1 + \Delta x \cdot f(x_1, y_1) \\ &= \frac{1{,}098}{1{,}000} + \frac{1}{2} \cdot f\left(\frac{1}{2}, \frac{1{,}098}{1{,}000}\right) \\ &= \frac{1{,}098}{1{,}000} + \frac{1}{2} \cdot \frac{1}{5}\left[\left(\frac{1{,}098}{1{,}000}\right)^2 - \left(\frac{1}{2}\right)^2\right] = \frac{11{,}935{,}604}{10{,}000{,}000} \end{aligned}$$

and

$$\begin{aligned} y_2 = y_{1+1} &= y_1 + \Delta x \cdot \frac{1}{2}[f(x_1, y_1) + f(x_2, p)] \\ &= \frac{1{,}098}{1{,}000} + \frac{1}{2} \cdot \frac{1}{2}\left[f\left(\frac{1}{2}, \frac{1{,}098}{1{,}000}\right) + f\left(1, \frac{11{,}935{,}604}{10{,}000{,}000}\right)\right] \\ &= \frac{1{,}098}{1{,}000} + \frac{1}{2} \cdot \frac{1}{2} \cdot \frac{1}{5}\left[\left(\frac{1{,}098}{1{,}000}\right)^2 - \left(\frac{1}{2}\right)^2 + \left(\frac{11{,}935{,}604}{10{,}000{,}000}\right)^2 - 1^2\right] \\ &= \frac{1{,}167{,}009{,}521}{1{,}000{,}000{,}000} = 1.167009\ldots \quad . \end{aligned}$$

k	t_k	y_k
0	0	1
1	0.5	1.0980
2	1.0	1.1670
3	1.5	1.1450
4	2.0	0.9533
5	2.5	0.5070
6	3.0	−0.2423

(a)

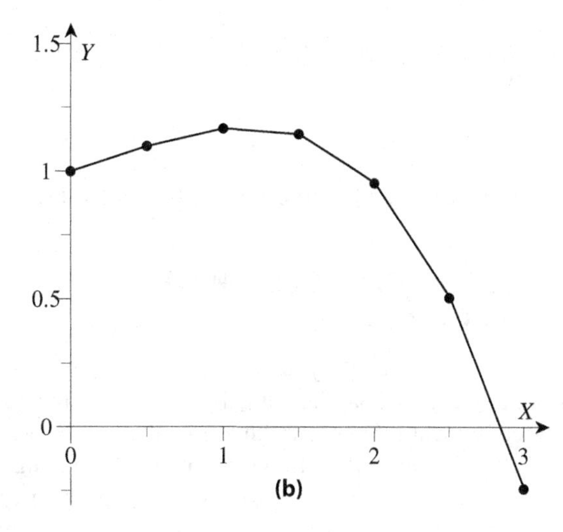

Figure 12.2: Results of using the improved Euler method to numerically solve initial-value problem (12.7) with $\Delta x = {}^1\!/_2$ and $N = 6$: **(a)** The numerical solution in which $y_k \approx y(x_k)$ (for $k \geq 2$, the values of y_k are to the nearest 0.0001). **(b)** The piecewise linear graph based on this numerical solution.

Step 3. With $k = 2$*:*

$$x_3 = x_{2+1} = x_2 + \Delta x = 1 + \frac{1}{2} = \frac{3}{2} \quad ,$$

$$\begin{aligned} p &= y_2 + \Delta x \cdot f(x_2, y_2) \\ &= \frac{1{,}167{,}009{,}521}{1{,}000{,}000{,}000} + \frac{1}{2} f\left(1, \frac{1{,}167{,}009{,}521}{1{,}000{,}000{,}000}\right) \\ &= \frac{1{,}167{,}009{,}521}{1{,}000{,}000{,}000} + \frac{1}{2} \cdot \frac{1}{5}\left[\left(\frac{1{,}167{,}009{,}521}{1{,}000{,}000{,}000}\right)^2 - 1^2\right] = \frac{1{,}203{,}200{,}643}{1{,}000{,}000{,}000} \end{aligned}$$

and

$$\begin{aligned} y_3 = y_{2+1} &= y_2 + \Delta x \cdot \frac{1}{2}\left[f(x_2, y_2) + f(x_3, p)\right] \\ &= \frac{1{,}167{,}009{,}521}{1{,}000{,}000{,}000} + \frac{1}{2} \cdot \frac{1}{2}\left[f\left(1, \frac{1{,}167{,}009{,}521}{1{,}000{,}000{,}000}\right) + f\left(\frac{3}{2}, \frac{1{,}203{,}200{,}643}{1{,}000{,}000{,}000}\right)\right] \\ &= \frac{1{,}167{,}009{,}521}{1{,}000{,}000{,}000} + \frac{1}{2} \cdot \frac{1}{2} \cdot \frac{1}{5}\left[\left(\frac{1{,}167{,}009{,}521}{1{,}000{,}000{,}000}\right)^2 - 1^2 \right. \\ &\qquad\qquad \left. + \left(\frac{1{,}203{,}200{,}643}{1{,}000{,}000{,}000}\right)^2 - \left(\frac{3}{2}\right)^2\right] \\ &= \frac{1{,}144{,}989{,}671}{1{,}000{,}000{,}000} = 1.144989\ldots \quad . \end{aligned}$$

For obvious reasons, the remaining steps were carried out using a computer program employing the above iterative procedure. The results can be seen in the table in Figure 12.2a.

As just indicated, the use of a computer to carry out the above computations is highly recommended, especially since most applications require much smaller values of Δx (and correspondingly larger values of N) than used in our example to reduce the errors in the computations. Either write your own program or set up a worksheet using your favorite computer math package. Be sure to be able to easily change

$$\Delta x \quad , \quad N \text{ or } x_{\max} \quad , \quad f(x,y) \quad , \quad x_0 \quad \text{and} \quad y_0$$

so you can easily change the parameters and consider different initial-value problems.[2]

Using the Results of the Method

Tables and Graphs

What you do with the results of your computations depends on why you are doing these computations. If N is not too large, it may be a good idea to write the obtained values of the x_k's and y_k's in a table for convenient reference (with a note that $y_k \approx y(x_k)$ for each k) as done in the table in Figure 12.2a for our example. It may also be enlightening to sketch (or have a computer sketch) the corresponding piecewise linear approximation of the graph of $y = y(x)$ as x goes from x_0 to x_N. Simply plot each (x_k, y_k) on an XY–plane and then connect the successive points with straight lines as in Figure 12.2b for our example.

General Observations on Accuracy

Keep in mind that each y_k generated by the improved Euler method is an approximation to the solution at x_k to some initial-value problem

$$\frac{dy}{dx} = f(x,y) \qquad \text{with} \quad y(x_0) = y_0 \quad . \tag{12.8}$$

There are two obvious questions regarding the accuracy of this method:

1. How much better is this method than the original Euler method developed in Chapter 10?

and

2. To what extent does shrinking the step size improve the result?

The answer to both questions is "significantly", and is illustrated in Figure 12.3.

It should be noted that, not only does the improved Euler method usually generate a more accurate numerical solution to our initial-value problem than the original Euler method, the accuracy of the improved Euler method increases faster than does the accuracy of the original Euler method as the step size, Δx, is decreased. In particular, by expanding the proof of Theorem 10.1 on page 188, it can be shown that:

If $y = y(x)$ is the true solution to initial-value problem (12.8) on an interval $(x_0, x_0 + L)$, and if f is "reasonably smooth", then there is a constant M_{IE} such that

$$|y(x_k) - y_k| < M_{\text{IE}} \cdot (\Delta x)^2$$

whenever $k = 0, 1, \ldots, N$ and

$$\{ x_0 \,,\, x_1 \,,\, x_2 \,,\, x_3 \,,\, \ldots \,,\, x_N \} \qquad \text{and} \qquad \{ y_0 \,,\, y_1 \,,\, y_2 \,,\, y_3 \,,\, \ldots \,,\, y_N \}$$

is a numerical solution to initial-value problem (12.8) obtained using the improved Euler method with step size Δx and total number of steps N satisfying

$$0 < \Delta x \cdot N \leq L \quad .$$

[2] In fact, your favorite computer math package may well already have some version of the above in its repertoire.

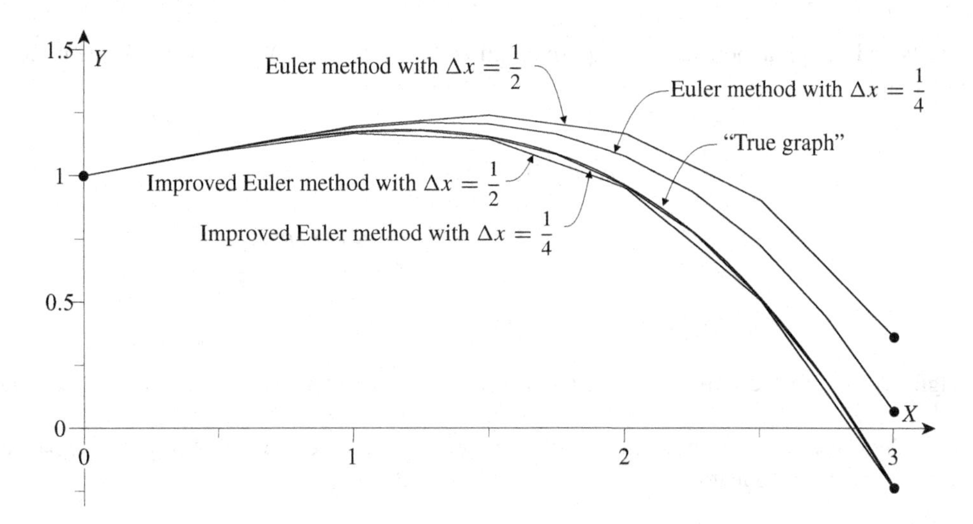

Figure 12.3: Four piecewise linear approximations to the true graph of the solution to initial-value problem (12.7) generated by the Euler and the improved Euler methods with step sizes $\Delta x = 1/2$ and $\Delta x = 1/4$. Also drawn is the graph of the true solution (generated by an extremely accurate approximation). Note how much better the graphs generated by the improved Euler method match the graph of the true solution — especially the improved Euler method with $\Delta x = 1/4$.

Recall that a similar statement was made for the original Euler method. There, however, the error bound was

$$|y(x_k) - y_k| < M_E \cdot \Delta x$$

for some constant M_E. So, with the original Euler method, we can expect to cut the error in half by reducing the step size by half. On the other hand, reducing the step size by half using the improved Euler method can be expected to reduce the error by $(\text{half})^2$ (i.e., one fourth).

The Improved Euler Method as a Single-Stage Predictor-Corrector Method, and Corresponding Multi-Stage Methods

Take another look at the iterative computations in the improved Euler method (equation set (12.6)):

$$x_{k+1} = x_k + \Delta x \quad ,$$

$$p = y_k + \Delta x \cdot f(x_k, y_k)$$

and

$$y_{k+1} = y_k + \Delta x \cdot \frac{1}{2}\left[f(x_k, y_k) + f(x_{k+1}, p)\right] \quad .$$

In this, p is a "prediction" of the value of $y(x_{k+1})$ based on the original Euler method, and this value is used to provide a "corrected" estimate of $y(x_{k+1})$ in the computation for y_{k+1} in the last line. Because of this, the improved Euler method is sometimes referred to as a *predictor-corrector* scheme. Occasionally, for reasons soon to be made clear, we might even refer to this as a *single-stage* improved Euler method (or a single-stage predictor-corrector method).

Of course, the above y_{k+1} is simply a second prediction of the value of $y(x_{k+1})$. So, instead of using the value computed above for y_{k+1}, we could use that value as a second prediction to provide what we hope is an even more accurate approximation y_{k+1} of $y(x_{k+1})$. This gives us the iterative computations:

$$x_{k+1} = x_k + \Delta x \quad ,$$

$$p_1 = y_k + \Delta x \cdot f(x_k, y_k) \quad ,$$

$$p_2 = y_k + \Delta x \cdot \frac{1}{2}\left[f(x_k, y_k) + f(x_{k+1}, p_1)\right]$$

and

$$y_{k+1} = y_k + \Delta x \cdot \frac{1}{2}\left[f(x_k, y_k) + f(x_{k+1}, p_2)\right] \quad .$$

We might refer to the resulting improved Euler/predictor-corrector method as being a *two-stage* method.

But why stop at just two "predictors"? The above set of iterative computations can be expanded to any number M of predictors:

$$x_{k+1} = x_k + \Delta x \quad ,$$

$$p_1 = y_k + \Delta x \cdot f(x_k, y_k) \quad ,$$

$$p_2 = y_k + \Delta x \cdot \frac{1}{2}\left[f(x_k, y_k) + f(x_{k+1}, p_1)\right]$$

$$\vdots$$

$$p_M = y_k + \Delta x \cdot \frac{1}{2}\left[f(x_k, y_k) + f(x_{k+1}, p_{M-1})\right]$$

and

$$y_{k+1} = y_k + \Delta x \cdot \frac{1}{2}\left[f(x_k, y_k) + f(x_{k+1}, p_M)\right] \quad .$$

This, as you surely now realize, would yield an *M-stage* improved Euler method.

In some applications, the value of M is not predetermined. Instead, at each step, y_{k+1} is recomputed using increasing values of M until the difference between y_{k+1} computed using M predictors and y_{k+1} computed using $M + 1$ predictors are within some very small, predetermined amount. The last computed value for y_{k+1} is then saved and used in the next step to find y_{k+2}.

12.3 A Few Other Methods Worth Brief Discussion

Because they play small roles in the next section on the Runge-Kutta method, let us briefly describe two more schemes for approximating $y(\bar{x} + \Delta x)$ where Δx is some small distance and $y = y(x)$ is the solution to

$$\frac{dy}{dx} = f(x, y) \qquad \text{with} \quad y(\bar{x}) = \bar{y} \quad .$$

Euler-Point Slope

This approach is illustrated in Figure 12.4a. In it, we first compute the "Euler point" $(\bar{x} + \Delta x, y_E)$ where y_E is the standard Euler method approximation to $y(\bar{x} + \Delta x)$. This is used as a prediction and the final approximation y_{EPS} to $y(\bar{x} + \Delta x)$ is computed using the straight line through $(\bar{x}, \bar{y})$

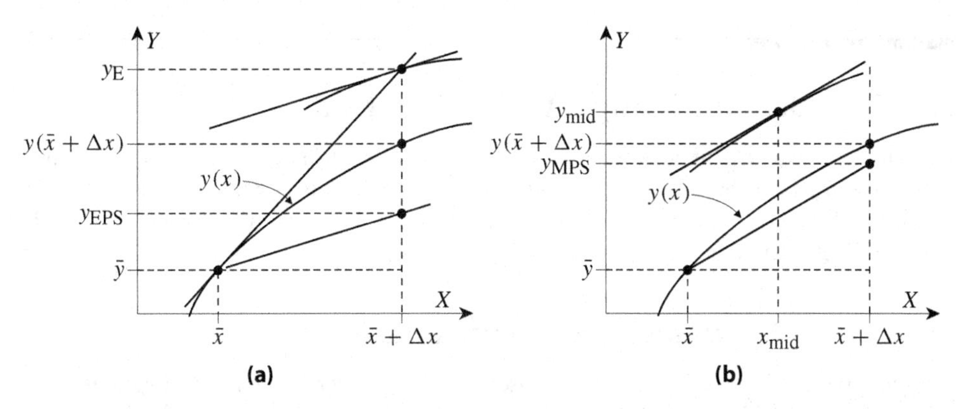

Figure 12.4: Two more approximations, y_{EPS} and y_{MPS}, of $y(\bar{x}+\Delta x)$ where y is the solution to $y' = f(x, y)$ passing through $(\bar{x}, \bar{y})$: In **(a)** (Euler-point slope approximation) y_{EPS} is on the line through $(\bar{x}, \bar{y})$ with slope $f(\bar{x}+\Delta x, y_E)$. In **(b)** (mid-point slope approximation) y_{MPS} is on the line through $(\bar{x}, \bar{y})$ whose slope is $f(x_{mid}, y_{mid})$ where $x_{mid} = \bar{x} + {}^{\Delta x}/_{2}$ and y_{mid} is some approximation to $y(x_{mid})$.

with slope equal to $f(x, y)$ at the Euler point. In other words,

$$y(\bar{x}+\Delta x) \approx y_{EPS}$$

where

$$y_{EPS} = \bar{y} + \Delta x \cdot f(\bar{x}+\Delta x, y_E)$$

with

$$y_E = \bar{y} + \Delta x \cdot f(\bar{x}, \bar{y}) \quad .$$

By itself, there is no reason to believe that this scheme is any better than the Euler method. Do observe, however, that the average of y_E and y_{EPS},

$$\frac{1}{2}[y_E + y_{EPS}] = \cdots = \bar{y} + \Delta x \cdot \frac{1}{2}[f(\bar{x}, \bar{y}) + f(\bar{x}+\Delta x, y_E)] \quad ,$$

is the approximation for $y(\bar{x}+\Delta x)$ used for the improved Euler method (see equation (12.6b) on page 224).

Mid-Point Slope Methods

The basic scheme for this is illustrated in Figure 12.4b. Here, the approximation y_{MPS} to $y(\bar{x}+\Delta x)$ is computed using a straight line through $(\bar{x}, \bar{y})$ whose slope is given by $f(x_{mid}, y_{mid})$ where x_{mid} is the midpoint of the interval $(\bar{x}, \bar{x}+\Delta x)$,

$$x_{mid} = \bar{x} + \frac{1}{2}\Delta x \quad ,$$

and y_{mid} is some approximation to $y(x_{mid})$. So,

$$y(\bar{x}+\Delta x) \approx y_{MPS}$$

where

$$y_{MPS} = \bar{y} + \Delta x \cdot f(x_{mid}, y_{mid}) \qquad \text{with} \quad x_{mid} = x + \frac{1}{2}\Delta x \quad .$$

The approximation y_{mid} to $y(x_{mid})$ can be determined using the approximations for either the Euler or improved Euler method with step size ${}^{\Delta x}/_{2}$, or by using some other method.

12.4 The Classic Runge-Kutta Method

There are, in fact, several "Runge-Kutta methods" for solving first-order initial-value problems. The one we will discuss is the best known one, and is the one usually meant when discussing *the* Runge-Kutta method. It is also the one originally developed by the German mathematicians Carl Runge and Martin Wilhelm Kutta in the late 1800s and early 1900s.[3]

Refining Our Fundamental Approximations

To get to the Runge-Kutta method, it helps to use integrals to rederive the approximations on which the Euler and improved Euler methods are based. So, assume we have a solution $y = y(x)$ to a first-order differential equation

$$\frac{dy}{dx} = f(x, y) \quad .$$

Then, of course,

$$y'(x) = f(x, y(x)) \quad ,$$

which, to simplify our discussion, we will write as

$$y'(x) = g(x) \qquad \text{with} \quad g(x) = f(x, y(x)) \quad .$$

Integrating this from $\bar{x}$ to $\bar{x} + \Delta x$ gives us

$$y(\bar{x} + \Delta x) - y(\bar{x}) = \int_{\bar{x}}^{\bar{x}+\Delta x} y'(x)\,dx = \int_{\bar{x}}^{\bar{x}+\Delta x} g(x)\,dx \quad ,$$

which we will rewrite as

$$y(\bar{x} + \Delta x) = y(\bar{x}) + \int_{\bar{x}}^{\bar{x}+\Delta x} g(x)\,dx \qquad \text{with} \quad g(x) = f(x, y(x)) \quad . \tag{12.9}$$

We can now rederive the fundamental approximations for the Euler and improved Euler by using different ways of approximating the integral in equation (12.9). By approximating the integral with the area of a rectangle of height $g(\bar{x})$ and width Δx (see Figure 12.5b), we get

$$y(\bar{x} + \Delta x) \approx y(\bar{x}) + \Delta x \cdot g(\bar{x}) = y(\bar{x}) + \Delta x \cdot f(x, y(\bar{x})) \quad ,$$

the fundamental approximation for each step of the Euler method.

By approximating the integral with the area of the trapezoid indicated in Figure 12.5b, we get

$$\begin{aligned} y(\bar{x} + \Delta x) &\approx y(\bar{x}) + \Delta x \cdot \frac{1}{2}[g(\bar{x}) + g(\bar{x} + \Delta x)] \\ &= \bar{y} + \Delta x \cdot \frac{1}{2}[f(\bar{x}, \bar{y}) + f(\bar{x} + \Delta x, y(\bar{x} + \Delta x))] \quad , \end{aligned}$$

the same as approximation (12.4) on page 224 for the improved Euler method.

If you recall "approximating integrals" from calculus, you will realize that the next step is "Simpson's rule", that is, approximating the integral of g using the integral over $(\bar{x}, \bar{x} + \Delta x)$ of the parabola $P = P(x)$ that passes through three particular points on the graph of g, namely,

$$(\bar{x}, g(\bar{x})) \quad , \quad (\bar{x}_{\text{m}}, g(\bar{x}_{\text{m}})) \quad \text{and} \quad (\bar{x}_{\text{r}}, g(\bar{x}_{\text{r}})) \quad ,$$

where $\bar{x}_{\text{m}}$ is the midpoint of the interval $(\bar{x}, \bar{x} + \Delta x)$ and $\bar{x}_{\text{r}}$ is the right-hand endpoint,

$$\bar{x}_{\text{m}} = \bar{x} + \frac{1}{2}\Delta x \qquad \text{and} \qquad \bar{x}_{\text{r}} = \bar{x} + \Delta x \quad .$$

(See Figure 12.5c).

[3] To learn more about the "family" of Runge-Kutta methods (which, technically, includes the Euler and improved Euler methods), the interested reader should consult a more advanced text on numerical methods.

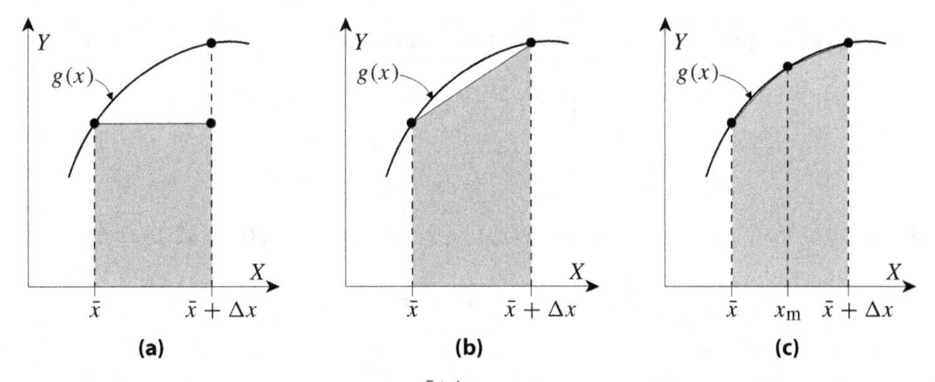

Figure 12.5: Approximating the integral $\int_{\bar{x}}^{\bar{x}+\Delta x} g(x)\,dx$ using **(a)** a rectangle, **(b)** a trapezoid and **(c)** a parabola.

To find the formula for this parabola, it helps to be clever and write its formula as

$$P(x) = A + B(x-\bar{x}) + C(x-\bar{x})(x-\bar{x}_{\mathrm{m}})$$

where A, B and C are constants to be determined. And since this parabola is supposed to pass through the three points listed above, we must have

$$\begin{aligned} g(\bar{x}) &= P(\bar{x}) \\ &= A + B(\bar{x}-\bar{x}) + C(\bar{x}-\bar{x})(x-\bar{x}_{\mathrm{m}}) = A \quad , \end{aligned}$$

$$\begin{aligned} g(\bar{x}_{\mathrm{m}}) &= P(\bar{x}_{\mathrm{m}}) \\ &= A + B(\bar{x}_{\mathrm{m}}-\bar{x}) + C(\bar{x}_{\mathrm{m}}-\bar{x})(\bar{x}_{\mathrm{m}}-\bar{x}_{\mathrm{m}}) = A + B\cdot\frac{1}{2}\Delta x \end{aligned}$$

and

$$\begin{aligned} g(\bar{x}_{\mathrm{r}}) &= P(\bar{x}_{\mathrm{r}}) \\ &= A + B(\bar{x}_{\mathrm{r}}-\bar{x}) + C(\bar{x}_{\mathrm{r}}-\bar{x})(x-\bar{x}_{\mathrm{m}}) = A + B\cdot\Delta x + C\cdot\frac{1}{2}(\Delta x)^2 \quad . \end{aligned}$$

This gives us an easily solved system, yielding

$$A = g(\bar{x}) \quad ,$$

$$B = \frac{2\left[g(\bar{x}_{\mathrm{m}}) - g(\bar{x})\right]}{\Delta x}$$

and

$$C = \frac{2\left[g(\bar{x}_{\mathrm{r}}) - 2g(\bar{x}_{\mathrm{m}}) + g(\bar{x})\right]}{(\Delta x)^2} \quad .$$

Computing the desired integral of the resulting $P(x)$ is also straightforward. As you can verify,

$$\int_{\bar{x}}^{\bar{x}+\Delta x} P(x)\,dx = \frac{\Delta x}{6}\left[g(\bar{x}) + 4g\left(\bar{x}+\frac{1}{2}\Delta x\right) + g(\bar{x}+\Delta x)\right] \quad .$$

Remember, we are interested in this integral because it is supposed to be a good approximation of the integral in equation (12.9). Using it as such, we get

$$\int_{\bar{x}}^{\bar{x}+\Delta x} g(x)\,dx \approx \int_{\bar{x}}^{\bar{x}+\Delta x} P(x)\,dx = \frac{\Delta x}{6}\left[g(\bar{x}) + 4g\left(\bar{x}+\frac{1}{2}\Delta x\right) + g(\bar{x}+\Delta x)\right] \quad .$$

This, combined with equation (12.9) becomes our first approximation for the Runge-Kutta method:

$$y(\bar{x} + \Delta x) \approx y(\bar{x}) + \frac{\Delta x}{6}\left[g(\bar{x}) + 4g\left(\bar{x} + \tfrac{1}{2}\Delta x\right) + g(\bar{x} + \Delta x)\right] \tag{12.10a}$$

where

$$g(x) = f(x, y(x)) \quad . \tag{12.10b}$$

It should be noted that, since $g(x) = f(x, y(x)) = y'(x)$, approximation (12.10a) can be written as

$$y(\bar{x} + \Delta x) \approx y(\bar{x}) + \Delta x \cdot S$$

where

$$S = \frac{1}{6}\left[y'(\bar{x}) + 4y'\left(\bar{x} + \tfrac{1}{2}\Delta x\right) + y'(\bar{x} + \Delta x)\right]$$

$$= \text{an "average" of slopes of } y = y(x) \text{ between } x = \bar{x} \text{ and } x = \bar{x} + \Delta x \quad .$$

The Iterative Formulas

Now let's consider developing an iterative formula for computing y_{k+1} from y_k in a numerical solution

$$\{x_0, x_1, x_2, x_3, \ldots, x_N\} \qquad \text{and} \qquad \{y_0, y_1, y_2, y_3, \ldots, y_N\}$$

with step size Δx and using approximation (12.10) to generate an iterative formula for computing $y_{k+1} \approx y(x_k + \Delta x)$ from an already computed $y_k \approx y(x_k)$.

Letting $\bar{x} = x_k$, approximation (12.10) becomes

$$y(x_k + \Delta x) \approx y(x_k) + \frac{\Delta x}{6}\left[g(x_k) + 4g\left(x_k + \tfrac{1}{2}\Delta x\right) + g(x_k + \Delta x)\right] \tag{12.11a}$$

with

$$g(x) = f(x, y(x)) \quad . \tag{12.11b}$$

Since we don't know the formula for $y(x)$, we must use what seem to be reasonable approximations based on previous discussions to approximate

$$y(x_k) \quad , \quad g(x_k) \quad , \quad g\left(x_k + \tfrac{1}{2}\Delta x\right) \quad \text{and} \quad g(x_k + \Delta x) \quad .$$

There are many possible approximations that we could use. Here are the ones used in the classic Runge-Kutta method:

1. For $y(x_k)$ and $g(x_k)$, the choice is pretty obvious since we already have $y_k \approx y(x_k)$:

$$y(x_k) \approx y_k$$

and

$$g(x_k) = f(x_k, y(x_k)) \approx f(x_k, y_k) \quad .$$

2. For $g\left(x_k + {}^{\Delta x}/_2\right)$, we take the average of the values obtained by approximating $y\left(x_k + {}^{\Delta x}/_2\right)$ using y_k with the basic Euler and the Euler-point slope method from Section 12.3, *both with step size* ${}^{\Delta x}/_2$:

$$g\left(x_k + \tfrac{1}{2}\Delta x\right) = \frac{1}{2}\left[g\left(x_k + \tfrac{1}{2}\Delta x\right) + g\left(x_k + \tfrac{1}{2}\Delta x\right)\right]$$

$$= \frac{1}{2}\left[f\left(x_k + \tfrac{1}{2}\Delta x, y\left(x_k + \tfrac{1}{2}\Delta x\right)\right) + f\left(x_k + \tfrac{1}{2}\Delta x, y\left(x_k + \tfrac{1}{2}\Delta x\right)\right)\right]$$

$$\approx \frac{1}{2}\left[f\left(x_k + \tfrac{1}{2}\Delta x, p_1\right) + f\left(x_k + \tfrac{1}{2}\Delta x, p_2\right)\right]$$

where

$$p_1 = y_k + \frac{1}{2}\Delta x \cdot f(x_k, y_k)$$

and

$$p_2 = y_k + \frac{1}{2}\Delta x \cdot f(x_k, p_1) \quad .$$

(Note that this is using the basic ideas behind the improved Euler method.)

3. For $g(x_k + \Delta x)$, we approximate $y(x_k + \Delta x)$ using the midpoint slope method from Section 12.3 with p_2 approximating the y–coordinate of the midpoint,

$$g(x_k + \Delta x) = f(x_k + \Delta x, y(x_k + \Delta x)) \approx f(x_k + \Delta x, p_3)$$

where

$$p_3 = y_k + \Delta x \cdot f\left(x_k + \frac{1}{2}\Delta x, p_2\right) \quad .$$

Combining these approximations with approximation (12.11) then gives us

$$y(x_k + \Delta x) = y(x_{k+1}) \approx y_{k+1}$$

where

$$y_{k+1} = y_k + \frac{\Delta x}{6}\Big[f(x_k, y_k) + 2f\left(x_k + \frac{1}{2}\Delta x, p_1\right) + 2f\left(x_k + \frac{1}{2}\Delta x, p_2\right) + f(x_k + \Delta x, p_3)\Big]$$

with

$$p_1 = y_k + \frac{1}{2}\Delta x \cdot f(x_k, y_k) \quad ,$$

$$p_2 = y_k + \frac{1}{2}\Delta x \cdot f(x_k, p_1)$$

and

$$p_3 = y_k + \Delta x \cdot f\left(x_k + \frac{1}{2}\Delta x, p_2\right) \quad .$$

For computational purposes, it turns out to be more convenient (or, at least, more traditional) to rewrite the above in terms of an "average of slopes":

$$y_{k+1} = y_k + \frac{\Delta x}{6}\left[s_1 + 2s_2 + 2s_3 + s_4\right] \tag{12.12a}$$

where

$$s_1 = f(x_k, y_k) \quad , \tag{12.12b}$$

$$s_2 = f\left(x_k + \frac{1}{2}\Delta x\,,\, y_k + \frac{1}{2}\Delta x \cdot s_1\right) \quad , \tag{12.12c}$$

$$s_3 = f\left(x_k + \frac{1}{2}\Delta x\,,\, y_k + \frac{1}{2}\Delta x \cdot s_2\right) \tag{12.12d}$$

and

$$s_4 = f(x_{k+1}, y_k + \Delta x \cdot s_3) \quad . \tag{12.12e}$$

This is the set used for the iterative computations in the classic Runge-Kutta method.

Implementing the Classic Runge-Kutta Method (Illustrated)

Let us outline the steps in using the Runge-Kutta Method to numerically solve some first-order differential equation with initial data $y(x_0) = y_0$, say,

$$5\frac{dy}{dx} - y^2 = -x^2 \qquad \text{with} \quad y(0) = 1 \quad . \tag{12.13}$$

(Yes, this is the same problem we used to illustrate both the basic Euler method and the improved Euler method.)

As with the basic Euler and improved Euler methods, the implementation consists of two preliminary steps in which some constants and the formula for f are determined for the iterative steps, followed by the actual iterative steps in which the x_k's and y_k's are computed.

Preliminary Steps

1. Rewrite the initial-value problem with the differential equation in derivative formula form,

$$\frac{dy}{dx} = f(x,y) \qquad \text{with} \quad y(x_0) = y_0 \quad ,$$

and, from this, determine the formula for $f(x,y)$ and the values of x_0 and y_0.

Rewriting the differential equation in our example in derivative formula form, this initial-value problem is

$$\frac{dy}{dx} = \frac{1}{5}\left[y^2 - x^2\right] \qquad \textit{with} \quad y(0) = 1 \quad .$$

So,

$$f(x,y) = \frac{1}{5}\left[y^2 - x^2\right] \quad , \qquad x_0 = 0 \qquad \textit{and} \qquad y_0 = 1 \quad .$$

2. Decide on a distance Δx for the step size, a positive integer N for the maximum number of steps, and a maximum value desired for x, x_{max}. Choose these quantities so that

$$x_{max} = x_N = x_0 + N\Delta x \quad .$$

For Δx and N, let us, again, pick

$$\Delta x = \frac{1}{2} \qquad \textit{and} \qquad N = 6 \quad .$$

Then

$$x_{max} = x_0 + N\Delta x = 0 + 6 \cdot \frac{1}{2} = 3 \quad .$$

Iterative Steps

1, 2, …. For each integer $k \geq 0$, successively compute x_{k+1}, s_1, s_2, s_3, s_4 and y_{k+1} using the values of x_k and y_k found in a previous step and equation set (12.12); that is,

$$x_{k+1} = x_k + \Delta x \quad ,$$

$$s_1 = f(x_k, y_k) \quad ,$$

$$s_2 = f\left(x_k + \tfrac{1}{2}\Delta x \,,\, y_k + \tfrac{1}{2}\Delta x \cdot s_1\right) \quad ,$$

$$s_3 = f\left(x_k + \frac{1}{2}\Delta x \,,\, y_k + \frac{1}{2}\Delta x \cdot s_2\right) \quad ,$$

$$s_4 = f(x_{k+1} \,,\, y_k + \Delta x \cdot s_3)$$

and

$$y_{k+1} = y_k + \frac{\Delta x}{6}[s_1 + 2s_2 + 2s_3 + s_4] \quad ,$$

where $f(x, y)$, Δx, x_0 and y_0 are as determined in the preliminary steps.

Continue repeating these computations until x_N and y_N are computed. (Be sure to save the x_k and y_k values computed!)

For our example, we will only attempt the first iterative step and part of the second before going to the computer for the rest of the computations:

Step 1. With $k = 0$:

$$x_1 = x_{0+1} = x_0 + \Delta x = 0 + \frac{1}{2} = \frac{1}{2} \quad ,$$

$$s_1 = f(x_0, y_0) = f(0, 1) = \frac{1}{5}[1^2 - 0^2] = \frac{1}{5} \quad ,$$

$$\begin{aligned} s_2 &= f\left(x_0 + \frac{1}{2}\Delta x \,,\, y_0 + \frac{1}{2}\Delta x \cdot s_1\right) \\ &= f\left(0 + \frac{1}{2} \cdot \frac{1}{2} \,,\, 1 + \frac{1}{2} \cdot \frac{1}{2} \cdot \frac{1}{5}\right) \\ &= f\left(\frac{1}{4}, \frac{21}{20}\right) = \frac{1}{5}\left[\left(\frac{21}{20}\right)^2 - \left(\frac{1}{4}\right)^2\right] = \frac{26}{125} \quad , \end{aligned}$$

$$\begin{aligned} s_3 &= f\left(x_0 + \frac{1}{2}\Delta x \,,\, y_0 + \frac{1}{2}\Delta x \cdot s_2\right) \\ &= f\left(0 + \frac{1}{2} \cdot \frac{1}{2} \,,\, 1 + \frac{1}{2} \cdot \frac{1}{2} \cdot \frac{26}{125}\right) \\ &= f\left(\frac{1}{4}, \frac{263}{250}\right) = \frac{1}{5}\left[\left(\frac{263}{250}\right)^2 - \left(\frac{1}{4}\right)^2\right] = \frac{261{,}051}{1{,}250{,}000} \quad , \end{aligned}$$

$$\begin{aligned} s_4 &= f(x_1 \,,\, y_0 + \Delta x \cdot s_3) \\ &= f\left(\frac{1}{2} \,,\, 1 + \frac{1}{2} \cdot \frac{261{,}051}{1{,}250{,}000}\right) \\ &= f\left(\frac{1}{2}, \frac{2{,}761{,}051}{2{,}500{,}000}\right) = \frac{1}{5}\left[\left(\frac{2{,}761{,}051}{2{,}500{,}000}\right)^2 - \left(\frac{1}{2}\right)^2\right] = \frac{6{,}060{,}902{,}624{,}601}{31{,}250{,}000{,}000{,}000} \end{aligned}$$

and

$$\begin{aligned} y_1 = y_{0+1} &= y_0 + \frac{\Delta x}{6}[s_1 + 2s_2 + 2s_3 + s_4] \\ &= 1 + \frac{1}{2} \cdot \frac{1}{6}\left[\frac{1}{5} + 2 \cdot \frac{26}{125} + 2 \cdot \frac{261{,}051}{1{,}250{,}000} + \frac{6{,}060{,}902{,}624{,}601}{31{,}250{,}000{,}000{,}000}\right] \\ &= \frac{413{,}363{,}452{,}624{,}601}{375{,}000{,}000{,}000{,}000} = 1.102302\ldots \quad . \end{aligned}$$

Step 2. With $k = 1$ *(and using a computer):*

$$x_2 = x_{1+1} = x_1 + \Delta x = \frac{1}{2} + \frac{1}{2} = 1 \quad ,$$

$$\begin{aligned}
s_1 &= f(x_1, y_1) = f\left(\frac{1}{2}, \frac{413{,}363{,}452{,}624{,}601}{375{,}000{,}000{,}000{,}000}\right) \\
&= \frac{1}{5}\left[\left(\frac{413{,}363{,}452{,}624{,}601}{375{,}000{,}000{,}000{,}000}\right)^2 - \left(\frac{1}{2}\right)^2\right] \\
&= \frac{135{,}713{,}093{,}965{,}730{,}755{,}355{,}430{,}409{,}201}{703{,}125{,}000{,}000{,}000{,}000{,}000{,}000{,}000{,}000} = 0.193014\ldots \quad ,
\end{aligned}$$

$$\begin{aligned}
s_2 &= f\left(x_1 + \frac{1}{2}\Delta x \,,\, y_1 + \frac{1}{2}\Delta x \cdot s_1\right) \\
&= f\left(\frac{1}{2} + \frac{1}{2}\cdot\frac{1}{2} \,,\, y_1 + \frac{1}{2}\cdot\frac{1}{2}\cdot s_1\right) \\
&= f\left(\frac{3}{4}, \frac{413{,}363{,}452{,}624{,}601}{375{,}000{,}000{,}000{,}000} + \frac{1}{4}\cdot\frac{135{,}713{,}093{,}965{,}730{,}755{,}355{,}430{,}409{,}201}{703{,}125{,}000{,}000{,}000{,}000{,}000{,}000{,}000{,}000}\right) \\
&= f\left(\frac{3}{4}, \frac{3{,}235{,}938{,}988{,}650{,}238{,}255{,}355{,}430{,}409{,}201}{2{,}812{,}500{,}000{,}000{,}000{,}000{,}000{,}000{,}000{,}000}\right) \\
&= \cdots = 0.152255\ldots \quad ,
\end{aligned}$$

$$s_3 = f\left(x_1 + \frac{1}{2}\Delta x \,,\, y_1 + \frac{1}{2}\Delta x \cdot s_2\right) = \cdots = 0.147587\ldots \quad ,$$

$$s_4 = f(x_2 \,,\, y_1 + \Delta x \cdot s_3) = \cdots = 0.076640\ldots$$

and

$$y_2 = y_{1+1} = y_1 + \frac{\Delta x}{6}[s_1 + 2s_2 + 2s_3 + s_4] = \cdots = 1.174747\ldots \quad .$$

For $k \geq 2$, the values for x_k and y_k have been computed by a computer using the above described iterative steps. The result is given in the table in Figure 12.6a.

As with the other numerical methods (and as is obvious from our rather simple example, above), the use of a computer to carry out the above computations is highly recommended. Either write your own program or set up a worksheet using your favorite computer math package. And, again, be sure to be able to easily change

$$\Delta x \quad , \quad N \text{ or } x_{max} \quad , \quad f(x,y) \quad , \quad x_0 \quad \text{and} \quad y_0$$

so you can easily change the computational parameters and consider different initial-value problems.[4]

Using the Results of the Method Tables and Graphs

The comments made regarding the use of the Runge-Kutta method for generating tables and piecewise linear graphs remain the same as were made for the improved Euler method.

General Observations on Accuracy

The Runge-Kutta method does require a few more computations per step than the improved Euler method (though it is still quite easy to write a computer program to carry out those steps). Typically, however, these extra computations are well worth it, and result in more accurate approximations which improve very rapidly as the step size is decreased. By greatly expanding on the proof of Theorem 10.1 on page 188, it can be shown that:

[4] In fact, your favorite computer math package may well already have some version of the above in its repertoire.

k	t_k	y_k
0	0	1
1	0.5	1.1023
2	1.0	1.1747
3	1.5	1.1546
4	2.0	0.9625
5	2.5	0.5142
6	3.0	−0.2366

(a)

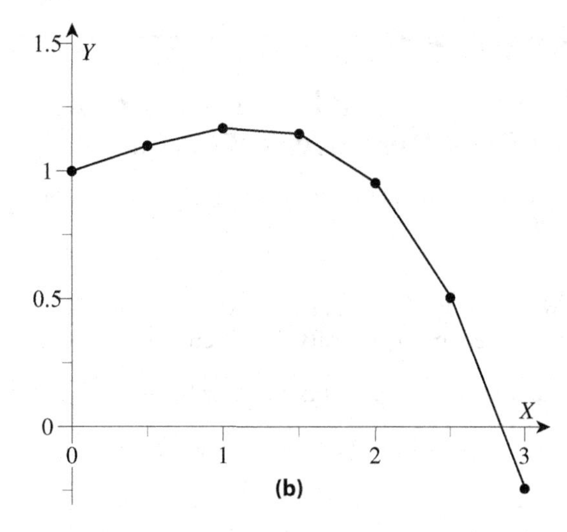

Figure 12.6: Results of using the classic Runge-Kutta method to numerically solve initial-value problem (12.13) with $\Delta t = 1/2$ and $N = 6$: **(a)** The numerical solution in which $y_k \approx y(x_k)$ (for $k \geq 1$, the values of y_k are to the nearest 0.0001). **(b)** The piecewise linear graph based on this numerical solution.

If $y = y(x)$ is the true solution to initial-value problem

$$\frac{dy}{dx} = f(x, y) \qquad \textit{with} \quad y(x_0) = y_0$$

on an interval $(x_0, x_0 + L)$, and if f is "reasonably smooth", then there is a constant M_{RG} such that

$$|y(x_k) - y_k| < M_{\text{RG}} \cdot (\Delta x)^4$$

whenever $k = 0, 1, \ldots, N$ and

$$\{x_0, x_1, x_2, x_3, \ldots, x_N\} \qquad \textit{and} \qquad \{y_0, y_1, y_2, y_3, \ldots, y_N\}$$

is a numerical solution to the above initial-value problem obtained using the classic Runge-Kutta method with step size Δx and total number of steps N satisfying

$$0 < \Delta x \cdot N \leq L \quad .$$

Recall that similar statements were made for the original Euler method and the improved Euler method. For these methods, however, the error bounds were, respectively,

$$|y(x_k) - y_k| < M_{\text{E}} \cdot \Delta x$$

and

$$|y(x_k) - y_k| < M_{\text{IE}} \cdot (\Delta x)^2 \quad ,$$

for constants M_{E} and M_{IE}. So, by reducing the step size by half we can expect to cut the error in the approximations

- by one half when using the original Euler method,
- by one fourth when using the improved Euler method

and

- by one sixteenth when using the Runge-Kutta method.

12.5 Some Additional Comments

Order of the Numerical Method

The error estimate for a numerical method for solving a first-order initial-value problem is often of the form

$$|y(x_k) - y_k| < M \cdot (\Delta x)^m$$

where M and m are constants and Δx is the step size. The exponent on Δx, m, is called the *order* and is used to help classify the different methods. Checking back, you can see that

- the basic Euler method is a first-order method,
- the improved Euler method is a second-order method, and
- the classic Runge-Kutta method is a fourth-order method.

As should be clear by now, higher-order methods tend to yield more accurate results.

Catastrophic Failure?

Because of their better accuracy, catastrophic failures (such as illustrated in Figure 10.3 on page 182) are less likely to occur using the improved Euler and the Runge-Kutta methods, especially when using small step sizes. Still, these catastrophic failures can occur, and it is always a good idea to do a reality check by graphing the corresponding approximation on top of the slope field for the differential equation to see if the approximations are reasonable.

Other Methods

Be aware that he subject of "numerical analysis of differential equations" goes far beyond what can be covered in this text. What we've discussed (and will discuss later in Chapter 40) should be viewed as an introduction. For a more complete development of numerical analysis and, in particular, the numerical analysis of differential equations, the interested reader should at least start with a good text and/or a good course on the subject.

Additional Exercises

12.1. *Several initial-value problems are given below, along with values for two of the following parameters: step size Δx, number of steps N and maximum variable of interest x_{max}. For each, find the corresponding numerical solution using the improved Euler method with the indicated parameter values. Do these problems without a calculator or computer.*

a. $\dfrac{dy}{dx} = \dfrac{y}{x}$ *with* $y(1) = -1$; $\Delta x = \dfrac{1}{3}$ *and* $N = 3$

b. $\dfrac{dy}{dx} = -8xy$ *with* $y(0) = 10$; $x_{max} = 1$ *and* $N = 4$

c. $2x + 5\dfrac{dy}{dx} = y$ *with* $y(0) = 2$; $x_{max} = 2$ *and* $\Delta x = \dfrac{1}{2}$

d. $\dfrac{dy}{dx} + \dfrac{y}{x} = 4$ *with* $y(1) = 8$; $\Delta x = \dfrac{1}{2}$ *and* $N = 6$

12.2 a. *Using your favorite computer language or computer math package, write a program or worksheet for finding the numerical solution to an arbitrary first-order initial-value problem using the improved Euler method. Make it easy to change the differential equation and the computational parameters (step size, number of steps, etc.).*[5]

b. *Test your program/worksheet by using it to re-compute the numerical solutions for the problems in Exercise 12.1, above.*

12.3. *Using your program/worksheet from Exercise 12.2 a with each of the following step sizes, find an approximation for* $y(5)$ *where* $y = y(x)$ *is the solution to*

$$\frac{dy}{dx} = \sqrt[3]{x^2 + y^2 + 1} \qquad \text{with} \quad y(0) = 0 \quad .$$

a. $\Delta x = 1$ **b.** $\Delta x = 0.1$ **c.** $\Delta x = 0.01$ **d.** $\Delta x = 0.001$

12.4. *The last exercise was the same as Exercise 10.4, but with the "Euler method" replaced by the "improved Euler method". Compare the results of the two exercises.*

12.5. *Consider the initial-value problem*

$$\frac{dy}{dx} = (y-1)^2 \qquad \text{with} \quad y(0) = -\frac{13}{10} \quad .$$

This is the problem discussed in Section 10.3 in the illustration of a "catastrophic failure" of Euler's method. It was also the subject of Exercise 10.6. There, you found that the solution to this initial-value problem is

$$y(x) = 1 - \frac{23}{23x + 10} \qquad \text{and that} \qquad y(4) = \frac{79}{102} \quad .$$

a. *Using your program/worksheet for the improved Euler method from Exercise 12.2 a, find the numerical solution to the above initial-value problem with* $x_{max} = 4$ *and step size* $\Delta x = 1/2$ *. (Compare this solution to the corresponding one for Exercise 10.6.)*

b. *Find the approximation to* $y(4)$ *generated by the improved Euler method with each of the following step sizes (use your answer to the previous part or your program/worksheet from Exercise 12.2 a). Also, compute the magnitude of the error in using this approximation for the exact value given above.*

i. $\Delta x = 1$ **ii.** $\Delta x = \dfrac{1}{2}$ **iii.** $\Delta x = \dfrac{1}{4}$ **iv.** $\Delta x = \dfrac{1}{10}$

c. *Compare your answers to the above with your answers to Exercise 10.6 c. Note that the approximate values for* $y(4)$ *obtained via the improved Euler method are much more accurate than those obtained by the Euler method, except for the case with* $\Delta x = 1$ *. Why did the improved Euler method fail so badly in this one case? (Hint: Consider the slope field.)*

12.6. *Several initial-value problems are given below, along with values for two of the following parameters: step size* Δx *, number of steps* N *, and maximum variable of interest* x_{max} *. For each, find the corresponding numerical solution using the classic Runge-Kutta method with the indicated parameter values. Do these problems without a programable calculator or computer.*

---

[5] If your computer math package uses infinite precision or symbolic arithmetic, you may have to include commands to ensure your results are given as decimal approximations.

a. $\dfrac{dy}{dx} = -8xy$ *with* $y(0) = 10$; $x_{max} = 1$ *and* $N = 2$

b. $2x + 5\dfrac{dy}{dx} = y$ *with* $y(0) = 2$; $x_{max} = 1$ *and* $\Delta x = \dfrac{1}{2}$

c. $\dfrac{dy}{dx} + \dfrac{y}{x} = 4$ *with* $y(1) = 8$; $\Delta x = \dfrac{1}{2}$ *and* $N = 2$

d. $\dfrac{dy}{dx} + \dfrac{y}{x} = 4$ *with* $y(1) = 8$; $\Delta x = \dfrac{1}{3}$ *and* $N = 3$

12.7 a. *Using your favorite computer language or computer math package, write a program or worksheet for finding the numerical solution to an arbitrary first-order initial-value problem using the classic Runge-Kutta method. Make it easy to change the differential equation and the computational parameters (step size, number of steps, etc.).*[6]

b. *Test your program/worksheet by using it to re-compute the numerical solutions for the problems in Exercise 12.6, above.*

12.8. *Using your program/worksheet from Exercise 12.7 a with each of the following step sizes, find an approximation for* $y(5)$ *where* $y = y(x)$ *is the solution to*

$$\frac{dy}{dx} = \sqrt[3]{x^2 + y^2 + 1} \qquad \text{with} \quad y(0) = 0 \quad .$$

a. $\Delta x = 1$ **b.** $\Delta x = 0.1$ **c.** $\Delta x = 0.01$ **d.** $\Delta x = 0.001$

12.9. *The last exercise was the same as Exercises 10.4 and 12.3, but with the "Euler method" and "improved Euler method" replaced by the "classic Runge-Kutta method". Compare the results of these three exercises.*

12.10. *Consider the initial-value problem*

$$\frac{dy}{dx} = (y - 1)^2 \qquad \text{with} \quad y(0) = -\frac{13}{10} \quad .$$

This is the problem discussed in Section 10.3 in the illustration of a "catastrophic failure" of Euler's method. It was also the subject of Exercises 10.6 and 12.5. In Exercise 10.6, you found that the solution to this initial-value problem is

$$y(x) = 1 - \frac{23}{23x + 10} \qquad \text{and that} \qquad y(4) = \frac{79}{102} \quad .$$

a. *Using your program/worksheet for the classic Runge-Kutta method from Exercise 12.7 a, find the numerical solution to the above initial-value problem with* $x_{max} = 4$ *and step size* $\Delta x = 1/2$. *(Compare this solution to the corresponding ones for Exercises 10.6 and 12.5.)*

b. *Find the approximation to* $y(4)$ *generated by the classic Runge-Kutta method with each of the following step sizes (use your answer to the previous part or your program/worksheet from Exercise 12.7 a). Also, compute the magnitude of the error in using this approximation for the exact value given above.*

i. $\Delta x = 1$ **ii.** $\Delta x = \dfrac{1}{2}$ **iii.** $\Delta x = \dfrac{1}{4}$ **iv.** $\Delta x = \dfrac{1}{10}$

c. *Compare your answers to the above with your answers to Exercise 12.5 b.*

---

[6] If your computer math package uses infinite precision or symbolic arithmetic, you may have to include commands to ensure your results are given as decimal approximations.

Part III

Second- and Higher-Order Equations

13

Higher-Order Equations: Extending First-Order Concepts

Let us switch our attention from first-order differential equations to differential equations of order two or higher. Our main interest will be with second-order differential equations, both because it is natural to look at second-order equations after studying first-order equations, and because second-order equations arise in applications much more often than do third-, or fourth- or eighty-third-order equations. Some examples of second-order differential equations are[1]

$$y'' + y = 0 \quad ,$$

$$y'' + 2xy' - 5\sin(x)\,y = 30e^{3x} \quad ,$$

and

$$(y+1)y'' = (y')^2 \quad .$$

Still, differential equations of even higher orders, such as

$$8y''' + 4y'' + 3y' - 83y = 2e^{4x} \quad ,$$

$$x^3 y^{(iv)} + 6x^2 y'' + 3xy' - 83\sin(x)y = 2e^{4x} \quad ,$$

and

$$y^{(83)} + 2y^3\, y^{(53)} - x^2 y'' = 18 \quad ,$$

can arise in applications, at least on occasion. Fortunately, many of the ideas used in solving these are straightforward extensions of those used to solve second-order equations. We will make use of this fact extensively in the following chapters.

Unfortunately, though, the methods we developed to solve first-order differential equations are of limited direct use in solving higher-order equations. Remember, most of those methods were based on integrating the differential equation after rearranging it into a form that could be legitimately integrated. This rarely is possible with higher-order equations, and that makes solving higher-order equations more of a challenge. This does not mean that those ideas developed in previous chapters are useless in solving higher-order equations, only that their use will tend to be subtle rather than obvious.

Still, there are higher-order differential equations that, after the application of a simple substitution, can be treated and solved as first-order equations. While our knowledge of first-order equations is still fresh, let us consider some of the more important situations in which this is possible. We will also take a quick look at how the basic ideas regarding first-order initial-value problems extend to

[1] For notational brevity, we will start using the 'prime' notation for derivatives a bit more. It is still recommended, however, that you use the '$^d/_{dx}$' notation when finding solutions just to help keep track of the variables involved.

higher-order initial-value problems. And finally, to cap off this chapter, we will briefly discuss the higher-order extensions of the existence and uniqueness theorems from Section 3.3.

13.1 Treating Some Second-Order Equations as First-Order

Suppose we have a second-order differential equation (with y being the yet unknown function and x being the variable). With luck, it is possible to convert the given equation to a first-order differential equation for another function v via the substitution $v = y'$. With a little more luck, that first-order equation can then be solved for v using methods discussed in previous chapters, and y can then be obtained from v by solving the first-order differential equation given by the original substitution, $y' = v$.

This approach requires some luck because, typically, setting $v = y'$ does not lead to a differential equation for just the one unknown function v. Instead, it usually results in a differential equation with *two* unknown functions, y and v, along with the variable x. This does not simplify our equation at all! So, being lucky here means that the conversion does yield a differential equation just involving v and one variable.

It turns out that we get lucky with two types of second-order differential equations: those that do not explicitly contain a y and those that do not explicitly contain an x. The first type will be especially important to us since solving this type of equation is part of an important method for solving more general differential equations (the "reduction of order" method in Chapter 14). It is also, typically, the easier type of equation to solve. So let's now spend a few moments discussing how to solve these equations. (We'll say more about the second type in a few pages.)

Solving Second-Order Differential Equations Not Explicitly Containing y

If the equation explicitly involves x, ${}^{dy}/_{dx}$, and ${}^{d^2y}/_{dx^2}$ — but not y — then we can naturally view the differential equation as a "first-order equation for ${}^{dy}/_{dx}$". For convenience, we usually set

$$\frac{dy}{dx} = v \quad .$$

Consequently,

$$\frac{d^2y}{dx^2} = \frac{d}{dx}\left[\frac{dy}{dx}\right] = \frac{d}{dx}[v] = \frac{dv}{dx} \quad .$$

Under these substitutions, the equation becomes a first-order differential equation for v. And since the original equation did not have a y, neither does the differential equation for v. This means we have a reasonable chance of solving this equation for $v = v(x)$ using methods developed in previous chapters. Then, assuming $v(x)$ can be so obtained,

$$y(x) = \int \frac{dy}{dx}\,dx = \int v(x)\,dx \quad .$$

When solving these equations, you normally end up with a formula involving two distinct arbitrary constants: one from the general solution to the first-order differential equation for v and the other arising from the integration of v to get y. Don't forget them, and be sure to label them as *different* arbitrary constants.

!►Example 13.1: *Consider the second-order differential equation*

$$\frac{d^2y}{dx^2} + 2\frac{dy}{dx} = 30e^{3x} \quad .$$

Setting

$$\frac{dy}{dx} = v \qquad \text{and} \qquad \frac{d^2y}{dx^2} = \frac{dv}{dx} \quad ,$$

as suggested above, the differential equation becomes

$$\frac{dv}{dx} + 2v = 30e^{3x} \quad .$$

This is a linear first-order differential equation with integrating factor

$$\mu = e^{\int 2\,dx} = e^{2x} \quad .$$

Proceeding as normal with linear first-order equations,

$$e^{2x}\left[\frac{dv}{dx} + 2v = 30e^{3x}\right]$$

$$\hookrightarrow \qquad e^{2x}\frac{dv}{dx} + 2e^{2x}v = 30e^{3x}e^{2x}$$

$$\hookrightarrow \qquad \frac{d}{dx}\left[e^{2x}v\right] = 30e^{5x}$$

$$\hookrightarrow \qquad \int \frac{d}{dx}\left[e^{2x}v\right]dx = \int 30e^{5x}\,dx$$

$$\hookrightarrow \qquad e^{2x}v = 6e^{5x} + c_0 \quad .$$

Hence,

$$v = e^{-2x}\left[6e^{5x} + c_0\right] = 6e^{3x} + c_0e^{-2x} \quad .$$

But $v = {}^{dy}/_{dx}$*, so the last equation can be rewritten as*

$$\frac{dy}{dx} = 6e^{3x} + c_0e^{-2x} \quad ,$$

which is easily integrated,

$$y = \int\left[6e^{3x} + c_0e^{-2x}\right]dx = 2e^{3x} - \frac{c_0}{2}e^{-2x} + c_2 \quad .$$

Thus (letting $c_1 = -{}^{c_0}/_2$*), the solution to our original differential equation is*

$$y(x) = 2e^{3x} - c_1e^{-2x} + c_2 \quad .$$

If your differential equation for v is separable and you are solving it as such, don't forget to check for the constant solutions to this differential equation and to take these "constant-v" solutions into account when integrating $y' = v$.

!▶Example 13.2: *Consider the second-order differential equation*

$$\frac{d^2y}{dx^2} = -\left(\frac{dy}{dx} - 3\right)^2 \quad . \tag{13.1}$$

Letting

$$\frac{dy}{dx} = v \qquad \text{and} \qquad \frac{d^2y}{dx^2} = \frac{dv}{dx} \quad ,$$

the differential equation becomes

$$\frac{dv}{dx} = -(v-3)^2 \quad . \tag{13.2}$$

This equation has a constant solution,

$$v = 3 \quad ,$$

which we can rewrite as

$$\frac{dy}{dx} = 3 \quad .$$

Integrating then gives us

$$y(x) = 3x + c_0 \quad .$$

This describes all the "constant-v" solutions to our original differential equation.

To find the nonconstant solutions to equation (13.2), divide through by $(v-3)^2$ and integrate:

$$\frac{dv}{dx} = -(v-3)^2$$

$$\hookrightarrow \qquad (v-3)^{-2}\frac{dv}{dx} = -1$$

$$\hookrightarrow \qquad \int (v-3)^{-2}\frac{dv}{dx}\,dx = -\int 1\,dx$$

$$\hookrightarrow \qquad -(v-3)^{-1} = -x + c_1$$

$$\hookrightarrow \qquad v = 3 + \frac{1}{x-c_1} \quad .$$

But, since $v = y'$, this last equation is the same as

$$\frac{dy}{dx} = 3 + \frac{1}{x-c_1} \quad ,$$

which is easily integrated, yielding

$$y(x) = 3x + \ln|x-c_1| + c_2 \quad .$$

Gathering all the solutions we've found gives us the set consisting of

$$y = 3x + \ln|x-c_1| + c_2 \qquad \text{and} \qquad y(x) = 3x + c_0 \tag{13.3}$$

describing all possible solutions to our original differential equation.

Equations of Even Higher Orders

With just a little imagination, the basic ideas discussed above can be applied to a few differential equations of even higher order. Here is an example:

!► *Example 13.3:* *Consider the third-order equation*

$$3\frac{d^3y}{dx^3} = \left(\frac{d^2y}{dx^2}\right)^{-2} \quad .$$

Set

$$v = \frac{d^2y}{dx^2} \quad .$$

Then

$$\frac{dv}{dx} = \frac{d}{dx}\left[\frac{d^2y}{dx^2}\right] = \frac{d^3y}{dx^3} \quad ,$$

and the original differential equation reduces to a simple separable first-order equation,

$$3\frac{dv}{dx} = v^{-2} \quad .$$

Multiplying both sides by v^2 *and proceeding as usual with such equations:*

$$3v^2\frac{dv}{dx} = 1$$

$$\hookrightarrow \qquad \int 3v^2\frac{dv}{dx}\,dx = \int 1\,dx$$

$$\hookrightarrow \qquad v^3 = x + c_1$$

$$\hookrightarrow \qquad v = (x + c_1)^{1/3} \quad .$$

So

$$\frac{d^2y}{dx^2} = v = (x + c_1)^{1/3} \quad .$$

Integrating once:

$$\frac{dy}{dx} = \int \frac{d^2y}{dx^2}\,dx = \int (x + c_1)^{1/3}\,dx = \frac{3}{4}(x + c_1)^{4/3} + c_2 \quad .$$

And once again:

$$y = \int \frac{dy}{dx}\,dx = \int \left[\frac{3}{4}(x + c_1)^{4/3} + c_2\right]dx = \frac{9}{28}(x + c_1)^{7/3} + c_2 x + c_3 \quad .$$

Converting a Differential Equation to a System*

Consider what we actually have after taking a second-order differential equation (with y being the yet unknown function and x being the variable) and converting it to a first-order equation for

* These comments relate the material in this chapter to more advanced concepts and methods that will be developed much later in this text. This discussion won't help you solve any differential equations now, but will give a little hint of an approach to dealing with higher-order equations we could take (and will explore in the distant future). You might find them interesting. Then, again, you may just want to ignore this discussion for now.

v through the substitution $v = y'$. We actually have a pair of first-order differential equations involving the two unknown functions y and v. The first equation in the pair is simply the equation for the substitution, $y' = v$, and the second is what we obtain after using the substitution with the original second-order equation. If we are lucky, we can directly solve the second equation of this pair. But, as the next example illustrates, we have this pair whether or not either of the above cases applies. Together, this pair forms a "system" of first-order differential equations" whose solution is the pair $y(x)$ and $v(x)$.

!►Example 13.4: *Suppose our original differential equation is*

$$\frac{d^2y}{dx^2} + 2x\frac{dy}{dx} = 5\sin(x)\,y \quad .$$

Setting

$$v = \frac{dy}{dx} \quad \text{and} \quad \frac{dv}{dx} = \frac{d^2y}{dx^2} \quad ,$$

the original differential equation reduces to

$$\frac{dv}{dx} + 2xv = 5\sin(x)\,y \quad .$$

Thus, $v = v(x)$ *and* $y = y(x)$*, together, must satisfy both*

$$v = \frac{dy}{dx} \quad \text{and} \quad \frac{dv}{dx} + 2xv = 5\sin(x)\,y \quad .$$

This is a system of two first-order equations. Traditionally, each equation is rewritten in derivative formula form, and the system is then written as

$$\frac{dy}{dx} = v$$

$$\frac{dv}{dx} = 5\sin(x)\,y - 2xv \quad .$$

As just illustrated, almost any second-order differential equation encountered in practice can be converted to a system of two first-order equations involving two unknown functions. In fact, almost any N^{th}-order differential equation can be converted using similar ideas to a system of N first-order differential equations involving N unknown functions. This is significant because methods exist for dealing with such systems. In many cases, these methods are analogous to methods we used with first-order differential equations. We may discuss some of these methods in the future (the distant future).

13.2 The Other Class of Second-Order Equations "Easily Reduced" to First-Order†

As noted a few pages ago, the substitution $v = y'$ can also be useful when no formulas of x explicitly appear in the given second-order differential equation. Such second-order differential equations are said to be *autonomous* (this extends the definition of "autonomous" given for first-order differential equations in Chapter 3).

† The material in this section is interesting and occasionally useful, but not essential to the rest of this text. At least give this section a quick skim before going to the discussion of initial-value problems in the next section (and promise to return to this section when convenient or necessary).

Solving Second-Order Autonomous Equations

Unfortunately, if you have an autonomous differential equation (i.e., one possibly containing y, $^{dy}/_{dx}$, and $^{d^2y}/_{dx^2}$ — but not x), then simply letting

$$v = \frac{dy}{dx}$$

and proceeding as in the previous section leads to a differential equation involving three entities — $^{dv}/_{dx}$, y and v. Unless that equation is *very* simple (say, $^{dv}/_{dx} = 0$), there won't be much you can do with it.

To avoid that impasse, take a second route: Eliminate all references to x by viewing the v as a function of y instead of x. This works because our equation contains no explicit formulas of x. This also means that the substitution for $^{d^2y}/_{dx^2}$ must be converted, via the above substitution and the chain rule, to a formula involving v and its derivative with respect to y as follows:

$$\frac{d^2y}{dx^2} = \frac{d}{dx}\left[\frac{dy}{dx}\right] = \frac{d}{dx}[v] = \frac{dv}{dx} \underset{\substack{\text{chain}\\\text{rule}}}{=} \frac{dv}{dy}\frac{dy}{dx} = \frac{dv}{dy}v \quad .$$

The resulting equation will be a first-order differential equation for $v(y)$. Solving that equation, and plugging the resulting formula for $v(y)$ into our original substitution,

$$\frac{dy}{dx} = v(y) \quad ,$$

gives us another first-order differential equation, this time for y. Solving this yields $y(x)$.

Since this use of $v = y'$ is a little less natural than that in the previous section, let us outline the steps more explicitly while doing an example. For our example, we will solve the apparently simple equation

$$\frac{d^2y}{dx^2} + y = 0 \quad .$$

(By the way, this equation will turn out to be rather important to us. We will return to it several times in the next few chapters.)

1. Set

$$\frac{dy}{dx} = v$$

and remember that, by this and the chain rule,

$$\frac{d^2y}{dx^2} = \frac{dv}{dx} = \frac{dv}{dy}\frac{dy}{dx} = \frac{dv}{dy}v \quad .$$

Using these substitutions with our example,

$$\frac{d^2y}{dx^2} + y = 0$$

becomes

$$v\frac{dv}{dy} + y = 0 \quad .$$

2. Solve the resulting differential equation for v as a function of y.

For our example:

$$v\frac{dv}{dy} + y = 0$$

$$\hookrightarrow \qquad v\frac{dv}{dy} = -y$$

$$\hookrightarrow \qquad \int v \frac{dv}{dy}\, dy = -\int y\, dy$$

$$\hookrightarrow \qquad \frac{1}{2}v^2 = -\frac{1}{2}y^2 + c_0$$

$$\hookrightarrow \qquad v = \pm\sqrt{2c_0 - y^2} \quad .$$

3. Rewrite the original substitution,
$$\frac{dy}{dx} = v \quad ,$$
replacing the v with the formula just found for v. Then observe that this is another first-order differential equation. In fact, you should notice that it is a separable first-order equation.

Replacing the v in
$$\frac{dy}{dx} = v$$
with the formula just obtained in our example for v, we get
$$\frac{dy}{dx} = \pm\sqrt{2c_0 - y^2} \quad .$$
Why, yes, this is a separable first-order differential equation for y!

4. Solve the first-order differential equation just obtained. This gives the general solution to the original second-order differential equation.

For our example,
$$\frac{dy}{dx} = \pm\sqrt{2c_0 - y^2}$$
$$\hookrightarrow \qquad \frac{1}{\sqrt{2c_0 - y^2}}\frac{dy}{dx} = \pm 1$$
$$\hookrightarrow \qquad \int \frac{1}{\sqrt{2c_0 - y^2}}\frac{dy}{dx}\, dx = \pm\int 1\, dx \quad .$$
Evaluating these integrals (after, perhaps, consulting our old calculus text or a handy table of integrals) yields
$$\arcsin\left(\frac{y}{a}\right) = \pm x + b \quad ,$$
where a and b are arbitrary constants with $a^2 = 2c_0$. (Note that c_0 had to be positive for square root to make sense.)

Taking the sine of both sides and recalling that sine is an odd function, we see that
$$\frac{y}{a} = \sin(\pm x + b) = \pm\sin(x \pm b) \quad .$$
Thus, letting $c_1 = \pm a$ and $c_2 = \pm b$, we have
$$y(x) = c_1 \sin(x + c_2)$$
as a general solution to our original differential equation,
$$\frac{d^2y}{dx^2} + y = 0 \quad .$$
(Later, after developing more theory, we will find easier ways to solve this and certain similar 'linear' equations.)

A Few Comments

Most of these should be pretty obvious:

1. Again, remember to check for the "constant-v" solutions of any separable differential equation for v.

2. It may be that the original differential equation does not explicitly contain either x or y. If so, then the approach just described and the approach described in the previous section may both be appropriate. Which one you choose is up to you. Often, it makes little difference (though the first is usually at least a little more straightforward). But occasionally (as illustrated in Exercise 13.10) one approach may be much easier to carry out than the other.

3. To be honest, we won't be using the method just outlined in later sections or chapters. Many of the second-order autonomous equations arising in applications are also "linear", and we will develop better methods for dealing with these equations over the next few chapters (where we will also learn just what it means to say that a differential equation is "linear"). It should also be mentioned that, much later, we will develop clever ways to analyze the possible solutions to fairly arbitrary autonomous differential equations after rewriting these equations as systems.

   Still, the method described here is invaluable for completely solving certain autonomous differential equations that are not "linear".

13.3 Initial-Value Problems

Initial Values with Higher-Order Equations

Remember, an Nth-order initial-value problem consists of an Nth-order differential equation along with the set of assignments

$$y(x_0) = y_0 \quad , \quad y'(x_0) = y_1 \quad , \quad y''(x_0) = y_2 \quad , \quad \ldots \quad \text{and} \quad y^{(N-1)}(x_0) = y_{N-1}$$

where x_0 is some single fixed number and the y_k's are the desired values of the function and its first few derivatives at position x_0.

In particular, a first-order initial-value problem consists of a first-order differential equation with a $y(x_0) = y_0$ initial condition. For example,

$$x\frac{dy}{dx} + 4y = x^3 \qquad \text{with} \quad y(1) = 3 \quad .$$

A second-order initial-value problem consists of a second-order differential equation along with $y(x_0) = y_0$ and $y'(x_0) = y_1$ initial conditions. For example,

$$\frac{d^2y}{dx^2} + 2\frac{dy}{dx} = 30e^{3x} \qquad \text{with} \quad y(0) = 5 \quad \text{and} \quad y'(0) = 14 \quad .$$

A third-order initial-value problem consists of a third-order differential equation along with $y(x_0) = y_0$, $y'(x_0) = y_1$ and $y''(x_0) = y_2$ initial conditions. For example,

$$3\frac{d^3y}{dx^3} = \left(\frac{d^2y}{dx^2}\right)^{-2} \qquad \text{with} \quad y(0) = 4 \quad , \quad y'(0) = 6 \quad \text{and} \quad y''(0) = 8 \quad .$$

And so on.

Solving Higher-Order Initial-Value Problems
The Basic Approach

The basic procedure for solving a typical higher-order initial-value problem is just about the same as the procedure for solving a first-order initial-value problem. You just need to account for the additional initial values.

First find the general solution to the differential equation. (Though we haven't proven it, you should expect the formula for the general solution to have as many arbitrary/undetermined constants as you have initial conditions.) Use the formula found for the general solution with each initial condition. This creates a system of algebraic equations for the yet-undetermined constants which can be solved for those constants. Solve the system by whatever means you can, and use those values for the constants in the formula for the differential equation's general solution. The resulting formula is the solution to the initial-value problem.

One example should suffice.

!►Example 13.5: *Consider the second-order initial-value problem*

$$\frac{d^2y}{dx^2} + 2\frac{dy}{dx} = 30e^{3x} \qquad \text{with} \quad y(0) = 5 \quad \text{and} \quad y'(0) = 14 \quad .$$

From Example 13.1 we already know that

$$y(x) = 2e^{3x} - c_1 e^{-2x} + c_2$$

is the general solution to the differential equation. Since the initial conditions include the value of $y'(x)$ at $x = 0$, we will also need the formula for y',

$$y'(x) = \frac{dy}{dx} = 6e^{3x} + 2c_1 e^{-2x} \quad ,$$

which we can obtain by either differentiating the above formula for y or copying the formula for y' from the work done in the example. Combining these formulas with the given initial conditions yields

$$5 = y(0) = 2e^{3\cdot 0} - c_1 e^{-2\cdot 0} + c_2 = 2 - c_1 + c_2$$

and

$$14 = y'(0) = 6e^{3\cdot 0} + 2c_1 e^{-2\cdot 0} = 6 + 2c_1 \quad .$$

That is,

$$5 = 2 - c_1 + c_2$$

and

$$14 = 6 + 2c_1 \quad .$$

Doing the obvious arithmetic, we get the system

$$\begin{aligned} -c_1 + c_2 &= 3 \\ 2c_1 \quad\;\; &= 8 \end{aligned}$$

of two equations and two unknowns. This is an easy system to solve. From the second equation we immediately see that

$$c_1 = \frac{8}{2} = 4 \quad .$$

Then, solving the first equation for c_2 and using the value just found for c_1, we see that

$$c_2 = c_1 + 3 = 4 + 3 = 7 \quad .$$

Thus, for $y(x)$ to satisfy the given differential equation and the two given initial conditions, we must have

$$y(x) = 2e^{3x} - c_1 e^{-2x} + c_2 \quad \text{with} \quad c_1 = 4 \quad \text{and} \quad c_2 = 7 \quad .$$

That is,

$$y(x) = 2e^{3x} - 4e^{-2x} + 7$$

is the solution to our initial-value problem.

An Alternative Approach

At times, it may be a little easier to determine the values of the arbitrary/undetermined constants "as they arise" in solving the differential equation. This is especially true when using the methods discussed in Sections 13.1 and 13.2, where we used the substitution $v = y'$ to convert the differential equation to a first-order differential equation for v. For the sort of equation considered in Section 13.1, this substitution immediately gives a first-order initial-value problem with

$$v(x_0) = y'(x_0) = y_1 \quad .$$

For the type of equation considered in Section 13.2 (the autonomous differential equations), the initial condition for $v = v(y)$ comes from combining $v = y'$ and the original initial conditions

$$y(x_0) = y_0 \quad \text{and} \quad y'(x_0) = y_1$$

into

$$v(y_0) = y'(x_0) = y_1 \quad .$$

Let's do one simple example.

!►*Example 13.6:* *Consider the initial-value problem*

$$\frac{d^2y}{dx^2} + 2\frac{dy}{dx} = 30e^{3x} \quad \text{with} \quad y(0) = 9 \quad \text{and} \quad y'(0) = 2 \quad .$$

The differential equation is the same as in Example 13.1 on page 247. Letting $v(x) = y'(x)$ yields the first-order initial-value problem

$$\frac{dv}{dx} + 2v = 30e^{3x} \quad \text{with} \quad v(0) = y'(0) = 2 \quad .$$

As shown in Example 13.1, the general solution to the differential equation for v is

$$v = 6e^{3x} + c_0 e^{-2x} \quad .$$

Combining this with the initial value yields

$$2 = v(0) = 6e^{3 \cdot 0} + c_0 e^{-\cdot 0} = 6 + c_0 \quad .$$

So, $c_0 = 2 - 6 = -4$, and

$$\frac{dy}{dx} = v = 6e^{3x} - 4e^{-2x} \quad .$$

Integrating this,

$$y(x) = \int \left[6e^{3x} - 4e^{-2x}\right] dx = 2e^{3x} + 2e^{-2x} + c_1 \quad ,$$

and using this formula for y *with the initial condition* $y(0) = 9$ *gives us*

$$9 = y(0) = 2e^{3 \cdot 0} + 2e^{-2 \cdot 0} + c_1 = 2 + 2 + c_1 = 4 + c_1 \quad .$$

Thus, $c_1 = 9 - 4 = 5$, *and*

$$y(x) = 2e^{3x} + 2e^{-2x} + 5$$

is the solution to our initial-value problem.

As the example illustrates, one advantage of this approach is that you only deal with one unknown constant at a time. This approach also bypasses obtaining the general solution to the original differential equation. Consequently, if the general solution is also desired, then the slight advantages of this method are considerably reduced.

13.4 On the Existence and Uniqueness of Solutions

Second-Order Problems

When we were dealing with first-order differential equations, we often found it useful to rewrite a given first-order equation in the derivative formula form,

$$\frac{dy}{dx} = F(x, y) \quad .$$

Extending this form to higher-order differential equations is straightforward. In particular, if we algebraically solve a second-order differential equation for the second derivative, y'', in terms of x, y and the first derivative, y', then we will have rewritten our differential equation in the form

$$y'' = F\left(x, y, y'\right)$$

for some function F of three variables. We will call this the *second-derivative formula form* for the second-order differential equation.

!►Example 13.7: *Solving the equation*

$$y'' + 2xy' = 5\sin(x)\, y$$

for the second derivative yields

$$y'' = \underbrace{5\sin(x)\, y - 2xy'}_{F(x,y,y')} \quad .$$

Replacing the derivative on the right with the symbol z, *we see that the formula for* F *is*

$$F(x, y, z) = 5\sin(x)\, y - 2xz \quad .$$

To be honest, we won't find the second-derivative formula form particularly useful in solving second-order differential equations until we seriously start dealing with systems of differential equations. It is mentioned here because it is the form used in the following basic theorems on the existence and uniqueness of solutions to second-order initial-value problems:

Theorem 13.1 (existence and uniqueness for second-order initial-value problems)
Consider a second-order initial-value problem

$$y'' = F\left(x, y, y'\right) \qquad \textit{with} \quad y(x_0) = y_0 \quad \textit{and} \quad y'(x_0) = z_0$$

in which $F = F(x, y, z)$ *and the corresponding partial derivatives* ${}^{\partial F}/_{\partial y}$ *and* ${}^{\partial F}/_{\partial z}$ *are all continuous in some open region containing the point* (x_0, y_0, z_0) *. This initial-value problem then has exactly one solution over some open interval* (α, β) *containing* x_0 *. Moreover, this solution and its first and second derivatives are continuous over that interval.*

Theorem 13.2 (existence and uniqueness for second-order initial-value problems)
Consider a second-order initial-value problem

$$y'' = F\left(x, y, y'\right) \qquad \textit{with} \quad y(x_0) = y_0 \quad \textit{and} \quad y'(x_0) = z_0$$

over an interval (α, β) *containing the point* x_0 *, and with* $F = F(x, y, z)$ *being a continuous function on an infinite slab*

$$\mathcal{R} = \{(x, y, z) : \alpha < x < \beta \ , \ -\infty < y < \infty \ \text{ and } \ -\infty < z < \infty\} \quad .$$

Further suppose that, on $\mathcal{R}$ *, the partial derivatives* ${}^{\partial F}/_{\partial y}$ *and* ${}^{\partial F}/_{\partial z}$ *are continuous and are functions of* x *only. Then the initial-value problem has exactly one solution on* (α, β) *. Moreover, the solution and its first and second derivatives are all continuous on this interval.*

The above theorems are the second-order analogs of the theorems on existence and uniqueness for first-order differential equations given in Section 3.3, and are also special cases of the two theorems we'll discuss next. They assure us that most of the second-order initial-value problems encountered in practice are, in theory, solvable. We will find the second theorem particularly important in rigorously establishing some useful results concerning the general solutions to an important class of second-order differential equations.

Problems of Any Order

The biggest difficulty in extending the above existence and uniqueness results for second-order problems to problems of arbitrary order N is that we quickly run out of letters to denote the variables and constants. So we will use subscripts.

Extending the idea of the "derivative formula form" for a first-order differential equation remains trivial. If, given an Nth-order differential equation, we can algebraically solve for the Nth-order derivative $y^{(N)}$ in terms of x , y , and the other derivatives of y , then we will say that we've gotten the differential equation into the *derivative formula form* for the differential equation,

$$y^{(N)} = F\left(x, y, y', \ldots, y^{(N-1)}\right) \quad .$$

Note that F will be a function of $N + 1$ variables, which we will denote by x , s_1 , s_2 ,..., s_N .

!▶Example 13.8: *Solving the equation*

$$y^{(4)} - x^2 y''' - yy'' + 2xy^3 y' = 0$$

for the fourth derivative yields

$$y^{(4)} = \underbrace{x^2 y''' + yy'' - 2xy^3 y'}_{F(x,y,y',y'',y''')} \quad .$$

The formula for F is then

$$F(x, s_1, s_2, s_3, s_4) = x^2 s_4 + s_1 s_3 - 2x(s_1)^3 s_2 \quad .$$

Again, the main reason to mention this form is that it is used in the Nth-order analogs of the existence and uniqueness theorems given in Section 3.3. These analogous theorems are:

Theorem 13.3 (existence and uniqueness for N^{th}-order initial-value problems)
Consider an Nth-order initial-value problem

$$y^{(N)} = F\left(x, y, y', \ldots, y^{(N-1)}\right)$$

with

$$y(x_0) = \sigma_1 \quad , \quad y'(x_0) = \sigma_2 \quad , \quad \ldots \quad \text{and} \quad y^{(N-1)}(x_0) = \sigma_N$$

in which $F = F(x, s_1, s_2, \ldots, s_N)$ and the corresponding partial derivatives

$$\frac{\partial F}{\partial s_1} \quad , \quad \frac{\partial F}{\partial s_2} \quad , \quad \ldots \quad \text{and} \quad \frac{\partial F}{\partial s_n}$$

are all continuous functions in some open region containing the point $(x_0, \sigma_1, \sigma_2, \ldots, \sigma_N)$. This initial-value problem then has exactly one solution over some open interval (α, β) containing x_0. Moreover, this solution and its derivatives up to order N are continuous over that interval.

Theorem 13.4 (existence and uniqueness for N^{th}-order initial-value problems)
Consider an Nth-order initial-value problem

$$y^{(N)} = F\left(x, y, y', \ldots, y^{(N-1)}\right)$$

with

$$y(x_0) = \sigma_1 \quad , \quad y'(x_0) = \sigma_2 \quad , \quad \ldots \quad \text{and} \quad y^{(N-1)}(x_0) = \sigma_N$$

over an interval (α, β) containing the point x_0, and with $F = F(x, s_1, s_2, \ldots, s_{N+1})$ being a continuous function on

$$\mathcal{R} = \{(x, s_1, s_2, \ldots, s_N) : \alpha < x < \beta \text{ and } -\infty < s_k < \infty \text{ for } k = 1, 2, \ldots, N\} \quad .$$

Further suppose that, on $\mathcal{R}$, the partial derivatives

$$\frac{\partial F}{\partial s_1} \quad , \quad \frac{\partial F}{\partial s_2} \quad , \quad \ldots \quad \text{and} \quad \frac{\partial F}{\partial s_N}$$

are all continuous and are functions of x only. Then the initial-value problem has exactly one solution on (α, β). Moreover, the solution and its first and second derivatives are all continuous on this interval.

A good way to prove the four theorems above is to use a "systems" approach. We'll discuss this further when we get to "systems of differential equations" in Chapter 38.

Additional Exercises

13.1. *None of the following second-order equations explicitly contains* y. *Solve each using the substitution* $v = y'$ *as described in Section 13.1.*

a. $xy'' + 4y' = 18x^2$

b. $xy'' = 2y'$

c. $y'' = y'$

d. $y'' + 2y' = 8e^{2x}$

e. $xy'' = y' - 2x^2y'$

f. $(x^2 + 1)y'' + 2xy' = 0$

13.2. *For each of the following, determine if the given differential equation explicitly contains* y. *If it does not, solve it using the substitution* $v = y'$ *as described in Section 13.1.*

a. $y'' = 4x\sqrt{y'}$

b. $y'y'' = 1$

c. $yy'' = -(y')^2$

d. $xy'' = (y')^2 - y'$

e. $xy'' - y' = 6x^5$

f. $yy'' - (y')^2 = y'$

g. $y'' = 2y' - 6$

h. $(y - 3)y'' = (y')^2$

i. $y'' + 4y' = 9e^{-3x}$

j. $y'' = y'\left(y' - 2\right)$

13.3. *Solve the following higher-order differential equations using the basic ideas from Section 13.1 (as done in Example 13.3 on page 249):*

a. $y''' = y''$

b. $xy''' + 2y'' = 6x$

c. $y''' = 2\sqrt{y''}$

d. $y^{(4)} = -2y'''$

13.4. *The following second-order equations are all autonomous. Solve each using the substitution* $v = y'$ *as described in Section 13.2.*

a. $yy'' = (y')^2$

b. $3yy'' = 2(y')^2$

c. $\sin(y)y'' + \cos(y)(y')^2 = 0$

d. $y'' = y'$

e. $(y')^2 + yy'' = 2yy'$

f. $y^2y'' + y'' + 2y(y')^2 = 0$

13.5. *For each of the following, determine if the given differential equation is autonomous. If it is, then solve it using the substitution* $v = y'$ *as described in Section 13.2.*

a. $y'' = 4x\sqrt{y'}$

b. $yy'' = -(y')^2$

c. $y'y'' = 1$

d. $xy'' = (y')^2 - y'$

e. $xy'' - y' = 6x^5$

f. $yy'' - (y')^2 = y'$

g. $yy'' = 2(y')^2$

h. $(y - 3)y'' = (y')^2$

i. $y'' + 4y' = 9e^{-3x}$

j. $y'' = y'\left(y' - 2\right)$

13.6. *Solve the following initial-value problems. In several cases you can use the general solutions already found for the corresponding differential equations in exercise sets 13.1 and 13.3:*

a. $xy'' + 4y' = 18x^2$ *with* $y(1) = 8$ *and* $y'(1) = -3$

b. $xy'' = 2y'$ with $y(-1) = 4$ and $y'(-1) = 12$

c. $y'' = y'$ with $y(0) = 8$ and $y'(0) = 5$

d. $y'' + 2y' = 8e^{2x}$ with $y(0) = 0$ and $y'(0) = 0$

e. $y''' = y''$ with $y(0) = 10$, $y'(0) = 5$ and $y''(0) = 2$

f. $xy''' + 2y'' = 6x$ with $y(1) = 2$, $y'(1) = 1$ and $y''(1) = 4$

g. $xy'' + 2y' = 6$ with $y(1) = 4$ and $y'(1) = 5$

h. $2xy'y'' = (y')^2 - 1$ with $y(1) = 0$ and $y'(1) = \sqrt{3}$

13.7. *Solve the following initial-value problems. In several cases you can use the general solutions already found for the corresponding differential equations in exercise set 13.4:*

a. $yy'' = (y')^2$ with $y(0) = 5$ and $y'(0) = 15$

b. $3yy'' = 2(y')^2$ with $y(0) = 8$ and $y'(0) = 6$

c. $3yy'' = 2(y')^2$ with $y(1) = 1$ and $y'(1) = 9$

d. $yy'' + 2(y')^2 = 3yy'$ with $y(0) = 2$ and $y'(0) = \frac{3}{4}$

e. $y'' = -y'e^{-y}$ with $y(0) = 0$ and $y'(0) = 2$

13.8. *In solving a second-order differential equation using the methods described in this chapter, we first solved a first-order differential equation for $v = y'$, obtaining a formula for $v = y'$ involving an arbitrary constant. Sometimes the value of the first 'arbitrary' constant affects how we solve $v = y'$ for y. You will illustrate this using*

$$y'' = -2x(y')^2 \tag{13.4}$$

in the following set of problems.

a. *Using the "alternative approach" to solving initial-value problems (as illustrated in Example 13.6 on page 255), find the solution to differential equation (13.4) satisfying each of the following sets of initial values:*

i. $y(0) = 3$ and $y'(0) = 4$ **ii.** $y(0) = 3$ and $y'(0) = 0$

iii. $y(1) = 0$ and $y'(1) = 1$ **iv.** $y(0) = -\frac{1}{4}$ and $y'(1) = 5$

(Observe how different the solutions to these different initial-value problems are, even though they all involve the same differential equation.)

b. *Find the set of all possible solutions to differential equation (13.4).*

13.9. *We will again illustrate the issue raised at the beginning of Exercise 13.8, but using differential equation*

$$y'' = 2yy' \quad . \tag{13.5}$$

a. *Using the "alternative approach" to solving initial-value problems (as illustrated in Example 13.6 on page 255), find the solution to differential equation (13.5) satisfying each of the following sets of initial values:*

i. $y(0) = 0$ and $y'(0) = 1$ **ii.** $y(0) = 1$ and $y'(0) = 1$

iii. $y(0) = 1$ and $y'(0) = 0$ **iv.** $y(0) = 0$ and $y'(0) = -1$

(Again, observe how different the solutions to these different initial-value problems are, even though they all involve the same differential equation.)

b. *Find the set of all possible solutions to differential equation (13.5).*

13.10. In Exercise 13.2 g, you showed that the differential equation

$$y'' = 2y' - 6$$

is easily solved using the substitution

$$v = \frac{dy}{dx} \quad \text{with} \quad \frac{d^2y}{dx^2} = \frac{dv}{dx} \quad .$$

(If you didn't do Exercise 13.2 g, go back and do it.) Since the differential equation is also autonomous, the other substitution discussed in this chapter,

$$v = \frac{dy}{dx} \quad \text{with} \quad \frac{d^2y}{dx^2} = \frac{dv}{dy}v \quad ,$$

should also be considered as appropriate. Try it and see what you get for v as a function of y. Can you, personally, actually go further and replace v with $^{dy}/_{dx}$ and solve the resulting differential equation? What is the moral of this exercise?

14

Higher-Order Linear Equations and the Reduction of Order Method

We have just seen that some higher-order differential equations can be solved using methods for first-order equations after applying the substitution $v = {}^{dy}/_{dx}$. Unfortunately, this approach has its limitations. Moreover, as we will later see, many of those differential equations that can be so solved can also be solved more easily using the theory and methods that will be developed in the next few chapters. This theory and methodology apply to the class of "linear" differential equations. This is a rather large class that includes a great many differential equations arising in applications. In fact, this class of equations is so important and the theory for dealing with these equations is so extensive that we will not again seriously consider higher-order nonlinear differential equations for many, many chapters.

14.1 Linear Differential Equations of All Orders

The Equations

Recall that a first-order differential equation is said to be linear if and only it can be written as

$$\frac{dy}{dx} + py = f \tag{14.1}$$

where $p = p(x)$ and $f = f(x)$ are known functions. Observe that this is the same as saying that a first-order differential equation is linear if and only if it can be written as

$$a\frac{dy}{dx} + by = g \tag{14.2}$$

where a, b and g are known functions of x. After all, the first equation is equation (14.2) with $a = 1$, $b = p$ and $f = g$, and any equation in the form of equation (14.2) can be converted to one looking like equation (14.1) by simply dividing through by a (so $p = {}^{b}/_{a}$ and $f = {}^{g}/_{a}$).

Higher-order analogs of either equation (14.1) or equation (14.2) can be used to define when a higher-order differential equation is "linear". We will find it slightly more convenient to use analogs of equation (14.2) (which was the reason for the above observations). Second- and third-order linear equations will first be described so you can start seeing the pattern. Then the general definition will be given. For convenience (and because there are only so many letters in the alphabet), we may start denoting different functions with subscripts.

A second-order differential equation is said to be *linear* if and only if it can be written as

$$a_0\frac{d^2y}{dx^2} + a_1\frac{dy}{dx} + a_2y = g \tag{14.3}$$

where a_0, a_1, a_2 and g are known functions of x. (In practice, generic second-order differential equations are often denoted by

$$a\frac{d^2y}{dx^2} + b\frac{dy}{dx} + cy = g \quad ,$$

instead.) For example,

$$\frac{d^2y}{dx^2} + x^2\frac{dy}{dx} - 6x^4y = \sqrt{x+1} \qquad \text{and} \qquad 3\frac{d^2y}{dx^2} + 8\frac{dy}{dx} - 6y = 0$$

are second-order linear differential equations, while

$$\frac{d^2y}{dx^2} + y^2\frac{dy}{dx} = \sqrt{x+1} \qquad \text{and} \qquad \frac{d^2y}{dx^2} = \left(\frac{dy}{dx}\right)^2$$

are not.

A third-order differential equation is said to be *linear* if and only if it can be written as

$$a_0\frac{d^3y}{dx^3} + a_1\frac{d^2y}{dx^2} + a_2\frac{dy}{dx} + a_3y = g$$

where a_0, a_1, a_2, a_3 and g are known functions of x. For example,

$$x^3\frac{d^3y}{dx^3} + x^2\frac{d^2y}{dx^2} + x\frac{dy}{dx} - 6y = e^x \qquad \text{and} \qquad \frac{d^3y}{dx^3} - y = 0$$

are third-order linear differential equations, while

$$\frac{d^3y}{dx^3} - y^2 = 0 \qquad \text{and} \qquad \frac{d^3y}{dx^3} + y\frac{dy}{dx} = 0$$

are not.

Getting the idea?

In general, for any positive integer N, we refer to an Nth-order differential equation as being *linear* if and only if it can be written as

$$a_0\frac{d^Ny}{dx^N} + a_1\frac{d^{N-1}y}{dx^{N-1}} + \cdots + a_{N-2}\frac{d^2y}{dx^2} + a_{N-1}\frac{dy}{dx} + a_Ny = g \tag{14.4}$$

where a_0, a_1, ..., a_N and g are known functions of x. For convenience, this equation will often be written using the prime notation for derivatives,

$$a_0y^{(N)} + a_1y^{(N-1)} + \cdots + a_{N-2}y'' + a_{N-1}y' + a_Ny = g \quad .$$

The function g on the right side of the above equation is often called the *forcing function* for the differential equation (because it often describes a force affecting whatever phenomenon the equation is modeling). If $g = 0$ (i.e., $g(x) = 0$ for every x in the interval of interest), then the equation is said to be *homogeneous*.[1] Conversely, if g is nonzero somewhere on the interval of interest, then we say the differential equation is *nonhomogeneous*.

As we will later see, solving a nonhomogeneous equation

$$a_0y^{(N)} + a_1y^{(N-1)} + \cdots + a_{N-2}y'' + a_{N-1}y' + a_Ny = g$$

[1] You may recall the term "homogeneous" from Chapter 6. If you compare what "homogeneous' meant there with what it means here, you will find absolutely no connection. The same term is being used for two completely different concepts.

is usually best done after first solving the homogeneous equation generated from the original equation by simply replacing g with 0,

$$a_0 y^{(N)} + a_1 y^{(N-1)} + \cdots + a_{N-2} y'' + a_{N-1} y' + a_N y = 0 \quad .$$

This corresponding homogeneous equation is officially called either the *corresponding homogeneous equation* or the *associated homogeneous equation*, depending on the author (we will use whichever phrase we feel like at the time). Do observe that the zero function,

$$y(x) = 0 \qquad \text{for all} \quad x \quad ,$$

is always a solution to a homogeneous linear differential equation (verify this for yourself). This is called the *trivial solution* and is not a very exciting solution. Invariably, the interest is in finding the *non*trivial solutions.

Intervals of Interest for Linear Equations

When attempting to solve a linear differential equation, be it second order,

$$a_0 y'' + a_1 y' + a_2 y = g \quad , \tag{14.5}$$

or of arbitrary order,

$$a_0 y^{(N)} + a_1 y^{(N-1)} + \cdots + a_{N-2} y'' + a_{N-1} y' + a_N y = g \quad , \tag{14.6}$$

we should keep in mind that we are seeking a solution (or general solution) valid over some "interval of interest", (α, β). To ensure that the solutions exist and are reasonably well behaved on the interval, we will usually require that

> *the g and the a_k's in equation (14.5) or equation (14.6) (depending on the equation of interest) must all be continuous functions on the interval (α, β) with a_0 never being zero at any point in this interval.*

Often, in practice, this assumption is not explicitly stated, or the interval (α, β) is not explicitly given. In these cases, you should usually assume that the interval of interest (α, β) is any interval over which the above assumption holds for the given differential equation.

Why is this assumption important? It is because of the following two theorems:

Theorem 14.1 (existence and uniqueness for second-order linear initial-value problems)
Consider the initial-value problem

$$ay'' + by' + cy = g \qquad \text{with} \quad y(x_0) = A \quad \text{and} \quad y'(x_0) = B$$

over an interval (α, β) containing the point x_0. Assume, further, that a, b, c and g are continuous functions on (α, β) with a never being zero at any point in this interval.. Then the initial-value problem has exactly one solution on (α, β). Moreover, the solution and its first and second derivatives are all continuous on this interval.

Theorem 14.2 (existence and uniqueness for N^th^-order linear initial-value problems)
Consider the initial-value problem

$$a_0 y^{(N)} + a_1 y^{(N-1)} + \cdots + a_{N-2} y'' + a_{N-1} y' + a_N y = g$$

with

$$y(x_0) = A_1 \quad , \quad y'(x_0) = A_2 \quad , \quad \ldots \quad \text{and} \quad y^{(N-1)}(x_0) = A_N$$

over an interval (α, β) *containing the point* x_0 *. Assume, further, that* g *and the* a_k *'s are continuous functions on* (α, β) *with* a_0 *never being zero at any point in this interval. Then the initial-value problem has exactly one solution on* (α, β) *. Moreover, the solution and its derivatives up to order* N *are all continuous on this interval.*

These theorems assure us that the initial-value problems we'll encounter are "completely solvable", at least, in theory.

Both of the above theorems are actually corollaries of theorems from the previous chapter (Theorems 13.2 on page 257 and 13.4 on page 258). For those interested, here is the proof of one:

PROOF (of Theorem 14.1): Start by rewriting the differential equation,

$$ay'' + by' + cy = g$$

as

$$y'' = \frac{g}{a} - \frac{b}{a}y' - \frac{c}{a}y \quad ,$$

and observe that this is

$$y'' = F(x, y, y') \qquad \text{where} \quad F(x, y, z) = \frac{g(x)}{a(x)} - \frac{b(x)}{a(x)}z - \frac{c(x)}{a(x)}y \quad .$$

The partial derivatives $\partial F/\partial y$ and $\partial F/\partial z$ are easily computed:

$$\frac{\partial F}{\partial y} = -\frac{c(x)}{a(x)} \qquad \text{and} \qquad \frac{\partial F}{\partial z} = -\frac{b(x)}{a(x)} \quad .$$

Note that these partial derivatives are functions of x only. In addition, since a, b and c are continuous on (α, β), and a is never zero on (α, β), it is easy to see that F, $\partial F/\partial y$ and $\partial F/\partial z$ are all continuous on

$$\mathcal{R} = \{\, (x, y, z) : \alpha < x < \beta \ , \ -\infty < y < \infty \ \text{and} \ -\infty < z < \infty \,\} \quad .$$

The claims of our theorem now follow immediately from Theorem 13.2 (using $y_0 = A$ and $z_0 = B$). ∎

14.2 Introduction to the Reduction of Order Method

The rest of this chapter is devoted to the *reduction of order method.* This is a method for converting any linear differential equation to another linear differential equation of lower order, and then constructing the general solution to the original differential equation using the general solution to the lower-order equation. In general, to use this method with an Nth-order linear differential equation

$$a_0 y^{(N)} + a_1 y^{(N-1)} + \cdots + a_{N-2} y'' + a_{N-1} y' + a_N y = g \quad ,$$

we need to already have one known nontrivial solution $y_1 = y_1(x)$ to the corresponding homogeneous differential equation. We then try a substitution of the form

$$y = y_1 u$$

where $u = u(x)$ is a yet unknown function (and $y_1 = y_1(x)$ is the aforementioned known solution). Plugging this substitution into the differential equation then leads to a linear differential equation for u. As we will see, because y_1 satisfies the corresponding homogeneous equation, the differential equation for u ends up being of the form

$$A_0 u^{(N)} + A_1 u^{(N-1)} + \cdots + A_{N-2} u'' + A_{N-1} u' = g$$

— remarkably, there is no "$A_N u$" term. This means we can use the substitution

$$v = u' \quad ,$$

as discussed in Chapter 13, to rewrite the differential equation for u as an $(N-1)^{\text{th}}$-order differential equation for v,

$$A_0 v^{(N-1)} + A_1 v^{(N-2)} + \cdots + A_{N-2} v' + A_{N-1} v = g \quad .$$

So we have reduced the order of the equation to be solved. If a general solution $v = v(x)$ for this equation can be found, then the most general formula for u can be obtained from v by integration (since $u' = v$). Finally then, by going back to the original substitution formula $y = y_1 u$, we can obtain a general solution to the original differential equation.

This method is especially useful for solving second-order, homogeneous linear differential equations since (as we will see) it reduces the problem to one of solving relatively simple first-order differential equations. Accordingly, we will first concentrate on its use in finding general solutions to second-order, homogeneous linear differential equations. Then we will briefly discuss using reduction of order with second-order, nonhomogeneous linear equations, and with both homogeneous and nonhomogeneous linear differential equations of higher orders.

14.3 Reduction of Order for Homogeneous Linear Second-Order Equations

The Method

Here we lay out the details of the reduction of order method for second-order, homogeneous linear differential equations. To illustrate the method, we'll use the differential equation

$$x^2 y'' - 3xy' + 4y = 0 \quad .$$

Note that the first coefficient, x^2, vanishes when $x = 0$. From comments made earlier (see Theorem 14.1 on page 265), we should suspect that $x = 0$ ought not be in any interval of interest for this equation. So we will be solving over the intervals $(0, \infty)$ and $(-\infty, 0)$.

Before starting the reduction of order method, we need one nontrivial solution y_1 to our differential equation. Ways for finding that first solution will be discussed in later chapters. For now let us just observe that if

$$y_1(x) = x^2 \quad ,$$

then

$$\begin{aligned} x^2 y_1'' - 3xy_1' + 4y_1 &= x^2 \frac{d^2}{dx^2}\big[x^2\big] - 3x\frac{d}{dx}\big[x^2\big] + 4\big[x^2\big] \\ &= x^2[2 \cdot 1] - 3x[2x] + 4x^2 \\ &= x^2 \underbrace{[2 - (3 \cdot 2) + 4]}_{0!} = 0 \quad . \end{aligned}$$

Thus, one solution to the above differential equation is $y_1(x) = x^2$.

As already stated, this method is for finding a general solution to some homogeneous linear second-order differential equation

$$ay'' + by' + cy = 0$$

(where a, b, and c are known functions with $a(x)$ never being zero on the interval of interest). We will assume that we already have one nontrivial particular solution $y_1(x)$ to this generic differential equation.

For our example (as already noted), we will seek a general solution to

$$x^2 y'' - 3xy' + 4y = 0 \quad . \tag{14.7}$$

The one (nontrivial) solution we know is $y_1(x) = x^2$.

Here are the details in using the reduction of order method to solve the above:

1. Let

$$y = y_1 u$$

where $u = u(x)$ is a function yet to be determined. To simplify "plugging into the differential equation", go ahead and compute the corresponding formulas for the derivatives y' and y'' using the product rule:

$$y' = (y_1 u)' = y_1' u + y_1 u'$$

and

$$\begin{aligned} y'' = (y')' &= (y_1' u + y_1 u')' \\ &= (y_1' u)' + (y_1 u')' \\ &= (y_1'' u + y_1' u') + (y_1' u' + y_1 u'') \\ &= y_1'' u + 2y_1' u' + y_1 u'' \quad . \end{aligned}$$

For our example,

$$y = y_1 u = x^2 u$$

where $u = u(x)$ *is the function yet to be determined. The derivatives of* y *are*

$$y' = (x^2 u)' = 2xu + x^2 u'$$

and

$$\begin{aligned} y'' = (y')' &= (2xu + x^2 u')' \\ &= (2xu)' + (x^2 u')' \\ &= (2u + 2xu') + (2xu' + x^2 u'') \\ &= 2u + 4xu' + x^2 u'' \quad . \end{aligned}$$

2. Plug the formulas just computed for y, y' and y'' into the differential equation, group together the coefficients for u and each of its derivatives, and simplify as far as possible. (We'll do this with the example first and then look at the general case.)

Plugging the formulas just computed above for y, y' and y'' into equation (14.7), we get

$$\begin{aligned} 0 &= x^2y'' - 3xy' + 4y \\ &= x^2\left[2u + 4xu' + x^2u''\right] - 3x\left[2xu + x^2u'\right] + 4\left[x^2u\right] \\ &= 2x^2u + 4x^3u' + x^4u'' - 6x^2u - 3x^3u' + 4x^2u \\ &= x^4u'' + \left[4x^3 - 3x^3\right]u' + \left[2x^2 - 6x^2 + 4x^2\right]u \\ &= x^4u'' + x^3u' + 0 \cdot u \quad . \end{aligned}$$

Notice that the u term drops out! So the resulting differential equation for u is simply

$$x^4u'' + x^3u' = 0 \quad ,$$

which we can further simplify by dividing out x^4,

$$u'' + \frac{1}{x}u' = 0 \quad .$$

In general, plugging the formulas for y and its derivatives into the given differential equation yields

$$\begin{aligned} 0 &= ay'' + by' + cy \\ &= a\left[y_1''u + 2y_1'u' + y_1u''\right] + b\left[y_1'u + y_1u'\right] + c\left[y_1u\right] \\ &= ay_1''u + 2ay_1'u' + ay_1u'' + by_1'u + by_1u' + cy_1u \\ &= ay_1u'' + \left[2ay_1' + by_1\right]u' + \left[ay_1'' + by_1' + cy_1\right]u \quad . \end{aligned}$$

That is, the differential equation becomes

$$Au'' + Bu' + Cu = 0$$

where

$$A = ay_1 \quad , \quad B = 2ay_1' + by_1 \quad \text{and} \quad C = ay_1'' + by_1' + cy_1 \quad .$$

But remember, y_1 is a solution to the homogeneous equation

$$ay'' + by' + cy = 0 \quad .$$

Consequently,

$$C = ay_1'' + by_1' + cy_1 = 0 \quad ,$$

and the differential equation for u automatically reduces to

$$Au'' + Bu' = 0 \quad .$$

The u term always drops out!

3. Now find the general solution to the second-order differential equation just obtained for u,

$$Au'' + Bu' = 0 \quad ,$$

via the substitution method discussed in Section 13.1:

(a) Let $u' = v$ (and, thus, $u'' = v' = {}^{dv}/_{dx}$) to convert the second-order differential equation for u to the first-order differential equation for v,

$$A\frac{dv}{dx} + Bv = 0 \quad .$$

(It is worth noting that this first-order differential equation is both linear and separable.)

(b) Find the general solution $v(x)$ to this first-order equation. (Since it is both linear and separable, you can solve it using either the procedure developed for first-order linear equations or the approach developed for first-order separable equations.)

(c) Using the formula just found for v, integrate the substitution formula $u' = v$ to obtain the formula for u,

$$u(x) = \int v(x)\,dx \quad .$$

Don't forget all the arbitrary constants.

In our example, we obtained

$$u'' + \frac{1}{x}u' = 0 \quad .$$

Letting $v = u'$ *and, thus,* $v' = u''$ *this becomes*

$$\frac{dv}{dx} + \frac{1}{x}v = 0 \quad .$$

Equivalently,

$$\frac{dv}{dx} = -\frac{1}{x}v \quad .$$

This is a relatively simple separable first-order equation. It has one constant solution, $v = 0$. *To find the others, we divide through by* v *and proceed as usual with such equations:*

$$\frac{1}{v}\frac{dv}{dx} = -\frac{1}{x}$$

$$\hookrightarrow \quad \int \frac{1}{v}\frac{dv}{dx}\,dx = -\int \frac{1}{x}\,dx$$

$$\hookrightarrow \quad \ln|v| = -\ln|x| + c_0$$

$$\hookrightarrow \quad v = \pm e^{-\ln|x| + c_0}$$

$$\hookrightarrow \quad v = \pm e^{c_0}x^{-1} \quad .$$

Letting $A = \pm e^{c_0}$, *this simplifies to*

$$v = \frac{A}{x} \quad ,$$

which also accounts for the constant solution (when $A = 0$ *).*

Since $u' = v$, it then follows that

$$u(x) = \int v(x)\,dx = \int \frac{A}{x}\,dx = A\ln|x| + B \quad .$$

4. Finally, plug the formula just obtained for $u(x)$ into the first substitution,

$$y = y_1 u \quad ,$$

used to convert the original differential equation for y to a differential equation for u. The resulting formula for $y(x)$ will be a general solution for that original differential equation. (Sometimes that formula can be simplified a little. Feel free to do so.)

In our example, the solution we started with was $y_1(x) = x^2$. Combined with the $u(x)$ just found, we have

$$y = y_1 u = x^2[A\ln|x| + B] \quad .$$

That is,

$$y(x) = Ax^2\ln|x| + Bx^2$$

is the general solution to equation (14.7).

An Observation About the General Solution

In the above example, the general solution obtained was

$$y(x) = Ax^2\ln|x| + Bx^2 \quad .$$

It will be worth noting that, after renaming the arbitrary constants A and B as c_2 and c_1, respectively, we can rewrite this as

$$y(x) = c_1 y_1(x) + c_2 y_2(x)$$

where

$$y_1(x) = x^2$$

is the one solution we already knew, and

$$y_2(x) = x^2\ln|x|$$

is another function that arose in the course of our procedure, and which, we should also note, can be written as

$$y_2(x) = y_1(x)u_0(x) \qquad \text{with} \qquad u_0(x) = \ln|x| \quad .$$

Moreover:

1. Since $y(x) = c_1 y_1(x) + c_2 y_2(x)$ is a solution to our differential equation for any choice of constants c_1 and c_2, it follows (by taking $c_1 = 0$ and $c_2 = 1$) that y_2 is also a particular solution to our differential equation.

2. Since $u_0(x)$ is NOT a constant, y_2 is NOT a constant multiple of y_1.

It turns out that the above observation does not just hold for this one example. If you look carefully at the reduction of order method for solving any second-order linear homogeneous differential equation,

$$ay'' + by' + cy = 0 \quad ,$$

you will find that this method always results in a general solution of the form

$$y(x) = c_1 y_1(x) + c_2 y_2(x)$$

where

1. c_1 and c_2 are arbitrary constants,
2. y_1 is the one known particular solution needed to start the method, and
3. y_2 is another particular solution to the given differential equation which is NOT a constant multiple of y_1.

As noted, the above can be easily verified by looking at the general formulas that arise in using the reduction of order method.[2] For now though, let us take the above as something you should observe in the solutions you obtain in doing the exercises for this section.

And why are the above observations important? Because they will serve as the starting point for a more general discussion of "general solutions" in the next chapter, and which, in turn, will lead to other methods for solving our differential equations, often (but not always) without the need to go through the full reduction of order method.

14.4 Reduction of Order for Nonhomogeneous Linear Second-Order Equations

If you look back over our discussion in Section 14.3, you will see that the reduction of order method applies almost as well in solving a nonhomogeneous equation

$$ay'' + by' + cy = g \quad ,$$

provided that "one solution y_1" is a solution to the corresponding *homogeneous* equation

$$ay'' + by' + cy = 0 \quad .$$

Then, letting $y = y_1 u$ in the nonhomogeneous equation and then replacing u' with v leads to an equation of the form

$$Av' + Bv = g$$

instead of

$$Av' + Bv = 0 \quad .$$

So the resulting first-order equation for v is certainly linear, but is probably not separable. Still, we know how to solve such equations. Solving that first-order linear differential equation for v and continuing with the method already described finally yields the general solution to the desired nonhomogeneous differential equation.

We will do one example. Then I'll tell you why the method is rarely used in practice.

[2] Actually, there are technical issues arising at points where y_1 is zero.

!► Example 14.1: Let us try to solve the second-order nonhomogeneous linear differential equation

$$x^2y'' - 3xy' + 4y = \sqrt{x} \tag{14.8}$$

over the interval $(0, \infty)$.

As we saw in our main example in Section 14.3, the corresponding homogeneous equation

$$x^2y'' - 3xy' + 4y = 0$$

has $y_1(x) = x^2$ as one solution (in fact, from that example, we know the entire general solution to this homogeneous equation, but we only need this one particular solution for the method). Let

$$y = y_1 u = x^2 u$$

where $u = u(x)$ is the function yet to be determined. The derivatives of y are

$$y' = \left(x^2u\right)' = 2xu + x^2u'$$

and

$$\begin{aligned} y'' = (y')' &= \left(2xu + x^2u'\right)' \\ &= 2u + 2xu' + 2xu' + x^2u'' \\ &= 2u + 4xu' + x^2u'' \quad . \end{aligned}$$

Plugging these into equation (14.8) yields

$$\begin{aligned} \sqrt{x} &= x^2y'' - 3xy' + 4y \\ &= x^2\left[2u + 4xu' + x^2u''\right] - 3x\left[2xu + x^2u'\right] + 4\left[x^2u\right] \\ &= 2x^2u + 4x^3u' + x^4u'' - 6x^2u - 3x^3u' + 4x^2u \\ &= x^4u'' + \left[4x^3 - 3x^3\right]u' + \left[2x^2 - 6x^2 + 4x^2\right]u \\ &= x^4u'' + x^3u' + 0 \cdot u \quad . \end{aligned}$$

As before, the u term drops out. In this case, we are left with

$$x^4u'' + x^3u' = \sqrt{x} \quad .$$

That is,

$$x^4v' + x^3v = x^{1/2} \qquad \text{with} \quad v = u' \quad .$$

This is a relatively simple first-order linear equation. To help find the integrating factor, we now divide through by x^4, obtaining

$$\frac{dv}{dx} + \frac{1}{x}v = x^{-7/2} \quad .$$

So the integrating factor is

$$\mu = e^{\int 1/x \, dx} = e^{\ln|x|} = |x| \quad .$$

Since we are just attempting to solve over the interval $(0, \infty)$, we really just have

$$\mu = x \quad .$$

Multiplying the last differential equation for v and proceeding as usual when solving first-order linear differential equations:

$$x\left[\frac{dv}{dx} + \frac{1}{x}v\right] = x\left[x^{-7/2}\right]$$

$$\hookrightarrow \quad x\frac{dv}{dx} + v = x^{-5/2}$$

$$\hookrightarrow \quad \frac{d}{dx}\left[xv\right] = x^{-5/2}$$

$$\hookrightarrow \quad \int \frac{d}{dx}\left[xv\right]dx = \int x^{-5/2}\,dx$$

$$\hookrightarrow \quad xv = -\frac{2}{3}x^{-3/2} + c_1$$

$$\hookrightarrow \quad v = -\frac{2}{3}x^{-5/2} + \frac{c_1}{x} \quad .$$

Recalling that $v = u'$, we can rewrite the last line as

$$\frac{du}{dx} = -\frac{2}{3}x^{-5/2} + \frac{c_1}{x} \quad .$$

Thus,

$$u = \int \frac{du}{dx}\,dx = \int \left[-\frac{2}{3}x^{-5/2} + \frac{c_1}{x}\right]dx$$

$$= \left(\frac{2}{3}\right)^2 x^{-3/2} + c_1\ln(x) + c_2$$

$$= \frac{4}{9}x^{-3/2} + c_1\ln(x) + c_2 \quad ,$$

and the general solution to our nonhomogeneous equation is

$$y(x) = x^2u(x) = x^2\left[\frac{4}{9}x^{-3/2} + c_1\ln(x) + c_2\right]$$

$$= \frac{4}{9}x^{1/2} + c_1x^2\ln(x) + c_2x^2 \quad .$$

For no obvious reason at this point, let's observe that we can write this solution as

$$y(x) = c_1x^2\ln(x) + c_2x^2 + \frac{4}{9}\sqrt{x} \quad . \tag{14.9}$$

It should be observed that, in the above example, we only used one particular solution, $y_1(x) = x^2$, to the homogeneous differential equation

$$x^2y'' - 3xy' + 4y = 0$$

even though we had already found the general solution

$$Ax^2\ln|x| + Bx^2 \quad .$$

Later, in Chapter 24, we will develop a refinement of the reduction of order method for solving second-order, nonhomogeneous linear differential equations that makes use of the entire general solution to the corresponding homogeneous equation. This refinement (the "variation of parameters" method) has two distinct advantages over the reduction of order method when solving *non*homogeneous differential equations:

1. The computations required for the refined procedure tend to be simpler and more easily carried out.

2. With a few straightforward modifications, the refined procedure readily extends to being a useful method for dealing with nonhomogeneous linear differential equations of any order. For the reasons discussed in the next section, the same cannot be said about the basic reduction of order method.

That is why, in practice, the basic reduction of order method is rarely used with nonhomogeneous equations.

14.5 Reduction of Order in General

In theory, reduction of order can be applied to any linear equation of any order, homogeneous or not. Whether its application is useful is another issue.

!►Example 14.2: *Consider the third-order homogeneous linear differential equation*

$$y''' - 8y = 0 \quad . \tag{14.10}$$

If you rewrite this equation as

$$y''' = 8y \quad ,$$

and think about what happens when you differentiate exponentials, you will realize that

$$y_1(x) = e^{2x}$$

is 'obviously' a solution to our differential equation (verify it yourself). Letting

$$y = y_1 u = e^{2x} u$$

and repeatedly using the product rule, we get

$$y' = \left(e^{2x}u\right)' = 2e^{2x}u + e^{2x}u' \quad ,$$

$$\begin{aligned} y'' = \left(e^{2x}u\right)'' &= \left(2e^{2x}u + e^{2x}u'\right)' \\ &= 4e^{2x}u + 2e^{2x}u' + 2e^{2x}u' + e^{2x}u'' \\ &= 4e^{2x}u + 4e^{2x}u' + e^{2x}u'' \end{aligned}$$

and

$$\begin{aligned} y''' = \left(e^{2x}u\right)''' &= \left(4e^{2x}u + 4e^{2x}u' + e^{2x}u''\right)' \\ &= 8e^{2x}u + 4e^{2x}u' + 8e^{2x}u' + 4e^{2x}u'' + 2e^{2x}u'' + e^{2x}u''' \\ &= 8e^{2x}u + 12e^{2x}u' + 6e^{2x}u'' + e^{2x}u''' \quad . \end{aligned}$$

So, using $y = e^{2x}u$,

$$y''' - 8y = 0$$

$$\hookrightarrow \quad \left[8e^{2x}u + 12e^{2x}u' + 6e^{2x}u'' + e^{2x}u'''\right] - 8\left[e^{2x}u\right] = 0$$

$$\hookrightarrow \quad e^{2x}u''' + 6e^{2x}u'' + 12e^{2x}u' + \left[8e^{2x} - 8e^{2x}\right]u = 0 \quad .$$

Again, the u *term vanishes, leaving us with*

$$e^{2x}u''' + 6e^{2x}u'' + 12e^{2x}u' = 0 \quad .$$

Letting $v = u'$ *and dividing out the exponential, this becomes the second-order differential equation*

$$v'' + 6v' + 12v = 0 \quad . \tag{14.11}$$

Thus we have changed the problem of solving the third-order differential equation to one of solving a second-order differential equation. If we can now correctly guess a particular solution v_1 *to that second-order differential equation, we could again use reduction of order to get the general solution* $v(x)$ *to that second-order equation, and then use that and the fact that* $y = e^{2x}u$ *with* $v = u'$ *to obtain the general solution to our original differential equation. Unfortunately, even though the order is less, "guessing" a solution to equation (14.11) is a good deal more difficult than was guessing a particular solution to the original differential equation, equation (14.10).*

As the example illustrates, even if we can, somehow, obtain one particular solution to a given N^{th}-order linear homogeneous linear differential equation, and then use it to reduce the problem to solving an $(N-1)^{\text{th}}$-order differential equation, that lower-order differential equation may be just as hard to solve as the original differential equation (unless $N = 2$). In fact, we will learn how to solve differential equations such as equation (14.11), but those methods can also be used to find the general solution to the original differential equation, equation (14.10), as well.

Still it does no harm to know that the problem of solving an Nth-order linear homogeneous linear differential equation can be reduced to that of solving an $(N-1)^{\text{th}}$-order differential equation, at least when we have one solution to the original equation. For the record, here is a theorem to that effect:

Theorem 14.3 (reduction of order in homogeneous equations)
Let y *be any solution to some* N^{th}*-order homogeneous differential equation*

$$a_0y^{(N)} + a_1y^{(N-1)} + \cdots + a_{N-2}y'' + a_{N-1}y' + a_Ny = g \tag{14.12}$$

where g *and the* a_k*'s are known functions on some interval* (α, β) *, and let* y_1 *be a nontrivial particular solution to the corresponding homogeneous equation*

$$a_0y^{(N)} + a_1y^{(N-1)} + \cdots + a_{N-2}y'' + a_{N-1}y' + a_Ny = 0 \quad .$$

Set

$$u = \frac{y}{y_1} \qquad (\text{so that } y = y_1 u) \quad .$$

Then $v = u'$ *satisfies an* $(N-1)^{th}$*-order differential equation*

$$A_0v^{(N-1)} + A_1v^{(N-2)} + \cdots + A_{N-2}v' + A_{N-1}v = g \quad ,$$

where the A_k's are functions on the interval (α, β) that can be determined from the a_k's along with y_1 and its derivatives.

The proof is relatively straightforward: You see what happens when you repeatedly use the product rule with $y = y_1 u$, and plug the results into the equation (14.12). You can fill in the details yourself (see Exercise 14.4).

Additional Exercises

14.1. *For each of the following differential equations, identify*

i. *the order of the equation,*

ii. *whether the equation is linear or not, and,*

iii. *if it is linear, whether the equation is homogeneous or not.*

a. $y'' + x^2 y' - 4y = x^3$

b. $y'' + x^2 y' - 4y = 0$

c. $y'' + x^2 y' = 4y$

d. $y'' + x^2 y' + 4y = y^3$

e. $xy' + 3y = e^{2x}$

f. $y''' + y = 0$

g. $(y+1)y'' = (y')^3$

h. $y'' = 2y' - 5y + 30e^{3x}$

i. $y^{(iv)} + 6y'' + 3y' - 83y - 25 = 0$

j. $yy''' + 6y'' + 3y' = y$

k. $y''' + 3y' = x^2 y$

l. $y^{(55)} = \sin(x)$

14.2. *For each of the following, first verify that the given y_1 is a solution to the given differential equation, and then find the general solution to the differential equation using the given y_1 with the method of reduction of order.*

a. $y'' - 5y' + 6y = 0 \quad , \quad y_1(x) = e^{2x}$

b. $y'' - 10y' + 25y = 0 \quad , \quad y_1(x) = e^{5x}$

c. $x^2 y'' - 6xy' + 12y = 0 \quad \text{on} \quad x > 0 \quad , \quad y_1(x) = x^3$

d. $2x^2 y'' - xy' + y = 0 \quad \text{on} \quad x > 0 \quad , \quad y_1(x) = x$

e. $4x^2 y'' + y = 0 \quad \text{on} \quad x > 0 \quad , \quad y_1(x) = \sqrt{x}$

f. $y'' - \left(4 + \frac{2}{x}\right)y' + \left(4 + \frac{4}{x}\right)y = 0 \quad \text{on} \quad x > 0 \quad , \quad y_1(x) = e^{2x}$

g. $(x+1)y'' + xy' - y = 0 \quad , \quad y_1(x) = e^{-x}$

h. $y'' - \frac{1}{x}y' - 4x^2 y = 0 \quad , \quad y_1(x) = e^{-x^2}$

i. $y'' + y = 0 \quad , \quad y_1(x) = \sin(x)$

j. $xy'' + (2+2x)y' + 2y = 0 \quad \text{on} \quad x > 0 \quad , \quad y_1(x) = x^{-1}$

k. $\sin^2(x)y'' - 2\cos(x)\sin(x)y' + \left(1 + \cos^2(x)\right)y = 0 \quad , \quad y_1(x) = \sin(x)$

l. $x^2y'' - 2xy' + \left(x^2+2\right)y = 0$ on $x > 0$, $y_1(x) = x\sin(x)$

m. $x^2y'' + xy' + y = 0$ on $x > 0$, $y_1(x) = \sin(\ln|x|)$

n. $x^2y'' + xy' + \left(x^2 - \frac{1}{4}\right)y = 0$ on $x > 0$, $y_1(x) = x^{-1/2}\cos(x)$

14.3. *Several nonhomogeneous differential equations are given below. For each, first verify that the given y_1 is a solution to the corresponding homogeneous differential equation, and then find the general solution to the given nonhomogeneous differential equation using reduction of order with the given y_1.*

a. $y'' - 4y' + 3y = 9e^{2x}$, $y_1(x) = e^{3x}$

b. $y'' - 6y' + 8y = e^{4x}$, $y_1(x) = e^{2x}$

c. $x^2y'' + xy' - y = \sqrt{x}$ on $x > 0$, $y_1(x) = x$

d. $x^2y'' - 20y = 27x^5$ on $x > 0$, $y_1(x) = x^5$

e. $xy'' + (2+2x)y' + 2y = 8e^{2x}$ on $x > 0$, $y_1(x) = x^{-1}$

f. $(x+1)y'' + xy' - y = (x+1)^2$, $y_1(x) = e^{-x}$

14.4. *Prove the claims in Theorem 14.3 assuming:*

a. $N = 3$ **b.** $N = 4$ **c.** N *is any positive integer*

14.5. *Each of the following is one of the relatively few third- and fourth-order differential equations that can be easily solved via reduction of order. For each, first verify that the given y_1 is a solution to the given differential equation or to the corresponding homogeneous equation (as appropriate), and then find the general solution to the differential equation using the given y_1 with the method of reduction of order.*

a. $y''' - 9y'' + 27y' - 27y = 0$, $y_1(x) = e^{3x}$

b. $y''' - 9y'' + 27y' - 27y = e^{3x}\sin(x)$, $y_1(x) = e^{3x}$

c. $y^{(4)} - 8y''' + 24y'' - 32y' + 16y = 0$, $y_1(x) = e^{2x}$

d. $x^3y''' - 4y'' + 10y' - 12y = 0$ on $x > 0$, $y_1(x) = x^2$

15

General Solutions to Homogeneous Linear Differential Equations

The reduction of order method is, admittedly, limited. First of all, you must already have one solution to the given differential equation before you can start the method. Moreover, it's just not that helpful when the order of the equation is greater than two. We will be able to get around these limitations, at least when dealing with certain important classes of differential equations, by using methods that will be developed in later chapters. An important element of those methods is the construction of general solutions from a "suitably chosen" collection of particular solutions. That element is the topic of this chapter. We will discover how to choose that "suitably chosen" set and how to (easily) construct a general solution from that set. Once you know that, you can proceed straight to Chapter 17 (skipping, if you wish, Chapter 16) and learn how to fully solve some of the more important equations found in applications.

By the way, we will not abandon the reduction of order method. It will still be needed, and the material in this chapter will help tell us when it is needed.

15.1 Second-Order Equations (Mainly)

The Problem and a Basic Question

Throughout this section, our interest will be in "finding" a general solution over some interval (α, β) to a fairly arbitrary second-order homogeneous linear differential equation

$$ay'' + by' + cy = 0 \quad .$$

As suggested in the last chapter, we will limit the interval (α, β) to being one over which the functions a, b and c are all continuous with a never being zero.

At the end of Section 14.3, we observed that the reduction of order method always seemed to yield a general solution to the above differential equation in the form

$$y(x) = c_1 y_1(x) + c_2 y_2(x)$$

where c_1 and c_2 are arbitrary constants, and y_1 and y_2 are a pair of particular solutions that are not constant multiples of each other. Then, y_1 was a previously known solution to the differential equation, and y_2 was a solution that arose from the order of reduction method. You were not, however, given much advice on how to find that first particular solution. We will discover (in later chapters) that there are relatively simple methods for finding "that one solution" at least for certain common types of differential equations. We will further discover that, while these methods do not, themselves, yield general solutions, they often do yield sets of particular solutions. Here is a simple example:

!►Example 15.1: *Consider the homogeneous differential equation*

$$y'' + y = 0$$

over the entire real line, $(-\infty, \infty)$*. We can find at least two solutions by rewriting this as*

$$y'' = -y \quad ,$$

and then asking ourselves if we know of any basic functions (powers, exponentials, trigonometric functions, etc.) that satisfy this. It should not take long to recall that

$$y(x) = \cos(x) \qquad \text{and} \qquad y(x) = \sin(x)$$

are two such functions: If $y(x) = \cos(x)$*, then*

$$y''(x) = \frac{d^2}{dx^2}[\cos(x)] = \frac{d}{dx}[-\sin(x)] = -\cos(x) = -y(x) \quad ,$$

and if $y(x) = \sin(x)$*, then*

$$y''(x) = \frac{d^2}{dx^2}[\sin(x)] = \frac{d}{dx}[\cos(x)] = -\sin(x) = -y(x) \quad .$$

Thus, both

$$y(x) = \cos(x) \qquad \text{and} \qquad y(x) = \sin(x)$$

are solutions to our differential equation. Moreover, it should be clear that neither is a constant multiple of the other. So, to find the general solution of our differential equation, should we use the reduction of order method with $y_1(x) = \cos(x)$ *as the known solution or the reduction of order method with* $y_1(x) = \sin(x)$ *as the known solution — or can we simply say*

$$y(x) = c_1 \cos(x) + c_2 \sin(x)$$

is the general solution, skipping the reduction of order method altogether?

Thus, we are led to the basic question:

Can the general solution to any second-order homogeneous linear differential equation

$$ay'' + by' + cy = 0 \quad ,$$

be given by

$$y(x) = c_1 y_1(x) + c_2 y_2(x)$$

where c_1 *and* c_2 *are arbitrary constants, and* y_1 *and* y_2 *are any two solutions that are not constant multiples of each other?*

Our goal is to answer this question by the end of this section.

Linear Combinations and the Principle of Superposition

Time to introduce a little terminology to simplify future discussion: Given any finite collection of functions — y_1, y_2, ... and y_N — a *linear combination* of these y_k's is any expression of the form

$$c_1 y_1 + c_2 y_2 + \cdots + c_N y_N$$

where the c_k's are constants. If these constants are all arbitrary, then the expression is, unsurprisingly, an *arbitrary linear combination.* In practice, we will often refer to some function y as a linear combination of the y_k's (over some interval (α, β)) to indicate there are constants c_1, c_2, ... and c_N such that

$$y(x) = c_1 y_1(x) + c_2 y_2(x) + \cdots + c_N y_N(x) \qquad \text{for all } x \text{ in } (\alpha, \beta) \quad .$$

!►*Example 15.2:* *Here are some linear combinations of* $\cos(x)$ *and* $\sin(x)$ *:*

$$4\cos(x) + 2\sin(x) \quad ,$$

$$3\cos(x) - 5\sin(x)$$

and

$$\sin(x) \quad (\text{i.e., } 0\cos(x) + 1\sin(x)) \quad .$$

And

$$c_1\cos(x) + c_2\sin(x) + c_3 e^x$$

is an arbitrary linear combination of the three functions $\cos(x)$, $\sin(x)$ *and* e^x .

Let us now assume that we have found two solutions y_1 and y_2 to our homogeneous differential equation

$$ay'' + by' + cy = 0$$

on (α, β). This means that

$$a y_1'' + b y_1' + c y_1 = 0 \qquad \text{and} \qquad a y_2'' + b y_2' + c y_2 = 0 \quad .$$

Let's see what happens when we plug into our differential equation some linear combination of these two solutions, say,

$$y = 2y_1 + 6y_2 \quad .$$

By the fundamental properties of differentiation, we know that

$$y' = [2y_1 + 6y_2]' = [2y_1]' + [6y_2]' = 2{y_1}' + 6{y_2}'$$

and

$$y'' = [2y_1 + 6y_2]'' = [2y_1]'' + [6y_2]'' = 2{y_1}'' + 6{y_2}'' \quad .$$

So,

$$\begin{aligned} ay'' + by' + cy &= a[2y_1 + 6y_2]'' + b[2y_1 + 6y_2]' + c[2y_1 + 6y_2] \\ &= 2a{y_1}'' + 6a{y_2}'' + 2b{y_1}' + 6b{y_2}' + 2cy_1 + 6cy_2 \\ &= 2\left[a{y_1}'' + b{y_1}' + cy_1\right] + 6\left[a{y_2}'' + b{y_2}' + cy_2\right] \\ &= 2[0] + 6[0] \\ &= 0 \quad . \end{aligned}$$

Thus, the linear combination $2y_1 + 6y_2$ is another solution to our homogeneous linear differential equation.

Of course, there was nothing special about the constants 2 and 6. If we had used any linear combination of y_1 and y_2

$$y = c_1 y_1 + c_2 y_2 \quad ,$$

then the above computations would have yielded

$$\begin{aligned} ay'' + by' + cy & \\ &= a\left[c_1 y_1 + c_2 y_2\right]'' + b\left[c_1 y_1 + c_2 y_2\right]' + c\left[c_1 y_1 + c_2 y_2(x)\right] \\ &= \cdots \\ &= c_1\left[a y_1{}'' + b y_1{}' + c y_1\right] + c_2\left[a y_2{}'' + b y_2{}' + c y_2\right] \\ &= c_1\,[0] + c_2\,[0] \\ &= 0 \quad . \end{aligned}$$

Nor is there any reason to stop with two solutions. If y had been any linear combination

$$y = c_1 y_1 + c_2 y_2 + \cdots + c_K y_K$$

with each y_k being a solution to our homogeneous differential equation,

$$a y_k{}'' + b y_k{}' + c y_k = 0 \quad ,$$

then the above computations — expanded to account for the N solutions — clearly would have yielded

$$ay'' + by' + cy = c_1[0] + c_2[0] + \cdots + c_K[0] = 0 \quad .$$

This is a major result, often called the "principle of superposition".[1] Being a major result, it deserves its own theorem:

Theorem 15.1 (principle of superposition [for second-order equations])
Any linear combination of solutions to a second-order homogeneous linear differential equation is, itself, a solution to that homogeneous linear equation.

Note that this partially answers our basic question on page 280 by assuring us that

$$y = c_1 y_1 + c_2 y_2$$

is a solution to our differential equation for every choice of constants c_1 and c_2. What remains is to see whether this describes *all* possible solutions.

Linear Independence and Fundamental Sets

We now know that if we have, say, three solutions y_1, y_2 and y_3 to our homogeneous differential equation, then any linear combination of these functions

$$y = c_1 y_1 + c_2 y_2 + c_3 y_3$$

[1] The name comes from the fact that, geometrically, the graph of a linear combination of functions can be viewed as a "superposition" of the graphs of the individual functions.

is also a solution. But what if one of these y_k's is also a linear combination of the other y_k's, say,

$$y_3 = 4y_1 + 2y_2 \quad .$$

Then we can simplify our expression for y by noting that

$$\begin{aligned} y &= c_1y_1 + c_2y_2 + c_3y_3 \\ &= c_1y_1 + c_2y_2 + c_3[4y_1 + 2y_2] \\ &= [c_1 + 4c_3]y_1 + [c_2 + 2c_3]y_2 \quad . \end{aligned}$$

Since $c_1 + 4c_3$ and $c_2 + 2c_3$ are, themselves, just constants — call them a_1 and a_2 — our formula for y reduces to

$$y = a_1y_1 + a_2y_2 \quad .$$

Thus, our original formula for y did not require y_3 at all. In fact, including this redundant function gives us a formula with more constants than necessary. Not only is this a waste of ink, it will cause difficulties when we use these formulas in solving initial-value problems.

This prompts even more terminology to simplify future discussion. Suppose

$$\{ y_1, y_2, \ldots, y_M \}$$

is a set of functions defined on some interval. We will say this set is *linearly independent* (over the given interval) if none of the y_k's can be written as a linear combination of any of the others (over the given interval). On the other hand, if at least one y_k in the set can be written as a linear combination of some of the others, then we will say the set is *linearly dependent* (over the given interval).

!▶Example 15.3: *The set of functions*

$$\{ y_1(x), y_2(x), y_3(x) \} = \{ \cos(x), \sin(x), 4\cos(x) + 2\sin(x) \} \quad .$$

is linearly dependent (over any interval) since the last function is clearly a linear combination of the first two.

By the way, we should observe the almost trivial fact that, whatever functions y_1, y_2, ... and y_M may be,

$$0 = 0 \cdot y_1 + 0 \cdot y_2 + \cdots + 0 \cdot y_M \quad .$$

So the zero function can always be treated as a linear combination of other functions, and, hence, cannot be one of the functions in any linearly *in*dependent set.

Linear Independence for Function Pairs

Matters simplify greatly when our set is just a pair of functions

$$\{ y_1, y_2 \} \quad .$$

In this case, the statement that one of these y_k's is a linear combination of the other over the interval (α, β) is just the statement that either $y_1 = c_2y_2$ for some constant c_2, that is,

$$y_1(x) = c_2y_2(x) \qquad \text{for all } x \text{ in } (\alpha, \beta) \quad ,$$

or that $y_2 = c_1y_1$ for some constant c_1, that is,

$$y_2(x) = c_1y_1(x) \qquad \text{for all } x \text{ in } (\alpha, \beta) \quad .$$

Either way, one function is simply a constant multiple of the other over the interval of interest. In fact, unless $c_1 = 0$ or $c_2 = 0$, then each function is a constant multiple of the other with $c_1 \cdot c_2 = 1$. Thus, for a pair of functions, the concepts of linear independence and dependence reduce to the following:

The set $\{ y_1, y_2 \}$ is linearly *in*dependent.

$\iff$ Neither y_1 nor y_2 is a constant multiple of the other.

and

The set $\{ y_1, y_2 \}$ is linearly dependent.

$\iff$ Either y_1 or y_2 is a constant multiple of the other.

In practice, this makes it relatively easy to determine when a pair of functions is linearly independent.

!►Example 15.4: *In Example 15.3 we obtained*

$$\{ y_1(x), y_2(x) \} = \{ \cos(x), \sin(x) \}$$

as a pair of solutions for the homogeneous second-order linear differential equation

$$y'' + y = 0 \quad .$$

It should be clear that there is no constant c_1 *or* c_2 *such that*

$$\cos(x) = c_2 \sin(x) \qquad \text{for all } x \text{ in } (\alpha, \beta)$$

or

$$\sin(x) = c_1 \cos(x) \qquad \text{for all } x \text{ in } (\alpha, \beta) \quad .$$

After all, if we were to believe, say, that $\sin(x) = c_1 \cos(x)$, *then we would have to believe that there is a constant* c_1 *such that*

$$\frac{\sin(x)}{\cos(x)} = c_1 \qquad \text{for all } x \text{ in } (\alpha, \beta) \quad ,$$

which, in turn, would require that we believe

$$0 = \frac{0}{1} = \frac{\sin(0)}{\cos(0)} = c_1 = \frac{\sin(\pi/4)}{\cos(\pi/4)} = \frac{\sqrt{2}/2}{\sqrt{2}/2} = 1 \quad !$$

So, clearly, neither $\sin(x)$ *nor* $\cos(x)$ *is a constant multiple of the other over the real line. Hence,* $\{\cos(x), \sin(x)\}$ *is a linearly independent set (over the entire real line), and*

$$y(x) = c_1 \sin(x) + c_2 \sin(x)$$

not only describes many possible solutions to our differential equation, it cannot be simplified to an expression with fewer arbitrary constants.

Fundamental Solution Sets and General Solutions

One final bit of terminology: Given any homogeneous linear differential equation

$$a_0 y^{(N)} + a_1 y^{(N-1)} + \cdots + a_{N-2} y'' + a_{N-1} y' + a_N y = 0$$

(over an interval (α, β)), a *fundamental set of solutions* (for the given differential equation) is simply any linearly independent set of solutions

$$\{y_1, y_2, \ldots, y_M\}$$

such that the arbitrary linear combination of these solutions,

$$y = c_1 y_1 + c_2 y_2 + \cdots + c_M y_M$$

is a general solution to the differential equation.

While the above definition holds for any homogeneous linear differential equation, our current interest is focused on the case where $N = 2$, and, at this point, you are probably suspecting that any linearly independent pair $\{y_1, y_2\}$ of solutions to our second-order differential equation is a fundamental set. This suspicion can be confirmed by applying the principle of superposition and the following lemma on the existence and uniqueness of solutions to the problems we are dealing with:

Lemma 15.2 (existence and uniqueness for second-order homogeneous linear equations)
Consider the initial-value problem

$$ay'' + by' + cy = 0 \quad \text{with} \quad y(x_0) = A \quad \text{and} \quad y'(x_0) = B$$

over an interval (α, β) containing the point x_0. Assume, further, that a, b, c and g are continuous functions on (α, β) with a never being zero at any point in this interval.. Then the initial-value problem has exactly one solution on (α, β). Moreover, the solution and its first and second derivatives are all continuous on this interval.

This lemma is nothing more than Theorem 14.1 on page 265 with "$g = 0$". To see how it is relevant to us, let's go back to the differential equation in an earlier example.

!►Example 15.5: *We know that one linearly independent pair of solutions to*

$$y'' + y = 0$$

on $(-\infty, \infty)$ is

$$\{y_1, y_2\} = \{\cos(x), \sin(x)\} \quad .$$

In addition, the principle of superposition assures us that any linear combination of these solutions

$$y(x) = c_1 \cos(x) + c_2 \sin(x)$$

is also a solution to the differential equation. To go further and verify that this pair is a fundamental set of solutions for the above differential equation, and that the above expression for y is a general solution, we need to show that every solution to this differential equation can be written as the above linear combination for some choice of constants c_1 and c_2.

So let $\hat{y}$ be any single solution to the above differential equation, and consider the problem of solving the initial-value problem

$$y'' + y = 0 \quad \text{with} \quad y(0) = A \quad \text{and} \quad y'(0) = B \tag{15.1a}$$

where

$$A = \widehat{y}(0) \qquad \text{and} \qquad B = \widehat{y}'(0) \quad . \tag{15.1b}$$

Obviously, $\widehat{y}$ is a solution, but so is

$$y(x) = c_1 \cos(x) + c_2 \sin(x)$$

provided we can find constants c_1 and c_2 such that the above and its derivative,

$$y'(x) = \frac{d}{dx}[c_1 \cos(x) + c_2 \sin(x)] = -c_1 \sin(x) + c_2 \cos(x) \quad ,$$

equal A and B, respectively, when $x = 0$. This means we want to find c_1 and c_2 so that

$$y(0) = c_1 \cos(0) + c_2 \sin(0) = A$$

and

$$y'(0) = -c_1 \sin(0) + c_2 \cos(0) = B \quad .$$

Since $\cos(0) = 1$ and $\sin(0) = 0$, this system reduces to

$$c_1 \cdot 1 + c_2 \cdot 0 = A$$

and

$$-c_1 \cdot 0 + c_2 \cdot 1 = B \quad ,$$

immediately telling us that $c_1 = A$ and $c_2 = B$. Thus, both

$$\widehat{y}(x) \qquad \text{and} \qquad A\cos(x) + B\sin(x)$$

are solutions over $(-\infty, \infty)$ to initial-value problem (15.1). But Lemma 15.2 tells us that there is only one solution. So our two solutions must be the same; that is, we must have

$$\widehat{y}(x) = A\cos(x) + B\sin(x) \qquad \text{for all } x \text{ in } (-\infty, \infty) \quad .$$

Thus, not only is

$$y(x) = c_1 \cos(x) + c_2 \sin(x)$$

a solution to

$$y'' + y = 0$$

for every pair of constants c_1 and c_2, this arbitrary linear combination describes every possible solution to this differential equation. In other words,

$$y(x) = c_1 \cos(x) + c_2 \sin(x)$$

is a general solution to

$$y'' + y = 0 \quad ,$$

and

$$\{\cos(x), \sin(x)\}$$

is a fundamental set of solutions to the above differential equation.

Before leaving this example, let us make one more observation; namely that, if x_0, A and B are any three real numbers, then Lemma 15.2 tells us that there is exactly one solution to the initial-value problem

$$y'' + y = 0 \qquad \text{with} \quad y(x_0) = A \quad \text{and} \quad y'(x_0) = B \quad .$$

Moreover, by the above analysis, we know that a solution is given by

$$y(x) = c_1 \cos(x) + c_2 \sin(x)$$

for some single pair of constants c_1 *and* c_2 *. Hence, there is one and only one choice of constants* c_1 *and* c_2 *such that*

$$A = y(x_0) = c_1 \cos(x_0) + c_2 \sin(x_0)$$

and

$$B = y'(x_0) = -c_1 \sin(x_0) + c_2 \cos(x_0) \quad .$$

It turns out that much of the analysis just done in the last example for

$$y'' + y = 0 \qquad \text{on} \quad (-\infty, \infty)$$

can be repeated for any given

$$ay'' + by' + cy = 0 \qquad \text{on} \quad (\alpha, \beta)$$

provided a, b and c are all continuous functions on (α, β) with a never being zero. More precisely, if you are given such a differential equation, along with a linearly independent pair of solutions $\{y_1, y_2\}$, then by applying the principle of superposition and Lemma 15.2 as done in the above example, you can show:

1. The arbitrary linear combination

$$y(x) = c_1 y_1(x) + c_2 y_2(x) \qquad \text{for all } x \text{ in } (\alpha, \beta)$$

is a general solution to the differential equation (and, hence, $\{y_1, y_2\}$ is a fundamental set of solutions).

2. Given any point x_0 in (α, β), and any two real numbers A and B, then there is exactly one choice of constants c_1 and c_2 such that

$$y(x) = c_1 y_1(x) + c_2 y_2(x) \qquad \text{for all } x \text{ in } (\alpha, \beta)$$

is the solution to the initial-value problem

$$ay'' + by' + cy = 0 \qquad \text{with} \quad y(x_0) = A \quad \text{and} \quad y'(x_0) = B \quad .$$

Try it yourself:

?►Exercise 15.1: *Consider the homogeneous linear differential equation*

$$y'' - y = 0 \quad .$$

a: *Verify that*

$$\{e^x, e^{-x}\}$$

is a linearly independent pair of solutions to the above differential equation on $(-\infty, \infty)$ *.*

b: *Verify that*

$$y(x) = c_1 e^x + c_2 e^{-x}$$

satisfies the given differential equation for any choice of constants c_1 *and* c_2 *.*

c: *Let A and B be any two real numbers, and find the values of c_1 and c_2 (in terms of A and B such that*

$$y(x) = c_1 e^x + c_2 e^{-x}$$

satisfies

$$y'' - y = 0 \quad \text{with} \quad y(0) = A \quad \text{and} \quad y'(0) = B \quad .$$

(Answer: $c_1 = (A+B)/2$ and $c_2 = (A-B)/2$)

d: *Let $\widehat{y}$ be any solution to the given differential equation, and, using Lemma 15.2 and the results from the last exercise (with $A = \widehat{y}(0)$ and $B = \widehat{y}'(0)$), show that*

$$\widehat{y}(x) = c_1 e^x + c_2 e^{-x} \qquad \text{for all } x \text{ in } (-\infty, \infty)$$

for some choice of constants c_1 and c_2 .

e: *Note that, by the above, it follows that*

$$y(x) = c_1 e^x + c_2 e^{-x}$$

is a general solution to

$$y'' - y = 0 \qquad \text{on} \quad (-\infty, \infty) \quad .$$

Hence, $\{e^x, e^{-x}\}$ is a fundamental set of solutions to this differential equation. Now convince yourself (preferably using Lemma 15.2) that every initial-value problem

$$y'' - y = 0 \qquad \text{with} \quad y(x_0) = A \quad \text{and} \quad y'(x_0) = B$$

has a solution of the form

$$y(x) = c_1 e^x + c_2 e^{-x}$$

(assuming, of course, that x_0 , A and B are real numbers).

The Big Theorem on Second-Order Homogeneous Equations

With the principle of superposition and existence/uniqueness Lemma 15.2, you should be able to verify that any given linearly independent pair $\{y_1, y_2\}$ of solutions to any given reasonable second-order homogeneous linear differential equation

$$ay'' + by' + cy = 0$$

is a fundamental set of solutions to the differential equation. In fact, it should seem reasonable that we could prove a general theorem to this effect, allowing us to simply invoke that theorem without going through the details we went through in the above example and exercise, and even without knowing a particular set of solutions. That theorem would look something like the following:

Theorem 15.3 (general solutions to second-order homogenous linear differential equations)
Let (α, β) be some open interval, and suppose we have a second-order homogeneous linear differential equation

$$ay'' + by' + cy = 0$$

where, on (α, β) , the functions a , b and c are continuous, and a is never zero. Then the following statements all hold:

1. *Fundamental sets of solutions for this differential equation (over (α, β)) exist.*

2. *Every fundamental solution set consists of a pair of solutions.*

3. *If* $\{y_1, y_2\}$ *is any linearly independent pair of particular solutions over* (α, β) *, then:*

   (a) $\{y_1, y_2\}$ *is a fundamental set of solutions.*

   (b) *A general solution to the differential equation is given by*

   $$y(x) = c_1 y_1(x) + c_2 y_2(x)$$

   where c_1 *and* c_2 *are arbitrary constants.*

   (c) *Given any point* x_0 *in* (α, β) *and any two fixed values* A *and* B *, there is exactly one ordered pair of constants* $\{c_1, c_2\}$ *such that*

   $$y(x) = c_1 y_1(x) + c_2 y_2(x)$$

   also satisfies the initial conditions

   $$y(x_0) = A \quad \text{and} \quad y'(x_0) = B \quad .$$

This theorem can be considered as the "Big Theorem on Second-Order, Homogeneous Linear Differential Equations". It will be used repeatedly, often without comment, in the chapters that follow. A full proof of this theorem is given in the next chapter for the dedicated reader. That proof is, basically, an expansion of the discussion given in Example 15.5.

By the way, the statement about "initial conditions" in the above theorem further assures us that second-order sets of initial conditions are appropriate for second-order homogeneous linear differential equations. It also assures us that no linearly independent pair of solutions for a second-order homogeneous linear differential equation can become "degenerate" at any point in the interval (α, β) . (To see why we might be worried about "degeneracy", see Exercises 15.3 and 15.4 at the end of the chapter.)

!▶Example 15.6: *Consider the differential equation*

$$x^2 y'' + xy' - 4y = 0 \quad .$$

This is

$$ay'' + by' + cy = 0$$

with

$$a(x) = x^2 \quad , \quad b(x) = x \quad \text{and} \quad c = -4 \quad .$$

These are all continuous functions everywhere, but

$$a(0) = 0^2 = 0 \quad .$$

So, to apply the above theorem, our interval of interest (α, β) *must not include* $x = 0$ *. Accordingly, let's consider solving*

$$x^2 y'' + xy' - 4y = 0 \quad \text{on} \quad (0, \infty) \quad .$$

You can easily verify that one pair of solutions[2] *is*

$$y_1(x) = x^{-2} \quad \text{and} \quad y_2(x) = x^2 \quad .$$

[2] We'll discuss a method for finding these particular solutions in Chapter 20.

It should be obvious that neither is a constant multiple of the other. Theorem 15.3 now immediately tells us that this linearly independent pair

$$\{x^{-2}, x^2\}$$

is a fundamental set of solutions for our differential equation, and that

$$y(x) = c_1 x^{-2} + c_2 x^2$$

is a general solution to this differential equation.

(By the way, note that one of the solutions, specifically $y_1(x) = x^{-2}$, "blows up" at the point where $a(x) = x^2$ is zero. This is something that often happens with solutions of

$$ay'' + by' + cy = 0$$

at a point where a is zero.)

15.2 Homogeneous Linear Equations of Arbitrary Order

The discussion in the previous section can, naturally, be extended to an analogous discussion concerning solutions to homogeneous linear equations of any order. Two results of this discussion that should be noted are the generalizations of the principle of superposition and the big theorem on second-order homogeneous linear differential equations (Theorem 15.3). Here are those generalizations:

Theorem 15.4 (principle of superposition)
Any linear combination of solutions to a homogeneous linear differential equation is, itself, a solution to that homogeneous linear equation.

Theorem 15.5 (general solutions to homogenous linear differential equations)
Let (α, β) be some open interval, and suppose we have an Nth-order homogeneous linear differential equation

$$a_0 y^{(N)} + a_1 y^{(N-1)} + \cdots + a_{N-2} y'' + a_{N-1} y' + a_N y = 0$$

where, on (α, β), the a_k's are all continuous functions with a_0 never being zero. Then the following statements all hold:

1. *Fundamental sets of solutions for this differential equation (over (α, β)) exist.*
2. *Every fundamental solution set consists of exactly N solutions.*
3. *If $\{y_1, y_2, \ldots, y_N\}$ is any linearly independent set of N particular solutions over (α, β), then:*
   (a) *$\{y_1, y_2, \ldots, y_N\}$ is a fundamental set of solutions.*
   (b) *A general solution to the differential equation is given by*

$$y(x) = c_1 y_1(x) + c_2 y_2(x) + \cdots + c_N y_N(x)$$

where c_1, c_2, ... and c_N are arbitrary constants.

(c) *Given any point x_0 in (α, β) and any N fixed values A_1, A_2, ... and A_N, there is exactly one ordered set of constants $\{c_1, c_2, \ldots, c_N\}$ such that*

$$y(x) = c_1 y_1(x) + c_2 y_2(x) + \cdots + c_N y_N(x)$$

also satisfies the initial conditions

$$y(x_0) = A_1 \quad , \quad y'(x_0) = A_2 \quad ,$$
$$y''(x_0) = A_3 \quad , \quad \ldots \quad \text{and} \quad y^{(N-1)}(x_0) = A_N \quad .$$

The proof of the more general version of the principle of superposition is a straightforward extension of the derivation of the original version. Proving Theorem 15.5 is more challenging and will be discussed in the next chapter after proving the theorem it generalizes.

15.3 Linear Independence and Wronskians

As we saw in Section 15.1, determining whether a set of just two solutions $\{y_1, y_2\}$ is linearly independent or not is simply a matter of checking whether one function is a constant multiple of the other. However, when the set has three or more solutions, $\{y_1, y_2, y_3, \ldots\}$, then the basic method of determining linear independence requires checking to see if any of the y_k's is a linear combination of the others. This can be a difficult task. Fortunately, this task can be simplified by the use of "Wronskians".

Definition of Wronskians

Let $\{y_1, y_2, \ldots, y_N\}$ be a set of N sufficiently differentiable functions on an interval (α, β). The corresponding *Wronskian*, denoted by either W or $W[y_1, y_2, \ldots, y_N]$, is the function on (α, β) generated by the following determinant of a matrix of derivatives of the y_k's:

$$W = W[y_1, y_2, \ldots, y_N] = \begin{vmatrix} y_1 & y_2 & y_3 & \cdots & y_N \\ y_1' & y_2' & y_3' & \cdots & y_N' \\ y_1'' & y_2'' & y_3'' & \cdots & y_N'' \\ \vdots & \vdots & \vdots & \vdots & \vdots \\ y_1^{(N-1)} & y_2^{(N-1)} & y_3^{(N-1)} & \cdots & y_N^{(N-1)} \end{vmatrix} \quad .$$

In particular, if $N = 2$,

$$W = W[y_1, y_2] = \begin{vmatrix} y_1 & y_2 \\ y_1' & y_2' \end{vmatrix} = y_1 y_2' - y_1' y_2 \quad .$$

!▶Example 15.7: *Let's find $W[y_1, y_2]$ on the real line when*

$$y_1(x) = x^2 \quad \text{and} \quad y_2(x) = x^3 \quad .$$

In this case,

$$y_1'(x) = 2x \quad \text{and} \quad y_2'(x) = 3x^2 \quad ,$$

and

$$W[y_1, y_2] = \begin{vmatrix} y_1(x) & y_2(x) \\ {y_1}'(x) & {y_2}'(x) \end{vmatrix} = \begin{vmatrix} x^2 & x^3 \\ 2x & 3x^2 \end{vmatrix} = x^2 3x^2 - 2xx^3 = x^4 \quad .$$

Applications of Wronskians

One reason for our interest in Wronskians is that they naturally arise when solving with initial-value problems. For example, suppose we have a pair of functions y_1 and y_2, and we want to find constants c_1 and c_2 such that

$$y(x) = c_1 y_1(x) + c_2 y_2(x)$$

satisfies

$$y(x_0) = 2 \quad \text{and} \quad y'(x_0) = 5$$

for some given point x_0 in our interval of interest. In solving for c_1 and c_2, you can easily show (and we will do it in Section 16.1) that

$$c_1 W(x_0) = 2{y_2}'(x_0) - 5y_2(x_0) \quad \text{and} \quad c_2 W(x_0) = 5y_1(x_0) - 2{y_1}'(x_0) \quad .$$

Thus, if $W(x_0) \neq 0$, then there is exactly one possible value for c_1 and one possible value for c_2, namely,

$$c_1 = \frac{2{y_2}'(x_0) - 5y_2(x_0)}{W(x_0)} \quad \text{and} \quad c_2 = \frac{5y_1(x_0) - 2{y_1}'(x_0)}{W(x_0)} \quad .$$

However, if $W(x_0) = 0$, then the system reduces to

$$0 = 2{y_2}'(x_0) - 5y_2(x_0) \quad \text{and} \quad 0 = 5y_1(x_0) - 2{y_1}'(x_0)$$

which cannot be solved for c_1 and c_2. (In practice, you probably don't even notice that your formulas involve the Wronskian.)

Another reason for our interest — a possibly more important reason — is that the vanishing of a Wronskian of a set of solutions signals that the given set is not a good choice in constructing solutions to initial-value problems. The value of this fact is enhanced by the following remarkable theorem:

Theorem 15.6 (Wronskians and fundamental solution sets)
Let W be the Wronskian of any set $\{y_1, y_2, \ldots, y_N\}$ of N particular solutions to an Nth-order homogeneous linear differential equation

$$a_0 y^{(N)} + a_1 y^{(N-1)} + \cdots + a_{N-2} y'' + a_{N-1} y' + a_N y = 0$$

on some open interval (α, β). Assume further that the a_k's are continuous functions with a_0 never being zero on (α, β). Then:

1. *If $W(x_0) = 0$ for any single point x_0 in (α, β), then $W(x) = 0$ for every point x in (α, β), and the set $\{y_1, y_2, \ldots, y_N\}$ is not linearly independent (and, hence, is not a fundamental solution set) on (α, β).*

2. *If $W(x_0) \neq 0$ for any single point x_0 in (α, β), then $W(x) \neq 0$ for every point x in (α, β), and $\{y_1, y_2, \ldots, y_N\}$ is a fundamental set of solutions for the given differential equation on (α, β).*

This theorem (whose proof will be discussed in the next chapter) gives us a relatively easy way to determine if a set of solutions to a homogeneous linear differential equation is a fundamental set of solutions. This test is especially useful when the order of the differential equation is 3 or higher.

!▶Example 15.8: *Consider the functions*

$$y_1(x) = 1 \quad , \quad y_2(x) = \cos(2x) \quad \text{and} \quad y_3(x) = \sin^2(x) \quad .$$

You can easily verify that all are solutions (over the entire real line) to the homogeneous third-order linear differential equation

$$y''' + 4y' = 0 \quad .$$

So, is

$$\left\{ 1, \cos(2x), \sin^2(x) \right\}$$

a fundamental set of solutions for this differential equation? To check we compute the first-order derivatives

$$y_1{}'(x) = 0 \quad , \quad y_2{}'(x) = -2\sin(2x) \quad \text{and} \quad y_3{}'(x) = 2\sin(x)\cos(x) \quad ,$$

the second-order derivatives

$$y_1{}''(x) = 0 \quad , \quad y_2{}''(x) = -4\cos(2x) \quad \text{and} \quad y_3{}''(x) = 2\cos^2(x) - 2\sin^2(x) \quad ,$$

and form the corresponding Wronskian,

$$W(x) = W[1, \cos(2x), \sin^2(x)] = \begin{vmatrix} 1 & \cos(2x) & \sin^2(x) \\ 0 & -2\sin(2x) & 2\sin(x)\cos(x) \\ 0 & -4\cos(2x) & 2\cos^2(x) - 2\sin^2(x) \end{vmatrix} \quad .$$

Rather than compute this determinant for all values of x (which would be very tedious), let us simply pick a convenient value for x, say $x = 0$, and compute the Wronskian at that point:

$$W(0) = \begin{vmatrix} 1 & \cos(2\cdot 0) & \sin^2(0) \\ 0 & -2\sin(2\cdot 0) & 2\sin(0)\cos(0) \\ 0 & -4\cos(2\cdot 0) & 2\cos^2(0) - 2\sin^2(0) \end{vmatrix} = \begin{vmatrix} 1 & 1 & 0 \\ 0 & 0 & 0 \\ 0 & -4 & 2 \end{vmatrix} = 0 \quad .$$

Theorem 15.6 now assures us that, because this Wronskian vanishes at that one point, it must vanish everywhere. More importantly for us, this theorem also tells us that $\{1, \cos(2x), \sin^2(x)\}$ is not a fundamental set of solutions for our differential equation.

!▶Example 15.9: *Now consider the functions*

$$y_1(x) = 1 \quad , \quad y_2(x) = \cos(2x) \quad \text{and} \quad y_3(x) = \sin(2x) \quad .$$

Again, you can easily verify that all are solutions (over the entire real line) to the homogeneous third-order linear differential equation

$$y''' + 4y' = 0 \quad .$$

So, is

$$\left\{ 1, \cos(2x), \sin(2x) \right\}$$

a fundamental set of solutions for our differential equation, above? To check we compute the appropriate derivatives and form the corresponding Wronskian,

$$W(x) = W[1, \cos(2x), \sin(2x)]$$

$$= \begin{vmatrix} y_1 & y_2 & y_3 \\ y_1{}' & y_2{}' & y_3{}' \\ y_1{}'' & y_2{}'' & y_3{}'' \end{vmatrix} = \begin{vmatrix} 1 & \cos(2x) & \sin(2x) \\ 0 & -2\sin(2x) & 2\cos(2x) \\ 0 & -4\cos(2x) & -2\sin(2x) \end{vmatrix} .$$

Letting $x = 0$, we get

$$W(0) = \begin{vmatrix} 1 & \cos(2\cdot 0) & \sin(2\cdot 0) \\ 0 & -2\sin(2\cdot 0) & 2\cos(2\cdot 0) \\ 0 & -4\cos(2\cdot 0) & -2\sin(2\cdot 0) \end{vmatrix} = \begin{vmatrix} 1 & 1 & 0 \\ 0 & 0 & 2 \\ 0 & -4 & 0 \end{vmatrix} = 8 \neq 0 .$$

Theorem 15.6 assures us that, since this Wronskian is nonzero at one point, it is nonzero everywhere, and that $\{1, \cos(2x), \sin(2x)\}$ is a fundamental set of solutions for our differential equation. Hence,

$$y(x) = c_1 \cdot 1 + c_2 \cos(2x) + c_3 \sin(2x)$$

is a general solution to our third-order differential equation.

Additional Exercises

15.2. Several initial-value problems are given below, each involving a second-order homogeneous linear differential equation, and each with a pair of functions $y_1(x)$ and $y_2(x)$. Verify that each pair $\{y_1, y_2\}$ is a fundamental set of solutions to the given differential equation (verify both that y_1 and y_2 are solutions and that the pair is linearly independent), and then find a linear combination of these solutions that satisfies the given initial-value problem.

a. I.v. problem: $y'' + 4y = 0$ with $y(0) = 2$ and $y'(0) = 6$.

Functions: $y_1(x) = \cos(2x)$ and $y_2(x) = \sin(2x)$.

b. I.v. problem: $y'' - 4y = 0$ with $y(0) = 0$ and $y'(0) = 12$.

Functions: $y_1(x) = e^{2x}$ and $y_2(x) = e^{-2x}$.

c. I.v. problem: $y'' + y' - 6y = 0$ with $y(0) = 8$ and $y'(0) = -9$.

Functions: $y_1(x) = e^{2x}$ and $y_2(x) = e^{-3x}$.

d. I.v. problem: $y'' - 4y' + 4y = 0$ with $y(0) = 1$ and $y'(0) = 6$.

Functions: $y_1(x) = e^{2x}$ and $y_2(x) = xe^{2x}$.

e. I.v. problem: $x^2y'' - 4xy' + 6y = 0$ with $y(1) = 0$ and $y'(1) = 4$.

Functions: $y_1(x) = x^2$ and $y_2(x) = x^3$.

f. I.v. problem: $4x^2y'' + 4xy' - y = 0$ with $y(1) = 8$ and $y'(1) = 1$.

Functions: $y_1(x) = \sqrt{x}$ and $y_2(x) = \frac{1}{\sqrt{x}}$.

g. *I.v. problem:* $x^2y'' - xy' + y = 0$ *with* $y(1) = 5$ *and* $y'(1) = 3$.

Functions: $y_1(x) = x$ *and* $y_2(x) = x \ln|x|$.

h. *I.v. problem:* $xy'' - y' + 4x^3y = 0$
with $y(\sqrt{\pi}) = 3$ *and* $y'(\sqrt{\pi}) = 4$.

Functions: $y_1(x) = \cos\left(x^2\right)$ *and* $y_2(x) = \sin\left(x^2\right)$.

i. *I.v. problem:* $(x+1)^2y'' - 2(x+1)y' + 2y = 0$
with $y(0) = 0$ *and* $y'(0) = 4$.

Functions: $y_1(x) = x^2 - 1$ *and* $y_2(x) = x + 1$.

15.3. *In Exercise 15.2 e, above, you found that* $\{x^2, x^3\}$ *is a fundamental set of solutions to*

$$x^2y'' - 4xy' + 6y = 0 \quad ,$$

at least over some interval containing $x_0 = 1$.

a. *What is the largest interval containing* $x_0 = 1$ *for which Theorem 15.3 on page 288 assures us that* $\left\{x^2, x^3\right\}$ *is a fundamental set of solutions to the above differential equation?*

b. *Attempt to find constants* c_1 *and* c_2 *so that*

$$y(x) = c_1x^2 + c_2x^3$$

satisfies the initial conditions

$$y(0) = 0 \quad \text{and} \quad y'(0) = -4 \quad .$$

What goes wrong? Why does this not violate the claim in Theorem 15.3 about initial-value problems being solvable?

15.4. *In Exercise 15.2 h, above, you found that* $\left\{\cos\left(x^2\right), \sin\left(x^2\right)\right\}$ *is a fundamental set of solutions to*

$$xy'' - y' + 4x^3y = 0 \quad ,$$

at least over some interval containing $x_0 = \pi$.

a. *What is the largest interval containing* $x_0 = \pi$ *that Theorem 15.3 on page 288 assures us* $\left\{\cos\left(x^2\right), \sin\left(x^2\right)\right\}$ *is a fundamental set of solutions to the above differential equation?*

b. *Attempt to find constants* c_1 *and* c_2 *so that*

$$y(x) = c_1 \cos\left(x^2\right) + c_2 \sin\left(x^2\right)$$

satisfies the initial conditions

$$y(0) = 1 \quad \text{and} \quad y'(0) = 4 \quad .$$

What goes wrong? Why does this not violate the claim in Theorem 15.3 about initial-value problems being solvable?

15.5. *Some third- and fourth-order initial-value problems are given below, each involving a homogeneous linear differential equation, and each with a set of three or four functions* $y_1(x)$, $y_2(x)$, ... *Verify that these functions form a fundamental set of solutions to the given differential equation, and then find a linear combination of these solutions that satisfies the given initial-value problem.*

a. *I.v. problem:* $y''' + 4y' = 0$
with $y(0) = 3$, $y'(0) = 8$ *and* $y''(0) = 4$.

Functions: $y_1(x) = 1$, $y_2(x) = \cos(2x)$ *and* $y_3(x) = \sin(2x)$.

b. *I.v. problem:* $y''' + 4y' = 0$
with $y(0) = 3$, $y'(0) = 8$ *and* $y''(0) = 4$.

Functions: $y_1(x) = 1$, $y_2(x) = \sin^2(x)$ *and* $y_3(x) = \sin(x)\cos(x)$.

c. *I.v. problem:* $y^{(4)} - y = 0$
with $y(0) = 0$, $y'(0) = 4$, $y'''(0) = 0$ *and* $y''(0) = 0$.

Functions: $y_1(x) = \cos(x)$, $y_2(x) = \sin(x)$, $y_3(x) = \cosh(x)$
and $y_4(x) = \sinh(x)$.

15.6. *Particular solutions to the differential equation in each of the following initial-value problems can found by assuming*

$$y(x) = e^{rx}$$

where r is a constant to be determined. To determine these constants, plug this formula for y into the differential equation, observe that the resulting equation miraculously simplifies to a simple algebraic equation for r, and solve for the possible values of r.

Do that with each equation and use those solutions (with the big theorem on general solutions to second-order homogeneous linear equations — Theorem 15.3 on page 288) to construct a general solution to the differential equation. Then, finally, solve the given initial-value problem.

a. $y'' - 4y = 0$ *with* $y(0) = 1$ *and* $y'(0) = 0$

b. $y'' + 2y' - 3y = 0$ *with* $y(0) = 0$ *and* $y'(0) = 1$

c. $y'' - 10y' + 9y = 0$ *with* $y(0) = 8$ *and* $y'(0) = -24$

d. $y'' + 5y' = 0$ *with* $y(0) = 1$ *and* $y'(0) = 0$

15.7. *Find solutions of the form*

$$y(x) = e^{rx}$$

where r is a constant (as in the previous exercise) and use the solutions found (along with the results given in Theorem 15.5 on page 290) to construct general solutions to the following differential equations:

a. $y''' - 9y' = 0$ **b.** $y^{(4)} - 10y'' + 9y = 0$

15.8. *Thus far, we've derived two general solutions to*

$$y'' + y = 0 \quad .$$

In Chapter 13 we obtained

$$y(x) = A\sin(x + B) \quad ,$$

and in this chapter, we got

$$y(x) = c_1\sin(x) + c_2\cos(x) \quad .$$

Using a well-known trigonometric identity, verify that these two solutions are equivalent, and find how c_1 and c_2 are related to A and B.

15.9. Let y_1 and y_2 be the following functions on the entire real line:

$$y_1(x) = \begin{cases} -x^2 & \text{if } x < 0 \\ x^2 & \text{if } 0 \le x \end{cases} \quad \text{and} \quad y_2(x) = \begin{cases} x^2 & \text{if } x < 0 \\ 3x^2 & \text{if } 0 \le x \end{cases} .$$

a. Verify that

i. $\{y_1, y_2\}$ is not linearly dependent on the entire real line, but

ii. the Wronskian for $\{y_1, y_2\}$ is zero over the entire real line (even at $x = 0$).

b. Why do the results in the previous part not violate Theorem 15.6 on page 292?

c. Is there an interval (α, β) on which $\{y_1, y_2\}$ is linearly dependent?

15.10. Let $\{y_1, y_2\}$ be a linearly independent pair of solutions over an interval (α, β) to some second-order homogeneous linear differential equation

$$ay'' + by' + cy = 0 .$$

As usual, assume a, b and c are continuous functions on (α, β) with a never being zero over that interval. Also, as usual, let

$$W = W[y_1, y_2] = y_1 y_2' - y_1' y_2 .$$

Do the following, using the fact that W is never zero on (α, β).

a. Show that, if $y_1(x_0) = 0$ for some x_0 in (α, β), then $y_1'(x_0) \neq 0$ and $y_2(x_0) \neq 0$.

b. Show that, if $y_1'(x_0) = 0$ for some x_0 in (α, β), then $y_1(x_0) \neq 0$ and $y_2'(x_0) \neq 0$.

c. Why can we not have $W(x) > 0$ for some x in (α, β) and $W(x) < 0$ for other x in (α, β)? That is, explain (briefly) why we must have either

$$W(x) > 0 \quad \text{for all } x \text{ in } (\alpha, \beta)$$

or

$$W(x) < 0 \quad \text{for all } x \text{ in } (\alpha, \beta) .$$

d. For the following, assume $W(x) > 0$ for all x in (α, β).[3] Let $[\alpha_0, \beta_0]$ be a closed subinterval of (α, β) such that

$$y_1(\alpha_0) = 0 \quad , \quad y_1(\beta_0) = 0$$

and

$$y_1(x) > 0 \quad \text{whenever} \quad \alpha_0 < x < \beta_0 .$$

i. How do we know that neither $y_1'(\alpha_0)$ nor $y_1'(\beta_0)$ are zero? Which one is positive? Which one is negative? (It may help to draw a rough sketch of the graph of y_1 based on the above information.)

ii. Using the Wronskian, determine if $y_2(\alpha_0)$ is positive or negative. Then determine if $y_2(\beta_0)$ is positive or negative.

[3] Similar results can be derived assuming $W(x) < 0$ for all x in (α, β).

iii. *Now show that there must be a point x_0 in the open interval (α_0, β_0) at which y_2 is zero.*

(What you've just shown is that there must be a zero of y_2 between any two zeroes α_0 and β_0 of y_1. You can easily expand this to the following statement:

> *Between any two zeroes of y_1 is a zero of y_2, and, likewise, between any two zeroes of y_2 is a zero of y_1.*

This tells us something about the graphs of linearly independent pairs of solutions to second-order homogeneous differential equations. It turns out to be an important property of these solution pairs when considering a type of differential equation problem involving the values of solutions at pairs of points, instead of at single points.)

16

Verifying the Big Theorems and an Introduction to Differential Operators

There are two parts to this chapter. The first (Sections 16.1 and 16.2) is a discussion of the proofs of the main theorems in the last chapter. To be precise, Section 16.1 contains a fairly complete verification of Theorem 15.3 (the big theorem on second-order homogeneous linear differential equations), along with a partial verification of Theorem 15.6 on Wronskians, while Section 16.2 contains a rather brief discussion on generalizing the results just verified in Section 16.1.

The rest of the chapter is devoted to a fairly elementary development of "linear differential operators". This material provides a slightly different perspective on linear differential equations, and can be enlightening to the careful reader (and confusing to the less careful reader). In addition, this material will make it easier to prove a few more advanced results later on in this text.

To be honest, most beginning students of differential equations can probably skip this chapter and go straight to the next chapter where we actually develop methods for completely solving a large number of important differential equations. In fact, it may be better to do so, promising yourself to return to this chapter as the need or interest arises.

16.1 Verifying the Big Theorem on Second-Order, Homogeneous Equations

Our main goal is to verify Theorem 15.3 on page 288. Accordingly, throughout this section we will assume the basic assumptions of that theorem, namely, that we have an open interval (α, β) and functions a, b and c that are continuous on (α, β) with a never being zero on this interval. With these assumptions, we will explore what we can about solutions to the second-order homogeneous linear differential equation

$$ay'' + by' + cy = 0 \quad .$$

Incidentally, along the way, we will also verify the claim concerning Wronskians in Theorem 15.6 for the case where $N = 2$.

Linear Independence, Initial-Value Problems and Wronskians

Much of our analysis will reduce to being able to solve any initial-value problem

$$ay'' + by' + cy = 0 \quad \text{with} \quad y(x_0) = A \quad \text{and} \quad y'(x_0) = B \tag{16.1}$$

where x_0 is some point in (α, β), and A and B are any two real numbers. Recall that Lemma 15.2 on page 285 assures us that any such initial-value problem has one and only one solution.

Solving Initial-Value Problems

Until further notice, we will assume that we have a pair of solutions over (α, β)

$$\{ y_1 , y_2 \}$$

to our differential equation

$$ay'' + by' + cy = 0 \quad .$$

By the principle of superposition (Theorem 15.1 on page 282), we know that any linear combination of these two solutions

$$y(x) = c_1 y_1(x) + c_2 y_2(x) \qquad \text{for all } x \text{ in } (\alpha, \beta)$$

is also a solution to our differential equation. Now let us consider solving initial-value problem (16.1) using this linear combination.

As already noted, this linear combination satisfies our differential equation. So all we need to do is to find constants c_1 and c_2 such that

$$A = y(x_0) = c_1 y_1(x_0) + c_2 y_2(x_0)$$

and

$$B = y'(x_0) = c_1 {y_1}'(x_0) + c_2 {y_2}'(x_0) \quad .$$

That is, we need to solve the linear algebraic system of two equations

$$\begin{aligned} c_1 y_1(x_0) + c_2 y_2(x_0) &= A \\ c_1 {y_1}'(x_0) + c_2 {y_2}'(x_0) &= B \end{aligned}$$

for c_1 and c_2. But this is easy. Start by multiplying each equation by ${y_2}'(x_0)$ or $y_2(x_0)$, as appropriate:

$$\begin{matrix} \big[c_1 y_1(x_0) + c_2 y_2(x_0) = A\big] {y_2}'(x_0) & & c_1 y_1(x_0) {y_2}'(x_0) + c_2 y_2(x_0) {y_2}'(x_0) = A {y_2}'(x_0) \\ & \Longrightarrow & \\ \big[c_1 {y_1}'(x_0) + c_2 {y_2}'(x_0) = B\big] y_2(x_0) & & c_1 {y_1}'(x_0) y_2(x_0) + c_2 {y_2}'(x_0) y_2(x_0) = B y_2(x_0) \end{matrix}$$

Subtracting the second equation from the first (and looking carefully at the results) yields

$$c_1 \big[\underbrace{y_1(x_0) {y_2}'(x_0) - {y_1}'(x_0) y_2(x_0)}_{W(x_0)} \big] + c_2 \big[\underbrace{y_2(x_0) {y_2}'(x_0) - {y_2}'(x_0) y_2(x_0)}_{0} \big]$$
$$= A {y_2}'(x_0) - B y_2(x_0) \quad .$$

That is,

$$c_1 W(x_0) = A {y_2}'(x_0) - B y_2(x_0) \tag{16.2}$$

where W is the *Wronskian* of function pair $\{y_1, y_2\}$, which (as you may recall from Section 15.3), is the function on (α, β) given by

$$W = W[y_1, y_2] = y_1 {y_2}' - {y_1}' y_2 \quad .$$

Similar computations yield

$$c_2 W(x_0) = B y_1(x_0) - A {y_1}'(x_0) \quad . \tag{16.3}$$

Observe that, if $W(x_0) \neq 0$, then there is exactly one possible value for c_1 and one possible value for c_2, namely,

$$c_1 = \frac{Ay_2'(x_0) - By_2(x_0)}{W(x_0)} \quad \text{and} \quad c_2 = \frac{By_1(x_0) - Ay_1'(x_0)}{W(x_0)} \quad .$$

On the other hand, if $W(x_0) = 0$, then equations (16.2) and (16.3) reduce to the pair

$$0 = Ay_2'(x_0) - By_2(x_0) \quad \text{and} \quad 0 = By_1(x_0) - Ay_1'(x_0)$$

which cannot be solved for c_1 and c_2. If the right sides of these last two equations just happen to both be 0, then the values of c_1 and c_2 are irrelevant — any values work. And if either right-hand side is nonzero, then no values for c_1 and c_2 will work.

The fact that the solvability of the above initial-value problem depends entirely on whether $W(x_0)$ is zero or not is an important fact. Let us enshrine this fact in a lemma for later reference.

Lemma 16.1 (Wronskians and initial-value problems)
Let x_0, A and B be any three fixed real values with x_0 in (α, β), and assume $\{y_1, y_2\}$ is a pair of functions differentiable at x_0. Also, let W be the Wronskian of $\{y_1, y_2\}$,

$$W = y_1 y_2' - y_1' y_2 \quad .$$

Then the problem of finding constants c_1 and c_2 such that

$$y = c_1 y_1 + c_2 y_2$$

satisfies the system

$$y(x_0) = A \quad \text{and} \quad y'(x_0) = B$$

has exactly one solution (i.e., exactly one choice for c_1 and exactly one choice for c_2) if and only if $W(x_0) \neq 0$.

Wronskians and Linear Independence

It turns out that the Wronskian of $\{y_1, y_2\}$ provides an alternative test for determining whether this pair of solutions is linearly dependent or linearly independent. To start, suppose $\{y_1, y_2\}$ is linearly dependent. Then either at least one of these functions, say, y_1, is a constant multiple of the other function; that is, for some constant κ,

$$y_1(x) = \kappa y_2(x) \quad \text{for all } x \text{ in } (\alpha, \beta) \quad .$$

This, of course, means that

$$y_1'(x) = \kappa y_2'(x) \quad \text{for all } x \text{ in } (\alpha, \beta) \quad .$$

Thus,

$$W = y_1 y_2' - y_1' y_2 = \kappa y_2 y_2' - \kappa y_2' y_2 = 0 \quad .$$

So,

$$\begin{aligned} &\{y_1, y_2\} \text{ is linearly dependent over } (\alpha, \beta) \\ &\qquad \Longrightarrow \quad W = 0 \text{ everywhere on } (\alpha, \beta) . \end{aligned} \tag{16.4a}$$

With a little thought, you'll realize that this is equivalent to

$$\begin{aligned} &W \neq 0 \text{ somewhere on } (\alpha, \beta) \\ &\qquad \Longrightarrow \quad \{y_1, y_2\} \text{ is linearly independent over } (\alpha, \beta) . \end{aligned} \tag{16.4b}$$

Showing that the above implications still hold with the arrows reversed requires a bit more work. To simplify that work, let's first prove the following two corollaries of Lemma 15.2 on page 285:

Corollary 16.2
Let x_0 be a point in an interval (α, β), and assume a, b and c are continuous functions on (α, β) with a never being zero in (α, β). Then the only solution to

$$ay'' + by' + cy = 0 \quad \text{with} \quad y(x_0) = 0 \quad \text{and} \quad y'(x_0) = 0 \quad .$$

is the trivial solution,

$$y(x) = 0 \quad \text{for all } x \text{ in } (\alpha, \beta) \quad .$$

PROOF: The trivial solution is certainly a solution to the given initial-value problem, and Lemma 15.2 assures us that this solution is the only solution. ∎

Corollary 16.3
Let x_0 be a point in an interval (α, β), and assume a, b and c are continuous functions on (α, β) with a never being zero in (α, β). Assume that $\{y_1, y_2\}$ is a pair of solutions on (α, β) to

$$ay'' + by' + cy = 0 \quad .$$

If there is a point x_0 in (α, β) and a constant κ such that

$$y_1(x_0) = \kappa y_2(x_0) \quad \text{and} \quad y_1{}'(x_0) = \kappa y_2{}'(x_0) \quad ,$$

then

$$y_1(x) = \kappa y_2(x) \quad \text{for all } x \text{ in } (\alpha, \beta)$$

and, hence, the pair $\{y_1, y_2\}$ is linearly dependent on (α, β).

PROOF: Observe that both $y = y_1$ and $y = \kappa y_2$ satisfy the same initial-value problem,

$$ay'' + by' + cy = 0 \quad \text{with} \quad y(x_0) = A \quad \text{and} \quad y'(x_0) = B$$

where

$$A = y_1(x_0) \quad \text{and} \quad B = y_1{}'(x_0) \quad .$$

Again, Lemma 15.2 tells us that this initial-value problem only has one solution. Hence $y = y_1$ and $y = \kappa y_2$ must be the same, and the claims in the corollary follow immediately. ∎

Now suppose the Wronskian of our pair $\{y_1, y_2\}$ of solutions to our differential equation is zero at some point x_0 in (α, β). That is,

$$W(x_0) = y_1(x_0)y_2{}'(x_0) - y_2(x_0)y_1{}'(x_0) = 0 \quad \text{for some } x_0 \text{ in } (\alpha, \beta) \quad .$$

This, of course, means that

$$y_1(x_0)y_2{}'(x_0) = y_2(x_0)y_1{}'(x_0) \quad . \tag{16.5}$$

Let us consider all the possibilities. For simplicity, we will start by assuming $y_2(x_0) \neq 0$:

1. If $y_2(x_0) \neq 0$ and $y_1'(x_0) \neq 0$, then equation (16.5) yields
$$y_1(x_0)y_2'(x_0) = y_2(x_0)y_1'(x_0) \neq 0 \quad ,$$
implying that $y_1(x_0) \neq 0$ and $y_2'(x_0) \neq 0$. Consequently, we can divide both sides of this equation by these two values and set
$$\kappa = \frac{y_1(x_0)}{y_2(x_0)} = \frac{y_1'(x_0)}{y_2'(x_0)} \quad .$$
Thus, κ is a constant such that
$$y_1(x_0) = \kappa y_2(x_0) \qquad \text{and} \qquad y_1'(x_0) = \kappa y_2'(x_0) \quad ,$$
and Corollary 16.3 tells us that $\{y_1, y_2\}$ is linearly dependent.

2. If $y_2(x_0) \neq 0$ and $y_1'(x_0) = 0$, then equation (16.5) yields
$$y_1(x_0)y_2'(x_0) = y_2(x_0)y_1'(x_0) = y_2(x_0) \cdot 0 = 0 \quad ,$$
telling us that $y_1(x_0) = 0$ or $y_2'(x_0) = 0$.
   (a) But if $y_1(x_0) = 0$, then y_1 is the solution to
$$ay'' + by' + cy = 0 \qquad \text{with} \quad y(x_0) = 0 \quad \text{and} \quad y'(x_0) = 0 \quad ,$$
which, as noted in Corollary 16.2 means that y_1 is the trivial solution, automatically making $\{y_1, y_2\}$ linearly dependent.
   (b) On the other hand, if $y_2'(x_0) = 0$, then we can set
$$\kappa = \frac{y_1(x_0)}{y_2(x_0)} \quad .$$
This, along with the fact that $y_1'(x_0) = 0$ and $y_2'(x_0) = 0$ gives us
$$y_1(x_0) = \kappa y_2(x_0) \qquad \text{and} \qquad y_1'(x_0) = \kappa y_2'(x_0) \quad ,$$
and Corollary 16.3 tells us that $\{y_1, y_2\}$ is linearly dependent.

From the above it follows that $\{y_1, y_2\}$ must be linearly dependent if we have both $W(x_0) = 0$ and $y_2(x_0) \neq 0$. Now let's assume $W(x_0) = 0$ and $y_2(x_0) = 0$. Equation (16.5) still holds, but, this time, yields
$$y_1(x_0)y_2'(x_0) = y_2(x_0)y_1'(x_0) = 0 \cdot y_1'(x_0) = 0 \quad ,$$
which means that $y_1(x_0) = 0$ or $y_2'(x_0) = 0$.

1. If $y_2(x_0) = 0$ and $y_2'(x_0) = 0$, then y_2 satisfies
$$ay'' + by' + cy = 0 \qquad \text{with} \quad y(x_0) = 0 \quad \text{and} \quad y'(x_0) = 0 \quad .$$
Again, Corollary 16.2 tells us that y_1 is the trivial solution, automatically making $\{y_1, y_2\}$ linearly dependent.

2. On the other hand, if $y_2(x_0) = 0$, $y_2'(x_0) \neq 0$ and $y_1(x_0) = 0$, then we can set
$$\kappa = \frac{y_1'(x_0)}{y_2'(x_0)}$$
and observe that, from this and the fact that $y_1(x_0) = 0 = y_2(x_0)$, we have
$$y_1(x_0) = \kappa y_2(x_0) \qquad \text{and} \qquad y_1'(x_0) = \kappa y_2'(x_0) \quad ,$$
which, according to Corollary 16.3, means that $\{y_1, y_2\}$ is linearly dependent.

Thus, we have that $\{y_1, y_2\}$ is linearly dependent whenever $W(x_0) = 0$ whether or not $y_2(x_0) = 0$. That is,

$$W(x_0) = 0 \text{ for some } x_0 \text{ in } (\alpha, \beta) \implies \{y_1, y_2\} \text{ is linearly dependent over } (\alpha, \beta) \quad . \tag{16.6a}$$

Equivalently,

$$\{y_1, y_2\} \text{ is linearly independent over } (\alpha, \beta) \implies W(x) \neq 0 \text{ for every } x \text{ in } (\alpha, \beta) \quad . \tag{16.6b}$$

Linearly Independent Pairs and General Solutions

Now assume that our pair of solutions $\{y_1, y_2\}$ is linearly independent over (α, β), and let $\widehat{y}$ be any single solution to our differential equation. Pick some x_0 in (α, β), and consider the initial-value problem

$$ay'' + by' + cy = 0 \quad \text{with} \quad y(x_0) = A \quad \text{and} \quad y'(x_0) = B$$

where

$$A = \widehat{y}(x_0) \quad \text{and} \quad B = \widehat{y}'(x_0) \quad .$$

Clearly, $y = \widehat{y}$ is a solution to the initial-value problem.

On the other hand, implication (16.6b) tells us that, since $\{y_1, y_2\}$ is linearly independent, the corresponding Wronskian is nonzero at x_0, and Lemma 16.1 and the principle of superposition then assure us that there are constants c_1 and c_2 such that

$$y = c_1 y_1 + c_2 y_2$$

is also a solution to this same initial-value problem. So we have "two" solutions to our initial-value problem,

$$\widehat{y} \quad \text{and} \quad y = c_1 y_1 + c_2 y_2 \quad .$$

But (by Lemma 15.2 on page 285) we know this initial-value problem only has one solution. This means our two solutions must be the same,

$$\widehat{y}(x) = c_1 y_1(x) + c_2 y_2(x) \quad \text{for every } x \text{ in } (\alpha, \beta) \quad .$$

Hence, any solution to our differential equation can be written as a linear combination of y_1 and y_2. This shows that

$$\{y_1, y_2\} \text{ is linearly independent over } (\alpha, \beta) \implies y = c_1 y_1 + c_2 y_2 \text{ is a general solution to our differential equation} \quad , \tag{16.7}$$

which, incidentally, means that the linearly independent pair $\{y_1, y_2\}$ is a fundament set of solutions for our differential equation.

Summarizing the Results So Far

Combining the results given in Lemma 16.1 and implications (16.4), (16.6) and (16.7) gives us the following lemma.

Lemma 16.4

Assume a, b and c are continuous functions on an interval (α, β) with a never being zero in (α, β), and let $\{y_1, y_2\}$ be a pair of solutions on (α, β) to

$$ay'' + by' + cy = 0 \quad .$$

Then all of the following statements are equivalent (i.e., if one holds, they all hold):

1. *The pair* $\{y_1, y_2\}$ *is linearly independent on* (α, β) *.*

2. *The Wronskian of* $\{y_1, y_2\}$ *,*
$$W = y_1 y_2{}' - y_2 y_1{}' \quad ,$$
is nonzero at one point in (α, β) *.*

3. *The Wronskian of* $\{y_1, y_2\}$ *,*
$$W = y_1 y_2{}' - y_2 y_1{}' \quad ,$$
is nonzero at every point in (α, β) *.*

4. *Given any point* x_0 *in* (α, β) *and any two fixed values* A *and* B *, there is exactly one ordered pair of constants* $\{c_1, c_2\}$ *such that*
$$y(x) = c_1 y_1(x) + c_2 y_2(x)$$
also satisfies the initial conditions
$$y(x_0) = A \qquad \text{and} \qquad y'(x_0) = B \quad .$$

5. *The arbitrary linear combination of* y_1 *and* y_2 *,*
$$y = c_1 y_1 + c_2 y_2 \quad ,$$
is a general solution for the above differential equation.

6. *The pair* $\{y_1, y_2\}$ *is a fundamental set of solutions for the above differential equation.*

If you check, you will see that most of our "big theorem", Theorem 15.3, follows immediately from this lemma, as does the theorem on Wronskians, Theorem 15.6, for the case where $N = 2$.

Proving the Rest of Theorem 15.3

All that remains to achieving our goal of verifying Theorem 15.3 is to verify that fundamental sets of solutions exist, and that a fundamental set of solutions must contain exactly two solutions. Verifying these two facts will be easy.

Existence of Fundamental Sets

Let x_0 be some point in (α, β) , and let y_1 and y_2 be, respectively, the solutions to the initial-value problems
$$ay'' + by' + cy = 0 \qquad \text{with} \quad y(x_0) = 1 \quad \text{and} \quad y'(x_0) = 0$$
and
$$ay'' + by' + cy = 0 \qquad \text{with} \quad y(x_0) = 0 \quad \text{and} \quad y'(x_0) = 1 \quad .$$

(By Lemma 15.2, we know these solutions exist.) Computing their Wronskian at x_0 , we get
$$W(x_0) = y_1(x_0) y_2{}'(x_0) - y_2(x_0) y_1{}'(x_0) = 1 \cdot 1 - 0 \cdot 0 \neq 0 \quad ,$$
which, according to Lemma 16.4, means that $\{y_1, y_2\}$ is a fundamental set of solutions to our differential equation.

Size of Fundamental Sets of Solutions

It should be clear that we cannot solve every initial-value problem

$$ay'' + by' + cy = 0 \quad \text{with} \quad y(x_0) = A \quad \text{and} \quad y'(x_0) = B$$

using a linear combination of a single particular solution y_1 of the differential equation. After all, just try finding a constant c_1 so that

$$y(x) = c_1 y_1(x)$$

satisfies

$$ay'' + by' + cy = 0 \quad \text{with} \quad y(x_0) = A \quad \text{and} \quad y'(x_0) = B$$

when

$$A = y_1(x_0) \quad \text{and} \quad B = 1 + y_1{}'(x_0) \quad .$$

So a fundamental set of solutions for our differential equation must contain at least two solutions.

On the other hand, if anyone were to propose the existence of a fundamental solution set of more than two solutions

$$\{y_1, y_2, y_3, \ldots\} \quad ,$$

then the required linear independence of the set (i.e., that no solution in this set is a linear combination of the others) would automatically imply that the set of just the first two solutions,

$$\{y_1, y_2\} \quad ,$$

is also linearly independent. But then Lemma 16.4 tells us that this smaller set is a fundamental set of solutions for our differential equation, and that every other solution, including the y_3 in the set originally proposed, is a linear combination of y_1 and y_2. Hence, the originally proposed set of three or more solutions cannot be linearly independent, and, hence, is not a fundamental set of solutions for our differential equation.

So, a fundamental set of solutions for a second-order, homogeneous linear differential equation cannot contain less than two solutions or more than two solutions. It must contain exactly two solutions.

And that completes our proof of Theorem 15.3.

16.2 Proving the More General Theorems on General Solutions and Wronskians

Extending the discussion in the previous section into proofs of the more general theorems in Chapter 15 (Theorem 15.5 on page 290 and Theorem 15.6 on page 292) is relatively straightforward provided you make use of some basic facts normally developed in a good introductory course on linear algebra. We will discuss this further (and briefly) in Section 38.6 in the context of "systems of differential equations".

16.3 Linear Differential Operators
The Operator Associated with a Linear Differential Equation

Sometimes, when given some N^{th}-order linear differential equation

$$a_0 \frac{d^N y}{dx^N} + a_1 \frac{d^{N-1} y}{dx^{N-1}} + \cdots + a_{N-2} \frac{d^2 y}{dx^2} + a_{N-1} \frac{dy}{dx} + a_N y = g \quad ,$$

it is convenient to let $L[y]$ denote the expression on the left side, whether or not y is a solution to the differential equation. That is, for any sufficiently differentiable function y,

$$L[y] = a_0 \frac{d^N y}{dx^N} + a_1 \frac{d^{N-1} y}{dx^{N-1}} + \cdots + a_{N-2} \frac{d^2 y}{dx^2} + a_{N-1} \frac{dy}{dx} + a_N y \quad .$$

To emphasize that y is a function of x, we may also use $L[y(x)]$ instead of $L[y]$. For much of what follows, y need not be a solution to the given differential equation, but it does need to be sufficiently differentiable on the interval of interest for all the derivatives in the formula for $L[y]$ to make sense.

While we defined $L[y]$ as the left side of the above differential equation, the expression for $L[y]$ is completely independent of the equation's right side. Because of this and the fact that the choice of y is largely irrelevant to the basic definition, we will often just define " L " by stating

$$L = a_0 \frac{d^N}{dx^N} + a_1 \frac{d^{N-1}}{dx^{N-1}} + \cdots + a_{N-2} \frac{d^2}{dx^2} + a_{N-1} \frac{d}{dx} + a_N$$

where the a_k's are functions of x on the interval of interest.[1]

!▶ *Example 16.1:* *If our differential equation is*

$$\frac{d^2 y}{dx^2} + x^2 \frac{dy}{dx} - 6y = \sqrt{x+1} \quad ,$$

then

$$L = \frac{d^2}{dx^2} + x^2 \frac{d}{dx} - 6 \quad ,$$

and, for any twice-differentiable function $y = y(x)$,

$$L[y(x)] = L[y] = \frac{d^2 y}{dx^2} + x^2 \frac{dy}{dx} - 6y \quad .$$

In particular, if $y = \sin(2x)$, then

$$\begin{aligned} L[y] = L\big[\sin(2x)\big] &= \frac{d^2}{dx^2}\big[\sin(2x)\big] + x^2 \frac{d}{dx}\big[\sin(2x)\big] - 6\big[\sin(2x)\big] \\ &= -4\sin(2x) + x^2 \cdot 2\cos(2x) - 6\sin(2x) \\ &= 2x^2\cos(2x) - 10\sin(2x) \quad . \end{aligned}$$

[1] If using " L " is just too much shorthand for you, observe that the formulas for L can be written in summation form:

$$L[y] = \sum_{k=0}^{N} a_k \frac{d^{N-k} y}{dx^{N-k}} \qquad \text{and} \qquad L = \sum_{k=0}^{N} a_k \frac{d^{N-k}}{dx^{N-k}} \quad .$$

You can use these summation formulas instead of " L " if you wish.

Observe that L is something into which we plug a function (such as the $\sin(2x)$ in the above example) and out of which pops another function (which, in the above example, ended up being $2x^2\cos(2x) - 10\sin(2x)$). Anything that so converts one function into another is often called an *operator* (on functions), and since the general formula for computing $L[y]$ looks like a linear combination of differentiations up to order N,

$$L[y] = a_0\frac{d^N y}{dx^N} + a_1\frac{d^{N-1}y}{dx^{N-1}} + \cdots + a_{N-2}\frac{d^2 y}{dx^2} + a_{N-1}\frac{dy}{dx} + a_N y \quad ,$$

it is standard to refer to L as a *linear differential operator (of order N).*

We should also note that our linear differential operators are "linear" in the sense normally defined in linear algebra:

Lemma 16.5

Assume L is a linear differentiable operator

$$L = a_0\frac{d^N}{dx^N} + a_1\frac{d^{N-1}}{dx^{N-1}} + \cdots + a_{N-2}\frac{d^2}{dx^2} + a_{N-1}\frac{d}{dx} + a_N$$

where the a_k's are functions on some interval (α, β). If y_1 and y_2 are any two sufficiently differentiable functions on (α, β), and c_1 and c_2 are any two constants, then

$$L[c_1 y_1 + c_2 y_2] = c_1 L[y_1] + c_2 L[y_2] \quad .$$

To prove this lemma, you basically go through the same computations as used to derive the principle of superposition (see the derivations just before Theorem 15.1 on page 282).

The Composition Product

Definition and Notation

The *(composition) product* $L_2 L_1$ of two linear differential operators L_1 and L_2 is the differential operator given by

$$L_2 L_1[\phi] = L_2\big[L_1[\phi]\big]$$

for every sufficiently differentiable function $\phi = \phi(x)$.[2]

!►Example 16.2: Let

$$L_1 = \frac{d}{dx} + x^2 \qquad \text{and} \qquad L_2 = \frac{d}{dx} + 4 \quad .$$

For any twice-differentiable function $\phi = \phi(x)$, we have

$$\begin{aligned}
L_2 L_1[\phi] = L_2\left[L_1[\phi]\right] &= L_2\left[\frac{d\phi}{dx} + x^2\phi\right] \\
&= \frac{d}{dx}\left[\frac{d\phi}{dx} + x^2\phi\right] + 4\left[\frac{d\phi}{dx} + x^2\phi\right] \\
&= \frac{d^2\phi}{dx^2} + \frac{d}{dx}\left[x^2\phi\right] + 4\frac{d\phi}{dx} + 4x^2\phi \\
&= \frac{d^2\phi}{dx^2} + 2x\phi + x^2\frac{d\phi}{dx} + 4\frac{d\phi}{dx} + 4x^2\phi \\
&= \frac{d^2\phi}{dx^2} + \left(4 + x^2\right)\frac{d\phi}{dx} + \left(2x + 4x^2\right)\phi \quad .
\end{aligned}$$

[2] The notation $L_2 \circ L_1$, instead of $L_2 L_1$ would also be correct.

Cutting out the middle yields

$$L_2L_1[\phi] = \frac{d^2\phi}{dx^2} + \left(4+x^2\right)\frac{d\phi}{dx} + \left(2x+4x^2\right)\phi$$

for every sufficiently differentiable function ϕ *. Thus*

$$L_2L_1 = \frac{d^2}{dx^2} + \left(4+x^2\right)\frac{d}{dx} + \left(2x+4x^2\right) \quad .$$

When we have formulas for our operators L_1 and L_2, it will often be convenient to replace the symbols " L_1 " and " L_2 " with their formulas enclosed in parentheses. We will also enclose any function ϕ being "plugged into" the operators with square brackets, " $[\phi]$ ". This will be called the *product notation.*[3]

!►Example 16.3: *Using the product notation, let us recompute* L_2L_1 *for*

$$L_1 = \frac{d}{dx} + x^2 \qquad \text{and} \qquad L_2 = \frac{d}{dx} + 4 \quad .$$

Letting $\phi = \phi(x)$ *be any twice-differentiable function,*

$$\begin{aligned}
\left(\frac{d}{dx} + 4\right)\left(\frac{d}{dx} + x^2\right)[\phi] &= \left(\frac{d}{dx} + 4\right)\left[\frac{d\phi}{dx} + x^2\phi\right] \\
&= \frac{d}{dx}\left[\frac{d\phi}{dx} + x^2\phi\right] + 4\left[\frac{d\phi}{dx} + x^2\phi\right] \\
&= \frac{d^2\phi}{dx^2} + \frac{d}{dx}\left[x^2\phi\right] + 4\frac{d\phi}{dx} + 4x^2\phi \\
&= \frac{d^2\phi}{dx^2} + 2x\phi + x^2\frac{d\phi}{dx} + 4\frac{d\phi}{dx} + 4x^2\phi \\
&= \frac{d^2\phi}{dx^2} + \left(4+x^2\right)\frac{d\phi}{dx} + \left(2x+4x^2\right)\phi \quad .
\end{aligned}$$

So,

$$L_2L_1 = \left(\frac{d}{dx} + 4\right)\left(\frac{d}{dx} + x^2\right) = \frac{d^2}{dx^2} + \left(4+x^2\right)\frac{d}{dx} + \left(2x+4x^2\right) \quad ,$$

just as derived in the previous example.

Algebra of the Composite Product

The notation $L_2L_1[\phi]$ is convenient, but it is important to remember that it is shorthand for

compute $L_1[\phi]$ *and plug the result into* L_2 .

[3] Many authors do not enclose "the function being plugged in" in square brackets, and just write $L_2L_1\phi$. We are avoiding that because it does not explicitly distinguish between " ϕ as a function being plugged in" and " ϕ as an operator, itself". For the first, $L_2L_1\phi$ means the function you get from computing $L_2\left[L_1[\phi]\right]$. For the second, $L_2L_1\phi$ means the operator such that, for any sufficiently differentiable function ψ,

$$L_2\left[L_1\left[\phi[\psi]\right]\right] = L_2\left[L_1[\phi\psi]\right] \quad .$$

The two possible interpretations for $L_2L_1\phi$ are not the same.

The result of this can be quite different from

compute $L_2[\phi]$ and plug the result into L_1 ,

which is what $L_1 L_2[\phi]$ means. Thus, in general,

$$L_2 L_1 \neq L_1 L_2 \quad .$$

In other words, the composition product of differential operators is generally *not* commutative.

!►Example 16.4: *In the previous two examples, we saw that*

$$\left(\frac{d}{dx} + 4\right)\left(\frac{d}{dx} + x^2\right) = \frac{d^2}{dx^2} + \left(4 + x^2\right)\frac{d}{dx} + \left(2x + 4x^2\right) \quad .$$

On the other hand, switching the order of the two operators, and letting ϕ be any sufficiently differentiable function gives

$$\begin{aligned}
\left(\frac{d}{dx} + x^2\right)\left(\frac{d}{dx} + 4\right)[\phi] &= \left(\frac{d}{dx} + x^2\right)\left[\frac{d\phi}{dx} + 4\phi\right] \\
&= \frac{d}{dx}\left[\frac{d\phi}{dx} + 4\phi\right] + x^2\left[\frac{d\phi}{dx} + 4\phi\right] \\
&= \frac{d^2\phi}{dx^2} + 4\frac{d\phi}{dx} + x^2\frac{d\phi}{dx} + 4x^2\phi \\
&= \frac{d^2\phi}{dx^2} + \left(4 + x^2\right)\frac{d\phi}{dx} + 4x^2\phi \quad .
\end{aligned}$$

Thus,

$$\left(\frac{d}{dx} + x^2\right)\left(\frac{d}{dx} + 4\right) = \frac{d^2}{dx^2} + \left(4 + x^2\right)\frac{d}{dx} + 4x^2 \quad .$$

After comparing this with the first equation in this example, we clearly see that

$$\left(\frac{d}{dx} + x^2\right)\left(\frac{d}{dx} + 4\right) \neq \left(\frac{d}{dx} + 4\right)\left(\frac{d}{dx} + x^2\right) \quad .$$

?►Exercise 16.1: *Let*

$$L_1 = \frac{d}{dx} \quad \text{and} \quad L_2 = x \quad ,$$

and verify that

$$L_2 L_1 = x\frac{d}{dx} \quad \text{while} \quad L_1 L_2 = x\frac{d}{dx} + 1 \quad .$$

Later (in Chapters 19 and 22) we will be dealing with special situations in which the composition product is commutative. In fact, the material we are now developing will be most useful in verifying certain theorems involving those situations. In the meantime, just remember that, in general,

$$L_2 L_1 \neq L_1 L_2 \quad .$$

Here are a few other short and easily verified notes about the composition product:

1. In the above examples, the operators L_2 and L_1 were all first-order differential operators. This was not necessary. We could have used, say,

$$L_2 = x^3\frac{d^3}{dx^3} + \sin(x)\frac{d^2}{dx^2} - xe^{3x}\frac{d}{dx} + 87\sqrt{x}$$

and

$$L_1 = \frac{d^{26}}{dx^{26}} - x^3 \frac{d^3}{dx^3} \quad ,$$

though we would have certainly needed many more pages for the calculations.

2. There is no need to limit ourselves to composition products of just two operators. Given any number of linear differential operators — L_1, L_2, L_3, ... — the composition products $L_3L_2L_1$, $L_4L_3L_2L_1$, etc. are defined to be the differential operators satisfying, for each and every sufficiently differentiable function ϕ,

$$L_3L_2L_1[\phi] = L_3\big[L_2\big[L_1[\phi]\big]\big] \quad ,$$

$$L_4L_3L_2L_1[\phi] = L_4\big[L_3\big[L_2\big[L_1[\phi]\big]\big]\big] \quad ,$$

$$\vdots$$

Naturally, the order of the operators is still important.

3. Any composition product of linear differential operators is, itself, a linear differential operator. Moreover, the order of the product

$$L_K \cdots L_2L_1$$

is the sum

$$(\text{the order of } L_K) + \cdots + (\text{the order of } L_2) + (\text{the order of } L_1) \quad .$$

4. Though not commutative, the composition product is associative. That is, if L_1, L_2 and L_3 are three linear differential operators, and we ‘precompute’ the products L_2L_1 and L_3L_2, and then compute

$$(L_3L_2)L_1 \quad , \quad L_3(L_2L_1) \quad \text{and} \quad L_3L_2L_1 \quad ,$$

we will discover that

$$(L_3L_2)L_1 = L_3(L_2L_1) = L_3L_2L_1 \quad .$$

5. Keep in mind that we are dealing with linear differential operators and that their products are linear differential operators. In particular, if α is some constant and ϕ is any sufficiently differentiable function, then

$$L_K \cdots L_2L_1[\alpha\phi] = \alpha L_K \cdots L_2L_1[\phi] \quad .$$

And, of course,

$$L_K \cdots L_2L_1[0] = 0 \quad .$$

Factoring

Now suppose we have some linear differential operator L. If we can find other linear differential operators L_1, L_2, L_3 ..., and L_K such that

$$L = L_K \cdots L_2L_1 \quad ,$$

then, in analogy with the classical concept of factoring, we will say that we have *factored* the operator L. The product $L_N \cdots L_2 L_1$ will be called a *factoring* of L, and we may even refer to the individual operators L_1, L_2, L_3, ... and L_N as *factors* of L. Keep in mind that, because composition multiplication is order dependent, it is not usually enough to simply specify the factors. The order must also be given.

!►Example 16.5: *In Example 16.3, we saw that*

$$\frac{d^2}{dx^2} + \left(4+x^2\right)\frac{d}{dx} + \left(2x+4x^2\right) = \left(\frac{d}{dx} + 4\right)\left(\frac{d}{dx} + x^2\right) \quad .$$

So

$$\left(\frac{d}{dx} + 4\right)\left(\frac{d}{dx} + x^2\right)$$

is a factoring of

$$\frac{d^2}{dx^2} + \left(4+x^2\right)\frac{d}{dx} + \left(2x+4x^2\right)$$

with factors

$$\frac{d}{dx} + 4 \quad \text{and} \quad \frac{d}{dx} + x^2 \quad .$$

In addition, from Example 16.4 we know

$$\frac{d^2}{dx^2} + \left(4+x^2\right)\frac{d}{dx} + 4x^2 = \left(\frac{d}{dx} + x^2\right)\left(\frac{d}{dx} + 4\right) \quad .$$

Thus

$$\frac{d}{dx} + x^2 \quad \text{and} \quad \frac{d}{dx} + 4$$

are also factors for

$$\frac{d^2}{dx^2} + \left(4+x^2\right)\frac{d}{dx} + 4x^2 \quad ,$$

but the factoring here is

$$\left(\frac{d}{dx} + x^2\right)\left(\frac{d}{dx} + 4\right) \quad .$$

Let's make a simple observation. Assume a given linear differential operator L can be factored as $L = L_K \cdots L_2 L_1$. Assume, also, that $y_1 = y_1(x)$ is a function satisfying

$$L_1[y_1] = 0 \quad .$$

Then

$$L[y_1] = L_K \cdots L_2 L_1[y_1] = L_K \cdots L_2\big[L_1[y_1]\big] = L_K \cdots L_2[0] = 0 \quad .$$

This proves the following theorem:

Theorem 16.6
Let L be a linear differential operator with factoring $L = L_K \cdots L_2 L_1$. Then any solution to

$$L_1[y] = 0$$

is also a solution to

$$L[y] = 0 \quad .$$

Warning: On the other hand, if, say, $L = L_2 L_1$, then solutions to $L_2[y] = 0$ will usually *not* be solutions to $L[y] = 0$.

!►Example 16.6: Consider

$$\frac{d^2y}{dx^2} + \left(4 + x^2\right)\frac{dy}{dx} + 4x^2 y = 0 \quad .$$

As derived in Example 16.4,

$$\frac{d^2}{dx^2} + \left(4 + x^2\right)\frac{d}{dx} + 4x^2 = \left(\frac{d}{dx} + x^2\right)\left(\frac{d}{dx} + 4\right) \quad .$$

So our differential equation can be written as

$$\left(\frac{d}{dx} + x^2\right)\left(\frac{d}{dx} + 4\right)[y] = 0 \quad .$$

That is,

$$\left(\frac{d}{dx} + x^2\right)\left[\frac{dy}{dx} + 4y\right] = 0 \quad . \tag{16.8}$$

Now consider

$$\frac{dy}{dx} + 4y = 0 \quad .$$

This is a simple first-order linear and separable differential equation, whose general solution is easily found to be $y = c_1 e^{-4x}$ *. In particular,* e^{-4x} *is a solution. According to the above theorem,* e^{-4x} *is also a solution to our original differential equation. Let's check to be sure:*

$$\begin{aligned}
\frac{d^2}{dx^2}\left[e^{-4x}\right] + \left(4 + x^2\right)\frac{d}{dx}\left[e^{-4x}\right] + 4x^2 e^{-4x} &= \left(\frac{d}{dx} + x^2\right)\left(\frac{d}{dx} + 4\right)\left[e^{-4x}\right] \\
&= \left(\frac{d}{dx} + x^2\right)\left[\frac{d}{dx}\left[e^{-4x}\right] + 4e^{-4x}\right] \\
&= \left(\frac{d}{dx} + x^2\right)\left[-4e^{-4x} + 4e^{-4x}\right] \\
&= \left(\frac{d}{dx} + x^2\right)[0] \\
&= \frac{d0}{dx} + x^2 \cdot 0 \\
&= 0 \quad .
\end{aligned}$$

Keep in mind, though, that e^{-4x} *is simply one of the possible solutions, and that there will be solutions not given by* $c_1 e^{-4x}$ *.*

Unfortunately, unless it is of an exceptionally simple type (such as considered in Chapter 19), factoring a linear differential operator is a very nontrivial problem. And even with those simple types that we will be able to factor, we will find the main value of the above to be in deriving even simpler methods for finding solutions. Consequently, in practice, you should not expect to be solving many differential equations via "operator factoring".

Additional Exercises

16.2 a. *State the linear differential operator* L *corresponding to the left side of*

$$\frac{d^2y}{dx^2} + 5\frac{dy}{dx} + 6y = 0 \quad .$$

b. *Using this* L *, compute each of the following:*

i. $L[\sin(x)]$ **ii.** $L\left[e^{4x}\right]$ **iii.** $L\left[e^{-3x}\right]$ **iv.** $L\left[x^2\right]$

c. *Based on the answers to the last part, what is one solution to the differential equation in part* **a***?*

16.3 a. *State the linear differential operator* L *corresponding to the left side of*

$$\frac{d^2y}{dx^2} - 5\frac{dy}{dx} + 9y = 0 \quad .$$

b. *Using this* L *, compute each of the following:*

i. $L[\sin(x)]$ **ii.** $L[\sin(3x)]$ **iii.** $L\left[e^{2x}\right]$ **iv.** $L\left[e^{2x}\sin(x)\right]$

16.4 a. *State the linear differential operator* L *corresponding to the left side of*

$$x^2\frac{d^2y}{dx^2} + 5x\frac{dy}{dx} + 6y = 0 \quad .$$

b. *Using this* L *, compute each of the following:*

i. $L[\sin(x)]$ **ii.** $L\left[e^{4x}\right]$ **iii.** $L\left[x^3\right]$

16.5 a. *State the linear differential operator* L *corresponding to the left side of*

$$\frac{d^3y}{dx^3} - \sin(x)\frac{dy}{dx} + \cos(x)\,y = x^2 + 1 \quad ,$$

b. *and then, using this* L *, compute each of the following:*

i. $L[\sin(x)]$ **ii.** $L[\cos(x)]$ **iii.** $L\left[x^2\right]$

16.6. *Several choices for linear differential operators* L_1 *and* L_2 *are given below. For each choice, compute* L_2L_1 *and* L_1L_2 *.*

a. $L_1 = \frac{d}{dx} + x$ and $L_2 = \frac{d}{dx} - x$

b. $L_1 = \frac{d}{dx} + x^2$ and $L_2 = \frac{d}{dx} + x^3$

c. $L_1 = x\frac{d}{dx} + 3$ and $L_2 = \frac{d}{dx} + 2x$

d. $L_1 = \frac{d^2}{dx^2}$ and $L_2 = x$

e. $L_1 = \frac{d^2}{dx^2}$ and $L_2 = x^3$

f. $L_1 = \frac{d^2}{dx^2}$ and $L_2 = \sin(x)$

16.7. Compute the following composition products:

a. $\left(\frac{d}{dx} + 2\right)\left(\frac{d}{dx} + 3\right)$

b. $\left(x\frac{d}{dx} + 2\right)\left(x\frac{d}{dx} + 3\right)$

c. $\left(x\frac{d}{dx} + 4\right)\left(\frac{d}{dx} + \frac{1}{x}\right)$

d. $\left(\frac{d}{dx} + 4x\right)\left(\frac{d}{dx} + \frac{1}{x}\right)$

e. $\left(\frac{d}{dx} + \frac{1}{x}\right)\left(\frac{d}{dx} + 4x\right)$

f. $\left(\frac{d}{dx} + 5x^2\right)^2$

g. $\left(\frac{d}{dx} + x^2\right)\left(\frac{d^2}{dx^2} + \frac{d}{dx}\right)$

h. $\left(\frac{d^2}{dx^2} + \frac{d}{dx}\right)\left(\frac{d}{dx} + x^2\right)$

16.8. Verify that

$$\frac{d^2}{dx^2} + \left(\sin(x) - 3\right)\frac{d}{dx} - 3\sin(x) = \left(\frac{d}{dx} + \sin(x)\right)\left(\frac{d}{dx} - 3\right) \quad ,$$

and, using this factorization, find one solution to

$$\frac{d^2y}{dx^2} + \left(\sin(x) - 3\right)\frac{dy}{dx} - 3\sin(x)y = 0 \quad .$$

16.9. Verify that

$$\frac{d^2}{dx^2} + x\frac{d}{dx} + \left(2 - 2x^2\right) = \left(\frac{d}{dx} - x\right)\left(\frac{d}{dx} + 2x\right) \quad ,$$

and, using this factorization, find one solution to

$$\frac{d^2y}{dx^2} + x\frac{dy}{dx} + \left(2 - 2x^2\right)y = 0 \quad .$$

16.10. Verify that

$$x^2\frac{d^2}{dx^2} - 7x\frac{d}{dx} + 16 = \left(x\frac{d}{dx} - 4\right)^2 \quad ,$$

and, using this factorization, find one solution to

$$x^2\frac{d^2y}{dx^2} - 7x\frac{dy}{dx} + 16y = 0 \quad .$$

17

Second-Order Homogeneous Linear Equations with Constant Coefficients

A very important class of second-order homogeneous linear equations consists of those with constant coefficients; that is, those that can be written as

$$ay'' + by' + cy = 0$$

where a, b and c are real-valued constants (with $a \neq 0$). Some examples are

$$y'' - 5y' + 6y = 0 \quad ,$$

$$y'' - 6y' + 9y = 0$$

and

$$y'' - 6y' + 13y = 0 \quad .$$

There are two reasons these sorts of differential equations are important: First of all, they often arise in applications. Secondly, as we will see, it is relatively easy to find fundamental sets of solutions for these equations.

Do note that, because the coefficients are constants, they are, trivially, continuous functions on the entire real line. Consequently, we can take the entire real line as the interval of interest, and be confident that any solutions derived will be valid on all of $(-\infty, \infty)$.

IMPORTANT: What we will derive and define here (e.g., "the characteristic equation") is based on the assumption that the coefficients in our differential equation — the a, b and c above — are constants. Some of the results will even require that these constants be real valued. Do not, later, try to blindly apply what we develop here to differential equations in which the coefficients are not real-valued constants.

17.1 Deriving the Basic Approach

Seeking Inspiration

Let us look for clues on how to solve our second-order equations by first looking at solving a first-order, homogeneous linear differential equation with constant coefficients, say,

$$2\frac{dy}{dx} + 6y = 0 \quad .$$

Since we are considering 'linear' equations, let's solve it using the method developed for first-order linear equations: First divide through by the first coefficient, 2, to get

$$\frac{dy}{dx} + 3y = 0 \quad .$$

The integrating factor is then

$$\mu = e^{\int 3\,dx} = e^{3x} \quad .$$

Multiplying through and proceeding as usual with first-order linear equations:

$$e^{3x}\left[\frac{dy}{dx} + 3y\right] = e^{3x} \cdot 0$$

$$\hookrightarrow \qquad e^{3x}\frac{dy}{dx} + 3e^{3x}y = 0$$

$$\hookrightarrow \qquad \frac{d}{dx}\left[e^{3x}y\right] = 0$$

$$\hookrightarrow \qquad e^{3x}y = c$$

$$\hookrightarrow \qquad y = ce^{-3x} \quad .$$

So a general solution to

$$2\frac{dy}{dx} + 6y = 0$$

is

$$y = ce^{-3x} \quad .$$

Clearly, there is nothing special about the numbers used here. Replacing 2 and 6 with constants a and b in the above would just as easily have given us the fact that a general solution to

$$a\frac{dy}{dx} + by = 0$$

is

$$y = ce^{rx} \qquad \text{where} \quad r = -\frac{b}{a} \quad .$$

Thus we see that all solutions to first-order homogeneous linear equations with constant coefficients are given by constant multiples of exponential functions.

Exponential Solutions with Second-Order Equations

Now consider the second-order case. For convenience, we will use

$$y'' - 5y' + 6y = 0$$

as an example, keeping in mind that our main interest is in finding all possible solutions to an arbitrary second-order homogeneous differential equation

$$ay'' + by' + cy = 0$$

where a, b and c are constants.

From our experience with the first-order case, it seems reasonable to expect at least some of the solutions to be exponentials. So let us find all such solutions by setting

$$y = e^{rx}$$

where r is a constant to be determined, plugging this formula into our differential equation, and seeing if a constant r can be determined.

For our example,

$$y'' - 5y' + 6y = 0 \quad .$$

Letting $y = e^{rx}$ yields

$$\frac{d^2}{dx^2}\left[e^{rx}\right] - 5\frac{d}{dx}\left[e^{rx}\right] + 6\left[e^{rx}\right] = 0$$

$$\hookrightarrow \qquad r^2 e^{rx} - 5re^{rx} + 6e^{rx} = 0$$

$$\hookrightarrow \qquad e^{rx}\left[r^2 - 5r + 6\right] = 0 \quad .$$

Since e^{rx} can never be zero, we can divide it out, leaving the algebraic equation

$$r^2 - 5r + 6 = 0 \quad .$$

Before solving this for r, let us pause and consider the more general case.

More generally, letting $y = e^{rx}$ in

$$ay'' + by' + cy = 0 \tag{17.1}$$

yields

$$a\frac{d^2}{dx^2}\left[e^{rx}\right] + b\frac{d}{dx}\left[e^{rx}\right] + c\left[e^{rx}\right] = 0$$

$$\hookrightarrow \qquad ar^2 e^{rx} + bre^{rx} + ce^{rx} = 0$$

$$\hookrightarrow \qquad e^{rx}\left[ar^2 + br + c\right] = 0 \quad .$$

Since e^{rx} can never be zero, we can divide it out, leaving us with the algebraic equation

$$ar^2 + br + c = 0 \tag{17.2}$$

(remember: a, b and c are constants). Equation (17.2) is called the *characteristic equation* for differential equation (17.1). Note the similarity between the original differential equation and its characteristic equation. The characteristic equation is nothing more that the algebraic equation obtained by replacing the various derivatives of y with corresponding powers of r (treating y as being the zeroth derivative of y):

$$ay'' + by' + cy = 0 \qquad \text{(original differential equation)}$$

$$\hookrightarrow \qquad ar^2 + br + c = 0 \qquad \text{(characteristic equation)} \quad .$$

The nice thing is that the characteristic equation is easily solved for r by either factoring the polynomial or using the quadratic formula. These values for r must then be the values of r for which $y = e^{rx}$ are (particular) solutions to our original differential equation. Using what we developed in previous chapters, we may then be able to construct a general solution to the differential equation.

In our example, letting $y = e^{rx}$ *in*

$$y'' - 5y' + 6y = 0$$

led to the characteristic equation

$$r^2 - 5r + 6 = 0 \quad ,$$

which factors to

$$(r-2)(r-3) = 0 \quad .$$

Hence,

$$r - 2 = 0 \quad \text{or} \quad r - 3 = 0 \quad .$$

So the possible values of r *are*

$$r = 2 \quad \text{and} \quad r = 3 \quad ,$$

which, in turn, means

$$y_1 = e^{2x} \quad \text{and} \quad y_2 = e^{3x}$$

are solutions to our original differential equation. Clearly, neither of these functions is a constant multiple of the other; so, after recalling the big theorem on solutions to second-order, homogeneous linear differential equations, Theorem 15.3 on page 288, we know that

$$\{ e^{2x}, e^{3x} \}$$

is a fundamental set of solutions and

$$y(x) = c_1 e^{2x} + c_2 e^{3x}$$

is a general solution to our differential equation.

We will discover that we can always construct a general solution to any given homogeneous linear differential equation with constant coefficients using the solutions to its characteristic equation. But first, let us restate what we have just derived in a somewhat more concise and authoritative form, and briefly consider the nature of the possible solutions to the characteristic equation.

17.2 The Basic Approach, Summarized

To solve a second-order homogeneous linear differential equation

$$ay'' + by' + cy = 0$$

in which a, b and c are constants, start with the assumption that

$$y(x) = e^{rx}$$

where r is a constant to be determined. Plugging this formula for y into the differential equation yields, after a little computation and simplification, the differential equation's *characteristic equation* for r,

$$ar^2 + br + c = 0 \quad .$$

Alternatively, the characteristic equation can simply be constructed by replacing the derivatives of y in the original differential equation with the corresponding powers of r.

By the way, the polynomial on the left side of the characteristic equation,

$$ar^2 + br + c \quad ,$$

is called the *characteristic polynomial* for the differential equation. Recall from algebra that a "root" of a polynomial $p(r)$ is the same as a solution to $p(r) = 0$. So we can — and will — use the terms "solution to the characteristic equation" and "root of the characteristic polynomial" interchangeably.

Since the characteristic polynomial is only of degree two, solving the characteristic equation for r should present no problem. If this equation is simple enough, we can factor the polynomial and find the values of r by inspection. At worst, we must recall that the solution to the polynomial equation

$$ar^2 + br + c = 0$$

can always be obtained via the quadratic formula,

$$r = \frac{-b \pm \sqrt{b^2 - 4ac}}{2a} \quad .$$

Notice how the nature of the value r depends strongly on the value under the square root, $b^2 - 4ac$. There are three possibilities:

1. If $b^2 - 4ac > 0$, then $\sqrt{b^2 - 4ac}$ is some positive value, and we have two distinct real values for r,
$$r_- = \frac{-b - \sqrt{b^2 - 4ac}}{2a} \quad \text{and} \quad r_+ = \frac{-b + \sqrt{b^2 - 4ac}}{2a} \quad .$$
(In practice, we may denote these solutions by r_1 and r_2, instead.)

2. If $b^2 - 4ac = 0$, then
$$r = \frac{-b \pm \sqrt{b^2 - 4ac}}{2a} = \frac{-b \pm \sqrt{0}}{2a} \quad ,$$
and we only have one real root for our characteristic equation, namely,
$$r = -\frac{b}{2a} \quad .$$

3. If $b^2 - 4ac < 0$, then the quantity under the square root is negative, and, thus, this square root gives rise to an *imaginary number*. To be explicit,
$$\sqrt{b^2 - 4ac} = \sqrt{-1 \cdot \left|b^2 - 4ac\right|} = i\sqrt{\left|b^2 - 4ac\right|}$$
where "$i = \sqrt{-1}$".[1] Thus, in this case, we will get two distinct complex roots, r_+ and r_- with
$$r_\pm = \frac{-b \pm \sqrt{b^2 - 4ac}}{2a} = \frac{-b \pm i\sqrt{|b^2 - 4ac|}}{2a} = -\frac{b}{2a} \pm i\frac{\sqrt{|b^2 - 4ac|}}{2a} \quad .$$

Whatever the case, if we find r_0 to be a root of the characteristic polynomial, then, by the very steps leading to the characteristic equation, it follows that

$$y_0(x) = e^{r_0 x}$$

is a solution to our original differential equation. As you can imagine, though, the nature of the corresponding general solution to this differential equation depends strongly on which of the above three cases we are dealing with. Let us consider each case.

[1] Some people prefer to use j instead of i for $\sqrt{-1}$. Clearly, these people don't know how to spell 'imaginary'.

17.3 Case 1: Two Distinct Real Roots

Suppose the characteristic equation for

$$ay'' + by' + cy = 0$$

has two distinct (i.e., different) real solutions r_1 and r_2. Then we have that both

$$y_1 = e^{r_1 x} \qquad \text{and} \qquad y_2 = e^{r_2 x}$$

are solutions to the differential equation. Since we are assuming r_1 and r_2 are not the same, it should be clear that neither y_1 nor y_2 is a constant multiple of the other. Hence

$$\left\{ e^{r_1 x}, e^{r_2 x} \right\}$$

is a linearly independent set of solutions to our second-order homogeneous linear differential equation. The big theorem on solutions to second-order, homogenous linear differential equations, Theorem 15.3 on page 288, then tells us that

$$y(x) = c_1 e^{r_1 x} + c_2 e^{r_2 x}$$

is a general solution to our differential equation.

We will later find it convenient to have a way to refer back to the results of the above observations. That is why those results are now restated in the following lemma:

Lemma 17.1
Let a, b and c be constants with $a \neq 0$. If the characteristic equation for

$$ay'' + by' + cy = 0$$

has two distinct real solutions r_1 and r_2, then

$$y_1(x) = e^{r_1 x} \qquad \text{and} \qquad y_2(x) = e^{r_2 x}$$

are two solutions to this differential equation. Moreover, $\{e^{r_1 x}, e^{r_2 x}\}$ is a fundamental set for the differential equation, and

$$y(x) = c_1 e^{r_1 x} + c_2 e^{r_2 x}$$

is a general solution.

The example done while deriving the basic approach illustrated this case. Another example, however, may not hurt.

!▶Example 17.1: *Consider the differential equation*

$$y'' + 7y' = 0 \quad .$$

Assuming $y = e^{rx}$ in this equation gives

$$\frac{d^2}{dx^2}\left[e^{rx}\right] + \frac{d}{dx}\left[e^{rx}\right] = 0$$

$$\hookrightarrow \qquad r^2 e^{rx} + 7re^{rx} = 0 \quad .$$

Dividing out e^{rx} *gives us the characteristic equation*

$$r^2 + 7r = 0 \quad .$$

In factored form, this is

$$r(r+7) = 0 \quad ,$$

which means that

$$r = 0 \qquad \text{or} \qquad r + 7 = 0 \quad .$$

Consequently, the solutions to the characteristic equation are

$$r = 0 \qquad \text{and} \qquad r = -7 \quad .$$

The two corresponding solutions to the differential equation are

$$y_1 = e^{0 \cdot x} = 1 \qquad \text{and} \qquad y_2 = e^{-7x} \quad .$$

Thus, our fundamental set of solutions is

$$\left\{ 1, e^{-7x} \right\} \quad ,$$

and the corresponding general solution to our differential equation is

$$y(x) = c_1 \cdot 1 + c_2 e^{-7x} \quad ,$$

which is slightly more simply written as

$$y(x) = c_1 + c_2 e^{-7x} \quad .$$

17.4 Case 2: Only One Root Using Reduction of Order

If the characteristic polynomial only has one root r, then

$$y_1(x) = e^{rx}$$

is one solution to our differential equation. This, alone, is not enough for a general solution, but we can use this one solution with the reduction of order method to get the full general solution. Let us do one example this way.

!▶Example 17.2: *Consider the differential equation*

$$y'' - 6y' + 9y = 0 \quad .$$

The characteristic equation is

$$r^2 - 6r + 9 = 0 \quad ,$$

which factors nicely to

$$(r-3)^2 = 0 \quad ,$$

giving us $r = 3$ as the only root. Consequently, we have

$$y_1(x) = e^{3x}$$

as one solution to our differential equation.

To find the general solution, we start the reduction of order method as usual by letting

$$y(x) = y_1(x)u(x) = e^{3x}u(x) \quad .$$

The derivatives are then computed,

$$y'(x) = \left[e^{3x}u\right]' = 3e^{3x}u + e^{3x}u'$$

and

$$\begin{aligned} y''(x) &= \left[3e^{3x}u + e^{3x}u'\right]' \\ &= 3 \cdot 3e^{3x}u + 3e^{3x}u' + 3e^{3x}u' + e^{3x}u'' \\ &= 9e^{3x}u + 6e^{3x}u' + e^{3x}u'' \quad , \end{aligned}$$

and plugged into the differential equation,

$$\begin{aligned} 0 &= y'' - 6y' + 9y \\ &= \left[9e^{3x}u + 6e^{3x}u' + e^{3x}u''\right] - 6\left[3e^{3x}u + e^{3x}u'\right] + 9\left[e^{3x}u\right] \\ &= e^{3x}\left\{9u + 6u' + u'' - 18u - 6u' + 9u\right\} \quad . \end{aligned}$$

Dividing out the exponential and grouping together the coefficients for u, u' and u'' yield

$$0 = u'' + [6 - 6]u' + [9 - 18 + 9]u = u'' \quad .$$

As expected, the "u term" drops out. Even nicer, though, is that the "u' term" also drops out, leaving us with $u'' = 0$; that is, to be a little more explicit,

$$\frac{d^2u}{dx^2} = 0 \quad .$$

No need to do anything fancy here — just integrate twice. The first time yields

$$\frac{du}{dx} = \int \frac{d^2u}{dx^2}\,dx = \int 0\,dx = A \quad .$$

Integrating again,

$$u(x) = \int \frac{du}{dx}\,dx = \int A\,dx = Ax + B \quad .$$

Thus,

$$y(x) = e^{3x}u(x) = e^{3x}[Ax + B] = Axe^{3x} + Be^{3x}$$

is the general solution to the differential equation being considered here.

Most of the labor in the last example was in carrying out the reduction of order method. That labor was greatly simplified by the fact that the differential equation for u simplified to

$$u'' = 0 \quad ,$$

which, in turn, meant that

$$u(x) = Ax + B \quad ,$$

and so

$$y(x) = e^{3x}u(x) = Axe^{3x} + Be^{3x} \quad .$$

Will we always be this lucky? To see, let us consider the most general case where the characteristic equation

$$ar^2 + br + c = 0$$

has only one root. As noted when we discussed the possible solutions to the characteristic polynomial (see page 321), this means

$$r = -\frac{b}{2a} \quad .$$

Let us go through the reduction of order method, keeping this fact in mind.

Start with the one known solution,

$$y_1(x) = e^{rx} \qquad \text{where} \quad r = -\frac{b}{2a} \quad .$$

Set

$$y(x) = y_1(x)u(x) = e^{rx}u(x) \quad ,$$

compute the derivatives,

$$y'(x) = \left[e^{rx}u\right]' = re^{rx}u + e^{rx}u'$$

and

$$\begin{aligned} y''(x) &= \left[re^{rx}u + e^{rx}u'\right]' \\ &= r \cdot re^{rx}u + re^{rx}u' + re^{rx}u' + e^{rx}u'' \\ &= r^2e^{rx}u + 2re^{rx}u' + e^{rx}u'' \quad , \end{aligned}$$

and plug these into the differential equation,

$$\begin{aligned} 0 &= ay'' + by' + cy \\ &= a\left[r^2e^{rx}u + 2re^{rx}u' + e^{rx}u''\right] + b\left[re^{rx}u + e^{rx}u'\right] + c\left[e^{rx}u\right] \\ &= e^{rx}\left\{ar^2u + 2aru' + au'' + bru + bu' + cu\right\} \quad . \end{aligned}$$

Dividing out the exponential and grouping together the coefficients for u, u' and u'', we get

$$0 = au'' + [2ar + b]u' + [ar^2 + br + c]u \quad .$$

Since r satisfies the characteristic equation,

$$ar^2 + br + c = 0 \quad ,$$

the "u term" drops out, as it should. Moreover, because $r = -{}^{b}/_{2a}$,

$$2ar + b = 2a\left[-\frac{b}{2a}\right] + b = -b + b = 0 \quad ,$$

and the "u' term" also drops out, just as in the example. Dividing out the a (which, remember, is a nonzero constant), the differential equation for u simplifies to

$$u'' = 0 \quad .$$

Integrating twice yields

$$u(x) = Ax + B \quad ,$$

and, thus,

$$y(x) = y_1(x)u(x) = e^{rx}[Ax + B] = Axe^{rx} + Be^{rx} \quad . \tag{17.3}$$

Skipping Reduction of Order

Let us stop and reflect on what the last formula, equation (17.3), tells us. It tells us that, whenever the characteristic polynomial has only one root r, then the general solution of the differential equation is a linear combination of the two functions

$$e^{rx} \quad \text{and} \quad xe^{rx} \quad .$$

If we remember this, we don't need to go through the reduction of order method when solving these sorts of equations. This is a nice shortcut for solving these differential equations. And since these equations arise relatively often in applications, it is a shortcut worth remembering. To aid remembrance, here is the summary of what we have derived:

Lemma 17.2
Let a, b and c be constants with $a \neq 0$. If the characteristic equation for

$$ay'' + by' + cy = 0$$

has only one solution r, then

$$y_1(x) = e^{rx} \quad \text{and} \quad y_2(x) = xe^{rx}$$

are two solutions to this differential equation. Moreover, $\{e^{rx}, xe^{rx}\}$ is a fundamental set for the differential equation, and

$$y(x) = c_1 e^{rx} + c_2 x e^{rx}$$

is a general solution.

Let's redo Example 17.2 using this lemma:

!►Example 17.3: *Consider the differential equation*

$$y'' - 6y' + 9y = 0 \quad .$$

The characteristic equation is

$$r^2 - 6r + 9 = 0 \quad ,$$

which factors nicely to

$$(r - 3)^2 = 0 \quad ,$$

giving us $r = 3$ as the only root. Consequently,

$$y_1(x) = e^{3x}$$

is one solution to our differential equation. By our work above, summarized in Lemma 17.2, we know a second solution is

$$y_2(x) = xe^{3x}$$

and a general solution is

$$y(x) = c_1 e^{3x} + c_2 x e^{3x} \quad .$$

17.5 Case 3: Complex Roots

Blindly Using Complex Roots

Let us start with an example.

!►Example 17.4: *Consider solving*

$$y'' - 6y' + 13y = 0 \quad .$$

The characteristic equation is

$$r^2 - 6r + 13 = 0 \quad .$$

Factoring this is not easy for most, so we will resort to the quadratic formula for finding the possible values of r :

$$\begin{aligned} r &= \frac{-b \pm \sqrt{b^2 - 4ac}}{2a} \\ &= \frac{-(-6) \pm \sqrt{(-6)^2 - 4 \cdot 1 \cdot 13}}{2 \cdot 1} \\ &= \frac{6 \pm \sqrt{-16}}{2} \\ &= \frac{6 \pm i4}{2} = 3 \pm i2 \quad . \end{aligned}$$

So we get

$$y_+(x) = e^{(3+i2)x} \qquad \text{and} \qquad y_-(x) = e^{(3-i2)x}$$

as two solutions to the differential equation. Since the factors in the exponents are different, we can reasonably conclude that neither $e^{(3+i2)x}$ nor $e^{(3-i2)x}$ is a constant multiple of the other, and so

$$y(x) = c_+ e^{(3+i2)x} + c_- e^{(3-i2)x}$$

should be a general solution to our differential equation.

In general, if the roots to the characteristic equation for

$$ay'' + by' + cy = 0$$

are not real valued, then, as noted earlier, we will get a pair of complex-valued roots

$$r = \frac{-b \pm \sqrt{b^2 - 4ac}}{2a} = \frac{-b \pm i\sqrt{|b^2 - 4ac|}}{2a} = -\frac{b}{2a} \pm i\frac{\sqrt{|b^2 - 4ac|}}{2a} \quad .$$

For convenience, let's write these roots generically as

$$r_+ = \lambda + i\omega \qquad \text{and} \qquad r_- = \lambda - i\omega$$

where λ and ω are the real numbers

$$\lambda = -\frac{b}{2a} \qquad \text{and} \qquad \omega = \frac{\sqrt{|b^2 - 4ac|}}{2a} \quad .$$

Don't bother memorizing these formulas for λ and ω, but do observe that these two values for r, $\lambda + i\omega$ and $\lambda - i\omega$, form a "conjugate pair;" that is, they differ only by the sign on the imaginary part.

It should also be noted that ω cannot be zero; otherwise we would be back in case 2 (one real root). So $r_- = \lambda + i\omega$ and $r_+ = \lambda - i\omega$ are two distinct roots. As we will verify, this means

$$y_+(x) = e^{r_+ x} = e^{(\lambda+i\omega)x} \quad \text{and} \quad y_-(x) = e^{r_- x} = e^{(\lambda-i\omega)x}$$

are two independent solutions, and, from a purely mathematical point of view, there is nothing wrong with using

$$y(x) = c_+ e^{(\lambda+i\omega)x} + c_- e^{(\lambda-i\omega)x}$$

as a general solution to our differential equation.

There are problems, however, with using e^{rx} when r is a complex number. For one thing, it introduces complex numbers into problems that, most likely, should be entirely describable using just real-valued functions. More importantly, you might not yet know what e^{rx} means when r is a complex number! (Quick, graph $e^{(3+i2)x}$!) To deal with these problems, you have several choices:

1. Pray you never have to deal with complex roots when solving differential equations. (A very bad choice since such equations turn out to be among the most important ones in applications.)
2. Blindly use $e^{(\lambda \pm i\omega)x}$, hoping that no one questions your results and that your results are never used in a system the failure of which could result in death, injury, or the loss of substantial sums of money and/or prestige.
3. Derive an alternative formula for the general solution, and hope that no one else ever springs a $e^{(\lambda \pm i\omega)x}$ on you. (Another bad choice since these exponentials are used throughout the sciences and engineering.)
4. Spend a little time learning what $e^{(\lambda \pm i\omega)x}$ means.

Guess which choice we take.

The Complex Exponential Function

Let us now consider the quantity e^z where z is any complex value. We, like everyone else, will call this the *complex exponential function*. In our work,

$$z = rx = (\lambda \pm i\omega)x = \lambda x \pm i\omega x \quad ,$$

where λ and ω are real constants. Keep in mind that our r value came from a characteristic equation that, in turn, came from plugging $y(x) = e^{rx}$ into a differential equation. In getting the characteristic equation, we did our computations assuming the e^{rx} behaved just as it would behave if r were a real value. So, in determining just what e^{rx} means when r is complex, we should assume the complex exponential satisfies the rules satisfied by the exponential we already know. In particular, we must have, for any constant r (real or complex),

$$\frac{d}{dx}\left[e^{rx}\right] = re^{rx} \quad ,$$

and, for any two constants A and B,

$$e^{A+B} = e^A e^B \quad .$$

Thus,

$$e^{rx} = e^{(\lambda \pm i\omega)x} = e^{\lambda x \pm i\omega x} = e^{\lambda x} e^{\pm i\omega x} \quad .$$

The first factor, $e^{\lambda x}$ is just the ordinary, 'real' exponential — a function you should be able to graph in your sleep. It is the second factor, $e^{\pm i\omega x}$, that we need to come to grips with.

To better understand $e^{\pm i\omega x}$, let us first examine

$$\phi(t) = e^{it} \quad .$$

Later, we'll replace t with $\pm\omega x$.

One thing we can do with this $\phi(t)$ is to compute it at $t = 0$,

$$\phi(0) = e^{i \cdot 0} = e^0 = 1 \quad .$$

We can also differentiate ϕ, getting

$$\phi'(t) = \frac{d}{dt}e^{it} = ie^{it} \quad .$$

Letting $t = 0$ here gives

$$\phi'(0) = ie^{i \cdot 0} = ie^0 = i \quad .$$

Notice that the right side of the formula for $\phi'(t)$ is just i times ϕ. Differentiating $\phi'(t)$ again, we get

$$\phi''(t) = \frac{d}{dt}\left[ie^{it}\right] = i^2 e^{it} = -1 \cdot \phi(t) \quad ,$$

giving us the differential equation

$$\phi'' = -\phi \quad ,$$

which is better written as

$$\phi'' + \phi = 0$$

simply because we've seen this equation before (using y instead of ϕ for the unknown function).

What we have just derived is that $\phi(t) = e^{it}$ satisfies the initial-value problem

$$\phi'' + \phi = 0 \qquad \text{with} \quad \phi(0) = 1 \quad \text{and} \quad \phi'(0) = i \quad .$$

We can solve this! From Example 15.5 on page 285, we know

$$\phi(t) = A\cos(t) + B\sin(t)$$

is a general solution to the differential equation. Using this formula for ϕ we also have

$$\phi'(t) = \frac{d}{dt}[A\cos(t) + B\sin(t)] = -A\sin(t) + B\cos(t) \quad .$$

Applying the initial conditions gives

$$1 = \phi(0) = A\cos(0) + B\sin(0) = A \cdot 1 + B \cdot 0 = A$$

and

$$i = \phi'(0) = -A\sin(0) + B\cos(0) = -a \cdot 0 + B \cdot 1 = B \quad .$$

Thus,

$$e^{it} = \phi(t) = 1 \cdot \cos(t) + i \cdot \sin(t) = \cos(t) + i\sin(t) \quad .$$

The last formula is so important that we should write it again with the middle cut out and give it a reference number:

$$e^{it} = \cos(t) + i\sin(t) \quad . \tag{17.4}$$

Before replacing t with $\pm\omega x$, it is worth noting what happens when t is replaced by $-t$ in the above formula. In doing this, remember that the cosine is an even function and that the sine is an odd function, that is,

$$\cos(-t) = \cos(t) \qquad \text{while} \qquad \sin(-t) = -\sin(t) \quad .$$

So
$$e^{-it} = e^{i(-t)} = \cos(-t) + i\sin(-t) = \cos(t) - i\sin(t) \quad .$$
Cutting out the middle leaves us with
$$e^{-it} = \cos(t) - i\sin(t) \quad . \tag{17.5}$$

Formulas (17.4) and (17.5) are the (famous) Euler formulas for e^{it} and e^{-it}. They really should be written as a pair:
$$e^{it} = \cos(t) + i\sin(t) \tag{17.6a}$$
$$e^{-it} = \cos(t) - i\sin(t) \tag{17.6b}$$
That way, it is clear that when you add these two equations together, the sines cancel out and we get
$$e^{it} + e^{-it} = 2\cos(t) \quad .$$
On the other hand, subtracting equation (17.6b) from equation (17.6a) yields
$$e^{it} - e^{-it} = 2i\sin(t) \quad .$$
So, the sine and cosine functions can be rewritten in terms of complex exponentials,
$$\cos(t) = \frac{e^{it} + e^{-it}}{2} \quad \text{and} \quad \sin(t) = \frac{e^{it} - e^{-it}}{2i} \quad . \tag{17.7}$$
This is nice. This means that many computations involving trigonometric functions can be done using exponentials instead, and this can greatly simplify those computations.

!►Example 17.5: *Consider deriving the trigonometric identity involving the product of the sine and the cosine functions. Using equation set (17.7) and basic algebra,*
$$\begin{aligned}\sin(t)\cos(t) &= \frac{e^{it} + e^{-it}}{2} \cdot \frac{e^{it} - e^{-it}}{2i} \\ &= \frac{\left(e^{it}\right)^2 - \left(e^{-it}\right)^2}{2 \cdot 2i} \\ &= \frac{e^{i2t} - e^{-i2t}}{2 \cdot 2i} \\ &= \frac{1}{2} \cdot \frac{e^{i(2t)} - e^{-i(2t)}}{2i} = \frac{1}{2}\sin(2t) \quad .\end{aligned}$$
Thus, we have (re)derived the trigonometric identity
$$2\sin(t)\cos(t) = \sin(2t) \quad .$$
(Compare this to the derivation you did years ago without complex exponentials!)

But we digress — our interest is not in rederiving trigonometric identities, but in figuring out what to do when we get $e^{(\lambda \pm i\omega)x}$ as solutions to a differential equation. Using the law of exponents and formulas (17.6a) and (17.6b) (with ωx replacing t), we see that
$$e^{(\lambda + i\omega)x} = e^{\lambda x + i\omega x} = e^{\lambda x}e^{i\omega x} = e^{\lambda x}[\cos(\omega x) + i\sin(\omega x)]$$
and
$$e^{(\lambda - i\omega)x} = e^{\lambda x - i\omega x} = e^{\lambda x}e^{-i\omega x} = e^{\lambda x}[\cos(\omega x) - i\sin(\omega x)] \quad .$$
So now we know how to interpret $e^{(\lambda \pm i\omega)x}$, and now we can get back to solving our differential equations.

Intelligently Using Complex Roots

Recall, we are interested in solving

$$ay'' + by' + cy = 0$$

when a, b and c are real-valued constants, and the solutions to the corresponding characteristic equation are complex. Remember, also, that these complex roots will form a conjugate pair,

$$r_+ = \lambda + i\omega \qquad \text{and} \qquad r_- = \lambda - i\omega$$

where λ and ω are real numbers with $\omega \neq 0$. This gave us

$$y_+(x) = e^{r_+ x} = e^{(\lambda+i\omega)x} \qquad \text{and} \qquad y_-(x) = e^{r_- x} = e^{(\lambda-i\omega)x}$$

as two solutions to our differential equation. From our discussion of the complex exponential, we now know that

$$y_+(x) = e^{(\lambda+i\omega)x} = e^{\lambda x}[\cos(\omega x) + i\sin(\omega x)] \tag{17.8a}$$

and

$$y_-(x) = e^{(\lambda-i\omega)x} = e^{\lambda x}[\cos(\omega x) - i\sin(\omega x)] \quad . \tag{17.8b}$$

Clearly, neither y_+ nor y_- is a constant multiple of the other. So each of the equivalent formulas

$$y(x) = c_+ y_+(x) + c_- y_-(x) \quad ,$$

$$y(x) = c_+ e^{(\lambda+i\omega)x} + c_- e^{(\lambda-i\omega)x}$$

and

$$y(x) = c_+ e^{\lambda x}[\cos(\omega x) + i\sin(\omega x)] + c_- e^{\lambda x}[\cos(\omega x) - i\sin(\omega x)] \quad ,$$

can, legitimately, be used as a general solution to our differential equation.

Still, however, these solutions introduce complex numbers into formulas that, in applications, should probably just involve real numbers. To avoid that, let us derive an alternative fundamental pair of solutions by choosing the constants c_+ and c_- appropriately. The basic idea is the same as used to derive the formulas (17.7) for $\sin(t)$ and $\cos(t)$ in terms of complex exponentials. First add equations (17.8a) and (17.8b) together, noting how the sine terms cancel out:

$$\begin{array}{rl} y_+(x) &= e^{\lambda x}[\cos(\omega x) + i\sin(\omega x)] \\ +\{y_-(x) &= e^{\lambda x}[\cos(\omega x) - i\sin(\omega x)]\} \\ \hline y_+(x) + y_-(x) &= 2e^{\lambda x}\cos(\omega x) \end{array} \quad .$$

So

$$y_1(x) = \frac{1}{2}y_+(x) + \frac{1}{2}y_-(x) = e^{\lambda x}\cos(\omega x)$$

is a solution to our differential equation. On the other hand, the cosine terms cancel out when we subtract equation (17.8b) from equation (17.8a), leaving us with

$$y_+(x) - y_-(x) = 2ie^{\lambda x}\sin(\omega x) \quad .$$

So

$$y_2(x) = \frac{1}{2i}y_+(x) - \frac{1}{2i}y_-(x) = e^{\lambda x}\sin(\omega x)$$

is another solution to our differential equation. Again, it should be clear that our latest solutions are not constant multiples of each other. Consequently,

$$y_1(x) = e^{\lambda x}\cos(\omega x) \qquad \text{and} \qquad y_2(x) = e^{\lambda x}\sin(\omega x)$$

form a fundamental set, and

$$y(x) = c_1 e^{\lambda x}\cos(\omega x) + c_2 e^{\lambda x}\sin(\omega x)$$

is a general solution to our differential equation.

All this work should be summarized so we won't forget:

Lemma 17.3

Let a, b and c be real-valued constants with $a \neq 0$. If the characteristic equation for

$$ay'' + by' + cy = 0$$

does not have one or two distinct real solutions, then it will have a conjugate pair of solutions

$$r_+ = \lambda + i\omega \qquad \text{and} \qquad r_- = \lambda - i\omega$$

where λ and ω are real numbers with $\omega \neq 0$.

Moreover, both

$$\left\{ e^{(\lambda+i\omega)x}, e^{(\lambda-i\omega)x} \right\} \qquad \text{and} \qquad \left\{ e^{\lambda x}\cos(\omega x), e^{\lambda x}\sin(\omega x) \right\}$$

are fundamental sets of solutions for the differential equation, and the general solution can be written as either

$$y(x) = c_+ e^{(\lambda+i\omega)x} + c_- e^{(\lambda-i\omega)x} \tag{17.9a}$$

or

$$y(x) = c_1 e^{\lambda x}\cos(\omega x) + c_2 e^{\lambda x}\sin(\omega x) \tag{17.9b}$$

as desired.

In practice, formula (17.9b) is often preferred since it is a formula entirely in terms of real-valued functions. On the other hand, if you are going to do a bunch of calculations with $y(x)$ involving differentiation and/or integration, you may prefer using formula (17.9a), since calculus with exponentials — even complex exponentials — is much easier that calculus with products of exponentials and trigonometric functions.

By the way, instead of "memorizing" the above theorem, it may be better to just remember that you can get the $e^{\lambda x}\cos(\omega x)$ and $e^{\lambda x}\sin(\omega x)$ solutions from the real and imaginary parts of

$$\begin{aligned} e^{(\lambda \pm i\omega)x} &= e^{\lambda x} e^{\pm i\omega x} \\ &= e^{\lambda x}[\cos(\omega x) \pm i\sin(\omega x)] = e^{\lambda x}\cos(\omega x) \pm ie^{\lambda x}\sin(\omega x) \quad . \end{aligned}$$

!▶Example 17.6: *Again, consider solving*

$$y'' - 6y' + 13y = 0 \quad .$$

From Example 17.4, we already know

$$r_+ = 3 + i2 \qquad \text{and} \qquad r_- = 3 - i2$$

are the two solutions to the characteristic equation,

$$r^2 - 6r + 13 = 0 \quad .$$

Thus, one fundamental set of solutions consists of

$$e^{(3+i2)x} \qquad \text{and} \qquad e^{(3-i2)x}$$

with

$$e^{(3\pm i2)x} = e^{3x}e^{\pm i2x}$$

$$= e^{3x}\left[\cos(2x) \pm i\sin(2x)\right] = e^{3x}\cos(2x) \pm ie^{3x}\sin(2x) \quad .$$

Alternatively, we can use the pair of real-valued functions

$$e^{3x}\cos(2x) \qquad \text{and} \qquad e^{3x}\sin(2x)$$

as a fundamental set. Using this, we have

$$y(x) = c_1 e^{3x}\cos(2x) + c_2 e^{3x}\sin(2x)$$

as the general solution to the differential equation in terms of just real-valued functions.

17.6 Summary

Combining Lemma 17.1 on page 322, Lemma 17.2 on page 326, and Lemma 17.3 on page 332, we get the big theorem on solving second-order homogeneous linear differential equations with constant coefficients:

Theorem 17.4
Let a, b and c be real-valued constants with $a \neq 0$. Then the characteristic polynomial for

$$ay'' + by' + cy = 0$$

will have either one or two distinct real roots or will have two complex roots that are complex conjugates of each other. Moreover:

1. *If there are two distinct real roots r_1 and r_2, then*

$$\left\{ e^{r_1 x} , e^{r_2 x} \right\}$$

is a fundamental set of solutions to the differential equation, and

$$y(x) = c_1 e^{r_1 x} + c_2 e^{r_2 x}$$

is a general solution.

2. *If there is only one real root r, then*

$$\left\{ e^{rx} , xe^{rx} \right\}$$

is a fundamental set of solutions to the differential equation, and

$$y(x) = c_1 e^{rx} + c_2 xe^{rx}$$

is a general solution.

3. *If there is a conjugate pair of roots* $r = \lambda \pm i\omega$*, then both*

$$\left\{ e^{(\lambda+i\omega)x} , e^{(\lambda-i\omega)x} \right\} \quad \text{and} \quad \left\{ e^{\lambda x}\cos(\omega x) , e^{\lambda x}\sin(\omega x) \right\}$$

are fundamental sets of solutions to the differential equation, and either

$$y(x) = c_{+}e^{(\lambda+i\omega)x} + c_{-}e^{(\lambda-i\omega)x}$$

or

$$y(x) = c_1 e^{\lambda x}\cos(\omega x) + c_2 e^{\lambda x}\sin(\omega x)$$

can be used as a general solution.

Additional Exercises

17.1. *Find the general solution to each of the following:*

a. $y'' - 7y' + 10y = 0$

b. $y'' + 2y' - 24y = 0$

c. $y'' - 25y = 0$

d. $y'' + 3y' = 0$

e. $4y'' - y = 0$

f. $3y'' + 7y' - 6y = 0$

17.2. *Solve the following initial-value problems:*

a. $y'' - 8y' + 15y = 0$ *with* $y(0) = 1$ *and* $y'(0) = 0$

b. $y'' - 8y' + 15y = 0$ *with* $y(0) = 0$ *and* $y'(0) = 1$

c. $y'' - 8y' + 15y = 0$ *with* $y(0) = 5$ *and* $y'(0) = 19$

d. $y'' - 9y = 0$ *with* $y(0) = 1$ *and* $y'(0) = 0$

e. $y'' - 9y = 0$ *with* $y(0) = 0$ *and* $y'(0) = 1$

f. $y'' - 9y = 0$ *with* $y(0) = 3$ *and* $y'(0) = -3$

17.3. *Find the general solution to each of the following:*

a. $y'' - 10y' + 25y = 0$

b. $y'' + 2y' + y = 0$

c. $4y'' - 4y' + y = 0$

d. $25y'' - 10y' + y = 0$

e. $16y'' - 24y' + 9y = 0$

f. $9y'' + 12y' + 4y = 0$

17.4. *Solve the following initial-value problems:*

a. $y'' - 8y' + 16y = 0$ *with* $y(0) = 1$ *and* $y'(0) = 0$

b. $y'' - 8y' + 16y = 0$ *with* $y(0) = 0$ *and* $y'(0) = 1$

c. $y'' - 8y' + 16y = 0$ *with* $y(0) = 3$ *and* $y'(0) = 14$

d. $4y'' + 4y' + y = 0$ *with* $y(0) = 1$ *and* $y'(0) = 0$

e. $4y'' + 4y' + y = 0$ *with* $y(0) = 0$ *and* $y'(0) = 1$

f. $4y'' + 4y' + y = 0$ *with* $y(0) = 6$ *and* $y'(0) = -5$

17.5. *Find the general solution — expressed in terms of real-valued functions only — to each of the following:*

a. $y'' + 25y = 0$

b. $y'' + 2y' + 5y = 0$

c. $y'' - 2y' + 5y = 0$

d. $y'' - 4y' + 29y = 0$

e. $9y'' + 18y' + 10y = 0$

f. $4y'' + y = 0$

17.6. *Solve the following initial-value problems (again, your final answers should be in terms of real-valued functions only):*

a. $y'' + 16y = 0$ *with* $y(0) = 1$ *and* $y'(0) = 0$

b. $y'' + 16y = 0$ *with* $y(0) = 0$ *and* $y'(0) = 1$

c. $y'' + 16y = 0$ *with* $y(0) = 4$ *and* $y'(0) = 12$

d. $y'' - 4y' + 13y = 0$ *with* $y(0) = 1$ *and* $y'(0) = 0$

e. $y'' - 4y' + 13y = 0$ *with* $y(0) = 0$ *and* $y'(0) = 1$

f. $y'' - 4y' + 13y = 0$ *with* $y(0) = 5$ *and* $y'(0) = 31$

17.7. *For each of the following initial-value problems, find the solution and then sketch the solution's graph over the interval* $(0, 4^1\!/_2)$ *:*

a. $y'' - y' + \left(\frac{1}{4} + 4\pi^2\right) y = 0$ *with* $y(0) = 1$ *and* $y'(0) = \frac{1}{2}$

b. $y'' - y' + \left(\frac{1}{4} + 4\pi^2\right) y = 0$ *with* $y(0) = 1$ *and* $y'(0) = -\frac{1}{2}$

17.8. *Find the general solution to each of the following. Express your answers in terms of real-valued functions only.*

a. $y'' - 9y = 0$

b. $y'' + 9y = 0$

c. $y'' + 6y' + 9y = 0$

d. $y'' + 6y' - 9y = 0$

e. $9y'' - 6y' + y = 0$

f. $y'' + 6y' + 10y = 0$

g. $y'' - 4y' + 40y = 0$

h. $2y'' - 5y' + 2y = 0$

i. $y'' + 10y' + 25y = 0$

j. $9y'' - y = 0$

k. $9y'' + y = 0$

l. $9y'' + y' = 0$

m. $y'' + 4y' + 7y = 0$

n. $y'' + 4y' + 5y = 0$

o. $y'' + 4y' + 4y = 0$

p. $y'' - 2y' - 15y = 0$

q. $y'' - 4y' = 0$

r. $y'' + 8y' + 16y = 0$

s. $4y'' + 3y = 0$

t. $4y'' - 4y' + 5y = 0$

17.9. *Use the complex exponential (as in Example 17.5 on page 330) to verify each of the following trigonometric identities:*

a. $\sin^2(t) + \cos^2(t) = 1$

b. $\sin^2(t) = \frac{1}{2} - \frac{1}{2}\cos(2t)$

c. $\cos(A + B) = \cos(A)\cos(B) - \sin(A)\sin(B)$

d. $\sin(A)\sin(B) = \cos(A - B) - \cos(A + B)$

18

Springs: Part I

Second-order differential equations arise in many applications. We saw one involving a falling object at the beginning of this text (the falling frozen duck example in Section 1.2). In fact, since acceleration is given by the second derivative of position, any application involving Newton's equation $F = ma$ has the potential to be modeled by a second-order differential equation.

In this chapter we will consider a class of applications involving masses moving at the ends of springs. This is a particularly good class of examples for us. For one thing, the basic model is relatively easy to derive and is given by a second-order differential equation with constant coefficients. So we will be able to apply what we learned in the last chapter to derive reasonably accurate descriptions of the motion under a variety of situations. Moreover, most of us already have an intuitive idea of how these "mass/spring systems" behave. Hopefully, what we derive will correspond to what we expect, and may even refine our intuitive understanding.

It should also be noted that much of the analysis we will develop here carries over to the analysis of other applications involving things that vibrate or oscillate. For example, the analysis of current in basic electric circuits is completely analogous to the analysis we'll carry out here.

18.1 Modeling the Action

The Mass/Spring System

Imagine a horizontal spring with one end attached to an immobile wall and the other end attached to some object of interest (say, a box of frozen ducks) which can slide along the floor, as in Figure 18.1. For brevity, this entire assemblage of spring, object, wall, etc. will be called a *mass/spring system*. Let us assume that:

1. The object can only move back and forth in the one horizontal direction.
2. Newtonian physics apply.
3. The total force acting on the object is the sum of:
   (a) The force from the spring responding to the spring being compressed and stretched.
   (b) The forces resisting motion because of air resistance and friction between the box and the floor.
   (c) Any other forces acting on the object. (This term will usually be zero in this chapter. We include it here for use in later chapters, so we don't have to re-derive the equation for the spring to include other forces.)

   (All forces are directed parallel to the direction of the object's motion.)

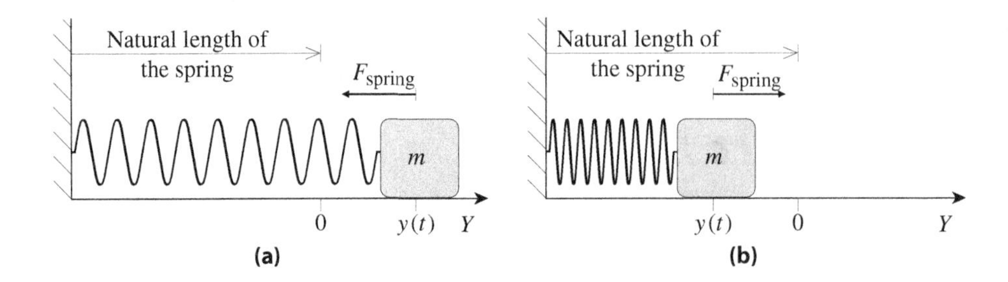

Figure 18.1: The mass/spring system with the direction of the spring force F_{spring} on the mass **(a)** when the spring is extended ($y(t) > 0$), and **(b)** when the spring is compressed ($y(t) < 0$).

4. The spring is an "ideal spring" with no mass. It has some natural length at which it is neither compressed nor stretched, and it can be both stretched and compressed. (So the coils are not so tightly wound that they are pressed against each other, making compression impossible.)

Our goal is to describe how the position of the object varies with time, and to see how this object's motion depends on the different parameters of our mass/spring system (the object's mass, the strength of the spring, the slipperiness of the floor, etc.).

To set up the general formulas and equations, we'll first make the following traditional symbolic assignments:

$$m = \text{the mass (in kilograms) of the object} \quad ,$$

$$t = \text{the time (in seconds) since the mass/spring system was set into motion} \quad ,$$

and

$$y = \text{the displacement (in meters) of the object from its position when the spring is at its natural length} \quad .$$

This means our Y–axis is horizontal (nontraditional, maybe, but convenient for this application), and positioned so that $y = 0$ is the "equilibrium position" of the object. Let us also direct the Y–axis so that the spring is stretched when $y > 0$, and compressed when $y < 0$ (again, see Figure 18.1).

Modeling the Forces

The motion of the object is governed by Newton's law $F = ma$ with F being the force acting on the box and

$$a = a(t) = \text{acceleration of the box at time } t = \frac{d^2y}{dt^2} \quad .$$

By our assumptions,

$$F = F_{\text{resist}} + F_{\text{spring}} + F_{\text{other}}$$

where

$$F_{\text{resist}} = \text{force due to the air resistance and friction} \quad ,$$

$$F_{\text{spring}} = \text{force from the spring due to it being compressed or stretched} \quad ,$$

and

$$F_{\text{other}} = \text{any other forces acting on the object} \quad .$$

Thus,

$$F_{\text{resist}} + F_{\text{spring}} + F_{\text{other}} = F = ma = m\frac{d^2y}{dt^2} \quad ,$$

which, for convenient reference later, we will rewrite as

$$m\frac{d^2y}{dt^2} - F_{\text{resist}} - F_{\text{spring}} = F_{\text{other}} \quad . \tag{18.1}$$

The resistive force, F_{resist}, is basically the same as the force due to air resistance discussed in the *Better Falling Object Model* in Chapter 1 (see page 11) — we are just including friction from the floor along with the friction with the air (or whatever medium surrounds our mass/spring system). So let us model the total resistive force here the same way we modeled the force of air resistance in Chapter 1:

$$F_{\text{resist}} = -\gamma \times \text{velocity of the box} = -\gamma\frac{dy}{dt} \tag{18.2}$$

where γ is some nonnegative constant. Because of the role it will play in determining how much the resistive forces "dampen" the motion, we call γ the *damping constant.* It will be large if the air resistance is substantial (possibly because the mass/spring system is submerged in water instead of air) or if the object does not slide easily on the floor. It will be small if there is little air resistance and the floor is very slippery. And it will be zero if there is no air resistance and no friction with the floor (a very idealized situation).

Now consider what we know about the spring force, F_{spring}. At any given time t, this force depends only on how much the spring is stretched or compressed at that time, and that, in turn, is completely described by $y(t)$. Hence, we can describe the spring force as a function of y, $F_{\text{spring}} = F_{\text{spring}}(y)$. Moreover:

1. If $y = 0$, then the spring is at its natural length, neither stretched nor compressed, and exerts no force on the box. So $F_{\text{spring}}(0) = 0$.
2. If $y > 0$, then the spring is stretched and exerts a force on the box pulling it backwards. So $F_{\text{spring}}(y) < 0$ whenever $y > 0$.
3. Conversely, if $y < 0$, then the spring is compressed and exerts a force on the box pushing it forwards. So $F_{\text{spring}}(y) > 0$ whenever $y < 0$.

Knowing nothing more about the spring force, we might as well model it using the simplest mathematical formula satisfying the above:

$$F_{\text{spring}}(y) = -\kappa y \tag{18.3}$$

where κ is some positive constant.

Formula (18.3) is the famous *Hooke's law* for springs. Experiments have shown it to be a good model for the spring force, provided the spring is not stretched or compressed too much. The constant κ in this formula is called the *spring constant.* It describes the "stiffness" of the spring (i.e., how strongly it resists being stretched), and can be determined by compressing or stretching the spring by some amount y_0, and then measuring the corresponding force F_0 at the end of the spring. Hooke's law then says that

$$\kappa = -\frac{F_0}{y_0} \quad .$$

And because κ is a positive constant, we can simplify things a little bit more to

$$\kappa = \frac{|F_0|}{|y_0|} \quad . \tag{18.4}$$

!▶Example 18.1: *Assume that, when our spring is pulled out* 2 *meters beyond its natural length, we measure that the spring is pulling back with a force* F_0 *of magnitude*

$$|F_0| = 18 \quad \left(\frac{\text{kg}\cdot\text{meter}}{\text{sec}^2}\right) \quad .$$

Then,

$$\kappa = \frac{|F_0|}{|y_0|} = \frac{18}{2} \quad \left(\frac{\text{kg}\cdot\text{meter/sec}^2}{\text{meter}}\right) \quad .$$

That is,

$$\kappa = 9 \quad \left(\frac{\text{kg}}{\text{sec}^2}\right) \quad .$$

(This is a pretty weak spring.)

A Note on Units

We defined m, t, and y to be numerical values describing mass, position, and time in terms of kilograms, seconds, and meters. Consequently, everything derived from these quantities — velocity, acceleration, the resistance coefficient γ, and the spring constant κ — are numerical values describing physical parameters in terms of corresponding units. Of course, velocity and acceleration are in terms of meters/second and meters/second^2, respectively. And, because "$F = ma$", any value for force should be interpreted as being in terms of kilogram·meter/second^2 (also called *newtons*). As indicated in the example above, the corresponding units associated with the spring constant are kilogram/second^2, and as you can readily verify, the resistance coefficient γ is in terms of kilograms/second.

In the above example, the units involved in every calculation were explicitly given in parentheses. In the future, we will not explicitly state the units in every calculation and trust that you, the reader, can determine the units appropriate for the final results from the information given.

Indeed, for our purposes, the actual choice of the units is not important. The formulas we have developed and illustrated (along with those we will later develop and illustrate) remain just as valid if m, t, and y are in terms of, say, grams, weeks, and miles, respectively, provided it is understood that the values of the corresponding velocities, accelerations, etc. are in terms of miles/week, miles/week^2, etc.

!▶Example 18.2: *Pretend that, when our spring is pulled out* 2 *miles beyond its natural length, we measure that the spring is pulling back with a force* F_0 *of magnitude*

$$|F_0| = 18 \quad \left(\frac{\text{gram}\cdot\text{mile}}{\text{week}^2}\right) \quad .$$

Then,

$$\kappa = \frac{|F_0|}{|y_0|} = \frac{18}{2} \quad \left(\frac{\text{gram}\cdot\text{mile/week}^2}{\text{meter}}\right) \quad .$$

That is,

$$\kappa = 9 \quad \left(\frac{\text{gram}}{\text{week}^2}\right) \quad .$$

18.2 The Mass/Spring Equation and Its Solutions

The Differential Equation

Replacing F_{resist} and F_{spring} in equation (18.1) with the formulas for these forces from equations (18.2) and (18.3), we get

$$m\frac{d^2y}{dt^2} + \gamma\frac{dy}{dt} + \kappa y = F_{\text{other}} \quad . \tag{18.5}$$

This is the differential equation for $y(t)$, the position y of the object in the system at time t.

For the rest of this chapter, let us assume the object is moving "freely" under the influence of no forces except those from friction and from the spring's compression and expansion.[1] Thus, for the rest of this chapter, we will restrict our interest to the above differential equation with $F_{\text{other}} = 0$,

$$m\frac{d^2y}{dt^2} + \gamma\frac{dy}{dt} + \kappa y = 0 \quad . \tag{18.6}$$

This is a second-order homogeneous linear differential equation with constant coefficients; so we can solve it by the methods discussed in the previous chapter. The precise functions in these solutions (sine/cosines, exponentials, etc.) will depend on the coefficients. We will go through all the possible cases soon.

Keep in mind that the mass, m, and the spring constant, κ, are positive constants for a real spring. On the other hand, the damping constant, γ, can be positive or zero. This is significant. Because $\gamma = 0$ when there is no resistive force to dampen the motion, we say the mass/spring system is *undamped* when $\gamma = 0$. We will see that the motion of the mass in this case is relatively simple.

If, however, there is a nonzero resistive force to dampen the motion, then $\gamma > 0$. Accordingly, in this case, we say the mass/spring system is *damped.* We will see that there are three subcases to consider, according to whether $\gamma^2 - 4\kappa m$ is negative, zero or positive.

Let's now carefully examine, case by case, the solutions that can arise.

Undamped Systems

If $\gamma = 0$, differential equation (18.6) reduces to

$$m\frac{d^2y}{dt^2} + \kappa y = 0 \quad . \tag{18.7}$$

The corresponding characteristic equation,

$$mr^2 + \kappa = 0 \quad ,$$

has roots

$$r_{\pm} = \pm\frac{\sqrt{-\kappa m}}{m} = \pm i\omega_0 \qquad \text{where} \quad \omega_0 = \sqrt{\frac{\kappa}{m}} \quad .$$

From our discussions in the previous chapter, we know the general solution to our differential equation is given by

$$y(t) = c_1\cos(\omega_0 t) + c_2\sin(\omega_0 t)$$

where c_1 and c_2 are arbitrary constants. However, for graphing purposes (and a few other purposes) it is convenient to write our general solution in yet another form. To derive this form, plot (c_1, c_2) as

[1] We will introduce other forces in later chapters.

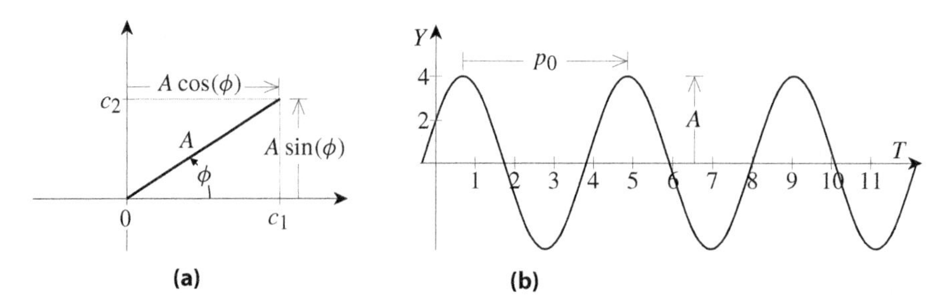

Figure 18.2: **(a)** Expressing c_1 and c_2 as $A\cos(\phi)$ and $A\sin(\phi)$. **(b)** The graph of $y(t)$ for the undamped mass/spring system of Example 18.3.

a point on a Cartesian coordinate system, and let A and ϕ be the corresponding polar coordinates of this point (see Figure 18.2a). That is, let

$$A = \sqrt{(c_1)^2 + (c_2)^2}$$

and let ϕ be the angle in the range $[0, 2\pi)$ with

$$c_1 = A\cos(\phi) \qquad \text{and} \qquad c_2 = A\sin(\phi) \quad .$$

Using this and the well-known trigonometric identity

$$\cos(\theta \pm \phi) = \cos(\theta)\cos(\phi) \mp \sin(\theta)\sin(\phi) \quad ,$$

we get

$$\begin{aligned} c_1\cos(\omega_0 t) + c_2\sin(\omega_0 t) &= [A\cos(\phi)]\cos(\omega_0 t) + [A\sin(\phi)]\sin(\omega_0 t) \\ &= A\big[\cos(\omega_0 t)\cos(\phi) + \sin(\omega_0 t)\sin(\phi)\big] \\ &= A\cos(\omega_0 t - \phi) \quad . \end{aligned}$$

Thus, our general solution is given by either

$$y(t) = c_1\cos(\omega_0 t) + c_2\sin(\omega_0 t) \tag{18.8a}$$

or, equivalently,

$$y(t) = A\cos(\omega_0 t - \phi) \tag{18.8b}$$

where

$$\omega_0 = \sqrt{\frac{\kappa}{m}}$$

and the other constants are related by

$$A = \sqrt{(c_1)^2 + (c_2)^2} \quad , \qquad \cos(\phi) = \frac{c_1}{A} \qquad \text{and} \qquad \sin(\phi) = \frac{c_2}{A} \quad .$$

It is worth noting that ω_0 depends only on the spring constant and the mass of the attached object. The other constants are "arbitrary" and are determined by the initial position and velocity of the attached object.

Either of the formulas from set (18.8) can be used for the position y of the box at time t. One advantage of using formula (18.8a) is that the constants c_1 and c_2 are fairly easily determined in many of initial-value problems involving this differential equation. However, formula (18.8b) gives an even simpler description of how the position varies with time. It tells us that the position is completely described by a single shifted cosine function multiplied by the positive constant A and shifted so that

$$y(0) = A\cos(\phi) \quad .$$

You should be well acquainted with such functions. The graph of one is sketched in Figure 18.2b. Take a look at it, and then read on.

Formula (18.8b) tells us that the object is oscillating back and forth from $y = A$ to $y = -A$. Accordingly, we call A the *amplitude* of the oscillation. The *(natural) period* p_0 of the oscillation is the time it takes the mass to go through one complete "cycle" of oscillation. Using formula (18.8b), rewritten as

$$y(t) = A\cos(X) \qquad \text{with} \quad X = \omega_0 t - \phi \quad ,$$

we see that *our system going through one cycle as t varies from $t = t_0$ to $t = t_0 + p_0$* is the same as $\cos(X)$ *going through one complete cycle as X varies from*

$$X = X_0 = \omega_0 t_0 - \phi$$

to

$$X = \omega_0(t_0 + p_0) - \phi = X_0 + \omega_0 p_0 \quad .$$

But, as we well know, $\cos(X)$ goes through one complete cycle as X goes from $X = X_0$ to $X = X_0 + 2\pi$. Thus,

$$\omega_0 p_0 = 2\pi \quad , \tag{18.9}$$

and the natural period of our system is

$$p_0 = \frac{2\pi}{\omega_0} \quad .$$

This is the "time per cycle" for the oscillations in the mass/spring system. Its reciprocal,

$$\nu_0 = \frac{1}{p_0} = \frac{\omega_0}{2\pi} \quad ,$$

then gives the "cycles per unit time" for the system (typically measured in terms of *hertz*, with one hertz equaling one cycle per second). We call ν_0 the *(natural) frequency* for the system. The closely related quantity originally computed, ω_0 (which can be viewed as describing "radians per unit time"), will be called the *(natural) angular frequency* for the system.[2] Because the natural frequency ν_0 is usually more easily measured than the natural circular frequency ω_0, it is sometimes more convenient to express the formulas for position (formula set 18.8) with $2\pi\nu_0$ replacing ω_0,

$$y(t) = c_1\cos(2\pi\nu_0 t) + c_2\sin(2\pi\nu_0 t) \quad , \tag{18.8a$'$}$$

and, equivalently,

$$y(t) = A\cos(2\pi\nu_0 t - \phi) \quad . \tag{18.8b$'$}$$

By the way, the angle ϕ in all the formulas above is called the *phase angle* of the oscillations, and any motion described by these formulas is referred to as *simple harmonic motion.*

!►Example 18.3: *Assume we have an undamped mass/spring system in which the spring's spring constant κ and the attached object's mass m are*

$$\kappa = 9 \left(\frac{\text{kg}}{\text{sec}^2}\right) \qquad \textit{and} \qquad m = 4 \ \text{(kg)}$$

[2] Many authors refer to ω_0 as a *circular frequency* instead of an angular frequency.

(as in Example 18.1). Let us try to find and graph the position y at time t of the attached object, assuming the object's initial position and velocity are

$$y(0) = 2 \quad \text{and} \quad y'(0) = 3\sqrt{3} \quad .$$

With the above values for γ, m and κ, the differential equation for y, equation (18.6), becomes

$$4\frac{d^2y}{dx^2} + 9y = 0 \quad .$$

As noted in our discussion, the general solution can be given by either

$$y(t) = c_1 \cos(\omega_0 t) + c_2 \sin(\omega_0 t)$$

or

$$y(t) = A\cos(\omega_0 t - \phi)$$

where the natural angular frequency is

$$\omega_0 = \sqrt{\frac{\kappa}{m}} = \sqrt{\frac{9}{4}} = \frac{3}{2} \quad .$$

This means the natural period p_0 and the natural frequency ν_0 of the system are

$$p_0 = \frac{2\pi}{\omega_0} = \frac{2\pi}{3/2} = \frac{4\pi}{3} \quad \text{and} \quad \nu_0 = \frac{1}{p_0} = \frac{3}{4\pi} \quad .$$

To determine the other constants in the above formulas for $y(t)$, we need to consider the given initial conditions. Using the first formula for y, we have

$$y(t) = c_1 \cos\left(\frac{3}{2}t\right) + c_2 \sin\left(\frac{3}{2}t\right)$$

and

$$y'(t) = -\frac{3}{2}c_1 \sin\left(\frac{3}{2}t\right) + c_2\frac{3}{2}\cos\left(\frac{3}{2}t\right) \quad .$$

Plugging these into the initial conditions yields

$$2 = y(0) = c_1 \cos\left(\frac{3}{2}\cdot 0\right) + c_2 \sin\left(\frac{3}{2}\cdot 0\right) = c_1$$

and

$$3\sqrt{3} = y'(0) = -\frac{3}{2}c_1 \sin\left(\frac{3}{2}\cdot 0\right) + c_2\frac{3}{2}\cos\left(\frac{3}{2}\cdot 0\right) = \frac{3}{2}c_2 \quad .$$

So

$$c_1 = 2 \quad , \quad c_2 = \frac{2}{3}\cdot 3\sqrt{3} = 2\sqrt{3} \quad ,$$

and

$$A = \sqrt{(c_1)^2 + (c_2)^2} = \sqrt{(2)^2 + (2\sqrt{3})^2} = \sqrt{4+12} = 4 \quad .$$

This gives us enough information to graph $y(t)$. From our computations, we know this graph is a cosine shaped curve with amplitude $A = 4$ and period $p_0 = {}^{4\pi}/_{3}$. It is shifted horizontally so that the initial conditions

$$y(0) = 2 \quad \text{and} \quad y'(0) = 3\sqrt{3} > 0$$

are satisfied. In other words, the graph must cross the Y–axis at $y = 2$, and the graph's slope at that crossing point must be positive. That is how the graph in Figure 18.2b was constructed.

To find the phase angle ϕ, we must solve the pair of trigonometric equations

$$\cos(\phi) = \frac{c_1}{A} = \frac{2}{4} = \frac{1}{2} \quad \text{and} \quad \sin(\phi) = \frac{c_2}{A} = \frac{2\sqrt{3}}{4} = \frac{\sqrt{3}}{2}$$

for $0 \leq \phi < 2\pi$. From our knowledge of trigonometry, we know the first of these equations is satisfied if and only if

$$\phi = \frac{\pi}{3} \quad \text{or} \quad \phi = \frac{5\pi}{3} \quad ,$$

while the second is satisfied if and only if

$$\phi = \frac{\pi}{3} \quad \text{or} \quad \phi = \frac{2\pi}{3} \quad .$$

Hence, for both of the above trigonometric equations to hold, we must have

$$\phi = \frac{\pi}{3} \quad .$$

Finally, using the values just obtained, we can completely write out two equivalent formulas for our solution:

$$y(t) = 2\cos\left(\frac{3}{2}t\right) + 2\sqrt{3}\sin\left(\frac{3}{2}t\right)$$

and

$$y(t) = 4\cos\left(\frac{3}{2}t - \frac{\pi}{3}\right) \quad .$$

Damped Systems

If $\gamma > 0$, then all coefficients in our differential equation

$$m\frac{d^2y}{dt^2} + \gamma\frac{dy}{dt} + \kappa y = 0$$

are positive. The corresponding characteristic equation is

$$mr^2 + \gamma r + \kappa = 0 \quad ,$$

and its solutions are given by

$$r = r_{\pm} = \frac{-\gamma \pm \sqrt{\gamma^2 - 4\kappa m}}{2m} \quad . \tag{18.10}$$

As we saw in the last chapter, the nature of the differential equation's solution, $y = y(t)$, depends on whether $\gamma^2 - 4\kappa m$ is positive, negative or zero. And this, in turn, depends on the positive constants γ, κ and m as follows:

$$\gamma < 2\sqrt{\kappa m} \iff \gamma^2 - 4\kappa m < 0 \quad ,$$

$$\gamma = 2\sqrt{\kappa m} \iff \gamma^2 - 4\kappa m = 0 \quad ,$$

and

$$2\sqrt{\kappa m} < \gamma \iff \gamma^2 - 4\kappa m > 0 \quad .$$

For reasons that may (or may not) be clear by the end of this section, we say that a mass/spring system is, respectively,

$$\textit{underdamped} \quad , \quad \textit{critically damped} \quad \text{or} \quad \textit{overdamped}$$

if and only if

$$0 < \gamma < 2\sqrt{\kappa m} \quad , \quad \gamma = 2\sqrt{\kappa m} \quad \text{or} \quad 2\sqrt{\kappa m} < \gamma \quad .$$

Since we've already considered the case where $\gamma = 0$, the first damped cases considered will be the underdamped mass/spring systems (where $0 < \gamma < 2\sqrt{\kappa m}$).

Underdamped Systems ($0 < \gamma < 2\sqrt{\kappa m}$)

In this case,

$$\sqrt{\gamma^2 - 4\kappa m} = \sqrt{-\left|\gamma^2 - 4\kappa m\right|} = i\sqrt{\left|\gamma^2 - 4\kappa m\right|} = i\sqrt{4\kappa m - \gamma^2} \quad ,$$

and formula (18.10) for the $r_\pm$'s can be written as

$$r_\pm = -\alpha \pm i\omega \qquad \text{where} \qquad \alpha = \frac{\gamma}{2m} \qquad \text{and} \qquad \omega = \frac{\sqrt{4\kappa m - \gamma^2}}{2m} \quad .$$

Both α and ω are positive real values, and, from the discussion in the previous chapter, we know a corresponding general solution to our differential equation is

$$y(t) = c_1 e^{-\alpha t}\cos(\omega t) + c_2 e^{-\alpha t}\sin(\omega t) \quad .$$

Factoring out the exponential and applying the same analysis to the linear combination of sines and cosines as was done for the undamped case, we get that the position y of the box at time t is given by any of the following:

$$y(t) = e^{-\alpha t}\left[c_1\cos(\omega t) + c_2\sin(\omega t)\right] \quad , \tag{18.11a}$$

$$y(t) = Ae^{-\alpha t}\cos(\omega t - \phi) \tag{18.11b}$$

and even

$$y(t) = Ae^{-\alpha t}\cos\left(\omega\left[t - \frac{\phi}{\omega}\right]\right) \quad . \tag{18.11c}$$

These three formulas are equivalent, and the arbitrary constants are related, as before, by

$$A = \sqrt{(c_1)^2 + (c_2)^2} \quad , \quad \cos(\phi) = \frac{c_1}{A} \quad \text{and} \quad \sin(\phi) = \frac{c_2}{A} \quad .$$

Note the similarities and differences in the motion of the undamped system and the underdamped system. In both cases, a shifted cosine function plays a major role in describing the position of the mass. In the underdamped system this cosine function has angular frequency ω and, hence, corresponding period and frequency

$$p = \frac{2\pi}{\omega} \qquad \text{and} \qquad \nu = \frac{\omega}{2\pi} \quad .$$

However, in the underdamped system, this shifted cosine function is also multiplied by a decreasing exponential, reflecting the fact that the motion is being damped, but not so damped as to completely prevent oscillations in the box's position. (You will further analyze how p and ν vary with γ in Exercise 18.7 b.)

Because the α in the formula set (18.11) determines the rate at which the maximum values of $y(t)$ are decreasing as t increases, let us call α the *decay coefficient* for our system. It is also tempting to call ω, p and ν the angular frequency, period and frequency of the system, but, because $y(t)$ is not truly periodic, this terminology is not truly appropriate. Instead, let's refer to these quantities the *angular quasi-frequency*, *quasi-period* and *quasi-frequency* of the system.[3] And, if you must give them names, call A the *quasi-amplitude* and $Ae^{-\alpha t}$ the *time-varying amplitude* of the system.

And, again, it is sometimes more convenient to express our formulas in terms of the quasi-frequency ν instead of the angular quasi-frequency ω, with, for example, formulas (18.11a) and (18.11b) being rewritten as

$$y(t) = e^{-\alpha t}\left[c_1\cos(2\pi\nu t) + c_2\sin(2\pi\nu t)\right] \tag{18.11a$'$}$$

and

$$y(t) = Ae^{-\alpha t}\cos(2\pi\nu t - \phi) \quad . \tag{18.11b$'$}$$

!►Example 18.4: *Again, assume the spring constant κ and the mass m in our mass/spring system are*

$$\kappa = 9 \left(\frac{\text{kg}}{\text{sec}^2}\right) \qquad \text{and} \qquad m = 4 \quad (\text{kg}) \quad .$$

For the system to be underdamped, the resistance coefficient γ must satisfy

$$0 < \gamma < 2\sqrt{\kappa m} = 2\sqrt{9\cdot 4} = 12 \quad .$$

For this example, assume $\gamma = 2$. Then the position y at time t of the object is given by

$$y(t) = Ae^{-\alpha t}\cos(\omega t - \phi)$$

where

$$\alpha = \frac{\gamma}{2m} = \frac{2}{2\times 4} = \frac{1}{4}$$

and

$$\omega = \frac{\sqrt{4\kappa m - \gamma^2}}{2m} = \frac{\sqrt{(4\cdot 9\cdot 3) - 2^2}}{2\cdot 4} = \frac{\sqrt{35}}{4} \quad .$$

The corresponding quasi-period for the system is

$$p = \frac{2\pi}{\omega} = \frac{2\pi}{\sqrt{35}/4} = \frac{8\pi}{\sqrt{35}} \approx 4.25 \quad .$$

To keep this example short, we won't solve for A and ϕ from some set of initial conditions. Instead, we'll just set

$$A = 4 \qquad \text{and} \qquad \phi = \frac{\pi}{3} \quad ,$$

and note that the resulting graph of $y(t)$ is sketched in Figure 18.3.

[3] Some authors prefer using "pseudo" instead of "quasi".

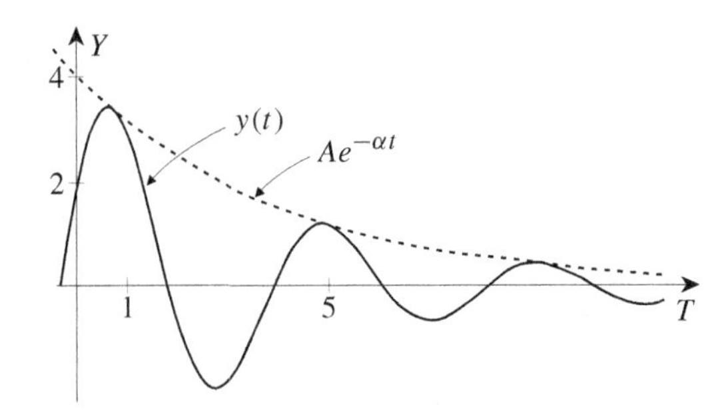

Figure 18.3: Graph of $y(t)$ for the underdamped mass/spring system of Example 18.4.

Critically Damped Systems ($\gamma = 2\sqrt{\kappa m}$)

In this case,

$$\sqrt{\gamma^2 - 4\kappa m} = 0$$

and

$$r_{\pm} = \frac{-\gamma \pm \sqrt{\gamma^2 - 4\kappa m}}{2m} = \frac{-2\sqrt{\kappa m} \pm \sqrt{0}}{2m} = -\frac{\sqrt{\kappa m}}{m} = -\sqrt{\frac{\kappa}{m}} \quad .$$

So the corresponding general solution to our differential equation is

$$y(t) = c_1 e^{-\alpha t} + c_2 t e^{-\alpha t} \qquad \text{where} \qquad \alpha = \sqrt{\frac{\kappa}{m}} \quad .$$

Factoring out the exponential yields

$$y(t) = [c_1 + c_2 t]e^{-\alpha t} \quad . \tag{18.12}$$

The cosine factor in the underdamped case has now been replaced with a formula for a straight line, $c_1 + c_2 t$. If $y'(0)$ is positive, then $y(t)$ will initially increase as t increases. However, at some point, the decaying exponential will force the graph of $y(t)$ back down towards 0 as $t \to \infty$. This is illustrated in Figure 18.4a.

If, on the other hand, $y(0)$ is positive and $y'(0)$ is negative, then the slope of the straight line is negative, and the graph will initially head downward as t increases. Eventually, $c_1 + c_2 t$ will be negative. And, again, the decaying exponential will eventually force $y(t)$ back (upward, this time) towards 0 as $t \to \infty$. This is illustrated in Figure 18.4b.

!►Example 18.5: *Once again, assume the spring constant κ and the mass m in our mass/spring system are*

$$\kappa = 9 \left(\frac{\text{kg}}{\text{sec}^2}\right) \qquad \textit{and} \qquad m = 4 \quad (\text{kg}) \quad .$$

For the system to be critically damped, the resistance coefficient γ must satisfy

$$\gamma = 2\sqrt{\kappa m} = 2\sqrt{9 \cdot 4} = 12 \quad .$$

Assuming this, the position y at time t of the object in this mass/spring system is given by

$$y(t) = [c_1 + c_2 t]e^{-\alpha t} \qquad \textit{where} \quad \alpha = \sqrt{\frac{\kappa}{m}} = \sqrt{\frac{9}{4}} = \frac{3}{2} \quad .$$

The graph of this with $(c_1, c_2) = (2, 8)$ is sketched in Figure 18.4a; the graph of this with $(c_1, c_2) = (2, -4)$ is sketched in Figure 18.4b.

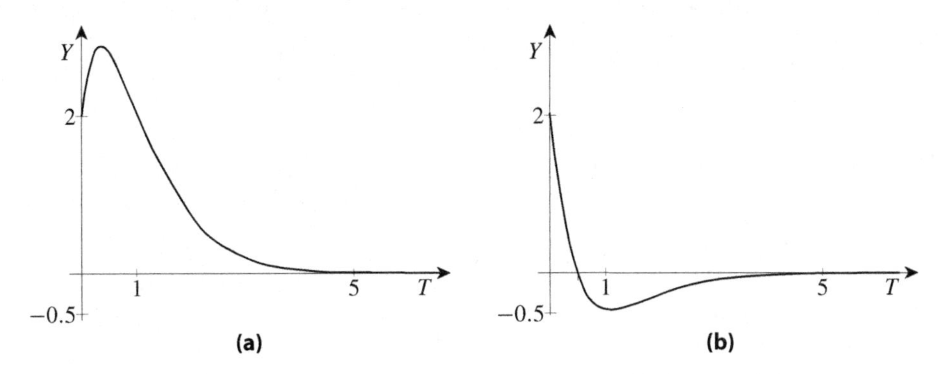

Figure 18.4: Graph of $y(t)$ for the critically damped mass/spring system of Example 18.5 **(a)** with $y(0) = 2$ and $y'(0) > 0$ **(b)** with $y(0) = 2$ and $y'(0) < 0$.

Overdamped Systems ($2\sqrt{\kappa m} < \gamma$)

In this case, it is first worth observing that

$$\gamma > \sqrt{\gamma^2 - 4\kappa m} > 0 \quad .$$

Consequently, the formula for the $r_\pm$'s (equation (18.10)),

$$r_\pm = \frac{-\gamma \pm \sqrt{\gamma^2 - 4\kappa m}}{2m}$$

can be written as

$$r_+ = -\alpha \quad \text{and} \quad r_- = -\beta$$

where α and β are the *positive* values

$$\alpha = \frac{\gamma - \sqrt{\gamma^2 - 4\kappa m}}{2m} \quad \text{and} \quad \beta = \frac{\gamma + \sqrt{\gamma^2 - 4\kappa m}}{2m} \quad .$$

Hence, the corresponding general solution to the differential equation is

$$y(t) = c_1 e^{-\alpha t} + c_2 e^{-\beta t} \quad ,$$

a linear combination of two decaying exponentials.

Some of the possible graphs for y are illustrated in Figure 18.5.

!▶ Example 18.6: *Once again, assume the spring constant and the mass in our mass/spring system are, respectively,*

$$\kappa = 9 \left(\frac{\text{kg}}{\text{sec}^2}\right) \quad \text{and} \quad m = 4 \text{ (kg)} \quad .$$

For the system to be overdamped, the resistance coefficient γ must satisfy

$$\gamma > 2\sqrt{\kappa m} = 2\sqrt{9 \cdot 4} = 12 \quad .$$

In particular, the system is overdamped if the resistance coefficient γ is 15. Assuming this, the general position y at time t of the object in our system is given by

$$y(t) = c_1 e^{-\alpha t} + c_2 e^{-\beta t}$$

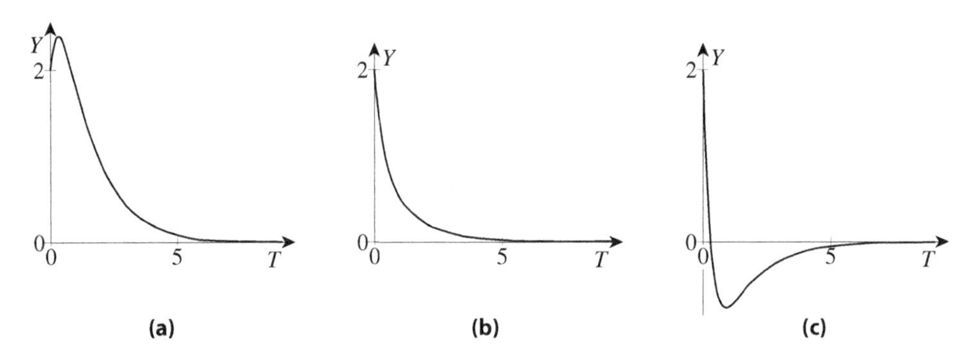

Figure 18.5: Graphs of $y(t)$ (with $y(0) = 2$) for the overdamped mass/spring system of Example 18.6. In **(a)** $y'(0) > 0$. In **(b)** and **(c)** $y'(0) < 0$ with the magnitude of $y'(0)$ in **(c)** being significantly larger than in **(b)**.

where

$$\alpha = \frac{\gamma - \sqrt{\gamma^2 - 4\kappa m}}{2m} = \frac{15 - \sqrt{15^2 - 4 \cdot 9 \cdot 4}}{2 \cdot 4} = \cdots = \frac{3}{4}$$

and

$$\beta = \frac{\gamma + \sqrt{\gamma^2 - 4\kappa m}}{2m} = \frac{15 + \sqrt{15^2 - 4 \cdot 9 \cdot 4}}{2 \cdot 4} = \cdots = 3 \quad .$$

Figures 18.5a, 18.5b, and 18.5c were drawn using this formula with, respectively,

$$(c_1, c_2) = (4, -2) \quad , \quad (c_1, c_2) = (1, 1) \quad \text{and} \quad (c_1, c_2) = (-2, 4) \quad .$$

Additional Exercises

18.1. *Assume we have a single undamped mass/spring system, and do the following:*

a. *Find the spring constant κ for the spring given that, when pulled out $^1/_2$ meter beyond its natural length, the spring pulls back with a force F_0 of magnitude*

$$|F_0| = 2 \left(\frac{\text{kg}\cdot\text{meter}}{\text{sec}^2} \right) \quad .$$

b. *Find the natural angular frequency ω_0, the natural frequency ν_0 and the natural period p_0 of this system assuming the mass of the attached object is 16 kilograms.*

c. *Four different sets of initial conditions are given below for this mass/spring system. For each, determine the corresponding amplitude A and phase angle ϕ for the system, and sketch the graph of the position over time $y = y(t)$. (Use the values of κ and ω_0 derived above, and assume position and time are given in meters and seconds, respectively.):*

i. $y(0) = 2$ *and* $y'(0) = 0$ ***ii.*** $y(0) = 0$ *and* $y'(0) = 2$

iii. $y(0) = 0$ *and* $y'(0) = -2$ ***iv.*** $y(0) = 2$ *and* $y'(0) = \sqrt{3}$

18.2. *Assume we have a single undamped mass/spring system, and do the following:*

a. *Find the spring constant* κ *for the spring given that, when pulled out* $1/4$ *meter beyond its natural length, the spring pulls back with a force* F_0 *of magnitude*

$$|F_0| = 72 \left(\frac{\text{kg}\cdot\text{meter}}{\text{sec}^2}\right) .$$

b. *Find the natural angular frequency* ω_0 *, the natural frequency* ν_0 *and the natural period* p_0 *of this system when the mass of the attached object is* 2 *kilograms.*

c. *Three different sets of initial conditions are given below for this mass/spring system. For each, determine the corresponding amplitude* A *, and sketch the graph of the position over time* $y = y(t)$ *:*

i. $y(0) = 1$ *and* $y'(0) = 0$ **ii.** $y(0) = 0$ *and* $y'(0) = 1$

iii. $y(0) = 1$ *and* $y'(0) = 3$

18.3. *Suppose we have an undamped mass/spring system with natural angular frequency* ω_0 *. Let* y_0 *and* v_0 *be, respectively, the position and velocity of the object at* $t = 0$ *. Show that the corresponding amplitude* A *is given by*

$$A = \sqrt{{y_0}^2 + \left[\frac{v_0}{\omega_0}\right]^2} ,$$

and that the phase angle satisfies

$$\tan(\phi) = \frac{v_0}{y_0\omega_0} .$$

18.4. *Suppose that a particular undamped mass/spring system has natural period* $p_0 = 3$ *seconds. What is the spring constant* κ *of the spring if the mass* m *of the object is (in kilograms)*

a. $m = 1$ **b.** $m = 2$ **c.** $m = \frac{1}{2}$

18.5. *Suppose we have an underdamped mass/spring system with decay coefficient* α *and angular quasi-frequency* ω *. Let* y_0 *and* v_0 *be, respectively, the position and velocity of the object at* $t = 0$ *. Show that the corresponding amplitude* A *is given by*

$$A = \sqrt{{y_0}^2 + \left[\frac{v_0 + \alpha y_0}{\omega}\right]^2} ,$$

while the phase angle satisfies

$$\tan(\phi) = \frac{v_0 + \alpha y_0}{y_0\omega} .$$

18.6. *Consider a damped mass/spring system with spring constant, mass and damping coefficient being, respectively,*

$$\kappa = 37 \quad , \quad m = 4 \quad \text{and} \quad \gamma = 4 .$$

a. *Verify that this is an underdamped system.*

b. *Find the decay coefficient* α *, the angular quasi-frequency* ω *, the quasi-period* p *and the quasi-frequency* ν *of this system.*

c. *Three different sets of initial conditions are given below for this mass/spring system. For each, determine the corresponding quasi-amplitude A for the system and then roughly sketch the graph of the position over time $y = y(t)$.*

i. $y(0) = 1$ and $y'(0) = 0$ **ii.** $y(0) = 4$ and $y'(0) = -2$

iii. $y(0) = 0$ and $y'(0) = 1$ **iv.** $y(0) = 2$ and $y'(0) = 2$

18.7 a. *Let ω be the angular quasi-frequency of some underdamped mass/spring system. Show that*

$$\omega = \sqrt{(\omega_0)^2 - \left(\frac{\gamma}{2m}\right)^2}$$

where m is the mass of the object in the system, γ is the damping constant and ω_0 is the natural frequency of the corresponding undamped system.

b. *Suppose we have a mass/spring system in which we can adjust the damping coefficient γ (the mass m and the spring constant κ remain constant). How do the quasi-frequency ν and the quasi-period p vary as γ varies from $\gamma = 0$ up to $\gamma = 2\sqrt{\kappa m}$? (Compare ν and p to the natural frequency and period of the corresponding undamped system, ν_0 and p_0.)*

18.8. *Consider a damped mass/spring system with spring constant, mass and damping coefficient being, respectively,*

$$\kappa = 4 \quad , \quad m = 1 \quad \text{and} \quad \gamma = 4 \quad .$$

a. *Verify that this system is critically damped.*

b. *Find and roughly graph the position of the object over time, $y(t)$, assuming that*

$$y(0) = 2 \quad \text{and} \quad y'(0) = 0 \quad .$$

c. *Find and roughly graph the position of the object over time, $y(t)$, assuming that*

$$y(0) = 0 \quad \text{and} \quad y'(0) = 2 \quad .$$

18.9. *Consider a damped mass/spring system with spring constant, mass and damping coefficient being, respectively,*

$$\kappa = 4 \quad , \quad m = 1 \quad \text{and} \quad \gamma = 5 \quad .$$

a. *Verify that this system is overdamped.*

b. *Find and roughly graph the position of the object over time, $y(t)$, assuming that*

$$y(0) = 2 \quad \text{and} \quad y'(0) = 0 \quad .$$

c. *Find and roughly graph the position of the object over time, $y(t)$, assuming that*

$$y(0) = 0 \quad \text{and} \quad y'(0) = 2 \quad .$$

19

Arbitrary Homogeneous Linear Equations with Constant Coefficients

In Chapter 17, we saw how to solve any equation of the form

$$ay'' + by' + cy = 0$$

when a, b and c are real constants. Unsurprisingly, the same basic ideas apply when dealing with any equation of the form

$$a_0 y^{(N)} + a_1 y^{(N-1)} + \cdots + a_{N-2} y'' + a_{N-1} y' + a_N y = 0$$

when N is some positive integer and the a_k's are all real constants. Assuming $y = e^{rx}$ still leads to the corresponding "characteristic equation"

$$a_0 r^N + a_1 r^{N-1} + \cdots + a_{N-2} r^2 + a_{N-1} r + a_N = 0 \quad ,$$

and a general solution to the differential equation can then be obtained using the solutions to the characteristic equation, much as we did in Chapter 17. Computationally, the only significant difficulty is in the algebra needed to find the roots of the characteristic polynomial.

So let us look at that algebra, first.

19.1 Some Algebra

A basic fact of algebra is that any second-degree polynomial

$$p(r) = ar^2 + br + c$$

can be factored to

$$p(r) = a(r - r_1)(r - r_2)$$

where r_1 and r_2 are the roots of the polynomial (i.e., the solutions to $p(r) = 0$). These roots may be complex, in which case r_1 and r_2 are complex conjugates of each other (assuming a, b and c are real numbers). It is also possible that $r_1 = r_2$, in which case the factored form of the polynomial is more concisely written as

$$p(r) = a(r - r_1)^2 \quad .$$

The idea of "factoring", of course, extends to polynomials of higher degree. And to use this idea with these polynomials, it will help to introduce the "completely factored form" for an arbitrary N^{th}-degree polynomial

$$p(r) = a_0 r^N + a_1 r^{N-1} + \cdots + a_{N-2} r^2 + a_{N-1} r + a_N \quad .$$

We will say that we've (re)written this polynomial into its *completely factored form* if and only if we've factored it to an expression of the form

$$p(r) = a_0 (r - r_1)^{m_1} (r - r_2)^{m_2} \cdots (r - r_K)^{m_K} \tag{19.1}$$

where

$$\{ r_1, r_2, \ldots, r_K \}$$

is the set of all different (possibly complex) roots of the polynomial (i.e., values of r satisfying $p(r) = 0$), and

$$\{ m_1, m_2, \ldots, m_K \}$$

is some corresponding set of positive integers.

Let's make a few simple observations regarding the above, and then look at a few examples.

1. It will be important for our discussion that

$$\{ r_1, r_2, \ldots, r_K \}$$

is the set of all *different* roots of the polynomial. If $j \neq k$, then $r_j \neq r_k$.

2. Each m_k is the largest integer such that $(r - r_k)^{m_k}$ is a factor of the original polynomial. Consequently, for each r_k, there is only one possible value for m_k. We call m_k the *multiplicity* of r_k.

3. As shorthand, we often say that r_k is a *simple root* if its multiplicity is 1, a *double root* if its multiplicity is 2, a *triple root* if its multiplicity is 3, and so on.

4. If you multiply out all the factors in the completely factored form in line (19.1), you get a polynomial of degree

$$m_1 + m_2 + \cdots + m_K \quad .$$

Since this polynomial is supposed to be $p(r)$, an Nth-degree polynomial, we must have

$$m_1 + m_2 + \cdots + m_K = N \quad .$$

!► Example 19.1: *By straightforward multiplication, you can verify that*

$$2(r-4)^3(r+5) = 2r^4 - 14r^3 - 24r^2 + 352r - 640 \quad .$$

This means

$$p(r) = 2r^4 - 14r^3 - 24r^2 + 352r - 640$$

can be written in completely factored form

$$p(r) = 2(r-4)^3(r-[-5]) \quad .$$

This polynomial has two distinct real roots, 4 and -5. The root 4 has multiplicity 3, and -5 is a simple root.

!►Example 19.2: *Straightforward multiplication also verifies that*

$$(r-3)^5 = r^5 - 15r^4 + 90r^3 - 270r^2 + 405r - 243 \quad .$$

Thus,

$$r^5 - 15r^4 + 90r^3 - 270r^2 + 405r - 243$$

has the completely factored form

$$(r-3)^5 \quad .$$

Here, 3 is the only distinct root, and this root has multiplicity 5 .

!►Example 19.3: *As the last example, for now, you can show that*

$$\left(r - [3+4i]\right)^2 \left(r - [3-4i]\right)^2 = r^4 - 12r^3 + 86r^2 - 300r + 625 \quad .$$

Hence,

$$\left(r - [3+4i]\right)^2 \left(r - [3-4i]\right)^2$$

is the completely factored form for

$$r^4 - 12r^3 + 86r^2 - 300r + 625 \quad .$$

This time we have two complex roots, $3+4i$ and $3-4i$, with each being a double root.

Can every polynomial be written in completely factored form? The next theorem says "yes".

Theorem 19.1 (complete factorization theorem)
Every polynomial can be written in completely factored form.

Note that we are not requiring the coefficients of the polynomial be real. This theorem is valid for every polynomial with real or complex coefficients. Unfortunately, you will have to accept this theorem on faith. Its proof requires developing more theory than is appropriate in this text.[1]

Unfortunately, also, this theorem does not tell us how to find the completely factored form. Of course, if the polynomial is of degree 2 ,

$$ar^2 + br + c \quad ,$$

then we can find the roots via the quadratic formula,

$$r = r_{\pm} = \frac{-b \pm \sqrt{b^2 - 4ac}}{2a} \quad .$$

Analogs of this formula do exist for polynomials of degrees 3 and 4 , but these analogs are rather complicated and not often used unless the user is driven by great need. For polynomials of degrees greater than 4 , it has been shown that no such analogs exist.

This means that finding the completely factored form may require some of those "tricks for factoring" you learned long ago in your old algebra classes. We'll review a few of those tricks later in examples involving differential equations.

[1] Those who are interested may want to look up the "Fundamental Theorem of Algebra" in a text on complex variables. The complete factorization theorem given here is a corollary of that theorem.

19.2 Solving the Differential Equation

The Characteristic Equation

Suppose we have some N^{th}-order differential equation of the form

$$a_0 y^{(N)} + a_1 y^{(N-1)} + \cdots + a_{N-2} y'' + a_{N-1} y' + a_N y = 0 \tag{19.2}$$

where the a_k's are all constants (and $a_0 \neq 0$). Since

$$\begin{aligned}
\left(e^{rx}\right)' &= re^{rx} \quad , \\
\left(e^{rx}\right)'' &= \left(re^{rx}\right)' = r \cdot re^{rx} = r^2 e^{rx} \quad , \\
\left(e^{rx}\right)''' &= \left(r^2 e^{rx}\right)' = r^2 \cdot re^{rx} = r^3 e^{rx} \quad , \\
&\vdots
\end{aligned}$$

for any constant r, it is easy to see that plugging $y = e^{rx}$ into the differential equation yields

$$a_0 r^N e^{rx} + a_1 r^{N-1} e^{rx} + \cdots + a_{N-2} r^2 e^{rx} + a_{N-1} r e^{rx} + a_N e^{rx} = 0 \quad ,$$

which, after dividing out e^{rx}, gives us the corresponding *characteristic equation*

$$a_0 r^N + a_1 r^{N-1} + \cdots + a_{N-2} r^2 + a_{N-1} r + a_N = 0 \quad . \tag{19.3}$$

As before, we refer to the polynomial on the left,

$$p(r) = a_0 r^N + a_1 r^{N-1} + \cdots + a_{N-2} r^2 + a_{N-1} r + a_N \quad ,$$

as the *characteristic polynomial* for the differential equation. Also, as in a previous chapter, it should be observed that the characteristic equation can be obtained from the original differential equation by simply replacing the derivatives of y with the corresponding powers of r.

According to the complete factorization theorem, the above characteristic equation can be rewritten in completely factored form,

$$a_0 (r - r_1)^{m_1} (r - r_2)^{m_2} \cdots (r - r_K)^{m_K} = 0 \tag{19.4}$$

where the r_k's are all the different roots of the characteristic polynomial, and the m_k's are the multiplicities of the corresponding roots. It turns out that, for each root r_k with multiplicity m_k, we can identify a corresponding linearly independent set of m_k particular solutions to the original differential equation. It will be obvious (once you see them) that no solution generated from one root can be written as a linear combination of solutions generated from the other roots. Hence, the set of all these particular solutions generated from all the r_k's will be a linearly independent set containing

$$m_1 + m_2 + \cdots + m_K = N$$

solutions. From the big theorem on solutions to homogeneous equations (Theorem 15.5 on page 290), we then know that this big set is a fundamental set of solutions for the differential equation, and that the general solution is given by an arbitrary linear combination of these particular solutions.

Exactly which particular solutions are generated from each individual root depends on the multiplicity and whether the root is real valued or not.

Particular Solutions Corresponding to One Root

In the following, we will assume r_k is a root of multiplicity m_k to our characteristic polynomial. That is,

$$(r - r_k)^{m_k}$$

is one factor in equation (19.4). However, since the choice of k will usually be irrelevant in this discussion, we will, unless needed, drop the subscripts.

The Basic Result

Assume r is a root of multiplicity m to our characteristic polynomial. Then, as before,

$$e^{rx}$$

is one particular solution to the differential equation, and if $m = 1$, it is the only solution corresponding to this root that we need to find.

So now assume $m > 1$. In the previous chapter, we found that

$$x\, e^{rx}$$

is a second solution to the differential equation when r is a repeated root and $N = 2$. This was obtained via reduction of order. For the more general case being considered here, it can be shown that $x\, e^{rx}$ is still a solution. In fact, it can be shown that the m particular solutions to the differential equation corresponding to root r can be generated one after the other by simply multiplying the previously found solution by x. That is, we have the following theorem:

Theorem 19.2
Let r be a root of multiplicity m to the characteristic polynomial for

$$a_0 y^{(N)} + a_1 y^{(N-1)} + \cdots + a_{N-2} y'' + a_{N-1} y' + a_N y = 0$$

where the a_k's are all constants. Then

$$\left\{ e^{rx} , x\, e^{rx} , x^2 e^{rx} , \ldots , x^{m-1} e^{rx} \right\}$$

is a linearly independent set of m solutions to the differential equation.

The proof of this theorem will be discussed later, in Section 19.4. (And it probably should be noted that $x^m e^{rx}$ ends up *not* being a solution to the differential equation.)

Particular Solutions Corresponding to a Real Root

If r is a real root of multiplicity m to our characteristic polynomial, then Theorem 19.2, above, tells us that

$$\left\{ e^{rx} , x\, e^{rx} , x^2 e^{rx} , \ldots , x^{m-1} e^{rx} \right\}$$

is the linearly independent set of m real-valued solutions to the differential equation corresponding to that root. No more need be said.

!►*Example 19.4:* *Consider the homogeneous differential equation*

$$y^{(5)} - 15y^{(4)} + 90y''' - 270y'' + 405y' - 243y = 0 \quad .$$

Its characteristic equation is

$$r^5 - 15r^4 + 90r^3 - 270r^2 + 405r - 243 = 0 \quad .$$

The left side of this equation is the polynomial from Example 19.2. Going back to that example, we discover that this characteristic equation can be factored to

$$(r-3)^5 = 0 \quad .$$

So 3 is the only root, and it has multiplicity 5. Theorem 19.2 then tells us that the linearly independent set of 5 corresponding particular solutions to the differential equation is

$$\left\{ e^{3x} , xe^{3x} , x^2e^{3x} , x^3e^{3x} , x^4e^{3x} \right\} \quad .$$

Since 5 is also the order of the differential equation, we know (via Theorem 15.5 on page 290) that the above set is a fundamental set of solutions to our homogeneous differential equation, and, thus,

$$y(x) = c_1e^{3x} + c_2xe^{3x} + c_3x^2e^{3x} + c_4x^3e^{3x} + c_5x^4e^{3x}$$

is a general solution for our differential equation.

Particular Solutions Corresponding to a Complex Root

Let λ and ω be two real numbers with $\omega \neq 0$. In Chapter 17 we observed that if $e^{(\lambda+i\omega)x}$ is one solution to a given second-order linear differential equation with constant coefficients, then so is $e^{(\lambda-i\omega)x}$. With a little bit of work (see Section 19.5), we can extend that observation to the following:

Theorem 19.3

Let r_+ be a root of multiplicity m to the characteristic polynomial for

$$a_0y^{(N)} + a_1y^{(N-1)} + \cdots + a_{N-2}y'' + a_{N-1}y' + a_Ny = 0$$

where the a_k's are all real constants. Also assume that $r_+ = \lambda + i\omega$ where λ and ω are real with $\omega \neq 0$. Then $r_- = \lambda - i\omega$ is also a root of multiplicity m to the characteristic polynomial, and the union of

$$\left\{ e^{(\lambda+i\omega)x} , x\,e^{(\lambda+i\omega)x} , x^2e^{(\lambda+i\omega)x} , \ldots , x^{m-1}e^{(\lambda+i\omega)x} \right\}$$

with

$$\left\{ e^{(\lambda-i\omega)x} , x\,e^{(\lambda-i\omega)x} , x^2e^{(\lambda-i\omega)x} , \ldots , x^{m-1}e^{(\lambda-i\omega)x} \right\}$$

is a linearly independent set of $2m$ solutions to the differential equation.

For many problems, the solution sets in this theorem are not particularly desirable sets of solutions because they introduce complex values into computations that we expect to yield real values. But recall how we dealt with complex-exponential solutions for second-order equations, constructing alternative pairs of solutions via linear combinations. Unsurprisingly, the same idea works here, and we can construct an alternative pair of solutions $\{y_{k,1}, y_{k,2}\}$ from each pair

$$\left\{ x^ke^{(\lambda+i\omega)x} , x^ke^{(\lambda-i\omega)x} \right\}$$

via the linear combinations

$$y_{k,1}(x) = \frac{1}{2}x^ke^{(\lambda+i\omega)x} + \frac{1}{2}x^ke^{(\lambda-i\omega)x}$$

and

$$y_{k,2}(x) = \frac{1}{2i}x^k e^{(\lambda+i\omega)x} - \frac{1}{2i}x^k e^{(\lambda-i\omega)x} \quad .$$

Since

$$e^{(\lambda\pm i\omega)x} = e^{\lambda x}\left[\cos(\omega x) \mp i\sin(\omega x)\right] \quad ,$$

you can easily verify that

$$y_{k,1} = x^k e^{\lambda x}\cos(\omega x) \qquad \text{and} \qquad y_{k,2} = x^k e^{\lambda x}\sin(\omega x) \quad .$$

It is also "easily" verified that the set of these functions, with $k = 0, 1, 2, \ldots, m-1$, is linearly independent.

Thus, instead of using

$$\left\{ e^{(\lambda+i\omega)x}, x\,e^{(\lambda+i\omega)x}, x^2 e^{(\lambda+i\omega)x}, \ldots, x^{m-1}e^{(\lambda+i\omega)x} \right\}$$

and

$$\left\{ e^{(\lambda-i\omega)x}, x\,e^{(\lambda-i\omega)x}, x^2 e^{(\lambda-i\omega)x}, \ldots, x^{m-1}e^{(\lambda-i\omega)x} \right\}$$

as the two linearly independent sets corresponding to roots $\lambda + i\omega$ and $\lambda - i\omega$, we can use the sets of real-valued functions

$$\left\{ e^{\lambda x}\cos(\omega x), x\,e^{\lambda x}\cos(\omega x), x^2 e^{\lambda x}\cos(\omega x), \ldots, x^{m-1}e^{\lambda x}\cos(\omega x) \right\}$$

and

$$\left\{ e^{\lambda x}\sin(\omega x), x\,e^{\lambda x}\sin(\omega x), x^2 e^{\lambda x}\sin(\omega x), \ldots, x^{m-1}e^{\lambda x}\sin(\omega x) \right\} \quad .$$

!►Example 19.5: *Consider the differential equation*

$$y^{(4)} - 12y''' + 86y'' - 300y' + 625y = 0 \quad .$$

Its characteristic equation is

$$r^4 - 12r^3 + 86r^2 - 300r + 625 = 0 \quad ,$$

which, as we saw in Example 19.3, can be factored to

$$\left(r - [3+4i]\right)^2 \left(r - [3-4i]\right)^2 = 0 \quad .$$

Here, we have a conjugate pair of roots, $3+4i$ *and* $3-4i$, *each with multiplicity* 2. *So the corresponding particular real-valued solutions to the differential equation are*

$$e^{3x}\cos(4x) \quad , \quad xe^{3x}\cos(4x) \quad , \quad e^{3x}\sin(4x) \quad \text{and} \quad xe^{3x}\sin(4x) \quad .$$

And since our homogeneous, linear differential equation is of order 4, *its general solution is given by an arbitrary linear combination of these four solutions,*

$$y(x) = c_1 e^{3x}\cos(4x) + c_2 xe^{3x}\cos(4x) + c_3 e^{3x}\sin(4x) + c_4 xe^{3x}\sin(4x) \quad ,$$

which, to save space, might also be written as

$$y(x) = [c_1 + c_2 x]e^{3x}\cos(4x) + [c_3 + c_4 x]e^{3x}\sin(4x) \quad .$$

By the way, we might as well note that, as an immediate corollary of Theorem 19.3, we have:

Corollary 19.4

The complex roots to any polynomial equation occur as conjugate pairs with the same multiplicities. That is, for any given polynomial equation, and for any two real numbers λ *and* ω, $\lambda + i\omega$ *is a root of multiplicity* m *if and only if* $\lambda - i\omega$ *is a root of multiplicity* m.

19.3 More Examples

The most difficult part of solving a high-order homogeneous linear differential equation with constant coefficients is the factoring of its characteristic polynomial. Unfortunately, the methods commonly used to factor second-degree polynomials do not nicely generalize to methods for factoring polynomials of higher degree. So we have to use whatever algebraic tricks we can think of. And if all else fails, we can run to the computer and let our favorite math package attempt the factoring.

Here are a few examples to help you recall some of the useful tricks for factoring polynomials of order three or above.

!►Example 19.6: *Consider the seventh-order homogeneous differential equation*

$$y^{(7)} - 625y''' = 0 \quad .$$

The characteristic equation is

$$r^7 - 625r^3 = 0 \quad .$$

An obvious choice of action would be to first factor out r^3,

$$r^3\left(r^4 - 625\right) = 0 \quad .$$

Cleverly noting that $r^4 = [r^2]^2$ *and* $625 = 25^2$, *and then applying well-known algebraic formulas, we have*

$$r^3\left(r^4 - 625\right) = 0$$

$$\hookrightarrow \qquad r^3\left([r^2]^2 - [25]^2\right) = 0$$

$$\hookrightarrow \qquad r^3\left(r^2 - 25\right)\left(r^2 + 25\right) = 0$$

$$\hookrightarrow \qquad r^3(r-5)(r+5)\left(r^2 + 25\right) = 0 \quad .$$

Now

$$r^2 + 25 = 0 \quad \Longrightarrow \quad r^2 = -25 \quad \Longrightarrow \quad r = \pm\sqrt{-25} = \pm 5i \quad .$$

So our characteristic equation can be written as

$$r^3(r-5)(r+5)(r-5i)(r+5i) = 0 \quad .$$

To be a little more explicit,

$$(r-0)^3(r-5)\big(r-[-5]\big)(r-5i)\big(r-[-5i]\big) = 0 \quad .$$

Thus, our characteristic polynomial has 5 *different roots:*

$$0 \quad , \quad 5 \quad , \quad -5 \quad , \quad 5i \quad \text{and} \quad -5i \quad .$$

The root 0 *has multiplicity* 3, *and the differential equation has corresponding particular solutions*

$$e^{0\cdot x} \quad , \quad xe^{0\cdot x} \quad \text{and} \quad x^2e^{0\cdot x} \quad ,$$

which most of us would rather write as

$$1 \quad , \quad x \quad \text{and} \quad x^2 \quad .$$

The roots 5 *and* -5 *each have multiplicity* 1. *So the differential equation has corresponding particular solutions*

$$e^{5x} \quad \text{and} \quad e^{-5x} \quad .$$

Finally, we have a pair of complex roots $5i$ *and* $-5i$, *each with multiplicity* 1. *Since these are of the form* $\lambda \pm i\omega$ *with* $\lambda = 0$ *and* $\omega = 5$, *the corresponding real-valued particular solutions to our differential equation are*

$$\cos(5x) \quad \text{and} \quad \sin(5x) \quad .$$

Taking an arbitrary linear combination of the above seven particular solutions, we get

$$y(x) = c_1 \cdot 1 + c_2 x + c_3 x^2 + c_4 e^{5x} + c_5 e^{-5x} + c_6 \cos(5x) + c_7 \sin(5x)$$

as a general solution to our differential equation.

!► **Example 19.7:** Consider

$$y''' - 19y' + 30y = 0 \quad .$$

The characteristic equation is

$$r^3 - 19r + 30 = 0 \quad .$$

Few people can find a first factoring of this characteristic polynomial,

$$p(r) = r^3 - 19r + 30 \quad ,$$

by inspection. But remember,

$$(r - r_1) \text{ is a factor of } p(r) \iff p(r_1) = 0 \quad .$$

This means we can test candidates for r_1 *by just seeing if* $p(r_1) = 0$. *Good candidates here would be the integer factors of* 30 (± 1, ± 2, ± 3, ± 5, ± 6, ± 10 *and* ± 15).

Trying $r_1 = 1$, *we get*

$$p(1) = 1^3 - 19 \cdot 1 + 30 = 12 \neq 0 \quad .$$

So $r_1 \neq 1$.

Trying $r_1 = -1$, *we get*

$$p(-1) = (-1)^3 - 19 \cdot (-1) + 30 = -1 + 19 + 30 \neq 0 \quad .$$

So $r_1 \neq -1$.

Trying $r_1 = 2$, *we get*

$$p(2) = (2)^3 - 19 \cdot (2) + 30 = 8 - 38 + 30 = 0 \quad .$$

Success! One root is $r_1 = 2$ *and one factor of our characteristic polynomial is* $(r - 2)$. *To get our first factoring, we then divide* $r - 2$ *into the characteristic polynomial:*

$$\begin{array}{r|l} & r^2 + 2r - 15 \\ \hline r - 2 & r^3 \qquad\quad - 19r + 30 \\ & -r^3 + 2r^2 \\ \hline & \quad 2r^2 - 19r \\ & -2r^2 + 4r \\ \hline & \qquad -15r + 30 \\ & \qquad 15r - 30 \\ \hline & \qquad\qquad 0 \end{array} \quad .$$

Thus,

$$r^3 - 19r + 30 = (r-2)(r^2+2r-15) \quad .$$

By inspection, we see that

$$r^2 + 2r - 15 = (r+5)(r-3) \quad .$$

So, our characteristic equation

$$r^3 - 19r + 30 = 0$$

factors to

$$(r-2)(r-[-5])(r-3) = 0 \quad ,$$

and, thus,

$$y(x) = c_1 e^{2x} + c_2 e^{-5x} + c_3 e^{3x}$$

is a general solution to our differential equation.

19.4 On Verifying Theorem 19.2

Theorem 19.2 claims to give a linearly independent set of solutions to a linear homogeneous differential equation with constant coefficients corresponding to a repeated root for the equation's characteristic polynomial. Our task of verifying this claim will be greatly simplified if we slightly expand our discussion of "factoring" linear differential operators from Section 16.3. (You may want to go back and quickly review that section.)

Linear Differential Operators with Constant Coefficients

First, we need to expand our terminology a little. When we refer to L as being an *N^{th}-order linear differential operator with constant coefficients*, we just mean that L is an Nth-order linear differential operator

$$L = a_0 \frac{d^N}{dx^N} + a_1 \frac{d^{N-1}}{dx^{N-1}} + \cdots + a_{N-2} \frac{d^2}{dx^2} + a_{N-1} \frac{d}{dx} + a_N$$

in which all the a_k's are constants. Its *characteristic polynomial* $p(r)$ is simply the polynomial

$$p(r) = a_0 r^N + a_1 r^{N-1} + \cdots + a_{N-2} r^2 + a_{N-1} r + a_N \quad .$$

It turns out that factoring a linear differential operator with constant coefficients is remarkably easy if you already have the factorization for its characteristic polynomial.

!►Example 19.8: *Consider the linear differential operator*

$$L = \frac{d^2}{dx^2} - 5\frac{d}{dx} + 6 \quad .$$

Its characteristic polynomial is

$$r^2 - 5r + 6 \quad ,$$

which factors to

$$(r-2)(r-3) \quad .$$

Now, consider the analogous composition product

$$\left(\frac{d}{dx} - 2\right)\left(\frac{d}{dx} - 3\right) \quad .$$

Letting ϕ be any suitably differentiable function, we see that

$$\begin{aligned}
\left(\frac{d}{dx} - 2\right)\left(\frac{d}{dx} - 3\right)[\phi] &= \left(\frac{d}{dx} - 2\right)\left[\frac{d\phi}{dx} - 3\phi\right] \\
&= \frac{d}{dx}\left[\frac{d\phi}{dx} - 3\phi\right] - 2\left[\frac{d\phi}{dx} - 3\phi\right] \\
&= \frac{d^2\phi}{dx^2} - 3\frac{d\phi}{dx} - 2\frac{d\phi}{dx} + 6\phi \\
&= \frac{d^2\phi}{dx^2} - 5\frac{d\phi}{dx} + 6\phi \\
&= L[\phi] \quad .
\end{aligned}$$

Thus,

$$L = \left(\frac{d}{dx} - 2\right)\left(\frac{d}{dx} - 3\right) \quad .$$

Redoing this last example with the numbers 2 and 3 replaced by constants r_1 and r_2 leads to the first result of this section:

Lemma 19.5

Let L be a second-order linear differential operator with constant coefficients,

$$L = a\frac{d^2}{dx^2} + b\frac{d}{dx} + c \quad ,$$

and let

$$p(r) = a(r - r_1)(r - r_2)$$

be a factorization of the characteristic polynomial for L (the roots r_1 and r_2 need not be different, nor must they be real). Then the operator L has factorization

$$L = a\left(\frac{d}{dx} - r_1\right)\left(\frac{d}{dx} - r_2\right) \quad .$$

PROOF: First of all, by the definition of p and elementary algebra,

$$ar^2 + br + c = p(r) = a(r - r_1)(r - r_2) = ar^2 - a[r_1 + r_2]r + ar_1r_2 \quad .$$

So,

$$b = -a[r_1 + r_2] \qquad \text{and} \qquad c = ar_1r_2 \quad .$$

Now, let ϕ be any sufficiently differentiable function. By the above,

$$\begin{aligned}
a\left(\frac{d}{dx} - r_1\right)\left(\frac{d}{dx} - r_2\right)[\phi] &= a\left(\frac{d}{dx} - r_1\right)\left[\frac{d\phi}{dx} - r_2\phi\right] \\
&= a\left(\frac{d}{dx}\left[\frac{d\phi}{dx} - r_2\phi\right] - r_1\left[\frac{d\phi}{dx} - r_2\phi\right]\right) \\
&= a\left(\frac{d^2\phi}{dx^2} - r_2\frac{d\phi}{dx} - r_1\frac{d\phi}{dx} + r_1r_2\phi\right)
\end{aligned}$$

$$
\begin{aligned}
&= a\frac{d^2\phi}{dx^2} - a\left[r_1 + r_2\right]\frac{d\phi}{dx} + ar_1r_2\phi \\
&= a\frac{d^2\phi}{dx^2} + b\frac{d\phi}{dx} + c\phi \\
&= L[\phi] \quad .
\end{aligned}
$$

Clearly, straightforward extensions of these arguments will show that, for any factorization of the characteristic polynomial of any linear differential operator with constant coefficients, there is a corresponding factorization of that operator. To simplify writing the factors of the operator when r_k is a multiple root of the characteristic polynomial, let us agree that

$$
\begin{aligned}
\left(\frac{d}{dx} - r_1\right)^0 &= 1 \quad , \\
\left(\frac{d}{dx} - r_1\right)^1 &= \left(\frac{d}{dx} - r_1\right) \quad , \\
\left(\frac{d}{dx} - r_1\right)^2 &= \left(\frac{d}{dx} - r_1\right)\left(\frac{d}{dx} - r_1\right) \quad , \\
\left(\frac{d}{dx} - r_1\right)^3 &= \left(\frac{d}{dx} - r_1\right)\left(\frac{d}{dx} - r_1\right)\left(\frac{d}{dx} - r_1\right) \quad , \\
&\vdots
\end{aligned}
$$

Using this notation along with the obvious extension of the above proof yields the next theorem.

Theorem 19.6 (factorization of constant coefficient operators)

Let L be a linear differential operator with constant coefficients, and let

$$
p(r) = a(r - r_1)^{m_1}(r - r_2)^{m_2} \cdots (r - r_K)^{m_K}
$$

be the completely factored form for the characteristic polynomial for L. Then

$$
L = a\left(\frac{d}{dx} - r_1\right)^{m_1}\left(\frac{d}{dx} - r_2\right)^{m_2} \cdots \left(\frac{d}{dx} - r_K\right)^{m_K} \quad .
$$

Let us make two observations regarding the polynomial p, one of the roots r_j of this polynomial, and the operator L from the last theorem:

1. Because the order in which we write the factors of a polynomial is irrelevant, we have

$$
p(r) = a(r - r_1)^{m_1}(r - r_2)^{m_2} \cdots (r - r_K)^{m_K} = ap_j(r)(r - r_j)^{m_j}
$$

where $p_j(r)$ is the product of all the $(r - r_k)$'s with $r_k \neq r_j$. Hence, L can be factored by

$$
L = aL_j\left(\frac{d}{dx} - r_j\right)^{m_j}
$$

where L_j is the composition product of all the $\left({}^{d}\!/\!_{dx} - r_k\right)$'s with $r_k \neq r_j$.

2. If y is a solution to

$$
\left(\frac{d}{dx} - r_j\right)^{m_j}[y] = 0 \quad ,
$$

then

$$
L[y] = aL_j\left(\frac{d}{dx} - r_j\right)^{m_j}[y] = aL_j\left[\left(\frac{d}{dx} - r_j\right)^{m_j}[y]\right] = aL_j[0] = 0 \quad .
$$

Together, these observations give us the actual result we will use.

Corollary 19.7
Let L be a linear differential operator with constant coefficients; let

$$p(r) = a(r - r_1)^{m_1}(r - r_2)^{m_2} \cdots (r - r_K)^{m_K}$$

be the completely factored form for the characteristic polynomial for L, and let r_j be any one of the roots of p. Suppose, further, that y is a solution to

$$\left(\frac{d}{dx} - r_j\right)^{m_j} [y] = 0 \quad .$$

Then y is also a solution to

$$L[y] = 0 \quad .$$

Proof of Theorem 19.2

Theorem 19.2 claims that, if r is a root of multiplicity m for the characteristic polynomial of some linear homogeneous differential equation with constant coefficients, then

$$\left\{ e^{rx} , x\,e^{rx} , x^2 e^{rx} , \ldots , x^{m-1} e^{rx} \right\}$$

is a linearly independent set of solutions to that differential equation. First we will verify that each of these $x^k e^{rx}$'s is a solution to the differential equation. Then we will confirm the linear independence of this set.

Verifying the Solutions

If you look back at Corollary 19.7, you will see that we need only show that

$$\left(\frac{d}{dx} - r\right)^m \left[x^k e^{rx}\right] = 0 \tag{19.5}$$

whenever k is a nonnegative integer less than m. To expedite our main computations, we'll do two preliminary computations. And, since at least one may be useful in a later chapter, we'll describe the results in an easily referenced lemma.

Lemma 19.8
Let r, α and β be constants with α being a positive integer and β being real valued. Then

$$\left(\frac{d}{dx} - r\right)^\alpha \left[e^{rx}\right] = 0 \quad \text{and} \quad \left(\frac{d}{dx} - r\right)^\alpha \left[x^\beta e^{rx}\right] = \beta \left(\frac{d}{dx} - r\right)^{\alpha-1} \left[x^{\beta-1} e^{rx}\right] \quad .$$

PROOF: For the first:

$$\begin{aligned}
\left(\frac{d}{dx} - r\right)^\alpha \left[e^{rx}\right] &= \left(\frac{d}{dx} - r\right)^{\alpha-1} \left(\frac{d}{dx} - r\right)\left[e^{rx}\right] \\
&= \left(\frac{d}{dx} - r\right)^{\alpha-1} \left[re^{rx} - re^{rx}\right] = \left(\frac{d}{dx} - r\right)^{\alpha-1} [0] = 0 \quad .
\end{aligned}$$

For the second:

$$
\begin{aligned}
\left(\frac{d}{dx} - r\right)^{\alpha} \left[x^{\beta} e^{rx}\right] &= \left(\frac{d}{dx} - r\right)^{\alpha-1} \left(\frac{d}{dx} - r\right) \left[x^{\beta} e^{rx}\right] \\
&= \left(\frac{d}{dx} - r\right)^{\alpha-1} \left[\frac{d}{dx}\left[x^{\beta} e^{rx}\right] - rx^{\beta} e^{rx}\right] \\
&= \left(\frac{d}{dx} - r\right)^{\alpha-1} \left[\beta x^{\beta-1} e^{rx} + x^{\beta} r e^{rx} - rx^{\beta} e^{rx}\right] \\
&= \beta \left(\frac{d}{dx} - r\right)^{\alpha-1} \left[x^{\beta-1} e^{rx}\right] \quad .
\end{aligned}
$$

∎

Now let k be an positive integer less than m. Using the above lemma, we see that

$$
\begin{aligned}
\left(\frac{d}{dx} - r\right)^{m} \left[x^{k} e^{rx}\right] &= k \left(\frac{d}{dx} - r\right)^{m-1} \left[x^{k-1} e^{rx}\right] \\
&= k(k-1) \left(\frac{d}{dx} - r\right)^{m-2} \left[x^{k-2} e^{rx}\right] \\
&= k(k-1)(k-2) \left(\frac{d}{dx} - r\right)^{m-3} \left[x^{k-3} e^{rx}\right] \\
&\ \ \vdots \\
&= k(k-1)(k-2)\cdots(k-[k-1]) \left(\frac{d}{dx} - r\right)^{m-k} \left[x^{k-k} e^{rx}\right] \\
&= k! \left(\frac{d}{dx} - r\right)^{m-k} \left[e^{rx}\right] \\
&= 0 \quad ,
\end{aligned}
$$

verifying equation (19.5).

Verifying Linear Independence

To finish verifying the claim of Theorem 19.2, we need only confirm that

$$
\left\{ e^{rx} \, , \, x e^{rx} \, , \, x^2 e^{rx} \, , \, \ldots \, , \, x^{m-1} e^{rx} \right\}
$$

is a linearly independent set of functions on the real line. Well, let's ask if this set could be, instead, a linearly *dependent* set of functions on the real line. Then one of these functions, say, $x^{\kappa} e^{rx}$, would be a linear combination of the others,

$$
x^{\kappa} e^{rx} = \text{linear combination of the other } x^{k} e^{rx}\text{'s} \quad .
$$

Subtract $x^{\kappa} e^{rx}$ from both sides, and you get

$$
0 = \text{linear combination of the other } x^{k} e^{rx}\text{'s} \; - \; 1 \cdot x^{\kappa} e^{rx} \quad ,
$$

which we can rewrite as

$$
0 = c_0 e^{rx} + c_1 x e^{rx} + c_2 x^2 e^{rx} + \cdots + c_m x^{m-1} e^{rx}
$$

where the c_k's are constants with

$$
c_{\kappa} = -1 \quad .
$$

Dividing out e^{rx} reduces the above to

$$0 = c_0 + c_1 x + c_2 x^2 + \cdots + c_{m-1} x^{m-1} \quad . \tag{19.6}$$

Since this is supposed to hold for all x, it should hold for $x = 0$, giving us

$$0 = c_0 + c_1 \cdot 0 + c_2 \cdot 0^2 + \cdots + c_{m-1} \cdot 0^{m-1} = c_0 \quad .$$

Now differentiate both sides of equation (19.6) and plug in $x = 0$:

$$\frac{d}{dx}[0] = \frac{d}{dx}\left[c_0 + c_1 x + c_2 x^2 + \cdots + c_{m-1} x^{m-1}\right]$$

$$\hookrightarrow \qquad 0 = 0 + c_1 + 2c_2 x + \cdots + (m-1)c_{m-1} x^{m-2}$$

$$\hookrightarrow \qquad 0 = 0 + c_1 + 2c_2 \cdot 0 + \cdots + (m-1)c_{m-1} \cdot 0^{m-2}$$

$$\hookrightarrow \qquad 0 = c_1 \quad .$$

Differentiating both sides of equation (19.6) twice and plugging in $x = 0$:

$$\frac{d^2}{dx^2}[0] = \frac{d^2}{dx^2}\left[c_0 + c_1 x + c_2 x^2 + c_3 x^3 \cdots + c_{m-1} x^{m-1}\right]$$

$$\hookrightarrow \qquad 0 = \frac{d}{dx}\left[0 + c_1 + 2c_2 x + 3c_3 x^2 \cdots + (m-1)c_{m-1} x^{m-2}\right]$$

$$\hookrightarrow \qquad 0 = 0 + 0 + 2c_2 + 6c_3 x \cdots + (m-1)c_{m-1}(m-2)x^{m-2}$$

$$\hookrightarrow \qquad 0 = 0 + 0 + 2c_2 + 6c_3 \cdot 0 + \cdots + (m-1)(m-2)c_{m-1} \cdot 0^{m-2}$$

$$\hookrightarrow \qquad 0 = c_2 \quad .$$

Clearly, we can differentiate equation (19.6) again and again, plug in $x = 0$, and, eventually, obtain

$$0 = c_k \qquad \text{for} \quad k = 0, 1, 2, \ldots, m-1 \quad .$$

But one of these c_k's is c_κ which we know is -1 (assuming our set of $x^k e^{rx}$'s is linearly *dependent*). In other words, for our set of $x^k e^{rx}$'s to be linearly *dependent*, we must have

$$0 = c_\kappa = -1 \quad ,$$

which is impossible. So our set of $x^k e^{rx}$'s cannot be linearly dependent. It must be linearly independent, just as Theorem 19.2 claimed.

19.5 On Verifying Theorem 19.3

It will help to begin with a brief discussion of complex conjugation in the algebra of complex numbers.

Algebra with Complex Conjugates

Recall that a complex number z is something that can be written as

$$z = x + iy$$

where x and y are real numbers, which we generally refer to as, respectively, the *real* and the *imaginary parts* of z. Along these lines, we say z is *real* if and only if $z = x$ (i.e., $y = 0$), and we say z is *imaginary* if and only if $z = iy$ (i.e., $x = 0$).

The corresponding *complex conjugate* of z — denoted z^* — is z with the sign of its imaginary part switched,

$$z = x + iy \quad \Longrightarrow \quad z^* = x + i(-y) = x - iy \quad .$$

Note that

$$\left(z^*\right)^* = (x - iy)^* = x + iy = z \quad ,$$

and that

$$z^* = z \qquad \text{if } z \text{ is real} \quad .$$

We will use these facts in a moment. We will also use formulas involving the complex conjugates of sums and products. To derive them, let

$$z = x + iy \qquad \text{and} \qquad c = a + ib$$

where x, y, a and b are all real, and compute out

$$(c + z)^* \quad , \quad c^* + z^* \quad , \quad (cz)^* \quad \text{and} \quad c^*z^*$$

in terms of a, b, x and y. You'll quickly discover that

$$(c + z)^* = c^* + z^* \qquad \text{and} \qquad (cz)^* = c^*z^* \quad .$$

In particular, if c is real,

$$(cz)^* = c^*z^* = cz^* \quad .$$

Proving Theorem 19.3

First of all, the theorem's claim that

$$\left\{ e^{(\lambda+i\omega)x} \, , \, x e^{(\lambda+i\omega)x} \, , \, x^2 e^{(\lambda+i\omega)x} \, , \, \ldots \, , \, x^{m-1} e^{(\lambda+i\omega)x} \right\}$$

is a linearly independent set of solutions to the differential equation follows immediately from Theorem 19.2. With a little thought, it should be clear that the other claims in Theorem 19.3 will follow from the following lemma, along with the relationship between the solutions to the given differential equation and the roots of the differential equation's characteristic polynomial.

Lemma 19.9
Let k, λ and ω be real numbers with k being an integer and $\omega \neq 0$, and assume

$$y_+(x) = x^k e^{(\lambda+i\omega)x}$$

is a solution to the Nth-order homogeneous differential equation with constant coefficients

$$a_0 y^{(N)} + a_1 y^{(N-1)} + \cdots + a_{N-1} y' + a_N y = 0$$

where the a_m's are all real. Then

$$y_-(x) = x^k e^{(\lambda - i\omega)x}$$

is also a solution to this differential equation.

PROOF: For convenience, let u and v be the real-valued functions given by

$$u(x) = x^k e^{\lambda x} \cos(\omega x) \qquad \text{and} \qquad v(x) = x^k e^{\lambda x} \sin(\omega x) \quad .$$

Observe that

$$y_\pm(x) = x^k e^{(\lambda \pm i\omega)x} = x^k [\cos(\omega x) \pm i \sin(\omega x)] = u(x) \pm i v(x) = (y_\mp(x))^*$$

and that, for any positive integer n,

$$y_\pm{}^{(n)} = \frac{d^n}{dx^n}[u(x) \pm i v(x)] = \frac{d^n u}{dx^n} \pm i \frac{d^n v}{dx^n} \quad .$$

Hence,

$$\left(y_\pm{}^{(n)}\right)^* = \left(\frac{d^n u}{dx^n} \pm i \frac{d^n v}{dx^n}\right)^* = \frac{d^n u}{dx^n} \mp i \frac{d^n v}{dx^n} = y_\mp{}^{(n)} \quad .$$

Now, using all the above and the fact that y_+ is a solution to the given differential equation, we have

$$\begin{aligned}
0 = 0^* &= \left(a_0 y_+{}^{(N)} + a_1 y_+{}^{(N-1)} + \cdots + a_{N-1} y_+{}' + a_N y_+\right)^* \\
&= \left(a_0 y_+{}^{(N)}\right)^* + \left(a_1 y_+{}^{(N-1)}\right)^* + \cdots + \left(a_{N-1} y_+{}'\right)^* + \left(a_N y_+\right)^* \\
&= a_0 \left(y_+{}^{(N)}\right)^* + a_1 \left(y_+{}^{(N-1)}\right)^* + \cdots + a_{N-1} \left(y_+{}'\right)^* + a_N \left(y_+\right)^* \\
&= a_0 y_-{}^{(N)} + a_1 y_-{}^{(N-1)} + \cdots + a_{N-1} y_-{}' + a_N y_- \quad ,
\end{aligned}$$

showing that y_- is also a solution to the differential equation. ∎

Additional Exercises

19.1. *Using clever factoring of the characteristic polynomials (such as was done in Example 19.6 on page 360), find the general solution to each of the following:*

a. $y^{(4)} - 4y''' = 0$

b. $y^{(4)} + 4y'' = 0$

c. $y^{(4)} - 34y'' + 225y = 0$

d. $y^{(4)} - 81y = 0$

e. $y^{(4)} - 18y'' + 81y = 0$

f. $y^{(5)} + 18y''' + 81y' = 0$

19.2. *For each of the following differential equations, one or more roots to the corresponding characteristic polynomial can be found by "testing candidates" (as illustrated in Example 19.7 on page 361). Using this fact, find the general solution to each.*

a. $y''' - y'' + y' - y = 0$

b. $y''' - 6y'' + 11y' - 6y = 0$

c. $y''' - 8y'' + 37y' - 50y = 0$

d. $y''' - 9y'' + 31y' - 39y = 0$

e. $y^{(4)} + y''' + 2y'' + 4y' - 8y = 0$

f. $y^{(4)} + 2y''' + 10y'' + 18y' + 9y = 0$

19.3. *Find the solution to each of the following initial-value problems:*

a. $y''' + 4y' = 0$ with $y(0) = 4$, $y'(0) = 6$ and $y''(0) = 8$

b. $y''' - 6y'' + 12y' - 8y = 0$
with $y(0) = 5$, $y'(0) = 13$ and $y''(0) = 86$

c. $y^{(4)} + 26y'' + 25y = 0$
with $y(0) = 6$, $y'(0) = -28$, $y''(0) = -102$ and $y'''(0) = 622$

d. $y^{(4)} + y''' + 9y'' + 9y' = 0$
with $y(0) = 10$, $y'(0) = 0$, $y''(0) = 6$ and $y'''(0) = -60$

19.4. *Find the general solution to each of the following:*

a. $y''' - 8y = 0$

b. $y''' + 216y = 0$

c. $y^{(4)} - 3y'' - 4y = 0$

d. $y^{(4)} + 13y'' + 36y = 0$

e. $y^{(6)} - 3y^{(4)} + 3y'' - y = 0$

f. $y^{(6)} - 2y''' + y = 0$

g. $16y^{(4)} - y = 0$

h. $4y^{(4)} + 15y'' - 4y = 0$

i. $y^{(4)} - 4y''' + 16y' - 16y = 0$

j. $y^{(6)} + 16y''' + 64y = 0$

20

Euler Equations

We now know how to completely solve any equation of the form

$$ay'' + by' + cy = 0$$

or even

$$a_0 y^{(N)} + a_1 y^{(N-1)} + \cdots + a_{N-2} y'' + a_{N-1} y' + a_N y = 0$$

in which the coefficients are all real constants (provided we can completely factor the corresponding characteristic polynomial).

Let us now consider some equations of the form

$$ay'' + by' + cy = 0$$

or even

$$a_0 y^{(N)} + a_1 y^{(N-1)} + \cdots + a_{N-2} y'' + a_{N-1} y' + a_N y = 0$$

when the coefficients are not all constants. In particular, let us consider the "Euler equations", described more completely in the next section, in which the coefficients happen to be particularly simple polynomials.[1]

As with the constant-coefficient equations, we will discuss the second-order Euler equations (and their solutions) first, and then note how those ideas extend to corresponding higher-order Euler equations.

20.1 Second-Order Euler Equations

Basics

A second-order differential equation is called an *Euler equation* if it can be written as

$$\alpha x^2 y'' + \beta x y' + \gamma y = 0$$

where α, β and γ are constants (in fact, we will assume they are real-valued constants). For example,

$$x^2 y'' - 6xy' + 10y = 0 \quad ,$$

$$x^2 y'' - 9xy' + 25y = 0$$

[1] These differential equations are also called *Cauchy–Euler equations*, *Euler–Cauchy equations* and *Cauchy equations*. By the way, "Euler" is pronounced "oi′ler".

and

$$x^2y'' - 3xy' + 20y = 0$$

are the Euler equations we'll solve to illustrate the methods we'll develop below. In these equations, the coefficients are not constants but are constants times the variable raised to the power equaling the order of the corresponding derivative. Notice, too, that the first coefficient, αx^2, vanishes at $x = 0$. This means we should not attempt to solve these equations over intervals containing 0. For convenience, we will use $(0, \infty)$ as the interval of interest. You can easily verify that the same formulas derived using this interval also work using the interval $(-\infty, 0)$ after replacing the x in these formulas with either $-x$ or $|x|$.

Euler equations are important for two or three good reasons:

1. They are easily solved.
2. They occasionally arise in applications, though not nearly as often as equations with constant coefficients.
3. They are especially simple cases of a broad class of differential equations for which infinite series solutions can be obtained using the "method of Frobenius".[2]

The basic approach to solving Euler equations is similar to the approach used to solve constant-coefficient equations: Assume a simple formula for the solution y involving one constant "to be determined", plug that formula for y into the differential equation, simplify and solve the resulting equation for the constant, and then construct the general solution using the constants found and the basic theory already developed.

The appropriate form for the solution to an Euler equation is not the exponential assumed for a constant-coefficient equation. Instead, it is

$$y(x) = x^r$$

where r is a constant to be determined. This choice for $y(x)$ can be motivated by either first considering the solutions to the corresponding first-order Euler equations

$$\alpha x \frac{dy}{dx} + \beta y = 0 \quad ,$$

or by just thinking about what happens when you compute

$$x^m \frac{d^m}{dx^m}\left[x^r\right] \quad .$$

We will outline the details of the method in a moment. Do not, however, bother memorizing anything except for the first assumption about the form of the solution and general outline of the method. The precise formulas that arise are not as easily memorized as the corresponding formulas for differential equations with constant coefficients. Moreover, you won't probably be using them enough later on to justify memorizing these formulas.

The Steps in Solving Second-Order Euler Equations

Here are the basic steps for finding a general solution to any second-order Euler equation

$$\alpha x^2y'' + \beta xy' + \gamma y = 0 \qquad \text{for} \quad x > 0 \quad .$$

Remember α, β and γ are real numbers. To illustrate the basic method, we will solve

$$x^2y'' - 6xy' + 10y = 0 \qquad \text{for} \quad x > 0 \quad .$$

---

[2] We won't start discussing the method of Frobenius until Chapter 35.

1. Assume a solution of the form
$$y = y(x) = x^r$$
where r is a constant to be determined.

2. Plug the assumed formula for y into the differential equation and simplify. Let's do the example first:

Replacing y with x^r gives
$$\begin{aligned} 0 &= x^2 y'' - 6xy' + 10y \\ &= x^2\left[x^r\right]'' - 6x\left[x^r\right]' + 10\left[x^r\right] \\ &= x^2\left[r(r-1)x^{r-2}\right] - 6x\left[rx^{r-1}\right] + 10\left[x^r\right] \\ &= (r^2 - r)x^r - 6rx^r + 10x^r \\ &= \left[r^2 - r - 6r + 10\right]x^r \\ &= \left[r^2 - 7r + 10\right]x^r \quad . \end{aligned}$$

Since we are solving on an interval where $x \neq 0$, we can divide out the x^r, leaving us with the algebraic equation
$$r^2 - 7r + 10 = 0 \quad .$$

In general, replacing y with x^r gives
$$\begin{aligned} 0 &= \alpha x^2 y'' + \beta xy' + \gamma y \\ &= \alpha x^2\left[x^r\right]'' + \beta x\left[x^r\right]' + \gamma\left[x^r\right] \\ &= \alpha x^2\left[r(r-1)x^{r-2}\right] + \beta x\left[rx^{r-1}\right] + \gamma\left[x^r\right] \\ &= \alpha(r^2 - r)x^r + \beta rx^r + \gamma x^r \\ &= \left[\alpha r^2 - \alpha r + \beta r + \gamma\right]x^r \\ &= \left[\alpha r^2 + (\beta - \alpha)r + \gamma\right]x^r \quad . \end{aligned}$$

Dividing out the x^r leaves us with the second-degree polynomial equation
$$\alpha r^2 + (\beta - \alpha)r + \gamma = 0 \quad .$$

This equation, known as the *indicial equation* corresponding to the given Euler equation,[3] is analogous to the characteristic equation for a second-order homogeneous linear differential equation with constant coefficients. (Don't memorize this equation — it is easy enough to simply rederive it each time. Besides, analogous equations for higher-order Euler equations are significantly different.)[4]

3. Solve the polynomial equation for r.

In our example, we obtained the indicial equation
$$r^2 - 7r + 10 = 0 \quad ,$$

[3] Often, though, it's just called "the equation for r".

[4] However, there is a shortcut for finding the indicial equations which may be useful if you are solving large numbers of Euler equations of different orders. See Exercise 20.5 at the end of this chapter.

which factors to

$$(r-2)(r-5) = 0 \quad .$$

So $r = 2$ *and* $r = 5$ *are the possible values of* r *.*

4. Remember that, for each value of r obtained, x^r is a solution to the original Euler equation. If there are two distinct real values r_1 and r_2 for r, then

$$\{ x^{r_1} , x^{r_2} \}$$

is clearly a fundamental set of solutions to the differential equation, and

$$y(x) = c_1 x^{r_1} + c_2 x^{r_2}$$

is a general solution. If there is only one value for r, then

$$y_1(x) = x^r$$

is one solution to the differential equation, and the general solution can be obtained via reduction of order. (The cases where there is only one value of r and where the two values of r are complex will be examined more closely in the next section.)

In our example, we obtained two values for r *,* 2 *and* 5 *. So*

$$\{ x^2 , x^5 \}$$

is a fundamental set of solutions to the differential equation, and

$$y(x) = c_1 x^2 + c_2 x^5$$

is a general solution.

20.2 The Special Cases

A Single Value for r

Let's do an example and then discuss what happens in general.

!►Example 20.1: *Consider*

$$x^2 y'' - 9xy' + 25y = 0 \qquad \text{for} \quad x > 0 \quad .$$

Letting $y = x^r$ *, we get*

$$\begin{aligned} 0 &= x^2 y'' - 9xy' + 25y \\ &= x^2 [x^r]'' - 9x [x^r]' + 10 [x^r] \\ &= x^2 [r(r-1)x^{r-2}] - 9x [rx^{r-1}] + 25 [x^r] \\ &= (r^2 - r)x^r - 9rx^r + 25x^r \end{aligned}$$

$$= \left[r^2 - r - 9r + 25\right]x^r$$
$$= \left[r^2 - 10r + 25\right]x^r \quad .$$

Dividing out the x^r *, this becomes*

$$r^2 - 10r + 25 = 0 \quad ,$$

which factors to

$$(r - 5)^2 = 0 \quad .$$

So $r = 5$ *, and the corresponding solution to the differential equation is*

$$y_1(x) = x^5 \quad .$$

Since we only have one solution, we cannot just write out the general solution as we did in the previous example. But we can still use the reduction of order method. So let

$$y(x) = x^5 u(x) \quad .$$

Computing the derivatives,

$$y'(x) = \left[x^5 u\right]' = 5x^4 u + x^5 u'$$

and

$$y''(x) = \left[5x^4 u + x^5 u'\right]' = 20x^3 u + 10x^4 u' + x^5 u'' \quad ,$$

and plugging into the differential equation yields

$$\begin{aligned} 0 &= x^2 y'' - 9xy' + 25y \\ &= x^2\left[20x^3 u + 10x^4 u' + x^5 u''\right] - 9x\left[5x^4 u + x^5 u'\right] + 25\left[x^5 u\right] \\ &= 20x^5 u + 10x^6 u' + x^7 u'' - 45x^5 u - 9x^6 u' + 25x^5 u \\ &= x^7 u'' + \left[10x^6 - 9x^6\right]u' + \left[20x^5 - 45x^5 + 25x^5\right]u \\ &= x^7 u'' + x^6 u' \quad . \end{aligned}$$

Letting $v = u'$ *, this becomes*

$$x^7 v' + x^6 v = 0 \quad ,$$

a simple separable first-order equation. Solving it:

$$x^7 \frac{dv}{dx} + x^6 v = 0$$

$$\hookrightarrow \qquad \frac{1}{v}\frac{dv}{dx} = -\frac{x^6}{x^7} = -\frac{1}{x}$$

$$\hookrightarrow \qquad \int \frac{1}{v}\frac{dv}{dx}\,dx = -\int \frac{1}{x}\,dx$$

$$\hookrightarrow \qquad \ln|v| = -\ln|x| + c_0$$

$$\hookrightarrow \qquad v = \pm e^{-\ln|x| + c_0} = \frac{c_2}{x} \quad .$$

Thus,

$$u(x) = \int u'(x)\,dx = \int v(x)\,dx = \int \frac{c_2}{x}\,dx = c_2 \ln|x| + c_1 \quad ,$$

and the general solution to the differential equation is

$$y(x) = x^5 u(x) = x^5[c_2 \ln|x| + c_1] = c_1 x^5 + c_2 x^5 \ln|x| \quad .$$

While just using reduction of order is recommended, you can show that, if your indicial equation only has one solution r, then

$$y_1(x) = x^r \qquad \text{and} \qquad y_2(x) = x^r \ln|x|$$

will always be solutions to the differential equation. Since they are clearly not constant multiples of each other, they form a fundamental set for the differential equation. Thus, in this case,

$$y(x) = c_1 x^r + c_2 x^r \ln|x|$$

will always be a general solution to the given Euler equation. This may be worth remembering, if you expect to be solving many Euler equations (which you probably won't). Otherwise just remember how to use reduction of order.

Verifying this claim is left to the interested reader (see Exercise 20.3 on page 383).

Complex Values for r

Again, we start with an example.

!►Example 20.2: *Consider*

$$x^2 y'' - 3xy' + 20y = 0 \qquad \text{for} \quad x > 0 \quad .$$

Using $y = x^r$, we get

$$\begin{aligned} 0 &= x^2 y'' - 3xy' + 20y \\ &= x^2\left[x^r\right]'' - 3x\left[x^r\right]' + 20\left[x^r\right] \\ &= x^2\left[r(r-1)x^{r-2}\right] - 3x\left[rx^{r-1}\right] + 20\left[x^r\right] \\ &= x^r\left[r^2 - r - 3r + 20\right] \quad , \end{aligned}$$

which simplifies to

$$r^2 - 4r + 20 = 0 \quad .$$

The solution to this is

$$r = \frac{-(-4) \pm \sqrt{(-4)^2 - 4(20)}}{2} = \frac{4 \pm \sqrt{-64}}{2} = 2 \pm i4 \quad .$$

Thus, we have two distinct values for r, $2+i4$ and $2-i4$. Presumably, then, we could construct a general solution from

$$x^{2+i4} \qquad \text{and} \qquad x^{2-i4} \quad ,$$

provided we had some idea as to just what "x to a complex power" means.

So let's figure out what "x to a complex power" means.

For exactly the same reasons as when we were solving constant coefficient equations, the complex solutions to the indicial equation will occur as complex conjugate pairs

$$r_+ = \lambda + i\omega \quad \text{and} \quad r_- = \lambda - i\omega \quad ,$$

which, formally at least, yield

$$y_+(x) = x^{r_+} = x^{\lambda + i\omega} \quad \text{and} \quad y_-(x) = x^{r_-} = x^{\lambda - i\omega}$$

as solutions to the original Euler equation. Now, assuming the standard algebraic rules remain valid for complex powers,[5]

$$x^{\lambda \pm i\omega} = x^\lambda x^{\pm i\omega} \quad ,$$

and, for $x > 0$,

$$x^{\pm i\omega} = e^{\ln|x|^{\pm i\omega}} = e^{\pm i\omega \ln|x|} = \cos(\omega \ln|x|) \pm i \sin(\omega \ln|x|) \quad .$$

So our two solutions can be written as

$$y_+(x) = x^\lambda \left[\cos(\omega \ln|x|) + i \sin(\omega \ln|x|) \right]$$

and

$$y_-(x) = x^\lambda \left[\cos(\omega \ln|x|) - i \sin(\omega \ln|x|) \right] \quad .$$

To get solutions in terms of only real-valued functions, essentially do what was done when we had complex-valued roots to characteristic equations for constant-coefficient equations, namely, use the fundamental set

$$\{ y_1, y_2 \}$$

where

$$y_1(x) = \frac{1}{2} y_+(x) + \frac{1}{2} y_-(x) = \cdots = x^\lambda \cos(\omega \ln|x|)$$

and

$$y_2(x) = \frac{1}{2i} y_+(x) - \frac{1}{2i} y_-(x) = \cdots = x^\lambda \sin(\omega \ln|x|) \quad .$$

Note that these are just the real and the imaginary parts of the formulas for $y_\pm = x^{\lambda \pm i\omega}$.

If you really wish, you can memorize what we just derived, namely:

> *If you get*
>
> $$r = \lambda \pm i\omega$$
>
> *when assuming $y = x^r$ is a solution to an Euler equation, then*
>
> $$y_1(x) = x^\lambda \cos(\omega \ln|x|) \quad \textit{and} \quad y_2(x) = x^\lambda \sin(\omega \ln|x|)$$
>
> *form a corresponding linearly independent pair of real-valued solutions to the differential equation, and*
>
> $$y(x) = c_1 x^\lambda \cos(\omega \ln|x|) + c_2 x^\lambda \sin(\omega \ln|x|)$$
>
> *is a general solution in terms of just real-valued functions.*

Memorizing these formulas is not recommended. It's easy enough (and safer) to simply re-derive the formulas for $x^{\lambda \pm i\omega}$ as needed, and then just take the real and imaginary parts as our the two real-valued solutions.

---

[5] They do.

!▶ Example 20.3: *Let us finish solving*

$$x^2y'' - 3xy' + 20y = 0 \quad \text{for} \quad x > 0 \quad .$$

From above, we got the complex-power solutions

$$y_\pm(x) = x^{2 \pm i4} \quad .$$

Rewriting this using the corresponding complex exponential, we get

$$y_\pm(x) = x^2 x^{\pm i4} = x^2 e^{\ln|x|^{\pm i4}}$$

$$= x^2 e^{\pm i4 \ln|x|} = x^2\left[\cos(4\ln|x|) \pm i\sin(4\ln|x|)\right] \quad .$$

Taking the real and imaginary parts of this then yields the corresponding linearly independent pair of real-valued solutions to the differential equation,

$$y_1(x) = x^2\cos(4\ln|x|) \quad \text{and} \quad y_2(x) = x^2\sin(4\ln|x|) \quad .$$

Thus,

$$y(x) = c_1x^2\cos(4\ln|x|) + c_2x^2\sin(4\ln|x|)$$

is a general solution in terms of just real-valued functions.

20.3 Euler Equations of Any Order

The definitions and ideas just described for second-order Euler equations are easily extended to analogous differential equations of any order. The natural extension of the concept of a second-order Euler differential equation is that of an Nth-order Euler equation, which is any differential equation that can be written as

$$\alpha_0 x^N y^{(N)} + \alpha_1 x^{N-1} y^{(N-1)} + \cdots + \alpha_{N-2}x^2y'' + \alpha_{N-1}xy' + \alpha_N y = 0$$

where the α_k's are all constants (and $\alpha_0 \neq 0$). We will further assume they are all real constants.

The basic ideas used to find the general solution to an Nth-order Euler equation over $(0, \infty)$ are pretty much the same as used to solve the second-order Euler equations:

1. Assume a solution of the form
$$y = y(x) = x^r$$
where r is a constant to be determined.

2. Plug the assumed formula for y into the differential equation and simplify. The result will be an Nth-degree polynomial equation
$$A_0r^N + A_1r^{N-1} + \cdots + A_{N-1}r + A_N = 0 \quad .$$
This is the *indicial equation* for the given Euler equation, and the polynomial on the left is the *indicial polynomial.* It is easily shown that the A_k's are all real (assuming the α_k's are real) and that $A_0 = \alpha_0$. However, the relation between the other A_k's and α_k's will depend on the order of the original differential equation.

3. Solve the indicial equation. The same tricks used to help solve the characteristic equations in Chapter 19 can be used here. And, as with those characteristic equations, we will obtain a list of all the different roots of the indicial polynomial,

$$r_1 \quad , \quad r_2 \quad , \quad r_3 \quad , \quad \dots \quad \text{and} \quad r_K \quad ,$$

along with their corresponding multiplicities,

$$m_1 \quad , \quad m_2 \quad , \quad m_3 \quad , \quad \dots \quad \text{and} \quad m_K \quad .$$

As noted in Chapter 19,

$$m_1 + m_2 + m_3 + \cdots + m_K = N \quad .$$

What you do next with each r_k depends on whether r_k is real or complex, and on the multiplicity m_k of r_k.

4. If $r = r_k$ is real, then there will be a corresponding linearly independent set of $m = m_k$ solutions to the differential equation. One of these, of course, will be $y = x^r$. If this root's multiplicity m is greater than 1, then a second corresponding solution to the Euler equation is obtained by multiplying the first, x^r, by $\ln|x|$, just as in the second-order case. This — multiplying the last solution found by $\ln|x|$ — turns out to be the pattern for generating the other solutions when $m = m_k > 2$. That is, the set of solutions to the differential equation corresponding to $r = r_k$ is

$$\left\{ x^r \, , \, x^r \ln|x| \, , \, x^r (\ln|x|)^2 \, , \, \dots \, , \, x^r (\ln|x|)^{m-1} \right\}$$

with $m = m_k$. (We'll verify this rigorously in the next section.)

5. If a root is complex, say, $r = \lambda + i\omega$, and has multiplicity m, then (as noted in Corollary 19.4 on page 359) this root's complex conjugate $r^* = \lambda - i\omega$ is another root of multiplicity m. By the same arguments given for real roots, we have that the functions

$$x^{\lambda+i\omega} \quad , \quad x^{\lambda+i\omega}\ln|x| \quad , \quad x^{\lambda+i\omega}(\ln|x|)^2 \quad , \quad \dots \quad \text{and} \quad x^{\lambda+i\omega}(\ln|x|)^{m-1}$$

along with

$$x^{\lambda-i\omega} \quad , \quad x^{\lambda-i\omega}\ln|x| \quad , \quad x^{\lambda-i\omega}(\ln|x|)^2 \quad , \quad \dots \quad \text{and} \quad x^{\lambda-i\omega}(\ln|x|)^{m-1}$$

make up a linearly independent set of $2m$ solutions to the Euler equation. To obtain the corresponding set of real-valued solutions, we again use the fact that, for $x > 0$,

$$x^{\lambda \pm i\omega} = x^\lambda x^{\pm i\omega} = x^\lambda e^{\pm i\omega \ln|x|} = x^\lambda \big[\cos(\omega \ln|x|) \pm i \sin(\omega \ln|x|) \big] \qquad (20.1)$$

to obtain the alternative set of $2m$ solutions

$$\begin{aligned}\big\{ & x^\lambda \cos(\omega\ln|x|) \, , \, x^\lambda \sin(\omega\ln|x|) \, , \\ & x^\lambda \cos(\omega\ln|x|) \ln|x| \, , \, x^\lambda \sin(\omega\ln|x|) \ln|x| \, , \\ & x^\lambda \cos(\omega\ln|x|) (\ln|x|)^2 \, , \, x^\lambda \sin(\omega\ln|x|) (\ln|x|)^2 \, , \\ & \dots \, , \, x^\lambda \cos(\omega\ln|x|) (\ln|x|)^{m-1} \, , \, x^\lambda \sin(\omega\ln|x|) (\ln|x|)^{m-1} \big\}\end{aligned}$$

for the Euler equation.

6. Now form the set of solutions to the Euler equation consisting of the m_k solutions described above for each real root r_k, and the $2m_k$ real-valued solutions described above for each conjugate pair of roots r_k and $r_k{}^*$. Since (as we saw in Chapter 19) the sum of the multiplicities equals N, and since the r_k's are distinct, it will follow that this set will be a fundamental set of solutions for our Euler equation. Thus, finally, a general solution to the given Euler equation can be written out as an arbitrary linear combination of the functions in this set.

We will do two examples (skipping some of the tedious algebra).

!► Example 20.4: *Consider the third-order Euler equation*

$$x^3y''' - 6x^2y'' + 19xy' - 27y = 0 \quad \text{for} \quad x > 0 \quad .$$

Plugging in $y = x^r$, we get

$$x^3r(r-1)(r-2)x^{r-3} - 6x^2r(r-1)x^{r-2} + 19xrx^{r-1} - 27x^r = 0 \quad ,$$

which, after a bit of algebra, reduces to

$$r^3 - 9r^2 + 27r - 27 = 0 \quad .$$

This is the indicial equation for our Euler equation. You can verify that its factored form is

$$(r-3)^3 = 0 \quad .$$

So the only root to our indicial polynomial is $r = 3$, and it has multiplicity 3. As discussed above, the corresponding fundamental set of solutions to the Euler equation is

$$\left\{ x^3 , x^3 \ln|x| , x^3(\ln|x|)^2 \right\} \quad ,$$

and the corresponding general solution is

$$y = c_1x^3 + c_2x^3\ln|x| + c_3x^3(\ln|x|)^2 \quad .$$

!► Example 20.5: *Consider the fourth-order Euler equation*

$$x^4y^{(4)} + 6x^3y''' + 25x^2y'' + 19xy' + 81y = 0 \quad \text{for} \quad x > 0 \quad .$$

Plugging in $y = x^r$, we get

$$x^4r(r-1)(r-2)(r-3)x^{r-4} + 6x^3r(r-1)(r-2)x^{r-3} + 25x^2r(r-1)x^{r-2} + 19xrx^{r-1} + 81x^r = 0 \quad ,$$

which simplifies to

$$r^4 + 18r^2 + 81 = 0 \quad .$$

Solving this yields

$$r = \pm 3i \quad \text{with multiplicity } 2 \quad ,$$

and the four corresponding real-valued solutions to our Euler equation are

$$\cos(3\ln|x|) \quad , \quad \sin(3\ln|x|) \quad , \quad \cos(3\ln|x|)\ln|x| \quad \text{and} \quad \sin(3\ln|x|)\ln|x| \quad .$$

The general solution, then, is

$$y = c_1\cos(3\ln|x|) + c_2\sin(3\ln|x|) + c_4\cos(3\ln|x|)\ln|x| + c_4\sin(3\ln|x|)\ln|x| \quad .$$

20.4 The Relation Between Euler and Constant Coefficient Equations

Let us suppose that

$$A_0 r^N + A_1 r^{N-1} + \cdots + A_{N-1} r + A_N = 0 \tag{20.2}$$

is the indicial equation for some N^{th}-order Euler equation

$$\alpha_0 x^N \frac{d^N y}{dx^N} + \alpha_1 x^{N-1} \frac{d^{N-1} y}{dx^{N-1}} + \cdots + \alpha_{N-2} x^2 \frac{d^2 y}{dx^2} + \alpha_N y = 0 \quad . \tag{20.3}$$

Observe that polynomial equation (20.2) is also the characteristic equation for the Nth-order constant coefficient equation

$$A_0 \frac{d^N Y}{dt^N} + A_1 \frac{d^{N-1} Y}{dt^{N-1}} + \cdots + A_{N-1} \frac{dY}{dt} + A_N Y = 0 \quad . \tag{20.4}$$

This means that, if r is a solution to polynomial equation (20.2), then

$$x^r \qquad \text{and} \qquad e^{rt}$$

are solutions, respectively, to the above Euler equation and the above constant coefficient equation. This suggests that these two differential equations are related to each other, possibly through a substitution of the form

$$x^r = e^{rt} \quad .$$

Taking the rth root of both sides, this simplifies to

$$x = e^t \qquad \text{or, equivalently,} \qquad \ln|x| = t \quad .$$

Exploring this possibility further eventually leads to the following lemma about the solutions to the above differential equations:

Lemma 20.1

Assume that two homogeneous linear differential equations of equal order are given, with one being an Euler equation and the other having constant coefficients. Further assume that the indicial equation of the Euler equation is the same as the characteristic equation of the other. Also, let $y(x)$ and $Y(t)$ be two functions, with y defined on $(0,\infty)$, and $Y(t)$ defined on $(-\infty,\infty)$, and related by the substitution $x = e^t$ (equivalently, $\ln|x| = t$); that is,

$$y(x) = Y(t) \qquad \text{where} \quad x = e^t \quad \text{and} \quad t = \ln|x| \quad .$$

Then y is a solution to the given Euler equation if and only if Y is a solution to the given constant-coefficient equation.

The proof of this lemma involves repeated chain rule computations such as

$$\frac{dy}{dx} = \frac{d}{dx} Y(t) = \frac{dt}{dx}\frac{d}{dt} Y(t) = \frac{d\ln|x|}{dx}\frac{d}{dt} Y(t) = \frac{1}{x}\frac{dY}{dt} = e^{-t}\frac{dY}{dt} \quad . \tag{20.5}$$

We'll leave the details to the adventurous (see Exercises 20.6, 20.7 and 20.8).

There are two noteworthy consequences of this lemma:

1. It gives us another way to solve Euler equations. To be specific: we can use the substitution in the lemma to convert the Euler equation into a constant coefficient equation (with t as the variable); solve that coefficient equation for its general solution (in terms of functions of t), and then use the substitution backwards to get the general solution to the Euler equation (in terms of functions of x).[6]

2. We can now confirm the claim made (and used) in the previous section about solutions to the Euler equation corresponding to a root r of multiplicity m to the indicial equation. After all, if r is a solution of multiplicity m to equation (20.2), then we know that

$$\left\{ e^{rt} \, , \, te^{rt} \, , \, t^2 e^{rt} \, , \, \ldots \, , \, t^{m-1} e^{rt} \right\}$$

is a set of solutions to constant coefficient equation (20.4). The lemma then assures us that this set, with $t = \ln|x|$, is the corresponding set of solutions to Euler equation (20.3). But, using this substitution,

$$t^k e^{rt} = \left(e^t\right)^r t^k = x^r \left(\ln|x|\right)^k \quad .$$

So the set of solutions obtained to the Euler equation is

$$\left\{ x^r \, , \, x^r \ln|x| \, , \, x^r (\ln|x|)^2 \, , \, \ldots \, , \, x^r (\ln|x|)^{m-1} \right\} \quad ,$$

just as claimed in the previous section.

Additional Exercises

20.1. *Find the general solution (in terms of real-valued functions) to each of the following Euler equations on* $(0, \infty)$ *:*

a. $x^2 y'' - 5xy' + 8y = 0$

b. $x^2 y'' - 2y = 0$

c. $x^2 y'' - 2xy' = 0$

d. $2x^2 y'' - xy' + y = 0$

e. $x^2 y'' - 5xy' + 9y = 0$

f. $x^2 y'' + 5xy' + 4y = 0$

g. $4x^2 y'' + y = 0$

h. $x^2 y'' - 19xy' + 100y = 0$

i. $x^2 y'' - 5xy' + 13y = 0$

j. $x^2 y'' - xy' + 10y = 0$

k. $x^2 y'' + 5xy' + 29y = 0$

l. $x^2 y'' + xy' + y = 0$

m. $2x^2 y'' + 5xy' + y = 0$

n. $4x^2 y'' + 37y = 0$

o. $x^2 y'' + xy' = 0$

p. $x^2 y'' + xy' - 25y = 0$

q. $4x^2 y'' + 8xy' + 5y = 0$

r. $3x^2 y'' - 7xy' + 3y = 0$

[6] It may be argued that this method, requiring the repeated use of the chain rule, is more tedious and error-prone than the one developed earlier, which only requires algebra and differentiation of x^r. That would be a good argument.

20.2. *Solve the following initial-value problems:*

a. $x^2y'' - 2xy' - 10y = 0$ *with* $y(1) = 5$ and $y'(1) = 4$

b. $4x^2y'' + 4xy' - y = 0$ *with* $y(4) = 0$ and $y'(4) = 2$

c. $x^2y'' - 11xy' + 36y = 0$ *with* $y(1) = 1/2$ and $y'(1) = 2$

d. $x^2y'' - xy' + y = 0$ *with* $y(1) = 3$ and $y'(1) = 0$

e. $x^2y'' - xy' + 2y = 0$ *with* $y(1) = 3$ and $y'(1) = 0$

f. $x^2y'' - 3xy' + 13y = 0$ *with* $y(1) = 9$ and $y'(1) = 3$

20.3. *Suppose that the indicial equation for a second-order Euler equation only has one solution* r*. Using reduction of order (or any other approach you think appropriate) show that both*

$$y_1(x) = x^r \qquad \text{and} \qquad y_2(x) = x^r \ln|x|$$

are solutions to the differential equation on $(0, \infty)$*.*

20.4. *Find the general solution to each of the following third- and fourth-order Euler equations on* $(0, \infty)$*:*

a. $x^3y''' + 2x^2y'' - 4xy' + 4y = 0$

b. $x^3y''' + 2x^2y'' + xy' - y = 0$

c. $x^3y''' - 5x^2y'' + 14xy' - 18y = 0$

d. $x^3y''' - 3x^2y'' + 7xy' - 8y = 0$

e. $x^4y^{(4)} + 6x^3y''' + 15x^2y'' + 9xy' + 16y = 0$

f. $x^4y^{(4)} + 6x^3y''' - 3x^2y'' - 9xy' + 9y = 0$

g. $x^4y^{(4)} + 2x^3y''' + x^2y'' - xy' + y = 0$

h. $x^4y^{(4)} + 6x^3y''' + 7x^2y'' + xy' - y = 0$

20.5. *While memorizing the indicial equations is not recommended, it must be admitted that there is a simple, easily derived shortcut to finding these equations.*

a. *Show that the indicial equation for the second-order Euler equation*

$$\alpha x^2y'' + \beta xy' + \gamma y = 0$$

is given by

$$\alpha r(r-1) + \beta r + \gamma = 0 \quad .$$

b. *Show that the indicial equation for the third-order Euler equation*

$$\alpha_0 x^3y''' + \alpha_1 x^2y'' + \alpha_2 xy' + \alpha_3 y = 0$$

is given by

$$\alpha_0 r(r-1)(r-2) + \alpha_1 r(r-1) + \alpha_2 r + \alpha_3 = 0 \quad .$$

c. *So what do you suspect is the general shortcut for finding the indicial equation of any Euler equation?*

20.6. *Confirm that the claim of Lemma 20.1 holds when $N = 2$ by considering the general second-order Euler equation*

$$\alpha x^2 y'' + \beta x y' + \gamma y = 0$$

and doing the following:

a. *Find the corresponding indicial equation.*

b. *Using the substitution $x = e^t$, convert the above Euler equation to a second-order constant coefficient differential equation, and write out the corresponding characteristic equation. Remember, $x = e^t$ is equivalent to $t = \ln|x|$. (You may want to glance back at the chain rule computations in line (20.5).)*

c. *Confirm (by inspection!) that the characteristic equation for the constant coefficient equation just obtained is identical to the indicial equation for the above Euler equation.*

20.7. *Confirm that the claim of Lemma 20.1 holds when $N = 3$ by considering the general third-order Euler equation*

$$\alpha_0 x^3 y''' + \alpha_1 x^2 y'' + \alpha_2 x y' + \alpha_3 y = 0$$

and doing the following:

a. *Find the corresponding indicial equation.*

b. *Convert the above Euler equation to a third-order constant coefficient differential equation using the substitution $x = e^t$.*

c. *Confirm that the characteristic equation for the constant coefficient equation just obtained is identical to the indicial equation for the above Euler equation.*

20.8. *Confirm that the claim of Lemma 20.1 holds when N is any positive integer.*

21

Nonhomogeneous Equations in General

Now that we are proficient at solving many homogeneous linear differential equations, including

$$y'' - 4y = 0 \quad ,$$

it is time to expand our skills to solving nonhomogeneous linear equations, such as

$$y'' - 4y = 5e^{3x} \quad .$$

21.1 General Solutions to Nonhomogeneous Equations

In Chapter 15, we saw that any linear combination

$$y = c_1 y_1 + c_2 y_2$$

of two solutions y_1 and y_2 to a second-order *homogeneous* linear differential equation

$$ay'' + by' + cy = 0$$

(on some interval) is another solution to that differential equation. However, for a second-order *non*homogeneous linear differential equation

$$ay'' + by' + cy = g \quad ,$$

the situation is not as simple. To see this, let us compute

$$ay'' + by' + cy$$

assuming y is some linear combination of any two sufficiently differentiable functions y_1 an
say,

$$y = 2y_1 + 6y_2 \quad .$$

By the fundamental properties of differentiation, we know that[1]

$$y' = [2y_1 + 6y_2]' = [2y_1]' + [6y_2]' = 2y_1{}' + 6y_2{}'$$

[1] If the computations look familiar, it's because we did very similar computations in deriving the princi
in Chapter 15 (see page 281).

and

$$y'' = [2y_1 + 6y_2]'' = [2y_1]'' + [6y_2]'' = 2y_1{}'' + 6y_2{}'' \quad .$$

So,

$$\begin{aligned} ay'' + by' + cy &= a\,[2y_1 + 6y_2]'' + b\,[2y_1 + 6y_2]' + c\,[2y_1 + 6y_2(x)] \\ &= 2ay_1{}'' + 6ay_2{}'' + 2by_1{}' + 6by_2{}' + 2cy_1(x) + 6cy_2(x) \\ &= 2\left[ay_1{}'' + by_1{}' + cy_1\right] + 6\left[ay_2{}'' + by_2{}' + cy_2\right] \quad . \end{aligned}$$

Of course, there was nothing special about the constants 2 and 6. If we had used any linear combination of y_1 and y_2

$$y = c_1y_1 + c_2y_2 \quad , \tag{21.1a}$$

then the above computations would have yielded

$$ay'' + by' + cy = c_1\left[ay_1{}'' + by_1{}' + cy_1\right] + c_2\left[ay_2{}'' + by_2{}' + cy_2\right] \quad . \tag{21.1b}$$

Now, suppose we have two particular solutions y_p and y_q to the nonhomogeneous equation

$$ay'' + by' + cy = g \quad .$$

This means

$$ay_p{}'' + by_p{}' + cy_p = g \qquad \text{and} \qquad ay_q{}'' + by_q{}' + cy_q = g \quad .$$

From equation set (21.1) we see that if

$$y = 2y_p + 6y_q \quad ,$$

then

$$\begin{aligned} ay'' + by' + cy &= 2\left[ay_p{}'' + by_p{}' + cy_p\right] + 6\left[ay_q{}'' + by_q{}' + cy_q\right] \\ &= 2[g] + 6[g] \\ &= 8g \\ &\neq g \quad , \end{aligned}$$

showing that this linear combination of solutions to our nonhomogeneous differential equation is *not* a solution to our original nonhomogeneous equation. So it is *NOT* true that, in general, a linear combination of solutions to a nonhomogeneous differential equation is another solution to that nonhomogeneous differential equation.

Notice, however, what happens when we use the difference between these two particular solutions

$$y = y_q - y_p = 1 \cdot y_q + (-1)y_p \quad .$$

Then

$$\begin{aligned} ay'' + by' + cy &= 1\left[ay_p{}'' + by_p{}' + cy_p\right] + (-1)\left[ay_q{}'' + by_q{}' + cy_q\right] \\ &= g - g \\ &= 0 \quad , \end{aligned}$$

which means that $y = y_q - y_p$ is a solution to the corresponding homogeneous equation

$$ay'' + by' + cy = 0 \quad .$$

Let me rephrase this:

If y_p and y_q are any two solutions to a given second-order nonhomogeneous linear differential equation, then

$$y_q = y_p + \text{a solution to the corresponding homogeneous equation} \quad .$$

On the other hand, if

$$y = y_p + y_h$$

where y_p is any particular solution to the nonhomogeneous equation and y_h is any solution to the corresponding homogeneous equation (so that

$$ay_p'' + by_p' + cy_p = g \qquad \text{and} \qquad ay_h'' + by_h' + cy_h = 0 \qquad),$$

then equation set (21.1) yields

$$\begin{aligned} ay'' + by' + cy &= \left[ay_p'' + by_p' + cy_p\right] + \left[ay_0'' + by_0' + cy_0\right] \\ &= g + 0 \\ &= g \quad . \end{aligned}$$

Thus:

If y_p is a particular solution to a given second-order nonhomogeneous linear differential equation, and

$$y = y_p + \text{any solution to the corresponding homogeneous equation} \quad ,$$

then y is also a solution to the nonhomogeneous differential equation.

If you think about it, you will realize that we've just derived the form for a general solution to any nonhomogeneous linear differential equation order two; namely,

$$y = y_p + y_h$$

where y_p is any one particular solution to that nonhomogeneous differential equation and y_h a general solution to the corresponding homogeneous linear differential equation. And if you think about it a little more, you will realize that analogous computations can be done for any nonhomogeneous linear differential equation, no matter what its order is. That gives us the following theorem:

Theorem 21.1 (general solutions to nonhomogeneous equations)
A general solution to any given nonhomogeneous linear differential equation is given by

$$y = y_p + y_h$$

where y_p is any particular solution to the given nonhomogeneous equation, and y_h is a general solution to the corresponding homogeneous differential equation.[2]

[2] Many texts refer to the general solution of the corresponding homogeneous differential equation as "the complementary solution" and denote it by y_c instead of y_h. We are using y_h to help remind us that this is the general solution to the corresponding *h*omogeneous differential equation.

!▶Example 21.1: *Consider the nonhomogeneous differential equation*

$$y'' - 4y = 5e^{3x} \quad . \tag{21.2}$$

Observe that

$$\left[e^{3x}\right]'' - 4\left[e^{3x}\right] = 3^2 e^{3x} - 4e^{3x} = 5e^{3x} \quad .$$

So one particular solution to our nonhomogeneous equation is

$$y_p(x) = e^{3x} \quad .$$

The corresponding homogeneous equation is

$$y'' - 4y = 0 \quad ,$$

a linear equation with constant coefficients. Its characteristic equation,

$$r^2 - 4 = 0 \quad ,$$

has solutions $r = 2$ and $r = -2$. So this homogeneous equation has

$$\{\, y_1(x) \,,\, y_2(x) \,\} = \{\, e^{2x} \,,\, e^{-2x} \,\}$$

as a fundamental set of solutions, and

$$y_h(x) = c_1 e^{2x} + c_2 e^{-2x}$$

as a general solution.

As we saw in deriving Theorem 21.1, the general solution to nonhomogeneous equation (21.2) is then

$$\begin{aligned} y(x) &= y_p(x) + y_h(x) \\ &= e^{3x} + c_1 e^{2x} + c_2 e^{-2x} \quad . \end{aligned}$$

(Note that there are only two arbitrary constants, and that they are only in the formula for y_h. There is no arbitrary constant corresponding to y_p !)

This last example illustrates what happens when we limit ourselves to second-order equations. More generally, if we recall how we construct general solutions to the corresponding homogeneous equations, then we get the following corollary of Theorem 21.1:

Corollary 21.2 (general solutions to second-order nonhomogeneous linear equations)

A general solution to a second-order nonhomogeneous linear differential equation

$$ay'' + by' + cy = g$$

is given by

$$y(x) = y_p(x) + c_1 y_1(x) + c_2 y_2(x) \tag{21.3}$$

where y_p is any particular solution to the nonhomogeneous equation, and $\{y_1, y_2\}$ is any fundamental set of solutions for the corresponding homogeneous equation

$$ay'' + by' + cy = 0 \quad .$$

Do note that there are only two arbitrary constants c_1 and c_2 in formula (21.3), and that they are multiplying only particular solutions to the corresponding homogeneous equation. The particular solution to the nonhomogeneous equation, y_p, is NOT multiplied by an arbitrary constant!

Of course, if we don't limit ourselves to second-order equations, but still recall how to construct general solutions to homogeneous equations from a fundamental set of solutions to that homogeneous equation, then we get the Nth-order analog of the last corollary:

Corollary 21.3 (general solutions to N^{th}-order nonhomogeneous linear equations)
A general solution to an Nth-order nonhomogeneous linear differential equation

$$a_0 y^{(N)} + a_1 y^{(N-1)} + \cdots + a_{N-2} y'' + a_{N-1} y' + a_N y = g$$

is given by

$$y(x) = y_p(x) + c_1 y_1(x) + c_2 y_2(x) + \cdots + c_N y_N(x) \tag{21.4}$$

where y_p is any particular solution to the nonhomogeneous equation, and $\{y_1, y_2, \ldots, y_N\}$ is any fundamental set of solutions for the corresponding homogeneous equation

$$a_0 y^{(N)} + a_1 y^{(N-1)} + \cdots + a_{N-2} y'' + a_{N-1} y' + a_N y = 0 \quad .$$

21.2 Superposition for Nonhomogeneous Equations

Before discussing methods for finding particular solutions, we should note that equation (21.1) on page 386 is describing a "principle of superposition for nonhomogeneous equations"; namely, that if y_1, y_2, g_1 and g_2 are functions satisfying

$$a{y_1}'' + b{y_1}' + cy_1 = g_1 \qquad \text{and} \qquad a{y_2}'' + b{y_2}' + cy_2 = g_2$$

over some interval, and y is some linear combination of the two solutions

$$y = c_1 y_2 + c_2 y_2 \quad ,$$

then, over that interval,

$$\begin{aligned} ay'' + by' + cy &= c_1 \left[a{y_1}'' + b{y_1}' + cy_1\right] + c_2 \left[a{y_2}'' + b{y_2}' + cy_2\right] \\ &= c_1 g_1 + c_2 g_2 \quad , \end{aligned}$$

showing that $y = c_1 y_2 + c_2 y_2$ is a solution to

$$ay'' + by' + cy = c_1 g_1 + c_2 g_2 \quad .$$

Obviously, similar computations will yield similar results involving any number of "(y_j, g_j) pairs", and using comparable nonhomogeneous linear differential equations of any order. This gives us:

Theorem 21.4 (principle of superposition for nonhomogeneous linear equations)
Let a_0, a_1, ... and a_N be functions on some interval (α, β), and let K be a positive integer. Assume $\{y_1, y_2, \ldots, y_K\}$ and $\{g_1, g_2, \ldots, g_K\}$ are two sets of K functions related over (α, β) by

$$a_0 {y_k}^{(N)} + \cdots + a_{N-2} {y_k}'' + a_{N-1} {y_k}' + a_N y_k = g_k \qquad \text{for} \quad k = 1, 2, \ldots, K \quad .$$

Then, for any set of K constants $\{c_1, c_2, \ldots, c_K\}$, a particular solution to

$$a_0 y^{(N)} + \cdots + a_{N-2} y'' + a_{N-1} y' + a_N y = G$$

where

$$G = c_1 g_1 + c_2 g_2 + \cdots + c_K g_K$$

is given by

$$y_p = c_1 y_1 + c_2 y_2 + \cdots + c_K y_K \quad .$$

This principle gives us a means for constructing solutions to certain nonhomogeneous equations as linear combinations of solutions to simpler nonhomogeneous equations, provided, of course, we have the solutions to those simpler equations.

!▶Example 21.2: *From our last example, we know that*

$$y_1(x) = e^{3x} \quad \text{satisfies} \quad {y_1}'' - 4y_1 = 5e^{3x} \quad .$$

The principle of superposition (with $K = 1$) then assures us that, for any constant a_1,

$$y_p(x) = a_1 y_1(x) = a_1 e^{3x} \quad \text{satisfies} \quad {y_1}'' - 4y_1 = a_1\left[5e^{3x}\right] \quad .$$

For example, a particular solution to

$$y'' - 4y = e^{3x} \quad ,$$

which we will rewrite as

$$y'' - 4y = \frac{1}{5}\left[5e^{3x}\right]$$

is given by

$$y_p(x) = \frac{1}{5} y_1(x) = \frac{1}{5} e^{3x} \quad .$$

And for the general solution, we simply add the general solution to the corresponding homogeneous equation found in the previous example:

$$y(x) = y_p(x) + y_h(x) = \frac{1}{5} e^{3x} + c_1 e^{2x} + c_2 e^{-2x} \quad .$$

The basic use of superposition requires that we already know the appropriate "y_k's". At times, we may not already know them, but, with luck, we can make good "guesses" as to appropriate y_k's and then, after computing the corresponding g_k's, use the principle of superposition.

!▶Example 21.3: *Consider solving*

$$y'' - 4y = 2x^2 - 8x + 3 \quad . \tag{21.5}$$

Let us "guess" that a particular solution can be given by a linear combination of

$$y_1(x) = x^2 \quad , \quad y_2(x) = x \quad \text{and} \quad y_3(x) = 1 \quad .$$

Plugging these into the left-hand side of equation (21.5), we get

$$g_1(x) = {y_1}'' - 4y_1 = \left[x^2\right]'' - 4\left[x^2\right] = 2 - 4x^2 \quad ,$$

$$g_2(x) = {y_2}'' - 4y_2 = [x]'' - 4[x] = -4x$$

and

$$g_3(x) = {y_3}'' - 4y_3 = [1]'' - 4[1] = -4 \quad .$$

Now set

$$y_p(x) = a_1 y_1(x) + a_2 y_2(x) + a_3 y_3(x) \quad .$$

By the principle of superposition,

$$\begin{aligned} y_p{}'' - 4y_p &= a_1 g_1(x) + a_2 g_2(x) + a_3 g_3(x) \\ &= a_1\left[2 - 4x^2\right] + a_2[-4x] + a_3[-4] \\ &= -4a_1 x^2 - 4a_2 x + [2a_1 - 4a_3] \quad . \end{aligned}$$

This means y_p *is a solution to our differential equation,*

$$y'' - 4y = 2x^2 - 8x + 3 \quad ,$$

if and only if

$$-4a_1 = 2 \quad , \quad -4a_2 = -8 \quad \text{and} \quad 2a_1 - 4a_3 = 3 \quad .$$

Solving for the a_k*'s yields*

$$a_1 = -\frac{1}{2} \quad , \quad a_2 = 2 \quad \text{and} \quad a_1 = -1 \quad .$$

Thus, a particular solution to our differential equation is given by

$$y_p(x) = a_1 y_1(x) + a_2 y_2(x) + a_3 y_3(x) = -\frac{1}{2}x^2 + 2x - 1 \quad ,$$

and a general solution is

$$y(x) = y_p(x) + y_h(x) = -\frac{1}{2}x^2 + 2x - 1 + c_1 e^{2x} + c_2 e^{-2x} \quad .$$

By the way, we'll further discuss the "art of making good guesses" in the next chapter, and develop a somewhat more systematic method that uses superposition in a slightly more subtle way.

21.3 Reduction of Order

In practice, finding a particular solution to a nonhomogeneous linear differential equation can be a challenge. One method, the basic reduction of order method for second-order nonhomogeneous linear differential equations was briefly discussed in Section 14.4. If you wish, you can go back and skim that section. Or don't. Truth is, better methods will be developed in the next few chapters.

Additional Exercises

21.1. What should $g(x)$ be so that $y(x) = e^{3x}$ is a solution to

a. $y'' + y = g(x)$?

b. $x^2 y'' - 4y = g(x)$?

c. $y^{(3)} - 4y' + 5y = g(x)$?

21.2. What should $g(x)$ be so that $y(x) = x^3$ is a solution to

a. $y'' + 4y' + 4y = g(x)$?

b. $x^2y'' + 4xy' + 4y = g(x)$?

c. $y^{(4)} + xy^{(3)} + 4y'' - \frac{3}{x}y' = g(x)$?

21.3 a. Can $y(x) = \sin(x)$ be a solution to

$$y'' + y = g(x)$$

for some nonzero function g ? (Give a reason for your answer.)

b. What should $g(x)$ be so that $y(x) = x\sin(x)$ is a solution to

$$y'' + y = g(x) \quad ?$$

21.4 a. Can $y(x) = x^4$ be a solution on $(0, \infty)$ to

$$x^2y'' - 6xy' + 12y = g(x)$$

for some nonzero function g ? (Give a reason for your answer.)

b. What should $g(x)$ be so that $y(x) = x^4 \ln|x|$ is a solution to

$$x^2y'' - 6xy' + 12y = g(x) \qquad \text{for} \quad x > 0 \quad ?$$

21.5. Consider the nonhomogeneous linear differential equation

$$y'' + 4y = 24e^{2x} \quad .$$

a. Verify that $y_p(x) = 3e^{2x}$ is one particular solution to this differential equation.

b. What is y_h, the general solution to the corresponding homogeneous equation?

c. What is the general solution to the above nonhomogeneous equation?

d. Find the solution to the above nonhomogeneous equation that also satisfies each of the following sets of initial conditions:

i. $y(0) = 6$ and $y'(0) = 6$

ii. $y(0) = -2$ and $y'(0) = 2$

21.6. Consider the nonhomogeneous linear differential equation

$$y'' + 2y' - 8y = 8x^2 - 3 \quad .$$

a. Verify that one particular solution to this equation is

$$y_p(x) = -x^2 - \frac{1}{2}x \quad .$$

b. What is y_h, the general solution to the corresponding homogeneous equation?

c. What is the general solution to the above nonhomogeneous equation?

d. Find the solution to the above nonhomogeneous equation that also satisfies each of the following sets of initial conditions:

i. $y(0) = 0$ and $y'(0) = 0$

ii. $y(0) = 1$ and $y'(0) = -3$

21.7. *Consider the nonhomogeneous linear differential equation*

$$y'' - 9y = 36 \quad .$$

a. *Verify that one particular solution to this equation is* $y_p(x) = -4$.

b. *Find the general solution to this nonhomogeneous equation.*

c. *Find the solution to the above nonhomogeneous equation that also satisfies*

$$y(0) = 8 \quad \text{and} \quad y'(0) = 6 \quad .$$

21.8. *Consider the nonhomogeneous linear differential equation*

$$y'' - 3y' - 10y = -6e^{4x} \quad .$$

a. *Verify that one particular solution to this equation is* $y_p(x) = e^{4x}$.

b. *Find the general solution to this nonhomogeneous equation.*

c. *Find the solution to the above nonhomogeneous equation that also satisfies*

$$y(0) = 6 \quad \text{and} \quad y'(0) = 8 \quad .$$

21.9. *Consider the nonhomogeneous linear differential equation*

$$y'' - 3y' - 10y = 7e^{5x} \quad .$$

a. *Verify that one particular solution to this equation is* $y_p(x) = xe^{5x}$.

b. *Find the general solution to this nonhomogeneous equation.*

c. *Find the solution to the above nonhomogeneous equation that also satisfies*

$$y(0) = 12 \quad \text{and} \quad y'(0) = -2 \quad .$$

21.10. *Consider the nonhomogeneous linear differential equation*

$$y'' + 6y' + 9y = 169\sin(2x) \quad .$$

a. *Verify that one particular solution to this equation is* $y_p(x) = 5\sin(2x) - 12\cos(2x)$.

b. *Find the general solution to this nonhomogeneous equation.*

c. *Find the solution to the above nonhomogeneous equation that also satisfies*

$$y(0) = -10 \quad \text{and} \quad y'(0) = 9 \quad .$$

21.11. *Consider the nonhomogeneous linear differential equation*

$$x^2y'' - 4xy' + 6y = 10x + 12 \quad \text{for} \quad x > 0 \quad .$$

a. *Verify that one particular solution to this equation is* $y_p(x) = 5x + 2$.

b. *Find the general solution to this nonhomogeneous equation.*

c. *Find the solution to the above nonhomogeneous equation that also satisfies*

$$y(1) = 6 \quad \text{and} \quad y'(1) = 8 \quad .$$

21.12. *Consider the nonhomogeneous linear differential equation*

$$y^{(4)} + y'' = 1 \quad .$$

a. *Verify that one particular solution to this equation is* $y_p(x) = \frac{1}{2}x^2$.

b. *Find the general solution to this nonhomogeneous equation.*

c. *Find the solution to the above nonhomogeneous equation that also satisfies*

$$y(0) = 4 \quad , \quad y'(0) = 3 \quad , \quad y''(0) = 0 \qquad \text{and} \qquad y^{(3)}(0) = 2 \quad .$$

21.13. *In Exercises 21.8 and 21.9, you saw that* $y_1(x) = e^{4x}$ *is a particular solution to*

$$y'' - 3y' - 10y = -6e^{4x} \quad ,$$

and that $y_2(x) = xe^{5x}$ *is a particular solution to*

$$y'' - 3y' - 10y = 7e^{5x} \quad .$$

Using this and the principle of superposition, find a particular solution y_p *to each of the following:*

a. $y'' - 3y' - 10y = e^{4x}$

b. $y'' - 3y' - 10y = e^{5x}$

c. $y'' - 3y' - 10y = -18e^{4x} + 14e^{5x}$

d. $y'' - 3y' - 10y = 35e^{5x} + 12e^{4x}$

21.14. *In Exercise 21.11, you verified that* $y_1(x) = 5x + 2$ *is a particular solution to*

$$x^2y'' - 4xy' + 6y = 10x + 12 \qquad \text{for} \quad x > 0 \quad .$$

It should also be clear that $y_2(x) = 1$ *is a particular solution to*

$$x^2y'' - 4xy' + 6y = 6 \qquad \text{for} \quad x > 0 \quad .$$

Using these facts and the principle of superposition, find a particular solution y_p *to each of the following on* $(0, \infty)$:

a. $x^2y'' - 4xy' + 6y = 1$

b. $x^2y'' - 4xy' + 6y = x$

c. $x^2y'' - 4xy' + 6y = 22x + 24$

21.15 a. *What should* $g(x)$ *be so that* $y(x)$ *is a solution to*

$$x^2y'' - 7xy' + 15y = g(x) \qquad \text{for} \quad x > 0$$

i. *when* $y(x) = x^2$?

ii. *when* $y(x) = x$?

iii. *when* $y(x) = 1$?

b. *Using the results just derived and the principle of superposition, find a particular solution* y_p *to*

$$x^2y'' - 7xy' + 15y = x^2 \qquad \text{for} \quad x > 0 \quad .$$

c. *Using the results derived above and the principle of superposition, find a particular solution* y_p *to*

$$x^2y'' - 7xy' + 15y = 4x^2 + 2x + 3 \qquad \text{for} \quad x > 0 \quad .$$

22

Method of Undetermined Coefficients (aka: Method of Educated Guess)

In this chapter, we will discuss one particularly simple-minded, yet often effective, method for finding particular solutions to nonhomogeneous differential equations. As the above title suggests, the method is based on making "good guesses" regarding these particular solutions. And, as always, "good guessing" is usually aided by a thorough understanding of the problem (being 'educated'), and usually works best if the problem is not too complicated. Fortunately, you have had the necessary education, and a great many nonhomogeneous differential equations of interest are sufficiently simple.

As usual, we will start with second-order equations, and then observe that everything developed also applies, with little modification, to similar nonhomogeneous differential equations of any order.

22.1 Basic Ideas

Suppose we wish to find a particular solution to a nonhomogeneous second-order differential equation

$$ay'' + by' + cy = g \quad .$$

If g is a relatively simple function and the coefficients — a, b and c — are constants, then, after recalling what the derivatives of various basic functions look like, we might be able to make a good guess as to what sort of function $y_p(x)$ yields $g(x)$ after being plugged into the left side of the above equation. Typically, we won't be able to guess exactly what $y_p(x)$ should be, but we can often guess a formula for $y_p(x)$ involving specific functions and some constants that can then be determined by plugging the guessed formula for $y_p(x)$ into the differential equation and solving the resulting algebraic equation(s) for those constants (provided the initial 'guess' was good).

!►*Example 22.1:* *Consider*

$$y'' - 2y' - 3y = 36e^{5x} \quad .$$

Since all derivatives of e^{5x} *equal some constant multiple of* e^{5x} *, it should be clear that, if we let*

$$y(x) = \text{some multiple of } e^{5x} \quad ,$$

then

$$y'' - 2y' - 3y = \text{some other multiple of } e^{5x} \quad .$$

So let us let A be some constant "to be determined", and try

$$y_p(x) = Ae^{5x}$$

as a particular solution to our differential equation:

$$y_p'' - 2y_p' - 3y_p = 36e^{5x}$$

$$\hookrightarrow \quad [Ae^{5x}]'' - 2[Ae^{5x}]' - 3[Ae^{5x}] = 36e^{5x}$$

$$\hookrightarrow \quad [25Ae^{5x}] - 2[5Ae^{5x}] - 3[Ae^{5x}] = 36e^{5x}$$

$$\hookrightarrow \quad 25Ae^{5x} - 10Ae^{5x} - 3Ae^{5x} = 36e^{5x}$$

$$\hookrightarrow \quad 12Ae^{5x} = 36e^{5x}$$

$$\hookrightarrow \quad A = 3 \quad .$$

So our "guess", $y_p(x) = Ae^{5x}$, satisfies the differential equation only if $A = 3$. Thus,

$$y_p(x) = 3e^{5x}$$

is a particular solution to our nonhomogeneous differential equation.

In the next section, we will determine the appropriate "first guesses" for particular solutions corresponding to different choices of g in our differential equation. These guesses will involve specific functions and initially unknown constants that can be determined as we determined A in the last example. Unfortunately, as we will see, the first guesses will sometimes fail. So we will discuss appropriate second (and, when necessary, third) guesses, as well as when to expect the first (and second) guesses to fail.

Because all of the guesses will be linear combinations of functions in which the coefficients are "constants to be determined", this whole approach to finding particular solutions is formally called the *method of undetermined coefficients*. Less formally, it is also called the *method of (educated) guess*.

Keep in mind that this method only finds a particular solution for a differential equation. In practice, we usually need the general solution, which (as we know from our discussion in the previous chapter) can be constructed from any particular solution along the general solution to the corresponding homogeneous equation (see Theorem 21.1 and Corollary 21.2 on page 387).

!▶Example 22.2: *Consider finding the general solution to*

$$y'' - 2y' - 3y = 36e^{5x} \quad .$$

From the last example, we know

$$y_p(x) = 3e^{5x}$$

is a particular solution to the differential equation. The corresponding homogeneous equation is

$$y'' - 2y' - 3y = 0 \quad .$$

Its characteristic equation is

$$r^2 - 2r - 3 = 0 \quad ,$$

which factors as

$$(r+1)(r-3) = 0 \quad .$$

So $r = -1$ *and* $r = 3$ *are the possible values of* r *, and*

$$y_h(x) = c_1 e^{-x} + c_2 e^{3x}$$

is the general solution to the corresponding homogeneous differential equation.

As noted in Corollary 21.2, it then follows that

$$y(x) = y_p(x) + y_h(x) = 3e^{5x} + c_1 e^{-x} + c_2 e^{3x} \quad .$$

is a general solution to our nonhomogeneous differential equation.

Also keep in mind that you may not just want the general solution but also the one solution that satisfies some particular initial conditions.

!►Example 22.3: *Consider the initial-value problem*

$$y'' - 2y' - 3y = 36e^{5x} \qquad \text{with} \quad y(0) = 9 \quad \text{and} \quad y'(0) = 25 \quad .$$

From above, we know the general solution to the differential equation is

$$y(x) = 3e^{5x} + c_1 e^{-x} + c_2 e^{3x} \quad .$$

Its derivative is

$$y'(x) = \left[3e^{5x} + c_1 e^{-x} + c_2 e^{3x}\right]' = 15e^{5x} - c_1 e^{-x} + 3c_2 e^{3x} \quad .$$

This, with our initial conditions, gives us

$$9 = y(0) = 3e^{5\cdot 0} + c_1 e^{-0} + c_2 e^{3\cdot 0} = 3 + c_1 + c_2$$

and

$$25 = y'(0) = 15e^{5\cdot 0} - c_1 e^{-0} + 3c_2 e^{3\cdot 0} = 15 - c_1 + 3c_2 \quad ,$$

which, after a little arithmetic, becomes the system

$$\begin{aligned} c_1 + c_2 &= 6 \\ -c_1 + 3c_2 &= 10 \end{aligned} \quad .$$

Solving this system by whatever means you prefer yields

$$c_1 = 2 \qquad \text{and} \qquad c_2 = 4 \quad .$$

So the solution to the given differential equation that also satisfies the given initial conditions is

$$y(x) = 3e^{5x} + c_1 e^{-x} + c_2 e^{3x} = 3e^{5x} + 2e^{-x} + 4e^{3x} \quad .$$

22.2 Good First Guesses for Various Choices of g

In all of the following, we are interested in finding a particular solution $y_p(x)$ to

$$ay'' + by' + cy = g \tag{22.1}$$

where a, b and c are constants and g is the indicated type of function. In each subsection, we will describe a class of functions for g and the corresponding 'first guess' as to the formula for a particular solution y_p. In each case, the formula will involve constants "to be determined". These constants are then determined by plugging the guessed formula for y_p into the differential equation and solving the system of algebraic equations that results. Of course, if the resulting equations are not solvable for those constants, then the first guess is not adequate, and you'll have to read the next section to learn a good 'second guess'.

Exponentials

As already illustrated in Example 22.1:

If, for some constants C and α,

$$g(x) = Ce^{\alpha x} \quad ,$$

then a good first guess for a particular solution to differential equation (22.1) is

$$y_p(x) = Ae^{\alpha x}$$

where A is a constant to be determined.

Sines and Cosines

!►Example 22.4: *Consider*

$$y'' - 2y' - 3y = 65\cos(2x) \quad .$$

A naive first guess for a particular solution might be

$$y_p(x) = A\cos(2x) \quad ,$$

where A is some constant to be determined. Unfortunately, here is what we get when we plug this guess into the differential equation:

$$y_p{}'' - 2y_p{}' - 3y_p = 65\cos(2x)$$

$$\hookrightarrow \quad [A\cos(2x)]'' - 2[A\cos(2x)]' - 3[A\cos(2x)] = 65\cos(2x)$$

$$\hookrightarrow \quad -4A\cos(2x) + 4A\sin(2x) - 3A\cos(2x) = 65\cos(2x)$$

$$\hookrightarrow \quad A[-7\cos(2x) + 4\sin(2x)] = 65\cos(2x) \quad .$$

But there is no constant A satisfying this last equation for all values of x. So our naive first guess will not work.

Since our naive first guess resulted in an equation involving both sines and cosines, let us add a sine term to the guess and see if we can get all the resulting sines and cosines in the resulting equation to balance. That is, assume

$$y_p(x) = A\cos(2x) + B\sin(2x)$$

where A and B are constants to be determined. Plugging this into the differential equation:

$$y_p'' - 2y_p' - 3y_p = 65\cos(2x)$$

$$\hookrightarrow \quad [A\cos(2x) + B\sin(2x)]'' - 2[A\cos(2x) + B\sin(2x)]'$$
$$- 3[A\cos(2x) + B\sin(2x)] = 65\cos(2x)$$

$$\hookrightarrow \quad -4A\cos(2x) - 4B\sin(2x) - 2[-2A\sin(2x) + 2B\cos(2x)]$$
$$- 3[A\cos(2x) + B\sin(2x)] = 65\cos(2x)$$

$$\hookrightarrow \quad (-7A - 4B)\cos(2x) + (4A - 7B)\sin(2x) = 65\cos(2x) \quad .$$

For the cosine terms on the two sides of the last equation to balance, we need

$$-7A - 4B = 65 \quad ,$$

and for the sine terms to balance, we need

$$4A - 7B = 0 \quad .$$

This gives us a relatively simple system of two equations in two unknowns. Its solution is easily found. From the second equation, we have

$$B = \frac{4}{7}A \quad .$$

Combining this with the first equation yields

$$65 = -7A - 4\left[\frac{4}{7}A\right] = \left[-\frac{49}{7} - \frac{16}{7}\right]A = -\frac{65}{7}A \quad .$$

Thus,

$$A = -7 \quad \text{and} \quad B = \frac{4}{7}A = \frac{4}{7}(-7) = -4 \quad ,$$

and a particular solution to the differential equation is given by

$$y_p(x) = A\cos(2x) + B\sin(2x) = -7\cos(2x) - 4\sin(2x) \quad .$$

The last example illustrates the fact that, typically, if $g(x)$ is a sine or cosine function (or a linear combination of a sine and cosine function with the same frequency) then a linear combination of both the sine and cosine can be used for $y_p(x)$. Thus, we have the following rule:

If, for some constants C_c, C_s and ω,

$$g(x) = C_c\cos(\omega x) + C_s\sin(\omega x)$$

then a good first guess for a particular solution to differential equation (22.1) is

$$y_p(x) = A\cos(\omega x) + B\sin(\omega x)$$

where A and B are constants to be determined.

Polynomials

!► Example 22.5: *Let us find a particular solution to*

$$y'' - 2y' - 3y = 9x^2 + 1 \quad .$$

Now consider, if y is any polynomial of degree N, then y, y' and y'' are also polynomials of degree N or less. So the expression "$y'' - 2y' - 3y$" would then be a polynomial of degree N. Since we want this to match the right side of the above differential equation, which is a polynomial of degree 2, it seems reasonable to try a polynomial of degree N with $N = 2$. So we "guess"

$$y_p(x) = Ax^2 + Bx + C \quad .$$

In this case

$$y_p'(x) = 2Ax + B \quad \text{and} \quad y_p''(x) = 2A \quad .$$

Plugging these into the differential equation:

$$y_p'' - 2y_p' - 3y_p = 9x^2 + 1$$

$$\hookrightarrow \quad 2A - 2[2Ax + B] - 3[Ax^2 + Bx + C] = 9x^2 + 1$$

$$\hookrightarrow \quad -3Ax^2 + [-4A - 3B]x + [2A - 2B - 3C] = 9x^2 + 1 \quad .$$

For the last equation to hold, the corresponding coefficients to the polynomials on the two sides must equal, giving us the following system:

$$\begin{aligned} x^2 \text{ terms:} &\quad -3A = 9 \\ x \text{ terms:} &\quad -4A - 3B = 0 \\ \text{constant terms:} &\quad 2A - 2B - 3C = 1 \end{aligned}$$

So,

$$A = -\frac{9}{3} = -3 \quad ,$$

$$B = -\frac{4A}{3} = -\frac{4(-3)}{3} = 4$$

and

$$C = \frac{1 - 2A + 2B}{-3} = \frac{1 - 2(-3) + 2(4)}{-3} = \frac{15}{-3} = -5 \quad .$$

And the particular solution is

$$y_p(x) = Ax^2 + Bx + C = -3x^2 + 4x - 5 \quad .$$

Generalizing from this example, we can see that the rule for the first guess for $y_p(x)$ when g is a polynomial is:

If

$$g(x) = \text{a polynomial of degree } K \quad ,$$

then a good first guess for a particular solution to differential equation (22.1) is a Kth-degree polynomial

$$y_p(x) = A_0 x^K + A_1 x^{K-1} + \cdots + A_{K-1}x + A_K$$

where the A_k's are constants to be determined.

Products of Exponentials, Polynomials, and Sines and Cosines

If g is a product of the simple functions discussed above, then the guess for y_p must take into account everything discussed above. That leads to the following rule:

If, for some pair of polynomials $P(x)$ and $Q(x)$, and some pair of constants α and ω,

$$g(x) = P(x)e^{\alpha x}\cos(\omega x) + Q(x)e^{\alpha x}\sin(\omega x)$$

then a good first guess for a particular solution to differential equation (22.1) is

$$\begin{aligned} y_p(x) = {} & \left[A_0x^K + A_1x^{K-1} + \cdots + A_{K-1}x + A_K\right]e^{\alpha x}\cos(\omega x) \\ & + \left[B_0x^K + B_1x^{K-1} + \cdots + B_{K-1}x + B_K\right]e^{\alpha x}\sin(\omega x) \end{aligned}$$

where the A_k's and B_k's are constants to be determined and K is the highest power of x appearing in polynomial $P(x)$ or $Q(x)$.

(Note that the above include the cases where $\alpha = 0$ or $\omega = 0$. In these cases the formula for y_p simplifies a bit.)

!►Example 22.6: *To find a particular solution to*

$$y'' - 2y' - 3y = 65x\cos(2x) \quad ,$$

we should start by assuming it is of the form

$$y_p(x) = [A_0x + A_1]\cos(2x) + [B_0x + B_1]\sin(2x) \quad .$$

With a bit of work, you can verify yourself that, with $y = y_p(x)$, the above differential equation reduces to

$$\begin{aligned} & [-2A_0 - 7A_1 + 4B_0 - 4B_1]\cos(2x) + [-7A_0 - 4B_0]x\cos(2x) \\ & \quad + [-4A_0 + 4A_1 - 2B_0 - 7B_1]\sin(2x) + [4A_0 - 7B_0]x\sin(2x) = 65x\cos(2x) \quad . \end{aligned}$$

Comparing the terms on either side of the last equation, we get the following system:

$$\begin{array}{rl} \cos(2x) \text{ terms:} & -2A_0 - 7A_1 + 4B_0 - 4B_1 = 0 \\ x\cos(2x) \text{ terms:} & -7A_0 \qquad\quad - 4B_0 \qquad\quad = 65 \\ \sin(2x) \text{ terms:} & -4A_0 + 4A_1 - 2B_0 - 7B_1 = 0 \\ x\sin(2x) \text{ terms:} & 4A_0 \qquad\quad - 7B_0 \qquad\quad = 0 \end{array}$$

Solving this system yields

$$A_0 = -7 \quad , \quad A_1 = -\frac{158}{65} \quad , \quad B_0 = -4 \quad . \quad \text{and} \quad B_1 = \frac{244}{65} \quad .$$

So a particular solution to the differential equation is given by

$$y_p(x) = \left[-7x - \frac{158}{65}\right]\cos(2x) + \left[-4x + \frac{244}{65}\right]\sin(2x) \quad .$$

22.3 When the First Guess Fails

!►*Example 22.7:* *Consider*

$$y'' - 2y' - 3y = 28e^{3x} \quad .$$

Our first guess is

$$y_p(x) = Ae^{3x} \quad .$$

Plugging it into the differential equation:

$$y_p{}'' - 2y_p{}' - 3y_p = 28e^{3x}$$

$$\hookrightarrow \quad \left[Ae^{3x}\right]'' - 2\left[Ae^{3x}\right]' - 3\left[Ae^{3x}\right] = 28e^{3x}$$

$$\hookrightarrow \quad \left[9Ae^{3x}\right] - 2\left[3Ae^{3x}\right] - 3\left[Ae^{3x}\right] = 28e^{3x}$$

$$\hookrightarrow \quad 9Ae^{3x} - 6Ae^{3x} - 3Ae^{3x} = 28e^{3x} \quad .$$

But when we add up the left side of the last equation, we get the impossible equation

$$0 = 28e^{3x} \quad !$$

No value for A can make this equation true. So our first guess fails.

Why did it fail? Because the guess, Ae^{3x}, was already a solution to the corresponding homogeneous equation

$$y'' - 2y' - 3y = 0 \quad ,$$

which we would have realized if we had recalled the general solution to this homogeneous differential equation. So the left side of our differential equation will have to vanish when we plug in this guess, leaving us with an 'impossible' equation.

In general, whenever our first guess for a particular solution contains a term that is also a solution to the corresponding homogeneous differential equation, then the contribution of that term to

$$ay_p{}'' + by_p{}' + cy_p = g$$

vanishes, and we are left with an equation or a system of equations with no possible solution. In these cases, we can still attempt to solve the problem using the first guess with the reduction of order method mentioned in the previous chapter. To save time, though, I will tell you what would happen. You would discover that, if the first guess fails, then there is a particular solution of the form

$$x \times \text{"the first guess"}$$

unless this formula also contains a term satisfying the corresponding homogeneous differential equation, in which case there is a particular solution of the form

$$x^2 \times \text{"the first guess"} \quad .$$

Thus, instead of using reduction of order (or the method we'll learn in the next chapter), we can apply the following rules for generating the *appropriate guess* for the form for a particular solution $y_p(x)$ (given that we've already figured out the first guess using the rules in the previous section):

If the first guess for $y_p(x)$ contains a term that is also a solution to the corresponding homogeneous differential equation, then consider

$$x \times \text{"the first guess"}$$

as a "second guess". If this (after multiplying through by the x) does not contain a term satisfying the corresponding homogeneous differential equation, then set

$$y_p(x) = \text{"second guess"} = x \times \text{"the first guess"} \quad .$$

If, however, the second guess also contains a term satisfying the corresponding homogeneous differential equation, then set

$$y_p(x) = \text{"the third guess"}$$

where

$$\text{"third guess"} = x \times \text{"the second guess"} = x^2 \times \text{"the first guess"} \quad .$$

It must be emphasized that the second guess is used only if the first fails (i.e., has a term that satisfies the homogeneous equation). If the first guess works, then the second (and third) guesses will not work. Likewise, if the second guess works, then the third guess is not only unnecessary, it will not work. If, however the first and second guesses fail, you can be sure that the third guess will work.

!▶Example 22.8: Again, consider

$$y'' - 2y' - 3y = 28e^{3x} \quad .$$

Our first guess

$$Ae^{3x}$$

was a solution to the corresponding homogeneous differential equation. So we try a second guess of the form

$$x \times \text{"first guess"} = x \times Ae^{3x} = Axe^{3x} \quad .$$

Comparing this (our second guess) to the general solution

$$y_h(x) = c_1 e^{-x} + c_1 e^{3x}$$

of the corresponding homogeneous equation (see Exercise 22.2), we see that our second guess is not a solution to the corresponding homogeneous differential equation, and, so, we can find a particular solution to our nonhomogeneous differential equation by setting

$$y_p(x) = \text{"second guess"} = Axe^{3x} \quad .$$

The first two derivatives of this are

$$y_p{}'(x) = Ae^{3x} + 3Axe^{3x}$$

and

$$y_p{}''(x) = 3Ae^{3x} + 3Ae^{3x} + 9Axe^{3x} = 6Ae^{3x} + 9Axe^{3x} \quad .$$

Using this:

$$y_p{}'' - 2y_p{}' - 3y_p = 28e^{3x}$$

$$\hookrightarrow \quad \left[Axe^{3x}\right]'' - 2\left[Axe^{3x}\right]' - 3\left[Axe^{3x}\right] = 28e^{3x}$$

$$\hookrightarrow \quad \left[6Ae^{3x} + 9Axe^{3x}\right] - 2\left[Ae^{3x} + 3Axe^{3x}\right] - 3Axe^{3x} = 28e^{3x}$$

$$\hookrightarrow \quad \underbrace{[9 - 2(3) - 3]}_{0} Axe^{3x} + [6 - 2]Ae^{3x} = 28e^{3x}$$

$$\hookrightarrow \quad 4Ae^{3x} = 28e^{3x} \quad .$$

Thus,

$$A = \frac{28}{4} = 7$$

and

$$y_p(x) = 7xe^{3x} \quad .$$

22.4 Method of Guess in General

If you think about why the method of (educated) guess works with second-order equations, you will realize that this basic approach will work just as well with any linear differential equation with constant coefficients,

$$a_0 y^{(N)} + a_1 y^{(N-1)} + \cdots + a_{N-1} y' + a_N y = g \quad ,$$

provided the $g(x)$ is any of the types of functions already discussed. The appropriate first guesses are exactly the same, and, if a term in one 'guess' happens to already satisfy the corresponding homogeneous differential equation, then x times that guess will be an appropriate 'next guess'. The only modification in our method is that, with higher-order equations, we may have to go to a fourth guess or a fifth guess or $\ldots$.

!►Example 22.9: *Consider the seventh-order nonhomogeneous differential equation*

$$y^{(7)} - 625y^{(3)} = 6e^{2x} \quad .$$

An appropriate first guess for a particular solution is still

$$y_p(x) = Ae^{2x} \quad .$$

Plugging this guess into the differential equation:

$$y_p{}^{(7)} - 625y_p{}^{(3)} = 6e^{2x}$$

$$\hookrightarrow \quad \left[Ae^{2x}\right]^{(7)} - 625\left[Ae^{2x}\right]^{(3)} = 6e^{2x}$$

$$\hookrightarrow \quad 2^7 Ae^{2x} - 625 \cdot 2^3 Ae^{2x} = 6e^{2x}$$

$$\hookrightarrow \quad 128Ae^{2x} - 5{,}000Ae^{2x} = 6e^{2x}$$

$$\hookrightarrow \quad -4{,}872Ae^{2x} = 6e^{2x}$$

$$\hookrightarrow \quad A = -\frac{6}{4{,}872} = -\frac{1}{812} \quad .$$

So a particular solution to our differential equation is

$$y_p(x) = -\frac{1}{812}e^{2x} \quad .$$

Fortunately, we dealt with the corresponding homogeneous equation,

$$y^{(7)} - 625y^{(3)} = 0 \quad ,$$

in Example 19.6 on page 360. Looking back at that example, we see that the general solution to this homogeneous differential equation is

$$y_h(x) = c_1 + c_2x + c_3x^2 + c_4e^{5x} + c_5e^{-5x} + c_6\cos(5x) + c_7\sin(5x) \quad . \tag{22.2}$$

Thus, the general solution to our nonhomogeneous equation,

$$y^{(7)} - 625y^{(3)} = 6e^{2x} \quad ,$$

is

$$\begin{aligned} y(x) &= y_p(x) + y_h(x) \\ &= -\frac{1}{812}e^{2x} + c_1 + c_2x + c_3x^2 + c_4e^{5x} + c_5e^{-5x} \\ &\qquad + c_6\cos(5x) + c_7\sin(5x) \quad . \end{aligned}$$

!►Example 22.10: Now consider the nonhomogeneous equation

$$y^{(7)} - 625y^{(3)} = 300x + 50 \quad .$$

Since the right side is a polynomial of degree one, the appropriate first guess for a particular solution is

$$y_p(x) = Ax + B \quad .$$

However, the general solution to the corresponding homogeneous equation (formula (22.2), above) contains both a constant term and a cx term. So plugging this guess into the nonhomogeneous differential equation will yield the impossible equation

$$0 = 300x + 50 \quad .$$

Likewise, both terms of the second guess,

$$x \times \text{"first guess"} = x \times (Ax + B) = Ax^2 + Bx \quad ,$$

and the last term of the third guess,

$$x \times \text{"second guess"} = x \times (Ax^2 + Bx) = Ax^3 + Bx^2 \quad ,$$

satisfy the corresponding homogeneous differential equation, and, thus, would fail. The fourth guess,

$$x \times \text{"third guess"} = x \times (Ax^3 + Bx^2) = Ax^4 + Bx^3 \quad ,$$

has no terms in common with the general solution to the corresponding homogeneous equation (formula (22.2), above). So the appropriate "guess" here is

$$y_p(x) = \text{"fourth guess"} = Ax^4 + Bx^3 \quad .$$

Using this:

$$y_p^{(7)} - 625y_p^{(3)} = 300x + 50$$

$$\hookrightarrow \quad \left[Ax^4 + Bx^3\right]^{(7)} - 625\left[Ax^4 + Bx^3\right]^{(3)} = 300x + 50$$

$$\hookrightarrow \quad 0 - 625[A \cdot 4 \cdot 3 \cdot 2x + B \cdot 3 \cdot 2 \cdot 1] = 300x + 50$$

$$\hookrightarrow \quad -15{,}000Ax - 3{,}750B = 300x + 50 \quad .$$

Thus,

$$A = -\frac{300}{15{,}000} = -\frac{1}{50} \quad \text{and} \quad B = -\frac{50}{3{,}750} = -\frac{1}{75} \quad ,$$

and a particular solution to our nonhomogeneous differential equation is given by

$$y_p(x) = -\frac{x^4}{50} - \frac{x^3}{75} \quad .$$

For the sake of completeness, let us end our development of the method of (educated) guess (more properly called the method of undetermined coefficients) with a theorem that does two things:

1. It concisely summarizes the rules we've developed in this chapter. (But its conciseness may make it too dense to be easily used — so just remember the rules we've developed instead of memorizing this theorem.)

2. It assures us that the method we've just developed will always work.

Theorem 22.1

Suppose we have a nonhomogeneous linear differential equation with constant coefficients

$$a_0 y^{(N)} + a_1 y^{(N-1)} + \cdots + a_{N-1}y' + a_N y = g$$

where

$$g(x) = P(x)e^{\alpha x}\cos(\omega x) + Q(x)e^{\alpha x}\sin(\omega x)$$

for some pair of polynomials $P(x)$ and $Q(x)$, and some pair of constants α and ω. Let K be the highest power of x appearing in $P(x)$ or $Q(x)$, and let M be the smallest nonnegative integer such that

$$x^M e^{\alpha x}\cos(\omega x)$$

is not a solution to the corresponding homogeneous differential equation.

Then there are constants A_0, A_1, ... and A_K, and constants B_0, B_1, ... and B_K such that

$$\begin{aligned} y_p(x) = {} & x^M\left[A_0x^K + A_1x^{K-1} + \cdots + A_{K-1}x + A_K\right]e^{\alpha x}\cos(\omega x) \\ & + x^M\left[B_0x^K + B_1x^{K-1} + \cdots + B_{K-1}x + B_K\right]e^{\alpha x}\sin(\omega x) \end{aligned} \tag{22.3}$$

is a particular solution to the given nonhomogeneous differential equation.

Proving this theorem is not that difficult, provided you have the right tools. Those who are interested can turn to Section 22.7 for the details.

22.5 Common Mistakes

A Bad Alternative to Formula (22.3)

One common mistake is to use

$$x^M\big[A_0x^K + A_1x^{K-1} + \cdots + A_{K-1}x + A_K\big]e^{\alpha x}\big[C_1\cos(\omega x) + C_2\sin(\omega x)\big]$$

for $y_p(x)$ instead of formula (22.3). These two formulas are not equivalent.

!►Example 22.11: *Let us suppose that the particular solution we are seeking is actually*

$$y_p(x) = [2x+3]\cos(2x) + [4x-5]\sin(2x) \quad ,$$

and that we are (incorrectly) trying to use the "guess"

$$y_p(x) = [A_0x + A_1][C_1\cos(2x) + C_2\sin(2x)]$$

to find it. Setting the guess equal to the correct answer, and multiplying things out, we get

$$\begin{aligned}
&[2x+3]\cos(2x) + [4x-5]\sin(2x) \\
&\qquad = [A_0x + A_1][C_1\cos(2x) + C_2\sin(2x)] \\
&\qquad = [A_0C_1x + A_1C_1]\cos(2x) + [A_0C_2x + A_1C_2]\sin(2x)] \quad .
\end{aligned}$$

Thus, we must have

$$A_0C_1 = 2 \quad , \quad A_1C_1 = 3 \quad , \quad A_0C_2 = 4 \quad \text{and} \quad A_1C_2 = -5 \quad .$$

But then,

$$\frac{2}{3} = \frac{A_0C_1}{A_1C_1} = \frac{A_0}{A_1} = \frac{A_0C_2}{A_1C_2} = \frac{4}{-5} \quad ,$$

which is impossible. So we cannot find the correct formula for y_p *using*

$$y_p(x) = [A_0x + A_1][C_1\cos(2x) + C_2\sin(2x)]$$

instead of formula (22.3).

The problem here is that, while the products A_0C_1 *,* A_1C_1 *,* A_0C_2 *and* A_1C_2 *do define four constants*

$$A_0C_1 = D_1 \quad , \quad A_1C_1 = D_2 \quad , \quad A_0C_2 = D_3 \quad \text{and} \quad A_1C_2 = D_4 \quad ,$$

these constants are not independent constants — given any three of those constants, the fourth is related to the other three by

$$\frac{D_1}{D_2} = \frac{A_0C_1}{A_1C_1} = \frac{A_0C_2}{A_1C_2} = \frac{D_3}{D_4} \quad .$$

Using Too Many Undetermined Coefficients

It may be argued that there is no harm in using expressions with extra undetermined coefficients, say,

$$y_p(x) = \left[A_0x^3 + A_1x^2 + A_2x + A_0\right]\cos(2x) + \left[B_0x^3 + B_1x^2 + B_2x + B_3\right]\sin(2x)$$

when Theorem 22.1 assures you that

$$y_p(x) = \left[A_0x + A_1\right]\cos(2x) + \left[B_0x + B_1\right]\sin(2x)$$

will suffice. After all, won't the extra coefficients just end up being zero? Well, yes, *IF* you do all your calculations correctly. But, by including those extra terms, you have greatly increased the difficulty and length of your calculations, thereby increasing your chances of making errors in your calculations. And why complicate your calculation so much when you should already know that those extra terms will all be zero?

So, make sure your "guess" contains the right number of coefficients to be determined — not too many, and not too few.

22.6 Using the Principle of Superposition

Suppose we have a nonhomogeneous linear differential equation with constant coefficients

$$a_0y^{(N)} + a_1y^{(N-1)} + \cdots + a_{N-1}y' + a_Ny = g$$

where g is the sum of functions

$$g(x) = g_1(x) + g_2(x) + \cdots$$

with each of the g_k's requiring a different 'guess' for y_p. One approach to finding a particular solution $y_p(x)$ to this is to construct a big guess by adding together all the guesses suggested by the g_k's. This typically leads to rather lengthy formulas and requires keeping track of many undetermined constants, and that often leads to errors in computations — errors that, themselves, may be difficult to recognize or track down.

Another approach is to break down the differential equation to a collection of slightly simpler differential equations,

$$a_0y^{(N)} + a_1y^{(N-1)} + \cdots + a_{N-1}y' + a_Ny = g_1 \quad ,$$

$$a_0y^{(N)} + a_1y^{(N-1)} + \cdots + a_{N-1}y' + a_Ny = g_2 \quad ,$$

$$\vdots$$

and, for each g_k, find a particular solution $y = y_{pk}$ to

$$a_0y^{(N)} + a_1y^{(N-1)} + \cdots + a_{N-1}y' + a_Ny = g_k \quad .$$

By the principle of superposition for nonhomogeneous equations discussed in Section 21.2, we know that a particular solution to the differential equation of interest,

$$a_0y^{(N)} + a_1y^{(N-1)} + \cdots + a_{N-1}y' + a_Ny = g_1 + g_2 + \cdots \quad ,$$

can then be constructed by simply adding up the y_{pk}'s,

$$y_p(x) = y_{p1}(x) + y_{p2}(x) + \cdots \quad .$$

Typically, the total amount of computational work is essentially the same for either approach. Still the approach of breaking the problem into simpler problems and using superposition is usually considered to be easier to actually carry out since we are dealing with smaller formulas and fewer variables at each step.

!►Example 22.12: *Consider*

$$y'' - 2y' - 3y = 65\cos(2x) + 9x^2 + 1 \quad .$$

Because

$$g_1(x) = 65\cos(2x) \qquad \text{and} \qquad g_2(x) = 9x^2 + 1$$

lead to different initial guesses for $y_p(x)$, we will break this problem into the separate problems of finding particular solutions to

$$y'' - 2y' - 3y = 65\cos(2x)$$

and

$$y'' - 2y' - 3y = 9x^2 + 1 \quad .$$

Fortunately, these happen to be differential equations considered in previous examples. From Example 22.4 we know that a particular solution to the first of these two equations is

$$y_{p1}(x) = -7\cos(2x) - 4\sin(2x) \quad ,$$

and from Example 22.5 we know that a particular solution to the second of these two equations is

$$y_{p2}(x) = -3x^2 + 4x - 5 \quad .$$

So, by the principle of superposition, we have that a particular solution to

$$y'' - 2y' - 3y = 65\cos(2x) + 9x^2 + 1$$

is given by

$$\begin{aligned} y_p(x) &= y_{p1}(x) + y_{p2}(x) \\ &= -7\cos(2x) - 4\sin(2x) - 3x^2 + 4x - 5 \quad . \end{aligned}$$

22.7 On Verifying Theorem 22.1

Theorem 22.1, which confirms our "method of guess", is the main theorem of this chapter. Its proof follows relatively easily using some of the ideas developed in Sections 16.3 and 19.4 on multiplying and factoring linear differential operators.

A Useful Lemma

Rather than tackle the proof of Theorem 22.1 directly, we will first prove the following lemma. This lemma contains much of our theorem, and its proof nicely illustrates the main ideas in the proof of the main theorem. After this lemma's proof, we'll see about proving our main theorem.

Lemma 22.2

Let L be a linear differential operator with constant coefficients, and assume y_p is a function satisfying

$$L[y_p] = g$$

where, for some nonnegative integer K and constants ρ, b_0, b_1, ... and b_K,

$$g(x) = b_0 x^K e^{\rho x} + b_1 x^{K-1} e^{\rho x} + \cdots + b_{K-1} x e^{\rho x} + b_K e^{\rho x} \quad .$$

Then there are constants A_0, A_1, ... and A_K such that

$$y_p(x) = x^M \left[A_0 x^K + A_1 x^{K-1} + \cdots + A_{K-1} x + A_K \right] e^{\rho x}$$

where

$$M = \begin{cases} \text{multiplicity of } \rho & \text{if } \rho \text{ is a root of } L\text{'s characteristic polynomial} \\ 0 & \text{if } \rho \text{ is not a root of } L\text{'s characteristic polynomial} \end{cases} \quad .$$

PROOF: Let y_q be any function on the real line satisfying

$$L[y_q] = g \quad .$$

(Theorem 13.4 on page 258 assures us that such a function exists.) Applying an equality from Lemma 19.8 on page 365, you can easily verify that

$$\left(\frac{d}{dx} - \rho\right)^{K+1} [g] = 0 \quad .$$

Hence,

$$\left(\frac{d}{dx} - \rho\right)^{K+1} \left[L[y_q]\right] = \left(\frac{d}{dx} - \rho\right)^{K+1} [g] = 0 \quad .$$

That is, y_q is a solution to the homogeneous linear differential equation with constant coefficients

$$\left(\frac{d}{dx} - \rho\right)^{K+1} L[y] = 0 \quad .$$

Letting r_1, r_2, ... and r_L be all the roots other than ρ to the characteristic polynomial for L, we can factor the characteristic equation for the last differential equation above to

$$a(r - \rho)^{K+1} (r - r_1)^{m_1} (r - r_2)^{m_2} \cdots (r - r_L)^{m_L} (r - \rho)^M = 0 \quad .$$

Equivalently,

$$a(r - r_1)^{m_1} (r - r_2)^{m_2} \cdots (r - r_L)^{m_L} (r - \rho)^{M+K+1} = 0 \quad .$$

From what we learned about the general solutions to homogeneous linear differential equations with constant coefficients in Chapter 19, we know that

$$y_q(x) = Y_1(x) + Y_\rho(x)$$

where Y_1 is a linear combination of the $x^k e^{rx}$'s arising from the roots other than ρ, and

$$Y_\rho(x) = C_0 e^{\rho x} + C_1 x e^{\rho x} + C_2 x^2 e^{\rho} + \cdots + C_{M+K} x^{M+K} e^{\rho x} \quad .$$

Now let $Y_{\rho,0}$ consist of the first M terms of Y_ρ, and set

$$y_p = y_q - Y_1 - Y_{\rho,0} \quad .$$

Observe that, while y_q is a solution to the nonhomogeneous differential equation $L[y] = g$, every term in $Y_1(x)$ and $Y_{\rho,0}(x)$ is a solution to the corresponding homogeneous differential equation $L[y] = 0$. Hence,

$$L[y_p] = L[y_q - Y_1 - Y_{\rho,0}] = L[y_q] - L[Y_1] - L[Y_{\rho,0}] = g - 0 - 0 \quad .$$

So y_p is a solution to the nonhomogeneous differential equation in the lemma. Moreover,

$$\begin{aligned} y_p(x) &= y_q - Y_1 - Y_{\rho,0} \\ &= Y_\rho(x) - \text{the first } M \text{ terms of } Y_\rho(x) \\ &= C_M x^M e^{\rho x} + C_{M+1} x^{M+1} x e^{\rho x} + C_{M+2} x^{M+2} e^{\rho} + \cdots + C_{M+K} x^{M+K} e^{\rho x} \\ &= x^M \left[C_M + C_{M+1} x + C_{M+2} x^2 + \cdots + C_{M+K} x^K \right] e^{\rho x} \quad , \end{aligned}$$

which, except for minor cosmetic differences, is the formula for y_p claimed in the lemma. ∎

Proving the Main Theorem

Take a look at Theorem 22.1 on page 406. Observe that, if you set $\rho = \alpha$, then our lemma is just a restatement of that theorem with the additional assumption that $\omega = 0$. So the claims of Theorem 22.1 follow immediately from our lemma when $\omega = 0$.

Verifying the claims of Theorem 22.1 when $\omega \neq 0$ requires just a little more work. Simply let

$$\rho = \alpha + i\omega$$

and redo the lemma's proof (making the obvious modifications) with the double factor

$$\left(\frac{d}{dx} - \rho\right)\left(\frac{d}{dx} - \rho^*\right)$$

replacing the single factor

$$\left(\frac{d}{dx} - \rho\right) \quad ,$$

and keeping in mind what you know about the solutions to a homogeneous differential equation with constant coefficients corresponding to the complex roots of the characteristic polynomial. The details will be left to the interested reader.

Additional Exercises

22.1. *Find both a particular solution* y_p *(via the method of educated guess) and a general solution* y *to each of the following:*

a. $y'' + 9y = 52e^{2x}$ **b.** $y'' - 6y' + 9y = 27e^{6x}$

c. $y'' + 4y' - 5y = 30e^{-4x}$ **d.** $y'' + 3y' = e^{x/2}$

22.2. *Solve the initial-value problem*

$$y'' - 3y' - 10y = -5e^{3x} \quad \text{with} \quad y(0) = 5 \quad \text{and} \quad y'(0) = 3 \quad .$$

22.3. *Find both a particular solution* y_p *(via the method of educated guess) and a general solution* y *to each of the following:*

a. $y'' + 9y = 10\cos(2x) + 15\sin(2x)$ **b.** $y'' - 6y' + 9y = 25\sin(6x)$

c. $y'' + 3y' = 26\cos\left(\frac{x}{3}\right) - 12\sin\left(\frac{x}{3}\right)$ **d.** $y'' + 4y' - 5y = \cos(x)$

22.4. *Solve the initial-value problem*

$$y'' - 3y' - 10y = -4\cos(x) + 7\sin(x) \quad \text{with} \quad y(0) = 8 \quad \text{and} \quad y'(0) = -5 \quad .$$

22.5. *Find both a particular solution* y_p *(via the method of educated guess) and a general solution* y *to each of the following:*

a. $y'' - 3y' - 10y = -200$ **b.** $y'' + 4y' - 5y = x^3$

c. $y'' - 6y' + 9y = 18x^2 + 3x + 4$ **d.** $y'' + 9y = 9x^4 - 9$

22.6. *Solve the initial-value problem*

$$y'' + 9y = x^3 \quad \text{with} \quad y(0) = 0 \quad \text{and} \quad y'(0) = 0 \quad .$$

22.7. *Find both a particular solution* y_p *(via the method of educated guess) and a general solution* y *to each of the following:*

a. $y'' + 9y = 25x\cos(2x)$ **b.** $y'' - 6y' + 9y = e^{2x}\sin(x)$

c. $y'' + 9y = 54x^2e^{3x}$ **d.** $y'' = 6xe^x\sin(x)$

e. $y'' - 2y' + y = [-6x - 8]\cos(2x) + [8x - 11]\sin(2x)$

f. $y'' - 2y' + y = [12x - 4]e^{-5x}$

22.8. *Solve the initial-value problem*

$$y'' + 9y = 39xe^{2x} \quad \text{with} \quad y(0) = 1 \quad \text{and} \quad y'(0) = 0 \quad .$$

22.9. *Find both a particular solution* y_p *(via the method of educated guess) and a general solution* y *to each of the following:*

a. $y'' - 3y' - 10y = -3e^{-2x}$ **b.** $y'' + 4y' = 20$

c. $y'' + 4y' = x^2$

d. $y'' + 9y = 3\sin(3x)$

e. $y'' - 6y' + 9y = 10e^{3x}$

f. $y'' + 4y' = 4xe^{-4x}$

22.10. Find a general solution to each of the following, using the method of educated guess to find a particular solution.

a. $y'' - 3y' - 10y = \left[72x^2 - 1\right]e^{2x}$

b. $y'' - 3y' - 10y = 4xe^{6x}$

c. $y'' - 10y' + 25y = 6e^{5x}$

d. $y'' - 10y' + 25y = 6e^{-5x}$

e. $y'' + 4y' + 5y = 24\sin(3x)$

f. $y'' + 4y' + 5y = 8e^{-3x}$

g. $y'' - 4y' + 5y = e^{2x}\sin(x)$

h. $y'' - 4y' + 5y = e^{-x}\sin(x)$

i. $y'' - 4y' + 5y = 100$

j. $y'' - 4y' + 5y = e^{-x}$

k. $y'' - 4y' + 5y = 10x^2 + 4x + 8$

l. $y'' + 9y = e^{2x}\sin(x)$

m. $y'' + y = 6\cos(x) - 3\sin(x)$

n. $y'' + y = 6\cos(2x) - 3\sin(2x)$

22.11. For each of the following, state the appropriate guess for the form for a particular solution $y_p(x)$. This 'guess' should be the one that works, not necessarily the first. Leave the coefficients 'undetermined'; that is, do NOT actually determine the values of the coefficients.

a. $y'' - 4y' + 5y = x^3e^{-x}\sin(x)$

b. $y'' - 4y' + 5y = x^3e^{2x}\sin(x)$

c. $y'' - 5y' + 6y = x^2e^{-7x} + 2e^{-7x}$

d. $y'' - 5y' + 6y = x^2$

e. $y'' - 5y' + 6y = 4e^{-8x}$

f. $y'' - 5y' + 6y = 4e^{3x}$

g. $y'' - 5y' + 6y = x^2e^{3x}$

h. $y'' - 5y' + 6y = x^2\cos(2x)$

i. $y'' - 5y' + 6y = x^2e^{3x}\sin(2x)$

j. $y'' - 4y' + 20y = e^{4x}\sin(2x)$

k. $y'' - 4y' + 20y = e^{2x}\sin(4x)$

l. $y'' - 4y' + 20y = x^3\sin(4x)$

m. $y'' - 10y' + 25y = 3x^2e^{5x}$

n. $y'' - 10y' + 25y = 3x^4$

22.12. Find particular solutions to the following differential equations. For your convenience, y_h, the solution to the corresponding homogeneous equation (which you found in Chapter 19) is also given for each differential equation.

a. $y^{(4)} - 4y^{(3)} = 12e^{-2x}$, $y_h(x) = c_1 + c_2x + c_3x^2 + c_4e^{4x}$

b. $y^{(4)} - 4y^{(3)} = 10\sin(2x)$, $y_h(x) = c_1 + c_2x + c_3x^2 + c_4e^{4x}$

c. $y^{(4)} - 4y^{(3)} = 32e^{4x}$, $y_h(x) = c_1 + c_2x + c_3x^2 + c_4e^{4x}$

d. $y^{(4)} - 4y^{(3)} = 32x$, $y_h(x) = c_1 + c_2x + c_3x^2 + c_4e^{4x}$

e. $y^{(3)} - y'' + y' - y = x^2$, $y_h(x) = c_1e^x + c_2\cos(x) + c_3\sin(x)$

f. $y^{(3)} - y'' + y' - y = 30\cos(2x)$, $y_h(x) = c_1e^x + c_2\cos(x) + c_3\sin(x)$

g. $y^{(3)} - y'' + y' - y = 6e^x$, $y_h(x) = c_1e^x + c_2\cos(x) + c_3\sin(x)$

22.13. *For each of the following, state the appropriate guess for the form for a particular solution y_p. Leave the coefficients undetermined; do NOT actually determine the values of the coefficients. Again, for your convenience, y_h, the solution to the corresponding homogeneous equation (which you found in Chapter 19) is also given for each differential equation.*

a. $y^{(5)} + 18y^{(3)} + 81y' = x^2e^{3x}$,
$y_h(x) = c_1 + [c_2 + c_3x]\cos(3x) + [c_4 + c_5x]\sin(3x)$

b. $y^{(5)} + 18y^{(3)} + 81y' = x^2\sin(3x)$,
$y_h(x) = c_1 + [c_2 + c_3x]\cos(3x) + [c_4 + c_5x]\sin(3x)$

c. $y^{(5)} + 18y^{(3)} + 81y' = x^2e^{3x}\sin(3x)$,
$y_h(x) = c_1 + [c_2 + c_3x]\cos(3x) + [c_4 + c_5x]\sin(3x)$

d. $y^{(3)} - y'' + y' - y = 30x\cos(2x)$, $y_h(x) = c_1e^x + c_2\cos(x) + c_3\sin(x)$

e. $y^{(3)} - y'' + y' - y = 3x\cos(x)$, $y_h(x) = c_1e^x + c_2\cos(x) + c_3\sin(x)$

f. $y^{(3)} - y'' + y' - y = 3xe^x\cos(x)$, $y_h(x) = c_1e^x + c_2\cos(x) + c_3\sin(x)$

g. $y^{(3)} - y'' + y' - y = 3x^5e^{2x}$, $y_h(x) = c_1e^x + c_2\cos(x) + c_3\sin(x)$

22.14. *Find particular solutions to the following. Use superposition and/or answers to previous exercises, if practical.*

a. $y'' - 6y' + 9y = 27e^{6x} + 25\sin(6x)$

b. $y'' + 9y = 25x\cos(2x) + 3\sin(3x)$

c. $y'' - 4y' + 5y = 5\sin^2(x)$ *(Hint: Use a trig. identity to rewrite the* $\sin^2(x)$ *in a form we've already discussed.)*

d. $y'' - 4y' + 5y = 20\sinh(x)$

22.15. *With obvious modifications, the method for finding particular solutions to nonhomogeneous equations with constant coefficients discussed in this chapter can be applied to finding particular solutions to other nonhomogeneous equations. Below is a list of several nonhomogeneous Euler equations along with the solution to the corresponding homogeneous equation (found in Chapter 19). For each, apply the "appropriately modified method of educated guess" to find a particular solution:*

a. $x^2y'' - 5xy' + 8y = 5x^{-3}$, $y_h(x) = = c_1x^2 + c_2x^4$

b. $2x^2y'' - xy' + y = 50x^3$, $y_h(x) = c_1x + c_2\sqrt{x}$

c. $2x^2y'' + 5xy' + y = 85\cos(2\ln|x|)$, $y_h(x) = c_1x^{-1/2} + c_2x^{-1}$

d. $x^2y'' - 2y = 15\cos(3\ln|x|) - 10\sin(3\ln|x|)$, $y(x) = c_1x^2 + c_2x^{-1}$

e. $3x^2y'' - 7xy' + 3y = 4x^3$, $y_h(x) = c_1x^3 + c_2\sqrt[3]{x}$

f. $2x^2y'' + 5xy' + y = 10x^{-1}$, $y_h(x) = c_1x^{-1/2} + c_2x^{-1}$

g. $x^2y'' - 5xy' + 9y = 6x^3$, $y(x) = c_1x^3 + c_2x^3\ln|x|$

h. $x^2y'' + 5xy' + 4y = 64x^2\ln|x|$, $y(x) = c_1x^{-2} + c_2x^{-2}\ln|x|$

23

Springs: Part II (Forced Vibrations)

Let us, again, look at those mass/spring systems discussed in Chapter 18. Remember, in such a system we have a spring with one end attached to an immobile wall and the other end attached to some object that can move back and forth under the influences of the spring and whatever friction may be in the system. Now that we have methods for dealing with nonhomogeneous differential equations (in particular, the method of educated guess), we can expand our investigations to mass/spring systems that are under the influence of outside forces such as gravity or of someone pushing and pulling the object. Of course, the limitations of the method of guess will limit the forces we can consider. Still, these forces happen to be particularly relevant to mass/spring systems, and our analysis will lead to some very interesting results — results that can be extremely useful not just when considering springs but also when considering other systems in which things vibrate or oscillate.

23.1 The Mass/Spring System

In Chapter 18, we derived

$$m\frac{d^2y}{dt^2} + \gamma\frac{dy}{dt} + \kappa y = F$$

to model the mass/spring system. In this differential equation:

1. $y = y(t)$ is the position (in meters) at time t (in seconds) of the object attached to the spring. As before, the Y–axis is positioned so that

   (a) $y = 0$ is the location of the object when the spring is at its natural length. (This is the "equilibrium point" of the object, at least when $F = 0$.)

   (b) $y > 0$ when the spring is stretched.

   (c) $y < 0$ when the spring is compressed.

   In Chapter 18 we visualized the spring as laying horizontally as in Figure 23.1a, but that was just to keep us from thinking about the effect of gravity on this mass/spring system. Now, we can allow the spring (and Y–axis) to be either horizontal or vertical or even at some other angle. All that is important is that the motion of the object only be along the Y–axis. (Do note, however, that if the spring is hanging vertically, as in Figure 23.1c, then the Y–axis is actually pointing *downward*.)

2. m is the mass (in kilograms) of the object attached to the spring (assumed to be positive, of course).

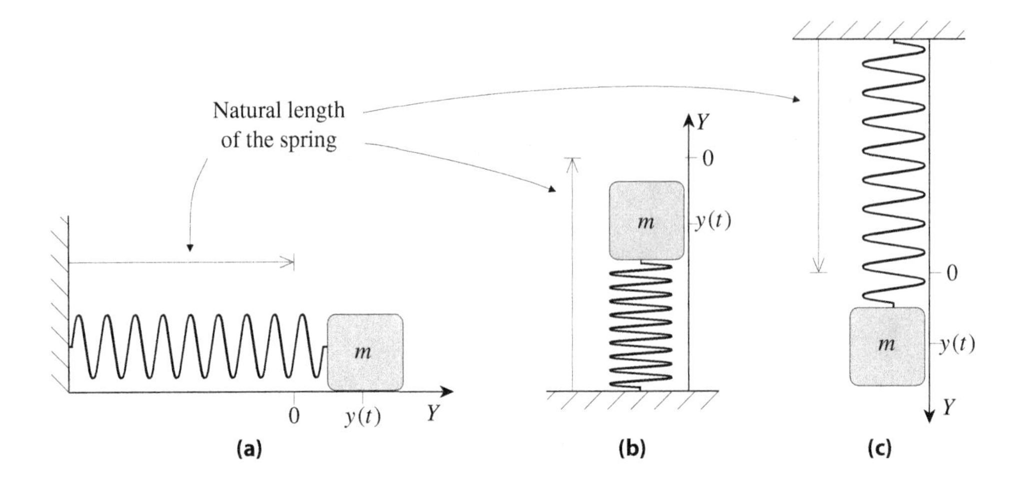

Figure 23.1: Three equivalent mass/spring systems with slightly different orientations.

3. κ is the spring constant, a positive quantity describing the "stiffness" of the spring (with "stiffer" springs having larger values for κ).

4. γ is the damping constant, a nonnegative quantity describing how much friction is in the system resisting the motion (with $\gamma = 0$ corresponding to an ideal system with no friction whatsoever).

5. F is the sum of all forces acting on the spring other than those due to the spring responding to being compressed and stretched, and the frictional forces in the system resisting motion.

Since we are expanding on the results from Chapter 18, let us recall some of the major results derived there regarding the general solution y_h to the corresponding homogeneous equation

$$m\frac{d^2y_h}{dt^2} + \gamma\frac{dy_h}{dt} + \kappa y = 0 \quad . \tag{23.1}$$

If there is no friction in the system, then we say the system is undamped, and the solution to equation (23.1) is

$$y_h(t) = c_1 \cos(\omega_0 t) + c_2 \sin(\omega_0 t)$$

or, equivalently,

$$y_h(t) = A\cos(\omega_0 t - \phi)$$

where

$$\omega_0 = \sqrt{\frac{\kappa}{m}}$$

is the natural angular frequency of the system, and the other constants are related by

$$A = \sqrt{(c_1)^2 + (c_2)^2} \quad , \qquad \cos(\phi) = \frac{c_1}{A} \qquad \text{and} \qquad \sin(\phi) = \frac{c_2}{A} \quad .$$

When convenient, we can rewrite the above formulas for y_h in terms of the system's natural frequency ν_0 by simply replacing each ω_0 with $2\pi\nu_0$.

If there is friction resisting the object's motion (i.e., if $0 < \gamma$), then we say the system is damped, and we can further classify the system as being underdamped, critically damped and overdamped, according to the precise relation between γ, κ and m. In these cases, different solutions to equation (23.1) arise, but in each of these cases, every term of the solution $y_h(t)$ has an exponentially decreasing factor. This factor ensures that

$$y_h(t) \rightarrow 0 \quad \text{as} \quad t \rightarrow \infty \quad .$$

That is what will be particularly relevant in this chapter.

(At this point, you may want to go back and quickly review Chapter 18 yourself, verifying the above and filling in some of the details glossed over. In particular, you may want to glance back over the brief note on 'units' starting on page 340.)

23.2 Constant Force

Let us first consider the case where the external force is constant. For example, the spring might be hanging vertically and the external force is the force of gravity on the object. Letting F_0 be that constant, the differential equation for $y = y(t)$ is

$$m\frac{d^2y}{dt^2} + \gamma\frac{dy}{dt} + \kappa y = F_0 \quad .$$

From our development of the method of guess, we know the general solution is

$$y(t) = y_h(t) + y_p(t)$$

where y_h is as described in the previous section, and the particular solution, y_p, is some constant,

$$y_p(t) = y_0 \qquad \text{for all} \quad t \quad .$$

Plugging this constant solution into the differential equation, we get

$$m \cdot 0 + \gamma \cdot 0 + \kappa y_0 = F_0 \quad .$$

Hence,

$$y_0 = \frac{F_0}{\kappa} \quad . \tag{23.2}$$

If the system is undamped, then

$$y(t) = y_h(t) + y_0 = c_1\cos(\omega_0 t) + c_2\sin(\omega_0 t) + y_0 \quad ,$$

which tells us that the object is oscillating about $y = y_0$. On the other hand, if the system is damped, then

$$\lim_{t\rightarrow\infty} y(t) = \lim_{t\rightarrow\infty} [y_h(t) + y_0] = 0 + y_0 \quad .$$

In this case, $y = y_0$ is where the object finally ends up. Either way, the effect of this constant force is to change the object's equilibrium point from $y = 0$ to $y = y_0$. Accordingly, if L is the natural length of the spring, then we call $L + y_0$ the *equilibrium length* of the spring in this mass/spring system under the constant force F_0.

It's worth noting that, in practice, y_0 is a quantity that can often be measured. If we also know the force, then relation (23.2) can be used to determine the spring constant κ.

!▶Example 23.1: *Suppose we have a spring whose natural length is 1 meter. We attach a 2 kilogram mass to its end and hang it vertically (as in Figure 23.1c), letting the force of gravity (near the Earth's surface) act on the mass. After the mass stops bobbing up and down, we measure the spring and find that its length is now 1.4 meters, 0.4 meters longer than its natural length. This gives us y_0 (as defined above), and since we are near the Earth's surface,*

$$F_0 = \text{force of gravity on the mass} = mg = 2 \times 9.8 \quad \left(\frac{\text{kg·meter}}{\text{sec}^2}\right) \quad .$$

Solving equation (23.2) for the spring constant and plugging in the above values, we get

$$\kappa = \frac{F_0}{y_0} = \frac{2 \times 9.8}{0.4} = 49 \quad \left(\frac{\text{kg}}{\text{sec}^2}\right) \quad .$$

We should note that the sign of F_0 and y_0 in the calculations can both be positive or negative, depending on the orientation of the system relative to the positive direction of the Y–axis. Still, κ must be positive. So, to simply avoid having to keep track of the signs, let us rewrite the above relation between κ, F_0 and y_0 as

$$\kappa = \left|\frac{F_0}{y_0}\right| \quad . \tag{23.3}$$

23.3 Resonance and Sinusoidal Forces

The mass/spring systems being considered here are but a small subset of all the things that naturally vibrate or oscillate at or around fixed frequencies — consider the swinging of a pendulum after being pushed, the vibrations of a guitar string or a steel beam after being plucked or struck — even an ordinary drinking glass may vibrate when lightly struck. And if these vibrating/oscillating systems are somehow forced to move using a force that, itself, varies periodically, then we may see *resonance*. This is the tendency of the system's vibrations or oscillations to become very large when the force periodically fluctuates at certain frequencies. Sometimes, these oscillations can be so large that the system breaks. Because of resonance, bridges have collapsed, singers have shattered glass, and small but vital parts of motors have broken off at inconvenient moments. (On the other hand, if you are in a swing, you use resonance in pumping the swing to swing as high as possible, and if you are a musician, your instrument may well use resonance to amplify the mellow tones you want amplified. So resonance is not always destructive.)

We can investigate the phenomenon of resonance in our mass/spring system by looking at the solutions to

$$m\frac{d^2y}{dt^2} + \gamma\frac{dy}{dt} + \kappa y = F(t)$$

when $F(t)$ is a sinusoidal, that is,

$$F = F(t) = a\cos(\eta t) + b\sin(\eta t)$$

where a, b and η are constants with $\eta > 0$. Naturally, we call η the *forcing angular frequency*, and the corresponding frequency, $\mu = {}^{\eta}/_{2\pi}$, the *forcing frequency*. To simplify our imagery, let us use an appropriate trigonometric identity (see page 342), and rewrite this function as a shifted cosine function,

$$F(t) = F_0\cos(\eta t - \phi)$$

where

$$F_0 = \sqrt{a^2 + b^2} \quad , \quad \cos(\phi) = \frac{a}{F_0} \quad \text{and} \quad \sin(\phi) = \frac{b}{F_0} \quad .$$

(Such a force can be generated by an unbalanced flywheel on the object spinning with angular velocity η about an axis perpendicular to the Y–axis. Alternatively, one could use a very well-trained flapping bird.)

The value of ϕ is relatively unimportant to our investigations, so let's set $\phi = 0$ and just consider the system modeled by

$$m\frac{d^2y}{dt^2} + \gamma\frac{dy}{dt} + \kappa y = F_0 \cos(\eta t) \quad . \tag{23.4}$$

You can easily verify, at your leisure, that completely analogous results are obtained using $\phi \neq 0$. The only change is that each particular solution y_p will have a corresponding nonzero shift.

In all that follows, keep in mind that F_0 and η are positive constants. You might even want to observe that letting $\eta \to 0$ leads to the constant force case just considered in the previous section.

It is convenient to consider the undamped and damped systems separately. We'll start with an ideal mass/spring system in which there is no friction to dampen the motion.

Sinusoidal Force in Undamped Systems

If the system is undamped, equation (23.4) reduces to

$$m\frac{d^2y}{dt^2} + \kappa y = F_0 \cos(\eta t) \quad ,$$

and the general solution to the corresponding homogeneous equation is

$$y_h(t) = c_1 \cos(\omega_0 t) + c_2 \sin(\omega_0 t) \qquad \text{with} \quad \omega_0 = \sqrt{\frac{\kappa}{m}} \quad .$$

To save a little effort later, let's observe that the equation for the natural angular frequency ω_0 can be rewritten as $\kappa = m(\omega_0)^2$. This and a little algebra allow us to rewrite the above differential equation as

$$\frac{d^2y}{dt^2} + (\omega_0)^2 y = \frac{F_0}{m}\cos(\eta t) \quad . \tag{23.5}$$

As discussed in the previous two chapters, the general solution to this is

$$y(t) = y_h(t) + y_p(t) = c_1 \cos(\omega_0 t) + c_2 \sin(\omega_0 t) + y_p(t)$$

where y_p is of the form

$$y_p(t) = \begin{cases} A\cos(\eta t) + B\sin(\eta t) & \text{if} \quad \eta \neq \omega_0 \\ At\cos(\omega_0 t) + Bt\sin(\omega_0 t) & \text{if} \quad \eta = \omega_0 \end{cases} \quad .$$

We now have two cases to consider: the case where $\eta = \omega_0$, and the case where $\eta \neq \omega_0$. Let's start with the most interesting of these two cases.

The Case Where $\eta = \omega_0$

If the forcing angular frequency η is the same as the natural angular frequency ω_0 of our mass/spring system, then

$$y_p(t) = At\cos(\omega_0 t) + Bt\sin(\omega_0 t) \quad .$$

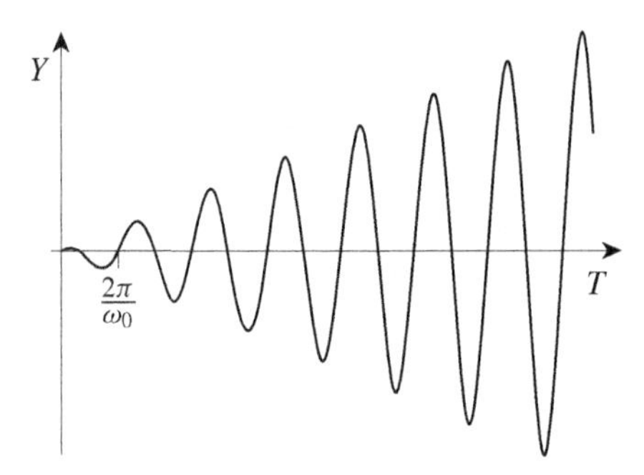

Figure 23.2: Graph of a particular solution exhibiting the "runaway" resonance in an undamped mass/spring system having natural angular frequency ω_0.

Right off, you can see that this is describing oscillations of larger and larger amplitude as time goes on. To get a more precise picture of the motion, plug the above formula for $y = y_p$ into differential equation (23.5). You can easily verify that the result is

$$\left[2B\omega_0 - At(\omega_0)^2\right]\cos(\omega_0 t) + \left[-2A\omega_0 - (Bt\omega_0)^2\right]\sin(\omega_0 t)$$
$$+ (\omega_0)^2\left[At\cos(\omega_0 t) + Bt\sin(\omega_0 t)\right] = \frac{F_0}{m}\cos(\omega_0 t) \quad ,$$

which simplifies to

$$2B\omega_0\cos(\omega_0 t) - 2A\omega_0\sin(\omega_0 t) = \frac{F_0}{m}\cos(\omega_0 t) \quad .$$

Comparing the cosine terms and the sine terms on either side of this equation then gives us the pair

$$2B\omega_0 = \frac{F_0}{m} \quad \text{and} \quad -2A\omega_0 = 0 \quad .$$

Thus,

$$B = \frac{F_0}{2m\omega_0} \quad , \quad A = 0 \quad ,$$

the particular solution is

$$y_p(t) = \frac{F_0}{2m\omega_0}t\sin(\omega_0 t) \quad , \tag{23.6}$$

and the general solution is

$$y(t) = y_h(t) + y_p(t) = c_1\cos(\omega_0 t) + c_2\sin(\omega_0 t) + \frac{F_0}{2m\omega_0}t\sin(\omega_0 t) \quad . \tag{23.7}$$

The graph of y_p is sketched in Figure 23.2. Clearly we have true, "runaway" resonance here. As time increases, the size of the oscillations are becoming steadily larger, dwarfing those in the y_h term. With each oscillation, the object moves further and further from its equilibrium point, stretching and compressing the spring more and more (try visualizing that motion!). Wait long enough, and, according to our model, the magnitude of the oscillations will exceed any size desired ... unless the spring breaks.

The Case Where $\eta \neq \omega_0$

Plugging

$$y_p(t) = A\cos(\eta t) + B\sin(\eta t)$$

into equation (23.5) yields

$$-\eta^2 [A\cos(\eta t) + B\sin(\eta t)] + (\omega_0)^2 [A\cos(\eta t) + B\sin(\eta t)] = \frac{F_0}{m}\cos(\eta t) \quad ,$$

which simplifies to

$$\left[(\omega_0)^2 - \eta^2\right] A\cos(\eta t) + \left[(\omega_0)^2 - \eta^2\right] B\sin(\eta t) = \frac{F_0}{m}\cos(\eta t) \quad .$$

Comparing the cosine terms and the sine terms on either side of this equation then gives us the pair

$$\left[(\omega_0)^2 - \eta^2\right] A = \frac{F_0}{m} \quad \text{and} \quad \left[(\omega_0)^2 - \eta^2\right] B = 0 \quad .$$

Thus,

$$A = \frac{F_0}{m\left[(\omega_0)^2 - \eta^2\right]} \quad \text{and} \quad B = 0 \quad ,$$

the particular solution is

$$y_p(t) = \frac{F_0}{m\left[(\omega_0)^2 - \eta^2\right]}\cos(\eta t) \quad , \tag{23.8}$$

and the general solution is

$$\begin{aligned} y(t) &= y_h(t) + y_p(t) \\ &= c_1\cos(\omega_0 t) + c_2\sin(\omega_0 t) + \frac{F_0}{m\left[(\omega_0)^2 - \eta^2\right]}\cos(\eta t) \quad . \end{aligned} \tag{23.9}$$

Here, the oscillations in the y_p term are not increasing with time. However, if the forcing angular frequency η is close to the natural angular frequency ω_0 of the system (and $F_0 \neq 0$), then

$$(\omega_0)^2 - \eta^2 \approx 0$$

and, so, the amplitude of the oscillations in y_p,

$$\left|\frac{F_0}{m\left[(\omega_0)^2 - \eta^2\right]}\right| \quad ,$$

will be very large. If we can adjust the forcing angular frequency η (but keeping F_0 constant), then we can make the amplitude of the oscillations in y_p as large as we could wish. So, again, our solutions are exhibiting "resonance" (perhaps we should call this "near resonance").

Some Comments About What We've Just Derived

1. *Relevance of the y_h term*: Because the oscillations in the y_p term are not increasing with time when $\eta \neq \omega_0$, every term in formula (23.9) can play a relatively significant role in the long-term motion of the object in an undamped mass/spring system. In addition, the oscillations in the y_h term can "interfere" with the y_p term to prevent $y(t)$ from reaching its maximum value within the first oscillation from when the object is initially still. In fact, the interaction of the y_h terms with the y_p term can lead to some very interesting motion. However, exploring how y_h and y_p can interact goes a little outside of our current discussions of "resonance". Accordingly, we will delay a more complete discussion of this interaction to Section 23.4, after finishing our discussion of resonance.

2. *The limit as near resonance approaches true resonance*: The *resonant frequency* of a system is the forcing frequency at which resonance is most pronounced for that system. The above analysis tells us that the resonant frequency for an undamped mass/spring system is the same as the system's natural frequency. At least, it tells us that when the forcing function is given by a cosine function. It turns out that, using more advanced tools, we will show (in Chapter 30) that we get those ever-increasing oscillations whenever the force is given by a periodic function having the same frequency as the natural frequency of that undamped mass/spring system.

   Something you might expect is that, as η gets closer and closer to the natural angular frequency ω_0, the corresponding solution y of equation 23.5 satisfying some given initial values will approach that obtained when $\eta = \omega_0$. This is, indeed, the case, and its verification will be left as an exercise (Exercise 23.5 on page 428).

3. *Limitations in our model*: Keep in mind that our model for the mass/spring system was based on certain assumptions regarding the behavior of springs. In particular, the κ term in our differential equation came from Hooke's law,

$$F_{\text{spring}}(y) = -\kappa y \quad ,$$

   relating the spring's force to the object's position. As we noted after deriving Hooke's law (page 339), this is a good model for the spring force, *provided the spring is not stretched or compressed too much.* So if our formulas for $y(t)$ have $|y(t)|$ becoming too large for Hooke's law to remain valid, then these formulas are probably are not that accurate after $|y(t)|$ becomes that large. Precisely what happens after the oscillations become so large that our model is no longer valid will depend on the spring and the force.

Sinusoidal Force in Damped Systems

If the system is damped, then we need to consider equation (23.4),

$$m\frac{d^2y}{dt^2} + \gamma\frac{dy}{dt} + \kappa y = F_0\cos(\eta t) \quad , \tag{23.4$'$}$$

assuming $0 < \gamma$. As noted a few pages ago, the terms in y_h, the solution to the corresponding homogeneous differential equation, all contain decaying exponential factors. Hence,

$$y_h(t) \rightarrow 0 \quad \text{as} \quad t \rightarrow \infty \quad ,$$

and we can assume a particular solution of the form

$$y_p(t) = A\cos(\eta t) + B\sin(\eta t) \quad .$$

We can then write the general solution to our nonhomogeneous differential equation as

$$y(t) = y_h(t) + y_p(t) = y_h(t) + A\cos(\eta t) + B\sin(\eta t) \quad ,$$

and observe that, as $t \rightarrow \infty$,

$$y(t) = y_h(t) + y_p(t) \rightarrow 0 + y_p(t) = A\cos(\eta t) + B\sin(\eta t) \quad .$$

This tells us that any long-term behavior of y depends only on y_p, and may explain why, in these cases, we refer to $y_h(t)$ as the *transient* part of the solution and $y_p(t)$ as the *steady-state* part of the solution.

The analysis of the particular solution,

$$y_p(t) = A\cos(\eta t) + B\sin(\eta t) \quad ,$$

is relatively straightforward, but a little tedious. We'll leave the computational details to the interested reader (Exercise 23.6 on page 428), and quickly summarize the high points.

Plugging in the above formula for y_p into our differential equation and solving for A and B yield

$$y_p(t) = A\cos(\eta t) + B\sin(\eta t) \tag{23.10a}$$

with

$$A = \frac{\eta\gamma F_0}{\left[\kappa - m\eta^2\right]^2 + \eta^2\gamma^2} \quad \text{and} \quad B = -\frac{\left[\kappa - m\eta^2\right] F_0}{\left[\kappa - m\eta^2\right]^2 + \eta^2\gamma^2} \quad . \tag{23.10b}$$

Using a little trigonometry, we can rewrite this as

$$y_p(t) = C\cos(\eta t - \phi) \tag{23.11a}$$

where the amplitude of these forced vibrations is

$$C = \frac{F_0}{\sqrt{\left[\kappa - m\eta^2\right]^2 + \eta^2\gamma^2}} \tag{23.11b}$$

and ϕ satisfies

$$\cos(\phi) = \frac{A}{C} \quad \text{and} \quad \sin(\phi) = \frac{B}{C} \quad . \tag{23.11c}$$

Note that the amplitude, C, does not blow up with time, nor does it become infinite for any forcing angular frequency η. So we do not have the "runaway" resonance exhibited by an undamped mass/spring system. Still this amplitude does vary with the forcing frequency. With a little work, you can show that the amplitude of the forced vibrations has a maximum value provided the friction is not too great. To be specific, if

$$\gamma < \sqrt{2\kappa m} \quad ,$$

then the maximum amplitude occurs when the forcing angular frequency is

$$\eta_0 = \sqrt{\frac{\kappa}{m} - \frac{\gamma^2}{2m^2}} \quad . \tag{23.12}$$

This is the resonant angular frequency for the given damped mass/spring system. Plugging this value for η back into formula (23.11b) then yields (after a little algebra) the maximum amplitude

$$C_{\max} = \frac{2mF_0}{\gamma\sqrt{4\kappa m - \gamma^2}} \quad . \tag{23.13}$$

If, on the other hand, $\gamma > \sqrt{2\kappa m}$, then there is no maximum value for the amplitude as η varies over $(0, \infty)$. Instead, the amplitude steadily decreases as η increases.

You might recall that a damped mass/spring system given by

$$m\frac{d^2y}{dt^2} + \gamma\frac{dy}{dt} + \kappa y = F_0\cos(\eta t)$$

is further classified, respectively, as

underdamped , critically damped or overdamped

according to whether

$$0 < \gamma < 2\sqrt{\kappa m} \quad , \quad \gamma = 2\sqrt{\kappa m} \quad \text{or} \quad 2\sqrt{\kappa m} < \gamma \quad .$$

Since "resonance" can only occur if $\gamma < \sqrt{2\kappa m}$ (and since $\sqrt{2} < 2$), it should be clear that it makes no sense to talk about resonant frequencies for critically or overdamped systems. Indeed, we can even further subdivide the underdamped systems into those having resonant frequencies and those that do not.

Finally, let us also observe that the formula for η_0 can be rewritten as

$$\eta_0 = \sqrt{(\omega_0)^2 - \frac{1}{2}\left(\frac{\gamma}{m}\right)^2}$$

where, as you should recall,

$$\omega_0 = \sqrt{\frac{\kappa}{m}}$$

is the natural angular frequency of the corresponding undamped system. From this we see that the resonant frequency of the damped system is always less than the natural frequency of the undamped system. In fact, as γ increases from 0 to $m\omega_0\sqrt{2}$, the damped system's resonant angular frequency shrinks from ω_0 to 0.

23.4 More on Undamped Motion under Nonresonant Sinusoidal Forces

When two or more sinusoidal functions of different frequencies are added together, they can alternatively amplify and interfere with each other to produce a graph that looks somewhat like a single sinusoidal function whose amplitude varies in some regular fashion. This is illustrated in Figure 23.3 in which graphs of

$$\cos(\eta t) - \cos(\omega_0 t)$$

have been sketched using one value for ω_0 and two values for η. The first figure (Figure 23.3a) illustrates what is commonly called the *beat phenomenon*, in which we appear to have a fairly high frequency sinusoidal whose amplitude seems to be given by another, more slowly varying sinusoidal. This slowly varying sinusoidal gives us the individual "beats" in which the high-frequency function intensifies and fades (Figure 23.3a shows three beats).

This beat phenomenon is typical of the sum (or difference) of two sinusoidal functions of almost the same frequency, and can be analyzed somewhat using trigonometric identities. For the functions graphed in Figure 23.3a we can use basic trigonometric identities to show that

$$\cos(\eta t) - \cos(\omega_0 t) = -2\sin\left(\frac{\eta + \omega_0}{2}t\right)\sin\left(\frac{\eta - \omega_0}{2}t\right) \quad .$$

Thus, we have

$$\cos(\eta t) - \cos(\omega_0 t) = A(t)\sin(\omega_{\text{high}}t) \qquad \text{with} \quad \omega_{\text{high}} = \frac{\eta + \omega_0}{2}$$

where

$$A(t) = \pm 2\sin(\omega_{\text{low}}t) \qquad \text{with} \quad \omega_{\text{low}} = \left|\frac{\omega_0 - \eta}{2}\right| \quad .$$

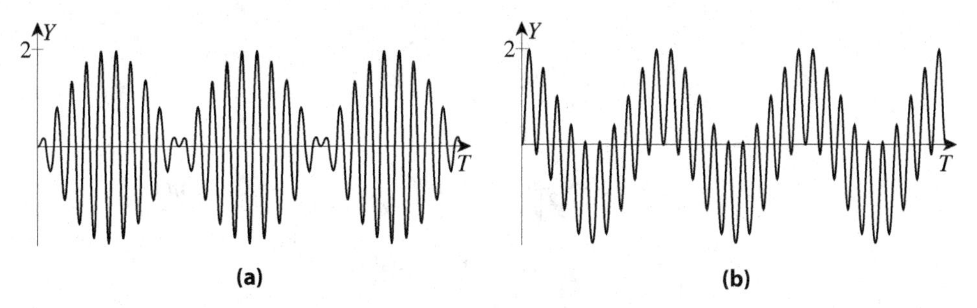

Figure 23.3: Graph of $\cos(\eta t) - \cos(\omega_0 t)$ with $\omega_0 = 2$ and with **(a)** $\eta = 0.9\,\omega_0$ and **(b)** $\eta = 0.1\,\omega_0$. (Drawn using the same horizontal scales for both graphs).

The angular frequency of the high-frequency wiggles in Figure 23.3a are approximately ω_{high}, while ω_{low} corresponds to the angular frequency of pairs of beats. (Visualizing $A(t)$ as a slowly varying amplitude only makes sense if $A(t)$ varies much more slowly than $\sin(\omega_{\text{high}} t)$. And, if you think about it, you will realize that, if $\eta \approx \omega_0$, then

$$\omega_{\text{high}} = \frac{\eta + \omega_0}{2} \approx \omega_0 \quad \text{and} \quad \omega_{\text{low}} = \left|\frac{\omega_0 - \eta}{2}\right| \approx 0 \quad .$$

So this analysis is justified if the forcing frequency is close, but not equal, to the resonant frequency.)

The general phenomenon just described (with or without "beats") occurs whenever we have a linear combination of sinusoidal functions. In particular, it becomes relevant whenever describing the behavior of an undamped mass/spring system with a sinusoidal forcing function not at resonant frequency. Let's do one general example:

!▶ Example 23.2: *Consider an undamped mass/spring system having resonant angular frequency ω_0 under the influence of a force given by*

$$F(t) = F_0 \cos(\eta t)$$

where $\eta \neq \omega_0$. Assume further that the object in the system (with mass m) is initially at rest. In other words, we want to find the solution to the initial-value problem

$$\frac{d^2y}{dt^2} + (\omega_0)^2 y = \frac{F_0}{m}\cos(\eta t) \quad \text{with} \quad y(0) = 0 \quad \text{and} \quad y'(0) = 0 \quad .$$

From our work a few pages ago (see equation (23.9)), we know

$$y(t) = c_1 \cos(\omega_0 t) + c_2 \sin(\omega_0 t) + \frac{F_0}{m\left[(\omega_0)^2 - \eta^2\right]}\cos(\eta t) \quad .$$

To satisfy the initial conditions, we must then have

$$0 = y(0) = c_1 \cos(0) + c_2 \sin(0) + \frac{F_0}{m\left[(\omega_0)^2 - \eta^2\right]}\cos(0)$$

and

$$0 = y'(0) = -c_1\omega_0 \sin(0) + c_2\omega_0 \cos(0) - \frac{F_0\eta}{m\left[(\omega_0)^2 - \eta^2\right]}\sin(0) \quad ,$$

which simplifies to the pair

$$0 = c_1 + \frac{F_0}{m\left[(\omega_0)^2 - \eta^2\right]} \quad \text{and} \quad 0 = c_2\omega_0 \quad .$$

So

$$c_1 = -\frac{F_0}{m\left[(\omega_0)^2 - \eta^2\right]} \quad , \quad c_2 = 0 \quad ,$$

and

$$\begin{aligned} y(t) &= -\frac{F_0}{m\left[(\omega_0)^2 - \eta^2\right]}\cos(\omega_0 t) + \frac{F_0}{m\left[(\omega_0)^2 - \eta^2\right]}\cos(\eta t) \\ &= \frac{F_0}{m\left[(\omega_0)^2 - \eta^2\right]}\left[\cos(\eta t) - \cos(\omega_0 t)\right] \quad . \end{aligned}$$

If $\eta = 0.9\,\omega_0$ *, the last formula for* y *reduces to*

$$y(t) = \frac{100}{19} \cdot \frac{F_0}{m(\omega_0)^2}\left[\cos(\eta t) - \cos(\omega_0 t)\right] \quad ,$$

and the graph of the object's position at time t *is the same as the graph in Figure 23.3a with the amplitude multiplied by*

$$\frac{100}{19} \cdot \frac{F_0}{m(\omega_0)^2} \quad .$$

If $\eta = 0.1\,\omega_0$ *, then*

$$y(t) = \frac{100}{99} \cdot \frac{F_0}{m(\omega_0)^2}\left[\cos(\eta t) - \cos(\omega_0 t)\right] \quad ,$$

and the graph of the object's position at time t *is the same as the graph in Figure 23.3b with the amplitude multiplied by*

$$\frac{100}{99} \cdot \frac{F_0}{m(\omega_0)^2}$$

(which, it should be noted, is approximately $1/5$ *the amplitude when* $\eta = 0.9\omega_0$ *).*

?►Exercise 23.1: *Consider the mass/spring system just discussed in the last example. Using the graphs in Figure 23.3, try to visualize the motion of the object in this system*

a: *when the forcing frequency is* 0.9 *the natural frequency.*

b: *when the forcing frequency is* 0.1 *the natural frequency.*

Additional Exercises

23.2. *A spring, whose natural length is* 0.1 *meter, is stretched to an equilibrium length of* 0.12 *meter when suspended vertically (near the Earth's surface) with a* 0.01 *kilogram mass at the end.*

a. *Find the spring constant* κ *for this spring.*

b. *Find the natural angular frequency* ω_0 *and the natural frequency* ν_0 *for this mass/spring system, assuming the system is undamped.*

23.3. *All of the following concern a single spring of natural length* 1 *meter mounted vertically with one end attached to the floor (as in Figure 23.1b on page 415).*

a. *Suppose we place a 25 kilogram box of frozen ducks on top of the spring, and, after moving the box down to its equilibrium point, we find that the length of the spring is now 0.9 meter.*

i. *What is the spring constant for this spring?*

ii. *What is the natural angular frequency of the mass/spring system assuming the system is undamped?*

iii. *Approximately how many times per second will this box bob up and down assuming the system is undamped and the box is moved from its equilibrium point and released? (In other words, what is the natural frequency,* ν_0 *?)*

b. *Suppose we replace the box of frozen ducks with a single 2 kilogram chicken.*

i. *Now what is the equilibrium length of the spring?*

ii. *What is the natural angular frequency of the undamped chicken/spring system?*

iii. *Assuming the system is undamped, not initially at equilibrium, and the chicken is not flapping its wings, how many times per second does this bird bob up and down?*

c. *Next, the chicken is replaced with a box of imported fruit. After the box stops bobbing up and down, we find that the length of the spring is 0.85 meter. What is the mass of this box of fruit?*

d. *Finally, everything is taken off the spring, and a bunch of red, helium filled balloons is tied onto the end of the spring, stretching it to an new equilibrium length of 1.02 meters. What is the buoyant force of this bunch of balloons?*

23.4. *A live 2 kilogram chicken is securely attached to the top of the the floor-mounted spring of natural length 1 meter (similar to that described in Exercise 23.3, above). Nothing else is on the spring. Knowing that the spring will break if it is stretched or compressed by half its natural length, and hoping to use the resonance of the system to stretch or compress the spring to its breaking point, the chicken starts flapping its wings. The force generated by the chicken's flapping wings* t *seconds after it starts to flap is*

$$F(t) = F_0 \cos(2\pi \mu t)$$

where μ *is the frequency of the wing flapping (flaps/second) and*

$$F_0 = 3 \left(\frac{\text{kg·meter}}{\text{sec}^2} \right) .$$

For the following exercises, also assume the following:

1. *This chicken/spring system is undamped and has natural frequency* $\nu_0 = 6$ *(hertz).*

2. *The model given by differential equation (23.5) on page 419 is valid for this chicken/spring system right up to the point where the spring breaks.*

3. *The chicken's position at time* t*,* $y(t)$ *is just given by the particular solution* y_p *found by the method of educated guess (formula (23.6) on page 420 or formula (23.8) on page 421, depending on* μ *).*

a. *Suppose the chicken flaps at the natural frequency of the system.*

i. *What is the formula for the chicken's position at time* t *?*

ii. *When does the amplitude of the oscillations become large enough to break the spring?*

b. *Suppose that the chicken manages to consistently flap its wings 3 times per second.*

i. *What is the formula for the chicken's position at time t ?*

ii. *Does the chicken break the spring? If so, when?*

c. *What is the range of values for μ , the flap frequency, that the chicken can flap at, eventually breaking the spring? (That is, find the minimum and maximum values of μ so that the corresponding near resonance will stretch or compress the spring enough to break it.)*

23.5. For each $\eta > 0$, let y_η be the solution to

$$\frac{d^2 y_\eta}{dt^2} + (\omega_0)^2 y_\eta = \frac{F_0}{m}\cos(\eta t) \qquad \text{with} \quad y_\eta(0) = 0 \quad \text{and} \quad {y_\eta}'(0) = 0$$

where ω_0 , m and F_0 are all positive constants. (Note that this describes an undamped mass/spring system in which the mass is initially at rest.)

a. Find $y_\eta(t)$ assuming $\eta \neq \omega_0$.

b. Find $y_\eta(t)$ assuming $\eta = \omega_0$.

c. Verify that, for each t ,

$$\lim_{\eta \to \omega_0} y_\eta(t) = y_{\omega_0}(t) \quad .$$

d. Using a computer math package, sketch the graph of y_η from $t = 0$ to $t = 100$ when $\omega_0 = 5$, ${}^{F_0}\!/_m = 1$, and

i. $\eta = 0.1\,\omega_0$ **ii.** $\eta = 0.5\,\omega_0$ **iii.** $\eta = 0.75\,\omega_0$ **iv.** $\eta = 0.9\,\omega_0$

v. $\eta = 0.99\,\omega_0$ **vi.** $\eta = \omega_0$ **vii.** $\eta = 1.1\,\omega_0$ **viii.** $\eta = 2\,\omega_0$

In particular, observe what happens when $\eta \approx \omega_0$, and how these graphs illustrate the result given in part c of this exercise.

23.6. Consider a damped mass/spring system given by

$$m\frac{d^2y}{dt^2} + \gamma\frac{dy}{dt} + \kappa y = F_0\cos(\eta t)$$

where m , γ , κ and F_0 are all positive constants. (This is the same as equation (23.4).)

a. Using the method of educated guess, derive the particular solution given by equation set (23.10) on page 423.

b. Then show that the solution in the previous part can be rewritten as described by equation set (23.11) on page 423; that is, verify that the solution can be written as

$$y_p(t) = C\cos(\eta t - \phi)$$

where the amplitude of these forced vibrations is

$$C = \frac{F_0}{\sqrt{\left[\kappa - m\eta^2\right]^2 + \eta^2\gamma^2}}$$

and ϕ satisfies

$$\cos(\phi) = \frac{A}{C} \qquad \text{and} \qquad \sin(\phi) = \frac{B}{C} \quad .$$

c. *Next, by "finding the maximum of C with respect to η", show that the angular resonant frequency of the system is*

$$\eta_0 = \sqrt{\frac{\kappa}{m} - \frac{\gamma^2}{2m^2}} \quad ,$$

and that the corresponding maximum amplitude is

$$C_{max} = \frac{2mF_0}{\gamma\sqrt{4\kappa m - \gamma^2}}$$

provided $\gamma < \sqrt{2\kappa m}$. What happens if, instead, $\gamma \geq \sqrt{2\kappa m}$?

d. *Assume that $\gamma < \sqrt{2\kappa m}$. Then the system is underdamped and, as noted in Chapter 18, the general solution to the corresponding homogeneous equation is*

$$y_h(t) = c_1 e^{-\alpha t}\cos(\omega t) + c_2 e^{-\alpha t}\sin(\omega t)$$

where α is the decay coefficient and ω is the angular quasi-frequency. Verify that this ω, the resonant angular frequency η_0, and the natural angular frequency ω_0 of the corresponding undamped system are related by

$$(\eta_0)^2 + \left(\frac{\gamma}{2m}\right)^2 = \omega^2 = (\omega_0)^2 - \left(\frac{\gamma}{2m}\right)^2 \quad ,$$

and, from this, conclude that $\eta_0 < \omega < \omega_0$.

24

Variation of Parameters (A Better Reduction of Order Method)

"Variation of parameters" is another way to solve nonhomogeneous linear differential equations, be they second order,

$$ay'' + by' + cy = g \quad ,$$

or even higher order,

$$a_0 y^{(N)} + a_1 y^{(N-1)} + \cdots + a_{N-1} y' + a_N y = g \quad .$$

One advantage of this method over the method of undetermined coefficients from Chapter 22 is that the differential equation does not have to be simple enough that we can 'guess' the form for a particular solution. In theory, the method of variation of parameters will work whenever g and the coefficients are reasonably continuous functions. As you may expect, though, it is not quite as simple a method as the method of guess. So, for 'sufficiently simple' differential equations, you may still prefer using the guess method instead of what we'll develop here.

We will first develop the variation of parameters method for second-order equations. Then we will see how to extend it to deal with differential equations of even higher order.[1] As you will see, the method is really just a very clever improvement on the reduction of order method for solving nonhomogeneous equations.

24.1 Second-Order Variation of Parameters

Derivation of the Method

Suppose we want to solve a second-order nonhomogeneous differential equation

$$ay'' + by' + cy = g$$

over some interval of interest, say,

$$x^2 y'' - 2xy' + 2y = 3x^2 \qquad \text{for} \quad x > 0 \quad .$$

Let us also assume that the corresponding homogeneous equation,

$$ay'' + by' + cy = 0 \quad ,$$

[1] It is possible to use a "variation of parameters" method to solve first-order nonhomogeneous linear equations, but that would be just plain silly.

has already been solved. That is, we already have an independent pair of functions $y_1 = y_1(x)$ and $y_2 = y_2(x)$ for which

$$y_h(x) = c_1 y_1(x) + c_2 y_2(x)$$

is a general solution to the homogeneous equation.

For our example,

$$x^2 y'' - 2xy' + 2y = 3x^2 \quad ,$$

the corresponding homogeneous equation is the Euler equation

$$x^2 y'' - 2xy' + 2y = 0 \quad .$$

You can easily verify that this homogeneous equation is satisfied if y is either

$$y_1 = x \qquad \text{or} \qquad y_2 = x^2 \quad .$$

Clearly, the set $\{x, x^2\}$ is linearly independent, and, so, the general solution to the corresponding homogeneous equation is

$$y_h = c_1 x + c_2 x^2 \quad .$$

Now, in using reduction of order to solve our nonhomogeneous equation

$$ay'' + by' + cy = g \quad ,$$

we would first assume a solution of the form

$$y = y_0 u$$

where $u = u(x)$ is an unknown function 'to be determined', and $y_0 = y_0(x)$ is any single solution to the corresponding homogeneous equation. However, we do not just have a single solution to the corresponding homogeneous equation — we have two: y_1 and y_2 (along with all linear combinations of these two). So why don't we use both of these solutions and assume, instead, a solution of the form

$$y = y_1 u + y_2 v$$

where y_1 and y_2 are the two solutions to the corresponding homogeneous equation already found, and $u = u(x)$ and $v = v(x)$ are two unknown functions to be determined.

For our example,

$$x^2 y'' - 2xy' + 2y = 3x^2 \quad ,$$

we already have that

$$y_1 = x \qquad \text{and} \qquad y_2 = x^2$$

form a fundamental pair of solutions to the corresponding homogeneous differential equation. So, in this case, the assumption that

$$y = y_1 u + y_2 v$$

is

$$y = xu + x^2 v$$

where $u = u(x)$ and $v = v(x)$ are two functions to be determined.

To determine the two unknown functions $u(x)$ and $v(x)$, we will need two equations. One, of course, must be the original differential equation that we are trying to solve. The other equation can be chosen at our convenience (provided it doesn't contradict or simply repeat the original differential equation). Here is a remarkably clever choice for that other equation:

$$y_1 u' + y_2 v' = 0 \quad . \tag{24.1}$$

For our example,

$$y_1 = x \qquad \text{and} \qquad y_2 = x^2 \quad .$$

So we will require that

$$xu' + x^2 v' = 0 \quad .$$

To see why this is such a clever choice, let us now compute y' and y'', and see what the differential equation becomes in terms of u and v. We'll do this for the example first.

For our example,

$$y = xu + x^2 v \quad ,$$

and we required that

$$xu' + x^2 v' = 0 \quad .$$

Computing the first derivative, rearranging a little, and applying the above requirement:

$$\begin{aligned} y' &= \left[xu + x^2 v\right]' \\ &= u + xu' + 2xv + x^2 v' \\ &= u + 2xv + \underbrace{xu' + x^2 v'}_{0} \quad . \end{aligned}$$

So

$$y' = u + 2xv \quad ,$$

and

$$y'' = [u + 2xv]' = u' + 2v + 2xv' \quad .$$

Notice that the formula for y'' does not involve any second derivatives of u and v. Plugging the above formulas for y, y' and y'' into the left side of our original differential equation, we see that

$$x^2 y'' - 2xy' + 2y = 3x^2$$

$$\hookrightarrow \quad x^2\left[u' + 2v + 2xv'\right] - 2x\left[u + 2xv\right] + 2\left[xu + x^2 v\right] = 3x^2$$

$$\hookrightarrow \quad x^2 u' + 2x^2 v + 2x^3 v' - 2xu - 4x^2 v + 2xu + 2x^2 v = 3x^2$$

$$\hookrightarrow \quad x^2 u' + 2x^3 v' + \big[\underbrace{2x^2 - 4x^2 + 2x^2}_{0}\big]v + \big[\underbrace{-2x + 2x}_{0}\big]u = 3x^2 \quad .$$

Hence, our original differential equation,

$$x^2 y'' - 2xy' + 2y = 3x^2 \quad ,$$

reduces to

$$x^2 u' + 2x^3 v' = 3x^2 \quad .$$

For reasons that will be clear in by the end of this section, let us divide this equation through by x^2 *, giving us*

$$u' + 2xv' = 3 \quad . \tag{24.2}$$

Keep in mind that this is what our differential equation reduces to if we start by letting

$$y = xu + x^2 v$$

and requiring that

$$xu' + x^2 v' = 0 \quad .$$

Now back to the general case, where our differential equation is

$$ay'' + by' + cy = g \quad .$$

If we set

$$y = y_1 u + y_2 v$$

where y_1 and y_2 are solutions to the corresponding homogeneous equation, and require that

$$y_1 u' + y_2 v' = 0 \quad ,$$

then

$$\begin{aligned} y' &= [y_1 u + y_2 v]' \\ &= [y_1 u]' + [y_2 v]' \\ &= y_1' u + y_1 u' + y_2' v + y_2 v' \\ &= y_1' u + y_2' v + \underbrace{y_1 u' + y_2 v'}_{0} \quad . \end{aligned}$$

So

$$y' = y_1' u + y_2' v \quad ,$$

and

$$\begin{aligned} y'' &= [y_1' u + y_2' v]' \\ &= y_1'' u + y_1' u' + y_2'' v + y_2' v' \\ &= y_1' u' + y_2' v' + y_1'' u + y_2'' v \quad . \end{aligned}$$

Remember, y_1 and y_2, being solutions to the corresponding homogeneous equation, satisfy

$$ay_1'' + by_1' + cy_1 = 0 \qquad \text{and} \qquad ay_2'' + by_2' + cy_2 = 0 \quad .$$

Using all the above, we have

$$\begin{aligned} & ay'' + by' + cy = g \\ \hookrightarrow \quad & a[y_1' u' + y_2' v' + y_1'' u + y_2'' v] + b[y_1' u + y_2' v] + c[y_1 u + y_2 v] = g \\ \hookrightarrow \quad & a[y_1' u' + y_2' v'] + [\underbrace{ay_1'' + by_1' + cy_1}_{0}]u + [\underbrace{ay_2'' + by_2' + cy_2}_{0}]v = g \quad . \end{aligned}$$

The vanishing of the u and v terms should not be surprising. A similar vanishing occurred in the original reduction of order method. What we also have here, thanks to the 'other equation' that we chose, is that no second-order derivatives of u or v occur either. Consequently, our original differential equation,

$$ay'' + by' + cy = g \quad ,$$

reduces to

$$a\left[y_1{}'u' + y_2{}'v'\right] = g \quad .$$

Dividing this by a then yields

$$y_1{}'u' + y_2{}'v' = \frac{g}{a} \quad .$$

Keep in mind what the last equation is. It is what our original differential equation reduces to after setting

$$y = y_1 u + y_2 v \tag{24.3}$$

(where y_1 and y_2 are solutions to the corresponding homogeneous equation), and requiring that

$$y_1 u' + y_2 v' = 0 \quad .$$

This means that the derivatives u' and v' of the unknown functions in formula (24.3) must satisfy the pair (or system) of equations

$$\begin{aligned} y_1 u' + y_2 v' &= 0 \\ y_1{}'u' + y_2{}'v' &= \frac{g}{a} \end{aligned} \quad .$$

This system can be easily solved for u' and v'. Integrating what we get for u' and v' then gives us the formulas for u and v which we can plug back into formula (24.3) for y, the solution to our nonhomogeneous differential equation.

Let's finish our example:

We have

$$y = xu + x^2 v$$

where u' and v' satisfy the system

$$\begin{aligned} xu' + x^2 v' &= 0 \\ u' + 2xv' &= 3 \end{aligned} \quad .$$

(The first was the equation we chose to require; the second was what the differential equation reduced to.) From the first equation in this system, we have that

$$u' = -xv' \quad .$$

Combining this with the second equation:

$$u' + 2xv' = 3$$

$$\hookrightarrow \quad -xv' + 2xv' = 3$$

$$\hookrightarrow \quad xv' = 3$$

$$\hookrightarrow \quad v' = \frac{3}{x} \quad .$$

Hence, also,

$$u' = -xv' = -x \cdot \frac{3}{x} = -3 \quad .$$

Remember, the primes denote differentiation with respect to x *. So we have*

$$\frac{du}{dx} = -3 \qquad \text{and} \qquad \frac{dv}{dx} = \frac{3}{x} \quad .$$

Integrating yields

$$u = \int \frac{du}{dx}\,dx = \int -3\,dx = -3x + c_1$$

and

$$v = \int \frac{dv}{dx}\,dx = \int \frac{3}{x}\,dx = 3\ln|x| + c_2 \quad .$$

Plugging these into the formula for y *, we get*

$$\begin{aligned} y &= xu + x^2 v \\ &= x\big[-3x + c_1\big] + x^2\big[3\ln|x| + c_2\big] \\ &= -3x^2 + c_1 x + 3x^2\ln|x| + c_2 x^2 \\ &= 3x^2\ln|x| + c_1 x + (c_2 - 3)x^2 \quad , \end{aligned}$$

which simplifies a little to

$$y = 3x^2\ln|x| + C_1 x + C_2 x^2 \quad .$$

This, at long last, is our solution to

$$x^2 y'' - 2xy' + 2y = 3x^2 \quad .$$

Before summarizing our work (and reducing it to a fairly simple procedure) let us make two observations based on the above:

1. If we keep the arbitrary constants arising from the indefinite integrals of u' and v', then the resulting formula for y is a general solution to the nonhomogeneous equation. If we drop the arbitrary constants (or use definite integrals), then we will end up with a particular solution.

2. After plugging the formulas for u and v into $y = y_1 u + y_2 v$, some of the resulting terms can often be absorbed into other terms. (In the above, for example, we absorbed the $-3x^2$ and $c_2 x^2$ terms into one $C_2 x^2$ term.)

Summary: How to Do It

If you look back over our derivation, you will see that we have the following:

To solve

$$ay'' + by' + cy = g \quad ,$$

first find a fundamental pair $\{y_1, y_2\}$ *of solutions to the corresponding homogeneous equation*

$$ay'' + by' + cy = 0 \quad .$$

Then set

$$y = y_1 u + y_2 v \tag{24.4}$$

assuming that $u = u(x)$ and $v = v(x)$ are unknown functions whose derivatives satisfy the system

$$y_1 u' + y_2 v' = 0 \tag{24.5a}$$

$$y_1{}' u' + y_2{}' v' = \frac{g}{a} \tag{24.5b}$$

Solve the system for u' and v'; integrate to get the formulas for u and v, and plug the results back into (24.4). That formula for y is your solution.

The above procedure is what we call (the method of) *variation of parameters* (for solving a second-order nonhomogeneous differential equation). Notice the similarity between the two equations in the system. That makes it relatively easy to remember the system of equations. This similarity is carried over to problems with equations of even higher order.

It is possible to reduce the procedure further to a single (not-so-simple) formula. We will discuss that in Section 24.3. However, as noted in that section, it is probably best to use the above method for most of your work, instead of that equation.

And remember:

1. To get a general solution to the nonhomogeneous equation, find u and v using indefinite integrals and keep the arbitrary constants arising from that integration. Otherwise, you get a particular solution.

2. After plugging the formulas for u and v into $y = y_1 u + y_2 v$, some of the resulting terms can often be absorbed into other terms. Go ahead and do so; it simplifies your final result.

!►Example 24.1: *Using the above procedure, let us find the general solution to*

$$y'' + y = \tan(x) \quad .$$

The corresponding homogeneous equation is

$$y'' + y = 0 \quad .$$

We've already solved this equation a couple of times; its general solution is

$$y_h = c_1 \cos(x) + c_2 \sin(x) \quad .$$

So we will take

$$y_1 = \cos(x) \qquad \text{and} \qquad y_2 = \sin(x)$$

as the independent pair of solutions to the corresponding homogeneous equation in the solution formula

$$y = y_1 u + y_2 v \quad .$$

That is, as the formula for the general solution to our nonhomogeneous differential equation, we will use

$$y = \cos(x) u + \sin(x) v$$

where $u = u(x)$ and $v = v(x)$ are functions to be determined.

For our differential equation, $a = 1$ and $g = \tan(x)$. Thus, with our choice of y_1 and y_2, the system

$$y_1 u' + y_2 v' = 0$$

$$y_1{}' u' + y_2{}' v' = \frac{g}{a}$$

is

$$\cos(x)u' + \sin(x)v' = 0$$
$$-\sin(x)u' + \cos(x)v' = \tan(x)$$

This system can be solved several different ways. Why don't we just observe that, if we solve the first equation for v'*, we get*

$$v' = -\frac{\cos(x)}{\sin(x)}u' \quad .$$

Combining this with the second equation (and recalling a little trigonometry) then yields

$$-\sin(x)u' + \cos(x)\left[-\frac{\cos(x)}{\sin(x)}u'\right] = \tan(x)$$

$$\hookrightarrow \qquad \left(-\sin(x) - \frac{\cos^2(x)}{\sin(x)}\right)u' = \tan(x)$$

$$\hookrightarrow \qquad -\left(\frac{\sin^2(x) + \cos^2(x)}{\sin(x)}\right)u' = \tan(x)$$

$$\hookrightarrow \qquad -\left(\frac{1}{\sin(x)}\right)u' = \tan(x) \quad .$$

Thus,

$$u' = -\tan(x)\sin(x) = -\frac{\sin^2(x)}{\cos(x)}$$

and

$$v' = -\frac{\cos(x)}{\sin(x)}u' = -\frac{\cos(x)}{\sin(x)} \times \left[-\frac{\sin^2(x)}{\cos(x)}\right] = \sin(x) \quad .$$

To integrate the formula for u'*, it may help to first observe that*

$$u'(x) = -\frac{\sin^2(x)}{\cos(x)} = -\frac{1-\cos^2(x)}{\cos(x)} = -\sec(x) + \cos(x) \quad .$$

It may also help to review the integration of the secant function in your old calculus text. After doing so, you'll see that

$$\begin{aligned} u(x) &= \int u'(x)\,dx \\ &= \int \big[\cos(x) - \sec(x)\big]\,dx \\ &= \sin(x) - \ln|\sec(x) + \tan(x)| + c_1 \end{aligned}$$

and

$$v(x) = \int \sin(x)\,dx = -\cos(x) + c_2 \quad .$$

Plugging these formulas for u *and* v *back into our solution formula*

$$y = y_1 u + y_2 v = \cos(x)u + \sin(x)v \quad ,$$

we get

$$\begin{aligned} y &= \cos(x)\big[\sin(x) - \ln|\sec(x) + \tan(x)| + c_1\big] + \sin(x)\big[-\cos(x) + c_2\big] \\ &= -\cos(x)\ln|\sec(x) + \tan(x)| + c_1\cos(x) + c_2\sin(x) \end{aligned}$$

as the general solution to the nonhomogeneous equation

$$y'' + y = \tan(x) \quad .$$

Possible Difficulties

The main 'work' in carrying out the variation of parameters method to solve

$$ay'' + by' + cy = g$$

is in solving the system

$$y_1 u' + y_2 v' = 0$$

$$y_1' u' + y_2' v' = \frac{g}{a}$$

for u' and v', and then integrating the results to get the formulas for u and v. One can foresee two possible issues: (1) the 'solvability' of the system, and (2) the 'integrability' of the solutions to the system (assuming the solutions can be obtained).

Remember, some systems of equations are "degenerate" and not truly solvable, either having no solution or infinitely many solutions. For example,

$$u' + 3v' = 0$$

$$u' + 3v' = 1$$

clearly has no solution. Fortunately, this is not an issue in the variation of parameters method. As we will see in Section 24.3 (when discussing a formula for this method), the requirement that $\{y_1, y_2\}$ be a fundamental set for the corresponding homogeneous differential equation ensures that the above system is nondegenerate and can be solved uniquely for u' and v'.

The issue of the "integrability" of the formulas obtained for u' and v' is much more significant. In our discussion of a variation of parameters formula (again, in Section 24.3), it will be noted that, in theory, the functions obtained for u' and v' will be integrable for any reasonable choice of a, b, c and g. In practice, though, it might not be possible to find usable formulas for the integrals of u' and v'. In these cases it may be necessary to use definite integrals instead of indefinite, numerically evaluating these integrals to obtain approximate values for specific choices of $u(x)$ and $v(x)$. We will discuss this further in Section 24.3.

24.2 Variation of Parameters for Even Higher Order Equations

Take another quick look at part of our derivation in the previous section. In setting

$$y = y_1 u + y_2 v$$

and then requiring

$$y_1 u' + y_2 v' = 0 \quad ,$$

we ensured that the formula for y',

$$y' = [y_1 u + y_2 v]' = y_1{}' u + y_1 u' + y_2{}' v + y_2 v'$$
$$= y_1{}' u + y_2{}' v + y_1 u' + y_2 v' = y_1{}' u + y_2{}' v + 0 \quad ,$$

contains no derivatives of the unknown functions u and v.

Suppose, instead, that we have three known functions y_1, y_2 and y_3, and we set

$$y = y_1 u + y_2 v + y_3 w$$

where u, v and w are unknown functions. For the same reasons as before, requiring that

$$y_1 u' + y_2 v' + y_3 w' = 0 \tag{24.6}$$

will insure that the formula for y' contains no derivative of u, v and w, but will simply be

$$y' = y_1{}' u + y_2{}' v + y_3{}' w \quad .$$

Differentiating this yields

$$y'' = y_1{}'' u + y_1{}' u' + y_2{}'' v + y_2{}' v' + y_3{}'' w + y_3{}' w'$$
$$= \left[y_1{}'' u + y_2{}'' v + y_3{}'' w \right] + \left[y_1{}' u' + y_2{}' v' + y_3{}' w' \right] \quad ,$$

which reduces to

$$y'' = y_1{}'' u + y_2{}'' v + y_3{}'' w$$

provided we require that

$$y_1{}' u' + y_2{}' v' + y_3{}' w' = 0 \quad . \tag{24.7}$$

Thus, requiring equations (24.6) and (24.7) prevents derivatives of the unknown functions from appearing in the formulas for either y' or y''. As you can easily verify, differentiating the last formula for y'' and plugging the above formulas for y, y' and y'' into a third-order differential equation

$$a_0 y''' + a_1 y'' + a_2 y' + a_3 y = g$$

then yield

$$a_0 \left[y_1{}'' u' + y_2{}'' v' + y_3{}'' w' \right] + [\cdots] u + [\cdots] v + [\cdots] w = g \tag{24.8}$$

where the coefficients in the u, v and w terms will vanish if y_1, y_2 and y_3 are solutions to the corresponding homogeneous differential equation.

Together equations (24.6), (24.7) and (24.8) form a system of three equations in three unknown functions. If you look at this system, and recall the original formula for y, you'll see that we've derived the variation of parameters method for solving third-order nonhomogeneous linear differential equations:

To solve the nonhomogeneous differential equation

$$a_0 y''' + a_1 y'' + a_2 y' + a_3 y = g \quad ,$$

first find a fundamental set of solutions $\{y_1, y_2, y_3\}$ *to the corresponding homogeneous equation*

$$a_0 y''' + a_1 y'' + a_2 y' + a_3 y = 0 \quad .$$

Then set

$$y = y_1 u + y_2 v + y_3 w \tag{24.9}$$

assuming that $u = u(x)$, $v = v(x)$ and $w = w(x)$ are unknown functions whose derivatives satisfy the system

$$y_1 u' + y_2 v' + y_3 w' = 0 \tag{24.10a}$$

$$y_1' u' + y_2' v' + y_3' w' = 0 \tag{24.10b}$$

$$y_1'' u' + y_2'' v' + y_3'' w' = \frac{g}{a_0} \tag{24.10c}$$

Solve the system for u', v' and w'; integrate to get the formulas for u, v and w, and plug the results back into formula (24.9) for y. That formula is the solution to the original nonhomogeneous differential equation.

Extending the method to nonhomogeneous linear equations of even higher order is straightforward. We simply continue to let y be given by formulas similar to formula (24.9) using fundamental sets of solutions to the corresponding homogeneous equations. Repeatedly imposing requirements patterned after equations (24.10a) and (24.10b) to ensure that no derivatives of unknown functions remain until we compute the highest order derivative in the differential equation, we eventually get the variation of parameters method for solving any nonhomogeneous linear differential equation:

To solve the Nth-order nonhomogeneous differential equation

$$a_0 y^{(N)} + a_1 y^{(N-1)} + \cdots + a_{N-1} y' + a_N y = g \quad ,$$

first find a fundamental set of solutions $\{y_1, y_2, \ldots, y_N\}$ to the corresponding homogeneous equation

$$a_0 y^{(N)} + a_1 y^{(N-1)} + \cdots + a_{N-1} y' + a_N y = 0 \quad .$$

Then set

$$y = y_1 u_1 + y_2 u_2 + \cdots + y_N u_N \tag{24.11}$$

assuming that the u_k's are unknown functions whose derivatives satisfy the system

$$y_1 u_1' + y_2 u_2' + \cdots + y_N u_N' = 0 \tag{24.12a}$$

$$y_1' u_1' + y_2' u_2' + \cdots + y_N' u_N' = 0 \tag{24.12b}$$

$$y_1'' u_1' + y_2'' u_2' + \cdots + y_N'' u_N' = 0 \tag{24.12c}$$

$$\vdots$$

$$y_1^{(N-2)} u_1' + y_2^{(N-2)} u_2' + \cdots + y_N^{(N-2)} u_N' = 0 \tag{24.12d}$$

$$y_1^{(N-1)} u_1' + y_2^{(N-1)} u_2' + \cdots + y_N^{(N-1)} u_N' = \frac{g}{a_0} \tag{24.12e}$$

Solve the system for u_1', u_2', … and u_N'; integrate to get the formula for each u_k, and then plug the results back into formula (24.11) for y. That formula is the solution to the original nonhomogeneous differential equation.

As with the second-order case, the above system can be shown to be nondegenerate, and the resulting formula for each u_k' can be shown to be integrable, at least in some theoretical sense, as long as g and the a_k's are continuous functions with a_0 never being zero on the interval of interest.

24.3 The Variation of Parameters Formula Second-Order Version with Indefinite Integrals

By solving system (24.5) for u' and v' using generic y_1 and y_2, integrating, and then plugging the result back into formula (24.4)

$$y = y_1 u + y_2 v \quad ,$$

you can show that the solution to

$$ay'' + by' + cy = g \tag{24.13}$$

is given by

$$y(x) = -y_1(x) \int \frac{y_2(x) f(x)}{W(x)} \, dx + y_2(x) \int \frac{y_1(x) f(x)}{W(x)} \, dx \tag{24.14}$$

where

$$f(x) = \frac{g(x)}{a(x)} \quad , \quad W(x) = y_1(x) {y_2}'(x) - {y_1}'(x) y_2(x)$$

and $\{y_1, y_2\}$ is any fundamental set of solutions to the corresponding homogeneous equation. The details will be left as an exercise (Exercise 24.5 on page 445).

A few observations should be made about the elements in formula 24.14:

1. Back in Section 14.1, we saw that solutions to second-order linear differential equations are continuous and have continuous derivatives, at least over intervals on which a, b and c are continuous functions and a is never zero. Consequently, the above $W(x)$ will be a continuous function on any such interval.

2. Moreover, if you recall the discussion about the "Wronskian" corresponding to the function set $\{y_1, y_2\}$ (see Section 15.6), then you may have noticed that the $W(x)$ in the above formula is that very Wronskian. As noted in Theorem 15.6 on page 292, $W(x)$ will be nonzero at every point in our interval of interest, provided a, b and c are continuous functions and a is never zero on that interval.

Consequently, the integrands of the integrals in formula (24.14) will (theoretically at least) be nice integrable functions over our interval of interest as long as a, b, c and g are continuous functions and a is never zero over this interval.[2] And this verifies that, in theory, the variation of parameters method does always yield the solution to a nonhomogeneous linear second-order differential equation over appropriate intervals.

In practice, this author discourages the use of formula (24.14), at least at first. For most, trying to memorize and effectively use this formula is more difficult than remembering the basic system from which it was derived. And the small savings in computational time gained by using this formula is hardly worth the effort unless you are going to be solving many of equations of the form

$$ay'' + by' + cy = g$$

in which the left side remains the same, but you have several different choices for g.

[2] In fact, f does not have to even be continuous. It just cannot have particularly bad discontinuities.

Second-Order Version with Definite Integrals

If you fix a point x_0 in the interval of interest and rederive formula (24.14) using definite integrals instead of the indefinite ones used just above, you get that a particular solution to

$$ay'' + by' + cy = g$$

is given by

$$y_p(x) = -y_1(x) \int_{x_0}^{x} \frac{y_2(s)f(s)}{W(s)}\, ds + y_2(x) \int_{x_0}^{x} \frac{y_1(s)f(s)}{W(s)}\, ds \tag{24.15}$$

where, again,

$$f(s) = \frac{g(s)}{a(s)} \quad , \quad W(s) = y_1(s)y_2{}'(s) - y_1{}'(s)y_2(s)$$

and $\{y_1, y_2\}$ is any fundamental set of solutions to the corresponding homogeneous equation.

Then, of course,

$$y(x) = y_p(x) + c_1 y(x) + c_2(x) \tag{24.16}$$

is the corresponding general solution to the original nonhomogeneous differential equation.

There are two practical advantages to using definite integral formula (24.15) instead of the corresponding indefinite integral formula, formula (24.14):

1. Often, it is virtually impossible to find usable, explicit formulas for the integrals of

$$\frac{y_2(x)f(x)}{W(x)} \quad \text{and} \quad \frac{y_1(x)f(x)}{W(x)} \quad .$$

In these cases, formula (24.14), with its impossible-to-compute indefinite integrals, is of very little practical value. However, the definite integrals in formula (24.15) can still be accurately approximated for specific values of x using any decent numerical integration method. Thus, while we may not be able to obtain a nice formula for $y_p(x)$, we can still evaluate it at desired points on any reasonable interval of interest, possibly using these values to generate a table for $y_p(x)$ or to sketch its graph.

2. As you can easily show (Exercise 24.6), the y_p given by formula (24.15) satisfies the initial conditions

$$y(x_0) = 0 \quad \text{and} \quad y'(x_0) = 0 \quad .$$

This makes it a little easier to find the constants c_1 and c_2 such that

$$y(x) = y_p(x) + c_1 y(x) + c_2(x)$$

satisfies initial conditions

$$y(x_0) = A \quad \text{and} \quad y'(x_0) = B$$

for some values A and B (especially, if we cannot explicitly compute the integrals in formulas (24.14) and (24.15)).

For Arbitrary Orders

In using variation of parameters to solve the more general N^{th}-order nonhomogeneous differential equation

$$a_0 y^{(N)} + a_1 y^{(N-1)} + \cdots + a_{N-1} y' + a_N y = g \quad ,$$

we need to solve system (24.12) for $u_1{}'$, $u_2{}'$, ... and $u_N{}'$. If you carefully solve this system for arbitrary y_k's, or simply apply the procedure known as "Cramer's rule" (see almost any introductory text in linear algebra), you will discover that

$$u_k{}'(x) = (-1)^{N+k} \frac{W_k(x) f(x)}{W(x)} \qquad \text{for} \quad k = 1, 2, 3, \ldots, n$$

where $f = {}^g/_{a_0}$, W is the determinant of the matrix

$$\mathbf{M} = \begin{bmatrix} y_1 & y_2 & y_3 & \cdots & y_N \\ y_1{}' & y_2{}' & y_3{}' & \cdots & y_N{}' \\ y_1{}'' & y_2{}'' & y_3{}'' & \cdots & y_N{}'' \\ \vdots & \vdots & \vdots & \ddots & \vdots \\ y_1{}^{(N-1)} & y_2{}^{(N-1)} & y_3{}^{(N-1)} & \cdots & y_N{}^{(N-1)} \end{bmatrix}$$

and W_k is the determinant of the submatrix of $\mathbf{M}$ obtained by deleting the last row and k^{th} column.

Integrating and plugging into formula (24.11) for the differential equation's solution y, we obtain either

$$y(x) = \sum_{k=1}^{n} (-1)^{N+k} y_k(x) \int \frac{W_k(x) f(x)}{W(x)}\, dx \tag{24.17a}$$

or

$$y_p(x) = \sum_{k=1}^{n} (-1)^{N+k} y_k(x) \int_{x_0}^{x} \frac{W_k(s) f(s)}{W(s)}\, ds \tag{24.17b}$$

depending on whether indefinite or definite integrals are used.

Again, it should be noted that the W in these formulas is the Wronskian of the fundamental set $\{y_1, y_2, \ldots, y_N\}$, and, from the corresponding theory developed for these Wronskians (see, in particular, Theorem 15.6 on page 292), it follows that the above integrands will (theoretically at least) be nice integrable functions over our interval of interest as long as g and the a_k's are continuous functions and a_0 is never zero over this interval.[3]

Additional Exercises

24.1. *Find the general solution to each of the following nonhomogeneous differential equations. Use variation of parameters even if another method might seem easier. For your convenience, each equation is accompanied by a general solution to the corresponding homogeneous equation.*

a. $x^2 y'' - 2xy' + 2y = 3\sqrt{x} \quad , \quad y_h = c_1 x + c_2 x^2$

b. $y'' + y = \cot(x) \quad , \quad y_h = c_1 \cos(x) + c_2 \sin(x)$

c. $y'' + 4y = \csc(2x) \quad , \quad y_h = c_1 \cos(2x) + c_2 \sin(2x)$

d. $y'' - 7y' + 10y = 6e^{3x} \quad , \quad y_h = c_1 e^{2x} + c_2 e^{5x}$

[3] And, again, g does not have to even be continuous. It just cannot have particularly bad discontinuities.

e. $y'' - 4y' + 4y = \left[24x^2 + 2\right]e^{2x} \quad , \quad y_h = c_1 e^{2x} + c_2 x e^{2x}$

f. $y'' + 4y' + 4y = \dfrac{e^{-2x}}{1+x^2} \quad , \quad y_h = c_1 e^{-2x} + c_2 x e^{-2x}$

g. $x^2 y'' + xy' - y = \sqrt{x} \quad , \quad y_h = c_1 x + c_2 x^{-1}$

h. $x^2 y'' + xy' - 9y = 12x^3 \quad , \quad y_h = c_1 x^{-3} + c_2 x^3$

i. $x^2 y'' - 3xy' + 4y = x^2 \quad , \quad y_h = c_1 x^2 + c_2 x^2 \ln|x|$

j. $x^2 y'' + 5xy' + 4y = \ln|x| \quad , \quad y_h = c_1 x^{-2} + c_2 x^{-2} \ln|x|$

k. $x^2 y'' - 2y = \dfrac{1}{x-2} \quad , \quad y_h = c_1 x^2 + c_2 x^{-1}$

l. $xy'' - y' - 4x^3 y = x^3 e^{x^2} \quad , \quad y_h = c_1 e^{x^2} + c_2 e^{-x^2}$

m. $xy'' + (2+2x)y' + 2y = 8e^{2x} \quad , \quad y_h = c_1 x^{-1} + c_2 x^{-1} e^{-2x}$

n. $(x+1)y'' + xy' - y = (x+1)^2 \quad , \quad y_h = c_1 x + c_2 e^{-x}$

24.2. *Solve the following initial-value problems using variation of parameters to find the general solution to the given differential equations.*

a. $x^2 y'' - 2xy' - 4y = \dfrac{10}{x}$ *with* $y(1) = 3$ *and* $y'(1) = -15$

b. $y'' - y' - 6y = 12e^{2x}$ *with* $y(0) = 0$ *and* $y'(0) = 8$

24.3. *Find the general solution to each of the following nonhomogeneous differential equations. Use variation of parameters even if another method might seem easier. For your convenience, each equation is accompanied by a general solution to the corresponding homogeneous equation.*

a. $y''' - 4y' = 30e^{3x} \quad , \quad y_h = c_1 + c_2 e^{2x} + c_3 e^{-2x}$

b. $x^3 y''' - 3x^2 y'' + 6xy' - 6y = x^3 \quad , \quad y_h = c_1 x + c_2 x^2 + c_3 x^3$

24.4. *For each of the following, set up the system of equations (corresponding to system 24.5, 24.10 or 24.12) arising in solving the equations via variation of parameters.*

a. $x^3 y''' - 3x^2 y'' + 6xy' - 6y = e^{-x^2} \quad , \quad y_h = c_1 x + c_2 x^2 + c_3 x^3$

b. $y''' - y'' + y' - y = \tan(x) \quad , \quad y_h = c_1 e^x + c_2 \cos(x) + c_3 \sin(x)$

c. $y^{(4)} - 81y = \sinh(x) \quad , \quad y_h = c_1 e^{3x} + c_2 e^{-3x} + c_3 \cos(3x) + c_4 \sin(3x)$

d. $x^4 y^{(4)} + 6x^3 y''' - 3x^2 y'' - 9xy' + 9y = 12x \sin\left(x^2\right) \quad ,$

$$y_h = c_1 x + c_2 x^{-1} + c_3 x^3 + c_4 x^{-3}$$

24.5. *Derive integral formula (24.14) on page 442 for the solution* y *to*

$$ay'' + by' + cy = g$$

from the fact that $y = y_1 u + y_2 v$ *where*

$$y_1 u' + y_2 v' = 0 \qquad \text{and} \qquad y_1' u' + y_2' v' = \frac{g}{a} \quad .$$

(Hint: Start by solving the system for u' *and* v' *.)*

24.6. *Show that y_p given by formula (24.15) on page 443 satisfies the initial conditions*

$$y(x_0) = 0 \qquad \text{and} \qquad y'(x_0) = 0 \quad .$$

25

Review Exercises for Part III

Find the general solution — expressed in terms of real-valued functions only — to each of the following differential equations. In each case, the prime denotes differentiation with respect to x, and you may assume $x > 0$ when the equation is an Euler equation.

1. $y'' + 36y = 0$

2. $y'' - 12y' + 36y = 0$

3. $x^2y'' + xy' - 9y = 0$

4. $y'' - 36y = 0$

5. $y'' - 9y' + 14y = 0$

6. $x^2y'' - 7xy' + 16y = 0$

7. $2xy'' + y' = \sqrt{x}$

8. $y^{(4)} - 8y'' + 16y = 0$

9. $y'' + 6y' + 9y = 0$

10. $y'' + 3y = 0$

11. $x^2y'' + 7xy' + 9y = 0$

12. $x^2y'' + \frac{5}{2}y = 0$

13. $y^{(5)} - 6y^{(4)} + 13y^{(3)} = 0$

14. $x^2y'' - 6y = 0$

15. $y'' - 6y' + 25y = 0$

16. $y'' = (y')^2$

17. $x^2y'' + xy' + 9y = 0$

18. $y'' - 8y' + 25y = 0$

19. $x^2y'' + 2xy' - 30y = 0$

20. $y'' + y' - 30y = 0$

21. $16y'' - 8y' + y = 0$

22. $4x^2y'' + 8xy' + y = 0$

23. $y''' - 6y'' + 12y' - 8 = 0$

24. $2x^2y'' - 3xy' + 2y = 0$

25. $9x^2y'' + 3xy' + y = 0$

26. $y^{(4)} - 16y = 0$

27. $2y'' - 7y' + 3 = 0$

28. $y'' + 20y' + 100y = 0$

29. $xy'' = 3y'$

30. $y'' - 5y' = 0$

31. $y'' - 9y' + 14y = 98x^2$

32. $y'' - 12y' + 36y = 25\sin(3x)$

33. $y'' - 9y' + 14y = 576x^2e^{-x}$

34. $y'' - 12y' + 36y = 81e^{3x}$

35. $x^2y'' + xy' - 9y = 3\sqrt{x}$

36. $y'' - 12y' + 36y = 3xe^{6x} - 2e^{6x}$

37. $y'' + 36y = 6\sec(6x)$

38. $x^2y'' + 2xy' - 6y = 18\ln|x|$

39. $y'' + 6y' + 9y = 10e^{-3x}$

40. $2x^2y'' - xy' - 2y = 10x^2$

41. $y'' + 6y' + 9y = 2\cos(2x)$

42. $xy'' - y' = -3x(y')^3$

43. $x^2y'' + 3xy' + 2y = 6$

44. $x^2y'' + xy' - y = \dfrac{1}{x^2+1}$

45. $4y'' - 12y' + 9y = xe^{3x/2}$

46. $3y'' + 8y' - 3y = 123x\sin(3x)$

47. $y''' + 8y = e^{-2x}$

48. $y^{(6)} - 64y = e^{-2x}$

49. $x^2y'' + 3xy' + y = \dfrac{1}{(x+1)^2}$

50. $x^2y'' + 3xy' + y = \dfrac{1}{x}$

Part IV

The Laplace Transform

26

The Laplace Transform (Intro)

The Laplace transform is a mathematical tool based on integration and with many applications. In particular, it can simplify the solving of many differential equations. We will find it particularly useful when dealing with nonhomogeneous equations in which the forcing functions are not continuous. This makes it a valuable tool for engineers and scientists dealing with "real-world" applications.

By the way, the Laplace transform is just one of many "integral transforms" in general use. Conceptually and computationally, it is probably the simplest. If you understand the Laplace transform, then you will find it much easier to pick up the other transforms as needed.

26.1 Basic Definition and Examples

Definition, Notation and Other Basics

Let f be a 'suitable' function (more on that later). The *Laplace transform of* f, denoted by either F or $\mathcal{L}[f]$, is the function given by

$$F(s) = \mathcal{L}[f]|_s = \int_0^\infty f(t)e^{-st}\,dt \quad . \tag{26.1}$$

!►Example 26.1: *For our first example, let us use*

$$f(t) = \begin{cases} 1 & \text{if } t \leq 2 \\ 0 & \text{if } 2 < t \end{cases} \quad .$$

This is the relatively simple discontinuous function graphed in Figure 26.1a. To compute the Laplace transform of this function, we need to break the integral into two parts:

$$\begin{aligned} F(s) = \mathcal{L}[f]|_s &= \int_0^\infty f(t)e^{-st}\,dt \\ &= \int_0^2 \underbrace{f(t)}_{1} e^{-st}\,dt + \int_2^\infty \underbrace{f(t)}_{0} e^{-st}\,dt \\ &= \int_0^2 e^{-st}\,dt + \int_2^\infty 0\,dt = \int_0^2 e^{-st}\,dt \quad . \end{aligned}$$

So, if $s \neq 0$,

$$F(s) = \int_0^2 e^{-st}\,dt = \left.\frac{e^{-st}}{-s}\right|_{t=0}^{2} = -\frac{1}{s}\left[e^{-s\cdot 2} - e^{-s\cdot 0}\right] = \frac{1}{s}\left[1 - e^{-2s}\right] \quad .$$

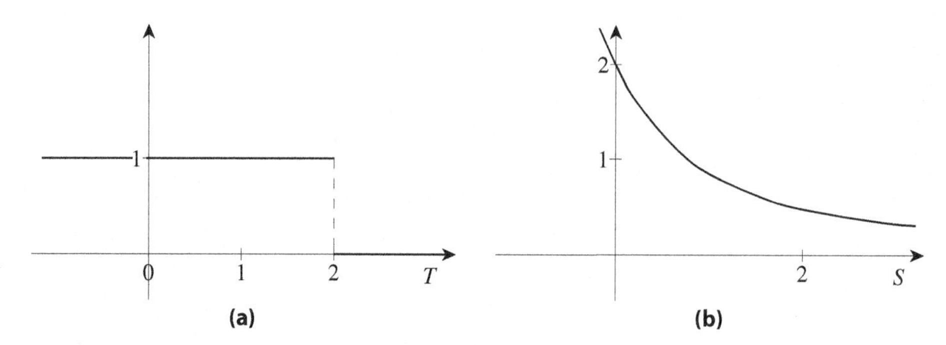

Figure 26.1: The graph of **(a)** the discontinuous function $f(t)$ from Example 26.1 and **(b)** its Laplace transform $F(s)$.

And if $s = 0$,

$$F(s) = F(0) = \int_0^2 e^{-0 \cdot t}\, dt = \int_0^2 1\, dt = 2 \quad .$$

This is the function sketched in Figure 26.1b. (Using L'Hôpital's rule, you can easily show that $F(s) \to F(0)$ *as* $s \to 0$. *So, despite our need to compute* $F(s)$ *separately when* $s = 0$, F *is a continuous function.)*

As the example just illustrated, we really are 'transforming' the function $f(t)$ into another function $F(s)$. This process of transforming $f(t)$ to $F(s)$ is also called the *Laplace transform* and, unsurprisingly, is denoted by $\mathcal{L}$. Thus, when we say "the Laplace transform", we can be referring to either the transformed function $F(s)$ or to the process of computing $F(s)$ from $f(t)$.

Some other quick notes:

1. There are standard notational conventions that simplify bookkeeping. The functions 'to be transformed' are (almost) always denoted by lowercase Roman letters — f, g, h, etc. — and t is (almost) always used as the variable in the formulas for these functions (because, in applications, these are typically functions of time). The corresponding 'transformed functions' are (almost) always denoted by the corresponding uppercase Roman letters — F, G, H, ETC. — and s is (almost) always used as the variable in the formulas for these functions.

   Thus, if we happen to refer to functions $f(t)$ and $F(s)$, it is a good bet that $F = \mathcal{L}[f]$.

2. Observe that, in the integral for the Laplace transform, we are integrating the inputted function $f(t)$ multiplied by the exponential e^{-st} over the positive T–axis. Because of the sign in the exponential, this exponential is a *rapidly decreasing* function of t when $s > 0$ and is a *rapidly increasing* function of t when $s < 0$. This will help determine both the sort of functions that are 'suitable' for the Laplace transform and the domains of the transformed functions.

3. It is also worth noting that, because the lower limit in the integral for the Laplace transform is $t = 0$, the formula for $f(t)$ when $t < 0$ is completely irrelevant. In fact, $f(t)$ need not even be defined for $t < 0$. For this reason, some authors explicitly limit the values for t to being nonnegative. We won't do this explicitly, but do keep in mind that the Laplace transform of a function $f(t)$ is only based on the values/formula for $f(t)$ with $t \geq 0$. This will become a little more relevant when we discuss inverting the Laplace transform (in Chapter 28).

4. As indicated by our discussion so far, we are treating the s in

$$F(s) = \int_0^\infty f(t)e^{-st}\,dt$$

as a real variable; that is, we are assuming s denotes a relatively arbitrary real value. Be aware, however, that in more advanced developments, s is often treated as a complex variable, $s = \sigma + i\xi$. This allows the use of results from the theory of analytic complex functions. But we won't need that theory (a theory which few readers of this text are likely to have yet seen). So, in this text (with one very brief exception in Chapter 28), s will always be assumed to be real.

Transforms of Some Common Functions

Before we can make much use of the Laplace transform, we need to build a repertoire of common functions whose transforms we know. It would also be a good idea to compute a number of transforms simply to get a better grasp of this whole 'Laplace transform' idea.

So let's get started.

!►Example 26.2 (transforms of favorite constants): *Let f be the zero function, that is,*

$$f(t) = 0 \qquad \text{for all} \quad t \quad .$$

Then its Laplace transform is

$$F(s) = \mathcal{L}[0]|_s = \int_0^\infty 0 \cdot e^{-st}\,dt = \int_0^\infty 0\,dt = 0 \quad . \tag{26.2}$$

Now let h be the unit constant function, that is,

$$h(t) = 1 \qquad \text{for all} \quad t \quad .$$

Then

$$H(s) = \mathcal{L}[1]|_s = \int_0^\infty 1 \cdot e^{-st}\,dt = \int_0^\infty e^{-st}\,dt \quad .$$

What comes out of this integral depends strongly on whether s is positive or not. If $s < 0$, then $0 < -s = |s|$ and

$$\int_0^\infty e^{-st}\,dt = \int_0^\infty e^{|s|t}\,dt = \frac{1}{|s|}e^{|s|t}\bigg|_{t=0}^\infty$$
$$= \lim_{t\to\infty} \frac{1}{|s|}e^{|s|t} - \frac{1}{|s|}e^{|s|\cdot 0} = \infty - \frac{1}{|s|} = \infty \quad .$$

If $s = 0$, then

$$\int_0^\infty e^{-st}\,dt = \int_0^\infty e^{0\cdot t}\,dt = \int_0^\infty 1\,dt = t\big|_{t=0}^\infty = \infty \quad .$$

Finally, if $s > 0$, then

$$\int_0^\infty e^{-st}\,dt = \frac{1}{-s}e^{-st}\bigg|_{t=0}^\infty = \lim_{t\to\infty} \frac{1}{-s}e^{-st} - \frac{1}{-s}e^{-s\cdot 0} = 0 + \frac{1}{s} = \frac{1}{s} \quad .$$

So,

$$\mathcal{L}[1]|_s = \int_0^\infty 1 \cdot e^{-st}\,dt = \begin{cases} \dfrac{1}{s} & \text{if} \quad 0 < s \\ \infty & \text{if} \quad s \le 0 \end{cases} \quad . \tag{26.3}$$

As illustrated in the last example, a Laplace transform $F(s)$ is often a well-defined (finite) function of s only when s is greater that some fixed number s_0 ($s_0 = 0$ in the example). This is a result of the fact that the larger s is, the faster e^{-st} goes to zero as $t \to \infty$ (provided $s > 0$). In practice, we will only give the formulas for the transforms over the intervals where these formulas are well-defined and finite. Thus, in place of equation (26.3), we will write

$$\mathcal{L}[1]|_s = \frac{1}{s} \quad \text{for} \quad s > 0 \quad . \tag{26.4}$$

As we compute Laplace transforms, we will note such restrictions on the values of s. To be honest, however, these restrictions will usually not be that important in our work. What will be important is that there is some finite value s_0 such that our formulas are valid whenever $s > s_0$.

Keeping this in mind, let's go back to computing transforms.

!►Example 26.3 (transforms of some powers of t): *We want to find*

$$\mathcal{L}[t^n]\big|_s = \int_0^\infty t^n e^{-st}\,dt = \int_0^\infty t^n e^{-st}\,dt \qquad \text{for} \quad n = 1, 2, 3, \ldots \quad .$$

With a little thought, you will realize this integral will not be finite if $s \leq 0$. So we will assume $s > 0$ in these computations. This, of course, means that

$$\lim_{t\to\infty} e^{-st} = 0 \quad .$$

It also means that, using L'Hôpital's rule, you can easily verify that

$$\lim_{t\to\infty} t^n e^{-st} = 0 \qquad \text{for} \quad n \geq 0 \quad .$$

Keeping the above in mind, consider the case where $n = 1$,

$$\mathcal{L}[t]|_s = \int_0^\infty t e^{-st}\,dt \quad .$$

This integral just cries out to be integrated "by parts":

$$\begin{aligned}
\mathcal{L}[t]|_s &= \int_0^\infty \underbrace{t}_{u}\, \underbrace{e^{-st}\,dt}_{dv} \\
&= uv\Big|_{t=0}^{\infty} - \int_0^\infty v\,du \\
&= t\left(\frac{1}{-s}\right)e^{-st}\Big|_{t=0}^{\infty} - \int_0^\infty \left(\frac{1}{-s}\right)e^{-st}\,dt \\
&= -\frac{1}{s}\Big[\underbrace{\lim_{t\to\infty} te^{-st}}_{0} - \underbrace{0\cdot e^{-s\cdot 0}}_{0} - \int_0^\infty e^{-st}\,dt\Big] = \frac{1}{s}\int_0^\infty e^{-st}\,dt \quad .
\end{aligned}$$

Admittedly, this last integral is easy to compute, but why bother since we computed it in the previous example! In fact, it is worth noting that combining the last computations with the computations for $\mathcal{L}[1]$ yields

$$\mathcal{L}[t]|_s = \frac{1}{s}\int_0^\infty e^{-st}\,dt = \frac{1}{s}\int_0^\infty 1\cdot e^{-st}\,dt = \frac{1}{s}\mathcal{L}[1]|_s = \frac{1}{s}\left[\frac{1}{s}\right] \quad .$$

So,

$$\mathcal{L}[t]|_s = \frac{1}{s^2} \quad \text{for} \quad s > 0 \quad . \tag{26.5}$$

Now consider the case where $n = 2$. Again, we start with an integration by parts:

$$\begin{aligned}
\mathcal{L}\left[t^2\right]\Big|_s &= \int_0^\infty t^2 e^{-st}\,dt \\
&= \int_0^\infty \underbrace{t^2}_{u} \underbrace{e^{-st}\,dt}_{dv} \\
&= uv\Big|_{t=0}^\infty - \int_0^\infty v\,du \\
&= t^2\left(\frac{1}{-s}\right)e^{-st}\Big|_{t=0}^\infty - \int_0^\infty \left(\frac{1}{-s}\right)e^{-st}2t\,dt \\
&= \frac{1}{-s}\Big[\underbrace{\lim_{t\to\infty} t^2 e^{-st}}_{0} - \underbrace{0^2 e^{-s\cdot 0}}_{0} - 2\int_0^\infty te^{-st}\,dt\Big] = \frac{2}{s}\int_0^\infty te^{-st}\,dt \quad .
\end{aligned}$$

But remember,

$$\int_0^\infty te^{-st}\,dt = \mathcal{L}[t]|_s \quad .$$

Combining the above computations with this (and referring back to equation (26.5)), we end up with

$$\mathcal{L}\left[t^2\right]\Big|_s = \frac{2}{s}\int_0^\infty te^{-st}\,dt = \frac{2}{s}\mathcal{L}[t]|_s = \frac{2}{s}\left[\frac{1}{s^2}\right] = \frac{2}{s^3} \quad . \tag{26.6}$$

Clearly, a pattern is emerging. I'll leave the computation of $\mathcal{L}\left[t^3\right]$ to you.

?►Exercise 26.1: *Assuming $s > 0$, verify (using integration by parts) that*

$$\mathcal{L}\left[t^3\right]\Big|_s = \frac{3}{s}\mathcal{L}\left[t^2\right]\Big|_s \quad ,$$

and from that and the formula for $\mathcal{L}\left[t^2\right]$ computed above, conclude that

$$\mathcal{L}\left[t^3\right]\Big|_s = \frac{3\cdot 2}{s^4} = \frac{3!}{s^4} \quad .$$

?►Exercise 26.2: *More generally, use integration by parts to show that, whenever $s > 0$ and n is a positive integer,*

$$\mathcal{L}\left[t^n\right]\Big|_s = \frac{n}{s}\mathcal{L}\left[t^{n-1}\right]\Big|_s \quad .$$

Using the results from the last two exercises, we have, for $s > 0$,

$$\begin{aligned}
\mathcal{L}\left[t^4\right]\Big|_s &= \frac{4}{s}\mathcal{L}\left[t^3\right]\Big|_s = \frac{4}{s}\cdot\frac{3\cdot 2}{s^4} = \frac{4\cdot 3\cdot 2}{s^5} = \frac{4!}{s^5} \quad , \\
\mathcal{L}\left[t^5\right]\Big|_s &= \frac{5}{s}\mathcal{L}\left[t^4\right]\Big|_s = \frac{5}{s}\cdot\frac{4\cdot 3\cdot 2}{s^5} = \frac{5\cdot 4\cdot 3\cdot 2}{s^6} = \frac{5!}{s^6} \quad , \\
&\;\;\vdots
\end{aligned}$$

In general, for $s > 0$ and $n = 1, 2, 3, \ldots$,

$$\mathcal{L}[t^n]\big|_s = \frac{n!}{s^{n+1}} \quad . \tag{26.7}$$

(If you check, you'll see that it even holds for $n = 0$.)

It turns out that a formula very similar to (26.7) also holds when n is not an integer. Of course, there is then the issue of just what $n!$ means if n is not an integer. Since the discussion of that issue may distract our attention away from the main issue at hand — that of getting a basic understanding of what the Laplace transform is by computing transforms of simple functions — let us hold off on that discussion for a few pages.

Instead, let's compute the transforms of some exponentials:

!►Example 26.4 (transform of a real exponential): *Consider computing the Laplace transform of e^{3t},*

$$\mathcal{L}[e^{3t}]\big|_s = \int_0^\infty e^{3t}e^{-st}\,dt = \int_0^\infty e^{3t-st}\,dt = \int_0^\infty e^{-(s-3)t}\,dt \quad .$$

If $s - 3$ is not positive, then $e^{-(s-3)t}$ is not a decreasing function of t, and, hence, the above integral will not be finite. So we must require $s - 3$ to be positive (that is, $s > 3$). Assuming this, we can continue our computations

$$\begin{aligned}\mathcal{L}[e^{3t}]\big|_s &= \int_0^\infty e^{-(s-3)t}\,dt \\ &= \frac{-1}{s-3}e^{-(s-3)t}\bigg|_{t=0}^\infty \\ &= \frac{-1}{s-3}\left[\lim_{t\to\infty} e^{-(s-3)t} - e^{-(s-3)0}\right] = \frac{-1}{s-3}[0 - 1] \quad .\end{aligned}$$

So

$$\mathcal{L}[e^{3t}]\big|_s = \frac{1}{s-3} \quad \text{for} \quad 3 < s \quad .$$

Replacing 3 with any other real number is trivial.

?►Exercise 26.3 (transforms of real exponentials): *Let α be any real number and show that*

$$\mathcal{L}[e^{\alpha t}]\big|_s = \frac{1}{s-\alpha} \quad \text{for} \quad \alpha < s \quad . \tag{26.8}$$

Complex exponentials are also easily done:

!►Example 26.5 (transform of a complex exponential): *Computing the Laplace transform of e^{i3t} leads to*

$$\begin{aligned}\mathcal{L}[e^{i3t}]\big|_s &= \int_0^\infty e^{i3t}e^{-st}\,dt \\ &= \int_0^\infty e^{-(s-i3)t}\,dt \\ &= \frac{-1}{s-i3}e^{-(s-i3)t}\bigg|_{t=0}^\infty = \frac{-1}{s-i3}\left[\lim_{t\to\infty} e^{-(s-i3)t} - e^{-(s-i3)0}\right] \quad .\end{aligned}$$

Now,

$$e^{-(s-i3)0} = e^0 = 1$$

and

$$\lim_{t\to\infty} e^{-(s-i3)t} = \lim_{t\to\infty} e^{-st+i3t} = \lim_{t\to\infty} e^{-st}\big[\cos(3t) + i\sin(3t)\big] \quad .$$

Since sines and cosines oscillate between 1 *and* -1 *as* $t \to \infty$*, the last limit does not exist unless*

$$\lim_{t\to\infty} e^{-st} = 0 \quad ,$$

and this occurs if and only if $s > 0$*. In this case,*

$$\lim_{t\to\infty} e^{-(s-i3)t} = \lim_{t\to\infty} e^{-st}\big[\cos(3t) + i\sin(3t)\big] = 0 \quad .$$

Thus, when $s > 0$*,*

$$\mathcal{L}\big[e^{i3t}\big]\big|_s = \frac{-1}{s-i3}\Big[\lim_{t\to\infty} e^{-(s-i3)t} - e^{-(s-i3)0}\Big] = \frac{-1}{s-i3}[0 - 1] = \frac{1}{s-i3} \quad .$$

Again, replacing 3 with any real number is trivial.

?►Exercise 26.4 (transforms of complex exponentials): *Let* α *be any real number and show that*

$$\mathcal{L}\big[e^{i\alpha t}\big]\big|_s = \frac{1}{s-i\alpha} \qquad \text{for} \quad 0 < s \quad . \tag{26.9}$$

26.2 Linearity and Some More Basic Transforms

Suppose we have already computed the Laplace transforms of two functions $f(t)$ and $g(t)$, and, thus, already know the formulas for

$$F(s) = \mathcal{L}[f]|_s \qquad \text{and} \qquad G(s) = \mathcal{L}[g]|_s \quad .$$

Now look at what happens if we compute the transform of any linear combination of f and g. Letting α and β be any two constants, we have

$$\begin{aligned}
\mathcal{L}[\alpha f(t) + \beta g(t)]|_s &= \int_0^\infty [\alpha f(t) + \beta g(t)]e^{-st}\,dt \\
&= \int_0^\infty \big[\alpha f(t)e^{-st} + \beta g(t)e^{-st}\big]\,dt \\
&= \alpha\int_0^\infty f(t)e^{-st}\,dt + \beta\int_0^\infty g(t)e^{-st}\,dt \\
&= \alpha\mathcal{L}[f(t)]|_s + \beta\mathcal{L}[g(t)]|_s = \alpha F(s) + \beta G(s) \quad .
\end{aligned}$$

Thus, the Laplace transform is a *linear transform*; that is, for any two constants α and β, and any two Laplace transformable functions f and g,

$$\mathcal{L}[\alpha f(t) + \beta g(t)] = \alpha\mathcal{L}[f] + \beta\mathcal{L}[g] \quad .$$

This fact will simplify many of our computations and is important enough to enshrine as a theorem. While we are at it, let's note that the above computations can be done with more functions than two, and that we, perhaps, should have noted the values of s for which the integrals are finite. Taking all that into account, we can prove:

Theorem 26.1 (linearity of the Laplace transform)
The Laplace transform is linear. That is,

$$\mathcal{L}[c_1 f_1(t) + c_2 f_2(t) + \cdots + c_n f_n(t)] = c_1\mathcal{L}[f_1(t)] + c_2\mathcal{L}[f_2(t)] + \cdots + c_n\mathcal{L}[f_n(t)]$$

where each c_k is a constant and each f_k is a "Laplace transformable" function.

Moreover, if, for each f_k we have a value s_k such that

$$F_k(s) = \mathcal{L}[f_k(t)]|_s \quad \text{for} \quad s_k < s \quad ,$$

then, letting s_{max} be the largest of these s_k's,

$$\mathcal{L}[c_1 f_1(t) + c_2 f_2(t) + \cdots + c_n f_n(t)]|_s$$
$$= c_1 F_1(s) + c_2 F_2(s) + \cdots + c_n F_n(s) \quad \text{for} \quad s_{max} < s \quad .$$

!► Example 26.6 (transform of the sine function): *Let us consider finding the Laplace transform of $\sin(\omega t)$ for any real value ω. There are several ways to compute this, but the easiest starts with using Euler's formula for the sine function along with the linearity of the Laplace transform:*

$$\mathcal{L}[\sin(\omega t)]|_s = \mathcal{L}\left[\frac{e^{i\omega t} - e^{-i\omega t}}{2i}\right]\Big|_s$$
$$= \frac{1}{2i}\mathcal{L}\left[e^{i\omega t} - e^{-i\omega t}\right]\Big|_s = \frac{1}{2i}\left[\mathcal{L}\left[e^{i\omega t}\right]\Big|_s - \mathcal{L}\left[e^{-i\omega t}\right]\Big|_s\right] \quad .$$

From Example 26.5 and Exercise 26.4, we know

$$\mathcal{L}\left[e^{i\omega t}\right]\Big|_s = \frac{1}{s - i\omega} \quad \text{for} \quad s > 0 \quad .$$

Thus, also,

$$\mathcal{L}\left[e^{-i\omega t}\right]\Big|_s = \mathcal{L}\left[e^{i(-\omega)t}\right]\Big|_s = \frac{1}{s - i(-\omega)} = \frac{1}{s + i\omega} \quad \text{for} \quad s > 0 \quad .$$

Plugging these into the computations for $\mathcal{L}[\sin(\omega t)]$ (and doing a little algebra) yields, for $s > 0$,

$$\mathcal{L}[\sin(\omega t)]|_s = \frac{1}{2i}\left[\mathcal{L}\left[e^{i\omega t}\right]\Big|_s - \mathcal{L}\left[e^{-i\omega t}\right]\Big|_s\right]$$
$$= \frac{1}{2i}\left[\frac{1}{s - i\omega} - \frac{1}{s + i\omega}\right]$$
$$= \frac{1}{2i}\left[\frac{s + i\omega}{(s - i\omega)(s + i\omega)} - \frac{s - i\omega}{(s + i\omega)(s - i\omega)}\right]$$
$$= \frac{1}{2i}\left[\frac{(s + i\omega) - (s - i\omega)}{s^2 - i^2\omega^2}\right]$$
$$= \frac{1}{2i}\left[\frac{2i\omega}{s^2 - i^2\omega^2}\right] \quad ,$$

which immediately simplifies to

$$\mathcal{L}[\sin(\omega t)]|_s = \frac{\omega}{s^2 + \omega^2} \quad \text{for} \quad s > 0 \quad . \tag{26.10}$$

?►Exercise 26.5 (transform of the cosine function): *Show that, for any real value* ω ,

$$\mathcal{L}[\cos(\omega t)]|_s = \frac{s}{s^2 + \omega^2} \quad \text{for} \quad s > 0 \quad . \tag{26.11}$$

26.3 Tables and a Few More Transforms

In practice, those using the Laplace transform in applications do not constantly recompute basic transforms. Instead, they refer to tables of transforms (or use software) to look up commonly used transforms, just as so many people use tables of integrals (or software) when computing integrals. We, too, can use tables (or software) *after*

1. you have computed enough transforms on your own to understand the basic principles,

and

2. we have computed the transforms appearing in the table so we know our table is correct.

The table we will use is Table 26.1, *Laplace Transforms of Common Functions (Version 1)*, on page 460. Checking that table, we see that we have already verified all but two or three of the entries, with those being the transforms of fairly arbitrary powers of t , t^α , and the "shifted step function", $\text{step}(t - \alpha)$. So let's compute them now.

Arbitrary Powers (and the Gamma Function)

Earlier, we saw that

$$\mathcal{L}[t^n]\big|_s = \int_0^\infty t^n e^{-st}\,dt = \frac{n!}{s^{n+1}} \quad \text{for} \quad s > 0 \tag{26.12}$$

when n is any nonnegative integer. Let us now consider computing

$$\mathcal{L}[t^\alpha]\big|_s = \int_0^\infty t^\alpha e^{-st}\,dt \quad \text{for} \quad s > 0$$

when α is any real number greater than -1 . (When $\alpha \leq -1$, you can show that t^α 'blows up' too quickly near $t = 0$ for the integral to be finite.)

The method we used to find $\mathcal{L}[t^n]$ becomes awkward when we try to apply it to find $\mathcal{L}[t^\alpha]$ when α is not an integer. Instead, we will 'cleverly' simplify the above integral for $\mathcal{L}[t^\alpha]$ by using the substitution $u = st$. Since t is the variable in the integral, this means

$$t = \frac{u}{s} \quad \text{and} \quad dt = \frac{1}{s}\,du \quad .$$

So, assuming $s > 0$ and $\alpha > -1$,

$$\begin{aligned}
\mathcal{L}[t^\alpha]\big|_s &= \int_0^\infty t^\alpha e^{-st}\,dt \\
&= \int_0^\infty \left(\frac{u}{s}\right)^\alpha e^{-u}\frac{1}{s}\,du \\
&= \int_0^\infty \frac{u^\alpha}{s^{\alpha+1}} e^{-u}\,du = \frac{1}{s^{\alpha+1}} \int_0^\infty u^\alpha e^{-u}\,du \quad .
\end{aligned}$$

Table 26.1: Laplace Transforms of Common Functions (Version 1)

In the following, α and ω are real-valued constants, and, unless otherwise noted, $s > 0$.

$f(t)$	$F(s) = \mathcal{L}[f(t)]\|_s$	Restrictions
1	$\dfrac{1}{s}$	
t	$\dfrac{1}{s^2}$	
t^n	$\dfrac{n!}{s^{n+1}}$	$n = 1, 2, 3, \ldots$
$\dfrac{1}{\sqrt{t}}$	$\dfrac{\sqrt{\pi}}{\sqrt{s}}$	
t^α	$\dfrac{\Gamma(\alpha+1)}{s^{\alpha+1}}$	$-1 < \alpha$
$e^{\alpha t}$	$\dfrac{1}{s-\alpha}$	$\alpha < s$
$e^{i\alpha t}$	$\dfrac{1}{s-i\alpha}$	
$\cos(\omega t)$	$\dfrac{s}{s^2+\omega^2}$	
$\sin(\omega t)$	$\dfrac{\omega}{s^2+\omega^2}$	
$\text{step}_\alpha(t)$, $\text{step}(t-\alpha)$	$\dfrac{e^{-\alpha s}}{s}$	$0 \leq \alpha$

Notice that the last integral depends only on the constant α — we've 'factored out' any dependence on the variable s. Thus, we can treat this integral as a constant (for each value of α) and write

$$\mathcal{L}[t^\alpha]\big|_s = \frac{C_\alpha}{s^{\alpha+1}} \quad \text{where} \quad C_\alpha = \int_0^\infty u^\alpha e^{-u}\,du \quad .$$

It just so happens that the above formula for C_α is very similar to the formula for something called the "Gamma function". This is a function that crops up in various applications (such as this) and, for $x > 0$, is given by

$$\Gamma(x) = \int_0^\infty u^{x-1} e^{-u}\,du \quad . \tag{26.13}$$

Comparing this with the formula for C_α, we see that

$$C_\alpha = \int_0^\infty u^\alpha e^{-u}\,du = \int_0^\infty u^{(\alpha+1)-1} e^{-u}\,du = \Gamma(\alpha+1) \quad .$$

So our formula for the Laplace transform of t^α (with $\alpha > -1$) can be written as

$$\mathcal{L}[t^\alpha]\big|_s = \frac{\Gamma(\alpha+1)}{s^{\alpha+1}} \quad \text{for} \quad s > 0 \quad . \tag{26.14}$$

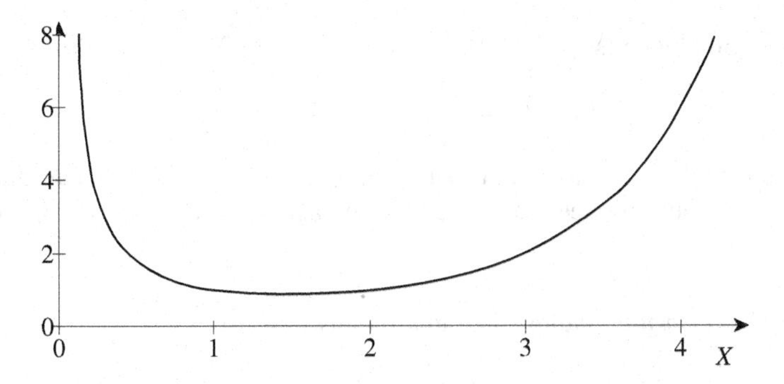

Figure 26.2: The graph of the Gamma function over the interval $(0,4)$. As $x \to 0^+$ or $x \to +\infty$, $\Gamma(x) \to +\infty$ very rapidly.

This is normally considered the preferred way to express $\mathcal{L}[t^\alpha]$ because the Gamma function is considered to be a "well-known" function. Perhaps you don't yet consider it "well known", but you can find tables for evaluating $\Gamma(x)$, and it is probably one of the functions already defined in your favorite computer math package. That makes graphing $\Gamma(x)$, as done in Figure 26.2, relatively easy.

As it is, we can readily determine the value of $\Gamma(x)$ when x is a positive integer by comparing our two formulas for $\mathcal{L}[t^n]$ when n is a nonnegative integer — the one mentioned at the start of our discussion (formula (26.12)), and the more general formula (formula (26.14)) just derived for $\mathcal{L}[t^\alpha]$ with $\alpha = n$:

$$\frac{n!}{s^{n+1}} = \mathcal{L}\big[t^n\big]\Big|_s = \frac{\Gamma(n+1)}{s^{n+1}} \qquad \text{when} \quad n = 0, 1, 2, 3, 4, \dots \quad .$$

Thus,

$$\Gamma(n+1) = n! \qquad \text{when} \quad n = 0, 1, 2, 3, 4, \dots \quad .$$

Letting $x = n+1$, this becomes

$$\Gamma(x) = (x-1)! \qquad \text{when} \quad x = 1, 2, 3, 4, \dots \quad . \tag{26.15}$$

In particular:

$$\Gamma(1) = (1-1)! = 0! = 1 \quad ,$$

$$\Gamma(2) = (2-1)! = 1! = 1 \quad ,$$

$$\Gamma(3) = (3-1)! = 2! = 2 \quad ,$$

$$\Gamma(4) = (4-1)! = 3! = 6 \quad ,$$

and

$$\Gamma(12) = (12-1)! = 11! = 39{,}916{,}800 \quad .$$

This shows that the Gamma function can be viewed as a generalization of the factorial. Indeed, you will find texts where the factorial is redefined for all positive numbers (not just integers) by

$$x! = \Gamma(x+1) \quad .$$

We won't do that.

Computing $\Gamma(x)$ when x is not an integer is not so simple. It can be shown that

$$\Gamma\left(\tfrac{1}{2}\right) = \sqrt{\pi} \quad . \tag{26.16}$$

Also, using integration by parts (just as you did in Exercise 26.2 on page 455), you can show that

$$\Gamma(x+1) = x\Gamma(x) \qquad \text{for} \quad x > 0 \quad , \tag{26.17}$$

which is analogous to the factorial identity $(n+1)! = (n+1)n!$. We will leave the verification of these to the more adventurous (see Exercise 26.13 on page 478), and go on to the computation of a few more transforms.

!►Example 26.7: *Consider finding the Laplace transforms of*

$$\frac{1}{\sqrt{t}} \quad , \quad \sqrt{t} \quad \text{and} \quad \sqrt[3]{t} \quad .$$

For the first, we use formulas (26.14) with $\alpha = -1/2$, *along with equation (26.16):*

$$\mathcal{L}\left[\frac{1}{\sqrt{t}}\right]\bigg|_s = \mathcal{L}\left[t^{-1/2}\right]\Big|_s = \frac{\Gamma\left(-\frac{1}{2}+1\right)}{s^{-1/2+1}} = \frac{\Gamma\left(\frac{1}{2}\right)}{s^{1/2}} = \frac{\sqrt{\pi}}{\sqrt{s}} \quad .$$

For the second, formula (26.14) with $\alpha = 1/2$ *gives*

$$\mathcal{L}\left[\sqrt{t}\right]\Big|_s = \mathcal{L}\left[t^{1/2}\right]\Big|_s = \frac{\Gamma\left(\frac{1}{2}+1\right)}{s^{1/2+1}} = \frac{\Gamma\left(\frac{3}{2}\right)}{s^{3/2}} \quad .$$

Using formulas (26.17) and (26.16), we see that

$$\Gamma\left(\frac{3}{2}\right) = \Gamma\left(\frac{1}{2}+1\right) = \frac{1}{2}\Gamma\left(\frac{1}{2}\right) = \frac{1}{2}\sqrt{\pi} \quad .$$

Thus

$$\mathcal{L}\left[\sqrt{t}\right]\Big|_s = \mathcal{L}\left[t^{1/2}\right]\Big|_s = \frac{\Gamma\left(\frac{3}{2}\right)}{s^{3/2}} = \frac{\sqrt{\pi}}{2s^{3/2}} \quad .$$

For the transform of $\sqrt[3]{t}$, *we simply have*

$$\mathcal{L}\left[\sqrt[3]{t}\right]\Big|_s = \mathcal{L}\left[t^{1/3}\right]\Big|_s = \frac{\Gamma\left(\frac{1}{3}+1\right)}{s^{1/3+1}} = \frac{\Gamma\left(\frac{4}{3}\right)}{s^{4/3}} \quad .$$

Unfortunately, there is not a formula analogous to (26.16) for $\Gamma\left(4/3\right)$ *or* $\Gamma\left(1/3\right)$. *There is the approximation*

$$\Gamma\left(\frac{4}{3}\right) \approx .8929795121 \quad ,$$

which can be found using either tables or a computer math package, but, since this is just an approximation, we might as well leave our answer as

$$\mathcal{L}\left[\sqrt[3]{t}\right]\Big|_s = \mathcal{L}\left[t^{1/3}\right]\Big|_s = \frac{\Gamma\left(\frac{4}{3}\right)}{s^{4/3}} \quad .$$

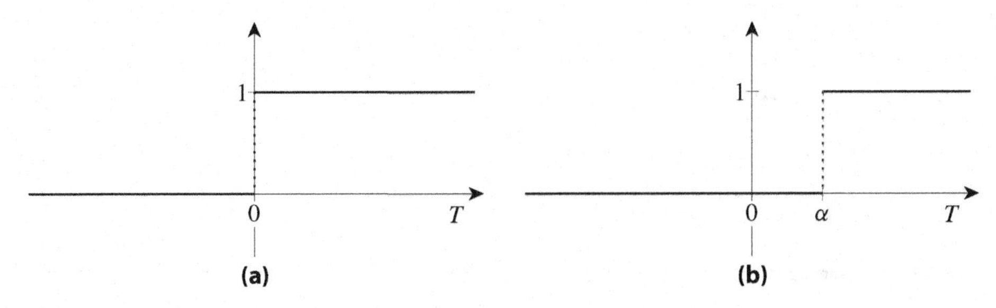

Figure 26.3: The graphs of **(a)** the basic step function $\text{step}(t)$ and **(b)** a shifted step function $\text{step}_\alpha(t)$ with $\alpha > 0$.

The Shifted Unit Step Function

Step functions are the simplest discontinuous functions we can have. The *(basic) unit step function,* which we will denote by $\text{step}(t)$, is defined by

$$\text{step}(t) = \begin{cases} 0 & \text{if } t < 0 \\ 1 & \text{if } 0 < t \end{cases} .$$

Its graph has been sketched in Figure 26.3a.[1]

For any real value α, the corresponding *shifted unit step function,* which we will denote by step_α, is given by

$$\text{step}_\alpha(t) = \text{step}(t-\alpha) = \begin{Bmatrix} 0 & \text{if } t-\alpha < 0 \\ 1 & \text{if } 0 < t-\alpha \end{Bmatrix} = \begin{cases} 0 & \text{if } t < \alpha \\ 1 & \text{if } \alpha < t \end{cases} .$$

Its graph, with $\alpha > 0$, has been sketched in Figure 26.3b. Do observe that the basic step function and the step function at zero are the same, $\text{step}(t) = \text{step}_0(t)$.

You may have noted that we've not defined the step functions at their points of discontinuity ($t = 0$ for $\text{step}(t)$, and $t = \alpha$ for $\text{step}_\alpha(t)$). That is because the value of a step function right at its single discontinuity will be completely irrelevant in any of our computations or applications. Observe this fact as we compute the Laplace transform of $\text{step}_\alpha(t)$ when $\alpha \geq 0$:

$$\begin{aligned} \mathcal{L}[\text{step}_\alpha(t)]\big|_s &= \int_0^\infty \text{step}_\alpha(t) e^{-st}\, dt \\ &= \int_0^\alpha \text{step}_\alpha(t) e^{-st}\, dt + \int_\alpha^\infty \text{step}_\alpha(t) e^{-st}\, dt \\ &= \int_0^\alpha 0 \cdot e^{-st}\, dt + \int_\alpha^\infty 1 \cdot e^{-st}\, dt = \int_\alpha^\infty e^{-st}\, dt \quad . \end{aligned}$$

You can easily show that the above integral is infinite if $s < 0$ or $s = 0$. But if $s > 0$, then the above becomes

$$\begin{aligned} \mathcal{L}[\text{step}_\alpha(t)]\big|_s &= \int_\alpha^\infty e^{-st}\, dt = \frac{1}{-s} e^{-st}\bigg|_{t=\alpha}^\infty \\ &= \lim_{t\to\infty} \frac{1}{-s} e^{-st} - \frac{1}{-s} e^{-s\alpha} = 0 + \frac{1}{s} e^{-\alpha s} \quad . \end{aligned}$$

[1] The unit step function is also called the *Heaviside step function*, and, in other texts, is often denoted by u and, occasionally, by h or H.

Thus,

$$\mathcal{L}\big[\text{step}_\alpha(t)\big]\big|_s = \frac{1}{s}e^{-\alpha s} \qquad \text{for} \quad s > 0 \quad \text{and} \quad \alpha \geq 0 \quad . \tag{26.18}$$

26.4 The First Translation Identity (and More Transforms)

The linearity of the Laplace transform allows us to construct transforms from linear combinations of known transforms. Other identities allow us to construct new transforms from other formulas involving known transforms. One particularly useful identity is the "first translation identity" (also called the "translation along the S–axis identity" for reasons that will soon be obvious). The derivation of this identity starts with the observation that, in the expression

$$F(s) = \mathcal{L}[f(t)]|_s = \int_0^\infty f(t)e^{-st}\,dt \qquad \text{for} \quad s > s_0 \quad ,$$

the s is simply a placeholder. It can be replaced with any symbol, say, X, that does not involve the integration variable, t,

$$F(X) = \mathcal{L}[f(t)]|_X = \int_0^\infty f(t)e^{-Xt}\,dt \qquad \text{for} \quad X > s_0 \quad .$$

In particular, let $X = s - \alpha$ where α is any real constant. Using this for X in the above gives us

$$F(s-\alpha) = \mathcal{L}[f(t)]|_{s-\alpha} = \int_0^\infty f(t)e^{-(s-\alpha)t}\,dt \qquad \text{for} \quad s-\alpha > s_0 \quad .$$

But

$$s - \alpha > s_0 \quad \Longleftrightarrow \quad s > s_0 + \alpha$$

and

$$\int_0^\infty f(t)e^{-(s-\alpha)t}\,dt = \int_0^\infty f(t)e^{-st}e^{\alpha t}\,dt$$
$$= \int_0^\infty e^{\alpha t}f(t)e^{-st}\,dt = \mathcal{L}\big[e^{\alpha t}f(t)\big]\big|_s \quad .$$

So the expression above for $F(s-\alpha)$ can be written as

$$F(s-\alpha) = \mathcal{L}\big[e^{\alpha t}f(t)\big]\big|_s \qquad \text{for} \quad s > s_0 + \alpha \quad .$$

This gives us the following identity:

Theorem 26.2 (first translation identity [translation along the S–axis])
If

$$\mathcal{L}[f(t)]|_s = F(s) \qquad \text{for} \quad s > s_0 \quad ,$$

then, for any real constant α *,*

$$\mathcal{L}\big[e^{\alpha t}f(t)\big]\big|_s = F(s-\alpha) \qquad \text{for} \quad s > s_0 + \alpha \quad . \tag{26.19}$$

This is called a 'translation identity' because the graph of the right side of the identity, equation (26.19), is the graph of $F(s)$ translated to the right by α.[2,3] (The second translation identity, in which f is the function shifted, will be developed later.)

!►Example 26.8: *Let us use this translation identity to find the transform of t^2e^{6t}. First we must identify the '$f(t)$' and the 'α' and then apply the identity:*

$$\mathcal{L}\left[t^2e^{6t}\right]\Big|_s = \mathcal{L}\Big[e^{6t}\underbrace{t^2}_{f(t)}\Big]\Big|_s = \mathcal{L}\left[e^{6t}f(t)\right]\Big|_s = F(s-6) \quad . \tag{26.20}$$

Here, $f(t) = t^2$ and $\alpha = 6$. From either formula (26.7) or Table 26.1, we know

$$F(s) = \mathcal{L}[f(t)]|_s = \mathcal{L}\left[t^2\right]\Big|_s = \frac{2!}{s^{2+1}} = \frac{2}{s^3} \quad \text{for} \quad s > 0 \quad .$$

So, for any $X > 0$,

$$F(X) = \frac{2}{X^3} \quad .$$

Using this with $X = s - 6$, the above computation of $\mathcal{L}\left[t^2e^{6t}\right]$ (equation set (26.20)) becomes

$$\mathcal{L}\left[t^2e^{6t}\right]\Big|_s = \cdots = F(\underbrace{s-6}_{X}) = F(X) = \frac{2}{X^3} = \frac{2}{(s-6)^3} \quad \text{for} \quad s-6 > 0 \quad .$$

That is,

$$\mathcal{L}\left[t^2e^{6t}\right]\Big|_s = \frac{2}{(s-6)^3} \quad \text{for} \quad s > 6 \quad .$$

Notice that, in the last example, we carefully rewrote the formula for $F(s)$ as a formula of another variable, X, and used that to get the formula for $F(s-3)$,

$$F(s) = \frac{2}{s^3} \implies F(X) = \frac{2}{X^3} \implies F(\underbrace{s-6}_{X}) = \frac{2}{(s-6)^3} \quad .$$

This helps to prevent dumb mistakes. It replaces the s with a generic placeholder X, which, in turn, is replaced with some formula of s. So long as you remember that the s in the first equation is, itself, simply a placeholder and can be replaced throughout the equation with another formula of s, you can go straight from the formula for $F(s)$ to the formula for $F(s-6)$. Unfortunately, this is often forgotten in the heat of computations, especially by those who are new to these sorts of computations. So it is strongly recommended that you include this intermediate step of replacing $F(s)$ with $F(X)$, and then use the formula for $F(X)$ with $X = s - 6$ (or $X =$ "whatever formula of s is appropriate").

Let's try another:

!►Example 26.9: *Find $\mathcal{L}\left[e^{3t}\sin(2t)\right]$. Here,*

$$\mathcal{L}\left[e^{3t}\sin(2t)\right]\Big|_s = \mathcal{L}\Big[e^{3t}\underbrace{\sin(2t)}_{f(t)}\Big]\Big|_s = \mathcal{L}\left[e^{3t}f(t)\right]\Big|_s = F(s-3) \quad .$$

In this case, $f(t) = \sin(2t)$. Recalling the formula for the transform of such a function (or

[2] More precisely, it's shifted to the right by α if $\alpha > 0$, and is shifted to the left by $|\alpha|$ if $\alpha < 0$.

[3] Some authors prefer to use the word "shifting" instead of "translation".

peeking back at formula (26.10) or Table 26.1), we have

$$F(s) = \mathcal{L}[f(t)]|_s = \mathcal{L}[\sin(2t)]|_s = \frac{2}{s^2+2^2} = \frac{2}{s^2+4} \quad \text{for} \quad s > 0 \quad .$$

So,

$$F(X) = \frac{2}{X^2+4} \quad \text{for} \quad X > 0 \quad .$$

In particular, using $X = s - 3$ *,*

$$F(s-3) = \frac{2}{(s-3)^2+4} \quad \text{for} \quad s - 3 > 0 \quad ,$$

and the above computation of $\mathcal{L}\left[e^{3t}\sin(2t)\right]$ *becomes*

$$\mathcal{L}\left[e^{3t}\sin(2t)\right]\Big|_s = \cdots = F(s-3) = \frac{2}{(s-3)^2+4} \quad \text{for} \quad s > 3 \quad .$$

In the homework, you'll derive the general formulas for

$$\mathcal{L}\left[t^n e^{\alpha t}\right]\Big|_s \quad , \quad \mathcal{L}\left[e^{\alpha t}\sin(\omega t)\right]\Big|_s \quad \text{and} \quad \mathcal{L}\left[e^{\alpha t}\cos(\omega t)\right]\Big|_s \quad .$$

These formulas are found in most tables of common transforms (but not ours).

26.5 What Is "Laplace Transformable"? (and Some Standard Terminology)

When we say a function f is "Laplace transformable", we simply mean that there is a finite value s_0 such that the integral for $\mathcal{L}[f(t)]|_s$,

$$\int_0^\infty f(t)e^{-st}\,dt \quad ,$$

exists and is finite for every value of s greater than s_0 . Not every function is Laplace transformable. For example, t^{-2} and e^{t^2} are not.

Unfortunately, further developing the theory of Laplace transforms assuming nothing more than the "Laplace transformability of our functions" is a bit difficult, and would lead to some rather ungainly wording in our theorems. To simplify our discussions, we will usually insist that our functions are, instead, "piecewise continuous" and "of exponential order". Together, these two conditions will ensure that a function is Laplace transformable, and they will allow us to develop some very general theory that can be applied using the functions that naturally arise in applications. Moreover, these two conditions will be relatively easy to visualize.

So let's find out just what these terms mean.

Jump Discontinuities and Piecewise Continuity

The phrase "piecewise continuity" suggests that we only require continuity on "pieces" of the function's interval of interest. That is a little misleading. For one thing, we want to limit the discontinuities between the pieces to "jump" discontinuities.

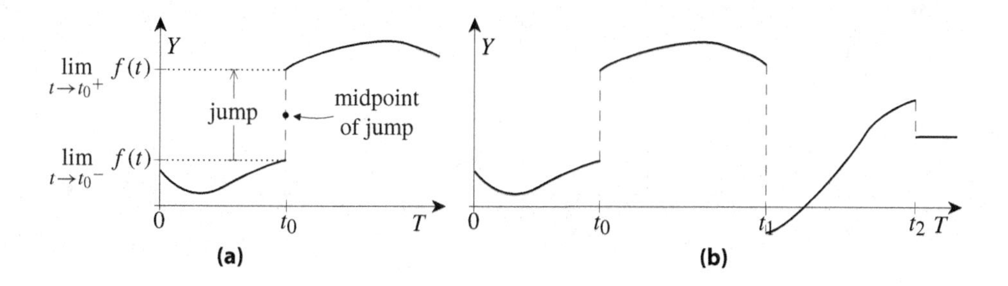

Figure 26.4: The graph of **(a)** a function with a jump discontinuity at t_0 and **(b)** a function with several jump discontinuities.

Jump Discontinuities

A function f is said to have a *jump discontinuity* at a point t_0 if the left- and right-hand limits

$$\lim_{t \to t_0^-} f(t) \qquad \text{and} \qquad \lim_{t \to t_0^+} f(t)$$

exist, but are different finite numbers. The *jump* at this discontinuity is the difference of the two limits,

$$\text{jump} = \lim_{t \to t_0^+} f(t) - \lim_{t \to t_0^-} f(t) \quad ,$$

and the average of the two limits is the *Y*-coordinate of the *midpoint* of the jump,

$$y_{\text{midpoint}} = \frac{1}{2}\left[\lim_{t \to t_0^+} f(t) + \lim_{t \to t_0^-} f(t)\right] \quad .$$

A generic example is sketched in Figure 26.4a. And right beside that figure (in Figure 26.4b) is the graph of a function with multiple jump discontinuities.

The simplest example of a function with a jump discontinuity is the basic step function, $\text{step}(t)$. Just looking at its graph (Figure 26.3a on page 462) you can see that it has a jump discontinuity at $t = 0$ with $\text{jump} = 1$, and $y = 1/2$ as the Y-coordinate of the midpoint.

On the other hand, consider the functions

$$f(t) = \frac{1}{(t-2)^2} \qquad \text{and} \qquad g(t) = \begin{cases} 0 & \text{if} \quad t < 2 \\ \dfrac{1}{(t-2)^2} & \text{if} \quad 2 < t \end{cases} \quad ,$$

sketched in Figures 26.5a and 26.5b, respectively. Both have discontinuities at $t = 2$. In each case, however, the limit of the function as $t \to 2$ from the right is infinite. Hence, we do not view these discontinuities as "jump" discontinuities.

Piecewise Continuity

We say that a function f is *piecewise continuous* on an *finite* open interval (a, b) if and only if both of the following hold:

1. f is continuous on the interval except for, at most, a finite number of jump discontinuities in (a, b).

2. The endpoint limits

$$\lim_{t \to a+} f(t) \qquad \text{and} \qquad \lim_{t \to b-} f(t)$$

exist and are finite.

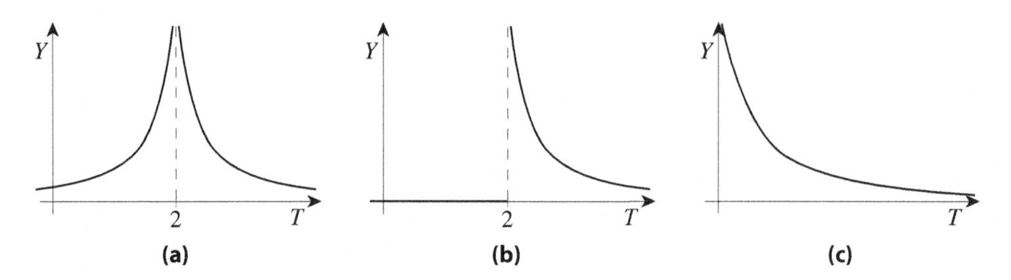

Figure 26.5: Functions having at least one point with an infinite left- or right-hand limit at some point.

We extend this concept to functions on infinite open intervals (such as $(0, \infty)$) by defining a function f to be *piecewise continuous* on an *infinite* open interval if and only if f is piecewise continuous on every finite open subinterval. In particular then, a function f being piecewise continuous on $(0, \infty)$ means that

$$\lim_{t \to 0^+} f(t)$$

is a finite value, and that, for every finite, positive value T, $f(t)$ has at most a finite number of discontinuities on the interval $(0, T)$, with each of those being a jump discontinuity.

For some of our discussions, we will only need our function f to be piecewise continuous on $(0, \infty)$. Strictly speaking, this says nothing about the possible value of $f(t)$ when $t = 0$. If, however, we are dealing with initial-value problems, then we may require our function f to be *piecewise continuous on* $[0, \infty)$, which simply means f is piecewise continuous on $(0, \infty)$, defined at $t = 0$, and

$$f(0) = \lim_{t \to 0^+} f(t) \quad .$$

Keep in mind that the "finite number" of jump discontinuities can be zero, in which case f has no discontinuities and is, in fact, continuous on that interval. What is important is that a piecewise continuous function cannot 'blow up' at any (finite) point in or at the ends of the interval. At worst, it has only 'a few' jump discontinuities in each finite subinterval.

The functions sketched in Figure 26.4 are piecewise continuous, at least over the intervals in the figures. And any step function is piecewise continuous on $(0, \infty)$. On the other hand, the functions sketched in Figures 26.5a and 26.5b, are not piecewise continuous on $(0, \infty)$ because they both "blow up" at $t = 2$. Consider even the function

$$f(t) = \frac{1}{t} \quad ,$$

sketched in Figure 26.5c. Even though this function is continuous on the interval $(0, \infty)$, we do not consider it to be piecewise continuous on $(0, \infty)$ because

$$\lim_{t \to 0^+} \frac{1}{t} = \infty \quad .$$

Two simple observations will soon be important to us:

1. If f is piecewise continuous on $(0, \infty)$, and T is any positive finite value, then the integral

$$\int_0^T f(t)\, dt$$

is well defined and evaluates to a finite number. Remember, geometrically, this integral is the "net area" between the graph of f and the T–axis over the interval $(0, T)$. The piecewise continuity of f assures us that f does not "blow up" at any point in $(0, T)$, and that we can divide the graph of f over $(0, T)$ into a finite number of fairly nicely behaved 'pieces' (see Figure 26.4b) with each piece enclosing a finite area.

2. The product of any two piecewise continuous functions f and g on $(0, \infty)$ will, itself, be piecewise continuous on $(0, \infty)$. You can easily verify this yourself using the fact that

$$\lim_{t \to t_0^{\pm}} f(t)g(t) = \lim_{t \to t_0^{\pm}} f(t) \times \lim_{t \to t_0^{\pm}} g(t) \quad .$$

Combining the above two observations with the obvious fact that, for any real value of s, $g(t) = e^{-st}$ is a piecewise continuous function of t on $(0, \infty)$ gives us:

Lemma 26.3
Let f be a piecewise continuous function on $(0, \infty)$, and let T be any finite positive number. Then the integral

$$\int_0^T f(t)e^{-st}\, dt$$

is a well-defined finite number for each real value s.

Because of our interest in the Laplace transform, we will want to ensure that the above integral converges to a finite number as $T \to \infty$. That is the next issue we will address.

Exponential Order*

Let f be a function on $(0, \infty)$, and let s_0 be some real number. We say that f is *of exponential order* s_0 if and only if there are finite constants M and T such that

$$|f(t)| \leq Me^{s_0 t} \qquad \text{whenever} \quad T \leq t \quad . \tag{26.21}$$

Often, the precise value of s_0 is not particularly important. In these cases we may just say that f is *of exponential order* to indicate that it is of exponential order s_0 for some value s_0.

Saying that f is of exponential order is just saying that the graph of $|f(t)|$ is bounded above by the graph of some constant multiple of some exponential function on some interval of the form $[T, \infty)$. Note that, if this is the case and s is any real number, then

$$\left|f(t)e^{-st}\right| = |f(t)|\, e^{-st} \leq Me^{s_0 t}e^{-st} = Me^{-(s-s_0)t} \qquad \text{whenever} \quad T \leq t \quad .$$

Moreover, if $s > s_0$, then $s - s_0$ is positive, and

$$\left|f(t)e^{-st}\right| \leq Me^{-(s-s_0)t} \to 0 \qquad \text{as} \quad t \to \infty \quad . \tag{26.22}$$

Thus, in the future, we will automatically know that

$$\lim_{t \to \infty} f(t)e^{-st} = 0$$

whenever f is of exponential order s_0 and $s > s_0$.

---

* More precisely: **Exponential Order as $t \to \infty$**. One can have "exponential order as $t \to -\infty$" and even "as $t \to 3$"). However, we are not interested in those cases, and it is silly to keep repeating "as $t \to \infty$".

Transforms of Piecewise Continuous Functions of Exponential Order

Now, suppose f is a piecewise continuous function of exponential order s_0 on the interval $(0, \infty)$. As already observed, the piecewise continuity of f assures us that

$$\int_0^T f(t)e^{-st}\,dt$$

is a well-defined finite number for each $T > 0$. And if $s > s_0$, then inequality (26.22), above, tells us that $f(t)e^{-st}$ is shrinking to 0 as $t \to \infty$ at least as fast as a constant multiple of some decreasing exponential. It is easy to verify that this is fast enough to ensure that

$$\lim_{T\to\infty}\int_0^T f(t)e^{-st}\,dt$$

converges to some finite value. And that gives us the following theorem on conditions ensuring the existence of Laplace transforms.

Theorem 26.4
If f is both piecewise continuous on $(0, \infty)$ and of exponential order s_0, then

$$F(s) = \mathcal{L}[f(t)]|_s = \int_0^\infty f(t)e^{-st}\,dt$$

is a well-defined function for $s > s_0$.

In the next several chapters, we will often assume that our functions of t are both piecewise continuous on $(0, \infty)$ and of exponential order. Mind you, not all Laplace transformable functions satisfy these conditions. For example, we've already seen that t^α with $-1 < \alpha$ is Laplace transformable. But

$$\lim_{t\to 0^+} t^\alpha = \infty \qquad \text{if} \quad \alpha < 0 \quad .$$

So those functions given by t^α with $-1 < \alpha < 0$ (such as $t^{-1/2}$) are not piecewise continuous on $(0, \infty)$, even though they are certainly Laplace transformable. Still, all the other functions on the left side of Table 26.1 on page 460 are piecewise continuous on $(0, \infty)$ and are of exponential order. More importantly, the functions that naturally arise in applications in which the Laplace transform may be useful are usually piecewise continuous on $(0, \infty)$ and of exponential order.

By the way, since you've probably just glanced at Table 26.1 on page 460, go back and look at the functions on the right side of the table. Observe that

1. these functions have no discontinuities in the intervals on which they are defined,

and

2. they all shrink to 0 as $s \to \infty$.

It turns out that you can extend the work used to obtain the above theorem to show that the above observations hold much more generally. More precisely, the above theorem can be extended to:

Theorem 26.5
If f is both piecewise continuous on $(0, \infty)$ and of exponential order s_0, then

$$F(s) = \mathcal{L}[f(t)]|_s = \int_0^\infty f(t)e^{-st}\,dt$$

is a continuous function on (s_0, ∞) *and*

$$\lim_{s \to \infty} F(s) = 0 \quad .$$

We will verify this theorem at the end of the chapter.

26.6 Further Notes on Piecewise Continuity and Exponential Order

Issues Regarding Piecewise Continuous Functions on $(0, \infty)$

In the next several chapters, we will be concerned mainly with functions that are piecewise continuous on $(0, \infty)$. There are a few small technical issues regarding these functions that could become significant later if we don't deal with them now. These issues concern the values of such functions at jumps.

On the Value of a Function at a Jump

Take a look at Figure 26.4b on page 466. Call the function sketched there f, and consider evaluating, say,

$$\int_0^{t_2} f(t)e^{-st}\,dt \quad .$$

The obvious approach is to break up the integral into three pieces,

$$\int_0^{t_2} f(t)e^{-st}\,dt = \int_0^{t_0} f(t)e^{-st}\,dt + \int_{t_0}^{t_1} f(t)e^{-st}\,dt + \int_{t_1}^{t_2} f(t)e^{-st}\,dt \quad ,$$

and use values/formulas for f over the intervals $(0, t_0)$, (t_0, t_1) and (t_1, t_2) to compute the individual integrals in the above sum. What you would not worry about would be the actual values of f at the points of discontinuity, t_0, t_1 and t_2. In particular, it would not matter if

$$f(t_0) = \lim_{t \to t_0^-} f(t) \qquad \text{or} \qquad f(t_0) = \lim_{t \to t_0^+} f(t)$$

or

$$f(t_0) = \text{the } Y\text{-coordinate of the midpoint of the jump} \quad .$$

This extends an observation made when we computed the Laplace transform of the shifted step function. There, we found that the precise value of $\text{step}_\alpha(t)$ at $t = \alpha$ was irrelevant to the computation of $\mathcal{L}[\text{step}_\alpha(t)]$. And the pseudo-computations in the previous paragraph point out that, in general, the value of any piecewise continuous function at a point of discontinuity will be irrelevant to the integral computations we will be doing with these functions.

Parallel to these observations are the observations of how we use functions with jump discontinuities in applications. Typically, a function with a jump discontinuity at $t = t_0$ is modeling something that changes so quickly around $t = t_0$ that we might as well pretend the change is instantaneous. Consider, for example, the output of a one-lumen old-fashioned incandescent light bulb switched on at $t = 2$: Until it is switched on, the bulb's light output is 0 lumen. For a brief period around $t = 2$ the filament is warming up and the light output increases from 0 to 1 lumen, and remains at 1 lumen thereafter. In practice, however, the warm-up time is so brief that we don't notice it, and we are content to describe the light output by

$$\text{light output at time } t = \begin{Bmatrix} 0 \text{ lumen} & \text{if} \quad t < 2 \\ 1 \text{ lumen} & \text{if} \quad 2 < t \end{Bmatrix} = \text{step}_2(t) \text{ lumen}$$

without giving any real thought as to the value of the light output the very instant we are turning on the bulb.[4]

What all this is getting to is that, for our work involving piecewise continuous functions on $(0, \infty)$,

the value of a function f *at any point of discontinuity* t_0 *in* $(0, \infty)$ *is irrelevant.*

What is important is not $f(t_0)$ but the one-sided limits

$$\lim_{t \to t_0^-} f(t) \qquad \text{and} \qquad \lim_{t \to t_0^+} f(t) \quad .$$

Because of this, we will not normally specify the value of a function at a discontinuity, at least not while developing Laplace transforms. If this disturbs you, go ahead and assume that, unless otherwise indicated, the value of a function at each jump discontinuity is given by the Y-coordinate of the jump's midpoint. It's as good as any other value.

Equality of Piecewise Continuous Functions

Because of the irrelevance of the value of a function at a discontinuity, we need to slightly modify what it means to say "$f = g$ on some interval". Henceforth, let us say that

$$f = g \text{ on some interval (as piecewise continuous functions)}$$

means

$$f(t) = g(t)$$

for every t in the interval at which f and g are continuous. We will not insist that f and g be equal at the relatively few points of discontinuity in the functions. But do note that we will still have

$$\lim_{t \to t_0^\pm} f(t) = \lim_{t \to t_0^+} g(t)$$

for every t_0 in the interval. Consequently, the graphs of f and g will have the same 'jumps' in the interval.

By the way, the phrase "as piecewise continuous functions" in the above definition is recommended but is often forgotten.

!►Example 26.10: *The functions*

$$\text{step}_2(t) \quad , \quad f(t) = \begin{cases} 0 & \text{if } t \leq 2 \\ 1 & \text{if } 2 < t \end{cases} \quad \text{and} \quad g(t) = \begin{cases} 0 & \text{if } t < 2 \\ 1 & \text{if } 2 \leq t \end{cases}$$

all satisfy

$$\text{step}_2(t) = f(t) = g(t)$$

[4] On the other hand, "What is the light output of a one-lumen light bulb the very instant the light is turned on?" may be a nice question to meditate upon if you are studying Zen.

for all values of t in $(0, \infty)$ except $t = 2$, at which each has a jump. So, as piecewise continuous functions,

$$\text{step}_2 = f = g \quad \text{on} \quad (0, \infty) \quad .$$

Conversely, if we know $h = \text{step}_2$ on $(0, \infty)$ (as piecewise continuous functions), then we know

$$h(t) = \begin{cases} 0 & \text{if} \quad 0 < t < 2 \\ 1 & \text{if} \quad 2 < t \end{cases} \quad .$$

We do not know (nor do we care about) the value of $h(t)$ when $t = 2$ (or when $t < 0$).

Testing for Exponential Order

Before deriving this test for exponential order, it should be noted that the "order" is not unique. After all, if

$$|f(t)| \leq Me^{s_0 t} \qquad \text{whenever} \quad T \leq t \quad ,$$

and $s_0 \leq \sigma$, then

$$|f(t)| \leq Me^{s_0 t} \leq Me^{\sigma t} \qquad \text{whenever} \quad T \leq t \quad ,$$

proving the following little lemma:

Lemma 26.6
If f is of exponential order s_0, then f is of exponential order σ for every $\sigma \geq s_0$.

Now here is the test:

Lemma 26.7 (test for exponential order)
Let f be a function on $(0, \infty)$.

1. *If there is a real value s_0 such that*

$$\lim_{t \to \infty} f(t)e^{-s_0 t} = 0 \quad ,$$

then f is of exponential order s_0.

2. *If*

$$\lim_{t \to \infty} f(t)e^{-st}$$

does not converge to 0 for any real value s, then f is not of exponential order.

PROOF: First, assume

$$\lim_{t \to \infty} f(t)e^{-s_0 t} = 0$$

for some real value s_0, and let M be any finite positive number you wish (it could be 1, $1/2$, 827, whatever). By the definition of "limits", the above assures us that, if t is large enough, then $f(t)e^{-s_0 t}$ is within M of 0. That is, whenever T is any single "large enough" value of t, then we must have

$$\left| f(t)e^{-s_0 t} - 0 \right| \leq M \qquad \text{whenever} \quad T \leq t \quad .$$

By elementary algebra, we can rewrite this as

$$|f(t)| \leq Me^{s_0 t} \qquad \text{whenever} \quad T \leq t \quad ,$$

which is exactly what we mean when we say "f is of exponential order s_0". That confirms the first part of the lemma.

To verify the second part of the lemma, assume

$$\lim_{t \to \infty} f(t)e^{-st}$$

does not converge to 0 for any real value s. If f were of exponential order, then it is of exponential order s_0 for some finite real number s_0, and, as noted in the discussion of expression (26.22) on page 469, we would then have that

$$\lim_{t \to \infty} f(t)e^{-st} = 0 \qquad \text{for} \quad s > s_0 \quad .$$

But we've assumed this is not possible; thus, it is not possible for f to be of exponential order. ∎

26.7 Proving Theorem 26.5
The Theorem and a Bad Proof

The basic premise of Theorem 26.5 is that we have a piecewise continuous function f on $(0, \infty)$ which is also of exponential order s_0. From the previous theorem, we know

$$F(s) = \mathcal{L}[f(t)]|_s = \int_0^\infty f(t)e^{-st}\,dt$$

is a well-defined function on (s_0, ∞). Theorem 26.5 further claims that

1. $F(s) = \mathcal{L}[f(t)]|_s$ is continuous on (s_0, ∞). That is,

$$\lim_{s \to s_1} F(s) = F(s_1) \qquad \text{for each} \quad s_1 > s_0$$

and

2. $\lim_{s \to \infty} F(s) = 0$.

In a naive attempt to verify these claims, you might try

$$\begin{aligned} \lim_{s \to s_1} F(s) &= \lim_{s \to s_1} \int_0^\infty f(t)e^{-st}\,dt \\ &= \int_0^\infty \lim_{s \to s_1} f(t)e^{-st}\,dt = \int_0^\infty f(t)e^{-s_1 t}\,dt = F(s_1) \quad , \end{aligned}$$

and

$$\begin{aligned} \lim_{s \to \infty} F(s) &= \lim_{s \to \infty} \int_0^\infty f(t)e^{-st}\,dt \\ &= \int_0^\infty \lim_{s \to \infty} f(t)e^{-st}\,dt = \int_0^\infty f(t) \cdot 0\,dt = 0 \quad . \end{aligned}$$

Unfortunately, these computations assume

$$\lim_{s \to \alpha} \int_0^\infty g(t, s)\,dt = \int_0^\infty \lim_{s \to \alpha} g(t, s)\,dt$$

which is NOT always true. Admittedly, it often is true, but there are exceptions (see, for example, Exercise 26.19 on page 480). And because there are exceptions, we cannot rely on this sort of "switching of limits with integrals" to prove our claims.

Preliminaries

There are two small observations that will prove helpful here and elsewhere.

The first concerns any function f which is piecewise continuous on $(0, \infty)$ and satisfies

$$|f(t)| \leq M_T e^{s_0 t} \qquad \text{whenever} \quad T \leq t \quad ,$$

for two positive values M_T and T. For convenience, let

$$g(t) = f(t)e^{-s_0 t} \qquad \text{for} \quad t > 0 \quad .$$

This is another piecewise continuous function on $(0, \infty)$, but it satisfies

$$|g(t)| = \left|f(t)e^{-s_0 t}\right| = |f(t)|\, e^{-s_0 t} \leq M e^{s_0 t} e^{-s_0 t} = M \qquad \text{for} \quad T < t \quad .$$

On the other hand, the piecewise continuity of g on $(0, \infty)$ means that g does not "blow up" anywhere in or at the endpoints of $(0, T)$. So it is easy to see (and to prove) that there is a constant B such that

$$|g(t)| \leq B \qquad \text{for} \quad 0 < t < T \quad .$$

Letting M_0 be the larger of B and M_T, we now have that

$$|g(t)| \leq M_0 \qquad \text{if} \quad 0 < t < T \quad \text{or} \quad T \leq t \quad .$$

So,

$$|f(t)|\, e^{-s_0 t} = \left|f(t)e^{-s_0 t}\right| = |g(t)| \leq M_0 \qquad \text{for all} \quad t > 0 \quad .$$

Multiply through by the exponential, and you've got:

Lemma 26.8
Assume f is a piecewise continuous function on $(0, \infty)$ which is also of exponential order s_0. Then there is a constant M_0 such that

$$|f(t)| \leq M_0 e^{s_0 t} \qquad \text{for all} \quad t > 0 \quad .$$

The above lemma will let us use the exponential bound $M_0 e^{s_0 t}$ over all of $(0, \infty)$, and not just (T, ∞). The next lemma is one you should either already be acquainted with or can easily confirm on your own.

Lemma 26.9
If g is an integrable function on the interval (a, b), then

$$\left|\int_a^b g(t)\,dt\right| \leq \int_a^b |g(t)|\,dt \quad .$$

Proof of Theorem 26.5

Now we will prove the two claims of Theorem 26.5. Keep in mind that f is a piecewise continuous function on $(0, \infty)$ of exponential order s_0, and that

$$F(s) = \int_0^\infty f(t)e^{-st}\,dt \qquad \text{for} \quad s > s_0 \quad .$$

We will make repeated use of the fact, stated in Lemma 26.8 just above, that there is a constant M_0 such that

$$|f(t)| \leq M_0 e^{s_0 t} \qquad \text{for} \quad 0 < t \quad . \tag{26.23}$$

Since the second claim is a little easier to verify, we will start with that.

Proof of the Second Claim

The second claim is that

$$\lim_{s\to\infty} F(s) = 0 \quad ,$$

which, of course, can be proven by showing

$$\lim_{s\to\infty} |F(s)| \leq 0 \quad .$$

Now let $s > s_0$. Using inequality (26.23) with the integral inequality from Lemma 26.9, we have

$$\begin{aligned} |F(s)| = \left|\int_0^\infty f(t)e^{-st}\,dt\right| &\leq \int_0^\infty \left|f(t)e^{-st}\right|\,dt \\ &= \int_0^\infty |f(t)|\,e^{-st}\,dt \\ &\leq \int_0^\infty M_0 e^{s_0 t}e^{-st}\,dt = M_0\mathcal{L}\big[e^{s_0 t}\big]\big|_s = \frac{M_0}{s-s_0} \quad . \end{aligned}$$

Thus,

$$\lim_{s\to\infty} |F(s)| \leq \lim_{s\to\infty} \frac{M_0}{s-s_0} = 0 \quad ,$$

confirming the claim.

Proof of the First Claim

The first claim is that F is continuous on (s_0, ∞). To prove this, we need to show that, for each $s_1 > s_0$,

$$\lim_{s\to s_1} F(s) = F(s_1) \quad .$$

Note that this limit can be verified by showing

$$\lim_{s\to s_1} |F(s) - F(s_1)| \leq 0 \quad .$$

Now let s and s_1 be two different points in (s_0, ∞) (Hence, $s - s_0 > 0$ and $s_1 - s_0 > 0$). Using the integral inequality from Lemma 26.9, we get

$$\begin{aligned} |F(s) - F(s_1)| &= \left|\int_0^\infty f(t)e^{-st}\,dt - \int_0^\infty f(t)e^{-s_1 t}\,dt\right| \\ &= \left|\int_0^\infty f(t)\left[e^{-st} - e^{-s_1 t}\right]dt\right| \leq \int_0^\infty |f(t)|\left|e^{-st} - e^{-s_1 t}\right|dt \quad . \end{aligned}$$

Applying inequality (26.23), we then have

$$|F(s) - F(s_1)| \leq M_0 \int_0^\infty e^{s_0 t}\left|e^{-st} - e^{-s_1 t}\right|dt \quad . \tag{26.24}$$

Now observe that, if $s \leq s_1$ and $t \geq 0$, then

$$e^{s_0 t}\left|e^{-st} - e^{-s_1 t}\right| = e^{s_0 t}\left[e^{-st} - e^{-s_1 t}\right] = e^{-(s-s_0)t} - e^{-(s_1-s_0)t} \quad .$$

Thus, when $s \leq s_1$,

$$\begin{aligned} \int_0^\infty e^{s_0 t}\left|e^{-st} - e^{-s_1 t}\right|dt &= \int_0^\infty \left[e^{-(s-s_0)t} - e^{-(s_1-s_0)t}\right]dt \\ &= \left[\frac{1}{s-s_0} - \frac{1}{s_1-s_0}\right] = \left|\frac{1}{s-s_0} - \frac{1}{s_1-s_0}\right| \quad . \end{aligned}$$

Almost identical computations also show that, when $s \geq s_1$,

$$\int_0^\infty e^{s_0 t} \left| e^{-st} - e^{-s_1 t} \right| dt = -\int_0^\infty \left[e^{-(s-s_0)t} - e^{-(s_1-s_0)t} \right] dt$$

$$= -\left[\frac{1}{s-s_0} - \frac{1}{s_1-s_0} \right] = \left| \frac{1}{s-s_0} - \frac{1}{s_1-s_0} \right| \quad .$$

Consequently, we can rewrite inequality (26.24) as

$$|F(s) - F(s_1)| \leq M_0 \left| \frac{1}{s-s_0} - \frac{1}{s_1-s_0} \right| \quad ,$$

and from this we have

$$\lim_{s \to s_1} |F(s) - F(s_1)| \leq \lim_{s \to s_1} M_0 \left| \frac{1}{s-s_0} - \frac{1}{s_1-s_0} \right| = M_0 \left| \frac{1}{s_1-s_0} - \frac{1}{s_1-s_0} \right| = 0 \quad ,$$

which is all we needed to show to confirm the first claim.

Additional Exercises

26.6. *Sketch the graph of each of the following choices of $f(t)$, and then find that function's Laplace transform by direct application of the definition, formula (26.1) on page 451 (i.e., compute the integral). Also, if there is a restriction on the values of s for which the formula of the transform is valid, state that restriction.*

a. $f(t) = 4$

b. $f(t) = 3e^{2t}$

c. $f(t) = \begin{cases} 2 & \text{if } t \leq 3 \\ 0 & \text{if } 3 < t \end{cases}$

d. $f(t) = \begin{cases} 0 & \text{if } t \leq 3 \\ 2 & \text{if } 3 < t \end{cases}$

e. $f(t) = \begin{cases} e^{2t} & \text{if } t \leq 4 \\ 0 & \text{if } 4 < t \end{cases}$

f. $f(t) = \begin{cases} e^{2t} & \text{if } 1 < t \leq 4 \\ 0 & \text{otherwise} \end{cases}$

g. $f(t) = \begin{cases} t & \text{if } 0 < t \leq 1 \\ 0 & \text{otherwise} \end{cases}$

h. $f(t) = \begin{cases} 0 & \text{if } 0 < t \leq 1 \\ t & \text{otherwise} \end{cases}$

26.7. *Find the Laplace transform of each using either formula (26.7) on page 456, formula (26.8) on page 456 or formula (26.9) on page 457, as appropriate:*

a. t^4 **b.** t^9 **c.** e^{7t} **d.** e^{i7t} **e.** e^{-7t} **f.** e^{-i7t}

26.8. *Find the Laplace transform of each of the following using Table 26.1 on page 460 (Transforms of Common Functions) and the linearity of the Laplace transform:*

a. $\sin(3t)$

b. $\cos(3t)$

c. 7

d. $\cosh(3t)$

e. $\sinh(4t)$

f. $3t^2 - 8t + 47$

g. $6e^{2t} + 8e^{-3t}$

h. $3\cos(2t) + 4\sin(6t)$

i. $3\cos(2t) - 4\sin(2t)$

26.9. *Compute the following Laplace transforms using Table 26.1 on page 460 (Transforms of Common Functions) and, where appropriate, properties of the Gamma function:*

a. $t^{3/2}$ **b.** $t^{5/2}$ **c.** $t^{-1/3}$ **d.** $\sqrt[4]{t}$ **e.** $\text{step}_2(t)$

26.10. *For the following, let*

$$f(t) = \begin{cases} 1 & \text{if } t < 2 \\ 0 & \text{if } 2 \leq t \end{cases} .$$

a. *Verify that* $f(t) = 1 - \text{step}_2(t)$ *, and using this and linearity,*

b. *compute* $\mathcal{L}[f(t)]|_s$.

26.11. *Find the Laplace transform of each of the following using Table 26.1 on page 460 (Transforms of Common Functions) and the first translation identity:*

a. te^{4t} **b.** t^4e^t **c.** $e^{2t}\sin(3t)$

d. $e^{2t}\cos(3t)$ **e.** $e^{3t}\sqrt{t}$ **f.** $e^{3t}\,\text{step}_2(t)$

26.12. *Verify each of the following using Table 26.1 on page 460 (Transforms of Common Functions) and the first translation identity (assume* α *and* ω *are real-valued constants and* n *is a positive integer):*

a. $\mathcal{L}\left[t^n e^{\alpha t}\right]\big|_s = \dfrac{n!}{(s-\alpha)^{n+1}}$ for $s > \alpha$

b. $\mathcal{L}\left[e^{\alpha t}\sin(\omega t)\right]\big|_s = \dfrac{\omega}{(s-\alpha)^2+\omega^2}$ for $s > \alpha$

c. $\mathcal{L}\left[e^{\alpha t}\cos(\omega t)\right]\big|_s = \dfrac{s-\alpha}{(s-\alpha)^2+\omega^2}$ for $s > \alpha$

d. $\mathcal{L}\left[e^{\alpha t}\,\text{step}_\omega(t)\right]\big|_s = \dfrac{1}{s-\alpha}e^{-\omega(s-\alpha)}$ for $s > \alpha$ and $\omega \geq 0$

26.13. *The following problems all concern the Gamma function,*

$$\Gamma(\sigma) = \int_0^\infty e^{-u}u^{\sigma-1}\,du .$$

a. *Using integration by parts, show that* $\Gamma(\sigma+1) = \sigma\Gamma(\sigma)$ *whenever* $\sigma > 0$.

b i. *By using an appropriate change of variables and symmetry, verify that*

$$\Gamma\left(\tfrac{1}{2}\right) = \int_{-\infty}^\infty e^{-\tau^2}\,d\tau .$$

ii. *Starting with the observation that, by the above,*

$$\Gamma\left(\tfrac{1}{2}\right)\Gamma\left(\tfrac{1}{2}\right) = \left(\int_{-\infty}^\infty e^{-x^2}\,dx\right)\left(\int_{-\infty}^\infty e^{-y^2}\,dy\right) = \int_{-\infty}^\infty\int_{-\infty}^\infty e^{-x^2-y^2}\,dx\,dy ,$$

show that

$$\Gamma\left(\tfrac{1}{2}\right) = \sqrt{\pi} .$$

(Hint: Use polar coordinates to integrate the double integral.)

26.14. *Several functions are given below. Sketch the graph of each over an appropriate interval, and decide whether each is or is not piecewise continuous on* $(0,\infty)$.

a. $f(t) = 2\,\text{step}_3(t)$

b. $g(t) = \text{step}_2(t) - \text{step}_3(t)$

c. $\sin(t)$

d. $\dfrac{\sin(t)}{t}$

e. $\tan(t)$

f. $\sqrt{t}$

g. $\dfrac{1}{\sqrt{t}}$

h. $t^2 - 1$

i. $\dfrac{1}{t^2 - 1}$

j. $\dfrac{1}{t^2 + 1}$

k. *The "ever increasing stair" function,*

$$\text{stair}(t) = \begin{cases} 0 & \text{if } t < 0 \\ 1 & \text{if } 0 < t < 1 \\ 2 & \text{if } 2 < t < 3 \\ 3 & \text{if } 3 < t < 4 \\ 4 & \text{if } 4 < t < 5 \\ \vdots & \quad\vdots \end{cases}$$

26.15. Assume f and g are two piecewise continuous functions on an interval (a, b) containing the point t_0. Assume further that f has a jump discontinuity at t_0 while g is continuous at t_0. Verify that the jump in the product fg at t_0 is given by

$$\text{"the jump in } f \text{ at } t_0\text{"} \times g(t_0) \quad .$$

26.16. Using either the basic definition or the test for exponential order (Lemma 26.7 on page 473), determine which of the following are of exponential order, and, for each function of exponential order, determine the possible values for the order.

a. e^{3t}

b. t^2

c. te^{3t}

d. e^{t^2}

e. $\sin(t)$

26.17. For the following, let α and σ be any two positive numbers.

a. Using basic calculus, show that $t^\alpha e^{-\sigma t}$ has a maximum value $M_{\alpha,\sigma}$ on the interval $[0,\infty)$. Also, find both where this maximum occurs and the value of $M_{\alpha,\sigma}$.

b. Explain why this confirms that

i. $t^\alpha \leq M_{\alpha,\sigma} e^{\sigma t}$ whenever $t > 0$, and that

ii. t^α is of exponential order σ for any $\sigma > 0$.

26.18. Assume f is a piecewise continuous function on $(0,\infty)$ of exponential order s_0, and let α and σ be any two positive numbers. Using the results of the last exercise, show that $t^\alpha f(t)$ is piecewise continuous on $(0,\infty)$ and of exponential order $s_0 + \sigma$.

26.19. *For this problem, let*

$$g(t,s) = 2ste^{-st^2} \quad .$$

Your goal, in the following, is to show that

$$\lim_{s\to\infty} \int_0^\infty g(t,s)\,dt \neq \int_0^\infty \lim_{s\to\infty} g(t,s)\,dt \quad .$$

a. *By simply computing the integral and then taking the limit, show that*

$$\lim_{s\to\infty} \int_0^\infty g(t,s)\,dt = 1 \quad .$$

b. *Then, using L'Hôpital's rule, verify that, for each $t \geq 0$,*

$$\lim_{s\to\infty} g(t,s) = 0 \quad ,$$

and observe that this means

$$\int_0^\infty \lim_{s\to\infty} g(t,s)\,dt = 0 \neq 1 = \lim_{s\to\infty} \int_0^\infty g(t,s)\,dt \quad .$$

27

Differentiation and the Laplace Transform

In this chapter, we will explore how the Laplace transform interacts with the basic operators of calculus: differentiation and integration. The greatest interest will be in the first identity that we will derive. This relates the transform of a derivative of a function to the transform of the original function, and will allow us to convert many initial-value problems to easily solved algebraic equations. But there are other useful relations involving the Laplace transform and either differentiation or integration. So we'll look at them, too.

27.1 Transforms of Derivatives

The Main Identity

To see how the Laplace transform can convert a differential equation to a simple algebraic equation, let us examine how the transform of a function's derivative,

$$\mathcal{L}\left[f'(t)\right]\Big|_s = \mathcal{L}\left[\frac{df}{dt}\right]\Big|_s = \int_0^\infty \frac{df}{dt}e^{-st}\,dt = \int_0^\infty e^{-st}\frac{df}{dt}\,dt \quad ,$$

is related to the corresponding transform of the original function,

$$F(s) = \mathcal{L}[f(t)]|_s = \int_0^\infty f(t)e^{-st}\,dt \quad .$$

The last formula above for $\mathcal{L}\left[f'(t)\right]$ clearly suggests using integration by parts, and to ensure that this integration by parts is valid, we need to assume f is continuous on $[0, \infty)$ and f' is at least piecewise continuous on $(0, \infty)$. Assuming this,

$$\begin{aligned}
\mathcal{L}\left[f'(t)\right]\Big|_s &= \int_0^\infty \underbrace{e^{-st}}_{u}\,\underbrace{\frac{df}{dt}\,dt}_{dv} \\
&= uv\Big|_{t=0}^\infty - \int_0^\infty v\,du \\
&= e^{-st}f(t)\Big|_{t=0}^\infty - \int_0^\infty f(t)\left[-se^{-st}\right]dt
\end{aligned}$$

$$= \lim_{t\to\infty} e^{-st} f(t) - e^{-s\cdot 0} f(0) - \int_0^\infty \left[-se^{-st} \right] f(t)\, dt$$

$$= \lim_{t\to\infty} e^{-st} f(t) - f(0) + s\int_0^\infty f(t)e^{-st}\, dt \quad .$$

Now, if f is of exponential order s_0, then

$$\lim_{t\to\infty} e^{-st} f(t) = 0 \qquad \text{whenever} \quad s > s_0$$

and

$$F(s) = \mathcal{L}[f(t)]|_s = \int_0^\infty f(t)e^{-st}\, dt \qquad \text{exists for} \quad s > s_0 \quad .$$

Thus, continuing the above computations for $\mathcal{L}\big[f'(t)\big]$ with $s > s_0$, we find that

$$\mathcal{L}\big[f'(t)\big]\big|_s = \lim_{t\to\infty} e^{-st} f(t) - f(0) + s\int_0^\infty f(t)e^{-st}\, dt$$

$$= 0 - f(0) + s\mathcal{L}[f(t)]|_s \quad ,$$

which is a little more conveniently written as

$$\mathcal{L}\big[f'(t)\big]\big|_s = s\mathcal{L}[f(t)]|_s - f(0) \tag{27.1a}$$

or even as

$$\mathcal{L}\big[f'(t)\big]\big|_s = sF(s) - f(0) \quad . \tag{27.1b}$$

This will be a very useful result, well worth preserving in a theorem.

Theorem 27.1 (transform of a derivative)
Let $F = \mathcal{L}[f]$ where f is a continuous function of exponential order s_0 on $[0, \infty)$. If f' is at least piecewise continuous on $(0, \infty)$, then

$$\mathcal{L}\big[f'(t)\big]\big|_s = sF(s) - f(0) \qquad \textit{for} \quad s > s_0 \quad .$$

Extending these identities to formulas for the transforms of higher derivatives is easy. First, for convenience, rewrite equation (27.1a) as

$$\mathcal{L}\big[g'(t)\big]\big|_s = s\mathcal{L}[g(t)]|_s - g(0)$$

or, equivalently, as

$$\mathcal{L}\left[\frac{dg}{dt}\right]\bigg|_s = s\mathcal{L}[g(t)]|_s - g(0) \quad .$$

(Keep in mind that this assumes g is a continuous function of exponential order, g' is piecewise continuous and s is larger than the order of g.) Now we simply apply this equation with $g = f'$, $g = f''$, etc. Assuming all the functions are sufficiently continuous and are of exponential order, we see that

$$\mathcal{L}\big[f''(t)\big]\big|_s = \mathcal{L}\left[\frac{df'}{dt}\right]\bigg|_s = s\mathcal{L}\big[f'(t)\big]\big|_s - f'(0)$$

$$= s\left[sF(s) - f(0)\right] - f'(0)$$

$$= s^2F(s) - sf(0) - f'(0) \quad .$$

Using this, we then see that

$$\begin{aligned}\mathcal{L}[f'''(t)]\big|_s = \mathcal{L}\left[\frac{df''}{dt}\right]\bigg|_s &= s\mathcal{L}[f''(t)]\big|_s - f''(0)\\ &= s\left[s^2F(s) - sf(0) - f'(0)\right] - f''(0)\\ &= s^3F(s) - s^2f(0) - sf'(0) - f''(0) \quad .\end{aligned}$$

Clearly, if we continue, we will end up with the following corollary to Theorem 27.1:

Corollary 27.2 (transforms of derivatives)
Let $F = \mathcal{L}[f]$ where f is a continuous function of exponential order s_0 on $[0,\infty)$. If f' is at least piecewise continuous on $(0,\infty)$, then

$$\mathcal{L}[f'(t)]\big|_s = sF(s) - f(0) \qquad \text{for} \quad s > s_0 \quad .$$

If, in addition, f' is a continuous function of exponential order s_0, and f'' is at least piecewise continuous, then

$$\mathcal{L}[f''(t)]\big|_s = s^2F(s) - sf(0) - f'(0) \qquad \text{for} \quad s > s_0 \quad .$$

More generally, if f, f', f'', ... and $f^{(n-1)}$ are all continuous functions of exponential order s_0 on $[0,\infty)$ for some positive integer n, and $f^{(n)}$ is at least piecewise continuous on $(0,\infty)$, then

$$\begin{aligned}\mathcal{L}[f^{(n)}(t)]\big|_s = s^nF(s) &- s^{n-1}f(0) - s^{n-2}f'(0)\\ &- s^{n-3}f''(0) - \cdots - sf^{(n-2)}(0) - f^{(n-1)}(0) \qquad \text{for} \quad s > s_0 \quad .\end{aligned}$$

Using the Main Identity

Let us now see how these identities can be used in solving initial-value problems. We'll start with something simple:

!►Example 27.1: *Consider the initial-value problem*

$$\frac{dy}{dt} - 3y = 0 \qquad \text{with} \quad y(0) = 4 \quad .$$

Observe what happens when we take the Laplace transform of the differential equation (i.e., we take the transform of both sides). Initially, we just have

$$\mathcal{L}\left[\frac{dy}{dt} - 3y\right]\bigg|_s = \mathcal{L}[0]|_s \quad .$$

By the linearity of the transform and fact that $\mathcal{L}[0] = 0$, this is the same as

$$\mathcal{L}\left[\frac{dy}{dt}\right]\bigg|_s - 3\mathcal{L}[y]|_s = 0 \quad .$$

Letting $Y = \mathcal{L}[y]$ and applying the "transform of the derivative identity" (Theorem 27.1, above), our equation becomes

$$\left[sY(s) - y(0)\right] - 3Y(s) = 0 \quad ,$$

which, since the initial condition is $y(0) = 4$, *can be rewritten as*

$$sY(s) - 4 - 3Y(s) = 0 \quad .$$

This is a simple algebraic equation that we can easily solve for $Y(s)$. *First, gather the* $Y(s)$ *terms together and add* 4 *to both sides,*

$$[s - 3]Y(s) = 4 \quad ,$$

and then divide through by $s - 3$,

$$Y(s) = \frac{4}{s - 3} \quad .$$

Thus, we have the Laplace transform Y *of the solution* y *to the original initial-value problem. Of course, it would be nice if we can recover the formula for* $y(t)$ *from* $Y(s)$. *And this is fairly easily done, provided we remember that*

$$\frac{4}{s - 3} = 4 \cdot \frac{1}{s - 3} = 4\,\mathcal{L}\big[e^{3t}\big]\big|_s = \mathcal{L}\big[4e^{3t}\big]\big|_s \quad .$$

Combining the last two equations with the definition of Y, *we now have*

$$\mathcal{L}[y(t)]|_s = Y(s) = \frac{4}{s - 3} = \mathcal{L}\big[4e^{3t}\big]\big|_s \quad .$$

That is,

$$\mathcal{L}[y(t)] = \mathcal{L}\big[4e^{3t}\big] \quad ,$$

from which it seems reasonable to expect

$$y(t) = 4e^{3t} \quad .$$

We will confirm that this is valid reasoning when we discuss the "inverse Laplace transform" in the next chapter.

In general, it is fairly easy to find the Laplace transform of the solution to an initial-value problem involving a linear differential equation with constant coefficients and a 'reasonable' forcing function.[1] Simply take the transform of both sides of the differential equation involved, apply the basic identities, avoid getting lost in the bookkeeping, and solve the resulting simple algebraic equation for the unknown function of s. But keep in mind that this is just the Laplace transform $Y(s)$ of the solution $y(t)$ to the original problem. Recovering $y(t)$ from the $Y(s)$ found will usually not be as simple as in the last example. We'll discuss this (the recovering of $y(t)$ from $Y(s)$) in greater detail in the next chapter. For now, let us just practice finding the "$Y(s)$".

!►Example 27.2: *Let's find the Laplace transform* $Y(s) = \mathcal{L}[y(t)]|_s$ *when* y *is the solution to the initial-value problem*

$$y'' - 7y' + 12y = 16e^{2t} \qquad \text{with} \quad y(0) = 6 \quad \text{and} \quad y'(0) = 4 \quad .$$

Taking the transform of the equation and proceeding as in the last example:

$$\mathcal{L}\big[y'' - 7y' + 12y\big]\big|_s = \mathcal{L}\big[16e^{2t}\big]\big|_s$$

$$\hookrightarrow \qquad \mathcal{L}\big[y''\big]\big|_s - 7\mathcal{L}\big[y'\big]\big|_s + 12\mathcal{L}[y]|_s = 16\mathcal{L}\big[e^{2t}\big]\big|_s$$

[1] i.e., a forcing function whose transform is easily computed

$$\hookrightarrow \quad \left[s^2 Y(s) - sy(0) - y'(0)\right] - 7\left[sY(s) - y(0)\right] + 12Y(s) = \frac{16}{s-2}$$

$$\hookrightarrow \quad \left[s^2 Y(s) - s6 - 4\right] - 7\left[sY(s) - 6\right] + 12Y(s) = \frac{16}{s-2}$$

$$\hookrightarrow \quad s^2 Y(s) - 6s - 4 - 7sY(s) + 7 \cdot 6 + 12Y(s) = \frac{16}{s-2}$$

$$\hookrightarrow \quad \left[s^2 - 7s + 12\right]Y(s) - 6s + 38 = \frac{16}{s-2}$$

$$\hookrightarrow \quad \left[s^2 - 7s + 12\right]Y(s) = \frac{16}{s-2} + 6s - 38 \quad .$$

Thus,

$$Y(s) = \frac{16}{(s-2)(s^2 - 7s + 12)} + \frac{6s - 38}{s^2 - 7s + 12} \quad . \tag{27.2}$$

If desired, we can obtain a slightly more concise expression for $Y(s)$ by finding the common denominator and adding the two terms on the right,

$$Y(s) = \frac{16}{(s-2)(s^2 - 7s + 12)} + \frac{(s-2)(6s-38)}{(s-2)\left(s^2 - 7s + 12\right)} \quad ,$$

obtaining

$$Y(s) = \frac{6s^2 - 50s + 92}{(s-2)\left(s^2 - 7s + 12\right)} \quad . \tag{27.3}$$

We will finish solving the above initial-value problem in Example 28.6 on page 504. At that time, we will find the second expression for $Y(s)$ to be more convenient. At this point, though, there is no significant advantage gained by reducing expression (27.2) to (27.3). When doing similar problems in the exercises, go ahead and "find the common denominator and add" if the algebra is relatively simple. Otherwise, leave your answers as the sum of two terms.

However, do observe that we did NOT multiply out the factors in the denominator, but left them as

$$(s-2)\left(s^2 - 7s + 12\right) \quad .$$

Do the same in your own work. In the next chapter, we will see that leaving the denominator in factored form will simplify the task of recovering $y(t)$ from $Y(s)$.

27.2 Derivatives of Transforms

In addition to the "transforms of derivatives" identities just discussed, there are some "derivatives of transforms" identities worth discussing. To derive the basic identity, we start with a generic transform,

$$F(s) = \mathcal{L}[f(t)]|_s = \int_0^\infty f(t)e^{-st}\,dt \quad ,$$

and (naively) look at its derivative,

$$\begin{aligned} F'(s) = \frac{dF}{ds} &= \frac{d}{ds}\int_0^\infty f(t)e^{-st}\,dt \\ &= \int_0^\infty \frac{\partial}{\partial s}e^{-st} f(t)\,dt \\ &= \int_0^\infty (-t)e^{-st} f(t)\,dt = -\underbrace{\int_0^\infty tf(t)e^{-st}\,dt}_{\mathcal{L}[tf(t)]|_s} \quad . \end{aligned}$$

Cutting out the middle of the above set of equalities gives us the identity

$$\frac{dF}{ds} = -\mathcal{L}[tf(t)]|_s \quad .$$

Since we will often use this identity to compute transforms of functions multiplied by t, let's move the negative sign to the other side and rewrite this identity as

$$\mathcal{L}[tf(t)]|_s = -\frac{dF}{ds} \tag{27.4a}$$

or, equivalently, as

$$\mathcal{L}[tf(t)]|_s = -\frac{d}{ds}\mathcal{L}[f(t)]|_s \quad . \tag{27.4b}$$

The cautious reader may be concerned about the validity of

$$\frac{d}{ds}\int_0^\infty g(t,s)\,dt = \int_0^\infty \frac{\partial}{\partial s}[g(t,s)]\,dt \quad ,$$

blithely used (with $g(t,s) = e^{-st} f(t)$) in the above derivation This is a legitimate concern, and is why we must consider the above a somewhat "naive" derivation, instead of a truly rigorous one. Fortunately, the above derivations can be rigorously verified whenever f is of exponential order s_0 and we restrict s to being greater than s_0. This gives the following theorem:

Theorem 27.3 (derivatives of transforms)
Let $F = \mathcal{L}[f]$ where f is a piecewise continuous function of exponential order s_0 on $(0,\infty)$. Then $F(s)$ is differentiable on $s > s_0$, and

$$\mathcal{L}[tf(t)]|_s = -\frac{dF}{ds} \quad \text{for} \quad s > s_0 \quad . \tag{27.5}$$

A rigorous proof of this theorem is not hard, but is a bit longer than our naive derivation. The interested reader can find it in the appendix starting on page 493.

Now let's try using our new identity.

!▸Example 27.3: *Find the Laplace transform of $t\sin(3t)$. Here, we have*

$$\mathcal{L}[t\sin(3t)]|_s = \mathcal{L}[tf(t)]|_s = -\frac{dF}{ds}$$

with $f(t) = \sin(3t)$. From the tables (or memory), we find that

$$F(s) = \mathcal{L}[f(t)]|_s = \mathcal{L}[\sin(3t)]|_s = \frac{3}{s^2+9} \quad .$$

Applying the identity just derived (identity (27.5)) yields

$$\begin{aligned}\mathcal{L}[t\sin(3t)]|_s = \mathcal{L}[tf(t)]|_s &= -\frac{dF}{ds}\\ &= -\frac{d}{ds}\left[\frac{3}{s^2+9}\right] = -\frac{-3\cdot 2s}{\left(s^2+9\right)^2} = \frac{6s}{\left(s^2+9\right)^2} \quad .\end{aligned}$$

Deriving corresponding identities involving higher-order derivatives and higher powers of t is straightforward. Simply use the identity in Theorem 27.3 repeatedly, replacing $f(t)$ with $tf(t)$, $t^2 f(t)$, etc.:

$$\begin{aligned}\mathcal{L}\left[t^2 f(t)\right]\Big|_s = \mathcal{L}\left[t[tf(t)]\right]\Big|_s &= -\frac{d}{ds}\mathcal{L}[tf(t)]|_s\\ &= -\frac{d}{ds}\left[-\frac{dF}{ds}\right] = (-1)^2\frac{d^2F}{ds^2} \quad ,\end{aligned}$$

$$\begin{aligned}\mathcal{L}\left[t^3 f(t)\right]\Big|_s = \mathcal{L}\left[t[t^2 f(t)]\right]\Big|_s &= -\frac{d}{ds}\mathcal{L}\left[t^2 f(t)\right]\Big|_s\\ &= -\frac{d}{ds}\left[(-1)^2\frac{d^2F}{ds^2}\right] = (-1)^3\frac{d^3F}{ds^3} \quad ,\end{aligned}$$

$$\begin{aligned}\mathcal{L}\left[t^4 f(t)\right]\Big|_s = \mathcal{L}\left[t[t^3 f(t)]\right]\Big|_s &= -\frac{d}{ds}\mathcal{L}\left[t^3 f(t)\right]\Big|_s\\ &= -\frac{d}{ds}\left[(-1)^3\frac{d^3F}{ds^3}\right] = (-1)^4\frac{d^4F}{ds^4} \quad ,\end{aligned}$$

and so on. Clearly, then, as a corollary to Theorem 27.3, we have:

Corollary 27.4 (derivatives of transforms)
Let $F = \mathcal{L}[f]$ where f is a piecewise continuous function of exponential order s_0. Then $F(s)$ is infinitely differentiable for $s > s_0$, and

$$\mathcal{L}\left[t^n f(t)\right]\Big|_s = (-1)^n\frac{d^nF}{ds^n} \qquad \text{for} \quad n = 1, 2, 3, \ldots \quad .$$

For easy reference, all the Laplace transform identities we've derived so far are listed in Table 27.1. Also in the table are two identities that will be derived in the next section.

Table 27.1: Commonly Used Identities (Version 1)

In the following, $F(s) = \mathcal{L}[f(t)]|_s$.

$h(t)$	$H(s) = \mathcal{L}[h(t)]\|_s$	Restrictions
$f(t)$	$\int_0^\infty f(t)e^{-st}\,dt$	
$e^{\alpha t} f(t)$	$F(s-\alpha)$	α is real
$\frac{df}{dt}$	$sF(s) - f(0)$	
$\frac{d^2 f}{dt^2}$	$s^2 F(s) - sf(0) - f'(0)$	
$\frac{d^n f}{dt^n}$	$s^n F(s) - s^{n-1} f(0) - s^{n-2} f'(0) - s^{n-3} f''(0) - \cdots - f^{(n-1)}(0)$	$n = 1, 2, 3, \ldots$
$t\,f(t)$	$-\frac{dF}{ds}$	
$t^n\,f(t)$	$(-1)^n \frac{d^n F}{ds^n}$	$n = 1, 2, 3, \ldots$
$\int_0^t f(\tau)\,d\tau$	$\frac{F(s)}{s}$	
$\frac{f(t)}{t}$	$\int_s^\infty F(\sigma)\,d\sigma$	

27.3 Transforms of Integrals and Integrals of Transforms

Analogous to the differentiation identities

$$\mathcal{L}\big[f'(t)\big]\big|_s = sF(s) - f(0) \qquad \text{and} \qquad \mathcal{L}[tf(t)]|_s = -F'(s)$$

are a pair of identities concerning transforms of integrals and integrals of transforms. These identities will not be nearly as important to us as the differentiation identities, but they do have their uses and are considered to be part of the standard set of identities for the Laplace transform.

Before we start, however, take another look at the above differentiation identities. They show that, under the Laplace transform, the differentiation of one of the functions, $f(t)$ or $F(s)$, corresponds to the *multiplication* of the other by the appropriate variable. This may lead you to suspect that the analogous integration identities show that, under the Laplace transform, integration of one of the functions, $f(t)$ or $F(s)$, corresponds to the *division* of the other by the appropriate variable. Be suspicious. We will confirm (and use) this suspicion.

Transform of an Integral

Let

$$g(t) = \int_0^t f(\tau)\,d\tau$$

where f is piecewise continuous on $(0,\infty)$ and of exponential order s_0. From calculus, we (should) know the following:

1. g is continuous on $[0,\infty)$.[2]

2. g is differentiable at every point in $(0,\infty)$ at which f is continuous, and

$$\frac{dg}{dt} = \frac{d}{dt}\int_0^t f(\tau)\,d\tau = f(t) \quad .$$

3. $g(0) = \int_0^0 f(\tau)\,d\tau = 0 \quad .$

In addition, it is not that difficult to show (see Lemma 27.7 on page 492) that g is also of exponential order s_1 where s_1 is any *positive* value greater than or equal to s_0. So both f and g have Laplace transforms, which, as usual, will be denoted by F and G, respectively. Letting $s > s_1$, and using the second and third facts listed above, along with our first differentiation identity, we have

$$\frac{dg}{dt} = f(t)$$

$$\hookrightarrow \qquad \mathcal{L}\left[\frac{dg}{dt}\right]\bigg|_s = \mathcal{L}[f(t)]|_s$$

$$\hookrightarrow \qquad sG(s) - \underbrace{g(0)}_{0} = F(s)$$

$$\hookrightarrow \qquad sG(s) = F(s) \quad .$$

Dividing through by s and recalling what G and g represent then give us the following theorem:

Theorem 27.5 (transform of an integral)
Let $F = \mathcal{L}[f]$ where f is any piecewise continuous function on $(0,\infty)$ of exponential order s_0, and let s_1 be any positive value greater than or equal to s_0. Then

$$\int_0^t f(\tau)\,d\tau$$

is a continuous function of t on $[0,\infty)$ of exponential order s_1, and

$$\mathcal{L}\left[\int_0^t f(\tau)\,d\tau\right]\bigg|_s = \frac{F(s)}{s} \qquad \text{for} \quad s > s_1 \quad .$$

!►Example 27.4: *Let α be any nonnegative real number. The "ramp at α function" can be defined by*

$$\text{ramp}_\alpha(t) = \int_0^t \text{step}_\alpha(\tau)\,d\tau \quad .$$

[2] If the continuity of g is not obvious, take a look at the discussion of Theorem 2.1 on page 30.

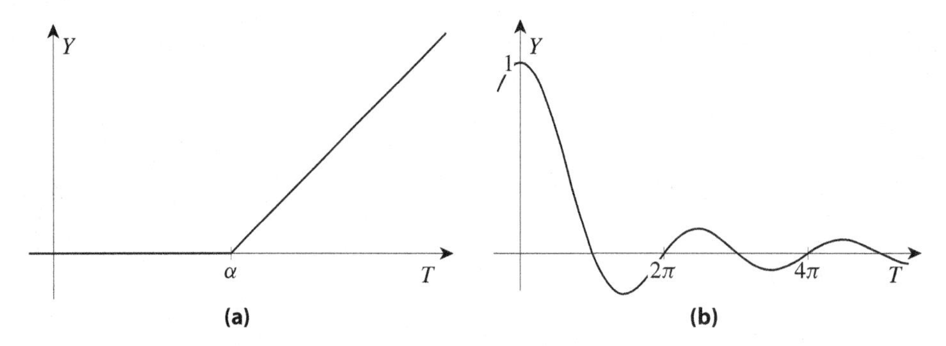

Figure 27.1: The graphs of **(a)** the ramp function at α from Example 27.4 and **(b)** the sinc function from Example 27.5.

If $t \geq \alpha$,

$$\begin{aligned} \text{ramp}_\alpha(t) &= \int_0^t \text{step}_\alpha(\tau)\,d\tau \\ &= \int_0^\alpha \text{step}_\alpha(\tau)\,d\tau + \int_\alpha^t \text{step}_\alpha(\tau)\,d\tau \\ &= \int_0^\alpha 0\,d\tau + \int_\alpha^t 1\,d\tau = 0 + t - \alpha \quad . \end{aligned}$$

If $t < \alpha$,

$$\text{ramp}_\alpha(t) = \int_0^t \text{step}_\alpha(\tau)\,d\tau = \int_0^t 0\,d\tau = 0 \quad .$$

So

$$\text{ramp}_\alpha(t) = \begin{cases} 0 & \text{if } t < \alpha \\ t - \alpha & \text{if } \alpha \leq t \end{cases} \quad ,$$

which is the function sketched in Figure 27.1a.

By the integral formula we gave for $\text{ramp}_\alpha(t)$ *and the identity in Theorem 27.5,*

$$\mathcal{L}\big[\text{ramp}_\alpha(t)\big]\big|_s = \mathcal{L}\left[\int_0^t \text{step}_\alpha(\tau)\,d\tau\right]\bigg|_s = \mathcal{L}\left[\int_0^t f(\tau)\,d\tau\right]\bigg|_s = \frac{F(s)}{s}$$

where $f(t) = \text{step}_\alpha(t)$, $s_0 = 0$ *and*

$$F(s) = \mathcal{L}[f(t)]|_s = \mathcal{L}\big[\text{step}_\alpha(t)\big]\big|_s = \frac{e^{-\alpha s}}{s} \qquad \text{for} \quad s > 0 \quad .$$

So,

$$\mathcal{L}\big[\text{ramp}_\alpha(t)\big]\big|_s = \frac{F(s)}{s} = \frac{e^{-\alpha s}}{s \cdot s} = \frac{e^{-\alpha s}}{s^2} \qquad \text{for} \quad s > 0 \quad .$$

Integral of a Transform

The identity just derived should reinforce our suspicion that, under the Laplace transform, the *division* of $f(t)$ by t should correspond to some integral of F. To confirm this suspicion and derive that integral, let's assume

$$g(t) = \frac{f(t)}{t} \quad .$$

where f is some piecewise continuous function on $(0, \infty)$ of exponential order s_0. Let us further assume that

$$\lim_{t \to 0^+} g(t) = \lim_{t \to 0^+} \frac{f(t)}{t}$$

converges to some finite value. Clearly then, g is also piecewise continuous on $(0, \infty)$ and of exponential order s_0.

Using an old trick along with the "derivative of a transform" identity (in Theorem 27.3), we have

$$F(s) = \mathcal{L}[f(t)]|_s = \mathcal{L}\left[t \cdot \frac{f(t)}{t}\right]\bigg|_s = \mathcal{L}[tg(t)]|_s = -\frac{dG}{ds} \quad .$$

Cutting out the middle and renaming the variable as σ,

$$F(\sigma) = -\frac{dG}{d\sigma} \quad ,$$

allows us to use s as a limit when we integrate both sides,

$$\int_a^s F(\sigma)\, d\sigma = -\int_a^s \frac{dG}{d\sigma}\, d\sigma = -G(s) + G(a) \quad ,$$

which we can then solve for $G(s)$:

$$G(s) = G(a) - \int_a^s F(\sigma)\, d\sigma = G(a) + \int_s^a F(\sigma)\, d\sigma \quad .$$

All this, of course, assumes a and s are any two real numbers greater than s_0. Since we are trying to get a formula for G, we don't know $G(a)$. What we do know (from Theorem 26.5 on page 470) is that $G(a) \to 0$ as $a \to \infty$. This, along with the fact that s and a are independent of each other, means that

$$G(s) = \lim_{a \to \infty} G(s) = \lim_{a \to \infty} \left[G(a) + \int_s^a F(\sigma)\, d\sigma\right] = 0 + \int_s^\infty F(\sigma)\, d\sigma \quad .$$

After recalling what G and g originally denoted, we discover that we have verified:

Theorem 27.6 (integral of a transform)
Let $F = \mathcal{L}[f]$ where f is piecewise continuous on $(0, \infty)$ and of exponential order s_0. Assume further that

$$\lim_{t \to 0^+} \frac{f(t)}{t}$$

converges to some finite value. Then

$$\frac{f(t)}{t}$$

is also piecewise continuous on $(0, \infty)$ and of exponential order s_0. Moreover,

$$\mathcal{L}\left[\frac{f(t)}{t}\right]\bigg|_s = \int_s^\infty F(\sigma)\, d\sigma \qquad \text{for} \quad s > s_0 \quad .$$

!►Example 27.5: *The sinc function (pronounced "sink") is defined by*

$$\operatorname{sinc}(t) = \frac{\sin(t)}{t} \qquad \text{for} \quad t \neq 0 \quad .$$

Its limit as $t \to 0$ is easily computed using L'Hôpital's rule, and defines the value of $\operatorname{sinc}(t)$ when $t = 0$,

$$\operatorname{sinc}(0) = \lim_{t \to 0} \operatorname{sinc}(t) = \lim_{t \to 0} \frac{\sin(t)}{t} = \lim_{t \to 0} \frac{\frac{d}{dt}[\sin(t)]}{\frac{d}{dt}[t]} = \lim_{t \to 0} \frac{\cos(t)}{1} = 1 \quad .$$

The graph of the sinc function is sketched in Figure 27.1b.

Now, by definition,

$$\text{sinc}(t) = \frac{f(t)}{t} \quad \textit{with} \quad f(t) = \sin(t) \quad \textit{for} \quad t > 0 \quad .$$

Clearly, this f satisfies all the requirements for f given in Theorem 27.6 (with $s_0 = 0$). Thus, for $s > 0$, we have

$$\mathcal{L}[\text{sinc}(t)]|_s = \mathcal{L}\left[\frac{f(t)}{t}\right]\Big|_s = \int_s^\infty F(\sigma)\, d\sigma$$

with

$$F(\sigma) = \mathcal{L}[f(t)]|_\sigma = \mathcal{L}[\sin(t)]|_\sigma = \frac{1}{\sigma^2 + 1} \quad .$$

So, for $s > 0$,

$$\begin{aligned}
\mathcal{L}[\text{sinc}(t)]|_s &= \int_s^\infty \frac{1}{\sigma^2 + 1}\, d\sigma \\
&= \arctan(\sigma)\Big|_s^\infty \\
&= \lim_{\sigma \to \infty} \arctan(\sigma) - \arctan(s) = \frac{\pi}{2} - \arctan(s) \quad .
\end{aligned}$$

(By the way, you can derive the equivalent formula

$$\mathcal{L}[\text{sinc}(t)]|_s = \arctan\left(\frac{1}{s}\right)$$

using either arctangent identities or the substitution $\sigma = {}^1\!/_u$ in the last integral above.)

Addendum

Here's a little fact used in deriving the "transform of an integral" identity in Theorem 27.5. We prove it here because the proof could distract from the more important part of that derivation.

Lemma 27.7

Let f be any piecewise continuous function on $(0, \infty)$ of exponential order s_0. Then the function g, given by

$$g(t) = \int_0^t f(\tau)\, d\tau \quad ,$$

is of exponential order s_1 where s_1 is any positive value greater than or equal to s_0.

PROOF: Since f is piecewise continuous on $(0, \infty)$ and of exponential order s_0, Lemma 26.8 on page 475 assures us that there is a constant M such that

$$|f(t)| \leq Me^{s_0 t} \quad \text{whenever} \quad 0 < t \quad .$$

Now let s_1 be any positive value greater than or equal to s_0, and let $t > 0$. Using the above and the integral inequality from Lemma 26.9 on page 475, we have

$$\begin{aligned}
|g(t)| = \left|\int_0^t f(\tau)\, d\tau\right| &\leq \int_0^t |f(\tau)|\, d\tau \\
&\leq \int_0^t Me^{s_0 \tau}\, d\tau \\
&\leq \int_0^t Me^{s_1 \tau}\, d\tau = \frac{M}{s_1}\left[e^{s_1 t} - e^{s_1 \cdot 0}\right] \leq \frac{M}{s_1} e^{s_1 t} \quad .
\end{aligned}$$

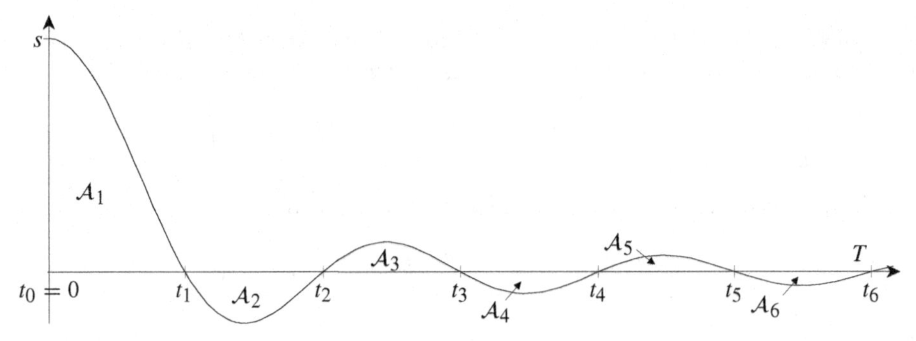

Figure 27.2: The graph of $\frac{\sin(st)}{t}$ for some $s > 0$ with $t_k = {}^{k\pi}/_{s}$.

So, letting $M_1 = {}^{M}/_{s_1}$,

$$|g(t)| < M_1 e^{s_1 t} \qquad \text{for} \quad t > 0 \quad ,$$

verifying that g is of exponential order s_1. ∎

27.4 Appendix: Differentiating the Transform

The Main Issue

On page 486, we derived the "derivative of a transform" identity, $F' = -\mathcal{L}[tf(t)]$, naively using the "fact" that

$$\frac{d}{ds}\int_0^\infty g(t,s)\,dt = \int_0^\infty \frac{\partial}{\partial s}[g(t,s)]\,dt \quad . \tag{27.6}$$

The problem is that this "fact", while *often* true, is not *always* true.

!►Example 27.6: Let

$$g(t,s) = \frac{\sin(st)}{t} \qquad \text{for} \quad t > 0 \quad \text{and} \quad s > 0 \quad .$$

(Compare this function to the sinc function in Example 27.5 on page 491.) It is easily verified that the graph of this function over the positive T–axis is as sketched in Figure 27.2. Recalling the relation between integration and area, we also see that, for each $s > 0$,

$$\int_0^\infty g(t,s)\,dt = \lim_{T\to\infty}\int_0^T \frac{\sin(st)}{t}\,dt$$

$$= \mathcal{A}_1 - \mathcal{A}_2 + \mathcal{A}_3 - \mathcal{A}_4 + \mathcal{A}_5 - \mathcal{A}_6 + \cdots = \sum_{k=1}^{\infty}(-1)^{k+1}\mathcal{A}_k \quad .$$

where each $\mathcal{A}_k$ is the area enclosed by the graph of g and the T–axis interval (t_{k-1}, t_k) described in Figure 27.2. Notice that this last summation is an alternating series whose terms are steadily decreasing to zero. As you surely recall from your calculus course, any such summation is convergent. Hence, so is the above integral of g. That is, this integral is well defined and finite *for each* $s > 0$ *. Unfortunately, this integral cannot be evaluated by elementary means. Still, using the substitution* $t = {}^{\tau}/_{s}$ *, we can reduce this integral to a slightly simpler form:*

$$\int_0^\infty g(t,s)\,dt = \int_0^\infty \frac{\sin(st)}{t}\,dt = \int_0^\infty \frac{\sin(\tau)}{\tau/s}\frac{d\tau}{s} = \int_0^\infty \frac{\sin(\tau)}{\tau}\,d\tau \quad .$$

Thus, in fact, this integral does not depend on s *. Consequently,*

$$\frac{d}{ds}\int_0^\infty g(t,s)\,dt = \frac{d}{ds}\int_0^\infty \frac{\sin(\tau)}{\tau}\,d\tau = 0 \quad .$$

On the other hand,

$$\begin{aligned}\int_0^\infty \frac{\partial}{\partial s}[g(t,s)]\,dt &= \int_0^\infty \frac{\partial}{\partial s}\left[\frac{\sin(st)}{t}\right]dt \\ &= \int_0^\infty \frac{\cos(st)\cdot t}{t}\,dt \\ &= \int_0^\infty \cos(st)\,dt = \lim_{t\to\infty}\frac{\sin(st)}{s} \quad ,\end{aligned}$$

which is not 0 *— it does not even converge! Thus, at least for this choice of* $g(t,s)$ *,*

$$\frac{d}{ds}\int_0^\infty g(t,s)\,dt \neq \int_0^\infty \frac{\partial}{\partial s}[g(t,s)]\,dt \quad .$$

There are fairly reasonable conditions ensuring that equation (27.6) holds, and our use of it on page 486 in deriving the "derivative of the transform" identity can be justified once we know those "reasonable conditions". But instead, let's see if we can rigorously verify our identity just using basic facts from elementary calculus.

The Rigorous Derivation

Our goal is to prove Theorem 27.3 on page 486. That is, we want to rigorously derive the identity

$$F'(s) = -\mathcal{L}[tf(t)]|_s \qquad (27.7)$$

assuming $F = \mathcal{L}[f]$ with f being a piecewise continuous function of exponential order s_0 . We will also assume $s > s_0$.

First, you should verify that the results of Exercise 26.18 on page 479 and Lemma 26.8 on page 475 give us:

Lemma 27.8
If $f(t)$ *is of exponential order* s_0 *, and* n *is any positive integer, then* $t^n f(t)$ *is a piecewise continuous function on* $(0,\infty)$ *of exponential order* $s_0 + \sigma$ *for any* $\sigma > 0$ *. Moreover, there is a constant* M_σ *such that*

$$\left|t^n f(t)\right| \leq M_\sigma e^{(s_0+\sigma)t} \qquad \text{for all} \quad t > 0 \quad .$$

Since we can always find a positive σ such that $s_0 < s_0 + \sigma < s$, this lemma assures us that $\mathcal{L}\big[t^n f(t)\big]\big|_s$ is well defined for $s > s_0$.

Now let's consider $F'(s)$. By definition

$$F'(s) = \lim_{\Delta s\to 0}\frac{F(s+\Delta s) - F(s)}{\Delta s}$$

provided the limit exists. Taking $|\Delta s|$ small enough so that $s + \Delta s$ is also greater than s_0 (even if $\Delta s < 0$), we have

$$\begin{aligned}\frac{F(s+\Delta s) - F(s)}{\Delta s} &= \frac{1}{\Delta s}[F(s+\Delta s) \; - \; F(s)] \\ &= \frac{1}{\Delta s}\left[\int_0^\infty f(t)\underbrace{e^{-(s+\Delta s)t}}_{e^{-st}e^{-\Delta st}} dt \; - \; \int_0^\infty f(t)e^{-st}\, dt\right] \quad ,\end{aligned}$$

which simplifies to

$$\frac{F(s+\Delta s) - F(s)}{\Delta s} = \int_0^\infty f(t)\left(\frac{1}{\Delta s}\right)\left[e^{-(\Delta s)t} - 1\right]e^{-st}\, dt \quad . \tag{27.8}$$

To deal with the integral in the last equation, we will use the fact that, for any value x, the exponential of x is given by its Taylor series,

$$e^x = \sum_{k=0}^{\infty}\frac{1}{k!}x^k = 1 + x + \frac{1}{2!}x^2 + \frac{1}{3!}x^3 + \frac{1}{4!}x^4 + \cdots \quad .$$

So

$$\begin{aligned}e^x - 1 &= x + \frac{1}{2!}x^2 + \frac{1}{3!}x^3 + \frac{1}{4!}x^4 + \cdots \\ &= x + x^2 E(x)\end{aligned}$$

where

$$E(x) = \frac{1}{2!} + \frac{1}{3!}x + \frac{1}{4!}x^2 + \cdots \quad .$$

Consequently (using $x = -(\Delta s)t$),

$$\begin{aligned}\frac{1}{\Delta s}\left[e^{-(\Delta s)t} - 1\right] &= \frac{1}{\Delta s}\left[-(\Delta s)t + [-(\Delta s)t]^2 E(-(\Delta s)t)\right] \\ &= -t + (\Delta s)t^2 E(-(\Delta s)t) \quad .\end{aligned}$$

Combined with equation (27.8), this yields

$$\begin{aligned}\frac{F(s+\Delta s) - F(s)}{\Delta s} &= \int_0^\infty f(t)\left[-t + (\Delta s)t^2 E(-(\Delta s)t)\right]e^{-st}\, dt \\ &= \int_0^\infty f(t)[-t]e^{-st}\, dt + \int_0^\infty f(t)\left[(\Delta s)t^2 E(-(\Delta s)t)\right]e^{-st}\, dt \quad .\end{aligned}$$

That is,

$$\frac{F(s+\Delta s) - F(s)}{\Delta s} = -\mathcal{L}[tf(t)]\big|_s + \Delta s\int_0^\infty t^2 f(t)E(-(\Delta s)t)e^{-st}\, dt \quad . \tag{27.9}$$

Obviously, the question now is *What happens to the second term on the right when* $\Delta s \to 0$*?* To help answer that, let us observe that, for all x,

$$\begin{aligned}|E(x)| &= \left|\frac{1}{2!} + \frac{1}{3!}x + \frac{1}{4!}x^2 + \frac{1}{5!}x^3 + \cdots\right| \\ &\leq \frac{1}{2!} + \frac{1}{3!}|x| + \frac{1}{4!}|x|^2 + \frac{1}{5!}|x|^3 + \cdots \\ &< 1 + |x| + \frac{1}{2!}|x|^2 + \frac{1}{3!}|x|^3 + \cdots = e^{|x|} \quad .\end{aligned}$$

Moreover, as noted in Lemma 27.8, for each $\sigma > 0$, there is a positive constant M_σ such that

$$\left|t^2 f(t)\right| \leq M_\sigma e^{(s_0+\sigma)t} \qquad \text{for all} \quad t > 0 \quad .$$

Remember, $s > s_0$. And since σ can be chosen as close to 0 as desired, and we are taking the limit as $\Delta s \to 0$, we may assume that $s + [\sigma + |\Delta s|] > s_0$. Doing so and applying the above then yields

$$\begin{aligned}
\left|\Delta s \int_0^\infty t^2 f(t) E(-(\Delta s)t) e^{-st}\, dt\right| &\leq |\Delta s| \int_0^\infty \left|t^2 f(t)\right| |E(-(\Delta s)t)|\, e^{-st}\, dt \\
&\leq |\Delta s| \int_0^\infty M_\sigma e^{(s_0+\sigma)t} e^{|-(\Delta s)t|} e^{-st}\, dt \\
&= |\Delta s|\, M_\sigma \int_0^\infty e^{-(s-s_0-\sigma-|\Delta s|)t}\, dt \\
&\leq |\Delta s| \frac{M_\sigma}{s - s_0 - \sigma - |\Delta s|} \quad .
\end{aligned}$$

Thus,

$$\begin{aligned}
0 \leq \lim_{\Delta s \to 0} \left|\Delta s \int_0^\infty t^2 f(t) E(-(\Delta s)t) e^{-st}\, dt\right| &\leq \lim_{\Delta s \to 0} |\Delta s| \frac{M_\sigma}{s - s_0 - \sigma - |\Delta s|} \\
&= 0 \cdot \frac{M_\sigma}{s - s_0 - \sigma - 0} = 0 \quad ,
\end{aligned}$$

which, of course, means that

$$\lim_{\Delta s \to 0} \left|\Delta s \int_0^\infty t^2 f(t) E(-(\Delta s)t) e^{-st}\, dt\right| = 0 \quad .$$

Combining this with equation (27.9), we finally obtain

$$\lim_{\Delta s \to 0} \frac{F(s + \Delta s) - F(s)}{\Delta s} = -\mathcal{L}[tf(t)]|_s + 0 \quad ,$$

verifying both the differentiability of F at s, and equation (27.7).

Additional Exercises

27.1. *Find the Laplace transform $Y(s)$ of the solution to each of the following initial-value problems. Just find $Y(s)$ using the ideas illustrated in Examples 27.1 and 27.2. Do NOT solve the problem using methods developed before we started discussing Laplace transforms and then computing the transform! Also, do not attempt to recover $y(t)$ from each $Y(s)$ you obtain.*

a. $y' + 4y = 0$ *with* $y(0) = 3$

b. $y' - 2y = t^3$ *with* $y(0) = 4$

c. $y' + 3y = \text{step}_4(t)$ *with* $y(0) = 0$

d. $y'' - 4y = t^3$ *with* $y(0) = 1$ *and* $y'(0) = 3$

e. $y'' + 4y = 20e^{4t}$ with $y(0) = 3$ and $y'(0) = 12$

f. $y'' + 4y = \sin(2t)$ with $y(0) = 3$ and $y'(0) = 5$

g. $y'' + 4y = 3\,\text{step}_2(t)$ with $y(0) = 0$ and $y'(0) = 5$

h. $y'' + 5y' + 6y = e^{4t}$ with $y(0) = 1$ and $y'(0) = 0$

i. $y'' - 5y' + 6y = t^2 e^{4t}$ with $y(0) = 0$ and $y'(0) = 2$

j. $y'' - 5y' + 6y = 7$ with $y(0) = 2$ and $y'(0) = 4$

k. $y'' - 4y' + 13y = e^{2t}\sin(3t)$ with $y(0) = 4$ and $y'(0) = 3$

l. $y'' + 4y' + 13y = 4t + 2e^{2t}\sin(3t)$ with $y(0) = 4$ and $y'(0) = 3$

m. $y''' - 27y = e^{-3t}$ with $y(0) = 2$, $y'(0) = 3$ and $y''(0) = 4$

27.2. Compute the Laplace transforms of the following functions using the given tables and the 'derivative of the transforms identities' from Theorem 27.3 (and its corollary).

a. $t\cos(3t)$ **b.** $t^2\sin(3t)$ **c.** te^{-7t}

d. $t^3 e^{-7t}$ **e.** $t\,\text{step}(t-3)$ **f.** $t^2\,\text{step}_4(t)$

27.3. Verify the following identities using the 'derivative of the transforms identities' from Theorem 27.3.

a. $\mathcal{L}[t\sin(\omega t)]|_s = \dfrac{2\omega s}{\left(s^2+\omega^2\right)^2}$ **b.** $\mathcal{L}[t\cos(\omega t)]|_s = \dfrac{s^2-\omega^2}{\left(s^2+\omega^2\right)^2}$

27.4. For the following, let y be the solution to

$$t\frac{d^2y}{dt^2} + \frac{dy}{dt} + ty = 0 \quad \text{with} \quad y(0) = 1 \quad \text{and} \quad y'(0) = 0 \quad .$$

The above differential equation, known as Bessel's equation of order zero, is important in many two-dimensional problems involving circular symmetry. The solution to this equation with the above initial values turns out to be particularly important. It is called the Bessel function (of the first kind) of order zero, and is universally denoted by J_0. Thus, in the following,

$$y(t) = J_0(t) \quad \text{and} \quad Y(s) = \mathcal{L}[y(t)]|_s = \mathcal{L}[J_0(t)]|_s \quad .$$

a. Using the differentiation identities from this chapter, show that

$$\left(s^2+1\right)\frac{dY}{ds} + sY = 0 \quad .$$

b. The above differential equation for Y is a simple first-order differential equation. Find its general solution.

c. It can be shown (trust the author on this) that

$$\int_0^\infty J_0(t) = 1 \quad .$$

What does this tell you about $Y(0)$?

d. Using what you now know about $Y(s)$, find $= \mathcal{L}[J_0(t)]|_s$.

27.5. *Compute the Laplace transforms using the tables provided. You will have to apply two different identities.*

a. $te^{4t}\sin(3t)$ **b.** $te^{4t}\cos(3t)$ **c.** $te^{4t}\,\text{step}(t-3)$ **d.** $e^{3t}t^2\,\text{step}_1(t)$

27.6. *The following concern the ramp at α function whose transform was computed in Example 27.4 on page 489. Assume $\alpha > 0$.*

a. *Verify that the "ramp-squared at α" function,*

$$\text{ramp}_\alpha^2(t) = \left(\text{ramp}_\alpha(t)\right)^2 \quad ,$$

satisfies

$$\text{ramp}_\alpha^2(t) = \int_0^t 2\,\text{ramp}_\alpha(\tau)\,d\tau \quad .$$

b. *Using the above and the "transform of an integral" identity, find $\mathcal{L}\left[\text{ramp}_\alpha^2(t)\right]\big|_s$.*

27.7. *The sine-integral function, Si, is given by*

$$\text{Si}(t) = \int_0^t \frac{\sin(\tau)}{\tau}\,d\tau \quad .$$

In Example 27.5, it is shown that

$$\mathcal{L}\left[\frac{\sin(t)}{t}\right]\bigg|_s = \arctan\left(\frac{1}{s}\right) \quad .$$

What is $\mathcal{L}[\text{Si}(t)]|_s$?

27.8. *Verify that the limit of each of the following functions as $t \to 0$ is a finite number, and then find the Laplace transform of that function using the "integral of a transform" identity.*

a. $\dfrac{1-e^{-t}}{t}$ **b.** $\dfrac{e^{2t}-1}{t}$ **c.** $\dfrac{e^{-2t}-e^{3t}}{t}$

d. $\dfrac{1-\cos(t)}{t}$ **e.** $\dfrac{1-\cosh(t)}{t}$ **f.** $\dfrac{\sin(3t)}{t}$

28

The Inverse Laplace Transform

In the last chapter, we saw how to find Laplace transforms of "unknown" functions satisfying various initial-value problems. Of course, it's not the transforms of those unknown functions which are usually of interest. It's the functions, themselves, that are of interest. So let us turn to the general issue of finding a function $y(t)$ when all we know is its Laplace transform $Y(s)$.

28.1 Basic Notions

On Recovering a Function from Its Transform

In attempting to solve the differential equation in Example 27.1, we got

$$Y(s) = \frac{4}{s-3} \quad ,$$

which, since

$$Y(s) = \mathcal{L}[y(t)]|_s \qquad \text{and} \qquad \frac{4}{s-3} = \mathcal{L}\left[4e^{3t}\right]\Big|_s \quad ,$$

we rewrote as

$$\mathcal{L}[y(t)] = \mathcal{L}\left[4e^{3t}\right] \quad .$$

From this, it seemed reasonable to conclude that

$$y(t) = 4e^{3t} \quad .$$

But, what if there were another function $f(t)$ with the same transform as $4e^{3t}$? Then we could not be sure whether the above $y(t)$ should be $4e^{3t}$ or that other function $f(t)$. Fortunately, someone has managed to prove the following:

Theorem 28.1 (uniqueness of the transforms)
Suppose f and g are any two piecewise continuous functions on $[0,\infty)$ of exponential order and having the same Laplace transforms,

$$\mathcal{L}[f] = \mathcal{L}[g] \quad .$$

Then, as piecewise continuous functions,

$$f(t) = g(t) \qquad \text{on} \quad [0,\infty) \quad .$$

(You may want to quickly review the discussion of "equality of piecewise continuous functions" on page 472.)

This theorem actually follows from another result that will be briefly discussed at the end of this section. What is important, now, is that this theorem assures us that, if

$$\mathcal{L}[y(t)]|_s = \mathcal{L}\left[4e^{3t}\right]|_s \quad ,$$

then

$$y(t) = 4e^{3t} \quad ,$$

at least for $t \geq 0$.

What about for $t < 0$? Well, keep in mind that the Laplace transform of any function f,

$$F(s) = \mathcal{L}[f]|_s = \int_0^\infty f(t)e^{-st}\,dt \quad ,$$

involves integration only over the positive T–axis. The behavior of f on the negative T–axis has no effect on the formula for $F(s)$. In fact, $f(t)$ need not even be defined for $t < 0$. So, even if they exist, there can be no way to recover the values of $f(t)$ on the negative T–axis from $F(s)$. But that is not a real concern because we will just use the Laplace transform for problems over the positive T–axis — problems in which we have initial values at $t = 0$ and want to know what happens *later*.

What all this means is that we are only interested in functions of t with $t \geq 0$. That was hinted at when we began our discussions of the Laplace transform (see note 3 on page 452), but we did not make an issue of it to avoid getting too distracted by technical details. Now, with the inverse transform, requiring $t \geq 0$ becomes more of an issue. Still, there is no need to obsess about this any more than necessary, or to suddenly start including "for $t \geq 0$" with every formula of t. Let us just agree that the negative T–axis is irrelevant to our discussions, and that

in all formulas involving t, it is assumed that $t \geq 0$.

!►Example 28.1: *Somewhere above, we have*

$$y(t) = 4e^{3t} \quad .$$

What we really mean is that

$$y(t) = 4e^{3t} \qquad \text{for} \quad t \geq 0 \quad .$$

We have no idea what $y(t)$ is for $t < 0$. We don't even know whether, in whatever application this may have arisen, it makes sense to talk about $y(t)$ for $t < 0$, nor do we care.[1]

The Inverse Laplace Transform Defined

We can now officially define the inverse Laplace transform:

> Given a function $F(s)$, the *inverse Laplace transform of* F, denoted by $\mathcal{L}^{-1}[F]$, is that function f whose Laplace transform is F.

More succinctly:

$$f(t) = \mathcal{L}^{-1}[F(s)]|_t \quad \Longleftrightarrow \quad \mathcal{L}[f(t)]|_s = F(s) \quad .$$

Our theorem on uniqueness (Theorem 28.1) (along with our understanding about "always assuming $t \geq 0$") assures us that the above definition for $\mathcal{L}^{-1}[F]$ is unambiguous. In this definition, of course, we assume $F(s)$ can be given as $\mathcal{L}[f(t)]$ for some function f.

[1] For example: What if $y(t)$ denoted the temperature in a cup of coffee t minutes after being poured? Does it make sense to consider the temperature of the coffee before it exists? (Answer this assuming you are not a Zen master.)

!►Example 28.2: *We have*

$$\mathcal{L}^{-1}\left[\frac{4}{s-3}\right]\bigg|_t = 4e^{3t}$$

because

$$\frac{4}{s-3} = \mathcal{L}\left[4e^{3t}\right]\big|_s \quad .$$

Likewise, since

$$\mathcal{L}\left[t^3\right]\big|_s = \frac{6}{s^4} \quad ,$$

we have

$$t^3 = \mathcal{L}^{-1}\left[\frac{6}{s^4}\right]\bigg|_t \quad .$$

The fact that

$$f(t) = \mathcal{L}^{-1}[F(s)]|_t \iff \mathcal{L}[f(t)]|_s = F(s)$$

means that any table of Laplace transforms (such as Table 26.1 on page 460) is also a table of inverse Laplace transforms. Instead of reading off the $F(s)$ for each $f(t)$ found, read off the $f(t)$ for each $F(s)$.

As you may have already noticed, we take inverse transforms of "functions of s that are denoted by uppercase Roman letters" and obtain "functions of t that are denoted by the corresponding lowercase Roman letter". These notational conventions are consistent with the notational conventions laid down for the Laplace transform early in Chapter 26.

We should also note that the phrase "inverse Laplace transform" can refer to either the 'inverse transformed function' f or to the process of computing f from F.

By the way, there is a formula for computing inverse Laplace transforms. If you must know, it is

$$\mathcal{L}^{-1}[F(s)]|_t = \frac{1}{2\pi} \lim_{Y\to+\infty} \int_{-Y}^{Y} e^{t(\sigma+i\xi)} F(\sigma + i\xi)\, d\xi \quad .$$

The integral here is over a line in the complex plane, and σ is a suitably chosen positive value. In deriving this formula, you actually verify uniqueness Theorem 28.1. Unfortunately, deriving and verifying this formula go beyond our current abilities.[2]

Don't pretend to understand this formula, and don't try to use it until you've had a course in complex variables. Besides, it is not nearly as useful as a good table of transforms.

28.2 Linearity and Using Partial Fractions

Linearity of the Inverse Transform

The fact that the inverse Laplace transform is linear follows immediately from the linearity of the Laplace transform. To see this, consider $\mathcal{L}^{-1}[\alpha F(s) + \beta G(s)]$ where α and β are any two constants and F and G are any two functions for which inverse Laplace transforms exist. Following our conventions, we'll let

$$f(t) = \mathcal{L}^{-1}[F(s)]|_t \quad \text{and} \quad g(t) = \mathcal{L}^{-1}[G(s)]|_t \quad .$$

[2] Two derivations can be found in the third edition of *Transforms and Applications Handbook* (Ed: A. Poularikas, CRC Press). One, using Fourier transforms, is in Section 2.4.6 of the chapter on Fourier transforms by Howell. The other, using results from the theory of complex analytic functions, is in Section 5.6 of the chapter on Laplace transforms by Poularikas and Seely.

Remember, this is completely the same as stating that

$$\mathcal{L}[f(t)]|_s = F(s) \quad \text{and} \quad \mathcal{L}[g(t)]|_s = G(s) \quad .$$

Because we already know the Laplace transform is linear, we know

$$\alpha F(s) + \beta G(s) = \alpha\mathcal{L}[f(t)]|_s + \beta\mathcal{L}[g(t)]|_s = \mathcal{L}[\alpha f(t) + \beta g(t)]|_s \quad .$$

This, along the definition of the inverse transform and the above definitions of f and g, yields

$$\mathcal{L}^{-1}[\alpha F(s) + \beta G(s)]|_t = \alpha f(t) + \beta g(t) = \alpha\mathcal{L}^{-1}[F(s)]|_t + \beta\mathcal{L}^{-1}[G(s)]|_t \quad .$$

Redoing these little computations with as many functions and constants as desired then gives us the next theorem:

Theorem 28.2 (linearity of the inverse Laplace transform)
The inverse Laplace transform is linear. That is,

$$\mathcal{L}^{-1}[c_1F_1(s) + c_2F_2(s) + \cdots + c_nF_n(s)] = c_1\mathcal{L}^{-1}[F_1(s)] + c_2\mathcal{L}^{-1}[F_2(s)] + \cdots + c_n\mathcal{L}^{-1}[F_n(s)]$$

when each c_k is a constant and each F_k is a function having an inverse Laplace transform.

Let's now use the linearity to compute a few inverse transforms.

!►Example 28.3: *Let's find*

$$\mathcal{L}^{-1}\left[\frac{1}{s^2+9}\right]\bigg|_t \quad .$$

We know (or found in Table 26.1 on page 460) that

$$\mathcal{L}^{-1}\left[\frac{3}{s^2+9}\right]\bigg|_t = \sin(3t) \quad ,$$

which is almost what we want. To use this in computing our desired inverse transform, we will combine linearity with one of mathematics' oldest tricks — multiplying by 1 — with, in this case, $1 = 3/3$):

$$\mathcal{L}^{-1}\left[\frac{1}{s^2+9}\right]\bigg|_t = \mathcal{L}^{-1}\left[\frac{1}{3}\cdot\frac{3}{s^2+9}\right]\bigg|_t = \frac{1}{3}\mathcal{L}^{-1}\left[\frac{3}{s^2+9}\right]\bigg|_t = \frac{1}{3}\sin(3t) \quad .$$

The use of linearity along with 'multiplying by 1' will be used again and again. Get used to it.

!►Example 28.4: *Let's find the inverse Laplace transform of*

$$\frac{30}{s^7} + \frac{8}{s-4} \quad .$$

We know

$$\mathcal{L}^{-1}\left[\frac{6!}{s^7}\right]\bigg|_t = t^6 \quad \text{and} \quad \mathcal{L}^{-1}\left[\frac{1}{s-4}\right]\bigg|_t = e^{4t} \quad .$$

So,

$$\mathcal{L}^{-1}\left[\frac{30}{s^7} + \frac{8}{s-4}\right]\Big|_t = 30\mathcal{L}^{-1}\left[\frac{1}{s^7}\right]\Big|_t + 8\mathcal{L}^{-1}\left[\frac{1}{s-4}\right]\Big|_t$$

$$= 30\mathcal{L}^{-1}\left[\frac{1}{6!}\cdot\frac{6!}{s^7}\right]\Big|_t + 8e^{4t}$$

$$= \frac{30}{6!}\mathcal{L}^{-1}\left[\frac{6!}{s^7}\right]\Big|_t + 8e^{4t} = \frac{30}{6\cdot 5\cdot 4\cdot 3\cdot 2}t^6 + 8e^{4t} \quad ,$$

which, after a little arithmetic, reduces to

$$\mathcal{L}^{-1}\left[\frac{30}{s^7} + \frac{8}{s-4}\right]\Big|_t = \frac{1}{24}t^6 + 8e^{4t} \quad .$$

Partial Fractions

When using the Laplace transform with differential equations, we often get transforms that can be converted via 'partial fractions' to forms that are easily inverse transformed using the tables and linearity, as above. This means that the general method(s) of partial fractions are particularly important. By now, you should be well-acquainted with using partial fractions — remember, the basic idea is that, if we have a fraction of two polynomials

$$\frac{Q(s)}{P(s)}$$

and $P(s)$ can be factored into two smaller polynomials

$$P(s) = P_1(s)P_2(s) \quad ,$$

then two other polynomials $Q_1(s)$ and $Q_2(s)$ can be found so that

$$\frac{Q(s)}{P(s)} = \frac{Q(s)}{P_1(s)P_2(s)} = \frac{Q_1(s)}{P_1(s)} + \frac{Q_2(s)}{P_2(s)} \quad .$$

Moreover, if (as will usually be the case for us) the degree of $Q(s)$ is less than the degree of $P(s)$, then the degree of each $Q_k(s)$ will be less than the degree of the corresponding $P_k(s)$.

You probably used partial fractions to compute some of the integrals in the earlier chapters of this text. We'll go through a few examples to both refresh our memories of this technique and to see how it naturally arises in using the Laplace transform to solve differential equations.

!►Example 28.5: *In Exercise 27.1 e on page 497, you found that the Laplace transform of the solution to*

$$y'' + 4y = 20e^{4t} \quad \text{with} \quad y(0) = 3 \quad \text{and} \quad y'(0) = 12$$

is

$$Y(s) = \frac{3s^2 - 28}{(s-4)(s^2+4)} \quad .$$

The partial fraction expansion of this is

$$Y(s) = \frac{3s^2 - 28}{(s-4)(s^2+4)} = \frac{A}{s-4} + \frac{Bs + C}{s^2+4}$$

for some constants A, B and C. There are many ways to find these constants. The basic method is to "undo" the partial fraction expansion by getting a common denominator and adding up the fractions on the right:

$$\begin{aligned}\frac{3s^2 - 28}{(s-4)(s^2+4)} &= \frac{A}{s-4} + \frac{Bs+C}{s^2+4} \\ &= \frac{A(s^2+4)}{(s-4)(s^2+4)} + \frac{(s-4)(Bs+C)}{(s-4)(s^2+4)} \\ &= \cdots \\ &= \frac{(A+B)s^2 + (C-4B)s + 4(A-C)}{(s-4)(s^2+4)} \quad .\end{aligned}$$

Cutting out the middle and canceling out the common denominator lead to the equation

$$3 \cdot s^2 + 0 \cdot s - 28 = (A+B)s^2 + (C-4B)s + 4(A-C) \quad ,$$

which, in turn, means that our constants satisfy the three-by-three system

$$\begin{aligned} 3 &= A + B \\ 0 &= C - 4B \\ -28 &= 4A - 4C \end{aligned} \quad .$$

This is a relatively simple system. Solving it however you wish, you obtain

$$A = 1 \quad \text{and} \quad B = 2 \quad \text{and} \quad C = 8 \quad .$$

Hence

$$Y(s) = \frac{A}{s-4} + \frac{Bs+C}{s^2+4} = \frac{1}{s-4} + \frac{2s+8}{s^2+4} \quad ,$$

and

$$\begin{aligned} y(t) = \mathcal{L}^{-1}[Y(s)]\big|_t &= \mathcal{L}^{-1}\left[\frac{1}{s-4} + \frac{2s+8}{s^2+4}\right]\bigg|_t \\ &= \mathcal{L}^{-1}\left[\frac{1}{s-4}\right]\bigg|_t + 2\mathcal{L}^{-1}\left[\frac{s}{s^2+4}\right]\bigg|_t + 8\mathcal{L}^{-1}\left[\frac{1}{s^2+4}\right]\bigg|_t \\ &= e^{4t} + 2\mathcal{L}^{-1}\left[\frac{s}{s^2+2^2}\right]\bigg|_t + 8 \cdot \frac{1}{2}\mathcal{L}^{-1}\left[\frac{2}{s^2+2^2}\right]\bigg|_t \\ &= e^{4t} + 2\cos(2t) + 4\sin(2t) \quad . \end{aligned}$$

!► Example 28.6: *In Example 27.2 on page 484 we obtained*

$$Y(s) = \frac{16}{(s-2)(s^2-7s+12)} + \frac{6s-38}{s^2-7s+12}$$

and, equivalently,

$$Y(s) = \frac{6s^2 - 50s + 92}{(s-2)(s^2-7s+12)}$$

as the Laplace transform of the solution to some initial-value problem. While we could find partial fraction expansions for each term of the first expression above, it will certainly be more convenient to simply find the single partial fraction expansion for the second expression for $Y(s)$ *. But before attempting that, we should note that one factor in the denominator can be further factored,*

$$s^2 - 7s + 12 = (s-3)(s-4) \quad ,$$

giving us

$$Y(s) = \frac{6s^2 - 50s + 92}{(s-2)(s-3)(s-4)} \quad .$$

Now we can seek the partial fraction expansion of $Y(s)$ *:*

$$\begin{aligned}\frac{6s^2 - 50s + 92}{(s-2)(s-3)(s-4)} &= \frac{A}{s-2} + \frac{B}{s-3} + \frac{C}{s-4} \\ &= \cdots \\ &= \frac{A(s-3)(s-4) + B(s-2)(s-4) + C(s-2)(s-3)}{(s-2)(s-3)(s-4)} \quad .\end{aligned}$$

Cutting out the middle and canceling out the common denominator leave

$$\begin{aligned}6s^2 &- 50s + 92 \\ &= A(s-3)(s-4) + B(s-2)(s-4) + C(s-2)(s-3) \quad .\end{aligned} \tag{28.1}$$

Rather than multiplying out the right side of this equation and setting up the system that A *,* B *and* C *must satisfy for this equation to hold (as we did in the previous example), let's find these constants after making clever choices for the value of* s *in this last equation.*

Letting $s = 2$ *in equation (28.1):*

$$\begin{aligned}6 \cdot 2^2 &- 50 \cdot 2 + 92 \\ &= A(2-3)(2-4) + B(2-2)(2-4) + C(2-2)(2-3)\end{aligned}$$

$$\hookrightarrow \qquad 16 = 2A + 0B + 0C \quad \rightarrowtail \quad A = 8 \quad .$$

Letting $s = 3$ *in equation (28.1):*

$$\begin{aligned}6 \cdot 3^2 &- 50 \cdot 3 + 92 \\ &= A(3-3)(3-4) + B(3-2)(3-4) + C(3-2)(3-3)\end{aligned}$$

$$\hookrightarrow \qquad -4 = 0A - B + 0C \quad \rightarrowtail \quad B = 4 \quad .$$

Letting $s = 4$ *in equation (28.1):*

$$\begin{aligned}6 \cdot 4^2 &- 50 \cdot 4 + 92 \\ &= A(4-3)(4-4) + B(4-2)(4-4) + C(4-2)(4-3)\end{aligned}$$

$$\hookrightarrow \qquad -12 = 0A + 0B + 2C \quad \rightarrowtail \quad C = -6 \quad .$$

Combining the above results, we have

$$
\begin{aligned}
Y(s) &= \frac{6s^2 - 50s + 76}{(s-2)(s-3)(s-4)} \\
&= \frac{A}{s-2} + \frac{B}{s-3} + \frac{C}{s-4} = \frac{8}{s-2} + \frac{4}{s-3} - \frac{6}{s-4} \quad .
\end{aligned}
$$

Hence,

$$
\begin{aligned}
y(t) = \mathcal{L}^{-1}[Y(s)]|_t &= \mathcal{L}^{-1}\left[\frac{8}{s-2} + \frac{4}{s-3} - \frac{6}{s-4}\right]\Bigg|_t \\
&= 8\mathcal{L}^{-1}\left[\frac{1}{s-2}\right]\Bigg|_t + 4\mathcal{L}^{-1}\left[\frac{1}{s-3}\right]\Bigg|_t - 6\mathcal{L}^{-1}\left[\frac{1}{s-4}\right]\Bigg|_t \\
&= 8e^{2t} + 4e^{3t} - 6e^{4t} \quad .
\end{aligned}
$$

Do recall how to deal with repeated factors in the denominator. In particular, if your denominator has factors of the form

$$
(s+c)^n \qquad \text{or} \qquad \left(s^2 + bs + c\right)^n
$$

for some positive integer n and constants b and c, then the corresponding partial fraction expansions are

$$
\frac{A_1}{(s+c)^n} + \frac{A_2}{(s+c)^{n-1}} + \frac{A_3}{(s+c)^{n-2}} + \cdots + \frac{A_n}{s+c}
$$

and

$$
\frac{A_1 s + B_1}{\left(s^2+bs+c\right)^n} + \frac{A_2 s + B_2}{\left(s^2+bs+c\right)^{n-1}} + \frac{A_3 s + B_3}{\left(s^2+bs+c\right)^{n-2}} + \cdots + \frac{A_n s + B_n}{s^2+bs+c} \quad ,
$$

respectively.

!►Example 28.7: *The partial fraction expansion of*

$$
Y(s) = \frac{2s^2}{(s-6)^3}
$$

is of the form

$$
\frac{A}{(s-6)^3} + \frac{B}{(s-6)^2} + \frac{C}{s-6} \quad .
$$

To find the constants A, B and C, we proceed as in the previous examples:

$$
\begin{aligned}
\frac{2s^2}{(s-6)^3} &= \frac{A}{(s-6)^3} + \frac{B}{(s-6)^2} + \frac{C}{s-6} \\
&= \frac{A}{(s-6)^3} + \frac{B(s-6)}{(s-6)^2(s-6)} + \frac{C(s-6)^2}{(s-6)(s-6)^2} \\
&= \frac{A + B(s-6) + C(s-6)^2}{(s-6)^3} \quad .
\end{aligned}
$$

So we must have

$$
2s^2 = A + B(s-6) + C(s-6)^2 \quad .
$$

The value of A can be easily found by letting $s = 6$ in this equation, and the values of B and C can be found by letting $s = 6$ after taking derivatives of both sides of this equation. Or we can multiply out the right side and rewrite the left side more explicitly, obtaining

$$
2s^2 + 0s + 0 = Cs^2 + (B - 12C)s + (A - 6B + 36C) \quad .
$$

This tells us that the constants can be obtained by solving the system

$$\begin{aligned} C &= 2 \\ B - 12C &= 0 \\ A - 6B + 36C &= 0 \end{aligned} \quad .$$

In either case, you will discover that

$$A = 72 \quad , \quad B = 24 \quad \text{and} \quad C = 2 \quad .$$

Thus,

$$\begin{aligned} Y(s) &= \frac{2s^2}{(s-6)^3} \\ &= \frac{A}{(s-6)^3} + \frac{B}{(s-6)^2} + \frac{C}{s-6} \\ &= \frac{72}{(s-6)^3} + \frac{24}{(s-6)^2} + \frac{2}{s-6} \quad . \end{aligned}$$

In the next section, we will discuss an easy way to find the inverse transform of each of the terms in this partial fraction expansion.

28.3 Inverse Transforms of Shifted Functions

All the identities derived for the Laplace transform can be rewritten in terms of the inverse Laplace transform. Of particular value to us is the first shifting identity

$$\mathcal{L}\big[e^{at} f(t)\big]\big|_s = F(s-a)$$

where $F = \mathcal{L}[f(t)]$ and a is any fixed real number. In terms of the inverse transform, this is

$$\mathcal{L}^{-1}[F(s-a)]|_t = e^{at} f(t) \quad .$$

where $f = \mathcal{L}^{-1}[F(s)]$ and a is any fixed real number. Viewed this way, we have a nice way to find inverse transforms of functions that can be written as "shifts" of functions in our tables.

!►*Example 28.8:* *Consider*

$$\mathcal{L}^{-1}\left[\frac{1}{(s-6)^3}\right]\bigg|_t \quad .$$

Here, the 'shift' is clearly by $a = 6$, and we have, by the above identity,

$$\mathcal{L}^{-1}\left[\frac{1}{(s-6)^3}\right]\bigg|_t = \mathcal{F}^{-1}[F(s-6)]|_t = e^{6t} f(t) \quad . \tag{28.2}$$

We now need to figure out the $f(t)$ from the fact that

$$F(s-6) = \frac{1}{(s-6)^3} \quad .$$

Letting $X = s - 6$ in this equation, we have

$$F(X) = \frac{1}{X^3} \quad .$$

Thus,

$$F(s) = \frac{1}{s^3} \quad ,$$

and

$$f(t) = \mathcal{L}^{-1}[F(s)]\big|_t = \mathcal{L}^{-1}\left[\frac{1}{s^3}\right]\bigg|_t$$

$$= \mathcal{L}^{-1}\left[\frac{1}{2!} \cdot \frac{2!}{s^{2+1}}\right]\bigg|_t = \frac{1}{2!}\mathcal{L}^{-1}\left[\cdot \frac{2!}{s^{2+1}}\right]\bigg|_t = \frac{1}{2}t^2 \quad .$$

Plugging this back into equation (28.2), we obtain

$$\mathcal{L}^{-1}\left[\frac{1}{(s-6)^3}\right]\bigg|_t = \cdots = e^{6t} f(t) = e^{6t}\frac{1}{2}t^2 = \frac{1}{2}t^2 e^{6t} \quad .$$

In many cases, determining the shift is part of the problem.

!►Example 28.9: *Consider finding the inverse Laplace transform of*

$$\frac{1}{s^2 - 8s + 25} \quad .$$

If the denominator could be factored nicely, we would use partial fractions. This denominator does not factor nicely (unless we use complex numbers). When that happens, try "completing the square" to rewrite the denominator in terms of " $s - a$ *" for some constant* a *. Here,*

$$s^2 - 8s + 25 = s^2 - 2 \cdot 4s + \left[4^2 - 4^2\right] + 25$$

$$= \underbrace{s^2 - 2 \cdot 4s + 4^2}_{(s-4)^2} - 4^2 + 25 = (s-4)^2 + 9 \quad .$$

Hence,

$$\mathcal{L}^{-1}\left[\frac{1}{s^2 - 8s + 25}\right]\bigg|_t = \mathcal{L}^{-1}\left[\frac{1}{(s-4)^2 + 9}\right]\bigg|_t$$

$$= \mathcal{L}^{-1}[F(s-4)]\big|_t = e^{4t} f(t) \quad . \tag{28.3}$$

Again, we need to find $f(t)$ *from a shifted version of its transform. Here,*

$$F(s-4) = \frac{1}{(s-4)^2 + 9} \quad .$$

Letting $X = s - 4$ *in this equation, we have*

$$F(X) = \frac{1}{X^2 + 9} \quad ,$$

which means the formula for $F(s)$ *is*

$$F(s) = \frac{1}{s^2 + 9} \quad .$$

Thus,

$$f(t) = \mathcal{L}^{-1}[F(s)]\big|_t = \mathcal{L}^{-1}\left[\frac{1}{s^2 + 9}\right]\bigg|_t$$

$$= \mathcal{L}^{-1}\left[\frac{1}{3} \cdot \frac{3}{s^2 + 9}\right]\bigg|_t = \frac{1}{3}\mathcal{L}^{-1}\left[\frac{3}{s^2 + 3^2}\right]\bigg|_t = \frac{1}{3}\sin(3t) \quad .$$

Plugging this back into equation (28.3), we get

$$\mathcal{L}^{-1}\left[\frac{1}{s^2 - 8s + 25}\right]\bigg|_t = \cdots = e^{4t} f(t) = e^{4t}\frac{1}{3}\sin(3t) = \frac{1}{3}e^{4t}\sin(3t) \quad .$$

Additional Exercises

28.1. *Using the tables (mainly, Table 26.1 on page 460) or your own memory, find the inverse Laplace transform for each of the following:*

a. $\dfrac{1}{s-6}$ **b.** $\dfrac{1}{s+2}$ **c.** $\dfrac{1}{s^2}$

d. $\dfrac{6}{s^4}$ **e.** $\dfrac{5}{s^2+25}$ **f.** $\dfrac{s}{s^2+3\pi^2}$

28.2. *Using the tables and linearity, find the inverse Laplace transform for each of the following:*

a. $\dfrac{6}{s+2}$ **b.** $\dfrac{1}{s^4}$ **c.** $\dfrac{3}{\sqrt{s}} - \dfrac{8}{s-4}$

d. $\dfrac{4s^2-4}{s^5}$ **e.** $\dfrac{3s+1}{s^2+25}$ **f.** $\dfrac{1-e^{-4s}}{s}$

28.3. *In Exercise 27.3 on page 497, you found the transform of $t\sin(\omega t)$ and $t\cos(\omega t)$. Now verify the following inverse Laplace transforms assuming ω is any real constant:*

a. $\mathcal{L}^{-1}\left[\dfrac{s}{\left(s^2+\omega^2\right)^2}\right]\Bigg|_t = \dfrac{t}{2\omega}\sin(\omega t)$

b. $\mathcal{L}^{-1}\left[\dfrac{1}{\left(s^2+\omega^2\right)^2}\right]\Bigg|_t = \dfrac{1}{2\omega^3}[\sin(\omega t) - \omega t\cos(\omega t)]$

28.4. *Solve each of the following initial-value problems using the Laplace transform:*

a. $y' + 9y = 0$ *with* $y(0) = 4$

b. $y'' + 9y = 0$ *with* $y(0) = 4$ *and* $y'(0) = 6$

28.5. *Using the tables and partial fractions, find the inverse Laplace transform for each of the following:*

a. $\dfrac{7s+5}{(s+2)(s-1)}$ **b.** $\dfrac{s-1}{s^2-7s+12}$ **c.** $\dfrac{1}{s^2-4}$

d. $\dfrac{3s^2+6s+27}{s^3+9s}$ **e.** $\dfrac{1}{s^3-4s^2}$ **f.** $\dfrac{8s^3}{s^4-81}$

g. $\dfrac{5s^2+6s-40}{(s+6)\left(s^2+16\right)}$ **h.** $\dfrac{2s^3+3s^2+2s+27}{\left(s^2+9\right)\left(s^2+1\right)}$ **i.** $\dfrac{6s^2+62s+92}{(s+1)\left(s^2+10s+21\right)}$

28.6. *Solve each of the following initial-value problems using the Laplace transform (and partial fractions):*

a. $y'' - 9y = 0$ *with* $y(0) = 4$ *and* $y'(0) = 9$

b. $y'' + 9y = 27t^3$ *with* $y(0) = 0$ *and* $y'(0) = 0$

c. $y'' + 8y' + 7y = 165e^{4t}$ *with* $y(0) = 8$ *and* $y'(0) = 1$

28.7. *Using the translation identity (and the tables), find the inverse Laplace transform for each of the following:*

a. $\dfrac{1}{(s-7)^5}$ **b.** $\dfrac{1}{s^2-6s+45}$ **c.** $\dfrac{s}{s^2-6s+45}$ **d.** $\dfrac{1}{\sqrt{s+2}}$

e. $\dfrac{1}{s^2+8s+16}$ **f.** $\dfrac{s}{s^2-12s+40}$ **g.** $\dfrac{1}{s^2+12s+40}$ **h.** $\dfrac{s^2}{(s-3)^5}$

28.8. *Using the Laplace transform with the translation identity, solve the following initial-value problems:*

a. $y'' - 8y' + 17y = 0$ *with* $y(0) = 3$ *and* $y'(0) = 12$

b. $y'' - 6y' + 9y = e^{3t}t^2$ *with* $y(0) = 0$ *and* $y'(0) = 0$

c. $y'' + 6y' + 13y = 0$ *with* $y(0) = 2$ *and* $y'(0) = 8$

d. $y'' + 8y' + 17y = 0$ *with* $y(0) = 3$ *and* $y'(0) = -12$

28.9. *Using the Laplace transform, solve the following initial-value problems:*

a. $y'' = e^t \sin(t)$ *with* $y(0) = 0$ *and* $y'(0) = 0$

b. $y'' - 4y' + 40y = 122e^{-3t}$ *with* $y(0) = 0$ *and* $y'(0) = 8$

c. $y'' - 9y = 24e^{-3t}$ *with* $y(0) = 6$ *and* $y'(0) = 2$

d. $y'' - 4y' + 13y = e^{2t}\sin(3t)$ *with* $y(0) = 4$ *and* $y'(0) = 3$

28.10. *The inverse transforms of the following could be computed using partial fractions. Instead, find the inverse transform of each using the appropriate integration identity from Section 27.3.*

a. $\dfrac{1}{s(s^2+9)}$ **b.** $\dfrac{1}{s(s-4)}$ **c.** $\dfrac{1}{s(s-3)^2}$

29

Convolution

"Convolution" is an operation involving two functions that turns out to be rather useful in many applications. We have two reasons for introducing it here. First of all, convolution will give us a way to deal with inverse transforms of fairly arbitrary products of functions. Secondly, it will be a major element in some relatively simple formulas for solving a number of differential equations.

Let us start with just seeing what "convolution" is. After that, we'll discuss using it with the Laplace transform and in solving differential equations.

29.1 Convolution: The Basics

Definition and Notation

Let $f(t)$ and $g(t)$ be two functions. The *convolution* of f and g, denoted by $f * g$, is the function on $t \geq 0$ given by

$$f * g(t) = \int_0^t f(x)g(t-x)\,dx \quad .$$

!▶Example 29.1: Let

$$f(t) = e^{3t} \qquad \text{and} \qquad g(t) = e^{7t} \quad .$$

Since we will use $f(x)$ and $g(t-x)$ in computing the convolution, let us note that

$$f(x) = e^{3x} \qquad \text{and} \qquad g(t-x) = e^{7(t-x)} \quad .$$

So,

$$\begin{aligned}
f * g(t) &= \int_0^t f(x)g(t-x)\,dx \\
&= \int_0^t e^{3x} e^{7(t-x)}\,dx \\
&= \int_0^t e^{3x} e^{7t} e^{-7x}\,dx \\
&= e^{7t} \int_0^t e^{-4x}\,dx \\
&= e^{7t} \cdot \frac{-1}{4} e^{-4x} \Big|_{x=0}^{t} \\
&= \frac{-1}{4} e^{7t} e^{-4t} - \frac{-1}{4} e^{7t} e^{-4 \cdot 0} = \frac{-1}{4} e^{3t} + \frac{1}{4} e^{7t} \quad .
\end{aligned}$$

Simplifying this slightly, we have

$$f * g(t) = \frac{1}{4}\left[e^{7t} - e^{3t}\right] \qquad \textit{when} \quad f(t) = e^{3t} \quad \textit{and} \quad g(t) = e^{7t} \quad .$$

It is common practice to also denote the convolution $f * g(t)$ by $f(t) * g(t)$ where, here, $f(t)$ and $g(t)$ denote the formulas for f and g. Thus, instead of writing

$$f * g(t) = \frac{1}{4}\left[e^{7t} - e^{3t}\right] \qquad \text{when} \quad f(t) = e^{3t} \quad \text{and} \quad g(t) = e^{7t} \quad ,$$

we may just write

$$e^{3t} * e^{7t} = \frac{1}{4}\left[e^{7t} - e^{3t}\right] \quad .$$

This simplifies notation a little, but be careful — t is being used for two different things in this equation. On the left side, t is used to describe f and g; on the right side, t is the variable in the formula for the convolution. By convention, if we assign t a value, say, $t = 2$, then we are setting $t = 2$ in the final formula for the convolution. That is,

$$e^{3t} * e^{7t} \qquad \text{with} \quad t = 2$$

means *compute the convolution and replace the t in the resulting formula with* 2, which, by the above computations, is

$$\frac{1}{4}\left[e^{7\cdot 2} - e^{3\cdot 2}\right] = \frac{1}{4}\left[e^{14} - e^{6}\right] \quad .$$

It does NOT mean to compute

$$e^{3\cdot 2} * e^{7\cdot 2} \quad ,$$

which would give you a completely different result, namely,

$$e^{6} * e^{14} = \int_0^t e^6 e^{14}\, dx = e^{20} t \quad .$$

!▶Example 29.2: *Let us find*

$$\frac{1}{\sqrt{t}} * t^2 \qquad \textit{when} \quad t = 4 \quad .$$

Here,

$$f(t) = \frac{1}{\sqrt{t}} \qquad \textit{and} \qquad g(t) = t^2 \quad .$$

So

$$f(x) = \frac{1}{\sqrt{x}} \qquad \textit{and} \qquad g(t - x) = (t - x)^2 \quad ,$$

and

$$\begin{aligned}
\frac{1}{\sqrt{t}} * t^2 = f * g(t) &= \int_0^t \frac{1}{\sqrt{x}}(t - x)^2\, dx \\
&= \int_0^t x^{-1/2}\left[t^2 - 2tx + x^2\right] dx \\
&= \int_0^t \left[t^2 x^{-1/2} - 2tx^{1/2} + x^{3/2}\right] dx \\
&= t^2 2x^{1/2} - 2t\frac{2}{3}x^{3/2} + \frac{2}{5}x^{5/2}\Big|_{x=0}^{t} \\
&= 2t^2 \cdot t^{1/2} - \frac{4}{3}t \cdot t^{3/2} + \frac{2}{5}t^{5/2} \quad .
\end{aligned}$$

After a little algebra and arithmetic, this reduces to

$$\frac{1}{\sqrt{t}} * t^2 = \frac{16}{15} t^{5/2} \quad . \tag{29.1}$$

Thus, to compute

$$\frac{1}{\sqrt{t}} * t^2 \qquad \text{when} \quad t = 4 \quad ,$$

we actually compute

$$\frac{16}{15} t^{5/2} \qquad \text{with} \quad t = 4 \quad ,$$

obtaining

$$\frac{16}{15} \cdot 4^{5/2} = \frac{16}{15} \cdot 2^5 = \frac{512}{15} \quad .$$

Basic Identities

Let us quickly note a few easily verified identities that can simplify the computation of some convolutions.

The first identity is trivial to derive. Let α be a constant, and let f and g be two functions. Then, of course,

$$\int_0^t [\alpha f(x)] g(t-x)\, dx = \int_0^t f(x)[\alpha g(t-x)]\, dx = \alpha \int_0^t f(x) g(t-x)\, dx \quad ,$$

which we can rewrite as

$$[\alpha f] * g = f * [\alpha g] = \alpha [f * g] \quad .$$

In other words, we can "factor out constants".

A more substantial identity comes from looking at how switching the roles of f and g changes the convolution. That is, how does the result of computing

$$g * f(t) = \int_0^t g(x) f(t-x)\, dx$$

compare to what we get by computing

$$f * g(t) = \int_0^t f(x) g(t-x)\, dx \quad ?$$

Well, in the last integral, let's use the substitution $y = t - x$. Then $x = t - y$, $dx = -dy$ and

$$\begin{aligned} f * g(t) &= \int_{x=0}^{t} f(x) g(t-x)\, dx \\ &= \int_{y=t-0}^{t-t} f(t-y) g(y)(-1)\, dy \\ &= -\int_t^0 g(y) f(t-y)\, dy = \int_0^t g(y) f(t-y)\, dy \quad . \end{aligned}$$

The last integral is exactly the same as the integral for computing $g * f(t)$, except for the cosmetic change of denoting the variable of integration by y instead of x. So that integral is the formula for $g * f(t)$, and our computations just above reduce to

$$f * g(t) = g * f(t) \quad . \tag{29.2}$$

Thus we see that convolution is "commutative".

!►Example 29.3: *Let's consider the convolution*

$$t^2 * \frac{1}{\sqrt{t}} \quad .$$

Since we just showed that convolution is commutative, we know that

$$t^2 * \frac{1}{\sqrt{t}} = \frac{1}{\sqrt{t}} * t^2 \quad .$$

What an incredible stroke of luck! We've already computed the convolution on the right in Example 29.2. Checking back to equation (29.1), we find

$$\frac{1}{\sqrt{t}} * t^2 = \frac{16}{15} t^{5/2} \quad .$$

Hence,

$$t^2 * \frac{1}{\sqrt{t}} = \frac{1}{\sqrt{t}} * t^2 = \frac{16}{15} t^{5/2} \quad .$$

In addition to being commutative, convolution is "distributive" and "associative". That is, given three functions f, g and h,

$$[f+g] * h = [f * h] + [g * h] \quad , \tag{29.3}$$

$$f * [g+h] = [f * g] + [f * h] \tag{29.4}$$

and

$$f * [g * h] = [f * g] * h \quad . \tag{29.5}$$

The first and second equations are that "addition distributes over convolution". They are easily confirmed using the basic definition of convolution. For the first:

$$\begin{aligned} [f+g] * h(t) &= \int_0^t [f(x) + g(x)]h(t-x)\,dx \\ &= \int_0^t [f(x)h(t-x) + g(x)h(t-x)]\,dx \\ &= \int_0^t f(x)h(t-x)\,dx + \int_0^t g(x)h(t-x)]\,dx \\ &= [f * g] + [g * h] \quad . \end{aligned}$$

The second, equation (29.4), follows in a similar manner or by combining (29.3) with the commutativity of the convolution. The last equation in the list, equation (29.5), states that convolution is "associative", that is, when convolving three functions together, it does not matter which two you convolve first. Its verification requires showing that the two double integrals defining

$$f * [g * h] \qquad \text{and} \qquad [f * g] * h$$

are equivalent. This is a relatively straightforward exercise in substitution, and will be left as a challenge for the interested student (Exercise 29.3 on page 523).

Finally, just for fun, let's make three more simple observations:

$$0 * g(t) = g * 0(t) = \int_0^t 0 \cdot g(t-x)\,dx = 0 \quad .$$

$$f * 1(t) = 1 * f(t) = \int_0^t f(s) \cdot 1\,dx = \int_0^t f(s)\,dx \quad .$$

$$f * g(0) = \int_0^0 f(x)g(0-x)\,dx = 0 \quad .$$

Observations on the Existence of the Convolution

The observant reader will have noted that, if f and g are at least piecewise continuous on $(0,\infty)$, then, for any positive value t, the product $f(x)g(t-x)$ is a piecewise continuous function of x on $(0,t)$. It then follows that the integral in

$$f * g(t) = \int_0^t f(x)g(t-x)\,dx$$

is well defined and finite for every positive value of t. In other words, $f * g$ is a well-defined function on $(0,\infty)$, at least whenever f and g are both piecewise continuous on $(0,\infty)$. (In fact, it can then even be shown that $f * g(t)$ is a continuous function on $[0,\infty)$.)

But now observe that one of the functions in Example 29.2, namely $t^{-1/2}$, 'blows up' at $t = 0$ and, thus, is not piecewise continuous on $(0,\infty)$. So that example also demonstrates that, sometimes, $f * g$ is well defined on $(0,\infty)$ even though f or g is not piecewise continuous.

29.2 Convolution and Products of Transforms

To see one reason convolution is important in the study of Laplace transforms, let us examine the Laplace transform of the convolution of two functions $f(t)$ and $g(t)$. Our goal is a surprisingly simple formula of the corresponding transforms,

$$F(s) = \mathcal{L}[f(t)]|_s = \int_0^\infty f(t)e^{-st}\,dt$$

and

$$G(s) = \mathcal{L}[g(t)]|_s = \int_0^\infty g(t)e^{-st}\,dt \quad .$$

(The impatient can turn to Theorem 29.1 on page 517 for that formula.)

Keep in mind that we can rename the variable of integration in each of the above integrals. In particular, note (for future reference) that

$$F(s) = \int_0^\infty e^{-sx} f(x)\,dx \qquad \text{and} \qquad G(s) = \int_0^\infty e^{-sy} g(y)\,dy \quad .$$

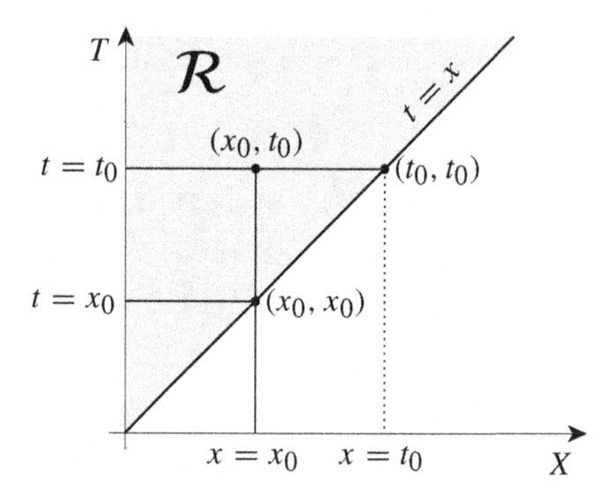

Figure 29.1: The region $\mathcal{R}$ for the transform of the convolution. Note that the coordinates of any point (x_0, t_0) in $\mathcal{R}$ must satisfy $0 < x_0 < t_0 < \infty$.

Now, simply writing out the integral formulas for the Laplace transform and for the convolution yields

$$\begin{aligned} \mathcal{L}[f * g(t)]|_s &= \int_0^\infty e^{-st} f * g(t)\, dt \\ &= \int_{t=0}^\infty e^{-st} \int_{x=0}^t f(x)g(t-x)\, dx\, dt \\ &= \int_{t=0}^\infty \int_{x=0}^t e^{-st} f(x)g(t-x)\, dx\, dt \quad . \end{aligned}$$

Combined with the observation that

$$e^{-st} = e^{-st+sx-sx} = e^{-s(t-x)}e^{-sx} \quad ,$$

the above sequence becomes

$$\begin{aligned} \mathcal{L}[f * g(t)]|_s &= \int_{t=0}^\infty \int_{x=0}^t e^{-sx} f(x)\, e^{-s(t-x)} g(t-x)\, dx\, dt \\ &= \int_{t=0}^\infty \int_{x=0}^t K(x,t)\, dx\, dt \end{aligned} \tag{29.6}$$

where, simply to simplify expressions in the next few lines, we've set

$$K(x,t) = e^{-sx} f(x)\, e^{-s(t-x)} g(t-x) \quad .$$

It is now convenient to switch the order of integration in the last double integral. According to the limits in that double integral, we are integrating over the region $\mathcal{R}$ in the XT–plane consisting of all (x,t) for which

$$0 < t < \infty$$

and, for each of these values of t,

$$0 < x < t \quad .$$

As illustrated in Figure 29.1, region $\mathcal{R}$ is the portion of the first quadrant of the XT–plane to the left of the line $t = x$. Equivalently, as can also be seen in this figure, $\mathcal{R}$ is the portion of the first quadrant above the line $t = x$. So $\mathcal{R}$ can also be described as the set of all (x,t) for which

$$0 < x < \infty$$

and, for each of these values of x,

$$x < t < \infty \quad .$$

Thus,

$$\int_{t=0}^{\infty} \int_{x=0}^{t} K(x,t)\,dx\,dt \;=\; \iint_{\mathcal{R}} K(x,t)\,dA \;=\; \int_{x=0}^{\infty} \int_{t=x}^{\infty} K(x,t)\,dt\,dx \quad .$$

Combining this with equation (29.6), and then continuing, we have

$$\begin{aligned}
\mathcal{L}[f * g(t)]|_s &= \int_{x=0}^{\infty} \int_{t=x}^{\infty} K(x,t)\,dt\,dx \\
&= \int_{x=0}^{\infty} \int_{t=x}^{\infty} e^{-sx} f(x)\, e^{-s(t-x)} g(t-x)\,dt\,dx \\
&= \int_{x=0}^{\infty} e^{-sx} f(x) \left[\int_{t=x}^{\infty} e^{-s(t-x)} g(t-x)\,dt \right] dx \quad .
\end{aligned}$$

Let us simplify the inner integral with the substitution $y = t - x$ (remembering that t is the variable of integration in this integral):

$$\int_{t=x}^{\infty} e^{-s(t-x)} g(t-x)\,dt \;=\; \int_{y=x-x}^{\infty - x} e^{-sy} g(y)\,dy \;=\; \int_{0}^{\infty} e^{-sy} g(y)\,dy \;=\; G(s) \quad !$$

Combining this with our last formula for $\mathcal{L}[f * g]$ then yields

$$\begin{aligned}
\mathcal{L}[f * g(t)]|_s &= \int_{0}^{\infty} f(x) e^{-sx} G(s)\,dx \\
&= \int_{0}^{\infty} e^{-sx} f(x)\,dx \cdot G(s) \;=\; F(s) \cdot G(s) \quad !
\end{aligned}$$

Thus,

$$\mathcal{L}[f * g(t)]|_s \;=\; F(s)G(s) \quad .$$

Equivalently,

$$f * g(t) \;=\; \mathcal{L}^{-1}[F(s)G(s)]|_t \quad .$$

If we had been a little more complete in our computations, we would have kept track of the exponential order of all the functions involved (see Exercise 29.9), and obtained all of the following theorem.

Theorem 29.1 (Laplace convolution identities)
Assume $f(t)$ and $g(t)$ are two functions of exponential order s_0, and with Laplace transforms

$$F(s) \;=\; \mathcal{L}[f(t)]|_s \qquad \text{and} \qquad G(s) \;=\; \mathcal{L}[g(t)]|_s \quad .$$

Then the convolution $f * g(t)$ is of exponential order s_1 for any $s_1 > s_0$. Moreover,

$$\mathcal{L}[f * g(t)]|_s \;=\; F(s)G(s) \qquad \text{for} \quad s > s_0 \tag{29.7}$$

and

$$\mathcal{L}^{-1}[F(s)G(s)]|_t \;=\; f * g(t) \quad . \tag{29.8}$$

Do remember that identities (29.7) and (29.8) are equivalent. It is also worthwhile to rewrite these identities as

$$\mathcal{L}[f * g(t)]|_s \;=\; \mathcal{L}[f(t)]|_s \cdot \mathcal{L}[g(t)]|_s \tag{29.7$'$}$$

and

$$\mathcal{L}^{-1}[F(s)G(s)]|_t = \mathcal{L}^{-1}[F(s)]|_t * \mathcal{L}^{-1}[G(s)]|_t \quad , \tag{29.8'}$$

respectively. These forms, especially the latter, are sometimes a little more convenient in practice.

!▶Example 29.4: *Consider finding the inverse Laplace transform of*

$$\frac{1}{s^2 - 10s + 21} \quad .$$

Factoring the denominator and applying the above, we get

$$\begin{aligned}
\mathcal{L}^{-1}\left[\frac{1}{s^2 - 10s + 21}\right]\bigg|_t &= \mathcal{L}^{-1}\left[\frac{1}{(s-3)(s-7)}\right]\bigg|_t \\
&= \mathcal{L}^{-1}\left[\frac{1}{s-3} \cdot \frac{1}{s-7}\right]\bigg|_t \\
&= \mathcal{L}^{-1}\left[\frac{1}{s-3}\right]\bigg|_t * \mathcal{L}^{-1}\left[\frac{1}{s-7}\right]\bigg|_t = e^{3t} * e^{7t} \quad .
\end{aligned}$$

As luck would have it, this convolution was computed in Example 29.1 on page 511),

$$e^{3t} * e^{7t} = \frac{1}{4}\left[e^{7t} - e^{3t}\right] \quad .$$

Thus,

$$\mathcal{L}^{-1}\left[\frac{1}{s^2 - 10s + 21}\right]\bigg|_t = e^{3t} * e^{7t} = \frac{1}{4}\left[e^{7t} - e^{3t}\right] \quad .$$

The inverse transform in the last example could also have been computed using partial fractions. Indeed, many of the inverse transforms we computed using partial fractions can also be computed using convolution. Whether one approach or the other is preferred depends on the opinion of the person doing the computing. However, as the next example shows, there are cases where convolution can be applied, but not partial fractions. We will also use this example to demonstrate how convolution naturally arises when solving differential equations.

!▶Example 29.5: *Consider solving the initial-value problem*

$$y'' + 9y = \frac{1}{\sqrt{t}} \qquad \text{with} \quad y(0) = 0 \quad \text{and} \quad y'(0) = 0 \quad .$$

Taking the Laplace transform of both sides:

$$\mathcal{L}[y'' + 9y]|_s = \mathcal{L}\left[\frac{1}{\sqrt{t}}\right]\bigg|_s$$

$$\hookrightarrow \qquad \mathcal{L}[y'']|_s + 9\mathcal{L}[y]|_s = \frac{\sqrt{\pi}}{\sqrt{s}}$$

$$\hookrightarrow \qquad s^2Y(s) - s\underbrace{y(0)}_{0} - \underbrace{y'(0)}_{0} + 9Y(s) = \frac{\sqrt{\pi}}{\sqrt{s}}$$

$$\hookrightarrow \qquad \left[s^2 + 9\right]Y(s) = \frac{\sqrt{\pi}}{\sqrt{s}}$$

$$\hookrightarrow \qquad Y(s) = \frac{\sqrt{\pi}}{\sqrt{s}(s^2 + 9)} \quad .$$

Thus, $y(t)$ is the inverse Laplace transform of

$$\frac{\sqrt{\pi}}{\sqrt{s}(s^2+9)} \quad .$$

Because the denominator does not factor into two polynomials, we cannot use partial fractions — we must use convolution,

$$\mathcal{L}^{-1}\left[\frac{\sqrt{\pi}}{\sqrt{s}(s^2+9)}\right]\bigg|_t = \mathcal{L}^{-1}\left[\frac{\sqrt{\pi}}{\sqrt{s}}\cdot\frac{1}{s^2+9}\right]\bigg|_t = \mathcal{L}^{-1}\left[\frac{\sqrt{\pi}}{\sqrt{s}}\right]\bigg|_t * \mathcal{L}^{-1}\left[\frac{1}{s^2+9}\right]\bigg|_t \quad .$$

Reversing the transform made on the right side of the above equations, we have

$$\mathcal{L}^{-1}\left[\frac{\sqrt{\pi}}{\sqrt{s}}\right]\bigg|_t = \frac{1}{\sqrt{t}} \quad .$$

Using our tables, we find that

$$\mathcal{L}^{-1}\left[\frac{1}{s^2+9}\right]\bigg|_t = \frac{1}{3}\mathcal{L}^{-1}\left[\frac{3}{s^2+3^2}\right]\bigg|_t = \frac{1}{3}\sin(3t) \quad .$$

Combining the above and recalling that "constants factor out", we then obtain

$$\begin{aligned} y(t) = \mathcal{L}^{-1}\left[\frac{\sqrt{\pi}}{\sqrt{s}(s^2+9)}\right]\bigg|_t &= \mathcal{L}^{-1}\left[\frac{\sqrt{\pi}}{\sqrt{s}}\right]\bigg|_t * \mathcal{L}^{-1}\left[\frac{1}{s^2+9}\right]\bigg|_t \\ &= \left[\frac{1}{\sqrt{t}}\right] * \left[\frac{1}{3}\sin(3t)\right] \\ &= \frac{1}{3}\frac{1}{\sqrt{t}} * \sin(3t) \quad . \end{aligned}$$

That is,

$$y(t) = \frac{1}{3}\int_0^t \frac{1}{\sqrt{x}}\sin(3[t-x])\,dx \quad .$$

Admittedly, this last integral is not easily evaluated by hand. But it is something that can be accurately approximated for any specific (nonnegative) value of t using routines found in many computer math packages. So it is still a usable formula.

29.3 Convolution and Differential Equations (Duhamel's Principle)

As illustrated in our last example, convolution has a natural role in solving differential equations when using the Laplace transform. However, if we look a little more carefully at the process of solving differential equations using the Laplace transform, we will find that convolution can play an even more significant role.

!►Example 29.6: *Let's consider solving the nonhomogeneous initial-value problem*

$$y'' - 10y' + 21y = f(t) \qquad \text{with} \quad y(0) = 0 \quad \text{and} \quad y'(0) = 0$$

where $f = f(t)$ is any Laplace transformable function. Naturally, we will use the Laplace transform. So let

$$Y(s) = \mathcal{L}[y(t)]|_s \qquad \text{and} \qquad F(s) = \mathcal{L}[f(t)]|_s \quad .$$

Because of our initial conditions, the "transform of the derivatives" identities simplify considerably:

$$\mathcal{L}\big[y''\big]\big|_s = s^2Y(s) - s\underbrace{y(0)}_{0} - \underbrace{y'(0)}_{0} = s^2Y(s)$$

and

$$\mathcal{L}\big[y'\big]\big|_s = sY(s) - \underbrace{y(0)}_{0} = sY(s) \quad .$$

Consequently,

$$\mathcal{L}\big[y'' - 10y' + 21y\big]\big|_s = \mathcal{L}[f(t)]|_s$$

$$\hookrightarrow \qquad \mathcal{L}\big[y''\big]\big|_s - 10\mathcal{L}\big[y'\big]\big|_s + 21\mathcal{L}[y]|_s = F(s)$$

$$\hookrightarrow \qquad s^2Y(s) - 10sY(s) + 21Y(s) = F(s)$$

$$\hookrightarrow \qquad \big[s^2 - 10s + 21\big]Y(s) = F(s) \quad .$$

Dividing through by the polynomial, we get

$$Y(s) = H(s)F(s) \qquad \text{where} \quad H(s) = \frac{1}{s^2 - 10s + 21} \quad .$$

Thus,

$$y(t) = \mathcal{L}^{-1}[Y(s)]|_t = \mathcal{L}^{-1}[H(s)F(s)]|_t \quad .$$

Applying the convolution identity then yields

$$y(t) = h * f(t) \tag{29.9a}$$

where

$$h(x) = \mathcal{L}^{-1}[H(s)]|_x = \mathcal{L}^{-1}\left[\frac{1}{s^2 - 10s + 21}\right]\bigg|_t \quad . \tag{29.9b}$$

The convolution $h * f$ can be computed using either

$$\int_0^t h(x)f(t-x)\,dx \qquad \text{or} \qquad \int_0^t h(t-x)f(x)\,dx \quad .$$

For no particular reason, we will choose the first integral formula.

To compute $h(x)$, we can use partial fractions or convolution. Or we can glance at Example 29.4 on page 518, discover that we've already computed $h(t)$, and just replace the t in that formula with x,

$$h(x) = \frac{1}{4}\left[e^{7x} - e^{3x}\right] \quad .$$

With this and our chosen integral formula for $h * f$, formula (29.9a), the solution to our initial-value problem, becomes

$$y(t) = \int_0^t \frac{1}{4}\left[e^{7x} - e^{3x}\right] f(t-x)\,dx \quad . \tag{29.10}$$

Formula (29.10) is a convenient way to describe the solutions to our initial-value problem, especially if we want to solve this problem for a number of different choices of $f(t)$. Using it, we can quickly write out a relatively simple integral formula for the solution corresponding to each $f(t)$. For example:

If $f(t) = e^{4t}$, then $f(t-x) = e^{4(t-x)}$ and formula (29.10) yields

$$y(t) = \int_0^t \frac{1}{4}\left[e^{7x} - e^{3x}\right] e^{4(t-x)}\,dx \quad .$$

If $f(t) = 1$, then $f(t-x) = 1$ and formula (29.10) yields

$$y(t) = \frac{1}{4}\int_0^t \left[e^{7x} - e^{3x}\right] \cdot 1\,dx \quad .$$

And finally, if $f(t) = \sqrt[3]{t}$, then $f(t-x) = \sqrt[3]{t-x}$ and formula (29.10) yields

$$y(t) = \frac{1}{4}\int_0^t \left[e^{7x} - e^{3x}\right] \sqrt[3]{t-x}\,dx \quad .$$

The first two integrals are easily evaluated, giving us

$$y(t) = \frac{1}{2}e^{7t} + \frac{1}{4}e^{3t} - \frac{1}{3}e^{4t} \qquad \text{and} \qquad y(t) = \frac{1}{28}e^{7t} - \frac{1}{12}e^{3t} - \frac{1}{21} \quad ,$$

respectively. The last integral,

$$y(t) = \frac{1}{4}\int_0^t \left[e^{7x} - e^{3x}\right] \sqrt[3]{t-x}\,dx \quad ,$$

is not easily evaluated by hand but can be accurately approximated for any value of t using routines found in our favorite computer math package.

?►Exercise 29.1: *Using the formula (29.10), find the solution to*

$$y'' - 10y' + 21y = e^{3t} \qquad \text{with} \quad y(0) = 0 \quad \text{and} \quad y'(0) = 0 \quad .$$

Generalizing what we just derived in the last example is easy. Suppose we have any second-order initial-value problem of the form

$$ay'' + by' + cy = f(t) \qquad \text{with} \quad y(0) = 0 \quad \text{and} \quad y'(0) = 0$$

where a, b and c are constants, and f is any Laplace transformable function. Then, taking the Laplace transform of both sides of the differential equation, letting

$$Y(s) = \mathcal{L}[y(t)]|_s \qquad \text{and} \qquad F(s) = \mathcal{L}[f(t)]|_s \quad ,$$

and noting that, because of our initial conditions, the "transform of the derivatives" identities simplify to

$$\mathcal{L}[y'']|_s = s^2Y(s) - s\underbrace{y(0)}_{0} - \underbrace{y'(0)}_{0} = s^2Y(s)$$

and

$$\mathcal{L}[y']|_s = sY(s) - \underbrace{y(0)}_{0} = sY(s) \quad ,$$

we see that

$$\mathcal{L}\big[ay'' + by' + cy\big]\big|_s = \mathcal{L}[f(t)]|_s$$

$$\hookrightarrow \quad a\mathcal{L}\big[y''\big]\big|_s + b\mathcal{L}\big[y'\big]\big|_s + c\mathcal{L}[y]|_s = F(s)$$

$$\hookrightarrow \quad as^2Y(s) + bsY(s) + cY(s) = F(s)$$

$$\hookrightarrow \quad \big[as^2 + bs + c\big]Y(s) = F(s) \quad .$$

Dividing through by the polynomial, we get

$$Y(s) = H(s)F(s) \qquad \text{where} \quad H(s) = \frac{1}{as^2 + bs + c} \quad .$$

So,

$$y(t) = \mathcal{L}^{-1}[Y(s)]|_t = \mathcal{L}^{-1}[H(s)F(s)]|_t \quad ,$$

and the convolution identity tells us that

$$y(t) = h * f(t) \tag{29.11a}$$

where

$$h(x) = \mathcal{L}^{-1}[H(s)]|_x = \mathcal{L}^{-1}\left[\frac{1}{as^2 + bs + c}\right]\bigg|_t \quad . \tag{29.11b}$$

The fact that the formula for y in equation set (29.11) is the solution to

$$ay'' + by' + cy = f(t) \qquad \text{with} \quad y(0) = 0 \quad \text{and} \quad y'(0) = 0$$

is often called *Duhamel's principle*. The function $H(s)$ is usually referred to as the *transfer function*, and its inverse transform, $h(t)$, is usually called the *impulse response function.*[1] Keep in mind that a, b and c are constants, and that we assumed f is Laplace transformable.

As illustrated in our example, Duhamel's principle makes it easy to write down solutions to the given initial-value problem once we have found h. This is especially useful if we need to find solutions to

$$ay'' + by' + cy = f(t) \qquad \text{with} \quad y(0) = 0 \quad \text{and} \quad y'(0) = 0$$

for a number of different choices of f.

But why stop at second-order problems? It should be clear that the above differential equation did not have to be second order. A completely analogous derivation can be done starting with any nonhomogeneous linear differential equation with constant coefficients, provided all the appropriate initial values are zero. Doing so leads to the following theorem:

Theorem 29.2 (Duhamel's principle)
Let N be any positive integer, let a_0, a_1, ... and a_N be any collection of real-valued constants, and let $f(t)$ be any Laplace transformable function. Then, the solution to

$$a_0y^{(N)} + a_1y^{(N-1)} + \cdots + a_{N-2}y'' + a_{N-1}y' + a_Ny = f(t)$$

satisfying the Nth-order "zero" initial conditions,

$$y(0) = 0 \quad , \quad y'(0) = 0 \quad , \quad y''(0) = 0 \quad , \quad \ldots \quad \text{and} \quad y^{N-1}(0) = 0 \quad ,$$

[1] The reason why h is called the "impulse response function" will be revealed in Chapter 31. A few authors also refer to h as a "weight" function.

is given by

$$y(t) = h * f(t)$$

where

$$h(x) = \mathcal{L}^{-1}[H(s)]|_x \qquad \text{and} \qquad H(s) = \frac{1}{a_0 s^n + a_1 s^{n-1} + \cdots + a_n} \quad .$$

Three quick notes:

1. As noted a few pages ago, the convolution $h * f$ can be computed using either
$$\int_0^t h(x) f(t-x)\,dx \qquad \text{or} \qquad \int_0^t h(t-x) f(x)\,dx \quad .$$
In practice, use whichever appears easier to compute given the h and f involved. In the examples here, we used the first. Later, when we re-examine "resonance" in mass/spring systems (Section 30.7), we will use the other integral formula.

2. It turns out that the $f(t)$ in Duhamel's principle (as described above) does not have to be Laplace transformable. By applying the above theorem with
$$f_T(t) = \begin{cases} f(t) & \text{if } t < T \\ 0 & \text{if } t \geq T \end{cases} \quad ,$$
and then letting $T \to \infty$, you can show that $y = h * f$ is well defined and satisfies the initial-value problem even when f is merely piecewise continuous on $(0, \infty)$.

3. It is not hard to show that, if you want the solution to
$$a_0 y^{(N)} + a_1 y^{(N-1)} + \cdots + a_{N-2} y'' + a_{N-1} y' + a_N y = f(t) \quad ,$$
but satisfying *non*zero initial conditions, then you simply need to add the solution obtained by Duhamel's principle to the solution to the corresponding homogeneous differential equation
$$a_0 y^{(N)} + a_1 y^{(N-1)} + \cdots + a_{N-2} y'' + a_{N-1} y' + a_N y = 0$$
that satisfies the desired initial conditions.

Additional Exercises

29.2. *Compute the convolution $f * g(t)$ of each of the following pairs of functions:*

a. $f(t) = e^{3t}$ and $g(t) = e^{5t}$ **b.** $f(t) = \frac{1}{\sqrt{t}}$ and $g(t) = t^2$

c. $f(t) = \sqrt{t}$ and $g(t) = 6$ **d.** $f(t) = t$ and $g(t) = e^{3t}$

e. $f(t) = t^2$ and $g(t) = t^2$ **f.** $f(t) = \sin(t)$ and $g(t) = t$

g. $f(t) = \sin(t)$ and $g(t) = \sin(t)$ **h.** $f(t) = \sin(t)$ and $g(t) = e^{-3t}$

29.3. *Verify the associative property of convolution. That is, verify equation (29.5) on page 514.*

29.4. *Using convolution, compute the inverse Laplace transform of each of the following:*

a. $\dfrac{1}{(s-4)(s-3)}$ **b.** $\dfrac{1}{s(s-3)}$ **c.** $\dfrac{1}{s(s^2+4)}$

d. $\dfrac{1}{(s-3)(s^2+1)}$ **e.** $\dfrac{1}{(s^2+9)^2}$ **f.** $\dfrac{s^2}{(s^2+4)^2}$

g. $\dfrac{1}{\sqrt{s}(s-3)}$ *(leave in integral form)* **h.** $\dfrac{1}{\sqrt{s}\,(s^2+4)}$ *(leave in integral form)*

29.5. *For each of the following initial-value problems, find the corresponding transfer function H and the impulse response function h, and write down the corresponding convolution integral formula for the solution:*

a. $y'' + 4y = f(t)$ *with* $y(0) = 0$ *and* $y'(0) = 0$

b. $y'' - 4y = f(t)$ *with* $y(0) = 0$ *and* $y'(0) = 0$

c. $y'' - 6y' + 9y = f(t)$ *with* $y(0) = 0$ *and* $y'(0) = 0$

d. $y'' - 6y' + 18y = f(t)$ *with* $y(0) = 0$ *and* $y'(0) = 0$

e. $y''' + 16y' = f(t)$ *with* $y(0) = 0$ *and* $y'(0) = 0$

29.6. *Using the results from Exercise 29.5 a, find the solution to*

$$y'' + 4y = f(t) \quad \text{with} \quad y(0) = 0 \quad \text{and} \quad y'(0) = 0$$

for each of the following choices of f:

a. $f(t) = 1$ **b.** $f(t) = t$ **c.** $f(t) = e^{3t}$

d. $f(t) = \sin(2t)$ **e.** $f(t) = \sin(\alpha t)$ *where* $\alpha \neq 2$

29.7. *Using the results from Exercise 29.5 c, find the solution to*

$$y'' - 6y' + 9y = f(t) \quad \text{with} \quad y(0) = 0 \quad \text{and} \quad y'(0) = 0$$

for each of the following choices of f:

a. $f(t) = 1$ **b.** $f(t) = t$ **c.** $f(t) = e^{3t}$

d. $f(t) = e^{-3t}$ **e.** $f(t) = e^{\alpha t}$ *where* $\alpha \neq 3$

29.8. *Using the results from Exercise 29.5 e, find the solution to*

$$y''' + 16y' = f(t) \quad \text{with} \quad y(0) = 0 \quad \text{and} \quad y'(0) = 0$$

for each of the following choices of f:

a. $f(t) = 1$ **b.** $f(t) = t$ **c.** $f(t) = e^{3t}$

d. $f(t) = \sin(4t)$ **e.** $f(t) = \sin(at)$ *where* $\alpha \neq 4$

29.9. *Let f and g be two piecewise continuous functions on the positive real line satisfying, for all $t > 0$,*

$$|f(t)| < M_f e^{s_0 t} \quad \text{and} \quad |g(t)| < M_g e^{s_0 t}$$

for some constants M_f, M_g and s_0.

a. *Show that $|f * g(t)| < M_f M_g e^{s_0 t} t$ whenever $t > 0$.*

b. *Why does this tell us that $f * g$ is of exponential order s_1 for any $s_1 > s_0$?*

30

Piecewise-Defined Functions and Periodic Functions

At the start of our study of the Laplace transform, it was claimed that the Laplace transform is "particularly useful when dealing with nonhomogeneous equations in which the forcing functions are not continuous". Thus far, however, we've done precious little with any discontinuous functions other than step functions. Let us now rectify the situation by looking at the sort of discontinuous functions (and, more generally, "piecewise-defined" functions) that often arise in applications, and develop tools and skills for dealing with these functions.

We will also take a brief look at transforms of periodic functions other than sines and cosines. As you will see, many of these functions are, themselves, piecewise defined. And finally, we will use some of the material we've recently developed to re-examine the issue of resonance in mass/spring systems.

30.1 Piecewise-Defined Functions

Piecewise-Defined Functions, Defined

When we talk about a "discontinuous function f" in the context of Laplace transforms, we usually mean f is a piecewise continuous function that is not continuous on the interval $(0, \infty)$. Such a function will have jump discontinuities at isolated points in this interval. Computationally, however, the real issue is often not so much whether there is a nonzero jump in the graph of f at a point t_0, but whether the formula for computing $f(t)$ is the same on either side of t_0. So we really should be looking at the more general class of "piecewise-defined" functions that, at worst, have jump discontinuities.

Just what is a *piecewise-defined* function? It is any function given by different formulas on different intervals. For example,

$$f(t) = \begin{cases} 0 & \text{if } t < 1 \\ 1 & \text{if } 1 < t < 2 \\ 0 & \text{if } 2 < t \end{cases} \qquad \text{and} \qquad g(t) = \begin{cases} 0 & \text{if } t \le 1 \\ t - 1 & \text{if } 1 < t < 2 \\ 1 & \text{if } 2 \le t \end{cases}$$

are two relatively simple piecewise-defined functions. The first (sketched in Figure 30.1a) is discontinuous because it has nontrivial jumps at $t = 1$ and $t = 2$. However, the second function (sketched in Figure 30.1b) is continuous because $t - 1$ goes from 0 to 1 as t goes from 1 to 2. There are no jumps in the graph of g.

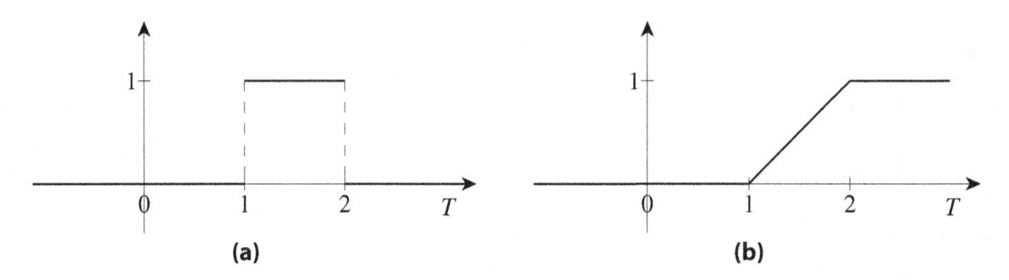

Figure 30.1: The graphs of two piecewise-defined functions.

By the way, we may occasionally refer to the sort of lists used above to define $f(t)$ and $g(t)$ as *sets of conditional formulas* for f and g, simply because these are sets of formulas with conditions stating when each formula is to be used.

Do note that, in the above formula set for f, we did not specify the values of $f(t)$ when $t = 1$ or $t = 2$. This was because f has jump discontinuities at these points and, as we agreed in Chapter 26 (see page 471), we are not concerned with the precise value of a function at its discontinuities. On the other hand, using the formula set given above for g, you can easily verify that

$$\lim_{t \to 1^-} g(t) = 0 = \lim_{t \to 2^+} g(t) \qquad \text{and} \qquad \lim_{t \to 1^-} g(t) = 1 = \lim_{t \to 2^+} g(t) \quad ;$$

so there is not a true jump in g at these points. That is why we went ahead and specified that

$$g(1) = 0 \qquad \text{and} \qquad g(2) = 1 \quad .$$

In the future, let us agree that, even if the value of a particular function f or g is not explicitly specified at a particular point t_0, as long as the left- and right-hand limits of the function at t_0 are defined and equal, then the function is defined at t_0 and is equal to those limits. That is, we'll assume

$$f(t_0) = \lim_{t \to t_0^-} f(t) = \lim_{t \to t_0^+} f(t)$$

whenever

$$\lim_{t \to t_0^-} f(t) = \lim_{t \to t_0^+} f(t) \quad .$$

This will simplify notation a little and may keep us from worrying about issues of continuity when those issues are not important.

Step Functions, Again

Most people would probably consider the step functions to be the simplest piecewise-defined functions. These include the basic step function,

$$\text{step}(t) = \begin{cases} 0 & \text{if} \quad t < 0 \\ 1 & \text{if} \quad 0 < t \end{cases}$$

(sketched in Figure 30.2a), as well as the step function at a point α,

$$\text{step}_\alpha(t) = \text{step}(t - \alpha) = \begin{cases} 0 & \text{if} \quad t < \alpha \\ 1 & \text{if} \quad \alpha < t \end{cases}$$

(sketched in Figure 30.2b).

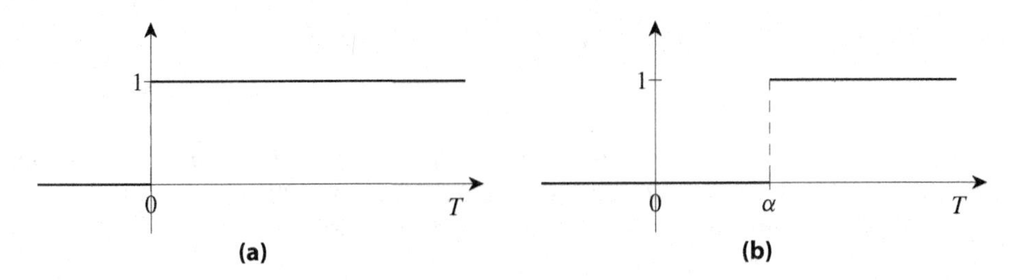

Figure 30.2: The graphs of **(a)** the basic step function $\text{step}(t)$ and **(b)** a shifted step function $\text{step}_\alpha(t)$ with $\alpha > 0$.

We will be dealing with other piecewise-defined functions, but, even with these other functions, we will find step functions useful. Step functions can be used as 'switches' — turning on and off the different formulas in our piecewise-defined functions. In this regard, let us quickly observe what we get when we multiply a step function by any function/formula $g(t)$:

$$g(t)\,\text{step}_\alpha(t) = \left\{\begin{matrix} g(t)\cdot 0 & \text{if} & t<\alpha \\ g(t)\cdot 1 & \text{if} & \alpha<t \end{matrix}\right\} = \begin{cases} 0 & \text{if} \quad t<\alpha \\ g(t) & \text{if} \quad \alpha<t \end{cases} .$$

Here, the step function at α 'switches on' $g(t)$ at $t = \alpha$. For example,

$$t^2\,\text{step}_3(t) = \begin{cases} 0 & \text{if} \quad t<3 \\ t^2 & \text{if} \quad 3<t \end{cases}$$

and

$$\sin(t-4)\,\text{step}_4(t) = \begin{cases} 0 & \text{if} \quad t<4 \\ \sin(t-4) & \text{if} \quad 4<t \end{cases} .$$

This fact will be very useful when applying Laplace transforms in problems involving piecewise-defined functions, and we will find ourselves especially interested in cases where the formula being multiplied by $\text{step}_\alpha(t)$ describes a function that is also translated by α (as in $\sin(t-4)\,\text{step}_4(t)$).

The Laplace transform of $\text{step}_\alpha(t)$ was computed in Chapter 26. If you don't recall how to compute this transform, it would be worth your while to go back to review that discussion. It is also worthwhile for us to look at a differential equation involving a step function.

!►*Example 30.1:* *Consider finding the solution to*

$$y'' + y = \text{step}_3 \qquad \textit{with} \quad y(0) = 0 \quad \textit{and} \quad y'(0) = 0 \quad .$$

Taking the Laplace transform of both sides:

$$\mathcal{L}[y'' + y]\big|_s = \mathcal{L}[\text{step}_3]\big|_s$$

$$\hookrightarrow \qquad \mathcal{L}[y'']\big|_s + \mathcal{L}[y]\big|_s = \frac{1}{s}e^{-3s}$$

$$\hookrightarrow \qquad \left[s^2Y(s) - sy(0) - y'(0)\right] + Y(s) = \frac{1}{s}e^{-3s}$$

$$\hookrightarrow \qquad \left[s^2+1\right]Y(s) = \frac{1}{s}e^{-3s}$$

$$\hookrightarrow \qquad Y(s) = \frac{1}{s(s^2+1)}e^{-3s} \quad .$$

Thus,

$$y(t) = \mathcal{L}^{-1}\left[\frac{1}{s(s^2+1)}e^{-3s}\right]\Big|_t \quad .$$

Here, we have the inverse transform of an exponential multiplied by a function whose inverse transform can easily be computed using, say, partial fractions. This would be a good point to pause and discuss, in general, what can be done in such situations.[1]

30.2 The "Translation Along the T-Axis" Identity

The Identity

As illustrated in the above example, we may often find ourselves with

$$\mathcal{L}^{-1}\big[e^{-\alpha s}F(s)\big]\big|_t$$

where α is some positive number and $F(s)$ is some function whose inverse Laplace transform, $f = \mathcal{L}^{-1}[F]$, is either known or can be found with relative ease. Remember, this means

$$F(s) = \mathcal{L}[f(t)]|_s = \int_0^\infty f(t)e^{-st}\,dt \quad .$$

Consequently,

$$e^{-\alpha s}F(s) = e^{-\alpha s}\int_0^\infty f(t)e^{-st}\,dt = \int_0^\infty f(t)e^{-\alpha s}e^{-st}\,dt = \int_0^\infty f(t)e^{-s(t+\alpha)}\,dt \quad .$$

Using the change of variables $\tau = t + \alpha$ (thus, $t = \tau - \alpha$), and being careful with the limits of integration, we see that

$$e^{-\alpha s}F(s) = \cdots = \int_{t=0}^\infty f(t)e^{-s(t+\alpha)}\,dt = \int_{\tau=\alpha}^\infty f(\tau-\alpha)e^{-s\tau}\,d\tau \quad . \qquad (30.1)$$

This last integral is almost, but not quite, the integral for the Laplace transform of $f(\tau - \alpha)$ (using τ instead of t as the symbol for the variable of integration). And the reason it is not is that this integral's limits start at α instead of 0. But that is where the limits would start if the function being transformed were 0 for $\tau < \alpha$. This, along with observations made a page or so ago, suggests viewing this integral as the transform of

$$f(t-\alpha)\,\text{step}_\alpha(t) = \begin{cases} 0 & \text{if } t < \alpha \\ f(t-\alpha) & \text{if } \alpha \le t \end{cases} \quad .$$

After all,

$$\begin{aligned}\int_{\tau=\alpha}^\infty f(\tau-\alpha)e^{-s\tau}\,d\tau &= \int_{t=\alpha}^\infty f(t-\alpha)e^{-st}\,dt \\ &= \int_{t=0}^\alpha f(t-\alpha)\cdot 0\cdot e^{-st}\,dt + \int_{t=\alpha}^\infty f(t-\alpha)\cdot 1\cdot e^{-st}\,dt\end{aligned}$$

[1] The observant reader will note that y can be found directly using convolution. However, beginners may find the computation of the needed convolution, $\sin(t) * \text{step}_3(t)$, a little tricky. The approach being developed here reduces the need for such convolutions and can be applied when convolution cannot be used. Still, convolutions with piecewise-defined functions can be useful and will be discussed in Section 30.4.

$$= \int_{t=0}^{\alpha} f(t-\alpha)\,\text{step}_\alpha(t) e^{-st}\,dt + \int_{t=\alpha}^{\infty} f(t-\alpha)\,\text{step}_\alpha(t) e^{-st}\,dt$$

$$= \int_{t=0}^{\infty} f(t-\alpha)\,\text{step}_\alpha(t) e^{-st}\,dt$$

$$= \mathcal{L}\big[f(t-\alpha)\,\text{step}_\alpha(t)\big]\big|_s \quad .$$

Combining the above computations with equation set (30.1) then gives us

$$e^{-\alpha s} F(s) = \cdots = \int_{\tau=\alpha}^{\infty} f(\tau-\alpha) e^{-s\tau}\,d\tau = \cdots = \mathcal{L}\big[f(t-\alpha)\,\text{step}_\alpha(t)\big]\big|_s \quad .$$

Cutting out the middle, we get our second translation identity:

Theorem 30.1 (second translation identity [translation along the T–axis])
Let

$$F(s) = \mathcal{L}[f(t)]|_s$$

where f is any Laplace transformable function. Then, for any positive constant α,

$$\mathcal{L}\big[f(t-\alpha)\,\text{step}_\alpha(t)\big]\big|_s = e^{-\alpha s} F(s) \quad . \tag{30.2a}$$

Equivalently,

$$\mathcal{L}^{-1}\big[e^{-\alpha s} F(s)\big]\big|_t = f(t-\alpha)\,\text{step}_\alpha(t) \quad . \tag{30.2b}$$

Computing Inverse Transforms

The Basic Computations

Computing inverse transforms using the translation along the T–axis identity is usually straightforward.

!►*Example 30.2:* *Consider finding the inverse Laplace transform of*

$$\frac{e^{-2s}}{s^2+1} \quad .$$

Applying the identity, we have

$$\mathcal{L}^{-1}\left[\frac{e^{-2s}}{s^2+1}\right]\Bigg|_t = \mathcal{L}^{-1}\Big[e^{-2s}\underbrace{\frac{1}{s^2+1}}_{F(s)}\Big]\Bigg|_t = \mathcal{L}^{-1}\big[e^{-2s}F(s)\big]\big|_t = f(t-2)\,\text{step}_2(t) \quad .$$

Here the inverse transform of F is easily read off the tables:

$$f(t) = \mathcal{L}^{-1}[F(s)]|_t = \mathcal{L}^{-1}\left[\frac{1}{s^2+1}\right]\Bigg|_t = \sin(t) \quad .$$

So, for any X,

$$f(X) = \sin(X) \quad .$$

Using this with $X = t-2$ in the above inverse transform computation then yields

$$\mathcal{L}^{-1}\left[\frac{e^{-2s}}{s^2+1}\right]\Bigg|_t = f(t-2)\,\text{step}_2(t) = \sin(t-2)\,\text{step}_2(t) \quad .$$

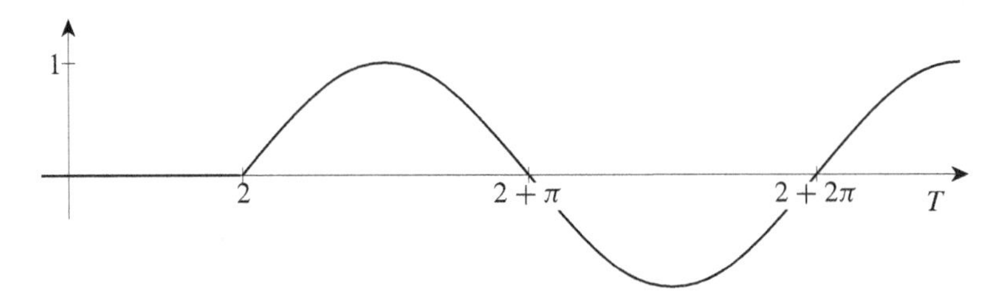

Figure 30.3: The graph of $\sin(t-2)\,\text{step}(t-2)$.

Keep in mind that

$$\sin(t-2)\,\text{step}_2(t) = \begin{cases} 0 & \text{if } t < 2 \\ \sin(t-2) & \text{if } 2 < t \end{cases} .$$

The graph of this function is sketched in Figure 30.3.

Observe that, as illustrated in Figure 30.3, the graph of

$$\mathcal{L}^{-1}\left[e^{-\alpha s}F(s)\right]\Big|_t = f(t-\alpha)\,\text{step}_\alpha(t)$$

is always zero for $t < \alpha$, and is the graph of $f(t)$ on $[0, \infty)$ shifted by α for $\alpha \leq t$. Remembering this can simplify graphing these types of functions.

Describing Piecewise-Defined Functions Arising From Inverse Transforms

Let us start with a simple, but illustrative, example.

!▶Example 30.3: *Consider computing the inverse Laplace transform of*

$$F(s) = \frac{1}{s^2}e^{-s} - \frac{1}{s^2}e^{-2s} .$$

Going to the tables, we see that

$$G(s) = \frac{1}{s^2} \implies g(t) = t .$$

Using this, along with linearity and the second translation identity, we get

$$\begin{aligned} f(t) = \mathcal{L}^{-1}[F(s)]|_t &= \mathcal{L}^{-1}\left[\frac{1}{s^2}e^{-s} - \frac{1}{s^2}e^{-2s}\right]\Big|_t \\ &= \mathcal{L}^{-1}\left[\frac{1}{s^2}e^{-1s}\right]\Big|_t - \mathcal{L}^{-1}\left[\frac{1}{s^2}e^{-2s}\right]\Big|_t \\ &= (t-1)\,\text{step}_1(t) - (t-2)\,\text{step}_2(t) . \end{aligned}$$

Note that the step functions tell us that 'significant changes' occur in $f(t)$ at the points $t = 1$ and $t = 2$.

While the above is a valid answer, it is not a particularly convenient answer. It would be much easier to graph and see what f really is if we go further and completely compute $f(t)$ on the intervals having $t = 1$ and $t = 2$ as endpoints:

For $t < 1$,

$$f(t) = (t-1)\underbrace{\text{step}_1(t)}_{0} - (t-2)\underbrace{\text{step}_2(t)}_{0} = 0 - 0 = 0 .$$

For $1 < t < 2$,

$$f(t) = (t-1)\underbrace{\text{step}_1(t)}_{1} - (t-2)\underbrace{\text{step}_2(t)}_{0} = (t-1) - 0 = t-1 \quad .$$

For $2 < t$,

$$f(t) = (t-1)\underbrace{\text{step}_1(t)}_{1} - (t-2)\underbrace{\text{step}_2(t)}_{1} = (t-1) - (t-2) = 1 \quad .$$

Thus,

$$f(t) = \begin{cases} 0 & \text{if } t < 1 \\ t-1 & \text{if } 1 < t < 2 \\ 1 & \text{if } 2 < t \end{cases} \quad .$$

(This is the function sketched in Figure 30.1b on page 525.)

As just illustrated, piecewise-defined functions naturally arise when computing inverse Laplace transforms using the second translation identity. Typically, use of this identity leads to an expression of the form

$$f(t) = g_0(t) + g_1(t)\,\text{step}_{\alpha_1}(t) + g_2(t)\,\text{step}_{\alpha_2}(t) + g_3(t)\,\text{step}_{\alpha_3}(t) + \cdots \tag{30.3}$$

where f is the function of interest, the $g_k(t)$'s are various formulas, and the α_k's are positive constants. This expression is a valid formula for f, and the step functions tell us that 'significant changes' occur in $f(t)$ at the points $t = \alpha_1$, $t = \alpha_2$, $t = \alpha_3$, ... Still, to get a better picture of the function $f(t)$, we will want to obtain the formulas for $f(t)$ over each of the intervals bounded by the α_k's. Assuming we were reasonably intelligent and indexed the α_k's so that

$$0 < \alpha_1 < \alpha_2 < \alpha_3 < \cdots \quad ,$$

we would have

For $t < \alpha_1$,

$$\begin{aligned} f(t) &= g_0(t) + g_1(t)\underbrace{\text{step}_{\alpha_1}(t)}_{0} + g_2(t)\underbrace{\text{step}_{\alpha_2}(t)}_{0} + g_3(t)\underbrace{\text{step}_{\alpha_3}(t)}_{0} + \cdots \\ &= g_0(t) + 0 + 0 + 0 + \cdots = g_0(t) \quad . \end{aligned}$$

For $\alpha_1 < t < \alpha_2$,

$$\begin{aligned} f(t) &= g_0(t) + g_1(t)\underbrace{\text{step}_{\alpha_1}(t)}_{1} + g_2(t)\underbrace{\text{step}_{\alpha_2}(t)}_{0} + g_3(t)\underbrace{\text{step}_{\alpha_3}(t)}_{0} + \cdots \\ &= g_0(t) + g_1(t) + 0 + 0 + \cdots = g_0(t) + g_1(t) \quad . \end{aligned}$$

For $\alpha_2 < t < \alpha_3$,

$$\begin{aligned} f(t) &= g_0(t) + g_1(t)\underbrace{\text{step}_{\alpha_1}(t)}_{1} + g_2(t)\underbrace{\text{step}_{\alpha_2}(t)}_{1} + g_3(t)\underbrace{\text{step}_{\alpha_3}(t)}_{0} + \cdots \\ &= g_0(t) + g_1(t) + g_2(t) + 0 + \cdots = g_0(t) + g_1(t) + g_2(t) \quad . \end{aligned}$$

And so on.

Thus, the function f described by formula (30.3), above, is also given by the conditional set of formulas

$$f(t) = \begin{cases} f_0(t) & \text{if } t < \alpha_1 \\ f_1(t) & \text{if } \alpha_1 < t < \alpha_2 \\ f_2(t) & \text{if } \alpha_2 < t < \alpha_3 \\ \vdots \end{cases}$$

where

$$f_0(t) = g_0(t) \quad ,$$

$$f_1(t) = g_0(t) + g_1(t) \quad ,$$

$$f_2(t) = g_0(t) + g_1(t) + g_2(t) \quad ,$$

$$\vdots$$

$$.$$

Computing Transforms with the Identity

The translation along the T–axis identity is also helpful in computing the transforms of piecewise-defined functions. Here, though, the computations typically require a little more care. We'll deal with fairly simple cases here, and develop this topic further in the next section.

!►Example 30.4: *Consider finding* $\mathcal{L}[g(t)]|_s$ *where*

$$g(t) = \begin{cases} 0 & \text{if } t < 3 \\ t^2 & \text{if } 3 < t \end{cases} \quad .$$

Remember, this function can also be written as

$$g(t) = t^2 \operatorname{step}_3(t) \quad .$$

Plugging this into the transform and applying our new translation identity gives

$$\mathcal{L}[g(t)]|_s = \mathcal{L}\big[t^2 \operatorname{step}_3(t)\big]\big|_s = \mathcal{L}\big[f(t-3) \operatorname{step}_3(t)\big]\big|_s = e^{-3s} F(s)$$

where

$$f(t-3) = t^2 \quad .$$

But we need the formula for $f(t)$, *not* $f(t-3)$, *to compute* $F(s)$. *To find that formula, let* $X = t - 3$ *(hence,* $t = X + 3$*) in the formula for* $f(t-3)$. *This gives*

$$f(X) = (X+3)^2 \quad .$$

Thus,

$$f(t) = (t+3)^2 = t^2 + 6t + 9 \quad ,$$

and

$$F(s) = \mathcal{L}[f(t)]|_s = \mathcal{L}\big[t^2 + 6t + 9\big]\big|_s$$

$$= \mathcal{L}\big[t^2\big]\big|_s + 6\mathcal{L}[t]|_s + 9\mathcal{L}[1]|_s = \frac{2}{s^3} + \frac{6}{s^2} + \frac{9}{s} \quad .$$

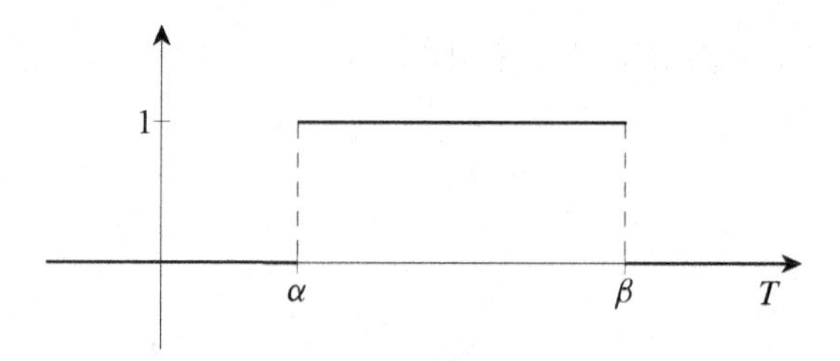

Figure 30.4: Graph of the rectangle function $\text{rect}_{(\alpha,\beta)}(t)$ with $-\infty < \alpha < \beta < \infty$.

Plugging this back into the above formula for $\mathcal{L}[g(t)]|_s$ *gives us*

$$\mathcal{L}[g(t)]|_s = e^{-3s}F(s) = e^{-3s}\left[\frac{2}{s^3} + \frac{6}{s^2} + \frac{9}{s}\right] .$$

30.3 Rectangle Functions and Transforms of More Piecewise-Defined Functions

Rectangle Functions

"Rectangle functions" are slight generalizations of step functions. Given any interval (α, β), the *rectangle function on* (α, β), denoted $\text{rect}_{(\alpha,\beta)}$, is the function given by

$$\text{rect}_{(\alpha,\beta)}(t) = \begin{cases} 0 & \text{if } t < \alpha \\ 1 & \text{if } \alpha < t < \beta \\ 0 & \text{if } \beta < t \end{cases} .$$

The graph of $\text{rect}_{(\alpha,\beta)}$ with $-\infty < \alpha < \beta < \infty$ has been sketched in Figure 30.4. You can see why it is called a rectangle function — it's graph looks rather "rectangular", at least when α and β are finite. If $\alpha = -\infty$ or $\beta = \infty$, the corresponding rectangle functions simplify to

$$\text{rect}_{(-\infty,\beta)}(t) = \begin{cases} 1 & \text{if } t < \beta \\ 0 & \text{if } \beta < t \end{cases}$$

and

$$\text{rect}_{(\alpha,\infty)}(t) = \begin{cases} 0 & \text{if } t < \alpha \\ 1 & \text{if } \alpha < t \end{cases} .$$

And if both $a = -\infty$ and $b = \infty$, then we have

$$\text{rect}_{(-\infty,\infty)}(t) = 1 \quad \text{for all } t .$$

All of these rectangle functions can be written as simple linear combinations of 1 and step functions at α and/or β, with, again, the step functions acting as 'switches' — switching the rectangle function 'on' (from 0 to 1 at α), and switching it 'off' (from 1 back to 0 at β). In particular, we clearly have

$$\text{rect}_{(-\infty,\infty)}(t) = 1 \quad \text{and} \quad \text{rect}_{(\alpha,\infty)}(t) = \text{step}_\alpha(t) .$$

Somewhat more importantly (for us), we should observe that, for $-\infty < \alpha < \beta < \infty$,

$$1 - \text{step}_\beta(t) = \begin{Bmatrix} 1-0 & \text{if} & t < \beta \\ 1-1 & \text{if} & \beta < t \end{Bmatrix} = \begin{Bmatrix} 1 & \text{if} & t < \beta \\ 0 & \text{if} & \beta < t \end{Bmatrix} = \text{rect}_{(-\infty,\beta)}(t) \quad ,$$

and

$$\text{step}_\alpha(t) - \text{step}_\beta(t) = \begin{Bmatrix} 0-0 & \text{if} & t < \alpha \\ 1-0 & \text{if} & \alpha < t < \beta \\ 1-1 & \text{if} & \beta < t \end{Bmatrix}$$

$$= \begin{Bmatrix} 0 & \text{if} & t < a \\ 1 & \text{if} & \alpha < t < \beta \\ 0 & \text{if} & \beta < t \end{Bmatrix} = \text{rect}_{(\alpha,\beta)}(t) \quad .$$

In summary, for $-\infty < \alpha < \beta < \infty$,

$$\text{rect}_{(\alpha,\beta)}(t) = \text{step}_\alpha(t) - \text{step}_\beta(t) \quad , \tag{30.4a}$$

$$\text{rect}_{(-\infty,\beta)}(t) = 1 - \text{step}_\beta(t) \tag{30.4b}$$

and

$$\text{rect}_{(\alpha,\infty)}(t) = \text{step}_\alpha(t) \quad . \tag{30.4c}$$

These formulas allow us to quickly compute the Laplace transforms of rectangle functions using the known transforms of 1 and the step functions.

!►*Example 30.5:*

$$\begin{aligned} \mathcal{L}\big[\text{rect}_{(3,4)}(t)\big]\big|_s &= \mathcal{L}\big[\text{rect}_{(3,4)}(t)\big]\big|_s \\ &= \mathcal{L}\big[\text{step}_3(t) - \text{step}_4(t)\big]\big|_s \\ &= \mathcal{L}\big[\text{step}_3(t)\big]\big|_s - \mathcal{L}\big[\text{step}_4(t)\big]\big|_s = \frac{1}{s}e^{-3s} - \frac{1}{s}e^{-4s} \quad . \end{aligned}$$

Transforming More General Piecewise-Defined Functions

To help us deal with more general piecewise-defined functions, let us make the simple observations that

$$g(t)\,\text{rect}_{(a,b)}(t) = \begin{Bmatrix} g(t)\cdot 0 & \text{if} & t < a \\ g(t)\cdot 1 & \text{if} & a < t < b \\ g(t)\cdot 0 & \text{if} & b < t \end{Bmatrix} = \begin{cases} 0 & \text{if} \quad t < a \\ g(t) & \text{if} \quad a < t < b \\ 0 & \text{if} \quad b < t \end{cases} ,$$

and

$$g(t)\,\text{rect}_{(-\infty,b)}(t) = \begin{Bmatrix} g(t)\cdot 1 & \text{if} & t < b \\ g(t)\cdot 0 & \text{if} & b < t \end{Bmatrix} = \begin{cases} g(t) & \text{if} \quad t < b \\ 0 & \text{if} \quad b < t \end{cases} .$$

So functions of the form

$$f(t) = \begin{cases} 0 & \text{if} \quad t < a \\ g(t) & \text{if} \quad a < t < b \\ 0 & \text{if} \quad b < t \end{cases} \qquad \text{and} \qquad h(t) = \begin{cases} g(t) & \text{if} \quad t < b \\ 0 & \text{if} \quad b < t \end{cases}$$

can be rewritten, respectively, as

$$f(t) = g(t)\,\mathrm{rect}_{(a,b)}(t) \qquad \text{and} \qquad h(t) = g(t)\,\mathrm{rect}_{(-\infty,b)}(t) \quad .$$

More generally, it should now be clear that anything of the form

$$f(t) = \begin{cases} g_0(t) & \text{if } t < \alpha_1 \\ g_1(t) & \text{if } \alpha_1 < t < \alpha_2 \\ g_2(t) & \text{if } \alpha_2 < t < \alpha_3 \\ \vdots & \end{cases} \tag{30.5a}$$

can be rewritten as

$$f(t) = g_0(t)\,\mathrm{rect}_{(-\infty,\alpha_1)}(t) + g_1(t)\,\mathrm{rect}_{(\alpha_1,\alpha_2)}(t) + g_2(t)\,\mathrm{rect}_{(\alpha_2,\alpha_3)}(t) + \cdots \quad . \tag{30.5b}$$

The second form (with the rectangle functions) is a bit more concise than the "conditional set of formulas" used in form (30.5a), and is generally preferred by typesetters. Of course, there is a more important advantage of form (30.5b): Assuming f is piecewise continuous and of exponential order, its Laplace transform can now be taken by expressing the rectangle functions in formula (30.5b) as the linear combinations of 1 and step functions given in equation set (30.4), and then using linearity and what we learned in the previous section about taking transforms of functions multiplied by step functions.

!▶Example 30.6: *Consider finding* $F(s) = \mathcal{L}[f(t)]|_s$ *when*

$$f(t) = \begin{cases} t^2 & \text{if } t < 3 \\ 0 & \text{if } 3 < t \end{cases} \quad .$$

From the above, we see that

$$\begin{aligned} f(t) &= t^2\,\mathrm{rect}_{(-\infty,3)}(t) \\ &= t^2\left[1 - \mathrm{step}_3(t)\right] = t^2 - t^2\,\mathrm{step}_3(t) \quad . \end{aligned}$$

So

$$F(s) = \mathcal{L}[f(t)]|_s = \mathcal{L}\left[t^2 - t^2\,\mathrm{step}_3(t)\right]\Big|_s = \mathcal{L}\left[t^2\right]\Big|_s - \mathcal{L}\left[t^2\,\mathrm{step}_3(t)\right]\Big|_s \quad .$$

The Laplace transform of t^2 *is in the tables, while the transform of* $t^2\,\mathrm{step}_3(t)$ *just happened to have been computed in Example 30.4 a few pages ago. Using these transforms, the above formula for* F *becomes*

$$F(s) = \frac{2}{s^3} - e^{-3s}\left[\frac{2}{s^3} + \frac{6}{s^2} + \frac{9}{s}\right] \quad .$$

!▶Example 30.7: *Consider finding* $F(s) = \mathcal{L}[f(t)]|_s$ *when*

$$f(t) = \begin{cases} 0 & \text{if } t < 2 \\ e^{3t} & \text{if } 2 < t < 4 \\ 0 & \text{if } 4 < t \end{cases} \quad .$$

From the above, we see that

$$\begin{aligned} f(t) &= e^{3t}\,\mathrm{rect}_{(2,4)}(t) \\ &= e^{3t}\left[\mathrm{step}_2(t) - \mathrm{step}_4(t)\right] = e^{3t}\,\mathrm{step}_2(t) - e^{3t}\,\mathrm{step}_4(t) \quad . \end{aligned}$$

Thus,

$$\mathcal{L}[f(t)]|_s = \mathcal{L}\big[e^{3t}\,\text{step}_2(t)\big]\big|_s - \mathcal{L}\big[e^{3t}\,\text{step}_4(t)\big]\big|_s \quad . \tag{30.6}$$

Both of the transforms on the right side of our last equation are easily computed via the translation identity developed in this chapter. For the first, we have

$$\mathcal{L}\big[e^{3t}\,\text{step}_2(t)\big]\big|_s = \mathcal{L}\big[g(t-2)\,\text{step}_2(t)\big]\big|_s = e^{-2s}G(s)$$

where

$$g(t-2) = e^{3t} \quad .$$

Letting $X = t - 2$ *(so* $t = X + 2$*), the last expression becomes*

$$g(X) = e^{3(X+2)} = e^{3X+6} = e^6 e^{3X} \quad .$$

So

$$g(t) = e^6 e^{3t}$$

and

$$G(s) = \mathcal{L}[g(t)]|_s = \mathcal{L}\big[e^6 e^{3t}\big]\big|_s = e^6\mathcal{L}\big[e^{3t}\big]\big|_s = e^6\frac{1}{s-3} \quad .$$

This, along with the first equation in this paragraph, gives us

$$\mathcal{L}\big[e^{3t}\,\text{step}_2(t)\big]\big|_s = e^{-2s}G(s) = e^{-2s}e^6\frac{1}{s-3} = \frac{e^{-2(s-3)}}{s-3} \quad .$$

The transform of $e^{3t}\,\text{step}_4(t)$ *can be computed in the same manner, yielding*

$$\mathcal{L}\big[e^{3t}\,\text{step}_4(t)\big]\big|_s = \frac{e^{-4(s-3)}}{s-3} \quad .$$

(The details of this computation are left to you.)

Finally, by combining the formulas we just obtained for the transforms of $e^{3t}\,\text{step}_2(t)$ *and* $e^{3t}\,\text{step}_4(t)$ *with equation (30.6), we have*

$$\begin{aligned}\mathcal{L}[f(t)]|_s &= \mathcal{L}\big[e^{3t}\,\text{step}_2(t)\big]\big|_s - \mathcal{L}\big[e^{3t}\,\text{step}_4(t)\big]\big|_s \\ &= \frac{e^{-2(s-3)}}{s-3} - \frac{e^{-4(s-3)}}{s-3} \quad .\end{aligned}$$

!► Example 30.8: *Let's find the Laplace transform* $F(s)$ *of*

$$f(t) = \begin{cases} 2 & \text{if } t < 1 \\ e^{3t} & \text{if } 1 < t < 3 \\ t^2 & \text{if } 3 < t \end{cases} \quad .$$

To apply the Laplace transform, we first convert the above to an equivalent expression involving step functions:

$$\begin{aligned} f(t) &= 2\,\text{rect}_{(-\infty,1)}(t) + e^{3t}\,\text{rect}_{(1,3)}(t) + t^2\,\text{rect}_{(3,\infty)}(t) \\ &= 2\big[1 - \text{step}_1(t)\big] + e^{3t}\big[\text{step}_1(t) - \text{step}_3(t)\big] + t^2\,\text{step}_3(t) \\ &= 2 - 2\,\text{step}_1(t) + e^{3t}\,\text{step}_1(t) - e^{3t}\,\text{step}_3(t) + t^2\,\text{step}_3(t) \quad . \end{aligned}$$

Using the tables and methods already discussed earlier in this chapter (as in Examples 30.7 and 30.4), we discover that

$$\mathcal{L}[2]|_s = \frac{2}{s} \quad , \quad \mathcal{L}[2\,\text{step}_1(t)]\big|_s = \frac{2e^{-s}}{s} \quad ,$$

$$\mathcal{L}\big[e^{3t}\,\text{step}_1(t)\big]\big|_s = \frac{e^{-(s-3)}}{s-3} \quad , \quad \mathcal{L}\big[e^{3t}\,\text{step}_3(t)\big]\big|_s = \frac{e^{-3(s-3)}}{s-3}$$

and

$$\mathcal{L}\big[t^2\,\text{step}_3(t)\big]\big|_s = e^{-3s}\left[\frac{2}{s^3} + \frac{6}{s^2} + \frac{9}{s}\right] \quad .$$

Combining the above and using the linearity of the Laplace transform, we obtain

$$\begin{aligned} F(s) &= \mathcal{L}[f(t)]|_s \\ &= \mathcal{L}\Big[2 - 2\,\text{step}_1(t) + e^{3t}\,\text{step}_1(t) - e^{3t}\,\text{step}_3(t) + t^2\,\text{step}_3(t)\Big]\Big|_s \\ &= \frac{2}{s} - \frac{2e^{-s}}{s} + \frac{e^{-(s-3)}}{s-3} - \frac{e^{-3(s-3)}}{s-3} + e^{-3s}\left[\frac{2}{s^3} + \frac{6}{s^2} + \frac{9}{s}\right] \quad . \end{aligned}$$

30.4 Convolution with Piecewise-Defined Functions

Take another look at Example 30.1 on page 527. As noted in the footnote, we could have by-passed much of the Laplace transform computation by simply observing that

$$y(t) = \sin(t) * \text{step}_3(t)$$

and computing that convolution. But in the footnote, it was claimed that computing such convolutions can be "a little tricky". Well, to be honest, it's not all that tricky. It's more an issue of careful bookkeeping.

When computing a convolution $h * f$ in which f is piecewise defined and h is not, you need to realize that the resulting convolution will also be piecewise defined, with (as you will see in the examples) the formula for $h * f$ changing at the same points where the formula for f changes. Hence, you should compute $h * f$ separately over the different intervals bounded by these points. Moreover, in computing the corresponding integrals, you will also need to account for the piecewise-defined nature of f, and break up the integral appropriately. To simplify all this, it is strongly recommended that you compute the convolution $h * f$ using the integral formula

$$h * f(t) = f * h(t) = \int_0^t f(x)h(t-x)\,dx$$

(and not with the integrand $h(x)f(t-x)$).

One or two examples should clarify matters.

!►Example 30.9: *Let's compute* $\sin(t) * \text{step}_3(t)$. *Since* step_3 *is piecewise defined, we will, as suggested, use the integral formula*

$$\sin(t) * \text{step}_3(t) = \int_0^t \text{step}_3(x)\sin(t-x)\,dx \quad .$$

First, we compute the integral assuming $t < 3$. This one is easy:

$$\sin(t) * \text{step}_3(t) = \int_0^t \underbrace{\text{step}_3(x)}_{\substack{0 \\ \text{since } x<t<3}} \sin(t-x)\,dx = \int_0^t 0 \cdot \sin(t-x)\,dx = 0 \quad .$$

So,

$$\sin(t) * \text{step}_3(t) = 0 \qquad \text{if} \quad t < 3 \quad . \tag{30.7}$$

On the other hand, if $3 < t$, then the interval of integration includes $x = 3$, the point at which the value of $\text{step}_3(x)$ radically changes from 0 to 1. Thus, we must break up our integral at the point $x = 3$ in computing $h * f$:

$$\begin{aligned}
\sin(t) * \text{step}_3(t) &= \int_0^t \text{step}_3(x)\sin(t-x)\,dx \\
&= \int_0^3 \underbrace{\text{step}_3(x)}_{\substack{0 \\ \text{since } x<3}}\sin(t-x)\,dx + \int_3^t \underbrace{\text{step}_3(x)}_{\substack{1 \\ \text{since } 3<x}}\sin(t-x)\,dx \\
&= \int_0^3 0 \cdot \sin(t-x)\,dx + \int_3^t 1 \cdot \sin(t-x)\,dx \\
&= 0 + \cos(t-t) - \cos(t-3) \\
&= 1 - \cos(t-3) \quad .
\end{aligned}$$

Thus,

$$\sin(t) * \text{step}_3(t) = 1 - \cos(t-3) \qquad \text{if} \quad 3 < t \quad . \tag{30.8}$$

Combining our two results (formulas (30.7) and (30.8)), we have the complete set of conditional formulas for our convolution,

$$\sin(t) * \text{step}_3(t) = \begin{cases} 0 & \text{if} \quad t < 3 \\ 1 - \cos(t-3) & \text{if} \quad 3 < t \end{cases} \quad .$$

Glance back at the above example and observe that, immediately after the computation of $\sin(t) * \text{step}_3(t)$ for each different case ($t < 3$ and $3 < t$), the resulting formula for the convolution was rewritten along with the values assumed for t (formulas (30.7) and (30.8), respectively). Do the same in your own computations! Always rewrite any derived formula for your convolution along with the values assumed for t. And write this someplace safe where you can easily find it. This is part of the bookkeeping, and helps ensure that you do not lose parts of your work when you compose the full set of conditional formulas for the convolution.

One more example should be quite enough.

!►Example 30.10: *Let's compute $e^{-3t} * f(t)$ where*

$$f(t) = \begin{cases} t & \text{if} \quad t < 2 \\ 2 & \text{if} \quad 2 < t < 4 \\ 0 & \text{if} \quad 4 < t \end{cases} \quad .$$

For this convolution, we can do a little "pre-computing" to simplify later steps:

$$\begin{aligned} e^{-3t} * f(t) &= \int_0^t f(x)e^{-3(t-x)}\,dx \\ &= \int_0^t f(x)e^{-3t+3x}\,dx = e^{-3t}\int_0^t f(x)e^{3x}\,dx \quad . \end{aligned}$$

Now, if $t < 2$,

$$e^{-3t} * f(t) = e^{-3t}\int_0^t \underbrace{f(x)}_{\substack{x \\ (\text{since } x < t < 2)}} e^{3x}\,dx = e^{-3t}\int_0^t xe^{3x}\,dx \quad .$$

This integral is easily computed using integration by parts, yielding

$$e^{-3t} * f(t) = e^{-3t}\left[\frac{t}{3}e^{3t} - \frac{1}{9}e^{3t} + \frac{1}{9}\right] = \frac{1}{9}\left[3t - 1 + e^{-3t}\right] \quad .$$

Thus,

$$e^{-3t} * f(t) = \frac{1}{9}\left[3t - 1 + e^{-3t}\right] \qquad \text{if} \quad t < 2 \quad . \tag{30.9}$$

On the other hand, when $2 < t < 4$,

$$\begin{aligned} e^{-3t} * f(t) &= e^{-3t}\int_0^t f(x)e^{3x}\,dx \\ &= e^{-3t}\left[\int_0^2 \underbrace{f(x)}_{\substack{x \\ (\text{since } x < 2)}} e^{3x}\,dx + \int_2^t \underbrace{f(x)}_{\substack{2 \\ (\text{since } 2 < x < t < 4)}} e^{3x}\,dx\right] \\ &= e^{-3t}\left[\int_0^2 xe^{3x}\,dx + \int_2^t 2\cdot e^{3x}\,dx\right] \\ &= \cdots \\ &= e^{-3t}\left[\frac{1}{9}\left(5e^6 + 1\right) + \frac{2}{3}\left(e^{3t} - e^6\right)\right] \\ &= \cdots \\ &= \frac{2}{3} + \frac{1}{9}\left[1 - e^6\right]e^{-3t} \quad . \end{aligned}$$

Thus,

$$e^{-3t} * f(t) = \frac{2}{3} + \frac{1}{9}\left[1 - e^6\right]e^{-3t} \qquad \text{for} \quad 2 < t < 4 \quad . \tag{30.10}$$

Finally, when $6 < t$,

$$\begin{aligned} e^{-3t} * f(t) &= e^{-3t}\int_0^t e^{3x} f(x)\,dx \\ &= e^{-3t}\left[\int_0^2 \underbrace{f(x)}_{\substack{x \\ (\text{since } x < 2)}} e^{3x}\,dx + \int_2^4 \underbrace{f(x)}_{\substack{2 \\ (\text{since } 2 < x < 4)}} e^{3x}\,dx + \int_4^t \underbrace{f(x)}_{\substack{0 \\ (\text{since } 4 < x)}} e^{3x}\,dx\right] \\ &= e^{-3t}\left[\int_0^2 xe^{3x}\,dx + \int_2^4 2\cdot e^{3x}\,dx + \int_4^t 0\cdot e^{3x}\,dx\right] \end{aligned}$$

$$= \cdots$$

$$= e^{-3t}\left[\frac{1}{9}\left(5e^6+1\right) + \frac{2}{3}\left(e^{12}-e^6\right) + 0\right]$$

$$= \cdots$$

$$= \frac{1}{9}\left[6e^{12}+1-e^6\right]e^{-3t} \quad .$$

Thus,

$$e^{-3t} * f(t) = \frac{1}{9}\left[6e^{12}+1-e^6\right]e^{-3t} \qquad \text{for} \quad 4 < t \quad . \tag{30.11}$$

Putting it all together, equations (30.9), (30.10) and (30.11) give us

$$e^{-3t} * f(t) = \begin{cases} \frac{1}{9}\left[3t-1+e^{-3t}\right] & \text{if} \quad t < 2 \\ \frac{2}{3} + \frac{1}{9}\left[1-e^6\right]e^{-3t} & \text{if} \quad 2 < t < 4 \\ \frac{1}{9}\left[6e^{12}+1-e^6\right]e^{-3t} & \text{if} \quad 4 < t \end{cases} \quad .$$

And what if both f and g are piecewise defined? Then you must keep track of where formulas of both $f(t)$ and $h(t-x)$ change. Fortunately, we will have little need to deal with such convolutions at this time.

30.5 Periodic Functions

Basics

Often, a function of interest f is *periodic with period* p for some positive value p. This means that the graph of the function remains unchanged when shifted to the left or right by p. This is equivalent to saying

$$f(t+p) = f(t) \qquad \text{for all} \quad t \quad . \tag{30.12}$$

You are well-acquainted with several periodic functions — the trigonometric functions, for example. In particular, the basic sine and cosine functions

$$\sin(t) \qquad \text{and} \qquad \cos(t)$$

are periodic with period $p = 2\pi$. But other periodic functions, such as the "saw" function sketched in Figure 30.5a and the "square-wave" function sketched in Figure 30.5b, can arise in applications.

Strictly speaking, a truly periodic function is defined on the entire real line, $(-\infty, \infty)$. For our purposes, though, it will suffice to have f "periodic on $(0, \infty)$" with period p. This simply means that f is that part of a periodic function along the positive T–axis. What $f(t)$ is for $t < 0$ is irrelevant. Accordingly, for functions *periodic on* $(0, \infty)$, we modify requirement (30.12) to

$$f(t+p) = f(t) \qquad \text{for all} \quad t > 0 \quad . \tag{30.13}$$

In what follows, however, it will usually be irrelevant as to whether a given function is truly periodic or merely periodic on $(0, \infty)$, In either case, we will refer to the function as "periodic", and specify whether it is defined on all of $(-\infty, \infty)$ or just $(0, -\infty)$ only if necessary.

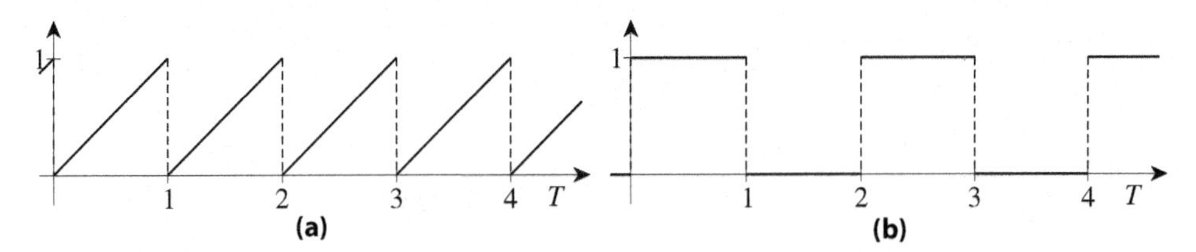

Figure 30.5: Two periodic functions: **(a)** a basic saw function, and **(b)** a basic square wave function.

A convenient way to describe a periodic function f with period p is by

$$f(t) = \begin{cases} f_0(t) & \text{if } 0 < t < p \\ f(t+p) & \text{in general} \end{cases} .$$

The $f_0(t)$ is the formula for f over the *base period interval* $(0, p)$. The second line is simply telling us that the function is periodic and that equation (30.12) or (30.13) holds and can be used to compute the function at points outside of the base period interval. (The value of $f(t)$ at $t = 0$ and integral multiples of p are determined — or ignored — following the conventions for piecewise-defined functions discussed in Section 30.1.)

!►Example 30.11: *Let* $\text{saw}(t)$ *denote the basic saw function sketched in Figure 30.5a. It clearly has period* $p = 1$, *has jump discontinuities at integer values of* t, *and is given on* $(0, \infty)$ *by*

$$\text{saw}(t) = \begin{cases} t & \textit{if } 0 < t < 1 \\ \text{saw}(t+1) & \textit{in general} \end{cases} .$$

In this case, the formula for computing $\text{saw}(\tau)$ *when* $0 < \tau < 1$ *is*

$$\text{saw}_0(\tau) = \tau \quad .$$

So, for example, $\text{saw}(3/4) = 3/4$.

On the other hand, to compute $\text{saw}(\tau)$ *when* $\tau > 1$ *(and not an integer), we must use*

$$\text{saw}(t+1) = \text{saw}(t)$$

repeatedly until we finally reach a value t *in the base period interval* $(0, 1)$. *For example,*

$$\begin{aligned} \text{saw}\left(\frac{8}{3}\right) &= \text{saw}\left(\frac{5}{3} + 1\right) = \text{saw}\left(\frac{5}{3}\right) \\ &= \text{saw}\left(\frac{2}{3} + 1\right) = \text{saw}\left(\frac{2}{3}\right) = \frac{2}{3} \quad . \end{aligned}$$

Often, the formula for the function over the base period interval is, itself, piecewise defined.

!►Example 30.12: *Let* $\text{sqwave}(t)$ *denote the square-wave function in Figure 30.5b. This function has period* $p = 2$, *and, over its base period interval* $(0, 2)$, *is given by*

$$\text{sqwave}(t) = \begin{cases} 1 & \textit{if } 0 < t < 1 \\ 0 & \textit{if } 1 < t < 2 \end{cases} .$$

So,

$$\text{sqwave}(t) = \begin{cases} 1 & \textit{if} \quad 0 < t < 1 \\ 0 & \textit{if} \quad 1 < t < 2 \\ \text{sqwave}(t-2) & \textit{in general} \end{cases} .$$

Before discussing Laplace transforms of periodic functions, let's make a couple of observations concerning a function f which is periodic with period p over $(0, \infty)$. We won't prove them. Instead, you should think about why these statements are "obviously true".

1. If f is piecewise continuous over $(0, p)$, then f is piecewise continuous over $(0, \infty)$.
2. If f is piecewise continuous over $(0, p)$, then f is of exponential order $s_0 = 0$.

Transforms of Periodic Functions

Suppose we want to find the Laplace transform

$$F(s) = \mathcal{L}[f(t)]|_s = \int_0^\infty f(t)e^{-st}\,dt$$

when f is piecewise continuous and periodic with period p. Because $f(t)$ satisfies

$$f(t) = f(t+p) \qquad \text{for} \quad t > 0 \quad ,$$

we should expect to (possibly) simplify our computations by partitioning the integral of the transform into integrals over subintervals of length p,

$$\begin{aligned} F(s) &= \int_0^\infty f(t)e^{-st}\,dt \\ &= \int_0^p f(t)e^{-st}\,dt + \int_p^{2p} f(t)e^{-st}\,dt + \int_{2p}^{3p} f(t)e^{-st}\,dt \\ &\quad + \int_{3p}^{4p} f(t)e^{-st}\,dt + \int_{4p}^{5p} f(t)e^{-st}\,dt + \cdots \quad . \end{aligned}$$

For brevity, let's rewrite this as

$$F(s) = \sum_{k=0}^{\infty} \int_{kp}^{(k+1)p} f(t)e^{-st}\,dt \quad . \tag{30.14}$$

Now consider using the substitution $\tau = t - kp$ in the kth term of this summation. Then $t = \tau + kp$,

$$e^{-st} = e^{-s(\tau+kp)} = e^{-kps}e^{-s\tau} \quad ,$$

and, by the periodicity of f,

$$\begin{aligned} f(\tau+p) &= f(\tau) \\ f(\tau+2p) &= f([\tau+p]+p) = f(\tau+p) = f(\tau) \\ f(\tau+3p) &= f([\tau+2p]+p) = f(\tau+2p) = f(\tau) \\ &\vdots \\ f(\tau+kp) &= \cdots = f(\tau) \quad . \end{aligned}$$

So,

$$\int_{t=kp}^{(k+1)p} f(t)e^{-st}\,dt = \int_{\tau=kp-kp}^{(k+1)p-kp} f(\tau+kp)e^{-s(\tau+kp)}\,dt$$
$$= \int_0^p f(\tau)e^{-kps}e^{-s\tau}\,dt = e^{-kps}\int_0^p f(\tau)e^{-s\tau}\,dt \quad .$$

Note that the last integral does not depend on k. Consequently, combining the last result with equation (30.14), we have

$$F(s) = \sum_{k=0}^{\infty} e^{-kps}\int_0^p f(\tau)e^{-s\tau}\,dt = \left[\sum_{k=0}^{\infty} e^{-kps}\right]\int_0^p f(\tau)e^{-s\tau}\,d\tau \quad .$$

Here we have an incredible stroke of luck, at least if you recall what a geometric series is and how to compute its sum. Assuming you do recall this, we have

$$\sum_{k=0}^{\infty} e^{-kps} = \sum_{k=0}^{\infty}\left[e^{-ps}\right]^k = \frac{1}{1-e^{-ps}} \quad . \tag{30.15}$$

We also have this if you do not recall about geometric series, but then you will certainly want to go to the brief review of geometric series in Section 32.1 to see how we get this equation.

Whether or not you recall about geometric series, equation (30.15) combined with the last formula for F (along with the observations made earlier regarding piecewise continuity and periodic functions) gives us the following theorem.

Theorem 30.2

Let f be a piecewise continuous and periodic function with period p. Then its Laplace transform F is given by

$$F(s) = \frac{F_0(s)}{1-e^{-ps}} \qquad \text{for} \quad s > 0$$

where

$$F_0(s) = \int_0^p f(t)e^{-st}\,dt \quad .$$

There are at least two alternative ways of describing F_0 in the above theorem. First of all, if f is given by

$$f(t) = \begin{cases} f_0(t) & \text{if} \quad 0 < t < p \\ f(t+p) & \text{in general} \end{cases} \quad ,$$

then, of course,

$$F_0(s) = \int_0^p f_0(t)e^{-st}\,dt \quad .$$

Also, using the fact that

$$\int_0^p f_0(t)e^{-st}\,dt = \int_0^\infty f_0(t)\,\text{rect}_{(0,p)}(t)\,e^{-st}\,dt \quad ,$$

we see that

$$F_0(s) = \mathcal{L}\big[f_0(t)\,\text{rect}_{(0,p)}(t)\big]\big|_s$$

or, equivalently, that

$$F_0(s) = \mathcal{L}\big[f(t)\,\text{rect}_{(0,p)}(t)\big]\big|_s \quad .$$

Whether any of the alternative descriptions of $F_0(s)$ is useful may depend on what transforms you have already computed.

!▶Example 30.13: *Let's find the Laplace transform of the saw function from Example 30.11 and sketched in Figure 30.5a,*

$$\operatorname{saw}(t) = \begin{cases} t & \text{if } 0 < t < 1 \\ \operatorname{saw}(t+1) & \text{in general} \end{cases} .$$

Here, $p = 1$, and the last theorem tells us that

$$\mathcal{L}[\operatorname{saw}(t)]|_s = \frac{F_0(s)}{1 - e^{-1 \cdot s}} = \frac{F_0(s)}{1 - e^{-s}} \qquad \text{for} \quad s > 0$$

where (using each of the formulas discussed for F_0)

$$F_0(s) = \int_0^1 \operatorname{saw}(t) e^{-st}\, dt \tag{30.16a}$$

$$= \int_0^1 t e^{-st}\, dt \tag{30.16b}$$

$$= \mathcal{L}\big[t \operatorname{rect}_{(0,1)}(t)\big]\big|_s \quad . \tag{30.16c}$$

Had the author been sufficiently clever, $\mathcal{L}\big[t \operatorname{rect}_{(0,1)}(t)\big]$ would have already been computed in a previous example, and we could write out the final result using formula (30.16c). But he wasn't, so let's just compute $F_0(s)$ using formula (30.16b) and integration by parts:

$$\begin{aligned} F_0(s) &= \int_0^1 t e^{-st}\, dt \\ &= -\frac{t}{s} e^{-st}\bigg|_{t=0}^{1} - \int_0^1 \left(-\frac{1}{s}\right) e^{-st}\, dt \\ &= -\frac{1}{s} e^{-s \cdot 1} + 0 - \frac{1}{s^2}\left[e^{-s \cdot 1} - e^{-s \cdot 0}\right] = \frac{1}{s^2}\left[1 - e^{-s} - s e^{-s}\right] \quad . \end{aligned}$$

Hence,

$$\begin{aligned} \mathcal{L}[\operatorname{saw}(t)]|_s &= \frac{F_0(s)}{1 - e^{-s}} \\ &= \frac{1}{s^2} \cdot \frac{1 - e^{-s} - s e^{-s}}{1 - e^{-s}} \\ &= \frac{1}{s^2}\left[1 - \frac{s e^{-s}}{1 - e^{-s}}\right] = \frac{1}{s^2} - \frac{1}{s} \cdot \frac{e^{-s}}{1 - e^{-s}} \quad . \end{aligned}$$

This is our transform. If you wish, you can apply a little algebra and 'simplify' it to

$$\mathcal{L}[\operatorname{saw}(t)]|_s = \frac{1}{s^2} - \frac{1}{s} \cdot \frac{1}{e^s - 1} \quad ,$$

though you may prefer to keep the formula with $1 - e^{-ps}$ in the denominator to remind you that this transform came from a periodic function with period p.

Just for fun, let's go even further using the fact that

$$\frac{e^{-s}}{1 - e^{-s}} = \frac{e^{-s}}{1 - e^{-s}} \cdot \frac{2e^{s/2}}{2e^{s/2}} = \frac{1}{2} \cdot \frac{2e^{-s/2}}{e^{s/2} - e^{-s/2}} = \frac{1}{2} \cdot \frac{e^{-s/2}}{\sinh(s/2)} \quad .$$

Thus, the above formula for the Laplace transform of the saw function can also be written as

$$\mathcal{L}[\operatorname{saw}(t)]|_s = \frac{1}{s^2} - \frac{1}{2s} \cdot \frac{e^{-s/2}}{\sinh(s/2)} \quad .$$

This is significant only in that it demonstrates why hyperbolic trigonometric functions are sometimes found in tables of transforms.

Table 30.1: Commonly Used Identities (Version 2)

In the following, $F(s) = \mathcal{L}[f(t)]|_s$.

$h(t)$	$H(s) = \mathcal{L}[h(t)]\|_s$	Restrictions
$f(t)$	$\int_0^\infty f(t)e^{-st}\,dt$	
$e^{\alpha t} f(t)$	$F(s-\alpha)$	α is real
$f(t-\alpha)\,\text{step}_\alpha(t)$	$e^{-\alpha s}F(s)$	$\alpha > 0$
$\frac{df}{dt}$	$sF(s) - f(0)$	
$\frac{d^2 f}{dt^2}$	$s^2F(s) - sf(0) - f'(0)$	
$\frac{d^n f}{dt^n}$	$s^nF(s) - s^{n-1}f(0) - s^{n-2}f'(0) - s^{n-3}f''(0) - \cdots - f^{(n-1)}(0)$	$n = 1, 2, 3, \ldots$
$t\,f(t)$	$-\frac{dF}{ds}$	
$t^n\,f(t)$	$(-1)^n \frac{d^n F}{ds^n}$	$n = 1, 2, 3, \ldots$
$\int_0^t f(\tau)\,d\tau$	$\frac{F(s)}{s}$	
$\frac{f(t)}{t}$	$\int_s^\infty F(\sigma)\,d\sigma$	
$f * g(t)$	$F(s)G(s)$	
f is periodic with period p	$\frac{\int_0^p f(t)e^{-st}\,dt}{1-e^{-ps}}$	

30.6 An Expanded Table of Identities

For reference, let us write out a new table of Laplace transform identities containing the identities listed in our first table of Laplace transform identities, Table 27.1 on page 488, along with some of the more important identities derived after making that table. Our new table is Table 30.1.

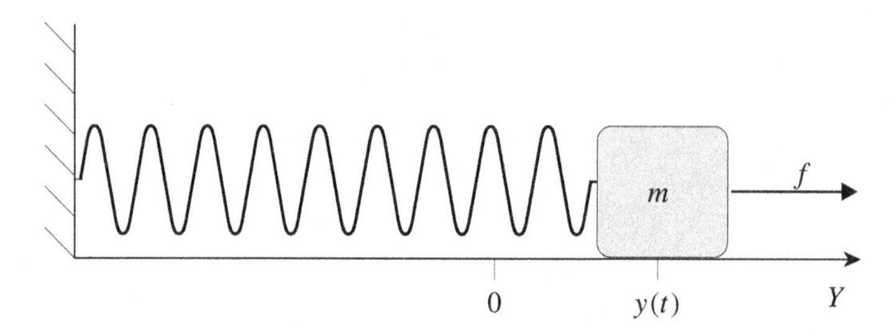

Figure 30.6: A mass/spring system with mass m and an outside force f acting on the mass.

30.7 Duhamel's Principle and Resonance

The Problem

Now is a good time to re-examine some of those "forced" mass/spring systems originally discussed in Chapter 23, and diagrammed in Figure 30.6. Recall that this system is modeled by

$$m\frac{d^2y}{dt^2} + \gamma\frac{dy}{dt} + \kappa y = f$$

where $y = y(t)$ is the position of the mass at time t (with $y = 0$ being the "equilibrium" position of the mass when $f = 0$), m is the mass of the object attached to the spring, κ is the spring constant, γ is the damping constant, and $f = f(t)$ is the sum of all forces acting on the spring other than the damping friction and the spring's reaction to being stretched and compressed (f was called F_{other} in Chapter 18 and F in Chapter 23). Remember, also, that m and κ are positive constants.

Our main interest will be in the phenomenon of resonance in an undamped system. Accordingly, we will assume $\gamma = 0$, and restrict our attention to solving

$$m\frac{d^2y}{dt^2} + \kappa y = f \quad . \tag{30.17}$$

Ultimately, we will further restrict our attention to cases in which f is periodic. But let's wait on that, and derive some basic formulas without assuming this periodicity.

Solutions Using Arbitrary f

The General Solution

As you know quite well by now, the general solution to our differential equation, equation (30.17), is

$$y(t) = y_p(t) + y_h(t)$$

where y_h is the general solution to the corresponding homogeneous differential equation, and y_p is any particular solution to the given nonhomogeneous differential equation.

The formula for y_h is already known. In Chapter 18, we found that

$$y_h(t) = c_1\cos(\omega_0 t) + c_2\sin(\omega_0 t) \qquad \text{where} \quad \omega_0 = \sqrt{\frac{\kappa}{m}} \quad .$$

Recall that ω_0 is the natural angular frequency of the mass/spring system, and is related to the system's natural frequency ν_0 and natural period p_0 via

$$\nu_0 = \frac{\omega_0}{2\pi} \qquad \text{and} \qquad p_0 = \frac{1}{\nu_0} = \frac{2\pi}{\omega_0} \quad .$$

For future use, note that y_h is a periodic function with period p_0 ; hence

$$y_h(t + p_0) - y_h(t) = 0 \quad \text{for all } t \quad .$$

That leaves finding a particular solution y_p . Let's take this to be the solution to the initial-value problem

$$m\frac{d^2y}{dt^2} + \kappa y = f \quad \text{with} \quad y(0) = 0 \quad \text{and} \quad y'(0) = 0 \quad .$$

This is easily found by either applying the Laplace transform and using the convolution identity in taking the inverse transform, or by appealing directly to Duhamel's principle. Either way, we get

$$y_p(t) = h * f(t) = \int_0^t h(t - x) f(x)\,dx$$

where

$$h(\tau) = \mathcal{L}^{-1}\left[\frac{1}{ms^2 + \kappa}\right]\bigg|_\tau = \frac{1}{m}\mathcal{L}^{-1}\left[\frac{1}{s^2 + \kappa/m}\right]\bigg|_\tau \quad .$$

Since $\omega_0 = \sqrt{\kappa/m}$,

$$h(\tau) = \frac{1}{m}\mathcal{L}^{-1}\left[\frac{1}{s^2 + (\omega_0)^2}\right]\bigg|_\tau = \frac{1}{\omega_0 m}\sin(\omega_0\tau) \quad .$$

Thus, the above integral formula for y_p can be written as

$$y_p(t) = \frac{1}{\omega_0 m}\int_0^t \sin(\omega_0[t - x])\, f(x)\,dx \quad . \tag{30.18}$$

The Difference Formula and First Theorem

For our studies, we will want to see how any solution y varies "over a cycle" (i.e., as t increases by p_0). This variance in y over a cycle is given by the difference $y(t + p_0) - y(t)$, and will be especially meaningful when the forcing function is periodic with period p_0 .

For now, let's consider the difference $y(t + p_0) - y(t)$ assuming $y = y_p + y_h$ is any solution to our differential equation. Of course, the y_h term is irrelevant because of its periodicity,

$$\begin{aligned} y(t + p_0) - y(t) &= \left[y_p(t + p_0) + y_h(t + p_0)\right] - \left[y_p(t) + y_h(t)\right] \\ &= y_p(t + p_0) - y_p(t) + \underbrace{y_h(t + p_0) - y_h(t)}_{0} \quad . \end{aligned}$$

Now, using formula (30.18) for y_p , we see that

$$\begin{aligned} y_p(t + p_0) &= \frac{1}{\omega_0 m}\int_0^{t+p_0} \sin(\omega_0[(t + p_0) - x])\, f(x)\,dx \\ &= \frac{1}{\omega_0 m}\int_0^{t+p_0} \sin(\omega_0[t - x] + \underbrace{\omega_0 p_0}_{2\pi})\, f(x)\,dx \\ &= \frac{1}{\omega_0 m}\int_0^{t+p_0} \sin(\omega_0[t - x])\, f(x)\,dx \\ &= \frac{1}{\omega_0 m}\int_0^{t} \sin(\omega_0[t - x])\, f(x)\,dx + \frac{1}{\omega_0 m}\int_t^{t+p_0} \sin(\omega_0[t - x])\, f(x)\,dx \quad . \end{aligned}$$

But the first integral in the last line is simply the integral formula for $y_p(t)$ given in equation (30.18). So the above reduces to

$$y(t+p_0) = y(t) + \frac{1}{\omega_0 m}\int_t^{t+p_0} \sin(\omega_0[t-x])\, f(x)\,dx \quad . \tag{30.19}$$

To further "reduce" our difference formula, let us use a well-known trigonometric identity:

$$\begin{aligned}
\int_t^{t+p_0} & \sin(\omega_0[t-x])\, f(x)\,dx \\
&= \int_t^{t+p_0} \sin(\omega_0 t - \omega_0 x)\, f(x)\,dx \\
&= \int_t^{t+p_0} [\sin(\omega_0 t)\cos(\omega_0 x) - \cos(\omega_0 t)\sin(\omega_0 x)]\, f(x)\,dx \\
&= \sin(\omega_0 t)\int_t^{t+p_0} \cos(\omega_0 x)\, f(x)\,dx - \cos(\omega_0 t)\int_t^{t+p_0} \sin(\omega_0 x)\, f(x)\,dx \quad .
\end{aligned}$$

Combining this result with the last equation for $y(t+p_0)$ and recalling the previous results derived in this section then yield:

Theorem 30.3
Let m and κ be positive constants, and let f be any piecewise continuous function of exponential order. Then, the general solution to

$$m\frac{d^2y}{dt^2} + \kappa y = f$$

is

$$y(t) = y_p(t) + c_1\cos(\omega_0 t) + c_2\sin(\omega_0 t)$$

where

$$\omega_0 = \sqrt{\frac{\kappa}{m}} \quad \text{and} \quad y_p(t) = \frac{1}{\omega_0 m}\int_0^t \sin(\omega_0[t-x])\, f(x)\,dx \quad .$$

Moreover,

$$y(t+p_0) - y(t) = \frac{1}{\omega_0 m}[\mathcal{I}_S(t)\sin(\omega_0 t) + \mathcal{I}_C(t)\cos(\omega t)] \quad \text{for} \quad t \geq 0$$

where

$$\mathcal{I}_S(t) = \int_t^{t+p_0} \cos(\omega_0 x)\, f(x)\,dx \quad \text{and} \quad \mathcal{I}_C(t) = -\int_t^{t+p_0} \sin(\omega_0 x)\, f(x)\,dx \quad .$$

Resonance from Periodic Forcing Functions

A Useful Fact

Take a look at Figure 30.7. It shows the graph of some periodic function g with period p_0, and with two regions of width p_0 "greyed in" in two shades of grey. The darker grey region is between the graph and the T–axis with $0 < t < p_0$. The lighter grey region is between the graph and the T–axis with $a < t < a + p_0$ for some real number a. Note the similarity in the shapes of the regions. In particular, note how the pieces of the lighter grey region can be rearranged to perfectly match the darker grey region. Consequently, the areas in each of these two regions, both above and

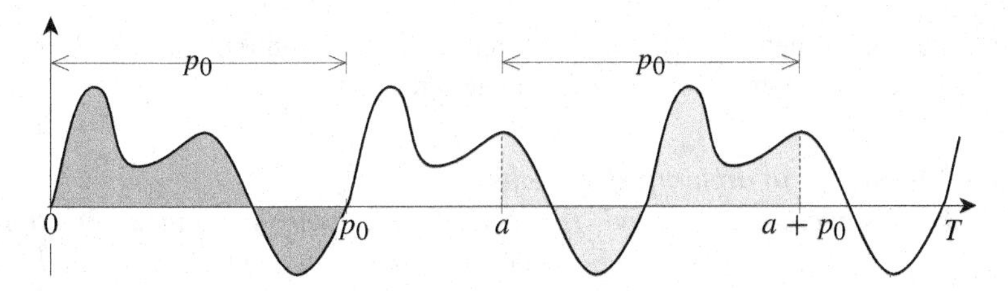

Figure 30.7: Illustration for Lemma 30.4.

below the T–axis, are the same. Add to this the relationship between "integrals" and "area", and you get the useful fact stated in the next lemma.

Lemma 30.4
Let g be a periodic, piecewise continuous function with period p_0. Then, for any t,

$$\int_t^{t+p_0} g(x)\,dx = \int_0^{p_0} g(x)\,dx \quad .$$

If you wish, you can rigorously prove this lemma using some basic theory from elementary calculus.

?►Exercise 30.1: *Prove Lemma 30.4. A good start would be to show that*

$$\frac{d}{dt}\int_t^{t+p_0} g(x)\,dx = 0 \quad .$$

Resonance

Now consider the formulas for $\mathcal{I}_S(t)$ and $\mathcal{I}_C(t)$ from Theorem 30.3,

$$\mathcal{I}_S(t) = \int_t^{t+p_0} \cos(\omega_0 x)\,f(x)\,dx \quad \text{and} \quad \mathcal{I}_C(t) = -\int_t^{t+p_0} \sin(\omega_0 x)\,f(x)\,dx \quad .$$

If f is also periodic with period p_0, then the products in these integrals are also periodic, each with period p_0. Lemma 30.4 then tells us that

$$\mathcal{I}_S(t) = \int_t^{t+p_0} \cos(\omega_0 x)\,f(x)\,dx = \int_0^{p_0} \cos(\omega_0 x)\,f(x)\,dx$$

and

$$\mathcal{I}_C(t) = -\int_t^{t+p_0} \sin(\omega_0 x)\,f(x)\,dx = -\int_0^{p_0} \sin(\omega_0 x)\,f(x)\,dx \quad .$$

Thus, if f is periodic with period p_0, the difference formula in Theorem 30.3 reduces to

$$y(t+p_0) - y(t) = \frac{1}{\omega_0 m}\left[\mathcal{I}_S \sin(\omega_0 t) + \mathcal{I}_C \cos(\omega t)\right]$$

where $\mathcal{I}_S$ and $\mathcal{I}_C$ are the *constants*

$$\mathcal{I}_S = \int_0^{p_0} \cos(\omega_0 x)\,f(x)\,dx \quad \text{and} \quad \mathcal{I}_C = -\int_0^{p_0} \sin(\omega_0 x)\,f(x)\,dx \quad .$$

Using a little more trigonometry (see the derivation of formula (18.8b) on page 342), we can reduce this to the even more convenient form given in the next theorem.

Theorem 30.5 (resonance in undamped systems)
Let m and p be positive constants, and f a periodic piecewise continuous function. Assume further that f has period p_0, the natural period of the mass/spring system modeled by

$$m\frac{d^2y}{dt^2} + \kappa y = f \quad .$$

That is,

$$\text{period of } f = p_0 = \frac{2\pi}{\omega_0} \quad \text{with} \quad \omega_0 = \sqrt{\frac{\kappa}{m}} \quad .$$

Also let

$$\mathcal{I}_S = \int_0^{p_0} \cos(\omega_0 x)\, f(x)\, dx \quad \text{and} \quad \mathcal{I}_C = -\int_0^{p_0} \sin(\omega_0 x)\, f(x)\, dx \quad .$$

Then, for any solution y to the above differential equation, and any $t > 0$,

$$y(t + p_0) - y(t) = A\cos(\omega_0 t - \phi) \tag{30.20}$$

where

$$A = \frac{1}{\omega_0 m}\sqrt{(\mathcal{I}_S)^2 + (\mathcal{I}_C)^2}$$

and with ϕ being the constant satisfying $0 \le \phi < 2\pi$,

$$\cos(\phi) = \frac{\mathcal{I}_C}{\sqrt{(\mathcal{I}_S)^2 + (\mathcal{I}_C)^2}} \quad \text{and} \quad \sin(\phi) = \frac{\mathcal{I}_S}{\sqrt{(\mathcal{I}_S)^2 + (\mathcal{I}_C)^2}} \quad .$$

To see what all this implies, assume f, y, etc. are as in the theorem, and look at what the difference formula tells us about $y(t_n)$ when t_0 is any fixed value in $[0, p_0)$, and

$$t_n = t_0 + np_0 \quad \text{for} \quad n = 1, 2, 3, \dots \quad .$$

The value of $y(t_0)$ can be computed using the integral formula for y_p in Theorem 30.3. To compute each $y(t_n)$, however, it is easier to use this computed value for $y(t_0)$ along with difference formula (30.20) and the fact that, for any integer k,

$$\cos(\omega_0 t_k - \phi) = \cos(\omega_0[t_0 + kp_0] - \phi) = \cos(\omega_0 t_0 - \phi + k\underbrace{\omega_0 p_0}_{2\pi}) = \cos(\omega_0 t_0 - \phi) \quad .$$

Doing so, we get

$$y(t_1) = y(t_0 + p_0) = y(t_0) + A\cos(\omega_0 t_0 - \phi) \quad ,$$

$$\begin{aligned} y(t_2) = y(t_0 + 2p_0) &= y(t_0 + p_0) + A\cos(\omega_0[t_0 + p_0] - \phi) \\ &= [y(t_0) + A\cos(\omega_0 t - \phi)] + A\cos(\omega_0 t_0 - \phi) \\ &= y(t_0) + 2A\cos(\omega_0 t_0 - \phi) \quad , \end{aligned}$$

$$\begin{aligned} y(t_3) = y(t_0 + 3p_0) &= y(t_0 + 2p_0) + A\cos(\omega_0[t_0 + 2p_0] - \phi) \\ &= [y(t_0) + 2A\cos(\omega_0 t - \phi)] + A\cos(\omega_0 t_0 - \phi) \\ &= y(t_0) + 3A\cos(\omega_0 t_0 - \phi) \quad , \end{aligned}$$

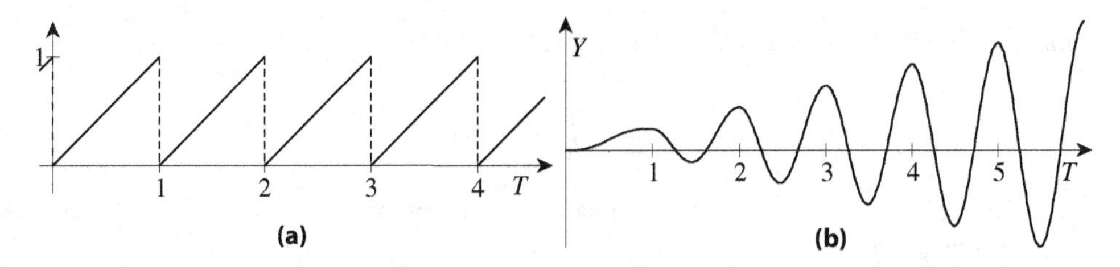

Figure 30.8: **(a)** A basic saw function, and **(b)** the corresponding response of an undamped mass/spring system with natural period 1 over 6 cycles.

and so on. In general,

$$y(t_n) = y(t_0) + nA\cos(\omega_0 t_0 - \phi) \quad . \tag{30.21}$$

Clearly, if $A \neq 0$ and $\omega_0 t_0 - \phi$ is neither $\pi/2$ or $3\pi/2$, then

$$y(t_n) \rightarrow \pm\infty \quad \text{as} \quad n \rightarrow \infty \quad .$$

This is certainly "runaway resonance".

Thus, it is the A in difference formula (30.20) that determines if we have "runaway resonance". If $A \neq 0$, the solution contains an oscillating term with a steadily increasing amplitude. On the other hand, if $A = 0$, then the solution y is periodic and does not "blow up".

By the way, for graphing purposes it may be convenient to use the periodicity of the cosine term and rewrite equation (30.21) as

$$y(t_n) = y(t_0) + nA\cos(\omega_0 t_n - \phi) \quad .$$

Replacing t_n with t, and recalling what n and τ represent, we see that this is the same as saying

$$y(t) = y(t_0) + nA\cos(\omega_0 t - \phi) \tag{30.22}$$

where n is the largest integer such that $np_0 \leq t$ and $t_0 = t - np_0$.

!►Example 30.14: *Let us use the theorems in this section to analyze the response of an undamped mass/spring system with natural period $p_0 = 1$ to a force f given by the basic saw function sketched in Figure 30.8a,*

$$f(t) = \text{saw}(t) = \begin{cases} t & \text{if } 0 < t < 1 \\ \text{saw}(t-1) & \text{if } 1 < t \end{cases} \quad .$$

The corresponding natural angular frequency is

$$\omega_0 = \frac{2\pi}{p_0} = 2\pi \quad .$$

The actual values of the mass m and spring constant κ are irrelevant provided they satisfy

$$2\pi = \omega_0 = \sqrt{\frac{\kappa}{m}} \quad .$$

Also, since the solution to the corresponding homogeneous differential equation was pretty much irrelevant in the discussion leading to our last theorem, let's assume our solution satisfies

$$y(0) = 0 \quad \text{and} \quad y'(0) = 0 \quad ,$$

so that the solution formula described in Theorem 30.3 becomes

$$y(t) = y_p(t) = \frac{1}{2\pi m}\int_0^t \sin(\omega_0[t-x])\,\mathrm{saw}(x)\,dx \quad .$$

In particular, if $0 \leq x \leq t < 1$ *, then* $\mathrm{saw}(x) = x$ *, and we can complete our computations of* $y(t)$ *using integration by parts:*

$$\begin{aligned} y(t) &= \frac{1}{2\pi m}\int_0^t \sin(2\pi[t-x])\,x\,dx \\ &= \frac{1}{2\pi m}\left[\frac{x}{2\pi}\cos(2\pi[t-x])\bigg|_{x=0}^{t} - \int_0^t \frac{1}{2\pi}\cos(2\pi[t-x])\,dx\right] \\ &= \frac{1}{2\pi m}\left[\frac{t}{2\pi}\cos(2\pi[t-t]) - \frac{0}{2\pi}\cos(2\pi[t-0]) \right. \\ &\qquad \left. + \frac{1}{(2\pi)^2}\sin(2\pi[t-t]) - \frac{1}{(2\pi)^2}\sin(2\pi[t-0])\right] \quad . \end{aligned}$$

This simplifies to

$$y(t) = \frac{1}{8\pi^3 m}[2\pi t - \sin(2\pi t)] \qquad \text{when} \quad 0 \leq t < 1 \quad . \tag{30.23}$$

In a similar manner, we find that

$$\mathcal{I}_S = \int_0^{p_0} \cos(\omega_0 x)\,f(x)\,dx = \int_0^1 \cos(2\pi x)\,x\,dx = \cdots = 0$$

and

$$\mathcal{I}_C = -\int_0^{p_0} \sin(\omega_0 x)\,f(x)\,dx = -\int_0^1 \sin(2\pi x)\,x\,dx = \cdots = \frac{1}{2\pi} \quad .$$

Thus,

$$A = \frac{1}{2\pi m}\sqrt{(\mathcal{I}_S)^2 + (\mathcal{I}_C)^2} = \frac{1}{4\pi^2 m} \quad .$$

Since $A \neq 0$ *, we have resonance. There is an oscillatory term whose amplitude steadily increases as* t *increases.*

To actually graph our solution, we still need to find the phase, ϕ *, which (according to our last theorem) is the value in* $[0, 2\pi)$ *such that*

$$\cos(\phi) = \frac{\mathcal{I}_C}{\sqrt{(\mathcal{I}_S)^2 + (\mathcal{I}_C)^2}} = 1 \qquad \text{and} \qquad \sin(\phi) = \frac{\mathcal{I}_S}{\sqrt{(\mathcal{I}_S)^2 + (\mathcal{I}_C)^2}} = 0 \quad .$$

Clearly $\phi = 0$ *.*

So let $t \geq 0$ *. Then, employing formula (30.22) (derived just before this example),*

$$\begin{aligned} y(t) &= y(t_0) + nA\cos(\omega_0 t - \phi) \\ &= \frac{1}{8\pi^3 m}[2\pi t_0 - \sin(2\pi t_0)] + \frac{n}{4\pi^2 m}\cos(2\pi t) \end{aligned}$$

where (since $p_0 = 1$ *)* n *is the largest integer with* $n \leq t$ *and* $t_0 = t - n$ *. This is the function graphed in Figure 30.8b.*

Additional Exercises

30.2. *Using the first translation identity or one of the differentiation identities, compute each of the following:*

a. $\mathcal{L}\left[e^{4t}\,\text{step}_6(t)\right]\Big|_s$ **b.** $\mathcal{L}\left[t\,\text{step}_6(t)\right]\Big|_s$

30.3. *Compute (using the translation along the T–axis identity) and then graph the inverse transforms of the following functions:*

a. $\dfrac{e^{-4s}}{s^3}$ **b.** $\dfrac{e^{-3s}}{s+2}$ **c.** $\sqrt{\pi}\,s^{-3/2}e^{-s}$

d. $\dfrac{\pi}{s^2+\pi^2}e^{-2s}$ **e.** $\dfrac{e^{-4s}}{(s-5)^3}$ **f.** $\dfrac{(s+2)e^{-5s}}{(s+2)^2+16}$

30.4. *Finish solving the differential equation in Example 30.1.*

30.5. *Compute and then graph the inverse transforms of the following functions (express your answers as sets of conditional formulas):*

a. $\dfrac{1-e^{-s}}{s^2}$ **b.** $\dfrac{e^{-s}+e^{-3s}}{s}$ **c.** $\dfrac{2}{s^3}-\dfrac{2+4s}{s^3}e^{-2s}$

d. $\dfrac{\pi\left(1+e^{-s}\right)}{s^2+\pi^2}$ **e.** $\dfrac{(s+4)e^{-12}-8e^{-3s}}{s^2-16}$ **f.** $\dfrac{e^{-2s}-2e^{-4s}+e^{-6s}}{s^2}$

30.6. *Find and graph the solution to each of the following initial-value problems:*

a. $y' = \text{step}_3(t)$ *with* $y(0) = 0$

b. $y' = \text{step}_3(t)$ *with* $y(0) = 4$

c. $y'' = \text{step}_2(t)$ *with* $y(0) = 0$ *and* $y'(0) = 0$

d. $y'' = \text{step}_2(t)$ *with* $y(0) = 4$ *and* $y'(0) = 6$

e. $y'' + 9y = \text{step}_{10}(t)$ *with* $y(0) = 0$ *and* $y'(0) = 0$

30.7. *Compute the Laplace transforms of the following functions using the translation along the T–axis identity. (Trigonometric identities may also be useful for some of these.)*

a. $f(t) = \begin{cases} 0 & \text{if } t < 6 \\ e^{4t} & \text{if } 6 < t \end{cases}$ **b.** $g(t) = \begin{cases} 0 & \text{if } t < 4 \\ \dfrac{1}{\sqrt{t-4}} & \text{if } 4 < t \end{cases}$

c. $t\,\text{step}_6(t)$ **d.** $te^{3t}\,\text{step}_2(t)$

e. $t^2\,\text{step}_6(t)$ **f.** $\sin(2(t-1))\,\text{step}_1(t)$

g. $\sin(2t)\,\text{step}_{\pi/2}(t)$ **h.** $\sin(2t)\,\text{step}_{\pi/4}(t)$

i. $\sin(2t)\,\text{step}_{\pi/6}(t)$

30.8. For each of the following choices of f :

i. *Graph the given function over the positive* T*–axis.*

ii. *Rewrite the function in terms of appropriate rectangle functions, and then rewrite that in terms of appropriate step functions.*

iii. *Then find the Laplace transform* $F(s) = \mathcal{L}[f(t)]|_s$.

a. $f(t) = \begin{cases} e^{-4t} & \text{if } t < 6 \\ 0 & \text{if } 6 < t \end{cases}$

b. $f(t) = \begin{cases} 2t - t^2 & \text{if } t < 2 \\ 0 & \text{if } 2 < t \end{cases}$

c. $f(t) = \begin{cases} 2 & \text{if } t < 3 \\ 2e^{-4(t-3)} & \text{if } 3 < t \end{cases}$

d. $f(t) = \begin{cases} \sin(\pi t) & \text{if } t < 1 \\ 0 & \text{if } 1 < t \end{cases}$

e. $f(t) = \begin{cases} t^2 & \text{if } t < 3 \\ 9 & \text{if } 3 < t \end{cases}$

f. $f(t) = \begin{cases} 0 & \text{if } t < 2 \\ 3 & \text{if } 2 < t < 4 \\ 0 & \text{if } 4 < t \end{cases}$

g. $f(t) = \begin{cases} 1 & \text{if } t < 1 \\ 2 & \text{if } 2 < t < 3 \\ 4 & \text{if } 3 < t \end{cases}$

h. $f(t) = \begin{cases} 0 & \text{if } t < 1 \\ \sin(\pi t) & \text{if } 1 < t < 2 \\ 0 & \text{if } 2 < t \end{cases}$

i. $f(t) = \begin{cases} 0 & \text{if } t \leq 1 \\ (t-1)^2 & \text{if } 1 < t < 3 \\ 4 & \text{if } 3 \leq t \end{cases}$

j. $f(t) = \begin{cases} t & \text{if } t \leq 2 \\ 4 - t & \text{if } 2 < t < 4 \\ 0 & \text{if } 3 \leq t \end{cases}$

30.9. *The infinite stair function,* $\text{stair}(t)$, *can be described in terms of rectangle functions by*

$$\begin{aligned} \text{stair}(t) &= \sum_{n=0}^{\infty} (n+1)\,\text{rect}_{(n,n+1)}(t) \\ &= 1\,\text{rect}_{(0,1)}(t) + 2\,\text{rect}_{(1,2)}(t) + 3\,\text{rect}_{(2,3)}(t) + 4\,\text{rect}_{(3,4)}(t) + \cdots \quad . \end{aligned}$$

Using this:

a. *Sketch the graph of* $\text{stair}(t)$ *over the positive* T*–axis, and rewrite the formula for* $\text{stair}(t)$ *in terms of step functions.*

b. *Assuming the linearity of the Laplace transform holds for infinite sums as well as finite sums, find an infinite sum formula for* $\mathcal{L}[\text{stair}(t)]|_s$.

c. *Recall the formula for the sum of a geometric series,*

$$\sum_{n=0}^{\infty} X^n = \frac{1}{1-X} \qquad \text{when} \quad |X| < 1 \quad .$$

Using this, simplify the infinite sum formula for $\mathcal{L}[\text{stair}(t)]|_s$ *which you obtained in the previous part of this exercise.*

30.10. *Find and graph the solution to each of the following initial-value problems:*

a. $y' = \text{rect}_{(1,3)}(t)$ *with* $y(0) = 0$

b. $y'' = \text{rect}_{(1,3)}(t)$ *with* $y(0) = 0$ *and* $y'(0) = 0$

c. $y'' + 9y = \text{rect}_{(1,3)}(t)$ *with* $y(0) = 0$ *and* $y'(0) = 0$

30.11. *Compute each of the following convolutions (assuming, in all, that* $t \geq 0$*):*

a. $t^2 * \text{step}_3(t)$

b. $1 * \text{step}_4(t)$

c. $\cos(t) * \text{rect}_{(0,\pi)}(t)$

d. $2e^{-2t} * \text{rect}_{(1,3)}(t)$

e. $e^{-2t} * \left[e^{5t}\,\text{rect}_{(1,3)}(t)\right]$

f. $\sin(t) * \left[\sin(t)\,\text{rect}_{(2\pi,3\pi)}(t)\right]$

g. $t * f(t)$ *where* $f(t) = \begin{cases} \sqrt{t} & \text{if } 0 \leq t < 4 \\ 2 & \text{if } 4 < t \end{cases}$

h. $\sin(t) * f(t)$ *where* $f(t) = \begin{cases} 1 & \text{if } t < 2\pi \\ \cos(t) & \text{if } 2\pi < t < 3\pi \\ -1 & \text{if } 3\pi < t \end{cases}$

30.12. *Each function listed below is at least periodic on* $(0, \infty)$. *Sketch the graph of each, and then find its Laplace transform using the methods developed in Section 30.5.*

a. $f(t) = \begin{cases} e^{-2t} & \text{if } 0 < t < 3 \\ f(t-3) & \text{if } t > 3 \end{cases}$

b. $f(t) = \text{sqwave}(t)$ *(from Example 30.12)*

c. $f(t) = \begin{cases} 1 & \text{if } 0 < t < 1 \\ -1 & \text{if } 1 < t < 2 \\ f(t-2) & \text{if } t > 2 \end{cases}$

d. $f(t) = \begin{cases} 2t - t^2 & \text{if } t < 2 \\ f(t-2) & \text{if } 2 < t \end{cases}$ *(see Exercise 30.8 b)*

e. $f(t) = \begin{cases} t & \text{if } 0 < t < 2 \\ 4 - t & \text{if } 2 < t < 4 \\ f(t-4) & \text{if } t > 4 \end{cases}$ *(see Exercise 30.8 j)*

f. $f(t) = |\sin(t)|$

30.13. *In each of the following exercises, you are given the natural period* p_0 *and a forcing function* f *for an undamped mass/spring system modeled by*

$$m\frac{d^2y}{dt^2} + \kappa y = f \quad .$$

Analyze the corresponding resonance occurring in each system. In particular, let y *be any solution to the modeling differential equation and:*

i. *Compute the difference* $y(t + p_0) - y(t)$.

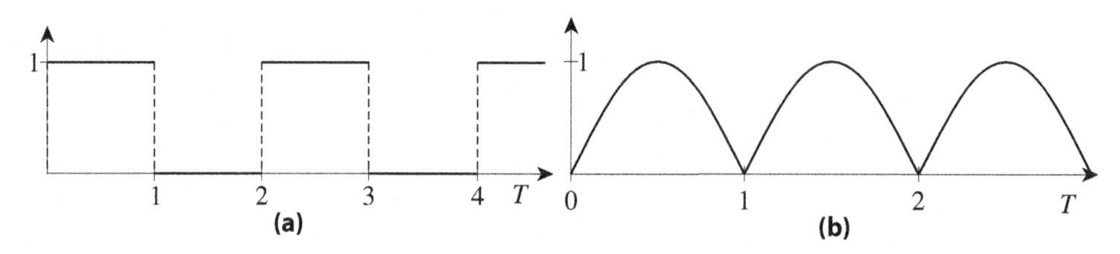

Figure 30.9: **(a)** The square wave function for Exercise 30.13 a, and **(b)** the rectified sine function for Exercise 30.13 b.

ii. *Compute the formula for* $y(t)$ *assuming* $y(0) = 0$ *and* $y'(0) = 0$. *(Express your answer using* t_0 *and* n *where* n *is the largest integer such that* $np_0 \leq t$ *and* $t_0 = t - np_0$ *.)*

iii. *Using the formula just computed for part ii along with your favorite computer math package, sketch the graph of* y *over several cycles. (For convenience, assume* m *is a unit mass.)*

a. $p_0 = 2$ *and* f *is the basic square-wave sketched in Figure 30.9a. That is,*

$$f(t) = \begin{cases} 1 & \text{if } 0 < t < 1 \\ 0 & \text{if } 1 < t < 2 \\ f(t+2) & \text{in general} \end{cases} .$$

b. $p_0 = 1$ *and* $f(t) = |\sin(\pi t)|$ *, the "rectified sine function" sketched in Figure 30.9b.*

c. $p_0 = 1$ *and* $f(t) = \sin(4\pi t)$.

31

Delta Functions

This chapter introduces mathematical entities commonly known as "delta functions". As we will see, delta functions are not really functions, at least not in the classical sense. Nonetheless, with a modicum of care, they can be treated like functions. More importantly, they are useful. They are valuable in modeling both "strong forces of brief duration" (such as the force of a baseball bat striking a ball) and "point masses". Moreover, their mathematical properties turn out to be remarkable, making them some of the simplest "functions" to deal with. After a little practice, you may rank them with the constant functions as some of your favorite functions to deal with. Indeed, the basic delta function has a relation with the constant function $f = 1$ that will allow us to expand our discussion of Duhamel's principle.

31.1 Visualizing Delta Functions

What is commonly called "the delta function"— traditionally denoted by $\delta(t)$ — is best thought of as shorthand for a particular limiting process. One standard way to visualize $\delta(t)$ is as the limit

$$\delta(t) = \lim_{\epsilon \to 0^+} \frac{1}{\epsilon} \operatorname{rect}_{(0,\epsilon)}(t) \quad .$$

Look at the function we are taking the limit of,

$$\frac{1}{\epsilon} \operatorname{rect}_{(0,\epsilon)}(t) = \begin{cases} 0 & \text{if} \quad t < 0 \\ \dfrac{1}{\epsilon} & \text{if} \quad 0 < t < \epsilon \\ 0 & \text{if} \quad \epsilon < t \end{cases} \quad .$$

Graphs of this for various small positive values of ϵ have been sketched in Figure 31.1a. Notice that, for each ϵ, the nonzero part of the graph forms a rectangle of width ϵ and height $1/\epsilon$. Consequently, the area of this rectangle is $\epsilon \cdot 1/\epsilon = 1$. Keep in mind that we are taking a limit as $\epsilon \to 0$; so ϵ is "small", which means that this rectangle is very narrow and very high, starts at $t = 0$, and is of unit area. As we let $\epsilon \to 0$ this "very narrow and very high rectangle starting at $t = 0$ and of unit area" becomes (loosely speaking) an "infinitesimally narrow and infinitely high 'spike' at $t = 0$ enclosing unit area".

Strictly speaking, no function has such a spike as its graph. The closest we can come is the function that is zero everywhere except at $t = 0$, where we pretend the function is infinite. This sort of gives the infinite spike, but the "area enclosed" is not at all well defined. Still, the visualization

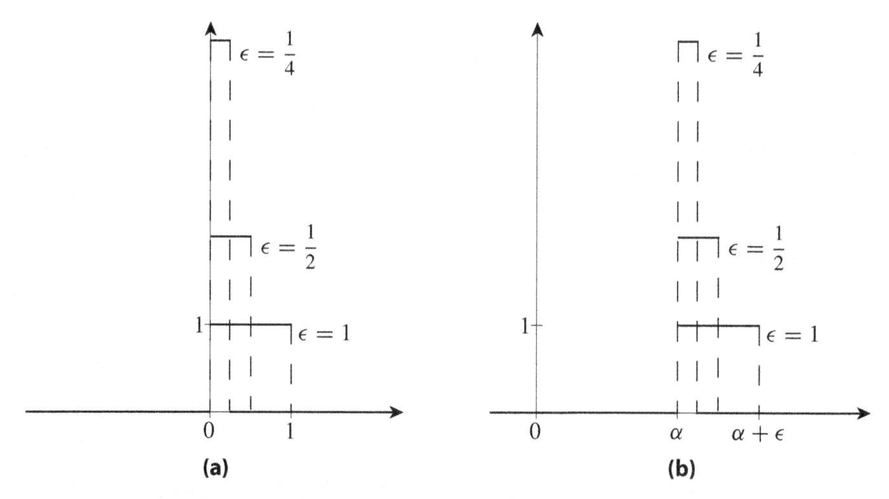

Figure 31.1: The graphs of **(a)** $\frac{1}{\epsilon}\,\text{rect}_{(0,\epsilon)}(t)$ and **(b)** $\frac{1}{\epsilon}\,\text{rect}_{(0,\epsilon)}(t-\alpha)$ (equivalently $\frac{1}{\epsilon}\,\text{rect}_{(\alpha,\alpha+\epsilon)}(t)$) for $\epsilon = 1$, $\epsilon = 1/2$ and $\epsilon = 1/4$.

of the delta function as "an infinite spike enclosing unit area" is useful, just as it is useful in physics to sometimes pretend that we can have a "point mass" (an infinitesimally small particle of nonzero mass).

The above describes "the" delta function. For any real number α, the delta function at α, $\delta_\alpha(t)$, is simply "the" delta function shifted by α,

$$\delta_\alpha(t) = \delta(t-\alpha) = \lim_{\epsilon \to 0} \frac{1}{\epsilon}\,\text{rect}_{(0,\epsilon)}(t-\alpha) \quad .$$

With a little thought (or a glance at Figure 31.1b), you can see that the nonzero part of $\text{rect}_{(0,\epsilon)}(t-\alpha)$ starts at $t = \alpha$ and ends at $t = \alpha + \epsilon$, and that

$$\delta_\alpha(t) = \delta(t-\alpha) = \lim_{\epsilon \to 0} \frac{1}{\epsilon}\,\text{rect}_{(\alpha,\alpha+\epsilon)}(t) \quad .$$

Do notice that

$$\delta_0(t) = \delta(t-0) = \delta(t) \quad .$$

This means that anything we derive concerning δ_α also holds for δ —just let $\alpha = 0$.

31.2 Delta Functions in Modeling

There are at least two general situations in which delta functions naturally arise when we attempt to describe "real-world" phenomena. One is when we attempt to model brief but strong forces. The other is when we imagine physical objects as "point masses". In both, the delta functions appear in integrals. This will be significant and is well worth observing in the models described below.

Since it will be especially useful to see how delta functions model "strong forces of brief duration", we'll start with that.

Strong Forces of Brief Duration

Consider the motion of some object under a force that varies with time. We will assume the object's motion is one dimensional (say, along some X–axis), and, as usual, we'll let

$$m = \text{the mass (in kilograms) of the object (assumed constant)} \quad ,$$

$$t = \text{time (in seconds)} \quad ,$$

$$v(t) = \text{velocity (in meters/second) of the object at time } t \quad ,$$

and

$$F(t) = \text{force (in kilogram}\cdot\text{meters/second}^2\text{) acting on the object at time } t \quad .$$

(Of course, any units for time, mass and distance can be used, as long as we are consistent.)

Newton's famous law of force gives

$$F(t) = m \times \text{acceleration} = m\frac{dv}{dt} \quad .$$

If we integrate this over an interval (t_0, t_1), we get

$$\int_{t_0}^{t_1} F(t)\,dt = \int_{t_0}^{t_1} m\frac{dv}{dt}\,dt = m\left[v(t_1) - v(t_0)\right] \quad .$$

So the integral of $F(t)$ from $t = t_0$ to $t = t_1$ is the object's mass times the change in the object's velocity over that period of time. This integral of F is sometimes called the *impulse* of the force over the interval (t_0, t_1) (with the total impulse being this integral with $t_0 = -\infty$ and $t_1 = \infty$).[1] Note that, following our above conventions for units, the units associated with the impulse is kilogram·meters/second.

Let's now restrict ourselves to situations in which the force is zero except for a very short period of time, during which the force is strong enough to significantly change the velocity of the object under question. We may be talking about the force of a baseball bat striking a baseball, or the force of some propellent (gunpowder, compressed air, etc.) forcing a bullet out of a gun, or even the force of a baseball in flight striking some unfortunate bat that fluttered out over the field to catch flies. For concreteness, let's pretend we are studying the force of a baseball bat hitting a baseball at "time $t = \alpha$". If we are very precise, we may let $t = \alpha$ be the first instant the bat comes into contact with the ball, and ϵ the length of time the bat remains in contact with the ball. Considering the situation, this length of time, ϵ , must be positive, but very small.

Before and after the bat touches the ball, this force is zero. So our $F(t)$ must be some function, such as,

$$\frac{1}{\epsilon}\operatorname{rect}_{(\alpha,\alpha+\epsilon)}(t) \quad ,$$

that satisfies

$$F(t) = 0 \qquad \text{if} \quad t < \alpha \quad \text{and if} \quad \alpha + \epsilon < t \quad .$$

Thus, if $t_0 < \alpha$ and $\alpha + \epsilon < t_1$, then

$$m\left[v(\alpha) - v(t_0)\right] = \int_{t_0}^{\alpha} F(t)\,dt = \int_{t_0}^{\alpha} 0\,dt = 0$$

and

$$m\left[v(t_1) - v(\alpha+\epsilon)\right] = \int_{\alpha+\epsilon}^{t_1} F(t)\,dt = \int_{\alpha+\epsilon}^{t_1} 0\,dt = 0 \quad .$$

---

[1] Students of physics will observe that the impulse is actually equal to the change in the *momentum*, mv .

Since t_0 can be any value less than α, and t_1 can be any value greater than $\alpha + \epsilon$, the last two equations tell us that the velocity is one constant v_{before} before the bat hits the ball, and another constant v_{after} afterwards, with

$$v_{\text{before}} = v(\alpha) \quad \text{and} \quad v_{\text{after}} = v(\alpha + \epsilon) \quad .$$

(We are using v_{before} and v_{after} because the expressions $v(\alpha)$ and $v(\alpha+\epsilon)$ will become problematic when we let $\epsilon \rightarrow 0$.)

The precise formula for $F(t)$ while the bat is in contact with the ball is typically both difficult to determine and of little interest. All we usually care about is describing $F(t)$ well enough to get the correct change in the velocity of the ball, $v_{\text{after}} - v_{\text{before}}$. So let us pick

$$F(t) = \frac{1}{\epsilon} \operatorname{rect}_{(\alpha,\alpha+\epsilon)}(t) \quad ,$$

and see what the resulting change of velocity is as t changes from t_0 to t_1 (with $t_0 < \alpha$ and $\alpha + \epsilon < t_1$):

$$\begin{aligned} m\left[v_{\text{after}} - v_{\text{before}}\right] &= m\left[v(t_1) - v(t_0)\right] \\ &= \int_{t_0}^{t_1} F(t)\, dt \\ &= \int_{t_0}^{t_1} \frac{1}{\epsilon} \operatorname{rect}_{(\alpha,\alpha+\epsilon)}(t)\, dt = \frac{1}{\epsilon} \int_{\alpha}^{\alpha+\epsilon} dt = 1 \quad . \end{aligned}$$

In other words,

$$F(t) = \frac{1}{\epsilon} \operatorname{rect}_{(\alpha,\alpha+\epsilon)}(t)$$

describes a force of duration ϵ starting at $t = \alpha$ with a total impulse of 1. Obviously, if we, instead, wanted a force of duration ϵ starting at $t = \alpha$ with a total impulse of $\mathcal{I}$, we could just multiply the above by $\mathcal{I}$. The corresponding velocity of the ball is then given by

$$v(t) = \begin{cases} v_{\text{before}} & \text{if} \quad t < \alpha \\ v_{\text{after}} & \text{if} \quad \alpha + \epsilon < t \end{cases} \quad ,$$

where

$$m\left[v_{\text{after}} - v_{\text{before}}\right] = \int_{t_0}^{t_1} \mathcal{I} \cdot \frac{1}{\epsilon} \operatorname{rect}_{(\alpha,\alpha+\epsilon)}(t)\, dt = \mathcal{I} \quad .$$

There is just one little complication: determining the length of time, ϵ, the bat is in contact with the ball. And, naturally, because this length of time is so close to being zero, we will simplify our computations by letting $\epsilon \rightarrow 0$. Thus, for some constant $\mathcal{I}$, we model the force by a *delta function force*

$$F(t) = \lim_{\epsilon \rightarrow 0^+} \mathcal{I} \cdot \frac{1}{\epsilon} \operatorname{rect}_{(\alpha,\alpha+\epsilon)}(t) = \mathcal{I}\, \delta_\alpha(t) \quad .$$

The resulting velocity of the ball $v(t)$ is then given by two constants v_{before} and v_{after}, with

$$v(t) = \begin{cases} v_{\text{before}} & \text{if} \quad t < \alpha \\ v_{\text{after}} & \text{if} \quad \alpha < t \end{cases} \quad ,$$

where

$$m\left[v_{\text{after}} - v_{\text{before}}\right] = \text{total impulse of } F = \mathcal{I} \quad .$$

Observe that using a delta function force leads to the velocity changing instantly from one constant to another. The velocity is no longer continuous, and the velocity right at $t = \alpha$ is no longer well defined. This is not physically possible but is still a very good approximation of what really happens.

We should also note that, because $F(t) = \delta_\alpha(t)$ corresponds to a force acting instantaneously at $t = \alpha$ with total impulse of 1, δ_α is also known as the *(instantaneous) unit impulse function at* α.

!►Example 31.1: *A baseball of mass* 0.145 *kilograms is thrown with a speed of* 35 *meters per second (about* 78 *miles per hour) towards a batter, who then hits the ball (with his bat), sending it back to the batter with a speed of 42 meters per second (about* 94 *miles per hour). We'll simplify matters slightly by assuming the ball travels along an* X*–axis both before and after it is hit, with the initial direction of travel in the negative direction and the final direction of travel in the positive direction along the* X*–axis. So, (in meters/second)*

$$v_{\text{before}} = -35 \qquad \text{and} \qquad v_{\text{after}} = 42 \quad .$$

Letting α *be the time the bat hits the ball, we can model the force of the bat on the ball by*

$$F(t) = \mathcal{I}\,\delta_\alpha(t) \qquad \left(\frac{\text{kg}\cdot\text{meter}}{\text{second}^2}\right)$$

where the impulse of the force is

$$\mathcal{I} = m\,[v_{\text{after}} - v_{\text{before}}] = 0.145\,[42+35] = 11.165 \qquad \left(\frac{\text{kg}\cdot\text{meter}}{\text{second}}\right) \quad .$$

As Density Functions for Point Masses

Suppose we have some material spread out along the X–axis. Recall that the linear density of the material at position x, $\rho(x)$, is the "mass per unit length" of the material at point x. More precisely, it is the function such that, if $x_0 < x_1$, then

$$\int_{x_0}^{x_1} \rho(x)\,dx$$

gives the mass of the material between positions $x = x_0$ and $x = x_1$.

Now think about what it means to have a density function

$$\rho(x) = \frac{m}{\epsilon}\,\text{rect}_{(\alpha,\alpha+\epsilon)}(x) = \begin{cases} 0 & \text{if} \quad t < 0 \\ \dfrac{m}{\epsilon} & \text{if} \quad 0 < t < \epsilon \\ 0 & \text{if} \quad \epsilon < t \end{cases}$$

where α, m and ϵ are real numbers with m and ϵ being positive. Here, all the mass is uniformly spread out in some object located between $x = \alpha$ and $x = \alpha+\epsilon$. Picking $x_0 < \alpha$ and $\alpha+\epsilon < x_1$, we see that

$$\begin{aligned} \text{total mass of the object} &= \int_{x_0}^{x_1} \rho(x)\,dx \\ &= \int_{x_0}^{x_1} \frac{m}{\epsilon}\,\text{rect}_{(\alpha,\alpha+\epsilon)}(x)\,dx = \frac{m}{\epsilon}\int_{\alpha}^{\alpha+\epsilon} 1\,dx = m \quad . \end{aligned}$$

So we have an object of mass m occupying the X–axis from $x = \alpha$ to $x = \alpha+\epsilon$.

In many applications, the width of the object, ϵ, is much smaller than the other dimensions involved, and taking account of this width complicates computations without significantly affecting the results of the computations. In these cases, it is common to simplify the mathematics by letting $\epsilon \to 0$ and thereby converting

our object of mass m occupying the region between $x = \alpha$ and $x = \alpha + \epsilon$

to

an object of mass m occupying the point $x = \alpha$.

In doing so, we see that

$$\rho(x) = \lim_{\epsilon \to 0^+} \frac{m}{\epsilon} \operatorname{rect}_{(\alpha,\alpha+\epsilon)}(x) = m \lim_{\epsilon \to 0^+} \frac{1}{\epsilon} \operatorname{rect}_{(\alpha,\alpha+\epsilon)}(x) = m\delta_\alpha(x) \quad .$$

Thus, the delta function at α multiplied by m describes the linear density of a "point mass" at α of mass m.

31.3 The Mathematics of Delta Functions

Integrals with Delta Functions

While we used

$$\delta_\alpha(t) = \delta(t - \alpha) = \lim_{\epsilon \to 0^+} \frac{1}{\epsilon} \operatorname{rect}_{(\alpha,\alpha+\epsilon)}(t) \tag{31.1}$$

to visualize the delta function at α, it is mathematically better to view δ_α through the integral equation

$$\int_{t_0}^{t_1} g(t)\delta_\alpha(t)\,dt = \lim_{\epsilon \to 0^+} \int_{t_0}^{t_1} g(t)\frac{1}{\epsilon} \operatorname{rect}_{(\alpha,\alpha+\epsilon)}(t)\,dt \tag{31.2}$$

where (t_0, t_1) can be any interval and g can be any function on (t_0, t_1) continuous at α. This means we are really viewing "$\delta_\alpha(t)$" as notation indicating a certain limiting process involving integration. Remember, that's how we actually used delta functions in modeling strong brief forces and point masses.

Since our interest is mainly in using delta functions with the Laplace transform, let us simplify matters a little and just consider the integral

$$\int_0^\infty g(t)\,\delta_\alpha(t)\,dt$$

when $\alpha \geq 0$ and g is any function continuous at α and piecewise continuous on $[0, \infty)$. Before applying equation (31.2), observe that, because $0 \leq \alpha$ and

$$g(t) \operatorname{rect}_{(\alpha,\alpha+\epsilon)}(t) = \begin{cases} 0 & \text{if} \quad t < \alpha \\ g(t) & \text{if} \quad \alpha < t < \alpha + \epsilon \\ 0 & \text{if} \quad \alpha + \epsilon < t \end{cases} \quad ,$$

we have

$$\int_0^\infty g(t) \cdot \frac{1}{\epsilon} \operatorname{rect}_{(\alpha,\alpha+\epsilon)}(t)\,dt = \frac{1}{\epsilon} \int_\alpha^{\alpha+\epsilon} g(t)\,dt \quad .$$

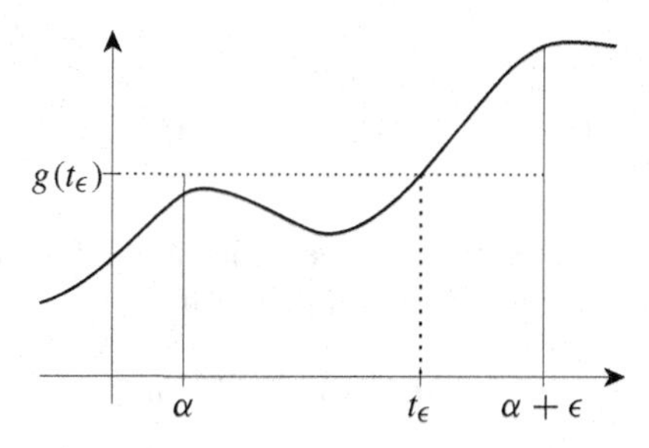

Figure 31.2: The rectangle with area equal to $\int_\alpha^{\alpha+\epsilon} g(t)\, dt$.

Because we will be taking the limit of the above as $\epsilon \to 0$, we can assume ϵ is small enough that g is continuous on the closed interval $[a, \alpha+\epsilon]$, and then apply the fact (illustrated in Figure 31.2) that

$$\begin{aligned}\int_\alpha^{\alpha+\epsilon} g(t)\, dt &= \text{"(net) area between } T\text{–axis and graph of } y = g(t) \text{ with } \alpha \leq t \leq \alpha+\epsilon\text{"} \\ &= \text{"(net) area of rectangle with base } [\alpha, \alpha+\epsilon] \text{ and (signed) height } g(t_\epsilon) \\ &\qquad\qquad \text{for some } t_\epsilon \text{ in the interval } [a, \alpha+\epsilon]\text{"} \\ &= \epsilon \times g(t_\epsilon) \qquad \text{for some } t_\epsilon \text{ in } [\alpha, \alpha+\epsilon] \quad .\end{aligned}$$

Combining the above and applying equation (31.2), we obtain

$$\begin{aligned}\int_0^\infty g(t)\, \delta_\alpha(t)\, dt &= \lim_{\epsilon \to 0} \int_0^\infty g(t) \cdot \frac{1}{\epsilon} \operatorname{rect}_{(\alpha,\alpha+\epsilon)}(t)\, dt \\ &= \lim_{\epsilon \to 0} \frac{1}{\epsilon} \int_\alpha^{\alpha+\epsilon} g(t)\, dt \\ &= \lim_{\epsilon \to 0} \frac{1}{\epsilon} \times \epsilon \times \left\{ g(t_\epsilon) \quad \text{for some } t_\epsilon \text{ in } [\alpha, \alpha+\epsilon] \right\} \\ &= \lim_{\epsilon \to 0} \left\{ g(t_\epsilon) \quad \text{for some } t_\epsilon \text{ in } [\alpha, \alpha+\epsilon] \right\} \\ &= g(t_\epsilon) \quad \text{for some } t_\epsilon \text{ in } [\alpha, \alpha+0] \quad .\end{aligned}$$

But, of course, the only t_ϵ in $[\alpha, \alpha+0]$ is $t_\epsilon = \alpha$. So the above reduces to a simple result that is important enough to place in a theorem.

Theorem 31.1

Let $\alpha \geq 0$ and let g be any piecewise continuous function on $[0, \infty)$ which is continuous at $t = \alpha$. Then

$$\int_0^\infty g(t)\, \delta_\alpha(t)\, dt = g(\alpha) \quad . \tag{31.3}$$

In particular, since $\delta = \delta_0$,

$$\int_0^\infty g(t)\, \delta(t)\, dt = g(0) \quad . \tag{31.4}$$

!►*Example 31.2:* *Actually, two examples:*

$$\int_0^\infty t^2\, \delta_3(t)\, dt = 3^2 = 9 \quad ,$$

and

$$\int_0^\infty (5-t)^3\,\delta(t)\,dt = (5-0)^3 = 125 \quad .$$

We derived the above theorem because it covers the cases of greatest interest to us. Still, it is worth noting that with just a little more work, you can verify that

$$\int_{t_0}^{t_1} g(t)\,\delta_\alpha(t)\,dt = \begin{cases} g(\alpha) & \text{if } t_0 \le \alpha < t_1 \\ 0 & \text{if } \alpha < t_0 \text{ or } t_1 \le \alpha \end{cases} \tag{31.5}$$

whenever g is a function continuous at α and piecewise continuous on $[t_0, t_1)$.

Equations (31.3) and (31.4) (and, more generally, equation (31.5)) are often used instead of equation (31.2) "fundamental descriptions" of the delta functions. Their simplicity belies their significance.

Laplace Transforms of Delta Functions

Finding the Laplace transform of a delta function is easy. Just use the integral formula for the Laplace transform along with an equation from Theorem 31.1. Assuming $\alpha \ge 0$, we have

$$\mathcal{L}[\delta_\alpha(t)]|_s = \int_0^\infty \delta_\alpha(t)e^{-st}\,dt = e^{-s\alpha} \quad ,$$

which we usually prefer to write as

$$\mathcal{L}[\delta_\alpha(t)]|_s = e^{-\alpha s} \quad .$$

In particular,

$$\mathcal{L}[\delta(t)]|_s = \mathcal{L}[\delta_0(t)]|_s = e^{-0s} = 1 \quad .$$

These transforms are important enough to add to our table of common transforms, giving us Table 31.1.

Differential Equations with Delta functions

Using the Laplace transform, it is relatively easy to solve many differential equations in which delta functions act as forcing functions. Let us look at two examples.

!►Example 31.3: *Let's find the solution to*

$$\frac{dy}{dt} = \delta_\alpha(t) \qquad \text{with} \quad y(0) = 0$$

where α is any positive real number.

Taking the Laplace transform of both sides:

$$\mathcal{L}\left[\frac{dy}{dt}\right]\Big|_s = \mathcal{L}[\delta_\alpha(t)]|_s$$

$$\hookrightarrow \qquad sY(s) - y(0) = e^{-\alpha s}$$

$$\hookrightarrow \qquad sY(s) - 0 = e^{-\alpha s}$$

$$\hookrightarrow \qquad Y(s) = \frac{e^{-\alpha s}}{s} \quad .$$

Table 31.1: Laplace Transforms of Common Functions (Version 2)

In the following, α and ω are real-valued constants, and, unless otherwise noted, $s > 0$.

$f(t)$	$F(s) = \mathcal{L}[f(t)]\|_s$	Restrictions
1	$\frac{1}{s}$	
t	$\frac{1}{s^2}$	
t^n	$\frac{n!}{s^{n+1}}$	$n = 1, 2, 3, \ldots$
$\frac{1}{\sqrt{t}}$	$\frac{\sqrt{\pi}}{\sqrt{s}}$	
t^α	$\frac{\Gamma(\alpha+1)}{s^{\alpha+1}}$	$-1 < \alpha$
$e^{\alpha t}$	$\frac{1}{s-\alpha}$	$\alpha < s$
$e^{i\alpha t}$	$\frac{1}{s-i\alpha}$	
$\cos(\omega t)$	$\frac{s}{s^2+\omega^2}$	
$\sin(\omega t)$	$\frac{\omega}{s^2+\omega^2}$	
$\text{step}_\alpha(t)$, $\text{step}(t-\alpha)$	$\frac{e^{-\alpha s}}{s}$	$0 \le \alpha$
$\delta(t)$	1	
$\delta_\alpha(t)$, $\delta(t-\alpha)$	$e^{-\alpha s}$	$0 \le \alpha$

Thus, the solution to our differential equation is

$$y(t) = \mathcal{L}^{-1}[Y(s)]\big|_t = \mathcal{L}^{-1}\left[\frac{e^{-\alpha s}}{s}\right]\bigg|_t = \text{step}_\alpha(t) \quad .$$

According to the last example, $y(t) = \text{step}_\alpha(t)$ is a solution to $y'(t) = \delta_\alpha(t)$. In other words,

$$\frac{d}{dt}\,\text{step}_\alpha(t) = \delta_\alpha(t) \quad .$$

This interesting fact may also be a disturbing fact for those of you who realize that step functions are not differentiable, at least not in the sense normally taught in calculus courses. The truth is that delta functions are somewhat exotic entities that are outside the classical theory of calculus. They are really examples of things better referred to as "generalized functions", and the above equation about the derivative of the step function, while not valid in a strict classical sense, is valid using a definition of differentiation appropriate for these generalized functions. (We will discuss this a little further in Section 31.5.)

But enough worrying about technicalities. Let's solve another differential equation with a delta function.

!►Example 31.4: *Now consider*

$$y'' - 10y' + 21y = \delta(t) \qquad \textit{with} \quad y(0) = 0 \quad \textit{and} \quad y'(0) = 0 \quad .$$

Taking the Laplace transform of both sides:

$$\mathcal{L}\left[y'' - 10y' + 21y\right]\big|_s = \mathcal{L}[\delta(t)]|_s$$

$$\hookrightarrow \qquad \mathcal{L}\left[y''\right]\big|_s - 10\mathcal{L}\left[y'\right]\big|_s + 21\mathcal{L}[y]|_s = 1$$

$$\hookrightarrow \qquad s^2Y(s) - 10sY(s) + 21Y(s) = 1$$

$$\hookrightarrow \qquad \left[s^2 - 10s + 21\right]Y(s) = 1 \quad .$$

So,

$$Y(s) = \frac{1}{s^2 - 10s + 21} \quad ,$$

which just happens to be the function whose inverse transform was found in Example 29.4 on page 518. Using the result of that example, we can just write out

$$y(t) = \mathcal{L}^{-1}[Y(s)]|_t = \frac{1}{4}\left[e^{7t} - e^{3t}\right] \quad .$$

31.4 Delta Functions and Duhamel's Principle

If you compare the results of the last example with the results of Example 29.6 on page 519, you'll notice that the solution $y(t)$ to

$$y'' - 10y' + 21y = \delta(t) \qquad \text{with} \quad y(0) = 0 \quad \text{and} \quad y'(0) = 0$$

and the impulse response function $h(t)$ for

$$y'' - 10y' + 21y = f(t)$$

are one and the same. Is this an amazing coincidence?

No.

In Section 29.3, we saw that, for any real constants a, b and c, and any Laplace transformable function f, the solution on $(0, \infty)$ to the generic initial-value problem

$$ay'' + by' + cy = f(t) \qquad \text{with} \quad y(0) = 0 \quad \text{and} \quad y'(0) = 0$$

is given by

$$y(t) = h * f(t)$$

where

$$h = \mathcal{L}^{-1}[H] \qquad \text{and} \qquad H(s) = \frac{1}{as^2 + bs + c} \quad .$$

Now consider the corresponding initial-value problem

$$ay'' + by' + cy = \delta(t) \qquad \text{with} \quad y(0) = 0 \quad \text{and} \quad y'(0) = 0 \quad ,$$

which is just the generic initial-value problem above with $f = \delta$. Taking the Laplace transform, we get

$$\mathcal{L}\big[ay'' + by' + cy\big]\big|_s = \mathcal{L}[\delta(t)]|_s$$

$$\hookrightarrow \qquad a\mathcal{L}\big[y''\big]\big|_s + b\mathcal{L}\big[y'\big]\big|_s + c\mathcal{L}[y]|_s = 1$$

$$\hookrightarrow \qquad as^2Y(s) + bsY(s) + cY(s) = 1$$

$$\hookrightarrow \qquad \big[as^2 + bs + c\big]Y(s) = 1 \quad .$$

Dividing by the polynomial and comparing the result with the above formula for H, we see that

$$Y(s) = \frac{1}{as^2 + bs + c} = H(s) \quad .$$

Thus,

$$y(t) = \mathcal{L}^{-1}[Y(s)]|_t = \mathcal{L}^{-1}[H(s)]|_t = h(t) \quad .$$

In other words, $h(t)$ is the solution to the particular initial-value problem

$$ah'' + bh' + ch = \delta(t) \qquad \text{with} \quad y(0) = 0 \quad \text{and} \quad y'(0) = 0 \quad .$$

This explains why h is commonly referred to as the "impulse response function" — well, almost explains. Here's a little background: In many applications, the solution to the initial-value problem

$$ay'' + by' + cy = f(t) \qquad \text{with} \quad y(0) = 0 \quad \text{and} \quad y'(0) = 0$$

describes how some physical system responds to an applied "force" f (actually, f might not be an actual force). With this interpretation, h does give the response of the system to a delta function force, and, as noted earlier, the delta function is also known as a unit impulse function. Hence the term "impulse response function" for h.

Of course, the generic computations just done can be done with higher-order differential equations. Combining this with Theorem 29.2 on page 522 yields:

Theorem 31.2 (Duhamel's principle, version 2)

Let N be any positive integer, let a_0, a_1, ... and a_N be any collection of real-valued constants, and let $f(t)$ be any Laplace transformable function. Then the solution to

$$a_0y^{(N)} + a_1y^{(N-1)} + \cdots + a_{N-2}y'' + a_{N-1}y' + a_Ny = f(t)$$

satisfying the Nth-order "zero" initial conditions,

$$y(0) = 0 \quad , \quad y'(0) = 0 \quad , \quad y''(0) = 0 \quad , \quad \ldots \quad \text{and} \quad y^{N-1}(0) = 0 \quad ,$$

is given by

$$y(t) = h * f(t) = \int_0^t h(x)f(t-x)\,dx$$

where $h(t)$ is the solution to

$$a_0 h^{(N)} + a_1 h^{(N-1)} + \cdots + a_{N-2} h'' + a_{N-1} h' + a_N h = \delta(t)$$

with

$$h(0) = 0 \quad , \quad h'(0) = 0 \quad , \quad h''(0) = 0 \quad , \quad \ldots \quad \text{and} \quad h^{N-1}(0) = 0 \quad .$$

There is a practical consequence to h being the impulse response function. Suppose you have a physical system in which you know the 'output' $y(t)$ is related to an 'input' $f(t)$ through a differential equation of the form given in the above theorem. Suppose, further, that you do not know exactly what that differential equation is. Maybe, for example, you have a mass/spring system some of whose basic parameters — mass, spring constant or damping constant — are unknown and cannot be easily measured. The above theorem tells us that, if we input the physical equivalent of a delta function (say, we provide a unit impulse to the mass/spring system by carefully hitting the mass with a hammer), then measuring the output over time will yield a description of the impulse response function, $h(t)$. Save those values for $h(t)$ in a computer, and you can then numerically evaluate the output $y(t)$ corresponding to any other input $f(t)$ through the formula

$$y(t) = f * h(t) \quad .$$

In practice, generating and inputting the physical equivalent of $\delta(t)$ is usually impossible. What is often possible is to generate and input a good approximation to the delta function, say,

$$\frac{1}{\epsilon} \operatorname{rect}_{(0,\epsilon)}(t)$$

for some small value of ϵ. The resulting measured output will not be $h(t)$ exactly, but, if the errors in measurement aren't too bad, it will be a close approximation.

31.5 Some "Issues" with Delta Functions

The astute reader may have noticed that we've glossed over a few troublesome issues in our discussion of delta functions. Let's deal with a few of these now.

Defining the Delta Functions

You may have noticed that we have not yet *defined* the delta function. In particular, I've not given you any formula for computing the values of $\delta(t)$ or $\delta_\alpha(t)$ for different values of t. Instead, you've only been told to *visualize* $\delta_\alpha(t)$ in terms of either the limit

$$\delta_\alpha(t) = \delta(t - \alpha) = \lim_{\epsilon \to 0} \frac{1}{\epsilon} \operatorname{rect}_{(\alpha,\alpha+\epsilon)}(t) \tag{31.6}$$

or the limit

$$\int_{t_0}^{t_1} g(t)\delta_\alpha(t)\,dt = \lim_{\epsilon \to 0^+} \int_{t_0}^{t_1} g(t)\frac{1}{\epsilon} \operatorname{rect}_{(\alpha,\alpha+\epsilon)}(t)\,dt \quad . \tag{31.7}$$

If you check other texts, you'll often find δ_a (with $\alpha \geq 0$) "defined" either as the limit in (31.6) or as the 'function' such that

$$\int_0^\infty g(t)\,\delta_\alpha(t)\,dt = g(\alpha) \tag{31.8}$$

whenever g is a function continuous at α. (This, recall, was something we derived from equation (31.7).) Both of these are good 'working' definitions in that, properly interpreted, they tell you how you should use the symbol δ_α in computations (provided you interpret the limit in (31.6) as really meaning the limit in (31.7)).

Unfortunately, if you treat either as a rigorous definition for a classical function δ_α, then you can rigorously derive

$$\delta_\alpha(t) = 0 \qquad \text{whenever} \quad t \neq \alpha \quad .$$

Rigorously applying the classical theory of integration normally developed in undergraduate mathematics, you then find that

$$\begin{aligned}\int_0^\infty g(t)\,\delta_\alpha(t)\,dt &= \int_0^\alpha g(t)\,\delta_\alpha(t)\,dt + \int_\alpha^\infty g(t)\,\delta_\alpha(t)\,dt \\ &= \int_0^\alpha g(t)\cdot 0\,dt + \int_\alpha^\infty g(t)\cdot 0\,dt = 0 \quad .\end{aligned}$$

In particular, using $g(t) = t^2$, $\alpha = 1$ and both equation (31.7) and the last equation above, we get

$$1 = 1^2 = \int_0^\infty t^2\,\delta_2(t)\,dt = 0 \quad !$$

The problem is that there is no classical function that satisfies either definition. Fortunately, there is a way to 'generalize' the classical notion of 'functions' yielding a class of things called "generalized functions". Delta functions are members of this class. Unfortunately, a proper development of "generalized functions" goes beyond the scope of this text. What can be said is that, if f is a generalized function, then, for every sufficiently smooth and integrable function g and suitable interval (t_0, t_1),

$$\int_{t_0}^{t_1} g(t)\,f(t)\,dt$$

"makes sense" in some generalized sense. For $f = \delta_\alpha$, this integral can be defined by equation (31.7).[2] Using the theory of generalized functions, along with the corresponding generalization of the theory of calculus, everything developed in this chapter can be rigorously defined or derived, including the observation that, "in a generalized sense",

$$\delta_\alpha(t) = \frac{d}{dt}\,\text{step}_\alpha(t) \quad .$$

For now, however, it may be best to view the computations we are doing with δ_α as shorthand for doing the same computations with

$$\frac{1}{\epsilon}\,\text{rect}_{(\alpha,\alpha+\epsilon)} \quad ,$$

and then letting $\epsilon \to 0^+$ in the final result.

!►Example 31.5: *Let's reconsider solving*

$$\frac{dy_\epsilon}{dt} = \delta_\alpha(t) \qquad \textit{with} \quad y(0) = 0$$

[2] If you must know, "generalized functions" are actually "continuous linear functionals on a suitable space of test functions", and if you want to find out what that means, see Part IV of the author's *Principles of Fourier Analysis*, or go to the library and look up books on either generalized functions or distributional theory, or just do an internet search for these terms.

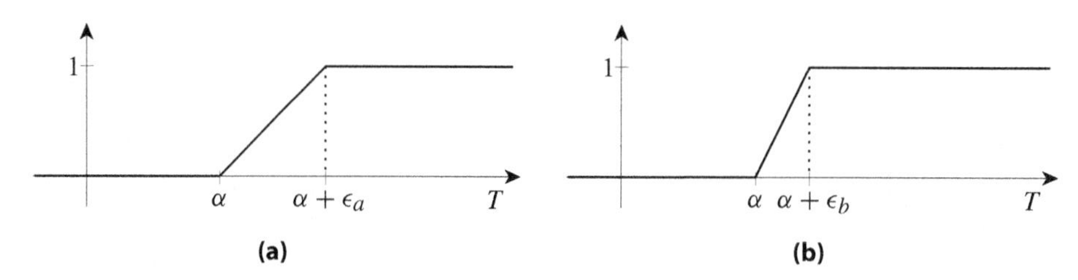

Figure 31.3: The graph of the solution y_ϵ to initial-value problem (31.9) **(a)** when $\epsilon = \epsilon_a$ and **(b)** when $\epsilon = \epsilon_b$ with $0 < \epsilon_b < \epsilon_a$.

where α is any positive real number. Doing the replacement suggested above, we'll first solve

$$\frac{dy_\epsilon}{dt} = \frac{1}{\epsilon}\operatorname{rect}_{(\alpha,\alpha+\epsilon)} \qquad \textit{with} \quad y(0) = 0 \quad , \tag{31.9}$$

assuming $\epsilon > 0$, and then take the limit of the result as $\epsilon \to 0$.

Taking the Laplace transform of both sides of the last equation:

$$\mathcal{L}\left[\frac{dy_\epsilon}{dt}\right]\Big|_s = \mathcal{L}\left[\frac{1}{\epsilon}\operatorname{rect}_{(\alpha,\alpha+\epsilon)}\right]\Big|_s$$

$$\hookrightarrow \qquad sY(s) - y(0) = \frac{1}{\epsilon}\mathcal{L}\left[\operatorname{rect}_{(\alpha,\alpha+\epsilon)}\right]\Big|_s$$

$$\hookrightarrow \qquad sY_\epsilon(s) - 0 = \frac{1}{\epsilon}\left[\frac{1}{s}e^{-\alpha s} - \frac{1}{s}e^{-(\alpha+\epsilon)s}\right]$$

$$\hookrightarrow \qquad Y_\epsilon(s) = \frac{1}{\epsilon}\left[\frac{1}{s^2}e^{-\alpha s} - \frac{1}{s^2}e^{-(\alpha+\epsilon)s}\right] \quad .$$

So,

$$\begin{aligned}
y_\epsilon(t) &= \mathcal{L}^{-1}\left[\frac{1}{\epsilon}\left[\frac{1}{s^2}e^{-\alpha s} - \frac{1}{s^2}e^{-(\alpha+\epsilon)s}\right]\right]\Big|_t \\
&= \frac{1}{\epsilon}\left\{\mathcal{L}^{-1}\left[\frac{1}{s^2}e^{-\alpha s}\right]\Big|_t - \mathcal{L}^{-1}\left[\frac{1}{s^2}e^{-(\alpha+\epsilon)}\right]\Big|_t\right\} \\
&= \frac{1}{\epsilon}\left\{[t-\alpha]\operatorname{step}_\alpha(t) - [t-(\alpha+\epsilon)]\operatorname{step}_{\alpha+\epsilon}(t)\right\} \\
&= \frac{1}{\epsilon}\begin{cases} 0 - 0 & \textit{if} \quad t < \alpha \\ [t-\alpha] - 0 & \textit{if} \quad \alpha < t < \alpha+\epsilon \\ [t-\alpha] - [t-(\alpha+\epsilon)] & \textit{if} \quad \alpha+\epsilon < t \end{cases} \\
&= \begin{cases} 0 & \textit{if} \quad t < \alpha \\ \dfrac{t-\alpha}{\epsilon} & \textit{if} \quad \alpha < t < \alpha+\epsilon \\ 1 & \textit{if} \quad \alpha+\epsilon < t \end{cases} \quad .
\end{aligned}$$

Graphs of y_ϵ for two different values of ϵ are sketched in Figure 31.3.

Finally, taking the limit, we get

$$y(t) = \lim_{\epsilon\to 0^+} y_\epsilon(t) = \lim_{\epsilon\to 0^+}\left\{\begin{array}{cl} 0 & \textit{if} \quad t < \alpha \\ \dfrac{t-\alpha}{\epsilon} & \textit{if} \quad \alpha < t < \alpha+\epsilon \\ 1 & \textit{if} \quad \alpha+\epsilon < t \end{array}\right\} = \begin{cases} 0 & \textit{if} \quad t < \alpha \\ 1 & \textit{if} \quad \alpha < t \end{cases} \quad .$$

That is,

$$y(t) = \text{step}_\alpha(t) \quad ,$$

just as obtained (with much less work!) in Example 31.3.

Continuity of Solutions and Problems with Initial Values

Early in this text, it was stated that solutions to first-order differential equations had to be continuous, and solutions to second-order differential equations had to be continuous and have continuous derivatives. But $y = \text{step}_\alpha(t)$, the solution to

$$\frac{dy}{dt} = \delta_\alpha(t) \qquad \text{with} \quad y(0) = 0$$

obtained in Exercises 31.3 and 31.5, is clearly not continuous. If you think about it, this may not be so surprising. Our original insistence on the continuity of solutions assumed we were using classical functions. The exotic nature of the delta functions takes us outside the classical theory to the idealized cases where instantaneous change can occur.

Normally, this is not a problem. Indeed, it can be desirable, especially if you are modeling "brief strong forces". One place where this can cause some confusion is when the discontinuities occur where initial data is given. In these cases, the confusion can be somewhat abated by remembering that a delta function really indicates a limiting process.

!►*Example 31.6:* *Letting $\alpha = 0$, we see that the solution to*

$$\frac{dy}{dt} = \delta(t) \qquad \text{with} \quad y(0) = 0$$

is

$$y(t) = \text{step}(t) \quad .$$

However, $\text{step}(t)$ has a jump at $t = 0$, and its limit from the right at this point is 1. So how can we say this step function satisfies the given initial condition, $y(0) = 0$? By going back to Exercise 31.5, which showed that the above solution should be viewed as the limit as $\epsilon \to 0$ of the function $y_\epsilon(t)$ graphed in Figure 31.3 with $\alpha = 0$. For each $\epsilon > 0$, $y_\epsilon(t)$ is continuous at $t = 0$ and satisfies $y_\epsilon(0) = 0$. As ϵ becomes smaller, the values of $y_\epsilon(t)$ increase more rapidly to 1 for positive values of t. So what we end up with after taking $\epsilon \to 0$ is that the left-hand limit of $y(t)$ at $t = 0$ is 0, but $y(t)$ "immediately" increases from 0 to 1 as t switches from negative values to positive values.

What this last example demonstrates is that, when the differential equation has a $\delta(t)$ in its forcing function, then initial conditions naively written as

$$y(0) = y_0 \quad , \quad y'(0) = y_1 \quad , \quad \dots$$

are, well, naive. What is really meant is that these values give the left-hand limits,

$$\lim_{t \to 0^-} y(t) = y_0 \quad , \quad \lim_{t \to 0^-} y'(0) = y_1 \quad , \quad \dots \quad .$$

Additional Exercises

31.1. *The speeds of the pitched and batted baseball given in Example 31.1 are close to 'typical' speeds. However, a really good professional pitcher can throw a fastball at a speed of* 45 *meters per second (a little over* 100 *miles per second), and a good batter can hit the ball back at* 49 *meters per second (almost* 110 *miles per hour). Assuming these values (and a* 0.145 *kilogram baseball):*

a. *Find the magnitude of the impulse the pitcher initially imparts to the thrown ball.*

b. *Find the magnitude of the impulse the batter imparts to the ball.*

31.2. *For the following, assume an object of mass m kilograms is initially moving along the X–axis with constant velocity v_{before} meters/second until its velocity is changed to v_{after} meters/second by a delta function force with impulse $\mathcal{I}$ kilogram·meters/second at time $t = \alpha$ seconds.*

a. *Find v_{after} assuming $m = 2$, $v_{\text{before}} = -10$ and*

i. $\mathcal{I} = 60$ **ii.** $\mathcal{I} = 100$ **iii.** $\mathcal{I} = 20$

b. *Assume $m = 0.2$ and $v_{\text{before}} = -40$. What impulse $\mathcal{I}$ is needed to obtain*

i. $v_{\text{after}} = 50$ **ii.** $v_{\text{after}} = 100$ **iii.** $v_{\text{after}} = 0$

c. *Assume $\mathcal{I} = 30$, and that the velocity of the object before and after $t = \alpha$ is determined by radar. What is the mass of the object if*

i. $v_{\text{before}} = -10$ *and* $v_{\text{after}} = 50$ **ii.** $v_{\text{before}} = 0$ *and* $v_{\text{after}} = 15$

31.3. *Using the results given in Theorem 31.1, compute the following integrals*

a. $\displaystyle\int_0^\infty t^2 \delta_4(t)\,dt$ **b.** $\displaystyle\int_0^\infty t^2 \delta(t)\,dt$

c. $\displaystyle\int_0^\infty \cos(t)\,\delta(t)\,dt$ **d.** $\displaystyle\int_0^\infty \sin(t)\,\delta_{\pi/6}(t)\,dt$

e. $\displaystyle\int_0^\infty t^2\,\text{rect}_{(1,4)}(t)\delta_3(t)\,dt$ **f.** $\displaystyle\int_0^\infty t^2\,\text{rect}_{(1,4)}(t)\delta_6(t)\,dt$

31.4. *Prove/derive equation (31.5) on page 564.*

31.5. *Show that*

$$g * \delta_\alpha(t) = g(t-\alpha)\,\text{step}_\alpha(t)$$

whenever $\alpha \geq 0$, $t > 0$ and g is a piecewise continuous function on $(0, \infty)$.

31.6. *Find and sketch the solution over $[0, \infty)$ to each of the following:*

a. $y' = 3\delta_2(t)$ *with* $y(0) = 0$

b. $y' = \delta_2(t) - \delta_4(t)$ *with* $y(0) = 0$

c. $y'' = \delta_3(t)$ *with* $y(0) = 0$ *and* $y'(0) = 0$

d. $y'' = \delta_1(t) - \delta_4(t)$ *with* $y(0) = 0$ *and* $y'(0) = 0$

e. $y' + 2y = 4\delta_1(t)$ with $y(0) = 0$

f. $y'' + y = \delta(t) + \delta_\pi(t)$ with $y(0) = 0$ and $y'(0) = 0$

g. $y'' + y = -2\delta_{\pi/2}(t)$ with $y(0) = 0$ and $y'(0) = 0$

31.7. *Find the solution on $t > 0$ to each of the following initial-value problems:*

a. $y' + 3y = \delta_2(t)$ with $y(0) = 2$

b. $y'' + 3y' = \delta(t)$ with $y(0) = 0$ and $y'(0) = 0$

c. $y'' + 3y' = \delta_1(t)$ with $y(0) = 0$ and $y'(0) = 1$

d. $y'' + 16y = \delta_2(t)$ with $y(0) = 0$ and $y'(0) = 0$

e. $y'' - 16y = \delta_{10}(t)$ with $y(0) = 0$ and $y'(0) = 0$

f. $y'' + y = \delta(t)$ with $y(0) = 0$ and $y'(0) = -1$

g. $y'' + 4y' - 12y = \delta(t)$ with $y(0) = 0$ and $y'(0) = 0$

h. $y'' + 4y' - 12y = \delta_3(t)$ with $y(0) = 0$ and $y'(0) = 0$

i. $y'' + 6y' + 9y = \delta_4(t)$ with $y(0) = 0$ and $y'(0) = 0$

j. $y'' - 12y' + 45y = \delta(t)$ with $y(0) = 0$ and $y'(0) = 0$

k. $y''' + 9y' = \delta_1(t)$ with $y(0) = 0$, $y'(0) = 0$ and $y''(0) = 0$

l. $y^{(4)} - 16y = \delta(t)$ with $y(0) = y'(0) = y''(0) = y'''(0) = 0$

Part V

Power Series and Modified Power Series Solutions

32

Series Solutions: Preliminaries (A Brief Review of Infinite Series, Power Series and a Little Complex Variables)

At this point, you should have no problem in solving any differential equation of the form

$$a\frac{d^2y}{dx^2} + b\frac{dy}{dx} + cy = 0 \qquad \text{or} \qquad ax^2\frac{d^2y}{dx^2} + bx\frac{dy}{dx} + cy = 0$$

when a, b and c are all constants. You've even solved a few (in Chapters 13 and 14) in which a, b and/or c were not constants. Unfortunately, the methods used in those chapters are somewhat limited. More general methods do exist, and, in the next few chapters, we will discuss some of the more important of these in which solutions are described in terms of "power series" and "modified power series".

Ideally, you are already well-enough acquainted with infinite series and power series to jump right into the discussion of the next chapter. As a precaution, though, you may want to skim through this chapter. It is a brief review of infinite series with an emphasis on power series, along with a brief discussion of using complex variables in these series. As much as possible, we'll limit our discussion to topics that will be needed in the next few chapters, including a few that probably were not emphasized during your first exposure to power series.

32.1 Infinite Series

Basic Basics

Recall that, in the language of mathematics, an *infinite series* is a summation with infinitely many terms. For example,

$$\sum_{k=1}^{\infty} \frac{1}{k} = 1 + \frac{1}{2} + \frac{1}{3} + \frac{1}{4} + \frac{1}{5} + \cdots$$

is the infamous *harmonic series*. More generally, an infinite series is anything of the form

$$\sum_{k=\gamma}^{\infty} \alpha_k \qquad \text{or (equivalently)} \qquad \alpha_\gamma + \alpha_{\gamma+1} + \alpha_{\gamma+2} + \alpha_{\gamma+3} + \alpha_{\gamma+4} + \cdots$$

where γ, the starting index, is some integer (often, it's 0 or 1), and the α_k's are things that can be added together. These α_k's may be numbers, functions or even matrices. For the moment, we will assume they are numbers (as in the harmonic series, above).

Given an arbitrary infinite series

$$\sum_{k=\gamma}^{\infty} \alpha_k = \alpha_\gamma + \alpha_{\gamma+1} + \alpha_{\gamma+2} + \alpha_{\gamma+3} + \alpha_{\gamma+4} + \cdots$$

and any integer $N \geq \gamma$, we define the corresponding N^{th} *partial sum* S_N by[1]

$$S_N = \text{sum of all terms from } a_\gamma \text{ to } a_N$$

$$= \sum_{k=\gamma}^{N} \alpha_k = \alpha_\gamma + \alpha_{\gamma+1} + \alpha_{\gamma+2} + \cdots + \alpha_N \quad .$$

Observe that

$$\lim_{N\to\infty} S_N = \lim_{N\to\infty} \sum_{k=\gamma}^{N} \alpha_k = \sum_{k=\gamma}^{\infty} \alpha_k = \alpha_\gamma + \alpha_{\gamma+1} + \alpha_{\gamma+2} + \alpha_{\gamma+3} + \alpha_{\gamma+4} + \cdots \quad .$$

Naturally, the usefulness of an infinite series usually depends on whether it actually adds up to some finite value; that is, whether

$$\lim_{N\to\infty} \sum_{k=\gamma}^{N} \alpha_k$$

is some finite value. If the above limit does exist as a finite value, then we say our series *converges*, and call that value the *sum* of that series (freely using $\sum_{k=\gamma}^{\infty} \alpha_k$ to denote both the series and its sum). On the other hand, if this limit does not exist as a finite value, then we say the series *diverges*.

Recall the following simple facts:

1. If $\sum_{k=\gamma}^{\infty} \alpha_k$ converges, then we can closely approximate its sum by any N^{th} partial sum, provided we choose N large enough.

2. If $\sum_{k=\gamma}^{\infty} \alpha_k$ converges, then its terms must shrink to zero as k gets large,

$$\alpha_k \to 0 \quad \text{as} \quad k \to \infty \quad .$$

Consequently, $\sum_{k=\gamma}^{\infty} \alpha_k$ can*not* converge (i.e., must diverge) if the terms do not shrink to zero as k gets large.

However, it is quite possible to have a series $\sum_{k=\gamma}^{\infty} \alpha_k$ that diverges even though

$$\alpha_k \to 0 \quad \text{as} \quad k \to \infty \quad .$$

The harmonic series, above, is one example. Even though

$$\alpha_k = \frac{1}{k} \to 0 \quad \text{as} \quad k \to \infty \quad ,$$

you can easily show that the series diverges (to infinity) using the integral test.

3. If $\sum_{k=\gamma}^{\infty} \alpha_k$ and $\sum_{k=\gamma}^{\infty} \beta_k$ are both convergent series, and A and B are any two finite numbers, then the series $\sum_{k=\gamma}^{\infty} [A\alpha_k]$ and $\sum_{k=\gamma}^{\infty} [A\alpha_k + B\beta_k]$ are also convergent. Moreover,

$$\sum_{k=\gamma}^{\infty} [A\alpha_k] = A \sum_{k=\gamma}^{\infty} \alpha_k \quad \text{and} \quad \sum_{k=\gamma}^{\infty} [A\alpha_k + B\beta_k] = A \sum_{k=\gamma}^{\infty} \alpha_k + B \sum_{k=\gamma}^{\infty} \beta_k \quad .$$

To illustrate some of the above concepts, and to give us a first glimpse of "power series", let's look at the "geometric series".

---

[1] It's also standard to define S_N to be the sum of the first N terms. Our choice will be slightly more convenient.

The Geometric Series

Let x be any finite value and let γ be any nonnegative integer. The corresponding *geometric series* is

$$\sum_{k=\gamma}^{\infty} x^k = x^\gamma + x^{\gamma+1} + x^{\gamma+2} + x^{\gamma+3} + x^{\gamma+4} + \cdots \quad .$$

If $\gamma = 0$, we may refer to this as a *basic* geometric series.[2]

Letting $\gamma = 0$ and using, respectively, $x = 0,\ 1/2,\ -1/2,\ 1,\ -1,\ 2$, and -2 gives us the following geometric series:

$$\sum_{k=0}^{\infty} 0^k = 1 + 0 + 0 + 0 + 0 + \cdots \quad ,$$

$$\sum_{k=0}^{\infty} \left(\frac{1}{2}\right)^k = 1 + \frac{1}{2} + \frac{1}{4} + \frac{1}{8} + \frac{1}{16} + \cdots \quad ,$$

$$\sum_{k=0}^{\infty} \left(-\frac{1}{2}\right)^k = 1 - \frac{1}{2} + \frac{1}{4} - \frac{1}{8} + \frac{1}{16} - \cdots \quad ,$$

$$\sum_{k=0}^{\infty} 1^k = 1 + 1 + 1 + 1 + 1 + \cdots \quad ,$$

$$\sum_{k=0}^{\infty} (-1)^k = 1 - 1 + 1 - 1 + 1 - \cdots \quad ,$$

$$\sum_{k=0}^{\infty} 2^k = 1 + 2 + 4 + 8 + 16 + \cdots \quad ,$$

and

$$\sum_{k=0}^{\infty} (-2)^k = 1 - 2 + 4 - 8 + 16 - \cdots \quad .$$

It should be obvious that a geometric series $\sum_{k=0}^{\infty} x^k$ converges if $x = 0$ and diverges whenever $|x| \geq 1$. It will also be worth noting that

$$\sum_{k=\gamma}^{\infty} x^k = x^\gamma + x^{\gamma+1} + x^{\gamma+2} + x^{\gamma+3} + x^{\gamma+4} + \cdots$$

$$= \sum_{k=0}^{\infty} x^{\gamma+k} = \sum_{k=0}^{\infty} x^\gamma x^k = x^\gamma \sum_{k=0}^{\infty} x^k \quad .$$

That is,

$$\sum_{k=\gamma}^{\infty} x^k = x^\gamma \sum_{k=0}^{\infty} x^k \quad . \tag{32.1}$$

[2] Since 0^0 is an indeterminant form, it may be argued that there is a problem with the x^0 term when $x = 0$. However, in every geometric series with $x \neq 0$, $x^0 = 1$. So, to be consistent, we automatically assume $x^0 = 1$ in a geometric series when $x = 0$.

Geometric series are unusual in that rather simple formulas can be derived for their partial sums. To see this, let

$$S_N = \sum_{k=0}^{N} x^k = x^0 + x^1 + x^2 + \cdots + x^N \quad .$$

If $x = 1$, then

$$S_N = \sum_{k=0}^{N} 1^k = \underbrace{1 + 1 + 1 + \cdots + 1}_{N+1 \text{ terms}} = N + 1 \quad .$$

If $x \neq 1$, then

$$\begin{aligned}(1-x)S_N &= S_N - xS_N \\ &= \left[x^0 + x^1 + x^2 + \cdots + x^N\right] \\ &\qquad - x\left[x^0 + x^1 + x^2 + \cdots + x^N\right] \\ &= \left[1 + x^1 + x^2 + \cdots + x^N\right] \\ &\qquad - \left[x^1 + x^2 + x^3 + \cdots + x^{N+1}\right] \\ &= 1 - x^{N+1} \quad .\end{aligned}$$

Dividing by $1 - x$ then gives us

$$\sum_{k=0}^{N} x^k = S_N = \frac{1 - x^{N+1}}{1 - x} \qquad \text{for} \quad x \neq 1 \quad . \tag{32.2}$$

!►Example 32.1: *With $x = 1/2$, the above formula for S_N becomes*

$$S_N = \frac{1 - \left(\frac{1}{2}\right)^{N+1}}{1 - \frac{1}{2}} = 2\left[1 - \left(\frac{1}{2}\right)^{N+1}\right] \quad .$$

In particular,

$$S_4 = 2\left[1 - \left(\frac{1}{2}\right)^{4+1}\right] = 2\left[1 - \frac{1}{32}\right] = \frac{31}{16} \quad .$$

Of greater interest is that

$$\lim_{N\to\infty} S_N = \lim_{N\to\infty} 2\left[1 - \left(\frac{1}{2}\right)^{N+1}\right] = 2\left[1 - \lim_{N\to\infty}\left(\frac{1}{2}\right)^{N+1}\right] = 2[1 - 0] = 2 \quad .$$

Thus, the geometric series with $x = 1/2$ converges, and

$$\sum_{k=0}^{\infty} \left(\frac{1}{2}\right)^k = \lim_{N\to\infty} S_N = 2 \quad .$$

?►Exercise 32.1: *Repeat the computations done in the last example, but using $x = -1/2$. Show that the corresponding geometric series converges with*

$$\sum_{k=0}^{\infty} \left(-\frac{1}{2}\right)^k = \frac{2}{3} \quad .$$

As you can easily verify for yourself (and as illustrated in the above example and exercise),

$$\lim_{N\to\infty} x^{N+1} = 0 \qquad \text{whenever} \quad |x| < 1 \quad .$$

This, along with equations (32.2) and (32.1) (and some of the other comments above), leads to the following:

Theorem 32.1 (geometric series)
The basic geometric series $\sum_{k=0}^{\infty} x^k$ converges if $|x| < 1$ and diverges if $|x| \geq 1$. Moreover,

$$\sum_{k=0}^{\infty} x^k = \frac{1}{1-x} \qquad \text{for} \quad |x| < 1 \quad .$$

More generally, for any nonnegative integer γ, the geometric series $\sum_{k=\gamma}^{\infty} x^k$ converges if and only if $|x| < 1$. Moreover,

$$\sum_{k=\gamma}^{\infty} x^k = \frac{x^\gamma}{1-x} \qquad \text{for} \quad |x| < 1 \quad .$$

Absolute Convergence and Convergence Tests

Absolute and Conditional Convergence

Recall that a series $\sum_{k=\gamma}^{\infty} \alpha_k$ can converge "absolutely" or "conditionally". It *converges absolutely* if and only if the corresponding series of absolute values

$$\sum_{k=\gamma}^{\infty} |\alpha_k|$$

converges. Basically, as the index increases in an absolutely convergent series, the terms shrink towards zero fast enough to ensure convergence. Consequently, it's easily verified that an absolutely convergent series is, as the terminology suggests, convergent. Moreover, by repeatedly using the triangle inequality,

$$|a+b| \leq |a| + |b| \quad ,$$

you can easily verify that

$$\left|\sum_{k=\gamma}^{\infty} \alpha_k\right| \leq \sum_{k=\gamma}^{\infty} |\alpha_k| \quad .$$

If a series converges but is not absolutely convergent, then it is converging because each term "cancels out" some of the previous terms, and the series is said to be *conditionally convergent*. Such a convergence is somewhat unstable and can be upset by, say, rearranging the terms of the series in a clever way. Because of this, we will much prefer series that converge absolutely.

Tests for Convergence and Divergence

In practice, it is rarely possible to determine the convergence of an infinite series by using its partial sums, simply because it is rarely possible to find usable formulas for these partial sums. That is why, in your calculus course, you were exposed to several "tests" for determining whether a given series converges or diverges. One, of course, is the basic comparison test.

Theorem 32.2 (the comparison test)
Let $\sum_{k=\gamma}^{\infty} \alpha_k$ and $\sum_{k=\mu}^{\infty} \beta_k$ be two infinite series of real numbers, and suppose that, for some integer K,

$$0 \leq \alpha_k \leq \beta_k \qquad \text{whenever} \quad K \leq k \quad .$$

Then

$$\sum_{k=\mu}^{\infty} \beta_k \text{ converges} \quad \Longrightarrow \quad \sum_{k=\gamma}^{\infty} \alpha_k \text{ converges absolutely} \quad ,$$

while

$$\sum_{k=\gamma}^{\infty} \alpha_k \text{ diverges} \quad \Longrightarrow \quad \sum_{k=\mu}^{\infty} \beta_k \text{ diverges} \quad .$$

We'll be using the above test in a few pages, and will briefly discuss two other well-known tests (the limit comparison and the limit ratio tests) later.

There are, of course, many other "tests for convergence", including the alternating series test, the integral test, the basic ratio test, and the root test. I'm sure you remember them all fondly and will be disappointed to learn that we will find little use for these other tests.

32.2 Power Series and Analytic Functions

Definition and Examples

A *power series* is any series of the form

$$\sum_{k=\gamma}^{\infty} a_k (x - x_0)^k$$

where x is a variable, x_0 and the a_k's are constants, and the starting index, γ, is some nonnegative integer. We'll often refer to x_0 as the *center* of the series, and say that the above power series is *centered at* or *about* x_0. We will also refer to the term $a_k(x - x_0)^k$ as the kth*-order term* of the series.[3]

In theory, γ can be any nonnegative integer; in practice, γ is often 0. Even when $\gamma \neq 0$, we can assume

$$\sum_{k=\gamma}^{\infty} a_k (x - x_0)^k = \sum_{k=0}^{\infty} a_k (x - x_0)^k$$

by simply setting

$$a_k = 0 \qquad \text{when} \quad k < \gamma \quad .$$

Also, in practice, many power series are centered at 0. And even if one isn't, we can convert it to one centered at 0 via a simple change of variables:

$$\sum_{k=\gamma}^{\infty} a_k (x - x_0)^k = \sum_{k=\gamma}^{\infty} a_k X^k \qquad \text{with} \quad X = x - x_0 \quad .$$

[3] Again, there is a minor issue with the 0th-order term appearing to be 'indeterminant' when $x = x_0$. But since

$$a_0 (x - x_0)^0 = a_0 \qquad \text{whenever} \quad x \neq x_0 \quad ,$$

we will, for consistency, automatically interpret $a_0(x - x_0)^0$ as a_0 even when $x = x_0$.

It is also worth noting that the same sort of computations leading to equation (32.1) also yield

$$\sum_{k=\gamma}^{\infty} a_k (x - x_0)^k = (x - x_0)^{\gamma} \sum_{k=0}^{\infty} a_{\gamma+k} (x - x_0)^k \quad . \tag{32.3}$$

The basic geometric series $\sum_{k=0}^{\infty} x^k$ is a power series. From Theorem 32.1, we know

$$\frac{1}{1-x} = \sum_{k=0}^{\infty} x^k \quad \text{for} \quad |x| < 1 \quad . \tag{32.4a}$$

Thus, the function $(1 - x)^{-1}$ can be represented by the above power series when $|x| < 1$. You may recall that many other functions can be represented by power series. Here are a few:

$$e^x = \sum_{k=0}^{\infty} \frac{1}{k!} x^k \quad \text{for} \quad -\infty < x < \infty \quad , \tag{32.4b}$$

$$\cos(x) = \sum_{k=0}^{\infty} \frac{(-1)^k}{(2k)!} x^{2k} \quad \text{for} \quad -\infty < x < \infty \quad , \tag{32.4c}$$

$$\sin(x) = \sum_{k=0}^{\infty} \frac{(-1)^k}{(2k+1)!} x^{2k+1} \quad \text{for} \quad -\infty < x < \infty \tag{32.4d}$$

and

$$\ln|x| = \sum_{k=1}^{\infty} \frac{(-1)^{k-1}}{k} (x - 1)^k \quad \text{for} \quad 0 < x < 2 \quad . \tag{32.4e}$$

Any function that can be represented on an open interval by a power series at a point x_0 in that interval is said to be *analytic* at x_0. It turns out that many functions of interest are analytic at most points in their domains. This fact will be vital in the next several chapters.

Convergence and the Radius of Convergence

If we are going to use a power series $\sum_{k=\gamma}^{\infty} a_k (x - x_0)^k$ as the formula for a function, it will be important to know the values of x for which this series is "makes sense" (i.e., is convergent). This set of x-values turns out to be an interval with x_0 as the midpoint. To see this, let us consider the series

$$\sum_{k=\gamma}^{\infty} a_k X^k \quad .$$

First of all, we clearly have convergence if $X = 0$ since every term with $k > 0$ is $a_k 0^k = 0$.

Now suppose we know $\sum_{k=\gamma}^{\infty} a_k r^k$ converges for some nonzero value r, and let X be any real value with $|X| < |r|$. Then $\left| X/r \right| < 1$ and, as noted in Theorem 32.3, the geometric series

$$\sum_{k=0}^{\infty} \left| \frac{X}{r} \right|^k$$

converges. Moreover, since $\sum_{k=\gamma}^{\infty} a_k r^k$ converges, we must have $\left| a_k r^k \right| \to 0$ as $k \to \infty$, which means there must be an integer K such that $\left| a_k r^k \right| < 1$ whenever $k > K$. And that means

$$\left| a_k X^k \right| = \left| a_k r^k \frac{X^k}{r^k} \right| = \left| a_k r^k \right| \cdot \left| \frac{X}{r} \right|^k < \left| \frac{X}{r} \right|^k \quad \text{for} \quad k > K \quad .$$

It then follows from the comparison test (Theorem 32.2) using the above convergent geometric series that $\sum_{k=\gamma}^{\infty} \left|a_k X^k\right|$ converges for this choice of X. In other words,

$$|X| < |r| \quad \text{and} \quad \sum_{k=\gamma}^{\infty} a_k r^k \text{ converges} \quad \Longrightarrow \quad \sum_{k=\gamma}^{\infty} a_k X^k \text{ converges absolutely} \quad .$$

On the other hand,

$$0 < |\rho| < |X| \quad \text{and} \quad \sum_{k=\gamma}^{\infty} a_k \rho^k \text{ diverges} \quad \Longrightarrow \quad \sum_{k=\gamma}^{\infty} a_k X^k \text{ diverges} \quad ,$$

because if $\sum_{k=\gamma}^{\infty} a_k X^k$ did not diverge, then the very arguments just used in the previous paragraph would falsely imply that $\sum_{k=\gamma}^{\infty} a_k \rho^k$ converges.

Letting $X = x - x_0$, and taking r as large as possible and/or ρ as small as possible then gives the existence of the value R (which may be 0 or $+\infty$) in the next theorem.

Theorem 32.3

For each power series $\sum_{k=\gamma}^{\infty} a_k (x - x_0)^k$, there is an R — which is either 0, a finite positive value or $+\infty$ — such that

$$|x - x_0| < R \quad \Longrightarrow \quad \sum_{k=\gamma}^{\infty} a_k (x - x_0)^k \text{ converges absolutely} \quad ,$$

while

$$R < |x - x_0| \quad \Longrightarrow \quad \sum_{k=\gamma}^{\infty} a_k (x - x_0)^k \text{ diverges} \quad .$$

The R in the above theorem is called the *radius of convergence* for the given power series. If $R = 0$, the power series only converges for $x = x_0$ (which means the series won't be of much use); if $R = +\infty$, the power series converges for all values of x (which is very nice). Otherwise, the series converges absolutely at every point in the interval $(x_0 - R, x_0 + R)$. Whether we have convergence when $x = x_0 \pm R$ depends on the particular series, and, frankly, will usually not be of great concern to us.

The radius of convergence for a given power series can sometimes be determined through careful use of the formulas in either the limit ratio test or the limit root test. You may recall doing this. We, however, will discover that the radii of convergence for the power series of interest to us can be determined much more easily from the "singularities" of whatever differential equation we will be trying to solve.

Algebra with Power Series and Analytic Functions

Addition

Adding two power series with the same center and starting index is trivial:

$$\sum_{k=\gamma}^{\infty} a_k (x - x_0)^k + \sum_{k=\gamma}^{\infty} b_k (x - x_0)^k = \sum_{k=\gamma}^{\infty} \left[a_k (x - x_0)^k + b_k (x - x_0)^k\right]$$

$$= \sum_{k=\gamma}^{\infty} [a_k + b_k] (x - x_0)^k \quad .$$

However, if (as will often happen in the next chapter) they have different starting indices,

$$\sum_{k=\gamma}^{\infty} a_k (x - x_0)^k + \sum_{k=\mu}^{\infty} b_k (x - x_0)^k \qquad \text{with} \quad \gamma \neq \mu \quad ,$$

then we will first convert the series with extra low-order terms to a finite sum with those extra terms added to an infinite series with the same starting index as the other.

!►Example 32.2: *Consider*

$$\sum_{k=0}^{\infty} a_k X^k + \sum_{k=2}^{\infty} b_k X^k \quad .$$

Now,

$$\sum_{k=0}^{\infty} a_k X^k = a_0 + a_1 X + a_2 X^2 + a_3 X^3 + a_4 X^4 + \cdots$$

$$= a_0 + a_1 X + \sum_{k=2}^{\infty} a_k X^k \quad .$$

So,

$$\sum_{k=0}^{\infty} a_k X^k + \sum_{k=2}^{\infty} b_k X^k = \left[a_0 + a_1 X + \sum_{k=2}^{\infty} a_k X^k \right] + \sum_{k=2}^{\infty} b_k X^k$$

$$= a_0 + a_1 X + \left[\sum_{k=2}^{\infty} a_k X^k + \sum_{k=2}^{\infty} b_k X^k \right]$$

$$= a_0 + a_1 X + \sum_{k=2}^{\infty} [a_k + b_k] X^k \quad .$$

Changing the Index

In the next few chapters, we will often find ourselves with expressions of the form

$$\sum_{k=\gamma}^{\infty} a_k X^{k+\omega}$$

where ω is some fixed integer. On those occasions, we will want to convert this summation formula involving $X^{k+\omega}$ to an equivalent formula involving X^n. We will do this using the index substitution $n = k + \omega$ (equivalently, $k = n - \omega$),

$$\sum_{k=\gamma}^{\infty} a_k X^{k+\omega} = \sum_{n-\omega=\gamma}^{\infty} a_{n-\omega} X^n = \sum_{n=\gamma+\omega}^{\infty} a_{n-\omega} X^n \quad .$$

The goal is to end up with a power series in which each term is a constant times X^n.

This sort of index manipulation is called a *change of index*, and is analogous to the "change of variables" often used to simplify integrals. Keep in mind that the index is an "internal variable" for each series. This means we can use different index substitutions on different summations.

!►Example 32.3: *Consider the sum of summations*

$$\sum_{k=0}^{\infty}(k+1)a_k X^{k+2} + \sum_{k=0}^{\infty} a_k X^k \quad .$$

Using $n = k+2$ (i.e., $k = n-2$) in the first summation,

$$\sum_{k=0}^{\infty}(k+1)a_k X^{k+2} = \sum_{n-2=0}^{\infty}([n-2]+1)a_{n-2}X^n = \sum_{n=2}^{\infty}(n-1)a_{n-2}X^n \quad .$$

For the second, we use $n = k$ and pull out the first two terms,

$$\sum_{k=0}^{\infty} a_k X^k = \sum_{n=0}^{\infty} a_n X^n = a_0 X^0 + a_1 X^1 + \sum_{n=2}^{\infty} a_n X^n \quad .$$

Thus,

$$\begin{aligned}\sum_{k=0}^{\infty}(k+1)a_k X^{k+2} + \sum_{k=0}^{\infty} a_k X^k &= \sum_{n=2}^{\infty}(n-1)a_{n-2}X^n + \left[a_0 + a_1 X + \sum_{n=2}^{\infty} a_n X^n\right] \\ &= a_0 + a_1 X + \left[\sum_{n=2}^{\infty}(n-1)a_{n-2}X^n + \sum_{n=2}^{\infty} a_n X^n\right] \\ &= a_0 + a_1 X + \sum_{n=2}^{\infty}\left[(n-1)a_{n-2} + a_n\right]X^n \quad .\end{aligned}$$

A Basic Equation

We will often find ourselves with the equation

$$\sum_{k=0}^{\infty} a_k (x-x_0)^k = 0 \qquad \text{for} \quad |x-x_0| < R \quad ,$$

which, in more explicit form (with $X = x - x_0$), is

$$a_0 + a_1 X + a_2 X^2 + a_3 X^3 + \cdots = 0 \qquad \text{for} \quad -R < X < R \quad .$$

Plugging in $X = 0$ gives

$$a_0 + \underbrace{a_1 0 + a_2 0^2 + a_3 0^3 + \cdots}_{0} = 0 \quad .$$

Hence,

$$a_0 = 0 \quad ,$$

and, for $|X| < R$,

$$\begin{aligned} & a_0 + a_1 X + a_2 X^2 + a_3 X^3 + \cdots = 0 \\ \hookrightarrow \quad & 0 + a_1 X + a_2 X^2 + a_3 X^3 + \cdots = 0 \\ \hookrightarrow \quad & X\left(a_1 + a_2 X + a_3 X^2 + \cdots\right) = 0 \quad .\end{aligned}$$

Assuming $R > 0$, the X factor can be divided out, leaving us with

$$a_1 + a_2X + a_3X^2 + \cdots = 0 \qquad \text{whenever} \qquad -R < X < R \quad .$$

Plugging $X = 0$ into this last equation then gives

$$a_1 = 0 \quad .$$

Continuing this process, we can show all the a_k's are 0, thus confirming the following:

Theorem 32.4

Let $\sum_{k=0}^{\infty} a_k(x - x_0)^k$ be a power series with a nonzero radius of convergence R. If

$$\sum_{k=0}^{\infty} a_k(x - x_0)^k = 0 \qquad \textit{for} \quad |x - x_0| < R \quad ,$$

then

$$a_k = 0 \qquad \textit{for} \quad k = 0, 1, 2, 3, \ldots \quad .$$

This simple theorem will be of fundamental importance for us.

By the way, we will refer to any power series $\sum_{k=0}^{\infty} a_k(x - x_0)^k$ as a *trivial power series* if and only if all the a_k's are zero. Along the same lines, we will say that a function f analytic at a point x_0 is *trivial* if and only if it is given by a trivial power series about x_0. An immediate corollary of the above is the following unsurprising result.

Corollary 32.5

Let f be a function analytic at x_0. Then f is trivial if and only if there is an open interval (a, b) containing x_0 such that

$$f(x) = 0 \qquad \textit{whenever} \quad a < x < b \quad .$$

Naturally, our main interest will be with nontrivial analytic functions; that is, analytic functions that are not trivial.

Calculus with Power Series and Analytic Functions

Differentiating Power Series

Suppose we have a function f given by some power series with a nonzero radius of convergence R,

$$f(x) = \sum_{k=0}^{\infty} a_k(x - x_0)^k \qquad \text{for} \quad |x - x_0| < R \quad .$$

To differentiate this, it seems reasonable to use the

$$\textit{derivative of a sum} = \textit{sum of the derivatives}$$

rule from elementary calculus:

$$\begin{aligned} f'(x) &= \frac{d}{dx} \sum_{k=0}^{\infty} a_k(x - x_0)^k \\ &= \frac{d}{dx}\big[a_0 + a_1(x - x_0) + a_2(x - x_0)^2 + a_3(x - x_0)^3 + \cdots \\ &\qquad\qquad + a_k(x - x_0)^k + \cdots\big] \end{aligned}$$

$$
\begin{aligned}
&= \frac{d}{dx}a_0 + \frac{d}{dx}a_1(x-x_0) + \frac{d}{dx}a_2(x-x_0)^2 + \frac{d}{dx}a_3(x-x_0)^3 + \cdots \\
&\qquad + \frac{d}{dx}a_k(x-x_0)^k + \cdots \\
&= 0 + a_1 + 2a_2(x-x_0) + 3a_3(x-x_0)^2 + \cdots + ka_k(x-x_0)^{k-1} + \cdots \\
&= \sum_{k=1}^{\infty} k\,a_k(x-x_0)^{k-1} \quad .
\end{aligned}
$$

Note that the derivative of the "$k = 0$ term for $f(x)$" is 0. That is why, in the last series above, we dropped the $k = 0$ term and started with $k = 1$. Strictly speaking, this is not necessary. Since

$$
ka_k(x-x_0)^{k-1} = 0 \qquad \text{when} \quad k = 0 \quad ,
$$

the above series formula for f' would still be valid if it started at $k = 0$ instead of $k = 1$. Still, dropping the $k = 0$ term in the above can help prevent some embarrassing mistakes in the sort of computations we'll be doing in the next chapter.

Repeating the above (in abbreviated form) with the series obtained for $f'(x)$, we get

$$
\begin{aligned}
f''(x) &= \frac{d}{dx}\sum_{k=1}^{\infty} k\,a_k(x-x_0)^{k-1} \\
&= \sum_{k=1}^{\infty} \frac{d}{dx}\left[k\,a_k(x-x_0)^{k-1}\right] = \sum_{k=2}^{\infty} k(k-1)\,a_k(x-x_0)^{k-2} \quad .
\end{aligned}
$$

Using this, we then have

$$
\begin{aligned}
f'''(x) &= \frac{d}{dx}\sum_{k=2}^{\infty} k(k-1)\,a_k(x-x_0)^{k-2} \\
&= \sum_{k=2}^{\infty} \frac{d}{dx}\left[k(k-1)\,a_k(x-x_0)^{k-2}\right] = \sum_{k=3}^{\infty} k(k-1)(k-2)\,a_k(x-x_0)^{k-3} \quad .
\end{aligned}
$$

Continuing these computations, you end up getting

$$
f^{(n)}(x) = \sum_{k=n}^{\infty} k(k-1)(k-2)\cdots(k-n+1)\,a_k(x-x_0)^{k-n}
$$

for any nonnegative integer n.

There is a technical issue with the above computations. The

derivative of a sum $=$ *sum of the derivatives*

rule from elementary calculus was only shown to be true when the sum had finitely many terms. Here we have infinitely many terms. In fact, there are infinite series of functions for which this rule fails. Fortunately, it does not fail for power series, and the following theorem can be rigorously confirmed (see any good calculus text).

Theorem 32.6 (differentiation of power series)
Suppose f is a function given by a power series with a nonzero radius of convergence R,

$$
f(x) = \sum_{k=0}^{\infty} a_k(x-x_0)^k \qquad \text{for} \quad |x-x_0| < R \quad .
$$

Then, for any positive integer n, the nth derivative of f exists. Moreover, R is also the radius of convergence of the differentiated series, with

$$f^{(n)}(x) = \sum_{k=n}^{\infty} k(k-1)(k-2)\cdots(k-n+1)\, a_k (x-x_0)^{k-n} \qquad \text{for} \quad |x-x_0| < R \quad .$$

In particular,

$$f'(x) = \sum_{k=1}^{\infty} k\, a_k (x-x_0)^{k-1} \qquad \text{for} \quad |x-x_0| < R \quad ,$$

and

$$f''(x) = \sum_{k=2}^{\infty} k(k-1)\, a_k (x-x_0)^{k-2} \qquad \text{for} \quad |x-x_0| < R \quad .$$

Integral analogs to the above theorem also hold. In particular, if

$$f(x) = \sum_{k=0}^{\infty} a_k (x-x_0)^k \qquad \text{for} \quad |x-x_0| < R \quad ,$$

then it can be verified that

$$\int_{x_0}^{x} f(t)\,dt = \sum_{k=0}^{\infty} \int_{x_0}^{x} a_k (t-x_0)^k\,dt = \sum_{k=0}^{\infty} \frac{a_k}{k+1}(x-x_0)^{k+1}$$

whenever $|x-x_0| < R$. This can be a useful fact, though we won't have much need for it.

Power Series for Analytic Functions

As already noted, any function f given by a power series centered at x_0 in some open interval containing x_0 is said to be *analytic* at x_0. If, in addition, f is analytic at every point in some interval, then we say f is *analytic on that interval.*

So suppose we have a function f analytic at x_0 with

$$f(x) = \sum_{k=0}^{\infty} a_k (x-x_0)^k \qquad \text{for} \quad |x-x_0| < R$$

for some $R > 0$. Our 'differentiation of power series' theorem (Theorem 32.6) tells us that f is, in fact, "infinitely differentiable" on the interval $(x_0 - R, x_0 + R)$.[4] That theorem also allows us to derive a simple relationship between the a_k's in the series and the derivatives of f at x_0.

Let's derive that relation. First, plugging $x = x_0$ into the above, we get

$$\begin{aligned} f(x_0) &= \sum_{k=0}^{\infty} a_k (x_0-x_0)^k \\ &= a_0 + a_1(x_0-x_0) + a_2(x_0-x_0)^2 + a_3(x_0-x_0)^3 + \cdots \\ &= a_0 + a_1 \cdot 0 + a_2 \cdot 0^2 + a_3 \cdot 0^3 + \cdots \\ &= a_0 \quad . \end{aligned}$$

[4] We say that a function f is *infinitely differentiable* at a point x if and only if $f^{(n)}(x)$ exists for every positive integer n, and is infinitely differentiable on a given interval if and only if it is infinitely differentiable at each point in the interval.

Then, using formulas from Theorem 32.6, we see that

$$\begin{aligned} f'(x_0) &= \sum_{k=1}^{\infty} k\, a_k (x_0 - x_0)^{k-1} \\ &= 1 a_1 + 2 a_2 \cdot 0 + 3 a_3 \cdot 0^2 + 4 a_4 \cdot 0^3 + \cdots \\ &= 1 a_1 \quad , \end{aligned}$$

and

$$\begin{aligned} f''(x_0) &= \sum_{k=2}^{\infty} k(k-1)\, a_k (x_0 - x_0)^{k-2} \\ &= 2 \cdot 1 a_2 + 3 \cdot 2 a_3 \cdot 0 + 4 \cdot 3 a_4 \cdot 0^2 + 5 \cdot 4 a_5 \cdot 0^3 + \cdots \\ &= (2 \cdot 1) a_2 \quad . \end{aligned}$$

More generally, for any positive integer n,

$$\begin{aligned} f^{(n)}(x_0) &= \sum_{k=n}^{\infty} k(k-1)(k-2)\cdots(k-n+1)\, a_k (x_0 - x_0)^{k-n} \\ &= n(n-1)(n-2)\cdots 1\, a_n \\ &\quad + (n+1)(n)(n-1)\cdots 2\, a_{n+1} \cdot 0 + (n+2)(n+1)(n)\cdots 3\, a_{n+2} \cdot 0^2 \\ &\quad + \cdots \\ &= n!\, a_n \quad . \end{aligned}$$

Dividing the last relation through by $n!$ and observing that the result also holds when $n = 0$ (interpreting $f^{(0)}$ as f) gives us the next theorem.

Theorem 32.7

Let f be analytic at x_0. Then, for every x in some open interval containing x_0,

$$f(x) = \sum_{k=0}^{\infty} a_k (x - x_0)^k \qquad \text{with} \quad a_k = \frac{f^{(k)}(x_0)}{k!} \quad .$$

As an immediate corollary, we have the following (which will be important when discussing "power series solutions to initial-value problems"):

Corollary 32.8

Assume

$$f(x) = \sum_{k=0}^{\infty} a_k (x - x_0)^k \qquad \text{for} \quad |x - x_0| < R$$

with $R > 0$. Then,

$$a_0 = f(x_0) \qquad \text{and} \qquad a_1 = f'(x_0) \quad .$$

You may recognize the series in Theorem 32.7, written a little more simply as

$$\sum_{k=0}^{\infty} \frac{f^{(k)}(x_0)}{k!} (x - x_0)^k \quad ,$$

as the *Taylor's series (formula)* for $f(x)$ about x_0, and you may recall having once computed Taylor series for such functions as e^x, $\sin(x)$, $\cos(x)$ and $\ln|x|$. In fact, the Taylor series about any point x_0 can be computed for any function f which is infinitely differentiable at that point. However, a function can be infinitely differentiable at a point x_0 without being analytic there — its Taylor series exists, but does not equal the function at any point other than x_0. With luck you saw an example, say,

$$f(x) = \begin{cases} 0 & \text{if } x = 0 \\ e^{-1/x^2} & \text{if } x \neq 0 \end{cases} ,$$

which can be shown to be infinitely differentiable but not analytic at $x_0 = 0$ (see Exercise 32.10).

Still, you probably saw that many functions are analytic at many points. You may well have already verified that such functions as

$$e^x \quad , \quad e^{-2x^2} \quad , \quad \sin(x) \quad \text{and} \quad \cos(x)$$

are analytic at every point on the real line, and that functions such as

$$\sqrt{x} \quad \text{and} \quad \ln x$$

are analytic at any point $x_0 > 0$. You may have even been given the impression that most functions typically encountered in "real life" are analytic at every point at which they are infinitely differentiable. In a sense, this is true, though very difficult to confirm using the methods normally developed in elementary calculus courses. (We'll return to this issue in Chapter 34.)

32.3 Elementary Complex Analysis

Up to now, we've acted as if we were only dealing with real numbers in our infinite series. In fact, just about everything said so far, up to the discussion of *Calculus with Power Series and Analytic Functions* (starting on page 587), holds even if the numbers are complex, provided we make some obvious changes in notation and phrasing.[5] In fact, we will later have particular interest in power series in which the variables are complex.

The Complex Plane

Recall that a *complex number* z is simply something that can be written as

$$z = x + iy$$

where x and y are real numbers, and i is a constant satisfying $i^2 = -1$. Because we'll be using such expressions so often, let us agree that, unless otherwise noted, in any expression of the form $z = x + iy$, both x and y are real numbers.

As you are probably aware, each complex number $z = x + iy$ can be identified with the point (x, y) in the XY–plane. When we do so, we generally refer to the plane as the *complex plane* and denote it by $\mathbb{C}$. The distance between any two points

$$z_1 = x_1 + iy_1 \quad \text{and} \quad z_2 = x_2 + iy_2$$

[5] And, after discussing "complex calculus" in Chapter 34, we'll see that everything said in *Calculus with Power Series ...* also holds for power series with complex variables.

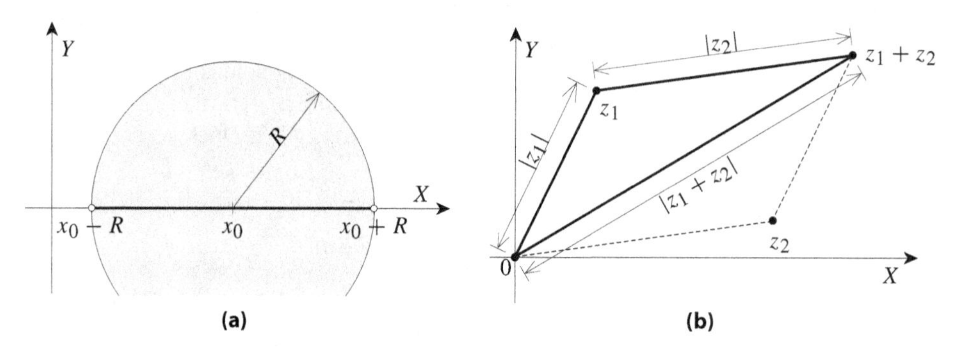

Figure 32.1: **(a)** The disk of radius R about x_0 in $\mathbb{C}$ with the interval $(x_0 - R, x_0 + R)$ on the X–axis, and **(b)** an illustration of the triangle inequality.

in the complex plane is simply their distance as points in the XY–plane,

$$|z_2 - z_1| = \sqrt{(x_2 - x_1)^2 + (y_2 - y_1)^2} \quad .$$

Note that, for any z_0 in $\mathbb{C}$ and $R > 0$, the set of all z satisfying $|z - z_0| < R$ is the disk of radius R centered at z_0. For comparison, recall that, when we were just considering real numbers, the set of x satisfying $|x - x_0| < R$ was the interval $(x_0 - R, x_0 + R)$ (see Figure 32.1a).

Power Series and Analytic Functions

If you look at the triangle having vertices 0, z_1, z_2 and $z_1 + z_2$ (as in Figure 32.1b), you'll realize that the triangle inequality,

$$|z_1 + z_2| \leq |z_1| + |z_2| \quad ,$$

holds for complex as well as real values. Using this fact, everything already stated regarding "absolute convergence" is easily shown to apply whether a series involves real or complex values. In particular, Theorem 32.3 on page 584 can be automatically expanded to:

Theorem 32.9
For each power series $\sum_{k=\gamma}^{\infty} a_k (z - z_0)^k$, there is an R — which is either 0, a finite positive value or $+\infty$ — such that

$$|z - z_0| < R \quad \Longrightarrow \quad \sum_{k=\gamma}^{\infty} a_k (z - z_0)^k \text{ converges absolutely} \quad ,$$

while

$$R < |z - z_0| \quad \Longrightarrow \quad \sum_{k=\gamma}^{\infty} a_k (z - z_0)^k \text{ diverges} \quad .$$

As before, we call the R in this theorem the *radius of convergence* for the power series, and we can refer to the point z_0 as the *center* for our series. This time, the terminology is truly appropriate, since the given power series does converge inside the disk of radius R about z_0, and diverges outside that disk.

Along these same lines, we extend the definition of analyticity to functions of complex variables by saying that a function f of a complex variable z is *analytic at a point z_0 in the complex plane* if and only if there is a power series $\sum_{k=0}^{\infty} a_k(z - z_0)^k$ and a $R > 0$ (possibly with $R = \infty$), such that

$$f(z) = \sum_{k=0}^{\infty} a_k(z - z_0)^k \qquad \text{for} \quad |z - z_0| < R \quad .$$

Naturally, we say that the function is analytic on a region in the complex plane if and only if it is analytic at each point in that plane.[6]

Let us note that, if f is a function of a real variable x given by, say,

$$f(x) = \sum_{k=0}^{\infty} a_k(x - x_0)^k \qquad \text{for} \quad |x - x_0| < R \quad ,$$

then f can be extended to a function of the complex variable by simply replacing x in the series with z,

$$f(z) = \sum_{k=0}^{\infty} a_k(z - x_0)^k \qquad \text{whenever} \quad |z - x_0| < R \quad .$$

What is more, it follows directly from Theorem 32.9 that the radii of convergence for both of the two series above are the same. In particular, the corresponding complex variable versions of the functions given by formulas 32.4 on page 583 are

$$\frac{1}{1-z} = \sum_{k=0}^{\infty} z^k \qquad \text{for} \quad |z| < 1 \quad , \tag{32.5a}$$

$$e^z = \sum_{k=0}^{\infty} \frac{1}{k!} z^k \qquad \text{for} \quad |z| < \infty \quad , \tag{32.5b}$$

$$\cos(z) = \sum_{k=0}^{\infty} \frac{(-1)^k}{(2k)!} z^{2k} \qquad \text{for} \quad |z| < \infty \quad , \tag{32.5c}$$

$$\sin(z) = \sum_{k=0}^{\infty} \frac{(-1)^k}{(2k+1)!} z^{2k+1} \qquad \text{for} \quad |z| < \infty \tag{32.5d}$$

and

$$\ln|z| = \sum_{k=1}^{\infty} \frac{(-1)^{k-1}}{k} (z-1)^k \qquad \text{for} \quad |z - 1| < 1 \quad . \tag{32.5e}$$

There is an issue here that may concern the thoughtful reader: some of the above functions have formulas other than the above power series for computing their values at complex points. For example, in Chapter 17, we learned of another formula for e^z when $z = x + iy$, namely,

$$e^z = e^{x+iy} = e^x [\cos(y) + i \sin(y)] \quad .$$

Can we be sure that this formula will give the same result as using the above power series for e^z,

$$\sum_{k=0}^{\infty} \frac{1}{k!} z^k = \sum_{k=0}^{\infty} \frac{1}{k!} (x + iy)^k \quad ?$$

Yes, we can. Trust the author on this. And if you don't feel that trust, turn ahead to Section 34.7 (starting on page 663) where we discuss the calculus of functions of a complex variable.

[6] If you've had a course in complex analysis, you may have seen a different definition for "analyticity". In Section 34.7, we will find that the two definitions are equivalent.

32.4 Additional Basic Material That May Be Useful

The material in the previous sections will be needed in the next chapter. But there are some additional facts about series and power series that will be useful later, especially when we get deeper into the rigorous theory behind the computations that we will be developing. For convenience, we'll provide some the more basic general facts here, and develop the more advanced material as needed. It won't hurt to skip this material initially, provided you return to it as needed.

Two More General Tests for Convergence

The well-known basic comparison test for determining if a given series converges or diverges was described in Theorem 32.2 on page 582 and was used in developing the radius of convergence for power series. In Chapter 34, we will find the next test, a clever refinement of the basic comparison test, to be useful.

Theorem 32.10 (the limit comparison test)
Let $\sum_{k=\gamma}^{\infty} \alpha_k$ and $\sum_{k=\mu}^{\infty} \beta_k$ be two infinite series, and suppose

$$\lim_{k\to\infty} \left|\frac{\alpha_k}{\beta_k}\right|$$

exists as either a finite number or as $+\infty$. Then

$$\lim_{k\to\infty} \left|\frac{\alpha_k}{\beta_k}\right| < \infty \quad \text{and} \quad \sum_{k=\mu}^{\infty} |\beta_k| \text{ converges} \quad \Longrightarrow \quad \sum_{k=\gamma}^{\infty} |\alpha_k| \text{ converges} \quad ,$$

while

$$\lim_{k\to\infty} \left|\frac{\alpha_k}{\beta_k}\right| > 0 \quad \text{and} \quad \sum_{k=\mu}^{\infty} |\beta_k| \text{ diverges} \quad \Longrightarrow \quad \sum_{k=\gamma}^{\infty} |\alpha_k| \text{ diverges} \quad .$$

Under certain conditions, you can use the limit of the ratio of the consecutive terms of a single series to construct a geometric series that can serve as a second series in the above limit comparison test. That leads to a third test, which will be used near the end of Chapter 37.

Theorem 32.11 (the limit ratio test)
Let $\sum_{k=\gamma}^{\infty} \alpha_k$ be an infinite series, and suppose

$$\lim_{k\to\infty} \left|\frac{\alpha_{k+1}}{\alpha_k}\right|$$

exists as either a finite number or as $+\infty$. Then

$$\lim_{k\to\infty} \left|\frac{\alpha_{k+1}}{\alpha_k}\right| < 1 \quad \Longrightarrow \quad \sum_{k=\gamma}^{\infty} \alpha_k \text{ converges absolutely} \quad ,$$

while

$$\lim_{k\to\infty} \left|\frac{\alpha_{k+1}}{\alpha_k}\right| > 1 \quad \Longrightarrow \quad \sum_{k=\gamma}^{\infty} \alpha_k \text{ diverges} \quad .$$

(If the limit is 1, there is no conclusion.)

The derivations of the above tests can be found in any reasonable elementary calculus text.

More on Algebra with Power Series and Analytic Functions

Multiplication

The following — a straightforward extension of a basic formula for computing products of polynomials — is worth a brief mention, especially since we will need it in Chapter 34.

Theorem 32.12

The product of two power series centered at the same point is another power series whose radius of convergence is at least as large as the smallest radius of convergence of the original two series. Moreover,

$$\left(\sum_{k=0}^{\infty} a_k(z-z_0)^k\right)\left(\sum_{k=0}^{\infty} b_k(z-z_0)^k\right) = \sum_{k=0}^{\infty} c_k(z-z_0)^k$$

with

$$c_k = a_0b_k + a_1b_{k-1} + \cdots + a_kb_0 = \sum_{j=0}^{k} a_jb_{k-j} \quad .$$

Factoring a Power Series/Analytic Function

On occasion, it will be convenient to "factor out" factors of the form $(z-z_0)^m$ from a function f analytic at z_0. Our ability to do this follows immediately from the fact that, being analytic at z_0, f is given by some power series with a nonzero radius of convergence R,

$$f(z) = \sum_{k=0}^{\infty} a_k(z-z_0)^k \qquad \text{for} \quad |z-z_0| < R \quad .$$

Note that

$$f(z_0) = a_0 + a_1(z_0-z_0) + a_2(z_0-z_0)^2 + \cdots = a_0 \quad ,$$

telling us that $f(z_0) = 0$ if and only if $a_0 = 0$.

Let's go a little further and assume $f(z_0) = 0$, Then, as just noted, $a_0 = 0$. Moreover,

$$\begin{aligned} f(z) &= a_0 + a_1(z-z_0) + a_2(z-z_0)^2 + a_3(z-z_0)^3 + \cdots \\ &= 0 + (z-z_0)\left[a_1 + a_2(z-z_0) + a_3(z-z_0)^2 + \cdots\right] \\ &= (z-z_0)\sum_{k=1}^{\infty} a_k(z-z_0)^{k-1} \qquad \text{for} \quad |z-z_0| < R \quad . \end{aligned}$$

Of course, we could also have $a_1 = 0$, in which case we can repeat the above to obtain

$$f(z) = (z-z_0)^2\sum_{k=2}^{\infty} a_k(z-z_0)^{k-2} \qquad \text{for} \quad |z-z_0| < R \quad .$$

Continuing until we finally reach a nonzero coefficient (assuming f in nontrivial), we get, for some positive integer m,

$$f(z) = (z-z_0)^m\sum_{k=m}^{\infty} a_k(z-z_0)^{k-m} \qquad \text{with} \quad a_m \neq 0 \quad ,$$

which, after letting $b_n = a_{n+m}$, can be written as

$$f(z) = (z - z_0)^m \sum_{n=0}^{\infty} b_n (z - z_0)^n \qquad \text{with} \quad b_0 \neq 0 \quad .$$

Thus, setting

$$f_0(z) = \sum_{n=0}^{\infty} b_n (z - z_0)^n$$

and noting that, even if $a_0 \neq 0$, it is trivially true that

$$f(z) = (z - z_0)^0 f(z) \quad ,$$

we get:

Lemma 32.13
Let f be a nontrivial function analytic at z_0. Then there is a nonnegative integer m such that

$$f(z) = (z - z_0)^m f_0(z_0)$$

where f_0 is a function analytic at z_0 with $f(z_0) \neq 0$. Moreover:

1. *$f(z_0) = 0$ if and only if $m > 0$.*
2. *The power series for f and f_0 about z_0 have the same radii of convergence.*

It is standard to refer to any point z_0 as a *zero* for an analytic function if $f(z_0) = 0$. It is also standard to refer to the m described in the last lemma as the *multiplicity* of the zero z_0.

Quotients of Analytic Functions

There is a technical issue that may arise when defining a function h as the quotient of two functions analytic at a given point.

!►Example 32.4: *Consider defining $h = {}^{f}/_{g}$ when f and g are the polynomials*

$$f(z) = z^2 - 1 \qquad \text{and} \qquad g(z) = z - 1 \quad .$$

For any value of z other than $z = 1$, we simply have

$$h(z) = \frac{f(z)}{g(z)} = \frac{z^2 - 1}{z - 1} \quad ,$$

which we can rewrite more simply by dividing out the common factor,

$$h(z) = \frac{z^2 - 1}{z - 1} = \frac{(z-1)(z+1)}{z - 1} = z + 1 \quad .$$

These two formulas for $h(z)$ give identical results whenever $z \neq 1$. However, we get two different results if we plug in $z = 1$:

$$h(1) = \frac{1^2 - 1}{1 - 1} = \frac{0}{0} \qquad \text{and} \qquad h(1) = 1 + 1 = 2 \quad .$$

The first expression is indeterminant, and is problematic in practical applications. The second is an finite number, and is clearly what we want to use for $h(1)$, especially since:

1. *It comes from a simpler formula for $h(z)$ when $z \neq 1$.*

2. *It is the same as the value we would obtain using the obvious limit,*

$$\lim_{z \to 1} h(z) = \lim_{z \to 1} \frac{f(z)}{g(z)} = \lim_{z \to 1} \frac{z^2 - 1}{z - 1} = \cdots = 2 \quad ,$$

computed either using L'Hôpital's rule or the formula obtained by dividing out the common factors. Hence, h is continuous at $z = 1$.

More generally, when we define a function h as the quotient of two other continuous functions, $h = {}^{f}/_{g}$, we automatically mean the function given by

$$h(z_0) = \frac{f(z_0)}{g(z_0)}$$

at each point z_0 in the common domain of f and g at which $g(z_0) \neq 0$; and, provided the limit exists as a finite number, by

$$h(z_0) = \lim_{z \to z_0} \frac{f(z)}{g(z)}$$

at each point z_0 in the common domain of f and g at which $g(z_0) = 0$. (If the above limit does not exist, we simply accept that h is not defined at that z_0.)

We should note that using the limit when f and g are analytic and zero at z_0 yields exactly the same result as if we were to divide out any common factors of $z - z_0$ in the quotient. After all, if f and g are analytic, then (as shown in the previous subsection) we can rewrite $f(z)$ and $g(z)$ as

$$f(z) = (z - z_0)^m f_0(z) \qquad \text{and} \qquad g(z) = (z - z_0)^n g_0(z)$$

where m and n are nonnegative integers, and f_0 and g_0 are functions analytic at z_0 with $f_0(z_0) \neq 0$ and $g_0(z_0) \neq 0$. Hence,

$$\frac{f(z)}{g(z)} = \frac{(z - z_0)^m f_0(z)}{(z - z_0)^n g_0(z)} = (z - z_0)^{m-n} \frac{f_0(z)}{g_0(z)} \quad , \tag{32.6}$$

and

$$\lim_{z \to z_0} \frac{f(z)}{g(z)} = \lim_{z \to z_0} (z - z_0)^{m-n} \frac{f_0(z_0)}{g_0(z_0)} \quad ,$$

which is finite if and only if $m \geq n$. Since this observation will later be used, let us make it a lemma.

Lemma 32.14

Let f and g be two functions analytic at z_0. Then

$$\lim_{z \to z_0} \frac{f(z)}{g(z)}$$

exists and is finite if and only if there is a nonnegative integer N and a $R > 0$ such that

$$\frac{f(z)}{g(z)} = (z - z_0)^N \frac{f_0(z)}{g_0(z)}$$

where f_0 and g_0 are functions analytic at z_0 with $f_0(z_0) \neq 0$ and $g_0(z_0) \neq 0$.

Partial Sum Approximations with Taylor Series

Another approach to deriving the Taylor series formula for function f analytic at a point x_0 starts with the well-known equality

$$f(x) - f(x_0) = \int_{x_0}^{x} f'(s)\,ds \quad .$$

Solving for $f(x)$ and using a "clever" integration by parts yields

$$\begin{aligned} f(x) &= f(x_0) + \int_{x_0}^{x} \underbrace{f'(s)}_{u}\,\underbrace{ds}_{dv} \\ &= f(x_0) + \left[\underbrace{f'(s)}_{u}\underbrace{(s-x_0)}_{v}\Big|_{s=x_0}^{x} - \int_{x_0}^{x} \underbrace{(s-x_0)}_{v}\,\underbrace{f''(s)\,ds}_{du} \right] \\ &= f(x_0) + f'(x)(x-x) - f'(x_0)(x_0-x) - \int_{x_0}^{x} (s-x_0)f''(s)\,ds \\ &= f(x_0) + 0 + f'(x_0)(x-x_0) - \int_{x_0}^{x} (s-x_0)f''(s)\,ds \quad . \end{aligned}$$

Repeating this again and again with similar "clever" uses of integration by parts ultimately leads to

$$f(x) = P_N(x) + E_N(x) \qquad \text{for} \quad N = 1, 2, 3, \ldots \tag{32.7a}$$

where $P_N(x)$ is the Nth*-degree Taylor polynomial,*

$$P_N(x) = \sum_{k=0}^{N} \frac{f^{(k)}(x_0)}{k!}(x-x_0)^k \quad , \tag{32.7b}$$

and $E_N(x)$ is the corresponding remainder term,

$$E_N(x) = (-1)^N \frac{1}{N!} \int_{x_0}^{x} f^{(N+1)}(s)(s-x)^N\,ds \quad . \tag{32.7c}$$

Note that $P_N(x)$ is the Nth partial sum for the power series for f about x_0, and $E_N(x)$ is the error in using this partial sum in place of $f(x)$.

Since we are assuming f is analytic at x_0, we know that

$$f(x) = \sum_{k=0}^{\infty} \frac{f^{(k)}(x_0)}{k!}(x-x_0)^k = \lim_{N\to\infty} P_N(x) \qquad \text{whenever} \quad |x-x_0| < R \quad ,$$

where R is the radius of convergence for the Taylor series. This, in turn, means that

$$\lim_{N\to\infty} E_N(x) = \lim_{N\to\infty} [f(x) - P_N(x)] = 0 \qquad \text{whenever} \quad |x-x_0| < R \quad .$$

Conversely, in theory, you can verify that f truly is analytic at x_0 and that its power series at that point has a radius of convergence of at least R by verifying that $E_N(x) \to 0$ as $N \to \infty$ whenever $x_0 - R < x < x_0 + R$.

In practice, computing $E_N(x)$ is rarely practical. Because of this, a slightly more usable error estimate in terms of upper bounds on the derivatives of f is often described in textbooks. For our purposes, let $[a, b]$ be some closed subinterval with

$$x_0 - R < a < x_0 < b < x_0 + R \quad .$$

By the analyticity of f, we know $\left|f^{(N+1)}\right|$ is continuous and, thus, has an upper bound on $[a,b]$; that is, there is a finite value M_N such that

$$\left|f^{(N+1)}(x)\right| \leq M_N \qquad \text{for} \quad a < x < b \quad .$$

Then, as you can easily verify,

$$\begin{aligned} |E_N(x)| &= \left|\frac{1}{N!}\int_{x_0}^{x} f^{(N+1)}(s)(s-x)^N \, ds\right| \\ &\leq \left|\frac{1}{N!}\int_{x_0}^{x} M_N (s-x)^N \, ds\right| = \frac{M_N}{(N+1)!}|x-x_0|^{N+1} \quad . \end{aligned}$$

The value of this estimate in showing that $\lim_{N\to\infty} E_N(x) \to 0$ depends on being able to find the appropriate upper bound M_N for each positive integer N. Unfortunately for us, finding these M_N's will not be practical. So we will not be using $E_N(x)$ to determine the analyticity of f or the radius of convergence for its power series.

So why go through this discussion of E_N? It is for this simple observation regarding a function f analytic at x_0: For any given positive integer N,

$$\frac{M_N}{(N+1)!}|x-x_0|^{N+1} \to 0 \quad \text{“quickly” as} \quad x \to x_0 \quad .$$

Hence, for any given positive integer N,

$$E_N(x) \to 0 \quad \text{“quickly” as} \quad x \to x_0 \quad ,$$

assuring us that the Nth partial sum

$$\sum_{k=0}^{N} \frac{f^{(k)}(x_0)}{k!}(x-x_0)^k$$

is a very good approximation to $f(x)$ on some (possibly small) open interval about x_0.

Additional Exercises

32.2. *Several expressions involving geometric series are given below. If the given expression is a partial sum or a convergent infinite series, compute its sum using the formulas developed for geometric series and their partial sums. If the given expression is a divergent series, say so.*

a. $\sum_{k=0}^{4}\left(\frac{1}{3}\right)^k$ **b.** $\sum_{k=0}^{8}\left(\frac{1}{3}\right)^k$ **c.** $\sum_{k=0}^{\infty}\left(\frac{1}{3}\right)^k$ **d.** $\sum_{k=5}^{\infty}\left(\frac{1}{3}\right)^k$

e. $\sum_{k=0}^{\infty}\left(-\frac{2}{3}\right)^k$ **f.** $\sum_{k=0}^{5}\left(\frac{3}{2}\right)^k$ **g.** $\sum_{k=0}^{\infty}\left(\frac{3}{2}\right)^k$ **h.** $\sum_{k=0}^{\infty}8\left(\frac{3}{7}\right)^k$

i. $\sum_{k=0}^{\infty}\left[3\left(\frac{2}{5}\right)^k - 4\left(\frac{3}{5}\right)^k\right]$ **j.** $\sum_{k=0}^{\infty}\left[3\left(\frac{1}{10}\right)^k + \frac{2}{3}\left(-\frac{3}{5}\right)^k\right]$

32.3. *Verify each of the following equations:*

a. $\displaystyle\sum_{k=0}^{\infty}\frac{1}{k+1}x^k + \sum_{k=2}^{\infty}\frac{1}{k-1}x^k = 1 + \frac{1}{2}x + \sum_{k=2}^{\infty}\frac{2k}{k^2-1}x^k$

b. $\displaystyle\sum_{k=0}^{\infty}\left(k^2+9\right)x^k - 6\sum_{k=1}^{\infty}kx^k = 9 + \sum_{k=1}^{\infty}(k-3)^2x^k$

c. $\displaystyle\sum_{k=2}^{\infty}(k-1)x^{k-2} = \sum_{n=0}^{\infty}(n+1)x^n$

d. $\displaystyle x\sum_{k=2}^{\infty}(k-1)x^{k-2} = \sum_{n=1}^{\infty}nx^n$

e. $\displaystyle\sum_{k=0}^{\infty}(k+1)x^{k+1} - \sum_{k=4}^{\infty}(k-1)x^{k-1} = x + 2x^2$

f. $\displaystyle x^3\sum_{k=0}^{\infty}a_kx^k = \sum_{n=3}^{\infty}a_{n-3}x^n$

g. $\displaystyle\left(x^2+5\right)\sum_{k=0}^{\infty}a_kx^k = 5a_0 + 5a_1x + \sum_{n=2}^{\infty}[a_{n-2}+5a_n]x^n$

h. $\displaystyle\sum_{k=2}^{\infty}k(k-1)a_kx^{k-2} - 3\sum_{k=0}^{\infty}a_kx^k = \sum_{n=0}^{\infty}\left[(n+2)(n+1)a_{n+2} - 3a_n\right]x^n$

32.4. *Rewrite each of the following expressions as a single power series centered at a point x_0, with the index being the order of each term. That is, if n is the index, then each term should be of the form*

$$[\textit{formula not involving } x] \times (x - x_0)^n \quad .$$

In most cases, $x_0 = 0$. And, in some cases, the first few terms will have to be written separately. Simplify your expressions as much as practical.

a. $\displaystyle\sum_{k=0}^{\infty}\frac{1}{k+1}x^k - \sum_{k=1}^{\infty}\frac{1}{k}x^k$

b. $\displaystyle x + \frac{1}{2}x^2 + \frac{1}{3}x^3 + \sum_{k=1}^{\infty}\frac{(-1)^k}{k}x^k$

c. $\displaystyle\sum_{k=1}^{\infty}3k^2(x-5)^{k+3}$

d. $\displaystyle x\sum_{k=2}^{\infty}k(k-1)x^{k-2}$

e. $\displaystyle(x-3)\sum_{k=2}^{\infty}k(k-1)x^{k-2}$

f. $\displaystyle x\sum_{k=2}^{\infty}k(k-1)(x-3)^{k-2}$

g. $\displaystyle\sum_{k=1}^{\infty}k^2a_kx^{k+3}$

h. $\displaystyle(x-1)^2\sum_{k=0}^{\infty}a_k(x-1)^k$

i. $\displaystyle\sum_{k=0}^{\infty}(k+1)a_kx^{k+1} - \sum_{k=4}^{\infty}(k-1)a_kx^{k-1}$

j. $\displaystyle\sum_{k=1}^{\infty}ka_kx^{k-1} + 5\sum_{k=0}^{\infty}a_kx^k$

k. $x^2 \sum_{k=2}^{\infty} k(k-1)a_k x^{k-2} \; - \; 4\sum_{k=0}^{\infty} a_k x^k$ **l.** $\sum_{k=2}^{\infty} k(k-1)a_k x^{k-2} \; - \; 3x^2 \sum_{k=0}^{\infty} a_k x^k$

32.5. *On page 581, we saw that*

$$\frac{1}{1-x} = \sum_{k=0}^{\infty} x^k \quad \text{for} \quad |x| < 1 \quad .$$

By differentiating this, find the power series about 0 *for each of the following:*

a. $\dfrac{1}{(1-x)^2}$ **b.** $\dfrac{1}{(1-x)^3}$

32.6. *Find the Taylor series about* x_0 *for each of the following:*

a. e^x *with* $x_0 = 0$ **b.** $\cos(x)$ *with* $x_0 = 0$

c. $\sin(x)$ *with* $x_0 = 0$ **d.** $\ln|x|$ *with* $x_0 = 1$

32.7 a. *Using the Taylor series formula from Theorem 32.7, find the fourth partial sum of the power series about* 0 *for*

$$f(x) = \sqrt{1+x} \quad .$$

b. *Using the results from the previous part along with a simple substitution, find the first five terms of the power series about* 0 *for*

$$g(x) = \sqrt{1-x^2} \quad .$$

c. *Let*

$$g(x) = \sqrt{1-x^2} \quad \text{and} \quad h(x) = \frac{1}{1-x^2} \quad .$$

Verify that

$$h(x) = -\frac{g'(x)}{x} \quad ,$$

and using this along with the results from the previous part, find the first four terms of the power series about 0 *for* $h(x)$.

32.8 a. *Using your favorite computer mathematics package (e.g., Maple or Mathematica), along with the Taylor series formula from Theorem 32.7, write a program/worksheet that will find the first* N *coefficients in the power series about* x_0 *for* f *where* x_0 *is any given point on the real line,* f *is any function analytic at* x_0, *and* N *is any given positive integer. Also, have your program/worksheet write out the corresponding* N^{th}*-degree partial sum of this power series. Be sure to write your program/worksheet so that* N, x_0 *and* f *are easily changed.*

b. *Use your program/worksheet with each of the following choices of* f, x_0 *and* N *to find the* N^{th}*-degree polynomial about* x_0 *for* f.

i. $f(x) = e^{2x}$ *with* $x_0 = 0$ *and* $N = 9$

ii. $f(x) = \dfrac{1}{\cos(x)}$ *with* $x_0 = 0$ *and* $N = 11$

iii. $f(x) = \sqrt{2x^2+1}$ *with* $x_0 = 2$ *and* $N = 7$

32.9. We saw that

$$\frac{1}{1-x} = \sum_{k=0}^{\infty} x^k \quad \text{for} \quad |x| < 1 \quad .$$

Replacing x with $-x$,

$$\frac{1}{1-(-x)} = \sum_{k=0}^{\infty} (-x)^k \quad \text{for} \quad |-x| < 1 \quad ,$$

gives us a power series formula for $(1+x)^{-1}$,

$$\frac{1}{1+x} = \sum_{k=0}^{\infty} (-x)^k \quad \text{for} \quad |x| < 1 \quad .$$

Find a power series representation (and its radius of convergence R) for each of the following by replacing the x in some of the "known" power series from Exercises 32.5 and 32.6, above, with a suitable formula of x, as just done above.

a. $\dfrac{1}{1-2x}$ **b.** $\dfrac{1}{1+x^2}$ **c.** $\dfrac{2}{2-x}$ **d.** $\dfrac{2}{(2-x)^2}$

e. e^{-x^2} **f.** $\sin\left(x^2\right)$

32.10. Let

$$f(x) = \begin{cases} 0 & \text{if} \quad x = 0 \\ e^{-1/x^2} & \text{if} \quad x \neq 0 \end{cases} \quad .$$

a. Verify that

$$f'(x) = \begin{cases} 0 & \text{if} \quad x = 0 \\ e^{-1/x^2} \dfrac{2}{x^3} & \text{if} \quad x \neq 0 \end{cases} \quad .$$

Note: The derivative at $x = 0$ should be computed using the basic definition

$$f'(0) = \lim_{\Delta x \to 0} \frac{f(0 + \Delta x) - f(0)}{\Delta x} \quad .$$

b. Verify that

$$f''(x) = \begin{cases} 0 & \text{if} \quad x = 0 \\ e^{-1/x^2} \dfrac{1}{x^6} \left[4 - 6x^2\right] & \text{if} \quad x \neq 0 \end{cases} \quad .$$

c. Continuing, it can be shown that, for any positive integer k,

$$f^{(k)}(x) = \begin{cases} 0 & \text{if} \quad x = 0 \\ e^{-1/x^2} \dfrac{1}{x^{3k}} p_k(x) & \text{if} \quad x \neq 0 \end{cases}$$

where $p_k(x)$ is some nonzero polynomial. Using this fact, write out the Taylor series for f about 0.

d. Why is f not analytic at 0 even though it is infinitely differentiable at 0 ?

33

Power Series Solutions I: Basic Computational Methods

When a solution to a differential equation is analytic at a point, then that solution can be represented by a power series about that point. In this and the next chapter, we will discuss when this can be expected, and how we might employ this fact to obtain usable power series formulas for the solutions to various differential equations. In this chapter, we will concentrate on two basic methods — an "algebraic method" and a "Taylor series method" — for computing our power series. Our main interest will be in the algebraic method. It is more commonly used and is the method we will extend in Chapter 35 to obtain "modified" power series solutions when we do not quite have the desired analyticity. But the algebraic method is not well suited for solving all types of differential equations, especially when the differential equations in question are not linear. For that reason (and others) we will also introduce the Taylor series method near the end of this chapter.

33.1 Basics

General Power Series Solutions

If it exists, a *power series solution* for a differential equation is just a power series formula

$$y(x) = \sum_{k=0}^{\infty} a_k (x - x_0)^k$$

for a solution y to the given differential equation in some open interval containing x_0. The series is a *general power series solution* if it describes all possible solutions in that interval.

As noted in the last chapter (Corollary 32.8 on page 590), if $y(x)$ is given by the above power series, then

$$a_0 = y(x_0) \qquad \text{and} \qquad a_1 = y'(x_0) \quad .$$

Because a general solution to a first-order differential equation normally has one arbitrary constant, we should expect a general power series solution to a first-order differential equation to also have a single arbitrary constant. And since that arbitrary constant can be determined by a given initial value $y(x_0)$, it makes sense to use a_0 as that arbitrary constant.

On the other hand, a general solution to a second-order differential equation usually has two arbitrary constants, and they are normally determined by initial values $y(x_0)$ and $y'(x_0)$. Consequently, we should expect the first two coefficients, a_0 and a_1, to assume the roles of the arbitrary constants in our general power series solutions for second-order differential equations.

The Two Methods, Briefly

The basic ideas of both the "algebraic method" and the "Taylor series method" are fairly simple.

The Algebraic Method

The *algebraic method* starts by assuming the solution y can be written as a power series

$$y(x) = \sum_{k=0}^{\infty} a_k (x - x_0)^k$$

with the a_k's being constants to be determined. This formula for y is then plugged into the differential equation. By using a lot of algebra and only a little calculus, we then "simplify" the resulting equation until it looks something like

$$\sum_{n=0}^{\infty} \left[n^{\text{th}} \text{ formula of the } a_k\text{'s} \right] x^n = 0 \quad .$$

As we saw in the last chapter, this means

$$n^{\text{th}} \text{ formula of } a_k\text{'s} = 0 \qquad \text{for} \quad n = 0, 1, 2, 3, \ldots \quad ,$$

which (as we will see) can be used to determine all the a_k's in terms of one or two arbitrary constants. Plugging these a_k's back into the series then gives the power series solution to our differential equation about the point x_0.

We will outline the details for this method in the next two sections for first- and second-order homogeneous linear differential equations

$$a(x)y' + b(x)y = 0 \qquad \text{and} \qquad a(x)y'' + b(x)y' + c(x)y = 0$$

in which the coefficients are rational functions. These are the equations for which the method is especially well suited.[1] For pedagogic reasons, we will deal with first-order equations first, and then expand our discussion to include second-order equations. It should then be clear that this approach can easily be extended to solve higher-order analogs of the equations discussed here.

The Taylor Series Method

The basic ideas behind the *Taylor series method* are even easier to describe. We simply use the given differential equation to find the values of all the derivatives of the solution $y(x)$ when $x = x_0$, and then plug these values into the formula for the Taylor series for y about x_0 (see Corollary 32.7 on page 590). Details will be laid out in Section 33.6.

[1] Recall that a rational function is a function that can be written as one polynomial divided by another polynomial. Actually, in theory at least, the algebraic method is "well suited" for a somewhat larger class of first- and second-order linear differential equations. We'll discuss this in the next chapter.

33.2 The Algebraic Method with First-Order Equations

Details of the Method

Here are the detailed steps of our algebraic method for finding a general power series solution to

$$a(x)y' + b(x)y = 0$$

assuming $a(x)$ and $b(x)$ are rational functions. To illustrate these steps, we'll find a general power series solution to

$$y' + \frac{2}{x-2}y = 0 \quad . \tag{33.1}$$

Admittedly you could solve this differential equation easier using methods from either Chapter 4 or 5, but it is a good equation for our first example.[2]

Before actually starting the method, there are two preliminary steps:

Pre-step 1: Rewrite your differential equation in *preferred form*, which is

$$A(x)y' + B(x)y = 0$$

where A and B are polynomials, preferably without common factors.

To get differential equation (33.1) into the form desired, we multiply the equation by $x - 2$. That gives us

$$(x-2)y' + 2y = 0 \quad . \tag{33.2}$$

Pre-step 2: If not already specified, choose a value for x_0. For reasons we will discuss later, x_0 should be chosen so that $A(x_0) \neq 0$. If initial conditions are given for $y(x)$ at some point, then use that point for x_0 (provided $A(x_0) \neq 0$). Otherwise, choose x_0 as convenient — which usually means choosing $x_0 = 0$.[3]

For our example, we have no initial values at any point, and the first coefficient, $x - 2$, is zero only if $x_0 = 2$. So let us choose x_0 as simply as possible; namely, $x_0 = 0$.

Now for the basic method:

Step 1: Assume a power series solution

$$y = y(x) = \sum_{k=0}^{\infty} a_k (x - x_0)^k \tag{33.3}$$

with a_0 being arbitrary and the other a_k's "to be determined", and then compute/write out the corresponding first derivative

$$y' = \frac{d}{dx}\sum_{k=0}^{\infty} a_k (x-x_0)^k$$

$$= \sum_{k=0}^{\infty} \frac{d}{dx}\left[a_k (x-x_0)^k\right] = \sum_{k=1}^{\infty} k a_k (x-x_0)^{k-1} \quad .$$

[2] Truth is, power series are rarely used to solve first-order differential equations because these equations are often more easily solved using the more direct methods developed earlier in this text. In fact, many texts don't even mention using power series with first-order equations. We're doing first-order equations here because this author likes to start as simple as possible.

[3] The requirement that $A(x_0) \neq 0$ is a slight simplification of requirements we'll develop in the next section. But "$A(x_0) \neq 0$" will suffice for now, especially if A and B are polynomials with no common factors.

(Remember that the derivative of the a_0 term is zero. Explicitly dropping this zero term in the series for y' is not necessary, but can simplify bookkeeping, later.)

Since we've already decided $x_0 = 0$, we assume

$$y = y(x) = \sum_{k=0}^{\infty} a_k x^k \quad , \tag{33.4}$$

and compute

$$y' = \frac{d}{dx}\sum_{k=0}^{\infty} a_k x^k = \sum_{k=0}^{\infty} \frac{d}{dx}\left[a_k x^k\right] = \sum_{k=1}^{\infty} k a_k x^{k-1} \quad .$$

Step 2: Plug the series for y and y' back into the differential equation and "multiply things out". (If $x_0 \neq 0$, see the comments on page 613.)

Some notes:

i. Absorb any x's from $A(x)$ and $B(x)$ into the series.

ii. Your goal is to get an equation in which zero equals the sum of a few power series about x_0.

Using the above series with the given differential equation, we get

$$\begin{aligned} 0 &= (x-2)y' + 2y \\ &= (x-2)\sum_{k=1}^{\infty} k a_k x^{k-1} + 2\sum_{k=0}^{\infty} a_k x^k \\ &= \left[x\sum_{k=1}^{\infty} k a_k x^{k-1} - 2\sum_{k=1}^{\infty} k a_k x^{k-1}\right] + 2\sum_{k=0}^{\infty} a_k x^k \\ &= \sum_{k=1}^{\infty} k a_k x^k + \sum_{k=1}^{\infty} (-2) k a_k x^{k-1} + \sum_{k=0}^{\infty} 2 a_k x^k \quad . \end{aligned}$$

Step 3: For each series in your last equation, do a change of index[4] so that each series looks like

$$\sum_{n=\text{something}}^{\infty} \big[\text{something not involving } x\big](x-x_0)^n \quad .$$

Be sure to appropriately adjust the lower limit in each series.

[4] see *Changing the Index* on page 585

In all but the second series in the example, the "change of index" is trivial ($n = k$). In the second series, we set $n = k - 1$ (equivalently, $k = n + 1$):

$$0 = \underbrace{\sum_{k=1}^{\infty} k a_k x^k}_{n=k} + \underbrace{\sum_{k=1}^{\infty} (-2) k a_k x^{k-1}}_{n=k-1} + \underbrace{\sum_{k=0}^{\infty} 2 a_k x^k}_{n=k}$$

$$= \sum_{n=1}^{\infty} n a_n x^n + \sum_{n+1=1}^{\infty} (-2)(n+1) a_{n+1} x^n + \sum_{n=0}^{\infty} 2 a_n x^n$$

$$= \sum_{n=1}^{\infty} n a_n x^n + \sum_{n=0}^{\infty} (-2)(n+1) a_{n+1} x^n + \sum_{n=0}^{\infty} 2 a_n x^n \quad .$$

Step 4: Convert the sum of series in your last equation into one big series. The first few terms will probably have to be written separately. Go ahead and simplify what can be simplified.

Since one of the series in the last equation begins with $n = 1$, we need to separate out the terms corresponding to $n = 0$ in the other series before combining series:

$$0 = \sum_{n=1}^{\infty} n a_n x^n + \sum_{n=0}^{\infty} (-2)(n+1) a_{n+1} x^n + \sum_{n=0}^{\infty} 2 a_n x^n$$

$$= \sum_{n=1}^{\infty} n a_n x^n + \left[(-2)(0+1) a_{0+1} x^0 + \sum_{n=1}^{\infty} (-2)(n+1) a_{n+1} x^n \right] + \left[2 a_0 x^0 + \sum_{n=1}^{\infty} 2 a_n x^n \right]$$

$$= [-2a_1 + 2a_0] x^0 + \sum_{n=1}^{\infty} [n a_n - 2(n+1) a_{n+1} + 2 a_n] x^n$$

$$= 2[a_0 - a_1] x^0 + \sum_{n=1}^{\infty} \big[(n+2) a_n - 2(n+1) a_{n+1} \big] x^n \quad .$$

Step 5: At this point, you have an equation basically of the form

$$\sum_{n=0}^{\infty} \big[n^{\text{th}} \text{ formula of the } a_k\text{'s} \big] (x - x_0)^n = 0 \quad ,$$

which is possible only if

$$n^{\text{th}} \text{ formula of the } a_k\text{'s} = 0 \qquad \text{for} \quad n = 0, 1, 2, 3, \ldots \quad .$$

Using this last equation:

(a) Solve for the a_k with the highest index, obtaining

$$a_{\text{highest index}} = \text{formula of } n \text{ and lower-indexed coefficients} \quad .$$

A few of these equations may need to be treated separately, but you should obtain one relatively simple formula that holds for all indices above some fixed value. This formula is a *recursion formula* for computing each coefficient from the previously computed coefficients.

(*b*) To simplify things just a little, do another change of index so that the recursion formula just derived is rewritten as

$$a_k = \text{formula of } k \text{ and lower-indexed coefficients} \quad .$$

From the last step in our example, we have

$$2[a_0 - a_1]x^0 + \sum_{n=1}^{\infty} \left[(n+2)a_n - 2(n+1)a_{n+1}\right]x^n = 0 \quad .$$

So,

$$2[a_0 - a_1] = 0 \tag{33.5a}$$

and, for $n = 1, 2, 3, 4, \ldots,$

$$(n+2)a_n - 2(n+1)a_{n+1} = 0 \quad . \tag{33.5b}$$

In equation (33.5a), a_1 *is the highest indexed* a_k *; solving for it in terms of the lower-indexed* a_k*'s (i.e.,* a_0 *) yields*

$$a_1 = a_0 \quad .$$

Equation (33.5b) also just contains two a_k*'s:* a_n *and* a_{n+1}*. Since* $n + 1 > n$*, we solve for* a_{n+1}*,*

$$a_{n+1} = \frac{n+2}{2(n+1)}a_n \quad \text{for} \quad n = 1, 2, 3, 4, \ldots \quad .$$

Letting $k = n + 1$ *(equivalently,* $n = k - 1$ *), this becomes*

$$a_k = \frac{k+1}{2k}a_{k-1} \quad \text{for} \quad k = 2, 3, 4, 5, \ldots \quad . \tag{33.6}$$

This is the recursion formula we will use.

Step 6: Use the recursion formula (and any corresponding formulas for the lower-order terms) to find all the a_k's in terms of a_0. Look for patterns!

In the last step, we saw that

$$a_1 = a_0 \quad .$$

Using this and recursion formula (33.6) with $k = 2, 3, 4, \ldots$ *(and looking for patterns), we obtain the following:*

$$a_2 = \frac{2+1}{2 \cdot 2}a_{2-1} = \frac{3}{2 \cdot 2}a_1 = \frac{3}{2 \cdot 2}a_0 \quad ,$$

$$a_3 = \frac{3+1}{2 \cdot 3}a_{3-1} = \frac{4}{2 \cdot 3}a_2 = \frac{4}{2 \cdot 3} \cdot \frac{3}{2 \cdot 2}a_0 = \frac{4}{2^3}a_0 \quad ,$$

$$a_4 = \frac{4+1}{2 \cdot 4}a_{4-1} = \frac{5}{2 \cdot 4}a_3 = \frac{5}{2 \cdot 4} \cdot \frac{4}{2^3}a_0 = \frac{5}{2^4}a_0 \quad ,$$

$$a_5 = \frac{5+1}{2 \cdot 5}a_{5-1} = \frac{6}{2 \cdot 5}a_4 = \frac{6}{2 \cdot 5} \cdot \frac{5}{2^5}a_0 = \frac{6}{2^5}a_0 \quad ,$$

$$\vdots$$

The pattern here is obvious:

$$a_k = \frac{k+1}{2^k} a_0 \qquad \text{for} \quad k = 2, 3, 4, \ldots \quad .$$

Note that this formula even gives us our $a_1 = a_0$ *equation,*

$$a_1 = \frac{1+1}{2^1} a_0 = \frac{2}{2} a_0 = a_0 \quad ,$$

and is even valid with $k = 0$*,*

$$a_0 = \frac{0+1}{2^2} a_0 = a_0 \quad .$$

So, in fact,

$$a_k = \frac{k+1}{2^k} a_0 \qquad \text{for} \quad k = 0, 1, 2, 3, 4, \ldots \quad . \tag{33.7}$$

Step 7: Using the formulas just derived for the coefficients, write out the resulting series for $y(x)$. Try to simplify it and factor out a_0.

Plugging the formula just derived for the a_k*'s into the power series assumed for* y *yields*

$$y(x) = \sum_{k=0}^{\infty} a_k x^k = \sum_{k=0}^{\infty} \frac{k+1}{2^k} a_0 x^k = a_0 \sum_{k=0}^{\infty} \frac{k+1}{2^k} x^k \quad .$$

So we have

$$\begin{aligned} y(x) &= a_0 \sum_{k=0}^{\infty} \frac{k+1}{2^k} x^k \\ &= a_0 \left[\frac{0+1}{2^0} x^0 + \frac{1+1}{2^1} x^1 + \frac{2+1}{2^2} x^2 + \frac{3+1}{2^3} x^3 + \cdots \right] \\ &= a_0 \left[1 + x + \frac{3}{4} x^2 + \frac{1}{2} x^3 + \cdots \right] \end{aligned}$$

as the series solution for our first-order differential equation (assuming it converges).

Last Step: See if you recognize the series derived as the series for some well-known function (you probably won't!).

By an amazing stroke of luck, in Exercise 32.9 d on page 602 we saw that

$$\frac{2}{(2-x)^2} = \frac{1}{2} \sum_{k=0}^{\infty} \frac{k+1}{2^k} x^k \quad .$$

So our formula for y *simplifies considerably:*

$$y(x) = a_0 \sum_{k=0}^{\infty} \frac{k+1}{2^k} x^k = a_0 \left[2 \cdot \frac{2}{(2-x)^2} \right] = \frac{4a_0}{(2-x)^2} \quad .$$

Practical Advice on Using the Method

General Comments

The method just described is a fairly straightforward procedure, at least up to the point where you are trying to "find a pattern" for the a_k's. The individual steps are, for the most part, simple and only involve elementary calculus and algebraic computations — but there are a lot of these elementary computations, and an error in just one can throw off the subsequent computations with disastrous consequences for your final answer. So be careful, write neatly, and avoid shortcuts and doing too many computations in your head. It may also be a good idea to do your work with your paper turned sideways, just to have enough room for each line of formulas.

On Finding Patterns

In computing the a_k's, we usually want to find some "pattern" described by some reasonably simple formula. In our above example, we found formula (33.7),

$$a_k = \frac{k+1}{2^k} a_0 \quad \text{for} \quad k = 0, 1, 2, 3, 4, \ldots \quad .$$

Using this formula, it was easy to write out the power series solution.

More generally, we will soon verify that the a_k's obtained by this method can all be simply related to a_0 by an expression of the form

$$a_k = \alpha_k a_0 \quad \text{for} \quad k = 0, 1, 2, 3, 4, \ldots$$

where $\alpha_0 = 1$ and the other α_k's are fixed numbers (hopefully given by some simple formula of k). In the example cited just above,

$$\alpha_k = \frac{k+1}{2^k} \quad \text{for} \quad k = 0, 1, 2, 3, 4, \ldots \quad .$$

Finding that pattern and its formula (i.e., the above mentioned α_k's) is something of an art and requires a skill that improves with practice. One suggestion is to avoid multiplying factors out. It was the author's experience that, in deriving formula (33.7), led him to leave 2^2 and 2^3 as 2^2 and 2^3, instead of as 4 and 8 — he suspected a pattern would emerge. Another suggestion is to compute "many" of the a_k's using the recursion formula before trying to identify the pattern. And once you believe you've found that pattern and derived that formula, say,

$$a_k = \frac{k+1}{2^k} a_0 \quad \text{for} \quad k = 0, 1, 2, 3, 4, \ldots \quad ,$$

test it by computing a few more a_k's using both the recursion formula directly and using your newly found formula. If the values computed using both methods don't agree, your formula is wrong. Better yet, if you are acquainted with the method of induction, use that to rigorously confirm your formula.[5]

Unfortunately, in practice, it may not be so easy to find such a pattern for your a_k's. In fact, it is quite possible to end up with a three-term (or more) recursion formula, say,

$$a_n = \frac{1}{n^2+1} a_{n-1} + \frac{2}{3n(n+3)} a_{n-2} \quad ,$$

which can make "finding patterns" quite difficult.

Even if you do see a pattern, it might be difficult to describe. In these cases, writing out a relatively simple formula for all the terms in the power series solution may not be practical. What we can still do, though, is to use the recursion formula to compute (or have a computer compute) as many terms as we think are needed for a reasonably accurate partial sum approximation.

[5] And to learn about using induction, see Section 33.7.

Terminating Series

It's worth checking your recursion formula

$$a_k = \text{formula of } k \text{ and lower-indexed coefficients}$$

to see if the right side becomes zero for some value K of k. Then

$$a_K = 0$$

and the computation of the subsequent a_k's may become especially simple. In fact, you may well have

$$a_k = 0 \qquad \text{for all} \quad k \geq K \quad .$$

This, essentially, "terminates" the series and gives you a polynomial solution — something that's usually easier to handle than a true infinite series solution.

!►Example 33.1: *Consider finding the power series solution about $x_0 = 0$ to*

$$\left(x^2 + 1\right) y' - 4xy = 0 \quad .$$

It is already in the right form. So, following the procedure, we let

$$y(x) = \sum_{k=0}^{\infty} a_k (x - x_0)^k = \sum_{k=0}^{\infty} a_k x^k \quad ,$$

and 'compute'

$$y'(x) = \frac{d}{dx} \sum_{k=0}^{\infty} a_k x^k = \sum_{k=0}^{\infty} \frac{d}{dx} \left[a_k x^k \right] = \sum_{k=1}^{\infty} a_k k x^{k-1} \quad .$$

Plugging this into the differential equation and carrying out the index manipulation and algebra of our method:

$$\begin{aligned}
0 &= \left(x^2 + 1\right) y' - 4xy \\
&= \left(x^2 + 1\right) \sum_{k=1}^{\infty} a_k k x^{k-1} - 4x \sum_{k=0}^{\infty} a_k x^k \\
&= x^2 \sum_{k=1}^{\infty} a_k k x^{k-1} + 1 \sum_{k=1}^{\infty} a_k k x^{k-1} - 4x \sum_{k=0}^{\infty} a_k x^k \\
&= \underbrace{\sum_{k=1}^{\infty} a_k k x^{k+1}}_{n = k+1} + \underbrace{\sum_{k=1}^{\infty} a_k k x^{k-1}}_{n = k-1} - \underbrace{\sum_{k=0}^{\infty} 4 a_k x^{k+1}}_{n = k+1} \\
&= \sum_{n=2}^{\infty} a_{n-1} (n-1) x^n + \sum_{n=0}^{\infty} a_{n+1} (n+1) x^n - \sum_{n=1}^{\infty} 4 a_{n-1} x^n
\end{aligned}$$

$$= \sum_{n=2}^{\infty} a_{n-1}(n-1)x^n + \left[a_{0+1}(0+1)x^0 + a_{1+1}(1+1)x^1 + \sum_{n=2}^{\infty} a_{n+1}(n+1)x^n \right]$$

$$- \left[4a_{1-1}x^1 + \sum_{n=2}^{\infty} 4a_{n-1}x^n \right]$$

$$= a_1x^0 + [2a_2 - 4a_0]\,x^1 + \sum_{n=2}^{\infty} \left[a_{n-1}(n-1) + a_{n+1}(n+1) - 4a_{n-1} \right] x^n$$

$$= a_1x^0 + [2a_2 - 4a_0]\,x^1 + \sum_{n=2}^{\infty} \left[(n+1)a_{n+1} + (n-5)a_{n-1} \right] x^n \quad .$$

Remember, the coefficient in each term must be zero. From the x^0 *term, we get*

$$a_1 = 0 \quad .$$

From the x^1 *term, we get*

$$2a_2 - 4a_0 = 0 \quad .$$

And for $n \geq 2$*, we have*

$$(n+1)a_{n+1} + (n-5)a_{n-1} = 0 \quad .$$

Solving each of the above for the a_k *with the highest index, we get*

$$a_1 = 0 \quad ,$$

$$a_2 = 2a_0$$

and

$$a_{n+1} = \frac{5-n}{n+1} a_{n-1} \qquad \text{for} \quad n = 2, 3, 4, \ldots \quad .$$

Letting $k = n+1$ *then converts the last equation to the recursion formula*

$$a_k = \frac{6-k}{k} a_{k-2} \qquad \text{for} \quad k = 3, 4, 5, \ldots \quad .$$

Now, using our recursion formula, we see that

$$a_3 = \frac{6-3}{3} a_{3-2} = \frac{3}{3} a_1 = \frac{1}{2} \cdot 0 = 0 \quad ,$$

$$a_4 = \frac{6-4}{4} a_{4-2} = \frac{2}{4} a_2 = \frac{1}{2} \cdot 2a_0 = a_0 \quad ,$$

$$a_5 = \frac{6-5}{5} a_{5-2} = \frac{1}{5} a_3 = \frac{1}{5} \cdot 0 = 0 \quad ,$$

$$a_6 = \frac{6-6}{6} a_{6-2} = \frac{0}{6} a_4 = 0 \quad ,$$

$$a_7 = \frac{6-7}{7} a_{7-2} = -\frac{1}{7} a_5 = -\frac{1}{7} \cdot 0 = 0 \quad ,$$

$$a_8 = \frac{6-8}{8} a_{8-2} = -\frac{2}{8} a_6 = -\frac{1}{4} \cdot 0 = 0 \quad ,$$

$$\vdots$$

Clearly, the vanishing of both a_5 *and* a_6 *means that the recursion formula will give us*

$$a_k = 0 \qquad \textit{whenever} \quad k > 4 \quad .$$

Thus,

$$\begin{aligned} y(x) &= \sum_{k=0}^{\infty} a_k x^k \\ &= a_0 + a_1 x + a_2 x^2 + a_3 x^3 + a_4 x^4 + a_5 x^5 + a_6 x^6 + a_7 x^7 + \cdots \\ &= a_0 + 0x + 2a_0 x^2 + 0x^3 + a_0 x^4 + 0x^5 + 0x^6 + 0x^7 + \cdots \\ &= a_0 \qquad + 2a_0 x^2 \qquad + a_0 x^4 \quad . \end{aligned}$$

That is, the power series for y *reduces to the polynomial*

$$y(x) = a_0 \left[1 + 2x^2 + x^4 \right] \quad .$$

If $x_0 \neq 0$

The computations in our procedure (and the others we'll develop) tend to get a little messier when $x_0 \neq 0$, and greater care needs to be taken. In particular, before you "multiply things out" in step 2, you should rewrite your polynomials $A(x)$ and $B(x)$ in terms of $(x - x_0)$ instead of x to better match the terms in the series. For example, if

$$A(x) = x^2 + 2 \qquad \text{and} \qquad x_0 = 1 \quad ,$$

then rewrite $A(x)$ as follows:

$$\begin{aligned} A(x) &= [(x-1)+1]^2 + 2 \\ &= \left[(x-1)^2 + 2(x-1) + 1 \right] + 2 = (x-1)^2 + 2(x-1) + 3 \quad . \end{aligned}$$

Alternatively (and probably better), you can just convert the differential equation

$$A(x)y' + B(x)y = 0 \tag{33.8a}$$

using the change of variables $X = x - x_0$. That is, first set

$$Y(X) = y(x) \qquad \text{with} \quad X = x - x_0$$

and then rewrite the differential equation for $y(x)$ in terms of Y and X. After noting that $x = X + x_0$ and that (via the chain rule)

$$y'(x) = \frac{d}{dx}[y(x)] = \frac{d}{dx}[Y(X)] = \frac{dY}{dX}\frac{dX}{dx} = \frac{dY}{dX}\frac{d}{dx}[x - x_0] = \frac{dY}{dX} = Y'(X) \quad ,$$

we see that this converted differential equation is simply

$$A(X + x_0)Y' + B(X + x_0)Y = 0 \quad . \tag{33.8b}$$

Consequently, if we can find a general power series solution

$$Y(X) = \sum_{k=0}^{\infty} a_k X^k$$

to the converted differential equation (equation (33.8b)), we can then generate the corresponding general power series to the original equation (equation (33.8a)) by rewriting X in terms of x,

$$y(x) = Y(X) = Y(x - x_0) = \sum_{k=0}^{\infty} a_k (x - x_0)^k \quad .$$

!►Example 33.2: *Consider the problem of finding the power series solution about $x_0 = 3$ for*

$$\left(x^2 - 6x + 10\right) y' + (12 - 4x)y = 0 \quad .$$

Proceeding as suggested, we let

$$Y(X) = y(x) \qquad \text{with} \quad X = x - 3 \quad .$$

Then $x = X + 3$, and

$$\left(x^2 - 6x + 10\right) y' + (12 - 4x)y = 0$$

$$\hookrightarrow \qquad \left([X+3]^2 - 6[X+3] + 10\right) Y' + (12 - 4[X+3])Y = 0 \quad .$$

After a bit of simple algebra, this last equation simplifies to

$$\left(X^2 + 1\right) Y' - 4XY = 0 \quad ,$$

which, by an amazing stroke of luck, is the differential equation we just dealt with in Example 33.1 (only now written using capital letters). From that example, we know

$$Y(X) = a_0 \left[1 + 2X^2 + X^4\right] \quad .$$

Thus,

$$y(x) = Y(X) = Y(x - 3) = a_0 \left[1 + 2(x-3)^2 + (x-3)^4\right] \quad .$$

Initial-Value Problems (and Finding Patterns, Again)

The method just described yields a power series solution

$$y(x) = \sum_{k=0}^{\infty} a_k (x - x_0)^k = a_0 + a_1(x - x_0) + a_2(x - x_0)^2 + a_3(x - x_0)^3 + \cdots$$

in which a_0 is an arbitrary constant. Remember,

$$y(x_0) = a_0 \quad .$$

So the above general series solution for

$$A(x)y' + B(x)y = 0$$

becomes the solution to the initial-value problem

$$A(x)y' + B(x)y = 0 \qquad \text{with} \quad y(x_0) = y_0$$

if we simply replace the arbitrary constant a_0 with the value y_0.

Along these lines, it is worth recalling that we are dealing with first-order homogeneous linear differential equations, and that the general solution to any such equation can be given as an arbitrary constant times any nontrivial solution. In particular, we can write the general solution y to any given first-order homogeneous linear differential equation as

$$y(x) = a_0 y_1(x)$$

where a_0 is an arbitrary constant and y_1 is the particular solution satisfying the initial condition $y(x_0) = 1$. So if our solutions can be written as power series about x_0, then there is a particular power series solution

$$y_1(x) = \sum_{k=0}^{\infty} \alpha_k (x - x_0)^k$$

where $\alpha_0 = y_1(x_0) = 1$ and the other α_k's are fixed numbers (hopefully given by some simple formula of k). It then follows that the general solution y is given by

$$y(x) = a_0 y_1(x_0) = a_0 \sum_{k=0}^{\infty} \alpha_k (x - x_0)^k = \sum_{k=0}^{\infty} a_k (x - x_0)^k$$

where

$$a_k = \alpha_k a_0 \qquad \text{for} \quad k = 0, 1, 2, 3, \ldots \quad ,$$

just as was claimed a few pages ago when we discussed "finding patterns". (This also confirms that we will always be able to factor out the a_0 in our series solutions.)

One consequence of these observations is that, instead of assuming a solution of the form

$$\sum_{k=0}^{\infty} a_k (x - x_0)^k \qquad \text{with} \quad a_0 \text{ arbitrary}$$

in the first step of our method, we could assume a solution of the form

$$\sum_{k=0}^{\infty} \alpha_k (x - x_0)^k \qquad \text{with} \quad \alpha_0 = 1 \quad ,$$

and then just multiply the series obtained by an arbitrary constant a_0. In practice, though, this approach is no simpler than that already outlined in the steps of our algebraic method.

33.3 Validity of the Algebraic Method for First-Order Equations

Our algebraic method will certainly lead to a general solution of the form

$$y(x) = \sum_{k=0}^{\infty} a_k (x - x_0)^k \qquad \text{with } a_0 \text{ arbitrary} \quad ,$$

provided such a general solution exists. But what assurance do we have that such solutions exist? And what about the radius of convergence? What good is a formula for a solution if we don't know the interval over which that formula is valid? And while we are asking these sorts of questions, why do we insist that $A(x_0) \neq 0$ in pre-step 2?

Let's see if we can at least partially answer these questions.

Non-Existence of Power Series Solutions

Let a and b be functions on an interval (α, β) containing some point x_0, and let y be any function on that interval satisfying

$$a(x)y' + b(x)y = 0 \quad \text{with} \quad y(x_0) \neq 0 \quad .$$

For the moment, assume this differential equation has a general power series solution about x_0 valid on (α, β). This means there are finite numbers a_0, a_1, a_2, ... such that

$$y(x) = \sum_{k=0}^{\infty} a_k (x - x_0)^k \quad \text{for} \quad \alpha < x < \beta \quad .$$

In particular, there are finite numbers a_0 and a_1 with

$$a_0 = y(x_0) \neq 0 \quad \text{and} \quad a_1 = y'(x_0) \quad .$$

Also observe that we can algebraically solve our differential equation for y', obtaining

$$y'(x) = -\frac{b(x)}{a(x)} y(x) \quad .$$

Thus,

$$a_1 = y'(x_0) = -\frac{b(x_0)}{a(x_0)} y(x_0) = -\frac{b(x_0)}{a(x_0)} a_0 \quad , \tag{33.9}$$

provided the above fraction is a finite number — which will certainly be the case if $a(x)$ and $b(x)$ are polynomials with $a(x_0) \neq 0$.

More generally, the fraction in equation (33.9) might be indeterminant. To get around this minor issue, we'll take limits:

$$\begin{aligned} a_1 = y'(x_0) = \lim_{x \to x_0} y'(x) &= \lim_{x \to x_0} \left[-\frac{b(x)}{a(x)} y(x) \right] \\ &= -\lim_{x \to x_0} \left[\frac{b(x)}{a(x)} \right] y(x_0) = -\lim_{x \to x_0} \left[\frac{b(x)}{a(x)} \right] a_0 \quad . \end{aligned}$$

Solving for the limit, we then have

$$\lim_{x \to x_0} \frac{b(x)}{a(x)} = -\frac{a_1}{a_0} \quad .$$

This means the above limit must exist and be a well-defined finite number *whenever* the solution y can be given by the above power series. And if you think about what this means when the above limit does *not* exist as a finite number, you get:

Lemma 33.1 (nonexistence of a power series solution)
Let a and b be two functions on some interval containing a point x_0. If

$$\lim_{x \to x_0} \frac{b(x)}{a(x)}$$

does not exist as a finite number, then

$$a(x)y' + b(x)y = 0$$

does not have a general power series solution about x_0 with an arbitrary constant term.

!▶Example 33.3: *Consider the differential equation*

$$(x-2)y' + 2y = 0 \quad .$$

This equation is

$$a(x)y' + b(x)y = 0$$

with

$$a(x) = (x-2) \quad \text{and} \quad b(x) = 2 \quad .$$

Note that these are polynomials without common factors but with $a(2) = 0$. *Consequently,*

$$\lim_{x \to 2} \frac{b(x)}{a(x)} = \lim_{x \to 2} \frac{2}{(x-2)} = \frac{2}{0} \quad ,$$

which is certainly not a finite number. Lemma 33.1 then tells us not to bother looking for a solution of the form

$$y(x) = \sum_{k=0}^{\infty} a_k (x-2)^k \quad \text{with } a_0 \text{ arbitrary} \quad .$$

No such solution exists.

Singular and Ordinary Points, and the Radius of Analyticity

Because of the 'singular' behavior just noted, we refer to any point z_0 for which

$$\lim_{z \to z_0} \frac{b(z)}{a(z)}$$

is *not* a well-defined finite number as a *singular point* for the differential equation

$$a(x)y' + b(x)y = 0 \quad .$$

Note that we used " z " in this definition, suggesting that we may be considering points on the complex plane as well. This can certainly be the case when a and b are rational functions. And if a and b are rational functions, then the nonsingular points (i.e., the points that are not singular points) are traditionally referred to as *ordinary points* for the above differential equation.

A related concept is that of the "radius of analyticity". If the differential equation has at least one singular point, then the *radius of analyticity* (for the differential equation about z_0) is the distance between z_0 and the singular point z_s closest to z_0 (which could be z_0 , itself). If the equation has no singular points, then we define the equation's *radius of analyticity* (about z_0) to be $+\infty$.

In using the above terminology, keep in mind that the singular point z_s of most interest (i.e., the one closest to z_0) might not be on the real line but some point in the complex plane off of the real line.

Validity of the Algebraic Method

We just saw that our algebraic method for finding power series solutions about x_0 will fail if x_0 is a singular point. On the other hand, there is a theorem assuring us that the method will succeed when x_0 is an ordinary point for our differential equation and even giving us a good idea of the interval over which the general power series solution is valid. Here is that theorem:

Theorem 33.2 (existence of power series solutions)
Let x_0 be an ordinary point on the real line for

$$a(x)y' + b(x)y = 0$$

where a and b are rational functions. Then this differential equation has a general power series solution

$$y(x) = \sum_{k=0}^{\infty} a_k (x - x_0)^k$$

in which a_0 is the arbitrary constant. Moreover, this solution is valid at least on the interval $(x_0 - R, x_0 + R)$ where R is the radius of analyticity about x_0 for the differential equation.

The proof of this theorem requires a good deal more work than did our derivation of the previous lemma. We will save that labor for the next chapter.

Identifying Singular and Ordinary Points

The basic approach to identifying a point z_0 as being either a singular or ordinary point for

$$a(x)y' + b(x)y = 0$$

is to look at the limit

$$\lim_{z \to z_0} \frac{b(z)}{a(z)} \quad .$$

If the limit is a finite number, x_0 is an ordinary point; otherwise, x_0 is a singular point. And if you think about how this limit is determined by the values of $a(z)$ and $b(z)$ as $z \to z_0$, you'll derive the shortcuts listed in the next lemma.

Lemma 33.3 (tests for ordinary/singular points)
Let z_0 be a point in the complex plane, and consider the differential equation

$$a(x)y' + b(x)y = 0$$

where a and b are rational functions. Then

1. *If $a(z_0)$ and $b(z_0)$ are both finite numbers with $a(z_0) \neq 0$, then z_0 is an ordinary point for the differential equation.*

2. *If $a(z_0)$ and $b(z_0)$ are both finite numbers with $a(z_0) = 0$ and $b(z_0) \neq 0$, then z_0 is a singular point for the differential equation.*

3. *If $a(z_0)$ is a finite nonzero number, and*

$$\lim_{z \to z_0} |b(z)| = \infty \quad ,$$

then z_0 is a singular point for the differential equation.

4. *If $b(z_0)$ is a finite number, and*

$$\lim_{z \to z_0} |a(z)| = \infty \quad ,$$

then z_0 is an ordinary point for the differential equation.

Applying the above to the differential equations of interest here (rewritten in the form recommended for the algebraic method) yields the following corollary.

Corollary 33.4

Let $A(x)$ and $B(x)$ be polynomials having no factors in common. Then a point z_0 on the complex plane is a singular point for

$$A(x)y' + B(x)y = 0$$

if and only if $A(z_0) = 0$.

!► Example 33.4: *To first illustrate the algebraic method, we used*

$$y' + \frac{2}{x-2}y = 0 \quad ,$$

which we rewrote as

$$(x-2)y' + 2y = 0 \quad .$$

Now

$$A(z_s) = z_s - 2 = 0 \quad \Longleftrightarrow \quad z_s = 2 \quad .$$

So this differential equation has just one singular point, $z_s = 2$. Any $x_0 \neq 2$ is then an ordinary point for the differential equation, and the corresponding radius of analyticity is

$$R_{x_0} = \text{distance from } x_0 \text{ to } z_s = |z_s - x_0| = |2 - x_0| \quad .$$

Theorem 33.2 then assures us that, about any $x_0 \neq 2$, the general solution to our differential equation has a power series formula, and its radius of convergence is at least equal to $|2 - x_0|$. In particular, the power series we found,

$$y = a_0 \sum_{k=0}^{\infty} \frac{k+1}{2^k} x^k \quad ,$$

is centered at $x_0 = 0$. So the corresponding radius of analyticity is

$$R = |2 - 0| = 2$$

and our theorems assure us that our series solution is valid at least on the interval $(x_0 - R, x_0 + R) = (-2, 2)$.

In this regard, let us note the following:

1. *If $|x| \geq 2$, then the terms of our power series solution*

$$y = a_0 \sum_{k=0}^{\infty} \frac{k+1}{2^k} x^k = a_0 \sum_{k=0}^{\infty} (k+1) \left(\frac{x}{2}\right)^k \quad ,$$

clearly increase in magnitude as k increases. Hence, this series diverges whenever $|x| \geq 2$. So, in fact, the radius of convergence is 2, and our power series solution is only valid on $(-2, 2)$.

2. *As we observed on page 609, the above power series is the power series about x_0 for*

$$y(x) = \frac{4a_0}{(2-x)^2} \quad .$$

But you can easily verify that this simple formula gives us a valid general solution to our differential equation on any interval not containing the singular point $x = 2$, not just $(-2, 2)$.

The last example shows that a power series for a solution may be valid over a smaller interval than the interval of validity for another formula for that solution. Of course, finding that more general formula may not be so easy, especially after we start dealing with higher-order differential equations.

The next example illustrates something slightly different, namely that the radius of convergence for a power series solution can, sometimes, be much larger than the corresponding radius of analyticity for the differential equation.

!►Example 33.5: *In Example 33.2, we considered the problem of finding the power series solution about $x_0 = 3$ for*

$$\left(x^2 - 6x + 10\right) y' + (12 - 4x)y = 0 \quad .$$

Any singular point z for this differential equation is given by

$$z^2 - 6z + 10 = 0 \quad .$$

Using the quadratic formula, we see that we have two singular points z_+ and z_- given by

$$z_{\pm} = \frac{6 \pm \sqrt{(-6)^2 - 4 \cdot 10}}{2} = 3 \pm 1i \quad .$$

The radius of analyticity about $x_0 = 3$ for our differential equation is the distance between each of these singular points and $x_0 = 3$,

$$|z_{\pm} - x_0| = |[3 \pm 1i] - 3| = |\pm i| = 1 \quad .$$

So the radius of convergence for our series is at least 1, which means that our series solution is valid on at least the interval

$$(x_0 - R, x_0 + R) = (3 - 1, 3 + 1) = (2, 4) \quad .$$

Recall, however, that Example 33.2 demonstrated the possibility of a "terminating series", and that our series solution to the above differential equation actually reduced to the polynomial

$$y(x) = a_0 \left[1 + 2x^2 + x^4\right] \quad ,$$

which is easily verified to be a valid solution on the entire real line $(-\infty, \infty)$, not just $(2, 4)$.

33.4 The Algebraic Method with Second-Order Equations

Extending the algebraic method to deal with second-order differential equations is straightforward. The only real complication (aside from the extra computations required) comes from the fact that our solutions will now involve two arbitrary constants instead of one, and that complication won't be particularly troublesome.

Details of the Method

Our goal, now, is to find a general power series solution to

$$a(x)y'' + b(x)y' + c(x)u = 0$$

assuming $a(x)$, $b(x)$ and $c(x)$ are rational functions. As hinted above, the procedure given here is very similar to that given in the previous section. Because of this, some of the steps will not be given in the same detail as before.

To illustrate the method, we will find a power series solution to

$$y'' - xy = 0 \quad . \tag{33.10}$$

This happens to be *Airy's equation.* It is a famous equation and *cannot* be easily solved by any method we've discussed earlier in this text.

Again, we have two preliminary steps:

Pre-step 1: Get the differential equation into *preferred form*, which is

$$A(x)y'' + B(x)y' + C(x)y = 0$$

where $A(x)$, $B(x)$ and $C(x)$ are polynomials, preferably with no factors shared by all three.

Our example is already in the desired form.

Pre-step 2: If not already specified, choose a value for x_0 such that $A(x_0) \neq 0$. If initial conditions are given for $y(x)$ at some point, then use that point for x_0 (provided $A(x_0) \neq 0$). Otherwise, choose x_0 as convenient — which usually means choosing $x_0 = 0$.[6]

For our example, we have no initial values at any point, so we choose x_0 as simply as possible; namely, $x_0 = 0$.

Now for the basic method:

Step 1: Assume

$$y = y(x) = \sum_{k=0}^{\infty} a_k(x - x_0)^k \tag{33.11}$$

with a_0 and a_1 being arbitrary and the other a_k's "to be determined", and then compute/write out the corresponding series for the first two derivatives,

$$y' = \sum_{k=0}^{\infty} \frac{d}{dx}\left[a_k(x - x_0)^k\right] = \sum_{k=1}^{\infty} ka_k(x - x_0)^{k-1}$$

and

$$y'' = \sum_{k=1}^{\infty} \frac{d}{dx}\left[ka_k(x - x_0)^{k-1}\right] = \sum_{k=2}^{\infty} k(k-1)a_k(x - x_0)^{k-2} \quad .$$

Step 2: Plug these series for y, y' and y'' back into the differential equation and "multiply things out" to get zero equaling the sum of a few power series about x_0.

[6] Again, the requirement that $A(x_0) \neq 0$ is a simplification of requirements we'll develop in the next section. But "$A(x_0) \neq 0$" will suffice for now, especially if A, B and C are polynomials with no factors shared by all three.

Step 3: For each series in your last equation, do a change of index so that each series looks like

$$\sum_{n=\text{something}}^{\infty} \big[\text{something not involving } x\big](x - x_0)^n \quad .$$

Step 4: Convert the sum of series in your last equation into one big series. The first few terms may have to be written separately. Simplify what can be simplified.

Since we've already decided $x_0 = 0$ in our example, we let

$$y = y(x) = \sum_{k=0}^{\infty} a_k x^k \quad , \tag{33.12}$$

and "compute"

$$y' = \sum_{k=0}^{\infty} \frac{d}{dx}\left[a_k x^k\right] = \sum_{k=1}^{\infty} k a_k x^{k-1}$$

and

$$y'' = \sum_{k=1}^{\infty} \frac{d}{dx}\left[k a_k x^{k-1}\right] = \sum_{k=2}^{\infty} k(k-1) a_k x^{k-2} \quad .$$

Plugging these into the given differential equation and carrying out the other steps stated above then yield the following sequence of equalities:

$$\begin{aligned}
0 &= y'' - xy \\
&= \sum_{k=2}^{\infty} k(k-1) a_k x^{k-2} - x \sum_{k=0}^{\infty} a_k x^k \\
&= \underbrace{\sum_{k=2}^{\infty} k(k-1) a_k x^{k-2}}_{n\,=\,k-2} + \underbrace{\sum_{k=0}^{\infty} (-1) a_k x^{k+1}}_{n\,=\,k+1} \\
&= \sum_{n=0}^{\infty} (n+2)(n+1) a_{n+2} x^n + \sum_{n=1}^{\infty} (-1) a_{n-1} x^n \\
&= (0+2)(0+1) a_{0+2} x^0 + \sum_{n=1}^{\infty} (n+2)(n+1) a_{n+2} x^n + \sum_{n=1}^{\infty} (-1) a_{n-1} x^n \\
&= 2a_2 x^0 + \sum_{n=1}^{\infty} [(n+2)(n+1) a_{n+2} - a_{n-1}] x^n \quad .
\end{aligned}$$

Step 5: At this point, you will have an equation of the basic form

$$\sum_{n=0}^{\infty} \big[n^{\text{th}} \text{ formula of the } a_k\text{'s}\big](x - x_0)^n = 0 \quad .$$

Now:

(a) Solve

$$n^{\text{th}} \text{ formula of the } a_k\text{'s} = 0 \qquad \text{for} \quad n = 0,\ 1,\ 2,\ 3,\ 4,\ \ldots$$

for the a_k with the highest index,

$$a_{\text{highest index}} = \text{formula of } n \text{ and lower-indexed coefficients} \quad .$$

Again, a few of these equations may need to be treated separately, but you will also obtain a relatively simple formula that holds for all indices above some fixed value. This is a *recursion formula* for computing each coefficient from previously computed coefficients.

(b) Using another change of index, rewrite the recursion formula just derived so that it looks like

$$a_k = \text{formula of } k \text{ and lower-indexed coefficients} \quad .$$

From the previous step in our example, we have

$$2a_2x^0 + \sum_{n=1}^{\infty}[(n+2)(n+1)a_{n+2} - a_{n-1}]x^n = 0 \quad .$$

So

$$2a_2 = 0 \quad ,$$

and, for $n = 1,\ 2,\ 3,\ 4,\ \ldots,$

$$(n+2)(n+1)a_{n+2} - a_{n-1} = 0 \quad .$$

The first tells us that

$$a_2 = 0 \quad .$$

Solving the second for a_{n+2} *yields the recursion formula*

$$a_{n+2} = \frac{1}{(n+2)(n+1)}a_{n-1} \qquad \text{for} \quad n = 1,\ 2,\ 3,\ 4,\ \ldots \quad .$$

Letting $k = n+2$ *(equivalently,* $n = k-2$ *), this becomes the recursion formula we will use,*

$$a_k = \frac{1}{k(k-1)}a_{k-3} \qquad \text{for} \quad k = 3,\ 4,\ 5,\ 6,\ \ldots \quad . \tag{33.13}$$

Step 6: Use the recursion formula (and any corresponding formulas for the lower-order terms) to find all the a_k's in terms of a_0 and a_1. Look for patterns!

We already saw that

$$a_2 = 0 \quad .$$

Using this and recursion formula (33.13) with $k = 3,\ 4,\ \ldots$ *(and looking for patterns), we see that*

$$a_3 = \frac{1}{3(3-1)}a_{3-3} = \frac{1}{3 \cdot 2}a_0 \quad ,$$

$$a_4 = \frac{1}{4(4-1)}a_{4-3} = \frac{1}{4 \cdot 3}a_1 \quad ,$$

$$a_5 = \frac{1}{5(5-1)}a_{5-3} = \frac{1}{5 \cdot 4}a_2 = \frac{1}{5 \cdot 4} \cdot 0 = 0 \quad ,$$

$$a_6 = \frac{1}{6(6-1)}a_{6-3} = \frac{1}{6 \cdot 5}a_3 = \frac{1}{6 \cdot 5} \cdot \frac{1}{3 \cdot 2}a_0 \quad ,$$

$$a_7 = \frac{1}{7(7-1)}a_{7-3} = \frac{1}{7 \cdot 6}a_4 = \frac{1}{7 \cdot 6} \cdot \frac{1}{4 \cdot 3}a_1 \quad ,$$

$$a_8 = \frac{1}{8(8-1)}a_{8-3} = \frac{1}{8 \cdot 7}a_5 = \frac{1}{8 \cdot 7} \cdot \frac{1}{5 \cdot 4} \cdot 0 = 0 \quad ,$$

$$a_9 = \frac{1}{9(9-1)}a_{9-3} = \frac{1}{9 \cdot 8}a_6 = \frac{1}{9 \cdot 8} \cdot \frac{1}{6 \cdot 5} \cdot \frac{1}{3 \cdot 2}a_0 \quad ,$$

$$a_{10} = \frac{1}{10(10-1)}a_{10-3} = \frac{1}{10 \cdot 9}a_7 = \frac{1}{10 \cdot 9} \cdot \frac{1}{7 \cdot 6} \cdot \frac{1}{4 \cdot 3}a_1 \quad ,$$

$$\vdots$$

There are three patterns here. The simplest is

$$a_k = 0 \qquad \text{when} \quad k = 2,\ 5,\ 8,\ 11,\ \dots$$

The other two are more difficult to describe. Look carefully and you'll see that the denominators are basically $k!$ with every third factor removed. If $k = 3$, 6, 9, ..., then

$$a_k = \frac{1}{(2 \cdot 3)(5 \cdot 6)(8 \cdot 9) \cdots ([k-1] \cdot k)}a_0 \quad .$$

If $k = 4$, 7, 10, ..., then

$$a_k = \frac{1}{(3 \cdot 4)(6 \cdot 7)(9 \cdot 10) \cdots ([k-1] \cdot k)}a_1 \quad .$$

Let us observe that we can use the change of indices $k = 3n$ and $k = 3n + 1$ to rewrite the last two expressions as

$$a_{3n} = \frac{1}{(2 \cdot 3)(5 \cdot 6) \cdots ([3n-1] \cdot 3n)}a_0 \qquad \text{for} \quad n = 1,\ 2,\ 3,\ \dots$$

and

$$a_{3n+1} = \frac{1}{(3 \cdot 4)(6 \cdot 7) \cdots (3n \cdot [3n+1])}a_1 \qquad \text{for} \quad n = 1,\ 2,\ 3,\ \dots \quad .$$

Step 7: Using the formulas just derived for the coefficients, write out the resulting series for $y(x)$. Try to simplify it to a linear combination of two power series, $y_1(x)$ and $y_2(x)$, with $y_1(x)$ multiplied by a_0 and $y_2(x)$ multiplied by a_1.

Plugging the formulas just derived for the a_k's into the power series assumed for y, we get

$$y(x) = \sum_{k=0}^{\infty} a_k x^k$$

$$= a_0 + a_1 x + a_2 x^2 + a_3 x^3 + a_4 x^4 + a_5 x^5 + a_6 x^6 + \cdots$$

$$= \left[a_0 + a_3x^3 + a_6x^6 + a_9x^9 + \cdots\right]$$
$$+ \left[a_1 + a_4x^4 + a_7x^7 + a_{10}x^{10} + \cdots\right]$$
$$+ \left[a_2 + a_5x^5 + a_8x^8 + a_{11}x^{11} + \cdots\right]$$

$$= \left[a_0 + a_3x^3 + a_6x^6 + \cdots + a_{3n}x^{3n} + \cdots\right]$$
$$+ \left[a_1 + a_4x^4 + a_7x^7 + \cdots + a_{3n+1}x^{3n+1} + \cdots\right]$$
$$+ \left[a_2 + a_5x^5 + a_8x^8 + \cdots + a_{3n+2}x^{3n+2} + \cdots\right]$$

$$= \left[a_0 + \frac{1}{3 \cdot 2}a_0x^3 + \cdots + \frac{1}{(2 \cdot 3)(5 \cdot 6) \cdots ([3n-1] \cdot 3n)}a_0x^{3n} + \cdots\right]$$
$$+ \left[a_1x + \frac{1}{4 \cdot 3}a_1x^4 + \cdots \right.$$
$$\left. + \frac{1}{(3 \cdot 4)(6 \cdot 7) \cdots (3n \cdot [3n+1])}a_1x^{3n+1} + \cdots\right]$$
$$+ \left[0 + 0x^5 + 0x^8 + 0x^{11} + \cdots\right]$$

$$= a_0\left[1 + \frac{1}{3 \cdot 2}x^3 + \cdots + \frac{1}{(2 \cdot 3)(5 \cdot 6) \cdots ([3n-1] \cdot 3n)}x^{3n} + \cdots\right]$$
$$+ a_1\left[x + \frac{1}{4 \cdot 3}x^4 + \cdots + \frac{1}{(3 \cdot 4)(6 \cdot 7) \cdots (3n \cdot [3n+1])}x^{3n+1} + \cdots\right]$$

So,

$$y(x) = a_0y_1(x) + a_1y_2(x) \tag{33.14a}$$

where

$$y_1(x) = 1 + \sum_{n=1}^{\infty} \frac{1}{(2 \cdot 3)(5 \cdot 6) \cdots ([3n-1] \cdot 3n)}x^{3n} \tag{33.14b}$$

and

$$y_2(x) = x + \sum_{n=1}^{\infty} \frac{1}{(3 \cdot 4)(6 \cdot 7) \cdots (3n \cdot [3n+1])}x^{3n+1} \quad . \tag{33.14c}$$

Last Step: See if you recognize either of the series derived as the series for some well-known function (you probably won't!).

It is unlikely that you have ever seen the above series before. So we cannot rewrite our power series solutions more simply in terms of better-known functions.

Practical Advice on Using the Method

The advice given for using this method with first-order equations certainly applies when using this method for second-order equations. All that can be added is that even greater diligence is needed in the individual computations. Typically, you have to deal with more power series terms when solving second-order differential equations, and that, naturally, provides more opportunities for error. That also leads to a greater probability that you will not succeed in finding "nice" formulas for the coefficients and may have to simply use the recursion formula to compute as many terms as you think necessary for a reasonably accurate partial sum approximation.

Initial-Value Problems (and Finding Patterns)

Observe that the solution obtained in our example (formula set 33.14) can be written as

$$y(x) = a_0 y_1(x) + a_1 y_2(x)$$

where $y_1(x)$ and $y_2(x)$ are power series about $x_0 = 0$ with

$$y_1(x) = 1 + \text{a summation of terms of order 2 or more}$$

and

$$y_2(x) = 1 \cdot (x - x_0) + \text{a summation of terms of order 2 or more} \quad .$$

In fact, we can derive this observation more generally after recalling that the general solution to a second-order, homogeneous linear differential equation is given by

$$y(x) = a_0 y_1(x) + a_1 y_2(x)$$

where a_0 and a_1 are arbitrary constants, and y_1 and y_2 form a linearly independent pair of particular solutions to the given differential equation. In particular, we can take y_1 to be the solution satisfying initial conditions

$$y_1(x_0) = 1 \quad \text{and} \quad y_1{}'(x_0) = 0 \quad ,$$

while y_2 is the solution satisfying initial conditions

$$y_2(x_0) = 0 \quad \text{and} \quad y_2{}'(x_0) = 1 \quad .$$

If our solutions can be written as power series about x_0, then y_1 and y_2 can be written as particular power series

$$y_1(x_0) = \sum_{k=0}^{\infty} \alpha_k (x - x_0)^k \quad \text{and} \quad y_2(x_0) = \sum_{k=0}^{\infty} \beta_k (x - x_0)^k$$

where

$$\alpha_0 = y_1(x_0) = 1 \quad , \quad \alpha_1 = y_1{}'(x_0) = 0 \quad ,$$

$$\beta_0 = y_2(x_0) = 0 \quad \text{and} \quad \beta_1 = y_2{}'(x_0) = 1 \quad ,$$

and the other α_k's and β_k's are fixed numbers (hopefully given by relatively simple formulas of k). Thus,

$$\begin{aligned} y_1(x) &= \alpha_0 + \alpha_1(x - x_0) + \alpha_2(x - x_0)^2 + \alpha_3(x - x_0)^3 + \cdots \\ &= 1 + 0 \cdot (x - x_0) + \alpha_2(x - x_0)^2 + \alpha_3(x - x_0)^3 + \cdots \\ &= 1 + \sum_{k=2}^{\infty} \alpha_k (x - x_0)^k \quad , \end{aligned}$$

while

$$\begin{aligned} y_2(x) &= \beta_0 + \beta_1(x-x_0) + \beta_2(x-x_0)^2 + \beta_3(x-x_0)^3 + \cdots \\ &= 0 + 1\cdot(x-x_0) + \beta_2(x-x_0)^2 + \beta_3(x-x_0)^3 + \cdots \\ &= 1\cdot(x-x_0) + \sum_{k=2}^{\infty}\beta_k(x-x_0)^k \quad , \end{aligned}$$

verifying that the observation made at the start of this subsection holds in general.

With regard to initial-value problems, we should note that, with these power series for y_1 and y_2,

$$y(x) = a_0 y_1(x) + a_1 y_2(x)$$

automatically satisfies the initial conditions

$$y(x_0) = a_0 \qquad \text{and} \qquad y'(x_0) = a_1$$

for any choice of constants a_0 and a_1.

Even and Odd Solutions (a Common Pattern)

Things simplify slightly when your recursion formula is of the form

$$a_k = \rho(k)\,a_{k-2} \qquad \text{for} \quad k = 2, 3, 4, \ldots$$

where $\rho(k)$ is some formula of k. This turns out to be a relatively common situation. When this happens,

$$\begin{aligned} a_2 &= \rho(2)\,a_0 & a_3 &= \rho(3)\,a_1 \\ a_4 &= \rho(4)\,a_2 = \rho(4)\rho(2)\,a_0 & a_5 &= \rho(5)\,a_3 = \rho(5)\rho(3)\,a_1 \\ a_6 &= \rho(6)\,a_4 = \rho(6)\rho(4)\rho(2)\,a_0 & a_7 &= \rho(7)\,a_5 = \rho(7)\rho(5)\rho(3)\,a_1 \\ a_8 &= \rho(8)\,a_6 = \rho(8)\rho(6)\rho(4)\rho(2)\,a_0 & a_9 &= \rho(9)\,a_7 = \rho(9)\rho(7)\rho(5)\rho(3)\,a_1 \\ &\vdots & &\vdots \end{aligned}$$

Clearly then, each even-indexed coefficient $a_k = a_{2m}$ with $k \geq 2$ is given by

$$a_{2m} = c_{2m}a_0 \qquad \text{where} \quad c_{2m} = \rho(2m)\,c_{2m-2} \quad ,$$

while each odd-indexed coefficient $a_k = a_{2m+1}$ with $k \geq 3$ is given by

$$a_{2m+1} = c_{2m+1}a_1 \qquad \text{where} \quad c_{2m+1} = \rho(2m+1)\,c_{2m-1} \quad .$$

If we further define $c_0 = c_1 = 1$ (simply so that $a_0 = c_0a_0$ and $a_1 = c_1a_1$), and let $x_0 = 0$, then

$$\begin{aligned} y(x) &= a_0 + a_1x + a_2x^2 + a_3x^3 + a_4x^4 + a_5x^5 + a_6x^6 + \cdots \\ &= c_0a_0 + c_1a_1x + c_2a_0x^2 + c_3a_1x^3 + c_4a_0x^4 + c_5a_1x^5 + c_6a_0x^6 + \cdots \\ &= a_0\left[c_0 + c_2x^2 + c_4x^4 + c_6x^6 + \cdots\right] + a_1\left[c_1x + c_3x^3 + c_5x^5 + c_7x^7 + \cdots\right] \\ &= a_0\sum_{m=0}^{\infty} c_{2m}x^{2m} + a_1\sum_{m=0}^{\infty} c_{2m+1}x^{2m+1} \quad . \end{aligned}$$

In other words, our power series solution is a linear combination of an even power series with an odd power series. Since these observations may prove useful in a few situations (and in a few exercises at the end of this chapter) let us summarize them in a little theorem (with $x - x_0$ replacing x).

Theorem 33.5

Let

$$y(x) = \sum_{k=0}^{\infty} a_k (x - x_0)^k$$

be a power series whose coefficients are related via a recursion formula

$$a_k = \rho(k)\, a_{k-2} \quad \text{for} \quad k = 2, 3, 4, \ldots$$

in which $\rho(k)$ *is some formula of* k *. Then*

$$y(x) = a_0 y_E(x) + a_1 y_O(x)$$

where y_E *and* y_O *are, respectively, the even- and odd-termed series*

$$y_E(x) = \sum_{m=0}^{\infty} c_{2m} (x - x_0)^{2m} \quad \text{and} \quad y_O(x) = \sum_{m=0}^{\infty} c_{2m+1} (x - x_0)^{2m+1}$$

with $c_0 = c_1 = 1$ *and*

$$c_k = \rho(k)\, c_{k-2} \quad \text{for} \quad k = 2, 3, 4, \ldots \quad .$$

33.5 Validity of the Algebraic Method for Second-Order Equations

Let's start by defining "ordinary" and "singular" points.

Ordinary and Singular Points, and the Radius of Analyticity

Given a differential equation

$$a(x)y'' + b(x)y' + c(x)y = 0 \tag{33.15}$$

we classify any point z_0 as a *singular point* if either of the limits

$$\lim_{z \to z_0} \frac{b(z)}{a(z)} \quad \text{or} \quad \lim_{z \to z_0} \frac{c(z)}{a(z)}$$

fails to exist as a finite number. Note that (just as with our definition of singular points for first-order differential equations) we used " z " in this definition, indicating that we may be considering points on the complex plane as well. This can certainly be the case when a , b and c are rational functions. And if a , b and c are rational functions, then the nonsingular points (i.e., the points that are not singular points) are traditionally referred to as *ordinary points* for the above differential equation.

The *radius of analyticity* for the above differential equation about any given point z_0 is defined just as before. It is the distance between z_0 and the singular point z_s closest to z_0 , provided the differential equation has at least one singular point. If the equation has no singular points, then we define the equation's *radius of analyticity* (about z_0) to be $+\infty$.

Again, it will be important to remember that the singular point z_s of most interest in a particular situation (i.e., the one closest to z_0) might not be on the real line, but some point in the complex plane off of the real line.

Nonexistence of Power Series Solutions

The above definitions are inspired by the same sort of computations as led to the analogous definitions for first-order differential equations in Section 33.3. I'll leave those computations to you. In particular, rewriting differential equation (33.15) as

$$y''(x) = -\frac{b(x)}{a(x)}y'(x) - \frac{c(x)}{a(x)}y(x) \quad ,$$

and using the relations between the values $y(x_0)$, $y'(x_0)$ and $y''(x_0)$, and the first three coefficients in

$$y(x) = \sum_{k=0}^{\infty} a_k(x - x_0)^k \quad ,$$

you should be able to prove the second-order analog to Lemma 33.1:

Lemma 33.6 (nonexistence of a power series solution)
If x_0 is a singular point for

$$a(x)y'' + b(x)y' + c(x)y = 0 \quad ,$$

then this differential equation does not have a power series solution $y(x) = \sum_{k=0}^{\infty} a_k(x - x_0)^k$ with a_0 and a_1 being arbitrary constants.

?►Exercise 33.1: *Verify Lemma 33.6.*

(By the way, a differential equation might have a "modified" power series solution about a singular point. We'll consider this possibility starting in Chapter 35.)

Validity of the Algebraic Method

Once again, we have a lemma telling us that our algebraic method for finding power series solutions about x_0 will fail if x_0 is a singular point (only now we are considering second-order equations). And, unsurprisingly, we also have a second-order analog of Theorem 33.7 assuring us that the method will succeed when x_0 is an ordinary point for our second-order differential equation, and even giving us a good idea of the interval over which the general power series solution is valid. That theorem is:

Theorem 33.7 (existence of power series solutions)
Let x_0 be an ordinary point on the real line for

$$a(x)y'' + b(x)y' + c(x)y = 0$$

where a , b and c are rational functions. Then this differential equation has a general power series solution

$$y(x) = \sum_{k=0}^{\infty} a_k(x - x_0)^k$$

with a_0 and a_1 being the arbitrary constants. Moreover, this solution is valid at least on the interval $(x_0 - R, x_0 + R)$ where R is the radius of analyticity about x_0 for the differential equation.

And again, we will wait until the next chapter to prove this theorem (or a slightly more general version of this theorem).

Identifying Singular and Ordinary Points

The basic approach to identifying a point z_0 as being either a singular or ordinary point for

$$a(x)y'' + b(x)y' + c(x)y = 0$$

is to look at the limits

$$\lim_{z \to z_0} \frac{b(z)}{a(z)} \quad \text{and} \quad \lim_{z \to z_0} \frac{c(z)}{a(z)} \quad .$$

If the limits are both finite numbers, x_0 is an ordinary point; otherwise x_0 is a singular point. And if you think about how these limits are determined by the values of $a(z)$ and $b(z)$ as $z \to z_0$, you'll derive the shortcuts listed in the next lemma.

Lemma 33.8 (tests for ordinary/singular points)
Let z_0 be a point on the complex plane, and consider a differential equation

$$a(x)y'' + b(x)y' + c(x)y = 0$$

in which a, b and c are all rational functions. Then:

1. *If $a(z_0)$, $b(z_0)$ and $c(z_0)$ are all finite values with $a(z_0) \neq 0$, then z_0 is an ordinary point for the differential equation.*

2. *If $a(z_0)$, $b(z_0)$ and $c(z_0)$ are all finite values with $a(z_0) = 0$, and either $b(z_0) \neq 0$ or $c(z_0) \neq 0$, then z_0 is a singular point for the differential equation.*

3. *If $a(z_0)$ is a finite value but either*

$$\lim_{z \to z_0} |b(z)| = \infty \quad \text{or} \quad \lim_{z \to z_0} |c(z)| = \infty \quad ,$$

   then z_0 is a singular point for the differential equation.

4. *If $b(z_0)$ and $c(z_0)$ are finite numbers, and*

$$\lim_{z \to z_0} |a(z)| = \infty \quad ,$$

   then z_0 is an ordinary point for the differential equation.

Applying the above to the differential equations of interest here (rewritten in the form recommended for the algebraic method) gives us:

Corollary 33.9
Let A, B and C be polynomials with no factors shared by all three. Then a point z_0 on the complex plane is a singular point for

$$A(x)y'' + B(x)y' + C(x)y = 0$$

if and only if $A(z_0) = 0$.

!►Example 33.6: *The coefficients in Airy's equation*

$$y'' - xy = 0$$

are polynomials with the first coefficient being

$$A(x) = 1 \quad .$$

Since there is no z_s in the complex plane such that $A(z_s) = 0$, Airy's equation has no singular point, and Theorem 33.7 assures us that the power series solution we obtained in solving Airy's equation (formula set (33.14) on page 625) is valid for all x.

33.6 The Taylor Series Method

The Basic Idea (Expanded)

In this approach to finding power series solutions, we compute the terms in the Taylor series for the solution,

$$y(x) = \sum_{k=0}^{\infty} \frac{y^{(k)}(x_0)}{k!}(x - x_0)^k \quad ,$$

in a manner reminiscent of the way you probably computed Taylor series in elementary calculus. Unfortunately, this approach often fails to yield a useful general formula for the coefficients of the power series solution (unless the original differential equation is very simple). Consequently, you typically end up with a partial sum of the power series solutions consisting of however many terms you've had the time or desire to compute. But there are two big advantages to this method over the general basic method:

1. The computation of the individual coefficients of the power series solution is a little more direct and may require a little less work than when using the algebraic method, at least for the first few terms (provided you are proficient with product and chain rules of differentiation).

2. The method can be used on a much more general class of differential equations than described so far. In fact, it can be used to formally find the Taylor series solution for any differential equation that can be rewritten in the form

$$y' = F_1(x, y) \qquad \text{or} \qquad y'' = F_2(x, y, y')$$

where F_1 and F_2 are known functions that are sufficiently differentiable with respect to all variables.

With regard to the last comment, observe that

$$a(x)y' + b(x)y = 0 \qquad \text{and} \qquad a(x)y'' + b(x)y' + c(x)y = 0$$

can be rewritten, respectively, as

$$y' = \underbrace{-\frac{b(x)}{a(x)}y}_{F_1(x,y)} \qquad \text{and} \qquad y'' = \underbrace{-\frac{b(x)}{a(x)}y' - \frac{c(x)}{a(x)}y}_{F_2(x,y,y')} \quad .$$

So this method can be used on the same differential equations we used the algebraic method on in the previous sections. Whether you would want to is a different matter.

The Steps in the Taylor Series Method

Here are the steps in our procedure for finding at least a partial sum for the Taylor series about a point x_0 for the solution to a fairly arbitrary first- or second-order differential equation. As an example, let us use the differential equation

$$y'' + \cos(x)y = 0 \quad .$$

As before we have two preliminary steps:

Pre-step 1: Depending on whether the original differential equation is first order or second order, respectively, rewrite it as

$$y' = F_1(x, y) \qquad \text{or} \qquad y'' = F_2(x, y, y') \quad .$$

For our example, we simply subtract $\cos(x)y$ *from both sides, obtaining*

$$y'' = -\cos(x)y \quad .$$

Pre-step 2: Choose a value for x_0. It should be an ordinary point if the differential equation is linear. (More generally, F_1 or F_2 must be "differentiable enough" to carry out all the subsequent steps in this procedure.) If initial values are given for $y(x)$ at some point, then use that point for x_0. Otherwise, choose x_0 as convenient — which usually means choosing $x_0 = 0$.

For our example, we choose $x_0 = 0$.

Now for the steps in the Taylor series method. In going through this method, note that the last step and some of the first few steps depend on whether your original differential equation is first or second order.

Step 1: Set

$$y(x_0) = a_0 \quad .$$

If an initial value for y at x_0 has already been given, you can use that value for a_0. Otherwise, treat a_0 as an arbitrary constant. It will be the first term in the power series solution.

For our example, we chose $x_0 = 0$*, and have no given initial values. So we set*

$$y(0) = a_0$$

with a_0 *being an arbitrary constant.*

Step 2: *(a)* If the original differential equation is first order, then compute $y'(x_0)$ by plugging x_0 and the initial value set above into the differential equation from the first pre-step,

$$y' = F_1(x, y) \quad .$$

Remember, to take into account the fact that y is shorthand for $y(x)$. Thus,

$$y'(x_0) = F_1(x_0, y(x_0)) = F_1(x_0, a_0) \quad .$$

Our sample differential equation is second order. So we do nothing here.

(b) If the original differential equation is second order, then just set

$$y'(x_0) = a_1 \quad .$$

If an initial value for y' at x_0 has already been given, then use that value for a_1. Otherwise, treat a_1 as an arbitrary constant.

Since our sample differential equation is second order, we set

$$y'(0) \;=\; a_1$$

with a_1 being an arbitrary constant.

Step 3 *(a)* If the original differential equation is first order, then differentiate both sides of

$$y' \;=\; F_1(x, y)$$

to obtain an expression of the form

$$y'' \;=\; F_2\left(x, y, y'\right) \quad .$$

Our sample differential equation is second order. So we do nothing here.

(b) Whether the original differential equation is first or second order, you now have a formula for y'',

$$y'' \;=\; F_2\left(x, y, y'\right) \quad .$$

Use this formula, along with the previous set of computed values for $y(x_0)$ and $y'(x_0)$, to compute $y''(x_0)$. Remember to take into account the fact that y and y' are shorthand for $y(x)$ and $y'(x)$. Thus,

$$y''(x_0) \;=\; F_2\left(x_0, y(x_0), y'(x_0)\right) \quad .$$

Save this newly computed value for future use.

For our example,

$$y'' \;=\; -\cos(x)y \quad , \quad x_0 = 0 \quad , \quad y(0) = a_0 \quad \text{and} \quad y'(0) = a_1 \quad .$$

So,

$$y''(0) \;=\; -\cos(0)y(0) \;=\; -1 \cdot a_0 \;=\; -a_0 \quad .$$

Step 4: Using the formula for y'' from the previous step, and the values for $y(x_0)$, $y'(x_0)$ and $y''(x_0)$ determined in the previous steps:

(a) Differentiate both sides of

$$y'' \;=\; F_2\left(x, y, y'\right)$$

to obtain an expression of the form

$$y''' \;=\; F_3\left(x, y, y', y''\right) \quad .$$

For our example,

$$y'' \;=\; -\cos(x)y \quad .$$

So, using the product rule,

$$y''' \;=\; \frac{d}{dx}[-\cos(x)y] \;=\; \underbrace{\sin(x)y \;-\; \cos(x)y'}_{F_3(x,y,y',y'')} \quad .$$

(*b*) Then compute $y'''(x_0)$ by plugging x_0 into the formula for $y'''(x)$ just derived,

$$y'''(x_0) = F_3\left(x_0, y(x_0), y'(x_0), y''(x_0)\right) \quad .$$

Save this newly computed value for future use.

For our example,

$$y'''(0) = \sin(0)y(0) - \cos(0)y'(0) = 0 \cdot a_0 - 1 \cdot a_1 = -a_1 \quad .$$

For steps 5, 6 and so on, simply repeat step 4 with derivatives of increasing order. In general:

Step k : Using the formula

$$y^{(k-1)} = F_{k-1}\left(x, y, y', y'', \ldots, y^{(k-2)}\right) \quad ,$$

obtained in the previous step, and values for $y(x_0)$, $y'(x_0)$, $y''(x_0)$, ... and $y^{(k-1)}(x_0)$ determined in the previous steps:

(*a*) Differentiate both sides of this formula for $y^{(k-1)}$ to obtain a corresponding formula for $y^{(k)}$,

$$y^{(k)} = F_k\left(x, y, y', y'', \ldots, y^{(k-1)}\right) \quad .$$

(*b*) Then compute $y^{(k)}(x_0)$ by plugging x_0 into the formula for $y^{(k)}(x)$ just derived,

$$y^{(k)}(x_0) = F_k\left(x_0, y(x_0), y'(x_0), \ldots, y^{(k-1)}(x_0)\right) \quad .$$

Save this newly computed value for future use.

Finally, you need to stop. Assuming you've computed $y^{(k)}(x_0)$ for k up to N, where N is either some predetermined integer or the order of the last derivative computed before you decided you've done enough:

Last Step: If you can determine a general formula for $y^{(k)}(x_0)$ in terms of a_0 or in terms of a_0 and a_1 (depending on whether the original differential equation is first or second order), then use that formula to write out the Taylor series for the solution,

$$y(x) = \sum_{k=0}^{\infty} \frac{y^{(k)}(x_0)}{k!}(x - x_0)^k \quad .$$

Otherwise, just use the computed values of the $y^{(k)}(x_0)$'s to write out the Nth partial sum for this Taylor series, obtaining the solution's N^{th} partial sum approximation

$$y(x) \approx \sum_{k=0}^{N} \frac{y^{(k)}(x_0)}{k!}(x - x_0)^k \quad .$$

In either case, try to simplify your result by gathering terms involving the arbitrary constants and, if possible, factoring out these constants.

Frankly, this method is easier to carry out than it is to describe.

!►Example 33.7: *Let us finish finding the sixth partial sum for the Taylor series about* 0 *for the solution to*

$$y'' + \cos(x)y = 0 \quad .$$

From the above, we already have

$$y(0) = a_0 \quad , \quad y'(0) = a_1 \quad , \quad y''(0) = -a_0 \quad , \quad y'''(0) = -a_1 \quad ,$$

and

$$y'''(x) = \sin(x)y - \cos(x)y' \quad .$$

Continuing:

$$\begin{aligned} y^{(4)}(x) &= \frac{d}{dx}\left[y^{(3)}(x)\right] \\ &= \frac{d}{dx}\left[\sin(x)y - \cos(x)y'\right] = \cos(x)y + 2\sin(x)y' - \cos(x)y'' \quad . \end{aligned}$$

And so,

$$\begin{aligned} y^{(4)}(0) &= \cos(0)y(0) + 2\sin(0)y'(0) - \cos(0)y''(0) \\ &= 1\cdot a_0 + 2\cdot 0\cdot a_1 - 1\cdot[-a_0] = 2a_0 \quad . \end{aligned}$$

Differentiating again:

$$\begin{aligned} y^{(5)}(x) &= \frac{d}{dx}\left[y^{(4)}(x)\right] \\ &= \frac{d}{dx}\left[\cos(x)y + 2\sin(x)y' - \cos(x)y''\right] \\ &= -\sin(x)y + \cos(x)y' + 2\cos(x)y' + 2\sin(x)y'' + \sin(x)y'' - \cos(x)y''' \\ &= -\sin(x)y + 3\cos(x)y' + 3\sin(x)y'' - \cos(x)y''' \quad . \end{aligned}$$

Hence,

$$\begin{aligned} y^{(5)}(0) &= -\sin(0)y(0) + 3\cos(0)y'(0) + 3\sin(0)y''(0) - \cos(0)y'''(0) \\ &= -0\cdot a_0 + 3\cdot 1\cdot a_1 + 3\cdot 0\cdot[-a_0] - 1\cdot[-a_1] \\ &= 4a_1 \quad . \end{aligned}$$

And again:

$$\begin{aligned} y^{(6)}(x) &= \frac{d}{dx}\left[y^{(5)}(x)\right] \\ &= \frac{d}{dx}\left[-\sin(x)y + 3\cos(x)y' + 3\sin(x)y'' - \cos(x)y'''\right] \\ &= \cdots \\ &= -\cos(x)y - 4\sin(x)y' + 6\cos(x)y'' + 4\sin(x)y''' - \cos(x)y^{(4)} \quad . \end{aligned}$$

So,

$$\begin{aligned} y^{(6)}(0) &= -\cos(0)y(0) - 4\sin(0)y'(0) + 6\cos(0)y''(0) \\ &\qquad + 4\sin(0)y'''(0) - \cos(0)y^{(4)}(0) \\ &= -1\cdot a_0 - 4\cdot 0\cdot a_1 + 6\cdot 1\cdot[-a_0] + 4\cdot 0\cdot[-a_1] - 1\cdot[2a_0] \\ &= -9a_0 \quad . \end{aligned}$$

Thus, the sixth partial sum of the Taylor series for y about 0 is

$$\begin{aligned}
S_6(x) &= \sum_{k=0}^{6} \frac{y^{(k)}(0)}{k!} x^k \\
&= y(0) + y'(0)x + \frac{y''(0)}{2!}x^2 + \frac{y'''(0)}{3!}x^3 + \frac{y^{(4)}(0)}{4!}x^4 + \frac{y^{(5)}(0)}{5!}x^5 + \frac{y^{(6)}(0)}{6!}x^6 \\
&= a_0 + a_1 x + \frac{-a_0}{2}x^2 + \frac{-a_1}{3!}x^3 + \frac{2a_0}{4!}x^4 + \frac{4a_1}{5!}x^5 + \frac{-9a_0}{6!}x^6 \\
&= a_0\left[1 - \frac{1}{2}x^2 + \frac{1}{12}x^4 - \frac{1}{80}x^6\right] + a_1\left[x - \frac{1}{6}x^3 + \frac{1}{30}x^5\right] \quad .
\end{aligned}$$

Validity of the Solutions

If the solutions to a given differential equation are analytic at a point x_0, then the above method will clearly find the Taylor series (i.e., the power series) for these solutions about x_0. Hence any valid theorem stating the existence and radius of convergence of power series solutions to a given type of differential equation about a point x_0 also assures us of the validity of the Taylor series method with these equations. In particular, we can appeal to Theorems 33.2 on page 618 and 33.7 on page 629, or even to the more general versions of these theorems in the next chapter (Theorems 34.9 and 34.10 starting on page 655). In particular, Theorem 34.9 will assure us that our above use of the Taylor series method in attempting to solve

$$y'' + \cos(x)y = 0$$

is valid. Hence, it will assure us that, at least when $x \approx 0$,

$$y(x) \approx a_0\left[1 - \frac{1}{2}x^2 + \frac{1}{12}x^4 - \frac{1}{80}x^6\right] + a_1\left[x - \frac{1}{6}x^3 + \frac{1}{30}x^5\right] \quad .$$

33.7 Appendix: Using Induction

The Basic Ideas

Suppose we have some sequence of numbers — $A_0, A_1, A_2, A_3, \ldots$ — with each of these numbers (other than A_0) related by some recursion formula to the previous number in the sequence. Further suppose we have some other formula $F(k)$, and we suspect that

$$A_k = F(k) \qquad \text{for} \quad k = 0, 1, 2, 3, \ldots \quad .$$

The obvious question is whether our suspicions are justified. Can we confirm that the above equality holds for every nonnegative integer k?

Well, let's make two more assumptions:

1. That we know

$$A_0 = F(0) \quad .$$

2. That we also know that, for every nonnegative integer N, we can rigorously derive the equality
$$A_{N+1} = F(N+1)$$
from
$$A_N = F(N)$$
using the recursion formula for the A_k's. For brevity, let's express this assumption as
$$A_N = F(N) \quad \Longrightarrow \quad A_{N+1} = F(N+1) \qquad \text{for} \quad N = 0,\ 1,\ 2,\ \ldots \quad . \tag{33.16}$$

These two assumptions are the "steps" in the basic principle of induction with a numeric sequence. The first — that $A_0 = F(0)$ — is the *base step* or *anchor*, and the second — implication (33.16) — is the *inductive step*. It is important to realize that, in implication (33.16), we are *not* assuming $A_N = F(N)$ is actually true, only that we could derive $A_{N+1} = F(N+1)$ *provided* $A_N = F(N)$.

However, if these assumptions hold, then we do know $A_0 = F(0)$, which is $A_N = F(N)$ with $N = 0$. That, combined with implication (33.16), assures us that we could then derive the equality
$$A_{0+1} = F(0+1) \quad .$$
So, in fact, we also know that
$$A_1 = F(1) \quad .$$
But this last equation, along with implication (33.16) (using $N = 1$) assures us that we could then obtain
$$A_{1+1} = F(1+1) \quad ,$$
assuring us that
$$A_2 = F(2) \quad .$$
And this, with implication (33.16) (this time using $N = 2$) tells us that we could then derive
$$A_{2+1} = F(2+1) \quad ,$$
thus confirming that
$$A_3 = F(3) \quad .$$
And, clearly, we can continue, successively confirming that
$$A_4 = F(4) \quad , \quad A_5 = F(5) \quad , \quad A_6 = F(6) \quad , \quad A_7 = F(7) \quad , \quad \ldots \quad .$$
Ultimately, we could confirm that
$$A_k = F(k)$$
for any given positive integer k, thus assuring us that our original suspicions were justified.

To summarize:

Theorem 33.10 (basic principle of induction for numeric sequences)
Let A_0, A_1, A_2, A_3, $\ldots$ be a sequence of numbers related to each other via some recursion formula, and let F be some function with $F(k)$ being defined and finite for each nonnegative integer k. Assume further that

1. *(base step) $A_0 = F(0)$,*

and that

2. *(inductive step) for each nonnegative integer N, the equality*

$$A_{N+1} = F(N+1)$$

can be obtained from the equality

$$A_N = F(N)$$

using the recursion formula for the A_k's.

Then

$$A_k = F(k) \qquad \text{for} \quad k = 0, 1, 2, 3, \ldots \quad .$$

To use the above, we must first verify that the two assumptions in the theorem actually hold for the case at hand. That is, we must verify both that

1. $A_0 = F(0)$

and that

2. for each positive integer N, $A_{N+1} = F(N+1)$ can be derived from $A_N = F(N)$ using the recursion formula.

In practice, the formula $F(k)$ comes from noting a "pattern" in computing the first few A_k's using a recursion formula. Consequently, the verification of the base step is typically contained in those computations. It is the verification of the inductive step that is most important. That step is where we confirm that the "pattern" observed in computing the first few A_k's continues and can be used to compute the rest of the A_k's.

!►Example 33.8: *For our A_k's, let us use the coefficients of the power series solution obtained in our first example of the algebraic method,*

$$A_k = a_k$$

where a_0 is an arbitrary constant, and the other a_k's are generated via recursion formula (33.6) on page 608,[7]

$$a_k = \frac{k+1}{2k} a_{k-1} \qquad \text{for} \quad k = 1, 2, 3, 4, \ldots \quad . \tag{33.17}$$

In particular, we obtained:

$$a_1 = \frac{2}{2}a_0 \quad , \quad a_2 = \frac{3}{2^2}a_0 \quad , \quad a_3 = \frac{4}{2^3}a_0 \quad , \quad a_4 = \frac{5}{2^4}a_0 \quad , \quad a_5 = \frac{6}{2^5}a_0 \quad ,$$

and "speculated" that, in general,

$$a_k = \frac{k+1}{2^k} a_0 \qquad \text{for} \quad k = 0, 1, 2, 3, 4, \ldots \quad . \tag{33.18}$$

So, we want to show that, indeed,

$$a_k = F(k) \quad \text{with} \quad F(k) = \frac{k+1}{2^k} a_0 \qquad \text{for} \quad k = 0, 1, 2, 3, 4, \ldots \quad .$$

To apply the principle of induction, we first must verify that

$$a_0 = F(0) \quad .$$

[7] Originally, recursion formula (33.6) was derived for $k \geq 2$. However, with $k = 1$, this recursion formula reduces to $a_1 = a_0$ which is true for this power series solution. So this formula is valid for all nonnegative integers.

But this is easily done by just computing $F(0)$,

$$F(0) = \frac{0+1}{2^0} a_0 = a_0 \quad .$$

Next comes the verification of the inductive step; that is, the verification that for any nonnegative integer N, the recursion formula for the a_k's yields the implication

$$a_N = F(N) \quad \Longrightarrow \quad a_{N+1} = F(N+1) \quad .$$

Now,

$$F(N) = \frac{N+1}{2^N} a_0 \qquad \text{and} \qquad F(N+1) = \frac{[N+1]+1}{2^{[N+1]}} a_0 = \frac{N+2}{2^{N+1}} a_0 \quad .$$

So what we really need to show is that the implication

$$a_N = \frac{N+1}{2^N} a_0 \quad \Longrightarrow \quad a_{N+1} = \frac{N+2}{2^{N+1}} a_0 \tag{33.19}$$

holds because of recursion formula (33.17), which, with $k = N+1$, becomes

$$a_{N+1} = \frac{[N+1]+1}{2[N+1]} a_{[N+1]-1} = \frac{N+2}{2[N+1]} a_N \quad .$$

Combining this with the first equation in implication (33.19) gives us

$$a_{N+1} = \frac{N+2}{2[N+1]} a_N = \frac{N+2}{2[N+1]} \cdot \frac{N+1}{2^N} a_0 = \frac{N+2}{2^{N+1}} a_0 \quad ,$$

which is the second equation in implication (33.19). This confirms that implication (33.19) holds for every nonnegative integer N.

With both steps of the inductive process successfully verified, the principle of induction (Theorem 33.10) assures us that yes, indeed,

$$a_k = \frac{k+1}{2^k} a_0 \qquad \text{for} \quad k = 0, 1, 2, 3, 4, \ldots$$

just as we originally speculated.

Usage Notes

In the above example, the A_k's were all the coefficients in a particular power series. In other examples, especially examples involving second-order equations, you may have to separately consider different subsequences of the coefficients.

!►Example 33.9: *In deriving the power series solution $\sum_{k=0}^{\infty} a_k x^k$ to Airy's equation, we obtained three patterns:*

$$a_{3n} = \frac{1}{(2 \cdot 3)(5 \cdot 6) \cdots ([3n-1] \cdot 3n)} a_0 \qquad \text{for} \quad n = 1, 2, 3, \ldots \quad ,$$

$$a_{3n+1} = \frac{1}{(3 \cdot 4)(6 \cdot 7) \cdots (3n \cdot [3n+1])} a_1 \qquad \text{for} \quad n = 1, 2, 3, \ldots$$

and

$$a_{3n+1} = 0 \qquad \text{for} \quad n = 1, 2, 3, \ldots \quad .$$

To verify each of these formulas, we would need to apply the method of induction three times:

1. *once with* $A_n = a_{3n}$ *and* $F(n) = \dfrac{1}{(2 \cdot 3)(5 \cdot 6) \cdots ([3n-1] \cdot 3n)} a_0$,

2. *once with* $A_n = a_{3n+1}$ *and* $F(n) = \dfrac{1}{(3 \cdot 4)(6 \cdot 7) \cdots (3n \cdot [3n+1])} a_1$, *and*

3. *once with* $A_n = a_{3n+2}$ *and* $F(n) = 0$.

It probably should be noted that there are slight variations on the method of induction. In particular, it should be clear that, in the inductive step, we need not base our derivation of $a_{N+1} = F(N+1)$ on just $a_N = F(N)$ — we could base it on any or all equations $a_k = F(k)$ with $k \leq N$. To be precise, a slight variation in our development of Theorem 33.10 would have given us the following:

Theorem 33.11 (alternative principle of induction for numeric sequences)
Let A_0 , A_1 , A_2 , A_3 , ... *be a sequence of numbers related to each other via some set of recursion formulas, and let* F *be some function with* $F(k)$ *being defined and finite for each nonnegative integer* k . *Assume further that*

1. *(base step)* $A_0 = F(0)$,

and that

2. *(inductive step) for each nonnegative integer* N , *the equality*

$$A_{N+1} = F(N+1)$$

   can be obtained from the equalities

$$A_k = F(k) \qquad \text{for} \quad k = 0, 1, 2, \ldots, N$$

   using the given set of recursion formulas.

Then

$$A_k = F(k) \qquad \text{for} \quad k = 0, 1, 2, 3, \ldots \quad .$$

This would be the version used with recursion formulas having two or more terms.

The General Principle of Induction

For the sake of completeness, it should be mentioned that the general principle of induction is a fundamental principle of logic. Here's the basic version:

Theorem 33.12 (basic general principle of induction)
Let S_0 , S_1 , S_2 , S_3 , ... *be a sequence of logical statements. Assume further that*

1. S_0 *is true,*

and that

2. *for each nonnegative integer* N , *it can be shown that* S_{N+1} *is true provided* S_N *is true.*

Then all the S_k *'s are true.*

You can find discussions of this principle in just about any "introduction to mathematical logic" or "introduction to abstract mathematics" textbook.

?►Exercise 33.2: *What is the logical statement* S_k *in Theorem 33.10?*

Additional Exercises

33.3. *Using the algebraic method from Section 33.2, find a general power series solution about x_0 to each of the first-order differential equations given below. In particular:*

i *Identify the recursion formula for the coefficients.*

ii *Use the recursion formula to find the corresponding series solution with at least four nonzero terms explicitly given (if the series terminates, just write out the corresponding polynomial solution).*

iii *Unless otherwise instructed, find a general formula for the coefficients and write out the resulting series.*

Also, then solve each of these equations using more elementary methods.

a. $y' - 2y = 0$ with $x_0 = 0$

b. $y' - 2xy = 0$ with $x_0 = 0$

c. $y' + \frac{2}{2x-1}y = 0$ with $x_0 = 0$

d. $(x-3)y' - 2y = 0$ with $x_0 = 0$

e. $\left(1+x^2\right)y' - 2xy = 0$ with $x_0 = 0$

f. $y' + \frac{1}{x-1}y = 0$ with $x_0 = 0$

g. $y' + \frac{1}{x-1}y = 0$ with $x_0 = 3$

h. $(1-x)y' - 2y = 0$ with $x_0 = 5$

i. $(2-x^3)y' - 3x^2y = 0$ with $x_0 = 0$

j. $(2-x^3)y' + 3x^2y = 0$ with $x_0 = 0$

k. $(1+x)y' - xy = 0$ with $x_0 = 0$
(Do not attempt finding a general formula for the coefficients.)

l. $(1+x)y' + (1-x)y = 0$ with $x_0 = 0$
(Do not attempt finding a general formula for the coefficients.)

33.4. *For each of the equations in the previous set, find each singular point z_s, the radius of analyticity R about the given x_0, and the interval I over which the power series solution is guaranteed to be valid according to Theorem 33.2.*

33.5. *Find the general power series solution about the given x_0 for each of the differential equations given below. In your work and answers:*

i *Identify the recursion formula for the coefficients.*

ii *Express your final answer as a linear combination of a linearly independent pair of power series solutions. If a series terminates, write out the resulting polynomial. Otherwise write out the series with at least four nonzero terms explicitly given.*

iii *Unless otherwise instructed, find a formula for the coefficients of each series, and write out the resulting series.*

a. $\left(1+x^2\right)y'' - 2y = 0$ *with* $x_0 = 0$

b. $y'' + xy' + y = 0$ *with* $x_0 = 0$

c. $\left(4+x^2\right)y'' + 2xy' = 0$ *with* $x_0 = 0$

d. $y'' - 3x^2y = 0$ *with* $x_0 = 0$
(Do not attempt finding a general formula for the coefficients.)

e. $\left(4-x^2\right)y'' - 5xy' - 3y = 0$ *with* $x_0 = 0$
(Do not attempt finding a general formula for the coefficients.)

f. $\left(1-x^2\right)y'' - xy' + 4y = 0$ *with* $x_0 = 0$ *(a Chebyshev equation)*
(Do not attempt finding a general formula for the coefficients.)

g. $y'' - 2xy' + 6y = 0$ *with* $x_0 = 0$ *(a Hermite equation)*
(Do not attempt finding a general formula for the coefficients.)

h. $\left(x^2-6x\right)y'' + 4(x-3)y' + 2y = 0$ *with* $x_0 = 3$

i. $y'' + (x+2)y' + 2y = 0$ *with* $x_0 = -2$

j. $(x^2-2x+2)y'' + (1-x)y' - 3y = 0$ *with* $x_0 = 1$
(Do not attempt finding a general formula for the coefficients.)

k. $y'' - 2y' - xy = 0$ *with* $x_0 = 0$
(Do not attempt finding a general formula for the coefficients.)

l. $y'' - xy' - 2xy = 0 = 0$ *with* $x_0 = 0$
(Do not attempt finding a general formula for the coefficients.)

33.6. For each of the equations in the previous set, find each singular point z_s, the radius of analyticity R about the given x_0, and the interval I over which the power series solution is guaranteed to be valid according to Theorem 33.7.

33.7. In describing the coefficients of power series solutions, experienced mathematicians often condense lengthy formulas through clever use of the factorial. Develop some of that experience by deriving the following expressions for the indicated products. Do not simply multiple all the factors and compare the two sides of each equation. Instead, use clever algebra to convert one side of the equation to the other.

a. Products of the first m even integers:

i. $6 \cdot 4 \cdot 2 = 2^3 3!$ (Hint: $6 \cdot 4 \cdot 2 = (2 \cdot 3)(2 \cdot 2)(2 \cdot 1)$)

ii. $8 \cdot 6 \cdot 4 \cdot 2 = 2^4 4!$

iii. $(2m)(2m-2)(2m-4)\cdots 6 \cdot 4 \cdot 2 = 2^m m!$

b. *Products of the first m odd integers:*

i. $5 \cdot 3 \cdot 1 \; = \; \dfrac{5!}{2^2 2!}$

ii. $7 \cdot 5 \cdot 3 \cdot 1 \; = \; \dfrac{7!}{2^3 3!}$

iii. $(2m-1)(2m-3)(2m-5)\cdots 5 \cdot 3 \cdot 1 \; = \; \dfrac{(2m-1)!}{2^{m-1}(m-1)!}$

33.8. The following exercises concern the Hermite equation, *which is*

$$y'' - 2xy' + \lambda y = 0$$

where λ is a constant (as in Exercise 33.5 g, above).

a. *Using the algebraic method, derive the general recursion formula (in terms of λ) for the general power series solution $y_\lambda(x) = \sum_{k=0}^{\infty} a_k x^k$ to the above Hermite equation.*

b. *Over what interval are these power series solutions guaranteed to be valid according to Theorem 33.7?*

c. *Using the recursion formula just found, along with Theorem 33.5 on page 628, verify that the general power series solution y_λ can be written as*

$$y_\lambda(x) = a_0 y_{\lambda,E}(x) + a_1 y_{\lambda,O}(x)$$

where $y_{\lambda,E}$ and $y_{\lambda,O}$ are, respectively, even- and odd-termed series

$$y_{\lambda,E}(x) = \sum_{m=0}^{\infty} c_{2m} x^{2m} \qquad \text{and} \qquad y_{\lambda,O}(x) = \sum_{m=0}^{\infty} c_{2m+1} x^{2m+1}$$

with $c_0 = c_1 = 1$ and the other c_k's satisfying some recursion formula. Write out that recursion formula.

d. *Assume N is an nonnegative integer, and find the one value λ_N for λ such that the above-found recursion formula yields $a_{N+2} = 0 \cdot a_N$.*

e. *Using the above, show that:*

i. *If $\lambda = \lambda_N$ for some nonnegative integer N , then exactly one of the two power series $y_{\lambda,E}(x)$ or $y_{\lambda,O}(x)$ reduces to an even or odd N^{th}-degree polynomial p_N , with*

$$p_N(x) = \begin{cases} y_{\lambda,E}(x) & \text{if } N \text{ is even} \\ y_{\lambda,O}(x) & \text{if } N \text{ is odd} \end{cases} ,$$

and with the other power series not reducing to a polynomial. (The polynomials, multiplied by suitable constants, are called the Hermite polynomials.)

ii. *If $\lambda \neq \lambda_N$ for any nonnegative integer N , then neither of the two power series $y_{\lambda,E}(x)$ or $y_{\lambda,O}(x)$ reduces to polynomial.*

f. *Find the polynomial solution $p_N(x)$ when*

i. $N = 0$ **ii.** $N = 1$ **iii.** $N = 2$ **iv.** $N = 3$

v. $N = 4$ **vi.** $N = 5$

33.9. *The following exercises concern the* Chebyshev[8] equation with parameter λ,

$$\left(1-x^2\right)y'' - xy' + \lambda y = 0$$

(as in Exercise 33.5 f, above). The parameter λ may be any constant.

a. *Using the algebraic method, derive the general recursion formula (in terms of λ) for the general power series solution $y_\lambda(x) = \sum_{k=0}^{\infty} a_k x^k$ to the above Chebyshev equation.*

b. *Using the recursion formula just found, along with Theorem 33.5 on page 628, verify that the general power series solution y_λ can be written as*

$$y_\lambda(x) = a_0 y_{\lambda,E}(x) + a_1 y_{\lambda,O}(x)$$

where $y_{\lambda,E}$ and $y_{\lambda,O}$ are, respectively, even- and odd-termed series

$$y_{\lambda,E}(x) = \sum_{m=0}^{\infty} c_{2m} x^{2m} \quad \text{and} \quad y_{\lambda,O}(x) = \sum_{m=0}^{\infty} c_{2m+1} x^{2m+1}$$

with $c_0 = c_1 = 1$ and the other c_k's satisfying some recursion formula. Write out that recursion formula.

c. *Assume N is an nonnegative integer, and find the one value λ_N for λ such that the above-found recursion formula yields $a_{N+2} = 0 \cdot a_N$.*

d. *Using the above, show that:*

i. *If $\lambda = \lambda_N$ for some nonnegative integer N, then exactly one of the two power series $y_{\lambda,E}(x)$ or $y_{\lambda,O}(x)$ reduces to an even or odd N^{th}-degree polynomial p_N, with*

$$p_N(x) = \begin{cases} y_{\lambda,E}(x) & \text{if } N \text{ is even} \\ y_{\lambda,O}(x) & \text{if } N \text{ is odd} \end{cases},$$

and with the other power series not reducing to a polynomial. (The polynomials, multiplied by suitable constants, are the Chebyshev polynomials of the first type.)

ii. *If $\lambda \neq \lambda_N$ for any nonnegative integer N, then neither of the two power series $y_{\lambda,E}(x)$ or $y_{\lambda,O}(x)$ reduces to a polynomial.*

e. *Now, find the following:*

i. λ_0 and $p_0(x)$ **ii.** λ_1 and $p_1(x)$ **iii.** λ_2 and $p_2(x)$

iv. λ_3 and $p_3(x)$ **v.** λ_4 and $p_4(x)$ **vi.** λ_5 and $p_5(x)$

f. *Now let λ be any constant (not necessarily λ_N).*

i. *What is the largest interval over which these power series solutions to the Chebyshev equation are guaranteed to be valid according to Theorem 33.7?*

ii. *Use the recursion formula along with the ratio test or limit ratio test to find the radius of convergence and largest interval of convergence for $y_{\lambda,E}(x)$ and for $y_{\lambda,O}(x)$, provided the series does not terminate as polynomials.*

g. *Verify each of the following using work already done above:*

i. *If $\lambda = N^2$ for some nonnegative integer N, then the Chebyshev equation with parameter λ has polynomial solutions, all of which are all constant multiples of $p_N(x)$.*

---

[8] also spelled *Tschebyscheff.*

ii. *If $\lambda \neq N^2$ for every nonnegative integer N, then the Chebyshev equation with parameter λ has no polynomial solutions (other than $y = 0$).*

iii. *If y_λ is a nonpolynomial solution to a Chebyshev equation on $(-1, 1)$, then it is given by a power series about $x_0 = 0$ with a radius of convergence of exactly 1.*

33.10. For any constant λ, the Legendre equation[9] with parameter λ is

$$(1 - x^2)y'' - 2xy' + \lambda y = 0 \quad .$$

This equation is the object of study in the following exercises.

a. *Using the algebraic method, derive the general recursion formula (in terms of λ) for the general power series solution $y_\lambda(x) = \sum_{k=0}^{\infty} a_k x^k$ to the above Legendre equation.*

b. *Using the recursion formula just found, along with Theorem 33.5 on page 628, verify that the general power series solution y_λ can be written as*

$$y_\lambda(x) = a_0 y_{\lambda,E}(x) + a_1 y_{\lambda,O}(x)$$

where $y_{\lambda,E}$ and $y_{\lambda,O}$ are, respectively, even- and odd-termed series

$$y_{\lambda,E}(x) = \sum_{m=0}^{\infty} c_{2m} x^{2m} \qquad \text{and} \qquad y_{\lambda,O}(x) = \sum_{m=0}^{\infty} c_{2m+1} x^{2m+1}$$

with $c_0 = c_1 = 1$ and the other c_k's satisfying some recursion formula. Write out that recursion formula.

c. *Assume N is a nonnegative integer, and find the one value λ_N for λ such that the above-found recursion formula yields $a_{N+2} = 0 \cdot a_N$.*

d. *Using the above, show that:*

i. *If $\lambda = \lambda_N$ for some nonnegative integer N, then exactly one of the two power series $y_{\lambda,E}(x)$ or $y_{\lambda,O}(x)$ reduces to an even or odd N^{th}-degree polynomial p_N, with*

$$p_N(x) = \begin{cases} y_{\lambda,E}(x) & \text{if } N \text{ is even} \\ y_{\lambda,O}(x) & \text{if } N \text{ is odd} \end{cases} \quad ,$$

and with the other power series not reducing to a polynomial. (The polynomials, multiplied by suitable constants, are called the Legendre polynomials.)

ii. *If $\lambda \neq \lambda_N$ for any nonnegative integer N, then neither of the two power series $y_{\lambda,E}(x)$ or $y_{\lambda,O}(x)$ reduces to a polynomial.*

e. *Based on the above, find the following:*

i. λ_0, $p_0(x)$, and $y_{0,O}(x)$

ii. λ_1, $p_1(x)$, and $y_{1,E}(x)$

iii. λ_2 and $p_2(x)$

iv. λ_3 and $p_3(x)$

v. λ_4 and $p_4(x)$

vi. λ_5 and $p_5(x)$

---

[9] The Legendre equations arise in problems involving three-dimensional spherical objects (such as the Earth). In Section 36.5, we will continue the analysis begun in this exercise, discovering that the "polynomial solutions" found here are of particular importance.

f. *Now let λ be any constant (not necessarily λ_N).*

i. *What is the largest interval over which these power series solutions to the Legendre equation are guaranteed to be valid according to Theorem 33.7?*

ii. *Use the recursion formula along with the ratio test or limit ratio test to find the radius of convergence and largest interval of convergence for $y_{\lambda,E}(x)$ and for $y_{\lambda,O}(x)$, provided the series does not terminate as polynomials.*

g. *Verify each of the following using work already done above:*

i. *If $\lambda = N(N+1)$ for some nonnegative integer N, then the Legendre equation with parameter λ has polynomial solutions, and they are all constant multiples of $p_N(x)$.*

ii. *If $\lambda \neq N(N+1)$ for every nonnegative integer N, then y_λ is not a polynomial.*

iii. *If y_λ is not a polynomial, then it is given by a power series about $x_0 = 0$ with a radius of convergence of exactly 1.*

33.11. For each of the following, use the Taylor series method to find the Nth-degree partial sum of the power series solution about x_0 to the given differential equation. If either applies, use Theorems 33.2 on page 618 or 33.7 on page 629 to determine an interval I over which the power series solutions are valid.

a. $y'' + 4y = 0$ with $x_0 = 0$ and $N = 5$

b. $y'' - x^2 y = 0$ with $x_0 = 0$ and $N = 5$

c. $y'' + e^{2x} y = 0$ with $x_0 = 0$ and $N = 4$

d. $\sin(x)\, y'' - y = 0$ with $x_0 = \frac{\pi}{2}$ and $N = 4$

e. $y'' + xy = \sin(x)$ with $x_0 = 0$ and $N = 5$

f. $y'' - \sin(x)\, y' - xy = 0$ with $x_0 = 0$ and $N = 4$

g. $y'' - y^2 = 0$ with $x_0 = 0$ and $N = 5$

h. $y' + \cos(y) = 0$ with $x_0 = 0$ and $N = 3$

34

Power Series Solutions II: Generalizations and Theory

A major goal in this chapter is to confirm the claims made in Theorems 33.2 and 33.7 regarding the validity of the algebraic method. Along the way, we will also expand both the set of differential equations for which this method can be considered and our definitions of "regular" and "singular" points. As a bonus, we'll also obtain formulas that, at least in some cases, can simplify the computation of the terms of the power series solutions.

34.1 Equations with Analytic Coefficients

In the previous chapter, we discussed an algebraic method for finding a general power series solution about a point x_0 to any differential equation of the form

$$A(x)y' + B(x)y = 0 \quad \text{or} \quad A(x)y'' + B(x)y' + C(x)y = 0$$

where $A(x)$, $B(x)$ and $C(x)$ are polynomials with $A(x_0) \neq 0$. Observe that these polynomials can be written as

$$A(x) = \sum_{k=0}^{N} a_k(x - x_0)^k \quad \text{with} \quad a_0 \neq 0 \quad ,$$

$$B(x) = \sum_{k=0}^{N} b_k(x - x_0)^k \quad \text{and} \quad C(x) = \sum_{k=0}^{N} c_k(x - x_0)^k$$

where N is the highest power appearing in these polynomials. Surely, you are now wondering: *Must N be finite? Or will our algebraic method still work if $N = \infty$?* That is, can we use our algebraic method to find power series solutions about x_0 to

$$A(x)y' + B(x)y = 0 \quad \text{and} \quad A(x)y'' + B(x)y' + C(x)y = 0$$

when $A(x)$, $B(x)$ and $C(x)$ are functions expressible as power series about x_0 (i.e., when A, B and C are functions analytic at x_0) and with $A(x_0) \neq 0$?

And the answer to this question is *yes*, at least in theory. Simply replace the coefficients in the differential equations with their power series about x_0, and follow the steps already outlined in Sections 33.2 and 33.4 (possibly using the formula from Theorem 32.12 on page 595 for multiplying infinite series).

There are, of course, some further questions you are bound to be asking regarding these power series solutions and the finding of them. In particular:

1. What about the radii of convergence for the resulting power series solutions?

and

2. Are there any shortcuts to what could clearly be a rather lengthy and tedious set of calculations?

For the answers, read on.

34.2 Ordinary and Singular Points, the Radius of Analyticity, and the Reduced Form

Introducing Complex Variables

To properly address at least one of our questions, and to simplify the statements of our theorems, it will help to start viewing the coefficients of our differential equations as functions of a complex variable z. We actually did this in the last chapter when we referred to a point z_s in the complex plane for which $A(z_s) = 0$. But A was a polynomial then, and viewing a polynomial as a function of a complex variable is so easy that we hardly noted doing so. Viewing other functions (such as exponentials, logarithms and trigonometric functions) as functions of a complex variable may be a bit more challenging.

Analyticity and Power Series

Let us start by recalling that we need not restrict the variable or the center in a power series to real values — they can be complex,

$$\sum_{k=0}^{\infty} a_k (z - z_0)^k \qquad \text{for} \quad |z - z_0| < R \quad ,$$

in which case the radius of convergence R is the radius of the largest open disk in the complex plane centered at z_0 on which the power series is convergent.[1]

Also recall that our definition of analyticity also applies to functions of a complex variable; that is, any function f of a complex variable is *analytic* at a point z_0 in the complex plane if and only if $f(z)$ can be expressed as a power series about z_0,

$$f(z) = \sum_{k=0}^{\infty} a_k (z - z_0)^k \qquad \text{for} \quad |z - z_0| < R$$

for some $R > 0$. Moreover, as also noted in Section 32.3, if f is any function of a real variable given by a power series on the interval $(x_0 - R, x_0 + R)$,

$$f(x) = \sum_{k=0}^{\infty} a_k (x - x_0)^k \quad ,$$

[1] If you don't recall this, quickly review Section 32.3.

then we can view this function as a function of the complex variable $z = x + iy$ on a disk of radius R about x_0 by simply replacing the real variable x with the complex variable z,

$$f(z) = \sum_{k=0}^{\infty} a_k (z - x_0)^k \quad .$$

We will do this automatically in all that follows.

By the way, do observe that, if

$$\lim_{z \to z_0} |f(z)| = \infty \quad ,$$

then f certainly is not analytic at z_0 !

Some Results from Complex Analysis

Useful insights regarding analytic functions can be gained from the theory normally developed in an introductory course on "complex analysis". Sadly, we do not have the time or space to properly develop that theory here. As an alternative, a brief overview of the relevant parts of that theory is given for the interested reader in an appendix near the end of this chapter (Section 34.7). From that appendix, we get the following two lemmas (both of which should seem reasonable):

Lemma 34.1
Assume F is a function analytic at z_0 with corresponding power series $\sum_{k=0}^{\infty} f_k (z - z_0)^k$, and let R be either some positive value or $+\infty$. Then

$$F(z) = \sum_{k=0}^{\infty} f_k (z - z_0)^k \qquad \text{whenever} \quad |z - z_0| < R$$

if and only if F is analytic at every complex point z satisfying

$$|z - z_0| < R \quad .$$

Lemma 34.2
Assume $F(z)$ and $A(z)$ are two functions analytic at a point z_0. Then the quotient F/A is also analytic at z_0 if and only if

$$\lim_{z \to z_0} \frac{F(z)}{A(z)}$$

is finite.

Let us note the following immediate corollary of the first lemma:

Corollary 34.3
Assume $F(x)$ is some function on the real line, and $\sum_{k=0}^{\infty} f_k (x - x_0)^k$ is a power series with a infinite radius of convergence . If

$$F(x) = \sum_{k=0}^{\infty} f_k (x - x_0)^k \qquad \text{for} \quad -\infty < x < \infty \quad ,$$

then $F(z)$ is analytic at every point in the complex plane.

From this corollary and the series in set (32.4) on page 583, it immediately follows that the sine and cosine functions, as well as the exponential functions, are all analytic on the entire complex plane.

Let us also note that the second lemma extends some observations regarding quotients of functions made in Section 32.4.[2]

Ordinary and Singular Points

Let z_0 be a point on the complex plane, and let a, b and c be functions on the complex plane. We will say that z_0 is an *ordinary point* for the first-order differential equation

$$a(x)y' + b(x)y = 0$$

if and only if the quotient

$$\frac{b(z)}{a(z)}$$

is analytic at z_0. And we will say that z_0 is an *ordinary point* for the second-order differential equation

$$a(x)y'' + b(x)y' + c(x)y = 0$$

if and only if the quotients

$$\frac{b(z)}{a(z)} \quad \text{and} \quad \frac{c(z)}{a(z)}$$

are both analytic at z_0.

Any point that is not an ordinary point (that is, any point at which the above quotients are not analytic) is called a *singular point* for the differential equation.

Using Lemma 34.2, you can easily verify the following shortcuts for determining whether a point is a singular or ordinary point for a given differential equation. You can then use these lemmas to verify that our new definitions reduce to those given in the last chapter when the coefficients of our differential equation are rational functions.

Lemma 34.4

Let z_0 be a point in the complex plane, and consider the differential equation

$$a(x)y' + b(x)y = 0$$

where a and b are functions analytic at z_0. Then:

1. *If $a(z_0) \neq 0$, then z_0 is an ordinary point for the differential equation.*
2. *If $a(z_0) = 0$ and $b(z_0) \neq 0$, then z_0 is a singular point for the differential equation.*
3. *The point z_0 is an ordinary point for this differential equation if and only if*

$$\lim_{z \to z_0} \frac{b(z)}{a(z)}$$

is finite.

[2] If you haven't already done so, now might be a good time to at least skim over the material in the subsection *More on Algebra with Power Series and Analytic Functions* starting on page 595.

Lemma 34.5
Let z_0 be a point in the complex plane, and consider the differential equation

$$a(x)y'' + b(x)y' + c(x)y = 0$$

where a, b and c are functions analytic at z_0. Then:

1. *If $a(z_0) \neq 0$, then z_0 is an ordinary point for the differential equation.*
2. *If $a(z_0) = 0$, and either $b(z_0) \neq 0$ or $c(z_0) \neq 0$, then z_0 is a singular point for the differential equation.*
3. *The point z_0 is an ordinary point for this differential equation if and only if*

$$\lim_{z \to z_0} \frac{b(z)}{a(z)} \quad \text{and} \quad \lim_{z \to z_0} \frac{c(z)}{a(z)}$$

are both finite.

!►Example 34.1: Consider the two differential equations

$$y'' + \sin(x)\, y = 0 \quad \text{and} \quad \sin(x)\, y'' + y = 0 \quad .$$

From Corollary 34.3, we know that the sine function is analytic at every point on the complex plane, and that

$$\sin(z) = 0 \quad \text{if} \quad z = n\pi \quad \text{with} \quad n = 0, \pm 1, \pm 2, \ldots \quad .$$

Moreover, it's not hard to show (see Exercise 34.4) that the above points are the only points in the complex plane at which the sine is zero.

What this means is that both coefficients of

$$y'' + \sin(x)\, y = 0$$

are analytic everywhere, with the first coefficient (which is simply the constant 1) never being zero. Thus, Lemma 34.5 assures us that every point in the complex plane is an ordinary point for this differential equation. It has no singular points.

On the other hand, while both coefficients of

$$\sin(x)\, y'' + 5y = 0$$

are analytic everywhere, the first coefficient is zero at $z_0 = 0$ (and at every other integral multiple of π). Since the second coefficient (again, the constant 1) is not zero at $z_0 = 0$, Lemma 34.5 tells us that $z_0 = 0$ (and every other integral multiple of π) is a singular point for this differential equation.

Radius of Analyticity
The Definition, Recycled

Why waste a perfectly good definition? Given

$$a(x)y' + b(x)y = 0 \quad \text{or} \quad a(x)y'' + b(x)y' + c(x)y = 0$$

we define the *radius of analyticity* (for the differential equation) about any given point z_0 to be the distance between z_0 and the singular point closest to z_0, unless the differential equation has no singular points, in which case we define the *radius of analyticity* to be $+\infty$.

This is precisely the same definition as given (twice) in the previous chapter.

Is the Radius Well Defined?

When the coefficients of our differential equations were just polynomials, it should have been obvious that there really was a "singular point closest to z_0" (provided the equation had singular points). But a cynical reader — especially one who has seen some advanced analysis — may wonder if such a singular point always exists with our more general equations, or if, instead, a devious mathematician could construct a differential equation with an infinite set of singular points, none of which are closest to the given ordinary point. Don't worry, no mathematician is devious enough.

Lemma 34.6
Let z_0 be an ordinary point for some first- or second-order linear homogeneous differential equation. Then, if the differential equation has singular points, there is at least one singular point z_s such that no other singular point is closer to z_0 than z_s.

The z_s in this lemma is a "singular point closest to z_0". There may, in fact, be other singular points at the same distance from z_0, but none closer. Anyway, this ensures that "the radius of analyticity" for a given differential equation about a given point is well defined.

The proof of Lemma 34.6 is subtle, and is discussed in an appendix (Section 34.8).

34.3 The Reduced Forms
A Standard Way to Rewrite Our Equations

There is some benefit in dividing a given differential equation

$$ay' + by = 0 \qquad \text{or} \qquad ay'' + by' + cy = 0$$

by the equation's leading coefficient, obtaining the equation's corresponding *reduced form*[3]

$$y' + Py = 0 \qquad \text{or} \qquad y'' + Py' + Qy = 0$$

(with $P = {}^{b}\!/_{a}$ and $Q = {}^{c}\!/_{a}$). For one thing, it may reduce the number of products of infinite series to be computed. In addition, it will allow us to use the generic recursion formulas that we will be deriving in a little bit. However, the advantages of using the reduced form depend somewhat on the ease in finding and using the power series for P (and, in the second-order case, for Q). In particular, if A, B and C are all relatively simple polynomials (with A not being a constant), then dividing

$$Ay'' + By' + Cy = 0$$

by A is unlikely to simplify your computations — don't do it unless ordered to do so in an exercise.

Ordinary Points and the Reduced Form

The next two lemmas will be important in deriving the general formulas for power series solutions. However, they follow almost immediately from Lemmas 34.1 and 34.2, along with the definitions of "reduced form", "regular and singular points", "radius of convergence" and "analyticity at z_0".

[3] also called the *normal form* by some authors

Lemma 34.7
Let

$$ay' + by = 0$$

have reduced form

$$y' + Py = 0 \quad .$$

Then z_0 is an ordinary point for this differential equation if and only if P is analytic at z_0. Moreover, if z_0 is an ordinary point, then P has a power series representation

$$P(z) = \sum_{k=0}^{\infty} p_k (z - z_0)^k \qquad \text{for} \quad |z - z_0| < R$$

where R is the radius of analyticity for this differential equation about z_0.

Lemma 34.8
Let

$$ay'' + by' + cy = 0$$

have reduced form

$$y'' + Py' + Qy = 0 \quad .$$

Then z_0 is an ordinary point for this differential equation if and only if both P and Q are analytic at z_0. Moreover, if z_0 is an ordinary point, then P and Q have power series representations

$$P(z) = \sum_{k=0}^{\infty} p_k (z - z_0)^k \qquad \text{for} \quad |z - z_0| < R$$

and

$$Q(z) = \sum_{k=0}^{\infty} q_k (z - z_0)^k \qquad \text{for} \quad |z - z_0| < R$$

where R is the radius of analyticity for this differential equation about z_0.

34.4 Existence of Power Series Solutions

Deriving the Generic Recursion Formulas

First-Order Case

Let us try to find the general power series solution to

$$y' + Py = 0$$

about $x_0 = 0$ when P is any function analytic at $x_0 = 0$. This analyticity means P has a power series representation

$$P(x) = \sum_{k=0}^{\infty} p_k x^k \qquad \text{for} \quad |x| < R$$

for some $R > 0$. We'll assume that this series and a value for R are known.

Our approach is to follow the method described in Section 33.2 as far as possible. We assume y is given by a yet unknown power series about $x_0 = 0$,

$$y(x) = \sum_{k=0}^{\infty} a_k x^k \quad ,$$

compute the corresponding series for y', plug that into the differential equation, and "compute" (using the above series for P and the formula for series multiplication from Theorem 32.12 on page 595):

$$y' + Py = 0$$

$$\hookrightarrow \quad \sum_{k=1}^{\infty} k a_k x^{k-1} + \left(\sum_{k=0}^{\infty} a_k x^k\right)\left(\sum_{k=0}^{\infty} p_k x^k\right) = 0$$

$$\hookrightarrow \quad \underbrace{\sum_{k=1}^{\infty} k a_k x^{k-1}}_{n=k-1} + \underbrace{\sum_{k=0}^{\infty} \left[\sum_{j=0}^{k} a_j p_{k-j}\right] x^k}_{n=k} = 0$$

$$\hookrightarrow \quad \sum_{n=0}^{\infty} (n+1) a_{n+1} x^n + \sum_{n=0}^{\infty} \left[\sum_{j=0}^{n} a_j p_{n-j}\right] x^n = 0$$

$$\hookrightarrow \quad \sum_{n=0}^{\infty} \left[(n+1) a_{n+1} + \sum_{j=0}^{n} a_j p_{n-j}\right] x^n = 0 \quad .$$

Thus,

$$(n+1) a_{n+1} + \sum_{j=0}^{n} a_j p_{n-j} = 0 \qquad \text{for} \quad n = 0, 1, 2, \dots \quad .$$

Solving for a_{n+1} and letting $k = n+1$ gives us

$$a_k = -\frac{1}{k} \sum_{j=0}^{k-1} a_j p_{k-1-j} \qquad \text{for} \quad k = 1, 2, 3, \dots \quad . \tag{34.1}$$

Of course, we would have obtained the same recursion formula with x_0 being any ordinary point for the given differential equation (just replace x in the above computations with $X = x - x_0$).

Second-Order Case

We will leave this derivation as an exercise.

?►Exercise 34.1: *Assume that, over some interval containing the point x_0, P and Q are functions given by power series*

$$P(x) = \sum_{k=0}^{\infty} p_k (x - x_0)^k \qquad \text{and} \qquad Q(x) = \sum_{k=0}^{\infty} q_k (x - x_0)^k \quad ,$$

and derive the recursion formula

$$a_k = -\frac{1}{k(k-1)} \sum_{j=0}^{k-2} \left[(j+1) a_{j+1} p_{k-2-j} + a_j q_{k-2-j}\right] \tag{34.2}$$

for the series solution

$$y(x) = \sum_{k=0}^{\infty} a_k(x - x_0)^k$$

to

$$y'' + Py' + Qy = 0 \quad .$$

(For simplicity, start with the case in which $x_0 = 0$.)

Validity of the Power Series Solutions

Here are the big theorems on the existence of power series solutions. They are also theorems on the computation of these solutions since they contain the recursion formulas just derived.

Theorem 34.9 (first-order series solutions)

Suppose x_0 is an ordinary point for a first-order homogeneous differential equation whose reduced form is

$$y' + Py = 0 \quad .$$

Then P has a power series representation

$$P(x) = \sum_{k=0}^{\infty} p_k(x - x_0)^k \qquad \text{for} \quad |x - x_0| < R$$

where R is the radius of analyticity about x_0 for this differential equation.

Moreover, a general solution to the differential equation is given by

$$y(x) = \sum_{k=0}^{\infty} a_k(x - x_0)^k \qquad \text{for} \quad |x - x_0| < R$$

where a_0 is arbitrary, and the other a_k's satisfy the recursion formula

$$a_k = -\frac{1}{k}\sum_{j=0}^{k-1} a_j p_{k-1-j} \quad . \tag{34.3}$$

Theorem 34.10 (second-order series solutions)

Suppose x_0 is an ordinary point for a second-order homogeneous differential equation whose reduced form is

$$y'' + Py' + Qy = 0 \quad .$$

Then P and Q have power series representations

$$P(x) = \sum_{k=0}^{\infty} p_k(x - x_0)^k \qquad \text{for} \quad |x - x_0| < R$$

and

$$Q(x) = \sum_{k=0}^{\infty} q_k(x - x_0)^k \qquad \text{for} \quad |x - x_0| < R$$

where R is the radius of analyticity about x_0 for this differential equation.

Moreover, a general solution to the differential equation is given by

$$y(x) = \sum_{k=0}^{\infty} a_k (x - x_0)^k \qquad \text{for} \quad |x - x_0| < R$$

where a_0 and a_1 are arbitrary, and the other a_k's satisfy the recursion formula

$$a_k = -\frac{1}{k(k-1)} \sum_{j=0}^{k-2} \left[(j+1) a_{j+1} p_{k-2-j} + a_j q_{k-2-j} \right] \quad . \tag{34.4}$$

There are four major parts to the proof of each of these theorems:

1. Deriving the recursion formula. (Done!)
2. Assuring ourselves that the coefficient functions in the reduced forms have the stated power series representations. (Done! See Lemmas 34.7 and 34.8.)
3. Verifying that the radius of convergence for the power series generated from the given recursion formula is at least R.
4. Noting that the calculations used to obtain each recursion formula also confirm that the resulting series satisfies the given differential equation over the interval $(x_0 - R, x_0 + R)$. (So noted!)

Thus, all that remains to proving these two major theorems is the verification of the claimed radii of convergence for the given series solutions. This verification is not difficult, but is a bit lengthy and technical, and may not be as exciting to the reader as was the derivation of the recursion formulas. Those who are interested should proceed to Section 34.5.

But now, let us try using our new theorems.

!►Example 34.2: *Consider, again, the differential equation from Example 33.7 on page 635,*

$$y'' + \cos(x) y = 0 \quad .$$

Again, let us try to find at least a partial sum of the general power series solution about $x_0 = 0$. This time, however, we will use the results from Theorem 34.10.

The equation is already in reduced form

$$y'' + Py' + Qy = 0$$

with $P(x) = 0$ and $Q(x) = \cos(x)$. Since both of these functions are analytic on the entire complex plane, the theorem assures us that there is a general power series solution

$$y(x) = \sum_{k=0}^{\infty} a_k x^k \qquad \text{for} \quad |x| < \infty$$

with a_0 and a_1 being arbitrary, and with the other a_k's being given through recursion formula (34.4). And to use this recursion formula, we need the corresponding power series representations for P and Q. The series for P, of course, is trivial,

$$P(x) = 0 \quad \Longleftrightarrow \quad P(x) = \sum_{k=0}^{\infty} p_k x^k \qquad \text{with} \quad p_k = 0 \qquad \text{for all} \quad k \quad .$$

Fortunately, the power series for Q is well known and only needs to be slightly rewritten for use in our recursion formula:

$$\begin{aligned} Q(x) &= \cos(x) \\ &= \sum_{m=0}^{\infty} (-1)^m \frac{1}{(2m)!} x^{2m} \\ &= 1 - \frac{1}{2!}x^2 + \frac{1}{4!}x^4 - \frac{1}{6!}x^6 + \cdots \\ &= (-1)^{0/2}x^0 + 0x^1 + (-1)^{2/2}\frac{1}{2!}x^2 + 0x^3 \\ &\quad + (-1)^{4/2}\frac{1}{4!}x^4 + 0x^5 + (-1)^{6/2}\frac{1}{6!}x^6 + 0x^7 + \cdots \quad . \end{aligned}$$

So,

$$q_0 = 1 \quad , \quad q_1 = 0 \quad , \quad q_2 = -\frac{1}{2!} \quad , \quad q_3 = 0 \quad , \quad q_4 = \frac{1}{4!} \quad , \quad \ldots \quad .$$

In general,

$$Q(x) = \sum_{k=0}^{\infty} q_k x^k \qquad \text{with} \quad q_k = \begin{cases} (-1)^{k/2}\frac{1}{k!} & \text{if } k \text{ is even} \\ 0 & \text{if } k \text{ is odd} \end{cases} \quad ,$$

and recursion formula (34.4) becomes, for $k \geq 2$,

$$\begin{aligned} a_k &= -\frac{1}{k(k-1)} \sum_{j=0}^{k-2} \Big[(j+1)a_{j+1} \underbrace{p_{k-2-j}}_{0} + a_j q_{k-2-j} \Big] \\ &= -\frac{1}{k(k-1)} \sum_{j=0}^{k-2} a_j \begin{Bmatrix} (-1)^{(k-2-j)/2}\frac{1}{(k-2-j)!} & \text{if } k-2-j \text{ is even} \\ 0 & \text{if } k-2-j \text{ is odd} \end{Bmatrix} \quad . \end{aligned} \tag{34.5}$$

However, since we are only attempting to find a partial sum and not the entire series, let us simply use the recursion formula

$$\begin{aligned} a_k &= -\frac{1}{k(k-1)} \sum_{j=0}^{k-2} \Big[(j+1)a_{j+1} \underbrace{p_{k-2-j}}_{0} + a_j q_{k-2-j} \Big] \\ &= -\frac{1}{k(k-1)} \sum_{j=0}^{k-2} a_j q_{k-2-j} \quad , \end{aligned}$$

with the particular q_n's given above, and with the factorials computed:

$$q_0 = 1 \quad , \quad q_1 = 0 \quad , \quad q_2 = -\frac{1}{2} \quad , \quad q_3 = 0 \quad , \quad q_4 = \frac{1}{24} \quad .$$

Doing so, we get

$$\begin{aligned} a_2 &= -\frac{1}{2(2-1)} \sum_{j=0}^{2-2} a_j q_{2-2-j} \\ &= -\frac{1}{2} \sum_{j=0}^{0} a_j q_{-j} = -\frac{1}{2} a_0 q_0 = -\frac{1}{2} a_0 \cdot 1 = -\frac{1}{2} a_0 \quad , \end{aligned}$$

$$
\begin{aligned}
a_3 &= -\frac{1}{3(3-1)}\sum_{j=0}^{3-2} a_j q_{3-2-j} \\
&= -\frac{1}{6}\sum_{j=0}^{1} a_j q_{1-j} = -\frac{1}{6}[a_0 q_1 + a_1 q_0] = -\frac{1}{6}[a_0 \cdot 0 + a_1 \cdot 1] = -\frac{1}{6}a_1 \quad ,
\end{aligned}
$$

$$
\begin{aligned}
a_4 &= -\frac{1}{4(4-1)}\sum_{j=0}^{4-2} a_j q_{4-2-j} \\
&= -\frac{1}{12}\sum_{j=0}^{2} a_j q_{2-j} \\
&= -\frac{1}{12}[a_0 q_2 + a_1 q_1 + a_2 q_0] \\
&= -\frac{1}{12}\left[a_0\left(-\frac{1}{2}\right) + a_1 \cdot 0 + \left(-\frac{1}{2}a_0\right)\cdot 1\right] = \frac{1}{12}a_0 \quad ,
\end{aligned}
$$

$$
\begin{aligned}
a_5 &= -\frac{1}{5(5-1)}\sum_{j=0}^{5-2} a_j q_{5-2-j} \\
&= -\frac{1}{20}\sum_{j=0}^{3} a_j q_{3-j} \\
&= -\frac{1}{20}[a_0 q_3 + a_1 q_2 + a_2 q_1 + a_3 q_0] \\
&= -\frac{1}{20}\left[a_0 \cdot 0 + a_1\left(-\frac{1}{2}\right) + a_2 \cdot 0 + \left(-\frac{1}{6}a_1\right)\cdot 1\right] = \frac{1}{30}a_1
\end{aligned}
$$

and

$$
\begin{aligned}
a_6 &= -\frac{1}{6(6-1)}\sum_{j=0}^{6-2} a_j q_{6-2-j} \\
&= -\frac{1}{30}\sum_{j=0}^{4} a_j q_{4-j} \\
&= -\frac{1}{30}[a_0 q_4 + a_1 q_3 + a_2 q_2 + a_3 q_1 + a_4 q_0] \\
&= -\frac{1}{30}\left[a_0\left(\frac{1}{24}\right) + a_1 \cdot 0 + \left(-\frac{1}{2}a_0\right)\left(-\frac{1}{2}\right) + a_3 \cdot 0 + \left(\frac{1}{12}a_0\right)\cdot 1\right] = -\frac{1}{80}a_0 \quad .
\end{aligned}
$$

Thus, the sixth partial sum of the power series for y about 0 is

$$
\begin{aligned}
S_6(x) &= a_0 + a_1 x + a_2 x^2 + a_3 x^3 + a_4 x^4 + a_5 x^5 + a_6 x^6 \\
&= a_0 + a_1 x - \frac{1}{2}a_0 x^2 - \frac{1}{6}a_1 x^3 + \frac{1}{12}a_0 x^4 + \frac{1}{30}a_1 x^5 - \frac{1}{80}a_0 x^6 \\
&= a_0\left[1 - \frac{1}{2}x^2 + \frac{1}{12}x^4 - \frac{1}{80}x^6\right] + a_1\left[x - \frac{1}{6}x^3 + \frac{1}{30}x^5\right] \quad ,
\end{aligned}
$$

just as we had found, using the Taylor series method, in Example 33.7 on page 635.

If you compare the work done in the last example with the work done in Example 33.7, it may appear that, while we obtained identical results, we may have expended more work in using the recursion formula from Theorem 34.10 than in using the Taylor series method. On the other hand, all the computations done in the last example were fairly simple arithmetic computations — computations we could easily program a computer to do, especially if we use recursion formula (34.5). So there can be computational advantages to using our new results.

34.5 Radius of Convergence for the Solution Series

To finish our proofs of Theorems 34.9 and 34.10, we need to verify that the radius of convergence for each of the given series solutions is at least the given value for R. We will do this for the solution series in Theorem 34.10, and leave the corresponding verification for Theorem 34.9 (which will be slightly easier) as an exercise.

What We Have, and What We Need to Show

Recall: We have a positive value R and two power series

$$\sum_{k=0}^{\infty} p_k X^k \qquad \text{and} \qquad \sum_{k=0}^{\infty} q_k X^k$$

that we know converge when $|X| < R$ (for simplicity, we're letting $X = x - x_0$). We also have a corresponding power series

$$\sum_{k=0}^{\infty} a_k X^k$$

where a_0 and a_1 are arbitrary, and the other coefficients are given by the recursion formula

$$a_k = -\frac{1}{k(k-1)} \sum_{j=0}^{k-2} \left[(j+1) a_{j+1} p_{k-2-j} + a_j q_{k-2-j} \right] \qquad \text{for} \quad k = 2, 3, 4, \ldots \quad .$$

We now only need to show that $\sum_{k=0}^{\infty} a_k X^k$ converges whenever $|X| < R$, and to do that, we will produce another power series $\sum_{k=0}^{\infty} b_k X^k$ whose convergence is "easily" shown using the limit ratio test, and which is related to our first series by

$$|a_k| \leq b_k \qquad \text{for} \quad k = 0, 1, 2, 3, \ldots \quad .$$

By the comparison test, it then follows that $\sum_{k=0}^{\infty} \left| a_k X^k \right|$, and hence also $\sum_{k=0}^{\infty} a_k X^k$, converges.

So let X be any value with $|X| < R$.

Constructing the Series for Comparison

Our first step in constructing $\sum_{k=0}^{\infty} b_k X^k$ is to pick some value r between $|X|$ and R,

$$0 \leq |X| < r < R \quad .$$

Since $|r| < R$, we know the series

$$\sum_{k=0}^{\infty} p_k r^k \qquad \text{and} \qquad \sum_{k=0}^{\infty} q_k r^k$$

both converge. But a series cannot converge if the terms in the series become arbitrarily large in magnitude. So the magnitudes of these terms — the $\left|p_k r^k\right|$'s and $\left|q_k r^k\right|$'s — must be bounded; that is, there must be a finite number M such that

$$\left|p_k r^k\right| < M \qquad \text{and} \qquad \left|q_k r^k\right| < M \qquad \text{for} \quad k = 0, 1, 2, 3, \ldots \quad .$$

Equivalently (since $r > 0$),

$$|p_k| < \frac{M}{r^k} \qquad \text{and} \qquad |q_k| < \frac{M}{r^k} \qquad \text{for} \quad k = 0, 1, 2, 3, \ldots \quad .$$

These inequalities, the triangle inequality and the recursion formula combine to give us, for $k = 2, 3, 4, \ldots$,

$$\begin{aligned}
|a_k| &= \left| -\frac{1}{k(k-1)} \sum_{j=0}^{k-2} \left[(j+1) a_{j+1} p_{k-2-j} \; + \; a_j q_{k-2-j} \right] \right| \\
&\leq \frac{1}{k(k-1)} \sum_{j=0}^{k-2} \left[(j+1) \left|a_{j+1}\right| \left|p_{k-2-j}\right| \; + \; \left|a_j\right| \left|q_{k-2-j}\right| \right] \\
&\leq \frac{1}{k(k-1)} \sum_{j=0}^{k-2} \left[(j+1) \left|a_{j+1}\right| \frac{M}{r^{k-2-j}} \; + \; \left|a_j\right| \frac{M}{r^{k-2-j}} \right] \quad ,
\end{aligned}$$

which we will rewrite as

$$|a_k| \leq \frac{1}{k(k-1)} \sum_{j=0}^{k-2} \frac{M \left[(j+1) \left|a_{j+1}\right| + \left|a_j\right| \right]}{r^{k-2-j}} \quad .$$

Now let $b_0 = |a_0|$, $b_1 = |a_1|$ and

$$b_k = \sum_{j=0}^{k-2} \frac{M \left[(j+1) \left|a_{j+1}\right| + \left|a_j\right| \right]}{r^{k-2-j}} \qquad \text{for} \quad k = 2, 3, 4, \ldots \quad .$$

From the preceding inequality, it is clear that we've chosen the b_k's so that

$$|a_k| \leq b_k \qquad \text{for} \quad k = 0, 1, 2, 3, \ldots \quad .$$

In fact, we even have

$$|a_k| \leq \frac{1}{k(k-1)} b_k \qquad \text{for} \quad k = 2, 3, \ldots \quad . \tag{34.6}$$

Thus,

$$\left|a_k X^k\right| \leq b_k |X|^k \qquad \text{for} \quad k = 2, 3, \ldots \quad ,$$

and (by the comparison test) we can confirm the convergence of $\sum_{k=0}^{\infty} a_k X^k$ by simply verifying the convergence of $\sum_{k=0}^{\infty} b_k |X|^k$.

Convergence of the Comparison Series

According to the limit convergence test (Theorem 32.2 on page 582), $\sum_{k=0}^{\infty} b_k |X|^k$ converges if

$$\lim_{k\to\infty} \left| \frac{b_{k+1} X^{k+1}}{b_k X^k} \right| < 1 \quad .$$

Well, let $k > 2$. Using the formula for the b_k's with k replaced with $k+1$, we get

$$\begin{aligned} b_{k+1} &= \sum_{j=0}^{[k+1]-2} \frac{M\left[(j+1)\left|a_{j+1}\right| + \left|a_j\right|\right]}{r^{[k+1]-2-j}} \\ &= \sum_{j=0}^{k-1} \frac{M\left[(j+1)\left|a_{j+1}\right| + \left|a_j\right|\right]}{r^{k-1-j}} \\ &= \sum_{j=0}^{k-2} \frac{M\left[(j+1)\left|a_{j+1}\right| + \left|a_j\right|\right]}{r^{k-1-j}} + \frac{M\left[([k-1]+1)\left|a_{[k-1]+1}\right| + \left|a_{[k-1]}\right|\right]}{r^{k-1-[k-1]}} \\ &= \frac{1}{r} \sum_{j=0}^{k-2} \frac{M\left[(j+1)\left|a_{j+1}\right| + \left|a_j\right|\right]}{r^{k-2-j}} + M\left[k\left|a_k\right| + \left|a_{k-1}\right|\right] \\ &= \frac{1}{r} b_k + kM\left|a_k\right| + M\left|a_{k-1}\right| \quad . \end{aligned}$$

But, by inequality (34.6),

$$kM\left|a_k\right| \le kM \frac{1}{k(k-1)} b_k = \frac{M}{k-1} b_k \quad . \tag{34.7}$$

Moreover, because the terms in the summation for b_k are all nonnegative real numbers,

$$\begin{aligned} b_k &= \sum_{j=0}^{k-2} \frac{M\left[(j+1)\left|a_{j+1}\right| + \left|a_j\right|\right]}{r^{k-2-j}} \\ &\ge \text{the last term in the summation} \\ &= \frac{M(j+1)\left|a_{j+1}\right|}{r^{k-2-j}} \quad \text{with} \quad j = k-2 \\ &= \frac{M([k-2]+1)\left|a_{[k-2]+1}\right|}{r^{k-2-[k-2]}} \\ &= M(k-1)\left|a_{k-1}\right| \quad . \end{aligned}$$

Thus,

$$M\left|a_{k-1}\right| \le \frac{1}{(k-1)} b_k \quad . \tag{34.8}$$

Combining inequalities (34.7) and (34.8) with the last formula above for b_{k+1} gives us

$$\begin{aligned} b_{k+1} &= \frac{1}{r} b_k + kM\left|a_k\right| + M\left|a_{k-1}\right| \\ &\le \frac{1}{r} b_k + \frac{M}{(k-1)} b_k + \frac{1}{(k-1)} b_k = \left[\frac{1}{r} + \frac{M+1}{k-1}\right] b_k \quad . \end{aligned}$$

That is,

$$\frac{b_{k+1}}{b_k} \le \frac{1}{r} + \frac{M+1}{k-1} \quad .$$

From this and the fact that $|X| < r$, we see that

$$\lim_{k\to\infty}\left|\frac{b_{k+1}X^{k+1}}{b_k X^k}\right| = \lim_{k\to\infty}\left[\frac{1}{r}+\frac{M+1}{k-1}\right]|X| = \left[\frac{1}{r}+0\right]|X| = \frac{|X|}{r} < 1 \quad ,$$

confirming (by the limit ratio test) that $\sum_{k=0}^{\infty} b_k X^k$ converges, and, thus, completing our proof of Theorem 34.10. ∎

To finish the proof of Theorem 34.9, do the following exercise:

?►Exercise 34.2: *Let $\sum_{k=0}^{\infty} p_k X^k$ be a power series that converges for $|X| < R$, and let $\sum_{k=0}^{\infty} a_k X^k$ be a power series where a_0 is arbitrary, and the other coefficients are given by the recursion formula*

$$a_k = -\frac{1}{k}\sum_{j=0}^{k-1} a_j p_{k-1-j} \qquad \text{for} \quad k = 1, 2, 3, \ldots \quad .$$

Show that $\sum_{k=0}^{\infty} a_k X^k$ converges also for $|X| < R$.

(Suggestion: Go back to the start of this section and "redo" the computations step by step, making the obvious modifications to deal with the given recursion formula.)

34.6 Singular Points and the Radius of Convergence

In the last section, we verified that the power series solutions obtained in this chapter are valid at least over $(x_0 - R, x_0 + R)$ where R is the distance from x_0 to the nearest singular point, provided the differential equation has singular points. This fact can be refined by the following theorem:

Theorem 34.11
Let

$$y(x) = \sum_{k=0}^{\infty} c_k (x - x_0)^k \qquad \text{for} \quad |x - x_0| < R$$

be a power series solution for some first- or second-order homogeneous linear differential equation. Assume, further, that R is finite and is the radius of convergence for the above power series. Then this differential equation has a singular point z_s with $|z_s - x_0| = R$.

The proof of this theorem, unfortunately, is nontrivial. The adventurous can read about it in an appendix, Section 34.9 (after reading Sections 34.7 and 34.8). By the way, the above theorem is actually a consequence of more general results obtained in the appendix, some of which will be useful in some of our more advanced work in Chapter 36.

34.7 Appendix: A Brief Overview of Complex Calculus

To properly address issues regarding the analyticity of our functions and the regions of convergence of their power series, we need to delve deeper into the theory of analytic functions — much deeper than normally presented in elementary calculus courses. Instead, we want the theory normally developed in introductory courses in complex analysis. That's because the complex-variable theory exposes a much closer relation between "differentiability" and "analyticity" than does the real-variable theory developed in elementary calculus. If you've had such a course, good; the following will be a review. If you've not had such a course, think about taking one, and read on. What follows is a brief synopsis of the relevant concepts and results from such a course, written assuming you have not had such a course (but have, at least, skimmed the introductory section on complex variables in Section 32.3).

Functions of a Complex Variable

In "complex analysis", the basic concepts and theories developed in elementary calculus are extended so that they apply to complex-valued functions of a complex variable. Thus, for example, where we may have considered the "real" polynomial and "real" exponential

$$p(x) = 3x^2 + 4x - 5 \qquad \text{and} \qquad h(x) = e^x \qquad \text{for all } x \text{ in } \mathbb{R}$$

in elementary calculus, in complex analysis we consider the "complex" polynomial and "complex" exponential

$$p(z) = 3z^2 + 4z - 5 \qquad \text{and} \qquad h(z) = e^z \qquad \text{for all } z = x + iy \text{ in } \mathbb{C} \quad .$$

Note that we treat z as a single entity. Still, the complex variable z is just $x + iy$. Consequently, much of complex analysis follows from what you already know about the calculus of functions of two variables. In particular, the partial derivatives with respect to x and y are defined just as they were defined back in your calculus course (and Section 3.7 of this text), and when we say f is *continuous at* z_0, we mean that

$$f(z_0) = \lim_{z \to z_0} f(z)$$

with the understanding that

$$\lim_{z \to z_0} f(z) = \lim_{\substack{x \to x_0 \\ y \to y_0}} f(x + iy) \qquad \text{with} \quad z_0 = x_0 + iy_0 \quad .$$

Along these lines, you should be aware that, in complex variables, we normally assume that functions are defined over subregions of the complex plane, instead of subintervals of the real line. In what follows, we will often require our region of interest to be *open* (as discussed in Section 3.7). For example, we will often refer to the disk of all z satisfying $|z - z_0| < R$ for some complex point z_0 and positive value R. Any such disk is an open region.

Complex Differentiability

Given a function f and a point $z_0 = x_0 + iy_0$ in the complex plane, the *complex derivative of* f *at* z_0 — denoted by $f'(z_0)$ or $\left.{}^{df}/_{dz}\right|_{z_0}$ — is given by

$$f'(z_0) = \left.\frac{df}{dz}\right|_{z_0} = \lim_{z \to z_0} \frac{f(z) - f(z_0)}{z - z_0} \quad .$$

If this limit exists as a finite complex number, we say that f is *differentiable with respect to the complex variable at* z_0 (*complex differentiable* for short). Remember, $z = x + iy$; so, for the above limit to make sense, the formula for f must be such that $f(x + iy)$ makes sense for every $x + iy$ in some open region about z_0 .

We further say that f is complex differentiable on a region of the complex plane if and only if it is complex differentiable at every point in the region.

Naturally, we can extend the complex derivative to higher orders:

$$f'' = \frac{d^2 f}{dz^2} = \frac{d}{dz}\frac{df}{dz} \quad , \quad f''' = \frac{d^3 f}{dz^3} = \frac{d}{dz}\frac{d^2 f}{dz^2} \quad , \quad \ldots \quad .$$

As with functions of a real variable, if $f^{(k)}$ exists for every positive integer k (at a point or in a region), then we say f is *infinitely differentiable* (at the point or in the region).

In many ways, the complex derivative is analogous to the derivative you learned in elementary calculus (the *real-variable derivative*). The same basic computational formulas apply, giving us, for example,

$$\frac{d}{dz}z^k = kz^{k-1} \quad , \quad \frac{d}{dz}e^{\alpha z} = \alpha e^{\alpha z} \quad \text{and} \quad \frac{d}{dz}[\alpha f(z) + \beta g(z)] = \alpha f'(z) + \beta g'(z) \quad .$$

In addition, the well-known product and quotient rules can easily be verified, and, in verifying these rules, you automatically verify the following:

Theorem 34.12
Assume f and g are complex differentiable on some open region of the complex plane. Then their product fg is also complex differentiable on that region. Moreover, so is their quotient $^f/_g$, provided $g(z) \neq 0$ for every z in this region.

Testing for Complex Differentiability

If f is complex differentiable in some open region of the complex plane, then, unsurprisingly, the chain rule can be shown to hold. In particular,

$$\frac{\partial}{\partial x}f(x+iy) = f'(x+iy)\cdot\frac{\partial}{\partial x}[x+iy] = f'(x+iy)\cdot 1 = f'(x+iy)$$

and

$$\frac{\partial}{\partial y}f(x+iy) = f'(x+iy)\cdot\frac{\partial}{\partial y}[x+iy] = f'(x+iy)\cdot i = if'(x+iy) \quad .$$

Combining these two equations, we get

$$\frac{\partial}{\partial y}f(x+iy) = if'(x+iy) = i\frac{\partial}{\partial x}f(x+iy) \quad .$$

Thus, if f is complex differentiable in some open region, then

$$\frac{\partial}{\partial y}f(x+iy) = i\frac{\partial}{\partial x}f(x+iy) \tag{34.9}$$

at every point $z = x + iy$ in that region.[4] Right off, this gives us a test for "nondifferentiability": If equation (34.9) does not hold throughout some region, then f is not complex differentiable on

[4] The two equations you get by splitting equation (34.9) into its real and imaginary parts are the famous *Cauchy-Riemann equations*.

that region. Remarkably, it can be shown that equation (34.9) can also be used to verify complex differentiability. More precisely, the following theorem can be verified using tools developed in a typical course in complex analysis.

Theorem 34.13
A function f is complex differentiable on an open region if and only if

$$\frac{\partial}{\partial y} f(x+iy) = i\frac{\partial}{\partial x} f(x+iy)$$

at every point $z = x+iy$ in the region. Moreover, in any open region on which f is complex differentiable,

$$f'(z) = \frac{d}{dz} f(z) = \frac{\partial}{\partial x} f(x+iy) \quad .$$

Differentiability of an Analytic Function

In the subsection starting on page 587 of Section 32.2, we discussed differentiating power series and analytic functions when the variable is real. That discussion remains true if we replace the real variable x with the complex variable z and use the complex derivative. In particular, we have:

Theorem 34.14 (differentiation of power series)
Suppose f is a function given by a power series,

$$f(z) = \sum_{k=0}^{\infty} a_k (z-z_0)^k \qquad \text{for} \quad |z-z_0| < R$$

for some $R > 0$. Then, for any positive integer n, the nth derivative of f exists. Moreover,

$$f^{(n)}(x) = \sum_{k=n}^{\infty} k(k-1)(k-2)\cdots(k-n+1)\, a_k (z-z_0)^{k-n} \qquad \text{for} \quad |z-z_0| < R \quad .$$

As an immediate corollary, we have:

Corollary 34.15
Let f be analytic at z_0 with power series representation

$$f(z) = \sum_{k=0}^{\infty} a_k (z-z_0)^k \qquad \text{whenever} \quad |z-z_0| < R \quad .$$

Then f is infinitely complex differentiable on the disk of all z with $|z-z_0| < R$. Moreover

$$a_k = \frac{f^{(k)}(z_0)}{k!} \qquad \text{for} \quad k = 0, 1, 2, \ldots \quad .$$

Complex Differentiability and Analyticity

Despite the similarity between complex differentiation and real-variable differentiation, complex differentiability is a much stronger condition on functions than is real-variable differentiability. The next theorem illustrates this.

Theorem 34.16
Assume $f(z)$ is complex differentiable in some open region $\mathcal{R}$. Then f is analytic at each point z_0 in $\mathcal{R}$, and is given by its Taylor series

$$f(z) = \sum_{k=0}^{\infty} \frac{f^{(k)}(z_0)}{k!}(z - z_0)^k \qquad \text{whenever} \quad |z - z_0| < R$$

where R is the radius of any open disk centered at z_0 and contained in region $\mathcal{R}$.

This remarkable theorem tells us that complex differentiability on an open region automatically implies analyticity on that region, and tells us the region over which a function's Taylor series converges and equals the function. Proving this theorem is beyond this text. It is, in fact, a summary of results normally derived over the course of many chapters of a typical text in complex analysis.

Keep in mind that we already saw that analyticity implied complex differentiability (Corollary 34.15). So as immediate corollaries to the above, we have:

Corollary 34.17
A function f is analytic at every point in an open region of the complex plane if and only if it is complex differentiable at every point in that region.

Corollary 34.18
Assume F is a function analytic at z_0 with corresponding power series $\sum_{k=0}^{\infty} f_k(z - z_0)^k$, and let R be either some positive value or $+\infty$. Then

$$F(z) = \sum_{k=0}^{\infty} f_k(z - z_0)^k \qquad \text{whenever} \quad |z - z_0| < R$$

if and only if F is analytic at every complex point z satisfying

$$|z - z_0| < R \quad .$$

The second corollary is especially of interest to us because it is the same as Lemma 34.1 on page 649, which we used extensively in this chapter. And the other lemma that we used, Lemma 34.2 on page 649? Well, using, in order,

1. Corollary 32.14 on page 597 on quotients of analytic functions,

2. Theorem 34.12 on page 664 on the differentiation of products and quotients

and

3. Corollary 34.17, above,

you can easily verify the following (which is the same as Lemma 34.2):

Corollary 34.19
Assume $F(z)$ and $A(z)$ are two functions analytic at a point z_0. Then the quotient ${}^F/_A$ is also analytic at z_0 if and only if

$$\lim_{z \to z_0} \frac{F(z)}{A(x)}$$

is finite.

The details are left to you.

?►Exercise 34.3: *Prove Corollary 34.19.*

34.8 Appendix: The "Closest Singular Point"

Here we want to answer a subtle question: Is it possible to have a first- or second-order linear homogeneous differential equation whose singular points are arranged in such a manner that none of them is the closest to some given ordinary point?

For example, could there be a differential equation having $z_0 = 0$ as an ordinary point, but whose singular points form an infinite sequence

$$z_1, z_2, z_3, \ldots \qquad \text{with} \quad |z_k| = 1 + \frac{1}{k} \quad ,$$

possibly located in the complex plane so that they "spiral around" the circle of radius 1 about $z_0 = 0$ without converging to some single point? Each of these singular points is closer to $z_0 = 0$ than the previous ones in the sequence, so not one of them can be called "a closest singular point".

Lemma 34.6 on page 652 claims that this situation can*not* happen. Let us see why we should believe this lemma.

The Problem and Fundamental Theorem

We are assuming that we have some first- or second-order linear homogeneous differential equation having singular points. We also assume z_0 is not one of these singular points — it is an ordinary point. Our goal is to show that

there is a singular point z_s such that no other singular point is closer to z_0.

If we can confirm such a z_s exists, then we've shown that the answer to this section's opening question is *No* (and proven Lemma 34.6).

We start our search for this z_s by rewriting our differential equation in reduced form

$$y' + P(x)y = 0 \qquad \text{or} \qquad y'' + P(x)y' + Q(x)y = 0$$

and recalling that a point z is an ordinary point for our differential equation if and only if the coefficient(s) (P for the first-order equation, and both P and Q for the second-order equation) are all analytic at z (see Lemmas 34.7 and 34.8). Consequently,

1. z_s is a closest singular point to z_0 for the first-order differential equation if and only if z_s is a point closest to z_0 at which P is not analytic,

and

2. z_s is a closest singular point to z_0 for the second-order differential equation if and only if z_s is a point closest to z_0 at which either P or Q is not analytic.

Either way, our problem of verifying the existence of a singular point z_s "closest to z_0" is reduced to the problem of verifying the existence of a point z_s "closest to z_0" at which a given function F is not analytic while still being analytic at z_0. That is, to prove Lemma 34.6, it will suffice to prove the following:

Theorem 34.20
Let F be a function that is analytic at some but not all points in the complex plane, and let z_0 be one of the points at which F is analytic. Then there is a positive value R_0 and a point z_s in the complex plane such that all the following hold:

1. $R_0 = |z_s - z_0|$.

2. *F is not analytic at z_s .*

3. *F is analytic at every z with $|z - z_0| < R_0$.*

Verifying Theorem 34.20
The Radius of Analyticity Function

Our proof of Theorem 34.20 will rely on properties of the *radius of analyticity function R_A for F*, which we define at each point z in the complex plane as follows:

- If F is not analytic at z, then $R_A(z) = 0$.
- If F is analytic at z, then $R_A(z)$ is the largest value of R such that

$$F \text{ is analytic on the open disk of radius } R \text{ about } z \quad . \tag{34.10}$$

(To see that this "largest value of R" exists when f is analytic at z, first note that the set of all positive values of R for which (34.10) holds forms an interval with 0 as the lower endpoint. Since we are assuming there are points at which F is not analytic, this interval must be finite, and, hence, has a finite upper endpoint. That endpoint is $R_A(z)$.)[5]

The properties of this function that will be used in our proof of Theorem 34.20 are summarized in the following lemmas.

Lemma 34.21
Let R_A be the radius of analyticity function corresponding to a function F analytic at some, but not all, points of the complex plane, and let z_0 be a point at which F is analytic. Then:

1. *If $|\zeta - z| < R_A(z)$, then F is analytic at ζ .*

2. *If F is not analytic at a point ζ, then $|\zeta - z| \geq R_A(z)$.*

3. *If $R > R_A(z)$, then there is a point in the open disk about z of radius R at which F is not analytic.*

[5] In practice, $R_A(z)$ is usually the radius of convergence R for the Taylor series for F about z. In theory, though, one can define F to not equal its Taylor series at some points in the disk of radius R about z. $R_A(z)$ is then the radius of the largest open disk about z on which F is given by its Taylor series about z.

Lemma 34.22
Let F be a function which is analytic at some but not all points of the complex plane, and let R_A be the radius of analyticity function corresponding to F. Then, for each complex point z,

$$F \text{ is analytic at } z \iff R_A(z) > 0 \quad .$$

Equivalently,

$$F \text{ is not analytic at } z \iff R_A(z) = 0 \quad .$$

Lemma 34.23
If F is a function analytic at some but not all points of the complex plane, then R_A, the radius of analyticity function corresponding to F, is a continuous function on the complex plane.

The claims in the first lemma follow immediately from the definition of R_A; so let us concentrate on proving the other two lemmas.

PROOF (Lemma 34.22): First of all, by definition

$$F \text{ is not analytic at } z \implies R_A(z) = 0 \quad .$$

Hence, we also have

$$R_A(z) > 0 \implies F \text{ is analytic at } z \quad .$$

On the other hand, if F is analytic at z, then there is a power series $\sum_{k=0}^{\infty} a_k(\varsigma - z)^k$ and a $R > 0$ such that

$$F(\varsigma) = \sum_{k=0}^{\infty} a_k(\varsigma - z)^k \qquad \text{whenever} \quad |\varsigma - z| < R \quad .$$

Corollary 34.18 immediately tells us that F is analytic on the open disk of radius R about z. Since $R_A(z)$ is the largest such R, $R \leq R_A(z)$. And since $0 < R$, we now have

$$F \text{ is analytic at } z \implies R_A(z) > 0 \quad .$$

This also means

$$R_A = 0 \implies F \text{ is not analytic at } z \quad .$$

Combining all the implications just listed yields the claims in the lemma. ∎

PROOF (Lemma 34.23): To verify the continuity of R_A, we need to show that

$$\lim_{z \to z_1} R_A(z) = R_A(z_1) \qquad \text{for each complex value } z_1 \quad .$$

There are two cases: the easy case with F not being analytic at z_1, and the less-easy case with F being analytic at z_1. For the second case, we will use pictures.

Consider the first case, where F is not analytic at z_1 (hence, $R_A(z_1) = 0$). Then, as noted in Lemma 34.21,

$$0 \leq R_A(z) \leq |z_1 - z| \qquad \text{for any } z \text{ in } \mathbb{C} \quad .$$

Taking limits, we see that

$$0 \leq \lim_{z \to z_1} R_A(z) \leq \lim_{z \to z_1} |z_1 - z| = 0 \quad ,$$

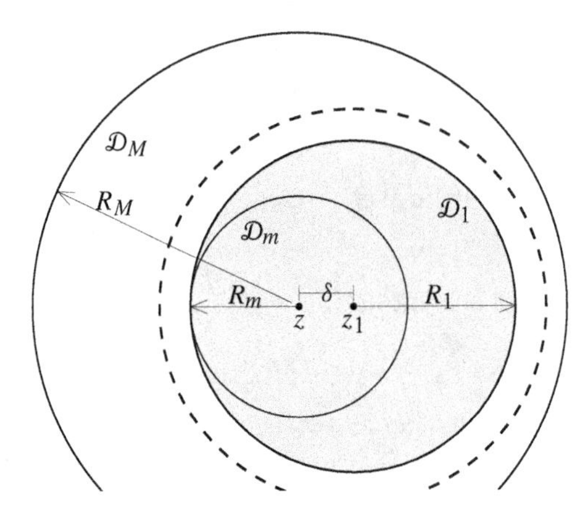

Figure 34.1: Limits on the radius of analyticity at z based on the radius of analyticity R_1 at z_1. Here $\delta = |z - z_1|$ $R_M = R_1 + 2\delta$ and $R_m = R_1 - \delta$.

which, since $R_A(z_1) = 0$, gives us

$$\lim_{z \to z_1} R_A(z) = R_A(z_1) \quad .$$

Next, assume F is analytic at z_1 (so that $R_A(z_1) > 0$), and let z be a point "close" to z_1. For notational convenience, let

$$R_1 = R_A(z_1)$$

and let $\mathcal{D}_1$ be the open disk centered at z_1 of radius R_1, as sketched in Figure 34.1. Note that, by the definition of R_1,

1. F is analytic at every point in $\mathcal{D}_1$, but
2. any open disk centered at z_1 larger than $\mathcal{D}_1$ (such as the one enclosed by the dashed-line circle in Figure 34.1) must contain a point at which F is not analytic.

Because we are just interested in limits as $z \to z_1$, we can assume z is close enough to z_1 that $|z - z_1| < R_1$. Let $\mathcal{D}_m$ and $\mathcal{D}_M$ be the open disks about z with radii

$$R_m = R_1 - |z - z_1| \qquad \text{and} \qquad R_M = R_1 + 2|z - z_1| \quad ,$$

as also illustrated in Figure 34.1. Now since $\mathcal{D}_m$ is contained in $\mathcal{D}_1$, F is analytic at every point in $\mathcal{D}_m$. Hence,

$$R_m \leq R_A(z) \quad .$$

On the other hand, inside $\mathcal{D}_M$ is another open disk that we had already noted contains a point at which F is not analytic. So F is not analytic at every point in $\mathcal{D}_M$. Thus,

$$R_A(z) < R_M \quad .$$

Combining the two inequalities above (and recalling what the notation means) gives us

$$R_A(z_1) - |z - z_1| \leq R_A(z) \leq R_A(z_1) + 2|z - z_1|$$

which, after letting $z \to z_1$, becomes

$$R_A(z_1) - 0 \leq \lim_{z \to z_1} R_A(z) \leq R_A(z_1) + 2 \cdot 0 \quad ,$$

clearly implying that

$$\lim_{z \to z_1} R_A(z) = R_A(z_1) \quad .$$

∎

Proof of Theorem 34.20

Remember: F is a function analytic at some but not all points in the complex plane, and z_0 is a point at which f is analytic.

Let

$$R_0 = R_A(z_0) \quad .$$

Because of the definition and properties of R_A, and the analyticity of F at z_0, we automatically have that $R_0 > 0$ and that F is analytic at every z with $|z - z_0| < R_0$. All that remains to proving our theorem is to show that there is a z_s at which F is not analytic and which satisfies $|z_s - z_0| = R_0$.

Now consider the possible values of $R_A(z)$ on the circle $|z - z_0| = R_0$. Since R_A is continuous, it must have a minimum value ρ at some point z_s in this circle. This means F must be analytic at each point closer than ρ to the circle $|z - z_0| = R_0$, as well at every point inside the disk enclosed by this circle. That is, F is analytic at every z on the open disk about z_0 of radius $R_0 + \rho$. And by the definition of R_A and R_0 and ρ, we must then have

$$R_A(z_0) = R_0 \leq R_0 + \rho \leq R_A(z_0) \quad ,$$

which is only possible if $\rho = 0$. But $\rho = R_A(z_s)$. So $R_A(z_s) = 0$, which, as Lemma 34.22 tells us, means that F is not analytic at z_s. ∎

34.9 Appendix: Singular Points and the Radius of Convergence for Solutions

Our goal in this section is to prove Theorem 34.11, which directly relates the radius of convergence for a power series solution for a given differential equation to a singular point for that differential equation. To do this, we must first expand on some of our discussion from the last few sections.

Analytic Continuation

Analytic continuation is any procedure that "continues" an analytic function defined on one region so that it becomes defined on a larger region. Perhaps it would be better to call it "analytic extension" because what we are really doing is extending the domain of our original function by creating an analytic function with a larger domain that equals the original function over the original domain.

We will "analytically extend" a power series solution to our differential equation on one disk to a solution on a larger disk using Taylor series. And to justify this, we will use the following theorem:

Theorem 34.24
Let $\mathcal{D}_A$ and $\mathcal{D}_B$ be two open disks in the complex plane that intersect each other. Assume that f_A is a function analytic on $\mathcal{D}_A$, and that f_B is a function analytic on $\mathcal{D}_B$. Further assume that there is an open disk $\mathcal{D}_0$ contained in both $\mathcal{D}_A$ and $\mathcal{D}_B$ (see Figure 34.2) such that

$$f_A(z) = f_B(z) \qquad \text{for every} \quad z \in \mathcal{D}_0 \quad .$$

Then

$$f_A(z) = f_B(z) \qquad \text{for every} \quad z \in \mathcal{D}_A \cap \mathcal{D}_B \quad .$$

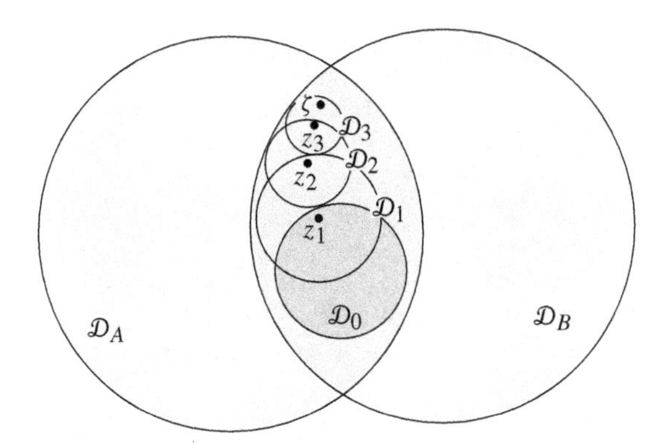

Figure 34.2: Disks in the complex plane for Theorem 34.24 and its proof.

Think of f_A as being the original function defined on $\mathcal{D}_A$, and f_B as some other analytic function that we constructed on $\mathcal{D}_B$ to match f_A on $\mathcal{D}_0$. This theorem tells us that we can define a "new" function f on $\mathcal{D}_A \cup \mathcal{D}_B$ by

$$f(z) = \begin{cases} f_A(z) & \text{if } z \text{ is in } \mathcal{D}_A \\ f_B(z) & \text{if } z \text{ is in } \mathcal{D}_B \end{cases} .$$

On the intersection, f is given both by f_A and f_B, but that is okay because the theorem assures us that f_A and f_B are the same on that intersection. And since f_A and f_B are, respectively, analytic at every point in $\mathcal{D}_A$ and $\mathcal{D}_B$, it follows that f is analytic on the union of $\mathcal{D}_A$ and $\mathcal{D}_B$, and satisfies

$$f(z) = f_A(z) \qquad \text{for each } z \text{ in } \mathcal{D}_A \quad .$$

That is, f is an "analytic extension" of f_A from the domain $\mathcal{D}_A$ to the domain $\mathcal{D}_A \cup \mathcal{D}_B$.

The proof of the above theorem is not difficult and is somewhat instructive.

PROOF (Theorem 34.24): We need to show that $f_A(\zeta) = f_B(\zeta)$ for every ζ in $\mathcal{D}_A \cap \mathcal{D}_B$. So let ζ be any point in $\mathcal{D}_A \cap \mathcal{D}_B$.

If $\zeta \in \mathcal{D}_0$ then, by our assumptions, we automatically have $f_A(\zeta) = f_B(\zeta)$.

On the other hand, if ζ in not in $\mathcal{D}_0$, then, as illustrated in Figure 34.2, we can clearly find a finite sequence of open disks $\mathcal{D}_1$, $\mathcal{D}_2$, ... and $\mathcal{D}_M$ with respective centers z_1, z_2, ... and z_M such that

1. each z_k is also in $\mathcal{D}_{k-1}$,
2. each $\mathcal{D}_k$ is in $\mathcal{D}_A \cap \mathcal{D}_B$, and
3. the last disk, $\mathcal{D}_M$, contains ζ.

Now, because f_A and f_B are the same on $\mathcal{D}_0$, so are all their derivatives. Consequently, the Taylor series for f_A and f_B about the point z_1 in $\mathcal{D}_0$ will be the same. And since $\mathcal{D}_1$ is a disk centered at z_1 and contained in both $\mathcal{D}_A$ and $\mathcal{D}_B$, we have

$$\begin{aligned} f_A(z) &= \sum_{k=0}^{\infty} \frac{{f_A}^{(k)}(z_1)}{k!}(z - z_1)^k \\ &= \sum_{k=0}^{\infty} \frac{{f_B}^{(k)}(z_1)}{k!}(z - z_1)^k = f_B(z) \qquad \text{for every } z \text{ in } \mathcal{D}_1 \quad . \end{aligned}$$

Repeating these arguments using the Taylor series for f_A and f_B at the points z_2, z_3, and so on, we eventually get

$$f_A(z) = f_B(z) \qquad \text{for every } z \text{ in } \mathcal{D}_M \quad .$$

In particular then,

$$f_A(\zeta) = f_B(\zeta) \quad ,$$

just as we wished to show. ∎

Ordinary and Singular Points for Power-Series Functions

It will help if we expand our notions of ordinary and singular points to any function given by a power series,

$$f(z) = \sum_{k=0}^{\infty} c_k(z - z_0)^k \qquad \text{for} \quad |z - z_0| < R \quad ,$$

assuming, of course, that $R > 0$. For convenience, let $\mathcal{D}$ be the open disk about z_0 of radius R. Then for each z_1 either in $\mathcal{D}$ or on its boundary, we will say:

1. z_1 an *ordinary point* for f if and only if there is a function f_1 analytic on a disk $\mathcal{D}_1$ of positive radius about z_1 and which equals f on the region where $\mathcal{D}$ and $\mathcal{D}_1$ overlap.

2. z_1 an *singular point* for f if and only if it is not an ordinary point for f.

Do note that Theorem 34.16 on page 666 assures us that every point in $\mathcal{D}$ is an ordinary point for f. So the only singular points must be on the boundary. And the next theorem tells us that there must be a singular point on the boundary of $\mathcal{D}$ when R is finite and the radius of convergence for the above power series.

Theorem 34.25
Let R be a positive finite number, and assume it is the radius of convergence for

$$f(z) = \sum_{k=0}^{\infty} c_k(z - z_0)^k \quad .$$

Then f must have a singular point on the circle $|z - z_0| = R$.

PROOF: For convenience, let $\mathcal{D}$ be the open disk of radius R about z_0,

$$\mathcal{D} = \{z : |z - z_0| < R\} \quad ,$$

and let $\overline{\mathcal{D}}$ be the union of $\mathcal{D}$ and its boundary,

$$\overline{\mathcal{D}} = \{z : |z - z_0| \leq R\} \quad .$$

Now, let's define a function R_C on $\overline{\mathcal{D}}$ as follows:

1. For each ordinary point ζ, there is a function f_1 analytic on a disk $\mathcal{D}_1$ of positive radius about ζ and which equals f on the region where $\mathcal{D}$ and $\mathcal{D}_1$ overlap. Since f_1 is analytic, it can be given by a power series about ζ. Let $R_C(\zeta)$ be the radius of convergence for that power series.

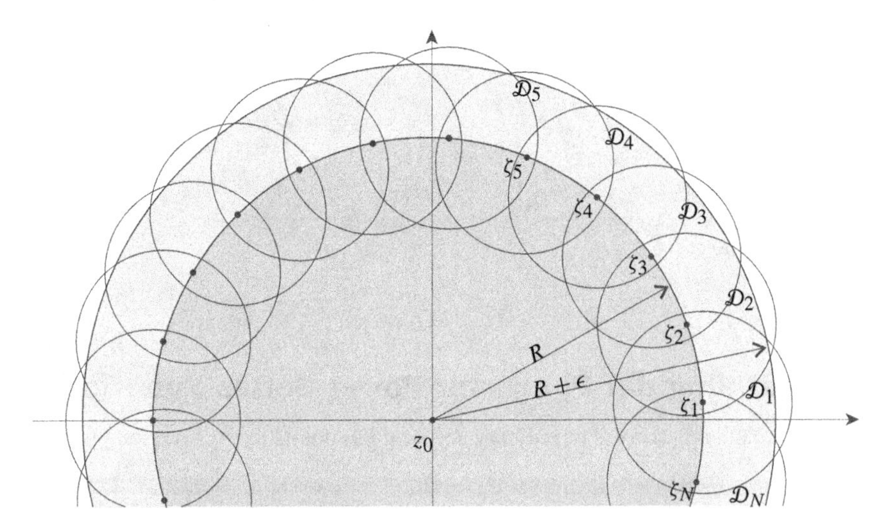

Figure 34.3: Disks for the proof of Theorem 34.25. The darker disk is the disk of radius R on which y is originally defined.

2. For each singular point ζ, let $R_C(\zeta) = 0$.

By definition, $R_C(\zeta) \geq 0$ for each ζ in $\overline{\mathcal{D}}$. We should also note that $R_C(\zeta)$ cannot be infinite for any ζ in $\overline{\mathcal{D}}$ because there would then be a function f_1 analytic on all of the complex plane and equaling f on the disk $\mathcal{D}$. And since f_1 is analytic everywhere, Theorem 34.16 assures us that the radius of convergence R_1 of the Taylor series of f_1 about z_0 would be infinite. But that Taylor series for f_1 about z_0 would have to be the same as the Taylor series for f about z_0 since $f = f_1$ on $\mathcal{D}$. And that, in turn, would mean the two power series have the same radius of convergence, giving us

$$\infty = R_1 = R < \infty \quad ,$$

which is impossible. Thus, it is not possible for $R_C(\zeta)$ to be infinite at any point ζ in $\overline{\mathcal{D}}$. In other words, R_C is a well-defined function on $\mathcal{D}$ with

$$0 \leq R_C(z) < \infty \qquad \text{for each } z \text{ in } \overline{\mathcal{D}} \quad .$$

The function R_C is very similar to the "radius of analyticity function" R_A discussed in Section 34.8, and, using arguments very similar to those used in that section for R_A, you can verify all the following:

1. $R_C(z) > 0$ if and only if z is an ordinary point for the differential equation.
2. $R_C(z) = 0$ if and only if z is a singular point for the differential equation.
3. R_C is a continuous function on the circle $|z - z_0| = R$.
4. $R_C(z)$ has a minimum value ρ at some point z_s on the circle $|z - z_0| = R$.

Now, if we can show $\rho = 0$, then the above tells us that the corresponding point z_s is a singular point for the given power series, and our theorem is proven. And to show that, it will suffice to show that $\rho > 0$ is impossible.

So, for the moment, assume $\rho > 0$. Then (as illustrated in Figure 34.3) we can choose a finite sequence of points ζ_1, ζ_2, ..., and ζ_N about the circle $|z - z_0| = R$ such that $|\zeta_N - \zeta_1| < \rho$ and

$$|\zeta_{n+1} - \zeta_n| < \rho \qquad \text{for} \quad n = 1, 2, 3, \ldots, n \quad .$$

For each ζ_n, let $\mathcal{D}_n$ be the open disk of radius ρ about ζ_n, and observe that, by basic geometry, the union of all the disks $\mathcal{D}_1$, $\mathcal{D}_2$, ..., and $\mathcal{D}_N$ contains not just the boundary of our original disk $\mathcal{D}$ but the annulus of all z satisfying

$$R \leq |z - z_0| \leq R + \epsilon$$

for some positive value ϵ.

By our choice of ζ_n's and $\mathcal{D}$'s, we know that, for each integer n from 1 to N, there is a power series function

$$f_n(z) = \sum_{k=0}^{\infty} c_{n,k}(z - \zeta_n)^k$$

defined on all of $\mathcal{D}_n$ and which equals our original function f in the overlap of $\mathcal{D}$ and $\mathcal{D}_n$. Repeated use of Theorem 34.24 then shows that any two functions from the set

$$\{f, f_1, f_2, \ldots, f_N\}$$

equal each other wherever both are defined. This allows us to define a "new" analytic function F on the union of all of our disks via

$$F(z) = \begin{cases} f(z) & \text{if } |z - z_0| < R \\ f_n(z) & \text{if } z \in \mathcal{D}_n \quad \text{for } n = 1, 2, \ldots, N \end{cases} .$$

Now, because the union of all of the disks contains the disk of radius $R+\epsilon$ about 0, Theorem 34.16 on page 666 on the radius of convergence for Taylor series assures us that the Taylor series for F about z_0 must have a radius of convergence of at least $R + \epsilon$. But, $f(z) = F(z)$ when $|z| < R$. So f and F have the same Taylor series at z_0, and, hence, these two power series share the same radius of convergence.

That is, if $\rho > 0$, then

$$\begin{aligned} R &= \text{radius of convergence for the power series of } f \text{ about } z_0 \\ &= \text{radius of convergence for the power series of } F \text{ about } z_0 > R \quad , \end{aligned}$$

which is clearly impossible. So, it is not possible to have $\rho > 0$. ∎

Complex Power Series Solutions

Throughout most of these chapters, we've been tacitly assuming that the derivatives in our differential equations are derivatives with respect to a real variable x,

$$y' = \frac{dy}{dx} \quad \text{and} \quad y'' = \frac{d^2y}{dx^2} \quad ,$$

as in elementary calculus. In fact, we can also have derivatives with respect to the complex variable z,

$$y' = \frac{dy}{dz} \quad \text{and} \quad y'' = \frac{d^2y}{dz^2} \quad ,$$

as described on page 663. Remember that, computationally, differentiation with respect to z is completely analogous to the differentiation with respect to x learned in basic calculus. In particular, if

$$y(z) = \sum_{k=0}^{\infty} a_k(z - z_0)^k \quad \text{for} \quad |z - z_0| < R$$

for some point z_0 in the complex plane and some $R > 0$, then

$$y'(z) = \frac{d}{dz}\sum_{k=0}^{\infty} a_k(z-z_0)^k$$
$$= \sum_{k=0}^{\infty}\frac{d}{dz}\left[a_k(z-z_0)^k\right] = \sum_{k=1}^{\infty} k a_k(z-z_0)^{k-1} \qquad \text{for} \quad |z-z_0| < R \quad .$$

Consequently, all of our computations in Chapters 33 and 34 can be carried out using the complex variable z instead of the real variable x, and using a point z_0 in the complex plane instead of a point x_0 on the real line. In particular, we have the following complex-variable analogs of Theorems 34.9 and 34.10:

Theorem 34.26 (first-order series solutions)
Suppose z_0 is an ordinary point for a first-order homogeneous differential equation whose reduced form is

$$y' + Py = 0 \quad .$$

Then P has a power series representation

$$P(z) = \sum_{k=0}^{\infty} p_k(z-z_0)^k \qquad \text{for} \quad |z-z_0| < R$$

where R is the radius of analyticity about z_0 for this differential equation.

Moreover, a general solution to the differential equation is given by

$$y(z) = \sum_{k=0}^{\infty} a_k(z-z_0)^k \qquad \text{for} \quad |z-z_0| < R$$

where a_0 is arbitrary, and the other a_k's satisfy the recursion formula

$$a_k = -\frac{1}{k}\sum_{j=0}^{k-1} a_j p_{k-1-j} \quad . \tag{34.11}$$

Theorem 34.27 (second-order series solutions)
Suppose z_0 is an ordinary point for a second-order homogeneous differential equation whose reduced form is

$$\frac{d^2y}{dz^2} + P\frac{dy}{dz} + Qy = 0 \quad .$$

Then P and Q have power series representations

$$P(z) = \sum_{k=0}^{\infty} p_k(z-z_0)^k \qquad \text{for} \quad |z-z_0| < R$$

and

$$Q(z) = \sum_{k=0}^{\infty} q_k(z-z_0)^k \qquad \text{for} \quad |z-z_0| < R$$

where R is the radius of analyticity about z_0 for this differential equation.

Moreover, a general solution to the differential equation is given by

$$y(z) = \sum_{k=0}^{\infty} a_k (z - z_0)^k \qquad \text{for} \quad |z - z_0| < R$$

where a_0 *and* a_1 *are arbitrary, and the other* a_k*'s satisfy the recursion formula*

$$a_k = -\frac{1}{k(k-1)} \sum_{j=0}^{k-2} \left[(j+1) a_{j+1} p_{k-2-j} + a_j q_{k-2-j} \right] \quad . \tag{34.12}$$

By the way, it should also be noted that, if

$$y(x) = \sum_{k=0}^{\infty} c_k (x - x_0)^k$$

is a solution to the real-variable differential equation for $|x - x_0| < R$, then

$$y(z) = \sum_{k=0}^{\infty} c_k (z - x_0)^k$$

is a solution to the corresponding complex-variable differential equation for $|z - x_0| < R$. This follows immediately from the last theorem and the relation between dy/dx and dy/dz (see the discussion of the complex derivative in Section 34.7).

Singular Points of Differential Equations and Solutions

So, let's suppose we have a power series solution

$$y(z) = \sum_{k=0}^{\infty} c_k (z - z_0)^k \qquad \text{for} \quad |z - x_0| < R$$

to some first- or second-order linear homogenous differential equation, and, as before, let $\mathcal{D}$ and $\overline{\mathcal{D}}$ be, respectively, the open and closed disks of radius R about z_0,

$$\mathcal{D} = \{z : |z - z_0| < R\} \qquad \text{and} \qquad \overline{\mathcal{D}} = \{z : |z - z_0| \leq R\} \quad .$$

Now consider any single point z_1 in $\overline{\mathcal{D}}$. If z_1 is an ordinary point for the given differential equation, then there is a disk $\mathcal{D}_1$ of some positive radius R_1 about z_1 such that, on that disk, general solutions to the differential equation exist and are given by power series about z_1. Using the material already developed in this section, it's easy to verify that, in particular, the above power series solution y is given by a power series about z_1 at least on the overlap of disks $\mathcal{D}$ and $\mathcal{D}_1$. And that means z_1 is also an ordinary point for the above power series solution. And, of course, this also means that a point z_1 in $\overline{\mathcal{D}}$ cannot be both an ordinary point for the differential equation and a singular point for y.

To summarize:

Lemma 34.28

Let R *be a positive finite number, and assume that, on the disk* $|z - z_0| < R$,

$$y(z) = \sum_{k=0}^{\infty} c_k (z - z_0)^k$$

is a power series solution to some first- or second-order linear homogeneous differential equation. Then, for each point ζ in the closed disk given by $|z - z_0| \le R$:

1. *If ζ is an ordinary point for the differential equation, then it is an ordinary point for the above power series solution.*
2. *If ζ is a singular point for the above power series solution, then it is a singular point for the differential equation.*

Combining the last lemma with Theorem 34.25 on singular points for power series functions, we get the main result of this appendix:

Theorem 34.29
Let

$$y(z) = \sum_{k=0}^{\infty} c_k (z - z_0)^k \qquad \text{for} \quad |z - z_0| < R$$

be a power series solution for some first- or second-order homogeneous linear differential equation. Assume, further, that R is finite and is the radius of convergence for the above power series. Then there is a point z_s with $|z_s - z_0| = R$ which is a singular point for both the above power series solution and the given differential equation.

Theorem 34.11 now follows as a corollary of the last theorem.

Additional Exercises

34.4. In the following, you will determine all the points in the complex plane at which certain common functions are zero. It may help to remember that, for a complex value $Z = X + iY$ to be zero, both its real part X and its imaginary part Y must be zero.

a. Using the fact that

$$e^{x+iy} = e^x [\cos(y) + i \sin(y)] \quad ,$$

show that e^z can never equal zero for any z in the complex plane.

b. In Chapter 17 we saw that

$$\sin(z) = \frac{e^{iz} - e^{-iz}}{2i} \qquad \text{and} \qquad \cos(z) = \frac{e^{iz} + e^{-iz}}{2} \quad ,$$

at least when z was a real value (see page 330). In fact, these formulas (along with the definition of the complex exponential) can define the sine and cosine functions at all values of z, real and complex. Using these formulas

i. Verify that

$$\sin(x + iy) = \frac{e^y + e^{-y}}{2} \sin(x) + i \frac{e^y - e^{-y}}{2} \cos(x)$$

and

$$\cos(x + iy) = \frac{e^y + e^{-y}}{2} \sin(x) - i \frac{e^y - e^{-y}}{2} \cos(x) \quad .$$

ii. *Using the above formulas, verify that*

$$\sin(z) = 0 \iff z = n\pi \quad \text{with} \quad n = 0, \pm 1, \pm 2, \ldots$$

and

$$\cos(z) = 0 \iff z = \left[n + \tfrac{1}{2}\right]\pi \quad \text{with} \quad n = 0, \pm 1, \pm 2, \ldots \quad .$$

c. *The hyperbolic sine and cosine are defined on the complex plane by*

$$\sinh(z) = \frac{e^z - e^{-z}}{2} \quad \text{and} \quad \cosh(z) = \frac{e^z + e^{-z}}{2} \quad .$$

Show that

$$\sinh(z) = 0 \iff z = in\pi \quad \text{with} \quad n = 0, \pm 1, \pm 2, \ldots$$

and

$$\cosh(z) = 0 \iff z = i\left[n + \tfrac{1}{2}\right]\pi \quad \text{with} \quad n = 0, \pm 1, \pm 2, \ldots \quad .$$

34.5. *For each of the following differential equations, find all singular points, as well as the radius of analyticity R about the given point x_0. You may have to use results from the previous problem. You may even have to expand on some of those results. And you may certainly need to rearrange a few equations.*

a. $y' - e^x y = 0$ with $x_0 = 0$

b. $y' - \tan(x)y = 0$ with $x_0 = 0$

c. $\sin(x)\, y'' + x^2 y' - e^x y = 0$ with $x_0 = 2$

d. $\sinh(x)\, y'' + x^2 y' - e^x y = 0$ with $x_0 = 2$

e. $\sinh(x)\, y'' + x^2 y' - \sin(x)\, y = 0$ with $x_0 = 2$

f. $e^{3x} y'' + \sin(x)\, y' + \dfrac{2}{(x^2+4)} y = 0$ with $x_0 = 0$

g. $y'' + \dfrac{1+e^x}{1-e^x} y = 0$ with $x_0 = 3$

h. $\left[x^2 - 4\right] y'' + \left[x^2 + x - 6\right] y = 0$ with $x_0 = 2$

i. $xy'' + \left[1 - e^x\right] y = 0$

j. $\sin\left(\pi x^2\right) y'' + x^2 y = 0$ with $x_0 = 0$

34.6. *Using recursion formula (34.3) on page 655, find the 4th-degree partial sum of the general power series solution for each of the following about the given point x_0. Also state the interval I over which you can be sure the full power series solution is valid.*

a. $y' - e^x y = 0$ with $x_0 = 0$

b. $y' + e^{2x} y = 0$ with $x_0 = 0$

c. $y' + \cos(x)\, y = 0$ with $x_0 = 0$

d. $y' + \ln|x|\, y = 0$ with $x_0 = 1$

34.7. *Using the recursion formula (34.4) on page 656, find the 4th-degree partial sum of the general power series solution for each of the following about the given point x_0. Also state the interval I over which you can be sure the full power series solution is valid.*

a. $y'' - e^x y = 0$ with $x_0 = 0$

b. $y'' + 3xy' - e^x y = 0$ with $x_0 = 0$

c. $xy'' - 3xy' + \sin(x)\, y = 0$ with $x_0 = 0$ and $N = 4$

d. $y'' + \ln|x|\, y = 0$ with $x_0 = 1$

e. $\sqrt{x}\, y'' + y = 0$ with $x_0 = 1$

f. $y'' + \left[1 + 2x + 6x^2\right]y' + [2 + 12x]\, y = 0$ with $x_0 = 0$

34.8 a. *Using your favorite computer mathematics package, along with recursion formula (34.3) on page 655, write a program/worksheet that will find the Nth partial sum of the power series solution about x_0 to*

$$y' + P(x)y = 0$$

for any given positive integer N, point x_0 and function $P(x)$ analytic at x_0. Finding the appropriate partial sum of the corresponding power series for P should be part of the program/worksheet (see Exercise 32.8 on page 601). Be sure to write your program/worksheet so that N, x_0 and P are easily changed.

b. *Use your program/worksheet to find the Nth-degree partial sum of the general power series solution about x_0 for each of the following differential equations and choices for N and x_0.*

i. $y' - e^x y = 0$ with $x_0 = 0$ and $N = 10$

ii. $y' + \sqrt{x^2 + 1}\, y = 0$ with $x_0 = 0$ and $N = 8$

iii. $\cos(x)y' + y = 0$ with $x_0 = 0$ and $N = 8$

iv. $y' + \sqrt{2x^2 + 1}\, y = 0$ with $x_0 = 2$ and $N = 5$

34.9 a. *Using your favorite computer mathematics package, along with recursion formula (34.4) on page 656, write a program/worksheet that will find the Nth partial sum of the power series solution about x_0 to*

$$y'' + P(x)y' + Q(x)y = 0$$

for any given positive integer N, point x_0, and functions $P(x)$ and $Q(x)$ analytic at x_0. Finding the appropriate partial sum of the corresponding power series for P and Q should be part of the program/worksheet (see Exercise 32.8 on page 601). Be sure to write your program/worksheet so that N, x_0, P and Q are easily changed.

b. *Use your program/worksheet to find the Nth-degree partial sum of the general power series solution about x_0 for each of the following differential equations and choices for N and x_0.*

i. $y'' - e^x y = 0$ with $x_0 = 0$ and $N = 8$

ii. $y'' + \cos(x)y = 0$ with $x_0 = 0$ and $N = 10$

iii. $y'' + \sin(x)y' + \cos(x)y = 0$ with $x_0 = 0$ and $N = 7$

iv. $\sqrt{x}\, y'' + y' + xy = 0$ with $x_0 = 1$ and $N = 5$

35

Modified Power Series Solutions and the Basic Method of Frobenius

The partial sums of a power series solution about an ordinary point x_0 of a differential equation provide fairly accurate approximations to the equation's solutions at any point x near x_0. This is true even if relatively low-order partial sums are used (provided you are just interested in the solutions at points very near x_0). However, these power series typically converge slower and slower as x moves away from x_0 towards a singular point, with more and more terms then being needed to obtain reasonably accurate partial sum approximations. As a result, the power series solutions derived in the previous two chapters usually tell us very little about the solutions near singular points. This is unfortunate because, in some applications, the behavior of the solutions near certain singular points can be a rather important issue.

Fortunately, in many of these applications, the singular point in question is not that "bad" a singular point. In these applications, the differential equation "is similar to" an easily solved Euler equation (which we discussed in Chapter 20), at least in some interval about that singular point. This fact will allow us to modify the algebraic method discussed in the previous chapters so as to obtain "modified" power series solutions about these points. The basic process for generating these modified power series solutions is typically called the *method of Frobenius*, and is what we will develop and use in this and the next two chapters.

By the way, we will only consider second-order homogeneous linear differential equations. One can extend the discussion here to first- and higher-order equations, but the important examples are second-order.

35.1 Euler Equations and Their Solutions

As already indicated, the Euler equations from Chapter 20 will play a fundamental role in our discussions and are, in fact, the simplest examples of the sort of equations of interest in this chapter. Let us quickly review them and their solutions, and take a look at what happens to their solutions about their singular points.

Recall that a standard second-order Euler equation is a differential equation that can be written as

$$\alpha_0 x^2 y'' + \beta_0 x y' + \gamma_0 y = 0$$

where α_0, β_0 and γ_0 are real constants with $\alpha_0 \neq 0$. Recall, also, that the basic method for solving such an equation begins with attempting a solution of the form $y = x^r$ where r is a constant to be determined. Plugging $y = x^r$ into the differential equation, we get

$$\alpha_0 x^2 \left[x^r\right]'' + \beta_0 x \left[x^r\right]' + \gamma_0 \left[x^r\right] = 0$$

$$\hookrightarrow \quad \alpha_0 x^2 \left[r(r-1)x^{r-2}\right] + \beta_0 x \left[r x^{r-1}\right] + \gamma_0 \left[x^r\right] = 0$$

$$\hookrightarrow \quad x^r \left[\alpha_0 r(r-1) + \beta_0 r + \gamma_0\right] = 0$$

$$\hookrightarrow \quad \alpha_0 r(r-1) + \beta_0 r + \gamma_0 = 0 \quad .$$

The last equation above is the indicial equation, which we typically rewrite as

$$\alpha_0 r^2 + (\beta_0 - \alpha_0) r + \gamma_0 = 0 \quad ,$$

and, from which, we can easily determine the possible values of r using basic algebra.

Generalizing slightly, we have the *shifted Euler equation*

$$\alpha_0 (x - x_0)^2 y'' + \beta_0 (x - x_0) y' + \gamma_0 y = 0$$

where x_0, α_0, β_0 and γ_0 are real constants with $\alpha_0 \neq 0$. Notice that x_0 is the one and only singular point of this equation. (Notice, also, that a standard Euler equation is a shifted Euler equation with $x_0 = 0$.)

To solve this slight generalization of a standard Euler equation, use the obvious slight generalization of the basic method for solving a standard Euler equation. First set

$$y = (x - x_0)^r \quad ,$$

where r is a yet unknown constant. Then plug this into the differential equation, compute, and simplify. Unsurprisingly, you end up with the corresponding indicial equation

$$\alpha_0 r(r-1) + \beta_0 r + \gamma_0 = 0 \quad ,$$

which you can rewrite as

$$\alpha_0 r^2 + (\beta_0 - \alpha_0) r + \gamma_0 = 0 \quad ,$$

and then solve for r, just as with the standard Euler equation. And, as with a standard Euler equation, there are then only three basic possibilities regarding the roots r_1 and r_2 to the indicial equation and the corresponding solutions to the Euler equations:

1. The two roots can be two different real numbers, $r_1 \neq r_2$.

   In this case, the general solution to the differential equation is

   $$y(x) = c_1 y_1(x) + c_2 y_2(x)$$

   with, at least when $x > x_0$,[1]

   $$y_1(x) = (x - x_0)^{r_1} \quad \text{and} \quad y_2(x) = (x - x_0)^{r_2} \quad .$$

   Observe that, for $j = 1$ or $j = 2$,

   $$\lim_{x \to x_0^+} \left|y_j(x)\right| = \lim_{x \to x_0} |x - x_0|^{r_j} = \begin{cases} 0 & \text{if } r_j > 0 \\ 1 & \text{if } r_j = 0 \\ +\infty & \text{if } r_j < 0 \end{cases} \quad .$$

[1] In all of these cases, the formulas and observations also hold when $x < x_0$, though we may wish to replace $x - x_0$ with $|x - x_0|$ to avoid minor issues with $(x - x_0)^r$ for certain values of r (such as $r = 1/2$).

2. The two roots can be the same real number, $r_2 = r_1$.

   In this case, we can use reduction of order and find that the general solution to the differential equation is (when $x > x_0$)

$$y(x) = c_1 y_1(x) + c_2 y_2(x)$$

with

$$y_1(x) = (x - x_0)^{r_1} \quad \text{and} \quad y_2(x) = (x - x_0)^{r_1} \ln|x - x_0| \quad .$$

After recalling how $\ln|x - x_0|$ behaves when $x \approx x_0$, we see that

$$\lim_{x \to x_0^+} |y_2(x)| = \lim_{x \to x_0} \left|(x - x_0)^{r_1} \ln|x - x_0|\right| = \begin{cases} 0 & \text{if } r_1 > 0 \\ +\infty & \text{if } r_1 \leq 0 \end{cases} \quad .$$

3. Finally, the two roots can be complex conjugates of each other,

$$r_1 = \lambda + i\omega \quad \text{and} \quad r_2 = \lambda - i\omega \quad \text{with} \quad \omega > 0 \quad .$$

   After recalling that

$$X^{\lambda \pm i\omega} = X^{\lambda} [\cos(\omega \ln|X|) \pm i \sin(\omega \ln|X|)] \quad \text{for} \quad X > 0$$

(see the discussion of complex exponents in Section 20.2), we find that the general solution to the differential equation for $x > x_0$ can be given by

$$y(x) = c_1 y_1(x) + c_2 y_2(x)$$

where y_1 and y_2 are the real-valued functions

$$y_1(x) = (x - x_0)^{\lambda} \cos(\omega \ln|x - x_0|)$$

and

$$y_2(x) = (x - x_0)^{\lambda} \sin(\omega \ln|x - x_0|) \quad .$$

The behavior of these solutions as $x \to x_0$ is a bit more complicated. First observe that, as X goes from 1 to 0, $\ln|X|$ goes from 0 to $-\infty$, which means that $\sin(\omega \ln|X|)$ and $\cos(\omega \ln|X|)$ then must oscillate infinitely many times between their maximum and minimum values of 1 and -1, as illustrated in Figure 35.1a. So $\sin(\omega \ln|X|)$ and $\cos(\omega \ln|X|)$ are bounded, but do not approach any single value as $X \to 0$. Taking into account how X^{λ} behaves (and replacing X with $x - x_0$), we see that

$$\lim_{x \to x_0^+} |y_i(x)| = 0 \quad \text{if} \quad \lambda > 0 \quad ,$$

and

$$\lim_{x \to x_0^+} |y_i(x)| \quad \text{does not exist if} \quad \lambda \leq 0 \quad .$$

Notice that the behavior of these solutions as $x \to x_0$ depends strongly on the values of r_1 and r_2. Notice, also, that you can rarely arbitrarily assign the initial values $y(x_0)$ and $y'(x_0)$ for these solutions.

!►Example 35.1: *Consider the shifted Euler equation*

$$(x - 3)^2 y'' - 2y = 0 \quad .$$

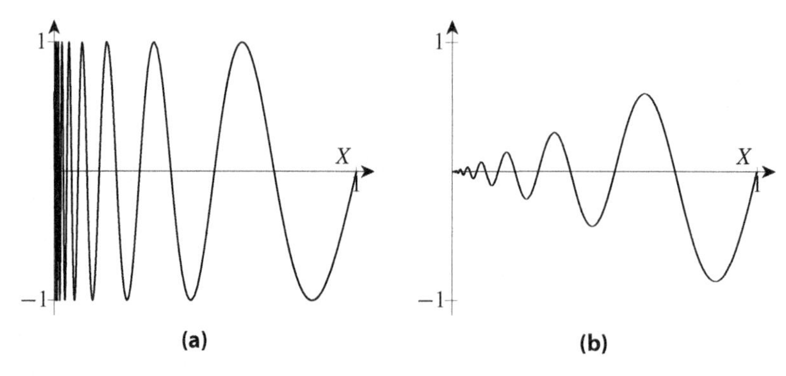

Figure 35.1: Graphs of **(a)** $\sin(\omega \ln |X|)$, and **(b)** $X^\lambda \sin(\omega \ln |X|)$ (with $\omega = 10$ and $\lambda = {}^{11}/_{10}$).

If $y = (x-3)^r$ *for any constant* r, *then*

$$(x-3)^2 y'' - 2y = 0$$

$$\hookrightarrow \quad (x-3)^2 \left[(x-3)^r\right]'' - 2\left[(x-3)^r\right] = 0$$

$$\hookrightarrow \quad (x-3)^2 r(r-1)(x-3)^{r-2} - 2(x-3)^r = 0$$

$$\hookrightarrow \quad (x-3)^r \left[r(r-1) - 2\right] = 0 \quad .$$

Thus, for $y = (x-3)^r$ *to be a solution to our differential equation,* r *must satisfy the indicial equation*

$$r(r-1) - 2 = 0 \quad .$$

After rewriting this as

$$r^2 - r - 2 = 0 \quad ,$$

and factoring,

$$(r-2)(r+1) = 0 \quad ,$$

we see that the two solutions to the indicial equation are $r = 2$ *and* $r = -1$. *Thus, the general solution to our differential equation is*

$$y = c_1(x-3)^2 + c_2(x-3)^{-1} \quad .$$

This has one term that vanishes as $x \to 3$ *and another that blows up as* $x \to 3$. *In particular, we cannot insist that* $y(3)$ *be any particular nonzero number, say, "*$y(3) = 2$*".*

!►Example 35.2: *Now consider*

$$x^2 y'' + xy' + y = 0 \quad ,$$

which is a shifted Euler equation, but with "shift" $x_0 = 0$.

The indicial equation for this is

$$r(r-1) + r + 1 = 0 \quad ,$$

which simplifies to

$$r^2 + 1 = 0 \quad .$$

So,

$$r = \pm\sqrt{-1} = \pm i \quad .$$

In other words, $r = \lambda \pm i\omega$ *with* $\lambda = 0$ *and* $\omega = 1$ *. Consequently, the general solution to the differential equation in this example is given by*

$$y = c_1 x^0 \cos(1 \ln|x|) + c_2 x^0 \sin(1 \ln|x|) \quad ,$$

which we naturally would prefer to write more simply as

$$y = c_1 \cos(\ln|x|) + c_2 \sin(\ln|x|) \quad .$$

Here, neither term vanishes or blows up. Instead, as x *goes from* 1 *to* 0 *, we have* $\ln|x|$ *going from* 0 *to* $-\infty$ *. This means that the sine and cosine terms oscillate infinitely many times as* x *goes from* 1 *to* 0 *, similar to the function illustrated in Figure 35.1a. Again, there is no way we can require that "*$y(0) = 2$*".*

The "blowing up" or "vanishing" of solutions illustrated above is typical behavior of solutions about the singular points of their differential equations. Sometimes it is important to know just how the solutions to a differential equation behave near a singular point x_0 . For example, if you know a solution describes some process that does not blow up as $x \to x_0$, then you know that those solutions that do blow up are irrelevant to your problem (this will become very important when we study boundary-value problems).

The tools we will develop in this chapter will yield "modified power series solutions" around singular points for at least some differential equations. The behavior of the solutions near these singular points can then be deduced from these modified power series. For which equations will this analysis be appropriate? Before answering that, I must first tell you the difference between a "regular" and an "irregular" singular point.

35.2 Regular and Irregular Singular Points (and the Frobenius Radius of Convergence)

Basic Terminology

Assume x_0 is singular point on the real line for some given homogeneous linear second-order differential equation

$$ay'' + by' + cy = 0 \quad . \tag{35.1}$$

We will say that x_0 is a *regular singular point* for this differential equation if and only if that differential equation can be rewritten as

$$(x - x_0)^2 \alpha(x) y'' + (x - x_0)\beta(x) y' + \gamma(x) y = 0 \tag{35.2}$$

where α , β and γ are 'suitable, well-behaved' functions about x_0 with $\alpha(x_0) \neq 0$. The above form will be called a *quasi-Euler form about* x_0 for the given differential equation, and the shifted Euler equation

$$(x - x_0)^2 \alpha_0 y'' + (x - x_0)\beta_0 y' + \gamma_0 y = 0 \tag{35.3a}$$

where

$$\alpha_0 = \alpha(x_0) \quad , \quad \beta_0 = \beta(x_0) \quad \text{and} \quad \gamma_0 = \gamma(x_0) \tag{35.3b}$$

will be called the *associated (shifted) Euler equation (about x_0).*

Precisely what we mean above by "'suitable, well-behaved' functions about x_0" depends, in practice, on the coefficients of the original differential equation. In general, it means that the functions α, β and γ in equation (35.2) are all analytic at x_0. However, if the coefficients of the original equation (the a, b and c in equation (35.1)) are rational functions, then we will be able to further insist that the α, β and γ in the quasi-Euler form be polynomials.

It is quite possible that our differential equation cannot be written in quasi-Euler form about x_0. Then we will say x_0 is an *irregular singular point* for our differential equation. Thus, every singular point on the real line for our differential equation is classified as being regular or irregular depending on whether the differential equation can be rewritten in quasi-Euler form about that singular point.

While we are at it, let's also define the *Frobenius radius of analyticity* about any point z_0 for our differential equation. It is simply the distance between z_0 and the closest singular point z_s other than z_0, provided such a point exists. If no such z_s exists, then the Frobenius radius of analyticity is ∞. The Frobenius radius of analyticity will play almost the same role as played by the radius of analyticity in the previous two chapters. In fact, it is the radius of analyticity if z_0 happens to be an ordinary point.

!►Example 35.3 (Bessel equations): *Let ν be any positive real constant. Bessel's equation of order ν is the differential equation* [2]

$$y'' + \frac{1}{x}y' + \left[1 - \left(\frac{\nu}{x}\right)^2\right]y = 0 \quad .$$

The coefficients of this,

$$a(x) = 1 \quad , \quad b(x) = \frac{1}{x} \quad \text{and} \quad c(x) = 1 - \left(\frac{\nu}{x}\right)^2 = \frac{x^2 - \nu^2}{x^2} \quad ,$$

are rational functions. Multiplying through by x^2 then gives us

$$x^2y'' + xy' + \left[x^2 - \nu^2\right]y = 0 \quad . \tag{35.4}$$

Clearly, $x_0 = 0$ is the one and only singular point for this differential equation. This, in turn, means that the Frobenius radius of analyticity about $x_0 = 0$ is ∞ (though the radius of analyticity, as defined in Chapter 33, is 0 since $x_0 = 0$ is a singular point.)

Now observe that the last differential equation is

$$(x-0)^2\alpha(x)y'' + (x-0)\beta(x)y' + \gamma(x)y = 0 \quad , \tag{35.5a}$$

where α, β and γ are the simple polynomials

$$\alpha(x) = 1 \quad , \quad \beta(x) = 1 \quad \text{and} \quad \gamma(x) = x^2 - \nu^2 \quad , \tag{35.5b}$$

which are certainly analytic about $x_0 = 0$ (and every other point on the complex plane). Moreover, since

$$\alpha(0) = 1 \neq 0 \quad , \quad \beta(0) = 1 \quad \text{and} \quad \gamma(0) = -\nu^2 \quad ,$$

we now have that:

[2] Bessel's equations and their solutions often arise in two-dimensional problems involving circular objects.

1. *The Bessel equation of order ν can be written in quasi-Euler form about $x_0 = 0$.*
2. *The singular point $x_0 = 0$ is a regular singular point.*

and

3. *The associated Euler equation about $x_0 = 0$ is*

$$(x - 0)^2 \cdot 1y'' + (x - 0) \cdot 1y' - \nu^2 y = 0 \quad ,$$

which, of course, is normally written

$$x^2 y'' + xy' - \nu^2 y = 0 \quad .$$

Two quick notes before going on:

1. Often, the point of interest is $x_0 = 0$, in which case we will write equation (35.2) more simply as

$$x^2 \alpha(x) y'' + x\beta(x) y' + \gamma(x) y = 0 \tag{35.6}$$

with the *associated Euler equation* being

$$x^2 \alpha_0 y'' + x\beta_0 y' + \gamma_0 y = 0 \tag{35.7a}$$

where

$$\alpha_0 = \alpha(0) \quad , \quad \beta_0 = \beta(0) \quad \text{and} \quad \gamma_0 = \gamma(0) \quad . \tag{35.7b}$$

2. In fact, any singular point on the complex plane can be classified as regular or irregular. However, this won't be particularly relevant to us. Our interest will only be in whether given singular points on the real line are regular or not.

Testing for Regularity

As illustrated in our last example, if the coefficients in our original differential equation are relatively simple rational functions, then it can be relatively straightforward to show that a given singular point x_0 is or is not regular by seeing if we can or cannot rewrite the equation in quasi-Euler form about x_0. An advantage of deriving this quasi-Euler form (if possible) is that we will want this quasi-Euler form in solving our differential equation. However, there are possible difficulties. If we cannot rewrite our equation in quasi-Euler form, then we may be left with the question of whether x_0 truly is an irregular singular point, or whether we just weren't clever enough to get the equation into quasi-Euler form. Also, if the coefficients in our original equation are not so simple, then the process of attempting to convert it to quasi-Euler form may be quite challenging.

A useful test for regularity is easily derived by first assuming that x_0 is a regular singular point for

$$a(x)y'' + b(x)y' + c(x)y = 0 \quad . \tag{35.8}$$

By definition, this means that this differential equation can be rewritten as

$$(x - x_0)^2 \alpha(x) y'' + (x - x_0)\beta(x) y' + \gamma(x) y = 0 \tag{35.9}$$

where α, β and γ are functions analytic at x_0 with $\alpha(x_0) \neq 0$. Dividing each of these two equations by its first coefficient converts our differential equation, respectively, to the two forms

$$y'' + \frac{b(x)}{a(x)} y' + \frac{c(x)}{a(x)} y = 0$$

and

$$y'' + \frac{\beta(x)}{(x - x_0)\alpha(x)}y' + \frac{\gamma(x)}{(x - x_0)^2\alpha(x)}y = 0 \quad .$$

But these last two equations describe the same differential equation, and have the same first coefficients. Clearly, the other coefficients must also be the same, giving us

$$\frac{b(x)}{a(x)} = \frac{\beta(x)}{(x - x_0)\alpha(x)} \quad \text{and} \quad \frac{c(x)}{a(x)} = \frac{\gamma(x)}{(x - x_0)^2\alpha(x)} \quad .$$

Equivalently,

$$(x - x_0)\frac{b(x)}{a(x)} = \frac{\beta(x)}{\alpha(x)} \quad \text{and} \quad (x - x_0)^2\frac{c(x)}{a(x)} = \frac{\gamma(x)}{\alpha(x)} \quad .$$

From this and the fact that α , β and γ are analytic at x_0 with $\alpha(x_0) \neq 0$ we get that the two limits

$$\lim_{x \to x_0} (x - x_0)\frac{b(x)}{a(x)} = \lim_{x \to x_0} \frac{\beta(x)}{\alpha(x)} = \frac{\beta(x_0)}{\alpha(x_0)}$$

and

$$\lim_{x \to x_0} (x - x_0)^2\frac{c(x)}{a(x)} = \lim_{x \to x_0} \frac{\gamma(x)}{\alpha(x)} = \frac{\gamma(x_0)}{\alpha(x_0)}$$

are finite.

So, x_0 being a regular singular point assures us that the above limits are finite. Consequently, if those limits are not finite, then x_0 cannot be a regular singular point for our differential equation. That gives us:

Lemma 35.1 (test for irregularity)
Assume x_0 is a singular point on the real line for

$$a(x)y'' + b(x)y' + c(x)y = 0 \quad .$$

If either of the two limits

$$\lim_{x \to x_0} (x - x_0)\frac{b(x)}{a(x)} \quad \text{and} \quad \lim_{x \to x_0} (x - x_0)^2\frac{c(x)}{a(x)}$$

is not finite, then x_0 is an irregular singular point for the differential equation.

This lemma is just a test for irregularity. It can be expanded to a more complete test if we make mild restrictions on the coefficients of the original differential equation. In particular, using properties of polynomials and rational functions, we can verify the following:

Theorem 35.2 (testing for regular singular points (ver.1))
Assume x_0 is a singular point on the real line for

$$a(x)y'' + b(x)y' + c(x)y = 0$$

where a , b and c are rational functions. Then x_0 is a regular singular point for this differential equation if and only if the two limits

$$\lim_{x \to x_0} (x - x_0)\frac{b(x)}{a(x)} \quad \text{and} \quad \lim_{x \to x_0} (x - x_0)^2\frac{c(x)}{a(x)}$$

are both finite values. Moreover, if x_0 is a regular singular point for the differential equation, then this differential equation can be written in quasi-Euler form

$$(x - x_0)^2 \alpha(x) y'' + (x - x_0)\beta(x) y' + \gamma(x) y = 0$$

where α, β and γ are polynomials with $\alpha(x_0) \neq 0$.

The full proof of this (along with a similar theorem applicable when the coefficients are merely quotients of analytic functions) is discussed in an appendix, Section 35.7.

!►Example 35.4: *Consider the differential equation*

$$y'' + \frac{1}{x^2} y' + \left[1 - \frac{1}{x^2}\right] y = 0 \quad .$$

Clearly, $x_0 = 0$ is a singular point for this equation. Writing out the limits given in the above theorem (with $x_0 = 0$) yields

$$\lim_{x \to 0} (x - 0) \frac{b(x)}{a(x)} = \lim_{x \to} x \cdot \frac{x^{-2}}{1} = \lim_{x \to 0} \frac{1}{x}$$

and

$$\lim_{x \to 0} (x - 0)^2 \frac{c(x)}{a(x)} = \lim_{x \to} x^2 \cdot \frac{1 - x^{-2}}{1} = \lim_{x \to 0} \left[x^2 - 1\right] \quad ,$$

The first limit is certainly not finite. So our test for regularity tells us that $x_0 = 0$ is an irregular singular point for this differential equation.

!►Example 35.5: *Consider*

$$2xy'' - 4y' - y = 0 \quad .$$

Again, $x_0 = 0$ is clearly the only singular point. Now,

$$\lim_{x \to 0} (x - 0) \frac{b(x)}{a(x)} = \lim_{x \to 0} x \cdot \frac{-4}{2x} = -2$$

and

$$\lim_{x \to 0} (x - 0)^2 \frac{c(x)}{a(x)} = \lim_{x \to 0} x^2 \cdot \frac{-1}{2x} = 0 \quad ,$$

both of which are finite. So $x_0 = 0$ is a regular singular point for our equation, and the given differential equation can be written in quasi-Euler form about 0. In fact, that form is obtained by simply multiplying the original differential equation by x,

$$2x^2 y'' - 4xy' - xy = 0 \quad .$$

35.3 The (Basic) Method of Frobenius

Motivation and Preliminary Notes

Let's suppose we have a second-order differential equation with x_0 as a regular singular point. Then, as we just discussed, this differential equation can be written as

$$(x - x_0)^2 \alpha(x) y'' + (x - x_0)\beta(x) y' + \gamma(x) y = 0$$

where α, β and γ are analytic at x_0 with $\alpha(x_0) \neq 0$. By continuity, when $x \approx x_0$, we have

$$\alpha(x) \approx \alpha_0 \quad , \quad \beta(x) \approx \beta_0 \quad \text{and} \quad \gamma(x) \approx \gamma_0$$

where

$$\alpha_0 = \alpha(x_0) \quad , \quad \beta_0 = \beta(x_0) \quad \text{and} \quad \gamma_0 = \gamma(x_0) \quad .$$

It should then seem reasonable that any solution $y = y(x)$ to

$$(x - x_0)^2 \alpha(x) y'' + (x - x_0)\beta(x) y' + \gamma(x) y = 0$$

can be approximated, at least when $x \approx x_0$, by a corresponding solution to the associated shifted Euler equation

$$(x - x_0)^2 \alpha_0 y'' + (x - x_0)\beta_0 y' + \gamma_0 y = 0 \quad .$$

And since some solutions to this shifted Euler equation are of the form $a_0(x - x_0)^r$ where a_0 is an arbitrary constant and r is a solution to the corresponding indicial equation,

$$\alpha_0 r(r-1) + \beta_0 r + \gamma_0 = 0 \quad ,$$

it seems reasonable to expect at least some solutions to our original differential equation to be approximated by this $a_0(x - x_0)^r$,

$$y(x) \approx a_0(x - x_0)^r \qquad \text{at least when} \qquad x \approx x_0 \quad .$$

Now, this is equivalent to saying

$$\frac{y(x)}{(x - x_0)^r} \approx a_0 \qquad \text{when} \quad x \approx x_0 \quad ,$$

which is more precisely stated as

$$\lim_{x \to x_0} \frac{y(x)}{(x - x_0)^r} = a_0 \quad . \tag{35.10}$$

At this point, there are a number of ways we might 'guess' at solutions satisfying the last approximation. Let us try a trick similar to one we've used before; namely, let's assume that y is the known approximate solution $(x - x_0)^r$ multiplied by some yet unknown function $a(x)$,

$$y(x) = (x - x_0)^r a(x) \quad .$$

To satisfy equation (35.10), we must have

$$a(x_0) = \lim_{x \to x_0} a(x) = \lim_{x \to x_0} \frac{y(x)}{(x - x_0)^r} = a_0 \quad ,$$

telling us that $a(x)$ is reasonably well behaved near x_0, perhaps even analytic. Well, let's hope it is analytic, because that means we can express $a(x)$ as a power series about x_0 with the arbitrary constant a_0 as the constant term,

$$a(x) = \sum_{k=0}^{\infty} a_k (x - x_0)^k \quad .$$

Then, with luck and skill, we might be able to use the methods developed in the previous chapters to find the a_k's in terms of a_0.

That is the starting point for what follows. We will assume a solution of the form

$$y(x) = (x - x_0)^r \sum_{k=0}^{\infty} a_k (x - x_0)^k$$

where r and the a_k's are constants to be determined, with a_0 being arbitrary. This will yield the "modified power series" solutions alluded to in the title of this chapter.

In the next subsection, we will describe a series of steps, generally called the *(basic) method of Frobenius*, for finding such solutions. You will discover that much of it is very similar to the algebraic method for finding power series solutions in Chapter 33.

Before we start that, however, there are a few things worth mentioning about this method:

1. We will see that the method of Frobenius always yields at least one solution of the form

$$y(x) = (x - x_0)^r \sum_{k=0}^{\infty} a_k (x - x_0)^k$$

where r is a solution to the appropriate indicial equation. If the indicial equation for the corresponding Euler equation

$$(x - x_0)^2 \alpha_0 y'' + (x - x_0) \beta_0 y' + \gamma_0 y = 0$$

has two distinct solutions, then we will see that the method often (but not always) leads to an independent pair of such solutions.

2. If, however, that indicial equation has only one solution r, then the fact that the corresponding shifted Euler equation has a 'second solution' in the form

$$(x - x_0)^r \ln |x - x_0|$$

may lead you to suspect that a 'second solution' to our original equation is of the form

$$(x - x_0)^r \ln |x - x_0| \sum_{k=0}^{\infty} a_k (x - x_0)^k \quad .$$

That turns out to be *almost* the case.

Just what can be done when the basic Frobenius method does not yield an independent pair of solutions will be discussed in the next chapter.

The (Basic) Method of Frobenius

Suppose we wish to find the solutions about some regular singular point x_0 to some second-order homogeneous linear differential equation

$$a(x)y'' + b(x)y' + c(x)y = 0$$

with $a(x)$, $b(x)$ and $c(x)$ being rational functions.[3] In particular, let us seek solutions to *Bessel's equation of order* $1/2$,

$$y'' + \frac{1}{x}y' + \left[1 - \frac{1}{4x^2}\right]y = 0 \qquad (35.11)$$

near the regular singular point $x_0 = 0$.

As with the algebraic method for finding power series solutions, there are two preliminary steps:

Pre-Step 1. If not already specified, choose the regular singular point x_0.

For our example, we choose $x_0 = 0$, which we know is the only regular singular point from the discussion in the previous section.

Pre-Step 2. Get the differential equation into quasi-Euler form; that is, into the form

$$(x - x_0)^2\alpha(x)y'' + (x - x_0)\beta(x)y' + \gamma(x)y = 0 \qquad (35.12)$$

where α, β and γ are polynomials, with $\alpha(x_0) \neq 0$, and with no factors shared by all three.[4]

To get the given differential equation into the form desired, we multiply equation (35.11) by $4x^2$. That gives us the differential equation

$$4x^2y'' + 4xy' + [4x^2 - 1]y = 0 \quad . \qquad (35.13)$$

(Yes, we could have just multiplied by x^2, but getting rid of any fractions will simplify computation.)

Now for the basic method of Frobenius:

Step 1. ***(a)*** Assume a solution of the form

$$y = y(x) = (x - x_0)^r \sum_{k=0}^{\infty} a_k(x - x_0)^k \qquad (35.14a)$$

with a_0 being an arbitrary *nonzero* constant.[5,6]

(b) Simplify this formula for the following computations by bringing the $(x - x_0)^r$ factor into the summation,

$$y = y(x) = \sum_{k=0}^{\infty} a_k(x - x_0)^{k+r} \quad . \qquad (35.14b)$$

---

[3] The "Frobenius method" for the more general equations is developed in Chapter 37.

[4] For computations, it is not necessary to explicitly factor out $x - x_0$ from the coefficients of the differential equation, only that the coefficients be polynomials that could be factored as in equation (35.12). Using this form will simplify bookkeeping, and will be convenient for discussions later.

[5] Insisting that $a_0 \neq 0$ will be important in determining the possible values of r.

[6] This procedure is valid whether or not $x - x_0$ is positive or negative. However, a few readers may have concerns about $(x - x_0)^r$ being imaginary if $x < x_0$ and r is, say, $1/2$. If you are one of those readers, go ahead and assume $x > x_0$ for now, and later read the short discussion of solutions on intervals with $x < x_0$ on page 702.

Step 2. Compute the corresponding modified power series for y' and y'' from the assumed series for y by differentiating "term-by-term".[7] This time, you can*not* drop the "$k = 0$" term in the summations because this term is not necessarily a constant.

Since we've already decided $x_0 = 0$, we assume

$$y = y(x) = x^r \sum_{k=0}^{\infty} a_k x^k = \sum_{k=0}^{\infty} a_k x^{k+r} \tag{35.15}$$

with $a_0 \neq 0$. Differentiating this term-by-term, we see that

$$\begin{aligned} y' &= \frac{d}{dx} \sum_{k=0}^{\infty} a_k x^{k+r} \\ &= \sum_{k=0}^{\infty} \frac{d}{dx} \left[a_k x^{k+r} \right] = \sum_{k=0}^{\infty} a_k (k+r) x^{k+r-1} \end{aligned}$$

and

$$\begin{aligned} y'' &= \frac{d}{dx} \sum_{k=0}^{\infty} a_k (k+r) x^{k+r-1} \\ &= \sum_{k=0}^{\infty} \frac{d}{dx} \left[a_k (k+r) x^{k+r-1} \right] = \sum_{k=0}^{\infty} a_k (k+r)(k+r-1) x^{k+r-2} \quad . \end{aligned}$$

Step 3. Plug these series for y, y' and y'' back into the differential equation, "multiply things out", and divide out the $(x - x_0)^r$ to get the left side of your equation in the form of the sum of a few power series about x_0.[8,9]

Combining the above series formulas for y, y' and y'' with our differential equation (equation (35.13)), we get

$$\begin{aligned} 0 &= 4x^2 y'' + 4xy' + [4x^2 - 1]y \\ &= 4x^2 \sum_{k=0}^{\infty} a_k (k+r)(k+r-1) x^{k+r-2} + 4x \sum_{k=0}^{\infty} a_k (k+r) x^{k+r-1} \\ &\quad + [4x^2 - 1] \sum_{k=0}^{\infty} a_k x^{k+r} \\ &= 4x^2 \sum_{k=0}^{\infty} a_k (k+r)(k+r-1) x^{k+r-2} + 4x \sum_{k=0}^{\infty} a_k (k+r) x^{k+r-1} \\ &\quad + 4x^2 \sum_{k=0}^{\infty} a_k x^{k+r} - 1 \sum_{k=0}^{\infty} a_k x^{k+r} \end{aligned}$$

---

[7] If you have any qualms about "term-by-term" differentiation here, see Exercise 35.6 at the end of the chapter.

[8] Dividing out the $(x - x_0)^r$ isn't necessary, but it simplifies the expressions slightly and reduces the chances of silly errors later.

[9] You may want to turn your paper sideways for more room!

$$= \sum_{k=0}^{\infty} a_k 4(k+r)(k+r-1)x^{k+r} + \sum_{k=0}^{\infty} a_k 4(k+r)x^{k+r}$$

$$+ \sum_{k=0}^{\infty} a_k 4x^{k+2+r} + \sum_{k=0}^{\infty} a_k(-1)x^{k+r} \quad .$$

Dividing out the x^r from each term then yields

$$0 = \sum_{k=0}^{\infty} a_k 4(k+r)(k+r-1)x^k + \sum_{k=0}^{\infty} a_k 4(k+r)x^k$$

$$+ \sum_{k=0}^{\infty} a_k 4x^{k+2} + \sum_{k=0}^{\infty} a_k(-1)x^k \quad .$$

Step 4. For each series in your last equation, do a change of index so that each series looks like

$$\sum_{n=\text{something}}^{\infty} \big[\text{something not involving } x\big](x-x_0)^n \quad .$$

Be sure to appropriately adjust the lower limit in each series.

In all but the third series in the example, the change of index is trivial, $n = k$. In the third series, we will set $n = k+2$ (equivalently, $n-2 = k$). This means, in the third series, replacing k with $n-2$, and replacing $k = 0$ with $n = 0+2 = 2$:

$$0 = \underbrace{\sum_{k=0}^{\infty} a_k 4(k+r)(k+r-1)x^k}_{n\,=\,k} + \underbrace{\sum_{k=0}^{\infty} a_k 4(k+r)x^k}_{n\,=\,k}$$

$$+ \underbrace{\sum_{k=0}^{\infty} a_k 4x^{k+2}}_{n\,=\,k+2} + \underbrace{\sum_{k=0}^{\infty} a_k(-1)x^k}_{n\,=\,k}$$

$$= \sum_{n=0}^{\infty} a_n 4(n+r)(n+r-1)x^n + \sum_{n=0}^{\infty} a_n 4(n+r)x^n$$

$$+ \sum_{n=2}^{\infty} a_{n-2} 4x^n + \sum_{n=0}^{\infty} a_n(-1)x^n \quad .$$

Step 5. Convert the sum of series in your last equation into one big series. The first few terms will probably have to be written separately. Simplify what can be simplified.

Since one of the series in the last equation begins with $n = 2$, we need to separate out the terms corresponding to $n = 0$ and $n = 1$ in the other series before combining the series:

$$0 = \sum_{n=0}^{\infty} a_n 4(n+r)(n+r-1)x^n + \sum_{n=0}^{\infty} a_n 4(n+r)x^n + \sum_{n=2}^{\infty} a_{n-2} 4x^n + \sum_{n=0}^{\infty} a_n(-1)x^n$$

$$= \left[a_0 \underbrace{4(0+r)(0+r-1)}_{4r(r-1)} x^0 + a_1 \underbrace{4(1+r)(1+r-1)}_{4(1+r)r} x^1 + \sum_{n=2}^{\infty} a_n 4(n+r)(n+r-1)x^n \right]$$

$$+ \left[a_0 \underbrace{4(0+r)}_{4r} x^0 + a_1 \underbrace{4(1+r)}_{4(1+r)} x^1 + \sum_{n=2}^{\infty} a_n 4(n+r)x^n \right]$$

$$+ \sum_{n=2}^{\infty} a_{n-2} 4x^n + \left[-a_0 x^0 - a_1 x^1 + \sum_{n=2}^{\infty} a_n(-1)x^n \right]$$

$$= a_0 \big[\underbrace{4r(r-1) + 4r - 1}_{4r^2 - 4r + 4r - 1} \big] x^0 + a_1 \big[\underbrace{4(1+r)r + 4(1+r) - 1}_{4r + 4r^2 + 4 + 4r - 1} \big] x^1$$

$$+ \sum_{n=2}^{\infty} \big[\underbrace{a_n 4(n+r)(n+r-1) + a_n 4(n+r) + a_{n-2} 4 + a_n(-1)}_{a_n[4(n+r)(n+r-1) + 4(n+r) - 1] + 4a_{n-2}} \big] x^n$$

$$= a_0 \big[4r^2 - 1 \big] x^0 + a_1 \big[4r^2 + 8r + 3 \big] x^1 + \sum_{n=2}^{\infty} \Big[a_n \big[4(n+r)(n+r) - 1 \big] + 4a_{n-2} \Big] x^n \quad .$$

So our differential equation reduces to

$$a_0 \big[4r^2 - 1 \big] x^0 + a_1 \big[4r^2 + 8r + 3 \big] x^1 + \sum_{n=2}^{\infty} \Big[a_n \big[4(n+r)^2 - 1 \big] + 4a_{n-2} \Big] x^n = 0 \quad . \tag{35.16}$$

Step 6. The first term in the last equation just derived will be of the form[10]

$$a_0 \big[\text{formula of } r \big] (x - x_0)^0 \quad .$$

Since each term in the series must vanish, we must have

$$a_0 \big[\text{formula of } r \big] = 0 \quad .$$

[10] If you did not originally get the differential equation into quasi-Euler form, the exponent on $(x - x_0)$ might not be zero. That's not important; the coefficient still must be zero.

Moreover, since $a_0 \neq 0$ (by assumption), the above must reduce to

$$\text{formula of } r \ = \ 0 \quad .$$

This is the *indicial equation* for the differential equation. It will be a quadratic equation (we'll see why later). Solve this equation for r. You will get two solutions (sometimes called either the *exponents* of the solution or the *exponents* of the singularity). Denote them by r_1 and r_2. If the exponents are real (which is common in applications), label the exponents so that $r_1 \geq r_2$. If the exponents are not real, then it does not matter which is labeled as r_1.[11]

In our example, the first term in the "big series" is the first term in equation (35.16),

$$a_0\left[4r^2 - 1\right]x^0 \quad .$$

Since this must be zero (and $a_0 \neq 0$ by assumption) the indicial equation is

$$4r^2 - 1 = 0 \quad . \tag{35.17}$$

Thus,

$$r = \pm\sqrt{\frac{1}{4}} = \pm\frac{1}{2} \quad .$$

Following the convention given above (that $r_1 \geq r_2$),

$$r_1 = \frac{1}{2} \qquad \text{and} \qquad r_2 = -\frac{1}{2} \quad .$$

Step 7. Using r_1 (the *largest* r if the exponents are real):

(a) Plug r_1 into the last series equation (and simplify, if possible). This will give you an equation of the form

$$\sum_{n=n_0}^{\infty} \left[n^{\text{th}} \text{ formula of } a_j\text{'s}\right](x - x_0)^n = 0 \quad .$$

Since each term must vanish, we must have

$$n^{\text{th}} \text{ formula of } a_j\text{'s} = 0 \qquad \text{for} \quad n_0 \leq n \quad .$$

(b) Solve this last set of equations for[12]

$$a_{\text{highest index}} = \text{formula of } n \text{ and lower indexed } a_j\text{'s} \quad .$$

A few of these equations may need to be treated separately, but you will also obtain a relatively simple formula that holds for all indices above some fixed value. This formula is the *recursion formula* for computing each coefficient a_n from the previously computed coefficients.

---

[11] We are assuming the coefficients of our differential equation— α, β and γ —are real-valued functions on the real line. In the very unlikely case they are not, then a more general convention should be used: If the solutions to the indicial equation differ by an integer, then label them so that $r_1 - r_2 \geq 0$. Otherwise, it does not matter which you call r_1 and which you call r_2. The reason for this convention will become apparent later (in Section 35.5) after we further discuss the formulas arising from the Frobenius method.

[12] If you did as suggested earlier and put the differential equation into quasi-Euler form, then n will be the "highest index" in this equation.

(c) (Optional) To simplify things just a little, do another change of indices so that the recursion formula just derived is rewritten as

$$a_k = \text{formula of } k \text{ and lower-indexed coefficients} \quad .$$

Letting $r = r_1 = 1/2$ in equation (35.16) yields

$$0 = a_0\big[4r^2 - 1\big]x^0 + a_1\big[4r^2 + 8r + 3\big]x^1 + \sum_{n=2}^{\infty}\Big[a_n\big[4(n+r)^2 - 1\big] + 4a_{n-2}\Big]x^n$$

$$= a_0\left[4\left(\frac{1}{2}\right)^2 - 1\right]x^0 + a_1\left[4\left(\frac{1}{2}\right)^2 + 8\left(\frac{1}{2}\right) + 3\right]x^1 + \sum_{n=2}^{\infty}\left[a_n\left[4\left(n+\frac{1}{2}\right)^2 - 1\right] + 4a_{n-2}\right]x^n$$

$$= a_0[0]x^0 + a_1 8x^1 + \sum_{n=2}^{\infty}\left[a_n\left[4n^2 + 4n\right] + 4a_{n-2}\right]x^n \quad .$$

The first term vanishes (as it should since $r = 1/2$ satisfies the indicial equation, which came from making the first term vanish). Doing a little more simple algebra, we see that, with $r = 1/2$, equation (35.16) reduces to

$$0a_0x^0 + 8a_1x^1 + \sum_{n=2}^{\infty} 4\big[n(n+1)a_n + a_{n-2}\big]x^n = 0 \quad . \qquad (35.18)$$

Since the individual terms in this series must vanish, we have

$$0a_0 = 0 \quad , \quad 8a_1 = 0$$

and

$$n(n+1)a_n + a_{n-2} = 0 \quad \text{for} \quad n = 2, 3, 4 \ldots \quad .$$

Solving for a_n gives us the recursion formula

$$a_n = \frac{-1}{n(n+1)}a_{n-2} \quad \text{for} \quad n = 2, 3, 4 \ldots \quad .$$

Using the trivial change of index, $k = n$, this is

$$a_k = \frac{-1}{k(k+1)}a_{k-2} \quad \text{for} \quad k = 2, 3, 4 \ldots \quad . \qquad (35.19)$$

(d) Use the recursion formula (and any corresponding formulas for the lower-order terms) to find all the a_k's in terms of a_0 and, possibly, one other a_m. Look for patterns!

From the first two terms in equation (35.18),

$$0a_0 = 0 \implies a_0 \text{ is arbitrary.}$$

$$8a_1 = 0 \implies a_1 = 0 \quad .$$

Using these values and recursion formula (35.19) with $k = 2, 3, 4, \ldots$ (and looking for patterns):

$$a_2 = \frac{-1}{2(2+1)}a_{2-2} = \frac{-1}{2\cdot 3}a_0 = \frac{-1}{3\cdot 2}a_0 \quad ,$$

$$a_3 = \frac{-1}{3(3+1)}a_{3-2} = \frac{-1}{3\cdot 4}a_1 = \frac{-1}{3\cdot 4}\cdot 0 = 0 \quad ,$$

$$a_4 = \frac{-1}{4(4+1)}a_{4-2} = \frac{-1}{4\cdot 5}a_2 = \frac{-1}{5\cdot 4}\cdot\frac{-1}{3\cdot 2}a_0 = \frac{(-1)^2}{5!}a_0 \quad ,$$

$$a_5 = \frac{-1}{5(5+1)}a_{5-2} = \frac{-1}{5\cdot 6}\cdot 0 = 0 \quad ,$$

$$a_6 = \frac{-1}{6(6+1)}a_{6-2} = \frac{-1}{6\cdot 7}a_4 = \frac{-1}{7\cdot 6}\cdot\frac{(-1)^2}{5!}a_0 = \frac{(-1)^3}{7!}a_0 \quad ,$$

$$\vdots$$

The patterns should be obvious here:

$$a_k = 0 \qquad \text{for} \quad k = 1, 3, 5, 7, \ldots \quad ,$$

and

$$a_k = \frac{(-1)^{k/2}}{(k+1)!}a_0 \qquad \text{for} \quad k = 2, 4, 6, 8, \ldots \quad .$$

Using $k = 2m$, this can be written more conveniently as

$$a_{2m} = (-1)^m \frac{a_0}{(2m+1)!} \qquad \text{for} \quad m = 1, 2, 3, 4, \ldots \quad .$$

Moreover, this last equation reduces to the trivially true statement "$a_0 = a_0$" if $m = 0$. So, in fact, it gives all the even-indexed coefficients,

$$a_{2m} = (-1)^m \frac{a_0}{(2m+1)!} \qquad \text{for} \quad m = 0, 1, 2, 3, 4, \ldots \quad .$$

(e) Using $r = r_1$ along with the formulas just derived for the coefficients, write out the resulting series for y. Try to simplify it and factor out the arbitrary constant(s).

Plugging $r = 1/2$ and the formulas just derived for the a_k's into the formula originally assumed for y (equation (35.15) on page 693), we get

$$y = x^r \sum_{k=0}^{\infty} a_k x^k$$

$$= x^r \left[\sum_{\substack{k=0 \\ k \text{ odd}}}^{\infty} a_k x^k + \sum_{\substack{k=0 \\ k \text{ even}}}^{\infty} a_k x^k \right]$$

$$= x^{1/2} \left[\sum_{\substack{k=0 \\ k \text{ odd}}}^{\infty} 0\cdot x^k + \sum_{m=0}^{\infty} (-1)^m \frac{a_0}{(2m+1)!} x^{2m} \right]$$

$$= x^{1/2} \left[0 + a_0 \sum_{m=0}^{\infty} (-1)^m \frac{1}{(2m+1)!} x^{2m} \right] \quad .$$

So one set of solutions to Bessel's equation of order $^1/_2$ *(equation (35.11) on page 692) is given by*

$$y = a_0 x^{1/2} \sum_{m=0}^{\infty} \frac{(-1)^m}{(2m+1)!} x^{2m} \tag{35.20}$$

with a_0 *being an arbitrary constant.*

Step 8. If $r_2 = r_1$, skip this step and just go to the next. If the exponents are complex (which is unlikely), see *Dealing with Complex Exponents* starting on page 712. But if the indicial equation has two distinct real solutions, now try to repeat step 7 with the other exponent, r_2, replacing r_1.

Before starting, however, you should be warned that, in attempting to redo step 7 with $r = r_2$, one of three things will happen:

i. With luck, this step will lead to a solution of the form

$$(x - x_0)^{r_2} \sum_{k=0}^{\infty} a_k (x - x_0)^k$$

with a_0 being the only arbitrary constant.

ii. This step can also lead to a series having two arbitrary constants, a_0 and one other coefficient a_M (our example will illustrate this). If you keep this second arbitrary constant, you will then end up with two particular series solutions — one multiplied by a_0 and one multiplied by a_M. However, it is easily shown (see Exercise 35.8) that the one multiplied by a_M is simply the series solution already obtained at the end of the previous step. Thus, in keeping a_M arbitrary, you will rederive the one solution you already have. Why bother? Go ahead and set $a_M = 0$ and continue. It will save you a lot of needless work.

iii. Finally, this step can lead to a contradiction. More precisely, the recursion formula might "blow up" for one of the coefficients. This tells you that there is no solution of the form

$$(x - x_0)^{r_2} \sum_{k=0}^{\infty} a_k (x - x_0)^k \qquad \text{with} \quad a_0 \neq 0 \quad .$$

If this happens, make note of it, and skip to the next step.

(Why we have these three possibilities will be further discussed in *Problems Possibly Arising in Step 8* starting on page 707.)

Letting $r = r_2 = -^1/_2$ *in equation (35.16) yields*

$$0 = a_0\big[4r^2 - 1\big]x^0 + a_1\big[4r^2 + 8r + 3\big]x^1$$
$$+ \sum_{n=2}^{\infty} \Big[a_n\big[4(n+r)^2 - 1\big] + 4a_{n-2}\Big]x^n$$

$$= a_0\left[4\left(-\tfrac{1}{2}\right)^2 - 1\right]x^0 + a_1\left[4\left(-\tfrac{1}{2}\right)^2 + 8\left(-\tfrac{1}{2}\right) + 3\right]x^1$$

$$+ \sum_{n=2}^{\infty}\left[a_n\left[4\left(n - \tfrac{1}{2}\right)^2 - 1\right] + 4a_{n-2}\right]x^n$$

$$= a_0 0x^0 + a_1 0x^1 + \sum_{n=2}^{\infty}\left[a_n\left[4n^2 - 4n\right] + 4a_{n-2}\right]x^n \quad .$$

That is,

$$0a_0x^0 + 0a_1x^1 + \sum_{n=2}^{\infty} 4\left[a_n n(n-1) + a_{n-2}\right]x^n = 0 \quad ,$$

which means that

$$0a_0 = 0 \quad , \quad 0a_1 = 0 \tag{35.21a}$$

and

$$a_n n(n-1) + a_{n-2} = 0 \quad \text{for} \quad n = 2, 3, 4, \ldots \quad . \tag{35.21b}$$

The first two equations hold for any values of a_0 and a_1. So, strictly speaking, both a_0 and a_1 are arbitrary constants. But, as noted above, keeping a_1 arbitrary in these calculations will simply lead to solution (35.20) with a_1 replacing a_0. Since we don't need to rederive this solution, we will simplify our work by setting

$$a_1 = 0 \quad .$$

From equation (35.21b) we get

$$a_n = \frac{-1}{n(n-1)}a_{n-2} \quad \text{for} \quad n = 2, 3, 4, \ldots \quad .$$

That is,

$$a_k = \frac{-1}{k(k-1)}a_{k-2} \quad \text{for} \quad k = 2, 3, 4, \ldots \quad . \tag{35.22}$$

This is the recursion formula for finding all the other a_k's in terms of a_0 (which is arbitrary) and a_0 (which we've set to 0).

Why don't you finish these computations as an exercise? You should have no trouble in obtaining

$$y = a_0 x^{-1/2} \sum_{m=0}^{\infty} \frac{(-1)^m}{(2m)!} x^{2m} \quad . \tag{35.23}$$

?►Exercise 35.1: *Do the computations left "as an exercise" in the last statement.*

Step 9. If the last step yielded a series solution, then the general solution to the original differential equation is the linear combination of the two series obtained at the end of steps 7 and 8. Write down this general solution (using different symbols for the two different arbitrary constants!). If step 8 did not yield a new series solution, then at least write down the one solution previously derived, noting that a second solution is still needed for the general solution to the differential equation. (Finding that second solution will be discussed in the next chapter.)

We are in luck. In the last step, we obtained series solution (35.23). Combining that with solution (35.20) (and renaming the two arbitrary constants as c_1 and c_2), gives us the general solution

$$y(x) = c_1 x^{1/2} \sum_{m=0}^{\infty} \frac{(-1)^m}{(2m+1)!} x^{2m} + c_2 x^{-1/2} \sum_{m=0}^{\infty} \frac{(-1)^m}{(2m)!} x^{2m} ,$$

for our original differential equation (equation (35.11) — Bessel's equation of order $1/2$).

Last Step. See if you recognize either of the series as the series for some well-known function. (This is possible occasionally, but not often.)

Our luck continues! Looking at the last formula above for y and recalling the power series for the sine and cosine, we see that

$$y(x) = c_1 y_1(x) + c_2 y_2(x)$$

where

$$y_2(x) = x^{-1/2} \sum_{m=0}^{\infty} \frac{(-1)^m}{(2m)!} x^{2m} = x^{-1/2} \sin(x)$$

and

$$y_1(x) = x^{1/2} \sum_{m=0}^{\infty} \frac{(-1)^m}{(2m+1)!} x^{2m}$$

$$= x^{1/2} x^{-1} \sum_{m=0}^{\infty} \frac{(-1)^m}{(2m+1)!} x^{2m+1} = x^{-1/2} \cos(x) .$$

That is, we can also write the general solution to Bessel's equation of order $1/2$ as

$$y = c_1 \frac{\cos(x)}{\sqrt{x}} + c_2 \frac{\sin(x)}{\sqrt{x}} . \qquad (35.24)$$

"First" and "Second" Solutions

For purposes of discussion, it is convenient to refer to the first solution found in the basic method of Frobenius,

$$y(x) = (x - x_0)^{r_1} \sum_{k=0}^{\infty} a_k (x - x_0)^k ,$$

as the *first solution*. If we then pick a particular nonzero value for a_0, then we have a *first particular solution*. In the above, our first solution was

$$y(x) = a_0 x^{1/2} \sum_{m=0}^{\infty} \frac{(-1)^m}{(2m+1)!} x^{2m} .$$

Taking $a_0 = 1$, we then have a first particular solution

$$y_1(x) = x^{1/2} \sum_{m=0}^{\infty} \frac{(-1)^m}{(2m+1)!} x^{2m} .$$

Naturally, if we also obtain a solution corresponding to r_2 in step 8,

$$y(x) = (x - x_0)^{r_2} \sum_{k=0}^{\infty} a_k (x - x_0)^k \quad ,$$

then we will refer to that as our *second solution*, with a *second particular solution* being this with a specific nonzero value chosen for a_0 and any specific value chosen for any other arbitrary constant. For example, in illustrating the basic method of Frobenius, we obtained

$$y(x) = a_0 x^{-1/2} \sum_{m=0}^{\infty} \frac{(-1)^m}{(2m)!} x^{2m} + a_1 x^{-1/2} \sum_{m=0}^{\infty} \frac{(-1)^m}{(2m+1)!} x^{2m+1}$$

as our second solution. Choosing $a_0 = 1$ and $a_1 = 0$, we get the second particular solution

$$y_2(x) = x^{-1/2} \sum_{m=0}^{\infty} \frac{(-1)^m}{(2m)!} x^{2m} \quad .$$

While we are at it, let us agree that, for the rest of this chapter, x_0 always denotes a regular singular point for some differential equation of interest, and that r_1 and r_2 always denote the corresponding exponents; that is, the solutions to the corresponding indicial equation. Let us further agree that these exponents are indexed according to the convention given in the basic Frobenius, with $r_1 \geq r_2$ if they are real.

35.4 Basic Notes on Using the Frobenius Method

The Obvious

One thing should be obvious: The method we've just outlined is even longer and more tedious than the algebraic method used in Chapter 33 to find power series solutions to similar equations about ordinary points. On the other hand, much of this method is based on that algebraic method, which, by now, you have surely mastered.

Naturally, all the 'practical advice' given regarding the algebraic method in Chapter 33 still holds, including the recommendation that you use

$$Y(X) = y(x) \quad \text{with} \quad X = x - x_0$$

to simplify your calculations when $x_0 \neq 0$.

But there are a number of other things you should be aware of:

Solutions on Intervals with $x < x_0$

On an interval with $x < x_0$, $(x - x_0)^r$ might be complex-valued (or even ambiguous) if r is not an integer. For example,

$$(x - x_0)^{1/2} = (-|x - x_0|)^{1/2} = (-1)^{1/2} |x - x_0|^{1/2} = \pm i \, |x - x_0|^{1/2} \quad .$$

More generally, we will have

$$(x - x_0)^r = (-|x - x_0|)^r = (-1)^r |x - x_0|^r \qquad \text{when} \quad x < x_0 \quad .$$

But this is not a significant issue because $(-1)^r$ can be viewed as a constant (possibly complex) and can be divided out of the final formula (or incorporated in the arbitrary constants). Thus, in our final formulas for $y(x)$, we can replace

$$(x - x_0)^r \quad \text{with} \quad |x - x_0|^r$$

to avoid having any explicitly complex-valued expressions (at least when r is not complex). Keep in mind that there is no reason to do this if r is an integer.

Convergence of the Series

It probably won't surprise you to learn that the Frobenius radius of analyticity serves as a lower bound on the radius of convergence for the power series found in the Frobenius method. To be precise, we have the following theorem:

Theorem 35.3

Let x_0 be a regular singular point for some second-order homogeneous linear differential equation

$$a(x)y'' + b(x)y' + c(x)y = 0$$

and let

$$y(x) = |x - x_0|^r \sum_{k=0}^{\infty} a_k (x - x_0)^k$$

be a modified power series solution to the differential equation found by the basic method of Frobenius. Then the radius of convergence for $\sum_{k=0}^{\infty} a_k (x - x_0)^k$ is at least equal to the Frobenius radius of convergence about x_0 for this differential equation.

We will verify this claim in Chapter 37. For now, let us simply note that this theorem assures us that the given solution y is valid at least on the intervals

$$(x_0 - R, x_0) \quad \text{and} \quad (x_0, x_0 + R)$$

where R is that Frobenius radius of convergence. Whether or not we can include the point x_0 depends on the value of the exponent r.

!►Example 35.6: *To illustrate the Frobenius method, we found modified power series solutions about $x_0 = 0$*

$$y_1(x) = x^{1/2} \sum_{m=0}^{\infty} \frac{(-1)^m}{(2m+1)!} x^{2m} \quad \text{and} \quad y_2(x) = x^{-1/2} \sum_{m=0}^{\infty} \frac{(-1)^m}{(2m)!} x^{2m}$$

for Bessel's equation of order $1/2$,

$$y'' + \frac{1}{x} y' + \left[1 - \frac{1}{4x^2}\right] y = 0 \quad .$$

Since there are no singular points for this differential equation other than $x_0 = 0$, the Frobenius radius of convergence for this differential equation about $x_0 = 0$ is $R = \infty$. That means the power series in the above formulas for y_1 and y_2 converge everywhere.

However, the $x^{1/2}$ and $x^{-1/2}$ factors multiplying these power series are not "well behaved" at $x_0 = 0$ — neither is differentiable there, and one becomes infinite as $x \to 0$. So, the above formulas for y_1 and y_2 are valid only on intervals not containing $x = 0$, the largest of which *are $(0, \infty)$ and $(-\infty, 0)$. Of course, on $(-\infty, 0)$ the square roots yield imaginary values, and we would prefer using the solutions*

$$y_1(x) = |x|^{1/2} \sum_{m=0}^{\infty} \frac{(-1)^m}{(2m+1)!} x^{2m} \quad \text{and} \quad y_2(x) = |x|^{-1/2} \sum_{m=0}^{\infty} \frac{(-1)^m}{(2m)!} x^{2m} \quad .$$

Variations of the Method

Naturally, there are several variations of "the basic method of Frobenius". The one just given is merely one the author finds convenient for initial discussion.

One variation you may want to consider is to find particular solutions without arbitrary constants by setting $a_0 = 1$. That is, in step 1, assume

$$y(x) = (x - x_0)^r \sum_{k=0}^{\infty} a_k (x - x_0)^k \quad \text{with} \quad a_0 = 1 \quad .$$

Assuming a_0 is 1, instead of an arbitrary nonzero constant, leads to a first particular solution

$$y_1(x) = (x - x_0)^{r_1} \sum_{k=0}^{\infty} a_k (x - x_0)^k \quad \text{with} \quad a_0 = 1 \quad .$$

With a little thought, you will realize that this is exactly the same as you would have obtained at the end of step 7, only not multiplied by an arbitrary constant. In particular, had we done this with the Bessel's equation used to illustrate the method, we would have obtained

$$y_1(x) = x^{1/2} \sum_{m=0}^{\infty} \frac{(-1)^m}{(2m+1)!} x^{2m}$$

instead of formula (35.20) on page 699.

If $r_2 \neq r_1$, then, with luck, doing step 8 with $a_0 = 1$ will yield a second particular solution

$$y_2(x) = (x - x_0)^{r_2} \sum_{k=0}^{\infty} a_k (x - x_0)^k \quad \text{with} \quad a_0 = 1 \quad .$$

In particular, doing this with the illustrating example would have yielded

$$y_2(x) = x^{-1/2} \sum_{m=0}^{\infty} \frac{(-1)^m}{(2m)!} x^{2m} \quad ,$$

instead of formula (35.23) on page 700.

Assuming the second particular solution can be found, this variant of the method yields a pair of particular solutions $\{y_1, y_2\}$ that, because $r_1 \neq r_1$, is easily seen to be linearly independent over any interval on which the formulas are valid. Thus,

$$y(x) = c_1 y_1(x) + c_2 y_2(x)$$

is the general solution to the differential equation over this interval.

Using the Method When x_0 Is an Ordinary Point

It should be noted that we can also find the power series solutions of a differential equation about an ordinary point x_0 using the basic Frobenius method (ignoring, of course, the first preliminary step). In practice, though, it would be silly to go through the extra work in the Frobenius method when you can use the shorter algebraic method from Chapter 33. After all, if x_0 is an ordinary point for our differential equation, then we already know the solution y can be written as

$$y(x) = a_0 y_1(x) + a_1 y_2(x)$$

where $y_1(x)$ and $y_2(x)$ are power series about x_0 with

$$\begin{aligned} y_1(x) &= 1 + \text{a summation of terms of order 2 or more} \\ &= (x - x_0)^0 \sum_{k=0}^{\infty} b_k (x - x_0)^k \qquad \text{with} \quad b_k = 1 \end{aligned}$$

and

$$\begin{aligned} y_2(x) &= 1 \cdot (x - x_0) + \text{a summation of terms of order 2 or more} \\ &= (x - x_0)^1 \sum_{k=0}^{\infty} c_k (x - x_0)^k \qquad \text{with} \quad c_k = 1 \end{aligned}$$

(see *Initial-Value Problems* on page 626). Clearly then, $r_2 = 0$ and $r_1 = 1$, and the indicial equation that would arise in using the Frobenius method would just be $r(r-1) = 0$.

So why bother solving for the exponents r_1 and r_2 of a differential equation at a point x_0 when you don't need to? It's easy enough to determine whether a point x_0 is an ordinary point or a regular singular point for your differential equation. Do so, and don't waste your time using the Frobenius method unless the point in question is a regular singular point.

35.5 About the Indicial and Recursion Formulas

In Chapter 37, we will closely examine the formulas involved in the basic Frobenius method. Here are a few things regarding the indicial equation and the recursion formulas that we will verify then (and which you should observe in your own computations now).

The Indicial Equation and the Exponents

Remember that one of the preliminary steps has us rewriting the differential equation as

$$(x - x_0)^2 \alpha(x) y'' + (x - x_0)\beta(x) y' + \gamma(x) y = 0$$

where α, β and γ are polynomials, with $\alpha(x_0) \neq 0$, and with no factors shared by all three. In practice, these polynomials are almost always real (i.e., their coefficients are real numbers). Let us assume this.

If you carefully follow the subsequent computations in the basic Frobenius method, you will discover that the indicial equation is just as we suspected at the start of Section 35.3, namely,

$$\alpha_0 r(r-1) + \beta_0 r + \gamma_0 = 0 \quad ,$$

where

$$\alpha_0 = \alpha(x_0) \quad , \quad \beta_0 = \beta(x_0) \quad \text{and} \quad \gamma_0 = \gamma(x_0) \quad .$$

The exponents for our differential equation (i.e., the solutions to the indicial equation) can then be found by rewriting the indicial equation as

$$\alpha_0 r^2 + (\beta_0 - \alpha_0)r + \gamma_0 = 0$$

and using basic algebra.

!►Example 35.7: *In describing the Frobenius method, we used Bessel's equation of order* $^1/_2$ *with* $x_0 = 0$. *That equation was (on page 692) rewritten as*

$$4x^2 y'' + 4xy' + [4x^2 - 1]y = 0 \quad ,$$

which is

$$(x-0)^2\alpha(x)y'' + (x-0)\beta(x)y' + \gamma(x)y = 0$$

with

$$\alpha(x) = 4 \quad , \quad \beta(x) = 4 \quad \text{and} \quad \gamma(x) = 4x^2 - 1 \quad .$$

So

$$\alpha_0 = \alpha(0) = 4 \quad , \quad \beta_0 = \beta(0) = 4 \quad \text{and} \quad \gamma_0 = \gamma(0) = -1 \quad .$$

According to the above, the corresponding indicial equation should be

$$4r(r-1) + 4r - 1 = 0 \quad ,$$

which simplifies to

$$4r^2 - 1 = 0 \quad ,$$

and which, indeed, is what we obtained (equation (35.17) on page 696) as the indicial equation for Bessel's equation of order $^1/_2$.

From basic algebra, we know that, if the coefficients of the indicial equation

$$\alpha_0 r^2 + (\beta_0 - \alpha_0)r + \gamma_0 = 0$$

are all real numbers, then the two solutions to the indicial equation are either both real numbers (possibly the same real numbers) or are complex conjugates of each other. This is usually the case in practice (and will always be the case in the examples and exercises of this chapter).

The Recursion Formulas

After finding the solutions r_1 and r_2 to the indicial equation, the basic method of Frobenius has us deriving the recursion formula corresponding to each of these r's. In Chapter 37, we'll discover that each of these recursion formulas can always be written as

$$a_k = \frac{F_k(r, a_0, a_1, \ldots, a_{k-1})}{(k+r-r_1)(k+r-r_2)} \quad \text{for} \quad k \geq \kappa \tag{35.25a}$$

where κ is some positive integer, and r is either r_1 or r_2, depending on whether this is the recursion formula corresponding to r_1 or r_2, respectively. It also turns out that this formula is derived from the seemingly equivalent relation

$$(k+r-r_1)(k+r-r_2)\,a_k = F_k(r, a_0, a_1, \ldots, a_{k-1}) \quad , \tag{35.25b}$$

which holds for every integer k greater than or equal to κ. (This later equation will be useful when we discuss possible "degeneracies" in the recursion formula).

At this point, all you need to know about each $F_k(r, a_0, a_1, \ldots, a_{k-1})$ is that it is a formula that yields a finite value for every possible choice of its variables. The actual formula is not that easily described and would not be all that useful for the differential equations being considered here.

Of greater interest is the fact that the denominator in the recursion formula factors so simply and predictably. The recursion formula obtained using $r = r_1$ will be

$$a_k = \frac{F_k(r_1, a_0, a_1, \ldots, a_{k-1})}{(k + r_1 - r_1)(k + r_1 - r_2)} = \frac{F_k(r_1, a_0, a_1, \ldots, a_{k-1})}{k(k + r_1 - r_2)} ,$$

while the recursion formula obtained using $r = r_2$ will be

$$a_k = \frac{F_k(r_2, a_0, a_1, \ldots, a_{k-1})}{(k + r_2 - r_1)(k + r_2 - r_2)} = \frac{F_k(r_2, a_0, a_1, \ldots, a_{k-1})}{(k - [r_1 - r_2])k} .$$

!►Example 35.8: *In solving Bessel's equation of order* $1/2$ *we obtained*

$$r_1 = \frac{1}{2} \quad \text{and} \quad r_2 = -\frac{1}{2} .$$

The recursion formulas corresponding to each of these were found to be

$$a_k = \frac{-1}{k(k+1)} a_{k-2} \quad \text{and} \quad a_k = \frac{-1}{k(k-1)} a_{k-2} \quad \text{for} \quad k = 2, 3, 4, \ldots ,$$

respectively (see formulas (35.19) and (35.22)). Looking at the recursion formula corresponding to $r = 1/2$ *we see that, indeed,*

$$a_k = \frac{-1}{k(k+1)} a_{k-2} = \frac{-1}{k\left(k + \frac{1}{2} - \left[-\frac{1}{2}\right]\right)} a_{k-2} = \frac{-1}{k\,(k + r_1 - r_2)} a_{k-2} .$$

Likewise, looking at the recursion formula corresponding to $r = 1/2$ *we see that*

$$a_k = \frac{-1}{(k-1)k} a_{k-2} = \frac{-1}{\left(k + \left[-\frac{1}{2}\right] - \frac{1}{2}\right) k} a_{k-2} = \frac{-1}{(k + r_2 - r_1)k} a_{k-2} .$$

Knowing that the denominator in your recursion formulas should factor into either

$$k(k + r_1 - r_2) \quad \text{or} \quad (k + r_2 - r_1)k$$

should certainly simplify your factoring of that denominator. And if your denominator does not so factor, then you know you made an error in your computations. So it provides a partial error check.

It also leads us to our next topic.

Problems Possibly Arising in Step 8

In discussing step 8 of the basic Frobenius method, we indicated that there may be "problems" in finding a modified series solution corresponding to r_2. Well, take a careful look at the recursion formula obtained using r_2:

$$a_k = \frac{F_k(r_2, a_0, a_1, \ldots, a_{k-1})}{(k - [r_1 - r_2])k} \quad \text{for} \quad k \geq \kappa$$

(remember, κ is some positive integer). See the problem? That's correct, the denominator may be zero for one of the k's. And it occurs if (and only if) r_1 and r_2 differ by some integer K greater than or equal to κ,

$$r_1 - r_2 = K \quad .$$

Then, when we attempt to compute a_K, we get

$$\begin{aligned} a_K &= \frac{F_K(r_2, a_0, a_1, \ldots, a_{k-1})}{(K - [r_1 - r_2])K} \\ &= \frac{F_K(r_2, a_0, a_1, \ldots, a_{k-1})}{(K - K)K} = \frac{F_K(r_2, a_0, a_1, \ldots, a_{k-1})}{0} \quad ! \end{aligned}$$

Precisely what we can say about a_K when this happens depends on whether the numerator in this expression is zero or not. To clarify matters slightly, let's rewrite the above using recursion relation (35.25b) which, in this case, reduces to

$$0 \cdot a_K = F_K(r_2, a_0, a_1, \ldots, a_{k-1}) \quad . \tag{35.26}$$

Now:

1. If $F_K(r_2, a_0, a_1, \ldots, a_{k-1}) \neq 0$, then the recursion formula "blows up" at $k = K$,

   $$a_K = \frac{F_K(r_2, a_0, a_1, \ldots, a_{k-1})}{0} \quad .$$

   More precisely, a_K must satisfy

   $$0 \cdot a_K = F_K(r_2, a_0, a_1, \ldots, a_{k-1}) \neq 0 \quad ,$$

   which is impossible. Hence, it is impossible for our differential equation to have a solution of the form $(x - x_0)^{r_2} \sum_{k=0}^{\infty} a_k (x - x_0)^k$ with $a_0 \neq 0$.

2. If $F_K(r_2, a_0, a_1, \ldots, a_{k-1}) = 0$, then the recursion equation for a_K reduces to

   $$a_K = \frac{0}{0} \quad ,$$

   an indeterminant expression telling us nothing about a_K. More precisely, (35.26) reduces to

   $$0 \cdot a_K = 0 \quad ,$$

   which is valid for any value of a_K; that is, a_K is another arbitrary constant (in addition to a_0). Moreover, a careful analysis will confirm that, in this case, we will re-derive the modified power series solution corresponding to r_1 with a_K being the arbitrary constant in that series solution (see Exercise 35.8).

But why don't we worry about the denominator in the recursion formula obtained using r_1,

$$a_k = \frac{F_k(r_1, a_0, a_1, \ldots, a_{k-1})}{k(k + r_1 - r_2)} \quad \text{for} \quad k \geq \kappa \quad ?$$

Because this denominator is zero only if $r_1 - r_2$ is some negative integer, contrary to our labeling convention of $r_1 \geq r_2$.

!►Example 35.9: *Consider*

$$2xy'' - 4y' - y = 0 \quad .$$

Multiplying this by x, we get it into the form

$$2x^2y'' - 4xy' - xy = 0 \quad .$$

As noted in Example 35.5 on page 689, this equation has one regular singular point, $x_0 = 0$. Setting

$$y = y(x) = x^r \sum_{k=0}^{\infty} a_k x^k = \sum_{k=0}^{\infty} a_k x^{k+r}$$

and differentiating, we get (as before)

$$y' = \frac{d}{dx} \sum_{k=0}^{\infty} a_k x^{k+r} = \sum_{k=0}^{\infty} a_k (k+r) x^{k+r-1}$$

and

$$y'' = \frac{d^2}{dx^2} \sum_{k=0}^{\infty} a_k x^{k+r} = \sum_{k=0}^{\infty} a_k (k+r)(k+r-1) x^{k+r-2} \quad .$$

Plugging these into our differential equation:

$$\begin{aligned} 0 &= 2x^2y'' - 4xy' - xy \\ &= 2x^2 \sum_{k=0}^{\infty} a_k (k+r)(k+r-1) x^{k+r-2} - 4 \sum_{k=0}^{\infty} a_k (k+r) x^{k+r-1} - x \sum_{k=0}^{\infty} a_k x^{k+r} \\ &= \sum_{k=0}^{\infty} 2a_k (k+r)(k+r-1) x^{k+r} - \sum_{k=0}^{\infty} 4a_k (k+r) x^{k+r} - \sum_{k=0}^{\infty} a_k x^{k+r+1} \quad . \end{aligned}$$

Dividing out the x^r, reindexing, and grouping like terms:

$$\begin{aligned} 0 &= \underbrace{\sum_{k=0}^{\infty} 2a_k (k+r)(k+r-1) x^k}_{n=k} - \underbrace{\sum_{k=0}^{\infty} 4a_k (k+r) x^k}_{n=k} - \underbrace{\sum_{k=0}^{\infty} a_k x^{k+1}}_{n=k+1} \\ &= \sum_{n=0}^{\infty} 2a_n (n+r)(n+r-1) x^n - \sum_{n=0}^{\infty} 4a_n (n+r) x^n - \sum_{n=1}^{\infty} a_n x^n \\ &= 2a_0 (0+r)(0+r-1) x^0 + \sum_{n=1}^{\infty} 2a_n (n+r)(n+r-1) x^n \\ &\quad - 4a_0 (0+r) x^0 - \sum_{n=1}^{\infty} 4a_{n-1} (n+r) x^n - \sum_{n=1}^{\infty} a_{n-1} x^n \\ &= a_0 \underbrace{[2r(r-1) - 4r]}_{2r^2-2r-4r} x^0 + \sum_{n=1}^{\infty} \{a_n \underbrace{[2(n+r)(n+r-1) - 4(n+r)]}_{2(n+r)(n+r-1-2)} - a_{n-1}\} x^n \quad . \end{aligned}$$

Thus, our differential equation reduces to

$$a_0\left[2\left(r^2-3r\right)\right]x^0 + \sum_{n=1}^{\infty}\{a_n\left[2(n+r)(n+r-3)\right]-a_{n-1}\}x^n = 0 \quad . \tag{35.27}$$

From the first term, we get the indicial equation:

$$\underbrace{2\left(r^2-3r\right)}_{2r(r-3)} = 0 \quad .$$

So r can equal 0 or 3. Since $3 > 0$, our convention has us identifying these solutions to the indicial equation as

$$r_1 = 3 \qquad \text{and} \qquad r_2 = 0 \quad .$$

Letting $r = r_1 = 3$ in equation (35.27), we see that

$$\begin{aligned} 0 &= a_0\left[2\left(3^2-3\cdot 3\right)\right]x^0 + \sum_{n=1}^{\infty}\{a_n\left[2(n+3)(n+3-3)\right]-a_{n-1}\}x^n \\ &= a_0\,[0]\,x^0 + \sum_{n=1}^{\infty}\{a_n\left[2(n+3)n\right]-a_{n-1}\}x^n \quad . \end{aligned}$$

So, for $n \geq 1$,

$$a_n\left[2(n+3)n\right]-a_{n-1} = 0 \quad .$$

Solving for a_n, we obtain the recursion formula

$$a_n = \frac{1}{2n(n+3)}a_{n-1} \qquad \text{for} \quad n = 1, 2, 3, \ldots \quad .$$

Equivalently,[13]

$$a_k = \frac{1}{2k(k+3)}a_{k-1} \qquad \text{for} \quad k = 1, 2, 3, \ldots \quad .$$

Applying this recursion formula:

$$a_1 = \frac{1}{2\cdot 1(1+3)}a_0 = \frac{1}{2\cdot 1(4)}a_0 \quad ,$$

$$\begin{aligned} a_2 &= \frac{1}{2\cdot 2(2+3)}a_1 = \frac{1}{2\cdot 2(5)}\cdot\frac{1}{2\cdot 1(4)}a_0 = \frac{1}{2^2(2\cdot 1)(5\cdot 4)}a_0 \\ &= \frac{3\cdot 2\cdot 1}{2^2(2\cdot 1)(5\cdot 4\cdot 3\cdot 2\cdot 1)}a_0 = \frac{6}{2^2(2\cdot 1)(5\cdot 4\cdot 3\cdot 2\cdot 1)}a_0 \quad , \end{aligned}$$

$$\begin{aligned} a_3 &= \frac{1}{2\cdot 3(3+3)}a_2 = \frac{1}{2\cdot 3(6)}\cdot\frac{6}{2^2(2\cdot 1)(5\cdot 4\cdot 3\cdot 2\cdot 1)}a_0 \\ &= \frac{6}{2^3(3\cdot 2\cdot 1)(6\cdot 5\cdot 4\cdot 3\cdot 2\cdot 1)}a_0 \quad , \end{aligned}$$

$$\vdots$$

$$a_k = \frac{6}{2^k\,k!(k+3)!}a_0 \quad .$$

---

[13] Observe that this is formula (35.25a) with $\kappa = 1$ and $F_k(r_1, a_0, a_1, \ldots, a_{k-1}) = \frac{1}{2}a_{k-1}$.

Notice that this last formula even holds for $k = 0$ and $k = 1$. So

$$a_k = \frac{6}{2^k\, k!(k+3)!}a_0 \qquad \text{for} \quad k = 0, 1, 2, 3, \dots \quad ,$$

and the corresponding modified power series solution is

$$y(x) = x^{r_1}\sum_{k=0}^{\infty} a_k x^k = a_0 x^3 \sum_{k=0}^{\infty} \frac{6}{2^k\, k!(k+3)!}x^k \quad .$$

Now let's try to find a similar solution corresponding to $r_2 = 0$. Letting $r = r_2 = 0$ in equation (35.27) yields

$$0 = a_0\left[2\left(0^2 - 3\cdot 0\right)\right]x^0 + \sum_{n=1}^{\infty}\{a_n\,[2(n+0)(n+0-3)] - a_{n-1}\}\,x^n$$

$$= a_0\,[0]\,x^0 + \sum_{n=1}^{\infty}\{a_n\,[2n(n-3)] - a_{n-1}\}\,x^n \quad .$$

So, for $n \geq 1$,

$$a_n\,[2n(n-3)] - a_{n-1} = 0 \quad .$$

Solving for a_n, we obtain the recursion formula

$$a_n = \frac{1}{2n(n-3)}a_{n-1} \qquad \text{for} \quad n = 1, 2, 3, \dots \quad .$$

Applying this, we get

$$a_1 = \frac{1}{2\cdot 1(1-3)}a_0 = \frac{1}{2\cdot 1(-2)}a_0 \quad ,$$

$$a_2 = \frac{1}{2\cdot 2(2-3)}a_1 = \frac{1}{2\cdot 2(-1)}\cdot\frac{1}{2\cdot 1(-2)}a_0 = \frac{1}{2^2 2\cdot 1(-1)(-2)}a_0 \quad ,$$

$$a_3 = \frac{1}{2\cdot 3(3-3)}a_1 = \frac{1}{2\cdot 2(0)}\cdot\frac{1}{2^2 2\cdot 1(-1)(-2)}a_0 = \frac{1}{0}\,a_0 \quad !$$

In this case, the a_3 "blows up" if $a_0 \neq 0$. In other words, for this last equation to be true, we must have

$$0\cdot a_3 = a_0 \quad ,$$

contrary to our requirement that $a_0 \neq 0$. So, for this differential equation, our search for a solution of the form

$$x^r\sum_{k=0}^{\infty} a_k x^k \qquad \text{with} \quad a_0 \neq 0$$

is doomed to failure when $r = r_2$. There is no such solution.

This leaves us with just the solutions found corresponding to $r = r_1$,

$$y(x) = a_0 x^3 \sum_{k=0}^{\infty} \frac{6}{2^k\, k!(k+3)!}x^k \quad .$$

In the next chapter, we will discuss how to find the complete general solution when, as in the last example, the basic method of Frobenius does not yield a second solution. We'll also finish solving the differential equation in the last example.

35.6 Dealing with Complex Exponents

In practice, the exponents r_1 and r_2 are usually real numbers. Still, complex exponents are possible. Fortunately, the differential equations arising from "real-world" problems usually have real-valued coefficients, which means that, if r_1 and r_2 are not real valued, they will probably be a conjugate pair

$$r_1 = r_+ = \lambda + i\omega \quad \text{and} \quad r_2 = r_- = \lambda - i\omega$$

with $\omega \neq 0$. In this case, step 7 of the Frobenius method (with $r_1 = r_+$, and renaming a_0 as c_+) leads to a solution of the form

$$y(x) = c_+ y_+(x)$$

where

$$y_+(x) = (x - x_0)^{\lambda + i\omega} \sum_{k=0}^{\infty} \alpha_k (x - x_0)^k$$

with $\alpha_0 = 1$. The other α_k's will be determined from the recursion formula and will simply be the a_k's described in the method with a_0 factored out. And they may well be complex valued since the recursion formula probably involves $r_+ = \lambda + i\omega$.

We could then continue with step 8 and compute the general series solution corresponding to $r_2 = r_-$. Fortunately, this lengthy set of computations is completely unnecessary because you can easily confirm that the complex conjugate y^* of any solution y to our differential equation is another solution to our differential equation (see Exercise 35.7 on page 718). Thus, as a second solution, we can just use

$$y_-(x) = \left[y_+(x)\right]^* = \left[(x - x_0)^{\lambda + i\omega} \sum_{k=0}^{\infty} \alpha_k (x - x_0)^k\right]^*$$

$$= (x - x_0)^{\lambda - i\omega} \sum_{k=0}^{\infty} {\alpha_k}^* (x - x_0)^k \quad .$$

It is also easy to verify that $\{y_+(x), y_-(x)\}$ is linearly independent. Hence,

$$y(x) = c_+ y_+(x) + c_- y_-(x)$$

is a general solution to our differential equation.

However, the two solutions y_+ and y_- will be complex valued. To avoid complex-valued solutions, we can employ the trick used before to derive a corresponding linearly independent pair $\{y_1, y_2\}$ of real-valued solutions to our differential equation by setting

$$y_1(x) = \frac{1}{2}\left[y_+(x) + y_-(x)\right] \quad \text{and} \quad y_2(x) = \frac{1}{2i}\left[y_+(x) - y_-(x)\right] \quad .$$

If the need arises, you can, with a little algebra, derive formulas for y_1 and y_2 in terms of the real and imaginary parts of $(x - x_0)^{\lambda + i\omega}$ and the α_k's. To be honest, though, that need is unlikely to ever arise.

35.7 Appendix: On Tests for Regular Singular Points

Proof of Theorem 35.2

The validity of Theorem 35.2 follows immediately from Lemma 35.1 (which we've already proven) and the next lemma.

Lemma 35.4

Let a, b and c be rational functions, and assume x_0 is a singular point on the real line for

$$a(x)y'' + b(x)y' + c(x)y = 0 \quad . \tag{35.28}$$

If the two limits

$$\lim_{x\to x_0}(x-x_0)\frac{b(x)}{a(x)} \quad \text{and} \quad \lim_{x\to x_0}(x-x_0)^2\frac{c(x)}{a(x)}$$

are both finite, then the differential equation can be written in quasi-Euler form

$$(x-x_0)^2\alpha(x)y'' + (x-x_0)\beta(x)y' + \gamma(x)y = 0$$

where α, β and γ are polynomials with $\alpha(x_0) \neq 0$.

PROOF: Because a, b and c are rational functions (i.e., quotients of polynomials), we can multiply equation (35.28) through by the common denominators, obtaining a differential equation in which all the coefficients are polynomials. Then factoring out the $(x-x_0)$-factors, and dividing out the factors common to all terms, we obtain

$$(x-x_0)^k A(x)y'' + (x-x_0)^m B(x)y' + (x-x_0)^n C(x)y = 0 \tag{35.29}$$

where A, B and C are polynomials with

$$A(x_0) \neq 0 \quad , \quad B(x_0) \neq 0 \quad \text{and} \quad C(x_0) \neq 0 \quad ,$$

and k, m and n are nonnegative integers, one of which must be zero. Moreover, since this is just a rewrite of equation (35.28), and x_0 is, by assumption, a singular point for this differential equation, we must have that $k \geq 1$. Hence $m = 0$ or $n = 0$.

Dividing equations (35.28) and (35.29) by their leading terms then yields, respectively,

$$y'' + \frac{b(x)}{a(x)}y' + \frac{c(x)}{a(x)}y = 0$$

and

$$y'' + (x-x_0)^{m-k}\frac{B(x)}{A(x)}y' + (x-x_0)^{n-k}\frac{C(x)}{A(x)}y = 0 \quad .$$

But these last two equations describe the same differential equation, and have the same first coefficients. Consequently, the other coefficients must be the same, giving us

$$\frac{b(x)}{a(x)} = (x-x_0)^{m-k}\frac{B(x)}{A(x)} \quad \text{and} \quad \frac{c(x)}{a(x)} = (x-x_0)^{n-k}\frac{C(x)}{A(x)} \quad .$$

Thus,

$$\lim_{x\to x_0}(x-x_0)\frac{b(x)}{a(x)} = \lim_{x\to x_0}(x-x_0)^{m+1-k}\frac{B(x)}{A(x)} = \lim_{x\to x_0}(x-x_0)^{m+1-k}\frac{B(x_0)}{A(x_0)}$$

and

$$\lim_{x\to x_0}(x-x_0)^2\frac{c(x)}{a(x)} = \lim_{x\to x_0}(x-x_0)^{n+2-k}\frac{C(x)}{A(x)} = \lim_{x\to x_0}(x-x_0)^{n+2-k}\frac{C(x_0)}{A(x_0)} \quad .$$

By assumption, though, these limits are finite, and for the limits on the right to be finite, the exponents must be zero or larger. This, combined with the fact that $k \geq 1$ means that k, m and l must satisfy

$$m+1 \geq k \geq 1 \qquad \text{and} \qquad n+2 \geq k \geq 1 \quad . \tag{35.30}$$

But remember, m, n and k are nonnegative integers with $m = 0$ or $n = 0$. That leaves us with only three possibilities:

1. If $m = 0$, then the first inequality in set (35.30) reduces to

$$1 \geq k \geq 1$$

telling us that $k = 1$. Equation (35.29) then becomes

$$(x-x_0)^1 A(x)y'' + B(x)y' + (x-x_0)^n C(x)y = 0 \quad .$$

Multiplying through by $x - x_0$, this becomes

$$(x-x_0)^2\alpha(x)y'' + (x-x_0)\beta(x)y' + \gamma(x)y = 0$$

where α, β and γ are the polynomials

$$\alpha(x) = A(x) \quad , \quad \beta(x) = B(x) \quad \text{and} \quad \gamma(x) = (x-x_0)^{n+1}C(x) \quad .$$

2. If $n = 0$, then the second inequality in set (35.30) reduces to

$$2 \geq k \geq 1 \quad .$$

So $k = 1$ or $k = 2$, giving us two options:

(a) If $k = 1$, then equation (35.29) becomes

$$(x-x_0)^1 A(x)y'' + (x-x_0)^m B(x)y' + C(x)y = 0 \quad ,$$

which can be rewritten as

$$(x-x_0)^2\alpha(x)y'' + (x-x_0)\beta(x)y' + \gamma(x)y = 0$$

where α, β and γ are the polynomials

$$\alpha(x) = A(x) \quad , \quad \beta(x) = (x-x_0)^m B(x) \quad \text{and} \quad \gamma(x) = (x-x_0)C(x) \quad .$$

(b) If $k = 2$, then the other inequality in set (35.30) becomes

$$m+1 \geq 2 \quad .$$

Hence $m \geq 1$. In this case, equation (35.29) becomes

$$(x-x_0)^2 A(x)y'' + (x-x_0)^m B(x)y' + C(x)y = 0$$

which is the same as

$$(x-x_0)^2\alpha(x)y'' + (x-x_0)\beta(x)y' + \gamma(x)y = 0$$

where α, β and γ are the polynomials

$$\alpha(x) = A(x) \quad , \quad \beta(x) = (x-x_0)^{m-1}B(x) \quad \text{and} \quad \gamma(x) = C(x) \quad .$$

Note that, in each case, we can rewrite our differential equation as

$$(x - x_0)^2\alpha(x)y'' + (x - x_0)\beta(x)y' + \gamma(x)y = 0$$

where α , β and γ are polynomials with

$$\alpha(x_0) = \alpha_0(x_0) \neq 0 \quad .$$

So, in each case, we have rewritten our differential equation in quasi-Euler form, as claimed in the lemma. ∎

Testing for Regularity When the Coefficients Are Not Rational

By using Lemma 32.13 on page 596 on factoring analytic functions, or Lemma 32.14 on page 597 on factoring quotients of analytic functions, you can easily modify the arguments in the last proof to obtain an analog of the above lemma appropriate to differential equations whose coefficients are not rational. Basically, just replace

"rational functions" with "quotients of functions analytic at x_0" ,

and replace

"polynomials" with "functions analytic at x_0" .

The resulting lemma, along with Lemma 35.1, then yields the following analog to Theorem 35.2.

Theorem 35.5 (testing for regular singular points (ver.2))
Assume x_0 is a singular point on the real line for the differential equation

$$a(x)y'' + b(x)y' + c(x)y = 0$$

where a , b and c are quotients of functions analytic at x_0 . Then x_0 is a regular singular point for this differential equation if and only if the two limits

$$\lim_{x \to x_0} (x - x_0)\frac{b(x)}{a(x)} \quad \text{and} \quad \lim_{x \to x_0} (x - x_0)^2 \frac{c(x)}{a(x)}$$

are both finite.

Additional Exercises

35.2. *Find a fundamental set of solutions $\{y_1, y_2\}$ for each of the following shifted Euler equations, and sketch the graphs of y_1 and y_2 near the singular point.*

a. $(x - 3)^2 y'' - 2(x - 3)y' + 2y = 0$

b. $2x^2 y'' + 5xy' + 1y = 0$

c. $(x - 1)^2 y'' - 5(x - 1)y' + 9y = 0$

d. $(x+2)^2y'' + (x+2)y' = 0$

e. $3(x-5)^2y'' - 4(x-5)y' + 2y = 0$

f. $(x-5)^2y'' + (x-5)y' + 4y = 0$

35.3. *Identify all of the singular points for each of the following differential equations, and determine which of those on the real line are regular singular points, and which are irregular singular points. Also, find the Frobenius radius of convergence R for the given differential equation about the given x_0.*

a. $x^2y'' + \frac{x}{x-2}y' + \frac{2}{x+2}y = 0 \quad , \quad x_0 = 0$

b. $x^3y'' + x^2y' + y = 0 \quad , \quad x_0 = 2$

c. $\left(x^3 - x^4\right)y'' + (3x-1)y' + 827y = 0 \quad , \quad x_0 = 1$

d. $y'' + \frac{1}{x-3}y' + \frac{1}{x-4}y = 0 \quad , \quad x_0 = 3$

e. $y'' + \frac{1}{(x-3)^2}y' + \frac{1}{(x-4)^2}y = 0 \quad , \quad x_0 = 4$

f. $y'' + \left(\frac{1}{x} - \frac{1}{3}\right)y' + \left(\frac{1}{x} - \frac{1}{4}\right)y = 0 \quad , \quad x_0 = 0$

g. $\left(4x^2 - 1\right)y'' + \left(4 - \frac{2}{x}\right)y' + \frac{1-x^2}{1+x^2}y = 0 \quad , \quad x_0 = 0$

h. $\left(4 + x^2\right)^2 y'' + y = 0 \quad , \quad x_0 = 0$

35.4. *Use the basic method of Frobenius to find modified power series solutions about $x_0 = 0$ for each of the following differential equations. In particular:*

i *Find, identify and solve the corresponding indicial equation for the equation's exponents r_1 and r_2.*

ii *Find the recursion formula corresponding to each exponent.*

iii *Find and explicitly write out at least the first four nonzero terms of all series solutions about $x_0 = 0$ that can be found by the basic Frobenius method (if a series terminates, find all the nonzero terms).*

iv *Try to find a general formula for all the coefficients in each series.*

v *Finally, either state the general solution or, when a second particular solution cannot be found by the basic method, give a reason that second solution cannot be found.*

(Note: $x_0 = 0$ is a regular singular point for each equation. You need not verify it.)

a. $x^2y'' - 2xy' + \left(x^2 + 2\right)y = 0$

b. $4x^2y'' + (1-4x)y = 0$

c. $x^2y'' + xy' + (4x-4)y = 0$

d. $\left(x^2 - 9x^4\right)y'' - 6xy' + 10y = 0$

e. $x^2y'' - xy' + \frac{1}{1-x}y = 0$

f. $y'' + \frac{1}{x}y' + y = 0$ *(Bessel's equation of order* 0 *)*

g. $y'' + \frac{1}{x}y' + \left[1 - \frac{1}{x^2}\right]y = 0$ *(Bessel's equation of order* 1 *)*

h. $2x^2y'' + \left(5x - 2x^3\right)y' + (1 - x^2)y = 0$

i. $x^2y'' - \left(5x + 2x^2\right)y' + (9 + 4x)y = 0$

j. $\left(3x^2 - 3x^3\right)y'' - \left(4x + 5x^2\right)y' + 2y = 0$

k. $x^2y'' - \left(x + x^2\right)y' + 4xy = 0$

l. $4x^2y'' + 8x^2y' + y = 0$

m. $x^2y'' + \left(x - x^4\right)y' + 3x^3y = 0$

n. $\left(9x^2 + 9x^3\right)y'' + \left(9x + 27x^2\right)y' + (8x - 1)y = 0$

35.5. *For each of the following, verify that the given* x_0 *is a regular singular point for the given differential equation, and then use the basic method of Frobenius to find modified power series solutions to that differential equation. In particular:*

- i *Find and solve the corresponding indicial equation for the equation's exponents* r_1 *and* r_2 *.*
- ii *Find the recursion formula corresponding to each exponent.*
- iii *Find and explicitly write out at least the first four nonzero terms of all series solutions about* $x_0 = 0$ *that can be found by the basic Frobenius method (if a series terminates, find all the nonzero terms).*
- iv *Try to find a general formula for all the coefficients in each series.*
- v *Finally, either state the general solution or, when a second particular solution cannot be found by the basic method, give a reason that second solution cannot be found.*

a. $(x - 3)y'' + (x - 3)y' + y = 0 \quad , \quad x_0 = 3$

b. $y'' + \frac{2}{x+2}y' + y = 0 \quad , \quad x_0 = -2$

c. $4y'' + \frac{4x - 3}{(x-1)^2}y = 0 \quad , \quad x_0 = 1$

d. $(x - 3)^2y'' + \left(x^2 - 3x\right)y' - 3y = 0 \quad , \quad x_0 = 3$

35.6. Let $\sum_{k=0}^{\infty} a_k x^k$ be a power series convergent on $(-R, R)$ for some $R > 0$. We already know that, on $(-R, R)$,

$$\frac{d}{dx}\sum_{k=0}^{\infty} a_k x^k = \sum_{k=0}^{\infty} \frac{d}{dx}\left[a_k x^k\right] \quad .$$

a. *Using the above and the product rule confirm that, on* $(-R, R)$,

$$\frac{d}{dx}\left[x^r \sum_{k=0}^{\infty} a_k x^k\right] = \sum_{k=0}^{\infty} a_k (k+r) x^{k+r-1} \quad .$$

for any real value r .

b. *How does this justify the term-by-term differentiation used in step 2 of the basic Frobenius method?*

35.7. For the following, assume u and v are real-valued functions on some interval $\mathcal{I}$, and let $y = u + iv$.

a. Verify that

$$\frac{dy^*}{dx} = \left(\frac{dy}{dx}\right)^* \qquad \text{and} \qquad \frac{d^2y^*}{dx^2} = \left(\frac{d^2y}{dx^2}\right)^*$$

where y^* is the complex conjugate of y .

b. Further assume that, on some interval $\mathcal{I}$, y satisfies

$$a(x)\frac{d^2y}{dx^2} + b(x)\frac{dy}{dx} + c(x)y = 0$$

where a , b and c are all real-valued functions. Using the results from the previous part of this exercise, show that

$$a(x)\frac{d^2y^*}{dx^2} + b(x)\frac{dy^*}{dx} + c(x)y^* = 0 \quad .$$

35.8. Let R , r_1 and r_2 be real numbers with $R > 0$ and $r_1 > r_2$, and let M be some positive integer. Suppose, further, that we have three modified power series solutions over $(-R, R)$

$$y_1(x) = x^{r_1} \sum_{k=0}^{\infty} \alpha_k x^k \qquad \text{with} \quad \alpha_0 = 1 \quad ,$$

$$y_2(x) = x^{r_2} \sum_{k=0}^{\infty} \beta_k x^k \qquad \text{with} \quad \beta_0 = 1$$

and

$$y_3(x) = x^{r_2} \sum_{k=M}^{\infty} \gamma_k x^k \qquad \text{with} \quad \gamma_M = 1$$

to a single second-order homogeneous linear differential equation.

a. Using the fact that $r_1 > r_2$, show that y_1 and y_2 cannot be constant multiples of each other, and, from this, conclude that $\{y_1, y_2\}$ must be a fundamental set of solutions to that one differential equation.

b. What does this say about solution y_3 ?

c. Show that, in fact, $y_3 = y_1$ and that $r_1 = r_2 + M$.

d. Why does the above verify the claim in step 8 on page 699 that we can set $a_M = 0$?

36

The Big Theorem on the Frobenius Method, with Applications

At this point, you may have a number of questions, including:

1. What do we do when the basic method does not yield the necessary linearly independent pair of solutions?
2. Are there any shortcuts?

To properly answer these questions requires a good bit of analysis — some straightforward and some, perhaps, not so straightforward. We will do that in the next chapter. Here, instead, we will present a few theorems summarizing the results of that analysis, and we will see how those results can, in turn, be applied to solve and otherwise gain useful information about solutions to some notable differential equations.

By the way, in the following, it does not matter whether we are restricting ourselves to differential equations with rational coefficients or are considering the more general case. The discussion holds for either.

36.1 The Big Theorems

The Theorems

The first theorem simply restates the definition of a "regular singular point", along with some results discussed earlier in Section 35.5.

Theorem 36.1 (the indicial equation and corresponding exponents)
Let x_0 be a point on the real line. Then x_0 is a regular singular point for a given second-order linear homogeneous differential equation if and only if that differential equation can be written as

$$(x - x_0)^2\alpha(x)y'' + (x - x_0)\beta(x)y' + \gamma(x)y = 0$$

where α, β and γ are all analytic at x_0 with $\alpha(x_0) \neq 0$. Moreover:

1. *The indicial equation arising in the method of Frobenius to solve this differential equation is*

$$\alpha_0 r(r - 1) + \beta_0 r + \gamma_0 = 0$$

where

$$\alpha_0 = \alpha(x_0) \quad , \quad \beta_0 = \beta(x_0) \quad \text{and} \quad \gamma_0 = \gamma(x_0) \quad .$$

2. *The indicial equation has exactly two solutions r_1 and r_2 (possibly identical). And, if $\alpha(x_0)$, $\beta(x_0)$ and $\gamma(x_0)$ are all real valued, then r_1 and r_2 are either both real valued or are complex conjugates of each other.*

The next theorem is "the big theorem" of the Frobenius method. It describes generic formulas for solutions about regular singular points and gives the intervals over which these formulas are valid. Proving it will be the major goal in the next chapter.

Theorem 36.2 (general solutions about regular singular points)
Assume x_0 is a regular singular point on the real line for some given second-order homogeneous linear differential equation with real coefficients. Let R be the corresponding Frobenius radius of convergence, and let r_1 and r_2 be the two solutions to the corresponding indicial equation, with $r_1 \geq r_2$ if they are real. Then, on the intervals $(x_0, x_0 + R)$ and $(x_0 - R, x_0)$, general solutions to the differential equation are given by

$$y(x) = c_1 y_1(x) + c_2 y_2(x)$$

where c_1 and c_2 are arbitrary constants, and y_1 and y_2 are solutions that can be written as follows[1]*:*

1. *In general,*

$$y_1(x) = |x - x_0|^{r_1} \sum_{k=0}^{\infty} a_k (x - x_0)^k \quad \text{with} \quad a_0 = 1 \quad . \tag{36.1}$$

2. *If $r_1 - r_2$ is not an integer, then*

$$y_2(x) = |x - x_0|^{r_2} \sum_{k=0}^{\infty} b_k (x - x_0)^k \quad \text{with} \quad b_0 = 1 \quad . \tag{36.2}$$

3. *If $r_1 - r_2 = 0$ (i.e., $r_1 = r_2$), then*

$$y_2(x) = y_1(x) \ln|x - x_0| + |x - x_0|^{1+r_1} \sum_{k=0}^{\infty} b_k (x - x_0)^k \quad . \tag{36.3}$$

4. *If $r_1 - r_2 = K$ for some positive integer K, then*

$$y_2(x) = \mu y_1(x) \ln|x - x_0| + |x - x_0|^{r_2} \sum_{k=0}^{\infty} b_k (x - x_0)^k \tag{36.4}$$

   where $b_0 = 1$, b_K is arbitrary and μ is some (nonarbitrary) constant (possibly zero). Moreover,

$$y_2(x) = y_{2,0}(x) + b_K y_1(x)$$

   where $y_{2,0}(x)$ is given by formula (36.4) with $b_K = 0$.

[1] In this theorem, we are assigning convenient values to the coefficients, such as a_0 and b_0, that could, in fact, be considered as arbitrary nonzero constants. Any coefficient not explicitly mentioned is not arbitrary.

Alternate Formulas
Solutions Corresponding to Integral Exponents

Remember that in developing the basic method of Frobenius, the first solution we obtained was actually of the form

$$y_1(x) = (x - x_0)^{r_1} \sum_{k=0}^{\infty} a_k (x - x_0)^k$$

and not

$$y_1(x) = |x - x_0|^{r_1} \sum_{k=0}^{\infty} a_k (x - x_0)^k \quad .$$

As noted on page 702, either solution is valid on both $(x_0, x_0 + R)$ and on $(x_0 - R, x_0)$. It's just that the second formula yields a real-valued solution even when $x < x_0$ and r_1 is a real number other than an integer.

Still, if r_1 is an integer — be it 8, 0, -2 or any other integer — then replacing

$$(x - x_0)^{r_1} \qquad \text{with} \qquad |x - x_0|^{r_1}$$

is completely unnecessary, and is usually undesirable. This is especially true if $r_1 = n$ for some nonnegative integer n, because then

$$y_1(x) = (x - x_0)^n \sum_{k=0}^{\infty} a_k (x - x_0)^k = \sum_{k=0}^{\infty} a_k (x - x_0)^{k+n}$$

is actually a power series about x_0, and the proof of Theorem 36.2 will show that this is a solution on the entire interval $(x_0 - R, x_0 + R)$. It might also be noted that, in this case, y_1 is analytic at x_0, a fact that might be useful in some applications.

Similar comments hold if the other exponent, r_2, is an integer. So let us make official:

Corollary 36.3 (solutions corresponding to integral exponents)
Let r_1 and r_2 be as in Theorem 36.2.

1. *If r_1 is an integer, then the solution given by formula (36.1) can be replaced by the solution given by*

$$y_1(x) = (x - x_0)^{r_1} \sum_{k=0}^{\infty} a_k (x - x_0)^k \qquad \text{with} \quad a_0 = 1 \quad . \tag{36.5}$$

2. *If r_2 is an integer while r_1 is not, then the solution given by formula (36.2) can be replaced by the solution given by*

$$y_2(x) = (x - x_0)^{r_2} \sum_{k=0}^{\infty} b_k (x - x_0)^k \qquad \text{with} \quad b_0 = 1 \quad . \tag{36.6}$$

3. *If $r_1 = r_2$ and are integers, then the solution given by formula (36.3) can be replaced by the solution given by*

$$y_2(x) = y_1(x) \ln|x - x_0| + (x - x_0)^{1+r_1} \sum_{k=0}^{\infty} b_k (x - x_0)^k \quad . \tag{36.7}$$

4. *If r_1 and r_2 are two different integers, then the solution given by formula (36.4) can be replaced by the solution given by*

$$y_2(x) = \mu y_1(x) \ln|x - x_0| + (x - x_0)^{r_2} \sum_{k=0}^{\infty} b_k (x - x_0)^k \quad . \tag{36.8}$$

Other Alternate Solutions Formulas

It should also be noted that alternative versions of formulas (36.3) and (36.4) can be found by simply factoring out $y_1(x)$ from both terms, giving us

$$y_2(x) = y_1(x)\left[\ln|x - x_0| + (x - x_0)^1 \sum_{k=0}^{\infty} c_k (x - x_0)^k\right] \tag{36.3$'$}$$

when $r_2 = r_1$, and

$$y_2(x) = y_1(x)\left[\mu \ln|x - x_0| + (x - x_0)^{-K} \sum_{k=0}^{\infty} c_k (x - x_0)^k\right] \tag{36.4$'$}$$

when $r_1 - r_2 = K$ for some positive integer K. In both cases, $c_0 = b_0$.

Theorem 36.2 and the Method of Frobenius

Remember that, using the basic method of Frobenius, we can find every solution of the form

$$y(x) = |x - x_0|^r \sum_{k=0}^{\infty} a_k (x - x_0)^k \qquad \text{with} \quad a_0 \neq 0 \quad ,$$

provided, of course, such solutions exist. Fortunately for us, statement 1 in the above theorem assures us that such solutions will exist corresponding to r_1. This means our basic method of Frobenius will successfully lead to a valid first solution (at least when the coefficients of the differential equation are rational). Whether there is a second solution y_2 of this form depends on the following:

1. If $r_1 - r_2$ is not an integer, then statement 2 of the theorem states that there is such a second solution. Consequently, step 8 of the method will successfully lead to the desired result.

2. If $r_1 - r_2$ is a positive integer, then there might be such a second solution, depending on whether or not $\mu = 0$ in formula (36.4). If $\mu = 0$, then formula (36.4) for y_2 reduces to the sort of modified power series we are seeking, and step 8 in the Frobenius method will give us this solution. What's more, as indicated in statement 4 of the above theorem, in carrying out step 8, we will also rederive the solution already obtained in steps 7 (unless we set $b_K = 0$)!

   On the other hand, if $\mu \neq 0$, then it follows from the above theorem that no such second solution exists. As a result, all the work carried out in step 8 of the basic Frobenius method will lead only to a disappointing end, namely, that the terms "blow up" (as discussed in the subsection *Problems Possibly Arising in Step 8* starting on page 707).[2]

3. Of course, if $r_2 = r_1$, then the basic method of Frobenius cannot yield a second solution different from the first. But our theorem does assure us that we can use formula (36.3) for y_2 (once we figure out the values of the b_k's).

So what can we do if the basic method of Frobenius does not lead to the second solution $y_2(x)$? At this point, we have two choices, both using the formula for $y_1(x)$ already found by the basic Frobenius method:

1. Use the reduction of order method.

[2] We will find a formula for μ in the next chapter. Unfortunately, it is not a simple formula and is not of much help in preventing us from attempting step 8 of the basic Frobenius method when that step does not lead to y_2.

2. Plug formula (36.3) or (36.4), as appropriate, into the differential equation, and solve for the b_k's (and, if appropriate, μ).

In practice — especially if all you have for $y_1(x)$ is the modified power series solution from the basic method of Frobenius — you will probably find the second of the above choices to be the better choice. It at least leads to a usable recursion formula for the b_k's. We will discuss this further in Section 36.6.

However, as we'll discuss in the next few sections, you might not need to find that y_2.

36.2 Local Behavior of Solutions: Issues

In many situations, we are interested in knowing something about the general behavior of a solution $y(x)$ to a given differential equation when x is at or near some point x_0. Certainly, for example, we were interested in what $y(x)$ and $y'(x)$ became as $x \to x_0$ when considering initial value problems.

In other problems (which we will see later), x_0 may be an endpoint of an interval of interest, and we may be interested in a solution y only if $y(x)$ and its derivatives remain well behaved (e.g., finite) as $x \to x_0$. This can become a very significant issue when x_0 is a singular point for the differential equation in question.

In fact, there are two closely related issues that will be of concern:

1. What does $y(x)$ approach as $x \to x_0$? Must it be zero? Can it be some nonzero finite value? Or does $y(x)$ "blow up" as $x \to x_0$? (And what about the derivatives of $y(x)$ as $x \to x_0$?)

2. Can we treat y as being analytic at x_0?

Of course, if y is analytic at x_0, then, for some $R > 0$, it can be represented by a power series

$$y(x) = \sum_{k=0}^{\infty} c_k (x - x_0)^k \qquad \text{for} \quad |x - x_0| < R \quad ,$$

and, as we already know,

$$\lim_{x \to x_0} y(x) = y(x_0) = c_0 \qquad \text{and} \qquad \lim_{x \to x_0} y'(x) = y'(x_0) = c_1 \quad .$$

Thus, when y is analytic at x_0, the limits in question are "well behaved", and the question of whether $y(x_0)$ must be zero or can be some nonzero value reduces to the question of whether c_0 is or is not zero.

In the next two sections, we will examine the possible behavior of solutions to a differential equation near a regular singular point for that equation. Our ultimate interest will be in determining when the solutions are "well behaved" and when they are not. Then, in Section 36.5, we will apply our results to an important set of equations from mathematical physics, the Legendre equations.

36.3 Local Behavior of Solutions: Limits at Regular Singular Points

As just noted, a major concern in many applications is the behavior of a solution $y(x)$ as x approaches a regular singular point x_0. In particular, it may be important to know whether

$$\lim_{x \to x_0} y(x)$$

is zero, some finite nonzero value, infinite or completely undefined.

We discussed this issue at the beginning of Chapter 33 for the shifted Euler equation

$$(x - x_0)^2 \alpha_0 y'' + (x - x_0)\beta_0 y' + \gamma_0 y = 0 \quad . \tag{36.9}$$

Let us try using what we know about those solutions after comparing them with the solutions given in our big theorem for

$$(x - x_0)^2 \alpha(x) y'' + (x - x_0)\beta(x) y' + \gamma(x) y = 0 \tag{36.10}$$

assuming

$$\alpha_0 = \alpha(x_0) \quad , \quad \beta_0 = \beta(x_0) \quad \text{and} \quad \gamma_0 = \gamma(x_0) \quad .$$

Remember, these two differential equations have the same indicial equation

$$\alpha_0 r(r - 1) + \beta_0 r + \gamma_0 = 0 \quad .$$

Naturally, we'll further assume α, β and γ are analytic at x_0 with $\alpha(x_0) \neq 0$ so we can use our theorems. Also, in the following, we will let r_1 and r_2 be the two solutions to the indicial equation, with $r_1 \geq r_2$ if they are real.

Preliminary Approximations

Observe that each of the solutions described in Theorem 36.2 involves one or more power series

$$\sum_{k=0}^{\infty} c_k (x - x_0)^k = c_0 + c_1(x - x_0) + c_2(x - x_0)^2 + c_3(x - x_0)^3 + \cdots$$

where $c_0 \neq 0$. As we've noted in earlier chapters,

$$\lim_{x \to x_0} \sum_{k=0}^{\infty} c_k (x - x_0)^k = \lim_{x \to x_0} \left[c_0 + c_1(x - x_0) + c_2(x - x_0)^2 + \cdots \right] = c_0 \quad .$$

This means

$$\sum_{k=0}^{\infty} c_k (x - x_0)^k \approx c_0 \qquad \text{when} \quad x \approx x_0 \quad .$$

Solutions Corresponding to r_1

Now, recall that the solutions corresponding to r_1 for the shifted Euler equation are constant multiples of

$$y_{\text{Euler},1}(x) = |x - x_0|^{r_1} \quad ,$$

while the corresponding solutions to equation (36.10) are constant multiples of something of the form

$$y_1(x) = |x - x_0|^{r_1} \sum_{k=0}^{\infty} a_k (x - x_0)^k \quad \text{with} \quad a_0 = 1 \quad .$$

Using the approximation just noted for the power series, we immediately have

$$y_1(x) \approx |x - x_0|^{r_1} a_0 = |x - x_0|^{r_1} = y_{\text{Euler},1}(x) \quad \text{when} \quad x \approx x_0 \quad ,$$

which is exactly what we suspected back the beginning of Section 35.3. In particular, if r_1 is real, then

$$\lim_{x \to x_0} |y_1(x)| = \lim_{x \to x_0} |x - x_0|^{r_1} = \begin{cases} 0 & \text{if } r_1 > 0 \\ 1 & \text{if } r_1 = 0 \\ +\infty & \text{if } r_1 < 0 \end{cases} \quad .$$

(For the case where r_1 is not real, see Exercise 36.3.)

Solutions Corresponding to r_2 When $r_2 \neq r_1$

In this case, all the corresponding solutions to the shifted Euler equation are given by constant multiples of

$$y_{\text{Euler},2} = |x - x_0|^{r_2} \quad .$$

If r_1 and r_2 do not differ by an integer, then the corresponding solutions to equation (36.10) are constant multiples of something of the form

$$y_2(x) = |x - x_0|^{r_2} \sum_{k=0}^{\infty} a_k (x - x_0)^k \quad \text{with} \quad a_0 = 1 \quad ,$$

and the same arguments given above with $r = r_1$ apply and confirm that

$$y_2(x) \approx |x - x_0|^{r_2} a_0 = |x - x_0|^{r_2} = y_{\text{Euler},2}(x) \quad \text{when} \quad x \approx x_0 \quad .$$

On the other hand, if $r_1 - r_2 = K$ for some positive integer K, then the corresponding solutions to (36.10) are constant multiples of something of the form

$$y_2(x) = \mu y_1(x) \ln|x - x_0| + |x - x_0|^{r_2} \sum_{k=0}^{\infty} b_k (x - x_0)^k \quad \text{with} \quad b_1 = 1 \quad .$$

Using approximations already discussed, we have, when $x \approx x_0$,

$$\begin{aligned} y_2(x) &\approx \mu |x - x_0|^{r_1} \ln|x - x_0| + |x - x_0|^{r_2} b_0 \\ &= \mu |x - x_0|^{r_2 + K} \ln|x - x_0| + |x - x_0|^{r_2} \\ &= |x - x_0|^{r_2} \left[\mu |x - x_0|^K \ln|x - x_0| + 1 \right] \quad . \end{aligned}$$

But K is a positive integer, and you can easily show, via L'Hôpital's rule, that

$$\lim_{x \to x_0} (x - x_0)^K \ln|x - x_0| = 0 \quad .$$

Thus, when $x \approx x_0$,

$$y_2(x) \approx |x - x_0|^{r_2} [\mu \cdot 0 + 1] = |x - x_0|^{r_2} \quad ,$$

confirming that, whenever $r_2 \neq r_1$

$$y_2(x) \approx y_{\text{Euler},2}(x) \qquad \text{when} \quad x \approx x_0 \quad .$$

In particular, if r_2 is real, then

$$\lim_{x \to x_0} |y_2(x)| = \lim_{x \to x_0} |x - x_0|^{r_1} = \begin{cases} 0 & \text{if} \quad r_2 > 0 \\ 1 & \text{if} \quad r_2 = 0 \\ +\infty & \text{if} \quad r_2 < 0 \end{cases} \quad .$$

Solutions Corresponding to r_2 When $r_2 = r_1$

In this case, all the second solutions to the shifted Euler equation are constant multiples of

$$y_{\text{Euler},2} = |x - x_0|^{r_2} \ln |x - x_0| \quad ,$$

and the corresponding solutions to our original differential equation are constant multiples of

$$y_2(x) = y_1(x) \ln |x - x_0| + |x - x_0|^{1+r_1} \sum_{k=0}^{\infty} b_k (x - x_0)^k \quad .$$

If $x \approx x_0$, then

$$\begin{aligned} y_2(x) &\approx |x - x_0|^{r_1} \ln |x - x_0| + |x - x_0|^{1+r_1} b_0 \\ &= |x - x_0|^{r_1} \ln |x - x_0| \left[1 + \frac{|x - x_0|}{\ln |x - x_0|} b_0 \right] \quad . \end{aligned}$$

But

$$\lim_{x \to x_0} \frac{|x - x_0|}{\ln |x - x_0|} = 0 \quad .$$

Consequently, when $x \approx x_0$,

$$y_2(x) \approx |x - x_0|^{r_1} \ln |x - x_0| \, [1 + 0 \cdot b_0] = |x - x_0|^{r_1} \ln |x - x_0| \quad ,$$

confirming that, again, we have

$$y_2(x) \approx y_{\text{Euler},2}(x) \qquad \text{when} \quad x \approx x_0 \quad .$$

Taking the limit (possibly using L'Hôpital's rule), you can then easily show that

$$\lim_{x \to x_0} |y_2(x)| = \lim_{x \to x_0} |a_0| \, |x - x_0|^{r_1} \, |\ln |x - x_0|| = \begin{cases} 0 & \text{if} \quad r_1 > 0 \\ +\infty & \text{if} \quad r_1 \leq 0 \end{cases} \quad .$$

!► Example 36.1 (Bessel's equation of order 1): *Suppose we are only interested in those solutions on $(0, \infty)$ to Bessel's equation of order 1,*

$$x^2 y'' + x y' + \left(x^2 - 1\right) y = 0 \quad ,$$

that do not "blow up" as $x \to 0$.

First of all, observe that this equation is already in the form

$$(x - x_0)^2 \alpha(x) y'' + (x - x_0) \beta(x) y' + \gamma(x) y = 0$$

with $x_0 = 0$, $\alpha(x) = 1$, $\beta(x) = 1$ and $\gamma(x) = x^2 - 1$. So $x_0 = 0$ is a regular singular point for this differential equation, and the indicial equation,

$$\alpha_0 r(r-1) + \beta_0 r + \gamma_0 = 0 \quad ,$$

is

$$1 \cdot r(r-1) + 1 \cdot r - 1 = 0 \quad .$$

This simplifies to

$$r^2 - 1 = 0 \quad ,$$

and has solutions $r = \pm 1$. Thus, the exponents for our differential equation are

$$r_1 = 1 \qquad \text{and} \qquad r_2 = -1 \quad .$$

By the analysis given above, we know that all solutions corresponding to r_1 are constant multiples of one particular solution y_1 satisfying

$$\lim_{x \to x_0} |y_1(x)| = \lim_{x \to x_0} |x - x_0|^1 = 0 \quad .$$

So none of the solutions corresponding to $r = 1$ blow up as $x \to 0$.

On the other hand, the analysis given above for solutions corresponding to r_2 when $r_2 \neq r_1$ tells us that all the (nontrivial) second solutions are (nonzero) constant multiples of one particular solution y_2 satisfying

$$\lim_{x \to x_0} |y_2(x)| = \lim_{x \to x_0} |x - x_0|^{-1} = \infty \quad .$$

So these do "blow up" as $x \to 0$.

Thus, since we are only interested in the solutions that do not blow up as $x \to 0$, we need only concern ourselves with those solutions corresponding to $r_1 = 1$. Those corresponding to $r_2 = -1$ are not relevant, and we need not spend time and effort finding their formulas.

Derivatives

To analyze the behavior of the derivative $y'(x)$ as $x \to x_0$, you simply differentiate the modified power series for $y(x)$, and then apply the ideas described above. Doing this is left for you (Exercise 36.4 at the end of the chapter).

36.4 Local Behavior: Analyticity and Singularities in Solutions

It is certainly easier to analyze how $y(x)$ or any of its derivatives behave as $x \to x_0$ when $y(x)$ is given by a basic power series about x_0

$$y(x) = \sum_{k=0}^{\infty} c_k (x - x_0)^k \quad .$$

Then, in fact, y is infinitely differentiable at x_0, and we know

$$\lim_{x \to x_0} y(x) = y(x_0) = c_0$$

and

$$\lim_{x \to x_0} y^{(k)}(x) = y^{(k)}(x_0) = k!\,c_k \qquad \text{for} \quad k = 1, 2, 3, \ldots \quad .$$

On the other hand, if $y(x)$ is given by one of those modified power series described in the big theorem, then, as we partly saw in the last section, computing the limits of $y(x)$ and its derivatives as $x \to x_0$ is more of a challenge. Indeed, unless that series is of the form

$$(x - x_0)^r \sum_{k=0}^{\infty} \alpha_k (x - x_0)^k$$

with r being zero or some positive integer, then $\lim_{x \to x_0} y(x)$ may blow up or otherwise fail to exist. And even if that limit does exist, you can easily show that the corresponding limit of the nth derivative, $y^{(n)}(x)$, will either blow up or fail to exist if $n > r$.

To simplify our discussion, let's slightly expand our "ordinary/singular point" terminology so that it applies to any function y. Basically, we want to refer to a point x_0 as an *ordinary point for a function* y if y is analytic at x_0, and we want to refer to x_0 as a *singular point for* y if y is not analytic at x_0.

There are, however, small technical issues that must be taken into account. Typically, our function y is defined on some interval (α, β), possibly by some power series or modified power series. So, initially at least, we must confine our definitions of ordinary and singular points for y to points in or at the ends of this interval. To be precise, for any such point x_0, we will say that x_0 is an *ordinary point* for y if and only if there is a power series $\sum_{k=0}^{\infty} a_k (x - x_0)^k$ with a nonzero radius of convergence R such that,

$$y(x) = \sum_{k=0}^{\infty} a_k (x - x_0)^k$$

for all x in (α, β) with $|x - x_0| < R$. If no such power series exists, then we'll say x_0 is a *singular point* for y.[3]

A rather general (and moderately advanced) treatment of 'singular points' for functions and solutions to differential equations was given in Section 34.9. One corollary of a result derived there (Lemma 34.28 on page 677) is the following unsurprising lemma:

Lemma 36.4

Let y be a solution on an interval (α, β) to some second-order homogeneous linear differential equation, and let x_0 be either in (α, β) or be one of the endpoints. Then

$$x_0 \text{ is an ordinary point for the differential equation} \quad \Longrightarrow \quad x_0 \text{ is an ordinary point for } y \quad .$$

Equivalently,

$$x_0 \text{ is a singular point for } y \quad \Longrightarrow \quad x_0 \text{ is a singular point for the differential equation} \quad .$$

This lemma does not say that y must have a singularity at each singular point of the differential equation. After all, if x_0 is a regular singular point, and the first solution r_1 of the indicial equation

[3] If x_0 is actually in (α, β), not an endpoint of (α, β), then

$$x_0 \text{ is an ordinary point for } y \quad \Longleftrightarrow \quad y \text{ is analytic at } x_0 \quad \Longleftrightarrow \quad x_0 \text{ is } \textit{point of analyticity} \text{ for } y \quad .$$

is a nonnegative integer, say, $r_1 = 3$, then our big theorem assures us of a solution y_1 given by

$$y_1(x) = (x - x_0)^3 \sum_{k=0}^{\infty} a_k (x - x_0)^k \quad ,$$

which is a true power series since

$$(x - x_0)^3 \sum_{k=0}^{\infty} a_k (x - x_0)^k = \sum_{k=0}^{\infty} a_k (x - x_0)^{k+3} = \sum_{n=0}^{\infty} c_k (x - x_0)^n$$

where

$$c_n = \begin{cases} 0 & \text{if } n < 3 \\ a_{n-3} & \text{if } n \geq 3 \end{cases} \quad .$$

Still, there are cases where one can be sure that a solution y has at least one singular point. In particular, consider the situation in which y is given by a power series

$$y(x) = \sum_{k=0}^{\infty} c_k (x - x_0)^k \quad \text{for} \quad |x - x_0| < R$$

when R is the radius of convergence and is finite. Replacing x with the complex variable z,

$$y(z) = \sum_{k=0}^{\infty} c_k (z - x_0)^k$$

yields a power series with radius of convergence R for a complex-variable function $y(z)$ analytic at least on the disk of radius R about z_0. Now, it can be shown that there must then be a point z_s on the edge of this disk at which $y(z)$ is not "well behaved". Unsurprisingly, it can also be shown that this z_s is a singular point for the differential equation. And if all the singular points of this differential equation happen to lie on the real line, then this singular point z_s must be one of the two points on the real line satisfying $|z_s - x_0| = R$, namely,

$$z_s = x_0 - R \quad \text{or} \quad z_s = x_0 + R \quad .$$

That gives us the following theorem (which will be useful in the following section and some exercises).

Theorem 36.5
Let x_0 be a point on the real line, and assume

$$y(x) = \sum_{k=0}^{\infty} c_k (x - x_0)^k \quad \text{for} \quad |x - x_0| < R$$

is a power series solution to some first- or second-order homogeneous linear differential equation. Suppose further, that both of the following hold:

1. *R is the radius of convergence for the above power series and is finite.*
2. *All the singular points for the differential equation are on the real axis.*

Then at least one of the two points $x_0 - R$ or $x_0 + R$ is a singular point for y.

Admittedly, our derivation of the above theorem was rather sketchy and involved claims that "can be shown". Fortunately, we can replace the "can be shown" with "has been shown". Turn back to Section 34.9 for a more satisfying development of a more general version of the above theorem (Theorem 34.29 on page 678). You will even find pictures there.

36.5 Case Study: The Legendre Equations

As an application of what we have just developed, let us analyze the behavior of the solutions to the set of Legendre equations. These equations often arise in physical problems in which some spherical symmetry can be expected, and in many applications, only those solutions that are "bounded" on $(-1, 1)$ are of interest. Our analysis here will allow us to readily find those solutions, and will save us from a lot of needless work dealing with those solutions that, ultimately, will not be of interest.

Let me remind you that a *Legendre equation* is any differential equation that can be written as

$$(1 - x^2)y'' - 2xy' + \lambda y = 0 \tag{36.11}$$

where λ, the equation's parameter, is some real constant. You may recall these equations from Exercise 33.10 on page 645, and we will begin our analysis here by recalling what we learned in that rather lengthy exercise.

What We Already Know

In Exercise 33.10 on page 645, you discovered the following:

1. The only singular points for each Legendre equation are $x = -1$ and $x = 1$.

2. For each λ, a general solution on $(-1, 1)$ of Legendre equation (36.11) is

$$y_\lambda(x) = a_0 y_{\lambda,E}(x) + a_1 y_{\lambda,O}(x)$$

where $y_{\lambda,E}(x)$ is a power series about 0 having just even-powered terms and whose first term is 1, and $y_{\lambda,O}(x)$ is a power series about 0 having just odd-powered terms and whose first term is x,

3. If $\lambda = m(m + 1)$ for some nonnegative integer m, and p_m is defined by

$$p_m(x) = \begin{cases} y_{\lambda,E}(x) & \text{if } m \text{ is even} \\ y_{\lambda,O}(x) & \text{if } m \text{ is odd} \end{cases} ,$$

then this $p_m(x)$ is a polynomial of degree m. In particular,

$$p_0(x) = y_{0,E}(x) = 1 ,$$

$$p_1(x) = y_{2,O}(x) = x ,$$

$$p_2(x) = y_{6,E}(x) = 1 - 3x^2 ,$$

$$p_3(x) = y_{12,O}(x) = x - \frac{5}{3}x^3 ,$$

$$p_4(x) = y_{20,E}(x) = 1 - 10x^2 + \frac{35}{3}x^4$$

and

$$p_5(x) = y_{30,O}(x) = x - \frac{14}{3}x^3 + \frac{21}{5}x^5 .$$

Moreover, any other polynomial solution to Legendre's equation of order $\lambda = m(m + 1)$ is a constant multiple of $p_m(x)$.

4. The Legendre equation with parameter λ has no polynomial solution on $(-1,1)$ if $\lambda \neq m(m+1)$ for every nonnegative integer m.

5. If y is a nonpolynomial solution to a Legendre equation on $(-1,1)$, then it is given by a power series about $x_0 = 0$ with a radius of convergence of exactly 1.

Let us now see what more we can determine about the solutions to the Legendre equations on $(-1,1)$ using the material developed in the last few sections. For convenience, we will record the noteworthy results as we derive them in a series of lemmas which, ultimately, will be summarized in a major theorem on Legendre equations, Theorem 36.11.

The Singular Points of the Solutions

First of all, we should note that, because polynomials are analytic everywhere, the polynomial solutions to Legendre's equation have no singular points.

Now, suppose y is a solution to a Legendre equation but is not a polynomial. As noted above, $y(x)$ can be given by a power series about $x_0 = 0$ with radius of convergence 1. This, and the fact that the only singular points for Legendre's equation are the points $x = -1$ and $x = 1$ on the real line, means that Lemma 36.5 applies and immediately gives us the following:

Lemma 36.6

Let y be a nonpolynomial solution to a Legendre equation on $(-1,1)$. Then y must have a singularity at either $x = -1$ or at $x = 1$ (or at both).

Solution Limits at $x = 1$

To determine the limits of the solutions at $x = 1$, we will first find the exponents of the Legendre equation at $x = 1$ by solving the appropriate indicial equation. To do this, it is convenient to first multiply the Legendre equation by -1 and factor the first coefficient, giving us

$$(x-1)(x+1)y'' + 2xy' - \lambda y = 0 \quad .$$

Multiplying this by $x-1$ converts the equation into quasi-Euler form

$$(x-1)^2\alpha(x)y'' + (x-1)\beta(x)y' + \gamma(x)y = 0$$

where

$$\alpha(x) = x + 1 \quad , \quad \beta(x) = 2x \quad \text{and} \quad \gamma(x) = -(x-1)\lambda \quad .$$

Thus, the corresponding indicial equation is

$$r(r-1)\alpha_0 + r\beta_0 + \gamma_0 = 0$$

with

$$\alpha_0 = \alpha(1) = 2 \quad , \quad \beta_0 = \beta(1) = 2 \quad \text{and} \quad \gamma_0 = \gamma(1) = 0 \quad .$$

That is, the exponents $r = r_1$ and $r = r_2$ must satisfy

$$r(r-1)2 + 2r = 0 \quad ,$$

which simplifies to $r^2 = 0$. So,

$$r_1 = r_2 = 0 \quad ,$$

and our big theorem on the Frobenius method, Theorem 36.2 on page 720, tells us that any solution y to a Legendre equation on $(-1, 1)$ can be written as

$$y(x) = c_1^+ y_1^+(x) + c_2^+ y_2^+$$

where

$$y_1^+(x) = |x - 1|^0 \sum_{k=0}^{\infty} a_k^+ (x-1)^k = \sum_{k=0}^{\infty} a_k^+ (x-1)^k \qquad \text{with} \quad a_0^+ = 1 \quad ,$$

and

$$y_2^+(x) = y_1^+(x) \ln|x-1| + (x-1) \sum_{k=0}^{\infty} b_k^+ (x-1)^k \quad .$$

Now, let's make some simple observations using these formulas:

1. Solution y_1^+ is analytic at $x = 1$ (i.e., $x = 1$ is an ordinary point for y_1^+). Moreover

$$\lim_{x \to 1} y_1^+(x) = a_0^+ = 1 \quad .$$

2. On the other hand, because of the logarithmic factor in y_2^+,

$$\lim_{x \to 1} y_2^+(x) = 1 \cdot \lim_{x \to 1} \ln|x-1| + (1-1) b_0^+ = -\infty \quad .$$

Hence, $x = 1$ is a singular point for solution y_2^+.

3. More generally, if $y(x) = c_1^+ y_1^+(x) + c_2^+ y_2^+(x)$ is any nontrivial solution to Legendre's equation on $(-1, 1)$, then the following must hold:

(a) If $c_2^+ \neq 0$, then $x = 1$ is a singular point for y, and

$$\lim_{x \to 1} |y(x)| = \left| c_1^+ \underbrace{\lim_{x \to 1} y_1^+(x)}_{1} + c_2^+ \underbrace{\lim_{x \to 1} y_2^+(x)}_{-\infty} \right| = \infty \quad .$$

(b) If $c_2^+ = 0$, then $c_1^+ \neq 0$, and $x = 1$ is an ordinary point for y. Moreover,

$$\lim_{x \to 1} y(x) = c_1^+ \lim_{x \to 1} y_1^+(x) = c_1^+ \neq 0 \quad .$$

(Hence, $x = 1$ is a singular point for y if and only if $c_2^+ \neq 0$.)

All together, these observations give us:

Lemma 36.7

Let y be a nontrivial solution on $(-1, 1)$ to Legendre's equation. Then $x = 1$ is a singular point for y if and only if

$$\lim_{x \to 1^-} |y(x)| = \infty \quad .$$

Moreover, if $x = 1$ is not a singular point for y, then

$$\lim_{x \to 1^-} y(x)$$

exists and is a finite nonzero value.

Solution Limits at $x = -1$

A very similar analysis using $x = -1$ instead of $x = 1$ leads to:

Lemma 36.8
Let y be a nontrivial solution on $(-1, 1)$ to Legendre's equation. Then $x = -1$ is a singular point for y if and only if

$$\lim_{x \to -1^+} |y(x)| = \infty \quad .$$

Moreover, if $x = -1$ is not a singular point for y, then

$$\lim_{x \to -1^+} y(x)$$

exists and is a finite nonzero value.

?►Exercise 36.1: *Verify the above lemma by redoing the analysis done in the previous subsection using $x = -1$ in place of $x = 1$.*

The Unboundedness of the Nonpolynomial Solutions

Recall that a function y is said to be *bounded* on an interval (a, b) if there is a finite number M which "bounds" the absolute value of $y(x)$ when x is in (a, b); that is,

$$|y(x)| \leq M \qquad \text{whenever} \quad a < x < b \quad .$$

Naturally, if a function is not bounded on the interval of interest, we say it is *unbounded.*

Now, if y happens to be one of those nonpolynomial solutions to a Legendre equation on $(-1, 1)$, then we know y has a singularity at either $x = -1$ or at $x = 1$ or at both (Lemma 36.6). Lemmas 36.7 and 36.8 then tell us that

$$\lim_{x \to 1^-} |y(x)| = \infty \qquad \text{or} \qquad \lim_{x \to -1^+} |y(x)| = \infty \quad ,$$

clearly telling us that $y(x)$ is not bounded on $(-1, 1)$!

Lemma 36.9
Let y be a nonzero solution to a Legendre equation on $(-1, 1)$. If y is not a polynomial, then it is not bounded on $(-1, 1)$.

The Polynomial Solutions and Legendre Polynomials

Now assume y is a nonzero polynomial solution to a Legendre equation.

First of all, because each polynomial is a continuous function on the real line, we know from a classic theorem of calculus that the polynomial has maximum and minimum values over any given closed subinterval. Thus, in particular, our polynomial solution y has a maximum and a minimum value over $[-1, 1]$, and, hence, is bounded on $(-1, 1)$. So

Lemma 36.10
Each polynomial solution to a Legendre equation is bounded on $(-1, 1)$.

But now remember that the Legendre equation with parameter λ has polynomial solutions if and only if $\lambda = m(m+1)$ for some nonnegative integer m, and those solutions are all constant multiples of

$$p_m(x) = \begin{cases} y_{\lambda,E}(x) & \text{if } m \text{ is even} \\ y_{\lambda,O}(x) & \text{if } m \text{ is odd} \end{cases} .$$

Since $x = 1$ is not a singular point for p_m, Lemma 36.7 tells us that

$$p_m(1) = \lim_{x\to 1} p_m(x) \neq 0 .$$

This allows us to define the mth *Legendre polynomial* P_m by

$$P_m(x) = \frac{1}{p_m(1)} p_m(x) \qquad \text{for} \quad m = 0, 1, 2, 3, \ldots .$$

Clearly, any constant multiple of p_m is also a constant multiple of P_m. So we can use the P_m's instead of the p_m's to describe all polynomial solutions to the Legendre equations.

In practice, it is more common to use the Legendre polynomials than the p_m's. In part, this is because

$$P_m(1) = \frac{1}{p_m(1)} p_m(1) = 1 \qquad \text{for} \quad m = 0, 1, 2, 3, \ldots .$$

With a little thought, you'll realize that this means that, for each nonnegative integer m, P_m is the polynomial solution to the Legendre equation with parameter $\lambda = m(m+1)$ that equals 1 when $x = 1$.

Summary

This is a good time to skim the lemmas in this section, along with the discussion of the Legendre polynomials, and verify for yourself that these results can be condensed into the following:

Theorem 36.11 (bounded solutions of the Legendre equations)
There are bounded, nontrivial solutions on $(-1, 1)$ to the Legendre equation

$$(1 - x^2)y'' - 2xy' + \lambda y = 0$$

if and only if $\lambda = m(m+1)$ for some nonnegative integer m. Moreover, y is such a solution if and only if y is a constant multiple of the mth Legendre polynomial.

We may find a use for this theorem later (much later).

36.6 Finding Second Solutions Using Theorem 36.2

Let's return to actually solving differential equations.

When the basic method of Frobenius fails to deliver a second solution, we can turn to the appropriate formulas given in Theorem 36.2 on page 720, namely, formula (36.3),

$$y_2(x) = y_1(x) \ln|x - x_0| + |x - x_0|^{1+r_1} \sum_{k=0}^{\infty} b_k (x - x_0)^k$$

or formula (36.4),

$$y_2(x) = \mu y_1(x) \ln|x - x_0| + |x - x_0|^{r_2} \sum_{k=0}^{\infty} b_k (x - x_0)^k \quad ,$$

depending, respectively, on whether the exponents r_1 and r_2 are equal or differ by a nonzero integer (the only cases for which the basic method might fail). The unknown constants in these formulas can then be found by fairly straightforward variations of the methods we've already developed. Plug formula (36.3) or (36.4), as appropriate, into the differential equation, derive the recursion formula for the b_k's (and the value of μ if using formula (36.4)), and compute as many of the b_k's as desired.

Because the procedures are straightforward modifications of what we've already done many times in the last few chapters, we won't describe the steps in detail. Instead, we'll illustrate the basic ideas with an example, and then comment on those basic ideas. As you've come to expect in these last few chapters, the computations are simple but lengthy. But do note how the linearity of the equations is used to break the computations into more digestible pieces.

The Second Solution When $r_1 - r_2$ Is a Positive Integer

!►Example 36.2: *In Example 35.9, starting on page 708, we attempted to find modified power series solutions about $x_0 = 0$ to*

$$2xy'' - 4y' - y = 0 \quad ,$$

which we rewrote as

$$2x^2y'' - 4xy' - xy = 0 \quad .$$

We found the exponents of this equation to be

$$r_1 = 3 \quad \text{and} \quad r_2 = 0 \quad ,$$

and obtained

$$y(x) = cx^3 \sum_{k=0}^{\infty} \frac{6}{2^k\, k!(k+3)!} x^k \tag{36.12}$$

as the solutions to the differential equation corresponding to r_1. Unfortunately, we found that there was not a similar solution corresponding to r_2.

To apply Theorem 36.2, we first need the particular solution corresponding to r_1

$$y_1(x) = x^3 \sum_{k=0}^{\infty} a_k x^k \quad \text{with} \quad a_0 = 1 \quad .$$

So we use formula (36.12) with c chosen so that

$$a_0 = c \cdot \frac{6}{2^0\, 0!(0+3)!} = 1 \quad .$$

Simple computations show that $c = 1$ and, so,

$$y_1(x) = x^3 \sum_{k=0}^{\infty} \frac{6}{2^k\, k!(k+3)!} x^k = \sum_{k=0}^{\infty} \frac{6}{2^k\, k!(k+3)!} x^{k+3} \quad .$$

According to Theorem 36.2, the general solution to our differential equation is

$$y(x) = c_1 y_1(x) + c_2 y_2(x)$$

where y_1 is as above, and (since $r_2 = 0$ and $K = r_1 - r_2 = 3$)

$$y_2(x) = \mu y_1(x) \ln |x - x_0| + \sum_{k=0}^{\infty} b_k x^k \quad ,$$

where $b_0 = 1$, b_3 is arbitrary (we'll take it to be zero), and μ is some constant. To simplify matters, let us rewrite the last formula as

$$y_2(x) = \mu Y_1(x) + Y_2(x)$$

where

$$Y_1(x) = y_1(x) \ln |x| \qquad \text{and} \qquad Y_2(x) = \sum_{k=0}^{\infty} b_k x^k \quad .$$

Thus,

$$\begin{aligned} 0 &= 2x^2 y_1'' - 4x y_1' - x y_1 \\ &= 2x^2 [\mu Y_1 + Y_2]'' - 4x[\mu Y_1 + Y_2]' - x[\mu Y_1 + Y_2] \quad . \end{aligned}$$

By the linearity of the derivatives, this can be rewritten as

$$0 = \mu \left\{ 2x^2 Y_1'' - 4x Y_1' - x Y_1 \right\} + \left\{ 2x^2 Y_2'' - 4x Y_2' - x Y_2 \right\} \quad . \tag{36.13}$$

Now, because y_1 is a solution to our differential equation, you can easily verify that

$$\begin{aligned} &2x^2 Y_1'' - 4x Y_1' - x Y_1 \\ &\quad = 2x^2 \frac{d^2}{dx^2} [y_1(x) \ln |x|] - 4x \frac{d}{dx} [y_1(x) \ln |x|] - x [y_1(x) \ln |x|] \\ &\quad = \cdots \\ &\quad = 4x y_1' - 6 y_1 \quad . \end{aligned}$$

Replacing the y_1 in the last line with its series formula then gives us

$$2x^2 Y_1'' - 4x Y_1' - x Y_1 = 4x \frac{d}{dx} \sum_{k=0}^{\infty} \frac{6}{2^k \, k!(k+3)!} x^{k+3} - 6 \sum_{k=0}^{\infty} \frac{6}{2^k \, k!(k+3)!} x^{k+3} \quad ,$$

which, after suitable computation and reindexing (you do it), becomes

$$2x^2 Y_1'' - 4x Y_1' - x Y_1 = \sum_{n=3}^{\infty} \frac{12(2n-3)}{2^{n-3}(n-3)!n!} x^n \quad . \tag{36.14}$$

Next, using the series formula for Y_2 we have

$$2x^2 Y_2'' - 4x Y_2' - x Y_2 = 2x^2 \frac{d^2}{dx^2} \sum_{k=0}^{\infty} b_k x^k - 4x \frac{d}{dx} \sum_{k=0}^{\infty} b_k x^k - x \sum_{k=0}^{\infty} b_k x^k \quad ,$$

which, after suitable computations and changes of indices (again, you do it!), reduces to

$$2x^2 Y_2'' - 4x Y_2' - x Y_2 = \sum_{n=1}^{\infty} \{2n(n-3) b_n - b_{n-1}\} x^n \quad . \tag{36.15}$$

Combining equations (36.13), (36.14) and (36.15):

$$0 = \mu\left\{2x^2Y_1'' - 4xY_1' - xY_1\right\} + \left\{2x^2Y_2'' - 4xY_2' - xY_2\right\}$$

$$= \mu\sum_{n=3}^{\infty}\frac{12(2n-3)}{2^{n-3}(n-3)!n!}x^n + \sum_{n=1}^{\infty}\{2n(n-3)b_n - b_{n-1}\}x^n$$

$$= \mu\sum_{n=3}^{\infty}\frac{12(2n-3)}{2^{n-3}(n-3)!n!}x^n + \left\{[-4b_1 - b_0]x^1 + [-4b_2 - b_1]x^2 + \sum_{n=3}^{\infty}\{2n(n-3)b_n - b_{n-1}\}x^n\right\}$$

$$= -[4b_1 + b_0]x^1 - [4b_2 + b_1]x^2 + \sum_{n=3}^{\infty}\left[2n(n-3)b_n - b_{n-1} + \mu\frac{12(2n-3)}{2^{n-3}(n-3)!n!}\right]x^n \quad .$$

Since each term in this last power series must be zero, we must have

$$-[4b_1 + b_0] = 0 \quad , \quad -[4b_2 + b_1] = 0$$

and

$$2n(n-3)b_n - b_{n-1} + \mu\frac{12(2n-3)}{2^{n-3}(n-3)!n!} = 0 \quad \text{for} \quad n = 3, 4, 5, \ldots \quad .$$

This (and the fact that we've set $b_0 = 1$) means that

$$b_1 = -\frac{1}{4}b_0 = -\frac{1}{4}\cdot 1 = -\frac{1}{4} \quad , \quad b_2 = -\frac{1}{4}b_1 = -\frac{1}{4}\left(\frac{-1}{4}\right) = \frac{1}{16}$$

and

$$2n(n-3)b_n = b_{n-1} - \mu\frac{12(2n-3)}{2^{n-3}(n-3)!n!} \quad \text{for} \quad n = 3, 4, 5, \ldots \quad . \tag{36.16}$$

Because of the $n-3$ factor in front of b_n, dividing the last equation by $2n(n-3)$ to get a recursion formula for the b_n's would result in a recursion formula that "blow ups" for $n = 3$. So we need to treat that case separately.

With $n = 3$ in equation (36.16), we get

$$2\cdot 3(3-3)b_3 = b_2 - \mu\frac{12(2\cdot 3-3)}{2^{3-3}(3-3)!3!}$$

$$\hookrightarrow \qquad 0 = \frac{1}{16} - \mu 6$$

$$\hookrightarrow \qquad \mu = \frac{1}{96} \quad .$$

Notice that we obtained the value for μ instead of b_3. As claimed in the theorem, b_3 is arbitrary. Because of this, and because we only need one second solution, let us set

$$b_3 = 0 \quad .$$

Now we can divide equation (36.16) by $2n(n-3)$ and use the value for μ just derived (with a little arithmetic) to obtain the recursion formula

$$b_n = \frac{1}{2n(n-3)}\left[b_{n-1} - \frac{(2n-3)}{2^n(n-3)!n!}\right] \quad \text{for} \quad n = 4, 5, 6, \ldots \quad .$$

So,

$$b_4 = \frac{1}{2\cdot 4(4-3)}\left[b_3 - \frac{(2\cdot 4-3)}{2^4(4-3)!4!}\right] = \frac{1}{8}\left[0 - \frac{5}{384}\right] = -\frac{5}{3{,}072} \quad ,$$

$$b_5 = \frac{1}{2\cdot 5(5-3)}\left[b_4 + \frac{(2\cdot 5-3)}{2^5(5-3)!5!}\right] = \frac{1}{20}\left[\frac{-5}{3{,}072} + \frac{7}{7{,}680}\right] = -\frac{11}{307{,}200} \quad ,$$

$$\vdots$$

We won't attempt to find a general formula for the b_n's here!

Thus, a second particular solution to our differential equation is

$$\begin{aligned} y_2(x) &= \mu y_1(x)\ln|x-x_0| + \sum_{k=0}^{\infty} b_k x^k \\ &= \frac{1}{96}y_1(x)\ln|x| + \left\{1 - \frac{1}{4}x + \frac{1}{16}x^2 + 0x^3 - \frac{5}{3{,}072}x^4 - \frac{11}{307{,}200}x^5 + \cdots\right\} \end{aligned}$$

where y_1 is our first particular solution,

$$y_1(x) = \sum_{k=0}^{\infty} \frac{6}{2^k\, k!(k+3)!}x^{k+3} \quad .$$

In general, when

$$r_1 - r_2 = K$$

for some positive integer K, the computations illustrated in the above example will yield a second particular solution. It will turn out that b_1, b_2, ... and b_{K-1} are all "easily computed" from b_0, just as in the example. You will also obtain a recursion relation for b_n in terms of lower-indexed b_k's and the coefficients from the series formula for y_1. This recursion formula (formula (36.16) in our example) will hold for $n \geq K$, but be degenerate when $n = K$ (just as in our example, where $K = 3$). From that degenerate case, the value of μ in formula (36.4) can be determined. The rest of the b_n's can then be computed using the recursion relation. Unfortunately, it is highly unlikely that you will be able to find a general formula for these b_n's in terms of just the index, n. So just compute as many as seem reasonable.

The Second Solution When $r_1 = r_2$

The basic ideas illustrated in the last example also apply when the exponents of the differential equation, r_1 and r_2, are equal. Of course, instead of using the formula used in the example for y_2, use formula (36.3),

$$y_2(x) = y_1(x)\ln|x-x_0| + |x-x_0|^{1+r_1}\sum_{k=0}^{\infty} b_k(x-x_0)^k \quad .$$

In this case, there is no "μ" to determine and none of the b_k's will be arbitrary. In some ways, that makes this a simpler case than considered in our example. You can work out the details in the exercises.

Additional Exercises

36.2. *For each differential equation and singular point x_0 given below, let r_1 and r_2 be the corresponding exponents (with $r_1 \geq r_2$ if they are real), and let y_1 and y_2 be the two modified power series solutions about the given x_0 described in the "big theorem on the Frobenius method", Theorem 36.2 on page 720, and do the following:*

i *If not already in the form*

$$(x - x_0)^2 \alpha(x) y'' + (x - x_0)\beta(x) y' + \gamma(x) y = 0$$

where α, β and γ are all analytic at x_0 with $\alpha(x_0) \neq 0$, then rewrite the differential equation in this form.

ii *Determine the corresponding indicial equation, and find r_1 and r_2.*

iii *Write out the corresponding shifted Euler equation*

$$(x - x_0)^2 \alpha(x_0) y'' + (x - x_0)\beta(x_0) y' + \gamma(x_0) y = 0 \quad ,$$

and find the solutions $y_{\text{Euler},1}$ and $y_{\text{Euler},2}$ which approximate, respectively, $y_1(x)$ and $y_2(x)$ when $x \approx x_0$.

iv *Determine the limits $\lim_{x \to x_0} |y_1(x)|$ and $\lim_{x \to x_0} |y_2(x)|$.*

Do not attempt to find the modified power series formulas for y_1 and y_2.

a. $x^2 y'' - 2xy' + \left(2 - x^2\right) y = 0 \quad , \quad x_0 = 0$

b. $x^2 y'' - 2x^2 y' + \left(x^2 - 2\right) y = 0 \quad , \quad x_0 = 0$

c. $y'' + \frac{1}{x} y' + y = 0 \quad , \quad x_0 = 0$ *(Bessel's equation of order 0)*

d. $x^2 \left(2 - x^2\right) y'' + \left(5x + 4x^2\right) y' + (1 + x^2) y = 0 \quad , \quad x_0 = 0$

e. $x^2 y'' - \left(5x + 2x^2\right) y' + 9y = 0 \quad , \quad x_0 = 0$

f. $x^2(1 + 2x) y'' + xy' + (4x^3 - 4) y = 0 \quad , \quad x_0 = 0$

g. $4x^2 y'' + 8xy' + (1 - 4x) y = 0 \quad , \quad x_0 = 0$

h. $x^2 y'' + xy' - (1 + 2x) y = 0 \quad , \quad x_0 = 0$

i. $xy'' + 4y' + \frac{12}{(x + 2)^2} y = 0 \quad , \quad x_0 = 0$

j. $xy'' + 4y' + \frac{12}{(x + 2)^2} y = 0 \quad , \quad x_0 = -2$

k. $(x - 3) y'' + (x - 3) y' + y = 0 \quad , \quad x_0 = 3$

l. $(1 - x^2) y'' - xy' + 3y = 0 \quad , \quad x_0 = 1$ *(Chebyshev equation with parameter 3)*

36.3. *Suppose x_0 is a regular singular point for some second-order homogeneous linear differential equation, and that the corresponding exponents are complex*

$$r_+ = \lambda + i\omega \quad \text{and} \quad r_- = \lambda - i\omega$$

(with $\omega \neq 0$). Let y be any nontrivial solution to this differential equation on an interval having x_0 as the left endpoint. Show that

$$\lim_{x \to x_0^+} y(x)$$

is zero if $\lambda > 0$, and does not exist if $\lambda \leq 0$.

36.4. *Assume x_0 is a regular singular point on the real line for*

$$(x - x_0)^2 \alpha(x) y'' + (x - x_0)\beta(x) y' + \gamma(x) y = 0$$

where, as usual, α, β and γ are all analytic at x_0 with $\alpha(x_0) \neq 0$. Assume the solutions r_1 and r_2 to the corresponding indicial equation are real, with $r_1 \geq r_2$. Let $y_1(x)$ and $y_2(x)$ be the corresponding solutions to the differential equation as described in the big theorem on the Frobenius method, Theorem 36.2.

a. *Compute the derivatives of y_1 and y_2 and show that, for $i = 1$ and $i = 2$,*

$$\lim_{x \to x_0} \left| y_i{}'(x) \right| = \begin{cases} 0 & \text{if } 1 < r_i \\ \infty & \text{if } 0 < r_i < 1 \\ \infty & \text{if } r_i < 0 \end{cases} .$$

Be sure to consider all cases.

b. *Compute $\lim_{x \to x_0} \left| y_2{}'(x) \right|$ when $r_1 = 1$ and when $r_1 = 0$.*

c. *What can be said about $\lim_{x \to x_0} \left| y_2{}'(x) \right|$ when $r_1 = 1$ and when $r_1 = 0$?*

36.5. *Recall that the Chebyshev equation with parameter λ is*

$$(1 - x^2) y'' - x y' + \lambda y = 0 \quad , \qquad (36.17)$$

where λ can be any constant. In Exercise 33.9 on page 644 you discovered that:

1. *The only singular points for each Chebyshev equation are $x = 1$ and $x = -1$.*

2. *For each λ, the general solution on $(-1, 1)$ to equation (36.17) is given by*

$$y_\lambda(x) = a_0 y_{\lambda,E}(x) + a_1 y_{\lambda,O}(x)$$

where $y_{\lambda,E}$ and $y_{\lambda,O}$ are, respectively, even- and odd-termed series

$$y_{\lambda,E}(x) = \sum_{\substack{k=0 \\ k \text{ is even}}}^{\infty} c_k x^k \quad \text{and} \quad y_{\lambda,O}(x) = \sum_{\substack{k=0 \\ k \text{ is odd}}}^{\infty} c_k x^k$$

with $c_0 = 1$, $c_1 = 1$ and the other c_k's determined from c_0 or c_1 via the recursion formula.

3. *Equation (36.17) has nontrivial polynomial solutions if and only if* $\lambda = m^2$ *for some nonnegative integer* m. *Moreover, for each such* m, *all the polynomial solutions are constant multiples of an* m^{th}*-degree polynomial* p_m *given by*

$$p_m(x) = \begin{cases} y_{\lambda,E}(x) & \text{if } m \text{ is even} \\ y_{\lambda,O}(x) & \text{if } m \text{ is odd} \end{cases} .$$

In particular,

$$p_0(x) = 1 \quad , \quad p_1(x) = x \quad , \quad p_2(x) = 1 - 2x^2 \quad ,$$

$$p_3(x) = x - \frac{4}{3}x^3 \quad , \quad p_4(x) = 1 - 8x^2 + 8x^4$$

and

$$p_5(x) = x - 4x^3 + \frac{16}{5}x^5 \quad .$$

4. *Each nonpolynomial solution to a Chebyshev equation on* $(-1, 1)$ *is given by a power series about* $x_0 = 0$ *whose radius of convergence is exactly* 1.

In the following, you will continue the analysis of the solutions to the Chebyshev equations in a manner analogous to our continuation of the analysis of the solutions to the Legendre equations in Section 36.5.

a. *Verify that* $x = 1$ *and* $x = -1$ *are regular singular points for each Chebyshev equation.*

b. *Find the exponents* r_1 *and* r_2 *at* $x = 1$ *and* $x = -1$ *of each Chebyshev equation.*

c. *Let* y *be a nonpolynomial solution to a Chebyshev equation on* $(-1, 1)$, *and show that either*

$$\lim_{x\to 1^-} |y'(x)| = \infty \qquad \text{or} \qquad \lim_{x\to -1^+} |y'(x)| = \infty$$

(or both limits are infinite).

d. *Verify that* $p_m(1) \neq 0$ *for each nonnegative integer* m.

e. *For each nonnegative integer* m, *the* m^{th} *Chebyshev polynomial* $T_m(x)$ *is the polynomial solution to the Chebyshev equation with parameter* $\lambda = m^2$ *satisfying* $T_m(1) = 1$. *Find* $T_m(x)$ *for* $m = 0, 1, 2, 3, 4$ *and* 5.

f. *Finish verifying that the Chebyshev equation*

$$(1 - x^2)y'' - 2xy' + \lambda y = 0$$

has nontrivial solutions with bounded first derivatives on $(-1, 1)$ *if and only if* $\lambda = m^2$ *for some nonnegative integer. Moreover,* y *is such a solution if and only if* y *is a constant multiple of the* m^{th} *Chebyshev polynomial.*

36.6. *The following differential equations all have* $x_0 = 0$ *as a regular singular point. For each, the corresponding exponents* r_1 *and* r_2 *are given, along with the solution* $y_1(x)$ *on* $x > 0$ *corresponding to* r_1, *as described in Theorem 36.2 on page 720. This solution can be found by the basic method of Frobenius. The second solution,* y_2, *cannot be found by the basic method, but, as stated in Theorem 36.2, it is of the form*

$$y_2(x) = y_1(x)\ln|x| + |x|^{1+r_1}\sum_{k=0}^{\infty} b_k x^k$$

or

$$y_2(x) = \mu y_1(x) \ln|x| + |x|^{r_2} \sum_{k=0}^{\infty} b_k x^k \quad ,$$

depending, respectively, on whether the exponents r_1 and r_2 are equal or differ by a nonzero integer. Do recall that, in the second formula, $b_0 = 1$, and b_K is arbitrary for $K = r_1 - r_2$.

"Find $y_2(x)$ for $x > 0$" for each of the following. More precisely, determine which of the above two formulas hold, and find at least the values of b_0, b_1, b_2, b_3 and b_4, along with the value for μ if appropriate. You may set any arbitrary constant equal to 0, and assume $x > 0$.

a. $4x^2 y'' + (1 - 4x)y = 0 \quad : \quad r_1 = r_2 = \frac{1}{2} \quad , \quad y_1(x) = \sqrt{x} \sum_{k=0}^{\infty} \frac{1}{(k!)^2} x^k$

b. $y'' + \frac{1}{x} y' + y = 0$ (Bessel's equation of order 0) $\quad : \quad r_1 = r_2 = 0 \quad ,$

$$y_1(x) = \sum_{m=0}^{\infty} \frac{(-1)^m}{(2^m m!)^2} x^{2m}$$

c. $x^2 y'' - \left(x + x^2\right) y' + 4xy = 0 \quad ; \quad r_1 = 2 \quad , \quad r_2 = 0 \quad ,$

$$y_1(x) = x^2 - \frac{2}{3} x^3 + \frac{1}{12} x^4$$

d. $x^2 y'' + xy' + (4x - 4)y = 0 \quad ; \quad r_1 = 2 \quad , \quad r_2 = -2 \quad ,$

$$y_1(x) = x^2 \sum_{k=0}^{\infty} \frac{(-4)^k 4!}{k!(k+4)!} x^k$$

37

Validating the Method of Frobenius

Let us now focus on verifying the claims made in the big theorems of Section 36.1: Theorem 36.1 on the indicial equation and Theorem 36.2 on solutions about regular singular points.

We will begin our work in a rather obvious manner — by applying the basic Frobenius method to a generic differential equation with a regular singular point (after rewriting the equation in a "reduced form") and then closely looking at the results of these computations. This, along with a theorem on convergence that we'll discuss, will tell us precisely when the basic method succeeds and why it fails for certain cases. After that, we will derive the alternative solution formulas (formulas (36.3) and (36.4) in Theorem 36.2 on page 720) and verify that they truly are valid solutions. Dealing with these later cases will be the challenging part.

37.1 Basic Assumptions and Symbology

Throughout this chapter, we are assuming that we have a second-order homogeneous linear differential equation having a point x_0 on the real line as a regular singular point, and having $R > 0$ as the Frobenius radius of convergence about x_0. For simplicity, we will further assume $x_0 = 0$, keeping in mind that corresponding results can be obtained when the regular singular point is nonzero by using the substitution $X = x - x_0$. Also (after recalling the comments made on page 702 about solutions when $x < x_0$), let us agree that we can restrict ourselves to analyzing the possible solutions on the interval $(0, R)$.

Since we are assuming $x_0 = 0$ is a regular singular point, our differential equation can be written as

$$x^2\alpha(x)y'' + x\beta(x)y' + \gamma(x)y = 0 \quad , \tag{37.1}$$

where α, β and γ are functions analytic at $x_0 = 0$ and with $\alpha(0) \neq 0$. Dividing through by α, we get the corresponding reduced form for our original differential equation

$$x^2y'' + xP(x)y' + Q(x)y = 0 \tag{37.2}$$

where

$$P(x) = \frac{\beta(x)}{\alpha(x)} \qquad \text{and} \qquad Q(x) = \frac{\gamma(x)}{\alpha(x)} \quad .$$

From Lemmas 34.1 and 34.2 on page 649, we know that we can express P and Q as power series

$$P(x) = \sum_{k=0}^{\infty} p_k x^k \qquad \text{for} \quad |x| < R \tag{37.3a}$$

and

$$Q(x) = \sum_{k=0}^{\infty} q_k x^k \qquad \text{for} \quad |x| < R \quad . \tag{37.3b}$$

For the rest of this chapter, we will be doing computations involving the above p_k's and q_k's. Don't forget this. And don't forget the relation between P and Q, and the coefficients of the first version of our differential equation. In particular, we might as well note here that

$$p_0 = P(0) = \frac{\beta(0)}{\alpha(0)} \qquad \text{and} \qquad q_0 = Q(0) = \frac{\gamma(0)}{\alpha(0)} \quad .$$

Finally, throughout this chapter, we will let $\mathcal{L}$ be the linear differential operator

$$\mathcal{L}[y] = x^2 y'' + xP(x)y' + Q(x)y \quad ,$$

so that we can write the differential equation we wish to solve, equation (37.2), in the very abbreviated form

$$\mathcal{L}[y] = 0 \quad .$$

This will make it easier to describe some of our computations.

37.2 The Indicial Equation and Basic Recursion Formula

Basic Derivations

First, let's see what we get from plugging the arbitrary modified power series

$$y(x) = x^r \sum_{k=0}^{\infty} c_k x^k = \sum_{k=0}^{\infty} c_k x^{k+r}$$

into $\mathcal{L}$ (using the formula from Theorem 32.12 on multiplying power series):

$$\begin{aligned}
\mathcal{L}[y] &= x^2 y'' + xP(x)y' + Q(x)y \\
&= x^2 \sum_{k=0}^{\infty} c_k (k+r)(k+r-1) x^{k+r-2} \\
&\quad + x \left(\sum_{k=0}^{\infty} p_k x^k \right) \left(\sum_{k=0}^{\infty} c_k (k+r) x^{k+r-1} \right) + \left(\sum_{k=0}^{\infty} q_k x^k \right) \left(\sum_{k=0}^{\infty} c_k x^{k+r} \right) \\
&= \sum_{k=0}^{\infty} c_k (k+r)(k+r-1) x^{k+r} \\
&\quad + \sum_{k=0}^{\infty} \sum_{j=0}^{k} c_j p_{k-j} (j+r) x^{k+r} + \sum_{k=0}^{\infty} \sum_{j=0}^{k} c_j q_{k-j} x^{k+r} \\
&= x^r \sum_{k=0}^{\infty} \left[c_k (k+r)(k+r-1) + \sum_{j=0}^{k} c_j \left[p_{k-j}(j+r) + q_{k-j} \right] \right] x^k \quad .
\end{aligned}$$

That is,

$$\mathcal{L}\left[x^r \sum_{k=0}^{\infty} c_k x^k\right] = x^r \sum_{k=0}^{\infty} L_k x^k$$

where

$$L_k = c_k(k+r)(k+r-1) + \sum_{j=0}^{k} c_j \left[p_{k-j}(j+r) + q_{k-j}\right] \quad .$$

Let's now look at the individual L_k's.

For $k = 0$,

$$\begin{aligned} L_0 &= c_0(0+r)(0+r-1) + \sum_{j=0}^{0} c_j \left[p_{0-j}(j+r) + q_{0-j}\right] \\ &= c_0 r(r-1) + c_0 \left[p_0 r + q_0\right] \\ &= c_0\left[r(r-1) + p_0 r + q_0\right] \quad . \end{aligned}$$

The expression in the last bracket will arise several more times in our computations. For convenience, we will let I be the corresponding polynomial function

$$I(\rho) = \rho(\rho-1) + p_0\rho + q_0 \quad .$$

Then

$$L_0 = c_0 I(r) \quad .$$

For $k > 0$,

$$\begin{aligned} L_k &= c_k(k+r)(k+r-1) + \sum_{j=0}^{k} c_j \left[p_{k-j}(j+r) + q_{k-j}\right] \\ &= c_k(k+r)(k+r-1) + \sum_{j=0}^{k-1} c_j \left[p_{k-j}(j+r) + q_{k-j}\right] + c_k \left[p_{k-k}(k+r) q_{k-k}\right] \\ &= c_k\Big[\underbrace{(k+r)(k+r-1) + p_0(k+r) + q_0}_{I(k+r)\,!}\Big] + \sum_{j=0}^{k-1} c_j \left[p_{k-j}(j+r) + q_{k-j}\right] \quad . \end{aligned}$$

We'll be repeating the above computations at least two more times in this chapter. To save time, let's summarize what we have.

Lemma 37.1

Let $\mathcal{L}$ be the differential operator

$$\mathcal{L}[y] = x^2 y'' + xP(x)y' + Q(x)y$$

where

$$P(x) = \sum_{k=0}^{\infty} p_k x^k \qquad \text{and} \qquad Q(x) = \sum_{k=0}^{\infty} q_k x^k \quad .$$

Then, for any modified power series $x^r \sum_k^{\infty} c_k x^k$,

$$\mathcal{L}\left[x^r \sum_{k=0}^{\infty} c_k x^k\right] = x^r\left[c_0 I(r) + \sum_{k=1}^{\infty}\left(c_k I(k+r) + \sum_{j=0}^{k-1} c_j \left[p_{k-j}(j+r) + q_{k-j}\right]\right)x^k\right]$$

where
$$I(\rho) = \rho(\rho - 1) + p_0\rho + q_0 \quad .$$

Our immediate interest is in finding a modified power series
$$y(x) = x^r \sum_{k=0}^{\infty} c_k x^k \qquad \text{with} \quad c_0 \neq 0$$
that satisfies our differential equation,
$$\mathcal{L}[y] = 0 \quad .$$

Applying the above lemma, we see that we must have
$$x^r \left[c_0 I(r) + \sum_{k=1}^{\infty} \left(c_k I(k+r) + \sum_{j=0}^{k-1} c_j \left[p_{k-j}(j+r) + q_{k-j} \right] \right) x^k \right] = 0 \quad ,$$
which means that each term in the above power series must be zero. That is,
$$I(r) = 0 \tag{37.4a}$$
and
$$c_k I(k+r) + \sum_{j=0}^{k-1} c_j \left[p_{k-j}(j+r) + q_{k-j} \right] = 0 \qquad \text{for} \quad k = 1, 2, 3, \ldots \quad . \tag{37.4b}$$

The Indicial Equation

The Equation and Its Solutions

You probably already recognized equation (37.4a) as the indicial equation from the basic method of Frobenius. In more explicit form, it's the polynomial equation
$$r(r-1) + p_0 r + q_0 = 0 \quad . \tag{37.5a}$$

Equivalently, we can write this equation as
$$r^2 + (p_0 - 1)r + q_0 = 0 \quad , \tag{37.5b}$$
or even
$$(r - r_1)(r - r_2) = 0 \tag{37.5c}$$
or
$$r^2 - (r_1 + r_2)r + r_1 r_2 = 0 \tag{37.5d}$$
where r_1 and r_2 are the solutions to the indicial equation,
$$r_1 = \frac{1 - p_0 + \sqrt{(p_0 - 1)^2 - 4q_0}}{2} \qquad \text{and} \qquad r_2 = \frac{1 - p_0 - \sqrt{(p_0 - 1)^2 - 4q_0}}{2} \quad .$$

This, of course, also means that we can write the formula for I in four different ways:
$$I(\rho) = \rho(\rho - 1) + p_0\rho + q_0 \quad , \tag{37.6a}$$
$$I(\rho) = \rho^2 + (p_0 - 1)\rho + q_0 \quad , \tag{37.6b}$$

$$I(\rho) = (\rho - r_1)(\rho - r_2) \tag{37.6c}$$

and

$$I(\rho) = \rho^2 - (r_1 + r_2)\rho + r_1 r_2 \quad . \tag{37.6d}$$

For the rest of this chapter, we will use whichever of the above formulas for I seems most convenient at the time. Also, r_1 and r_2 will always denote the two values given above. Do note that if both are real, then $r_1 \geq r_2$.

By the way, if you compare the second and last of the above formulas for $I(\rho)$, you'll see that

$$p_0 = 1 - (r_1 + r_2) \qquad \text{and} \qquad q_0 = r_1 r_2 \quad .$$

Later, we may find these observations useful.

Proof of Theorem 36.1

Recall that p_0 and q_0 are related to the coefficients in the equation we first started with,

$$x^2\alpha(x)y'' + x\beta(x)y' + \gamma(x)y = 0 \quad ,$$

via

$$p_0 = P(0) = \frac{\beta_0}{\alpha_0} \qquad \text{and} \qquad q_0 = Q(0) = \frac{\gamma_0}{\alpha_0}$$

where

$$\alpha_0 = \alpha(0) \quad , \qquad \beta_0 = \beta(0) \qquad \text{and} \qquad \gamma_0 = \gamma(0) \quad .$$

Using these relations, we can rewrite the first version of the indicial equation (equation (37.5a)) as

$$r(r-1) + \frac{\beta_0}{\alpha_0}r + \frac{\gamma_0}{\alpha_0} = 0 \quad ,$$

which, after multiplying through by α_0 is

$$\alpha_0 r(r-1) + \beta_0 r + \gamma_0 = 0 \quad .$$

This, along with the formulas for r_1 and r_2, completes the proof of Theorem 36.1 on page 719. ∎

Recursion Formulas

The Basic Recursion Formula

You probably also recognized that equation (37.4b) is, essentially, a recursion formula for any given value of r. Let us first rewrite it as

$$c_k I(k+r) = -\sum_{j=0}^{k-1} c_j \left[p_{k-j}(j+r) + q_{k-j}\right] \qquad \text{for} \quad k = 1, 2, 3, \ldots \quad . \tag{37.7}$$

If

$$I(k+r) \neq 0 \qquad \text{for} \quad k = 1, 2, 3, \ldots \quad ,$$

then we can solve the above for c_k, obtaining the generic recursion formula

$$c_k = \frac{-1}{I(k+r)} \sum_{j=0}^{k-1} c_j \left[p_{k-j}(j+r) + q_{k-j}\right] \qquad \text{for} \quad k = 1, 2, 3, \ldots \quad . \tag{37.8}$$

It will be worth noting that p_0 and q_0 do not explicitly appear in this recursion formula except in the formula for $I(k+r)$.

More General Recursion Formulas and a Convergence Theorem

Later, we will have to deal with recursion formulas of the form

$$c_k = \frac{1}{I(k+r)}\left(f_k - \sum_{j=0}^{k-1} c_j\left[p_{k-j}(j+r) + q_{k-j}\right]\right)$$

where the f_k's are coefficients of some power series convergent on $(-R, R)$. (Note that this reduces to recursion formula (37.8) if each f_k is 0.) To deal with the convergence of any power series based on any such a recursion formula, we have the following theorem:

Theorem 37.2
Let $R > 0$. Assume

$$\sum_{k=0}^{\infty} p_k x^k \quad , \quad \sum_{k=0}^{\infty} q_k x^k \quad \text{and} \quad \sum_{k=0}^{\infty} f_k x^k$$

are power series convergent for $|x| < R$, and

$$\sum_{k=0}^{\infty} c_k x^k$$

is a power series such that, for some value ω and some integer K_0,

$$c_k = \frac{1}{J(k)}\left(f_k - \sum_{j=0}^{k-1} c_j\left[p_{k-j}(j+\omega) + q_{k-j}\right]\right) \qquad \text{for} \quad k \geq K_0$$

where J is some second-degree polynomial function satisfying

$$J(k) \neq 0 \qquad \text{for} \quad k = K_0,\ K_0+1,\ K_0+2,\ \dots \quad .$$

Then $\sum_k^{\infty} c_k x^k$ is also convergent for $|x| < R$.

The proof of this convergence theorem will be given in Section 37.6. It is very similar to the convergence proofs developed in Chapter 34 for power series solutions.

37.3 The Easily Obtained Series Solutions

Now let r_j be either of the two solutions r_1 and r_2 to the indicial equation,

$$I(r) = 0 \quad .$$

To use recursion formula (37.8) with $r = r_j$, it suffices to have

$$I(k + r_j) \neq 0 \qquad \text{for} \quad k = 1, 2, 3, \dots \quad .$$

But r_1 and r_2 are the only solutions to $I(r) = 0$, so the last line tells us that, to use recursion formula (37.8) with $r = r_j$, it suffices to have

$$k + r_j \neq r_1 \quad \text{and} \quad k + r_j \neq r_2 \qquad \text{for} \quad k = 1, 2, 3, \dots \quad ;$$

that is, it suffices to have

$$r_1 - r_j \neq k \quad \text{and} \quad r_2 - r_j \neq k \qquad \text{for} \quad k = 1, 2, 3, \ldots \quad .$$

As long as this holds, we can start with any nonzero constant c_0 and generate subsequent c_k's via the basic recursion formula (37.8) to create a power series

$$\sum_{k=0}^{\infty} c_k x^k \quad .$$

Moreover, Theorem 37.2 assures us that this series is convergent for $|x| < R$. Consequently,

$$y(x) = x^{r_j} \sum_{k=0}^{\infty} c_k x^k$$

is a well-defined function, at least on $(0, R)$ (just what happens at $x = 0$ depends on the x^{r_j} factor in this formula). Plugging this formula back into our differential equation and basically repeating the computations leading to the indicial equation and the recursion formula would then confirm that this y is, indeed, a solution on $(0, R)$ to our differential equation. Let's record this:

Lemma 37.3
If r_j is either of the two solutions r_1 and r_2 to the indicial equation for the problem considered in this chapter, and

$$r_1 - r_j \neq k \qquad \text{and} \qquad r_2 - r_j \neq k \qquad \text{for} \quad k = 1, 2, 3, \ldots \quad , \tag{37.9}$$

then a solution on $(0, R)$ to the original differential equation is given by

$$y(x) = x^{r_j} \sum_{k=0}^{\infty} c_k x^k$$

where c_0 is any nonzero constant, and c_1, c_2, c_3, ... are given by recursion formula (37.8) with $r = r_j$.

Now let us consider the $r_j = r_1$ and $r_j = r_2$ cases separately, adding the assumption that the coefficients of our original differential equation are all real-valued in some interval about $x_0 = 0$. This means that the coefficients in the indicial equation are all real. Hence, we may assume that either both r_1 and r_2 are real with $r_1 \geq r_2$, or that r_1 and r_2 are complex conjugates of each other.

Solutions Corresponding to r_1

With $r_j = r_1$, condition (37.9) in the above lemma becomes

$$r_1 - r_1 \neq k \qquad \text{and} \qquad r_2 - r_1 \neq k \qquad \text{for} \quad k = 1, 2, 3, \ldots \quad .$$

Clearly, the only way this cannot be satisfied is if

$$r_2 - r_1 = K \qquad \text{for some positive integer } K \quad .$$

But, using the formulas for r_1 and r_2 from page 746, you can easily verify that

$$r_2 - r_1 = -\sqrt{(p_0 - 1)^2 - 4q_0} \quad ,$$

which cannot equal some positive integer. Thus, the above lemma assures us of the following:

One solution on $(0, R)$ *to our differential equation is given by*

$$y(x) = x^{r_1} \sum_{k=0}^{\infty} c_k x^k$$

where c_0 *is any nonzero constant, and* c_1 *,* c_2 *,* c_3 *,* $\ldots$ *are given by recursion formula (37.8) with* $r = r_1$ *.*

This confirms statement 1 in Theorem 36.2.

In particular, for the rest of our discussion, let us let y_1 be the solution

$$y_1(x) = x^{r_1} \sum_{k=0}^{\infty} a_k x^k \tag{37.10}$$

where $a_0 = 1$ and

$$a_k = \frac{-1}{I(k+r_1)} \sum_{j=0}^{k-1} a_j \left[p_{k-j}(j+r_1) + q_{k-j} \right] \qquad \text{for} \quad k = 1, 2, 3, \ldots \quad .$$

On occasion, we may call this our "first" solution.

"Unexceptional" Solutions Corresponding to r_2

With $r_j = r_2$, condition (37.9) in Lemma 37.3 becomes

$$r_1 - r_2 \neq k \qquad \text{and} \qquad r_2 - r_2 \neq k \qquad \text{for} \quad k = 1, 2, 3, \ldots \quad ,$$

which, obviously, is the same as

$$r_1 - r_2 \neq K \qquad \text{for each positive integer } K \quad .$$

Unfortunately, this requirement does not automatically hold. It is certainly possible that

$$r_1 - r_2 = K \qquad \text{for some positive integer } K \quad .$$

This is an "exceptional" case which we will have to examine further. For now, the lemma above simply assures us that:

If $r_1 - r_2$ *is not a positive integer, then a solution on* $(0, R)$ *to our differential equation is given by*

$$y(x) = x^{r_2} \sum_{k=0}^{\infty} c_k x^k$$

where c_0 *is any nonzero constant, and the other* c_k*'s are given by recursion formula (37.8) with* $r = r_2$ *.*

Of course, the solutions just described corresponding to r_2 will be the same as those corresponding to r_1 if $r_2 = r_1$ (i.e., $r_1 - r_2 = 0$). This is another exceptional case that we will have to examine later.

For the rest of this chapter, let us say that, if r_1 and r_2 are not equal and do not differ by an integer, then the "second solution" to our differential equation on $(0, R)$ is

$$y_2(x) = x^{r_2} \sum_{k=0}^{\infty} b_k x^k$$

where $b_0 = 1$ and

$$b_k = \frac{-1}{I(k+r_2)} \sum_{j=0}^{k-1} b_j \left[p_{k-j}(j+r_2) + q_{k-j} \right] \qquad \text{for} \quad k = 1, 2, 3, \ldots \quad .$$

We should note that, if r_1 and r_2 are two different values not differing by an integer, then the above y_1 and y_2 are clearly not constant multiples of each other (at least, it should be clear once you realize that the first terms of $y_1(x)$ and $y_2(x)$ are, respectively, x^{r_1} and x^{r_2}). Consequently $\{y_1, y_2\}$ is a fundamental set of solutions to our differential equation on $(0, R)$, and

$$y(x) = c_1 y_1(x) + c_2 y_2(x)$$

is a general solution to our differential equation over $(0, R)$. That finishes the proof of Theorem 36.2 up through statement 2.

Deriving the "Exceptional" Solutions

In the next two sections, we will derive formulas for the solutions corresponding to $r = r_2$ when r_1 and r_2 are equal or differ by a nonzero integer. In deriving these solutions, we could use the first solution, y_1 , with the reduction of order method from Chapter 14. Unfortunately, that gets somewhat messy and does not directly lead to useful recursion formulas. So, instead, we will take somewhat different approaches.

Since the approach we'll take when $r_2 = r_1$ is a bit more elementary (but still tedious) and somewhat less "clever" than the approach we'll take when $r_1 - r_2$ is a positive integer, we will consider the case where $r_2 = r_1$ first.

37.4 Second Solutions When $r_2 = r_1$

Recall that, based on what we learned from studying Euler equations, we suspected that a second solution to our differential equation when $r_2 = r_1$ will be of the form

$$y(x) = \ln|x|\, Y(x) \qquad \text{with} \quad Y(x) = x^{r_1} \sum_{k=0}^{\infty} b_k x^k \quad .$$

Unfortunately, this turns out not to be generally true. But since it seemed so reasonable at the time, let us still try using this, but with an added "error term". That is, let's try something of the form

$$y(x) = \ln|x|\, Y(x) + \epsilon(x) \quad . \tag{37.11}$$

Plugging this into the differential equation:

$$\begin{aligned} 0 &= \mathcal{L}[y] \\ &= x^2 y'' + xPy' + Qy \\ &= x^2 \left[\ln|x|\, Y'' + \frac{2}{x} Y' - \frac{1}{x^2} Y + \epsilon'' \right] + xP \left[\ln|x|\, Y' + \frac{1}{x} Y + \epsilon' \right] \\ &\quad + Q \left[\ln|x|\, Y(x) + \epsilon(x) \right] \\ &= \ln|x| \underbrace{\left[x^2 Y'' + xPY' + QY \right]}_{\mathcal{L}[Y]} + 2xY' - Y + PY + \underbrace{x^2 \epsilon'' + xP\epsilon' + Q\epsilon}_{\mathcal{L}[\epsilon]} \quad . \end{aligned}$$

Choosing Y to be our first solution,

$$Y(x) = y_1(x) = x^{r_1} \sum_{k=0}^{\infty} a_k x^k \quad ,$$

causes the natural log term to vanish, leaving us with

$$0 = 2xy_1{}' - y_1 + Py_1 + \mathcal{L}[\epsilon] \quad ,$$

which we can rewrite as

$$\mathcal{L}[\epsilon] = F(x) \tag{37.12a}$$

with

$$F(x) = y_1(x) - 2xy_1{}'(x) - P(x)y_1(x) \quad . \tag{37.12b}$$

It turns out that we will be seeing both the above differential equation and the function F when we deal with the case where r_2 and r_1 differ by a nonzero integer. So, for now, let's expand $F(x)$ using the series formulas for y_1 and P without assuming $r_2 = r_1$:

$$\begin{aligned}
F(x) &= y_1(x) - 2xy_1{}'(x) - P(x)y_1(x) \\
&= x^{r_1} \sum_{n=0}^{\infty} a_n x^n - 2x\left(x^{r_1} \sum_{n=0}^{\infty} a_n (r_1 + n) x^{n-1}\right) - \left(\sum_{n=0}^{\infty} p_n x^n\right)\left(x^{r_1} \sum_{m=0}^{\infty} a_m x^m\right) \\
&= x^{r_1}\left[\sum_{n=0}^{\infty} a_n x^n - \sum_{n=0}^{\infty} a_n (2r_1 + 2n) x^n - \left(\sum_{n=0}^{\infty} p_n x^n\right)\left(\sum_{m=0}^{\infty} a_m x^m\right)\right] \\
&= x^{r_1} \sum_{n=0}^{\infty}\left(a_n\left[1 - 2r_1 - 2n\right] - \sum_{j=0}^{n} a_j p_{n-j}\right) x^n \quad .
\end{aligned}$$

Recalling that $a_0 = 1$ and that, in general, $p_0 = 1 - r_1 - r_2$, we see that the first term in the series simplifies somewhat to

$$a_0\left[1 - 2r_1 - 2 \cdot 0)\right] - \sum_{j=0}^{0} a_j p_{0-j} = a_0[1 - 2r_1 - p_0] = r_2 - r_1 \quad .$$

For the other terms, we have

$$\begin{aligned}
a_n\left[1 - 2r_1 - 2n\right] - \sum_{j=0}^{n} a_j p_{n-j} &= a_n\left[1 - 2r_1 - 2n\right] - \sum_{j=0}^{n-1} a_j p_{n-j} - a_n p_0 \\
&= a_n\left[1 - 2r_1 - p_0 - 2n\right] - \sum_{j=0}^{n-1} a_j p_{n-j} \\
&= a_n\left[r_2 - r_1 - 2n\right] - \sum_{j=0}^{n-1} a_j p_{n-j} \quad .
\end{aligned}$$

So, in general,

$$F(x) = x^{r_1}\left[r_2 - r_1 + \sum_{n=1}^{\infty}\left(a_n\left[r_2 - r_1 - 2n\right] - \sum_{j=0}^{n-1} a_j p_{n-j}\right) x^n\right] \quad . \tag{37.13}$$

We should also note that, because of the way F was constructed from power series convergent for $|x| < R$, we automatically have that the power series factor in the above formula for $F(x)$ is convergent for $|x| < R$.

Now, let's again assume $r_2 = r_1$. With this assumption and the change of index $k = n - 1$, the above formula for $F(x)$ reduces further,

$$F(x) = x^{r_1}\left[0 + \sum_{n=1}^{\infty}\left(-2na_n - \sum_{j=0}^{n-1} a_j p_{n-j}\right)x^n\right]$$

$$= x^{r_1}\sum_{k=0}^{\infty}\left(-2(k+1)a_{k+1} - \sum_{j=0}^{k} a_j p_{k+1-j}\right)x^{k+1}\Bigg] \quad ,$$

which we can write more succinctly as

$$F(x) = x^{r_1+1}\sum_{k=0}^{\infty} f_k x^k \tag{37.14a}$$

with

$$f_k = -2(k+1)a_{k+1} - \sum_{j=0}^{k} a_j p_{k+1-j} \quad . \tag{37.14b}$$

Let us now consider a modified power series formula for our error term

$$\epsilon(x) = x^{\rho}\sum_{k=0}^{\infty}\epsilon_k x^k \quad .$$

From Lemma 37.2, we know that

$$\mathcal{L}[\epsilon(x)] = x^{\rho}\left[\epsilon_0 I(\rho) + \sum_{k=1}^{\infty}\left(\epsilon_k I(k+\rho) + \sum_{j=0}^{k-1}\epsilon_j\left[p_{k-j}(j+\rho) + q_{k-j}\right]\right)x^k\right] \quad .$$

Thus, the differential equation $\mathcal{L}[\epsilon] = F$ becomes

$$x^{\rho}\left[\epsilon_0 I(\rho) + \sum_{k=1}^{\infty}\left(\epsilon_k I(k+\rho) + \sum_{j=0}^{k-1}\epsilon_j\left[p_{k-j}(j+\rho) + q_{k-j}\right]\right)x^k\right] = x^{r_1+1}\sum_{k=0}^{\infty} f_k x^k \quad ,$$

which is satisfied when

$$\rho = r_1 + 1 \quad , \tag{37.15a}$$

$$\epsilon_0 I(\rho) = f_0 \tag{37.15b}$$

and, for $k = 1, 2, 3, \ldots$,

$$\epsilon_k I(k+\rho) + \sum_{j=0}^{k-1}\epsilon_j\left[p_{k-j}(j+r_1+1) + q_{k-j}\right] = f_k \quad . \tag{37.15c}$$

So let $\rho = r_1 + 1$ and observe that, because r_1 is a double root of $I(r)$,

$$I(k+\rho) = I(k+r_1+1) = ([r_1+k+1] - r_1)^2 = (k+1)^2 \quad .$$

System (37.15) now reduces to

$$\epsilon_0 = f_0$$

and, for $k = 1, 2, 3, \ldots,$

$$\epsilon_k(k+1)^2 + \sum_{j=0}^{k-1} \epsilon_j \left[p_{k-j}(j+r_1+1) + q_{k-j} \right] = f_k \quad .$$

That is,

$$\epsilon_0 = f_0 \tag{37.16a}$$

and, for $k = 1, 2, 3, \ldots,$

$$\epsilon_k = \frac{1}{(k+1)^2} \left(f_k - \sum_{j=0}^{k-1} \epsilon_j \left[p_{k-j}(j+r_1+1) + q_{k-j} \right] \right) \tag{37.16b}$$

where the f_k's are given by formula (37.14b).

Now recall just what we are looking for, namely, a function $\epsilon(x)$ such that

$$y_2(x) = y_1(x) \ln|x| + \epsilon(x)$$

is a solution to our original differential equation. We have obtained

$$\epsilon(x) = x^\rho \sum_{k=0}^{\infty} \epsilon_k x^k$$

where $\rho = r_1 + 1$ and the ϵ_k's are given by formula set (37.16). Plugging this formula back into the original differential equation and repeating the computations used to derive the above will confirm that y_2 is, indeed, a solution over $(0, R)$, provided the series for ϵ converges. Fortunately, using Theorem 37.2 on page 748 this convergence is easily confirmed.

Thus, the above y_2 is a solution to our original differential equation on the interval $(0, R)$. Moreover, y_2 is clearly not a constant multiple of y_1. So $\{y_1, y_2\}$ is a fundamental set of solutions,

$$y(x) = c_1 y_1(x) + c_2 y_2(x)$$

is a general solution to our differential equation over $(0, R)$, and we have verified statement 3 in Theorem 36.2 (with $b_k = \epsilon_k$).

37.5 Second Solutions When $r_1 - r_2 = K$

Preliminaries

Let's now assume r_1 and r_2 differ by some positive integer K, $r_1 - r_2 = K$. Setting

$$y(x) = x^{r_2} \sum_{k=0}^{\infty} b_k x^k$$

with $b_k = 1$ and using recursion formula (37.8) gives us

$$b_k = \frac{-1}{I(r_2+k)} \sum_{j=0}^{k-1} b_j \left[p_{k-j}(j+r) + q_{k-j} \right] \qquad \text{for} \quad k = 1, 2, 3, \ldots, K-1 \quad .$$

Unfortunately, $I(r_2 + K) = I(r_1) = 0$, giving us a "division by zero" when we attempt to compute b_K. This is the complication we will deal with for the rest of this section. It turns out that there are two subcases, depending on whether

$$\Gamma_K = \sum_{j=0}^{K-1} b_j \left[p_{K-j}(j + r_2) + q_{K-j} \right]$$

is zero or not. If it is zero, we get lucky.

The Case Where We Get Lucky

Recall that we actually derived our recursion formula from the requirement that, for

$$y(x) = x^{r_2} \sum_{k=0}^{\infty} b_k x^k$$

to be a solution to our differential equation, it suffices to have

$$b_k I(r_2 + k) = -\sum_{j=0}^{k-1} b_j \left[p_{k-j}(j + r) + q_{k-j} \right] \quad \text{for} \quad k = 1, 2, 3, \ldots, \quad . \tag{37.17}$$

As noted above, we can use this to find b_k for $k < K$. For $k = K$, we have $I(r_2 + K) = I(r_1) = 0$, and the above equation becomes

$$b_K \cdot 0 = \Gamma_K \quad .$$

If we are lucky, then $\Gamma_K = 0$ and the above equation is trivially true for any value of b_K. So b_K is arbitrary if $\Gamma_K = 0$. Pick any value you wish (say, $b_K = 0$), and use equation (37.17) to compute the rest of the b_k's for

$$y_2(x) = x^{r_2} \sum_{k=0}^{\infty} b_k x^k \quad .$$

Convergence Theorem 37.2 on page 748 now applies and assures us that the above power series converges for $|x| < R$. And then, again, the very computations leading to the indicial equation and recursion formulas verify that this $y(x)$ is a solution to our differential equation. Moreover, the leading term is x^{r_2}. Consequently, $y_1(x)$ and $y_2(x)$ are not constant multiples of each other. Hence, $\{y_1, y_2\}$ is a fundamental set of solutions, and

$$y(x) = c_1 y_1(x) + c_2 y_2(x)$$

is a general solution to our differential equation over $(0, R)$.

If you go back and check, you will see that the above y_2 is the solution claimed to exist in statement 4 of Theorem 36.2 when $\mu = 0$. Hence we've confirmed that part of the claim.

?►Exercise 37.1: *Show that, if we took $b_0 = 0$ and $b_K = 1$ in the above (instead of $b_0 = 1$ and $b_K = 0$), we would have obtained the first solution, $y_1(x)$. (Thus, if $r_1 - r_2 = K$ and $\Gamma_k = 0$, the Frobenius method will generate the complete general solution when using $r = r_2$.)*

The Other Case

Let us now assume

$$\Gamma_K = \sum_{j=0}^{K-1} b_j \left[p_{K-j}(j+r_2) + q_{K-j} \right] \neq 0 \quad .$$

Ultimately, we want to confirm that formula (36.4) in Theorem 36.2 does describe a solution to our differential equation. Before doing that, however, let us see how anyone could have come up with formula (36.4) in the first place.

Deriving a Solution as the Limit of Other Second Solutions

Suppose we have two differential equations that are very similar to each other. Does it not seem reasonable to expect one solution of one of these equations to also be very similar to some solution to the other differential equation? Of course it does, and this is what we will use to derive the second solution to our differential equation,

$$x^2 y'' + xPy' + Qy = 0 \quad . \tag{37.18}$$

Remember, this has the corresponding indicial equation

$$I(r) = 0 \qquad \text{with} \quad I(\rho) = (\rho - r_2)(\rho - r_1) \quad .$$

Also remember that we are assuming $r_1 = r_2 + K$ for some positive integer K, and that Γ_K (as defined above) is nonzero.

Now let r be any real value close to r_2 (say, with $|r - r_2| < 1$) and consider

$$x^2 y'' + xP_r y' + Q_r y = 0$$

where P_r and Q_r differ from P and Q only in having the first coefficients in their power series about 0 adjusted so that the corresponding indicial equation is

$$I_r(r) = 0 \qquad \text{with} \quad I_r(\rho) = (\rho - r)(\rho - r_1) \quad .$$

If $r = r_2$, this is our original equation. If $r \neq r_2$, this is our "approximating differential equation". From our discussion in Section 37.3 on the "easily obtained solutions", we know that, when $r \neq r_2$, a second solution to this equation is given by

$$y(x, r) = x^r \sum_{k=0}^{\infty} b_k(r) x^k$$

with $b_0(r) = 1$ and

$$b_k(r) = -\frac{1}{I_r(r+k)} \sum_{j=0}^{k-1} b_j(r)[P_{k-j}(j+r) + Q_{k-j}] \qquad \text{for} \quad k = 1, 2, \ldots \quad .$$

This, presumably, will approximate some second solution $y(x, r_1)$ to equation (37.18),

$$y(x, r_2) \approx y(x, r) \quad .$$

Presumably, also, this approximation improves as $r \to r_1$. So, let us go further and seek the $y(x, r_2)$ given by

$$y(x, r_2) = \lim_{r \to r_2} y(x, r) \quad .$$

Before going further, let us observe that

$$\begin{aligned} I_r(r+k) &= (r+k-r)(r+k-r_1) \\ &= k(k+r-[r_2+K]) = k(k-K+r-r_2) \quad . \end{aligned}$$

Thus,

$$b_k(r) = \frac{-1}{k(k-K+r-r_2)} \sum_{j=0}^{k-1} b_j(r)[P_{k-j}(j+r) + Q_{k-j}] \qquad \text{for} \quad k = 1, 2, \ldots \quad .$$

In particular,

$$b_K(r) = \frac{-1}{K(r-r_2)} \sum_{j=0}^{K-1} b_j(r)[P_{K-j}(j+r) + Q_{K-j}] \quad .$$

So, while we have

$$b_k(r_2) = \lim_{r \to r_2} b_k(r) \qquad \text{when} \quad k < K$$

being well-defined finite values, we also have

$$\lim_{r \to r_2} |b_K(r)| = \infty \quad ,$$

suggesting that

$$\lim_{r \to r_2} |b_k(r)| = \infty \qquad \text{for} \quad k > K$$

since the recursion formula for these $b_k(r)$'s all contain $b_K(r)$.

It must be noted, however, that we are assuming $\lim_{r \to r_2} y(x,r)$ exists despite the fact that individual terms in $y(x,r)$ behave badly as $r \to r_2$. Let's hold to this hope. Assuming this,

$$\begin{aligned} \lim_{r \to r_2} x^r \sum_{k=K}^{\infty} b_k(r)x^k &= \lim_{r \to r_2} \left[x^r \sum_{k=0}^{\infty} b_k(r)x^k - x^r \sum_{k=0}^{K-1} b_k(r)x^k \right] \\ &= \lim_{r \to r_2} \left[y(x,r) - x^r \sum_{k=0}^{K-1} b_k(r)x^k \right] = y(x,r_2) - x^{r_2} \sum_{k=0}^{K-1} b_k(r_2)x^k \quad , \end{aligned}$$

which is finite for each x in the interval of convergence. Consequently,

$$\lim_{r \to r_2} (r-r_2)x^r \sum_{k=K}^{\infty} b_k(r)x^k = 0 \quad ,$$

which we will rewrite as

$$\lim_{r \to r_2} x^r \sum_{k=K}^{\infty} \beta_k(r)x^k = 0 \tag{37.19}$$

by letting

$$\beta_k(r) = (r-r_2)b_k(r) \qquad \text{for} \quad r \neq r_2 \quad .$$

Now, let's start computing $y(x,r_2)$ as a limit using a simple, cheap trick:

$$\begin{aligned} y(x,r_2) &= \lim_{r \to r_2} y(x,r) \\ &= \lim_{r \to r_2} x^r \sum_{k=0}^{\infty} b_k(r)x^k \end{aligned}$$

$$= \lim_{r\to r_2} x^r \sum_{k=0}^{K-1} b_k(r)x^k \;+\; \lim_{r\to r_2} x^r \sum_{k=K}^{\infty} b_k(r)x^k$$

$$= x^{r_2} \sum_{k=0}^{K-1} b_k(r_2)x^k \;+\; \lim_{r\to r_2} \frac{r-r_2}{r-r_2} x^r \sum_{k=K}^{\infty} b_k(r)x^k$$

$$= x^{r_2} \sum_{k=0}^{K-1} b_k(r_2)x^k \;+\; \lim_{r\to r_2} \frac{x^r \sum_{k=K}^{\infty} \beta_k(r)x^k}{r-r_2} \quad .$$

Using L'Hôpital's rule, we see that

$$\lim_{r\to r_2} \frac{x^r \sum_{k=K}^{\infty} \beta_k(r)x^k}{r-r_2} = \lim_{r\to r_2} \frac{\dfrac{\partial}{\partial r}\left[x^r \sum_{k=K}^{\infty} \beta_k(r)x^k\right]}{\dfrac{\partial}{\partial r}[r-r_2]}$$

$$= \lim_{r\to r_2} \frac{x^r \ln|x| \sum_{k=K}^{\infty} \beta_k(r)x^k \;+\; x^r \sum_{k=K}^{\infty} \beta_k{}'(r)x^k}{1}$$

$$= x^{r_2} \ln|x| \sum_{k=K}^{\infty} \beta_k(r_2)x^k \;+\; x^{r_2} \sum_{k=K}^{\infty} \beta_k{}'(r_2)x^k \quad .$$

Combining the last two results gives

$$y(x,r_2) = x^{r_2} \ln|x| \sum_{k=K}^{\infty} \beta_k(r_2)x^k \;+\; x^{r_2} \sum_{k=0}^{\infty} \begin{Bmatrix} b_k(r_2) & \text{if} \quad k < K \\ \beta_k{}'(r_2) & \text{if} \quad K \leq k \end{Bmatrix} x^k \quad .$$

This is not a very "pretty" expression. To simplify it, let

$$\epsilon_k = \begin{cases} b_k(r_2) & \text{if} \quad k < K \\ \beta_k{}'(r_2) & \text{if} \quad K \leq k \end{cases} \quad ,$$

and observe that, letting $\alpha_k = \beta_{k+K}(r_2)$,

$$x^{r_2} \sum_{k=K}^{\infty} \beta_k(r_2)x^k = x^{r_1-K}\left[\alpha_0 x^K + \alpha_1 x^{K+1} + \alpha_2 x^{K+2} + \cdots\right] = x^{r_1} \sum_{k=0}^{\infty} \alpha_k x^k \quad .$$

Then

$$y(x,r_2) = \ln|x|\, Y(x) \;+\; \epsilon(x) \tag{37.20a}$$

where

$$Y(x) = x^{r_1} \sum_{k=0}^{\infty} \alpha_k x^k \qquad \text{and} \qquad \epsilon(x) = x^{r_2} \sum_{k=0}^{\infty} \epsilon_k x^k \quad . \tag{37.20b}$$

Admittedly, part of the derivation of formula (37.20) was based on "hope" and assumptions that seemed reasonable but were not rigorously justified. So we are not yet certain this formula does yield the desired solution. Moreover, the methods given in this derivation for computing the α_k's and ϵ_k's certainly appear to be rather difficult to carry out in practice. These are valid concerns that we will deal with by now ignoring just how we derived this formula. Instead, we will see about validating this formula and obtaining more usable recursion formulas for the α_k's and ϵ_k's via methods that, by now, should be familiar to the reader.

Verifying Our Solution

Notice how similar formula (37.20a) for $y(x, r_2)$ is to formula (37.11) on page 751 from which we derived the second solution $y(x)$ when $r_1 - r_2 = 0$ in Section 37.4. Let us be inspired by the work done in that section (and reuse as much of that work as possible) and try to find a solution of the form

$$y(x) = \ln|x|\, Y(x) + \epsilon(x)$$

where

$$\epsilon(x) = x^{r_2} \sum_{k=0}^{\infty} \epsilon_k x^k \quad .$$

Glancing back at the work near the beginning of Section 37.4, it should be clear that

$$y(x) = \ln|x|\, Y(x) + \epsilon(x)$$

will satisfy our differential equation $\mathcal{L}[y] = 0$ if

$$Y(x) = y_1(x) \qquad \text{and} \qquad \mathcal{L}[\epsilon] = F(x)$$

where, taking into account the facts that $I(r_2) = 0$ and $r_1 - r_2 = K$,

$$\begin{aligned}
\mathcal{L}[\epsilon] &= x^{r_2}\left[\epsilon_0 I(r_2) + \sum_{k=1}^{\infty}\left(\epsilon_k I(r_2 + k) + \sum_{j=0}^{k-1} \epsilon_j \left[p_{k-j}(r_2 + j) + q_{k-j}\right]\right)x^k\right] \\
&= x^{r_2} \sum_{k=1}^{\infty}\left(\epsilon_k I(r_2 + k) + \sum_{j=0}^{k-1} \epsilon_j \left[p_{k-j}(r_2 + j) + q_{k-j}\right]\right)x^k
\end{aligned}$$

and

$$\begin{aligned}
F(x) &= x^{r_1}\left[r_2 - r_1 + \sum_{n=1}^{\infty}\left(a_n\left[r_2 - r_1 - 2n\right] - \sum_{j=0}^{n-1} a_j p_{n-j}\right)x^n\right] \\
&= x^{r_2+K}\left[-K + \sum_{n=1}^{\infty}\left(a_n\left[-K - 2n\right] - \sum_{j=0}^{n-1} a_j p_{n-j}\right)x^n\right] \\
&= x^{r_2}\left[-Kx^K + \sum_{n=1}^{\infty}\left(a_n\left[-K - 2n\right] - \sum_{j=0}^{n-1} a_j p_{n-j}\right)x^{K+n}\right] \quad .
\end{aligned}$$

Using $k = n + K$, we can rewrite our last formula as

$$F(x) = x^{r_2}\left[-Kx^K + \sum_{k=K+1}^{\infty} f_k x^k\right]$$

with

$$f_k = a_{k-K}\left[K - 2k\right] - \sum_{j=0}^{k-K-1} a_j p_{k-K-j} \quad .$$

As in the previous section, we know the power series in the formula for $F(x)$ converges for $|x| < R$ because of the way it was constructed from power series already known to be convergent for these values of x .

So the differential equation, $\mathcal{L}[\epsilon] = F$, expands to

$$x^{r_2} \sum_{k=1}^{\infty} \left(\epsilon_k I(r_2 + k) \; + \; \sum_{j=0}^{k-1} \epsilon_j \left[p_{k-j}(r_2 + j) + q_{k-j} \right] \right) x^k$$

$$= \; x^{r_2} \left[-Kx^K \; + \sum_{k=K+1}^{\infty} f_k x^k \right] \quad ,$$

which means that we are seeking ϵ_k's satisfying the system

$$\epsilon_k I(r_2 + k) \; + \; \sum_{j=0}^{k-1} \epsilon_j \left[p_{k-j}(r_2 + j) + q_{k-j} \right] \; = \; \begin{cases} 0 & \text{if} \quad 1 \le k < K \\ -K & \text{if} \quad k = K \\ f_k & \text{if} \quad k > K \end{cases} \quad . \tag{37.21}$$

Solving for ϵ_k in the first few equations of this set yields

$$\epsilon_k \; = \; \frac{-1}{I(r_2 + k)} \sum_{j=0}^{k-1} \epsilon_j \left[p_{k-j}(r_2 + j) + q_{k-j} \right] \qquad \text{for} \quad k = 1, 2, \ldots, K - 1 \quad .$$

To simplify matters, let's recall that, at the start of this section, we had already obtained a set $\{b_0, b_1, \ldots, b_{K-1}\}$ satisfying $b_0 = 1$ and

$$b_k \; = \; \frac{-1}{I(r_2 + k)} \sum_{j=0}^{k-1} b_j \left[p_{k-j}(r_2 + j) + q_{k-j} \right] \qquad \text{for} \quad k = 1, 2, \ldots, K - 1 \quad .$$

It is then easily verified that, whatever value we have for ϵ_0,

$$\epsilon_k \; = \; \epsilon_0 b_k \qquad \text{for} \quad k = 1, 2, \ldots, K - 1 \quad .$$

Now, also recall that

$$\Gamma_K \; = \; \sum_{j=0}^{K-1} b_j \left[p_{K-j}(r_2 + j) + q_{K-j} \right] \; \neq \; 0 \quad ,$$

and take a look at the Kth equation in system (37.21):

$$\epsilon_K I(r_2 + K) \; + \; \sum_{j=0}^{K-1} \epsilon_j \left[p_{K-j}(r_2 + j) + q_{K-j} \right] \; = \; -K$$

$$\hookrightarrow \qquad \epsilon_K I(r_1) \; + \; \sum_{j=0}^{K-1} \epsilon_0 b_j \left[p_{K-j}(r_2 + j) + q_{K-j} \right] \; = \; -K$$

$$\hookrightarrow \qquad \epsilon_K \cdot 0 \; + \; \epsilon_0 \Gamma_K \; = \; -K \quad .$$

So ϵ_K can be any value, while

$$\epsilon_0 \; = \; -\frac{K}{\Gamma_K} \quad ,$$

and

$$\epsilon_k \; = \; \epsilon_0 \cdot b_k \; = \; -\frac{K b_k}{\Gamma_K} \qquad \text{for} \quad k = 1, 2, \ldots, K - 1 \quad .$$

For the remaining ϵ_k's, we simply solve each of the remaining equations in system (37.21) for ϵ_k (using whatever value of ϵ_K we choose), obtaining

$$\epsilon_k = \frac{1}{I(r_2+k)}\left[f_k - \sum_{j=0}^{k-1} \epsilon_j \left[p_{k-j}(r_2+j) + q_{k-j} \right] \right] \quad \text{for} \quad k > K \quad .$$

Theorem 37.2 tells us that the resulting $\sum_{k=0}^{\infty} \epsilon_k x^k$ converges for $|x| < R$, and that, along with all the computations above, tells us that

$$y(x) = y_1(x) \ln|x| + x^{r_2} \sum_{k=0}^{\infty} \epsilon_k x^k$$

is a solution to our original differential equation on $(0, R)$. Clearly, it is not a constant multiple of y_1, and so $\{y_1, \mu y\}$ is a fundamental set of solutions for any nonzero constant μ. In particular, the solution mentioned in Theorem 36.2 is the one with

$$\mu = \frac{1}{\epsilon_0} = -\frac{\Gamma_K}{K} \quad .$$

And that, except for verifying convergence Theorem 37.2, confirms statement 4 of Theorem 36.2, and completes the proof of Theorem 36.2, itself. ∎

37.6 Convergence of the Solution Series

Finally, let's verify Theorem 37.2 on page 748 on the convergence of our series. As you will see, the proof is very similar to (and a bit simpler than) the proofs of convergence in Chapter 34.

Assumptions and Claim

We are assuming that ω is some constant,

$$\sum_{k=0}^{\infty} f_k x^k \quad , \quad \sum_{k=0}^{\infty} p_k x^k \quad \text{and} \quad \sum_{k=0}^{\infty} q_k x^k$$

are power series convergent for $|x| < R$, J is a second-degree polynomial function, and K_0 is some nonnegative integer such that

$$J(k) \neq 0 \quad \text{for} \quad k = K_0,\ K_0+1,\ K_0+2,\ \dots \quad .$$

We are also assuming that we have a power series

$$\sum_{k=0}^{\infty} c_k x^k$$

whose coefficients satisfy

$$c_k = \frac{1}{J(k)}\left[f_k - \sum_{j=0}^{k-1} c_j \left[p_{k-j}(j+\omega) + q_{k-j} \right] \right] \quad \text{for} \quad k \geq K_0 \quad .$$

The claim of the theorem is that $\sum_k^{\infty} c_k x^k$ converges for all x satisfying $|x| < R$. This, of course, can be verified by showing $\sum_k^{\infty} |c_k|\,|x|^k$ converges for each x in $(-R, R)$.

The Proof

We start by letting x be any single value in $(-R, R)$. We then can (and do) choose X to be some value with $|x| < X < R$. Also, by the convergence of the series, we can (and do) choose M to be a positive value such that, for $k = 0, 1, 2, \ldots$,

$$\left|f_k X^k\right| < M \quad , \quad \left|p_k X^k\right| < M \quad \text{and} \quad \left|q_k X^k\right| < M \quad .$$

Now consider the power series $\sum_{k=0}^{\infty} C_k x^k$ with

$$C_k = |c_k| \quad \text{for} \quad k < K_0$$

and

$$C_k = \left|\frac{1}{J(k)}\right| \left[M X^{-k} + \sum_{j=0}^{k-1} C_j \left[M X^{-[k-j]}(j + |\omega|) + M X^{-[k-j]} \right] \right] \quad \text{for} \quad k \geq K_0 \quad .$$

Comparing the recursion formulas for c_k and C_k, it is obvious that

$$|c_k|\,|x|^k \leq C_k\,|x|^k \quad \text{for} \quad k = 0, 1, 2, \ldots \quad .$$

Consequently, the convergence of $\sum_k^{\infty} c_k x^k$ can be confirmed by showing $\sum_k^{\infty} C_k |x|^k$ converges, and, by the limit ratio test (Theorem 32.11 on page 594) that can be shown by verifying that

$$\lim_{k\to\infty} \left| \frac{C_{k+1} x^{k+1}}{C_k x^k} \right| \leq 1 \quad .$$

Fortunately, for $k > K_0$,

$$\begin{aligned}
C_{k+1} &= \left|\frac{1}{J(k+1)}\right| \left[M X^{-(k+1)} + \sum_{j=0}^{k} C_j \left[M X^{-[k+1-j]}(j + |\omega|) + M X^{-[k+1-j]} \right] \right] \\
&= \left|\frac{X^{-1}}{J(k+1)}\right| \left[\left(M X^{-k} + \sum_{j=0}^{k-1} C_j \left[M X^{-[k-j]}(j + |\omega|) + M X^{-[k-j]} \right] \right) \right. \\
&\qquad \left. + \; C_k \left[M X^{-[k-k]}(k + |\omega|) + M X^{-[k-k]} \right] \right] \\
&= \left|\frac{X^{-1}}{J(k+1)}\right| \left(|J(k)|\, C_k + C_k M[k + \omega + 1] \right) \\
&= \left| \frac{|J(k)| + M[k + \omega + 1]}{J(k+1)} \right| \cdot \frac{C_k}{X} \quad .
\end{aligned}$$

Thus,

$$\left| \frac{C_{k+1} x^{k+1}}{C_k x^k} \right| = \frac{C_{k+1}}{C_k} |x| = \left| \frac{|J(k)| + M[k + \omega + 1]}{J(k+1)} \right| \cdot \frac{|x|}{X} \quad .$$

Since J is a second-degree polynomial, you can easily verify that

$$\lim_{k\to\infty} \left| \frac{|J(k)| + M[k + \omega + 1]}{J(k+1)} \right| = 1 \quad .$$

Hence, since $|x| < X$,

$$\lim_{k\to\infty} \left| \frac{C_{k+1} x^{k+1}}{C_k x^k} \right| = \lim_{k\to\infty} \left| \frac{|J(k)| + M[k + \omega + 1]}{J(k+1)} \right| \cdot \frac{|x|}{X} = 1 \cdot \frac{|x|}{X} < 1 \quad . \qquad \blacksquare$$

Part VI

Systems of Differential Equations

(A Brief Introduction)

38

Systems of Differential Equations: A Starting Point

Thus far, we have been dealing with individual differential equations. But there are many applications that lead to sets of differential equations sharing common variables and solutions. These sets are generally referred to as "systems of differential equations".

In this chapter we will begin a brief discussion of these systems. Unfortunately, a complete discussion goes beyond the scope of this text and requires that the reader has had a decent course on linear algebra. So our discussion will be somewhat limited in scope and in the types of systems considered. The goal is for the reader to begin understanding the basics, including why these systems are important, and how some of the methods for analyzing their solutions can be especially illuminating, even when the system comes from a single differential equation. In fact, these methods are especially important in analyzing those differential equations that are not linear.

38.1 Basic Terminology and Notions

In general, a *k^{th}-order system of M differential equations with N unknowns* is simply a collection of M differential equations involving N unknown functions with k being the highest order of the derivatives explicitly appearing in the equations. For brevity, we may refer to such a system as a "k^{th}-order $M \times N$ system".

Our primary interest, however, will be in first-order $N \times N$ systems of differential equations that can be written as

$$\begin{aligned} x_1{}' &= f_1(t, x_1, x_2, \ldots, x_N) \\ x_2{}' &= f_2(t, x_1, x_2, \ldots, x_N) \\ &\vdots \\ x_N{}' &= f_N(t, x_1, x_2, \ldots, x_N) \end{aligned} \tag{38.1}$$

where the x_j's are (initially unknown) real-valued functions of t (hence $x' = {}^{dx}/_{dt}$), and the $f_k(t, x_1, x_2, \ldots, x_N)$'s — the *component functions* of the system — are known functions of $N + 1$ variables. Note that each equation is a first-order differential equation in which only one of the unknown functions is differentiated.[1] We will refer to such systems of differential equations as either *standard first-order systems* or, when extreme brevity is desired, as *standard systems*.

[1] Also note that we have gone back to using t as the independent variable.

While we will be consistent in using t as the variable in the initially unknown functions, we will feel free to use whatever symbols seem convenient for these functions of t. In fact, when $N = 2$ or $N = 3$ (which will be the case for almost all of our examples), we will abandon subscripts and denote our functions of t by x, y and, if needed, z, and we will write our generic systems as

$$\begin{aligned} x' &= f(t,x,y) \\ y' &= g(t,x,y) \end{aligned} \qquad \text{or} \qquad \begin{aligned} x' &= f(t,x,y,z) \\ y' &= g(t,x,y,z) \\ z' &= h(t,x,y,z) \end{aligned} \quad , \tag{38.2}$$

as appropriate.

!►Example 38.1: *Here is a simple standard first-order system:*

$$\begin{aligned} x' &= x + 2y \\ y' &= 5x - 2y \end{aligned} \quad .$$

Now suppose we have a standard first-order $N \times N$ system with unknown functions x_1, x_2, ... and x_N, along with some interval of interest (α, β). A *solution* to this system over this interval is any ordered set of N specific real-valued functions $\hat{x}_1$, $\hat{x}_2$, ... and $\hat{x}_N$ such that all the equations in the system are satisfied for all values of t in the interval (α, β) when we let[2]

$$x_1(t) = \hat{x}_1(t) \quad , \quad x_2(t) = \hat{x}_2(t) \quad , \quad \ldots \quad \text{and} \quad x_N(t) = \hat{x}_N(t) \quad .$$

A *general solution* to our system of differential equations (over (α, β)) is any ordered set of N formulas describing all possible such solutions. Typically, these formulas include arbitrary constants.[3]

!►Example 38.2: *Consider the system*

$$\begin{aligned} x' &= x + 2y \\ y' &= 5x - 2y \end{aligned}$$

over the entire real line, $(-\infty, \infty)$. *If we let*

$$x(t) = e^{3t} + 2e^{-4t} \qquad \text{and} \qquad y(t) = e^{3t} - 5e^{-4t} \quad ,$$

and plug these formulas for x *and* y *into the first differential equation in our system,*

$$x' = x + 2y \quad ,$$

we get

$$\frac{d}{dt}\left[e^{3t} + 2e^{-4t}\right] = \left[e^{3t} + 2e^{-4t}\right] + 2\left[e^{3t} - 5e^{-4t}\right]$$

$$\hookrightarrow \qquad 3e^{3t} - 2 \cdot 4e^{-4t} = [1+2]e^{3t} + [2 - 2 \cdot 5]e^{-4t}$$

$$\hookrightarrow \qquad 3e^{3t} - 8e^{-3t} = 3e^{3t} - 8e^{-3t} \quad ,$$

which is an equation valid for all values of t. *So these two functions,* x *and* y, *satisfy the first differential equation in the system over* $(-\infty, \infty)$.

---

[2] We could allow the $x_k(t)$'s to be complex valued, but this will not gain us anything with the systems of interest to us and would complicate the "graphing" techniques we'll later develop and employ.

[3] And it is also typical that the precise interval of interest, (α, β), is not explicitly stated, and may not even be precisely known.

Likewise, it is easily seen that these two functions also satisfy the second equation:

$$y' = 5x - 2y$$

$$\hookrightarrow \quad \frac{d}{dt}\left[e^{3t} - 5e^{-4t}\right] = 5\left[e^{3t} + 2e^{-4t}\right] - 2\left[e^{3t} - 5e^{-4t}\right]$$

$$\hookrightarrow \quad 3e^{3t} - 5(-4)e^{-4t} = [5-2]e^{3t} + [5 \cdot 22(-5)]e^{-4t}$$

$$\hookrightarrow \quad 3e^{3t} + 20e^{-4t} = 3e^{3t} + 20e^{-4t} \quad .$$

Thus, the pair

$$x(t) = e^{3t} + 2e^{-4t} \quad \text{and} \quad y(t) = e^{3t} - 5e^{-4t}$$

is a solution to our system over $(-\infty, \infty)$.

More generally, you can easily verify that, for any choice of constants c_1 *and* c_2,

$$x(t) = c_1 e^{3t} + 2c_2 e^{-4t} \quad \text{and} \quad y(t) = c_1 e^{3t} - 5c_2 e^{-4t}$$

satisfies the given system (see Exercise 38.4). In fact, we will later verify that the above is a general solution to the system. (Note that the two formulas in the above general solution share arbitrary constants. This will be typical.)

On the other hand, plugging the pair

$$x(t) = e^{3t} + 2e^{4t} \quad \text{and} \quad y(t) = 2e^{3t} + e^{4t}$$

into the first equation of our system yields

$$x' = x + 2y$$

$$\hookrightarrow \quad \frac{d}{dt}\left[e^{3t} + 2e^{4t}\right] = \left[e^{3t} + 2e^{4t}\right] + 2\left[2e^{3t} + e^{4t}\right]$$

$$\hookrightarrow \quad 3e^{3t} + 8e^{4t} = 5e^{3t} + 4e^{4t}$$

$$\hookrightarrow \quad 4e^{4t} = 2e^{3t} \quad ,$$

which is not true for every real value t. *So this last pair of functions is not a solution to our system.*

If, in addition to our standard first-order system, we have the value of every x_k specified at some single point t_0, then we have an *initial-value problem*, a solution of which is any solution to the system that also satisfies the given initial values. Unsurprisingly, we usually solve initial-value problems by first finding the general solution to the system, and then applying the initial conditions to the general solution to determine the values of the 'arbitrary' constants.

!►Example 38.3: *Consider the initial-value problem consisting of the system from the previous example,*

$$x' = x + 2y$$
$$y' = 5x - 2y \quad ,$$

along with the initial conditions

$$x(0) = 0 \quad \text{and} \quad y(0) = 1 \quad .$$

In the previous example, it was asserted that the pair

$$x(t) = c_1 e^{3t} + 2c_2 e^{-4t} \quad \text{and} \quad y(t) = c_1 e^{3t} - 5c_2 e^{-4t} \tag{38.3}$$

is a solution to our system for any choice of constants c_1 *and* c_2 *. Using these formulas with the initial conditions, we get*

$$0 = x(0) = c_1 e^{3 \cdot 0} + 2c_2 e^{-4 \cdot 0} = c_1 + 2c_2$$

and

$$1 = y(0) = c_1 e^{3 \cdot 0} - 5c_2 e^{-4 \cdot 0} = c_1 - 5c_2 \quad .$$

So, to find c_1 *and* c_2 *, we solve the simple algebraic linear system*

$$\begin{aligned} c_1 + 2c_2 &= 0 \\ c_1 - 5c_2 &= 1 \end{aligned} \quad .$$

Doing so however you wish, you should easily discover that

$$c_1 = \frac{2}{7} \quad \text{and} \quad c_2 = -\frac{1}{7} \quad ,$$

which, after plugging these values back into the general formulas for $x(t)$ *and* $y(t)$ *given in equation set (38.3), yields the solution to the given initial-value problem,*

$$x(t) = \frac{2}{7}e^{3t} - \frac{2}{7}e^{-4t} \quad \text{and} \quad y(t) = \frac{2}{7}e^{3t} + \frac{5}{7}e^{-4t} \quad .$$

By the way, you will occasionally hear the term "coupling" in describing the extent to which each equation of the system contains different unknown functions. A system is *completely uncoupled* if each equation involves just one of the unknown functions, as in

$$\begin{aligned} x' &= 5x + \sin(x) \\ y'' &= 4y \end{aligned} \quad ,$$

and is *weakly coupled* or *only partially coupled* if at least one of the equations only involves one unknown function, as in

$$\begin{aligned} x' &= 5x + 2y \\ y'' &= 4y \end{aligned} \quad .$$

Such systems can be solved in the obvious manner. First solve each equation involving a single unknown function, and then plug those solutions into the other equations, and deal with them.

!► Example 38.4: *Consider the system*

$$\begin{aligned} x' &= 5x + 2y \\ y' &= 4y \end{aligned} \quad .$$

The second equation, $y' = 4y$ *, is a simple linear and separable equation whose general solution you can readily find to be*

$$y(t) = c_1 e^{4t} \quad .$$

With this, the first equation in the system becomes

$$x' = 5x + 2c_1 e^{4t} \quad ,$$

another first-order linear equation that you should have little trouble solving (see Chapter 5). Its general solution is

$$x(t) = \frac{1}{2}c_1 e^{4t} + c_2 e^{5t} \quad .$$

So, the general solution to our system is the pair

$$x(t) = \frac{1}{2}c_1 e^{4t} + c_2 e^{5t} \qquad \text{and} \qquad y(t) = c_1 e^{4t} \quad .$$

For the most part, our interest will be in systems that are neither uncoupled nor weakly coupled.

38.2 A Few Illustrative Applications

Let us look at a few applications that naturally lead to standard 2×2 first-order systems.

A Falling Object

Way back in Section 1.2, we considered an object of mass m plummeting towards the ground under the influence of gravity. As we did there, let us set

$$t = \text{time (in seconds) since the object was dropped} \quad ,$$

$$y(t) = \text{vertical distance (in meters) between the object and the ground at time } t \quad ,$$

and

$$v(t) = \text{vertical velocity (in meters/second) of the object at time } t \quad .$$

We can view y and v as two unknown functions related by

$$\frac{dy}{dt} = v \quad .$$

Now, in developing a "better model" describing the fall (see the discussion starting on page 11), we took into account air resistance and obtained

$$\frac{dv}{dt} = -9.8 - \kappa v$$

where κ is a positive constant describing how strongly air resistance acts on the falling object. This gives us a system of two differential equations with two unknown functions,

$$\begin{aligned} y' &= v \\ v' &= -9.8 - \kappa v \end{aligned} \quad .$$

Fortunately, this is a very weakly coupled system whose second equation is a simple first-order equation involving only the function v. We've already solved it (in Example 4.7 on page 73), obtaining

$$v(t) = v_0 + c_1 e^{-\kappa t} \qquad \text{where} \qquad v_0 = -\frac{9.8}{\kappa} \quad .$$

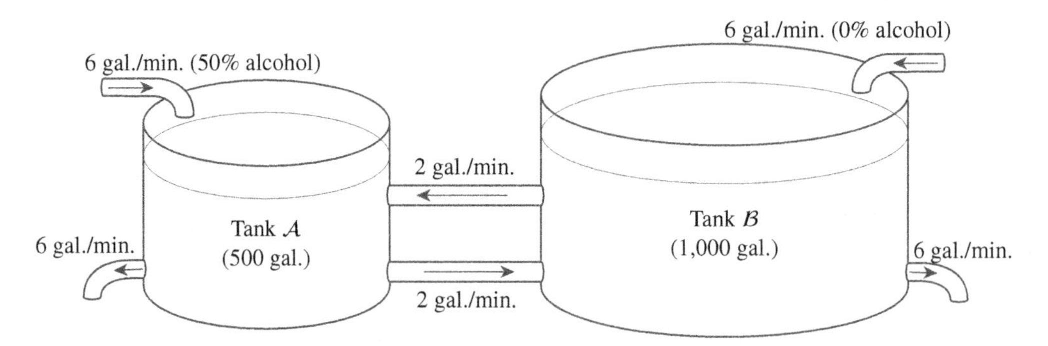

Figure 38.1: A simple system of two tanks containing water/alcohol mixtures.

Plugging this back into the first equation of the system yields

$$\frac{dy}{dt} = v_0 + c_1 e^{-\kappa t} \quad ,$$

which is easily integrated:

$$y(t) = \int \frac{dy}{dt}\,dt = \int \left[v_0 + c_1 e^{-\kappa t}\right] dt = v_0 t - \frac{c_1}{\kappa} e^{-\kappa t} + c_2 \quad .$$

So, the general solution to this system is the pair

$$y(t) = v_0 t - \frac{c_1}{\kappa} e^{-\kappa t} + c_2 \qquad \text{and} \qquad v(t) = v_0 + c_1 e^{-\kappa t} \quad .$$

Mixing Problems with Multiple Tanks

Let us expand, slightly, our discussion of "mixing" from Section 11.6 on page 211 by considering the situation illustrated in Figure 38.1. Here we have two tanks, $\mathcal{A}$ and $\mathcal{B}$. Each minute 6 gallons of a water/alcohol mix consisting of 50% alcohol is added to tank $\mathcal{A}$, and 6 gallons of the mix in tank $\mathcal{A}$ is drained out of the tank. At the same time, 6 gallons of pure water is added to tank $\mathcal{B}$ and 6 gallons of the mix in tank $\mathcal{B}$ is drained out of the tank. Meanwhile, the mix from each tank is pumped into the other tank at the rate of 2 gallons per minute.

Following standard conventions, we will let

$$t = \text{number of minutes since we started the mixing process} \quad ,$$

$$x = x(t) = \text{amount (in gallons) of alcohol in tank } \mathcal{A} \text{ at time } t$$

and

$$y = y(t) = \text{amount (in gallons) of alcohol in tank } \mathcal{B} \text{ at time } t \quad .$$

Let us assume that tank $\mathcal{A}$ initially contains 500 gallons of pure water, while tank $\mathcal{B}$ initially contains 1,000 gallons of an alcohol-water mix with 90 percent of that mix being alcohol. Note that the input and output flows for each tank cancel out, leaving the total amount of mix in each tank constant. So, our initial conditions are

$$x(0) = 0 \qquad \text{and} \qquad y(0) = \frac{90}{100} \times 1000 = 900 \quad ,$$

and the concentrations of the alcohol at time t in tanks $\mathcal{A}$ and $\mathcal{B}$ are, respectively,

$$\frac{x}{500} \qquad \text{and} \qquad \frac{y}{1000} \quad .$$

In this system we have six "flows" affecting the rate the amount of alcohol varies in each tank over time, each corresponding to one of the pipes in Figure 38.1. In each case the rate at which alcohol is flowing is simply the total flow rate of the mix in the pipe times the concentration of alcohol in that mix. Thus,

$$\begin{aligned} x' = \frac{dx}{dt} &= \text{change in the amount of alcohol in tank } \mathcal{A} \text{ per minute} \\ &= \text{rate alcohol is pumped into tank } \mathcal{A} \text{ from the outside} \\ &\quad + \text{rate alcohol is pumped into tank } \mathcal{A} \text{ from tank } \mathcal{B} \\ &\quad - \text{rate alcohol is pumped from tank } \mathcal{A} \text{ into tank } \mathcal{B} \\ &\quad - \text{rate alcohol is drained from tank } \mathcal{A} \\ &= \left(6 \times \frac{50}{100}\right) + \left(2 \times \frac{y}{1000}\right) - \left(2 \times \frac{x}{500}\right) - \left(6 \times \frac{x}{500}\right) \\ &= 3 + \frac{2}{1000}y - \frac{8}{500}x \end{aligned}$$

and

$$\begin{aligned} y' = \frac{dy}{dt} &= \text{change in the amount of alcohol in tank } \mathcal{B} \text{ per minute} \\ &= \text{rate alcohol is pumped into tank } \mathcal{B} \text{ from the outside} \\ &\quad + \text{rate alcohol is pumped into tank } \mathcal{B} \text{ from tank } \mathcal{A} \\ &\quad - \text{rate alcohol is pumped from tank } \mathcal{B} \text{ into tank } \mathcal{A} \\ &\quad - \text{rate alcohol is drained from tank } \mathcal{B} \\ &= (6 \times 0) + \left(2 \times \frac{x}{500}\right) - \left(2 \times \frac{y}{1000}\right) - \left(6 \times \frac{y}{1000}\right) \\ &= \frac{2}{500}x - \frac{8}{1000}y \quad . \end{aligned}$$

Thus, we have the system

$$\begin{aligned} x' &= -\frac{8}{500}x + \frac{2}{1000}y + 3 \\ y' &= \frac{2}{500}x - \frac{8}{1000}y \end{aligned} \quad . \tag{38.4}$$

Rabbits and Gerbils: A Competing Species Model

A Single Species Competing with Itself

Back in Chapter 11, we developed two models for population growth. Let us briefly review the "better" model, still assuming our population is a bunch of rabbits in an enclosed field. In that model

$$R(t) = \text{number of rabbits in the field after } t \text{ months}$$

and

$$R' = \frac{dR}{dt} = \text{change in the number of rabbits per month} = \beta R(t)$$

where β is the "net birth rate" (that is, 'the number of new rabbits normally born each month per rabbit' minus 'the fraction of the population that dies each month'). Under ideal conditions, β is a constant β_0, which can be determined from the natural reproductive rate for rabbits and the natural lifetime of a rabbit (see Section 11.2). But assuming β is constant led to a model that predicted an unrealistic number of rabbits in a short time. To take into account the decrease in the net birth rate that occurs when the number of rabbits increases, we added a correction term that decreases the net birth rate as the number of rabbits increases. Using the simplest practical correction term gave us

$$\beta = \beta_0 - \gamma R$$

where γ is some positive constant. Our differential equation for R, $R' = \beta R$, is then

$$R' = (\beta_0 - \gamma R)R$$

or, equivalently,

$$R' = \gamma(\kappa - R)R \qquad \text{where} \quad \kappa = \text{"the carrying capacity"} = \frac{\beta_0}{\gamma} \quad .$$

This is the "logistic equation", and we discussed it and its solutions in Section 11.4. In particular, we discovered that it has a stable equilibrium solution

$$R(t) = \kappa \qquad \text{for all} \quad t \quad .$$

Two Competing Species

Now suppose our field contains both rabbits and gerbils, all eating the same food and competing for the same holes in the ground. Then we should include an additional correction term to the net birth rate β to take into account the additional decrease in net birth rate for the rabbits that occurs as the number of gerbils increases, and the simplest way to add such a correction term is to simply subtract some positive constant times the number of gerbils. This gives us

$$\beta = \beta_0 - \gamma R - \alpha G$$

where α is some positive constant and

$$G = G(t) = \text{number of gerbils in the field at time } t \quad .$$

This means that our basic differential equation for the number of rabbits, $R' = \beta R$, becomes

$$R' = (\beta_0 - \gamma R - \alpha G)R \quad .$$

But, of course, there must be a similar differential equation describing the rate at which the gerbil population varies. So we actually have the system

$$\begin{aligned} R' &= (\beta_1 - \gamma_1 R - \alpha_1 G)R \\ G' &= (\beta_2 - \gamma_2 G - \alpha_2 R)G \end{aligned} \tag{38.5}$$

where β_1 and β_2 are the net birth rates per creature under ideal conditions for rabbits and gerbils, respectively, and the γ_k's and α_k's are positive constants that would probably have to be determined by experiment and measurement.

First-order system (38.5) is the classic *competing species model*. From our previous study of rabbit populations in Chapter 11, we know that

$$\beta_1 = \frac{5}{4} \quad .$$

Making somewhat uneducated but vaguely reasonable guesses for β_2 and the γ_k's and α_k's yields the system

$$\begin{aligned} R' &= \left(\frac{5}{4} - \frac{1}{160}R - \frac{3}{1000}G\right)R \\ G' &= \left(3 - \frac{3}{500}G - \frac{3}{160}R\right)G \end{aligned} \quad . \tag{38.6}$$

38.3 Converting Differential Equations to First-Order Systems

Converting a Single Second-Order Differential Equation

Let's start with an example.

!►Example 38.5: *Consider the second-order nonlinear differential equation*

$$y'' - 3y' + 8\cos(y) = 0 \quad ,$$

which we will rewrite as

$$y'' = 3y' - 8\cos(y) \quad .$$

Let us now introduce a second "unknown" function x *related to* y *by setting*

$$y' = x \quad .$$

Differentiating this and applying the last two equations yields

$$x' = y'' = 3y' - 8\cos(y) = 3x - 8\cos(y) \quad .$$

This (with the middle cut out), along with the definition of x *, then gives us the* 2×2 *standard first-order system*

$$\begin{aligned} x' &= 3x - 8\cos(y) \\ y' &= x \end{aligned} \quad .$$

Thus, we have converted the single second-order differential equation

$$y'' - 3y' + 8\cos(y) = 0$$

to the above 2×2 *standard first-order system. If we can solve this system, then we would also have the solution to the original single second-order differential equation. And even if we cannot truly solve the system, we may be able to use methods that we will later develop for systems to gain useful information about the desired solution* y *.*

In general, any second-order differential equation that can be written as

$$y'' = F\left(t, y, y'\right)$$

can be converted to a standard 2×2 first-order system by introducing a new unknown function x related to y via

$$y' = x \quad ,$$

and then observing that

$$x' = y'' = F\left(t, y, y'\right) = F(t, y, x) \quad .$$

This gives us the system

$$\begin{aligned} x' &= F(t, y, x) \\ y' &= x \end{aligned} \quad .$$

Let us cleverly call this system the *first-order system corresponding to the original differential equation.* If we can solve this system, then we automatically have the solution to our original differential equation. And even if we cannot easily solve the system, we will find that some of the tools we'll later develop for first-order systems will greatly aid us in analyzing the possible solutions, especially when the original equation is not linear.

Do note that, using this procedure, any second-order set of initial values

$$y(t_0) = a_1 \qquad \text{and} \qquad y'(t_0) = a_2$$

is converted to initial values

$$x(t_0) = a_2 \qquad \text{and} \qquad y(t_0) = a_1 \quad .$$

!▶Example 38.6: *Consider the second-order initial-value problem*

$$y'' - \sin(y)\,y' = 0 \qquad \text{with} \quad y(0) = 2 \quad \text{and} \quad y'(0) = 3 \quad .$$

Rewriting the differential equation as

$$y'' = \sin(y)\,y'$$

and letting $x = y'$ *(so that* $x' = y''$ *) lead to*

$$x' = y'' = \sin(y)\,y' = \sin(y)\,x = x\sin(y) \quad ,$$

giving us the standard first-order system

$$\begin{aligned} x' &= x\sin(y) \\ y' &= x \end{aligned}$$

with initial conditions

$$x(0) = y'(0) = 3 \qquad \text{and} \qquad y(0) = 2 \quad .$$

Converting Higher-Order Differential Equations and Systems

What we just did with a second-order differential equation can easily be extended to convert a third-order differential equation

$$y''' = F(t, y, y', y'')$$

to a standard first-order system with three differential equations. All we do is introduce two new functions x and z that are related to y and each other by

$$x = y' \qquad \text{and} \qquad z = x' = y'' \quad .$$

Then we have

$$z' = y''' = F(t, y, y', y'') = F(t, y, x, z) \quad ,$$

giving us the standard system

$$\begin{aligned} x' &= z \\ y' &= x \\ z' &= F(t, y, x, z) \end{aligned} \quad .$$

!▶Example 38.7: *Consider the third-order differential equation*

$$y''' - 3y'' + \sin(y)\,y' = 0 \quad ,$$

which we will rewrite as

$$y''' = 3y'' - \sin(y)\,y' \quad .$$

Introducing the functions x *and* z *related to each other and to* y *by*

$$x = y' \qquad \text{and} \qquad z = x' = y''$$

and observing that

$$z' = y''' = 3y'' - \sin(y)\,y' = 3z - \sin(y)\,x \quad ,$$

we see that the first-order system of three equations corresponding to our original third-order equation is

$$\begin{aligned} x' &= z \\ y' &= x \\ z' &= 3z - \sin(y)\,x \end{aligned} \quad .$$

Needless to say, the above can be extended to a simple process for converting any N^{th}-order differential equation that can be written as

$$y^{(N)} = F\left(t, y, y', y'', \ldots, y^{(N-1)}\right)$$

to an $N \times N$ standard first-order system. The biggest difficulty is that, if $N > 3$, you will probably want to start using subscripted functions. We'll see this later in Section 38.5.

It should also be clear that this process can be applied to convert higher-order systems of differential equations to standard first-order systems. The details are easy enough to figure out if you ever need to do so.

Why Do It?

It turns out that some of the useful methods we developed to deal with first-order differential equations can be modified to being useful methods for dealing with standard first-order systems. These methods include the use of slope fields (see Chapter 9) and the numerical methods discussed in Chapters 10 and 12. Consequently, the above process for converting a single second-order differential equation to a first-order system can be an important tool in analyzing solutions to second-order equations that cannot be easily handled by the methods previously discussed in this text. In particular, this conversion process is a particularly important element in the study of nonlinear second-order differential equations.

Another Example: The Pendulum

There is one more system that we will want for future use. This is the system describing the motion of the pendulum illustrated in Figure 38.2. This pendulum consists of a small weight of mass m attached at one end of a massless rigid rod, with the other end of the rod attached to a pivot so that the weight can swing around in a circle of radius L in a vertical plane. The forces acting on this pendulum are the downward force of gravity and, possibly, a frictional force from either friction at the pivot point or air resistance.

Let us describe the motion of this pendulum using the angle θ from the vertical downward line through the pivot to the rod, measured (in radians) in the counterclockwise direction. This means that $d\theta/dt$ is positive when the pendulum is moving counterclockwise, and is negative when the pendulum is moving clockwise.

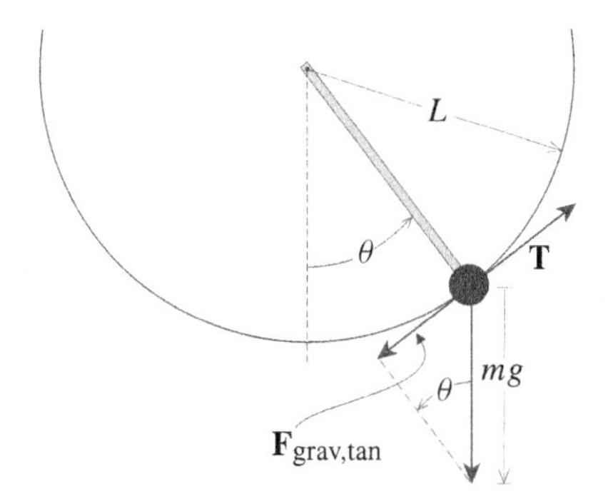

Figure 38.2: The pendulum system with a weight of mass m attached to a massless rod of length L swinging about a pivot point under the influence of gravity.

Since the motion is circular, and the 'positive' direction is counterclockwise, our interest is in the components of velocity, acceleration and force in the direction of vector $\mathbf{T}$ illustrated in Figure 38.2. This is the unit vector tangent to the circle of motion pointing in the counterclockwise direction from the current location of the weight. From basic physics and geometry, we know these tangential components of the weight's velocity and acceleration are, respectively,

$$v_{\text{tan}} = L\frac{d\theta}{dt} = L\theta' \qquad \text{and} \qquad a_{\text{tan}} = L\frac{d^2\theta}{dt^2} = L\theta'' \quad .$$

Using basic physics and trigonometry (and Figure 38.2), we see that the corresponding component of gravitational force is

$$F_{\text{grav,tan}} = -mg\sin(\theta) \quad .$$

For the frictional force, we'll use what we've used several times before,

$$F_{\text{fric,tan}} = -\gamma v_{\text{tan}} = -\gamma L\theta'$$

where γ is some nonnegative constant — either zero if this is an ideal pendulum having no friction, or a small to large positive value corresponding to a small to large frictional force acting on the pendulum.

Writing out the classic "$ma = F$" equation, we have

$$mL\theta'' = ma_{\text{tan}} = F_{\text{grav,tan}} + F_{\text{fric,tan}} = -mg\sin(\theta) - \gamma L\theta' \quad .$$

Cutting out the middle and dividing through by mL then gives us the slightly simpler equation

$$\theta'' = -\frac{g}{L}\sin(\theta) - \kappa\theta' \qquad \text{where} \quad \kappa = \frac{\gamma}{m} \quad . \tag{38.7}$$

Observe that, because of the $\sin(\theta)$ term, this second-order differential equation is *nonlinear*.

To convert this equation to a first-order system, let ω be the angular velocity, $^{d\theta}/_{dt}$. So

$$\theta' = \frac{d\theta}{dt} = \omega$$

and

$$\omega' = \theta'' = -\frac{g}{L}\sin(\theta) - \kappa\theta' = -\frac{g}{L}\sin(\theta) - \kappa\omega \quad ,$$

giving us the system

$$\begin{aligned} \theta' &= \omega \\ \omega' &= -\frac{g}{L}\sin(\theta) - \kappa\omega \end{aligned} \quad . \tag{38.8}$$

38.4 Using Laplace Transforms to Solve Systems

If this were a slightly more advanced treatment of systems (and if we had more space), we would discuss solving certain simple systems using "eigenvalues" and "eigenvectors". But that takes us beyond the scope of this text (in which an understanding of such terms is not assumed). So, as an alternative, let us observe that, if a system can be written as

$$\begin{aligned} x' &= ax + by + f(t) \\ y' &= cx + dy + g(t) \end{aligned}$$

where a, b, c and d are constants, and f and g are 'reasonable' functions on $(0, \infty)$, then we can find the solutions to this system by taking the Laplace transform of each equation, solving the resulting algebraic system for

$$X = \mathcal{L}[x] \qquad \text{and} \qquad Y = \mathcal{L}[y]$$

and then taking the inverse Laplace transform of each. If initial conditions $x(0)$ and $y(0)$ are known, then use them when taking the transforms of the derivatives. Otherwise, just view $x(0)$ and $y(0)$ as arbitrary constants, possibly renaming them x_0 and y_0 for convenience.

One example should suffice.

!►Example 38.8: *Consider solving the system*

$$\begin{aligned} x' &= x + 2y + 10e^{4t} \\ y' &= 2x + y \end{aligned}$$

with initial conditions

$$x(0) = 0 \qquad \text{and} \qquad y(0) = 2 \quad .$$

Taking the Laplace transform of the first equation:

$$\mathcal{L}\left[x'\right]\Big|_s = \mathcal{L}[x]|_s + 2\mathcal{L}[y]|_s + 10\mathcal{L}\left[e^{4t}\right]\Big|_s$$

$$\hookrightarrow \qquad sX(s) - x(0) = X(s) + 2Y(s) + \frac{10}{s-4}$$

$$\hookrightarrow \qquad sX(s) - 0 = X(s) + 2Y(s) + \frac{10}{s-4}$$

$$\hookrightarrow \qquad [s-1]X(s) - 2Y(s) = \frac{10}{s-4} \quad .$$

Doing the same with the second equation:

$$\mathcal{L}\left[y'\right]\Big|_s = 2\mathcal{L}[x]|_s + \mathcal{L}[y]|_s$$

$$\hookrightarrow \qquad sY(s) - y(0) = 2X(s) + Y(s)$$

$$\hookrightarrow \qquad sY(s) - 2 = 2X(s) + Y(s)$$

$$\hookrightarrow \qquad -2X(s) + [s-1]Y(s) = 2 \quad .$$

So the transforms $X = \mathcal{L}[x]$ *and* $Y = \mathcal{L}[y]$ *must satisfy the algebraic system*

$$\begin{aligned} [s-1]X(s) - 2Y(s) &= \frac{10}{s-4} \\ -2X(s) + [s-1]Y(s) &= 2 \end{aligned} \quad . \tag{38.9}$$

It is relatively easy to solve the above system. To find $X(s)$ *we can first add* $s-1$ *times the first equation to* 2 *times the second, and then apply a bit more algebra, to obtain*

$$\begin{aligned} [s-1]^2 X(s) - 4X(s) &= \frac{10(s-1)}{s-4} + 4 \\ \hookrightarrow \qquad \left(s^2 - 2s - 3\right) X(s) &= \frac{10(s-1) + 4(s-4)}{s-4} \\ \hookrightarrow \qquad (s+1)(s-3)X(s) &= \frac{14s - 26}{s-4} \\ \hookrightarrow \qquad X(s) &= \frac{14s - 26}{(s+1)(s-3)(s-4)} \quad . \end{aligned} \tag{38.10}$$

Similarly, to find $Y(s)$, *we can add* 2 *times the first equation in system (38.9) to* $s-1$ *times the second equation, obtaining*

$$-4Y(s) + [s-1]^2 Y(s) = \frac{20}{s-4} + 2(s-1) \quad .$$

Solving this for $Y(s)$ *then, eventually, yields*

$$Y(s) = \frac{2s^2 - 10s + 28}{(s+1)(s-3)(s-4)} \quad . \tag{38.11}$$

Finding the formulas for X *and* Y *was easy. Now we need to compute the formulas for* $x = \mathcal{L}^{-1}[X]$ *and* $y = \mathcal{L}^{-1}[Y]$ *from formulas (38.10) and (38.11) using the theory, techniques and tricks for finding inverse Laplace transforms developed in Part IV of this text:*

$$x(t) = \mathcal{L}^{-1}[X(s)]|_t = \mathcal{L}^{-1}\left[\frac{14s - 26}{(s+1)(s-3)(s-4)}\right]\Bigg|_t = \cdots = 6e^{4t} - 2e^{-t} - 4e^{3t}$$

and

$$y(t) = \mathcal{L}^{-1}[Y(s)]|_t = \mathcal{L}^{-1}\left[\frac{2s^2 - 10s + 28}{(s+1)(s-3)(s-4)}\right]\Bigg|_t = \cdots = 4e^{4t} + 2e^{-t} - 4e^{3t} \quad .$$

This is less easy, and, as you may have noted, the details are left to the reader "as a review of Laplace transforms" (and to save space here).

38.5 Existence, Uniqueness and General Solutions for Systems

Existence and Uniqueness of Solutions

In Section 3.3, two theorems were given — Theorems 3.1 and 3.2 — describing conditions ensuring the existence and uniqueness of solutions to first-order differential equations. Here are the "systems versions" of those theorems:

Theorem 38.1 (existence and uniqueness for general systems)
Consider an $N \times N$ standard first-order initial-value problem

$$\begin{aligned} x_1' &= f_1(t, x_1, x_2, \ldots, x_N) \\ x_2' &= f_2(t, x_1, x_2, \ldots, x_N) \\ &\vdots \\ x_N' &= f_N(t, x_1, x_2, \ldots, x_N) \end{aligned}$$

with

$$(x_1(t_0), x_2(t_0), \ldots, x_N(t_0)) = (a_1, a_2, \ldots, a_N) \quad .$$

Suppose each f_j and ${}^{\partial f_j}/_{\partial x_k}$ are continuous functions on an open region of the $TX_1X_2\cdots X_N$–space containing the point $(t_0, a_1, a_2, \ldots, a_N)$. This initial-value problem then has exactly one solution

$$(x_1, x_2, \ldots, x_N) = (x_1(t), x_2(t), \ldots, x_N(t))$$

over some open interval (α, β) containing t_0. Moreover, each x_k and its derivative are continuous over that interval.

Theorem 38.2
Consider an $N \times N$ standard first-order initial-value problem

$$\begin{aligned} x_1' &= f_1(t, x_1, x_2, \ldots, x_N) \\ x_2' &= f_2(t, x_1, x_2, \ldots, x_N) \\ &\vdots \\ x_N' &= f_N(t, x_1, x_2, \ldots, x_N) \end{aligned}$$

with

$$(x_1(t_0), x_2(t_0), \ldots, x_N(t_0)) = (a_1, a_2, \ldots, a_N)$$

over an interval (α, β) containing t_0, and with each f_j being a continuous function on the infinite strip

$$\mathcal{R} = \{(t, x_1, x_2, \ldots, x_N) : \alpha < t < \beta \text{ and } -\infty < x_k < \infty \text{ for } k = 1, 2, \ldots, N\} \quad .$$

Further suppose that, on $\mathcal{R}$, each ${}^{\partial f_j}/_{\partial x_k}$ is continuous and is a function of t only. Then the initial-value problem has exactly one solution

$$(x_1, x_2, \ldots, x_N) = (x_1(t), x_2(t), \ldots, x_N(t))$$

over (α, β). Moreover, each x_k and its derivative are continuous on that interval.

The proofs of the above two theorems are simply multidimensional versions of the proofs already discussed in Sections 3.4, 3.5 and 3.6 for Theorems 3.1 and 3.2.

General Solutions

In Exercise 38.4, you either have verified or will verify that, for any choice of constants c_1 and c_2, the pair

$$x(t) = c_1 e^{3t} + 2c_2 e^{-4t} \quad \text{and} \quad y(t) = c_1 e^{3t} - 5c_2 e^{-4t}$$

is a solution to the system

$$\begin{aligned} x' &= x + 2y \\ y' &= 5x - 2y \end{aligned} \quad .$$

But is the above a *general* solution, or could there be other solutions not described by the above?

Well, let us consider a more general situation; namely, let us consider "finding" the general solution to a 2×2 standard system

$$\begin{aligned} x' &= f(t, x, y) \\ y' &= g(t, x, y) \end{aligned}$$

over some interval (α, β) containing a point t_0. Let us further assume that the component functions are "reasonable". To be specific, we'll assume that f, g and their partial derivatives with respect to x and y are continuous on the entire XY–plane when $\alpha < t < \beta$. (This will allow us to apply Theorem 38.1 without getting bogged down in technicalities. Besides, all the 2×2 systems we'll see satisfy these conditions.)

Now suppose (as in the case we first started with) that we have found two formulas of t, c_1 and c_2

$$\hat{x}(t, c_1, c_2) \quad \text{and} \quad \hat{y}(t, c_1, c_2)$$

such that both of the following hold:

1. For each choice of real constants c_1 and c_2, the pair

$$x(t) = \hat{x}(t, c_1, c_2) \quad \text{and} \quad y(t) = \hat{y}(t, c_1, c_2)$$

is a solution to our standard system over (α, β).

2. For each ordered pair of real numbers A and B, there is a corresponding pair of real constants C_1 and C_2 such that

$$\hat{x}(t_0, C_1, C_2) = A \quad \text{and} \quad \hat{y}(t_0, C_1, C_2) = B \quad .$$

Now, let $x_1(t)$ and $y_1(t)$ be any particular solution pair to our system on (α, β), and let $A = x_1(t_0)$ and $B = y_1(t)$. Then, of course, x_1 and y_1, together, satisfy the initial-value problem

$$\begin{aligned} x' &= f(t, x, y) \\ y' &= g(t, x, y) \end{aligned}$$

with

$$(x(t_0), y(t_0)) = (A, B) \quad .$$

In addition, by our assumptions, we can also find constants C_1 and C_2 such that

$$x(t) = \hat{x}(t, C_1, C_2) \quad \text{and} \quad y(t) = \hat{y}(t, C_1, C_2)$$

is a solution to our standard system over (α, β) with

$$x(t_0) = \hat{x}(t_0, C_1, C_2) = A \quad \text{and} \quad y(t_0) = \hat{y}(t_0, C_1, C_2) = B \quad .$$

So, we have two solution pairs for the above initial-value problem, the pair

$$x_1(t) \qquad \text{and} \qquad y_1(t)$$

and the pair

$$x(t) = \hat{x}(t, C_1, C_2) \qquad \text{and} \qquad y(t) = \hat{y}(t, C_1, C_2) \quad .$$

But Theorem 38.1 tells us that there can only be one solution to the above initial-value problem. So our two solutions must be the same; that is, we must have

$$x_1(t) = \hat{x}(t, C_1, C_2) \qquad \text{and} \qquad y_1(t) = \hat{y}(t, C_1, C_2) \qquad \text{for} \quad \alpha < t < \beta \quad .$$

This means that every solution to our 2×2 standard system is given by

$$x(t) = \hat{x}(t, c_1, c_2) \qquad \text{and} \qquad y(t) = \hat{y}(t, c_1, c_2)$$

for some choice of constants c_1 and c_2. In other words, we have the following theorem:

Theorem 38.3

Let (α, β) be an interval containing a point t_0, and consider a 2×2 standard first-order system

$$\begin{aligned} x' &= f(t, x, y) \\ y' &= g(t, x, y) \end{aligned}$$

in which f, g and their partial derivatives with respect to x and y are continuous functions of all variables when t is in (α, β). Assume, further, that

$$\hat{x}(t, c_1, c_2) \qquad \text{and} \qquad \hat{y}(t, c_1, c_2)$$

are formulas of three variables satisfying both of the following:

1. *For each choice of real constants c_1 and c_2, the pair*

$$x(t) = \hat{x}(t, c_1, c_2) \qquad \text{and} \qquad y(t) = \hat{y}(t, c_1, c_2)$$

   satisfies the above first-order system over (α, β).

2. *For each ordered pair of real numbers (A, B), there is a corresponding ordered pair of real numbers (C_1, C_2) such that*

$$\hat{x}(t_0, C_1, C_2) = A \qquad \text{and} \qquad \hat{y}(t_0, C_1, C_2) = B \quad .$$

Then, treating c_1 and c_2 as arbitrary constants, the pair

$$\hat{x}(t, c_1, c_2) \quad \text{and} \quad \hat{y}(t, c_1, c_2)$$

is a general solution to the above system over (α, β).

!►Example 38.9: *As already noted, the pair*

$$x(t) = \underbrace{c_1 e^{3t} + 2c_2 e^{-4t}}_{\hat{x}(t, c_1, c_1)} \qquad \text{and} \qquad y(t) = \underbrace{c_1 e^{3t} - 5c_2 e^{-4t}}_{\hat{y}(t, c_1, c_1)}$$

satisfies

$$\begin{aligned} x' &= x + 2y \\ y' &= 5x - 2y \end{aligned}$$

over the entire real line, $(-\infty, \infty)$ for any choice of constants c_1 and c_2. Moreover, it is easily seen that this first-order system satisfies the conditions required in the last theorem with $(\alpha, \beta) = (-\infty, \infty)$. Hence, this theorem will assure us that the formula pair $(\hat{x}, \hat{y})$ is a general solution if, for each ordered pair of real numbers (A, B), there is a corresponding ordered pair of real numbers (C_1, C_2) such that

$$\hat{x}(t_0, C_1, C_2) = A \quad \text{and} \quad \hat{y}(t_0, C_1, C_2) = B \tag{38.12}$$

where t_0 is some fixed point, say, $t_0 = 0$. That is easily confirmed by simply solving this system for C_1 and C_2 in terms of A and B. Using our formulas for $\hat{x}$ and $\hat{y}$ with $t_0 = 0$, we see that system (38.12) reduces to the simple algebraic system

$$\begin{aligned} C_1 + 2C_2 &= A \\ C_1 - 5C_2 &= B \end{aligned} \quad .$$

Subtracting the second equation from the first yields

$$7C_2 = A - B \quad \rightarrowtail \quad C_2 = \frac{A - B}{7} \quad .$$

Plugging this back into the first equation gives

$$C_1 + 2\left[\frac{A - B}{7}\right] = A \quad \rightarrowtail \quad C_1 = \frac{5A + 2B}{7} \quad .$$

Thus, for any pair A and B, we can clearly find C_1 and C_2 such that system (38.12) holds when $t_0 = 0$. Theorem 38.3 then assures us that the pair

$$x(t) = c_1 e^{3t} + 2c_2 e^{-4t} \quad \text{and} \quad y(t) = c_1 e^{3t} - 5c_2 e^{-4t}$$

is, in fact, a general solution. That is, every solution is given by the above for some choice of c_1 and c_2.

The derivation leading to Theorem 38.3 can be easily redone starting with an $N \times N$ standard system, and using K arbitrary constants (where N and K are positive integers), giving us:

Theorem 38.4

Let (α, β) be an interval containing a point t_0, and consider an $N \times N$ standard first-order system

$$\begin{aligned} x_1' &= f_1(t, x_1, x_2, \ldots, x_N) \\ x_2' &= f_2(t, x_1, x_2, \ldots, x_N) \\ &\vdots \\ x_N' &= f_N(t, x_1, x_2, \ldots, x_N) \end{aligned}$$

in which each f_j and $\partial f_j / \partial x_k$ is a continuous function of all variables when t is in (α, β). Assume, further, that

$$\hat{x}_1(t, c_1, c_2, \ldots, c_K) \quad , \quad \hat{x}_2(t, c_1, c_2, \ldots, c_K) \quad , \quad \ldots \quad \text{and} \quad \hat{x}_N(t, c_1, c_2, \ldots, c_K)$$

are N formulas of $K + 1$ variables satisfying both of the following:

1. *For each choice of real constants* c_1 , c_2 , $\ldots$ *and* c_K , *the ordered set of* N *functions* x_1 , x_2 , $\ldots$ *and* x_N *given by*

$$x_n(t) = \hat{x}_n(t, c_1, c_2, \ldots, c_K) \qquad \text{for} \quad n = 1, 2, \ldots, N$$

satisfies the above first-order system over (α, β) .

2. *For each ordered* N*-tuple of real numbers* $(A_1, A_2, \ldots, A_N)$, *there is a corresponding* K*-tuple of real numbers* $(C_1, C_2, \ldots, C_K)$ *such that*

$$\hat{x}_n(t_0, C_1, C_2, \ldots, C_K) = A_n \qquad \text{for} \quad n = 1, 2, \ldots, N \quad .$$

Then, treating the c_k *'s as arbitrary constants,*

$$\hat{x}_1(t, c_1, c_2, \ldots, c_K) \quad , \quad \hat{x}_2(t, c_1, c_2, \ldots, c_K) \quad , \quad \ldots \quad \text{and} \quad \hat{x}_N(t, c_1, c_2, \ldots, c_K)$$

forms a general solution to the above system over (α, β) .

It should be noted that, using linear algebra, it can easily be shown that we must have $K \geq N$ for condition 2 in the last theorem to hold (with $K = N$ in most cases of interest).

38.6 Single N^{th}-Order Differential Equations

Several theorems regarding the existence and uniqueness of solutions to a single second- or higher-order differential equation were given near the end of Chapter 13. It is worth noting that all of these theorems can be derived from the results in the last section. To see this, let us consider a single N^{th}-order initial-value problem

$$y^{(N)} = F\left(t, y, y', \ldots, y^{(N-1)}\right) \tag{38.13a}$$

with

$$y(t_0) = a_1 \quad , \quad y'(t_0) = a_2 \quad , \quad \ldots \quad \text{and} \quad y^{(n-1)}(t_0) = a_N \quad . \tag{38.13b}$$

In this:

1. We are using t as the basic variable (so $y = y(t)$ and $y^{(k)} = {}^{d^k y}/_{dt^k}$).

2. t_0 , a_1 , a_1 , $\ldots$ and a_N are fixed real values.

3. $F(t, x_1, x_2, \ldots, x_N)$ is a function of $N + 1$ variables on some open region $\mathcal{R}$ of $N + 1$-dimensional space containing the point $(t_0, a_1, a_2, \ldots, a_N)$.

Our goal is to derive Theorem 13.3 on page 258. So assume, as in the theorem, that F and each ${}^{\partial F}/_{\partial x_k}$ is continuous on the region $\mathcal{R}$ (where $\mathcal{R}$ is as described in Theorem 13.3).

Let us now convert this single differential equation to a standard first-order system by introducing N new unknown functions x_1 , x_2 , $\ldots$ and x_N related to y via

$$x_1 = y \quad , \quad x_2 = y' \quad , \quad x_3 = y'' \quad , \quad \ldots \quad \text{and} \quad x_N = y^{(N-1)} \quad .$$

Then we have

$$\begin{aligned} {x_1}' &= y' = x_2 \quad , \\ {x_2}' &= \left(y'\right)' = y'' = x_3 \quad , \\ &\vdots \\ {x_{N-1}}' &= \left(y^{(N-2)}\right)' = y^{(N-1)} = x_N \end{aligned}$$

and, finally,

$$ {x_N}' = \left(y^{(N-1)}\right)' = y^{(N)} = F\left(t, y, y', \ldots, y^{(n-1)}\right) = F(t, x_1, x_2, x_3, \ldots, x_N) \quad . $$

Thus, our original initial-value problem can be rewritten as

$$\begin{aligned} {x_1}'(t) &= f_1(t, x_1, x_2, \ldots, x_N) \\ {x_2}'(t) &= f_2(t, x_1, x_2, \ldots, x_N) \\ &\vdots \\ {x_N}'(t) &= f_N(t, x_1, x_2, \ldots, x_N) \end{aligned}$$

with

$$ (x_1(t_0), x_2(t_0), \ldots, x_N(t_0)) = (a_1, a_2, \ldots, a_N) \quad , $$

where

$$\begin{aligned} f_1(t, x_1, x_2, \ldots, x_N) &= x_2 \quad , \\ f_2(t, x_1, x_2, \ldots, x_N) &= x_3 \quad , \\ &\vdots \end{aligned}$$

and

$$ f_{N-1}(t, x_1, x_2, \ldots, x_N) = x_N \quad , $$

while

$$ f_N(t, x_1, x_2, \ldots, x_N) = F(t, x_1, x_2, x_3, \ldots, x_N) \quad . $$

It is almost trivial to verify that each f_j and each $\partial f_j/\partial x_k$ with $j \neq N$ is a continuous function on the region $\mathcal{R}$. Moreover, the assumptions made on F ensure that f_N and the $\partial f_N/\partial x_k$'s are also continuous functions on $\mathcal{R}$. This means Theorem 38.1 on page 779 applies, telling us that

1. the above $N \times N$ initial-value problem has exactly one solution $(x_1, x_2, \ldots, x_N)$ over some open interval (α, β) containing t_0,

and, moreover,

2. each x_k and its derivative are continuous over the interval (α, β).

But, since

$$ \begin{bmatrix} x_1 \\ x_2 \\ \vdots \\ x_N \end{bmatrix} = \begin{bmatrix} y \\ y' \\ \vdots \\ y^{(N-1)} \end{bmatrix} \quad \text{and} \quad \begin{bmatrix} {x_1}' \\ {x_2}' \\ \vdots \\ {x_N}' \end{bmatrix} = \begin{bmatrix} y' \\ y'' \\ \vdots \\ y^{(N)} \end{bmatrix} \quad , $$

the above results concerning the x_k's tell us that there is exactly one y and that it and its derivatives up to order N are continuous over the interval (α, β). In other words, we've verified the following theorem:

Theorem 38.5 (existence and uniqueness for N^{th}-order initial-value problems)
Consider an Nth-order initial-value problem

$$y^{(N)} = F\left(t, y, y', \ldots, y^{(N-1)}\right)$$

with

$$y(t_0) = a_1 \quad , \quad y'(t_0) = a_2 \quad , \quad \ldots \quad \text{and} \quad y^{(N-1)}(t_0) = a_N$$

in which $F = F(t, x_1, x_2, \ldots, x_N)$ and the corresponding partial derivatives

$$\frac{\partial F}{\partial x_1} \quad , \quad \frac{\partial F}{\partial x_2} \quad , \quad \ldots \quad \text{and} \quad \frac{\partial F}{\partial x_N}$$

are all continuous functions in some open region containing the point $(t_0, a_1, a_2, \ldots, a_N)$. This initial-value problem then has exactly one solution y over some open interval (α, β) containing t_0. Moreover, this solution and its derivatives up to order N are continuous over that interval.

If you now go back and compare the above theorem with Theorem 13.3 on page 258, you will find that (except for cosmetic differences in notation) the two theorems are the same.

While we are at it, we should note that the above certainly applies if our differential equation is a homogeneous N^{th}-order linear differential equation

$$a_0 y^{(N)} + a_1 y^{(N-1)} + \cdots + a_{N-2} y'' + a_{N-1} y' + a_N y = 0$$

on an interval (α, β) on which the a_k's are all continuous functions with a_0 never being zero. In fact, it's easily verified that Theorem 38.2 applies after rewriting the differential equation as

$$y^{(N)} = \frac{-1}{a_0}\left[a_1 y^{(N-1)} + \cdots + a_{N-2} y'' + a_{N-1} y' + a_N y\right] \quad .$$

Moreover, if $\{y_1, y_2, \ldots, y_K\}$ is a linearly independent set of solutions to this differential equation on (α, β), then we can use Theorem 38.4 and the principle of superposition to show that

$$y(t) = c_1 y_1(t) + c_2 y_2(t) + \cdots + c_K y_K(t)$$

is a general solution if and only if, for each N-tuple of real numbers $(A_1, A_2, \ldots, A_N)$, there is a corresponding K-tuple of real numbers $(C_1, C_2, \ldots, C_K)$ such that, for $n = 1, 2, \ldots, N$,

$$C_1 {y_1}^{(n-1)}(t_0) + C_2 {y_2}^{(n-1)}(t_0) + \cdots + C_K {y_K}^{(n-1)}(t_0) = A_n$$

where t_0 is some fixed point in (α, β). Then, using Theorem 38.4 and basic results from linear algebra, you can, eventually, verify all the results regarding general solutions to N^{th}-order homogeneous equations described in Theorem 15.5 on page 290 and all the results on Wronskians in Theorem 15.6 on page 292.

Additional Exercises

38.1. Determine which of the following function pairs are solutions to the first-order system

$$x' = 2y$$
$$y' = 1 - 2x \quad .$$

a. $x(t) = \sin(2t) + \frac{1}{2}$ and $y(t) = \cos(2t)$

b. $x(t) = e^{2t} - 1$ and $y(t) = e^{2t}$

c. $x(t) = 3\cos(2t) + \frac{1}{2}$ and $y(t) = -3\sin(2t)$

38.2. Determine which of the following function pairs are solutions to the first-order system

$$x' = 4x - 3y$$
$$y' = 6x - 7y \quad .$$

a. $x(t) = 6e^{3t}$ and $y(t) = 2e^{3t}$

b. $x(t) = 3e^{2t} - e^{-5t}$ and $y(t) = 2e^{2t} - 3e^{-5t}$

c. $x(t) = 3e^{2t} + e^{-5t}$ and $y(t) = 2e^{2t} - 3e^{-5t}$

38.3. Determine which of the following function pairs are solutions on $(0, \infty)$ to the first-order system

$$tx' + 2x = 15y$$
$$ty' = x \quad .$$

a. $x(t) = 3t^3$ and $y(t) = -t^3$

b. $x(t) = 3t^3$ and $y(t) = t^3$

c. $x(t) = -5t^{-5}$ and $y(t) = t^{-5}$

38.4 a. Verify that, for any choice of constants c_1 and c_2,

$$x(t) = c_1e^{3t} + 2c_2e^{-4t} \quad \text{and} \quad y(t) = c_1e^{3t} - 5c_2e^{-4t}$$

satisfies the system

$$x' = x + 2y$$
$$y' = 5x - 2y \quad .$$

b. Then find the solution to this system that satisfies

$$x(0) = 7 \quad \text{and} \quad y(0) = -7 \quad .$$

38.5 a. Verify that, for any choice of constants c_1 and c_2,

$$x(t) = c_1e^{9t} - c_2e^{-3t} \quad \text{and} \quad y(t) = c_1e^{9t} + 2c_2e^{-3t}$$

satisfies the system

$$x' = 5x + 4y$$
$$y' = 8x + y \quad .$$

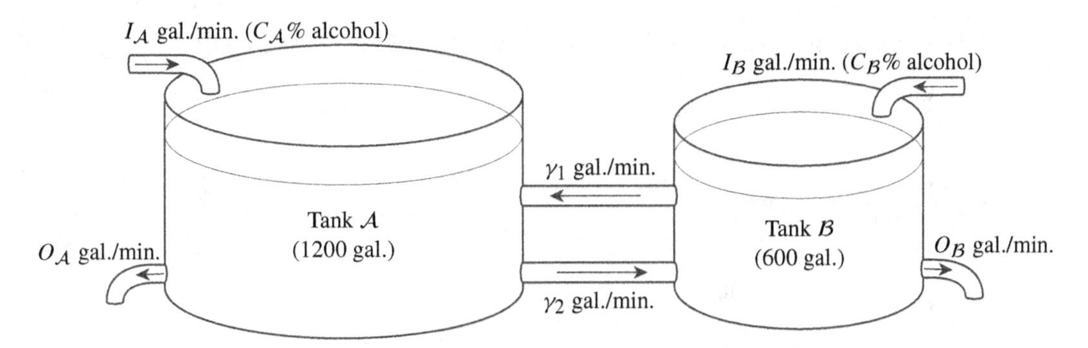

Figure 38.3: A simple system of two tanks containing water/alcohol mixtures for Exercise 38.8 .

b. *Then find the solution to this system that satisfies*

$$x(0) = 0 \quad \text{and} \quad y(0) = 9 \quad .$$

38.6 a. *Verify that, for any choice of constants c_1 and c_2,*

$$x(t) = c_1 e^{-2t} + 2c_2 e^{5t} \quad \text{and} \quad y(t) = -3c_1 e^{-2t} + c_2 e^{5t}$$

satisfies the system

$$\begin{aligned} x' &= 4x + 2y \\ y' &= 3x - y \end{aligned} \quad .$$

b. *Then find the solution to this system that satisfies*

$$x(0) = 0 \quad \text{and} \quad y(0) = -21 \quad .$$

38.7. *Solve each of the following weakly coupled systems:*

a. $$\begin{aligned} x'' + x &= 0 \\ y' &= x \end{aligned}$$

b. $$\begin{aligned} x' &= 2yx \\ ty' &= y \end{aligned}$$

c. $$\begin{aligned} x' + 2x &= 10z \\ zy' + 5zy &= 15x \\ z' - 3z &= 0 \end{aligned}$$

38.8. *Consider the tank system illustrated in Figure 38.3. Let x and y be, respectively, the amount of alcohol (measured in gallons) in tanks $\mathcal{A}$ and $\mathcal{B}$ at time t (measured in minutes), and find the standard first-order system describing how x and y vary over time when:*

a. $I_A = 5$, $C_A = 0$, $O_A = 5$, $I_B = 3$, $C_B = 100$, $O_B = 3$, $\gamma_1 = 1$ and $\gamma_2 = 1$

b. $I_A = 8$, $C_A = 25$, $O_A = 6$, $I_B = 2$, $C_B = 50$, $O_B = 4$, $\gamma_1 = 1$ and $\gamma_2 = 3$

38.9. *Rewrite each of the following differential equations as a standard first-order system.*

a. $y'' + 4y' + 2y = 0$

b. $y'' - 8t^2 y' - 32y = \sin(t)$

c. $y'' = 4 - y^2$

d. $t^2 y'' - 5ty' + 8y = 0$

e. $t^2 y'' - ty' + 10y = 0$

f. $y'' = 4t^2 - \sin(y')\, y$

g. $y''' + 2y'' - 3y' - 4y = 0$

h. $y''' + y'\left(t^2 + y^2\right) = 0$

38.10. Solve each of the following initial-value problems using the Laplace transform.

a. $\begin{aligned} x' &= x + 2y \\ y' &= 5x - 2y \end{aligned}$ with $\begin{aligned} x(0) &= 1 \\ y(0) &= 15 \end{aligned}$

b. $\begin{aligned} x' &= 2y \\ y' &= 2x \end{aligned}$ with $\begin{aligned} x(0) &= x_0 \\ y(0) &= y_0 \end{aligned}$

c. $\begin{aligned} x' &= 2y \\ y' &= -2x \end{aligned}$ with $\begin{aligned} x(0) &= x_0 \\ y(0) &= y_0 \end{aligned}$

d. $\begin{aligned} x' &= -2y \\ y' &= 8x \end{aligned}$ with $\begin{aligned} x(0) &= x_0 \\ y(0) &= y_0 \end{aligned}$

e. $\begin{aligned} x' &= 4x - 13y \\ y' &= x \end{aligned}$ with $\begin{aligned} x(0) &= 2 \\ y(0) &= 1 \end{aligned}$

f. $\begin{aligned} x' &= 3x + 2y \\ y' &= -2x + 3y \end{aligned}$ with $\begin{aligned} x(0) &= x_0 = a_1 \\ y(0) &= y_0 = a_2 \end{aligned}$

g. $\begin{aligned} x' &= 8x + 2y - 17 \\ y' &= 4x + y - 13 \end{aligned}$ with $\begin{aligned} x(0) &= 0 \\ y(0) &= 0 \end{aligned}$

h. $\begin{aligned} x' &= 8x + 2y + 7e^{2t} \\ y' &= 4x + y - 7e^{2t} \end{aligned}$ with $\begin{aligned} x(0) &= -1 \\ y(0) &= 1 \end{aligned}$

i. $\begin{aligned} x' &= 4x + 3y - 6e^{3t} \\ y' &= x + 6y + 2e^{3t} \end{aligned}$ with $\begin{aligned} x(0) &= 4 \\ y(0) &= 0 \end{aligned}$

j. $\begin{aligned} x' &= -y \\ y' &= 4x + 24t \end{aligned}$ with $\begin{aligned} x(0) &= 0 \\ y(0) &= 0 \end{aligned}$

k. $\begin{aligned} x' &= 4x - 13y \\ y' &= x + 19\cos(4t) - 13\sin(4t) \end{aligned}$ with $\begin{aligned} x(0) &= 13 \\ y(0) &= 0 \end{aligned}$

l. $\begin{aligned} x' &= 4x + 3y + 5\,\text{step}_2(t) \\ y' &= x + 6y + 17\,\text{step}_2(t) \end{aligned}$ with $\begin{aligned} x(0) &= 0 \\ y(0) &= 0 \end{aligned}$

38.11. Using Theorem 38.3 and results from Exercise 38.5, verify that

$$x(t) = c_1 e^{9t} - c_2 e^{-3t} \quad \text{and} \quad y(t) = c_1 e^{9t} + 2c_2 e^{-3t}$$

is a general solution to

$$\begin{aligned} x' &= 5x + 4y \\ y' &= 8x + y \end{aligned} \quad .$$

38.12. Using Theorem 38.3 and results from Exercise 38.6, verify that

$$x(t) = c_1 e^{-2t} + 2c_2 e^{5t} \quad \text{and} \quad y(t) = -3c_1 e^{-2t} + c_2 e^{5t}$$

is a general solution to

$$\begin{aligned} x' &= 4x + 2y \\ y' &= 3x - y \end{aligned} \quad .$$

39

Critical Points, Direction Fields and Trajectories

In this chapter, we will restrict our attention to certain standard first-order systems consisting of only two differential equations. Admittedly, just about everything we will discuss can be extended to larger systems, but, by staying with the smaller systems, we will be able to keep the notation and discussion more simple, and develop the main concepts more clearly. This will largely be because it is much easier to draw pictures with solutions consisting of just two real-valued functions, $x = x(t)$ and $y = y(t)$. And that will be important. Our goal is to find a way to "graphically represent" solutions in a useful manner. What we will develop, called "phase plane analysis", turns out to be a fundamental tool in the study of systems, especially for those many systems we cannot explicitly solve.

39.1 The Systems of Interest and Some Basic Notation

The Systems of Interest

As just noted above, the focus of this chapter concerns the standard 2×2 first-order system

$$\begin{aligned} x' &= f(t, x, y) \\ y' &= g(t, x, y) \end{aligned} \tag{39.1}$$

(though we may replace the symbols x and y with other symbols in some of the applications). In addition, we will often also require that our system be "regular" and/or "autonomous". So let us define these terms:

- System (39.1) is "regular" if the component functions, f and g, are "reasonably nice". More precisely, the system is *regular* if f, g and all their first partial derivatives exist and are continuous for all real values of their variables. Most of the systems we have seen are regular.
- System (39.1) is *autonomous* if and only if no component function explicitly depends on t. In this case, we will rewrite system (39.1) as

$$\begin{aligned} x' &= f(x, y) \\ y' &= g(x, y) \end{aligned} \quad . \tag{39.2}$$

  Autonomous systems naturally arise in applications. If you check, all the first-order systems derived in the previous chapter from applications are autonomous.

Some Convenient Notation

Remember that a solution over an interval (α, β) to a system

$$\begin{aligned} x' &= f(t, x, y) \\ y' &= g(t, x, y) \end{aligned} \tag{39.3}$$

consists of a pair of functions $x = x(t)$ and $y = y(t)$ that, together, satisfy the system for all t in the interval (α, β). In the last chapter, we occasionally wrote this pair explicitly as an ordered pair,

$$(x, y) = (x(t), y(t)) \quad .$$

We will find it even more convenient to do so in this chapter, especially when discussing generic systems. Let us also observe that we can write the above system, as well as the initial conditions

$$x(t_0) = x_0 \quad \text{and} \quad y(t_0) = y_0 \quad ,$$

in "vector form",

$$\begin{bmatrix} x' \\ y' \end{bmatrix} = \begin{bmatrix} f(t, x, y) \\ g(t, x, y) \end{bmatrix} \quad \text{with} \quad \begin{bmatrix} x(t_0) \\ y(t_0) \end{bmatrix} = \begin{bmatrix} x_0 \\ y_0 \end{bmatrix} \quad .$$

When convenient, we will further reduce the above line to

$$\mathbf{x}' = \mathbf{F}(t, \mathbf{x}) \quad \text{with} \quad \mathbf{x}(t_0) = \mathbf{x}_0$$

with the understanding that

$$\mathbf{x} = \mathbf{x}(t) = \begin{bmatrix} x(t) \\ y(t) \end{bmatrix} \quad , \quad \mathbf{x}' = \begin{bmatrix} x' \\ y' \end{bmatrix} \quad , \quad \mathbf{x}_0 = \begin{bmatrix} x_0 \\ y_0 \end{bmatrix}$$

and

$$\mathbf{F}(t, \mathbf{x}) = \begin{bmatrix} f(t, x, y) \\ g(t, x, y) \end{bmatrix} \quad .$$

Of course, if the system is autonomous, we will simply denote the system by $\mathbf{x}' = \mathbf{F}(\mathbf{x})$ with the understanding that

$$\mathbf{F}(\mathbf{x}) = \begin{bmatrix} f(x, y) \\ g(x, y) \end{bmatrix} \quad .$$

Along these same lines, we will let $\mathbf{0}$ denote the "zero vector"

$$\mathbf{0} = \begin{bmatrix} 0 \\ 0 \end{bmatrix} \quad .$$

!▶Example 39.1: *Consider the system*

$$\begin{aligned} x' &= x + 2y \\ y' &= 5x - 2y \end{aligned}$$

along with the initial conditions

$$x(0) = 0 \quad \text{and} \quad y(0) = 1 \quad .$$

In Example 38.3 we saw that a solution to this initial-value problem is (x, y) where

$$x = x(t) = \frac{2}{7}e^{3t} - \frac{2}{7}e^{-4t} \quad \text{and} \quad y = y(t) = \frac{2}{7}e^{3t} + \frac{5}{7}e^{-4t} \quad .$$

In matrix/vector form, this initial-value problem can be written as either

$$\begin{bmatrix} x' \\ y' \end{bmatrix} = \begin{bmatrix} x + 2y \\ 5x - 2y \end{bmatrix} \quad \textit{with} \quad \begin{bmatrix} x(0) \\ y(0) \end{bmatrix} = \begin{bmatrix} 0 \\ 1 \end{bmatrix}$$

or even as

$$\mathbf{x}' = \mathbf{F}(\mathbf{x}) \quad \textit{with} \quad \mathbf{x}(0) = \mathbf{x}_0$$

where

$$\mathbf{x} = \begin{bmatrix} x \\ y \end{bmatrix} \quad , \quad \mathbf{F}(\mathbf{x}) = \begin{bmatrix} x + 2y \\ 5x - 2y \end{bmatrix} \quad \textit{and} \quad \mathbf{x}_0 = \begin{bmatrix} 0 \\ 1 \end{bmatrix} \quad .$$

In the future, we will use, or not use, vector notation as is convenient.

39.2 Constant/Equilibrium Solutions

A solution (x, y) to a system $\mathbf{x}' = \mathbf{F}(t, \mathbf{x})$ is a *constant solution* if both x and y are constant functions; that is, for some pair of real numbers x_0 and y_0,

$$(x(t), y(t)) = (x_0, y_0) \qquad \text{for all} \quad t \quad .$$

But remember,

$$(x(t), y(t)) = (x_0, y_0) \quad \text{for all} \quad t \quad \iff \quad \big(x'(t), y'(t)\big) = (0, 0) \quad \text{for all} \quad t \quad .$$

Thus, (x, y) is a constant solution if and only if

$$\mathbf{x}'(t) = \begin{bmatrix} x'(t) \\ y'(t) \end{bmatrix} = \begin{bmatrix} 0 \\ 0 \end{bmatrix} = \mathbf{0} \qquad \text{for all} \quad t \quad .$$

Combining this with $\mathbf{x}' = \mathbf{F}(t, \mathbf{x})$, we get

$$\mathbf{0} = \mathbf{x}' = \mathbf{F}(t, \mathbf{x}) \quad .$$

Clearly then, for any pair of constants (x_0, y_0), we have that $(x(t), y(t)) = (x_0, y_0)$ is a constant solution of our system if and only if

$$\mathbf{0} = \mathbf{F}(t, \mathbf{x}_0) \qquad \text{for all} \quad t \qquad \text{where} \qquad \mathbf{x}_0 = \begin{bmatrix} x_0 \\ y_0 \end{bmatrix} \quad .$$

From this, you can find the constant solutions for a given system. (Do remember that we are insisting our solutions be real valued. So the constants must be real numbers.)

Constant solutions are especially important for autonomous systems. And, when the system is autonomous, it is traditional to call any constant solution an *equilibrium solution.* We will follow tradition.

!►Example 39.2: *Let's try to find every equilibrium solution for the autonomous system*

$$\begin{aligned} x' &= x(y^2 - 9) \\ y' &= (x - 1)(y^2 + 1) \end{aligned} \quad .$$

The constant/equilibrium solutions are all obtained by setting x' and y' both equal to 0, and then solving the resulting algebraic system,

$$\begin{aligned} 0 &= x(y^2 - 9) \\ 0 &= (x-1)(y^2+1) \end{aligned} \quad . \tag{39.4}$$

Consider the first equation, first:

$$0 = x(y^2 - 9)$$

$$\hookrightarrow \qquad x = 0 \quad \text{or} \quad y^2 = 9$$

$$\hookrightarrow \qquad x = 0 \quad \text{or} \quad y = 3 \quad \text{or} \quad y = -3 \quad .$$

If $x = 0$, then the second equation in system (39.4) reduces to

$$0 = (0-1)(y^2+1) \quad ,$$

which means that $y^2 = -1$, and, hence, $y = \pm i$. But these are not real numbers, as required. So we do not have an equilibrium solution

$$(x(t), y(t)) = (x_0, y_0) \qquad \text{with} \quad x_0 = 0 \quad .$$

On the other hand, if the first equation in system (39.4) is satisfied because $y = 3$, then the second equation in that system reduces to

$$0 = (x-1)\left(3^2+1\right) = (x-1)\cdot 10 \quad ,$$

telling us that $x = 1$. Thus, one equilibrium solution for our system of differential equations is

$$(x(t), y(t)) = (1, 3) \quad .$$

Finally, if the first equation in system (39.4) holds because $y = -3$, then the second equation in that system becomes

$$0 = (x-1)\cdot 10 \quad .$$

Hence, again, $x = 1$, and the corresponding equilibrium solution is

$$(x(t), y(t)) = (1, -3) \qquad \text{for all} \quad t \quad .$$

In summary, then, our system of differential equations has two equilibrium solutions,

$$(x(t), y(t)) = (1, 3) \qquad \text{and} \qquad (x(t), y(t)) = (1, -3) \quad .$$

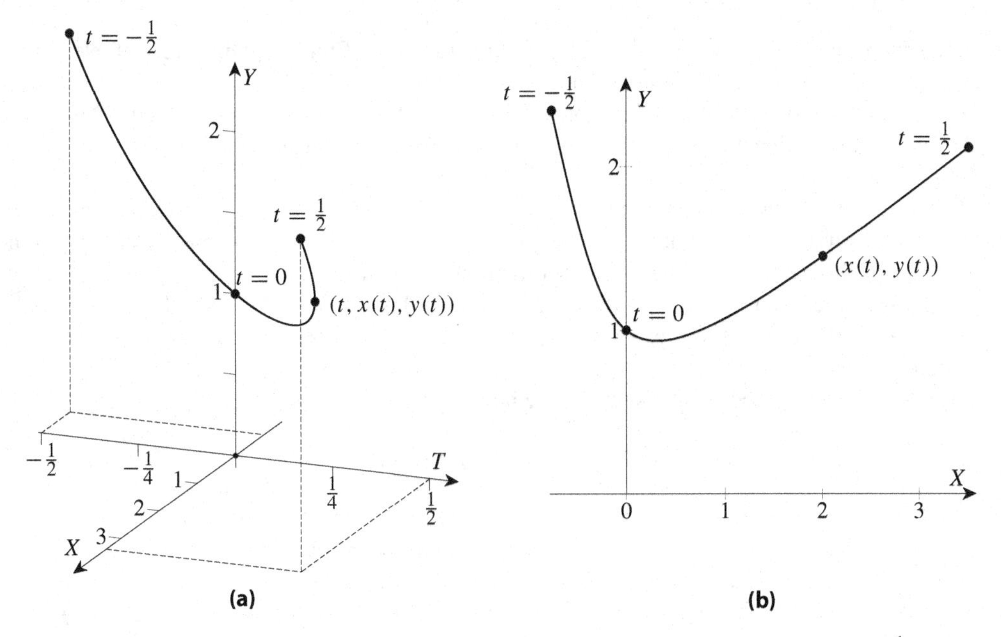

Figure 39.1: Two graphical representations of the solution from Example 39.3 with $|t| \leq 1/2$: **(a)** The actual graph, and **(b)** the curve traced out by $(x(t), y(t))$ in the XY–plane.

39.3 "Graphing" Standard Systems
True Graphs and Trajectories

Let us briefly discuss two ways of graphically representing solutions to standard first-order systems, starting with a simple example.

!►Example 39.3: *Consider "graphically representing" the solution to the initial-value problem*

$$\begin{bmatrix} x' \\ y' \end{bmatrix} = \begin{bmatrix} x + 2y \\ 5x - 2y \end{bmatrix} \quad \text{with} \quad \begin{bmatrix} x(0) \\ y(0) \end{bmatrix} = \begin{bmatrix} 0 \\ 1 \end{bmatrix} \quad .$$

From Example 38.3 we know that a solution to this initial-value problem is (x, y) *with*

$$x = x(t) = \frac{2}{7}e^{3t} - \frac{2}{7}e^{-4t} \quad \text{and} \quad y = y(t) = \frac{2}{7}e^{3t} + \frac{5}{7}e^{-4t} \quad .$$

To construct the actual graph of this solution, we need to plot each point $(t, x(t), y(t))$ *in* TXY*–space, using the above formulas for* $x(t)$ *and* $y(t)$*. This results in a curve in* TXY*–space. Part of this curve is sketched in Figure 39.1a.*

As an alternative to constructing the graph, we can sketch the curve in the XY*–plane that is traced out by* $(x(t), y(t))$ *as* t *varies. That is what is sketched in Figure 39.1b.*

Take a look at the two figures. Both were easily done using a computer math package.

In general, the graph of a solution to a standard 2×2 first-order system requires a coordinate system with 3 axes. Sketching such a graph is do-able, as in the above example, especially if you have a decent computer math package. Then you can even rotate the image to see the graph from different views. Unfortunately, when you are limited to just one view, as in Figure 39.1a, it may be somewhat difficult to properly interpret the figure. Because of this, we will rarely, if ever again, actually attempt to sketch true graphs of our solutions.

On the other hand, we will find the approach illustrated by Figure 39.1b quite useful. Moreover, the sketches that we will generate for 2×2 systems can give us insight as to the behavior of solutions to larger systems.

Observe that the curve in Figure 39.1b has a natural "direction of travel" corresponding to the way the curve is traced out as t increases. If you start at the point where $t = -1/2$ and travel the curve "in the positive direction" (that is, in the direction in which $(x(t), y(t))$ travels along the curve as t increases), then you would pass through the point where $t = 0$ and then through the point where $t = +1/2$. That makes this an *oriented* curve. We will call this oriented curve a *trajectory* for our system.

Just to be a bit more complete, suppose we have a solution

$$(x, y) = (x(t), y(t))$$

to a standard 2×2 first-order system $\mathbf{x}' = \mathbf{F}(t, \mathbf{x})$. For each t, we will view $(x(t), y(t))$ as a point on the XY–plane, and we will refer to the oriented curve traced out by this point as t increases as this solution's *trajectory*, with the orientation being the direction of travel along the curve as t increases. We will also refer to this oriented curve as a *trajectory for the system* $\mathbf{x}' = \mathbf{F}(t, \mathbf{x})$.

Velocity Vectors and the Direction of Travel

Do recall that

$$\mathbf{v}(t) = \mathbf{x}'(t) = \begin{bmatrix} x'(t) \\ y'(t) \end{bmatrix}$$

is the *velocity vector* $\mathbf{v}$ at time t of an object whose position at time t is $(x(t), y(t)$.[1] As you should recall from elementary multivariable calculus, this vector $\mathbf{v}(t)$ is an 'arrow' pointing in the direction the object is traveling at time t. So, if we pick some value t_0 for t and draw $\mathbf{v}^0 = \mathbf{v}(t_0)$ at position $(x(t_0), y(t_0))$, then $\mathbf{v}^0$ will be tangent to and pointing in the direction of travel of the trajectory of the object at $(x(t_0), y(t_0))$, thus giving us some idea of what that trajectory looks like near that point. And if $(x(t), y(t)$ is known to be a solution to

$$\mathbf{x}'(t) = \mathbf{F}(t, \mathbf{x}) \quad ,$$

then we can actually compute $\mathbf{v}^0 = \mathbf{x}'(t_0)$ for each choice of t_0 and $\mathbf{x}^0$ without even solving the system for $\mathbf{x}(t)$. All we need to do is to compute $\mathbf{F}(t_0, \mathbf{x}_0)$.

!►Example 39.4: *Consider the trajectories of two objects both of whose positions (x, y) at time t satisfy the nonautonomous system*

$$\begin{bmatrix} x' \\ y' \end{bmatrix} = \begin{bmatrix} t(x + y) \\ y - tx \end{bmatrix} \quad .$$

Assume the first object passes through the point $(x, y) = (1, 2)$ at time $t = 0$. At that time, its velocity is

$$\mathbf{v} = \begin{bmatrix} x' \\ y' \end{bmatrix} = \begin{bmatrix} 0(1 + 2) \\ 2 - 0 \cdot 1 \end{bmatrix} = \begin{bmatrix} 0 \\ 2 \end{bmatrix} \quad .$$

[1] This assumes we are using a Cartesian coordinate system. If the coordinate system is, say, polar, then the velocity is not so simply computed.

However, if the other object also passes through the point $(x, y) = (1, 2)$, *but at time* $t = 2$, *then its velocity at that time is*

$$\mathbf{v} = \begin{bmatrix} x' \\ y' \end{bmatrix} = \begin{bmatrix} 2(1+2) \\ 2 - 2 \cdot 1 \end{bmatrix} = \begin{bmatrix} 6 \\ 0 \end{bmatrix} .$$

Note the directions of travel for each of these objects as they pass through the point $(x, y) = (1, 2)$ *: The first is moving parallel to the* Y*–axis, while the second in moving parallel to the* X*–axis.*

As the last example illustrates, the direction of travel for a solution's trajectory through a given point may depend on "when" it passes through the point. However, this is only for *non*autonomous systems. If our first-order system $\mathbf{x}' = \mathbf{F}$ is autonomous, then $\mathbf{F}$ does not depend on t, only on the components of $\mathbf{x}$. Consequently, the "velocities" of

$$\mathbf{x}' = \mathbf{F}(\mathbf{x})$$

depend only on position, not t. We will use this fact for the rest of this chapter.

39.4 Sketching Trajectories for Autonomous Systems

Direction Fields

Suppose we are given a regular 2×2 autonomous first-order system of differential equations

$$\begin{aligned} x' &= f(x, y) \\ y' &= g(x, y) \end{aligned} .$$

Now, pick a point (x_0, y_0) on the XY–plane, and, using the given system, compute the 'velocity' at that point for the trajectory through that point,

$$\mathbf{v} = \begin{bmatrix} x' \\ y' \end{bmatrix} = \begin{bmatrix} f(x_0, y_0) \\ g(x_0, x_0) \end{bmatrix} .$$

This gives us a vector tangent at the point (x_0, x_0) to any trajectory passing through this point, and pointing in the direction of travel along the trajectory as t increases. So, if we draw a short arrow at (x_0, y_0) in the same direction as this velocity vector, we then have a short arrow tangent to the trajectory through this point and pointing in the "direction of travel" along this curve. We will call this short arrow a *direction arrow*. (This assumes $\mathbf{x}'$ is nonzero at the point. If it is zero, we have a "critical point". We'll discuss critical points in just a little bit.) In Figure 39.2a, we've sketched a few of these direction arrows at points along one particular trajectory.

If we sketch a corresponding direction arrow at every point (x_j, y_k) in a grid of points, then we have a *direction field*, as illustrated (along with a few trajectories) in Figure 39.2b. This direction field tells us the general behavior of the system's trajectories. To sketch a trajectory using a direction field, simply start at any chosen point in the plane, and sketch a curve following the directions indicated by the nearby direction arrows. "Go with the flow", and do not attempt to "connect the dots". The goal is to draw, as well as practical, a curve whose tangent at each point on the curve is lined up with the direction arrow that would be sketched at that point. That curve (if perfectly drawn), oriented in the direction given by the tangent direction arrows, is a trajectory of the system. (In practice, of course, it's as good an approximation to a trajectory as our drafting skills allow.)

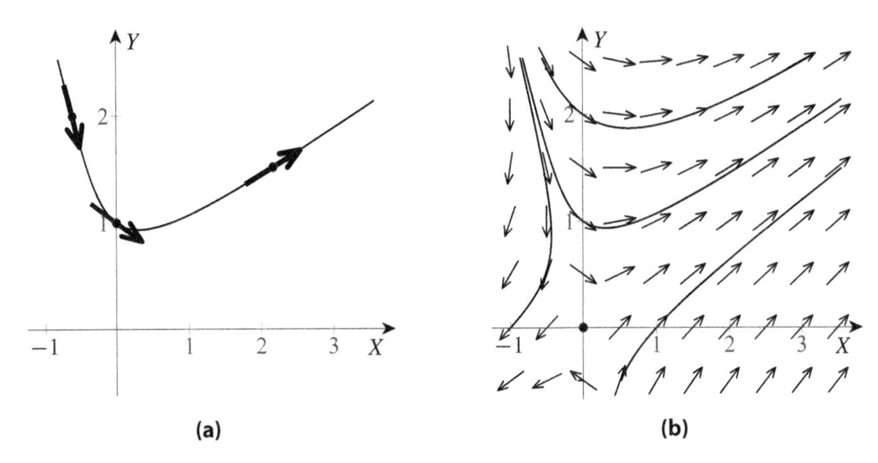

Figure 39.2: Direction arrows for trajectories for the system in Example 39.5, with **(a)** being a few direction arrows tangent to the trajectory passing through $(0, 1)$ (drawn oversized for clarity), and **(b)** being a more complete direction field, along with a few more trajectories.

The construction and use of a direction field for a 2×2 first-order autonomous system of differential equations is analogous to the construction and use of a slope field for a first-order differential equation (see Chapter 9). It's not exactly the same — we are now sketching trajectories instead of true graphs of solutions (which would require a T–axis), but the mechanics are very much the same. And, just as with slope lines for slope fields, it is good for your understanding to practice sketching the direction arrows for a few small direction fields, and, for the sake of your sanity, it is a good idea to learn how to construct direction fields (and trajectories) using a good computer math package.

Critical Points

When constructing a direction field, it is important to note each point (x_0, y_0) for which

$$\begin{bmatrix} f(x_0, y_0) \\ g(x_0, y_0) \end{bmatrix} = \begin{bmatrix} 0 \\ 0 \end{bmatrix} .$$

Any such point (x_0, y_0) is said to be a *critical point* for the system. Since the zero vector has no well-defined direction, we cannot sketch a direction arrow at a critical point. Instead, plot a clearly visible dot at this point. After all, if (x_0, y_0) is a critical point for our system, then

$$\begin{bmatrix} x' \\ y' \end{bmatrix} = \begin{bmatrix} f(x_0, y_0) \\ g(x_0, y_0) \end{bmatrix} = \begin{bmatrix} 0 \\ 0 \end{bmatrix} ,$$

which, in turn, means we have the constant/equilibrium solution

$$(x(t), y(t)) = (x_0, y_0) \qquad \text{for all} \quad t \quad ,$$

and, if you think about it for a moment, you'll realize that the point (x_0, y_0) is the entire trajectory of this solution.

In fact, this gives us an alternate definition for critical point, namely, that a *critical point* for an autonomous system of differential equations is the trajectory of an equilibrium solution for that system.

Finding critical points and determining the behavior of the trajectories in regions around them can be rather important. The mechanics of finding critical points is identical to the mechanics of finding equilibrium solutions (as illustrated in Example 39.2 on page 791). Issues regarding the behavior of near-by trajectories will be discussed in the next section.

!▶Example 39.5: *Consider, once again, the system*

$$\begin{bmatrix} x' \\ y' \end{bmatrix} = \begin{bmatrix} x + 2y \\ 5x - 2y \end{bmatrix} .$$

Plugging in $(x, y) = (0, 1)$ *, we get*

$$\begin{bmatrix} x' \\ y' \end{bmatrix} = \begin{bmatrix} 0 + 2 \cdot 1 \\ 5 \cdot 0 - 2 \cdot 1 \end{bmatrix} = \begin{bmatrix} 2 \\ -2 \end{bmatrix} .$$

Thus, the direction arrow sketched at position $(x, y) = (0, 1)$ *should be a short arrow (which we center at the point) parallel to and pointing in the same direction as the vector from the origin* $(0, 0)$ *to position* $(2, -2)$ *. That is what was sketched at point* $(0, 1)$ *in Figure 39.2a, along with the trajectory through that point.*

A more detailed direction field, along with three other trajectories, is illustrated in Figure 39.2b. Note the dot at $(0, 0)$ *. This is the critical point for the system and is the trajectory of the one equilibrium solution*

$$(x(t), y(t)) = (0, 0) \qquad \text{for all} \quad t \quad .$$

Trajectories and Solutions

Keep in mind that a trajectory through a point (a, b) for a regular autonomous system

$$\begin{bmatrix} x' \\ y' \end{bmatrix} = \begin{bmatrix} f(x, y) \\ g(x, y) \end{bmatrix}$$

is not a solution to the system; it is the path traced out by a solution to the initial-value problem

$$\begin{bmatrix} x' \\ y' \end{bmatrix} = \begin{bmatrix} f(x, y) \\ g(x, y) \end{bmatrix} \quad \text{with} \quad \begin{bmatrix} x(t_0) \\ y(t_0) \end{bmatrix} = \begin{bmatrix} a \\ b \end{bmatrix}$$

for any choice of t_0 in $(-\infty, \infty)$. And since there are infinitely many possible values of t_0, we should expect infinitely many solutions tracing out that one trajectory.

However, all the different solutions corresponding to a single trajectory are simply 'shifted' versions of each other. To see this, let $\big(\hat{x}(t), \hat{y}(t)\big)$ be a solution to

$$\begin{bmatrix} x' \\ y' \end{bmatrix} = \begin{bmatrix} f(x, y) \\ g(x, y) \end{bmatrix} \quad \text{with} \quad \begin{bmatrix} x(0) \\ y(0) \end{bmatrix} = \begin{bmatrix} a \\ b \end{bmatrix} .$$

You can then easily verify for yourself (see Exercise 39.11) that, for any real value t_0,

$$\begin{bmatrix} x \\ y \end{bmatrix} = \begin{bmatrix} x(t) \\ y(t) \end{bmatrix} = \begin{bmatrix} \hat{x}(t - t_0) \\ \hat{y}(t - t_0) \end{bmatrix}$$

is a solution to

$$\begin{bmatrix} x' \\ y' \end{bmatrix} = \begin{bmatrix} f(x, y) \\ g(x, y) \end{bmatrix} \quad \text{with} \quad \begin{bmatrix} x(t_0) \\ y(t_0) \end{bmatrix} = \begin{bmatrix} a \\ b \end{bmatrix} .$$

The uniqueness theorems briefly discussed in the last chapter then assure us that there are no other solutions to this initial-value problem.

This, by the way, leads to a minor technical point that should be discussed briefly. When specifying a solution $(x(t), y(t))$, we should also specify the domain of this solution, that is, the interval (α, β) over which t varies. In practice, we often don't mention that interval, simply assuming it to be "as large as possible" (which often means $(\alpha, \beta) = (-\infty, \infty)$). On the few occasions when it is particularly relevant, we will refer to a trajectory as being *complete* if it is the trajectory of a solution

$$\mathbf{x}(t) = \begin{bmatrix} x(t) \\ y(t) \end{bmatrix} \quad \text{with} \quad \alpha < t < \beta$$

where the interval (α, β) is "as large as possible". (That was the implicit assumption in the previous paragraph.)

Of course, while we may be most interested in the complete trajectories of our systems, in practice, the trajectories we sketch are often merely portions of complete trajectories simply because the complete trajectories often extend beyond the region over which we are making our drawings. One obvious exception, of course, is that a critical point (x_0, y_0) is a complete trajectory since it is the trajectory of the equilibrium solution

$$(x(t), y(t)) = (x_0, y_0) \quad \text{for} \quad -\infty < t < \infty \quad ,$$

and there certainly is no larger interval than $(-\infty, \infty)$.

Properties of Trajectories

It should be noted that the 'existence and uniqueness of solutions' implies a corresponding 'existence and uniqueness of (complete) trajectories', allowing us to say that "through each point there is one and only one (complete) trajectory". We'll say more about this in Section 39.7. Also in that section, we will verify the following facts about the trajectories of any regular autonomous standard system:

1. If a trajectory contains a critical point, then that critical point is the entire trajectory. Conversely, if a trajectory has nonzero length, then it does not contain a critical point of the system. (Hence, in sketching trajectories other than critical points, make sure your trajectories do not pass through any critical points.)

2. If a complete trajectory has an endpoint, then that endpoint must be a critical point for the system.

3. Any oriented curve not containing a critical point that "follows" the system's direction field (i.e., any oriented curve whose tangent vector at each point is in the same direction as the system's direction arrow at that point) is the trajectory of some solution to the system. (This is what we intuitively expect when "sketching trajectories" on a direction field.)

4. Any trajectory that is not a critical point is "smooth" in that no trajectory can have a "kink" or "corner" (as in Figure 39.3a).

5. No trajectory can "split" into two or more trajectories (as in Figure 39.3b), nor can two or more trajectories "merge" into one trajectory (as in Figure 39.3b with the arrows reversed).

Keeping the above in mind will help in both sketching trajectories from a direction field and in interpreting your results.

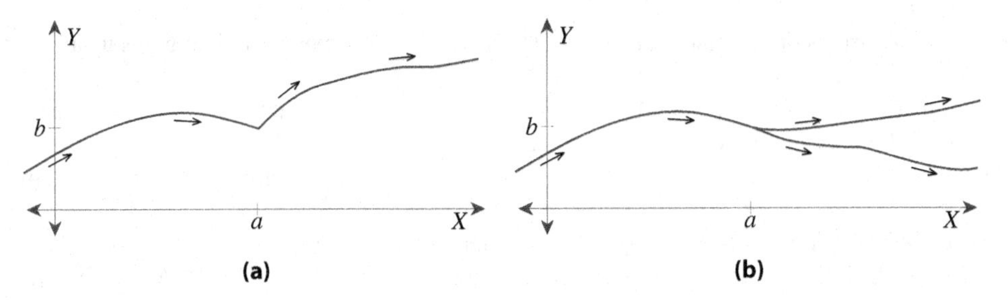

Figure 39.3: Two impossibilities for a trajectory of a regular autonomous system: **(a)** A sharp "kink" at some point, and **(b)** a "splitting" into two different trajectories at some point.

Phase Portraits and Planes

A little more terminology: When dealing with direction fields and trajectories for a standard 2×2 autonomous system, we refer to the plane on which we sketch the direction field and/or trajectories as the *phase plane* (as opposed to the "XY–plane" or "X_1X_2–plane" plane or ...). If we sketch an 'enlightening' representative sample of trajectories on the phase plane, then this sketch is said to be a *phase portrait* of the system. At this point, we are using direction fields to sketch phase portraits, so we are getting phase portraits superimposed on direction fields. If a phase portrait does not have an accompanying direction field to indicate direction of travel along the trajectories, then you should have little arrows on the trajectories to indicate the direction of travel for each trajectory.

Higher-Order Cases

The fundamental ideas just discussed and developed for any 2×2 regular autonomous standard system extend naturally to analogous ideas for any $N\times N$ regular autonomous standard system

$$\mathbf{x}' = \begin{bmatrix} x_1' \\ x_2' \\ \vdots \\ x_N' \end{bmatrix} = \begin{bmatrix} f_1(x_1,x_2,\ldots,x_N) \\ f_2(x_1,x_2,\ldots,x_N) \\ \vdots \\ f_N(x_1,x_2,\ldots,x_N) \end{bmatrix} = \mathbf{F}(\mathbf{x}) \quad .$$

As before, we define a *critical point* to be any point $\left(x_1^0, x_2^0, \ldots, x_N^0\right)$ in N-dimensional space for which

$$\begin{bmatrix} f_1\left(x_1^0,x_2^0,\ldots,x_N^0\right) \\ f_2\left(x_1^0,x_2^0,\ldots,x_N^0\right) \\ \vdots \\ f_N\left(x_1^0,x_2^0,\ldots,x_N^0\right) \end{bmatrix} = \begin{bmatrix} 0 \\ 0 \\ \vdots \\ 0 \end{bmatrix}$$

and, as before, any such point is the complete trajectory of the corresponding equilibrium solution

$$(x_1(t)\,,\, x_2(t)\,,\, \ldots x_N(t)) = \left(x_1^0\,,\, x_2^0\,,\, \ldots\,,\, x_N^0\right) \qquad \text{for all} \quad t$$

for our system $\mathbf{x}' = \mathbf{F}(\mathbf{x})$.

Likewise, at any point other than a critical point, we can, in theory, find a short vector pointing in the direction of travel of any trajectory through that point by just taking any short vector pointing in the same direction as $\mathbf{F}$ computed at that point. This gives a *direction arrow* at that point. Plotting these direction arrows on a suitable grid of points in N-dimensional space then gives us a *direction field* for the system.

Admittedly, few of us can actually sketch and use a direction field when $N > 2$ (especially if $N > 3$!). Still, we can find the critical points, and it turns out that much of what we will learn about the behavior of trajectories near critical points for 2×2 systems can apply when our systems are larger.

Extending our "phase" phraseology: Traditionally the N-dimensional space in which the trajectories would, in theory, be drawn is called the *phase space* (*phase plane* if $N = 2$), with any enlightening representative sample of trajectories in this space being called a *phase portrait*. Admittedly, visualizing a phase portrait when $N \geq 3$ requires a certain imagination.

39.5 Critical Points, Stability and Long-Term Behavior

A useful feature of a direction field for an autonomous system of differential equations is that it can give us some notion of the long-term behavior of the solutions to that system. All we need to do is to follow the sketched trajectories.

Critical Points and Stability

Stability of Equilibrium Solutions

Of particular interest will be the long-term behavior of solutions whose trajectories pass close to a critical point (x_0, y_0) of the system, and we will use this behavior to classify the 'stability' of that critical point and the corresponding equilibrium solution

$$(x(t), y(t)) = (x_0, y_0) \qquad \text{for all} \quad t \quad .$$

Loosely speaking we will classify this critical point and the corresponding equilibrium solution as being

- *stable* if and only if each trajectory that gets close to (x_0, y_0) stays close to (x_0, y_0) afterwards. That is, this critical point and equilibrium solution are stable if and only if each solution (x, y) to the system satisfying $(x(t_0), y(t_0)) \approx (x_0, y_0)$ for some t_0 also satisfies

$$(x(t), y(t)) \approx (x_0, y_0) \qquad \text{for all} \quad t > t_0 \quad ,$$

  as illustrated in Figures 39.4a and 39.4b.

- *asymptotically stable* if and only if each trajectory that gets close to (x_0, y_0) doesn't just stay close but converges to (x_0, y_0) as $t \to +\infty$. That is, this critical point and equilibrium solution are asymptotically stable if and only if each solution (x, y) to the system satisfying $(x(t_0), y(t_0)) \approx (x_0, y_0)$ for some t_0 also satisfies

$$\lim_{t \to +\infty} (x(t), y(t)) = (x_0, y_0) \quad ,$$

  as illustrated in Figure 39.4b. (Obviously, an asymptotically stable critical point is automatically stable.)

- *unstable* if and only if the equilibrium solution is not stable. Examples of unstable equilibrium solutions are illustrated in Figure 39.4c (in which the nearby trajectories all diverge away from the critical point) and in Figure 39.2b (in which trajectories approach the critical point $(0, 0)$ and then diverge away).

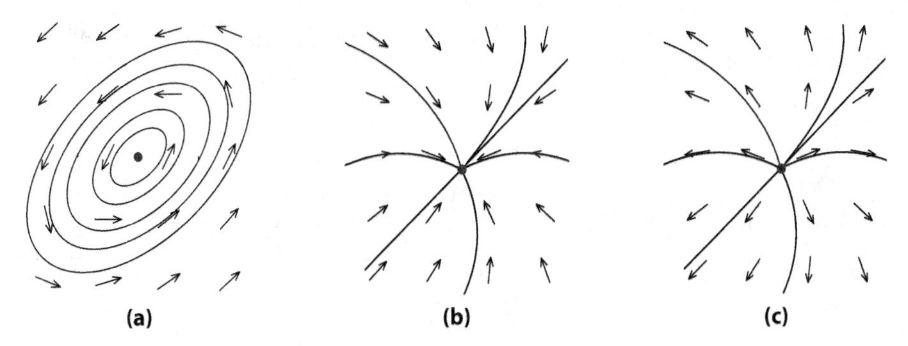

Figure 39.4: Two-dimensional direction fields and trajectories about critical points corresponding to equilibrium solutions that are **(a)** stable, but not asymptotically stable, **(b)** asymptotically stable and **(c)** unstable.

Be warned that the stability of an equilibrium solution is not always clear from just the direction field. The field may suggest that the nearby trajectories are loops circling the critical point (indicating that the equilibrium solution is stable) when, in fact, the nearby trajectories are either slowly spiraling into or away from the critical point (in which case the equilibrium solution is actually either asymptotically stable or is unstable).

Precise Definitions

Of course, precise mathematics requires precise definitions. So, to be precise, we classify our equilibrium solution

$$(x(t), y(t)) = (x_0, y_0) \qquad \text{for all} \quad t$$

and the corresponding critical point as being

- *stable* if and only if, for each $\epsilon > 0$, there is a corresponding $\delta > 0$ such that, if (x, y) is any solution to the system satisfying

  $$\sqrt{(x(t_0) - x_0)^2 + (y(t_0) - y_0)^2} < \delta \qquad \text{for some} \quad t_0 \quad ,$$

  then

  $$\sqrt{(x(t) - x_0)^2 + (y(t) - y_0)^2} < \epsilon \qquad \text{for all} \quad t > t_0 \quad .$$

- *asymptotically stable* if and only if there is a $\delta > 0$ such that, if (x, y) is any solution to the system satisfying

  $$\sqrt{(x(t_0) - x_0)^2 + (y(t_0) - y_0)^2} < \delta \qquad \text{for some} \quad t_0 \quad ,$$

  then

  $$\lim_{t \to +\infty} \sqrt{(x(t) - x_0)^2 + (y(t) - y_0)^2} = 0 \quad .$$

- *unstable* if the equilibrium solution is not stable.

While we are at it, we should give precise meanings to the 'convergence/divergence' of a trajectory to/from a critical point (x_0, y_0). So assume we have a trajectory and any solution (x, y) to the system that generates that trajectory. We'll say the trajectory

- *converges* to critical point (x_0, y_0) if and only if $\lim_{t \to +\infty}(x(t), y(t)) = (x_0, y_0)$,

and

- *diverges* from critical point (x_0, y_0) if and only if $\lim_{t \to -\infty}(x(t), y(t)) = (x_0, y_0)$.

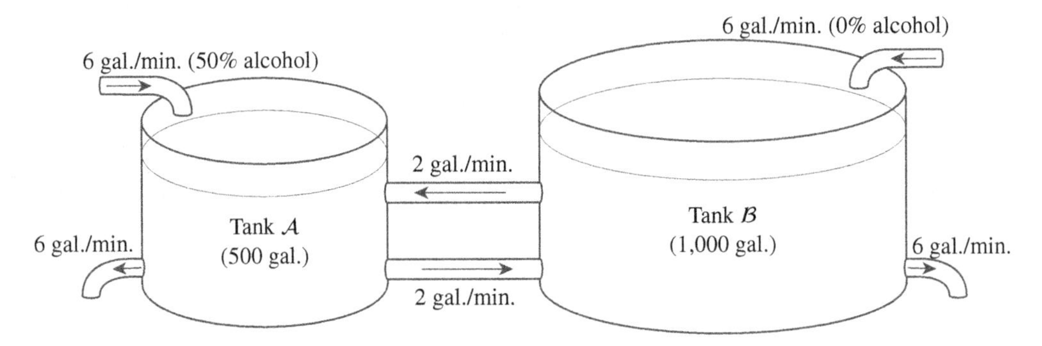

Figure 39.5: A simple system of two tanks containing water/alcohol mixtures.

Long-Term Behavior

A direction field may also give us some idea of the long-term behavior of the solutions to the given system at points far away from the critical points. Of course, this supposes that any patterns that appear to be evident in the direction field actually continue outside the region on which the direction field is drawn.

!▶ Example 39.6: *Consider the direction field and sample trajectories sketched in Figure 39.2b on page 795. In particular, look at the trajectory passing through the point $(1, 0)$, and follow it in the direction indicated by the direction field. The last part of this curve seems to be straightening out to a straight line proceeding further into the first quadrant at, very roughly, a 45 degree angle to both the positive X–axis and positive Y–axis. This suggests that, if $(x(t), y(t))$ is any solution to the direction field's system satisfying $(x(t_0), y(t_0)) = (1, 0)$ for some t_0, then*

$$\lim_{t\to\infty} (x(t), y(t)) = (\infty, \infty)$$

with

$$y(t) \approx x(t) \qquad \text{when} \quad t \text{ is "large"} \quad .$$

On the other hand, if you follow the trajectory passing through position $(-1, 0)$, then you probably get the impression that, as t increases, the trajectory is heading deeper into the third quadrant of the XY–plane, suggesting that if $(x(t), y(t))$ is any solution to the direction field's system satisfying $(x(t_0), y(t_0)) = (-1, 0)$ for some t_0, then

$$\lim_{t\to\infty} (x(t), y(t)) = (-\infty, -\infty) \quad .$$

You may even suspect that, for such a solution,

$$y(t) \approx x(t) \qquad \text{when} \quad t \text{ is "large"} \quad ,$$

though there is hardly enough of the trajectory sketched to be too confident of this suspicion.

39.6 Applications

Let us try to apply the above to three applications from the previous chapter; namely, the applications involving two-tank mixing, competing species and a swinging pendulum. In each case, we will find the critical points, attempt to use a (computer-generated) direction field to determine the stability of these points, and see what conclusions we can then derive.

A Mixing Problem

Our mixing problem is illustrated in Figure 39.5. In it, we have two tanks $\mathcal{A}$ and $\mathcal{B}$ containing, respectively, 500 and 1,000 gallons of a water/alcohol mix. Each minute 6 gallons of a water/alcohol mix containing 50% alcohol is added to tank $\mathcal{A}$, while 6 gallons of the mix is drained from that tank. At the same time, 6 gallons of pure water is added to tank $\mathcal{B}$, and 6 gallons of the mix in tank $\mathcal{B}$ is drained out. In addition, the two tanks are connected by two pipes, with one pumping liquid from tank $\mathcal{A}$ to tank $\mathcal{B}$ at a rate of 2 gallons per minute, and with the other pumping liquid in the opposite direction, from tank $\mathcal{B}$ to tank $\mathcal{A}$, at a rate of 2 gallons per minute.

In the previous chapter, we found that the system describing this mixing process is

$$\begin{aligned} x' &= -\frac{8}{500}x + \frac{2}{1000}y + 3 \\ y' &= \frac{2}{500}x - \frac{8}{1000}y \end{aligned} \quad , \tag{39.5}$$

where

$$t = \text{number of minutes since we started the mixing process} \quad ,$$

$$x = x(t) = \text{amount (in gallons) of alcohol in tank } \mathcal{A} \text{ at time } t \quad ,$$

and

$$y = y(t) = \text{amount (in gallons) of alcohol in tank } \mathcal{B} \text{ at time } t \quad .$$

To find any critical points of the system we first replace each derivative in system (39.5) with 0, obtaining

$$\begin{aligned} 0 &= -\frac{8}{500}x + \frac{2}{1000}y + 3 \\ 0 &= \frac{2}{500}x - \frac{8}{1000}y \end{aligned} \quad .$$

Then we solve this algebraic system. That is easily done. From the last equation we have

$$x = \frac{500}{2} \cdot \frac{8}{1000}y = 2y \quad .$$

Using this with the first equation, we get

$$0 = -\frac{8}{500} \cdot 2y + \frac{2}{1000}y + 3 = 3 - \frac{15}{500}y$$

$$\hookrightarrow \qquad y = 3 \cdot \frac{500}{15} = 100 \qquad \text{and} \qquad x = 2y = 2 \cdot 100 = 200 \quad .$$

So the one critical point for this system is

$$(x_0, y_0) = (200, 100) \quad .$$

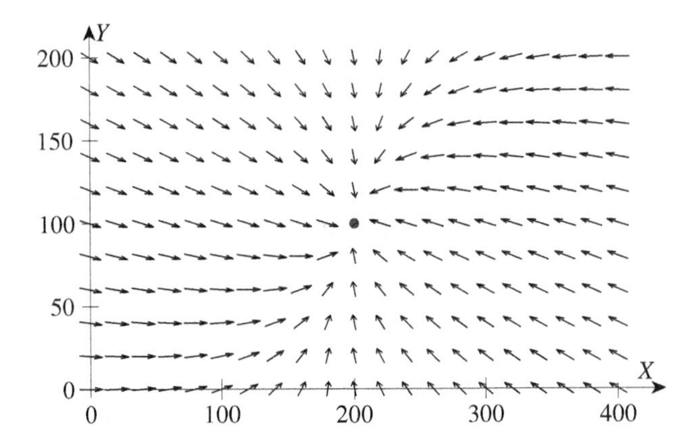

Figure 39.6: Direction field for the system of two tanks from Figure 39.5.

Considering the physical process being modeled, it should seem reasonable for this critical point to describe the long-term equilibrium of the system. That is, we should expect $(200, 100)$ to be an asymptotically stable equilibrium, and that

$$\lim_{t\to\infty} (x(t), y(t)) = (200, 100) \quad .$$

Checking the computer-generated direction field in Figure 39.6, we see that this is clearly the case. Around critical point $(200, 100)$, the direction arrows are all pointing toward this point. Thus, as $t \to \infty$ the concentrations of alcohol in tanks $\mathcal{A}$ and $\mathcal{B}$, respectively, approach

$$\frac{200}{500} \quad \text{(i.e., 40\%)} \qquad \text{and} \qquad \frac{100}{1000} \quad \text{(i.e., 10\%)} \quad .$$

Competing Species

The Model and Analysis

In our competing species example, we assumed that we have a large field containing rabbits and gerbils that are competing with each other for the resources in the field. Letting

$$R = R(t) = \text{number of rabbits in the field at time } t$$

and

$$G = G(t) = \text{number of gerbils in the field at time } t \quad ,$$

we derived the following system describing how the two populations vary over time:

$$\begin{aligned} R' &= (\beta_1 - \gamma_1 R - \alpha_1 G)R \\ G' &= (\beta_2 - \gamma_2 G - \alpha_2 R)G \end{aligned} \quad . \tag{39.6}$$

In this, β_1 and β_2 are the net birth rates per creature under ideal conditions for rabbits and gerbils, respectively, and the γ_k's and α_k's are positive constants that would probably have to be determined by experiment and measurement. In particular, the values we chose yielded the system

$$\begin{aligned} R' &= \left(\frac{5}{4} - \frac{1}{160}R - \frac{3}{1000}G\right)R \\ G' &= \left(3 - \frac{3}{500}G - \frac{3}{160}R\right)G \end{aligned} \quad . \tag{39.7}$$

Setting $R' = G' = 0$ in equation set (39.7) gives us the algebraic system we need to solve to find the critical points:

$$\begin{aligned} 0 &= \left(\frac{5}{4} - \frac{1}{160}R - \frac{3}{1000}G\right)R \\ 0 &= \left(3 - \frac{3}{500}G - \frac{3}{160}R\right)G \end{aligned} \quad . \tag{39.8}$$

The first equation in this algebraic system tells us that either

$$R = 0 \qquad \text{or} \qquad \frac{1}{160}R + \frac{3}{1000}G = \frac{5}{4} \quad .$$

If $R = 0$, the second equation reduces to

$$0 = \left(3 - \frac{3}{500}G\right)G \quad ,$$

which means that either

$$G = 0 \qquad \text{or} \qquad G = 500 \quad .$$

So two critical points are $(R, G) = (0, 0)$ and $(R, G) = (0, 500)$.

If, on the other hand, the first equation in algebraic system (39.8) holds because

$$\frac{1}{160}R + \frac{3}{1000}G = \frac{5}{4} \quad ,$$

then the system's second equation can only hold if either

$$G = 0 \qquad \text{or} \qquad \frac{3}{500}G + \frac{3}{160}R = 3 \quad .$$

If $G = 0$, then we can solve the first equation in the system, obtaining

$$R = \frac{5}{4} \cdot 160 = 200 \quad .$$

Thus, $(R, G) = (200, 0)$ is one critical point. Looking at what remains, we see that there is one more critical point, and it satisfies the simple algebraic linear system

$$\begin{aligned} \frac{1}{160}R + \frac{3}{1000}G &= \frac{5}{4} \\ \frac{3}{160}R + \frac{3}{500}G &= 3 \end{aligned} \quad .$$

You can easily verify that the solution to this is $(R, G) = (80, 250)$.

So the critical points for our system are (R, G) equaling

$$(0, 0) \quad , \quad (0, 500) \quad , \quad (200, 0) \quad \text{and} \quad (80, 250) \quad .$$

The first critical point tells us that, if we start with no rabbits and no gerbils, then the populations remain at 0 rabbits and 0 gerbils — no big surprise. The next two tell us that our populations can remain constant if either we have no rabbits with 500 gerbils, or we have 200 rabbits and no gerbils. The last critical point says that the two populations can coexist at 80 rabbits and 250 gerbils.

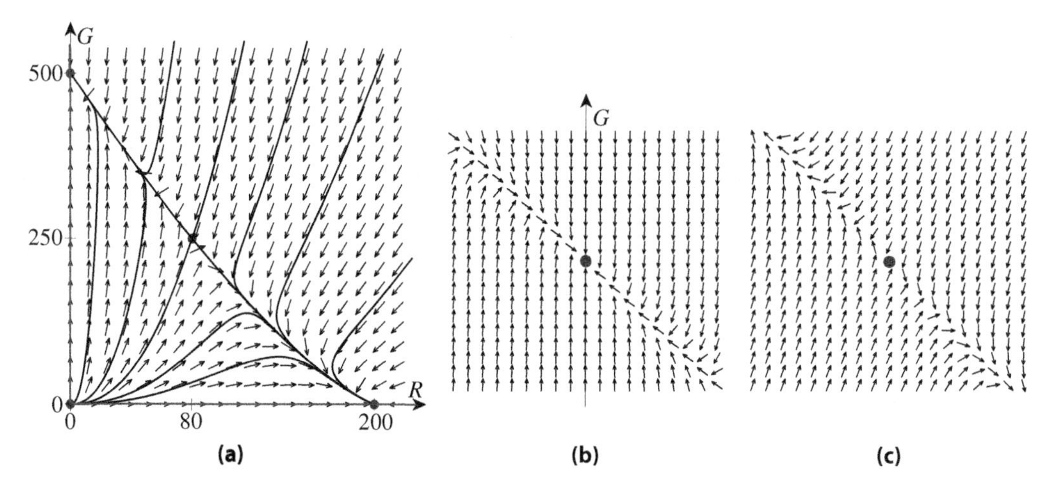

Figure 39.7: **(a)** A direction field and some trajectories for the competing species example with system (39.7), and detailed direction fields over very small regions about critical points **(b)** $(0, 500)$ and **(c)** $(80, 250)$.

But look at the direction field in Figure 39.7a for this system. From that, it is clear that critical point $(0, 0)$ is unstable. If we have at least a few rabbits and/or gerbils, then those populations do not, together, die out. We will always have at least some rabbits or gerbils.

On the other hand, critical point $(200, 0)$ certainly appears to be an asymptotically stable critical point. In fact, from the direction field it appears that, if we have, say $R(0) > 150$ and $G(0) < 250$, then the direction of the direction arrows "near" $(200, 0)$ indicate that

$$\lim_{t\to\infty} (R(t), G(t)) = (200, 0) \quad .$$

In other words, the gerbils die out and the number of rabbits stabilizes at 200.

Likewise, critical point $(500, 0)$ also appears to be asymptotically stable. Admittedly, a somewhat more detailed direction field about $(500, 0)$, such as in Figure 39.7b, may be desired to clarify this. Thus, if we start with enough gerbils and too few rabbits (say, $G(0) > 250$ and $R(0) < 80$), then

$$\lim_{t\to\infty} (R(t), G(t)) = (0, 500) \quad .$$

In other words, the rabbits die out and the number of gerbils approaches 500.

Finally, what about critical point $(80, 250)$? In the region about this critical point, we can see that a few direction arrows point towards this critical point, while others seem to lead the trajectories past the critical point. That suggests that $(80, 250)$ is an unstable critical point. Again, a more detailed direction field in a small area about critical point $(80, 250)$, such as in Figure 39.7c, is called for. This direction field shows more clearly that critical point $(80, 250)$ is unstable. Thus, while it is possible for the populations to stabilize at 80 rabbits and 250 gerbils, it is also extremely unlikely.

To summarize: It is possible for the two competing species to coexist, but, in the long run, it is much more likely that one or the other dies out, leaving us with either a field of 200 rabbits or a field of 500 gerbils, depending on the initial number of each.

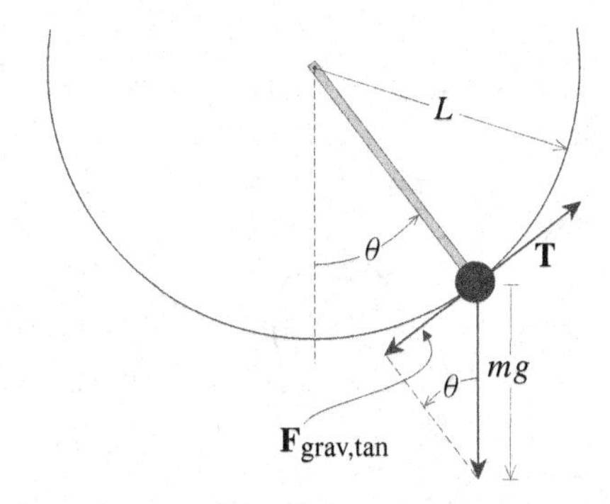

Figure 39.8: The pendulum system with a weight of mass m attached to a massless rod of length L swinging about a pivot point under the influence of gravity.

Some Notes

It turns out that different choices for the parameters in system (39.6) can lead to very different outcomes. For example, the system

$$\begin{aligned} R' &= (120 - 2R - 2G)R \\ G' &= (320 - 8G - 3R)G \end{aligned}$$

also has four critical points; namely,

$$(0,0) \quad , \quad (60,0) \quad , \quad (0,40) \quad \text{and} \quad (32,28) \quad .$$

In this case, however, the first three are unstable, and the last is asymptotically stable. Thus, if we start out with at least a few rabbits and a few gerbils, then

$$\lim_{t\to\infty} (R(t), G(t)) = (32, 28) \quad .$$

So neither population dies out. The rabbits and gerbils in this field are able to coexist, and there will eventually be approximately 32 rabbits and 28 gerbils.

It should also be noted that, sometimes, it can be difficult to determine the stability of a critical point from a given direction field. When this happens, a more detailed direction field about the critical point may be tried. A better approach involves a method using the eigenvalues of a certain matrix associated with the system and critical point. Unfortunately, this approach goes a little beyond the scope of this text. The reader is strongly encouraged to explore this method in the future.[2]

The (Damped) Pendulum

In the last chapter, we derived the system

$$\begin{aligned} \theta' &= \omega \\ \omega' &= -\gamma \sin(\theta) - \kappa\omega \end{aligned} \tag{39.9}$$

to describe the angular motion of the pendulum in Figure 39.8. Here

$\theta(t)$ = the angular position of pendulum at time t measured counterclockwise from the vertical line "below" the pivot point

[2] An exposition on this method can be found at the author's website for this text.

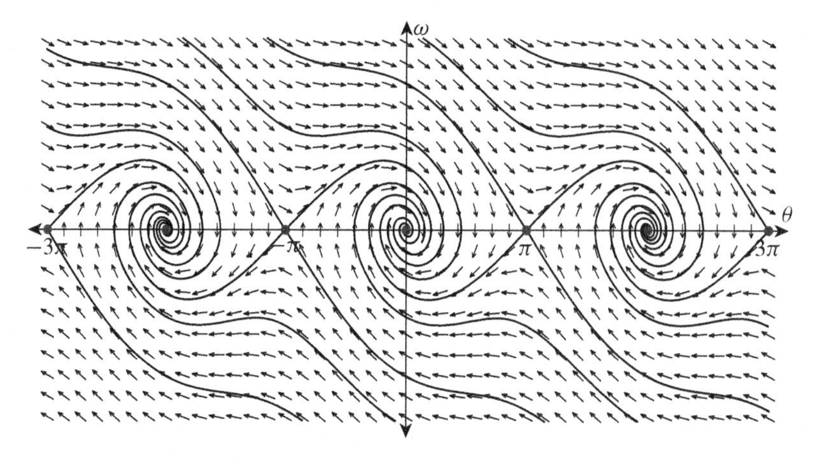

Figure 39.9: A phase portrait (with direction field) for pendulum system (39.10).

and

$$\omega(t) = \theta' = \text{the angular velocity of the pendulum at time } t \quad .$$

In addition, γ is the positive constant given by $\gamma = {}^{g}/_{L}$ where L is the length of the pendulum and g is the acceleration of gravity, and κ is the "drag coefficient", a nonnegative constant describing the effect friction has on the motion of the pendulum. The greater the effect of friction on the system, the larger the value of κ, with $\kappa = 0$ when there is no friction slowing down the pendulum. We will restrict our attention to the *damped* pendulum in which friction plays a role (i.e., $\kappa > 0$).[3] For simplicity, we will assume

$$\gamma = 8 \quad \text{and} \quad \kappa = 2 \quad ,$$

though the precise values for γ and κ will not play much of a role in our analysis. This gives us the system

$$\begin{aligned} \theta' &= \omega \\ \omega' &= -8\sin(\theta) - 2\omega \end{aligned} \quad . \tag{39.10}$$

Before going any further, observe that the right side of our system is periodic with period 2π with respect to θ. This means that, on the $\theta\omega$–plane, the pattern of the trajectories in any vertical strip of width 2π will be repeated in the next vertical strip of width 2π.

Setting $\theta' = 0$ and $\omega' = 0$ in system (39.10) yields the algebraic system

$$\begin{aligned} 0 &= \omega \\ 0 &= -8\sin(\theta) - 2\omega \end{aligned}$$

for the critical points. From this we get that there are infinitely many critical points, and they are given by

$$(\theta, \omega) = (n\pi, 0) \qquad \text{with} \quad n = 0, \pm 1, \pm 2, \ldots \quad .$$

A direction field with a few trajectories for this system is given in Figure 39.9. From it, we can see that the behavior of the trajectories near a critical point $(\theta, \omega) = (n\pi, 0)$ depends strongly on whether n is an even integer or an odd integer.

If n is an even integer, then the nearby trajectories are clearly spirals "spiraling" in towards the critical point $(\theta, \omega) = (n\pi, 0)$. Hence, every critical point $(n\pi, 0)$ with n even is asymptotically

[3] It turns out that the undamped pendulum, in which $\kappa = 0$, is more difficult to analyze.

stable. That is, if n is an even integer and

$$(\theta(t_0), \omega(t_0)) \approx (n\pi, 0)$$

for some t_0, then

$$(\theta(t), \omega(t)) \rightarrow (n\pi, 0) \qquad \text{as} \qquad t \rightarrow \infty \quad .$$

This makes sense. After all, if n is even, then $(\theta, \omega) = (n\pi, 0)$ describes a pendulum hanging straight down and not moving — certainly what most of us would call a 'stable equilibrium' position for the pendulum, and certainly the position we would expect a real-world pendulum (in which there is inevitably some friction slowing the pendulum) to eventually assume.

Now consider any critical point $(\theta, \omega) = (n\pi, 0)$ when n is an odd integer. From Figure 39.9 it is apparent that these critical points are unstable. This also makes sense. With n being an odd integer, $(\theta, \omega) = (n\pi, 0)$ describes a stationary pendulum balanced straight up from its pivot point, which is a physically unstable equilibrium. It may be possible to start the pendulum moving in such a manner that it approaches this configuration. But if the initial conditions are not just right, then the motion will be given by a trajectory that approaches and then goes away from that critical point. In particular, the trajectories near this critical point that pass through the horizontal axis (where the angular velocity ω is zero) are describing the pendulum slowing to a stop before reaching the upright position and then falling back down, while the trajectories near this critical point that pass over or below this point describe the pendulum traveling fast enough to reach and continue through the upright position.

From Figure 39.9, it is apparent that every trajectory eventually converges to one of the critical points, with most spiraling into one of the stable critical points. This tells us that, while the pendulum may initially have enough energy to spin in complete circles about the pivot point, it eventually stops spinning about the pivot and begins rocking back and forth in smaller and smaller arcs about its stable downward vertical position. Eventually, the arcs are so small that, for all practical purposes, the pendulum is motionless and hanging straight down.

By the way, it's fairly easy to redo the above using fairly arbitrary positive choices of γ and κ in pendulum system (39.9). As long as the friction is not too strong (i.e., as long as κ is not too large compared to γ), the resulting phase portrait will be quite similar to what we just obtained. However, if κ is too large compared to γ (to be precise, if $\kappa \geq 2\sqrt{\gamma}$), then, while the critical points and their stability remain the same as above, the trajectories no longer spiral about the stable critical points but approach them more directly. Interested readers are encouraged to redo the above with a relatively large κ to see for themselves.

39.7 Existence and Uniqueness of Trajectories

The existence and uniqueness of solutions to standard systems of differential equations were discussed in the last chapter. It is worth noting that the assumption of a system being regular immediately implies that the component functions all satisfy the conditions required in Theorem 38.1. So we automatically have:

Corollary 39.1

Suppose $\mathbf{x}' = \mathbf{F}(t, \mathbf{x})$ *is a regular* 2×2 *system. Then any initial-value problem involving this system,*

$$\mathbf{x}' = \mathbf{F}(t, \mathbf{x}) \qquad \textit{with} \quad \mathbf{x}(t_0) = \mathbf{a} \quad ,$$

has exactly one solution on some open interval (α, β) containing t_0. Moreover, the component functions of this solution and their first derivatives are continuous over that interval.

As almost immediate corollaries of Corollary 39.1 (along with a few observations made in Section 39.4), we have the following facts concerning trajectories of any given 2×2 standard first-order system that is both regular and autonomous:

1. Through each point of the plane there is exactly one complete trajectory.

2. If a trajectory contains a critical point for that system, then that entire trajectory is that single critical point. Conversely, if a trajectory for a regular autonomous system has nonzero length, then that trajectory does not contain a critical point.

3. Any trajectory that is not a critical point is "smooth" in that no trajectory can have a "kink" or "corner".

The other properties of trajectories that were claimed earlier all follow from the following theorem and its corollaries.

Theorem 39.2
Assume $\mathbf{x}' = \mathbf{F}(\mathbf{x})$ is a 2×2 standard first-order system that is both regular and autonomous, and let C be any oriented curve of nonzero length such that all the following hold at each point (x, y) in C:

1. *The point (x, y) is not a critical point for the system.*

2. *The curve C has a unique tangent line at (x, y), and that line is parallel to the vector $\mathbf{F}(\mathbf{x})$.*

3. *The direction of travel of C through (x, y) is in the same direction as given by the vector $\mathbf{F}(\mathbf{x})$.*

Then C (excluding any endpoints) is the trajectory for some solution to the system $\mathbf{x}' = \mathbf{F}(\mathbf{x})$.

This theorem assures us that, in theory, the curves drawn "following" a direction field will be trajectories of our system (in practice, of course, the curves we actually draw will be approximations). Combining this theorem with the existence and uniqueness results of Corollary 39.1 leads to the next two corollaries regarding complete trajectories.

Corollary 39.3
Two different complete trajectories of a regular autonomous system cannot intersect each other.

Corollary 39.4
If a complete trajectory of a regular autonomous system has an endpoint, that endpoint must be a critical point.

We'll discuss the proof of the above theorem in the next section for those who are interested. Verifying the two corollaries will be left as exercises (see Exercise 39.12).

39.8 Proving Theorem 39.2

The Assumptions

In all the following, let us assume we have some regular autonomous 2×2 standard system of differential equations

$$\mathbf{x}' = \mathbf{F}(\mathbf{x}) \quad ,$$

along with an oriented curve C of nonzero length such that all the following hold at each point (x, y) in C:

1. The point (x, y) is not a critical point for the system.
2. The curve C has a unique tangent line at (x, y), and that line is parallel to the vector $\mathbf{F}(\mathbf{x})$.
3. The direction of travel of C through (x, y) is in the same direction as given by the vector $\mathbf{F}(\mathbf{x})$.

Note that the requirement that C has a tangent line at each point in C means that we are excluding any endpoints of this curve.

Preliminaries

To verify our theorem, we will need some material concerning "parametric curves" that you should recall from your calculus course.

Norms and Normalizations

The *norm* (or length) of a column vector or vector-valued function

$$\mathbf{v} = \begin{bmatrix} v_1 \\ v_2 \end{bmatrix}$$

is

$$\|\mathbf{v}\| = \sqrt{[v_1]^2 + [v_2]^2} \quad .$$

If $\mathbf{v}$ is a nonzero vector, then we can *normalize* it by dividing it by its norm, obtaining a vector

$$\mathbf{n} = \frac{\mathbf{v}}{\|\mathbf{v}\|} = \frac{1}{\sqrt{[v_1]^2 + [v_2]^2}} \begin{bmatrix} v_1 \\ v_2 \end{bmatrix}$$

of unit length (i.e., $\|\mathbf{n}\| = 1$) and pointing in the same direction as $\mathbf{v}$.

Oriented Curves and Unit Tangents

If (x, y) is any point on any oriented curve at which the curve has a well-defined tangent line, then this curve has a *unit tangent vector* at (x, y), denoted by $\mathbf{T}(x, y)$, which is simply a unit vector tangent to the curve at that point, and pointing in the direction of travel along the curve at that point. For our oriented curve, C, that tangent line is parallel to $\mathbf{F}(\mathbf{x})$, and the direction of travel is given by $\mathbf{F}(\mathbf{x})$. So the unit tangent at (x, y) must be the normalization of $\mathbf{F}(\mathbf{x})$. That is

$$\mathbf{T}(x, y) = \frac{\mathbf{F}(\mathbf{x})}{\|\mathbf{F}(\mathbf{x})\|} \quad \text{for each } (x, y) \text{ in } C \quad . \tag{39.11}$$

Curve Parameterizations

A *parametrization* of an oriented curve C is an ordered pair of functions on some interval

$$(x(t), y(t)) \qquad \text{for} \quad t_S < t < t_E$$

that traces out the curve in the direction of travel along C as t varies from t_S to t_E. Given any such parametrization, we will automatically let

$$\mathbf{x} = \mathbf{x}(t) = \begin{bmatrix} x(t) \\ y(t) \end{bmatrix} \qquad \text{and} \qquad \mathbf{x}' = \mathbf{x}'(t) = \begin{bmatrix} x'(t) \\ y'(t) \end{bmatrix} \quad .$$

If we view our parametrization $(x(t), y(t))$ as giving the position at time t of some object traveling along C, then, provided the functions are suitably differentiable,

$$\mathbf{x}'(t) = \begin{bmatrix} x'(t) \\ y'(t) \end{bmatrix}$$

is the corresponding "velocity", of the object at time t. This is a vector pointing in the direction of travel of the object at time t, and whose length,

$$\left\|\mathbf{x}'(t)\right\| = \sqrt{[x'(t)]^2 + [y'(t)]^2} \quad ,$$

is the speed of the object at time t (i.e., as it goes through position $(x(t), y(t))$). Recall that the integral of this speed from $t = t_1$ to $t = t_2$,

$$\int_{t_1}^{t_2} \left\|\mathbf{x}'(t)\right\| \, dt \quad , \tag{39.12}$$

gives the signed distance one would travel along the curve in going from position $(x(t_1), y(t_1))$ to position $(x(t_2), y(t_2))$. This value is positive if $t_1 < t_2$ and negative if $t_1 > t_2$. Recall, also, that this distance (the *arclength*) is traditionally denoted by s.

The most fundamental parametrizations are the *arclength parametrizations*. To define one for our oriented curve C, first pick some point (x_0, y_0) on C. Then let s_S and s_E be, respectively, the negative and positive values such that s_E is the "maximal distance" that can be traveled in the positive direction along C from (x_0, y_0), and $|s_S|$ is the "maximal distance" that can be traveled in the negative direction along C from (x_0, y_0). These distances may be infinite.[4] Finally, define the arclength parametrization

$$(\tilde{x}(s), \tilde{y}(s)) \qquad \text{for} \quad s_S < s < s_E$$

as follows (and as indicated in Figure 39.10):

1. For $0 \leq s < s_E$ set $(\tilde{x}(s), \tilde{y}(s))$ equal to the point on C arrived at by traveling in the positive direction along C by a distance of s from (x_0, y_0).

2. For $s_S < s \leq 0$ set $(\tilde{x}(s), \tilde{y}(s))$ equal to the point on C arrived at by traveling in the negative direction along C by a distance of $|s|$ from (x_0, y_0).

We should note that if the curve intersects itself, then the same point $(\tilde{x}(s), \tilde{y}(s))$ may be given by more than one value of s. In particular, if C is a loop of length L, then $(\tilde{x}(s), \tilde{y}(s))$ will be periodic with $(\tilde{x}(s+L), \tilde{y}(s+L)) = (\tilde{x}(s), \tilde{y}(s))$ for every real value s.

[4] Better definitions for s_S and s_E are discussed in the 'technical note' at the end of this subsection.

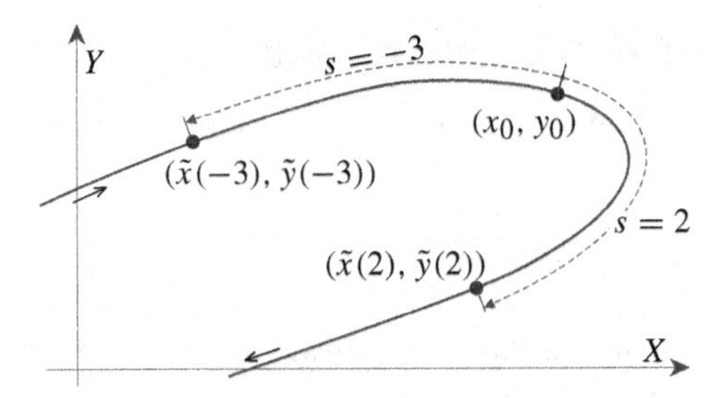

Figure 39.10: Two points given by an arclength parameterization $\tilde{\mathbf{x}}(s)$ of an oriented curve.

It should also be noted that, from arclength integral (39.12) and the fact that, by definition, s is the signed distance one would travel along the curve to go from $(\tilde{x}(0), \tilde{y}(0))$ to $(\tilde{x}(s), \tilde{y}(s))$, we automatically have

$$\int_0^s \left\| \tilde{\mathbf{x}}'(\sigma) \right\| \, d\sigma = s \quad .$$

Differentiating this yields

$$\left\| \tilde{\mathbf{x}}'(s) \right\| = \frac{d}{ds} \int_0^s \left\| \tilde{\mathbf{x}}'(\sigma) \right\| \, d\sigma = \frac{ds}{ds} = 1 \quad .$$

Hence, each $\tilde{\mathbf{x}}'(s)$ is a unit vector pointing in the direction of travel on C at $\tilde{\mathbf{x}}(s)$ — that is, $\tilde{\mathbf{x}}'(s)$ is the unit tangent vector for C at $(\tilde{x}(s), \tilde{y}(s))$. Combining this with equation (39.11) yields

$$\tilde{\mathbf{x}}'(s) = \mathbf{T}(\tilde{x}(s), \tilde{y}(s)) = \frac{\mathbf{F}(\tilde{\mathbf{x}}(s))}{\left\| \mathbf{F}(\tilde{\mathbf{x}}(s)) \right\|} \quad \text{for} \quad s_S < s < s_E \quad . \tag{39.13}$$

Technical Note on "Maximal Distances"

We set s_E equal to "the 'maximal distance' that can be traveled in the positive direction along C from (x_0, y_0)". Technically, this "maximal distance" may not exist because, technically, an endpoint of a trajectory need not actually be part of that trajectory.

To be more precise, let us define a subset S of the positive real numbers by specifying that

$$s \quad \text{is in} \quad S$$

if and only

> there is a point on C arrived at by traveling a distance of s in the positive direction along C from (x_0, y_0).

With a little thought, it should be clear that S must be a subinterval of $(0, \infty)$ (assuming some 'obvious facts' about the nature of curves). One end point of S must clearly be 0. The other endpoint gives us the value s_E. In particular, letting C^+ be that part of C containing all the points arrived at by traveling in the positive direction along C from (x_0, y_0):

1. If C^+ is a closed loop, then $s_E = \infty$ (because we keep going around the loop as s increases).
2. If C^+ is a curve that does not intersect itself, then s_E is the length of C^+ (which may be infinite).

Obviously, similar comments can be made regarding the definition of s_S.

Finishing the Proof of Theorem 39.2

Let us now use the arclength parameterization $(\tilde{x}, \tilde{y})$ to define a function $\tilde{t}$ of s by

$$\tilde{t}(s) = \int_0^s \frac{1}{\|\mathbf{F}(\tilde{\mathbf{x}}(\sigma))\|}\, d\sigma \qquad \text{for} \quad s_S < s < s_E \quad .$$

Since C contains no critical points, the integrand is always finite and positive, and the above function is a differentiable steadily increasing function with

$$\tilde{t}\,'(s) = \frac{1}{\|\mathbf{F}(\tilde{\mathbf{x}}(s))\|} \qquad \text{for} \quad s_S < s < s_E \quad .$$

Consequently, for each s in (s_E, s_E), there is exactly one corresponding t with $t = \tilde{t}(s)$. Thus, we can invert this relationship, defining a function $\tilde{s}$ by

$$\tilde{s}(t) = s \quad \Longleftrightarrow \quad t = \tilde{t}(s) \quad .$$

The function $\tilde{s}$ is defined on the interval (t_S, t_E) where

$$t_S = \lim_{s \to s_S{}^+} \tilde{t}(s) \qquad \text{and} \qquad t_E = \lim_{s \to s_E{}^-} \tilde{t}(s) \quad .$$

By definition,

$$s = \tilde{s}\left(\tilde{t}(s)\right) \qquad \text{for} \quad s_S < s < s_E \quad .$$

From this, the chain rule and the above formula for $\tilde{t}\,'$, we get

$$1 = \frac{ds}{ds} = \frac{d}{ds}\left[\tilde{s}\left(\tilde{t}(s)\right)\right] = \tilde{s}\,'\left(\tilde{t}(s)\right)\tilde{t}\,'(s) = \tilde{s}\,'\left(\tilde{t}(s)\right) \cdot \frac{1}{\|\mathbf{F}(\tilde{\mathbf{x}}(s))\|} \quad .$$

Hence,

$$\tilde{s}\,'\left(\tilde{t}(s)\right) = \|\mathbf{F}(\tilde{\mathbf{x}}(s))\| \quad . \tag{39.14}$$

Now let

$$(x(t), y(t)) = (\tilde{x}(\tilde{s}(t)), \tilde{y}(\tilde{s}(t))) \qquad \text{for} \quad t_S < t < t_E \quad .$$

Observe that $(x(t), y(t))$ will trace out C as t varies from t_S to t_E, and that

$$\mathbf{F}(\mathbf{x}(t)) = \mathbf{F}\left(\tilde{\mathbf{x}}(\tilde{s}(t))\right)$$

where

$$\mathbf{x}(t) = \begin{bmatrix} x(t) \\ y(t) \end{bmatrix} = \begin{bmatrix} \tilde{x}(\tilde{s}(t)) \\ \tilde{y}(\tilde{s}(t)) \end{bmatrix} = \tilde{\mathbf{x}}(\tilde{s}(t)) \quad .$$

The differentiation of this (using the chain rule applied to the components), along with equations (39.13) and (39.14), then yields

$$\mathbf{x}'(t) = \frac{d}{dt}\left[\tilde{\mathbf{x}}(\tilde{s}(t))\right] = \tilde{\mathbf{x}}\,'(\tilde{s}(t)) \cdot \tilde{s}\,'(t) = \frac{\mathbf{F}(\mathbf{x}(t))}{\|\mathbf{F}(\mathbf{x}(t))\|} \cdot \|\mathbf{F}(\mathbf{x}(t)\| = \mathbf{F}(\mathbf{x}(t)) \quad ,$$

completing our proof of Theorem 39.2. ∎

Additional Exercises

39.1. *Find all the constant/equilibrium solutions to each of the following systems:*

a. $\begin{bmatrix} x' \\ y' \end{bmatrix} = \begin{bmatrix} 2x - 5y \\ 3x - 7y \end{bmatrix}$

b. $\begin{bmatrix} x' \\ y' \end{bmatrix} = \begin{bmatrix} 2x - 5y + 4 \\ 3x - 7y + 5 \end{bmatrix}$

c. $\begin{bmatrix} x' \\ y' \end{bmatrix} = \begin{bmatrix} 3x + y \\ 6x + 2y \end{bmatrix}$

d. $\begin{bmatrix} x' \\ y' \end{bmatrix} = \begin{bmatrix} xy - 6y \\ x - y - 5 \end{bmatrix}$

e. $\begin{bmatrix} x' \\ y' \end{bmatrix} = \begin{bmatrix} x^2 - y^2 \\ x^2 - 6x + 8 \end{bmatrix}$

f. $\begin{bmatrix} x' \\ y' \end{bmatrix} = \begin{bmatrix} x\sin(y) \\ x^2 - 6x + 9 \end{bmatrix}$

g. $\begin{bmatrix} x' \\ y' \end{bmatrix} = \begin{bmatrix} 4x - xy \\ x^2y + y^3 - x^2 - y^2 \end{bmatrix}$

h. $\begin{bmatrix} x' \\ y' \end{bmatrix} = \begin{bmatrix} x^2 + y^2 + 4 \\ 2x - 6y \end{bmatrix}$

39.2. *Sketch the direction field for the system*

$$\begin{bmatrix} x' \\ y' \end{bmatrix} = \begin{bmatrix} -x + 2y \\ 2x - y \end{bmatrix}$$

a. on the 2×2 grid with $(x, y) = (0, 0)$, $(2, 0)$, $(0, 2)$ and $(2, 2)$.

b. on the 3×3 grid with $x = -1$, 0 and 1, and with $y = -1$, 0 and 1.

39.3. *Sketch the direction field for the system*

$$\begin{bmatrix} x' \\ y' \end{bmatrix} = \begin{bmatrix} (1 - 2x)(y + 1) \\ x - y \end{bmatrix}$$

on the 3×3 grid with $x = 0$, $1/2$ and 1, and with $y = 0$, $1/2$ and 1.

39.4. *A direction field for*

$$\begin{bmatrix} x' \\ y' \end{bmatrix} = \begin{bmatrix} -x + 2y \\ 2x - y \end{bmatrix}$$

has been sketched to the right. Using this system and direction field:

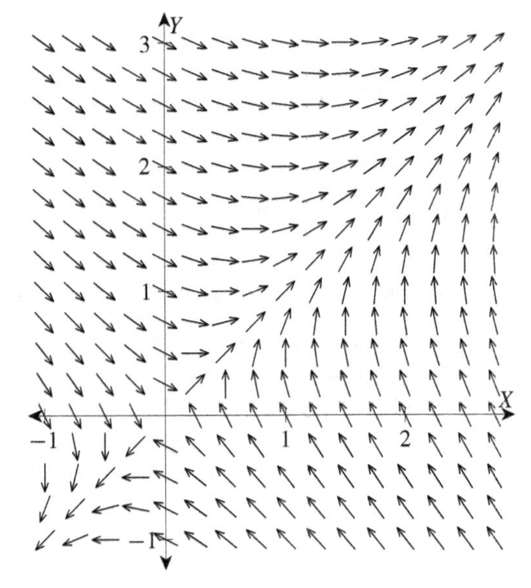

a. *Find and plot all the critical points.*

b. *Sketch the trajectories that go through points $(1, 0)$ and $(0, 1)$.*

c. *Sketch a phase portrait for this system.*

d. *Suppose $(x(t), y(t))$ is the solution to this system satisfying $(x(0), y(0)) = (1, 0)$. What apparently happens to $x(t)$ and $y(t)$ as t gets large?*

e. *As well as you can, decide whether the critical point found above is asymptotically stable, stable but not asymptotically stable, or unstable.*

39.5. A direction field for

$$\begin{bmatrix} x' \\ y' \end{bmatrix} = \begin{bmatrix} -x + 2y \\ -2x + 2 \end{bmatrix}$$

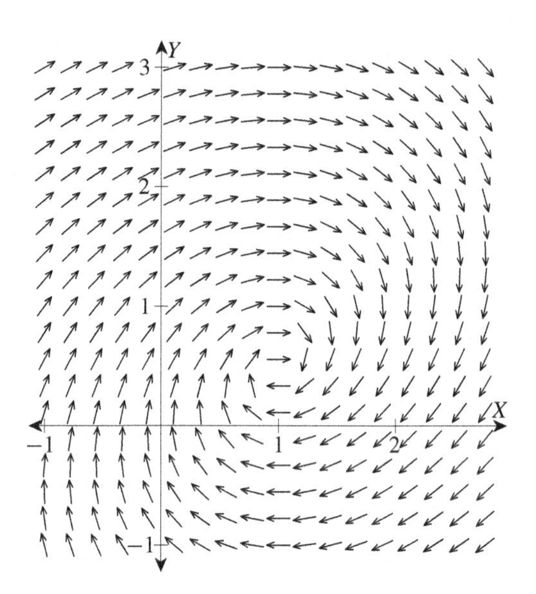

has been sketched to the right. Using this system and direction field:

a. Find and plot all the critical points.

b. Sketch the trajectories that go through points $(-1, 0)$ and $(0, 2)$.

c. Sketch a phase portrait for this system.

d. Suppose $(x(t), y(t))$ is the solution to this system satisfying $(x(0), y(0)) = (-1, 0)$. What apparently happens to $x(t)$ and $y(t)$ as t gets large?

e. As well as you can, decide whether the critical point found in part **a** is asymptotically stable, stable but not asymptotically stable, or unstable.

39.6. A direction field for

$$\begin{bmatrix} x' \\ y' \end{bmatrix} = \begin{bmatrix} y \\ -\sin(2x) \end{bmatrix}$$

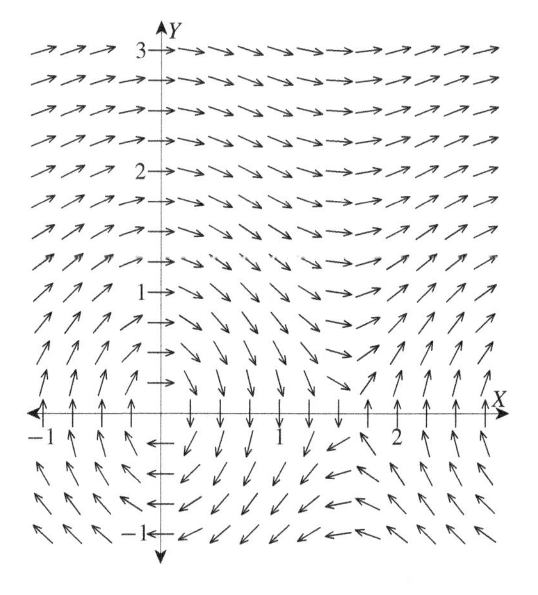

has been sketched to the right. Using this system and direction field:

a. Find and plot all the critical points.

b. All the critical points of this system are either stable (but not asymptotically stable) or unstable. Using this direction field, determine which are stable and which are unstable.

c. Sketch the trajectories that go through points $(1, 0)$ and $(0, 2)$.

d. Sketch a phase portrait for this system.

e. Suppose $(x(t), y(t))$ is the solution to this system satisfying $(x(0), y(0)) = (1, 0)$. What apparently happens to $x(t)$ and $y(t)$ as t gets large?

39.7. The system

$$\begin{bmatrix} x' \\ y' \end{bmatrix} = \begin{bmatrix} x - 2y - 1 \\ 2x - y - 2 \end{bmatrix}$$

has one critical point and that point is stable, but not asymptotically stable. A direction field for this system has been sketched to the right. Using this information:

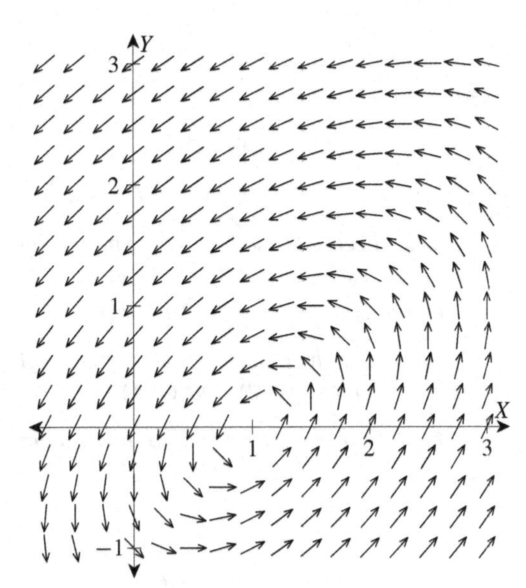

a. Find and plot the critical point.

b. Sketch the trajectories that go through points $(0,0)$ and $(0,1)$.

c. Sketch a phase portrait for this system.

d. Suppose $(x(t), y(t))$ is the solution to this system satisfying $(x(0), y(0)) = (0,0)$. What apparently happens to $x(t)$ and $y(t)$ as t gets large?

39.8. A direction field for

$$\begin{bmatrix} x' \\ y' \end{bmatrix} = \begin{bmatrix} -2x + y + 1 \\ -x - 4y + 5 \end{bmatrix}$$

has been sketched to the right. Using this system and direction field:

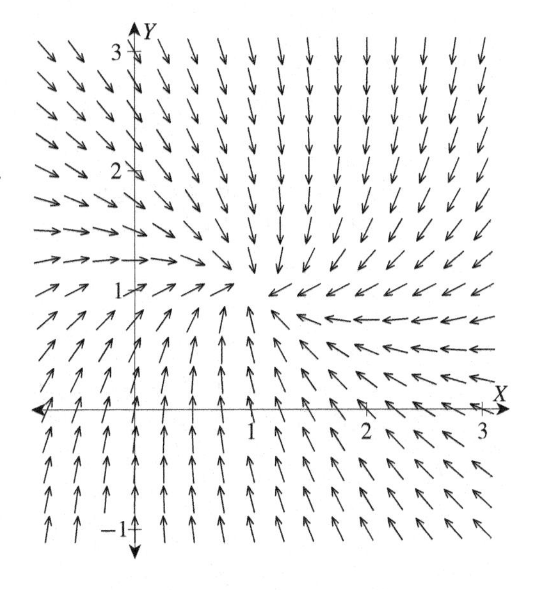

a. Find and plot the one critical point of this system.

b. Decide whether this critical point is asymptotically stable, stable but not asymptotically stable, or unstable.

c. Sketch the trajectories that go through points $(2,0)$ and $(0,2)$.

d. Sketch a phase portrait for this system.

e. Suppose $(x(t), y(t))$ is the solution to this system satisfying $(x(0), y(0)) = (1,0)$. What apparently happens to $x(t)$ and $y(t)$ as t gets large?

39.9. *A direction field for*

$$\begin{bmatrix} x' \\ y' \end{bmatrix} = \begin{bmatrix} x + 4y^2 - 1 \\ 2x - y - 2 \end{bmatrix}$$

has been sketched to the right. Using this system and direction field:

a. *Find and plot all the critical points.*

b. *Sketch the trajectories that go through the points* $(0, 0)$ *and* $(1, 1)$.

c. *Sketch a phase portrait for this system.*

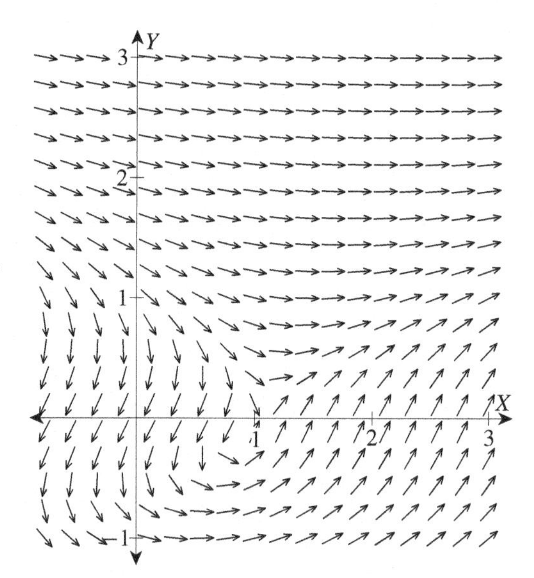

39.10. *Look up the commands for generating direction fields for systems of differential equations in your favorite computer math package. Then, use these commands to do the following for each problem below:*

i. *Sketch the indicated direction field for the given system.*

ii. *Use the resulting direction field to sketch (by hand) a phase portrait for the system.*

a. *The system is*

$$\begin{bmatrix} x' \\ y' \end{bmatrix} = \begin{bmatrix} -x + 2y \\ 2x - y \end{bmatrix} .$$

Use a 25×25 *grid on the region* $-1 \leq x \leq 3$ *and* $-1 \leq y \leq 3$. *(Compare the resulting direction field to the direction arrows computed in Exercise 39.2.)*

b. *The system is*

$$\begin{bmatrix} x' \\ y' \end{bmatrix} = \begin{bmatrix} (2x - 1)(y + 1) \\ y - x \end{bmatrix} .$$

Use a 25×25 *grid on the region* $-1 \leq x \leq 3$ *and* $-1 \leq y \leq 3$. *(Compare the resulting direction field to the direction field found in Exercise 39.3.)*

c. *The system is*

$$\begin{bmatrix} x' \\ y' \end{bmatrix} = \begin{bmatrix} x + 4y^2 - 1 \\ 2x - y - 2 \end{bmatrix} .$$

Use a 25×25 *grid on the region* $0 \leq x \leq 2$ *and* $-1 \leq y \leq 1$. *(This gives a 'close-up' view of the critical points of the system in Exercise 39.9.)*

d. *The system is*

$$\begin{bmatrix} x' \\ y' \end{bmatrix} = \begin{bmatrix} x + 4y^2 - 1 \\ 2x - y - 2 \end{bmatrix} .$$

Use a 25×25 *grid on the region* $3/4 \leq x \leq 5/4$ *and* $-1/4 \leq y \leq 1/4$. *(This gives an even more close-up view of the critical points of the system in Exercise 39.9.)*

39.11. *Consider the initial-value problem*

$$\begin{bmatrix} x' \\ y' \end{bmatrix} = \begin{bmatrix} f(x,y) \\ g(x,y) \end{bmatrix} \quad \text{with} \quad \begin{bmatrix} x(0) \\ y(0) \end{bmatrix} = \begin{bmatrix} a \\ b \end{bmatrix} \quad .$$

Assume the system is regular and autonomous, and that $(\tilde{x}, \tilde{y})$ *is a solution to the above on an interval* (α, β) *containing* 0. *Now let* t_0 *be any other point on the real line, and show that a solution to*

$$\begin{bmatrix} x' \\ y' \end{bmatrix} = \begin{bmatrix} f(x,y) \\ g(x,y) \end{bmatrix} \quad \text{with} \quad \begin{bmatrix} x(t_0) \\ y(t_0) \end{bmatrix} = \begin{bmatrix} a \\ b \end{bmatrix}$$

is then given by

$$(x(t), y(t)) = (\tilde{x}(t - t_0), \tilde{y}(t - t_0)) \quad \text{for} \quad \alpha + t_0 < t < \beta + t_0 \quad .$$

(Hint: Showing that the first-order system is satisfied is a simple chain rule problem.)

39.12. *Using Corollary 39.1 on page 809 on the existence and uniqueness of solutions to regular systems and Theorem 39.2 on page 810 on curves being trajectories, along with (possibly) the results of Exercise 39.11, verify the following:*

a. *Corollary 39.3 on page 810. (Hint: Start by assuming the two trajectories do intersect.)*

b. *Corollary 39.4 on page 810. (Hint: Start by assuming an endpoint of a given maximal trajectory is not a critical point.)*

40

Numerical Methods III: Systems and Higher-Order Equations

In Chapters 10 and 12 we developed methods for finding numerical solutions to a first-order initial-value problem involving a single differential equation [1]

$$\frac{dx}{dt} = f(t, x) \qquad \text{with} \quad x(t_0) = x_0 \quad .$$

As you should recall, a numerical solution to such a problem consists a pair of sequences

$$\{ t_0 , t_1 , t_2 , t_3 , \ldots , t_N \} \qquad \text{and} \qquad \{ x_0 , x_1 , x_2 , x_3 , \ldots , x_N \}$$

where t_0 and x_0 are given by the initial data, and with the other t_k's and x_k's computed so that $x_k \approx y(t_k)$ for each k, with $y = y(t)$ being the actual solution to the initial-value problem. Typically, of course, we seek a numerical solution when it is not practical to find the actual solution.

In this chapter, we will see how to extend those numerical methods to handle systems of differential equations and single higher-order differential equations. For brevity, we will limit our discussion to extending the basic Euler method from Chapter 10. Extending the improved Euler and Runge-Kutta methods from Chapter 12 is similar (though more involved), and can be carried out by the reader on their own (see Exercise 40.10). We will also simplify discussion by limiting our systems to systems of two equations, and limiting our higher-order equations to second order. This is a minor simplification. The reader should have no trouble figuring out how to deal with larger systems and equations of higher orders.

40.1 Brief Review of the Basic Euler Method

First, let us quickly rederive the basics of the Euler method previously described in Chapter 10, but using a slightly different derivation and notation to better fit the applications in the current chapter.

In Chapter 10, we were interested in a method for finding "approximate solutions" for a first-order initial-value problem involving a single differential equation,

$$\frac{dx}{dt} = f(t, x(t)) \qquad \text{with} \quad x(t_0) = x_0 \quad . \tag{40.1}$$

From this differential equation and the fact that

$$\lim_{\Delta t \to 0} \frac{x(t + \Delta t) - x(t)}{\Delta t} = \frac{dx}{dt} \quad ,$$

[1] We are modifying the notation slightly from that used in Chapters 10 and 12 to better handle the problems in this chapter.

we get the approximation

$$\frac{x(t+\Delta t) - x(t)}{\Delta t} \approx \frac{dx}{dt} = f(t, x(t))$$

where Δt is some chosen small positive value (the *step size*). Solving for $x(t+\Delta t)$ then yields the fundamental approximation for the Euler method:

$$x(t+\Delta t) \approx x(t) + \Delta t \cdot f(t, x(t)) \quad . \tag{40.2}$$

The Euler method is based on the above approximation for computing $x(t+\Delta t)$. It generates a numerical solution to problem (40.1) consisting of a pair of sequences

$$\{t_0, t_1, t_2, t_3, \ldots, t_N\} \quad \text{and} \quad \{x_0, x_1, x_2, x_3, \ldots, x_N\}$$

with

$$x_0 = x(t_0)$$

and the subsequent t_k's and x_k's computed successively by

$$t_{k+1} = t_k + \Delta t \quad \text{and} \quad x_{k+1} = x_k + \Delta t \cdot f(t_k, x_k) \quad ,$$

where the step size, Δt, is some prechosen distance. Note that, for $k = 1, \ldots, N$,

$$t_k = t_0 + k\Delta t$$

and, according to approximation (40.2)

$$x_1 = x_0 + \Delta t \cdot f(t_0, x_0) = x(t_0) + \Delta t \cdot f(t_0, x(t_0)) \approx x(t_1) \quad ,$$

$$x_2 = x_1 + \Delta t \cdot f(t_1, x_1) \approx x(t_1) + \Delta t \cdot f(t_1, x(t_1)) \approx x(t_2) \quad ,$$

$$x_3 = x_2 + \Delta t \cdot f(t_2, x_2) \approx x(t_2) + \Delta t \cdot f(t_2, x(t_2)) \approx x(t_3) \quad ,$$

and so on. Thus, each x_k is an approximation to $x(t_k)$. Just how good an approximation depends on the choice of Δt and just how well behaved the function f is. That was discussed in detail in Chapter 10.

40.2 The Euler Method for First-Order Systems

Generalities

Now let's suppose we have a first-order 2×2 system

$$\begin{aligned} x' &= f(t, x, y) \\ y' &= g(t, x, y) \end{aligned} \tag{40.3}$$

with initial conditions

$$x(t_0) = x_0 \quad \text{and} \quad y(t_0) = y_0 \quad .$$

(Remember $x' = {}^{dx}/_{dt}$ and $y' = {}^{dy}/_{dt}$.) Clearly, the arguments just given in the previous section can be repeated using both equations, yielding the system of approximations

$$\begin{aligned} x(t+\Delta t) &\approx x(t) + \Delta t \cdot f(t, x(t), y(t)) \\ y(t+\Delta t) &\approx y(t) + \Delta t \cdot g(t, x(t), y(t)) \end{aligned} \quad . \tag{40.4}$$

Accordingly, we can extend the Euler method from the previous section to a method for generating sequences

$$\{ t_0 , t_1 , t_2 , t_3 , \ldots , t_N \}$$

and

$$\{ (x_0, y_0) , (x_1, y_1) , (x_2, y_2) , (x_3, y_3) , \ldots , (x_N, y_N) \}$$

making up a *numerical solution* to system (40.3). In these sequences, t_0, x_0 and y_0 come straight from the initial data,

$$x(t_0) = x_0 \qquad \text{and} \qquad y(t_0) = y_0 \quad ,$$

and the subsequent t_k's, x_k's and y_k's are computed iteratively by

$$t_{k+1} = t_k + \Delta t \quad , \tag{40.5a}$$

$$x_{k+1} = x_k + \Delta t \cdot f(t_k, x_k, y_k) \tag{40.5b}$$

and

$$y_{k+1} = y_k + \Delta t \cdot g(t_k, x_k, y_k) \quad , \tag{40.5c}$$

where the step size, Δt, is some prechosen distance. From the definition of t_k it immediately follows that

$$t_k = t_0 + k\Delta t \quad ,$$

and from approximation (40.4) the definition of the x_k's and y_k's it should be clear that

$$x_k \approx x(t_k) \qquad \text{and} \qquad y_k \approx y(t_k) \qquad \text{for} \quad k = 1, 2, \ldots, N \quad .$$

Again, just how good these approximations are will depend on the choice of Δt and on how well behaved the functions f and g are. We'll discuss this more, later.

Implementing the Euler Method for Systems

In Section 10.2, we saw a step-by-step implementation of the Euler method to solve a first-order initial-value problem involving a single differential equation. The analog of this procedure for generating a numerical solution for a 2×2 first-order system

$$\begin{aligned} x' &= f(t, x, y) \\ y' &= g(t, x, y) \end{aligned} \tag{40.6}$$

with initial conditions

$$x(t_0) = x_0 \qquad \text{and} \qquad y(t_0) = y_0$$

is, unsurprisingly, quite similar. For completeness we'll describe this procedure in detail, illustrating its use with the system

$$\begin{aligned} x' &= x + \frac{1}{2}y \\ y' &= \frac{3}{4}x - \frac{1}{4}y \end{aligned} \tag{40.7a}$$

and initial conditions

$$x(0) = 0 \qquad \text{and} \qquad y(0) = 1 \quad . \tag{40.7b}$$

Preliminary Steps

1. Write out the initial-value problem as

$$x' = f(t, x, y)$$

$$y' = g(t, x, y)$$

with initial conditions

$$x(t_0) = x_0 \quad \text{and} \quad y(t_0) = y_0 \quad ,$$

and, from this, determine the formulas for $f(t, x, y)$ and $g(t, x, y)$, and the values of t_0, x_0 and y_0.

Our example is

$$x' = x + \frac{1}{2}y$$

$$y' = \frac{3}{4}x - \frac{1}{4}y$$

with initial conditions

$$x(0) = 0 \quad \textit{and} \quad y(0) = 1 \quad .$$

So,

$$f(t, x, y) = x + \frac{1}{2}y \quad , \quad g(t, x, y) = \frac{3}{4}x - \frac{1}{4}y \quad ,$$

$$t_0 = 0 \quad , \quad x_0 = 0 \quad \textit{and} \quad y_0 = 1 \quad .$$

2. Decide on a distance Δt for the step size, a positive integer N for the maximum number of steps, and a maximum value desired for t, $t_{\max}$. Choose these quantities so that

$$t_{\max} = t_N = t_0 + N\Delta t \quad .$$

For no good reason whatsoever, let us pick

$$\Delta t = \frac{1}{2} \quad \textit{and} \quad N = 6 \quad .$$

Then

$$t_{max} = t_0 + N\Delta t = 0 + 6 \cdot \frac{1}{2} = 3 \quad .$$

Iterative Steps

1, 2, ... For each integer $k \geq 0$, successively compute t_{k+1}. x_{k+1} and y_{k+1} using

$$t_{k+1} = t_k + \Delta t \quad ,$$

$$x_{k+1} = x_k + \Delta t \cdot f(t_k, x_k, y_k)$$

and

$$y_{k+1} = y_k + \Delta t \cdot g(t_k, x_k, y_k)$$

with the values of t_k, x_k and y_k found in a previous step. Continue repeating these computations until t_N, x_N and y_N are computed. (Be sure to save the values computed!)

For our example (omitting many computational details):

Step 1. With $k = 0$:

$$\begin{aligned} t_1 &= t_{0+1} = t_0 + \Delta t = 0 + \frac{1}{2} = \frac{1}{2} \quad , \\ x_1 &= x_{0+1} = x_0 + \Delta t \cdot f(t_0, x_0, y_0) \\ &= 0 + \frac{1}{2} \cdot f(0, 0, 1) \\ &= 0 + \frac{1}{2}\left[0 + \frac{1}{2} \cdot 1\right] = \frac{1}{4} = 0.25 \end{aligned}$$

and

$$\begin{aligned} y_1 &= y_{0+1} = y_0 + \Delta t \cdot g(t_0, x_0, y_0) \\ &= 0 + \frac{1}{2} \cdot g(0, 0, 1) \\ &= 1 + \frac{1}{2}\left[\frac{3}{4} \cdot 0 - \frac{1}{4} \cdot 1\right] = \frac{7}{8} = 0.875 \quad . \end{aligned}$$

Step 2. With $k = 1$:

$$\begin{aligned} t_2 &= t_{1+1} = t_1 + \Delta t = \frac{1}{2} + \frac{1}{2} = 1 \quad , \\ x_2 &= x_{1+1} = x_1 + \Delta t \cdot f(t_1, x_1, y_1) \\ &= \frac{1}{4} + \frac{1}{2} \cdot f\left(\frac{1}{2}, \frac{1}{4}, \frac{7}{8}\right) \\ &= \frac{1}{4} + \frac{1}{2}\left[\frac{1}{4} + \frac{1}{2} \cdot \frac{7}{8}\right] = \frac{19}{32} = 0.59375 \end{aligned}$$

and

$$\begin{aligned} y_2 &= y_{1+1} = y_2 + \Delta t \cdot g(t_1, x_1, y_1) \\ &= \frac{7}{8} + \frac{1}{2} \cdot g\left(\frac{1}{2}, \frac{1}{4}, \frac{7}{8}\right) \\ &= \frac{7}{8} + \frac{1}{2}\left[\frac{3}{4} \cdot \frac{1}{4} - \frac{1}{4} \cdot \frac{7}{8}\right] = \frac{55}{64} = 0.859375 \quad . \end{aligned}$$

Step 3. With $k = 2$,

$$\begin{aligned} t_3 &= t_{2+1} = t_2 + \Delta t = 1 + \frac{1}{2} = \frac{3}{2} \quad , \\ x_3 &= x_{2+1} = x_2 + \Delta t \cdot f(t_2, x_2, y_2) \\ &= \frac{19}{32} + \frac{1}{2} \cdot f\left(1, \frac{19}{32}, \frac{55}{64}\right) \\ &= \frac{19}{32} + \frac{1}{2}\left[\frac{19}{32} + \frac{1}{2} \cdot \frac{55}{64}\right] = \frac{283}{256} = 1.105468\ldots \end{aligned}$$

and

$$\begin{aligned} y_3 &= y_{2+1} = y_2 + \Delta t \cdot g(t_2, x_2, y_2) \\ &= \frac{55}{64} + \frac{1}{2} \cdot g\left(1, \frac{19}{32}, \frac{55}{64}\right) \\ &= \frac{55}{64} + \frac{1}{2}\left[\frac{3}{4} \cdot \frac{19}{32} - \frac{1}{4} \cdot \frac{55}{64}\right] = \frac{499}{512} = 0.974609\ldots \quad . \end{aligned}$$

k	t_k	x_k	y_k
0	0	0	1
1	0.5	0.2500	0.8750
2	1.0	0.5938	0.8594
3	1.5	1.1055	0.9746
4	2.0	1.9019	1.2673
5	2.5	3.1696	1.8221
6	3.0	5.2100	2.7830

(a)

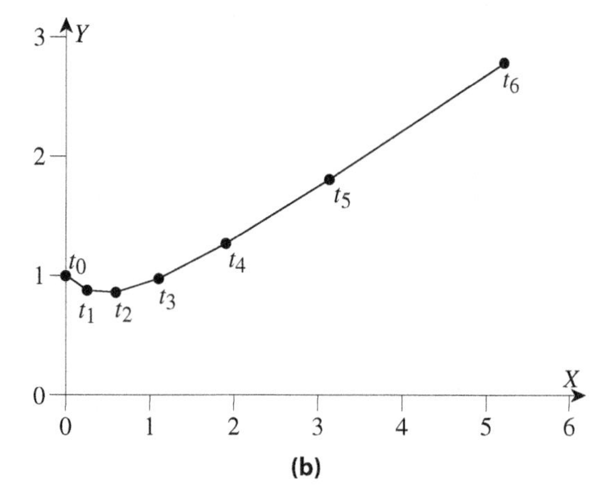

(b)

Figure 40.1: Results of Euler's method to numerically solve initial-value problem (40.7) on page 823 using $\Delta t = {}^1\!/_2$ and $N = 6$: **(a)** The numerical solution in which $x_k \approx x(t_k)$ and $y_k \approx y(t_k)$ (for $k \geq 2$, the values of x_k and y_k are to the nearest 0.0001). **(b)** The piecewise linear path for this approximate solution.

Step 4. With $k = 3$,

$$t_4 = t_{3+1} = t_3 + \Delta t = \frac{3}{2} + \frac{1}{2} = 2 \quad ,$$

$$\begin{aligned} x_4 = x_{3+1} &= x_3 + \Delta t \cdot f(t_3, x_3, y_3) \\ &= \frac{283}{256} + \frac{1}{2} \cdot f\left(1, \frac{283}{256}, \frac{499}{512}\right) = \cdots = 1.901855\ldots \end{aligned}$$

and

$$\begin{aligned} y_4 = y_{3+1} &= y_3 + \Delta t \cdot g(t_3, x_3, y_3) \\ &= \frac{499}{512} + \frac{1}{2} \cdot g\left(1, \frac{283}{256}, \frac{499}{512}\right) = \cdots = 1.267333\ldots \quad . \end{aligned}$$

Step 5. With $k = 4$,

$$t_5 = t_{4+1} = t_4 + \Delta t = 2 + \frac{1}{2} = \frac{5}{2} \quad ,$$

$$x_5 = x_{4+1} = x_4 + \Delta t \cdot f(t_4, x_4, y_4) = \cdots = 3.169616\ldots$$

and

$$y_5 = y_{4+1} = y_4 + \Delta t \cdot g(t_4, x_4, y_4) = \cdots = 1.822113\ldots \quad .$$

Step 6. With $k = 5$,

$$t_6 = t_{5+1} = t_5 + \Delta t = \frac{5}{2} + \frac{1}{2} = 3 \quad ,$$

$$x_6 = x_{5+1} = x_5 + \Delta t \cdot f(t_5, x_5, y_5) = \cdots = 5.209953\ldots$$

and

$$y_6 = y_{5+1} = y_5 + \Delta t \cdot g(t_5, x_5, y_5) = \cdots = 2.782955\ldots \quad .$$

Since we had chosen $N = 6$ and we've just computed t_6, x_6 and y_6, we are done with our computations.

A summary of our results are given in the table in Figure 40.1a.

As with the original Euler method described in Chapter 10, the use of a computer to carry out the above computations is highly recommended, especially since most applications require much smaller values of Δt (and correspondingly larger values of N) than used in our example to reduce the errors in the computations. You can either write your own program or set up a worksheet using your favorite computer math package. Be sure to be able to easily change

$$\Delta t \quad , \quad N \text{ or } t_{max} \quad , \quad f(x,y) \quad , \quad g(x,y) \quad , \quad t_0 \quad , \quad x_0 \quad \text{and} \quad y_0 \quad ,$$

so you can easily change the computational parameters and consider different initial-value problems.[2]

Using the Results of the Method

Tables and Graphs

What you do with the results of your computations depends on why you are doing these computations. If N is not too large, it may be a good idea to write the obtained values of the t_k's, x_k's and y_k's in a table for convenient reference (with a note that $x_k \approx x(t_k)$ and $y_k \approx y(t_k)$ for each k) as done in Figure 40.1a for our example. It may also be enlightening to sketch (or have a computer sketch) the corresponding piecewise linear approximation to the path of $(x(t), y(t))$ as t goes from t_0 to t_N . Simply plot each (x_k, y_k) on an XY–plane and then connect the successive points with straight lines as in Figure 40.1b for our example.

General Observations on Accuracy

Keep in mind that each (x_k, y_k) generated by Euler's method is an approximation to $(x(t_k), y(t_k))$, the solution at t_k to some initial-value problem

$$x' = f(t, x, y)$$

$$y' = g(t, x, y)$$

with

$$x(t_0) = x_0 \qquad \text{and} \qquad y(t_0) = y_0 \quad .$$

Just as with the Euler method discussed in Chapter 10, the error in using (x_k, y_k) for $(x(t_k), y(t_k))$ can increase significantly as k increases. If the system is autonomous (as in our example), it may be a good idea to draw that piecewise linear approximation to the path of $(x(t), y(t))$ on top of a direction field for the system, possibly along with a carefully sketched trajectory through (x_0, y_0) to get a visual idea of the accuracy of your numerical solution.

Also, as with the Euler method discussed in Chapter 10, we can typically reduce the errors in using the Euler method by taking smaller step sizes with appropriately larger numbers of steps. To be a little more precise, we will see that if f and g are "reasonably smooth" functions, then there is a corresponding constant M such that

$$|x(t_k) - x_k| < M \cdot \Delta t \qquad \text{and} \qquad |y(t_k) - y_k| < M \cdot \Delta t \tag{40.8}$$

for any numerical solution generated by the Euler method to the above initial-value problem on an interval $[t_0, t_0 + L]$. Just what is meant by "f and g are 'reasonably smooth' functions" will be discussed in the next subsection.

It also should be noted that, just as with the analogous error bound for the Euler method discussed in Chapter 10, the value of M in the above error bound may be large and very difficult to determine. In practice, it is common to simply find numerical solutions to a given problem by using smaller and smaller values of Δt , stopping only when the differences between the different solutions become "insignificant" .

[2] In fact, your favorite computer math package may well already have some version of the above in its repertoire.

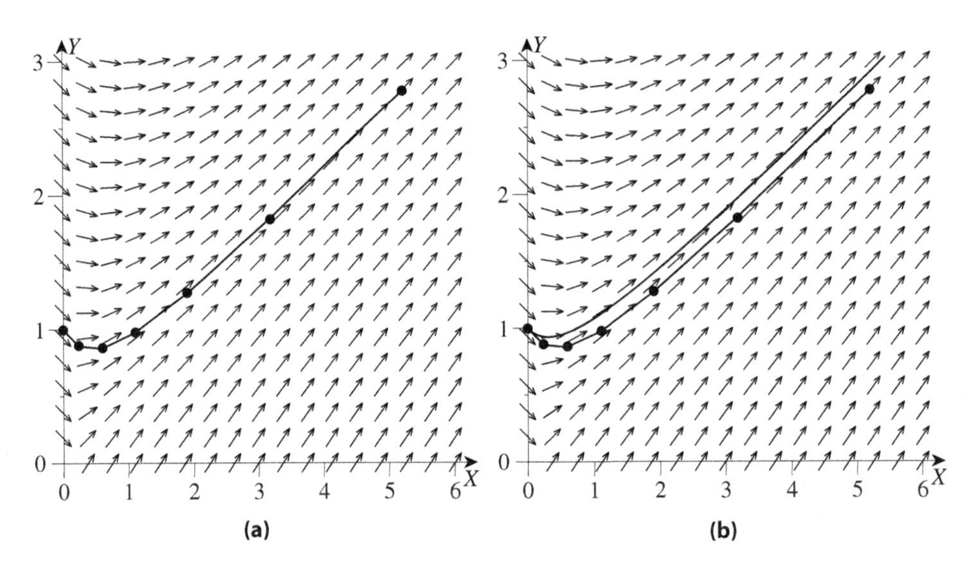

Figure 40.2: Comparison of the piecewise linear path for the approximate solution to initial-value problem (40.7) using Euler's method with $\Delta t = {}^1\!/_2$ and $N = 6$: **(a)** To a direction field for system (40.7a) **(b)** To a direction field and an actual trajectory for the solution to the initial-value problem.

!► *Example 40.1:* *In Figure 40.2a, the piecewise linear approximation sketched in Figure 40.1b is superimposed on a direction field for the system. Observe that this approximation follows the direction field fairly well except where $0 < x < {}^3\!/_2$. There, the approximation is clearly "undershooting" the path indicated by the direction arrows. Consequently, in this application of the Euler method, we should expect that the earlier computations are generating significant errors that, in turn, affect the later computations. This is confirmed in Figure 40.2b in which the actual trajectory through (x_0, y_0) is sketched. Clearly, a much smaller value of Δt than that used here should be used if a truly accurate approximation is desired.*

The Error Bound

So let us consider any numerical solution generated by the Euler method for the initial-value problem

$$\begin{aligned} x' &= f(t, x, y) \\ y' &= g(t, x, y) \end{aligned} \tag{40.9a}$$

with

$$x(t_0) = x_0 \quad \text{and} \quad y(t_0) = y_0 \quad . \tag{40.9b}$$

Let L be the length of the interval over which we are interested (so $t_0 \leq t \leq t_0 + L$ and let

$$\mathcal{R} = \{(t, x, y) : t_0 \leq t \leq t_0 + L \quad , \quad x_{\min} < x < x_{\max} \quad \text{and} \quad y_{\min} < y < y_{\max}\}$$

be a region on TXY–space such that all the following holds:

1. f , g and all first partial derivatives of f and g are continuous, bounded functions on $\mathcal{R}$.

2. There is a unique solution $(x(t), y(t))$ to the given initial-value problem valid for $t_0 \leq t \leq t_0 + L$, and each point $(t, x(t), y(t))$ is in $\mathcal{R}$ when $t_0 \leq t \leq t_0 + L$.

3. If
$$\{ t_0 \, , \, t_1 \, , \, t_2 \, , \, t_3 \, , \, \ldots \, , \, t_N \, \}$$
and
$$\{ (x_0, y_0) \, , \, (x_1, y_1) \, , \, (x_2, y_2) \, , \, (x_3, y_3) \, , \, \ldots \, , \, (x_N, y_N) \}$$
is any numerical solution generated by the Euler method to our initial-value problem with $t_N \leq t_0 + L$, then each point (t_k, x_k, y_k) is in $\mathcal{R}$.

We can now state the theorem on which error bound (40.8) is based.

Theorem 40.1 (Error bound for Euler's method)
Let f, g, t_0, x_0, y_0, L and $\mathcal{R}$ be as above, and let $(x, y) = (x(t), y(t))$ be the true solution to initial-value problem (40.6). Then there is a finite constant M such that, for $k = 0, 1, \ldots, N$,
$$|x(t_k) - x_k| \; < \; M \cdot \Delta t \qquad \text{and} \qquad |y(t_k) - y_k| \; < \; M \cdot \Delta t$$
whenever
$$\{ t_0 \, , \, t_1 \, , \, t_2 \, , \, t_3 \, , \, \ldots \, , \, t_N \, \}$$
and
$$\{ (x_0, y_0) \, , \, (x_1, y_1) \, , \, (x_2, y_2) \, , \, (x_3, y_3) \, , \, \ldots \, , \, (x_N, y_N) \}$$
is a numerical solution to initial-value problem (40.9) obtained from Euler's method with step size Δt and total number of steps N satisfying
$$0 \; < \; \Delta t \cdot N \; \leq \; L \quad .$$

The proof of this theorem can be obtained through a straightforward modification of the proof of Theorem 10.1 on page 188.

40.3 Extending Euler's Method to Second-Order Differential Equations

By combining the previous section's Euler method for numerically solving a 2×2 first-order system with the method for constructing a 2×2 first-order system from a single second-order differential equation from Section 38.3, we can easily develop a procedure for numerically solving a single second-order initial-value problem, such as
$$y'' \; - \; ty \; = \; 0 \qquad \text{with} \quad y(0) = 1 \quad \text{and} \quad y'(0) = 0 \quad .$$

You may recognize the above differential equation as Airy's equation, an equation first encountered in Chapter 33. There we constructed the power series solution for Airy's equation.

By the way, let us, for reasons soon to be obvious, observe that Airy's equation can be written as
$$y'' \; = \; ty \quad .$$

The Basic Extensions

Let's start with a generic second-order differential equation written in the form

$$y'' = h(t, y, y') \quad ,$$

and assume we want to find the solution satisfying some predetermined initial conditions at t_0,

$$y(t_0) = y_0 \qquad \text{and} \qquad y'(t_0) = y_1 \quad .$$

Following the discussion in Section 38.3, we convert this second-order problem to a first-order 2×2 system problem by introducing a new function, $x = x(t)$, with $x = y'$. Then

$$x' = y'' = h(t, y, y') = h(t, y, x) \quad ,$$

and we have the 2×2 first-order system

$$\begin{aligned} x' &= h(t, y, x) \\ y' &= x \end{aligned} \tag{40.10}$$

with initial conditions

$$x_0 = x(t_0) = y'(t_0) = y_1 \qquad \text{and} \qquad y_0 = y(t_0) \quad .$$

But system (40.10) is just system (40.3) on page 822 with

$$f(t, x, y) = h(t, y, x) \qquad \text{and} \qquad g(t, x, y) = x \quad .$$

So, for system (40.10), approximation (40.4) is

$$\begin{aligned} x(t + \Delta t) &\approx x(t) + \Delta t \cdot f(t, x(t), y(t)) = x(t) + \Delta t \cdot h(t, y(t), x(t)) \\ y(t + \Delta t) &\approx y(t) + \Delta t \cdot g(t, x(t), y(t)) = y(t) + \Delta t \cdot x(t) \end{aligned} \quad , \tag{40.11}$$

and the corresponding equations for the iterative step in the Euler method are

$$t_{k+1} = t_k + \Delta t \quad ,$$

$$x_{k+1} = x_k + \Delta t \cdot f(t_k, x_k, y_k) = x_k + \Delta t \cdot h(t_k, y_k, x_k)$$

and

$$y_{k+1} = y_k + \Delta t \cdot g(t_k, x_k, y_k) = y_k + \Delta t \cdot x_k \quad .$$

Warning: Observe that, in equation set (40.10), the symbols x and y in $h(t, y, x)$ are not in the order usually expected. Forgetting this fact and using $h(t, x, y)$ in your computations will lead to completely incorrect results.

The Procedure

For completeness and clarity, let us outline a procedure for numerically solving a second-order initial-value problem, using

$$y'' - ty = 0 \qquad \text{with} \quad y(0) = 1 \quad \text{and} \quad y'(0) = 0 \tag{40.12}$$

to illustrate each step.

Preliminary Steps

1. Set $x = y'$, rewrite the initial-value problem as

$$y'' = h(t, y, x) \qquad \text{with} \quad y(t_0) = y_0 \quad \text{and} \quad x(t_0) = y_1$$

and, from this, identify the formula for $h(t, y, x)$ and the initial values t_0, y_0 and $x_0 = x(t_0)$. *Careful: The x and y symbols in $h(t, y, x)$ are in reverse order from what you may normally expect.*

For our example, we have

$$y'' = ty \qquad \textit{with} \quad y(0) = 1 \quad \textit{and} \quad x(0) = y'(0) = 0 \quad .$$

So,

$$h(t, y, x) = ty \quad , \quad t_0 = 0 \quad , \quad x_0 = 0 \quad \textit{and} \quad y_0 = 1 \quad .$$

2. Decide on a distance Δt for the step size, a positive integer N for the maximum number of steps, and a maximum value desired for t, $t_{\max}$. Choose these quantities so that

$$t_{\max} = t_N = t_0 + N\Delta t \quad .$$

As usual, let us pick

$$\Delta t = \frac{1}{2} \qquad \textit{and} \qquad N = 6 \quad .$$

Then

$$t_{\max} = t_0 + N\Delta t = 0 + 6 \cdot \frac{1}{2} = 3 \quad .$$

Iterative Steps

1, 2, ... For each integer $k \geq 0$, successively compute t_{k+1}. x_{k+1} and y_{k+1} using

$$t_{k+1} = t_k + \Delta t \quad ,$$

$$x_{k+1} = x_k + \Delta t \cdot h(t_k, y_k, x_k)$$

and

$$y_{k+1} = y_k + \Delta t \cdot x_k$$

with the values of t_k, x_k and y_k found in a previous step. Continue repeating these computations until t_N, x_N and y_N are computed. (Be sure to save the values computed!)

For our example (omitting many computational details):

Step 1. *With* $k = 0$,

$$t_1 = t_{0+1} = t_0 + \Delta t = 0 + \frac{1}{2} = \frac{1}{2} \quad ,$$

$$\begin{aligned} x_1 = x_{0+1} &= x_0 + \Delta t \cdot h(t_0, y_0, x_0) \\ &= 0 + \frac{1}{2} \cdot h(0, 1, 0) \\ &= 0 + \frac{1}{2}[0 \cdot 1] = 0 \end{aligned}$$

and

$$y_1 = y_{0+1} = y_0 + \Delta t \cdot x_0$$
$$= 1 + \frac{1}{2} \cdot 0 = 1 \quad .$$

Step 2. With $k = 1$,

$$t_2 = t_{1+1} = t_1 + \Delta t = \frac{1}{2} + \frac{1}{2} = 1 \quad ,$$

$$x_2 = x_{1+1} = x_1 + \Delta t \cdot h(t_1, y_1, x_1)$$
$$= 0 + \frac{1}{2} \cdot h\left(\frac{1}{2}, 1, 0\right)$$
$$= 0 + \frac{1}{2}\left[\frac{1}{2} \cdot 1\right] = \frac{1}{4}$$

and

$$y_2 = y_{1+1} = y_1 + \Delta t \cdot x_1$$
$$= 1 + \frac{1}{2} \cdot 0 = 1 \quad .$$

Step 3. With $k = 2$,

$$t_3 = t_{2+1} = t_2 + \Delta t = 1 + \frac{1}{2} = \frac{3}{2} \quad ,$$

$$x_3 = x_{2+1} = x_2 + \Delta t \cdot h(t_2, y_2, x_2)$$
$$= \frac{1}{4} + \frac{1}{2} h\left(1, 1, \frac{1}{4}\right)$$
$$= \frac{1}{4} + \frac{1}{2}[1 \cdot 1] = \frac{3}{4}$$

and

$$y_3 = y_{2+1} = y_2 + \Delta t \cdot x_2$$
$$= 1 + \frac{1}{2} \cdot \frac{1}{4} = \frac{9}{8} = 1.125 \quad .$$

Step 4. With $k = 3$,

$$t_4 = t_{3+1} = t_3 + \Delta t = \frac{3}{2} + \frac{1}{2} = 2 \quad ,$$

$$x_4 = x_{3+1} = x_3 + \Delta t \cdot h(t_3, y_3, x_3)$$
$$= \frac{3}{4} + \frac{1}{2} h\left(\frac{3}{2}, \frac{9}{8}, \frac{3}{4}\right) = \cdots = \frac{51}{32}$$

and

$$y_4 = y_{3+1} = y_3 + \Delta t \cdot x_k$$
$$= \frac{9}{8} + \frac{1}{2} \cdot \frac{3}{4} = \frac{3}{2} = 1.5 \quad .$$

Step 5. With $k = 4$,

$$t_5 = t_{4+1} = t_4 + \Delta t = 2 + \frac{1}{2} = \frac{5}{2} \quad ,$$

$$x_5 = x_{4+1} = x_4 + \Delta t \cdot h(t_4, y_4, x_4) = \cdots = \frac{99}{32}$$

k	t_k	y_k
0	0	1.0000
1	0.5	1.0000
2	1.0	1.0000
3	1.5	1.1250
4	2.0	1.5000
5	2.5	2.2969
6	3.0	3.8438

(a)

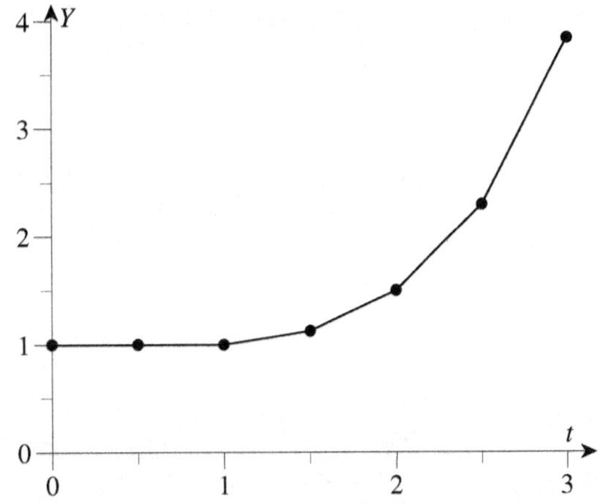

(b)

Figure 40.3: Results of using Euler's method to numerically solve Airy's equation with $y(0) = 1$ and $y'(0) = 0$ using $\Delta t = 1/2$ and $N = 6$: **(a)** The numerical solution in which $y_k \approx y(t_k)$ (for $k \geq 5$, the values of y_k are to the nearest 0.0001). **(b)** The piecewise linear graph based on this numerical solution.

and

$$y_5 = y_{4+1} = y_4 + \Delta t \cdot x_4 = \frac{3}{2} + \frac{1}{2} \cdot \frac{51}{32} = \frac{147}{64} = 2.296875 \quad .$$

Step 6. With $k = 5$,

$$t_6 = t_{5+1} = t_5 + \Delta t = \frac{5}{2} + \frac{1}{2} = 6 \quad ,$$

$$x_6 = x_{5+1} = x_5 + \Delta t \cdot h(t_5, y_5, x_5) = \cdots = \frac{1{,}527}{256}$$

and

$$y_6 = y_{5+1} = y_5 + \Delta t \cdot x_5 = \cdots = \frac{246}{99} = 3.84375 \quad .$$

Since we had chosen $N = 6$, *and we've just computed* t_6, x_6 *and* y_6, *we are done with our computations.*

A summary of our results are given in the table in Figure 40.3a.

Doing the Computations

To repeat the obvious: The use of a computer to carry out the above computations is highly recommended. Either write your own program or set up a worksheet using your favorite computer math package. Be sure to be able to easily change

$$\Delta t \quad , \quad N \text{ or } t_{max} \quad , \quad h(t, y, x) \quad , \quad t_0 \quad , \quad x_0 \quad \text{and} \quad y_0 \quad ,$$

so you can easily change the computational parameters and consider different initial-value problems.[3]

[3] In fact, your favorite computer math package may well already have some version of the above in its repertoire.

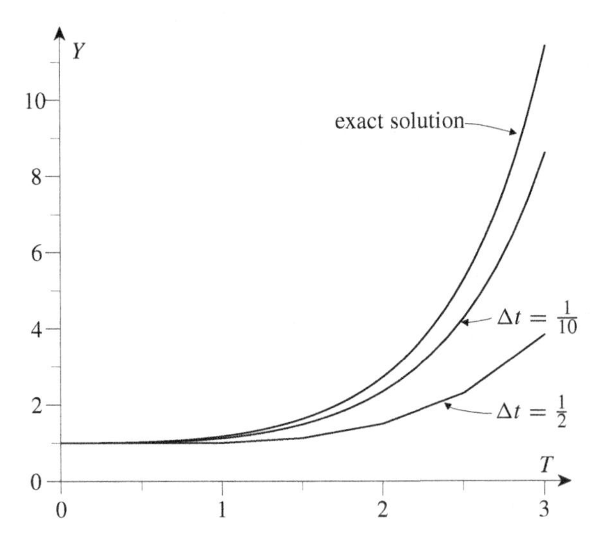

Figure 40.4: Graphs of the approximate solutions to initial-value problem (40.12) generated by Euler's method with $\Delta t = {}^1\!/_2$ and $\Delta t = {}^1\!/_{10}$, along with the graph of the exact solution.

Using the Results of the Method

The Numerical Solution

It should be noted that, strictly speaking, the actual numerical solution consists of the sequences of the t_k's and y_k's, and that $y_k \approx y(t_k)$. The x_k's (each of which approximates the $y'(t_k)$) are needed for the computation, but only need be saved at the end if there is an interest in the derivative of the solution beyond the computations.

Tables and Graphs

Again, if N is not too large, it may be a good idea to write the obtained values of the t_k's and y_k's in a table for convenient reference (with a note that $y_k \approx y(t_k)$ for each k) as done in Table 40.3a for our example. Whether you include the x_k's approximating the derivatives depends on your interests.

It may also be enlightening to sketch (or have a computer sketch) the corresponding piecewise linear approximation to the graph of $y(t)$ as t goes from t_0 to t_N, as done in Figure 40.3b.

General Observations on Accuracy

Because of the way we converted our second-order differential equation to a first-order system, the observations regarding the accuracy of using the Euler method to numerically solve initial-value problems involving systems translate directly to observations regarding the accuracy of using the Euler method to numerically solve initial-value problems involving a single second-order differential equation. In particular, we can typically reduce the errors in using the Euler method by taking smaller step sizes with appropriately larger numbers of steps, and if the h in $y'' = h(t, y, y')$ is "reasonably smooth", then there is a corresponding constant M such that

$$\left|y'(t_k) - x_k\right| < M \cdot \Delta t \qquad \text{and} \qquad |y(t_k) - y_k| < M \cdot \Delta t \tag{40.13}$$

for any numerical solution generated by the Euler method to the above initial-value problem on a suitable interval. Just what is meant by "h is 'reasonably smooth' " will be discussed in the next subsection.

As before, the value of M in the above error bound may be large and very difficult to determine. In practice, it is common to simply find numerical solutions to a given problem using smaller and smaller values of Δt, stopping only when the differences between the different solutions become "insignificant".

An illustration of the effect of shrinking the step size is given in Figure 40.4.

The Error Bound

So let us consider any numerical solution over an interval $[t_0, t_0 + L]$ generated by the Euler method for the initial-value problem

$$y'' = h(t, y, y') \qquad \text{with} \quad y(t_0) = y_0 \quad \text{and} \quad y'(t_0) = x_0 \quad . \tag{40.14}$$

Let

$$\mathcal{R} = \{(t, x, y) : t_0 \leq t \leq t_0 + L \quad , \quad x_{\min} < x < x_{\max} \quad \text{and} \quad y_{\min} < y < y_{\max}\}$$

be a region in TXY–space such that all the following holds:

1. h and all first partial derivatives of h are continuous, bounded functions on $\mathcal{R}$.

2. There is a unique solution $y = y(t)$ to the given initial-value problem valid for $t_0 \leq t \leq t_0 + L$, and each point $(t, y(t), y'(t))$ is in $\mathcal{R}$ when $t_0 \leq t \leq t_0 + L$.

3. If

$$\{ t_0 , t_1 , t_2 , t_3 , \ldots , t_N \}$$

and

$$\{ (x_0, y_0) , (x_1, y_1) , (x_2, y_2) , (x_3, y_3) , \ldots , (x_N, y_N) \}$$

is any numerical solution (including the approximations to $y'(t)$) to our initial-value problem with $t_N \leq t_0 + L$, then each point (t_k, x_k, y_k) is in $\mathcal{R}$.

We can now state the theorem on which error bound (40.13) is based. It is, in fact, a straightforward corollary of Theorem 40.1 on page 829.

Theorem 40.2 (Error bound for Euler's method)
Let h, t_0, x_0, y_0, L and $\mathcal{R}$ be as above, and let $y = y(t)$ be the true solution to initial-value problem (40.14). Then there is a finite constant M such that

$$\left|y'(t_k) - x_k\right| < M \cdot \Delta t \qquad \text{and} \qquad |y(t_k) - y_k| < M \cdot \Delta t$$

whenever $k = 0, 1, \ldots, N$ and

$$\{ t_0 , t_1 , t_2 , t_3 , \ldots , t_N \}$$

and

$$\{ (x_0, y_0) , (x_1, y_1) , (x_2, y_2) , (x_3, y_3) , \ldots , (x_N, y_N) \}$$

is a numerical solution to initial-value problem (40.14) obtained from Euler's method with step size Δt and total number of steps N satisfying

$$0 < \Delta t \cdot N \leq L \quad .$$

Additional Exercises

40.1. *Several initial-value problems are given below, along with values for two of the three parameters in Euler's method: step size Δt, number of steps N and maximum variable of interest x_{max}. For each, find the corresponding numerical solution using Euler's method with the indicated parameter values. Do these problems without a calculator or computer.*

a. $$\begin{aligned} x' &= -x + 2y \\ y' &= x - 2y \end{aligned}$$

with $x(0) = 4$ and $y(0) = 1$; $\Delta t = \frac{1}{2}$ and $N = 6$

b. $$\begin{aligned} x' &= -x + 2y \\ y' &= -2x + 2 \end{aligned}$$

with $x(0) = 2$ and $y(0) = 0$; $\Delta t = \frac{1}{2}$ and $N = 6$

c. $$\begin{aligned} tx' &= -x + 2y \\ ty' &= -2x + 2 \end{aligned}$$

with $x(1) = 2$ and $y(1) = 0$: $\Delta t = \frac{1}{3}$ and $N = 6$

d. $$\begin{aligned} tx' &= 2x - 3y \\ ty' &= 3x - 4y \end{aligned}$$

with $x(1) = 2$ and $y(1) = 1$; $t_{max} = 3$ and $N = 6$

40.2 a. *Using your favorite computer language or computer math package, write a program or worksheet for finding the numerical solution using Euler's method to an arbitrary initial-value problem involving a first-order 2×2 system. Make it easy to change the differential equations and the computational parameters (step size, number of steps, etc.).*[4]

b. *Test your program/worksheet by using it to re-compute the numerical solutions for the problems in Exercise 40.1, above.*

40.3. *Using your program/worksheet from Exercise 40.2 a with each of the following step sizes, find an approximation for $y(5)$ where $y = y(x)$ is the solution to*

$$\begin{aligned} x' &= (25 - 5x - 3y)x \\ y' &= (15 - 6x - 4y)y \end{aligned} \quad \text{with} \quad \begin{aligned} x(0) &= 1 \\ y(0) &= 1 \end{aligned} \quad .$$

a. $\Delta t = 0.5$ **b.** $\Delta t = 0.1$ **c.** $\Delta t = 0.01$

[4] If your computer math package uses infinite precision or symbolic arithmetic, you may have to include commands to ensure your results are given as decimal approximations.

40.4. A direction field for

$$x' = (25 - 5x - 3y)x$$
$$y' = (15 - 6x - 4y)y$$

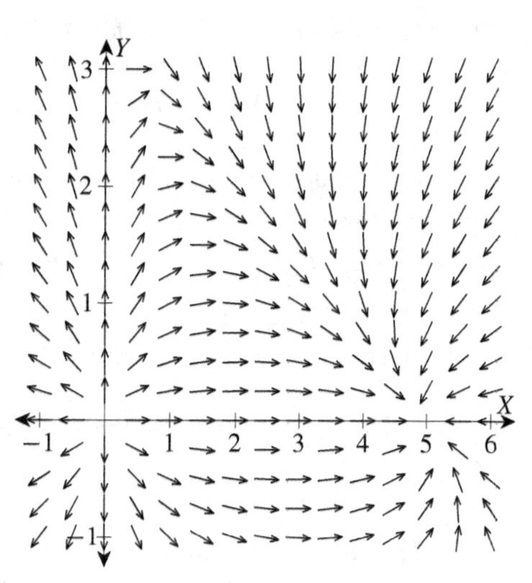

has been sketched to the right. Note that this is the same system that you found numerical solutions to in the last exercise. Using this system and direction field, as well as your answers to Exercise 40.3:

a. Find and plot all the critical points.

b. Sketch the trajectory that go through $(1, 1)$.

c. What can be said about the numerical solution found in Exercise 40.3 when

i. $\Delta t = .5$ **ii.** $\Delta t = .1$ **iii.** $\Delta t = .01$

40.5. Several second-order initial-value problems are given below, along with values for two of the three parameters in Euler's method: step size Δt, number of steps N and maximum variable of interest x_{max}. For each, find the corresponding numerical solution using Euler's method with the indicated parameter values. Do these problems without a calculator or computer.

a. $y'' - 2y' + y = 0$ with $y(0) = 1$ and $y'(0) = 1$;
$t_{max} = 2$ and $\Delta t = \frac{1}{2}$

b. $y'' + 2ty' + 2y = 0$ with $y(0) = 1$ and $y'(0) = 0$;
$N = 6$ and $\Delta t = \frac{1}{2}$

c. $t^2y'' - 6ty' + 10y = 0$ with $y(1) = 1$ and $y'(1) = 2$;
$N = 4$ and $\Delta t = \frac{1}{2}$

d. $ty'' - 2y' = 2t$ with $y(1) = 1$ and $y'(1) = 4$;
$N = 6$ and $\Delta t = \frac{1}{3}$

40.6 a. Using your favorite computer language or computer math package, write a program or worksheet for finding the numerical solution using Euler's method to an arbitrary second-order initial-value problem. Make it easy to change the differential equations and the computational parameters (step size, number of steps, etc.).[5]

b. Test your program/worksheet by using it to re-compute the numerical solutions for the problems in Exercise 40.5, above.

[5] If your computer math package uses infinite precision or symbolic arithmetic, you may have to include commands to ensure your results are given as decimal approximations.

40.7. Let $y = y(t)$ be the solution to the initial-value problem

$$\left(1 - t^2\right) y'' - ty' + 4y = 0 \quad \text{with} \quad y(0) = 1 \quad \text{and} \quad y'(0) = 0 \quad .$$

Find the approximation to $y(2)$ generated by the program/worksheet you wrote for Exercise 40.6 a using

a. $\Delta t = 0.1$ **b.** $\Delta t = 0.01$ **c.** $\Delta t = 0.001$

(Note: The above differential equation is the Chebyshev equation whose power series solution was found in Exercise 33.5 f on page 642.)

40.8. Let $y = y(t)$ be the solution to the initial-value problem

$$y'' - 2ty' + 6y = 0 \quad \text{with} \quad y(0) = 1 \quad \text{and} \quad y'(0) = 0 \quad .$$

Find the approximation to $y(.5)$ generated by the program/worksheet you wrote for Exercise 40.6 a using

a. $\Delta t = 0.1$ **b.** $\Delta t = 0.01$ **c.** $\Delta t = 0.001$

(Note: The above differential equation is the Hermite equation whose power series solution was found in Exercise 33.5 g on page 642.)

40.9. Let $y = y(t)$ be the solution to the initial-value problem

$$t^2 y'' + ty' + \left[t^2 - \frac{1}{4}\right] y = 0 \quad \text{with} \quad y(0) = 0 \quad \text{and} \quad y'(0) = 1 \quad .$$

Find the approximation to $y(5)$ generated by the program/worksheet you wrote for Exercise 40.6 a using

a. $\Delta t = 0.1$ **b.** $\Delta t = 0.01$ **c.** $\Delta t = 0.001$

(Note: The above differential equation is the Bessel's equation of order $1/2$ whose power series solution was found in Chapter 35.)

40.10. In finding a numerical solution to

$$\frac{dx}{dt} = f(t, x, y)$$

$$\frac{dy}{dt} = g(t, x, y)$$

with initial conditions

$$x(t_0) = x_0 \quad \text{and} \quad y(t_0) = y_0 \quad ,$$

we made the obvious extension of the basic Euler method of Chapter 10 to obtain equation set (40.5) on page 823 for iteratively computing the t_k's, x_k's and y_k's in the numerical solution. What would be the analogous set of equations for computing the t_k's, x_k's and y_k's if we were to, instead, make the obvious extension of the improved Euler method from Chapter 12?

Appendix

Author's Guide to Using This Text

The following comments are mainly directed to the instructors of introductory courses in differential equations. That said, the students in these courses and independent readers may also profit from this discussion.

This text was written to be as complete and rigorous as is practical for an introductory text on differential equations. Unfortunately, it is rarely practical for an instructor of an introductory course on differential equations to be as complete and rigorous as they would like. Constraints due to the time available, the mathematical background of the students, the imposed requirements on the topics to be covered, and, sometimes, meddling administrators, all force us to focus our instruction, abbreviate the development of some topics and even sacrifice interesting material. To a great extent that is why this text contains much more material than you, the instructor, can realistically cover — so that the interested student can, on their own time, go back and pick up some of the material that they will later find useful and/or interesting.

So let me, the author, tell you, the instructor, my opinion of what material in this text should be covered, what is optional, and what should probably be skipped in your course.

A.1 Overview

This text is divided into six parts:

I. The Basics

II. First-Order Equations

III. Second- and Higher-Order Equations

IV. The Laplace Transform

V. Power Series and Modified Power Series Solutions

VI. Systems of Differential Equations (A Brief Introduction)

The first three parts should be viewed as the core of your course and should be covered as completely as practical (subject to the further advice given in the chapter-by-chapter commentary that follows). The material in these three parts is needed for the next three parts. However, the next three parts are independent of each other (excluding one optional section in Part VI on the use of Laplace transforms to solve systems), and can be covered in any order. Feel free to choose whichever of these parts you want to cover, subject, of course, to the constraints of your situation.

One typical constraint is that "Laplace transforms" must be covered. Then I would suggest covering Part IV as completely as practical. Admittedly, most of the differential equations solved in this part using Laplace transforms can be solved just as easily by the more basic methods already covered in the previous parts. Still, I feel that the introduction to integral transform methods, convolution, Duhamel's principle and delta functions more than justifies the time spent on a decent development of "Laplace transforms". This is all material that should be in the repertoire of anyone doing applied mathematics in science or engineering. (A good example of the value of Duhamel's principle is in the more advanced development of resonance given in Section 27.7.)

If you do cover Part IV, you will probably have little or no time for anything else. If you do have a little time, you can start on an abbreviated development of power series solutions based on the material in Chapter 30, or you can introduce your students to systems of differential equations using as much of Part VI as you have time for. Given the antipathy towards "series" that so many students seem to develop in their calculus course, you and your students may prefer to end the course with "systems".

If you have time to cover Part V, possibly because you skipped all or much of Part IV on Laplace transforms, great! Part V is a fairly complete development of power series solutions and the modified power series solutions arising from the method of Frobenius. With luck, covering this material may counteract that antipathy towards series. Your students may even find the development of the Frobenius method as a natural extension and combination of methods they already knew to be illuminating. And don't neglect discussing the role of singularities in the behavior of the series solutions, especially when those solutions came from the Frobenius method. This oft overlooked topic is of singular importance in solving many of the boundary value problems arising from the separation of variables approach to solving partial differential equations.

As the title suggests, Part VI is only a brief introduction to systems of differential equations. It is not a complete development; it is, in fact, fairly elementary. For one thing, because this text is to be accessible by students who have not had a course in linear algebra, the use of eigenvalues and eigenvectors has been completely avoided. What is done is to introduce the concept of systems of differential equations, show how they can arise in applications (including applications involving a single nonlinear differential equation), develop the basics of phase plane analysis, and show how that analysis can yield useful information about the solutions to these systems, even though we might not be able to explicitly solve them. It may be a good way to end your course. There are lots of nice pictures.

A.2 Chapter-by-Chapter Guide

What follows are the author's suggestions of the chapters and sections to be covered (and not covered) in a "typical" introductory course on differential equations, along with some commentary on the material. Keep in mind that these are merely suggestions. Each instructor should use their own good judgment and adjust this schedule as appropriate so that their course best suits the backgrounds and needs of their students, the time available, and the instructor's own vision of how the course should be taught.

Part I: The Basics

1. The Starting Point

Cover all of Sections 1.1 and 1.2. Section 1.3 should be covered quickly, with the understanding that your students' understanding of and respect for this material will develop as they learn more about differential equations.

2. Integration and Differential Equations

Cover Sections 2.1 and 2.2 fairly quickly, emphasizing that this is stuff the students should have already seen and be familiar with. Let them know that much of homework is a review of the basic integration methods they will be using extensively for the rest of the course. Do not, however, skip these sections or skip on the homework. Many of your students will probably need the review.

It seems that the material in Sections 2.3 and 2.4 (on using definite integrals) is rarely part of introductory differential equation courses. It can be skipped. Still, it would not hurt to mention that using definite integrals makes it much easier to numerically solve those directly-integrable differential equations such as

$$\frac{dy}{dx} = e^{-x^2} \quad ,$$

which require integrating integrals that are not easily integrated.

Part II: First-Order Equations

3. Some Basics about First-Order Equations

Sections 3.1 and 3.2 are fundamental and should not be skipped.

Section 3.3 (on existence and uniqueness) should only be briefly discussed, and that discussion should probably be limited to Theorem 3.1. Most instructors will want to skip the rest of Chapter 3. (Still, you might tell the more mathematically inquisitive that the Picard iteration method developed in Section 3.4 is pretty cool, and that the discussion is fairly easy to follow since the boring parts have been removed and stuck in the Sections 3.5 and 3.6.)

4. Separable First-Order Equations

Cover Sections 4.1, 4.2, 4.3 and 4.4.

You can skip Sections 4.5 (*Existence, Uniqueness and False Solutions*), 4.6 (*On the Nature of Solutions to DEs*) and 4.7 (*Using and Graphing Implicit Solutions*). In an ideal world, the material in these sections would be recognized as important for understanding "solutions". But this is not the ideal world, and there isn't enough time to cover everything. Tell your students that they can read it on their own, and that understanding this material will help lead them to enlightenment.

You can also skip Section 4.8. It's on using definite integrals.

5. Linear First-Order Equations

Cover Sections 5.1 and 5.2. Sections 5.3 (on using definite integrals) and 5.4 (a bit of theory) can be ignored by most.

6. Simplifying Through Substitution

Cover the whole thing. Students should be able to recognize and use the "obvious" substitutions in Sections 6.2 and 6.3. They should also be able to use any other reasonable substitution suggested to them. On the other hand, I see little value in memorizing the substitution for the Bernoulli equations — being able to derive it (Exercise 6.8), however, is of great value.

7. The Exact Form and General Integrating Factors

Whether or not this chapter is covered depends on the background of your students. If they have had a course covering calculus of several variables — in particular, if they are acquainted with the multidimensional chain rule — then all of this chapter should be covered. After all, the methods for dealing with first-order differential equations discussed in Chapters 4 and 5 are all just special cases of the general methods discussed in this chapter.

On the other hand, many introductory courses in differential equations do not have multivariable calculus as a prerequisite, and the students in those courses cannot be expected to know the multidimensional chain rule on which almost everything in this chapter is based. That makes covering this material problematic. For that reason, this chapter and the rest of the text were written so that this chapter could be omitted. (But students in these courses should realize that they will want to read this chapter on their own after learning about the multidimensional chain rule.)

8. Review Exercises for Part of Part II

This chapter is simply a list of differential equations to be solved using methods from the previous chapters. Assign what seems reasonable.

One possible issue is that some of the equations are exact, which you may not have covered (see the comments for Chapter 7). You can either avoid assigning them, or let your students know that there may be problems that cannot be done using methods from the chapters covered.

9. Slope Fields: Graphing Solutions Without the Solutions

Cover Sections 9.1 and 9.2. These sections describe what slope fields are, why they can be useful, and how to construct and use them. I've also tried to make the exercises more meaningful than just "sketch a bunch of solution curves". If you want a computer assignment (highly recommended), require Exercise 9.8.

Sections 9.3, 9.4 and 9.5 cover useful stuff regarding slope fields that is rarely discussed in introductory differential equation courses. If time permits, a *brief* discussion of stability as approached in Section 9.3 may be enlightening, as might a discussion of "problem points", as done in Section 9.4, just to illustrate how solutions can go bad.

10. Numerical Methods I: The Euler Method

This development of Euler's numerical method for first-order differential equations follows naturally from the discussion of slope fields in the previous chapter. Cover Sections 10.1 and 10.2, and then briefly comment on the material in Sections 10.3 and 10.4 regarding the errors that can arise in using Euler's method. The detailed error analysis in Section 10.5 is only for the most dedicated.

(By the way, you can go straight from Chapter 9 to Chapter 11, and cover Chapter 10 later. If a chapter must be sacrificed because of time limitations, this would probably be one of the better candidates for the sacrifice.)

11. The Art and Science of Modeling with First-Order Equations

Cover Sections 11.1 through 11.6. Consider Section 11.7 on thermodynamics (Newton's law of heating and cooling) as optional.

Add few applications of your own if you want.

By the way, my approach to "applications" is a bit different from that found in many other texts. I prefer working on the reader's ability to derive and use the differential equations modeling any given situation, rather just than flashing a big list of applications.

12. Numerical Methods II: Beyond the Euler Method

Consider this a very optional chapter for most introductory courses in differential equations. It's the sort of chapter that the student should come back to in the future when, and if, that student really needs to know more than the basics of numerical methods discussed in Chapter 10.

Part III: Second- and Higher-Order Equations

13. Higher-Order Equations: Extending First-Order Concepts

Cover Section 13.1. That material is needed for future development. I'll leave it to you to decide whether the solution method in Section 13.2 is worth covering; it can be safely skipped.

Cover Section 13.3 on initial-value problems.

Section 13.4 is on the existence and uniqueness of solutions to higher-order differential equations, and should only be briefly discussed, if at all. At most, mention that the existence and uniqueness results for first-order equations have higher-order analogs.

14. Higher-Order Linear Equations and the Reduction of Order Method

Cover Sections 14.1, 14.2 and 14.3. All of this is needed for later development.

Extensions of the reduction of order method are discussed in Sections 14.4 and 14.5, along with explanations as to why these extensions are rarely used in practice. Consider these sections as very optional, but, to hint at future developments, you might mention that the variation of parameters method for solving nonhomogeneous equations, which will be developed later, is just an improvement on the reduction of order method for nonhomogeneous equations discussed in Section 14.4.

15. General Solutions to Homogeneous Linear Differential Equations

Cover Sections 15.1 and 15.2. This material is fundamental. However, if you want to concentrate on second-order equations at first, you can cover Section 15.1 now (which just concerns second-order equations), and then go on to Chapters 17 and 18 (and maybe the first two sections of Chapter 20 on Euler equations). You can then go back and cover Section 15.2 just before discussing the higher-order equations in Chapter 19.

Section 15.3 concerns Wronskians. Consider it optional. This section was written assuming the students had *not* yet taken a course in linear algebra. Under this assumption, I do not feel that Wronskians are worth any more than a brief mention here. It's later that Wronskians become truly of interest.

16. Verifying the Big Theorems and an Introduction to Differential Operators

You can, and probably should, completely skip this chapter.

The first two sections are there just to assure readers that I didn't make up the big theorems in the previous chapter. Of course, you can read it for your own personal enjoyment and so that you can tell me of all the typos in this chapter.

Section 16.3 can help the students better understand "differential operators" in the context of differential equations, but the material in this section is only used later in the text to prove a few theorems. Since you are not likely to be discussing these particular proofs, you can safely skip this section. Suggest this section as enrichment for your more inquisitive students.

17. Second-Order Homogeneous Linear Equations with Constant Coefficients

This may be the most important chapter for many of your students. By the end of the term, they should be able to solve these equations in their sleep. I also consider the introduction of the complex exponential as a practical tool for doing trigonometric function computations to be important.

Cover everything in this chapter except, possibly, Example 17.5 on page 330 (that example concerns using the complex exponential to derive trigonometric identities — a really nice thing, but you'll probably be a little pressed for time and will want to concentrate on solving differential equations.)

18. Springs: Part I

This is a fairly standard discussion of unforced springs (maybe with more emphasis than usual on the modeling process). Cover all of it quickly.

19. Arbitrary Homogeneous Linear Equations with Constant Coefficients

Cover all of Sections 19.1, 19.2 and 19.3.

Don't cover Sections 19.4 and 19.5. They contain rigorous proofs of theorems naively justified earlier in the chapter. Besides, you will need the material from Section 16.3 (which you probably skipped).

20. Euler Equations

Cover Sections 20.1 and 20.2. Section 20.3 is optional. Don't bother covering Section 20.4, though you may want to briefly comment on that material (which relates Euler equations to equations with constant coefficients via a substitution).

By the way, the approach to Euler equations taken here is taken to help reinforce the students' grasp of the theory developed for general linear equations. It will also help prepare them for the series methods (especially the method of Frobenius) that some of them will later see. That is why I downplay the substitution method commonly used by others.

21. Nonhomogeneous Equations in General

Cover Sections 21.1 and 21.2. Briefly mention that reduction of order can be used to solve nonhomogeneous equations (that's really all that's done in Section 21.3).

22. Method of Undetermined Coefficients

Cover everything in this chapter except Section 22.7 (verifying a theorem).

If you are running short on time, you might incorporate a brief discussion of resonance with the development of the material in Section 22.3, and then skip the next chapter.

23. Springs: Part II (Forced Vibrations)

This is the material on forced oscillations in mass/spring systems. If you have time, cover Sections 23.1, 23.2 and 23.3. Consider 23.4 (on the 'beat phenomenon') as very optional.

24. Variation of Parameters

Cover Sections 24.1 and 24.2. Do not cover Section 24.3 unless you really feel that the variation of parameters formula is worth memorizing (I don't).

25. Review Exercises for Part III

This chapter is simply a list of differential equations to be solved using methods from the previous chapters. Assign what seems reasonable.

Part IV: The Laplace Transform

26. The Laplace Transform (Intro)

Cover Sections 26.1, 26.2, 26.3, 26.4 and 26.5. (Don't waste much time on the gamma function — just about everything else in this chapter is much more important.)

The students should at least skim the first part of Section 26.6. This discusses some issues regarding piecewise continuous functions. In this text, I've adopted the view that "the value of a function at a point of discontinuity is 'irrelevant' ". Basically, I'm introducing the idea of functions being equal "almost everywhere" in the sense normally encountered in more advanced analysis courses. This is a practical and philosophical decision explained in the first part of Section 26.6. Point this out in class, but don't make a big deal of it.

Section 26.7 concerns two proofs — one bad and one good — on the existence of the Laplace transform for piecewise continuous functions of exponential order. Go ahead and skip it.

27. Differentiation and the Laplace Transform

Cover the material in Sections 27.1 and 27.2

Section 27.3, covering the integration identities, is optional. If you have time, do it; otherwise, just mention the existence of such identities as a counterpoint to the differentiation identities.

Section 27.4, which is mainly concerned with a rigorous derivation of the "derivative of a transform" identity, is very optional, though some may be intrigued by Example 27.6, which illustrates that, sometimes, that which seems obviously true is not true.

28. The Inverse Laplace Transform

Cover all of this chapter. And, yes, I really do wait until here to introduce the inverse transform. Trust me, it works.

29. Convolution

At least cover Sections 29.1 and 29.2. If time allows, cover 29.3 on Duhamel's principle — this is an introduction to a very important concept in a wide range of applications (including ordinary and partial differential equations, and the generic modeling of systems in physics, optics and engineering). Ultimately, it more justifies the development of convolution than does the formula $\mathcal{L}^{-1}[FG] = f*g$.

30. Piecewise-Defined Functions and Periodic Functions

At least cover the material in Sections 30.1 and 30.2. The material in Section 27.3 is the natural extension of that in the previous sections — at least discuss the rectangle function since it is used in the next chapter. Cover the rest of Section 30.3 if you have time. You might suggest that your students skim Section 30.4 if they suspect they might ever have to compute convolutions with piecewise-defined functions.

Section 30.5 is on periodic functions. Consider it optional.

Section 30.6 is just a table of identities for the Laplace transform.

Section 30.7 generalizes the discussion of resonance from Chapter 23. It is a nice application of Duhamel's principle which you certainly will not have time to cover. Maybe you will want to read it for yourself, or recommend it as enrichment for an interested student.

31. Delta Functions

Cover Sections 31.1, 31.2 (at least the material on "strong forces of brief duration") and 31.3. If you covered Duhamel's principle, you will want to discuss Section 31.4. Don't bother with Section 31.5 (on technical issues), other than to recommend it to the interested students.

Part V: Power Series and Modified Power Series Solutions

32. Series Solutions: Preliminaries

Much (but not all) of this chapter is a review of infinite series. Resist the temptation to simply tell your students to read it on their own. Quickly cover Sections 32.1, 32.2 and 32.3. Section 32.4, however, can be skipped initially and returned to if needed.

33. Power Series Solutions I: Basic Computational Methods

Cover all of Sections 33.1, 33.2, 33.3, 3.4 and 3.5. These cover the finding of basic power series solutions (and their radii of convergence) for first- and second-order linear equations with *rational* coefficients.

Section 33.6, on using Taylor series, can be enlightening. It can also be considered optional.

Section 33.7 is an appendix on using induction when deriving the formulas for the coefficients in a power series solutions. Theoretically, it should be covered. As a practical matter, it's probably best to simply tell your more mathematically mature students that this is a section they may want to read on their own.

34. Power Series Solutions II: Generalizations and Theory

In this chapter:

1. The ideas and solution methods discussed in the previous chapter are extended to corresponding ideas and solution methods for first- and second-order linear equations with *analytic* coefficients.

2. The validity of these methods and the claims of the relation between singular points and the radii of convergence for the power series solutions are rigorously verified.

It may be a good idea to quickly go through Sections 34.1 and 34.2 since some of the basic notions developed here will apply later in the next two chapters. However, the extension of solution method and the rigorous verification of the validity of the methods, no matter how skillfully done, may be something you would not want to deal with in an introductory course. I certainly wouldn't. So feel free to skip the rest of the chapter, telling your students that they can return to it if the need ever arises.

35. Modified Power Series Solutions and the Basic Method of Frobenius

Cover Sections 35.1, 35.2. 35.3, 35.4 and 35.5. This covers the development of the basic method of Frobenius as a natural extension of the power series method from Chapter 33 and the approach to solving Euler equations given in Chapter 20.

Section 35.6 discusses the case in which the exponents are complex. Skip it. It's included for the sake of completeness, but this case never arises in practice.

Section 35.7 is an appendix containing a proof and a generalization. It can be safely skipped.

36. The Big Theorem on the Frobenius Method, with Applications

Definitely cover Section 36.1. It contains the "Big Theorem" describing when the basic Frobenius method works, and describes the sort of modified power series solutions that arise in the 'exceptional cases' in which the basic method only works partially.

Sections 36.2, 36.3 and 36.4 contain a discussion of the behavior of modified power series solutions near singular points. While this seems to be rarely discussed, I consider knowing this behavior to be very important in many applications of the Frobenius method. For some solutions, this behavior near singularities is all you need to know. So I suggest covering these sections. If time allows, it may also be worthwhile to illustrate this material with the application in Section 36.5 concerning the Legendre equation and Legendre polynomials.

Adapting the basic Frobenius method to handle the 'exceptional cases' is discussed in Section 36.6. With any luck, your students will never need this. Let them read it if ever the need arises.

37. Validating the Method of Frobenius

You can, and probably should, completely skip this chapter.

Pity. I spent a lot of time and effort on this chapter, and I am rather proud of the result. It is straightforward, rigorous and understandable by students (and instructors). As far as I am aware, this chapter contains the only intelligible explanation of why you get the modified power series solutions that you do get for the 'exceptional cases.' But the chapter is mainly on rigorously proving the "Big Theorem" in the previous chapter and is intended to serve as a reference for those few who really would like to see a good proof of that theorem.

Part VI: Systems of Differential Equations (A Brief Introduction)

38. Systems of Differential Equations: A Starting Point

Cover Sections 38.1, 38.2 and 38.3. In these sections, the student is introduced to systems of differential equations and some interesting applications, one of which involves a single second-order nonlinear differential equation.

Section 38.4 discusses the use of Laplace transforms to solve certain systems. It's hardly the best way to solve these systems but is pretty well the only way that could be discussed within the parameters of this text. I would consider this section to be optional.

Section 38.5 on the existence and uniqueness of solutions and on identifying general solutions should also be considered optional.

In Section 38.6, some existence and uniqueness results for single N^{th}-order equations are discussed in context of the results given in Section 38.5 — very optional for most classes.

39. Critical Points, Direction Fields and Trajectories

Cover Sections 39.1, 39.2, 39.3, 39.4 and 39.5. The rudiments of phase plane analysis are developed in these sections.

Also cover Section 39.6, in which phase plane analysis is used to extract useful information regarding three applications (mixing in multiple tanks, two competing species and the motion of a pendulum). This could be a nice place to end the course, so consider the rest of the chapter (on verifying some properties of trajectories) as very optional.

40. Numerical Methods III: Systems and Higher-Order Equations

If you wish to discuss numerical methods, cover all of this chapter. Section 40.3 gives another good reason for studying systems even if your interest is in single higher-order differential equations. This, in my opinion, is the real reason for this chapter.

By the way, the numerical method being used in this chapter is the Euler method from Chapter 10. The more advanced methods from Chapter 12 are briefly mentioned, but are not needed for this chapter. So you can cover *Numerical Methods III* (Chapter 40) without having covered *Numerical Methods II* (Chapter 12).

Answers to Selected Exercises

Chapter 1

1. Second, fifth, fifth, forty-second **3a i.** Yes, it is **3a ii.** No, it is not **3a iii.** No **3b i.** No **3b ii.** Yes **3b iii.** No **3c i.** Yes **3c ii.** No **3c iii.** No **3d i.** No **3d ii.** No **3d iii.** Yes **3e i.** No **3e ii.** Yes **3e iii.** Yes **3f i.** No **3f ii.** No **3f iii.** Yes **3g i.** No **3g ii.** Yes **3g iii.** Yes **3h i.** Yes **3h ii.** Yes **3h iii.** Yes **4a i.** No **4a ii.** Yes **4a iii.** No **4b i.** No **4b ii.** No **4b iii.** Yes **4c i.** Yes **4c ii.** No **4c iii.** No **4d i.** No **4d ii.** No **4d iii.** Yes **5b i.** 9 **5b ii.** 5 **5c i.** $y(x) = \sqrt{x^2+9}$ **5c ii.** $y(x) = \sqrt{x^2+5}$ **6b i.** $y(x) = 4e^{x^2} - 3$ **6b ii.** $y(x) = 3e^{x^2-1} - 3$ **7b i.** $y(x) = 3\cos(2x) + 4\sin(2x)$ **7b ii.** $y(x) = \frac{1}{2}\sin(2x)$ **8a i.** 7 **8a ii.** No, because then $7 = y'(1) = 4$.
8a iii. $\frac{d^2y}{dx^2} = 3x\frac{dy}{dx} + 3y + 2x$, $y''(1) = 29$ **8a iv.** Of course.
8b i. $y''(0) = 4$ and $y'''(0) = 24$ **8b ii.** No. **9a.** 100/7 sec.
9b. -140 meters/second **9c i.** $y'(0) = 0$ is replaced with $y'(0) = 2$
9c ii. $-4.9t^2 + 2t + 1000$ **9c iii.** $-2\sqrt{4901}$ meters/second **10a.** $\frac{dv}{dt} = -9.8 - \kappa(v-2)$
10b i. 1000κ **10b ii.** $(v_{\text{hit}} + 9.8t_{\text{hit}})/1000$ **11b.** $y(x) = 3x + 5x\ln|x|$
11c. Because $y(x)$ is not continuous at $x = 0$.

Chapter 2

2a. Yes, it is directly integrable. **2b.** No, it is not directly integrable. **2c.** No
2d. Yes **2e.** No **2f.** Yes **2g.** Yes **2h.** No **2i.** No **2j.** No
3a. $y(x) = x^4 + c$ **3b.** $y(x) = -5e^{-4x} + c$ **3c.** $y(x) = 2\ln|x| - 2\sqrt{x} + c$
3d. $y(x) = 2\sqrt{x+4} + c$ **3e.** $y(x) = \frac{1}{2}\sin\left(x^2\right) + c$ **3f.** $y(x) = x\sin(x) + \cos(x) + c$
3g. $y(x) = \frac{1}{2}\ln\left|x^2-9\right| + c$ **3h.** $y(x) = \frac{1}{6}\ln\left|\frac{x-3}{x+3}\right| + c$ **3i.** $y(x) = \frac{1}{27}x^3 - \frac{1}{9}x + c$
3j. $y(x) = -\frac{1}{4}\sin(2x) + c_1x + c_2$ **3k.** $y(x) = \frac{1}{6}x^3 + \frac{3}{2}x^2 + c_1x + c_2$
3l. $y(x) = \frac{1}{24}x^4 + c_1x^3 + c_2x^2 + c_3x + c_4$ **4a.** $y(x) = 2x^2 + 5e^{2x} - 1$ for $-\infty < x < \infty$
4b. $y(x) = \frac{3}{2}(x+6)^{2/3} + 4$ for $-\infty < x < \infty$
4c. $y(x) = x - 2\ln|x+1| + 8$ for $-1 < x < \infty$
4d. $y(x) = 2\sqrt{x} - 2\ln|x| + 4$ for $0 < x < \infty$
4e. $y(x) = -\ln|\cos(x)| + 3$ for $-\frac{\pi}{2} < x < \frac{\pi}{2}$
4f. $y(x) = \arctan(x) + 3$ for $-\infty < x < \infty$
4g. $y(x) = \frac{4}{3}x^{3/2} - 2x\ln|x| + 6x + \frac{2}{3}$ for $0 < x < \infty$
5a. $y(x) = -2\cos\left(\frac{x}{2}\right) + 2 + y(0)$ **5b i.** 2 **5b ii.** 5 **5b iii.** 7

6a. $y(x) = 2(x+3)^{3/2} - 16 + y(1)$ **6b i.** 54 **6b ii.** 58 **6b iii.** -14

7a. $y(x) = \frac{7}{2} - \frac{1}{2}e^{-x^2}$ **7b.** $y(x) = 4 + \sqrt{x^2+5}$ **7c.** $y(x) = \arctan x - \frac{\pi}{2}$

7d. $y(x) = \frac{\sqrt{\pi}}{6}\operatorname{erf}(3x) + 1$ **7e.** $y(x) = \operatorname{Si}(x) + 4$ **7f.** $y(x) = \frac{1}{2}\operatorname{Si}\left(x^2\right)$

9a. $y = \begin{Bmatrix} 0 & \text{if} & x < 0 \\ x & \text{if} & 0 \le x \end{Bmatrix}$ (or ramp(x)) **9b.** $y(x) = \begin{Bmatrix} 2 & \text{if} & x < 1 \\ x+1 & \text{if} & 1 \le x \end{Bmatrix}$

9c. $y(x) = \begin{Bmatrix} 0 & \text{if} & x < 1 \\ x-1 & \text{if} & 1 \le x \le 2 \\ 1 & \text{if} & 2 < x \end{Bmatrix}$ **9d.** $y(x) = \begin{Bmatrix} 2x - \frac{1}{2}x^2 + c & \text{if} & x \le 2 \\ \frac{1}{2}x^2 - 2s + 4 + c & \text{if} & 2 < xs \end{Bmatrix}$

9e. $y(x) = \begin{Bmatrix} 0 & \text{if} & x < 0 \\ x & \text{if} & 0 \le x < 1 \\ 2\left(x - 1/2\right) & \text{if} & 1 \le x < 2\,x \\ 3\left(x - 2/2\right) & \text{if} & 2 \le x < 3 \\ 4\left(x - 3/2\right) & \text{if} & 3 \le x < 4 \end{Bmatrix}$

Chapter 3

4a. $\frac{dy}{dx} = 6x - 3xy$, const. solns.: $y = 2$ **4b.** $\frac{dy}{dx} = \frac{\sin(x+y)}{y}$, const. solns.: none

4c. $\frac{dy}{dx} = y^3 + 8$, const. solns.: $y = -2$

4d. $\frac{dy}{dx} = \frac{1-y^2}{x}$, const. solns.: $y = 1$ and $y = -1$

4e. $\frac{dy}{dx} = x + y^2$, const. solns.: none

4f. $\frac{dy}{dx} = 25y^2 - y^3$, const. solns.: $y = 0$, $y = 5$ and $y = -5$

4g. $\frac{dy}{dx} = \frac{y+3}{x-2}$, const. solns.: $y = -3$

4h. $\frac{dy}{dx} = \frac{x-3}{y-2}$, const. solns.: none

4i. $\frac{dy}{dx} = y^2 - 2y - 2$, const. solns.: $y = 1 + \sqrt{3}$ and $y = 1 - \sqrt{3}$

4j. $\frac{dy}{dx} = y^2 + (x-8)y - 8x$, const. solns.: $y = 8$

5. The equations in Exercises 3.4 c, 3.4 f and 3.4 i

6b. Because $\partial F/\partial y$ with $F(x,y) = 2\sqrt{y}$ is not continuous at the point $(1,0)$.

7a. $\psi_1(x) = 2 + x^2$ **7b.** $\psi_2(x) = 2 + x^2 + \frac{1}{4}x^4$ **7c.** $\psi_3(x) = 2 + x^2 + \frac{1}{4}x^4 + \frac{1}{24}x^6$

8a. $\psi_1(x) = 3 + 9x + 3x^2$ **8b.** $\psi_2(x) = 3 + 9x + 28x^2 + 29x^3 + \frac{9}{2}x^4 + \frac{1}{5}x^5$

Chapter 4

3a. $\frac{dy}{dx} = (3 - \sin(x))y^2$ **3b.** Not separable **3c.** Not separable **3d.** Separable

3e. $\frac{dy}{dx} = 4(2-y)$ **3f.** $\frac{dy}{dx} = x(4-y)$ **3g.** Not separable **3h.** $\frac{dy}{dx} = (x-2)(y-3)$

3i. Not separable **3j.** $\frac{dy}{dx} = e^x\,y^{-1}e^{-3y^2}$ **4a.** $y(x) = \pm\sqrt{x^2+c}$

4b. $y(x) = 3\tan(3x+c)$ **4c.** $y(x) = \pm\sqrt{Ax^2 - 9}$ **4d.** $y(x) = \tan(c + \arctan(x))$

4e. $y(x) = \arcsin(c - \cos(x))$ **4f.** $y(x) = \frac{1}{3}\ln\left|\frac{3}{2}e^{2x} + c\right|$ **5a.** $y(x) = \sqrt{x^2+8}$

5b. $y(x) = 3e^{x^2-x} - 1$ **5c.** $y(x) = -\sqrt{5e^{x^2} - 1}$ **5d.** $y(x) = \sqrt{\left(2x^{3/2}+3\right)^2 - 9}$

6a. $y = 4$ **6b.** $y = -1/2$ **6c.** $y = 3$ and $y = -3$ **6d.** $y = 0, \pm\pi, \pm 2\pi, \pm 3\pi, \ldots$
6e. No constant solution **6f.** $y = 0$ and $y = 100$ **7a.** $y(x) = 4 + A\exp\left(\frac{1}{2}x^2\right)$
7b. $y(x) = 3 + A\exp\left(\frac{1}{2}x^2 - 2x\right)$ **7c.** $y(x) = \left(c - 3x - \cos(x)\right)^{-1}$ and $y = 0$
7d. $y(x) = \arcsin\left(Ae^x\right)$ **7e.** $y(x) = Ax$ **7f.** $y^3 - 2y^2 = 2x^3 + 4x + c$
7g. $y(x) = \tan(\arctan(x) + c)$
7h. implicit: $y + y^{-1} = 2x^2 + c_1$, explicit: $y(x) = x^2 - c_2 \pm \sqrt{\left(c_2 - x^2\right)^2 - 1}$; also $y = 0$
7i. $y(x) = \ln(x + c)$ **7j.** $y(x) = \ln\left|Ae^x - 1\right|$ **7k.** $y(x) = \pm\left(c - 3x^2\right)^{-1/2}$ and $y = 0$
7l. $3y + y^{3/2} = 3x + x^{3/2} + c$ **7m.** $y(x) = \dfrac{1 + Ae^{2x^3}}{1 - Ae^{2x^3}}$ and $y = -1$
7n. $y(x) = \tan\left(x^3 + c\right)$ **7o.** $y(x) = \dfrac{100Ae^{200x}}{1 + Ae^{200x}}$ and $y = 100$ **8a.** $y(x) = 5 + 3e^{2x}$
8b. $y(x) = -\sqrt{18 - 2\cos(x)}$ **8c.** $y = -1$ **8d.** $y = 0$ **8e.** $y(x) = 2(2 - x)^{-1}$
8f. $y(x) = -\sqrt{1 + 3x^2}$ **8g.** $y^2 - 2\ln|y| = 4x^2 - 15$ **9a.** $y = 0$ and $y = \beta/\gamma$
9b. $y(x) = \dfrac{\beta}{\gamma - Ae^{-\beta x}}$ and $y = 0$ **10a.** $(-\infty, \infty)$ **10b.** $(-\infty, \infty)$ **10c.** $(-\infty, 2)$
10d. $(-e, \infty)$ **10e.** $\left(-\frac{2}{\sqrt{3}}, \frac{2}{\sqrt{3}}\right)$

Chapter 5

1a. Linear; $\dfrac{dy}{dx} + 3y = x^{-2}\sin(x)$ **1b.** Not linear **1c.** Not Linear **1d.** Not Linear
1e. Linear; $\dfrac{dy}{dx} - (3 + x)y = 1$ **1f.** Linear; $\dfrac{dy}{dx} - 4y = 8$ **1g.** Linear; $\dfrac{dy}{dx} + 0 \cdot y = e^{2x}$
1h. Linear; $\dfrac{dy}{dx} - \sin(x)\,y = 0$ **1i.** Not linear
1j. Linear; $\dfrac{dy}{dx} - 827x^{-1}y = -x^{-1}\cos\left(x^2\right)$ **2a.** $y(x) = 3 + ce^{-2x}$
2b. $y(x) = 4e^{3x} + ce^{-2x}$ **2c.** $y(x) = ce^{4x} - 4x - 1$ **2d.** $y(x) = ce^{x^2} - \frac{1}{2}$
2e. $y(x) = 2x^2 + cx^{-3}$ **2f.** $y(x) = [c - \cos x]x^{-2}$ **2g.** $y(x) = cx^3 - \frac{2}{5}\sqrt{x}$
2h. $y(x) = [x + c]\cos(x)$ **2i.** $y(x) = x^{-2}\left[4 + ce^{-5x}\right]$ **2j.** $y(x) = \left[\frac{2}{3}x^{3/2} + c\right]e^{-\sqrt{x}}$
3a. $y(x) = 7e^{3x} - 2$ **3b.** $y(x) = -2$ **3c.** $y(x) = \frac{1}{2}\left[e^{-3x} - e^{-5x}\right]$
3d. $y(x) = 4x^2 + 6x^{-3}$ **3e.** $y(x) = x[\sin(x) - 1]$
3f. $y(x) = 1 + x^2 + 3\sqrt{5}\left(1 + x^2\right)^{-1/2}$ **4a.** $y(x) = e^{-3x^2}\left[4 + \int_0^x e^{3s^2}\sin(s)\,ds\right]$
4b. $y(x) = \frac{1}{x}\left[10 + \int_2^x \frac{\sin(s)}{\sqrt{s}}\,ds\right]$ **4c.** $y(x) = x\left[\frac{8}{3} + \int_3^x e^{-s^2}\,ds\right]$
5. $\mu'(x) = \mu(x)p(x)$ and $y'(x) = f(x) - \dfrac{p(x)}{\mu(x)}\left[\mu(x_0)y_0 + \int_{x_0}^x \mu(s)f(s)\,ds\right]$

Chapter 6

1a. $3x + 3y + 2 = \tan(3y + C)$ **1b.** $y(x) = \frac{3}{2}x \pm \frac{1}{2}\sqrt{Ae^{-4x} - 1}$
1c. $y(x) = 2x - \frac{3}{4} + \frac{1}{4}\arcsin(8x + c)$ **2.** $y(x) = x - \dfrac{1}{x - 4}$
3a. $y(x) = -x(\ln|x| + c)^{-1}$ and $y = 0$ **3b.** $y = \pm x\sqrt{c + 2\ln|x|}$
3c. $y(x) = x\arcsin(Ax - 1)$ **4.** $y(x) = -x + \sqrt{2x^2 + 9}$

5a. $y(x) = \pm\left(1 + ce^{6x}\right)^{-1/2}$ and $y = 0$ **5b.** $y(x) = 2x^3\left(C - x^2\right)^{-1}$ and $y = 0$

5c. $y(x) = \left(\sin(x) + \frac{c}{\sin(x)}\right)^3$ and $y = 0$ **6.** $y(x) = \sqrt{11x^2 - 2x}$

7a. $u = \frac{y}{x}$; $y(x) = x\,(3\ln|x| + c)^{1/3}$

7b. $u = 2x + 3y + 4$; $y(x) = \frac{1}{3}\left(\frac{1}{2}x + c\right)^2 - \frac{2}{3}x - \frac{4}{3}$ and $y(x) = -\frac{1}{3}(2x + 4)$

7c. $u = y^{1/2}$; $y(x) = \left(x + \frac{c}{x}\right)^2$ and $y = 0$ **7d.** $u = 4x - y$; $y(x) = 4x - \arccos(x + c)$

7e. $u = y - x$; $y(x) = 1 + x - Ae^{-y}$ **7f.** $u = \frac{y}{x}$; $y\ln|y| - cy = x$ and $y = 0$

7g. $u = \frac{y}{x}$; $y(x) = -x \pm x\sqrt{\ln|x| + c}$

7h. $u = y^{-2}$; $y(x) = \pm\left(cx^2 - 2x^3\right)^{-1/2}$ and $y = 0$

7i. $u = 2x + y - 3$; $y(x) = (x + c)^2 - 2x + 3$ and $y(x) = 3 - 2x$

7j. $u = 2x + y - 3$; $\sqrt{2x + y - 3} - \ln\left(1 + \sqrt{2x + y - 3}\right) = x + c$

7k. $u = \frac{y}{x}$; $y(x) = x\left(\frac{1}{2}\ln|x| + c\right)^2 - x$ and $y = -x$

7l. $u = y^4$; $y(x) = \pm\left(8e^{2x} + ce^{-12x}\right)^{1/4}$

7m. $u = x - y + 3$; $y(x) = 3 + x + \left(Ae^{2x} - 1\right)\left(Ae^{2x} + 1\right)^{-1}$ and $y = x + 4$

7n. $u = y + x^2$; $y(x) = c^2 + 2cx$ and $y = -x^2$

7o. $u = \sin(y)$; $y(x) = \arcsin\big((c + x)e^{-x}\big)$ **7p.** $u = yx^{-2}$; $y(x) = x^2\tan(c + \ln|x|)$

8. $\frac{du}{dx} + (1 - n)p(x)u = (1 - n)f(x)$

Chapter 7

1a. $3y + 3x\frac{dy}{dx} = 0$, $y(x) = \frac{c}{x}$ **1b.** $-6x^2y + \left[2y - 2x^3\right]\frac{dy}{dx} = 0$, $y(x) = x^3 \pm \sqrt{c + x^6}$

1c. $\left[2xy - y^3\right] + \left[x^2 - 3xy^2\right]\frac{dy}{dx} = 0$, $x^2y - xy^3 = c$

1d. $\operatorname{Arctan}(y) + \frac{x}{1 + y^2}\frac{dy}{dx} = 0$, $y(x) = \tan\left(\frac{c}{x}\right)$ **2b.** $y(x) = \pm\sqrt{x + \frac{c}{x}}$

4a. $\phi(x, y) = x^2y + xy^2 + c_1$, $y(x) = -\frac{x}{2} \pm \frac{1}{2x}\sqrt{x^2 + C}$

4b. $\phi(x, y) = x^2y^3 + x^4 + c_1$, $y(x) = \left(cx^{-2} - x^2\right)^{1/3}$

4c. $\phi(x, y) = 2x - x^2 + y^3 + c_1$, $y(x) = \left(c + x^3 - 2x\right)^{1/3}$

4d. $\phi(x, y) = x^3y^2 + x + 3y^2 + c_1$, $y(x) = \pm\sqrt{\frac{c - x}{x^3 + 3}}$

4e. $\phi(x, y) = x^4y - \frac{1}{5}y^5 + c_1$, $x^4y - \frac{1}{5}y^5 = c$ **4f.** $\phi(x, y) = x\ln|xy| + c_1$, $y(x) = \frac{1}{x}e^{c/x}$

4g. $\phi(x, y) = x + xe^y + c_1$, $y(x) = \ln\left|\frac{c - x}{x}\right|$ **4h.** $\phi(x, y) = xe^y + y + c_1$, $xe^y + y = c$

5a. $\mu = x^3$, $y(x) = \pm\left(Cx^{-4} - 1\right)^{1/4}$ **5b.** $\mu = y^{-4}$, $xy^{-3} + y = c$

5c. $\mu = x^{-2}y^2$, $y^4 - x^{-2}y^3 = c$ **5d.** $\mu = \cos(y)$, $x\cos(y) + \sin(y) = c$

5e. $\mu = \sqrt{x}$, $y(x) = -\frac{1}{2} \pm \frac{1}{2}\sqrt{1 + Cx^{-3/2}}$ **5f.** $\mu = e^{-x^2}$, $y(x) = Ce^{x^2} - 1$

5g. $\mu = x^{-3}$, $y^4 - x^{-2}y^3 = c$ **5h.** $\mu = \sqrt{y}$, $y^{5/2} + x^2y^{3/2} = c$

5i. $\mu = xy^{1/3}$, $x^2y^{1/3} + x^4y^{7/3} = c$

Chapter 8

1. (A linear d.e.) $y(x) = cx^2 - 6x^3$ **2.** (A separable d.e.) $y = 0$ and $y(x) = \dfrac{3}{1 - Cx^3}$

3. (A separable d.e.) $y = 0$ and $y(x) = \dfrac{3}{12x - x^3 + C}$

4. (Linear substitution) $2\sqrt{x+y} - 2\ln\left|1 + \sqrt{x+y}\right| = x + c$

5. (A directly integrable d.e.) $y(x) = c - 3x^{-1} - 2x^{-1/2}$

6. (A homogeneous d.e.) $y(x) = \pm x\sqrt{(\ln|Ax|)^2 - 1}$

7. (Linear substitution) $y(x) = x + \dfrac{1 + Ae^{2x}}{1 - Ae^{2x}}$ **8.** (A linear d.e.) $y(x) = \dfrac{2}{x^2} + \dfrac{c}{x^4}$

9. (An exact d.e.) $y(x) = \pm\dfrac{\sqrt{12x + c}}{x}$

10. (A homogeneous and Bernoulli d.e.) $y(x) = x\sqrt[3]{3\ln|x| + c}$

11. (A linear d.e.) $y(x) = \dfrac{x^3}{6} + \dfrac{c}{x^3}$ **12.** (An exact d.e.) $y(x) = \pm\sqrt{\dfrac{c - x}{x^2 + 1}}$

13. (A Bernoulli equation) $y = 0$, $y(x) = \pm\dfrac{x}{\sqrt{2x^2 + c}}$

14. (A linear d.e.) $y(x) = \left(x^2 + 1\right)[c - 2\arctan(x)]$

15. (A separable d.e.) $y^2 - 8\ln|y| = 2x + C$ and $y = 0$

16. (A directly integrable d.e.) $y(x) = \dfrac{1}{2}\ln\left|x^2 - 4\right| + c$

17. (A separable d.e.) $y(x) = 3 \pm \sqrt{C + \ln x^2}$ **18.** (A Bernoulli d.e.) $y(x) = \dfrac{4(1+x)^3}{C + (1+x)^4}$

19. (An exact d.e.) $(x + y)\sin(y) + \cos(y) = C$

20. (The equation is both separable and exact.) $y(x) = \arcsin\left(\dfrac{1}{x+1}\right)$

21. (A directly integrable (and separable) d.e.) $y(x) = \dfrac{1}{2}\ln|\cos(x)| + c$

22. (A homogeneous and Bernoulli d.e.) $y(x) = \pm x\sqrt{Ax^2 - 2}$

23. (Linear substitution) $y + \ln|x + 2y + 1| = x + c$

24. (A homogeneous d.e.) $2\arctan\left(\dfrac{y}{x}\right) - \dfrac{1}{2}\ln\left|1 + \left[\dfrac{y}{x}\right]^2\right| = \ln|Ax|$

25. (A homogeneous d.e.) $y(x) = x\arcsin(Ax)$

26. (A separable d.e.) $y(x) = \tan\left(\dfrac{1}{2}x^2 + 3x + c\right)$

27. (Linear substitution) $y - \ln|x - 2y + 2| = C$ **28.** (An exact d.e.) $x\ln|y| + 3y = c$

29. (A separable d.e.) $y(x) = \tan(x + c)$ **30.** (A linear d.e.) $y(x) = ce^{3x} - 12e^{2x}$

31. (A homogeneous d.e.) $y = -x$ and $e^{y/x} = A[y + x]$

32. (A directly integrable d.e.) $y(x) = \dfrac{1}{3}x^3 - x^2 + 4x - 8\ln|x + 2|$

33. (A Bernoulli d.e.) $y(x) = \pm\sqrt[4]{2x^2 + Cx^4}$ **34.** (A Bernoulli d.e.) $y(x) = \sqrt[3]{6e^{4x} + ce^{12x}}$

35. (A linear d.e.) $y(x) = \dfrac{2x^3 + 3x^2 + c}{(x+1)^2}$ **36.** (An exact d.e.) $x^2y^2 + 4y^5 = c$

37. (A Bernoulli d.e.) $y(x) = \pm\sqrt{2e^{4x^2} + ce^{x^2}}$

38. (Linear substitution) $y(x) = x - 3 + \sqrt[3]{3x + C}$

39. (An exact d.e.) $y(x) = \dfrac{1}{x}\ln\left|c - \dfrac{1}{2}x^2\right|$ **40.** (A separable d.e.) $y(x) = \dfrac{1}{x - \sin(x) + C}$

41. (A linear d.e.) $y(x) = \dfrac{1}{5}[2\sin(x) - \cos(x)] + Ce^{-2x}$

42. (A directly integrable d.e.) $y(x) = c - \cos(x) - x^2$

43. (A separable d.e.) $y = 0$ and $y(x) = \pm\left(2[\sin(x) - x] + c\right)^{-1/2}$

44. (An exact d.e.) $y(x) = \pm\sqrt{\frac{1}{x}\ln\left|x^2+c\right|}$ **45.** (A linear d.e.) $y(x) = [x+c]e^{-3x}$

46. (Linear substitution) $y(x) = \frac{1}{3}\left[1+\arcsin\left(Ae^{3x}\right)\right] - 2x$ and $y(x) = \frac{1}{3}(n\pi - 1) - 2x$ with n being an integer **47.** (A separable d.e.) $y(x) = -\frac{1}{3}\ln\left|C - \frac{3}{4}e^{4x}\right|$

48. (A linear d.e.) $y(x) = Ce^{3x^2} - \frac{1}{4}e^{x^2}$ **49.** (An exact d.e.) $y(x) = x^2 \pm \sqrt{x^4 - x^2 - C}$

50. (A linear d.e.) $y(x) = x^{-3}\left[c - 3e^{-x^2}\right]$

Chapter 9

1b. $y(0)$ should be within $^1/_6$ of 2. **2a.**

2b.

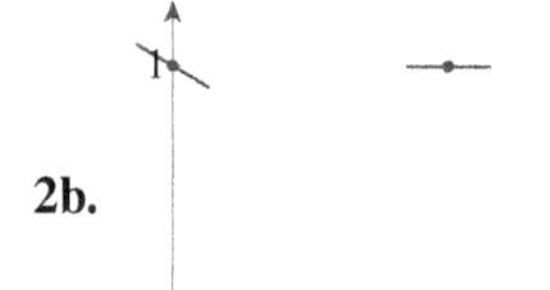

2c.

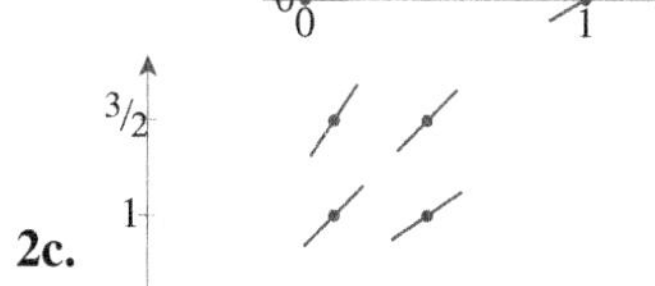

2d.

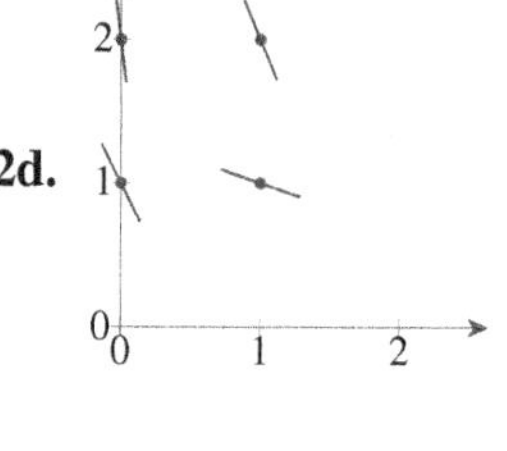

2e.

2f.

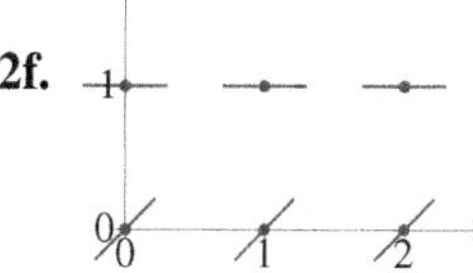

3. a.

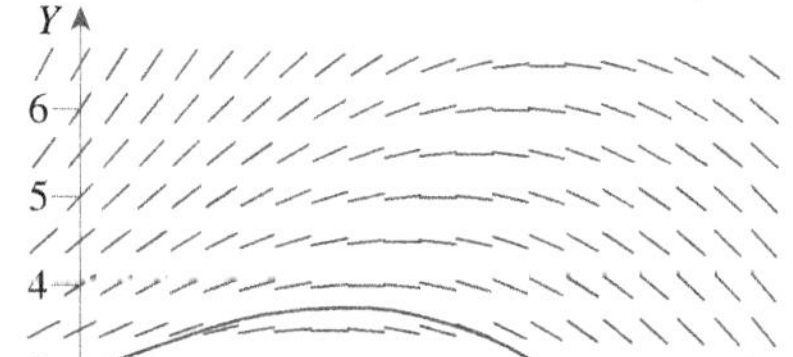

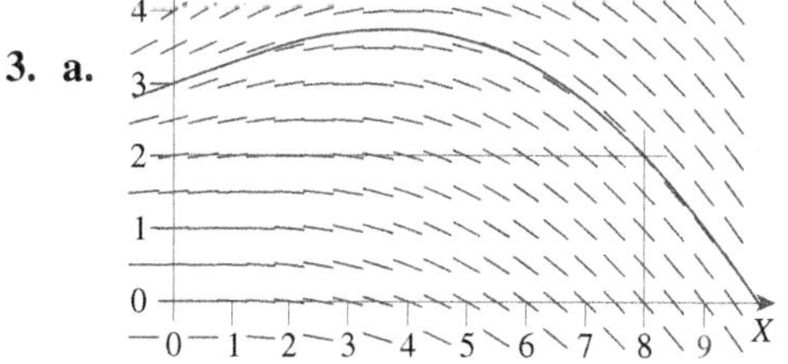

3b. $y(8) \approx 2$

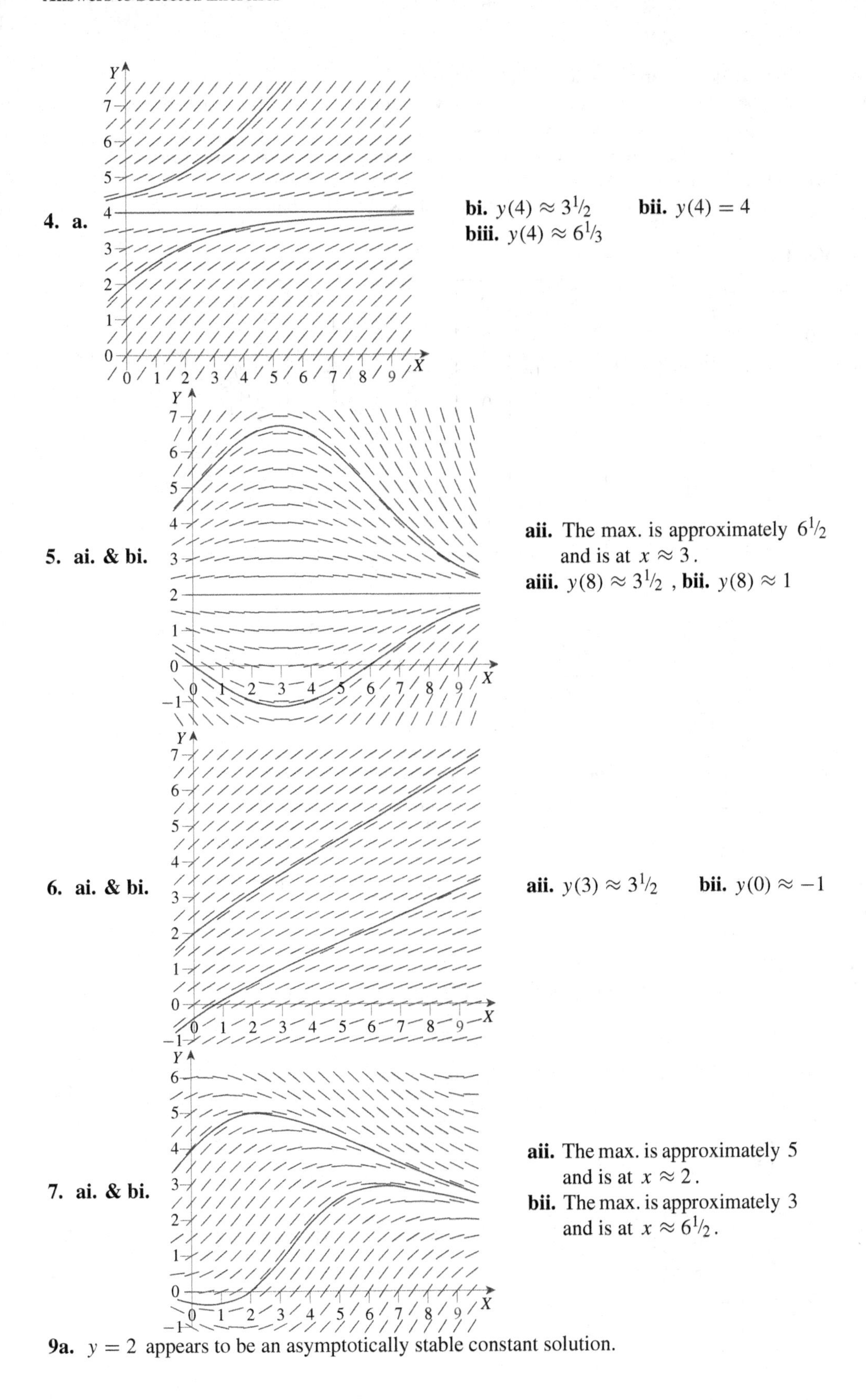

4. a.

bi. $y(4) \approx 3^1/_2$ **bii.** $y(4) = 4$
biii. $y(4) \approx 6^1/_3$

5. ai. & bi.

aii. The max. is approximately $6^1/_2$ and is at $x \approx 3$.
aiii. $y(8) \approx 3^1/_2$, **bii.** $y(8) \approx 1$

6. ai. & bi.

aii. $y(3) \approx 3^1/_2$ **bii.** $y(0) \approx -1$

7. ai. & bi.

aii. The max. is approximately 5 and is at $x \approx 2$.
bii. The max. is approximately 3 and is at $x \approx 6^1/_2$.

9a. $y = 2$ appears to be an asymptotically stable constant solution.

9b. $y = 2$ appears to be an unstable constant solution.

9c. $y = 2$ appears to be a stable (but maybe not asymptotically stable) constant solution.

9d. $y = 3$ appears to be an unstable constant solution.

9e. $y = 2$ appears to be an unstable constant solution, and $y = -2$ appears to be an asymptotically stable constant solution.

9f. $y = 1$ appears to be a stable (but not asymptotically stable) constant solution.

Chapter 10

1a.

k	x_k	y_k
0	1	-1
1	$4/3$	$-4/3$
2	$5/3$	$-5/3$
3	2	-2

1b.

k	x_k	y_k
0	0	10
1	$1/4$	10
2	$1/2$	5
3	$3/4$	0
4	1	0

1c.

k	x_k	y_k
0	0	2
1	$1/2$	$11/5$
2	1	$58/25$
3	$3/2$	$294/125$
4	2	$2{,}859/1{,}250$

1d.

k	x_k	y_k
0	1	8
1	$3/2$	6
2	2	6
3	$5/2$	$13/2$
4	3	$36/5$
5	$7/2$	8
6	4	$62/7$

2a.

k	x_k	y_k
0	0	0.0000
1	.5	0.0000
2	1.0	0.5000
3	1.5	1.2906
4	2.0	2.3263
5	2.5	3.5839
6	3.0	5.0488

2b.

k	x_k	y_k
0	0	1.0000
1	.3333	1.0000
2	0.6667	1.0556
3	1.0000	1.1637
4	1.3333	1.3178
5	1.6667	1.5095
6	2.0000	1.7309

2c.

k	x_k	y_k
0	1.0	2.0000
1	1.1	2.2000
2	1.2	2.4380
3	1.3	2.7294
4	1.4	3.0984
5	1.5	3.5855

2d.

k	x_k	y_k
0	0.0	0.0000
1	0.2	0.2000
2	0.4	0.3960
3	0.6	0.5805
4	0.8	0.7478
5	2.0	0.8944

4a. $y(5) \approx 12.1088$

4b. $y(5) \approx 15.7725$ **4c.** $y(5) \approx 16.2525$ **4d.** $y(5) \approx 16.3019$

5a. $y(3) \approx 0.81979$ with error $= 1.0568 \pm 0.00005$

5b. $y(3) \approx 0.36059$ with error $= .5976 \pm 0.00005$

5c. $y(3) \approx 0.07273$ with error $= .03097 \pm 0.00005$

5d. $y(3) \approx -0.08088$ with error $= .1561 \pm 0.00005$

5e. $y(3) \approx -0.22449$ with error $= .0125 \pm 0.00005$

5f. $y(3) \approx -0.23574$ with error $= .0013 \pm 0.00005$

5g. $y(3) \approx -0.23687$ with error $= .0001 \pm 0.00005$ **6a.** $y(x) = 1 - \frac{23}{23x + 10}$, $y(4) = \frac{79}{102}$

6b.

k	x_k	y_k
0	0.0	-1.3
1	0.5	1.345
2	1.0	1.4045125
3	1.5	1.486327681
4	2.0	1.604584988
5	2.5	1.787346492
6	3.0	2.097303741
7	3.5	2.699341491
8	4.0	4.143222243

6c i. $y(4) \approx 23950.5985$ with error $= 23949.824 \pm 0.0005$

6c ii. $y(4) \approx 4.1432$ with error $= 3.369 \pm 0.0005$
6c iii. $y(4) \approx 0.8081$ with error $= 0.03 \pm 0.0005$
6c iv. $y(4) \approx 0.7866$ with error $= 0.001 \pm 0.0005$ **7a i.** $y_k = (-1)^k 3$ **7a ii.** It does not.
7a iii. $y(x) \to 0$ as $x \to \infty$; $|y_k - y(x_k)| \approx 3$ **7b i.** $y_k = 3\left(\frac{3}{5}\right)^k$ **7b ii.** Yes.
8d. The solution becomes infinite.
8e. They are nonsense. The solution to the initial-value problem is not valid for $x > 7/3$.
9. It "blows up" at $k = 2$.

Chapter 11

2a. 6,538,035 **2b.** For each, the answer is $0.8 \times \ln 2 (\approx 0.55)$ month
2c. For each, the answer is $0.8 \times \ln 10 (\approx 1.84)$ months **2d.** 44.21 months
3a. $I(t) = e^{\beta t}$ with $\beta = \frac{1}{10}\ln(50)$ **3b.** 2,500 **3c.** 31.78 days **5a.** 0.023/year
5b i. 15.87 grams **5b ii.** 11.22 grams **5b iii.** 1.98 grams **6a.** $\delta \approx 0.00012$/year
6b i. 99.88 % **6b ii.** 98.8 % **6b iii.** 88.60 % **6b iv.** 54.61 % **6b v.** 29.82 %
6b vi. 0.24 % **6c.** 9,950 years ago **6d.** 4,222 years ago **6e.** $-\frac{5730}{\ln 2}\ln\left(\frac{A}{A_0}\right)$
7b i. 3,614.8 rabbits **7b ii.** 3,953,332.8 rabbits **7b iii.** 9,999,995.3 rabbits
7c. $\kappa = R_0 R(t) \dfrac{1 - e^{-\beta t}}{R_0 - R(t)e^{-\beta t}}$ **7d i.** 1,381.5 rabbits **7d ii.** 4,472.6 rabbits
8a. $\frac{dR}{dt} = \frac{5}{4}R - 500$
8b. $y = 400$ is the equilibrium solution. If we start with more, the population increases. If we start with less, the population rapidly decreases. (In fact, since $R' = -500$ when $R = 0$, this model has the number of rabbit becoming negative if we start with fewer than 400 rabbits!)
8c. $R(t) = 400 + (R_0 - 400)e^{5t/4}$ **9a.** $\frac{dR}{dt} = \frac{3}{4}R$
9b. There is no equilibrium solution; if we start with a positive number of rabbits, the population constantly increases. **9c.** $R(t) = R_0 e^{3t/4}$ **10a.** $\frac{dR}{dt} = \beta R - \gamma R^2 - h_0$
10b. $\frac{dR}{dt} = \left(\beta - \frac{1}{4}\right)R - \gamma R^2$ **11a.** $\frac{dy}{dt} = 50 - \frac{1}{20}y$ (initial condition: $y(0) = 200$)
11b. 1000 warriors **11c.** $y(t) = 1000 - 800e^{-t/20}$
11d. $20 \ln 8$ weeks (about 41.5 weeks) **12a.** $\frac{dy}{dt} = \frac{9}{4} - \frac{3}{1000}y$
12c. $y(t) = 750 - 750e^{-3t/1000}$ **12d i.** 22.2 **12d ii.** 123.6 **12d iii.** 712.7
12e. When $t = \frac{1000}{3}\ln 3 \approx 366.2$ **13a.** $\frac{dy}{dt} = \frac{9}{4} - \frac{3y}{1000}$ **13c.** $y(t) = 750 - 650e^{-3t/1000}$
13d i. 119.2 gallons **13d ii.** 207.1 gallons **13d iii.** 717.6 gallons
13e. When $t = \frac{1000}{3}\ln\frac{13}{5} \approx 318.5$ **14a.** $\frac{dy}{dt} = \frac{3}{2} - \frac{1}{400}y$ **14c.** $y(t) = 600 - 600e^{-t/400}$
14d i. 14.8 **14d ii.** 83.6 **14d iii.** 132.7 **14e.** 3 oz. salt/gal. water
14f. When $t = 400 \ln 3 \approx 439.4$ **15a i.** $\frac{dy}{dt} = \frac{3}{2} - \frac{1}{200}y$
15a iii. $y(t) = 300 - 300e^{-t/200}$ **15b i.** 14.6 **15b ii.** 77.8 **15b iii.** 118.0
15c. $3/2$ oz. salt/gal. water **15d.** Never **16a.** $500 - t$ **16b.** $\frac{dy}{dt} = 4 - \frac{3}{500 - t}y$
16c. $y(t) = 2(500 - t) - \left(10 - \frac{t}{50}\right)^3$ **16d i.** 38.808 **16d ii.** 198.528 **16d iii.** 288
16e i. $t = 499$ **16e ii.** $\frac{249{,}999}{125{,}000} = 1.999992$ ounces
17. Arrest the butler. The time of death was only about 3:20, over an hour after the butler reported finding the body. Besides, his fingerprints were on the bottle.

Chapter 12

1a.

k	x_k	y_k
0	1	-1
1	$4/3$	$-4/3$
2	$5/3$	$-5/3$
3	2	-2

1b.

k	x_k	y_k
0	0	10
1	$1/4$	$15/2$
2	$1/2$	$15/4$
3	$3/4$	$15/8$
4	1	$45/32$

1c.

k	x_k	y_k
0	0	2
1	$1/2$	$54/25$
2	1	$11{,}159/5{,}000$
3	$3/2$	$2{,}206{,}139/1{,}000{,}000$
4	2	$414{,}556{,}719/200{,}000{,}000$

1d.

k	x_k	y_k
0	1	8
1	$3/2$	7
2	2	7
3	$5/2$	$37/5$
4	3	8
5	$7/2$	$61/7$
6	4	$19/2$

3a. $y(5) \approx 15.988744$ **3b.** $y(5) \approx 16.303317$

3c. $y(5) \approx 16.307415$ **3d.** $y(5) \approx 16.307457$

5a.

k	x_k	y_k
0	0.0	-1.3
1	0.5	0.052256
2	1.0	0.338970
3	1.5	0.497173
4	2.0	0.595803
5	2.5	0.662650
6	3.0	0.710764
7	3.5	0.746981
8	4.0	0.775197

5b i. $y(4) \approx 1.908706 \times 10^{40}$ with error $= 1.908706 \times 10^{40} \pm 0.0000005 \times 10^{40}$
5b ii. $y(4) \approx 0.775197$ with error $= 0.000687 \pm 0.0000005$
5b iii. $y(4) \approx 0.770696$ with error $= 0.003813 \pm 0.0000005$

5b iv. $y(4) \approx 0.773919$ with error $= 0.000591 \pm 0.0000005$

6a.

k	x_k	y_k
0	0	10
1	$1/2$	$10/3$
2	1	$10/3$

6b.

k	x_k	y_k
0	0	2
1	$1/2$	$64{,}759/30{,}000$
2	1	$16{,}047{,}211{,}919/7{,}200{,}000{,}000$

6c.

k	x_k	y_k
0	1	8
1	$3/2$	7
2	2	7

6d.

k	x_k	y_k
0	1	8
1	$4/3$	$43/6$
2	$5/3$	$104/15$
3	2	7

8a. $y(5) \approx 16.30973295$ **8b.** $y(5) \approx 16.30745694$ **8c.** $y(5) \approx 16.30745728$

8d. $y(5) \approx 16.30745731$

10a.

k	x_k	y_k
0	0.0	−1.3
1	0.5	0.01070064
2	1.0	0.338090397
3	1.5	0.502658043
4	2.0	0.601696339
5	2.5	0.667843281
6	3.0	0.715149727
7	3.5	0.750661374
8	4.0	0.778300311

10b i. $y(4) \approx 41891.0998$ with error $= 41890.3253 \pm 0.00005$
10b ii. $y(4) \approx 0.778300311$ with error $= 0.003790507 \pm 0.0000000005$
10b iii. $y(4) \approx 0.774513033$ with error $= 0.000003229 \pm 0.0000000005$
10b iv. $y(4) \approx 0.774509045$ with error $= 0.00000076 \pm 0.000000005$

Chapter 13

1a. $y(x) = x^3 + \frac{c_1}{x^3} + c_2$ **1b.** $y(x) = Ax^3 + c$ **1c.** $y(x) = Ae^x + c$
1d. $y(x) = e^{2x} + c_1 e^{-2x} + c_2$ **1e.** $y(x) = Ae^{-x^2} + c$ **1f.** $y(x) = C_1 \arctan(x) + C_2$
2a. $y(x) = \frac{1}{5}x^5 + \frac{2}{3}c_1 x^3 + {c_1}^2 x + c_2$ and $y = C$ **2b.** $y(x) = c_2 \pm \frac{1}{3}(2x + c_1)^{3/2}$
2c. Equation contains y. **2d.** $y(x) = \frac{1}{A}\ln|1 + Ax| + C$, $y = c$ and $y(x) = x + c$
2e. $y(x) = \frac{1}{4}x^6 + c_1 x^2 + c_2$ **2f.** Equation contains y. **2g.** $y(x) = 3x + Ae^{2x} + C$
2h. Equation contains y. **2i.** $y(x) = Ae^{-4x} - 3e^{-3x} + B$ **2j.** $y(x) = B - \ln\left|A + e^{-2x}\right|$
3a. $y(x) = Ae^x + Bx + C$ **3b.** $y(x) = \frac{1}{3}x^3 - A\ln|x| + Bx + C$
3c. $y(x) = \frac{1}{12}(x + A)^4 + Bx + C$ and $y(x) = Bx + C$
3d. $y(x) = Ae^{-2x} + Bx^2 + Cx + D$ **4a.** $y(x) = Be^{ax}$ **4b.** $y(x) = (Ax + c)^3$
4c. $y(x) = \arccos(a - cx)$ **4d.** $y(x) = Ae^x + c$ **4e.** $y(x) = \pm\sqrt{Ae^{2x} + C}$
4f. $y^3 + 3y = Ax + C$ **5a.** Not autonomous **5b.** $y(x) = \pm\sqrt{Ax + c}$
5c. $y(x) = c_2 \pm \frac{1}{3}(2x + c_1)^{3/2}$ **5d.** Not autonomous **5e.** Not autonomous
5f. $y(x) = Be^{Ax} + \frac{1}{A}$ and $y = c$ **5g.** $y(x) = \frac{1}{Bx + C}$ **5h.** $y(x) = 3 + Be^{Ax}$
5i. Not autonomous **5j.** $y(x) = -\ln\left|A + Be^{-2x}\right|$ **6a.** $y(x) = x^3 + \frac{2}{x^3} + 5$
6b. $y(x) = 4x^3 + 8$ **6c.** $y(x) = 5e^x + 3$ **6d.** $y(x) = e^{2x} + e^{-2x} - 2$
6e. $y(x) = 2e^x + 3x + 8$ **6f.** $y(x) = \frac{1}{3}x^3 - 2\ln|x| + 2x + \frac{1}{3}$ **6g.** $y(x) = 3x - \frac{2}{x} + 3$
6h. $y(x) = \frac{1}{3}(2x + 1)^{3/2} - \sqrt{3}$ **7a.** $y(x) = 5e^{3x}$ **7b.** $y(x) = \frac{1}{8}(x + 4)^3$
7c. $y(x) = (3x - 2)^3$ **7d.** $y(x) = \sqrt[3]{3e^{3x} + 5}$ **7e.** $y(x) = \ln\left(2e^x - 1\right)$
8a i. $y(x) = 2\arctan(2x) + 3$ **8a ii.** $y = 3$ **8a iii.** $y(x) = 1 - \frac{1}{x}$
8a iv. $y(x) = \frac{1}{4}\ln\left|\frac{x - 2}{x + 2}\right| + 5$
8b. $y(x) = A\arctan(Ax) + c$, $y(x) = c - \frac{1}{x}$ and $y(x) = \frac{1}{2a}\ln\left|\frac{x - a}{x + a}\right| + c$
9a i. $y(x) = \tan(x)$ **9a ii.** $y(x) = \frac{1}{1 - x}$ **9a iii.** $y = 1$ **9a iv.** $y(x) = \frac{1 - e^{2x}}{1 + e^{2x}}$
9b. $y(x) = A\tan(Ax + B)$, $y(x) = \frac{1}{c - x}$ and $y(x) = c\frac{1 + Be^{2cx}}{1 - Be^{2cx}}$

Chapter 14

1a. Second-order, linear, nonhomogeneous **1b.** Second-order, linear, homogeneous
1c. Second-order, linear, homogeneous **1d.** Second-order, nonlinear
1e. First-order, linear, nonhomogeneous **1f.** Third-order, linear, homogeneous
1g. Second-order, nonlinear **1h.** Second-order, linear, nonhomogeneous
1i. Fourth-order, linear, nonhomogeneous **1j.** Third-order, nonlinear
1k. Third-order, linear, homogeneous **1l.** Fifty-fifth-order, linear, nonhomogeneous
2a. $y(x) = Ae^{3x} + Be^{2x}$ **2b.** $y(x) = Axe^{5x} + Be^{5x}$ **2c.** $y(x) = Ax^4 + Bx^3$
2d. $y(x) = A\sqrt{x} + Bx$ **2e.** $y(x) = A\sqrt{x}\ln|x| + B\sqrt{x}$ **2f.** $y(x) = Ax^3e^{2x} + Be^{2x}$
2g. $y(x) = Ax + Be^{-x}$ **2h.** $y(x) = Ae^{x^2} + Be^{-x^2}$ **2i.** $y(x) = A\cos(x) + B\sin(x)$
2j. $y(x) = Ax^{-1}e^{-2x} + Bx^{-1}$ **2k.** $y(x) = Ax\sin(x) + B\sin(x)$
2l. $y(x) = Ax\cos(x) + Bx\sin(x)$ **2m.** $y(x) = A\cos(\ln|x|) + B\sin(\ln|x|)$
2n. $y(x) = Ax^{-1/2}\cos(x) + Bx^{-1/2}\sin(x)$ **3a.** $y(x) = Ae^{3x} + Be^{x} - 9e^{2x}$
3b. $y(x) = Ae^{4x} + Be^{2x} - \frac{1}{2}xe^{2x}$ **3c.** $y(x) = Ax + Bx^{-1} - \frac{4}{3}\sqrt{x}$
3d. $y(x) = Ax^5 + Bx^{-4} + 3x^5\ln|x|$ **3e.** $y(x) = x^{-1}\left[e^{2x} + A + Be^{-2x}\right]$
3f. $y(x) = x^2 + 1 + Bx + Ae^{-x}$ **5a.** $y(x) = \left[A + Bx + Cx^2\right]e^{3x}$
5b. $y(x) = e^{3x}\cos(x) + Ae^{3x} + Bxe^{3x} + Cx^2e^{3x}$ **5c.** $y(x) = \left[A + Bx + Cx^2 + Dx^3\right]e^{2x}$
5d. $y(x) = Ax^2\ln|x| + Bx^2 + Cx^3$

Chapter 15

2a. $y(x) = 2\cos(2x) + 3\sin(2x)$ **2b.** $y(x) = 3e^{2x} - 3e^{-2x}$ **2c.** $y(x) = 3e^{2x} + 5e^{-3x}$
2d. $y(x) = e^{2x} + 4xe^{2x}$ **2e.** $y(x) = 4x^3 - 4x^2$ **2f.** $y(x) = 5x^{1/2} + 3x^{-1/2}$
2g. $y(x) = 5x - 2x\ln|x|$ **2h.** $y(x) = -3\cos\left(x^2\right) - \frac{2}{\sqrt{\pi}}\sin\left(x^2\right)$ **2i.** $y(x) = 4x^2 + 4x$
3a. $(0, \infty)$ **4a.** $(0, \infty)$ **5a.** $y(x) = 4 - \cos(2x) + 4\sin(2x)$
5b. $y(x) = 3 + 2\sin^2(x) + 8\sin(x)\cos(x)$ **5c.** $y(x) = 2\sin(x) + 2\sinh(x)$
6a. gen. soln.: $y(x) = c_1e^{2x} + c_2e^{-2x}$, i.v. soln.: $y(x) = \frac{1}{2}e^{2x} + \frac{1}{2}e^{-2x}$
6b. gen. soln.: $y(x) = c_1e^{x} + c_2e^{-3x}$, i.v. soln.: $y(x) = \frac{1}{4}e^{x} - \frac{1}{4}e^{-3x}$
6c. gen. soln.: $y(x) = c_1e^{x} + c_2e^{9x}$, i.v. soln.: $y(x) = 12e^{x} - 4e^{9x}$
6d. gen. soln.: $y(x) = c_1 + c_2e^{-5x}$, i.v. soln.: $y = 1$ **7a.** $y(x) = c_1 + c_2e^{3x} + c_3e^{-3x}$
7b. $y(x) = c_1e^{3x} + c_2e^{-3x} + c_3e^{x} + c_4e^{-x}$ **8.** $c_1 = A\cos(B)$ and $c_2 = A\sin(B)$
9c. Yes, $\{y_1, y_2\}$ is linearly dependent on any subinterval of either $(0, \infty)$ or $(-\infty, 0)$.

Chapter 16

2a. $L = \frac{d^2}{dx^2} + 5\frac{d}{dx} + 6$ **2b i.** $5\sin(x) + 5\cos(x)$ **2b ii.** $42e^{4x}$ **2b iii.** 0
2b iv. $6x^2 + 10x + 2$ **2c.** $y(x) = e^{-3x}$ **3a.** $L = \frac{d^2}{dx^2} - 5\frac{d}{dx} + 9$
3b i. $8\sin(x) - 5\cos(x)$ **3b ii.** $-15\cos(3x)$ **3b iii.** $3e^{2x}$
3b iv. $[2\sin(x) - \cos(x)]e^{2x}$ **4a.** $L = x^2\frac{d^2}{dx^2} + 5x\frac{d}{dx} + 6$
4b i. $\left(6 - x^2\right)\sin(x) + 5x\cos(x)$ **4b ii.** $\left[16x^2 + 20x + 6\right]e^{4x}$ **4b iii.** $27x^3$
5a. $L = \frac{d^3}{dx^3} - \sin(x)\frac{d}{dx} + \cos(x)$ **5b i.** $-\cos(x)$ **5b ii.** $1 + \sin(x)$
5b iii. $x^2\cos(x) - 2x\sin(x)$ **6a.** $L_2L_1 = \frac{d^2}{dx^2} + \left(1 - x^2\right)$, $L_1L_2 = \frac{d^2}{dx^2} - \left(x^2 + 1\right)$

6b. $L_2L_1 = \frac{d^2}{dx^2} + \left(x^2 + x^3\right)\frac{d}{dx} + \left(2x + x^5\right)$, $L_1L_2 = \frac{d^2}{dx^2} + \left(x^2 + x^3\right)\frac{d}{dx} + \left(3x^2 + x^5\right)$

6c. $L_2L_1 = x\frac{d^2}{dx^2} + \left(4 + 2x^2\right)\frac{d}{dx} + 6x$, $L_1L_2 = x\frac{d^2}{dx^2} + \left(3 + 2x^2\right)\frac{d}{dx} + 8x$

6d. $L_2L_1 = x\frac{d^2}{dx^2}$, $L_1L_2 = x\frac{d^2}{dx^2} + 2\frac{d}{dx}$

6e. $L_2L_1 = x^2\frac{d^2}{dx^2}$, $L_1L_2 = x^3\frac{d^2}{dx^2} + 6x^2\frac{d}{dx} + 6x$

6f. $L_2L_1 = \sin(x)\frac{d^2}{dx^2}$, $L_1L_2 = \sin(x)\frac{d^2}{dx^2} + 2\cos(x)\frac{d}{dx} - \sin(x)$ **7a.** $\frac{d^2}{dx^2} + 5\frac{d}{dx} + 6$

7b. $x^2\frac{d^2}{dx^2} + 6x\frac{d}{dx} + 6$ **7c.** $x\frac{d^2}{dx^2} + 5\frac{d}{dx} + \frac{3}{x}$ **7d.** $\frac{d^2}{dx^2} + \left(4x + \frac{1}{x}\right)\frac{d}{dx} + \left(4 - \frac{1}{x^2}\right)$

7e. $\frac{d^2}{dx^2} + \left(4x + \frac{1}{x}\right)\frac{d}{dx} + 8$ **7f.** $\frac{d^2}{dx^2} + 10x^2\frac{d}{dx} + \left(10x + 25x^4\right)$

7g. $\frac{d^3}{dx^3} + \left(1 + x^2\right)\frac{d^2}{dx^2} + x^2\frac{d}{dx}$ **7h.** $\frac{d^3}{dx^3} + \left(1 + x^2\right)\frac{d^2}{dx^2} + \left(4x + x^2\right)\frac{d}{dx} + 2(1 + x)$

8. $y(x) = ce^{3x}$ **9.** $y(x) = ce^{-x^2}$ **10.** $y(x) = cx^4$

Chapter 17

1a. $y(x) = c_1e^{2x} + c_2e^{5x}$ **1b.** $y(x) = c_1e^{4x} + c_2e^{-6x}$ **1c.** $y(x) = c_1e^{5x} + c_2e^{-5x}$
1d. $y(x) = c_1 + c_2e^{-3x}$ **1e.** $y(x) = c_1e^{x/2} + c_2e^{-x/2}$ **1f.** $y(x) = c_1e^{2x/3} + c_2e^{-3x}$
2a. $y(x) = \frac{5}{2}e^{3x} - \frac{3}{2}e^{5x}$ **2b.** $y(x) = -\frac{1}{2}e^{3x} + \frac{1}{2}e^{5x}$ **2c.** $y(x) = 3e^{3x} + 2e^{5x}$
2d. $y(x) = \frac{1}{2}e^{3x} + \frac{1}{2}e^{-3x}$ **2e.** $y(x) = \frac{1}{6}e^{3x} - \frac{1}{6}e^{-3x}$ **2f.** $y(x) = e^{3x} + 2e^{-3x}$
3a. $y(x) = c_1e^{5x} + c_2xe^{5x}$ **3b.** $y(x) = c_1e^{-x} + c_2xe^{-x}$
3c. $y(x) = c_1e^{x/2} + c_2xe^{x/2}$ **3d.** $y(x) = c_1e^{x/5} + c_2xe^{x/5}$
3e. $y(x) = c_1e^{3x/4} + c_2xe^{3x/4}$ **3f.** $y(x) = c_1e^{-2x/3} + c_2xe^{-2x/3}$
4a. $y(x) = e^{4x} - 4xe^{4x}$ **4b.** $y(x) = xe^{4x}$ **4c.** $y(x) = 3e^{4x} + 2xe^{4x}$
4d. $y(x) = e^{-x/2} + \frac{1}{2}xe^{-x/2}$ **4e.** $y(x) = xe^{-x/2}$ **4f.** $y(x) = 6e^{-x/2} - 2xe^{-x/2}$
5a. $y(x) = A\cos(5x) + B\sin(5x)$ **5b.** $y(x) = Ae^{-x}\cos(2x) + Be^{-x}\sin(2x)$
5c. $y(x) = Ae^{x}\cos(2x) + Be^{-x}\sin(2x)$ **5d.** $y(x) = Ae^{2x}\cos(5x) + Be^{2x}\sin(5x)$
5e. $y(x) = Ae^{-x}\cos\left(\frac{x}{3}\right) + Be^{-x}\sin\left(\frac{x}{3}\right)$ **5f.** $y(x) = A\cos\left(\frac{x}{2}\right) + Be^{2x}\sin\left(\frac{x}{2}\right)$
6a. $y(x) = \cos(4x)$ **6b.** $y(x) = \frac{1}{4}\sin(4x)$ **6c.** $y(x) = 4\cos(4x) + 3\sin(4x)$
6d. $y(x) = e^{2x}\cos(3x) - \frac{2}{3}e^{2x}\sin(3x)$ **6e.** $y(x) = \frac{1}{3}e^{2x}\sin(3x)$
6f. $y(x) = 5e^{2x}\cos(3x) + 7e^{2x}\sin(3x)$

7a. $y(x) = e^{x/2}\cos(2\pi x)$

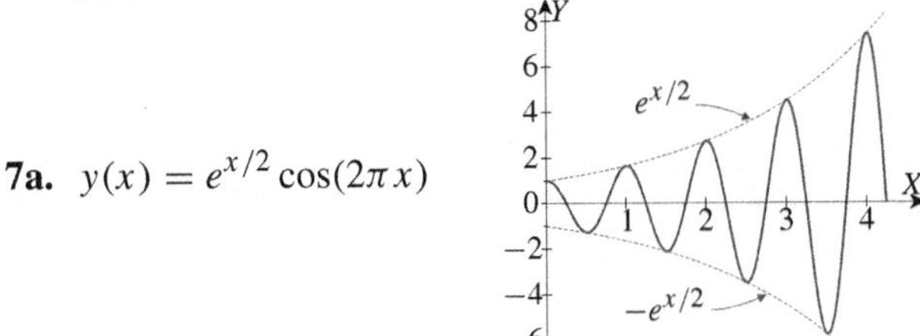

7b. $y(x) = e^{x/2}\cos(2\pi x)$

8a. $y(x) = c_1 e^{3x} + c_2 e^{-3x}$ **8b.** $y(x) = c_1\cos(3x) + c_2\sin(3x)$
8c. $y(x) = c_1 e^{-3x} + c_2 x e^{-3x}$ **8d.** $y(x) = c_1 e^{(-3+3\sqrt{2})x} + c_2 e^{(-3-3\sqrt{2})x}$
8e. $y(x) = c_1 e^{x/3} + c_2 x e^{x/3}$ **8f.** $y(x) = c_1 e^{-3x}\cos(x) + c_2 e^{-3x}\sin(x)$
8g. $y(x) = c_1 e^{2x}\cos(6x) + c_2 e^{2x}\sin(6x)$ **8h.** $y(x) = c_1 e^{2x} + c_2 e^{x/2}$
8i. $y(x) = c_1 e^{-5x} + c_2 x e^{-5x}$ **8j.** $y(x) = c_1 e^{x/3} + c_2 e^{-x/3}$
8k. $y(x) = c_1\cos\left(\frac{x}{3}\right) + c_2\sin\left(\frac{x}{3}\right)$ **8l.** $y(x) = c_1 + c_2 e^{-x/9}$
8m. $y(x) = c_1 e^{-2x}\cos\left(\sqrt{3}x\right) + c_2 e^{-2x}\sin\left(\sqrt{3}x\right)$
8n. $y(x) = c_1 e^{-2x}\cos(x) + c_2 e^{-2x}\sin(x)$ **8o.** $y(x) = c_1 e^{-2x} + c_2 x e^{-2x}$
8p. $y(x) = c_1 e^{-3x} + c_2 e^{5x}$ **8q.** $y(x) = c_1 + c_2 e^{4x}$ **8r.** $y(x) = c_1 e^{-4x} + c_2 x e^{-4x}$
8s. $y(x) = c_1\cos\left(\frac{\sqrt{3}}{2}x\right) + c_2\sin\left(\frac{\sqrt{3}}{2}x\right)$ **8t.** $y(x) = c_1 e^{x/2}\cos(x) + c_2 e^{x/2}\sin(x)$

Chapter 18

1a. $\kappa = 4$ (kg/sec^2) **1b.** $\omega_0 = \frac{1}{2}$ (sec^{-1}) , $\nu_0 = \frac{1}{4\pi}$ (sec^{-1}) , $p_0 = 4\pi$ (sec)
1c i. $A = 2\,, \phi = 0$ **1c ii.** $A = 4\,, \phi = \frac{\pi}{2}$ **1c iii.** $A = 4\,, \phi = \frac{3\pi}{2}$
1c iv. $A = 4\,, \phi = \frac{\pi}{3}$ **2a.** $\kappa = 288$ (kg/sec^2)
2b. $\omega_0 = 12$ (sec^{-1}) , $\nu_0 = \frac{6}{\pi}$ (sec^{-1}) , $p_0 = \frac{\pi}{6}$ (sec) **2c i.** $A = 1$ **2c ii.** $A = \frac{1}{12}$
2c iii. $A = \frac{\sqrt{17}}{4}$ **4a.** $\kappa = \frac{4\pi^2}{9}$ **4b.** $\kappa = \frac{8\pi^2}{9}$ **4c.** $\kappa = \frac{2\pi^2}{9}$
6b. $\alpha = \frac{1}{2}\,, \omega = 3\,, p = \frac{2\pi}{3}\,, \nu = \frac{3}{2\pi}$ **6c i.** $A = \frac{1}{6}\sqrt{37}$ **6c ii.** $A = 4$ **6c iii.** $A = \frac{1}{3}$
6c iv. $A = \frac{2}{3}\sqrt{10}$
7b. ν decreases from the natural frequency of the undamped system, ν_0, down to zero.
p increases from the natural period of the undamped system, p_0, up to ∞.

8b. $y(t) = [2 + 4t]e^{-2t}$

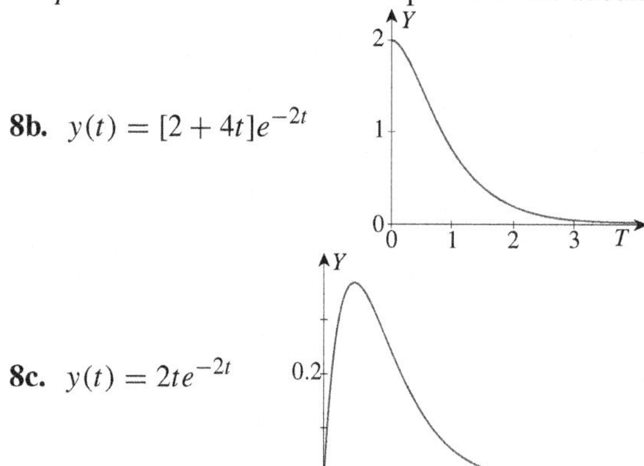

8c. $y(t) = 2te^{-2t}$

9b. $y(t) = \frac{8}{3}e^{-t} - \frac{2}{3}e^{-4t}$

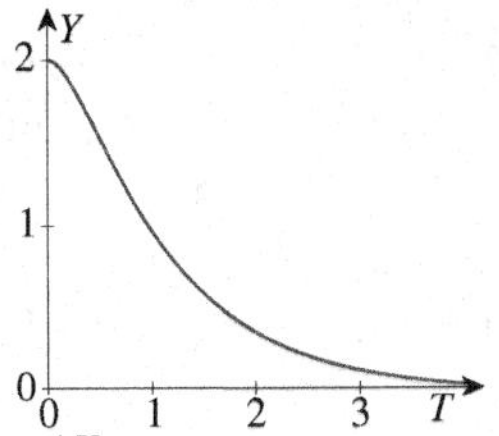

9c. $y(t) = \frac{2}{3}e^{-t} - \frac{2}{3}e^{-4t}$

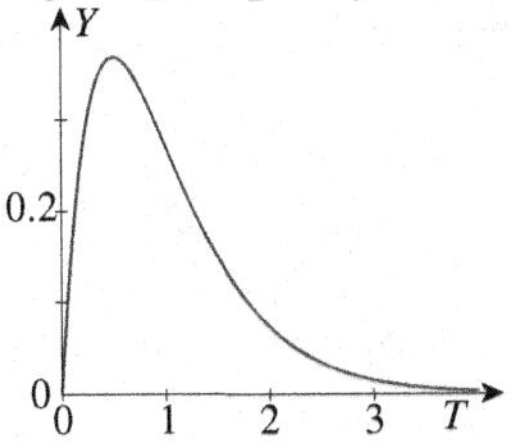

Chapter 19

1a. $y(x) = c_1 + c_2x + c_3x^2 + c_4e^{4x}$ **1b.** $y(x) = c_1 + c_2x + c_3\cos(2x) + c_4\sin(2x)$
1c. $y(x) = c_1e^{3x} + c_2e^{-3x} + c_3e^{5x} + c_4e^{-5x}$
1d. $y(x) = c_1e^{3x} + c_2e^{-3x} + c_3\cos(3x) + c_4\sin(3x)$
1e. $y(x) = [c_1 + c_2x]e^{3x} + [c_3 + c_4x]e^{-3x}$
1f. $y(x) = c_1 + [c_2 + c_3x]\cos(3x) + [c_4 + c_5x]\sin(3x)$
2a. $y(x) = c_1e^{x} + c_2\cos(x) + c_3\sin(x)$ **2b.** $y(x) = c_1e^{x} + c_2e^{2x} + c_3e^{3x}$
2c. $y(x) = c_1e^{2x} + c_2e^{3x}\cos(4x) + c_3e^{3x}\sin(4x)$
2d. $y(x) = [c_1 + c_2\cos(2x) + c_3\sin(2x)]\,e^{3x}$
2e. $y(x) = c_1e^{x} + c_2e^{-2x} + c_3\cos(2x) + c_4\sin(2x)$
2f. $y(x) = [c_1 + c_2x]e^{-x} + c_3\cos(3x) + c_4\sin(3x)$ **3a.** $y(x) = 6 - 2\cos(2x) + 3\sin(2x)$
3b. $y(x) = \left[5 + 3x + 27x^2\right]e^{2x}$ **3c.** $y(x) = 2\cos(x) - 3\sin(x) + 4\cos(5x) - 5\sin(5x)$
3d. $y(x) = 4 + 6e^{-x} + 2\sin(3x)$ **4a.** $y(x) = c_1e^{2x} + c_2e^{-x}\cos\left(\sqrt{3}x\right) + c_3e^{-x}\sin\left(\sqrt{3}x\right)$
4b. $y(x) = c_1e^{-6x} + c_2e^{3x}\cos\left(3\sqrt{3}x\right) + c_3e^{3x}\sin\left(3\sqrt{3}x\right)$
4c. $y(x) = c_1e^{2x} + c_2e^{-2x} + c_3\cos(x) + c_4\sin(x)$
4d. $y(x) = c_1\cos(2x) + c_2\sin(2x) + c_3\cos(3x) + c_4\sin(3x)$
4e. $y(x) = \left[c_1 + c_2x + c_3x^2\right]e^{-x} + \left[c_4 + c_5x + c_6x^2\right]e^{x}$
4f. $y(x) = [c_1 + c_2x]e^{x} + [c_3 + c_4x]e^{-x/2}\cos\left(\frac{\sqrt{3}}{2}x\right) + [c_5 + c_6x]e^{-x/2}\sin\left(\frac{\sqrt{3}}{2}x\right)$
4g. $y(x) = c_1e^{-x/2} + c_2e^{x/2} + c_3\cos\left(\frac{x}{2}\right) + c_4\sin\left(\frac{x}{2}\right)$
4h. $y(x) = c_1e^{-x/2} + c_2e^{x/2} + c_3\cos(2x) + c_4\sin(2x)$
4i. $y(x) = c_1e^{2x} + c_2xe^{2x} + c_3x^2e^{2x} + c_4e^{-2x}$
4j. $y(x) = [c_1 + c_2x]\,e^{-2x} + [c_3 + c_4x]\,e^{x}\cos\left(\sqrt{3}x\right) + [c_5 + c_6x]\,e^{x}\sin\left(\sqrt{3}x\right)$

Chapter 20

1a. $y(x) = c_1x^2 + c_2x^4$ **1b.** $y(x) = c_1x^2 + c_2x^{-1}$ **1c.** $y(x) = c_1 + c_2x^3$
1d. $y(x) = c_1x + c_2\sqrt{x}$ **1e.** $y(x) = c_1x^3 + c_2x^3\ln|x|$ **1f.** $y(x) = c_1x^{-2} + c_2x^{-2}\ln|x|$
1g. $y(x) = c_1\sqrt{x} + c_2\sqrt{x}\ln|x|$ **1h.** $y(x) = c_1x^{10} + c_2x^{10}\ln|x|$
1i. $y(x) = c_1x^3\cos(2\ln|x|) + c_2x^3\sin(2\ln|x|)$
1j. $y(x) = c_1x\cos(3\ln|x|) + c_2x\sin(3\ln|x|)$
1k. $y(x) = c_1x^{-2}\cos(5\ln|x|) + c_2x^{-2}\sin(5\ln|x|)$ **1l.** $y(x) = c_1\cos(\ln|x|) + c_2\sin(\ln|x|)$
1m. $y(x) = c_1x^{-1/2} + c_2x^{-1}$ **1n.** $y(x) = c_1\sqrt{x}\cos(3\ln|x|) + c_2\sqrt{x}\cos(3\ln|x|)$
1o. $y(x) = c_1 + c_2\ln|x|$ **1p.** $y(x) = c_1x^5 + c_2x^{-5}$

1q. $y(x) = c_1 x^{-1/2}\cos(\ln|x|) + c_2 x^{-1/2}\sin(\ln|x|)$ **1r.** $y(x) = c_1 x^3 + c_2\sqrt[3]{x}$
2a. $y(x) = 2x^5 + 3x^{-2}$ **2b.** $y(x) = 4x^{1/2} - 16x^{-1/2}$ **2c.** $y(x) = \frac{1}{2}x^6 - x^6\ln|x|$
2d. $y(x) = 3x - 3x\ln|x|$ **2e.** $y(x) = 3x\cos(\ln|x|) - 3x\sin(\ln|x|)$
2f. $y(x) = 9x^2\cos(3\ln|x|) - 5x^2\sin(3\ln|x|)$ **4a.** $y(x) = c_1 x + c_2 x^2 + c_3 x^{-2}$
4b. $y(x) = c_1 x + c_2\cos(\ln|x|) + c_3\sin(\ln|x|)$ **4c.** $y(x) = c_1 x^2 + c_2 x^3 + c_3 x^3\ln|x|$
4d. $y(x) = c_1 x^2 + c_2 x^2\ln|x| + c_3 x^2(\ln|x|)^2$
4e. $y(x) = [c_1 + c_3\ln|x|]\cos(2\ln|x|) + [c_2 + c_4\ln|x|]\sin(2\ln|x|)$
4f. $y(x) = c_1 x + c_2 x^{-1} + c_3 x^3 + c_4 x^{-3}$
4g. $y(x) = c_1 x + c_2 x\ln|x| + c_3 x(\ln|x|)^2 + c_4 x(\ln|x|)^3$
4h. $y(x) = c_1 x + c_2 x^{-1} + c_3\cos(\ln|x|) + c_4\sin(\ln|x|)$ **5c.** Replace each $x^k y^{(k)}$ in the differential equation with $r(r-1)(r-2)\cdots(r-[k-1])$

Chapter 21

1a. $g(x) = 10e^{3x}$ **1b.** $g(x) = \left(9x^2 - 4\right)e^{3x}$ **1c.** $g(x) = 20e^{3x}$
2a. $g(x) = 6x + 12x^2 + 4x^3$ **2b.** $g(x) = 22x^3$ **2c.** $g(x) = 21x$
3a. No, because $y'' + y = 0$ when $y(x) = \sin(x)$. **3b.** $g(x) = 2\cos(x)$
4a. No, because $x^2 y'' - 6xy' + 12y = 0$ when $y(x) = x^4$. **4b.** $g(x) = x^4$
5b. $y_h(x) = c_1\cos(2x) + c_2\sin(2x)$ **5c.** $y(x) = 3e^{2x} + c_1\cos(2x) + c_2\sin(2x)$
5d i. $y(x) = 3e^{2x} + 3\cos(2x)$ **5d ii.** $y(x) = 3e^{2x} - 5\cos(2x) - 2\sin(2x)$
6b. $y_h(x) = c_1 e^{2x} + c_2 e^{-4x}$ **6c.** $y(x) = -x^2 - \frac{1}{2}x + c_1 e^{2x} + c_2 e^{-4x}$
6d i. $y(x) = -x^2 - \frac{1}{2}x + \frac{1}{12}e^{2x} - \frac{1}{12}e^{-4x}$ **6d ii.** $y(x) = -x^2 - \frac{1}{2}x + \frac{1}{4}e^{2x} + \frac{3}{4}e^{-4x}$
7b. $y(x) = -4 + c_1 e^{3x} + c_2 e^{-3x}$ **7c.** $y(x) = -4 + 7e^{3x} + 5e^{-3x}$
8b. $y(x) = e^{4x} + c_1 e^{5x} + c_2 e^{-2x}$ **8c.** $y(x) = e^{4x} + 2e^{5x} + 3e^{-2x}$
9b. $y(x) = xe^{5x} + c_1 e^{5x} + c_2 e^{-2x}$ **9c.** $y(x) = xe^{5x} + 3e^{5x} + 9e^{-2x}$
10b. $y(x) = 5\sin(2x) - 12\cos(2x) + c_1 e^{-3x} + c_2 xe^{-3x}$
10c. $y(x) = 5\sin(2x) - 12\cos(2x) + 2e^{-3x} + 5xe^{-3x}$ **11b.** $y(x) = 5x + 2 + c_1 x^2 + c_2 x^3$
11c. $y(x) = 5x + 2 - 6x^2 + 5x^3$ **12b.** $y(x) = \frac{1}{2}x^2 + c_1 + c_2 x + c_3\cos(x) + c_4\sin(x)$
12c. $y(x) = \frac{1}{2}x^2 + 3 + 5x + \cos(x) - 2\sin(x)$ **13a.** $y_p(x) = \frac{1}{7}xe^{5x}$
13b. $y_p(x) = \frac{1}{7}xe^{5x}$ **13c.** $y_p(x) = 3e^{4x} + 2xe^{5x}$ **13d.** $y_p(x) = 5xe^{5x} - 2e^{4x}$
14a. $y_p(x) = \frac{1}{6}$ **14b.** $y_p(x) = \frac{1}{2}x$ **14c.** $y_p(x) = 11x + 4$ **15a i.** $g(x) - 3x^2$
15a ii. $g(x) = 8x$ **15a iii.** $g(x) = 15$ **15b.** $y_p(x) = \frac{1}{3}x^2$
15c. $y_p(x) = \frac{4}{3}x^2 + \frac{1}{4}x + \frac{1}{5}$

Chapter 22

1a. $y_p(x) = 4e^{2x}$, $y(x) = 4e^{2x} + c_1\cos(3x) + c_2\sin(3x)$
1b. $y_p(x) = 3e^{6x}$, $y(x) = 3e^{6x} + c_1 e^{3x} + c_2 xe^{3x}$
1c. $y_p(x) = -6e^{-4x}$, $y(x) = -6e^{-4x} + c_1 e^{x} + c_2 e^{-5x}$
1d. $y_p(x) = \frac{4}{7}e^{x/2}$, $y(x) = \frac{4}{7}e^{x/2} + c_1 + c_2 e^{-3x}$ **2.** $y(x) = \frac{1}{2}e^{3x} + 3e^{-2x} + \frac{3}{2}e^{5x}$
3a. $y_p(x) = 2\cos(2x) + 3\sin(2x)$, $y(x) = 2\cos(2x) + 3\sin(2x) + c_1\cos(3x) + c_2\sin(3x)$
3b. $y_p(x) = \frac{4}{9}\cos(6x) - \frac{1}{3}\sin(6x)$, $y(x) = \frac{4}{9}\cos(6x) - \frac{1}{3}\sin(6x) + c_1 e^{3x} + c_2 xe^{3x}$
3c. $y_p(x) = 9\cos\left(\frac{x}{3}\right) + 27\sin\left(\frac{x}{3}\right)$, $y(x) = 9\cos\left(\frac{x}{3}\right) + 27\sin\left(\frac{x}{3}\right) + c_1 + c_2 e^{-3x}$
3d. $y_p(x) = \frac{1}{13}\sin(x) - \frac{3}{26}\cos(x)$, $y(x) = \frac{1}{13}\sin(x) - \frac{3}{26}\cos(x) + c_1 e^{x} + c_2 e^{-5x}$
4. $y(x) = \frac{1}{2}\cos(x) - \frac{1}{2}\sin(x) + 6e^{-2x} + \frac{3}{2}e^{5x}$

5a. $y_p(x) = 20$, $y(x) = 20 + c_1e^{-2x} + c_2e^{5x}$

5b. $y_p(x) = -\frac{1}{5}x^3 - \frac{12}{25}x^2 - \frac{126}{125}x - \frac{624}{625}$, $y(x) = -\frac{1}{5}x^3 - \frac{12}{25}x^2 - \frac{126}{125}x - \frac{624}{625} + c_1e^x + c_2e^{-5x}$

5c. $y_p(x) = 2x^2 + 3x + 2$, $y(x) = 2x^2 + 3x + 2 + c_1e^{3x} + c_2xe^{3x}$

5d. $y_p(x) = x^4 - \frac{4}{3}x^2 - \frac{19}{27}$, $y(x) = x^4 - \frac{4}{3}x^2 - \frac{19}{27} + c_1\cos(3x) + c_2\sin(3x)$

6. $y(x) = \frac{1}{9}x^3 - \frac{2}{27}x + \frac{2}{81}\sin(3x)$

7a. $y_p(x) = 5x\cos(2x) + 4\sin(2x)$, $y(x) = 5x\cos(2x) + 4\sin(2x) + c_1\cos(3x) + c_2\sin(3x)$

7b. $y_p(x) = \frac{1}{2}e^{2x}\cos(x)$, $y(x) = \frac{1}{2}e^{2x}\cos(x) + c_1e^{3x} + c_2xe^{3x}$

7c. $y_p(x) = \left[3x^2 - 2x + \frac{1}{3}\right]e^{3x}$, $y(x) = \left[3x^2 - 2x + \frac{1}{3}\right]e^{3x} + c_1\cos(3x) + c_2\sin(3x)$

7d. $y_p(x) = [3\sin(x) - 3x\cos(x) + 3\cos(x)]e^x$,
$y(x) = [3\sin(x) - 3x\cos(x) + 3\cos(x)]e^x + c_1x + c_2$

7e. $y_p(x) = 2x\cos(2x) + \sin(2x)$, $y(x) = 2x\cos(2x) + \sin(2x) + c_1e^x + c_2xe^x$

7f. $y_p(x) = \frac{1}{3}xe^{5x}$, $y(x) = \frac{1}{3}xe^{5x} + c_1e^x + c_2xe^x$

8. $y(x) = 3xe^{2x} - \frac{12}{13}e^{2x} + \frac{25}{13}\cos(3x) - \frac{5}{13}\sin(3x)$

9a. $y_p(x) = \frac{3}{7}xe^{-2x}$, $y(x) = \frac{3}{7}xe^{-2x} + c_1e^{-2x} + c_2e^{5x}$

9b. $y_p(x) = 5x$, $y(x) = 5x + c_1 + c_2e^{-4x}$

9c. $y_p(x) = \frac{1}{12}x^3 - \frac{1}{16}x^2 + \frac{1}{32}x$, $y(x) = \frac{1}{12}x^3 - \frac{1}{16}x^2 + \frac{1}{32}x + c_1 + c_2e^{-4x}$

9d. $y_p(x) = -\frac{1}{2}x\cos(3x)$, $y(x) = -\frac{1}{2}x\cos(3x) + c_1\cos(3x) + c_2\sin(3x)$

9e. $y_p(x) = 5x^2e^{3x}$, $y(x) = 5x^2e^{3x} + c_1e^{3x} + c_2xe^{3x}$

9f. $y_p(x) = -\frac{1}{2}x^2e^{-4x} - \frac{1}{4}xe^{-4x}$, $y(x) = -\frac{1}{2}x^2e^{-4x} - \frac{1}{4}xe^{-4x} + c_1 + c_2e^{-4x}$

10a. $y(x) = \left(-6x^2 - x - 1\right)e^{2x} + c_1e^{-2x} + c_2e^{5x}$

10b. $y(x) = \left(\frac{1}{2}x - \frac{9}{16}\right)e^{6x} + c_1e^{-2x} + c_2e^{5x}$ **10c.** $y(x) = 3x^2e^{5x} + c_1e^{5x} + c_2xe^{5x}$

10d. $y(x) = \frac{3}{50}e^{-5x} + c_1e^{5x} + c_2xe^{5x}$

10e. $y(x) = -\frac{9}{5}\cos(3x) - \frac{3}{5}\sin(3x) + c_1e^{-2x}\cos(x) + c_2e^{-2x}\sin(x)$

10f. $y(x) = 4e^{-3x} + c_1e^{-2x}\cos(x) + c_2e^{-2x}\sin(x)$

10g. $y(x) = -\frac{1}{2}xe^{2x}\cos(x) + c_1e^{2x}\cos(x) + c_2e^{2x}\sin(x)$

10h. $y(x) = \frac{2}{39}e^{-x}\cos(x) + \frac{1}{13}e^{-x}\sin(x) + c_1e^{2x}\cos(x) + c_2e^{2x}\sin(x)$

10i. $y(x) = 20 + c_1e^{2x}\cos(x) + c_2e^{2x}\sin(x)$

10j. $y(x) = \frac{1}{10}e^{-x} + c_1e^{2x}\cos(x) + c_2e^{2x}\sin(x)$

10k. $y(x) = 2x^2 + 4x + 4 + c_1e^{2x}\cos(x) + c_2e^{2x}\sin(x)$

10l. $y(x) = \frac{3}{40}e^{2x}\sin(x) - \frac{1}{40}e^{2x}\cos(x) + c_1\cos(3x) + c_2\sin(3x)$

10m. $y(x) = \frac{3}{2}x\cos(x) + 3x\sin(x) + c_1\cos(x) + c_2\sin(x)$

10n. $y(x) = -2\cos(x) + \sin(x) + c_1\cos(x) + c_2\sin(x)$

11a. $y_p(x) = \left[A_0x^3 + A_1x^2 + A_2x + A_3\right]e^{-x}\sin(x) + \left[B_0x^3 + B_1x^2 + B_2x + B_3\right]e^{-x}\cos(x)$

11b. $y_p(x) = \left[A_0x^4 + A_1x^3 + A_2x^2 + A_3x\right]e^{2x}\sin(x)$
$+ \left[B_0x^4 + B_1x^3 + B_2x^2 + B_3x\right]e^{2x}\cos(x)$

11c. $y_p(x) = \left[Ax^2 + Bx + C\right]e^{-7x}$ **11d.** $y_p(x) = Ax^2 + Bx + C$

11e. $y_p(x) = Ae^{-8x}$ **11f.** $y_p(x) = Axe^{3x}$ **11g.** $y_p(x) = \left[Ax^3 + Bx^2 + Cx\right]e^{3x}$

11h. $y_p(x) = \left[Ax^2 + Bx + C\right]\cos(2x) + \left[Dx^2 + Ex + F\right]\sin(2x)$

11i. $y_p(x) = \left[Ax^2 + Bx + C\right]e^{3x}\cos(2x) + \left[Dx^2 + Ex + F\right]e^{3x}\sin(2x)$

11j. $y_p(x) = Ae^{4x}\sin(2x) + Be^{4x}\sin(2x)$ **11k.** $y_p(x) = Axe^{2x}\sin(4x) + Bxe^{2x}\cos(4x)$

11l. $y_p(x) = \left[Ax^3 + Bx^2 + Cx + D\right]\sin(4x) + \left[Ex^3 + Fx^2 + Gx + H\right]\cos(4x)$

11m. $y_p(x) = \left[Ax^4 + Bx^3 + Cx^2\right]e^{5x}$ **11n.** $y_p(x) = Ax^4 + Bx^3 + Cx^2 + Dx + C$

12a. $y_p(x) = \frac{1}{4}e^{-2x}$ **12b.** $y_p(x) = \frac{1}{8}\sin(2x) - \frac{1}{4}\cos(2x)$ **12c.** $y_p(x) = \frac{1}{2}xe^{4x}$

12d. $y_p(x) = -\frac{1}{3}x^4 - \frac{1}{3}x^3$ **12e.** $y_p(x) = -x^2 - 2x$ **12f.** $y_p(x) = 2\cos(2x) - 4\sin(2x)$

12g. $y_p(x) = 3xe^x$ **13a.** $y_p(x) = \left[Ax^2 + Bx + C\right]e^{3x}$

13b. $y_p(x) = \left[Ax^5 + Bx^3 + Cx^2\right]\cos(3x) + \left[Dx^5 + Ex^3 + Fx^2\right]\sin(3x)$

13c. $y_p(x) = \left[Ax^2 + Bx + C\right]e^{3x}\cos(3x) + \left[Dx^2 + Ex + F\right]e^{3x}\sin(3x)$

13d. $y_p(x) = [Ax + B]\cos(2x) + [Cx + D]\sin(2x)$

13e. $y_p(x) = \left[Ax^2 + Bx\right]\cos(x) + \left[Cx^2 + Dx\right]\sin(x)$

13f. $y_p(x) = [Ax + B]e^x\cos(x) + [Cx + D]e^x\sin(x)$

13g. $y_p(x) = \left[Ax^5 + Bx^4 + Cx^3 + Dx^2 + Ex + F\right]e^{2x}$

14a. $y_p(x) = 3e^{6x} + \frac{4}{9}\cos(6x) - \frac{1}{3}\sin(6x)$

14b. $y_p(x) = 5x\cos(2x) + 4\sin(2x) - \frac{1}{2}x\cos(3x)$

14c. $y_p(x) = \frac{1}{2} - \frac{1}{26}\cos(2x) + \frac{4}{13}\sin(2x)$

14d. $y_p(x) = 5e^x - e^{-x}$ (equiv., $y_p(x) = 6\sinh(x) + 4\cosh(x)$) **15a.** $y_p(x) = \frac{1}{7}x^{-3}$

15b. $y_p(x) = 5x^3$ **15c.** $y_p(x) = -7\cos(2\ln|x|) + 6\sin(2\ln|x|)$

15d. $y_p(x) = -\frac{3}{2}\cos(3\ln|x|) + \frac{1}{2}\sin(3\ln|x|)$ **15e.** $y_p(x) = \frac{1}{2}x^3\ln|x|$

15f. $y_p(x) = -10x^{-1}\ln|x|$ **15g.** $y_p(x) = 3x^3(\ln|x|)^2$ **15h.** $y_p(x) = 4x^2\ln|x| - 2x^2$

Chapter 23

2a. $\kappa = 4.9$ kg/sec^2 **2b.** $\omega_0 = 7\sqrt{10}$ /sec , $\nu_0 = \frac{7\sqrt{10}}{2\pi}$ hertz **3a i.** $\kappa = 2450$ kg/sec^2

3a ii. $\omega_0 = 7\sqrt{2}$ /sec) **3a iii.** $\frac{7\sqrt{2}}{2\pi} \approx 1.576$ times per second **3b i.** .992 meter

3b ii. $\omega_0 = 35$ /sec **3b iii.** $\frac{35}{2\pi} \approx 5.6$ times per second **3c.** 37.5 kg

3d. 49 kg·meter/sec^2 , **4a i.** $\frac{1}{16\pi}t\sin(12\pi t)$ **4a ii.** $t = 8\pi$ (≈ 25.1 seconds)

4b i. $\frac{1}{72\pi^2}\cos(6\pi t)$ **4b ii.** No. The amplitude of the oscillations is less than .002 meter.

4c. $\sqrt{36 - \delta} \le \mu \le \sqrt{36 + \delta}$ where $\delta = \frac{3}{4\pi^2}$

5a. $y_\eta(t) = \frac{F_0}{m\left[(\omega_0)^2 - \eta^2\right]}\left[\cos(\eta t) - \cos(\omega_0 t)\right]$ **5b.** $y_\eta(t) = \frac{F_0}{2m\omega_0}t\sin(\omega_0 t)$

Chapter 24

1a. $y(x) = 4\sqrt{x} + c_1x + c_2x^2$ **1b.** $y(x) = \sin(x)\ln|\csc(x) - \cot(x)| + c_1\cos(x) + c_2\sin(x)$

1c. $y(x) = -\frac{1}{2}x\cos(2x) + \frac{1}{4}\sin(2x)\ln|\sin(2x)| + c_1\cos(2x) + c_2\sin(2x)$

1d. $y(x) = -3e^{3x} + c_1e^{2x} + c_2e^{5x}$ **1e.** $y(x) = \left[2x^4 + x^2 + c_1 + c_2x\right]e^{2x}$

1f. $y(x) = \left[x \operatorname{Arctan}(x) - \frac{1}{2}\ln\left|1+x^2\right| + c_1 + c_2 x\right]e^{-2x}$ **1g.** $y(x) = c_1 x + c_2 x^{-1} - \frac{4}{3}\sqrt{x}$

1h. $y(x) = 2x^3 \ln|x| + c_1 x^{-3} + c_2 x^3$ **1i.** $y(x) = \frac{1}{2}x^2(\ln|x|)^2 + c_1 x^2 + c_2 x^2 \ln|x|$

1j. $y(x) = c_1 x^{-2} + c_2 x^{-2}\ln|x| + \frac{1}{4}[\ln|x| - 1]$

1k. $y(x) = c_1 x^2 + c_2 x^{-1} + \frac{1}{12}[1+x] + \frac{1}{24}x^2[\ln|x-2| - \ln|x|] - \frac{1}{3}x^{-1}\ln|x-2|$

1l. $y(x) = \frac{1}{8}x^2 e^{x^2} + c_1 e^{x^2} + c_2 e^{-x^2}$ **1m.** $y(x) = x^{-1}\left[e^{2x} + c_1 + c_2 e^{-2x}\right]$

1n. $y(x) = x^2 + 1 + c_1 x + c_2 e^{-x}$ **2a.** $y(x) = 5x^{-1} - 2x^4 - 2x^{-1}\ln|x|$

2b. $y(x) = -3e^{2x} - e^{-2x} + 4e^{3x}$ **3a.** $y(x) = 2e^{3x} + c_2 e^{2x} + c_3 e^{-2x}$

3b. $y(x) = \frac{1}{2}x^3 \ln|x| + c_1 x + c_2 x^2 + c_3 x^3$ **4a.**
$$\begin{aligned} xu' + x^2 v' + x^3 w' &= 0 \\ u' + 2xv' + 3x^2 w' &= 0 \\ 2v' + 6xw' &= x^{-3}e^{-x^2} \end{aligned}$$

4b.
$$\begin{aligned} e^x u' + \cos(x)v' + \sin(x)w' &= 0 \\ e^x u' - \sin(x)v' + \cos(x)w' &= 0 \\ e^x u' - \cos(x)v' - \sin(x)w' &= \tan(x) \end{aligned}$$

4c.
$$\begin{aligned} e^{3x}u_1' + e^{-3x}u_2' + \cos(3x)u_3' + \sin(3x)u_4' &= 0 \\ e^{3x}u_1' - e^{-3x}u_2' - \sin(3x)u_3' + \cos(3x)u_4' &= 0 \\ e^{3x}u_1' + e^{-3x}u_2' - \cos(3x)u_3' - \sin(3x)u_4' &= 0 \\ e^{3x}u_1' - e^{-3x}u_2' + \sin(3x)u_3' - \cos(3x)u_4' &= \frac{1}{27}\sinh(x) \end{aligned}$$

4d.
$$\begin{aligned} xu_1' + x^{-1}u_2' + x^3 u_3' + x^{-3}u_4' &= 0 \\ u_1' - x^{-2}u_2' + 3x^2 u_3' - 3x^{-4}u_4' &= 0 \\ x^{-3}u_2' + 3xu_3' + 6x^{-5}u_4' &= 0 \\ -x^{-4}u_2' + u_3' - 10x^{-6}u_4' &= 2x^{-3}\sin\left(x^2\right) \end{aligned}$$

Chapter 25

1. $y(x) = A\cos(6x) + B\sin(6x)$ **2.** $y(x) = c_1 e^{6x} + c_2 x e^{6x}$ **3.** $y(x) = c_1 x^3 + c_2 x^{-3}$

4. $y(x) = c_1 e^{6x} + c_2 e^{-6x}$ **5.** $y(x) = c_1 e^{2x} + c_2 e^{7x}$ **6.** $y(x) = c_1 x^4 + c_2 x^4 \ln|x|$

7. $y(x) = \frac{1}{3}x^{3/2} + C_1 x^{1/2} + C_2$ **8.** $y(x) = c_1 e^{2x} + c_2 x e^{2x} + c_3 e^{-2x} + c_4 x e^{-2x}$

9. $y(x) = c_1 e^{-3x} + c_2 x e^{-3x}$ **10.** $y(x) = A\cos\left(\sqrt{3}x\right) + B\sin\left(\sqrt{3}x\right)$

11. $y(x) = c_1 x^{-3} + c_2 x^{-3}\ln|x|$ **12.** $y(x) = A\sqrt{x}\cos\left(\frac{3}{2}\ln|x|\right) + B\sqrt{x}\sin\left(\frac{3}{2}\ln|x|\right)$

13. $y(x) = c_1 + c_2 x + c_3 x^2 + c_4 e^{3x}\cos(2x) + c_5 e^{3x}\sin(2x)$ **14.** $y(x) = c_1 x^3 + c_2 x^{-2}$

15. $y(x) = Ae^{3x}\cos(4x) + Be^{3x}\sin(4x)$ **16.** $y(x) = c_2 - \ln|x + c_1|$

17. $y(x) = A\cos(3\ln|x|) + B\sin(3\ln|x|)$ **18.** $y(x) = Ae^{4x}\cos(3x) + Be^{4x}\sin(4x)$

19. $y(x) = c_1 x^5 + c_2 x^{-6}$ **20.** $y(x) = c_1 e^{5x} + c_2 e^{-6x}$ **21.** $y(x) = c_1 e^{x/4} + c_2 x e^{x/4}$

22. $y(x) = c_1 x^{-1/2} + c_2 x^{-1/2}\ln|x|$ **23.** $y(x) = c_1 e^{2x} + c_2 x e^{2x} + c_3 x^2 e^{2x}$

24. $y(x) = c_1 x^2 + c_2\sqrt{x}$ **25.** $y(x) = c_1\sqrt[3]{x} + c_2\sqrt[3]{x}\ln|x|$

26. $y(x) = c_1 e^{2x} + c_2 e^{-2x} + c_3\cos(2x) + c_4\sin(2x)$ **27.** $y(x) = c_1 e^{x/2} + c_2 e^{3x}$

28. $y(x) = c_1 e^{-10x} + c_2 x e^{-10x}$ **29.** $y(x) = Ax^4 + C$ **30.** $y(x) = c_1 + c_2 e^{5x}$

31. $y(x) = 7x^2 + 9x + \frac{67}{14} + c_1 e^{2x} + c_2 e^{7x}$

32. $y(x) = \frac{4}{9}\cos(3x) + \frac{1}{3}\sin(3x) + c_1 e^{6x} + c_2 x e^{6x}$

33. $y(x) = \left(24x^2 + 22x + \frac{97}{12}\right)e^{-x} + c_1 e^{2x} + c_2 e^{7x}$

34. $y(x) = 9xe^{3x} + 6e^{3x} + c_1 e^{6x} + c_2 x e^{6x}$ **35.** $y(x) = c_1 x + c_2 x^2 - \frac{12}{35}\sqrt{x}$

36. $y(x) = \frac{1}{2}x^3 e^{6x} - x^2 e^{6x} + c_1 e^{6x} + c_2 x e^{6x}$

37. $y(x) = c_1\cos(6x) + c_2\sin(6x) + \frac{1}{6}\cos(6x)\ln|\cos(6x)| + x\sin(6x)$

38. $y(x) = c_1 x^2 + c_2 x^{-3} - \frac{1}{2} - 3\ln|x|$ **39.** $y(x) = \left[c_1 + c_2 x + 5x^2\right] e^{-3x}$

40. $y(x) = c_1 x^2 + c_2 x^{-1/2} + 2x^2 \ln|x|$

41. $y(x) = c_1 e^{-3x} + c_2 x e^{-3x} + \frac{10}{169}\cos(2x) + \frac{24}{169}\sin(2x)$

42. $y(x) = c_2$, $y = \frac{1}{2}\sqrt{2x^2 + c_1} + c_2$ **43.** $y(x) = c_1 x^{-1}\cos(\ln|x|) + c_2 x^{-1}\sin(\ln|x|) + 3$

44. $y(x) = c_1 x + c_2 x^{-1} - \frac{1}{2} - \frac{1}{2}\left[x + x^{-1}\right]\arctan(x)$ **45.** $y(x) = \left[c_1 + c_2 x + \frac{1}{24}x^3\right] e^{3x/2}$

46. $y(x) = c_1 e^{-3x} + c_2 e^{x/3} + \left(-2x - \frac{241}{246}\right)\cos(3x) + \left(-\frac{5}{2}x + \frac{54}{41}\right)\sin(3x)$

47. $y(x) = c_1 e^{-2x} + c_2 e^{x}\cos\left(\sqrt{3}x\right) + c_3 e^{x}\sin\left(\sqrt{3}x\right) + \frac{1}{12}xe^{-2x}$

48. $y(x) = c_1 e^{-2x} + c_2 e^{2x} + \left[c_3 e^{x} + c_4 e^{-x}\right]\cos\left(\sqrt{3}x\right) + \left[c_5 e^{x} + c_6 e^{-x}\right]\sin\left(\sqrt{3}x\right) - \frac{1}{192}xe^{-2x}$

49. $y(x) = c_1 x^{-1} + c_2 x^{-1}\ln|x| + x^{-1}\ln|x+1|$

50. $y(x) = c_1 x^{-1} + c_2 x^{-1}\ln|x| + \frac{1}{2}x^{-1}(\ln|x|)^2$

Chapter 26

6a. $F(s) = \frac{4}{s}$ for $s > 0$ **6b.** $F(s) = \frac{3}{s-2}$ for $s > 2$ **6c.** $F(s) = \frac{2}{s}\left[1 - e^{-3s}\right]$

6d. $F(s) = \frac{2}{s}e^{-3s}$ for $s > 0$ **6e.** $F(s) = \frac{1}{s-2}\left[1 - e^{4(2-s)}\right]$

6f. $F(s) = \frac{1}{s-2}\left[e^{2-s} - e^{4(2-s)}\right]$ **6g.** $F(s) = \frac{1}{s^2}\left[1 - e^{-s}\right] - \frac{1}{s}e^{-s}$

6h. $F(s) = \left[\frac{1}{s} + \frac{1}{s^2}\right]e^{-s}$ for $s > 0$ **7a.** $\frac{24}{s^5}$ for $s > 0$ **7b.** $\frac{9!}{s^{10}}$ for $s > 0$

7c. $\frac{1}{s-7}$ for $s > 7$ **7d.** $\frac{1}{s - i7}$ for $s > 0$ **7e.** $\frac{1}{s+7}$ for $s > -7$

7f. $\frac{1}{s + i7}$ for $s > -7$ **8a.** $\frac{3}{s^2+9}$ for $s > 0$ **8b.** $\frac{s}{s^2+9}$ for $s > 0$ **8c.** $\frac{7}{s}$ for $s > 0$

8d. $\frac{s}{s^2-9}$ for $s > 3$ **8e.** $\frac{4}{s^2-16}$ for $s > 4$ **8f.** $\frac{6}{s^3} - \frac{8}{s^2} + \frac{47}{s}$ for $s > 0$

8g. $\frac{6}{s-2} + \frac{8}{s+3}$ for $s > 2$ **8h.** $\frac{3s}{s^2+4} + \frac{24}{s^2+36}$ for $s > 0$ **8i.** $\frac{3s-8}{s^2+4}$ for $s > 0$

9a. $\frac{3\sqrt{\pi}}{4}s^{-5/2}$ for $s > 0$ **9b.** $\frac{15\sqrt{\pi}}{8}s^{-7/2}$ for $s > 0$ **9c.** $\Gamma\left(\frac{2}{3}\right)s^{-2/3}$ for $s > 0$

9d. $\Gamma\left(\frac{5}{4}\right)s^{-5/4}$ for $s > 0$ **9e.** $\frac{1}{s}e^{-2s}$ **10b.** $\frac{1}{s}\left[1 - e^{-2s}\right]$ **11a.** $\frac{1}{(s-4)^2}$ for $s > 4$

11b. $\frac{24}{(s-1)^5}$ for $s > 1$ **11c.** $\frac{3}{(s-2)^2+9}$ for $s > 2$ **11d.** $\frac{s-2}{(s-2)^2+9}$ for $s > 2$

11e. $\frac{\sqrt{\pi}}{2}(s-3)^{-3/2}$ for $s > 3$ **11f.** $\frac{1}{s-3}e^{-2(s-3)}$ for $s > 3$ **14a.** Piecewise continuous

14b. Piecewise continuous **14c.** Piecewise continuous **14d.** Piecewise continuous

14e. Not piecewise continuous **14f.** Piecewise continuous

14g. Not piecewise continuous **14h.** Piecewise continuous

14i. Not piecewise continuous **14j.** Piecewise continuous **14k.** Piecewise continuous

16a. Of exponential order $s_0 \geq 3$ **16b.** Of exponential order $s_0 > 0$

16c. Of exponential order $s_0 > 3$ **16d.** Not of exponential order

16e. Of exponential order $s_0 \geq 0$

17a. The maximum occurs at $t = \frac{\alpha}{\sigma}$ and is $M_{\alpha,\sigma} = \left(\frac{\alpha}{\sigma}\right)^{\alpha} e^{-\alpha}$.

Chapter 27

1a. $Y(s) = \frac{3}{s+4}$ **1b.** $Y(s) = \frac{4}{s-2} + \frac{6}{s^4(s-2)}$ **1c.** $Y(s) = \frac{e^{-4s}}{s(s+3)}$

1d. $Y(s) = \frac{s+3}{s^2-4} + \frac{6}{s^4(s^2-4)}$ **1e.** $Y(s) = \frac{3s^2-28}{(s-4)(s^2+4)}$

1f. $Y(s) = \frac{3s+5}{s^2+4} + \frac{2}{(s^2+4)^2}$ **1g.** $Y(s) = \frac{5}{s^2+4} + \frac{3e^{-2s}}{s(s^2+4)}$

1h. $Y(s) = \frac{s+5}{s^2+5s+6} + \frac{1}{(s-4)(s^2+5s+6)}$ **1i.** $Y(s) = \frac{2}{s^2-5s+6} + \frac{2}{(s-4)^3(s^2-5s+6)}$

1j. $Y(s) = \frac{2s-6}{s^2-5s+6} + \frac{7}{s(s^2-5s+6)}$ **1k.** $Y(s) = \frac{4s-13}{s^2-4s+13} + \frac{3}{(s^2-4s+13)^2}$

1l. $Y(s) = \frac{4s+19}{s^2+4s+13} + \frac{4}{s^2(s^2+4s+13)} + \frac{6}{(s^2-4s+13)(s^2+4s+13)}$

1m. $Y(s) = \frac{2s^2+3s+4}{s^3-27} + \frac{1}{(s+3)(s^3-27)}$ **2a.** $\frac{s^2-9}{(s^2+9)^2}$ **2b.** $\frac{18s^2-54}{(s^2+9)^3}$

2c. $\frac{1}{(s+7)^2}$ **2d.** $\frac{6}{(s+7)^4}$ **2e.** $\frac{1+3s}{s^2}e^{-3s}$ **2f.** $\frac{1+8s+16s^2}{s^3}e^{-4s}$

4b. $Y(s) = \frac{C}{\sqrt{s^2+1}}$ **4c.** $Y(0) = 1$ **4d.** $Y(s) = \frac{1}{\sqrt{s^2+1}}$ **5a.** $\frac{6(s-4)}{([s-4]^2+9)^2}$

5b. $\frac{(s-4)^2-9}{([s-4]^2+9)^2}$ **5c.** $\frac{3s-11}{(s-4)^2}e^{-3(s-4)}$ **5d.** $\frac{2+2(s-3)+(s-3)^2}{(s-3)^3}e^{-(s-3)}$

6b. $\frac{2}{s^3}e^{-\alpha s}$ for $s > 0$ **7.** $\frac{1}{s}\arctan\left(\frac{1}{s}\right)$ **8a.** $\ln\left(1+\frac{1}{s}\right)$ for $s > 0$

8b. $\ln\left(\frac{s}{s-2}\right)$ for $s > 2$ **8c.** $\ln\left(\frac{s-3}{s+2}\right)$ for $s > 3$ **8d.** $\frac{1}{2}\ln\left(1+\frac{1}{s^2}\right)$ for $s > 0$

8e. $\frac{1}{2}\ln\left(1-\frac{1}{s^2}\right)$ for $s > 1$ **8f.** $\arctan\left(\frac{3}{s}\right)$ for $s > 0$

Chapter 28

1a. e^{6t} **1b.** e^{-2t} **1c.** t **1d.** t^3 **1e.** $\sin(5t)$ **1f.** $\cos\left(\sqrt{3}\pi t\right)$ **2a.** $6e^{-2t}$

2b. $\frac{1}{6}t^3$ **2c.** $\frac{3}{\sqrt{\pi t}} - 8e^{4t}$ **2d.** $2t^2 - \frac{1}{6}t^4$ **2e.** $3\cos(5t) + \frac{1}{5}\sin(5t)$

2f. $1 - \text{step}_4(t)$ **4a.** $y(t) = 4e^{-9t}$ **4b.** $y(t) = 4\cos(3t) + 2\sin(3t)$ **5a.** $3e^{-2t} + 4e^t$

5b. $3e^{4t} - 2e^{3t}$ **5c.** $\frac{1}{4}e^{2t} - \frac{1}{4}e^{-2t}$ **5d.** $3 + 2\sin(3t)$ **5e.** $\frac{1}{16}e^{4t} - \frac{1}{4}t - \frac{1}{16}$

5f. $4\cos(3t) + 2e^{3t} + 2e^{-3t}$ **5g.** $2e^{-6t} + 3\cos(4t) - 3\sin(4t)$ **5h.** $2\cos(3t) + 3\sin(t)$

5i. $3e^{-t} + 5e^{-3t} - 2e^{-7t}$ **6a.** $y(t) = \frac{7}{2}e^{3t} + \frac{1}{2}e^{-3t}$ **6b.** $y(t) = 3t^3 - 2t + \frac{2}{3}\sin(3t)$

6c. $y(t) = 3e^{4t} + e^{-7t} + 4e^{-t}$ **7a.** $\frac{1}{24}t^4e^{7t}$ **7b.** $\frac{1}{6}e^{3t}\sin(6t)$

7c. $e^{3t}\left[\cos(6t) + \frac{1}{2}\sin(6t)\right]$ **7d.** $\frac{1}{\sqrt{\pi t}}e^{-2t}$ **7e.** te^{-4t} **7f.** $[\cos(2t) + 3\sin(2t)]e^{6t}$

7g. $\frac{1}{2}\sin(2t)e^{-6t}$ **7h.** $\left[\frac{1}{2}t^2 + t^3 + \frac{3}{8}t^4\right]e^{3t}$ **8a.** $y(t) = 3e^{4t}\cos(t)$

8b. $y(t) = \frac{1}{12}e^{3t}t^4$ **8c.** $y(t) = e^{-3t}[2\cos(2t) + 7\sin(2t)]$ **8d.** $y(t) = 3e^{-4t}\cos(t)$

9a. $y(t) = \frac{1}{2}\left(1 + t - \cos(t)e^t\right)$ **9b.** $y(t) = 2e^{-3t} + [3\sin(6t) - 2\cos(6t)]e^{2t}$

9c. $y(t) = 2e^{-3t} - 4te^{-3t} + 4e^{3t}$ **9d.** $y(t) = \left(\left[4 - \frac{t}{6}\right]\cos(3t) - \frac{29}{18}\sin(3t)\right)e^{2t}$

10a. $\frac{1}{9}[1 - \cos(3t)]$ **10b.** $\frac{1}{4}\left[e^{4t} - 1\right]$ **10c.** $\frac{1}{9}\left[1 + (3t-1)e^{3t}\right]$

Chapter 29

1. $\frac{1}{16}e^{7t} + \frac{1}{4}te^{3t} - \frac{1}{16}e^{3t}$ **2a.** $\frac{1}{2}[e^{5t} - e^{3t}]$ **2b.** $\frac{16}{15}t^{5/2}$ **2c.** $4t^{3/2}$

2d. $\frac{1}{9}[e^{3t} - 1 - 3t]$ **2e.** $\frac{1}{30}t^5$ **2f.** $t - \sin(t)$ **2g.** $\frac{1}{2}[\sin(t) - t\cos(t)]$

2h. $\frac{1}{10}\left[e^{-3t} + 3\sin(t) - \cos(t)\right]$ **4a.** $e^{4t} - e^{3t}$ **4b.** $\frac{1}{3}\left[e^{3t} - 1\right]$ **4c.** $\frac{1}{4}[1 - \cos(2t)]$

4d. $\frac{1}{10}\left[e^{3t} - 3\sin(t) - \cos(t)\right]$ **4e.** $\frac{1}{54}[\sin(3t) - 3t\cos(3t)]$ **4f.** $\frac{1}{4}[\sin(2t) + 2t\cos(2t)]$

4g. $\frac{1}{\sqrt{\pi}}e^{3t}\int_0^t \frac{1}{\sqrt{x}}e^{-3x}\,dx$ **4h.** $\frac{1}{2\sqrt{\pi}}\int_0^t \frac{1}{\sqrt{x}}\sin(2[t-x])\,dx$

5a. $h(x) = \frac{1}{2}\sin(2x)$, $y(t) = \frac{1}{2}\int_0^t \sin(2x)f(t-x)\,dx$

5b. $h(x) = \frac{1}{4}\left[e^{2x} - e^{-2x}\right]$, $y(t) = \frac{1}{4}\int_0^t \left[e^{2x} - e^{-2x}\right]f(t-x)\,dx$

5c. $h(x) = xe^{3x}$, $y(t) = \int_0^t xe^{3x}f(t-x)\,dx$

5d. $h(x) = \frac{1}{3}e^{3x}\sin(3x)$, $y(t) = \frac{1}{3}\int_0^t e^{3x}\sin(3x)f(t-x)\,dx$

5e. $h(x) = \frac{1}{16}[1 - \cos(4x)]$, $y(t) = \frac{1}{16}\int_0^t [1 - \cos(4x)]\,f(t-x)\,dx$

6a. $y(t) = \frac{1}{4}[1 - \cos(2t)]$ **6b.** $y(t) = \frac{1}{8}[2 - \sin(2t)]$

6c. $y(t) = \frac{1}{13}\left[e^{3t} - \cos(2t) - \frac{3}{2}\sin(2t)\right]$ **6d.** $y(t) = \frac{1}{8}[\sin(2t) - 2t\cos(2t)]$

6e. $y(t) = (2\alpha^2 - 8)^{-1}[\alpha\sin(2t) - 2\sin(\alpha t)]$ **7a.** $y(t) = \frac{1}{9}\left[3te^{3t} - e^{3t} + 1\right]$

7b. $y(t) = \frac{1}{27}\left[3te^{3t} - 2e^{3t} + 3t + 2\right]$ **7c.** $y = \frac{1}{2}t^2e^{3t}$ **7d.** $y = \frac{1}{36}\left[6te^{3t} - e^{3t} + e^{-3t}\right]$

7e. $y(t) = (3-\alpha)^{-2}\left[(3-\alpha)te^{3t} - e^{3t} + e^{\alpha t}\right]$ **8a.** $y(t) = \frac{1}{64}[4t - \sin(4t)]$

8b. $y(t) = \frac{1}{256}\left[8t^2 - 1 + \cos(4t)\right]$ **8c.** $y(t) = \frac{1}{1200}\left[16e^{3t} - 25 + 9\cos(4t) - 12\sin(4t)\right]$

8d. $y(t) = \frac{1}{64}[1 - 2t\sin(4t) - \cos(4t)]$

8e. $y(t) = \left[16\alpha\left(\alpha^2 - 16\right)\right]^{-1}\left[16\cos(\alpha t) + \alpha^2 - 16 - \alpha^2\cos(4t)\right]$

Chapter 30

2a. $\frac{1}{s-4}e^{-6(s-4)}$ **2b.** $\frac{1}{s^2}e^{-6s} + \frac{6}{s}e^{-6s}$ **3a.** $\frac{1}{2}(t-4)^2\,\text{step}_4(t)$ **3b.** $e^{-2(t-3)}\,\text{step}_3(t)$

3c. $2\sqrt{t-1}\,\text{step}_1(t)$ **3d.** $\sin(\pi t)\,\text{step}_2(t)$ **3e.** $\frac{1}{2}(t-4)^2e^{5(t-4)}\,\text{step}_4(t)$

3f. $e^{-2(t-5)}\cos(4(t-5))\,\text{step}_5(t)$ **4.** $[1 - \cos(t-3)]\,\text{step}_3(t)$ **5a.** $\begin{Bmatrix} t & \text{if} & t<1 \\ 1 & \text{if} & 1<t \end{Bmatrix}$

5b. $\begin{Bmatrix} 0 & \text{if} & t<1 \\ 1 & \text{if} & 1<t<3 \\ 2 & \text{if} & 3<t \end{Bmatrix}$ **5c.** $\begin{Bmatrix} t^2 & \text{if} & t<2 \\ 4 & \text{if} & 2<t \end{Bmatrix}$ **5d.** $\begin{Bmatrix} \sin(\pi t) & \text{if} & t<1 \\ 0 & \text{if} & 1<t \end{Bmatrix}$

5e. $\begin{Bmatrix} e^{4(t-3)} & \text{if} & t<3 \\ e^{-4(t-3)} & \text{if} & 3<t \end{Bmatrix}$ **5f.** $\begin{Bmatrix} 0 & \text{if} & t<2 \\ t-2 & \text{if} & 2<t<4 \\ 6-t & \text{if} & 4<t<6 \\ 0 & \text{if} & 6<t \end{Bmatrix}$ **6a.** $y(t) = (t-3)\,\text{step}_3(t)$

6b. $y(t) = 4 + (t-3)\,\text{step}_3(t)$ **6c.** $y(t) = \frac{1}{2}(t-2)^2\,\text{step}_2(t)$

6d. $y(t) = 6t + 4 + \frac{1}{2}(t-2)^2\,\text{step}_2(t)$ **6e.** $y(t) = \frac{1}{9}\left[1 - \cos(3[t-10])\right]\text{step}_{10}(t)$

7a. $\frac{1}{s-4}e^{-6(s-4)}$ **7b.** $\sqrt{\frac{\pi}{s}}e^{-4s}$ **7c.** $\frac{1}{s^2}e^{-6s} + \frac{6}{s}e^{-6s}$ **7d.** $\left[\frac{2}{s-3} + \frac{1}{(s-3)^2}\right]e^{6-2s}$

7e. $e^{-3s}\left[\frac{2}{s^3}+\frac{12}{s^2}+\frac{36}{s}\right]$ **7f.** $\frac{2e^{-s}}{s^2+4}$ **7g.** $-\frac{2}{s^2+4}e^{-\pi s/2}$ **7h.** $e^{-\pi s/4}\frac{s}{s^2+4}$

7i. $\frac{1}{2}\left[\frac{2}{s^2+4}+\frac{s\sqrt{3}}{s^2+4}\right]e^{-\pi s/6}$ **8a.** $F(s)=\frac{1}{s+4}\left[1-e^{-24-6s}\right]$

8b. $F(s)=\frac{2}{s^2}\left[1+e^{-2s}\right]-\frac{2}{s^3}\left[1-e^{-2s}\right]$ **8c.** $F(s)=\frac{2}{s}\left[1-e^{-3s}\right]+\frac{2}{s+4}e^{-3s}$

8d. $F(s)=\frac{\pi}{s^2+\pi^2}\left[1+e^{-s}\right]$ **8e.** $F(s)=\frac{2}{s^3}-\left[\frac{2}{s^3}+\frac{6}{s^2}\right]e^{-3s}$

8f. $F(s)=\frac{3}{s}\left[e^{-2s}-e^{-4s}\right]$ **8g.** $F(s)=\frac{1}{s}\left[1+e^{-2s}+2e^{-3s}\right]$

8h. $F(s)=-\frac{\pi}{s^2+\pi^2}\left[e^{-s}+e^{-2s}\right]$ **8i.** $F(s)=\frac{2}{s^3}e^{-s}-\left[\frac{2}{s^3}+\frac{4}{s^2}\right]e^{-3s}$

8j. $F(s)=\frac{1}{s^2}\left[1-2e^{-2s}+e^{-4s}\right]$ **9a.** $\sum_{n=0}^{\infty}\text{step}_n(t)$ **9b.** $\frac{1}{s}\sum_{n=0}^{\infty}e^{-ns}$ **9c.** $\frac{1}{s\left[1-e^{-s}\right]}$

10a. $y(t)=(t-1)\,\text{rect}_{(1,3)}(t)+2\,\text{step}_3(t)$ **10b.** $y(t)=\frac{1}{2}(t-1)^2\,\text{rect}_{(1,3)}+[2t-4]\,\text{step}_3(t)$

10c. $y(t)=\frac{1}{9}\left[\text{rect}_{(1,3)}(t)+\cos(3[t-3])\,\text{step}_3(t)-\cos(3[t-1])\,\text{step}_1(t)\right]$

11a. $\begin{cases}0 & \text{if } t<3\\ \frac{1}{3}t^3-3t^2+9t-9 & \text{if } 3<t\end{cases}$ **11b.** $\begin{cases}0 & \text{if } t<4\\ t-4 & \text{if } 4<t\end{cases}$

11c. $\begin{cases}\sin(t) & \text{if } 0\le t<\pi\\ 0 & \text{if } \pi<t\end{cases}$ **11d.** $\begin{cases}0 & \text{if } 0\le t<1\\ 1-e^{2-2t} & \text{if } 1<t<3\\ e^{6-2t}-e^{2-2t} & \text{if } 3<t\end{cases}$

11e. $\frac{1}{7}\begin{cases}0 & \text{if } t<1\\ e^{5t}-e^{7-2t} & \text{if } 1<t<3\\ e^{21-2t}-e^{7-2t} & \text{if } 3<t\end{cases}$

11f. $\frac{1}{2}\begin{cases}0 & \text{if } t<2\pi\\ \sin(t)+(2\pi-t)\cos(t) & \text{if } 2\pi<t<3\pi\\ -\pi\cos(t) & \text{if } 3\Pi<t\end{cases}$

11g. $\frac{1}{15}\begin{cases}4t^{5/2} & \text{if } t<4\\ 128+15(t-4)^2 & \text{if } 4<t\end{cases}$

11h. $\frac{1}{2}\begin{cases}2-2\cos(t) & \text{if } t<2\pi\\ (t-2\pi)\sin(t) & \text{if } 2\pi<t<3\pi\\ \pi\sin(t)-2-2\cos(t) & \text{if } 3\pi<t\end{cases}$ **12a.** $\frac{1-e^{-3(s+2)}}{(s+2)\left(1-e^{-3s}\right)}$

12b. $\frac{1-e^{-s}}{s\left(1-e^{-2s}\right)}$ $\left(\text{equivalently, } \frac{1}{s\left(1+e^{-s}\right)}\right)$

12c. $\frac{1-2e^{-s}+e^{-2s}}{s\left(1-e^{-2s}\right)}$ $\left(\text{equivalently, } \frac{1-e^{-s}}{s\left(1+e^{-s}\right)} \text{ or } \frac{1}{s}\tanh\left(\frac{s}{2}\right)\right)$

12d. $\frac{2\left(1+e^{-2s}\right)}{s^2\left(1-e^{-2s}\right)}-\frac{2}{s^3}$ $\left(\text{equivalently, } \frac{2}{s^2}\coth(s)-\frac{2}{s^3}\right)$

12e. $\frac{1-2e^{-2s}+e^{-4s}}{s^2\left(1-e^{-4s}\right)}$ $\left(\text{equivalently, } \frac{1-e^{-2s}}{s^2\left(1+e^{-2s}\right)} \text{ or } \frac{1}{s^2}\tanh(s)\right)$

12f. $\frac{1+e^{-\pi s}}{\left(s^2+1\right)\left(1-e^{-\pi s}\right)}$ $\left(\text{equivalently, } \frac{1}{s^2+1}\coth\left(\frac{\pi s}{2}\right)\right)$

13a. *(i)* $y(t+2\pi) - y(t) = \dfrac{-2}{\pi^2 m}\cos(\pi t)$

13a. *(ii)* $y(t) = \dfrac{1}{\pi^2 m}\begin{cases} 1-(2n+1)\cos(\pi t_0) & \text{if} \quad 0 \le t_0 < 1 \\ -2(n+1)\cos(\pi t_0) & \text{if} \quad 1 \le t_0 < 2 \end{cases}$

13a. *(iii)*

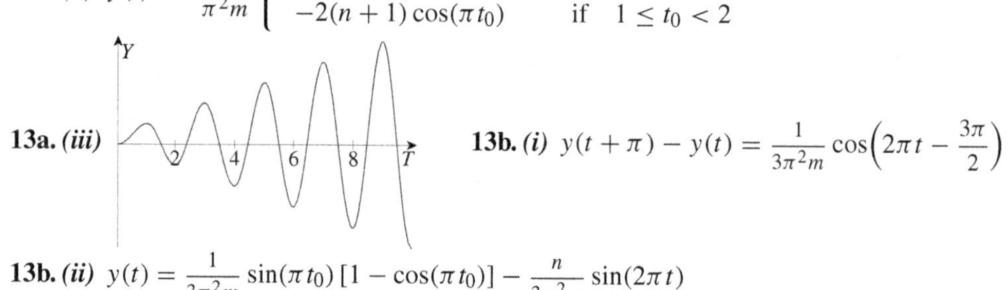

13b. *(i)* $y(t+\pi) - y(t) = \dfrac{1}{3\pi^2 m}\cos\left(2\pi t - \dfrac{3\pi}{2}\right)$

13b. *(ii)* $y(t) = \dfrac{1}{3\pi^2 m}\sin(\pi t_0)\left[1-\cos(\pi t_0)\right] - \dfrac{n}{3\pi^2 m}\sin(2\pi t)$

13b. *(iii)*

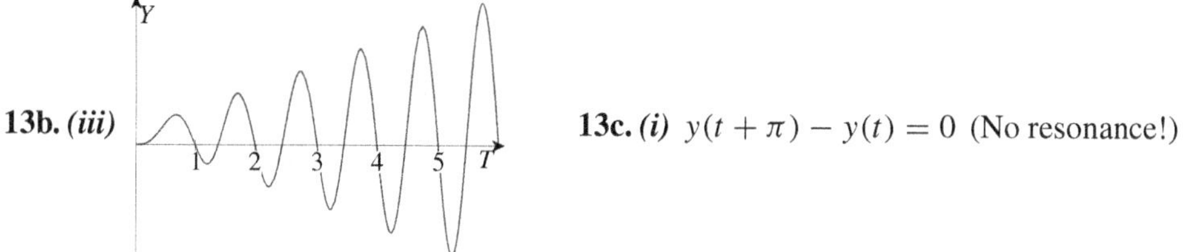

13c. *(i)* $y(t+\pi) - y(t) = 0$ (No resonance!)

13c. *(ii)* $y(t) = \dfrac{1}{12\pi^2 m}\left[2\sin(2\pi t) - \sin(4\pi t)\right]$

13c. *(iii)*

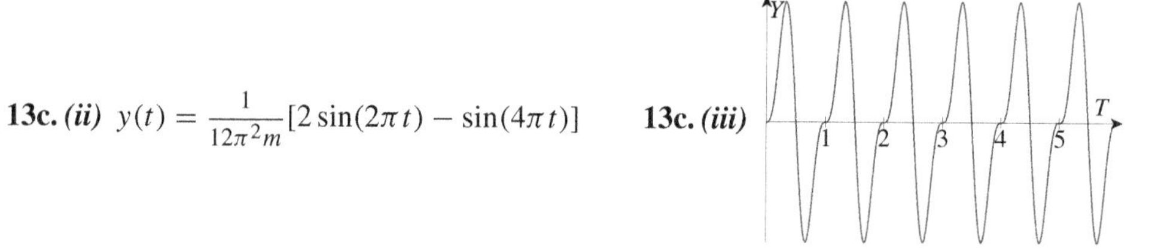

Chapter 31

1a. 6.525 kg·meter/sec **1b.** 13.63 kg·meter/sec **2a i.** 20 meter/sec
2a ii. 40 meter/sec **2a iii.** 0 meter/sec **2b i.** 18 kg·meter/sec **2b ii.** 28 kg·meter/sec
2b iii. 8 kg·meter/sec **2c i.** 0.5 kg **2c ii.** 2 kg **3a.** 16 **3b.** 0 **3c.** 1
3d. $\frac{1}{2}$ **3e.** 9 **3f.** 0 **6a.** $y(t) = 3\,\text{step}_2(t)$ **6b.** $y(t) = \text{rect}_{(2,4)}(t)$

6c. $y(t) = (t-3)\,\text{step}_3(t)$ **6d.** $y(t) = \begin{cases} 0 & \text{if} \quad t < 1 \\ t-1 & \text{if} \quad 1 < t < 4 \\ 3 & \text{if} \quad 4 < t \end{cases}$

6e. $y(t) = 4e^{-2(t-1)}\,\text{step}(t-1)$ **6f.** $y(t) = \begin{cases} \sin(t) & \text{if} \quad t < \pi \\ 0 & \text{if} \quad t < \pi \end{cases}$

6g. $y(t) = 2\cos(t)\,\text{step}\left(t - \dfrac{\pi}{2}\right)$ **7a.** $y(t) = 2e^{-3t} + e^{-3(t-2)}\,\text{step}(t-2)$

7b. $y(t) = \frac{1}{3}\left[1 - e^{-3t}\right]$ **7c.** $y(t) = \frac{2}{3}\left[1 - e^{-3t}\right] + \frac{1}{3}\left[1 - e^{-3(t-1)}\right]\text{step}_1(t)$

7d. $y(t) = \dfrac{1}{4}\sin(4(t-2))\,\text{step}_2(t)$ **7e.** $y(t) = \dfrac{1}{8}\left[e^{4(t-10)} - e^{-4(t-10)}\right]\text{step}_{10}(t)$

7f. $y(t) = 0$ **7g.** $y(t) = \frac{1}{8}\left[e^{2t} - e^{-6t}\right]$ **7h.** $y(t) = \frac{1}{8}\left[e^{2(t-3)} - e^{-6(t-3)}\right]\text{step}_3(t)$

7i. $y(t) = (t-4)e^{-3(t-4)}\,\text{step}_4(t)$ **7j.** $y(t) = \dfrac{1}{3}e^{6t}\sin(3t)$

7k. $y(t) = \frac{1}{9}\left[1 - \cos(3(t-1))\right]\text{step}_1(t)$

7l. $y(t) = \frac{1}{4}\left[e^{2(t-1)} - e^{-2(t-1)} + \frac{3}{2}\sin(2(t-1))\right]\text{step}_1(t)$

Chapter 32

2a. $\frac{121}{81}$ **2b.** $\frac{9{,}841}{6{,}561}$ **2c.** $\frac{3}{2}$ **2d.** $\frac{1}{162}$ **2e.** $\frac{3}{5}$ **2f.** $\frac{665}{32}$ **2g.** Diverges

2h. 14 **2i.** -5 **2j.** $\frac{15}{4}$ **4a.** $1 + \sum_{k=1}^{\infty} \frac{-1}{k(k+1)} x^k$ **4b.** $x^2 + \sum_{k=4}^{\infty} \frac{(-1)^k}{k} x^k$

4c. $\sum_{n=4}^{\infty} 3(n-3)^2(x-5)^n$ **4d.** $\sum_{n=1}^{\infty} (n+1)nx^n$ **4e.** $-6 + \sum_{n=1}^{\infty} [-2(n+1)(n+3)]x^n$

4f. $6 + \sum_{n=1}^{\infty} [2(n+1)(2n+3)]x^n$ **4g.** $\sum_{n=4}^{\infty} (n-3)^2 a_{n-3} x^n$ **4h.** $\sum_{n=2}^{\infty} a_{n-2}(x-1)^n$

4i. $a_0 x + 2a_1 x^2 + \sum_{n=3}^{\infty} n\left[a_{n-1} - a_{n+1}\right] x^n$ **4j.** $\sum_{n=0}^{\infty} \left[(n+1)a_{n+1} + 5a_n\right] x^n$

4k. $-4a_0 - 4a_1 x + \sum_{k=2}^{\infty} \left[k^2 - k - 4\right] a_k x^k$

4l. $2a_2 + 6a_3 x + \sum_{n=2}^{\infty} \left[(n+2)(n+1)a_{n+2} - 3a_{n-2}\right] x^n$ **5a.** $\sum_{k=0}^{\infty} (k+1)x^k$ for $|x| < 1$

5b. $\sum_{k=0}^{\infty} \frac{1}{2}(k+2)(k+1)x^k$ for $|x| < 1$ **6a.** $\sum_{k=0}^{\infty} \frac{1}{k!} x^k$ **6b.** $\sum_{k=0}^{\infty} \frac{(-1)^k}{(2k)!} x^{2k}$

6c. $\sum_{k=0}^{\infty} \frac{(-1)^k}{(2k+1)!} x^{2k+1}$ **6d.** $\sum_{k=1}^{\infty} \frac{(-1)^{k-1}}{k} (x-1)^k$ **7a.** $1 + \frac{1}{2}x - \frac{1}{8}x^2 + \frac{1}{16}x^3 - \frac{5}{128}x^4$

7b. $1 - \frac{1}{2}x^2 - \frac{1}{8}x^4 - \frac{1}{16}x^6 - \frac{5}{128}x^8$ **7c.** $1 + \frac{1}{2}x^2 + \frac{3}{8}x^4 + \frac{5}{16}x^6$

8b i. $1 + 2x + 2x^2 + \frac{4}{3}x^4 + \frac{4}{15}x^5 + \frac{4}{45}x^6 + \frac{8}{315}x^7 + \frac{2}{315}x^8 + \frac{4}{2835}x^9$

8b ii. $1 + \frac{1}{2}x^2 + \frac{5}{24}x^4 + \frac{61}{720}x^6 + \frac{277}{8064}x^8 + \frac{50521}{3628800}x^{10}$

8b iii. $3 - \frac{4}{3}(x-2) + \frac{1}{27}(x-2)^2 - \frac{4}{243}(x-2)^3 + \frac{31}{4374}(x-2)^4$
$- \frac{58}{19683}(x-2)^5 + \frac{139}{118098}(x-2)^6 - \frac{238}{531441}(x-2)^7$

9a. $\sum_{k=0}^{\infty} 2^k x^k$ with $R = \frac{1}{2}$ **9b.** $\sum_{k=0}^{\infty} (-1)^k x^{2k}$ with $R = 1$ **9c.** $\sum_{k=0}^{\infty} \frac{1}{2^k} x^k$ with $R = 2$

9d. $\sum_{k=0}^{\infty} \frac{k+1}{2^{k+1}} x^k$ with $R = 2$ **9e.** $\sum_{k=0}^{\infty} \frac{(-1)^k}{k!} x^{2k}$ with $R = \infty$

9f. $\sum_{k=0}^{\infty} \frac{(-1)^k}{(2k+1)!} x^{4k+2}$ with $R = \infty$ **10c.** Taylor series $= 0$

10d. Because $f(x)$ does not equal its Taylor series about 0 except right at $x = 0$.

Chapter 33

2. S_k is the statement: $a_k = F(k)$ for a given function F and a given sequence $A_0, A_1, A_2, \ldots$.

3a. $a_k = \frac{2}{k} a_{k-1}$ for $k \geq 1$, $y(x) = a_0 \sum_{k=0}^{\infty} \frac{2^k}{k!} x^k = a_0 \sum_{k=0}^{\infty} \frac{1}{k!}(2x)^k$

3b. $a_k = \frac{2}{k} a_{k-2}$ for $k \geq 2$ (with $a_1 = 0$) , $y(x) = a_0 \sum_{m=0}^{\infty} \frac{1}{m!} x^{2m} = a_0 e^{x^2}$

3c. $a_k = 2a_{k-1}$ for $k \geq 1$, $y(x) = a_0 \sum_{k=0}^{\infty} 2^k x^k = \frac{a_0}{1-2x}$

3d. $a_k = \frac{k-3}{3k} a_{k-1}$ for $k \geq 1$, $y(x) = a_0 \left[1 - \frac{2}{3}x + \frac{1}{9}x^2\right]$

3e. $a_k = \frac{4-k}{k} a_{k-2}$ for $k \geq 2$ (with $a_1 = 0$) , $y(x) = a_0 \left[1 + x^2\right]$

3f. $a_k = a_{k-1}$ for $k \geq 1$, $y(x) = a_0 \sum_{k=0}^{\infty} x^k = \frac{a_0}{1-x}$

3g. $a_k = -\frac{1}{2} a_{k-1}$ for $k \geq 1$, $y(x) = a_0 \sum_{k=0}^{\infty} \left(\frac{-1}{2}\right)^k (x-3)^k$

3h. $a_k = -\frac{1+k}{4k} a_{k-1}$ for $k \geq 1$, $y(x) = a_0 \sum_{k=0}^{\infty} (k+1) \left(\frac{-1}{4}\right)^k (x-5)^k$

3i. $a_k = \frac{1}{2} a_{k-3}$ for $k \geq 3$ (with $a_1 = a_2 = 0$) , $y(x) = a_0 \sum_{m=0}^{\infty} \frac{1}{2^m} x^{3m}$

3j. $a_k = \frac{k-6}{2k} a_{k-3}$ for $k \geq 3$ (with $a_1 = a_2 = 0$) , $y(x) = a_0 \left[1 - \frac{1}{2}x^3\right]$

3k. $a_k = \frac{1}{k} a_{k-2} - \frac{k-1}{k} a_{k-1}$ for $k \geq 2$ (with $a_1 = 0$) ,
$y(x) = a_0 \left[1 + \frac{1}{2}x^2 - \frac{1}{3}x^3 + \frac{3}{8}x^4 - \frac{11}{30}x^5 + \cdots\right]$

3l. $a_k = \frac{1}{k} a_{k-2} - a_{k-1}$ for $k \geq 2$ (with $a_1 = -a_0$) ,
$y(x) = a_0 \left[1 - x + \frac{3}{2}x^2 - \frac{11}{6}x^3 + \frac{53}{24}x^4 + \cdots\right]$

4a. No singular points, $R = \infty$, $I = (-\infty, \infty)$

4b. No singular points, $R = \infty$, $I = (-\infty, \infty)$ **4c.** $z_s = \frac{1}{2}$, $R = \frac{1}{2}$, $I = \left(-\frac{1}{2}, \frac{1}{2}\right)$

4d. $z_s = 3$, $R = 3$, $I = (-3, 3)$ **4e.** $z_s = \pm i$, $R = 1$, $I = (-1, 1)$

4f. $z_s = 1$, $R = 1$, $I = (-1, 1)$ **4g.** $z_s = 1$, $R = 2$, $I = (1, 5)$

4h. $z_s = 1$, $R = 4$, $I = (1, 9)$

4i. $z_s = \sqrt[3]{2},\ 2^{-2/3}\left(-1 \pm i\sqrt{3}\right)$; $R = \sqrt[3]{2}$, $I = \left(-\sqrt[3]{2}, \sqrt[3]{2}\right)$

4j. $z_s = \sqrt[3]{2},\ 2^{-2/3}\left(-1 \pm i\sqrt{3}\right)$; $R = \sqrt[3]{2}$, $I = \left(-\sqrt[3]{2}, \sqrt[3]{2}\right)$

4k. $z_s = -1$, $R = 1$, $I = (-1, 1)$ **4l.** $z_s = -1$, $R = 1$, $I = (-1, 1)$

5a. $a_k = \frac{4-k}{k} a_{k-2}$ for $k \geq 2$, $y(x) = a_0 y_1(x) + a_1 y_2(x)$ where $y_1(x) = 1 + x^2$ and

$$y_2(x) = x + \frac{1}{3}x^3 - \frac{1}{5 \cdot 3}x^5 + \frac{1}{7 \cdot 5}x^7 - \frac{1}{9 \cdot 7}x^9 + \cdots = \sum_{m=0}^{\infty} \frac{(-1)^{m+1}}{(1+2m)(2m-1)} x^{2m+1}$$

5b. $a_k = -\frac{1}{k} a_{k-2}$ for $k \geq 2$, $y(x) = a_0 y_1(x) + a_1 y_2(x)$ where

$$y_1(x) = 1 - \frac{1}{2}x^2 + \frac{1}{2^2 \cdot 2}x^4 - \frac{1}{2^3(3!)}x^6 + \cdots = \sum_{m=0}^{\infty} (-1)^m \frac{1}{2^m m!} x^{2m} \quad \text{and}$$

$$y_2(x) = x - \frac{1}{3}x^3 + \frac{1}{5 \cdot 3}x^5 - \frac{1}{7 \cdot 5 \cdot 3}x^7 + \cdots = \sum_{m=0}^{\infty} (-1)^m \frac{2^m m!}{(2m+1)!} x^{2m+1}$$

5c. $a_k = -\frac{k-2}{4k} a_{k-2}$ for $k \geq 2$, $y(x) = a_0 y_1(x) + a_1 y_2(x)$ where $y_1(x) = 1$ and

$$y_2(x) = x - \frac{1}{4 \cdot 3}x^3 + \frac{1}{4^2 \cdot 5}x^5 - \frac{1}{4^3 \cdot 7}x^7 + \cdots = \sum_{m=0}^{\infty} (-1)^m \frac{1}{4^m (2m+1)} x^{2m+1}$$

5d. $a_k = \frac{3}{k(k-1)} a_{k-4}$ for $k \geq 4$ (with $a_2 = a_3 = 0$) , $y(x) = a_0 y_1(x) + a_1 y_2(x)$ where

$y_1 = 1 + \frac{3}{4 \cdot 3} x^4 + \frac{3^2}{(8 \cdot 7)(4 \cdot 3)} x^8 + \frac{3^3}{(12 \cdot 11)(8 \cdot 7)(4 \cdot 3)} x^{12} + \cdots$ and

$y_2 = x + \frac{3}{5 \cdot 4} x^5 + \frac{3^2}{(9 \cdot 8)(5 \cdot 4)} x^9 + \frac{3^3}{(13 \cdot 12)(9 \cdot 8)(5 \cdot 4)} x^{13} + \cdots$

5e. $a_k = \frac{k+1}{4k} a_{k-2}$ for $k \geq 2$, $y(x) = a_0 y_1(x) + a_1 y_2(x)$ where

$y_1 = 1 + \frac{3}{4 \cdot 2} x^2 + \frac{5 \cdot 3}{4^2 (4 \cdot 2)} x^4 + \frac{7 \cdot 5 \cdot 3}{4^3 (6 \cdot 4 \cdot 2)} x^6 + \cdots \left(= \sum_{m=0}^{\infty} \frac{(2m+1)!}{4^{2m} (m!)^2} x^{2m} \right)$ and

$y_2 = x + \frac{4}{4 \cdot 3} x^3 + \frac{6 \cdot 4}{4^2 (5 \cdot 3)} x^5 + \frac{8 \cdot 6 \cdot 4}{4^3 (7 \cdot 5 \cdot 3)} x^7 + \cdots \left(= \sum_{m=0}^{\infty} \frac{(m+1)! m!}{(2m+1)!} x^{2m+1} \right)$

5f. $a_k = \frac{k-4}{k-1} a_{k-2}$ for $k \geq 2$, $y(x) = a_0 y_1(x) + a_1 y_2(x)$ where $y_1 = 1 - 2x^2$ and

$y_2 = x - \frac{1}{2} x^3 - \frac{1}{4 \cdot 2} x^5 - \frac{3}{6 \cdot 4 \cdot 2} x^7 - \frac{5 \cdot 3}{8 \cdot 6 \cdot 4 \cdot 2} x^7 + \cdots$

5g. $a_k = \frac{2(k-5)}{k(k-1)} a_{k-2}$ for $k \geq 2$, $y(x) = a_0 y_1(x) + a_1 y_2(x)$ where

$y_1(x) = 1 + \frac{2(-3)}{2!} x^2 + \frac{2^2 [(-1)(-3)]}{4!} x^4 + \frac{2^3 [(1)(-1)(-3)]}{6!} x^6 + \frac{2^3 [(3)(1)(-1)(-3)]}{8!} x^8 + \cdots$

and $y_2(x) = x - \frac{2}{3} x^3$

5h. $a_k = \frac{1}{9} a_{k-2}$ for $k \geq 2$, $y(x) = a_0 y_1(x) + a_1 y_2(x)$ where

$y_1(x) = \sum_{m=0}^{\infty} \frac{1}{9^m} (x-3)^{2m}$ and $y_2(x) = \sum_{m=0}^{\infty} \frac{1}{9^m} (x-3)^{2m+1}$

5i. $a_k = -\frac{1}{k-1} a_{k-2}$ for $k \geq 2$, $y(x) = a_0 y_1(x) + a_1 y_2(x)$ where

$y_1(x) = 1 - (x+2)^2 + \frac{1}{3}(x+2)^4 - \frac{1}{5 \cdot 3}(x+2)^6 + \frac{1}{7 \cdot 5 \cdot 3}(x+2)^8 + \cdots$

$\left(= \sum_{m=0}^{\infty} (-1)^m \frac{2^m m!}{(2m)!} (x+2)^{2m} \right)$

and $y_2 = (x+2) - \frac{1}{2}(x+2)^3 + \frac{1}{4 \cdot 2}(x+2)^5 - \frac{1}{6 \cdot 4 \cdot 2}(x+2)^7 + \cdots$

$\left(= \sum_{m=0}^{\infty} (-1)^m \frac{1}{2^m m!} (x+2)^{2m+1} \right)$

5j. $a_k = -\frac{k-5}{k} a_{k-2}$ for $k \geq 2$, $y(x) = a_0 y_1(x) + a_1 y_2(x)$ where

$y_1(x) = 1 - \frac{-3}{2}(x-1)^2 + \frac{(-1)(-3)}{4 \cdot 2}(x-1)^4 - \frac{(1)(-1)(-3)}{6 \cdot 4 \cdot 2}(x-1)^6 - \frac{(3)(1)(-1)(-3)}{8 \cdot 6 \cdot 4 \cdot 2}(x-1)^8$

and $y_2(x) = x - \frac{1}{3}$

5k. $a_k = \frac{2}{k} a_{k-1} + \frac{1}{k(k-1)} a_{k-3}$ for $k \geq 3$ (with $a_2 = a_1$) , $y(x) = a_0 y_1(x) + a_1 y_2(x)$ where

$y_1(x) = 1 + \frac{1}{6} x^3 + \frac{1}{12} x^4 + \frac{1}{30} x^5 + \cdots$ and $y_2(x) = x + x^2 + \frac{2}{3} x^3 + \frac{5}{12} x^4 + \frac{13}{60} x^5 + \cdots$

5l. $a_k = \frac{1}{k(k-1)} \left[(k-2) a_{k-2} + 2 a_{k-3} \right]$ for $k \geq 3$ (with $a_2 = 0$) , $y(x) = a_0 y_1(x) + a_1 y_2(x)$

where $y_1(x) = 1 + \frac{1}{3} x^3 + \frac{1}{20} x^5 + \frac{1}{45} x^6 + \cdots$

and $y_2(x) = x + \frac{1}{6} x^3 + \frac{1}{6} x^4 + \frac{1}{40} x^5 + \frac{1}{30} x^6 + \cdots$

6a. $z_s = \pm i$, $R = 1$, $I = (-1, 1)$ **6b.** No singular points, $R = \infty$, $I = (-\infty, \infty)$

6c. $z_s = \pm 2i$, $R = 2$, $I = (-2, 2)$ **6d.** No singular points, $R = \infty$, $I = (-\infty, \infty)$

6e. $z_s = \pm 2$, $R = 2$, $(-2, 2)$ **6f.** $z_s = \pm 1$, $R = 1$, $I = (-1, 1)$

6g. No singular points, $R = \infty$, $I = (-\infty, \infty)$ **6h.** $z_s = 0, 6$; $R = 3$, $I = (0, 6)$

6i. No singular points, $R = \infty$, $I = (-\infty, \infty)$ **6j.** $z_s = 1 \pm i$, $R = 1$, $I = (0, 2)$

6k. No singular points, $R = \infty$, $I = (-\infty, \infty)$

6l. No singular points, $R = \infty$, $I = (-\infty, \infty)$ **8a.** $a_k = \frac{2k - 4 - \lambda}{k(k-1)} a_{k-2}$ **8b.** $(-\infty, \infty)$

8c. $c_k = \frac{2k - 4 - \lambda}{k(k-1)} c_{k-2}$ **8d.** $\lambda_N = 2N$ **8f i.** $p_0(x) = 1$ **8f ii.** $p_1(x) = x$

8f iii. $p_2(x) = 1 - 2x^2$ **8f iv.** $p_3(x) = x - \frac{2}{3}x^3$ **8f v.** $p_4(x) = 1 - 4x^2 + \frac{4}{3}x^4$

8f vi. $p_5(x) = x - \frac{4}{3}x^3 + \frac{4}{15}x^5$ **9a.** $a_k = \frac{(k-2)^2 - \lambda}{k(k-1)} a_{k-2}$

9b. $c_k = \frac{(k-2)^2 - \lambda}{k(k-1)} c_{k-2}$ for $k \geq 2$ **9c.** $\lambda_N = N^2$ **9e i.** $\lambda_0 = 0$, $p_0(x) = 1$

9e ii. $\lambda_1 = 1$, $p_1(x) = x$ **9e iii.** $\lambda_2 = 4$ and $p_2(x) = 1 - 2x^2$

9e iv. $\lambda_3 = 9$ and $p_3(x) = x - \frac{4}{3}x^3$ **9e v.** $\lambda_4 = 16$ and $p_4(x) = 1 - 8x^2 + 8x^4$

9e vi. $\lambda_5 = 25$ and $p_5(x) = x - 4x^3 + \frac{16}{5}x^5$ **9f i.** $(-1, 1)$

9f ii. For both, $R = 1$ and $(-1, 1)$ **10a.** $a_k = \frac{k^2 - 3k - \lambda + 2}{k(k-1)} a_{k-2}$

10b. $c_k = \frac{k^2 - 3k - \lambda + 2}{k(k-1)} c_{k-2}$ for $k \geq 2$ **10c.** $\lambda = N(N+1)$

10e i. $\lambda_0 = 0$, $p_0(x) = 1$, and $y_{0,O}(x) = \sum_{n=0}^{\infty} \frac{1}{2n+1} x^{2n+1}$

10e ii. $\lambda_1 = 2$, $p_1(x) = x$, and $y_{1,E}(x) = \sum_{n=0}^{\infty} \frac{1}{1-2n} x^{2n}$

10e iii. $\lambda_2 = 6$ and $p_2(x) = 1 - 3x^2$ **10e iv.** $\lambda_3 = 12$ and $p_3(x) = x - \frac{5}{3}x^3$

10e v. $\lambda_4 = 20$ and $p_4(x) = 1 - 10x^2 + \frac{35}{3}x^4$

10e vi. $\lambda_5 = 30$ and $p_5(x) = x - \frac{14}{3}x^3 + \frac{21}{5}x^5$ **10f i.** $(-1, 1)$

10f ii. For both, $R = 1$ and $(-1, 1)$ **11a.** $a_0 \left[1 - \frac{2^2}{2!}x^2 + \frac{2^4}{4!}x^4 \right] + a_1 \left[x - \frac{2^2}{3!}x^3 + \frac{2^4}{5!}x^5 \right]$

11b. $a_0 \left[1 + \frac{2}{4!}x^4 \right] + a_1 \left[x + \frac{6}{5!}x^5 \right]$

11c. $a_0 \left[1 - \frac{1}{2!}x^2 - \frac{2}{3!}x^3 - \frac{3}{4!}x^4 \right] + a_1 \left[x - \frac{1}{3!}x^3 - \frac{4}{4!}x^4 \right]$

11d. $a_0 \left[1 + \frac{1}{2!}\left(x - \frac{\pi}{2}\right)^2 + \frac{2}{4!}\left(x - \frac{\pi}{2}\right)^4 \right] + a_1 \left[\left(x - \frac{\pi}{2}\right) + \frac{1}{3!}\left(x - \frac{\pi}{2}\right)^3 \right]$

11e. $a_0 + a_1 x + \frac{1 - a_0}{3!}x^3 - \frac{2a_1}{4!}x^4 - \frac{1}{5!}x^5$ **11f.** $a_0 \left[1 + \frac{1}{3}x^3 \right] + a_1 \left[x + \frac{1}{3!}x^3 + \frac{2}{4!}x^4 \right]$

11g. $a_0 + a_1 x + \frac{{a_0}^2}{2!}x^2 + \frac{2a_0 a_1}{3!}x^3 + \frac{2\left[{a_0}^3 + {a_1}^2\right]}{4!}x^4 + \frac{10{a_0}^2 a_1}{5!}x^5$

11h. $a_0 - \cos(a_0)\, x - \frac{\sin(a_0)\cos(a_0)}{2!}x^2 - \frac{\cos(a_0)\left[1 - 2\cos^2(a_0)\right]}{3!}x^3$

Chapter 34

5a. No singular points , $R = \infty$ **5b.** $z_s = \left[n + \frac{1}{2} \right]\pi$ with $n = 0, \pm 1, \pm 2, \ldots$, $R = 0$

5c. $z_s = 0, \pm\pi, \pm 2\pi, \pm 3\pi, \ldots$, $R = \pi - 2$

5d. $z_s = in\pi$ with $n = 0, \pm 1, \pm 2, \ldots$, $R = 2$

5e. $z_s = in\pi$ with $n = \pm 1, \pm 2, \ldots$, $R = \sqrt{4 + \pi^2}$ **5f.** $z_s = \pm 2i$, $R = 2$

5g. $z_s = ik2\pi$ with $k = 0, \pm 1, \pm 2, \ldots$, $R = 3$ **5h.** $z_s = -2$, $R = 4$

5i. No singular points , $R = \infty$ **5j.** $z_s = \pm\sqrt{n}$ with $n = 1, 2, 3, \ldots$, $R = 1$

6a. $a_0 \left[1 + x + x^2 + \frac{5}{6}x^3 + \frac{5}{8}x^4 \right]$, $I = (-\infty, \infty)$

6b. $a_0 \left[1 - x - \frac{1}{2}x^2 + \frac{1}{6}x^3 + \frac{3}{8}x^4 \right]$, $I = (-\infty, \infty)$

6c. $a_0 \left[1 - x - \frac{1}{2}x^2 + \frac{1}{8}x^4 \right]$, $I = (-\infty, \infty)$

6d. $a_0\left[1 - \frac{1}{2}(x-1)^2 + \frac{1}{6}(x-1)^3 + \frac{1}{24}(x-1)^4\right]$, $I = (0,2)$

7a. $a_0\left[1 + \frac{1}{2}x^2 + \frac{1}{6}x^3 + \frac{1}{12}x^4\right] + a_1\left[x + \frac{1}{6}x^3 + \frac{1}{12}x^4\right]$, $I = (-\infty,\infty)$

7b. $a_0\left[1 + \frac{1}{2}x^2 + \frac{1}{6}x^3 - \frac{1}{6}x^4\right] + a_1\left[x - \frac{1}{3}x^3 + \frac{1}{12}x^4\right]$, $(-\infty,\infty)$

7d. $a_0\left[1 - \frac{1}{6}(x-1)^3 + \frac{1}{24}(x-1)^4\right] + a_1\left[(x-1) - \frac{1}{12}(x-1)^4\right]$, $I = (0,2)$

7e. $a_0\left[1 - \frac{1}{2}(x-1)^2 + \frac{1}{12}(x-1)^3 + \frac{1}{96}(x-1)^4\right]$
$+ a_1\left[(x-1) - \frac{1}{6}(x-1)^3 + \frac{1}{24}(x-1)^4\right]$, $I = (0,2)$

7f. $a_0\left[1 - x^2 - \frac{5}{3}x^3 + \frac{11}{12}x^4\right] + a_1\left[x - \frac{1}{2}x^2 - \frac{1}{2}x^3 - \frac{9}{8}x^4\right]$, $I = (-\infty,\infty)$

8b i. $a_0\left[1 + x + x^2 + \frac{5}{6}x^3 + \frac{5}{8}x^4 + \frac{13}{30}x^5 + \frac{203}{720}x^6 + \frac{877}{5040}x^7 + \frac{23}{224}x^8 + \frac{1007}{17280}x^9 + \frac{4639}{145152}x^{10}\right]$

8b ii. $a_0\left[1 - x + \frac{1}{2}x^2 - \frac{1}{3}x^3 + \frac{5}{24}x^4 - \frac{1}{15}x^5 + \frac{13}{720}x^6 - \frac{11}{630}x^7 + \frac{361}{40320}x^8\right]$

8b iii. $a_0\left[1 - x + \frac{1}{2}x^2 - \frac{1}{3}x^3 + \frac{5}{24}x^4 - \frac{2}{15}x^5 + \frac{61}{720}x^6 - \frac{17}{315}x^7 + \frac{277}{8064}x^8\right]$

8b iv. $a_0\left[1 - 3(x-2) + \frac{23}{6}(x-2)^2 - \frac{407}{162}(x-2)^3 + \frac{1241}{1944}(x-2)^4 + \frac{21629}{87480}(x-2)^5\right]$

9b i. $a_0\left[1 + \frac{1}{2}x^2 + \frac{1}{6}x^3 + \frac{1}{12}x^4 + \frac{1}{24}x^5 + \frac{13}{720}x^6 + \frac{1}{140}x^7 + \frac{109}{40320}x^8\right]$
$+ a_1\left[x + \frac{1}{6}x^3 + \frac{1}{12}x^4 + \frac{1}{30}x^5 + \frac{1}{72}x^6 + \frac{29}{5040}x^7 + \frac{1}{448}x^8\right]$

9b ii. $a_0\left[1 - \frac{1}{2}x^2 + \frac{1}{12}x^4 - \frac{1}{80}x^6 + \frac{11}{8064}x^8 - \frac{17}{129600}x^{10}\right]$
$+ a_1\left[x - \frac{1}{6}x^3 + \frac{1}{30}x^5 - \frac{19}{5040}x^7 + \frac{29}{72576}x^9\right]$

9b iii. $a_0\left[1 - \frac{1}{2}x^2 + \frac{1}{6}x^4 - \frac{31}{720}x^6\right] + a_1\left[x - \frac{1}{3}x^3 + \frac{1}{10}x^5 - \frac{59}{2520}x^7\right]$

9b iv. $a_0\left[1 - \frac{1}{2}(x-1)^2 + \frac{1}{12}(x-1)^3 - \frac{1}{96}(x-1)^4 + \frac{31}{960}(x-1)^5\right]$
$+ a_1\left[(x-1) - \frac{1}{2}(x-1)^2 + \frac{1}{12}(x-2)^3 - \frac{3}{32}(x-1)^4 + \frac{71}{960}(x-4)^5\right]$

Chapter 35

2a. $y_1(x) = (x-3)^2$, $y_2(x) = x - 3$ **2b.** $y_1(x) = x^{-1/2}$, $y_2(x) = x^{-1}$

2c. $y_1(x) = (x-1)^3$, $y_2(x) = (x-1)^3 \ln|x-1|$ **2d.** $y_1(x) = 1$, $y_2(x) = \ln|x+2|$

2e. $y_1(x) = (x-5)^2$, $y_1(x) = \sqrt[3]{x-5}$

2f. $y_1(x) = \cos(2\ln|x-5|)$, $y_1(x) = \sin(2\ln|x-5|)$

3a. Reg. sing. pts.: $0,\ 2,\ -2$; No irreg. sing. pt. ; $R = 2$

3b. No Reg. sing. pt. ; Irreg. sing. pt.: 0 ; $R = 2$

3c. Reg. sing. pt.: 1 ; Irreg. sing. pt.: 0 ; $R = 1$

3d. Reg. sing. pt.: $3,\ 4$; No irreg. sing. pt. ; $R = 1$

3e. Reg. sing. pt.: 4 ; Irreg. sing. pt.: 3 ; $R = 1$

3f. Reg. sing. pt.: 0 ; No irreg. sing. pt. ; $R = \infty$

3g. Reg. sing. pts.: $0,\ \frac{1}{2},\ -\frac{1}{2},\ i,\ -i$; No irreg. sing. pts. ; $R = \frac{1}{2}$

3h. Reg. sing. pts.: $2i,\ -2i$; No irreg. sing. pt. ; $R = 2$

4a. $r^2 - 3r + 2 = 0$; $r_1 = 2$, $r_2 = 1$.

For $r = r_1$: $a_1 = 0$, $a_k = \frac{-1}{(k+1)k}a_{k-2}$; $y(x) = ay_1(x)$ with

$$y_1(x) = x^2\left[1 - \frac{1}{3!}x^2 + \frac{1}{5!}x^4 - \frac{1}{7!}x^6 - \cdots\right] = x^2\sum_{m=0}^{\infty}\frac{(-1)^m}{(2m+1)!}x^{2m} = x\sin(x) .$$

For $r = r_2$, $a_1 = 0$; $a_k = \frac{-1}{k(k-1)}a_{k-2}$; $y(x) = ay_2(x)$ with

$$y_2(x) = x\left[1 - \frac{1}{2!}x^2 + \frac{1}{4!}x^4 - \frac{1}{6!}x^6 + \cdots\right] = x\sum_{m=0}^{\infty}\frac{(-1)^m}{(2m)!}x^{2m} = x\cos(x) .$$

Gen. Soln.: $y(x) = c_1y_1(x) + c_2y_2(x)$

4b. $4r^2 - 4r + 1 = 0$; $r_1 = r_2 = \frac{1}{2}$; For $r = r_1$: $a_k = \frac{1}{k^2}a_{k-1}$;

$y(x) = ay_1(x)$ with $y_1(x) = \sqrt{x}\sum_{k=0}^{\infty}\frac{1}{(k!)^2}x^k$. No "second" value of r .

4c. $r^2 - 4 = 0$; $r_1 = 2$, $r_2 = -2$. For $r = r_1$: $a_k = \frac{-4}{k(k+4)}a_{k-1}$; $y(x) = ay_1(x)$ with

$$y_1(x) = x^2\left[1 - \frac{4}{5}x + \frac{4^2}{(2)(6\cdot 5)}x^2 - \frac{4^3}{(3\cdot 2)(7\cdot 6\cdot 5)}x^3 + \cdots\right] = x^2\sum_{k=0}^{\infty}\frac{(-4)^k 4!}{k!(k+4)!}x^k .$$

For $r = r_2$, the recursion formula blows up.

4d. $r^2 - 7r + 10 = 0$; $r_1 = 5$, $r_2 = 2$.

For $r = r_1$: $a_1 = 0$, $a_k = \frac{9(k+2)}{k}a_{k-2}$; $y(x) = ay_1(x)$ with

$$y_1(x) = x^2\left[1 + \frac{9\cdot 4}{2}x^2 + \frac{9^2\cdot 6}{2}x^4 + \frac{9^3\cdot 8}{2}x^6 - \cdots\right] = x^5\sum_{m=0}^{\infty}9^m(m+1)x^{2m} .$$

For $r = r_2$: $a_1 = 0$; $a_k = \frac{9(k-1)}{k-3}a_{k-2}$; $y(x) = ay_2(x)$ with

$$y_2(x) = x^2\left[1 - 9x^2 - 3\cdot 9^2x^4 - 5\cdot 9^3x^6 + \cdots\right] = x^2\sum_{m=0}^{\infty}9^m(1-2m)x^{2m} .$$

Gen. Soln.: $y(x) = c_1y_1(x) + c_2y_2(x)$

4e. $r^2 - 2r + 1 = 0$; $r_1 = r_2 = 1$. For $r = r_1$: $a_k = \frac{k-2}{k}a_{k-1}$;

$y(x) = ay_1(x)$ with $y_1(x) = x - x^2$. No "second" value of r .

4f. $r^2 = 0$; $r_1 = r_2 = 0$. For $r = r_1$: $a_1 = 0$, $a_k = \frac{-1}{k^2}a_{k-2}$; $y(x) = ay_1(x)$ with

$$y_1(x) = 1 - \frac{1}{2^2}x^2 + \frac{1}{(2\cdot 4)^2}x^4 - \frac{1}{(6\cdot 4\cdot 2)^2}x^6 + \cdots = \sum_{m=0}^{\infty}\frac{(-1)^m}{(2^m m!)^2}x^{2m} .$$

No "second value" for r .

4g. $r^2 - 1 = 0$; $r_1 = 1$, $r_2 = -1$. For $r = r_1$: $a_1 = 0$, $a_k = \frac{-1}{k(k+2)}a_{k-2}$;

$$y(x) = ay_1(x) \quad \text{with} \quad y_1(x) = x\left[1 - \frac{1}{2\cdot 4}x^2 + \frac{1}{(4\cdot 2)(6\cdot 4)}x^4 - \frac{1}{(6\cdot 4\cdot 2)(8\cdot 6\cdot 4)}x^6 + \cdots\right]$$

$$= x\sum_{m=0}^{\infty}\frac{(-1)^m}{2^{2m}m!(m+1)!}x^{2m} .$$ For $r = r_2$, the recursion formula blows up.

4h. $2r^2 + 3r + 1 = 0$; $r_1 = -1/2$, $r_2 = -1$.

For $r = r_1$: $a_1 = 0$, $a_k = \frac{2(k-2)}{k(2k+1)}a_{k-2}$; $y(x) = ay_1(x)$ with $y_1(x) = x^{-1/2}$.

For $r = r_2$: $a_1 = 0$, $a_k = \frac{2k-5}{k(2k-1)}a_{k-2}$; $y(x) = ay_2(x)$ with

$$y_2(x) = x^{-1}\left[1 - \frac{1}{2\cdot 3}x^2 - \frac{1}{4\cdot 2\cdot 7}x^4 - \frac{1}{6\cdot 4\cdot 2\cdot 11}x^6 + \cdots\right] = x^{-1}\sum_{m=0}^{\infty}\frac{-1}{2^m m!(4m-1)}x^{2m} .$$

Gen. Soln.: $y(x) = c_1y_1(x) + c_2y_2(x)$

4i. $r^2 - 6r + 9 = 0$; $r_1 = r_2 = 3$. For $r = r_1$: $a_k = \frac{2}{k}a_{k-1}$; $y(x) = ay_1(x)$ with

$$y_1(x) = x^3\left[1 + \frac{2}{1}x + \frac{2^2}{2}x^2 + \frac{2^3}{3\cdot 2}x^3 + \cdots\right] = x^3\sum_{k=0}^{\infty}\frac{2^k}{k!}x^k = x^3e^{2x} .$$

No "second" value of r .

4j. $3r^2 - 7r + 2 = 0$; $r_1 = 2$, $r_2 = 1/3$. For $r = r_1$: $a_k = \dfrac{k+1}{k} a_{k-1}$; $y(x) = ay_1(x)$ with

$$y_1(x) = x^2\left[1 + 2x + 3x^2 + 4x^3 + \cdots\right] = x^2 \sum_{k=0}^{\infty} (k+1)x^k .$$

For $r = r_2$: $a_k = \dfrac{3k-2}{3k-5} a_{k-1}$; $y(x) = ay_1(x)$ with

$$y_1(x) = \sqrt[3]{x}\left[1 - \frac{1}{2}x - \frac{4}{2}x^2 - \frac{7}{2}x^3 + \cdots\right] = \sqrt[3]{x} \sum_{k=0}^{\infty} \frac{2-3k}{2} x^k .$$

Gen. Soln.: $y(x) = c_1 y_1(x) + c_2 y_2(x)$

4k. $r^2 - 2r = 0$; $r_1 = 2$, $r_2 = 0$. For $r = r_1$: $a_k = \dfrac{k-3}{k(k+2)} a_{k-1}$;

$y(x) = ay_1(x)$ with $y_1(x) = x^2 - \frac{2}{3}x^3 + \frac{1}{12}x^4$.

For $r = r_2$, the recursion formula blows up.

4l. $4r^2 - 4r + 1 = 0$; $r_1 = r_2 = 1/2$. For $r = r_1$: $a_k = -\dfrac{2k-1}{k^2} a_{k-1}$;

$y(x) = ay_1(x)$ with $y_1(x) = \sqrt{x}\left[1 - x + \frac{3}{4}x^2 - \frac{5 \cdot 3}{9 \cdot 4}x^3 + \cdots\right]$.

$$= \sqrt{x}\left[1 + \sum_{k=1}^{\infty} (-1)^k \frac{(2k-1)(2k-3)\cdots 5 \cdot 3 \cdot 1}{(k!)^2} x^k\right] .$$ No "second" value of r .

4m. $r^2 = 0$; $r_1 = r_2 = 0$.

For $r = r_1$: $a_1 = 0$, $a_2 = 0$, $a_k = \dfrac{k-6}{k^2} a_{k-3}$; $y(x) = ay_1(x)$ with

$y_1(x) = 1 - \frac{1}{3}x^3$. No "second" value of r .

4n. $9r^2 - 1 = 0$; $r_1 = 1/3$, $r_2 = -1/3$.

For $r = r_1$: $a_k = -a_{k-1}$; $y(x) = ay_1(x)$ with

$$y_1(x) = x^{1/3}\left[1 - x + x^2 - x^3 + \cdots\right] = x^{1/3} \sum_{k=0}^{\infty} (-1)^k x^k = x^{1/3}(1+x)^{-1} .$$

For $r = r_2$: $a_k = -a_{k-1}$; $y(x) = ay_2(x)$ with

$$y_2(x) = x^{-1/3}\left[1 - x + x^2 - x^3 + \cdots\right] = x^{-1/3} \sum_{k=0}^{\infty} (-1)^k x^k = x^{-1/3}(1+x)^{-1} .$$

Gen. Soln.: $y(x) = c_1 y_1(x) + c_2 y_2(x)$

5a. $r^2 - r = 0$; $r_1 = 1$, $r_2 = 0$. For $r = r_1$: $a_k = -\dfrac{1}{k} a_{k-1}$; $y(x) = ay_1(x)$ with

$$y_1(x) = (x-3)\left[1 - \frac{1}{2}(x-3) + \frac{1}{2 \cdot 1}(x-3)^2 - \frac{1}{3 \cdot 2 \cdot 1}(x-3)^3 + \cdots\right]$$

$$= (x-3) \sum_{k=0}^{\infty} (-1)^k \frac{1}{k!}(x-3)^k = (x-3)e^{-(x-3)} .$$

For $r = r_2$, the recursion formula blows up.

5b. $r^2 + r = 0$; $r_1 = 0$, $r_2 = -1$.

For $r = r_1$: $a_k = -\dfrac{1}{(k+1)k} a_{k-2}$; $y(x) = ay_1(x)$ with

$$y_1(x) = 1 - \frac{1}{3!}(x+2)^2 + \frac{1}{5!}(x+2)^4 - \frac{1}{7!}(x+2)^6 + \cdots = \frac{\sin(x+2)}{x+2} .$$

For $r = r_2$: $a_k = -\dfrac{1}{k(k-1)} a_{k-2}$; $y(x) = ay_2(x)$ with

$$y_2(x) = \frac{1}{x+2}\left[1 - \frac{1}{2!}(x+2)^2 + \frac{1}{4!}(x+2)^4 - \frac{1}{6!}(x+2)^6 + \cdots\right] = \frac{\cos(x+2)}{x+2} .$$

Gen. Soln.: $y(x) = c_1 y_1(x) + c_2 y_2(x)$

5c. $4r^2 - 4r + 1 = 0$; $r_1 = r_2 = \frac{1}{2}$. For $r = r_1$: $a_k = -\frac{1}{k^2}a_{k-1}$;

$y(x) = ay_1(x)$ with $y_1(x) = \sqrt{x-1}\sum_{k=0}^{\infty}(-1)^k\frac{1}{(k!)^2}(x-1)^k$.

No "second" value of r .

5d. $r^2 + 2r - 3 = 0$; $r_1 = 1$, $r_2 = -3$.

For $r = r_1$: $a_k = \frac{-1}{k+4}a_{k-1}$; $y(x) = ay_1(x)$ with

$$y_1(x) = (x-3)\left[1 - \frac{1}{5}(x-3) + \frac{1}{6\cdot 5}(x-3)^2 - \frac{1}{7\cdot 6\cdot 5}(x-3)^3 + \cdots\right]$$

$$= (x-3)\sum_{k=0}^{\infty}\frac{(-1)^k 4!}{(k+4)!}(x-3)^k .$$

For $r = r_2$: $a_k = -\frac{1}{k}a_{k-1}$; $y(x) = ay_2(x)$ with

$$y_2(x) = (x-3)^{-3}\left[1 - (x-3) + \frac{1}{2}(x-3)^2 - \frac{1}{3\cdot 2}(x-3)^3 + \cdots\right]$$

$$= (x-3)^{-3}\sum_{k=0}^{\infty}\frac{(-1)^k}{k!}(x-3)^k = (x-3)^{-3}e^{-(x-3)} .$$

Gen. Soln.: $y(x) = c_1y_1(x) + c_2y_2(x)$

8b. y_3 must be a linear combination of y_1 and y_2.

Chapter 36

2a. $r^2 - 3r + 2 = 0$; $r_1 = 2$, $r_2 = 1$;
$y_{\text{Euler},1}(x) = x^2$, $y_{\text{Euler},2}(x) = x$; $\lim_{x\to 0}|y_1(x)| = 0$, $\lim_{x\to 0}|y_2(x)| = 0$

2b. $r^2 - r - 2 = 0$; $r_1 = 2$, $r_2 = -1$;
$y_{\text{Euler},1}(x) = x^2$, $y_{\text{Euler},2}(x) = x^{-1}$; $\lim_{x\to 0}|y_1(x)| = 0$, $\lim_{x\to 0}|y_2(x)| = \infty$

2c. $r^2 = 0$; $r_1 = r_2 = 0$;
$y_{\text{Euler},1}(x) = 1$, $y_{\text{Euler},2}(x) = \ln|x|$; $\lim_{x\to 0}|y_1(x)| = 1$, $\lim_{x\to 0}|y_2(x)| = \infty$

2d. $2r^2 + 3r + 1 = 0$; $r_1 = -1/2$, $r_2 = -1$;
$y_{\text{Euler},1}(x) = |x|^{-1/2}$, $y_{\text{Euler},2}(x) = x^{-1}$; $\lim_{x\to 0}|y_1(x)| = \infty$, $\lim_{x\to 0}|y_2(x)| = \infty$

2e. $r^2 - 6r + 9 = 0$; $r_1 = r_2 = 3$;
$y_{\text{Euler},1}(x) = x^3$, $y_{\text{Euler},2}(x) = x^3\ln|x|$; $\lim_{x\to 0}|y_1(x)| = 0$, $\lim_{x\to 0}|y_2(x)| = 0$

2f. $r^2 - 4 = 0$; $r_1 = 2$, $r_2 = -2$;
$y_{\text{Euler},1}(x) = x^2$, $y_{\text{Euler},2}(x) = x^{-2}$; $\lim_{x\to 0}|y_1(x)| = 0$, $\lim_{x\to 0}|y_2(x)| = \infty$

2g. $4r^2 + 4r + 1 = 0$; $r_1 = r_2 = -\frac{1}{2}$;
$y_{\text{Euler},1}(x) = |x|^{-1/2}$, $y_{\text{Euler},2}(x) = |x|^{-1/2}\ln|x|$; $\lim_{x\to 0}|y_1(x)| = \infty$, $\lim_{x\to 0}|y_2(x)| = \infty$

2h. $r^2 - 1 = 0$; $r_1 = 1$, $r_2 = -1$;
$y_{\text{Euler},1}(x) = x$, $y_{\text{Euler},2}(x) = x^{-1}$; $\lim_{x\to 0}|y_1(x)| = 0$, $\lim_{x\to 0}|y_2(x)| = \infty$

2i. $r^2 + 3r = 0$; $r_1 = 0$, $r_2 = -3$;
$y_{\text{Euler},1}(x) = 1$, $y_{\text{Euler},2}(x) = x^{-3}$; $\lim_{x\to 0}|y_1(x)| = 1$, $\lim_{x\to 0}|y_2(x)| = \infty$

2j. $r^2 - r - 6 = 0$; $r_1 = 3$, $r_2 = -2$;
$y_{\text{Euler},1}(x) = (x+2)^3$, $y_{\text{Euler},2}(x) = (x+2)^{-2}$; $\lim_{x\to 0}|y_1(x)| = 0$, $\lim_{x\to 0}|y_2(x)| = \infty$

2k. $r^2 - r = 0$; $r_1 = 1$, $r_2 = 0$;
$y_{\text{Euler},1}(x) = x - 3$, $y_{\text{Euler},2}(x) = 1$; $\lim_{x\to 3}|y_1(x)| = 0$, $\lim_{x\to 3}|y_2(x)| = 1$

2l. $2r^2 - r = 0$; $r_1 = \frac{1}{2}$, $r_2 = 0$;
$y_{\text{Euler},1}(x) = \sqrt{|x-1|}$, $y_{\text{Euler},2}(x) = 1$; $\lim_{x\to 1} |y_1(x)| = 0$, $\lim_{x\to 1} |y_2(x)| = 1$

4b. $\lim_{x\to x_0} |y_1{}'(x)| = 1$ when $r_1 = 1$; $\lim_{x\to x_0} |y_1{}'(x)| = a_1$ when $r_1 = 0$ **5b.** $r_1 = 1/2$, $r_2 = 0$

5e. $T_0(x) = 1$, $T_1(x) = x$, $T_2(x) = 2x^2 - 1$, $T_3(x) = 4x^3 - 3x$, $T_4(x) = 8x^4 - 8x^2 + 1$, $T_5(x) = 16x^5 - 20x^3 + 5x$

6a. $y_2(x) = y_1(x)\ln|x| + x^{3/2}\left[-2 - \frac{3}{4}x - \frac{11}{108}x^2 - \frac{25}{3{,}456}x^3 - \frac{137}{432{,}000}x^4 + \cdots\right]$

6b. $y_2(x) = y_1(x)\ln|x| + x\left[0 + \frac{1}{4}x + 0x^2 - \frac{3}{128}x^3 + 0x^4 + \cdots\right]$

6c. $y_2(x) = -6y_1(x)\ln|x| + 1 + 4x + 0x^2 - \frac{22}{3}x^3 + \frac{43}{24}x^4 + \cdots$

6d. $y_2(x) = -\frac{16}{9}y_1(x)\ln|x| + \frac{1}{x^2}\left[1 + \frac{4}{3}x + \frac{4}{3}x^2 + \frac{16}{9}x^3 + 0x^4 + \cdots\right]$

Chapter 38

1a. Yes, it is **1b.** No, it is not **1c.** Yes, it is **2a.** No, it is not **2b.** Yes, it is
2c. No, it is not **3a.** No, it is not **3b.** Yes, it is **3c.** Yes, it is

4b. $x(t) = 3e^{3t} + 4e^{-4t}$ and $y(t) = 3e^{3t} - 10e^{-4t}$

5b. $x(t) = 3e^{9t} - 3e^{-3t}$ and $y(t) = 3e^{9t} + 6e^{-3t}$

6b. $x(t) = 6e^{-2t} - 6e^{5t}$ and $y(t) = -18e^{-2t} - 3e^{5t}$

7a. $x(t) = c_1\cos(t) + c_2\sin(t)$ and $y(t) = c_1\sin(t) - c_2\cos(t) + c_3$

7b. $x(t) = Ae^{c_1 t^2}$ and $y(t) = c_1 t$

7c. $x(t) = 2Ae^{3t} + Be^{-2t}$, $y(t) = 6 + \left[\frac{15B}{A}t + C\right]e^{-5t}$ and $z(t) = Ae^{3t}$

8a. $x' = -\frac{6}{1200}x + \frac{1}{600}y$, $y' = 3 + \frac{1}{1200}x - \frac{4}{600}y$

8. $x' = 2 - \frac{9}{1200}x + \frac{1}{600}y$, $y' = 1 + \frac{3}{1200}x - \frac{5}{600}y$

9a. $x' = -2y - 4x$, $y' = x$

9b. $x' = 32y + 8t^2x + \sin(t)$, $y' = x$

9c. $x' = 4 - y^2$, $y' = x$

9d. $x' = 5t^{-1}x - 8t^{-2}y$, $y' = x$

9e. $x' = t^{-1}x - 10t^{-2}y$, $y' = x$

9f. $x' = 4t^2 - \sin(x)\,y$, $y' = x$

9g. $x' = z$, $y' = x$, $z' = 3x + 4y - 2z$

9h. $x' = z$, $y' = x$, $z' = -x\left(t^2 + y^2\right)$

10a. $x(t) = -4e^{-4t} + 5e^{3t}$ and $y(t) = 10e^{-4t} + 5e^{3t}$

10b. $x(t) = Ae^{2t} + Be^{-2t}$ and $y(t) = Ae^{2t} - Be^{-2t}$ where $A = \frac{1}{2}(x_0 + y_0)$ and $B = \frac{1}{2}(x_0 - y_0)$

10c. $x(t) = x_0\cos(2t) + y_0\sin(2t)$ and $y(t) = y_0\cos(2t) - x_0\sin(2t)$

10d. $x(t) = x_0\cos(4t) - \frac{y_0}{2}\sin(4t)$ and $y(t) = y_0\cos(4t) + 2x_0\sin(4t)$

10e. $x(t) = e^{2t}[2\cos(3t) - 3\sin(3t)]$ and $y(t) = e^{2t}\cos(3t)$

10f. $x(t) = e^{3t}\,[x_0\cos(2t) + y_0\sin(2t)]$ and $y(t) = e^{3t}\,[y_0\cos(2t) - x_0\sin(2t)]$

10g. $x(t) = -2e^{9t} + t + 2$ and $y(t) = -e^{9t} - 4t + 1$

10h. $x(t) = \frac{1}{2}\left[e^{2t} - 3\right]$ and $y(t) = -5e^{2t} + 6$

10i. $x(t) = e^{7t} + 3e^{3t} - 6te^{3t}$ and $y(t) = e^{7t} - e^{3t} + 2te^{3t}$

10j. $x(t) = 3\sin(2t) - 6t$ and $y(t) = -6\cos(2t) + 6$

10k. $x(t) = 13[\cos(4t) + \sin(4t)]$ and $y(t) = 8\sin(4t)$

10l. $x(t) = \left[2e^{7(t-2)} - 3e^{3(t-2)} + 1\right]\text{step}_2(t)$ and $y(t) = \left[2e^{7(t-2)} + e^{3(t-2)} - 3\right]\text{step}_2(t)$

Chapter 39

1a. $(0, 0)$ **1b.** $(3, 2)$ **1c.** Every (x_0, y_0) with $y_0 = -3x_0$ **1d.** $(6, 1)$ and $(5, 0)$
1e. $(2, 2)$, $(2, -2)$, $(4, 4)$ and $(4, -4)$ **1f.** $(3, n\pi)$ for $n = 0, \pm 1, \pm 2, \pm 3, \ldots$

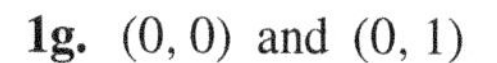

1g. $(0, 0)$ and $(0, 1)$ **1h.** No constant solutions. **2a.**

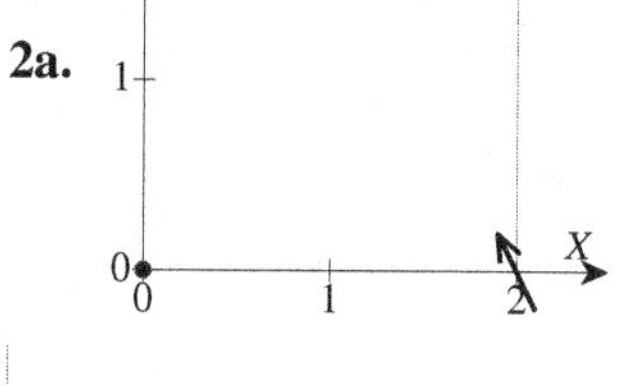

2b.

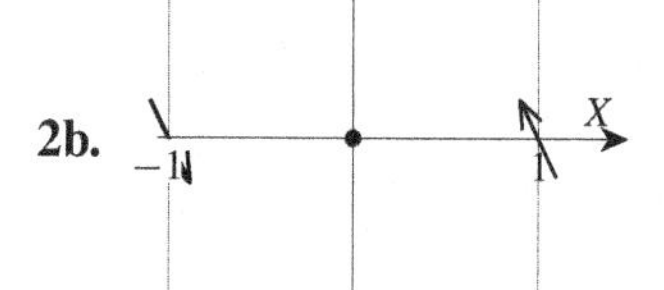

3.

4. a. $(0, 0)$

4b.

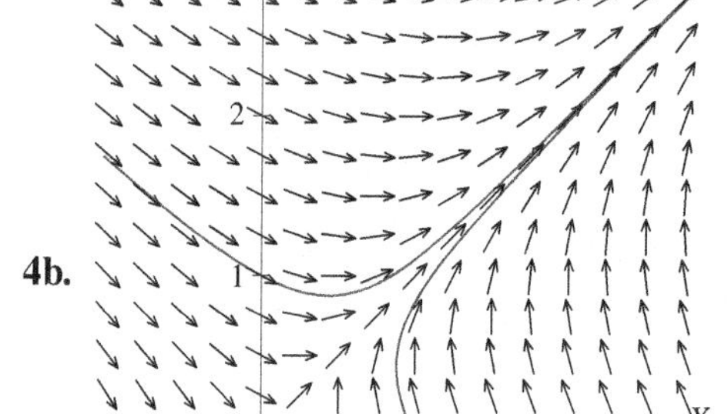

4c.

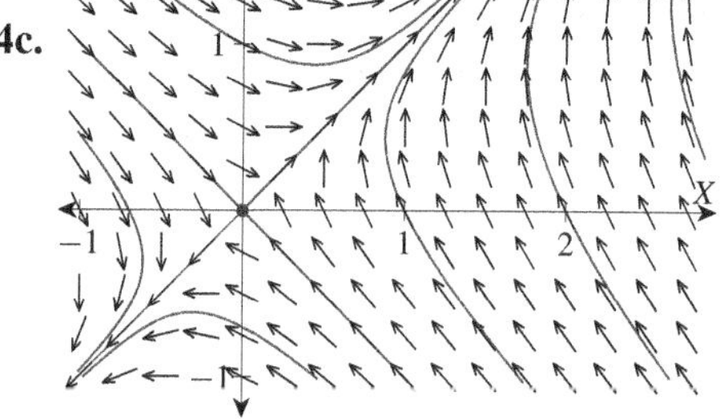

4d. They become large and nearly equal. **4e.** Unstable **5. a.** $(1, 1/2)$

5b.

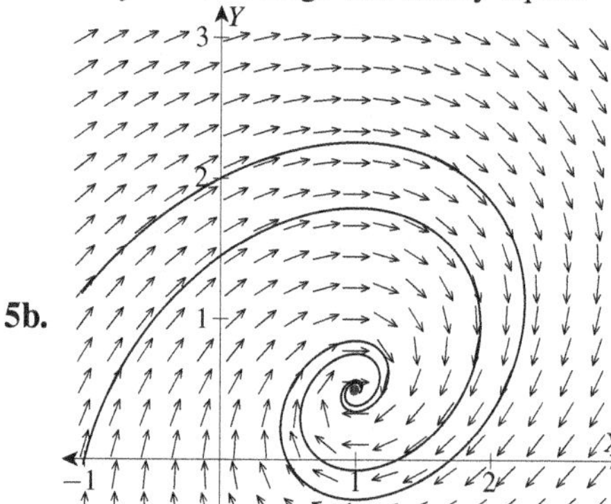

5c.

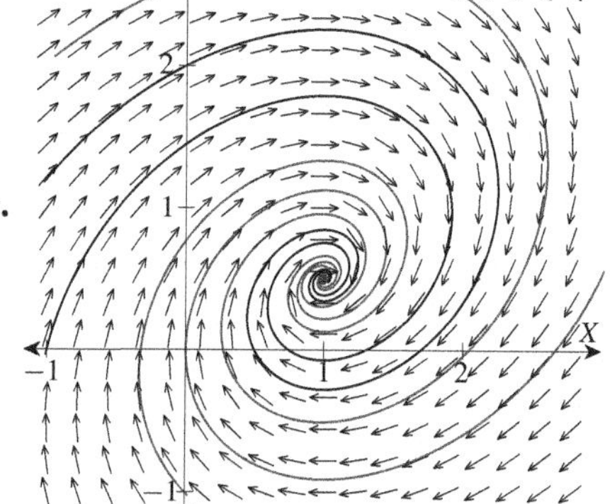

5d. $(x, y) \to (1, 1/2)$ **5e.** Asymptotically stable

6. a. $({}^{n\pi}/_{2}, 0)$ for $n = 0, \pm1, \pm2, \ldots$

6b. Stable: $(n\pi, 0)$ for $n = 0, \pm1, \pm2, \ldots$; unstable: $({}^{n\pi}/_{2}, 0)$ for $n = 1, \pm3, \pm5, \ldots$

6c.

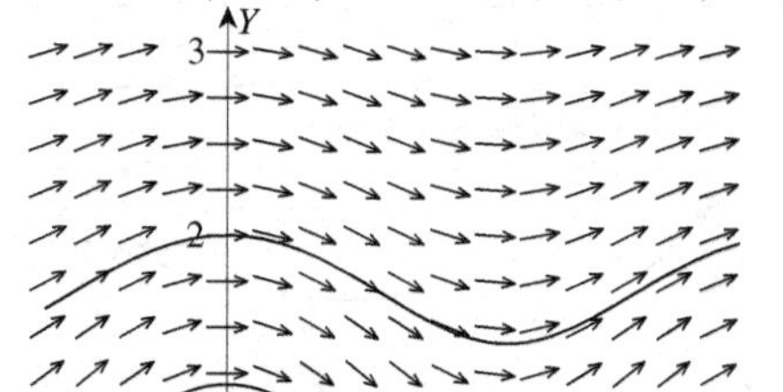

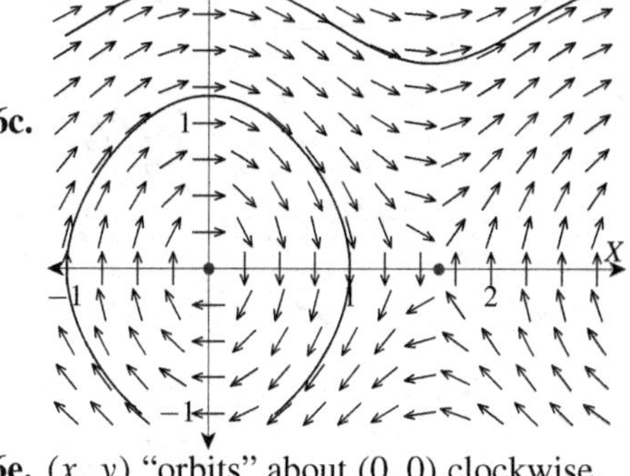

6d.

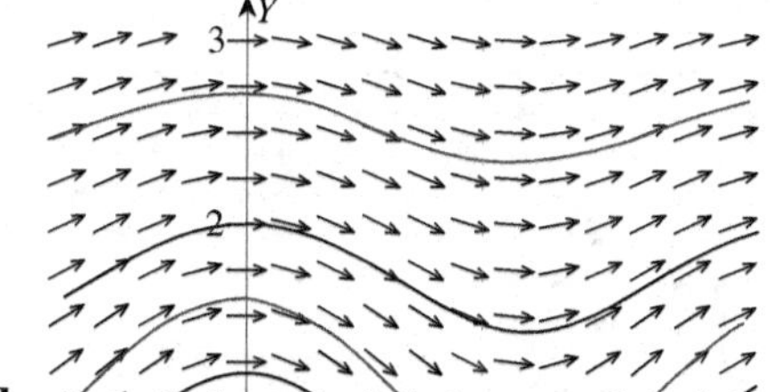

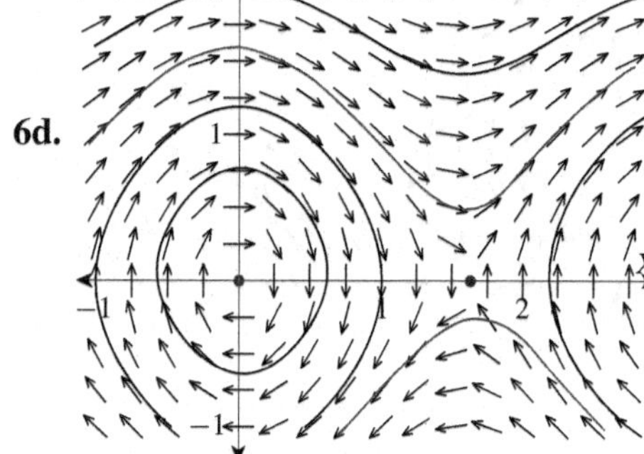

6e. (x, y) "orbits" about $(0, 0)$ clockwise.

7. a. $(1, 0)$

7b.

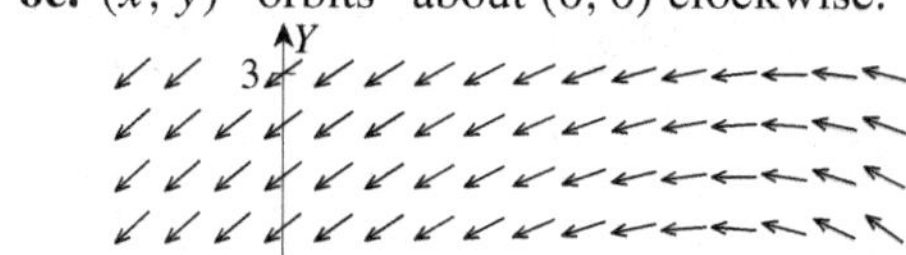

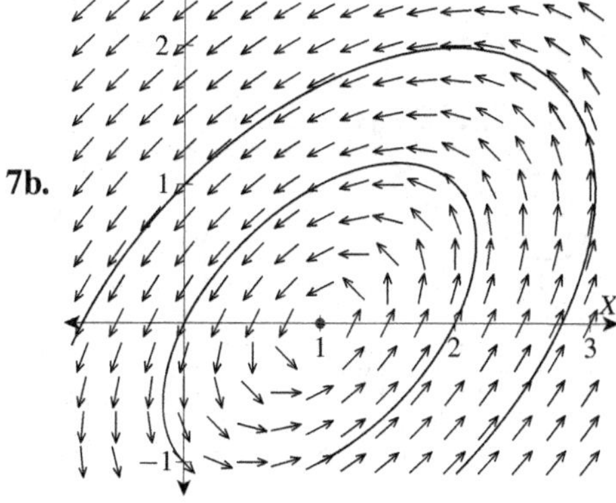

7c.

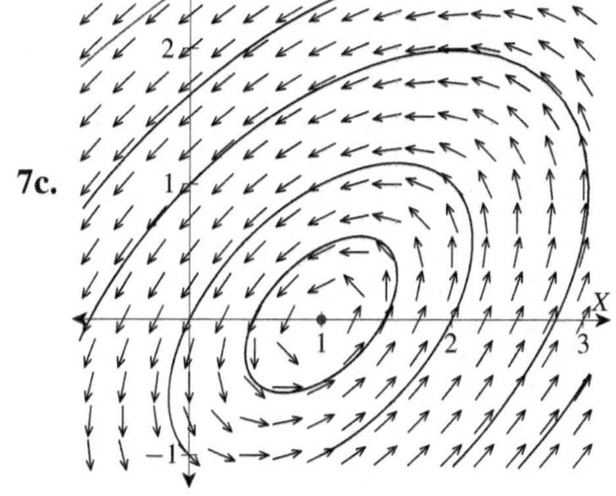

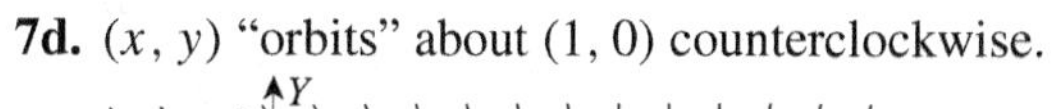

7d. (x, y) "orbits" about $(1, 0)$ counterclockwise.

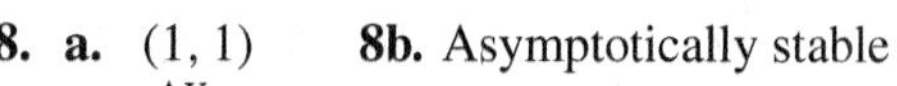

8. a. $(1, 1)$ **8b.** Asymptotically stable

8c.

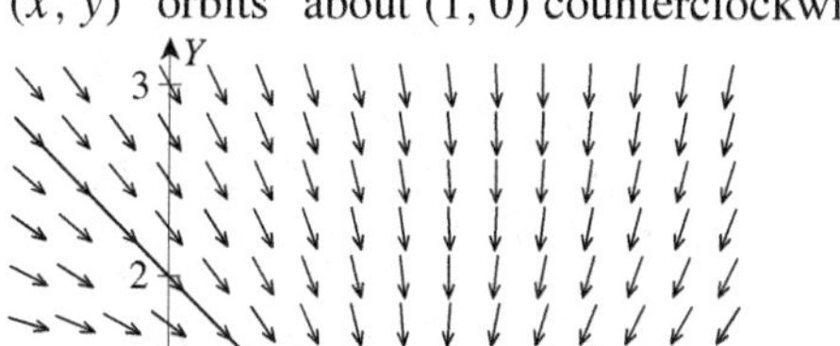

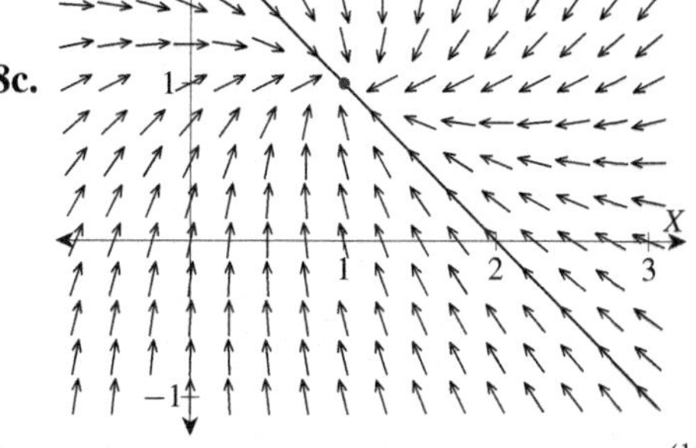

8d.

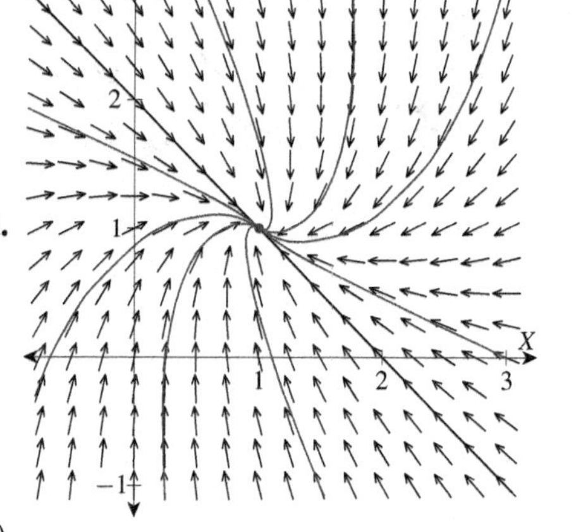

8e. $(x, y) \to (1, 1)$ **9. a.** $(1, 0)$ and $\left({}^{15}/_{16}, -{}^{1}/_{8}\right)$

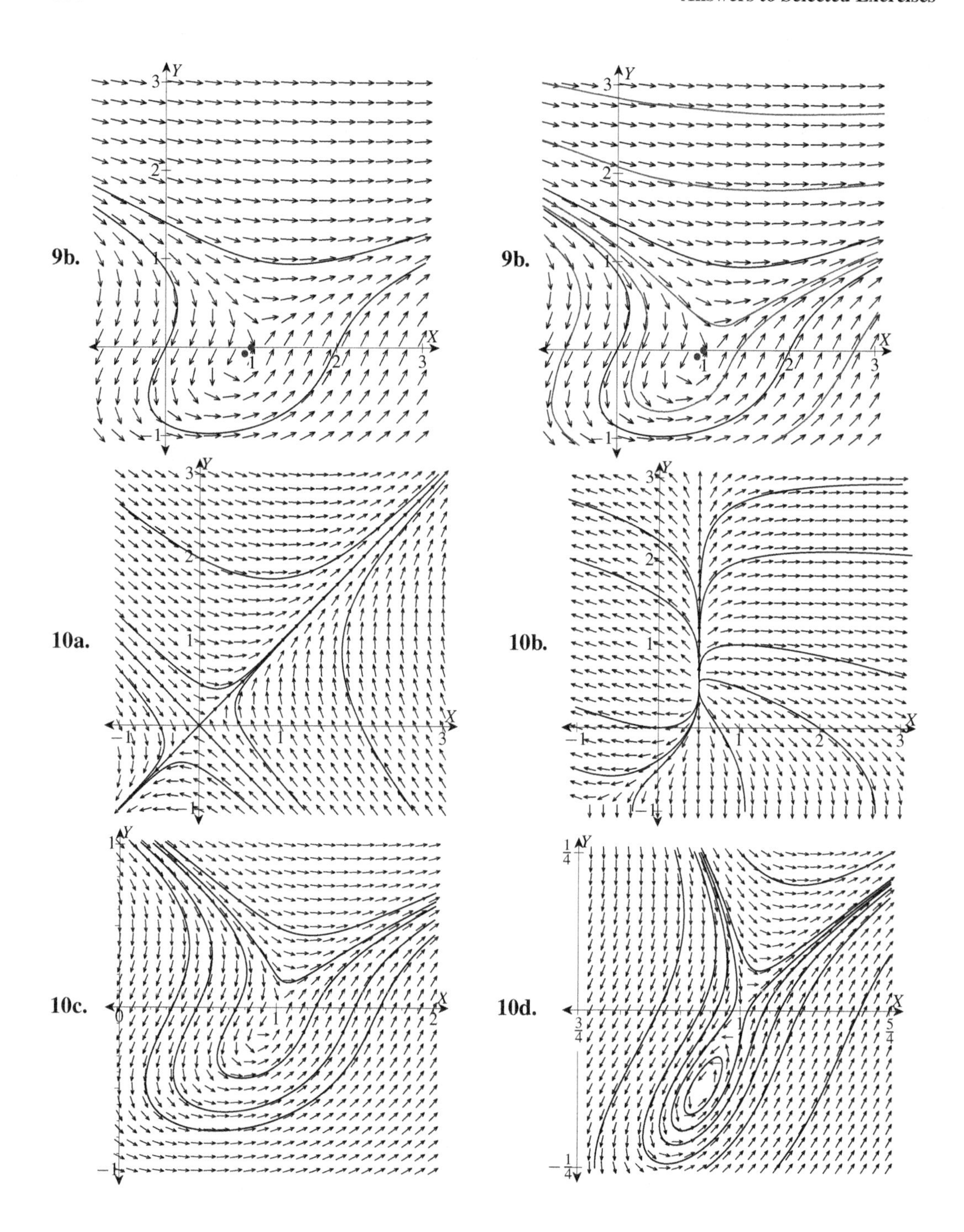
9b.
9b.
10a.
10b.
10c.
10d.

Chapter 40

1a.

k	t_k	x_k	y_k
0	0	4	1
1	$1/2$	3	2
2	1	$7/2$	$3/2$
3	$3/2$	$13/4$	$7/4$
4	2	$27/8$	$13/8$
5	$5/2$	$53/16$	$27/16$
6	3	$107/32$	$53/32$

1b.

k	t_k	x_k	y_k
0	0	2	0
1	$1/2$	1	-1
2	1	$-1/2$	-1
3	$3/2$	$-5/4$	$1/2$
4	2	$-1/8$	$11/4$
5	$5/2$	$43/16$	$31/8$
6	3	$167/32$	$35/162$

1c.

k	t_k	x_k	y_k
0	1	2	0
1	$4/3$	$4/3$	$-2/3$
2	$5/3$	$2/3$	$-5/6$
3	2	$1/5$	$-7/10$
4	$7/3$	$-1/15$	$-13/30$
5	$8/3$	$-19/105$	$-9/70$
6	3	$-4/21$	$1/6$

1d.

k	t_k	x_k	y_k
0	1	2	1
1	$4/3$	$7/3$	$5/3$
2	$5/3$	$9/4$	$7/4$
3	2	$21/10$	$17/10$
4	$7/3$	$39/20$	$97/60$
5	$8/3$	$127/70$	$107/70$
6	3	$949/560$	$809/560$

3a. $x(5) \approx -6.653 \times 10^{701}$ and $y(5) \approx -2.691 \times 10^{702}$
3b. $x(5) \approx 5.788583806$ and $y(5) \approx 2.464806994 \times 10^{-7}$
3c. $x(5) \approx 4.999999998$ and $y(5) \approx 4.397816374 \times 10^{-35}$
4. a. $(0,0)$, $(0, 15/4)$, $(0,5)$ and $(55/2, -75/2)$ **4ci.** It is obviously a catastrophic failure.
4cii. It is a fair approximation. **4ciii.** It is a much better approximation.

5a.

k	t_k	y_k
0	0	1
1	$1/2$	$3/2$
2	1	$9/4$
3	$3/2$	$27/8$
4	2	$81/16$

5b.

k	t_k	y_k
0	0	1
1	$1/2$	$3/2$
2	1	$9/4$
3	$3/2$	$3/4$
4	2	0
5	$5/2$	0
6	3	0

5c.

k	t_k	y_k
0	1	1
1	$3/2$	2
2	2	$7/2$
3	$5/2$	$52/9$
4	3	$1337/144$

5d.

k	t_k	y_k
0	1	4
1	$4/3$	$13/3$
2	$5/3$	$46/9$
3	2	$13/2$
4	$7/3$	$26/3$
5	$8/3$	$106/9$
6	3	16

7a. 0.5915247715 **7b.** 0.5083182131 **7c.** 0.5008224212

8a. -3.029881914 **8b.** 0.4054113025 **8c.** 0.6998790868 **9a.** $-.4185882949$
9b. $-.3459274604$ **9c.** $-.3391946824$

10.

$$t_{k+1} = t_k + \Delta t \quad ,$$

$$p_x = x_k + \Delta t \cdot f(t_k, x_k, y_k) \quad ,$$

$$p_y = y_k + \Delta t \cdot f(t_k, x_k, y_k) \quad ,$$

$$x_{k+1} = x_k + \Delta t \cdot \frac{1}{2}\left[f(t_k, x_k, y_k) + f(t_{k+1}, p_x, p_y)\right] \quad \text{and}$$

$$y_{k+1} = y_k + \Delta t \cdot \frac{1}{2}\left[g(t_k, x_k, y_k) + g(t_{k+1}, p_x, p_y)\right]$$

Index

Printed in the United States
by Baker & Taylor Publisher Services